THE DICTIONARY OF

Science

Edited by

Peter Lafferty

and

Julian Rowe

SIMON & SCHUSTER

A Paramount Communications Company

New York

First published in Great Britain in 1993 by Helicon Publishing Limited.

Published in the United States in 1994 by

Academic Reference Division
Simon & Schuster
15 Columbus Circle
New York, New York 10023

Typeset by Intype, London

Printed and bound in Great Britain by
The Bath Press Ltd, Bath, Avon

Printing number
1 2 3 4 5 6 7 8 9 10

Library of Congress Cataloging-in-Publication Data

A catalog record for this book is available from the Library of Congress.

ISBN 0–13–304718–0

THE DICTIONARY OF
Science

INTRODUCTION

Most dictionaries of science are aimed at students. Worthy as this aim is, it provides an excuse for publishers if the material is dull. As long as it is correct, the book has served its purpose; no matter that nobody would ever choose to open it for pleasure. *The Hutchinson Dictionary of Science* is rather different. It has been designed to be useful and appropriate for academic purposes, but at the same time to be interesting and accessible to the general reader. To this end we have included material covering a wider range of scientific topics than many other science dictionaries, as well as puzzles, quotations, and topical updates. Our intention is to show science not as an unchanging repository of facts but rather as a number of evolving and exciting areas of research. Our entries for super-conductivity, AIDS, and biodiversity, to name only three, should reflect this aim.

Arrangement of entries

Entries are ordered alphabetically, as if there were no spaces between words. Thus, entries for words beginning "sulphur" follow the order:

sulphur
sulphur dioxide
sulphuric acid
sulphurous acid
sulphur trioxide

Common or technical names?

Terms are usually placed under the better-known name rather than the technical name (thus, chloroform is placed under C and not under its technically correct name trichloromethane), but the technical term is also given. To aid comprehension for the non-specialist, terms are frequently explained when used within the text of an entry, even though they may have their own entry elsewhere.

Cross-references

These are shown by a ◊ symbol immediately preceding the reference. Cross-referencing is selective: a cross-reference is shown when another entry contains material directly relevant to the subject matter of an entry, in cases where the reader may not otherwise think of looking.

Units

SI and metric units are used throughout. Commonly used measurements of distances, temperatures, sizes, and so on include an approximate imperial equivalent.

Scientists

We have aimed to supply dates of birth and death for each of the many scientists whose work is described in the dictionary. Pre-eminent figures in each of the main fields of science are listed, with their dates, at the back of the book. The dates of other scientists are given within the body of the text.

CONTRIBUTORS

**Great Experiments
and Discoveries**
Peter Lafferty MSc
Julian Rowe PhD

Progress Reports
David Bradley PhD
John Broad MA MSc
Thomas Day PhD
Dougal Dixon MSc
Nigel Dudley
Barry Fox
James Le Fanu MD
Chris Pellant
Ian Ridpath FRAS
Peter Rodgers PhD
Jack Schofield MA
Professor Ian Stewart
Steve Smyth
Chris Stringer PhD DSc
Colin Tudge MA

Puzzles
R D Bagnall PhD
of the Centre for
Continuing Education,
University of Edinburgh

Subject Consultants
Dougal Dixon MSc
Nigel Dudley
Peter Lafferty MSc
Carol Lister PhD, FSS
Prof. Graham Littler MSc, FSS
Ian Ridpath FRAS
Julian Rowe PhD
Jack Schofield MA
Steve Smyth

EDITORS

Editors
Peter Lafferty MSc
Julian Rowe PhD

Project Editor
Sara Jenkins-Jones

Text Editors
Jane Anson
Lionel and Janet Browne
Ingrid von Essen
Edith Harkness
Ray Loughlin

Proofreader
Catherine Thompson

Design
Behram Kapadia

Production
Tony Ballsdon

Art Editor
Terence Caven

Additional Page Make-up
Helen Bird

A

A in physics, symbol for ◊ampere, a unit of electrical current.

abacus method of calculating with a handful of stones on 'a flat surface' (Latin *abacus*), familiar to the Greeks and Romans, and used by earlier peoples, possibly even in ancient Babylon; it still survives in the more sophisticated bead-frame form of the Russian *schoty* and the Japanese *soroban*. The abacus has been superseded by the electronic calculator.

The wires of a bead-frame abacus define place value (for example, in the decimal number system each successive wire, counting from right to left, would stand for units, tens, hundreds, thousands, and so on) and beads are slid to the top of each wire in order to represent the digits of a particular number. On a simple decimal abacus, for example, the number 8,493 would be entered by sliding three beads on the first wire (three units), nine beads on the second wire (nine tens), four beads on the third wire (four hundreds), and eight beads on the fourth wire (eight thousands).

abdomen in invertebrates, the part of the body below the ◊thorax, containing the digestive organs; in insects and other arthropods, it is the hind part of the body. In mammals, the abdomen is separated from the thorax by the diaphragm, a sheet of muscular tissue; in arthropods, commonly by a narrow constriction. In insects and spiders, the abdomen is characterized by the absence of limbs.

aberration of starlight apparent displacement of a star from its true position, due to the combined effects of the speed of light and the speed of the Earth in orbit around the Sun (about 30 km per second/18.5 mi per second).

Aberration, discovered 1728 by English astronomer James Bradley (1693–1762), was the first observational proof that the Earth orbits the Sun.

aberration, optical any of a number of defects that impair the image in an optical instrument. Aberration occurs because of minute variations in lenses and mirrors, and because different parts of the light ◊spectrum are reflected or refracted by varying amounts.

In *chromatic aberration* the image is surrounded by coloured fringes, because light of different colours is brought to different focal points by a lens. In *spherical aberration* the image is blurred because different parts of a spherical lens

rain falling past window of stationary train

rain falling past window of moving train

direction in which star seems to lie

true position of star

starlight enters telescope

starlight reaches eyepiece

movement of earth

aberration of starlight *The aberration of starlight is an optical illusion caused by the motion of the Earth. Rain falling appears vertical when seen from the window of a stationary train; when seen from the window of a moving train, the rain appears to follow a sloping path. In the same way, light from a star 'falling' down a telescope seems to follow a sloping path because the Earth is moving. This causes an apparent displacement, or aberration, in the position of the star.*

or mirror have different focal lengths. In *astigmatism* the image appears elliptical or cross-shaped because of an irregularity in the curvature of the lens. In *coma* the images appear progressively elongated towards the edge of the field of view.

The term was coined 1737 by French mathematician Alexis Clairaut (1713–1765).

abiotic factor nonorganic variable within the ecosystem, affecting the life of organisms. Examples include temperature, light, and soil structure. Abiotic factors can be harmful to the environment, as when sulphur dioxide emissions from power stations produce acid rain.

ablation in earth science, the loss of snow and ice from a ◊glacier by melting and evaporation. It is the opposite of ◊accumulation. Ablation is most significant near the snout, or foot, of a glacier and round its edges, since temperatures tend to be higher at lower altitudes. The rate of ablation also varies according to the time of year, being greatest during the summer. If total ablation exceeds total accumulation for a particular glacier, then the glacier will retreat, and vice versa.

abrasion in earth science, the effect of ◊corrasion, a type of erosion in which rock fragments scrape and grind away a surface. The rock fragments may be carried by rivers, wind, ice, or the sea. Striations, or grooves, on rock surfaces are common abrasions, caused by the scratching of rock debris embedded in glacier ice.

abrasive substance used for cutting and polishing or for removing small amounts of the surface of hard materials. There are two types: natural and artificial abrasives, and their hardness is measured using the ◊Mohs' scale. Natural abrasives include quartz, sandstone, pumice, diamond, and corundum; artificial abrasives include rouge, whiting, and carborundum.

abscissa in ◊coordinate geometry, the *x*-coordinate of a point—that is, the horizontal distance of that point from the vertical or *y*-axis. For example, a point with the coordinates (4, 3) has an abscissa of 3. The *y*-coordinate of a point is known as the ◊ordinate.

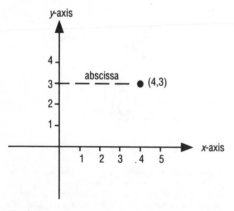

abscissa

abscissin or *abscissic acid* plant hormone found in all higher plants. It is involved in the process of ◊abscission and also inhibits stem elongation, germination of seeds, and the sprouting of buds.

abscission in botany, the controlled separation of part of a plant from the main plant body—most commonly, the falling of leaves or the dropping of fruit controlled by ◊abscissin. In ◊deciduous plants the leaves are shed before the winter or dry season, whereas ◊evergreen plants drop their leaves continually throughout the year. Fruitdrop, the abscission of fruit while still immature, is a naturally occurring process.

Abscission occurs after the formation of an abscission zone at the point of separation. Within this, a thin layer of cells, the abscission layer, becomes weakened and breaks down through the conversion of pectic acid to pectin. Consequently the leaf, fruit, or other part can easily be dislodged by wind or rain. The process is thought to be controlled by the amount of ◊auxin present. Fruitdrop is particularly common in fruit trees such as apples, and orchards are often sprayed with artificial auxin as a preventive measure.

absolute (of a value) in computing, real and unchanging. For example, an *absolute address* is a location in memory and an *absolute cell reference* is a single fixed cell in a spreadsheet display. The opposite of absolute is ◊relative.

absolute value or *modulus* in mathematics, the value, or magnitude, of a number irrespective of its sign. The absolute value of a number n is written $|n|$ (or sometimes as mod n), and is defined as the positive square root of n^2. For example, the numbers -5 and 5 have the same absolute value: $|5| = |-5| = 5$.

For a ◊complex number, the absolute value is its distance to the origin when it is plotted on an ◊Argand diagram, and can be calculated (without plotting) by applying the ◊Pythagoras' theorem. By definition, the absolute value of any complex number $a + bi$ is given by the expression: $|a + bi| = \sqrt{(a^2 + b^2)}$.

absolute zero lowest temperature theoretically possible, zero degrees Kelvin (0K), equivalent to $-273.15°C/-459.67°F$, at which molecules are motionless. Although the third law of ◊thermodynamics indicates the impossibility of reaching absolute zero exactly, a temperature of $2 \times 10^{-9}K$ (two billionths of a degree above absolute zero) was produced 1989 by Finnish scientists. Near absolute zero, the physical properties of some materials change substantially (see ◊cryogenics); for example, some metals lose their electrical resistance and become superconductive.

absorption in science, the taking up of one substance by another, such as a liquid by a solid (ink by blotting paper) or a gas by a liquid (ammonia by water). In biology, absorption describes the passing of nutrients or medication into and through tissues such as intestinal walls and blood vessels. In physics, absorption is the phenomenon by which a substance retains radiation of particular wavelengths; for example, a piece of blue glass absorbs all visible light except the wavelengths in the blue part of the spectrum; it also refers to the partial loss of energy resulting from light and other electromagnetic waves passing through a medium. In nuclear physics, absorption is the capture by elements, such as boron, of neutrons produced by fission in a reactor.

absorption spectroscopy or *absorptiometry* in analytical chemistry, a technique for determining the identity or amount present of a chemical substance by measuring the amount of electromagnetic radiation the substance absorbs at specific wavelengths; see ◊spectroscopy.

abyssal plain broad expanse of sea floor lying 3–6 km/2–4 mi below sea level. Abyssal plains are found in all the major oceans, and they extend from bordering continental rises to mid-oceanic ridges.

Underlain by outward-spreading, new oceanic crust extruded from ridges, abyssal plains are covered in deep-sea sediments derived from continental slopes and floating microscopic marine organisms. The plains often are interrupted by chains of volcanic islands where plates ride over hot spots in the mantle, and by sea mounts originally formed in oceanic ridge areas. In the Atlantic Ocean, for example, the abyssal plain extends about 930 mi/1,500 km off the east coast of the USA.

abyssal zone dark ocean region 2,000–6,000 m/ 6,500–19,500 ft deep; temperature 4°C/39°F. Three-quarters of the area of the deep-ocean floor lies in the abyssal zone, which is too far from the surface for ◊photosynthesis to take place. Some fish and crustaceans living there are blind or have their own light sources. The region above is the bathyal zone; the region below, the hadal zone.

abzyme in biotechnology, an artificially created antibody that can be used like an enzyme to accelerate reactions.

AC in physics, abbreviation for ◊alternating current.

accelerated freeze drying (AFD) common method of food preservation. See ◊food technology.

acceleration rate of change of the velocity of a moving body. It is usually measured in metres per second per second (m s^{-2}) or feet per second per second (ft s^{-2}). Because velocity is a ◊vector quantity (possessing both magnitude and direction) a body travelling at constant speed may be said to be accelerating if its direction of motion changes. According to Newton's second law of motion, a body will only accelerate if it is acted upon by an unbalanced, or resultant, ◊force.

Acceleration due to gravity is the acceleration of a body falling freely under the influence of the Earth's gravitational field; it varies slightly at different latitudes and altitudes. The value adopted internationally for gravitational acceleration is 9.806 m s^{-2}/32.174 ft s^{-2}.

The average acceleration a of an object travelling in a straight line over a period of time t may be calculated using the formula:

$$a = \text{change of velocity}/t$$

or, where u is its initial velocity and v its final velocity:

$$a = (v - u)/t.$$

A negative answer shows that the object is slowing down (decelerating). See also ◊equations of motion.

acceleration, secular in astronomy, the continuous and nonperiodic change in orbital velocity of one body around another, or the axial rotation period of a body.

An example is the axial rotation of the Earth. This is gradually slowing down owing to the gravitational effects of the Moon and the resulting production of tides, which have a frictional effect on the Earth. However, the angular ◊momentum of the Earth–Moon system is maintained, because the momentum lost by the Earth is passed to the Moon. This results in an increase in the Moon's orbital period and a consequential moving away from Earth. The overall effect is that the Earth's axial rotation period is increasing by about 15–millionths of a second a year, and the Moon is receding from the Earth at about 4 cm/1.5 in a year.

accelerator in physics, a device to bring charged particles (such as protons and electrons) up to high speeds and energies, at which they can be of use in industry, medicine, and pure physics. At low energies, accelerated particles can be used to produce the image on a television screen and generate X-rays (by means of a ◊cathode-ray tube), destroy tumour cells, or kill bacteria. When high-energy particles collide with other particles, the fragments formed reveal the nature of the fundamental forces of nature.

The first accelerators used high voltages (produced by ◊van de Graaff generators) to generate a strong, unvarying electric field. Charged particles were accelerated as they passed through the electric field. However, because the voltage produced by a generator is limited, these accelerators were replaced by machines where the particles passed through regions of alternating electric fields, receiving a succession of small pushes to accelerate them.

The first of these accelerators was the *linear accelerator* or *linac*. The linac consists of a line of metal tubes, called drift tubes, through which the particles travel. The particles are accelerated by electric fields in the gaps between the drift tubes.

Another way of making repeated use of an electric field is to bend the path of a particle into a circle so that it passes repeatedly through the same electric field. The first accelerator to use this idea was the *cyclotron* pioneered in the early 1930s by US physicist Ernest Lawrence. A cyclotron consists of an electromagnet with two hollow metal semi-circular structures, called dees, supported between the poles of an electromagnet. Particles such as protons are introduced at the centre of the machine and travel outwards in a spiral path, being accelerated by an oscillating electric field each time they pass through the gap between the dees. Cyclotrons can accelerate particles up to energies of 25 MeV (25 million electron volts); to produce higher energies, new techniques are needed.

In the ◊synchrotron, particles travel in a circular path of constant radius, guided by electromagnets. The strengths of the electromagnets are varied to keep the particles on an accurate path. Electric fields at points around the path accelerate the particles.

Early accelerators directed the particle beam onto a stationary target; large modern accelerators usually collide beams of particles that are travelling

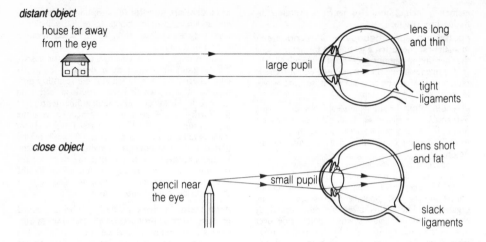

distant object

house far away
from the eye

large pupil

lens long
and thin

tight
ligaments

close object

pencil near
the eye

small pupil

lens short
and fat

slack
ligaments

accommodation *The process by which the shape of the lens in the eye is changed so that clear images of objects, whether distant or near, can be focused on the retina.*

in opposite directions. This arrangement doubles the effective energy of the collision.

The world's most powerful accelerator is the 2 km/1.25 mi diameter machine at ◊Fermilab near Batavia, Illinois, USA. This machine, the Tevatron, accelerates protons and antiprotons and then collides them at energies up to a thousand billion electron volts (or 1 TeV, hence the name of the machine). The largest accelerator is the ◊Large Electron Positron Collider at ◊CERN near Geneva, which has a circumference of 27 km/16.8 mi around which electrons and positrons are accelerated before being allowed to collide. The world's longest linac is also a colliding beam machine: the Stanford Linear Collider, in California, in which electrons and positrons are accelerated along a straight track, 3.2 km/2 mi long, and then steered to a head-on collision with other particles, such as protons and neutrons. Such experiments have been instrumental in revealing that protons and neutrons are made up of smaller elementary particles called ◊quarks.

accelerometer apparatus, either mechanical or electromechanical, for measuring ◊acceleration or deceleration—that is, the rate of increase or decrease in the ◊velocity of a moving object.

Accelerometers are used to measure the efficiency of the braking systems on road and rail vehicles; those used in aircraft and spacecraft can determine accelerations in several directions simultaneously. There are also accelerometers for detecting vibrations in machinery.

access time or *reaction time* in computing, the time taken by a computer, after an instruction has been given, to read from or write to ◊memory.

acclimation or *acclimatization* the physiological changes induced in an organism by exposure to new environmental conditions. When humans move to higher altitudes, for example, the number of red blood cells rises to increase the oxygen-carrying capacity of the blood in order to compensate for the lower levels of oxygen in the air.

accommodation in biology, the ability of the vertebrate ◊eye to focus on near or far objects by changing the shape of the lens.

For something to be viewed clearly the image must be precisely focused on the retina, the light-sensitive sheet of cells at the rear of the eye. Close objects can be seen when the lens takes up a more spherical shape, far objects when the lens is stretched and made thinner. These changes in shape are directed by the brain and by a ring of ciliary muscles lying beneath the iris.

From about the age of 40, the lens in the human eye becomes less flexible, causing the defect of vision known as *presbyopia* or lack of accommodation. People with this defect need different spectacles for reading and distance vision.

accumulation in earth science, the addition of snow and ice to a ◊glacier. It is the opposite of ◊ablation. Snow is added through snowfall and avalanches, and is gradually compressed to form ice as the glacier progresses. Although accumulation occurs at all parts of a glacier, it is most significant at higher altitudes near the glacier's start where temperatures are lower.

accumulator in electricity, a storage ◊battery—that is, a group of rechargeable secondary cells. A familiar example is the lead–acid car battery.

An ordinary 12–volt car battery consists of six lead–acid cells which are continually recharged by the car's alternator or dynamo. It has electrodes of lead and lead oxide in an electrolyte of sulphuric acid. Another common type of accumulator is the 'nife' or Ni Fe cell, which has electrodes of nickel and iron in a potassium hydroxide electrolyte.

accumulator in computing, a special register, memory location, in the ◊arithmetic and logic unit of the computer processor. It is used to hold the result of a calculation temporarily or to store data that is being transferred.

acesulfame-K non-carbohydrate sweetener that is up to 300 times as sweet as sugar. It is used in soft drinks and desserts.

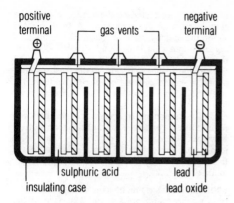

positive terminal ⊕ — gas vents — negative terminal ⊖

sulphuric acid | lead
insulating case | lead oxide

accumulator *The lead–acid car battery consists of connected cells with lead and lead oxide electrodes immersed in strong sulphuric acid. Chemical changes induced in the electrodes during charging are reversed when the battery discharges, releasing the stored electricity.*

acetaldehyde common name for ◊ethanal.

acetate common name for ◊ethanoate.

acetic acid common name for ◊ethanoic acid.

acetone common name for ◊propanone.

acetylcholine (ACh) chemical that serves as a ◊neurotransmitter, communicating nerve impulses between the cells of the nervous system. It is largely associated with the transmission of impulses across the ◊synapse (junction) between nerve cells and muscle cells.

ACh is produced in the synaptic knob (a swelling at the end of a nerve cell) and stored in vesicles until a nerve impulse triggers its discharge across the synapse. When the ACh reaches the membrane of the receiving cell, it binds with a specific site and brings about depolarization – a reversal of the electric charge on either side of the membrane, which brings about a new impulse (in nerve cells) or a contraction (in muscle cells). Its action is short-lived because it is quickly destroyed by the enzyme cholinesterase.

Anticholinergic drugs are used in medicine to block the action of ACh, thereby disrupting the passage of nerve impulses and relaxing certain muscles.

acetylene common name for ◊ethyne.

acetylsalicylic acid chemical name for the pain-killing drug ◊aspirin.

achene dry, one-seeded ◊fruit that develops from a single ◊ovary and does not split open to disperse the seed. Achenes commonly occur in groups, for example, the fruiting heads of buttercup *Ranunculus* and clematis. The outer surface may be smooth, spiny, ribbed, or tuberculate, depending on the species.

An achene with part of the fruit wall extended to form a membranous wing is called a **samara**; an example is the pendulous fruit of the ash *Fraxinus*. A ◊*caryopsis*, another type of achene, is formed when the ◊carpel wall becomes fused to the seed coat and is typical of grasses and cereals. A **cypsela** is derived from an inferior ovary and is characteristic of the daisy family (Compositae). It

often has a ◊pappus of hairs attached, which aids its dispersal by the wind, as in the dandelion.

Achernar or *Alpha Eridani* brightest star in the constellation Eridanus, and the ninth brightest star in the sky. It is a hot, luminous blue star with a true luminosity 250 times that of the Sun. It is 125 light years away.

Achilles tendon tendon pinning the calf muscle to the heel bone. It is one of the largest in the human body.

achromatic lens combination of lenses made from materials of different refractive indexes, constructed in such a way as to minimize chromatic aberration (which in a single lens causes coloured fringes around images because the lens diffracts the different wavelengths in white light to slightly different extents).

acid compound that, in solution in an ionizing solvent (usually water), gives rise to hydrogen ions (H^+ or protons). In modern chemistry, acids are defined as substances that are proton donors and accept electrons to form ◊ionic bonds. Acids react with ◊bases to form salts, and they act as solvents. Strong acids are corrosive; dilute acids have a sour or sharp taste, although in some organic acids this may be partially masked by other flavour characteristics.

Acids can be detected by using coloured indicators such as ◊litmus and methyl orange. The strength of an acid is measured by its hydrogen-ion concentration, indicated by the ◊pH value. Acids are classified as monobasic, dibasic, tribasic, and so forth, according to the number of hydrogen atoms, replaceable by bases, in a molecule. The first known acid was vinegar (ethanoic or acetic acid). Inorganic acids include boric, carbonic, hydrochloric, hydrofluoric, nitric, phosphoric, and sulphuric. Organic acids include acetic, benzoic, citric, formic, lactic, oxalic, and salicylic, as well as complex substances such as ◊nucleic acids and ◊amino acids.

acidic oxide oxide of a ◊nonmetal. Acidic oxides are covalent compounds. Those that dissolve in water, such as sulphur dioxide, give acidic solutions.

$$SO_2 + H_2O \rightleftharpoons H_2SO_{3(aq)} \rightleftharpoons H^+_{(aq)} + HSO^-_{3(aq)}$$

All acidic oxides react with alkalis to form salts.

$$CO_2 + NaOH \rightarrow NaHCO_3$$

acid rain acidic rainfall, thought to be caused principally by the release into the atmosphere of sulphur dioxide (SO_2) and oxides of nitrogen. Sulphur dioxide is formed from the burning of fossil fuels, such as coal, that contain high quantities of sulphur; nitrogen oxides are contributed from various industrial activities and from car exhaust fumes.

Acid rain is linked with damage to and death of forests and lake organisms in Scandinavia, Europe, and eastern North America. It also results in damage to buildings and statues. US and European power stations that burn fossil fuels release some 8 grams of sulphur dioxide and 3 grams of nitrogen oxides per kilowatt-hour. According to UK Depart-

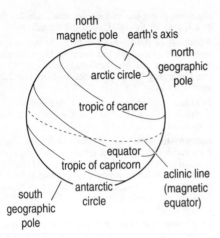

aclinic line *The magnetic equator, or the line at which the attraction of both magnetic poles is equal. Along the aclinic line, a compass needle swinging vertically will settle in a horizontal position.*

ment of Environment figures, emissions of sulphur dioxide from power stations will have to be decreased by 81% in order to arrest damage.

ACID RAIN

Rain fell in Scotland in 1974 with a pH of 2.4—nearly as acidic as lemon juice (pH 1–2).

---

acid salt chemical compound formed by the partial neutralization of a dibasic or tribasic ◊acid (one that contains two or three hydrogen atoms). Although a salt, it contains replaceable hydrogen, so it may undergo the typical reactions of an acid. Examples are sodium hydrogen sulphate ($NaHSO_4$) and acid phosphates.

aclinic line the magnetic equator, an imaginary line near the equator, where the compass needle balances horizontally, the attraction of the north and south magnetic poles being equal.

acoustic coupler device that enables computer data to be transmitted and received through a normal telephone handset; the handset rests on the coupler to make the connection. A small speaker within the device is used to convert the computer's digital output data into sound signals, which are then picked up by the handset and transmitted through the telephone system. At the receiving telephone, a second acoustic coupler or modem converts the sound signals back into digital data for input into a computer.

Unlike a ◊modem, an acoustic coupler does not require direct connection to the telephone system. However, interference from background noise means that the quality of transmission is poorer than with a modem, and more errors are likely to arise.

acoustic ohm cgs unit of acoustic impedance (the ratio of the sound pressure on a surface to the sound flux through the surface). It is analogous to the ohm as the unit of electrical ◊impedance.

acoustics in general, the experimental and theoretical science of sound and its transmission; in particular, that branch of the science that has to do with the phenomena of sound in a particular space such as a room or theatre.

Acoustical engineering is concerned with the technical control of sound, and involves architecture and construction, studying control of vibration, soundproofing, and the elimination of noise. It also includes all forms of sound recording and reinforcement, the hearing and perception of sounds, and hearing aids.

acquired characteristic feature of the body that develops during the lifetime of an individual, usually as a result of repeated use or disuse, such as the enlarged muscles of a weightlifter.

French naturalist Jean Baptiste Lamarck's theory of evolution assumed that acquired characteristics were passed from parent to offspring. Modern evolutionary theory does not recognize the inheritance of acquired characteristics because there is no reliable scientific evidence that it occurs, and because no mechanism is known whereby bodily changes can influence the genetic material. See also ◊central dogma.

acquired immune deficiency syndrome full name for the disease ◊AIDS.

acre traditional English land measure equal to

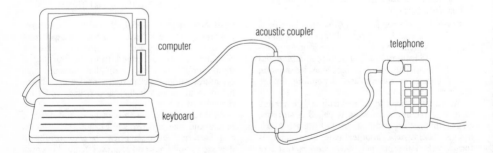

acoustic coupler *The acoustic coupler converts digital output from a computer to sound signals that can be sent down a telephone line.*

4,840 square yards (4,047 sq m/0.405 ha). Originally meaning a field, it was the size that a yoke of oxen could plough in a day.

As early as Edward I's reign (1272–1307) the acre was standardized by statute for official use, although local variations in Ireland, Scotland, and some English counties continued. It may be subdivided into 160 square rods (one square rod equalling 25.29 sq m/30.25 sq yd).

acre-foot unit sometimes used to measure large volumes of water, such as the capacity of a reservoir (equal to its area in acres multiplied by its average depth in feet). One acre-foot equals 1,233.5 cu m/43,560 cu ft or the amount of water covering one acre to a depth of one foot.

acridine $C_{13}H_9N$ organic compound that occurs in coal tar. It is extracted by dilute acids but can also be obtained synthetically. It is used to make dyes and drugs.

acrylic acid common name for ◊propenoic acid.

ACTH (adrenocorticotropic hormone) ◊hormone secreted by the anterior lobe of the ◊pituitary gland that controls the production of corticosteroid hormones by the ◊adrenal gland. It is commonly produced as a response to stress.

actinide any of a series of 15 radioactive metallic chemical elements with atomic numbers 89 (actinium) to 103 (lawrencium). Elements 89 to 95 occur in nature; the rest of the series are synthesized elements only. Actinides are grouped together because of their chemical similarities (for example, they are all bivalent), the properties differing only slightly with atomic number. The series is set out in a band in the ◊periodic table of the elements, as are the ◊lanthanides.

actinium (Greek *aktis* 'ray') white, radioactive, metallic element, the first of the actinide series, symbol Ac, atomic number 89, relative atomic mass 227; it is a weak emitter of high-energy alpha particles. Actinium occurs with uranium and radium in ◊pitchblende and other ores, and can be synthesized by bombarding radium with neutrons. The longest-lived isotope, Ac-227, has a half-life of 21.8 years (all the other isotopes have very short half-lives). Actinium was discovered 1899 by the French chemist André Debierne.

actinium K original name given 1939 to the radioactive element ◊francium by its discoverer, the French scientist Marguerite Perey (1909–1975).

action potential in biology, a change in the potential difference (voltage) across the membrane of a nerve cell when an impulse passes along it. A change in potential (from about –60 to +45 millivolts) accompanies the passage of sodium and potassium ions across the membrane.

activation analysis in analytical chemistry, a technique used to reveal the presence and amount of minute impurities in a substance or element. A sample of a material that may contain traces of a certain element is irradiated with ◊neutrons, as in a reactor. The gamma rays emitted by the material's radioisotope have unique energies and relative intensities, similar to the spectral lines from a luminous gas. Measurements and interpretation of the gamma-ray spectrum, using data from standard

samples for comparison, provide information on the amount of impurities present.

activation energy in chemistry, the energy required in order to start a chemical reaction. Some elements and compounds will react together merely by coming into contact (spontaneous reaction). For others it is necessary to supply energy in order to start the reaction, even if there is ultimately a net output of energy. This initial energy is the activation energy.

active transport in cells, the use of energy to move substances, usually molecules or ions, across a membrane.

Energy is needed because movement occurs against a concentration gradient, with substances being passed into a region where they are already present in significant quantities. Active transport thus differs from diffusion, the process by which substances move towards a region where they are in lower concentration, as when oxygen passes into the blood vessels of the lungs. Diffusion requires no input of energy.

activity in physics, the number of particles emitted in one second by a radioactive source. The term is used to describe the radioactivity or the potential danger of that source. The unit of activity is the becquerel (Bq), named after the French physicist Antoine Henri Becquerel.

activity series in chemistry, alternative name for ◊reactivity series.

acute angle an angle between 0° and 90°; that is, an amount of turn that is less than a quarter of a circle.

ACV abbreviation for *air-cushion vehicle*; see ◊hovercraft.

Ada high-level computer-programming language, developed and owned by the US Department of Defense, designed for use in situations in which a computer directly controls a process or machine, such as a military aircraft. The language took more than five years to specify, and became commercially available only in the late 1980s. It is named after English mathematician Ada Augusta Byron.

adaptation in biology, any change in the structure or function of an organism that allows it to survive and reproduce more effectively in its environment. In ◊evolution, adaptation is thought to occur as a result of random variation in the genetic make-up of organisms (produced by ◊mutation and ◊recombination) coupled with ◊natural selection.

adaptive radiation in evolution, the formation of several species, with ◊adaptations to different ways of life, from a single ancestral type. Adaptive radiation is likely to occur whenever members of a species migrate to a new habitat with unoccupied ecological niches. It is thought that the lack of competition in such niches allows sections of the migrant population to develop new adaptations, and eventually to become new species.

The colonization of newly formed volcanic islands has led to the development of many unique species. The 13 species of Darwin's finch on the Galápagos Islands, for example, are probably descended from a single species from the South American mainland. The parent stock evolved into

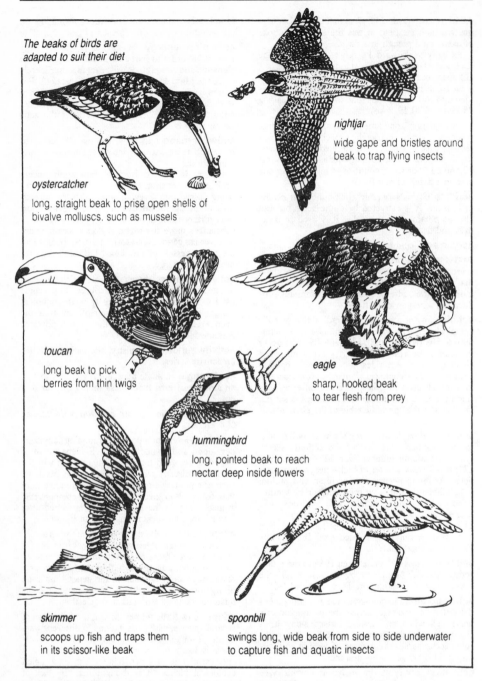

The beaks of birds are adapted to suit their diet

nightjar
wide gape and bristles around beak to trap flying insects

oystercatcher
long. straight beak to prise open shells of bivalve molluscs. such as mussels

toucan
long beak to pick berries from thin twigs

eagle
sharp, hooked beak to tear flesh from prey

hummingbird
long, pointed beak to reach nectar deep inside flowers

skimmer
scoops up fish and traps them in its scissor-like beak

spoonbill
swings long,. wide beak from side to side underwater to capture fish and aquatic insects

adaptation *The different types of beak found in birds are examples of adaptation.*

different species that now occupy a range of diverse niches.

ADC in electronics, abbreviation for ◊*analogue-to-digital converter.*

adder electronic circuit in a computer or calculator that carries out the process of adding two binary numbers. A separate adder is needed for each pair of binary ◊bits to be added. Such circuits are essential components of a computer's ◊arithmetic and logic unit (ALU).

adding machine device for adding (and usually subtracting, multiplying, and dividing) numbers, operated mechanically or electromechanically; now largely superseded by electronic ◊calculators.

addition reaction chemical reaction in which the atoms of an element or compound react with a double bond or triple bond in an organic compound by opening up one of the bonds and becoming attached to it, for example

$$CH_2=CH_2 + HCl \rightarrow CH_3CH_2Cl$$

An example is the addition of hydrogen atoms to ◊unsaturated compounds in vegetable oils to produce margarine.

additive in food, any natural or artificial chemical added to prolong the shelf life of processed foods (salt or nitrates), alter the colour or flavour of food, or improve its food value (vitamins or minerals). Many chemical additives are used and they are subject to regulation, since individuals may be affected by constant exposure even to traces of certain additives and may suffer side effects ranging from headaches and hyperactivity to cancer.

Within the European Community, approved additives are given an official ◊E number.

Flavours are said to increase the appeal of the food. They may be natural or artificial, and include artificial ◊sweeteners and monosodium glutamate (m.s.g.).

Colourings are used to enhance the visual appeal of certain foods.

Enhancers are used to increase or reduce the taste and smell of a food without imparting a flavour of their own.

Nutrients replace or enhance food value. Minerals and vitamins are added if the diet might otherwise be deficient, to prevent diseases such as beriberi and pellagra.

Preservatives are antioxidants and antimicrobials that control natural oxidation and the action of microorganisms. See ◊food technology.

Emulsifiers and *surfactants* regulate the consistency of fats in the food and on the surface of the food in contact with the air.

Thickeners, primarily vegetable gums, regulate the consistency of food. Pectin acts in this way on fruit products.

Leavening agents lighten the texture of baked goods without the use of yeasts. Sodium bicarbonate is an example.

Acidulants sharpen the taste of foods but may also perform a buffering function in the control of acidity.

Bleaching agents assist in the ageing of flours.

Anti-caking agents prevent powdered products coagulating into solid lumps.

additive: typical ingredients of orange squash

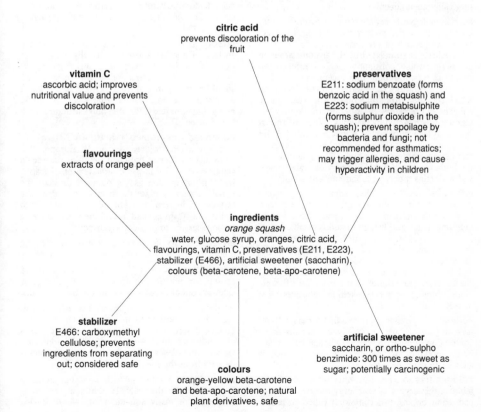

citric acid
prevents discoloration of the fruit

vitamin C
ascorbic acid; improves nutritional value and prevents discoloration

preservatives
E211: sodium benzoate (forms benzoic acid in the squash) and E223: sodium metabisulphite (forms sulphur dioxide in the squash); prevent spoilage by bacteria and fungi; not recommended for asthmatics; may trigger allergies, and cause hyperactivity in children

flavourings
extracts of orange peel

ingredients
orange squash
water, glucose syrup, oranges, citric acid, flavourings, vitamin C, preservatives (E211, E223), stabilizer (E466), artificial sweetener (saccharin), colours (beta-carotene, beta-apo-carotene)

stabilizer
E466: carboxymethyl cellulose; prevents ingredients from separating out; considered safe

colours
orange-yellow beta-carotene and beta-apo-carotene; natural plant derivatives, safe

artificial sweetener
saccharin, or ortho-sulpho benzimide: 300 times as sweet as sugar; potentially carcinogenic

Humectants control the humidity of the product by absorbing and retaining moisture.

Clarifying agents are used in fruit juices, vinegars, and other fermented liquids. Gelatin is the most common.

Firming agents restore the texture of vegetables that may be damaged during processing.

Foam regulators are used in beer to provide a controlled 'head' on top of the poured product.

address in a computer memory, a number indicating a specific location. At each address, a single piece of data can be stored. For microcomputers, this normally amounts to one ◊byte (enough to represent a single character, such as a letter or number).

The maximum capacity of a computer memory depends on how many memory addresses it can have. This is normally measured in units of 1,024 bytes (known as kilobytes, or K).

address bus in computing, the electrical pathway or ◊bus used to select the route for any particular data item as it is moved from one part of a computer to another.

adenoids masses of lymphoid tissue, similar to ◊tonsils, located in the upper part of the throat, behind the nose. They are part of a child's natural defences against the entry of germs but usually shrink and disappear by the age of ten.

Adenoids may swell and grow, particularly if infected, and block the breathing passages. If they become repeatedly infected, they may be removed surgically (adenoidectomy).

ADH in biology, abbreviation for *antidiuretic hormone*, part of the system maintaining a correct salt/water balance in vertebrates.

Its release is stimulated by the hypothalamus in the brain, which constantly receives information about salt concentration from receptors situated in the neck. In conditions of water shortage increased ADH secretion from the brain will cause more efficient conservation of water in the kidney, so that water is retained by the body. When an animal is able to take in plenty of water, decreased ADH secretion will cause the urine to become dilute so that more water leaves the body. The system allows the body to compensate for a varying water intake and maintains a correct blood concentration.

adhesive substance that sticks two surfaces together. Natural adhesives (glues) include gelatin in its crude industrial form (made from bones, hide fragments, and fish offal) and vegetable gums. Synthetic adhesives include thermoplastic and thermosetting resins, which are often stronger than the substances they join; mixtures of ◊epoxy resin and hardener that set by chemical reaction; and elastomeric (stretching) adhesives for flexible joints. Superglues are fast-setting adhesives used in very small quantities.

adiabatic in physics, a process that occurs without loss or gain of heat, especially the expansion or contraction of a gas in which a change takes place in the pressure or volume, although no heat is allowed to enter or leave.

adipose tissue type of ◊connective tissue of vertebrates that serves as an energy reserve, and also pads some organs. It is commonly called fat tissue,

and consists of large spherical cells filled with fat. In mammals, major layers are in the inner layer of skin and around the kidneys and heart.

adolescence in the human life cycle, the period between the beginning of puberty and adulthood.

ADP in biology, abbreviation for *adenosine diphosphate*, used in the manufacture of ◊ATP, the molecule used by all cells to drive their chemical reactions.

adrenal gland or *suprarenal gland* one of two ◊endocrine glands situated on top of the kidney. The adrenals are soft and yellow, and consist of two parts: the cortex and medulla. The *cortex* (outer part) secretes various ◊steroid hormones, controls salt and water metabolism, and regulates the use of carbohydrates, proteins, and fats. The *medulla* (inner part) secretes the hormones adrenaline and noradrenaline which, during times of stress, cause the heart to beat faster and harder, increase blood flow to the heart and muscle cells, and dilate airways in the lungs, thereby delivering more oxygen to cells throughout the body and in general preparing the body for 'fight or flight'.

adrenaline (US *epinephrine*) hormone secreted by the medulla of the ◊adrenal glands. Adrenaline's action on the ◊liver raises blood-sugar levels by stimulating glucose production; it also increases the heart rate, raises blood fatty-acid levels by its action on adipose tissue, and constricts and dilates blood vessels selectively.

adrenocorticotropic hormone hormone secreted by the anterior lobe of the ◊pituitary gland; see ◊ACTH.

adsorption taking up of a gas or liquid at the surface of another substance, usually a solid (for example, activated charcoal adsorbs gases). It involves molecular attraction at the surface, and should be distinguished from ◊absorption (in which a uniform solution results from a gas or liquid being incorporated into the bulk structure of a liquid or solid).

advanced gas-cooled reactor (AGR) type of ◊nuclear reactor widely used in W Europe. The AGR uses a fuel of enriched uranium dioxide in stainless-steel cladding and a moderator of graphite. Carbon dioxide gas is pumped through the reactor core to extract the heat produced by the ◊fission of the uranium. The heat is transferred to water in a steam generator, and the steam drives a turbogenerator to produce electricity.

advection fog ◊fog formed by warm air meeting a colder current or flowing over a cold surface.

adventitious root in plants, a root developing in an unusual position, as in ivy, where roots grow sideways out of the stem and cling to trees or walls.

aeolian referring to sediments carried, formed, eroded, or deposited by the wind. Such sediments include desert sands and dunes as well as deposits of windblown silt, called loess, carried long distances from deserts and from stream sediments derived from the melting of glaciers.

aerenchyma plant tissue with numerous air-filled spaces between the cells. It occurs in the stems and roots of many aquatic plants where it aids buoyancy and facilitates transport of oxygen around the plant.

aerial or *antenna* in radio and television broadcasting, a conducting device that radiates or receives electromagnetic waves. The design of an aerial depends principally on the wavelength of the signal. Long waves (hundreds of metres in wavelength) may employ long wire aerials; short waves (several centimetres in wavelength) may employ rods and dipoles; microwaves may also use dipoles—often with reflectors arranged like a toast rack—or highly directional parabolic dish aerials. Because microwaves travel in straight lines, giving line-of-sight communication, microwave aerials are usually located at the tops of tall masts or towers.

aerial oxidation in chemistry, a reaction in which air is used to oxidize another substance, as in the contact process for the manufacture of sulphuric acid, and in the ◊souring of wine.

$$2SO_2 + O_2 \rightleftharpoons 2SO_3$$

aerobic in biology, a description of those living organisms that require oxygen (usually dissolved in water) for the efficient release of energy contained in food molecules, such as glucose. They include almost all living organisms (plants as well as animals) with the exception of certain bacteria.

Aerobic reactions occur inside every cell and lead to the formation of energy-rich ◊ATP, subsequently used by the cell for driving its metabolic processes. Oxygen is used to convert glucose to carbon dioxide and water, thereby releasing energy. Most aerobic organisms die in the absence of oxygen, but certain organisms and cells, such as those found in muscle tissue, can function for short periods anaerobically (without oxygen). Other ◊anaerobic organisms can survive without oxygen.

aerodynamics branch of fluid physics that studies the forces exerted by air or other gases in motion—for example, the airflow around bodies (such as land vehicles, bullets, rockets, and aircraft) moving at speed through the atmosphere. For maximum efficiency, the aim is usually to design the shape of an object to produce a streamlined flow, with a minimum of turbulence in the moving air.

aerogenerator wind-powered electricity generator. These range from large models used in arrays on wind farms (see ◊wind turbine) to battery chargers used on yachts.

aeronautics science of travel through the Earth's atmosphere, including aerodynamics, aircraft structures, jet and rocket propulsion, and aerial navigation.

In *subsonic aeronautics* (below the speed of sound), aerodynamic forces increase at the rate of the square of the speed.

Transonic aeronautics covers the speed range from just below to just above the speed of sound and is crucial to aircraft design. Ordinary sound waves move at about 1,225 kph/760 mph at sea level, and air in front of an aircraft moving slower than this is 'warned' by the waves so that it can move aside. However, as the flying speed approaches that of the sound waves, the warning is too late for the air to escape, and the aircraft pushes the air aside, creating shock waves, which absorb much power and create design problems.

On the ground the shock waves give rise to a ◊sonic boom. It was once thought that the speed of sound was a speed limit to aircraft, and the term ◊sound barrier came into use.

Supersonic aeronautics concerns speeds above that of sound and in one sense may be considered a much older study than aeronautics itself, since the study of the flight of bullets, known as ◊ballistics, was undertaken soon after the introduction of firearms.

Hypersonics is the study of airflows and forces at speeds above five times that of sound (Mach 5); for example, for guided missiles, space rockets, and advanced concepts such as HOTOL (horizontal takeoff and landing). For all flight speeds streamlining is necessary to reduce the effects of air resistance.

Aeronautics is distinguished from astronautics, which is the science of travel through space. Astronavigation (navigation by reference to the stars) is used in aircraft as well as in ships and is a part of aeronautics.

aeroplane (US *airplane*) powered heavier-than-air craft supported in flight by fixed wings. Aeroplanes are propelled by the thrust of a jet engine or airscrew (propeller). They must be designed aerodynamically, since streamlining ensures maximum flight efficiency. The Wright brothers flew the first powered plane (a biplane) in Kitty Hawk, North Carolina, USA, 1903. For the history of aircraft and aviation, see ◊flight.

design Efficient streamlining prevents the formation of shock waves over the body surface and wings, which would cause instability and power loss. The wing of an aeroplane has the cross-sectional shape of an aerofoil, being broad and curved at the front, flat underneath, curved on top, and tapered to a sharp point at the rear. It is so shaped that air passing above it is speeded up, reducing pressure below atmospheric pressure. This follows from ◊Bernoulli's principle and results in a force acting vertically upwards, called lift, which counters the plane's weight. In level flight lift equals weight. The wings develop sufficient lift to support the plane when they move quickly through the air. The thrust that causes propulsion comes from the reaction to the air stream accelerated backwards by the propeller or the gases shooting backwards from the jet exhaust. In flight the engine thrust must overcome the air resistance, or ◊drag. Drag depends on frontal area (for example, large, airliner; small, fighter plane) and shape (drag coefficient); in level flight, drag equals thrust. The drag is reduced by streamlining the plane, resulting in higher speed and reduced fuel consumption for a given power. Less fuel need be carried for a given distance of travel, so a larger payload (cargo or passengers) can be carried.

The shape of a plane is dictated principally by

Science is all those things which are confirmed to such a degree that it would be unreasonable to withhold one's provisional consent.

Stephen Jay Gould *Lecture on Evolution* 1984

the speed at which it will operate (see ◊aeronautics). A low-speed plane operating at well below the speed of sound (about 965 kph/600 mph) need not be particularly well streamlined, and it can have its wings broad and projecting at right angles from the fuselage. An aircraft operating close to the speed of sound must be well streamlined and have swept-back wings. This prevents the formation of shock waves over the body surface and wings, which would result in instability and high power loss. Supersonic planes (faster than sound) need to be severely streamlined, and require a needle nose, extremely swept-back wings, and what is often termed a 'Coke-bottle' (narrow-waisted) fuselage, in order to pass through the sound barrier without suffering undue disturbance. To give great flexibility of operation at low as well as high speeds, some supersonic planes are designed with variable geometry, or ◊swing wings. For low-speed flight the wings are outstretched; for high-speed flight they are swung close to the fuselage to form an efficient ◊delta wing configuration.

Aircraft designers experiment with different designs in ◊wind tunnel tests, which indicate how their designs will behave in practice. Fighter jets in the 1990s are being deliberately designed to be aerodynamically unstable, to ensure greater agility; an example is the European Fighter Aircraft under development by the UK, Germany, Italy, and Spain. This is achieved by a main wing of continuously modifiable shape, the airflow over which is controlled by a smaller tilting foreplane. New aircraft are being made lighter and faster (to Mach 3) by the use of heat-resistant materials, some of which are also radar-absorbing, making the aircraft 'invisible' to enemy defences.

construction Planes are constructed using light but strong aluminium alloys such as duralumin (with copper, magnesium, and so on). For supersonic planes special stainless steel and titanium may be used in areas subjected to high heat loads. The structure of the plane, or the airframe (wings, fuselage, and so on) consists of a surface skin of alloy sheets supported at intervals by struts known as ribs and stringers. The structure is bonded together by riveting or by powerful adhesives such as ◊epoxy resins. In certain critical areas, which have to withstand very high stresses (such as the wing roots), body panels are machined from solid metal for extra strength. On the ground a plane rests on wheels, usually in a tricycle arrangement, with a nose wheel and two wheels behind, one under each wing. For all except some light planes the landing gear, or undercarriage, is retracted in flight to reduce drag. Seaplanes, which take off and land on water, are fitted with nonretractable hydrofoils.

flight control Wings by themselves are unstable in flight, and a plane requires a tail to provide stability. The tail comprises a horizontal tailplane and vertical tailfin, called the horizontal and vertical stabilizer respectively. The tailplane has hinged flaps at the rear called elevators to control pitch (attitude). Raising the elevators depresses the tail and inclines the wings upwards (increases the angle of attack). This speeds the airflow above the wings until lift exceeds weight and the plane climbs. However, the steeper attitude increases drag, so more power is needed to maintain speed and the engine throttle must be opened up. Moving the elevators in the opposite direction produces the reverse effect. The angle of attack is reduced, and the plane descends. Speed builds up rapidly if the engine is not throttled back. Turning (changing direction) is effected by moving the rudder hinged to the rear of the tailfin, and by banking (rolling) the plane. It is banked by moving the ailerons, interconnected flaps at the rear of the wings which move in opposite directions, one up, the other down. In planes with a delta wing, such as ◊Concorde, the ailerons and elevators are combined. Other movable control surfaces, called flaps, are fitted at the rear of the wings closer to the fuselage. They are extended to increase the width and camber (curve) of the wings during takeoff and landing, thereby creating extra lift, while movable sections at the front, or leading edges, of the wing, called slats, are also extended at these times to improve the airflow. To land, the nose of the plane is brought up so that the angle of attack of the wings exceeds a critical point and the airflow around them breaks down; lift is lost (a condition known as stalling), and the plane drops to the runway. A few planes (for example, the Harrier) have a novel method of takeoff and landing, rising and dropping vertically by swivelling nozzles to direct the exhaust of their jet engines downwards. The ◊helicopter and ◊convertiplane use rotating propellers (rotors) to obtain lift to take off vertically.

operation The control surfaces of a plane are operated by the pilot on the flight deck, by means of a control stick, or wheel, and by foot pedals (for the rudder). The controls are brought into action by hydraulic power systems. Advanced experimental high-speed craft known as control-configured vehicles use a sophisticated computer-controlled system. The pilot instructs the computer which manoeuvre the plane must perform, and the computer, informed by a series of sensors around the craft about the altitude, speed, and turning rate of the plane, sends signals to the control surface and throttle to enable the manoeuvre to be executed.

aerosol particles of liquid or solid suspended in a gas. Fog is a common natural example. Aerosol cans contain a substance such as scent or cleaner packed under pressure with a device for releasing it as a fine spray. Most aerosols used chlorofluorocarbons (CFCs) as propellants until these were found to cause destruction of the ◊ozone layer in the stratosphere.

The international community has agreed to phase out the use of CFCs, but most so-called 'ozone-friendly' aerosols use other ozone-depleting chemicals, although they are not as destructive as CFCs. Some of the products sprayed, such as pesticides, can be directly toxic to humans.

aestivation in zoology, a state of inactivity and reduced metabolic activity, similar to ◊hibernation, that occurs during the dry season in species such as lungfish and snails. In botany, the term is used to describe the way in which flower petals and sepals are folded in the buds. It is an important feature in ◊plant classification.

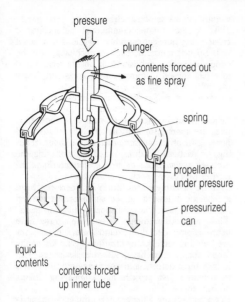

pressure

plunger

contents forced out as fine spray

spring

propellant under pressure

pressurized can

liquid contents

contents forced up inner tube

aerosol *The aerosol can produces a fine spray of liquid particles, called an aerosol. When the top button is pressed, a valve is opened, allowing the pressurized propellant in the can to force out a spray of the liquid contents. As the liquid sprays from the can, the small amount of propellant dissolved in the liquid vaporizes, producing a fine spray of small droplets.*

aether alternative form of ◊ether, the hypothetical medium once believed to permeate all of space.

AFD abbreviation for *accelerated freeze drying*, a common method of food preservation. See ◊food technology.

affine geometry geometry that preserves parallelism and the ratios between intervals on any line segment.

affinity in chemistry, the force of attraction (see ◊bond) between atoms that helps to keep them in combination in a molecule. The term is also applied to attraction between molecules, such as those of biochemical significance (for example, between ◊enzymes and substrate molecules). This is the basis for affinity ◊chromatography, by which biologically important compounds are separated.

The atoms of a given element may have a greater affinity for the atoms of one element than for another (for example, hydrogen has a great affinity for chlorine, with which it easily and rapidly combines to form hydrochloric acid, but has little or no affinity for argon).

afforestation planting of trees in areas that have not previously held forests. (*Reafforestation* is the planting of trees in deforested areas.) Trees may be planted (1) to provide timber and wood pulp; (2) to provide firewood in countries where this is an energy source; (3) to bind soil together and prevent soil erosion; and (4) to act as windbreaks.

Around 90% of Britain's timber needs are imported at an annual cost of over £4 billion. Between 1945 and 1980 the Forestry Commission doubled the forest area in the UK, planting, for example, Kielder Forest, Cumbria.

afterbirth in mammals, the placenta and other material, including blood and membranes, expelled from the uterus soon after birth. In the natural world it is often eaten.

afterburning method of increasing the thrust of a gas turbine (jet) aeroplane engine by spraying additional fuel into the hot exhaust duct between the turbojet and the tailpipe where it ignites. Used for short-term increase of power during takeoff, or during combat in military aircraft.

afterimage persistence of an image on the retina of the eye after the object producing it has been removed. This leads to persistence of vision, a necessary phenomenon for the illusion of continuous movement in films and television. The term is also used for the persistence of sensations other than vision.

after-ripening process undergone by the seeds of some plants before germination can occur. The length of the after-ripening period in different species may vary from a few weeks to many months.

It helps seeds to germinate at a time when conditions are most favourable for growth. In some cases the embryo is not fully mature at the time of dispersal and must develop further before germination can take place. Other seeds do not germinate even when the embryo is mature, probably owing to growth inhibitors within the seed that must be leached out or broken down before germination can begin.

agar jellylike carbohydrate, obtained from seaweeds. It is used mainly in microbiological experiments as a culture medium for growing bacteria and other microorganisms. The agar is resistant to breakdown by microorganisms, remaining a solid jelly throughout the course of the experiment.

It is also used in the food industry as a thickening agent in ice cream and other desserts, and in the canning of meat and fish.

agate banded or cloudy type of ◊chalcedony, a silica, SiO_2, that forms in rock cavities. Agates are used as ornamental stones and for art objects.

Agate stones, being hard, are also used to burnish and polish gold applied to glass and ceramics.

ageing in common usage, the period of deterioration of the physical condition of a living organism that leads to death; in biological terms, the entire life process.

Three current theories attempt to account for ageing. The first suggests that the process is genetically determined, to remove individuals that can no longer reproduce. The second suggests that it is due to the accumulation of mistakes during the replication of ◊DNA at cell division. The third suggests that it is actively induced by pieces of DNA that move between cells, or by cancer-causing viruses; these may become abundant in old cells and induce them to produce unwanted ◊proteins or interfere with the control functions of their DNA.

Agent Orange selective ◊weedkiller, notorious for its use in the 1960s during the Vietnam War by US forces to eliminate ground cover which could protect enemy forces. It was subsequently discovered to contain highly poisonous ◊dioxin. Thousands of US troops who had handled it later developed cancer or fathered deformed babies.

Agent Orange, named after the distinctive orange stripe on its packaging, combines equal parts of 2,4–D (2,4–trichlorophenoxyacetic acid) and 2,4,5–T (2,4,5–trichlorophenoxyacetic acid), both now banned in the USA. Companies that had manufactured the chemicals faced an increasing number of lawsuits in the 1970s. All the suits were settled out of court in a single class action (with individual claimants representing the larger group), resulting in the largest ever payment of its kind ($180 million) to claimants.

aggression in biology, behaviour used to intimidate or injure another organism (of the same or of a different species), usually for the purposes of gaining territory, a mate, or food. Aggression often involves an escalating series of threats aimed at intimidating an opponent without having to engage in potentially dangerous physical contact. Aggressive signals include roaring by red deer, snarling by dogs, the fluffing up of feathers by birds, and the raising of fins by some species of fish.

AGR abbreviation for ⟡*advanced gas-cooled reactor*, a type of nuclear reactor.

agriculture the practice of farming, including the cultivation of the soil (for raising crops) and the raising of domesticated animals. Crops are for human nourishment, animal fodder, or commodities such as cotton and sisal. Animals are raised for wool, milk, leather, dung (as fuel), or meat. The units for managing agricultural production vary from small holdings and individually owned farms to corporate-run farms and collective farms run by entire communities.

Agriculture developed in the Middle East and Egypt at least 10,000 years ago. Farming communities soon became the base for society in China, India, Europe, Mexico, and Peru, then spread throughout the world. Reorganization along more scientific and productive lines took place in Europe in the 18th century in response to dramatic population growth.

Mechanization made considerable progress in the USA and Europe during the 19th century. After World War II, there was an explosive growth in the use of agricultural chemicals: herbicides, insecticides, fungicides, and fertilizers. In the 1960s there was development of high-yielding species, especially in the **green revolution** of the Third World, and the industrialized countries began intensive farming of cattle, poultry, and pigs. In the 1980s, hybridization by genetic engineering methods and pest control by the use of chemicals plus ⟡pheromones were developed. However, there was also a reaction against some forms of intensive agriculture because of the pollution and habitat destruction caused. One result of this was a growth of alternative methods, including organic agriculture.

plants For plant products, the land must be prepared (ploughing, cultivating, harrowing, and rolling). Seed must be planted and the growing plant nurtured. This may involve fertilizers, irrigation, pest control by chemicals, and monitoring of acidity or nutrients. When the crop has grown, it must be harvested and, depending on the crop, processed in a variety of ways before it is stored or sold. Greenhouses allow cultivation of plants that

would otherwise find the climate too harsh. ⟡Hydroponics allows commercial cultivation of crops using nutrient-enriched solutions instead of soil. Special methods, such as terracing, may be adopted to allow cultivation in hostile terrain and to retain topsoil in mountainous areas with heavy rainfall.

livestock Animals may be semi-domesticated, such as reindeer, or fully domesticated but nomadic (where naturally growing or cultivated food supplies are sparse), or kept in one location. Animal farming involves accommodation (buildings, fencing, or pasture), feeding, breeding, gathering produce (eggs, milk, or wool), slaughtering, and further processing such as tanning.

organic farming From the 1970s there has been a movement towards more sophisticated natural methods without chemical sprays and fertilizers. These methods are desirable because nitrates have been seeping into the ground water, insecticides are found in lethal concentrations at the top of the ⟡food chain, some herbicides are associated with human birth defects, and hormones fed to animals to promote fast growth have damaging effects on humans.

overproduction The greater efficiency in agriculture achieved since the 19th century, coupled with post–World War II government subsidies for domestic production in the USA and the European Community (EC), have led to the development of high stocks, nicknamed 'lakes' (wine, milk) and 'mountains' (butter, beef, grain). There is no simple solution to this problem, as any large-scale dumping onto the market displaces regular merchandise. Increasing concern about the starving and the cost of storage has led the USA and the EC to develop measures for limiting production, such as letting arable land lie fallow to reduce grain crops. The USA had some success at selling surplus wheat to the USSR when the Soviet crop was poor, but the overall cost of bulk transport and the potential destabilization of other economies has acted against high producers exporting their excess on a regular basis to needy countries. Intensive farming methods also contribute to soil ⟡erosion and water pollution.

In the EC, a quota system for milk production coupled with price controls has reduced liquid milk and butter surpluses to manageable levels but has also driven out many small uneconomic producers who, by switching to other enterprises, risk upsetting the balance elsewhere. A voluntary 'set aside' scheme of this sort was proposed in the UK 1988.

Commerce, like industry, is merely a branch of agriculture. It is agriculture which furnishes the material of industry and commerce and which pays both . . .

On **agriculture** François Quesnay (1694–1774) in the *Encyclopédie*

agrochemical artificially produced chemical used in modern, intensive agricultural systems. Agrochemicals include nitrate and phosphate fertilizers, pesticides, some animal-feed additives, and pharmaceuticals. Many are responsible for

RECENT PROGRESS IN AGRICULTURE

Genetic engineering: curse or cure for agriculture

Over the next decade, the use of genetic engineering may initiate the greatest changes in agriculture since the large-scale introduction of artificial pesticides and fertilizers. Normally sober scientists have been making predictions usually associated with pulp science fiction: strains of wheat capturing their own nutrients from the air; cube-shaped tomatoes for easy packing; crops that fight off pests without pesticides; entirely new farm animals purpose-created for maximum productivity . . .

But does genetic engineering really offer such massive advantages? At present, there are sharp divisions of opinion between those who see biotechnology as an important step forward for farmers and growers, and those who see it as a largely irrelevant development which will bring with it a whole new set of health and environmental problems.

The word 'biotechnology' covers many different techniques. In its simplest form, biotechnology includes traditional crop breeding, which has been practised at least since the Sumerians developed settled agriculture in the Middle East 6,000 years ago. Modern developments in biotechnology have taken two main forms: (a) the use of **tissue culture** of plants and **embryo transfer** in animals to speed up conventional breeding techniques, allowing the creation of large numbers of individuals from a single plant and a rapid increase in rates of selective breeding, production of virus-free plants, export of livestock in embryo form, and so on; (b) the **transfer of genetic material** between cells of different individuals, or different species, usually by means of an infecting virus or plasmid, thus directly and permanently changing the nature of the organism.

Transfer of genetic material is fundamentally different from streamlined breeding techniques. In the latter case, the fundamentals of crop breeding still apply, with approximately the same constraints on which species can interbreed. In genetic transfer, genes and gene sequences can in theory be taken from any species and be placed in another to gain a specific result. So far, despite the existence of 500 companies and 150 research organizations, there is no commercially available product made by gene transfer. It is now tacitly admitted that making usable products by genetic manipulation has proved harder than was first hoped. There is also a growing concern about the health and safety aspects of releasing genetically manipulated organisms, which may have undetermined hazards for humans or the environment. Nonetheless, there are many products under development, aimed specifically at farmers and growers. These include crops resistant to specific diseases and viral attack; crops resistant to herbicides, so that these can be sprayed at higher concentrations for weed control; plants capable of capturing their own supplies of nitrogen from the air by means of genetically modified bacteria (something that occurs naturally in members of the pea and bean family); genetically manipulated animal feed to boost productivity; and artificial hormones to increase meat, milk, or wool production.

Supporters of biotechnology argue that genetic engineering offers the chance to break away from the current heavy reliance on polluting agrochemicals, and to make food production more efficient in an increasingly hungry world. However, there are questions hanging over the safety of several of these techniques. Using genetic manipulation to boost resistance to herbicides would *increase* the use of agrochemicals if it was widely adopted. Stimulating plants' innate abilities to fight off pests usually involves raising the concentration of 'natural pesticides' in the plant; in cases of severe pest attack, these toxic chemicals can rise to 300 times the background level and pose similar problems to those from pesticide residues in food. Changes to animals' hormone levels have seldom been achieved in the past without causing either suffering for the animals or attendant problems for people eating the products. Plans to introduce bovine somatotropin (BST) into the UK as a means of boosting milk production met with entrenched opposition from both farmers and consumers. We don't know the side effects of any new organism or characteristic and, once in use, the genetically manipulated organisms may well be impossible to recall.

Also, benefits may not be evenly distributed. The likelihood that companies will patent life forms for the first time inevitably means that an increasing cost will accrue for using the products of biotechnology. As in the case of the 'green revolution' of the 1960s, when improved strains of cereals were introduced into the Third World, it is likely to be the rich that benefit most, leaving the poorer farmers even worse off than before. Whatever the final outcome, the debate about biotechnology is sure to escalate over the next few years.

Steve Smyth

pollution and almost all are avoided by organic farmers.

AI abbreviation for ◊artificial intelligence.

AIDS (acronym for *a*cquired *i*mmune *d*eficiency *s*yndrome) the gravest of the sexually transmitted diseases, or STDs. It is caused by the human immunodeficiency virus (HIV), now known to be a ◊retrovirus, an organism first identified 1983. HIV is transmitted in body fluids, mainly blood and sexual secretions.

Sexual transmission of the AIDS virus endangers heterosexual men and women as well as high-risk groups, such as homosexual and bisexual men, prostitutes, intravenous drug-users sharing needles, and haemophiliacs and surgical patients treated with contaminated blood products. The virus itself is not selective, and infection is spreading to the population at large. The virus has a short life outside the body, which makes transmission of the infection by methods other than sexual contact, blood transfusion, and shared syringes extremely unlikely.

Infection with HIV is not synonymous with having AIDS; many people who have the virus in their blood are not ill, and only about half of those infected will develop AIDS within ten years. Some suffer AIDS-related illnesses but not the full-blown disease. However, there is no firm evidence to suggest that the proportion of those developing AIDS from being HIV-positive is less than 100%.

The effect of the virus in those who become ill is the devastation of the immune system, leaving the victim susceptible to (opportunistic) diseases that would not otherwise develop. In fact, diagnosis of AIDS is based on the appearance of rare tumours or opportunistic infections in unexpected candidates. Pneumocystis pneumonia, for instance, normally seen only in the malnourished or those whose immune systems have been deliberately suppressed, is common among AIDS victims and, for them, a leading cause of death.

The estimated incubation period is 9.8 years. Some AIDS victims die within a few months of the outbreak of symptoms, some survive for several years; roughly 50% are dead within three years. There is no cure for the disease, although the new drug zidovudine (◊AZT) was thought to delay the onset of AIDS and diminish its effects. The search continues for an effective vaccine.

The HIV virus originated in Africa, where the total number of cases up to Oct 1988 was 19,141. In Africa, the prevalence of AIDS among high-risk groups such as prostitutes may approach 30%. Previous reports of up to 80% of certain populations being affected are thought to have been grossly exaggerated by inaccurate testing methods. By Feb 1991, 323,378 AIDS cases in 159 countries had been reported to the World Health Organization (WHO), which estimated that over 1.3 million cases might have occurred worldwide, of which about 400,000 were at birth, during, or shortly after birth. WHO also estimated that at least 8–10 million individuals had been infected with HIV, and about half of these would develop AIDS within ten years of infection. By the year 2000 WHO expects that 15–20 million adults and 10 million children will have been infected with

HIV. The Global Aids Policy Coalition, Harvard University, USA, published a report *Aids in the World* 1993 in which it stated that 12.9 million people worldwide had been infected with HIV by the end of 1992. The same report predicts that 20 million people will be infected by 1995 and that 38–110 million adults and over 10 million children will be infected by the year 2000, with 25 million full-blown AIDS cases worldwide by the year 2000.

In the USA, attempts to carry out clinical trials of HIV vaccines on HIV-positive pregnant women were started 1991 in the hope that they might prevent the transmission of the virus to the fetus.

air see ◊atmosphere.

air conditioning system that controls the state of the air inside a building or vehicle. A complete air-conditioning unit controls the temperature and humidity of the air, removes dust and odours from it, and circulates it by means of a fan. US inventor W H Carrier developed the first effective air-conditioning unit 1902 for a New York printing plant.

The air in an air conditioner is cooled by a type of ◊refrigeration unit comprising a compressor and a condenser. The air is cleaned by means of filters and activated charcoal. Moisture is extracted by condensation on cool metal plates. The air can also be heated by electrical wires or, in large systems, pipes carrying hot water or steam; and cool, dry air may be humidified by circulating it over pans of water or through a water spray.

A specialized air-conditioning system is installed in spacecraft as part of the life-support system. This includes the provision of oxygen to breathe and the removal of exhaled carbon dioxide.

aircraft any aeronautical vehicle, which may be lighter than air (supported by buoyancy) or heavier than air (supported by the dynamic action of air on its surfaces). ◊Balloons and ◊airships are lighter-than-air craft. Heavier-than-air craft include the ◊aeroplane, glider, autogiro, and helicopter.

air-cushion vehicle (ACV) craft that is supported by a layer, or cushion, of high-pressure air. The ◊hovercraft is one form of ACV.

airglow faint and variable light in the Earth's atmosphere produced by chemical reactions (the recombination of ionized particles) in the ionosphere.

air mass large body of air with particular characteristics of temperature and humidity. An air mass forms when air rests over an area long enough to pick up the conditions of that area. For example, an air mass formed over the Sahara will be hot and dry. When an air mass moves to another area it affects the ◊weather of that area, but its own characteristics become modified in the process. For example, a Saharan air mass becomes cooler as it moves northwards.

air pollution contamination of the atmosphere caused by the discharge, accidental or deliberate, of a wide range of toxic airborne substances. Often the amount of the released substance is relatively high in a certain locality, so the harmful effects become more noticeable. The cost of preventing any discharge of pollutants into the air is prohibitive, so attempts are more usually made to reduce gradually the amount of discharge and to disperse

air pollution

pollutant	sources	effects
sulphur dioxide SO_2	oil, coal combustion in power stations	acid rain formed, which damages plants, trees, buildings, lakes
oxides of nitrogen NO and NO_2	high-temperature combustion in cars, and to some extent power stations	acid rain formed
lead compounds	from leaded petrol used by cars	nerve poison
carbon dioxide CO_2	oil, coal, petrol, diesel combustion	greenhouse effect
carbon monoxide CO	limited combustion of oil, coal, petrol, diesel fuels	poisonous, leads to photochemical smog in some areas
nuclear waste	nuclear power plants, nuclear weapon testing, war	radioactivity, contamination of locality, cancers, mutations, death

this as quickly as possible by using a very tall chimney, or by intermittent release.

California's Clean Air Act 1988 aims to reduce exhaust emissions of carbon monoxide, nitrogen oxides, hydrocarbons and other precursors of smog in increasingly stringent phases.

air sac in birds, a thin-walled extension of the lungs. There are nine of these and they extend into the abdomen and bones, effectively increasing lung capacity. In mammals, it is another name for the alveoli in the lungs, and in some insects, for widenings of the trachea.

airship or *dirigible* any aircraft that is lighter than air and power-driven, consisting of an elliptical balloon that forms the streamlined envelope or hull and has below it the propulsion system (propellers), steering mechanism, and space for crew, passengers and/or cargo. The balloon section is filled with lighter-than-air gas, either the nonflammable helium or, before helium was industrially available in large enough quantities, the easily ignited and flammable hydrogen. The envelope's form is maintained by internal pressure in the nonrigid (blimp) and semirigid types (in which the nose and tail sections have a metal framework connected by a rigid keel). The rigid type (zeppelin) maintains its form using an internal metal framework. Airships have been used for luxury travel, polar exploration, warfare, and advertising.

Rigid airships predominated from about 1900 until 1940. As the technology developed, the size of the envelope was increased from about 45 m/150 ft to more than 245 m/800 ft for the last two zeppelins built. In 1852 the first successful airship was designed and flown by Henri Giffard of France. In 1900 the first successful rigid type was designed by Count (Graf) Ferdinand von Zeppelin of Germany. In 1919 the first nonstop transatlantic roundtrip flight was completed by a rigid airship, the British R34. In the early 1920s a large source of helium was discovered in the USA and was substituted for hydrogen, reducing the danger of fire. The US military attempted to use zeppelins but abandoned the effort early on. In the 1920s and early 1930s luxury zeppelin services took passengers across the Atlantic faster and in greater comfort than the great ocean liners. The successful German airship *Graf Zeppelin*, completed 1927, was used for transatlantic, cruise, and round-the-world trips. In 1929 it travelled 32,000 km/20,000 mi around the world. It was retired and dismantled

after years of trouble-free service, and was replaced by the *Hindenburg* in 1936.

Several airship accidents were caused by structural break-up during storms and by fire. The last and best known was the *Hindenburg*, which had been forced to return to the use of flammable hydrogen by a US embargo on helium; it exploded and burned at the mooring mast at Lakehurst, New Jersey, USA, in 1937. The last and largest rigid airship was the German *Graf Zeppelin II*, completed just before World War II; it never saw commercial service but was used as a reconnaissance station off the English coast early in the war (it was the only zeppelin used in the war) and was soon retired and dismantled. Rigid airships, predominant from World War II, are no longer in use but blimps continued in use for coastal and antisubmarine patrol until the 1960s, and advertising blimps can be seen until this day. Recent interest in all types of airship has surfaced (including some with experimental and nontraditional shapes for the envelopes), since they are fuel-efficient, quiet, and capable of lifting enormous loads over great distances.

alabaster naturally occurring fine-grained white or light-coloured translucent form of ◊gypsum, often streaked or mottled. It is a soft material, used for carvings, and ranks second on the ◊Mohs' scale of hardness.

albedo the fraction of the incoming light reflected by a body such as a planet. A body with a high albedo, near 1, is very bright, while a body with a low albedo, near 0, is dark. The Moon has an average albedo of 0.12, Venus 0.65, Earth 0.37.

albumin or *albumen* any of a group of sulphur-containing ◊proteins. The best known is in the form of egg white; others occur in milk, and as a major component of serum. They are soluble in water and dilute salt solutions, and are coagulated by heat.

The presence of albumin in the urine, termed albuminuria or proteinuria, may be a symptom of a kidney disorder.

alchemy (Arabic *al-Kimya*) supposed technique of transmuting base metals, such as lead and mercury, into silver and gold by the philosopher's stone, a hypothetical substance, to which was also attributed the power to give eternal life.

This aspect of alchemy constituted much of the chemistry of the Middle Ages. More broadly, however, alchemy was a system of philosophy that dealt both with the mystery of life and the formation of

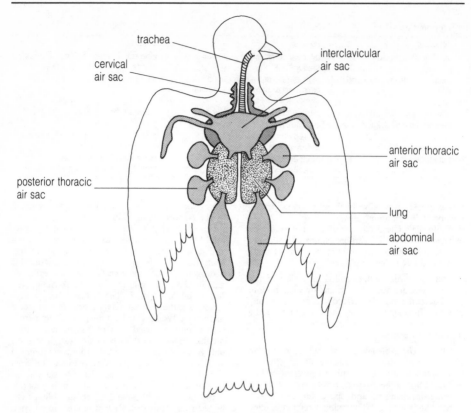

air sac *The air sacs of birds are thin-walled extensions of the lungs. They lie in the abdomen and thorax, and extend even into some of the bones. Compression and extension of the sacs assist ventilation of the lungs and respiratory system.*

inanimate substances. Alchemy was a complex and indefinite conglomeration of chemistry, astrology, occultism, and magic, blended with obscure and abstruse ideas derived from various religious systems and other sources. It was practised in Europe from ancient times to the Middle Ages but later fell into disrepute when ◊chemistry and ◊physics developed.

alcohol any member of a group of organic chemical compounds characterized by the presence of one or more aliphatic OH (hydroxyl) groups in the molecule, and which form ◊esters with acids. The main uses of alcohols are as solvents for gums, resins, lacquers, and varnishes; in the making of dyes; for essential oils in perfumery; and for medical substances in pharmacy. Alcohol (ethanol) is produced naturally in the ◊fermentation process and is consumed as part of alcoholic beverages.

Alcohols may be liquids or solids, according to the size and complexity of the molecule. The five simplest alcohols form a series in which the number of carbon and hydrogen atoms increases progressively, each one having an extra CH_2 (methylene) group in the molecule: methanol or wood spirit (methyl alcohol, CH_3OH); ethanol (ethyl alcohol, C_2H_5OH); propanol (propyl alcohol, C_3H_7OH); butanol (butyl alcohol, C_4H_9OH); and pentanol (amyl alcohol, $C_5H_{11}OH$). The lower alcohols are liquids that mix with water; the higher alcohols, such as pentanol, are oily liquids immiscible with water; and the highest are waxy solids—for example, hexadecanol (cetyl alcohol, $C_{16}H_{33}OH$) and melissyl alcohol ($C_{30}H_{61}OH$), which occur in sperm-whale oil and beeswax respectively. Alcohols containing the CH_2OH group are primary; those containing CHOH are secondary; while those containing COH are tertiary.

Aldebaran or *Alpha Tauri* brightest star in the constellation Taurus and the 14th brightest star in the sky; it marks the eye of the 'bull'. Aldebaran is a red giant 60 light years away, shining with a true luminosity of about 100 times that of the Sun.

aldehyde any of a group of organic chemical compounds prepared by oxidation of primary alcohols, so that the OH (hydroxyl) group loses its hydrogen to give an oxygen joined by a double bond to a carbon atom (the aldehyde group, with the formula CHO).

The name is made up from *al*cohol *dehyd*rogenation—that is, alcohol from which hydrogen has been removed. Aldehydes are usually liquids and include methanal (formaldehyde), ethanal (acetaldehyde), and benzaldehyde.

alexandrite rare gemstone variety of the mineral chrysoberyl (beryllium aluminium oxide $BeAl_2O_4$), which is green in daylight but appears red in artificial light.

algae (singular *alga*) diverse group of plants (including those commonly called seaweeds) that shows great variety of form, ranging from single-celled forms to multicellular seaweeds of considerable size and complexity.

Algae were formerly included within the division Thallophyta, together with fungi and bacteria. Their classification changed with increased awareness of the important differences existing between the algae and Thallophyta, and also between the groups of algae themselves; many botanists now place each algal group in a separate class or division of its own.

They can be classified into 12 divisions, largely to be distinguished by their pigmentation, including the *green algae* Chlorophyta, freshwater or terrestrial; *stoneworts* Charophyta; *golden-brown algae* Chrysophyta; *brown algae* Phaeophyta, mainly marine and including the *kelps Laminaria* and allies, the largest of all algae; *red algae* Rhodophyta, mainly marine and often living parasitically or as epiphytes on other algae; *diatoms* Bacillariophyta; *yellow-green algae* Xanthophyta, mostly freshwater and terrestrial; and *blue-green algae* Cyanophyta, of simple cell structure and without sexual reproduction, mostly freshwater or terrestrial.

ALGAE

fighting global warming Marine algae help combat global warming by removing carbon dioxide from the atmosphere during photosynthesis. They remove 10 billion tonnes of carbon dioxide each year—more than all the land plants combined.

---

algebra system of arithmetic applying to any set of non-numerical symbols (usually letters), and the axioms and rules by which they are combined or operated upon; sometimes known as *generalized arithmetic*.

The basics of algebra were familiar in Babylon 2000 BC, and were practised by the Arabs in the Middle Ages. In the 9th century, the Arab mathematician Muhammad ibn-Mūsā al-Khwārizmī first used the words *hisāb al-jabr* ('calculus of reduction') as part of the title of a treatise. Algebra is used in many branches of mathematics, for example, matrix algebra and Boolean algebra (the latter method was first devised in the 19th century by the British mathematician George Boole and used in working out the logic for computers).

alginate salt of alginic acid, $(C_6H_8O_6)n$, obtained from brown seaweeds and used in textiles, paper, food products, and pharmaceuticals.

ALGOL (acronym for *algo*rithmic *l*anguage) in computing, an early high-level programming language, developed in the 1950s and 1960s for scientific applications. A general-purpose language, ALGOL is best suited to mathematical work and has an algebraic style. Although no longer in common use, it has greatly influenced more recent languages, such as Ada and PASCAL.

Algol or *Beta Persei* ◊eclipsing binary, a pair of rotating stars in the constellation Perseus, one of which eclipses the other every 69 hours, causing its brightness to drop by two-thirds.

The brightness changes were first explained 1782 by English amateur astronomer John Goodricke (1764–1786).

Algonquin Radio Observatory site in Ontario, Canada, of the radio telescope, 46 m/150 ft in diameter, of the National Research Council of Canada, opened 1966.

algorithm procedure or series of steps that can be used to solve a problem. In computer science, it describes the logical sequence of operations to be performed by a program. A ◊flow chart is a visual representation of an algorithm.

The word derives from the name of the 9th-century Arab mathematician Muhammad ibn-Mūsā al-Khwārizmī.

alimentary canal in animals, the tube through which food passes; it extends from the mouth to the anus. It is a complex organ, adapted for ◊digestion. In human adults, it is about 9 m/30 ft long, consisting of the mouth cavity, pharynx, oesophagus, stomach, and the small and large intestines.

A constant stream of enzymes from the canal wall and from the pancreas assists the breakdown of food molecules into smaller, soluble nutrient molecules, which are absorbed through the canal wall into the bloodstream and carried to individual cells. The muscles of the alimentary canal keep the incoming food moving, mix it with the enzymes and other juices, and slowly push it in the direction of the anus, a process known as ◊peristalsis. The wall of the canal receives an excellent supply of blood and is folded so as to increase its surface area. These two adaptations ensure efficient absorption of nutrient molecules.

aliphatic compound any organic chemical compound in which the carbon atoms are joined in straight chains, as in hexane (C_6H_{14}), or in branched chains, as in 2–methylpentane ($CH_3CH(CH_3)CH_2CH_2CH_3$).

Aliphatic compounds have bonding electrons localized within the vicinity of the bonded atoms. ◊Cyclic compounds that do not have delocalized electrons are also aliphatic, as in the alicyclic compound cyclohexane (C_6H_{12}) or the heterocyclic piperidine ($C_5H_{11}N$). Compare ◊aromatic compound.

alkali in chemistry, a compound classed as a ◊base that is soluble in water. Alkalis neutralize acids and are soapy to the touch. The hydroxides of metals are alkalis; those of sodium and potassium being chemically powerful; both were historically derived from the ashes of plants.

The four main alkalis are sodium hydroxide (caustic soda, NaOH); potassium hydroxide (caustic potash, KOH); calcium hydroxide (slaked lime or limewater, $Ca(OH)_2$); and aqueous ammonia ($NH_{3(aq)}$). Their solutions all contain the hydroxide ion OH^-, which gives them a characteristic set of properties.

Alkalis react with acids to form a salt and water (neutralization).

$$KOH + HNO_3 \rightarrow KNO_3 + H_2O$$

$$OH^- + H^+ \rightarrow H_2O$$

They give a specific colour reaction with indicators; for example, litmus turns blue.

alkali metal any of a group of six metallic elements with similar chemical bonding properties: lithium, sodium, potassium, rubidium, caesium, and francium. They form a linked group in the ◊periodic table of the elements. They are univalent (have a valency of one) and of very low density (lithium, sodium, and potassium float on water); in general they are reactive, soft, low-melting-point metals. Because of their reactivity they are only found as compounds in nature.

alkaline-earth metal any of a group of six metallic elements with similar bonding properties: beryllium, magnesium, calcium, strontium, barium, and radium. They form a linked group in the ◊periodic table of the elements. They are strongly basic, bivalent (have a valency of two), and occur in nature only in compounds.

They and their compounds are used to make alloys, oxidizers, and drying agents.

alkaloid any of a number of physiologically active and frequently poisonous substances contained in some plants. They are usually organic bases and contain nitrogen. They form salts with acids and, when soluble, give alkaline solutions.

Substances in this group are included by custom rather than by scientific rules. Examples include morphine, cocaine, quinine, caffeine, strychnine, nicotine, and atropine.

In 1992, epibatidine, a chemical extracted from the skin of a Ecuadorian frog, was identified as a member of an entirely new class of alkaloid. It is an organochlorine compound, which is rarely found in animals, and a powerful pain killer, about 200 times as effective as morphine.

alkane member of a group of ◊hydrocarbons having the general formula C_nH_{2n+2}, commonly known as *paraffins*. Lighter alkanes, such as methane, ethane, propane, and butane, are colourless gases; heavier ones are liquids or solids. In nature they are found in natural gas and petroleum. As alkanes contain only single ◊covalent bonds, they are said to be saturated.

alkene member of the group of ◊hydrocarbons having the general formula C_nH_{2n}, formerly known as *olefins*. Lighter alkenes, such as ethene and propene, are gases, obtained from the ◊cracking of oil fractions. Alkenes are unsaturated compounds, characterized by one or more double bonds between adjacent carbon atoms. They react by addition, and many useful compounds, such as poly(ethene) and bromoethane, are made from them.

alkyne member of the group of ◊hydrocarbons with the general formula C_nH_{2n-2}, formerly known as the *acetylenes*. They are unsaturated compounds, characterized by one or more triple bonds between adjacent carbon atoms. Lighter alkynes, such as ethyne, are gases; heavier ones are liquids or solids.

allele one of two or more alternative forms of a ◊gene at a given position (locus) on a chromosome, caused by a difference in the ◊DNA. Blue and brown eyes in humans are determined by different alleles of the gene for eye colour.

Organisms with two sets of chromosomes (diploids) will have two copies of each gene. If the two alleles are identical the individual is said to be ◊homozygous at that locus; if different, the

name	molecular formula	structural formula
methane	CH_4	H | H–C–H | H

uses: domestic fuel (natural gas)

ethane	C_2H_6	H H | | H–C–C–H | | H H

uses: industrial fuel and chemical feedstock

propane	C_3H_8	H H H | | | H–C–C–C–H | | | H H H

uses: bottled gas (camping gas)

butane	C_4H_{10}	H H H H | | | | H–C–C–C–C–H | | | | H H H H

uses: bottled gas (lighter fuel, camping gas)

alkanes

individual is ◊heterozygous at that locus. Some alleles show ◊dominance over others.

allometry in biology, a regular relationship between a given feature (for example, the size of an organ) and the size of the body as a whole, when this relationship is not a simple proportion of body size. Thus, an organ may increase in size proportionately faster, or slower, than body size does. For example, a human baby's head is much larger in relation to its body than is an adult's.

The best-known allometric relationship is the *surface area law*: the ratio of body surface to total body volume decreases as body size gets larger. Large animals therefore lose less heat than small ones because they have proportionately less skin surface from which to radiate heat.

allotropy property whereby an element can exist in two or more forms (allotropes), each possessing different physical properties but the same state of matter (gas, liquid, or solid). The allotropes of carbon are diamond and graphite. Sulphur has several different forms (flowers of sulphur, plastic, rhombic, and monoclinic). These solids have different crystal structures, as do the white and grey forms of tin and the black, red, and white forms of phosphorus.

common alloys

name	approximate composition	uses
brass	20–45% zinc, 55–80% copper	decorative metalwork, plumbing fittings, industrial tubing
bronze		
common	2% zinc, 6% tin, 92% copper	machinery, decorative work
aluminium	10% aluminium, 90% copper	machinery castings
coinage	1% zinc, 4% tin, 95% copper	coins
cast iron	2–4% carbon, 96–98% iron	decorative metalwork, engine blocks, industrial machinery
dentist's amalgam	30% copper, 70% mercury	dental fillings
duralumin	0.5 % magnesium, 0.5% manganese, 4% copper, 95% aluminium	framework of aircraft
gold		
coinage	10% copper, 90% gold	coins
dental	14–28% silver, 14–28% copper, 58% gold	dental fillings
lead battery plate	6% antimony, 94% lead	car batteries
manganin	1.5% nickel, 16% manganese, 82.5% copper	resistance wire
nichrome	20% chromium, 80% nickel	heating elements
pewter	20% lead, 80% tin	utensils
silver		
coinage	10% copper, 90% silver	coins
solder	50% tin, 50% lead	joining iron surfaces
steel		
stainless	8–20% nickel, 11–20% chromium, 60–80% iron	kitchen utensils
tool	2–4% chromium, 6–7% molybdenum, 89–92% iron	tools

Oxygen exists as two gaseous allotropes: 'normal' oxygen (O_2) and ozone (O_3), which differ in their molecular configurations.

alloy metal blended with some other metallic or nonmetallic substance to give it special qualities, such as resistance to corrosion, greater hardness, or tensile strength. Useful alloys include bronze, brass, cupronickel, duralumin, German silver, gunmetal, pewter, solder, steel, and stainless steel.

Among the oldest alloys is bronze, whose widespread use ushered in the Bronze Age. Complex alloys are now common; for example, in dentistry, where a cheaper alternative to gold is made of chromium, cobalt, molybdenum, and titanium. Among the most recent alloys are superplastics: alloys that can stretch to double their length at specific temperatures, permitting, for example, their injection into moulds as easily as plastic.

alluvial deposit layer of broken rocky matter, or sediment, formed from material that has been carried in suspension by a river or stream and dropped as the velocity of the current changes. River plains and deltas are made entirely of alluvial deposits, but smaller pockets can be found in the beds of upland torrents.

Alluvial deposits can consist of a whole range of particle sizes, from boulders down through cobbles, pebbles, gravel, sand, silt, and clay. The raw materials are the rocks and soils of upland areas that are loosened by erosion and washed away by mountain streams. Much of the world's richest farmland lies on alluvial deposits. These deposits can also provide an economic source of minerals.

River currents produce a sorting action, with particles of heavy material deposited first while lighter materials are washed downstream. Hence heavy minerals such as gold and tin, present in the original rocks in small amounts, can be concentrated and deposited on stream beds in commercial quantities. Such deposits are called 'placer ores'.

alluvial fan roughly triangular sedimentary formation found at the base of slopes. An alluvial fan results when a sediment-laden stream or river rapidly deposits its load of gravel and silt as its speed is reduced on entering a plain.

The surface of such a fan slopes outward in a wide arc from an apex at the mouth of the steep valley. A small stream carrying a load of coarse particles builds a shorter, steeper fan than a large stream carrying a load of fine particles. Over time, the fan tends to become destroyed piecemeal by the continuing headward and downward erosion levelling the slope.

Almagest (Arabic *al* 'the' and Greek *majisti* 'greatest') book compiled by the Greek astronomer Ptolemy during the 2nd century AD, which included the idea of an Earth-centred universe. It survived in an Arabic translation. Some medieval books on astronomy, astrology, and alchemy were given the same title.

Each section of the book deals with a different branch of astronomy. The introduction describes the universe as spherical and contains arguments for the Earth being stationary at the centre. From this mistaken assumption, it goes on to describe the motions of the Sun, Moon, and planets;

eclipses; and the positions, brightness, and precession of the 'fixed stars'. The book drew on the work of earlier astronomers such as Hipparchus.

Alpha Centauri or *Rigil Kent* brightest star in the constellation Centaurus and the third brightest star in the sky. It is actually a triple star (see ◊binary star); the two brighter stars orbit each other every 80 years, and the third, Proxima Centauri, is the closest star to the Sun, 4.2 light years away, 0.1 light years closer than the other two.

alpha decay disintegration of the nucleus of an atom to produce an ◊alpha particle. See also ◊radioactivity.

alphanumeric data data made up of any of the letters of the alphabet and any digit from 0 to 9. The classification of data according to the type or types of character contained enables computer ◊validation systems to check the accuracy of data: a computer can be programmed to reject entries that contain the wrong type of character. For example, a person's name would be rejected if it contained any numeric data, and a bank-account number would be rejected if it contained any alphabetic data. A car's registration number, by comparison, would be expected to contain alphanumeric data but no punctuation marks.

alpha particle positively charged, high-energy particle emitted from the nucleus of a radioactive ◊atom. It is one of the products of the spontaneous disintegration of radioactive elements (see ◊radioactivity) such as radium and thorium, and is identical with the nucleus of a helium atom—that is, it consists of two protons and two neutrons. The process of emission, *alpha decay*, transforms one element into another, decreasing the atomic (or proton) number by two and the atomic mass (or nucleon number) by four.

Because of their large mass alpha particles have a short range of only a few centimetres in air, and can be stopped by a sheet of paper. They have a strongly ionizing effect (see ◊ionizing radiation) on the molecules that they strike, and are therefore capable of damaging living cells. Alpha particles travelling in a vacuum are deflected slightly by magnetic and electric fields.

Alps, Lunar mountain range on the Moon, NE of the Sea of Showers, cut by a valley 150 km/ 93 mi long.

Altair or *Alpha Aquilae* brightest star in the constellation Aquila and the 12th brightest star in the sky. It is a white star 16 light years away and forms the so-called Summer Triangle with the stars Deneb (in the constellation Cygnus) and Vega (in Lyra).

alternate angles a pair of angles that lie on opposite sides and at opposite ends of a transversal (a line that cuts two or more lines in the same plane). The alternate angles formed by a transversal of two parallel lines are equal.

alternating current (AC) electric current that flows for an interval of time in one direction and then in the opposite direction, that is, a current that flows in alternately reversed directions through or around a circuit. Electric energy is usually generated as alternating current in a power station, and

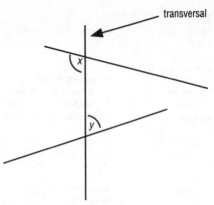

x and y are alternate angles

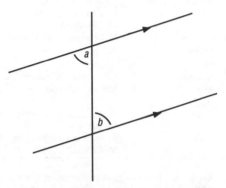

where a transversal cuts through a pair of parallel lines the alternate angles *a* and *b* are equal

alternate angles

alternating currents may be used for both power and lighting.

The advantage of alternating current over direct current (DC), as from a battery, is that its voltage can be raised or lowered economically by a transformer: high voltage for generation and transmission, and low voltage for safe utilization. Railways, factories, and domestic appliances, for example, use alternating current.

alternation of generations typical life cycle of terrestrial plants and some seaweeds, in which there are two distinct forms occurring alternately: *diploid* (having two sets of chromosomes) and *haploid* (one set of chromosomes). The diploid generation produces haploid spores by ◊meiosis, and is called the sporophyte, while the haploid generation produces gametes (sex cells), and is called the gametophyte. The gametes fuse to form a diploid ◊zygote which develops into a new sporophyte; thus the sporophyte and gametophyte alternate.

alternative energy see ◊energy, alternative

alternator electricity ◊generator that produces an alternating current.

altimeter instrument used in aircraft that measures altitude, or height above sea level. The

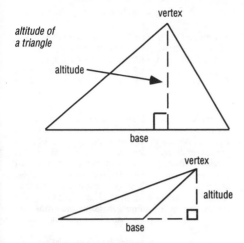

altitude of
a triangle

two altitudes of a quadrilateral

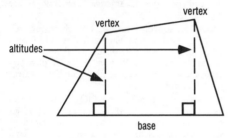

altitude
common type is a form of aneroid ◊barometer, which works by sensing the differences in air pressure at different altitudes. This must continually be recalibrated because of the change in air pressure with changing weather conditions. The ◊radar altimeter measures the height of the aircraft above the ground, measuring the time it takes for radio pulses emitted by the aircraft to be reflected. Radar altimeters are essential features of automatic and blind-landing systems.

altitude in geometry, the perpendicular distance from a ◊vertex (corner) of a figure, such as a triangle, to the base (the side opposite the vertex).

altitude measurement of height, usually given in metres above sea level.

altruism in biology, helping another individual of the same species to reproduce more effectively, as a direct result of which the altruist may leave fewer offspring itself. Female honey bees (workers) behave altruistically by rearing sisters in order to help their mother, the queen bee, reproduce, and forego any possibility of reproducing themselves.

ALU abbreviation for ◊*arithmetic and logic unit*.

alum any double sulphate of a monovalent metal or radical (such as sodium, potassium, or ammonium) and a trivalent metal (such as aluminium or iron). The commonest alum is the double sulphate of potassium and aluminium, $K_2Al_2(SO_4)_4.24H_2O$, a white crystalline powder that is readily soluble in water. It is used in curing animal skins. Other alums are used in papermaking and to fix dye in the textile industry.

alumina or *corundum* Al_2O_3 oxide of aluminium, widely distributed in clays, slates, and shales. It is formed by the decomposition of the feldspars in granite and used as an abrasive. Typically it is a white powder, soluble in most strong acids or caustic alkalis but not in water. Impure alumina is called 'emery'. Rubies and sapphires are corundum gemstones.

aluminium lightweight, silver-white, ductile and malleable, metallic element, symbol Al, atomic number 13, relative atomic mass 26.9815. It is the third most abundant element (and the most abundant metal) in the Earth's crust, of which it makes up about 8.1% by mass. It is an excellent conductor of electricity and oxidizes easily, the layer of oxide on its surface making it highly resistant to tarnish. In the USA the original name suggested by the British scientist Humphry Davy, 'aluminum', is retained.

Because of its rapid oxidation a great deal of energy is needed in order to separate aluminium from its ores, and the pure metal was not readily obtainable until the middle of the 19th century. Commercially, it is prepared by the electrolysis of ◊bauxite. In its pure state aluminium is a weak metal, but when combined with elements such as copper, silicon, or magnesium it forms alloys of great strength.

Because of its light weight (specific gravity 2.70), aluminium is widely used in the shipbuilding and aircraft industries. It is also used in making cooking utensils, cans for beer and soft drinks, and foil. It is much used in steel-cored overhead cables and for canning uranium slugs for nuclear reactors. Aluminium is an essential constituent in some magnetic materials; and, as a good conductor of electricity, is used as foil in electrical capacitors. A plastic form of aluminium, developed 1976, which moulds to any shape and extends to several times

*the extraction of aluminium from
bauxite by electrolysis*

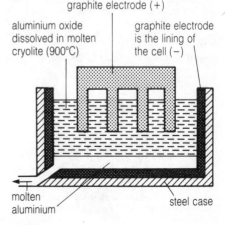

graphite electrode (+)

aluminium oxide
dissolved in molten
cryolite (900°C)

graphite electrode
is the lining of
the cell (−)

molten
aluminium

steel case

aluminium

its original length, has uses in electronics, cars, building construction, and so on.

Aluminium sulphate is the most widely used chemical in water treatment worldwide, but accidental excess (as at Camelford, N Cornwall, England, July 1989) makes drinking water highly toxic, and discharge into rivers kills all fish.

VALUABLE ALUMINIUM

In 1855 a Frenchman, Henri Sainte-Claire Deville, exhibited a shiny piece of aluminium in Paris. Napoleon III was so impressed that he asked Sainte-Claire to find a cheap way of producing the new metal. However, no easy way could be found and aluminium became more precious than gold. Napoleon is said to have had a set of aluminium cutlery made for his most honoured guests.

---

aluminium hydroxide or *alumina cream* Al(OH)$_3$ gelatinous ◊precipitate formed when a small amount of alkali solution is added to a solution of an aluminium salt.

It is an ◊amphoteric compound as it readily reacts with both acids and alkalis.

aluminium ore raw material from which aluminium is extracted. The main ore is bauxite, a mixture of minerals, found in economic quantities in Australia, Guinea, West Indies, and several other countries.

alveolus (plural *alveoli*) one of the many thousands of tiny air sacs in the ◊lungs in which exchange of oxygen and carbon dioxide takes place between air and the bloodstream.

AM abbreviation for ◊amplitude modulation.

amalgam any alloy of mercury with other metals. Most metals will form amalgams, except iron and platinum. Amalgam is used in dentistry for filling teeth, and usually contains copper, silver, and zinc as the main alloying ingredients. This amalgam is pliable when first mixed and then sets hard, but the mercury can leach out and may cause a type of heavy-metal poisoning.

Amalgamation, the process of forming an amalgam, is a technique sometimes used to extract gold and silver from their ores. The ores are treated with mercury, which combines with the precious metals.

amatol explosive consisting of ammonium nitrate

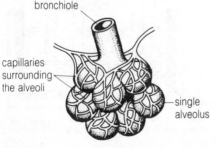

bronchiole

capillaries surrounding the alveoli

single alveolus

alveolus

and TNT (trinitrotoluene) in almost any proportions.

amber fossilized resin from coniferous trees of the Middle Tertiary period. It is often washed ashore on the Baltic coast with plant and animal specimens preserved in it; many extinct species have been found preserved in this way. It ranges in colour from red to yellow, and is used to make jewellery.

americium radioactive metallic element of the ◊actinide series, symbol Am, atomic number 95, relative atomic mass 243.13; it was first synthesized 1944. It occurs in nature in minute quantities in ◊pitchblende and other uranium ores, where it is produced from the decay of neutron-bombarded plutonium, and is the element with the highest atomic number that occurs in nature. It is synthesized in quantity only in nuclear reactors by the bombardment of plutonium with neutrons. Its longest-lived isotope is Am-243, with a half-life of 7,650 years.

The element was named by Glenn Seaborg, one of the team who first synthesized it in 1944, after the United States of America. Ten isotopes are known.

Ames Research Center US space-research (NASA) installation at Mountain View, California, USA, for the study of aeronautics and life sciences. It has managed the Pioneer series of planetary probes and is involved in the search for extra-terrestrial life.

amethyst variety of ◊quartz, SiO$_2$, coloured violet by the presence of small quantities of impurities such as manganese or iron; used as a semiprecious stone. Amethysts are found chiefly in the Ural Mountains, India, the USA, Uruguay, and Brazil.

amide any organic chemical derived from a fatty acid by the replacement of the hydroxyl group (–OH) by an amino group (–NH$_2$). One of the simplest amides is acetamide (CH$_3$CONH$_2$), which has a strong mousy odour.

amine any of a class of organic chemical compounds in which one or more of the hydrogen atoms of ammonia (NH$_3$) have been replaced by other groups of atoms.

Methyl amines have unpleasant ammonia odours and occur in decomposing fish. They are all gases at ordinary temperature. *Aromatic amine compounds* include aniline, which is used in dyeing.

amino acid water-soluble organic ◊molecule, mainly composed of carbon, oxygen, hydrogen, and nitrogen, containing both a basic amino group (NH$_2$) and an acidic carboxyl (COOH) group. When two or more amino acids are joined together, they are known as ◊peptides; ◊proteins are made up of interacting polypeptides (peptide chains consisting of more than three amino acids) and are folded or twisted in characteristic shapes.

Many different proteins are found in the cells of living organisms, but they are all made up of the same 20 amino acids, joined together in varying combinations (although other types of amino acid do occur infrequently in nature). Eight of these, the *essential amino acids*, cannot be synthesized by humans and must be obtained from the diet. Children need a further two amino acids that are

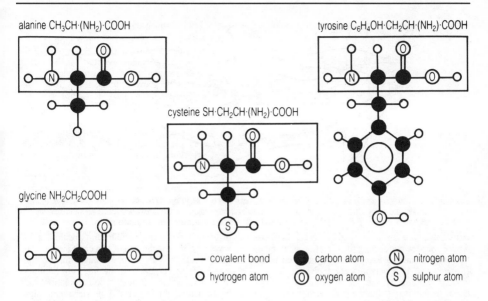

alanine $CH_3CH\cdot(NH_2)\cdot COOH$

tyrosine $C_6H_4OH\cdot CH_2CH\cdot(NH_2)\cdot COOH$

cysteine $SH\cdot CH_2CH\cdot(NH_2)\cdot COOH$

glycine NH_2CH_2COOH

— covalent bond   ● carbon atom   Ⓝ nitrogen atom

○ hydrogen atom   Ⓞ oxygen atom   Ⓢ sulphur atom

amino acid *Amino acids are natural organic compounds that make up proteins and can thus be considered the basic molecules of life. There are 20 different amino acids. They consist mainly of carbon, oxygen, hydrogen, and nitrogen. Each amino acid has a common core structure (consisting of two carbon atoms, two oxygen atoms, a nitrogen atom, and four hydrogen atoms) to which is attached a variable group, known as the R group. In glycine, the R group is a single hydrogen atom; in alanine, the R group consists of a carbon and three hydrogen atoms.*

not essential for adults. Other animals also need some preformed amino acids in their diet, but green plants can manufacture all the amino acids they need from simpler molecules, relying on energy mainly to produce the Sun and minerals (including nitrates) from the soil.

ammeter instrument that measures electric current, usually in ◊amperes.

ammonia NH_3 colourless pungent-smelling gas, lighter than air and very soluble in water. It is made on an industrial scale by the ◊Haber process, and used mainly to produce nitrogenous fertilizers, some explosives, and nitric acid.

In aquatic organisms and some insects, nitrogenous waste (from breakdown of amino acids and so on) is excreted in the form of ammonia, rather than urea as in mammals.

ammonite extinct marine ◊cephalopod mollusc of the order Ammonoidea, related to the modern nautilus. The shell was curled in a plane spiral and made up of numerous gas-filled chambers, the outermost containing the body of the animal. Many species flourished between 200 million and 65 million years ago, ranging in size from that of a small coin to 2 m/6 ft across.

ammonium NH_4^+ ion formed when ammonia accepts a proton H^+ from an acid. It is the only positive ion that is not a metal and that forms a series of salts. When heated with an alkali it produces ammonia gas.

Ammonium salts are a useful source of nitrogen for plants and are present in many fertilizer preparations.

ammonium carbonate $(NH_4)_2CO_3$ white, crystalline solid that readily sublimes at room temperature into its constituent gases: ammonia, carbon dioxide, and water. It was formerly used in ◊smelling salts.

ammonium chloride or *sal ammoniac* NH_4Cl a volatile salt that forms white crystals around volcanic craters. It is prepared synthetically for use in 'dry-cell' batteries, fertilizers, and dyes.

ammonium nitrate NH_4NO_3 colourless, crystalline solid, prepared by ◊neutralization of nitric acid with ammonia; the salt is crystallized from the solution. It sublimes on heating.

amnion innermost of three membranes that enclose the embryo within the egg (reptiles and birds) or within the uterus (mammals). It contains the amniotic fluid that helps to cushion the embryo.

amoeba (plural *amoebae*) one of the simplest living animals, consisting of a single cell and belonging to the ◊protozoa group. The body consists of colourless protoplasm. Its activities are controlled by the nucleus, and it feeds by flowing round and engulfing organic debris. It reproduces by ◊binary fission. Some species of amoeba are harmful parasites.

amp in physics, abbreviation for ampere, a unit of electrical current.

ampere SI unit (abbreviation amp, symbol A) of electrical current. Electrical current is measured in a similar way to water current, in terms of an amount per unit time; one ampere represents a flow of about 6.28×10^{18} ◊electrons per second, or a rate of flow of charge of one coulomb per second.

The ampere is defined as the current that produces a specific magnetic force between two long, straight, parallel conductors placed one metre (3.3 ft) apart in a vacuum. It is named after the French scientist André Ampère.

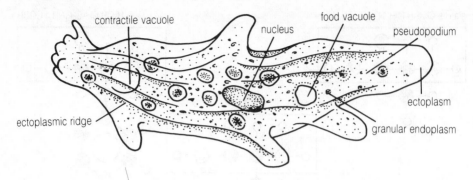

amoeba *The amoebae are among the simplest living organisms, consisting of a single cell. Within the cell, there is a nucleus, which controls cell activity, and many other microscopic bodies and vacuoles (fluid-filled spaces surrounded by a membrane) with specialized functions. Amoebae eat by flowing around food particles; engulfing the particle; a process called phagocytosis.*

Ampère's rule rule developed by French physicist André Ampère connecting the direction of an electric current and its associated magnetic currents. It states that if a person were travelling along a current-carrying wire in the direction of conventional current flow (from the positive to the negative terminal), and carrying a magnetic compass, then the north pole of the compass needle would be deflected to the left-hand side.

amphibian (Greek 'double life') member of the vertebrate class Amphibia, which generally spend their larval (tadpole) stage in fresh water, transferring to land at maturity (after ◊metamorphosis) and generally returning to water to breed. Like fish and reptiles, they continue to grow throughout life, and cannot maintain a temperature greatly differing from that of their environment. The class includes caecilians (wormlike in appearance), salamanders, frogs, and toads.

amphibole any one of a large group of rock-forming silicate minerals with an internal structure based on double chains of silicon and oxygen, and with a general formula $X_2Y_5Si_8O_{22}(OH)_2$; closely related to ◊pyroxene. Amphiboles form orthorhombic, monoclinic, and triclinic ◊crystals.

Amphiboles occur in a wide range of igneous and metamorphic rocks. Common examples are ◊hornblende (X = Ca, Y = Mg, Fe, Al) and tremolite (X = Ca, Y = Mg).

amphoteric term used to describe the ability of some chemical compounds to behave either as an ◊acid or as a ◊base depending on their environment. For example, the metals aluminium and zinc, and their oxides and hydroxides, act as bases in acidic solutions and as acids in alkaline solutions.

Amino acids and proteins are also amphoteric, as they contain both a basic (amino, $-NH_2$) and an acidic (carboxyl, $-COOH$) group.

amplifier electronic device that magnifies the strength of a signal, such as a radio signal. The ratio of output signal strength to input signal strength is called the *gain* of the amplifier. As well as achieving high gain, an amplifier should be free from distortion and able to operate over a range of frequencies. Practical amplifiers are usually complex circuits, although simple amplifiers can be built from single transistors or valves.

amplitude maximum displacement of an oscillation from the equilibrium position. For a wave motion, it is the height of a crest (or the depth of a trough). With a sound wave, for example, amplitude corresponds to the intensity (loudness) of the sound. In AM (amplitude modulation) radio broadcasting, the required audio-frequency signal is made to modulate (vary slightly) the amplitude of a continuously transmitted radio carrier wave.

amplitude or *argument* in mathematics, the angle in an ◊Argand diagram between the line that represents the complex number and the real (positive horizontal) axis. If the complex number is written in the form $r(\cos \theta + i \sin \theta)$, where r is radius and i = $\sqrt{-1}$, the amplitude is the angle $\tau\theta$).

amplitude modulation (AM) method by which radio waves are altered for the transmission of broadcasting signals. AM waves are constant in frequency, but the amplitude of the transmitting wave varies in accordance with the signal being broadcast.

ampulla in the inner ◊ear, a slight swelling at the end of each semicircular canal, able to sense the motion of the head. The sense of balance largely depends on sensitive hairs within the ampulla responding to movements of fluid within the canal.

amyl alcohol former name for ◊pentanol.

amylase one of a group of ◊enzymes that break down starches into their component molecules (sugars) for use in the body. It occurs widely in both plants and animals. In humans, it is found in saliva and in pancreatic juices.

anabatic wind warm wind that blows uphill in steep-sided valleys in the early morning. As the sides of a valley warm up in the morning the air above is also warmed and rises up the valley to give a gentle breeze. By contrast, a ◊katabatic wind is cool and blows down a valley at night.

anabolism process of building up body tissue, promoted by the influence of certain hormones. It is the constructive side of ◊metabolism, as opposed to catabolism.

anabranch (Greek *ana* 'again') stream that branches from a main river, then reunites with it. For example, the Great Anabranch in New South Wales, Australia, leaves the Darling near Menindee, and joins the Murray below the Darling–Murray confluence.

anaemia condition caused by a shortage of haemoglobin, the oxygen-carrying component of red blood cells. The main symptoms are fatigue, pallor, breathlessness, palpitations, and poor resistance to infection.

Anaemia arises either from abnormal loss or defective production of haemoglobin. Excessive loss occurs, for instance, with chronic slow bleeding or with accelerated destruction (◊haemolysis) of red blood cells. Defective production may be due to iron deficiency, vitamin B_{12} deficiency (pernicious anaemia), certain blood diseases (sickle-cell disease and thalassaemia), chronic infection, kidney disease, or certain kinds of poisoning. Untreated anaemia taxes the heart and may prove fatal.

anaerobic (of living organisms) not requiring oxygen for the release of energy from food molecules such as glucose. Anaerobic organisms include many bacteria, yeasts, and internal parasites.

Obligate anaerobes such as certain primitive bacteria cannot function in the presence of oxygen; but *facultative anaerobes*, like the fermenting yeasts and most bacteria, can function with or without oxygen. Anaerobic organisms release 19 times less of the available energy from their food than do ◊aerobic organisms.

In plants, yeasts, and bacteria, anaerobic respiration results in the production of alcohol and carbon dioxide, a process that is exploited by both the brewing and the baking industries (see ◊fermentation). Normally aerobic animal cells can respire anaerobically for short periods of time when oxygen levels are low, but are ultimately fatigued by the build-up of the lactic acid produced in the process. This is seen particularly in muscle cells during intense activity, when the demand for oxygen can outstrip supply (see ◊oxygen debt).

Although anaerobic respiration is a primitive and inefficient form of energy release, deriving from the period when oxygen was missing from the atmosphere, it can also be seen as an ◊adaptation. To survive in some habitats, such as the muddy bottom of a polluted river, an organism must be to a large extent independent of oxygen; such habitats are said to be *anoxic*.

anaesthetic drug that produces loss of sensation or consciousness; the resulting state is *anaesthesia*, in which the patient is insensitive to stimuli. Anaesthesia may also happen as a result of nerve disorder.

Ever since the first successful operation in 1846 on a patient rendered unconscious by ether, advances have been aimed at increasing safety and control. Sedatives may be given before the anaesthetic to make the process easier. The level and duration of unconsciousness are managed precisely. Where general anaesthesia may be inappropriate (for example, in childbirth, for a small procedure, or in the elderly), many other techniques are available. A topical substance may be applied to the skin or tissue surface; a local agent may be injected into the tissues under the skin in the area to be treated; or a regional block of sensation may be achieved by injection into a nerve. Spinal anaesthetic, such as epidural, is injected into the tissues surrounding the spinal cord, producing loss of feeling in the lower part of the body.

analgesic agent for relieving pain. Opiates alter the perception or appreciation of pain and are effective in controlling 'deep' visceral (internal) pain. Non-opiates, such as ◊aspirin, paracetamol, and NSAIDs (nonsteroidal anti-inflammatory drugs), relieve musculoskeletal pain and reduce inflammation in soft tissues.

Pain is felt when electrical stimuli travel along a nerve pathway, from peripheral nerve fibres to the brain via the spinal cord. An anaesthetic agent acts either by preventing stimuli from being sent (local), or by removing awareness of them (general). Analgesic drugs act on both.

Temporary or permanent analgesia may be achieved by injection of an anaesthetic agent into, or the severing of, a nerve. Implanted devices enable patients to deliver controlled electrical stimulation to block pain impulses. Production of the body's natural opiates, ◊endorphins, can be manipulated by techniques such as relaxation and biofeedback. However, for the severe pain of, for example, terminal cancer, opiate analgesics are required.

analogous in biology, term describing a structure that has a similar function to a structure in another organism, but not a similar evolutionary path. For example, the wings of bees and of birds have the same purpose—to give powered flight—but have different origins. Compare ◊homologous.

analogue (of a quantity or device) changing continuously; by contrast a ◊digital quantity or device varies in series of distinct steps. For example, an analogue clock measures time by means of a continuous movement of hands around a dial, whereas a digital clock measures time with a numerical display that changes in a series of discrete steps.

Most computers are digital devices. Therefore, any signals and data from a computer must be passed through a suitable ◊analogue-to-digital converter before they can be received and processed by computer. Similarly, output signals from digital computers must be passed through a digital-to-analogue converter before they can be received by an analogue device.

analogue computer computing device that performs calculations through the interaction of continuously varying physical quantities, such as voltages (as distinct from the more common ◊digital computer, which works with discrete quantities). An analogue computer is said to operate in real time (corresponding to time in the real world), and can therefore be used to monitor and control other events as they happen.

Although common in engineering since the 1920s, analogue computers are not general-purpose computers, but specialize in solving ◊differential calculus and similar mathematical problems. The earliest analogue computing device is thought to be the flat, or planispheric, astrolabe, which originated in about the 8th century.

analogue signal in electronics, current or voltage that conveys or stores information, and varies continuously in the same way as the information it represents (compare ◊digital signal). Analogue signals are prone to interference and distortion.

The bumps in the grooves of an LP (gramophone) record form a mechanical analogue of the

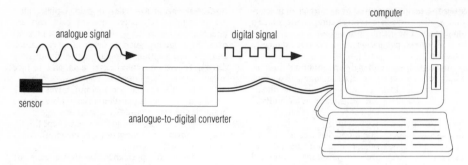

analogue signal

digital signal

computer

sensor

analogue-to-digital converter

analogue-to-digital converter

sound information stored, which is then is converted into an electrical analogue signal by the record player's pick-up device.

analogue-to-digital converter (ADC) electronic circuit that converts an analogue signal into a digital one. Such a circuit is needed to convert the signal from an analogue device into a digital signal for input into a computer. For example, many ◊sensors designed to measure physical quantities, such as temperature and pressure, produce an analogue signal in the form of voltage and this must be passed through an ADC before computer input and processing. A ◊digital-to-analogue converter performs the opposite process.

analysis in chemistry, the determination of the composition of substances; see ◊analytical chemistry.

analytical chemistry branch of chemistry that deals with the determination of the chemical composition of substances. *Qualitative analysis* determines the identities of the substances in a given sample; *quantitative analysis* determines how much of a particular substance is present.

Simple qualitative techniques exploit the specific, easily observable properties of elements or compounds—for example, the flame test makes use of the different flame-colours produced by metal cations when their compounds are held in a hot flame. More sophisticated methods, such as those of ◊spectroscopy, are required where substances are present in very low concentrations or where several substances have similar properties.

Most quantitative analyses involve initial stages in which the substance to be measured is extracted from the test sample, and purified. The final analytical stages (or 'finishes') may involve measurement of the substance's mass (gravimetry) or volume (volumetry, titrimetry), or a number of techniques initially developed for qualitative analysis, such as fluorescence and absorption spectroscopy, chromatography, electrophoresis, and polarography. Many modern methods enable quantification by means of a detecting device that is integrated into the extraction procedure (as in gas–liquid chromatography).

analytical engine programmable computing device designed by British mathematician Charles Babbage 1833. It was based on the ◊difference engine but was intended to automate the whole process of calculation. It introduced many of the

concepts of the digital computer but, because of limitations in manufacturing processes, was never built.

Among the concepts introduced were input and output, an arithmetic unit, memory, sequential operation, and the ability to make decisions based on data. It would have required at least 50,000 moving parts. The design was largely forgotten until some of Babbage's writings were rediscovered 1937.

analytical geometry another name for ◊coordinate geometry.

anatomy study of the structure of the body and its component parts, especially the ◊human body, as distinguished from physiology, which is the study of bodily functions.

Herophilus of Chalcedon (c. 330–c. 260 BC) is regarded as the founder of anatomy. In the 2nd century AD, the Graeco-Roman physician Galen produced an account of anatomy that was the only source of anatomical knowledge until *On the Working of the Human Body* 1543 by Belgian physician Andreas Vesalius. In 1628, English physician William Harvey published his demonstration of the circulation of the blood. Following the invention of the microscope, the Italian physiologist Marcello Malpighi and the Dutch microscopist Anton van Leeuwenhoek were able to found the study of histology (the laboratory study of cells and tissues). In 1747, Albinus (1697–1770), with the help of the artist Wandelaar (1691–1759), produced the most exact account of the bones and muscles, and in 1757–65 Swiss biologist Albrecht von Haller gave the most complete and exact description of the organs that had yet appeared. Among the anatomical writers of the early 19th century are the surgeon Charles Bell (1774–1842), Jonas Quain (1796–1865), and Henry Gray (1825–1861). Later in the century came stain techniques for microscopic examination, and the method of mechanically cutting very thin sections of stained tissues (using X-rays; see ◊radiography). Radiographic anatomy has been one of the triumphs of the 20th century, which has also been marked by immense activity in embryological investigation.

A human being: an ingenious assembly of portable plumbing.

On **anatomy** Christopher Morley (1890–1957)
Human Being 1932

andalusite aluminium silicate, Al_2SiO_5, a white to pinkish mineral crystallizing as square- or rhombus-based prisms. It is common in metamorphic rocks formed from clay sediments under low pressure conditions. Andalusite, kyanite, and sillimanite are all polymorphs of Al_2SiO_5.

andesite volcanic igneous rock, intermediate in silica content between rhyolite and basalt. It is characterized by a large quantity of feldspar ◊minerals, giving it a light colour. Andesite erupts from volcanoes at destructive plate margins (where one plate of the Earth's surface moves beneath another; see ◊plate tectonics), including the Andes, from which it gets its name.

AND gate in electronics, a type of ◊logic gate.

androecium male part of a flower, comprising a number of ◊stamens.

androgen general name for any male sex hormone, of which ◊testosterone is the most important. They are all ◊steroids and are principally involved in the production of male ◊secondary sexual characteristics (such as facial hair in humans).

Andromeda major constellation of the northern hemisphere, visible in autumn. Its main feature is the Andromeda galaxy. The star Alpha Andromedae forms one corner of the Square of Pegasus. It is named after the princess of Greek mythology.

Andromeda galaxy galaxy 2.2 million light years away from Earth in the constellation Andromeda, and the most distant object visible to the naked eye. It is the largest member of the ◊Local Group of galaxies. Like the Milky Way, it is a spiral galaxy, orbited by several companion galaxies but contains about twice as many stars. It is about 200,000 light years across.

ANDROMEDA GALAXY: A QUESTION OF SCALE

If our Sun were 2.5 cm/1 in across, the star nearest to Earth would be 716 km/445 mi away. On this scale, the most distant star in our galaxy would be over 13 million km/8 million mi away. The nearest large neighbouring galaxy, the Andromeda galaxy, would be 360 million km/227 million mi away.

---

anechoic chamber room designed to be of high sound absorbency. All surfaces inside the chamber are covered by sound-absorbent materials such as rubber. The walls are often covered with inward-facing pyramids of rubber, to minimize reflections. It is used for experiments in ◊acoustics and for testing audio equipment.

anemometer device for measuring wind speed and liquid flow. The most basic form, the *cup-type anemometer*, consists of cups at the ends of arms, which rotate when the wind blows. The speed of rotation indicates the wind speed.

Vane-type anemometers have vanes, like a small windmill or propeller, that rotate when the wind blows. *Pressure-tube anemometers* use the pressure generated by the wind to indicate speed. The wind blowing into or across a tube develops a pressure, proportional to the wind speed, that is measured by a manometer or pressure gauge. *Hot-wire anemometers* work on the principle that the rate at which heat is transferred from a hot wire to the surrounding air is a measure of the air speed. Wind speed is determined by measuring either the electric current required to maintain a hot wire at a constant temperature, or the variation of resistance while a constant current is maintained.

anemophily type of ◊pollination in which the pollen is carried on the wind. Anemophilous flowers are usually unscented, have either very reduced petals and sepals or lack them altogether, and do not produce nectar. In some species they are borne in ◊catkins. Male and female reproductive structures are commonly found in separate flowers. The male flowers have numerous exposed stamens, often on long filaments; the female flowers have long, often branched, feathery stigmas.

Many wind-pollinated plants, such as hazel *Corylus avellana*, bear their flowers before the leaves to facilitate the free transport of pollen. Since air movements are random, vast amounts of pollen are needed: a single birch catkin, for example, may produce over 5 million pollen grains.

aneroid barometer kind of ◊barometer.

angiosperm flowering plant in which the seeds are enclosed within an ovary, which ripens to a fruit. Angiosperms are divided into ◊monocotyledons (single seed leaf in the embryo) and ◊dicotyledons (two seed leaves in the embryo). They include the majority of flowers, herbs, grasses, and trees except conifers.

There are over 250,000 different species of angiosperm, found in a wide range of habitats. Like ◊gymnosperms, they are seed plants, but differ in that ovules and seeds are protected within the carpel. Fertilization occurs by male gametes passing into the ovary from a pollen tube. After fertilization the ovule develops into the seed while the ovary wall develops into the fruit.

angle in mathematics, the amount of turn or rotation; it may be defined by a pair of rays (half-lines) that share a common endpoint but do not lie on the same line. Angles are measured in ◊degrees (°) or ◊radians (rads)—a complete turn or circle being $360°$ or 2π rads.

Angles are classified generally by their degree measures: *acute angles* are less than $90°$; *right angles* are exactly $90°$ (a quarter turn); *obtuse angles* are greater than $90°$ but less than $180°$; *reflex angles* are greater than $180°$ but less than $360°$.

angle of declination angle at a particular point on the Earth's surface between the direction of the true or geographic North Pole and the magnetic north pole. The angle of declination has varied over time because of the slow drift in the position of the magnetic north pole.

angstrom unit (symbol Å) of length equal to 10^{-10} metre or one-ten-millionth of a millimetre, used for atomic measurements and the wavelengths of electromagnetic radiation. It is named after the Swedish scientist Anders Jonas Ångström.

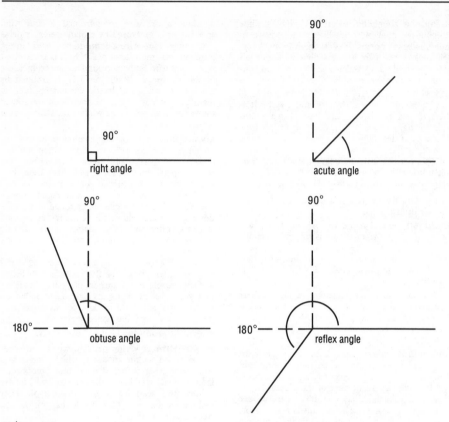

angle

angular momentum see ◊momentum.

anhydride chemical compound obtained by the removal of water from another compound; usually a dehydrated acid. For example, sulphur(VI) oxide (sulphur trioxide, SO_3) is the anhydride of sulphuric acid (H_2SO_4).

anhydrite naturally occurring anhydrous calcium sulphate ($CaSO_4$). It is used commercially for the manufacture of plaster of Paris and builders' plaster.

anhydrous of a chemical compound, containing no water. If the water of crystallization is removed

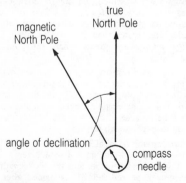

angle of declination

from blue crystals of copper(II) sulphate, a white powder (anhydrous copper sulphate) results. Liquids from which all traces of water have been removed are also described as being anhydrous.

aniline (Portuguese *anil* 'indigo') or *phenylamine* $C_6H_5NH_2$ one of the simplest aromatic chemicals (a substance related to benzene, with its carbon atoms joined in a ring). When pure, it is a colourless oily liquid; it has a characteristic odour, and turns brown on contact with air. It occurs in coal tar, and is used in the rubber industry and to make drugs and dyes. It is highly poisonous.

Aniline was discovered in 1826, and was originally prepared by the dry distillation of indigo, hence its name.

animal or *metazoan* member of the kingdom Animalia, one of the major categories of living things, the science of which is *zoology*. Animals are all ◊heterotrophs (they obtain their energy from organic substances produced by other organisms); they have eukaryotic cells (the genetic material is contained within a distinct nucleus) bounded by a thin cell membrane rather than the thick cell wall of plants. Most animals are capable of moving around for at least part of their life cycle.

The animal kingdom is divided into 18 major groups, or phyla. The large phylum Chordata (animals that have, at some time in their life, a notochord, or stiff rod of cells running along the length of their body) is subdivided into three subphyla, including the Vertebrata (animals with backbones).

In the past, it was common to include the single-celled ◊protozoa with the animals, but these are now classified as protists, together with single-celled plants. Thus all animals are multicellular. The oldest land animals known date back 440 million years. Their remains were found 1990 in a sandstone deposit near Ludlow, Shropshire, UK and included fragments of two centipedes a few centimetres long and a primitive spider measuring about 1 mm.

animal behaviour scientific study of the behaviour of animals, either by comparative psychologists (with an interest mainly in the psychological processes involved in the control of behaviour) or by ethologists (with an interest in the biological context and relevance of behaviour).

anion ion carrying a negative charge. An electrolyte, such as the salt zinc chloride ($ZnCl_2$), is dissociated in aqueous solution or in the molten state into doubly-charged Zn^{2+} zinc ◊cations and singly-charged Cl^- anions. During electrolysis, the zinc cations flow to the cathode (to become discharged and liberate zinc metal) and the chloride anions flow to the anode.

annealing process of heating a material (usually glass or metal) for a given time at a given temperature, followed by slow cooling, to increase ductility and strength. It is a common form of ◊heat treatment.

Ductile metals hardened by cold working may be softened by annealing. Thus thick wire may be annealed before being drawn into fine wire. Owing to internal stresses, glass objects made at high temperatures can break spontaneously as they cool unless they are annealed. Annealing releases the stresses in a controlled way and, for optical purposes, also improves the optical properties of the glass.

annelid any segmented worm of the phylum Annelida. Annelids include earthworms, leeches, and marine worms such as lugworms.

They have a distinct head and soft body, which is divided into a number of similar segments shut off from one another internally by membranous partitions, but there are no jointed appendages.

annihilation in nuclear physics, a process in which a particle and its 'mirror image' particle called an *antiparticle* collide and disappear, with the creation of a burst of energy. The energy created is equivalent to the mass of the colliding particles in accordance with the ◊mass–energy equation. For example, an electron and a positron annihilate to produce a burst of high-energy X-rays.

Not all particle-antiparticle interactions result in annihilation; the exception concerns the group called ◊mesons, which are composed of ◊quarks and their antiquarks. See ◊antimatter.

annual plant plant that completes its life cycle within one year, during which time it germinates, grows to maturity, bears flowers, produces seed, and then dies.

Examples include the common poppy *Papaver rhoeas* and groundsel *Senecio vulgaris*. Among garden plants, some that are described as 'annuals'

are actually ◊perennials, although usually cultivated as annuals because they cannot survive winter frosts. See also ◊ephemeral plant and ◊biennial plant.

annual rings or *growth rings* concentric rings visible on the wood of a cut tree trunk or other woody stem. Each ring represents a period of growth when new ◊xylem is laid down to replace tissue being converted into wood (secondary xylem). The wood formed from xylem produced in the spring and early summer has larger and more numerous vessels than the wood formed from xylem produced in autumn when growth is slowing down. The result is a clear boundary between the pale spring wood and the denser, darker autumn wood. Annual rings may be used to estimate the age of the plant (see ◊dendrochronology), although occasionally more than one growth ring is produced in a given year.

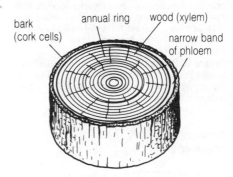

bark
(cork cells)   annual ring   wood (xylem)
                               narrow band
                               of phloem

annual rings

annular eclipse solar ◊eclipse in which the Moon does not completely obscure the Sun and a thin ring of sunlight remains visible. Annular eclipses occur when the Moon is at its furthest point from the Earth.

annulus (Latin 'ring') in geometry, the plane area between two concentric circles, making a flat ring.

anode in chemistry, the positive electrode of an electrolytic ◊cell, towards which negative particles (anions), usually in solution, are attracted. See ◊electrolysis.

An anode is given its positive charge by the application of an external electrical potential, unlike the positive electrode of an electrical (battery) cell, which acquires its charge in the course of a spontaneous chemical reaction taking place within the cell.

anodizing process that increases the resistance to ◊corrosion of a metal, such as aluminium, by building up a protective oxide layer on the surface. The natural corrosion resistance of aluminium is provided by a thin film of aluminium oxide; anodizing increases the thickness of this film and thus the corrosion protection.

It is so called because the metal becomes the ◊anode in an electrolytic bath containing a solution of, for example, sulphuric or chromic acid as the ◊electrolyte. During ◊electrolysis oxygen is produced at the anode, where it combines with the metal to form an oxide film.

anomalous expansion of water expansion of water as it is cooled from 4°C to 0°C. This behaviour is unusual because most substances contract when they are cooled. It means that water has a greater density at 4°C than at 0°C. Hence ice floats on water, and the water at the bottom of a pool in winter is warmer than at the surface. As a result large lakes freeze slowly in winter and aquatic life is more likely to survive.

anoxia in biology, deprivation of oxygen, a condition that rapidly leads to collapse or death, unless immediately reversed.

ANSI (abbreviation for *American National Standards Institute*) US national standards body. It sets official procedures in (among other areas) computing and electronics.

antagonistic muscles in the body, a pair of muscles allowing coordinated movement of the skeletal joints. The extension of the arm, for example, requires one set of muscles to relax, while another set contracts. The individual components of antagonistic pairs can be classified into extensors (muscles that straighten a limb) and flexors (muscles that bend a limb).

Antarctic Circle imaginary line that encircles the South Pole at latitude 66° 32′ S. The line encompasses the continent of Antarctica and the Antarctic Ocean.

The region south of this line experiences at least one night in the southern summer during which the Sun never sets, and at least one day in the southern winter during which the Sun never rises. See also ◊Arctic Circle.

Antares or *Alpha Scorpii* brightest star in the constellation Scorpius and the 15th brightest star in the sky. It is a red supergiant several hundred times larger than the Sun and perhaps 10,000 times as luminous, lies about 300 light years away, and fluctuates slightly in brightness.

antenna in zoology, an appendage ('feeler') on the head. Insects, centipedes, and millipedes each have one pair of antennae but there are two pairs in crustaceans, such as shrimps. In insects, the antennae are involved with the senses of smell and touch; they are frequently complex structures with large surface areas that increase the ability to detect scents.

antenna in radio and television, another name for ◊aerial.

anterior in biology, the front of an organism, usually the part that goes forward first when the animal is moving. The anterior end of the nervous system, over the course of evolution, has developed into a brain with associated receptor organs able to detect stimuli including light and chemicals.

human
arm raised

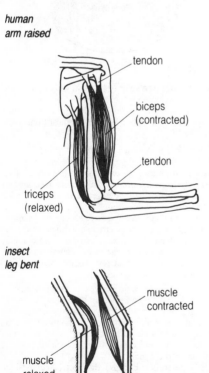

tendon

biceps
(contracted)

tendon

triceps
(relaxed)

arm lowered

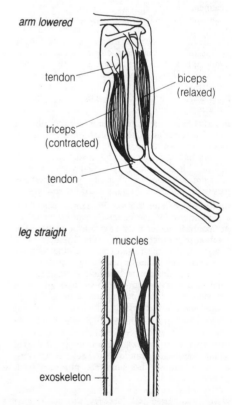

tendon

biceps
(relaxed)

triceps
(contracted)

tendon

insect
leg bent

muscle
contracted

muscle
relaxed

leg straight

muscles

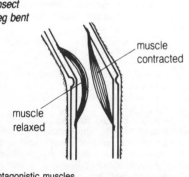

exoskeleton

antagonistic muscles

anther in a flower, the terminal part of a stamen in which the ◊pollen grains are produced. It is usually borne on a slender stalk or filament, and has two lobes, each containing two chambers, or pollen sacs, within which the pollen is formed.

antheridium organ producing the male gametes, ◊antherozoids, in algae, bryophytes (mosses and liverworts), and pteridophytes (ferns, club mosses, and horsetails). It may be either single-celled, as in most algae, or multicellular, as in bryophytes and pteridophytes.

antherozoid motile (or independently moving) male gamete produced by algae, bryophytes (mosses and liverworts), pteridophytes (ferns, club mosses, and horsetails), and some gymnosperms (notably the cycads). Antherozoids are formed in an antheridium and, after being released, swim by means of one or more ◊flagella, to the female gametes. Higher plants have nonmotile male gametes contained within ◊pollen grains.

anthracene white, glistening, crystalline, tricyclic, aromatic hydrocarbon with a faint blue fluorescence when pure. Its melting point is about 216°C/421°F and its boiling point 664°F/351°C. It occurs in the high-boiling-point fractions of coal tar, where it was discovered 1832 by the French chemists Auguste Laurent (1808–1853) and Jean Dumas (1800–1884).

anthracite (from Greek *anthrax*, 'coal') hard, dense, shiny variety of ◊coal, containing over 90% carbon and a low percentage of ash and impurities, which causes it to burn without flame, smoke, or smell.

Anthracite gives intense heat, but is slow-burning and slow to light; it is therefore unsuitable for open fires. Its characteristic composition is thought to be due to the action of bacteria in disintegrating the coal-forming material when it was laid down during the ◊Carboniferous period.

Among the chief sources of anthracite coal are Pennsylvania in the USA; S Wales, UK; the Donbas, Ukraine and Russia; and Shanxi province, China.

anthropic principle in science, the idea that 'the universe is the way it is because if it were different we would not be here to observe it'. The principle arises from the observation that if the laws of science were even slightly different, it would have been impossible for intelligent life to evolve. For example, if the electric charge on the electron were only slightly different, stars would have been unable to burn hydrogen and produce the chemical elements that make up our bodies. Scientists are undecided whether the principle is an insight into the nature of the universe or a piece of circular reasoning.

anthropology (Greek *anthropos* 'man' and *logos* 'discourse') study of humankind, which developed following 19th-century evolutionary theory to investigate the human species, past and present, physically, socially, and culturally.

The four subdisciplines are physical anthropology, linguistics, cultural anthropology, and archaeology.

anthropometry science dealing with the measurement of the human body, particularly stature, body weight, cranial capacity, and length of limbs, in samplings of the populations of living peoples, as well as the remains of buried and fossilized humans.

antibiotic drug that kills or inhibits the growth of bacteria and fungi. It is derived from living organisms such as fungi or bacteria, which distinguishes it from synthetic antimicrobials.

The earliest antibiotics, the ◊penicillins, came into use from 1941 and were quickly joined by chloramphenicol, the cephalosporins, erythromycins, tetracyclines, and aminoglycosides. A range of broad-spectrum antibiotics, the 4–quinolones, was developed 1989, of which ciprofloxacin was the first. Each class and individual antibiotic acts in a different way and may be effective against either a broad spectrum or a specific type of disease-causing agent. Use of antibiotics has become more selective as side effects, such as toxicity, allergy, and resistance, have become better understood. Bacteria have the ability to develop immunity following repeated or subclinical (insufficient) doses, so more advanced and synthetic antibiotics are continually required to overcome them.

antibody protein molecule produced in the blood by ◊lymphocytes in response to the presence of foreign or invading substances (◊antigens); such substances include the proteins carried on the surface of infecting microorganisms. Antibody production is only one aspect of ◊immunity in vertebrates.

Each antibody acts against only one kind of antigen, and combines with it to form a 'complex'. This action may render antigens harmless, or it may destroy microorganisms by setting off chemical changes that cause them to self-destruct. In other cases, the formation of a complex will cause antigens to form clumps that can then be detected and engulfed by white blood cells, such as ◊macrophages and ◊phagocytes.

Each bacterial or viral infection will bring about the manufacture of a specific antibody, which will then fight the disease. Many diseases can only be contracted once because antibodies remain in the blood after the infection has passed, preventing any further invasion. Vaccination boosts a person's resistance by causing the production of antibodies specific to particular infections.

Large quantities of specific antibodies can now be obtained by the monoclonal technique (see ◊monoclonal antibody). In 1989 a Cambridge University team developed genetically engineered bacteria to make a small part of an antibody (single domain antibodies) which bind to invaders such as toxins, bacteria, and viruses. Since they are smaller, they penetrate tissues more easily, and are potentially more effective in clearing organs of toxins. They can be produced more quickly, using fewer laboratory mice, and unlike conventional antibodies, they also disable viruses. In addition, single domain antibodies can be used to highlight other molecules, such as hormones in pregnancy testing.

anticholinergic any drug that blocks the passage of certain nerve impulses in the ◊central nervous system by inhibiting the production of ◊acetylcholine, a neurotransmitter.

anticline in geology, a fold in the rocks of the

Earth's crust in which the layers or beds bulge upwards to form an arch (seldom preserved intact).

The fold of an anticline may be undulating or steeply curved. A steplike bend in otherwise gently dipping or horizontal beds is a *monocline*. The opposite of an anticline is a *syncline*.

anticoagulant substance that suppresses the formation of blood clots. Common anticoagulants are heparin, produced by the liver and lungs, and derivatives of coumarin. Anticoagulants are used medically in treating heart attacks, for example. They are also produced by blood-feeding animals, such as mosquitoes, leeches, and vampire bats, to keep the victim's blood flowing.

Most anticoagulants prevent the production of thrombin, an enzyme that induces the formation from blood plasma of fibrinogen, to which blood platelets adhere and form clots.

anticyclone area of high atmospheric pressure caused by descending air, which becomes warm and dry. Winds radiate from a calm centre, taking a clockwise direction in the northern hemisphere and an anticlockwise direction in the southern hemisphere. Anticyclones are characterized by clear weather and the absence of rain and violent winds. In summer they bring hot, sunny days and in winter they bring fine, frosty spells, although fog and low cloud are not uncommon in the UK. *Blocking anticyclones*, which prevent the normal air circulation of an area, can cause summer droughts and severe winters.

For example, the summer drought in Britain 1976, and the severe winters of 1947 and 1963 were caused by blocking anticyclones.

antiferromagnetic material material with a very low magnetic ◊susceptibility that increases with temperature up to a certain temperature, called the ◊Néel temperature. Above the Néel temperature, the material becomes paramagnetic. See ◊magnetism.

Certain inorganic substances, such as manganese(II) oxide (MnO), iron(II) oxide (FeO), and manganese(II) sulphide (MnS), are antiferromagnetic.

antifreeze substance added to a water-cooling system (for example, that of a car) to prevent it freezing in cold weather.

The most common types of antifreeze contain the chemical ethylene ◊glycol, or (HOCH$_2$.CH$_2$OH), an organic alcohol with a freezing point of about $-15°C/5°F$.

The addition of this chemical depresses the freezing point of water significantly. A solution containing 33.5% by volume of ethylene glycol will not freeze until about $-20°C/-4°F$. A 50% solution will not freeze until $-35°C/-31°F$.

antigen any substance that causes the production of ◊antibodies by the body's immune system. Common antigens include the proteins carried on the surface of bacteria, viruses, and pollen grains. The proteins of incompatible blood groups or tissues also act as antigens, which has to be taken into account in medical procedures such as blood transfusions and organ transplants.

antiknock substance added to petrol to reduce knocking in car engines. It is a mixture of dibromoethane and tetraethyl lead.

Its use in leaded petrol has resulted in atmospheric pollution by lead compounds. Children exposed to this form of pollution over long periods of time can suffer impaired learning ability. Unleaded petrol has been used in the USA for some years, and is increasingly popular in the UK. Leaded petrol cannot be used in cars fitted with ◊catalytic converters.

antilogarithm or *antilog* the inverse of ◊logarithm, or the number whose logarithm to a given base is a given number. If $y = \log_a x$, then $x =$ antilog$_a$ y.

antimatter in physics, a form of matter in which most of the attributes (such as electrical charge, magnetic moment, and spin) of ◊elementary particles are reversed. Such particles (◊antiparticles) can be created in particle accelerators, such as those at ◊CERN in Geneva, Switzerland, and at ◊Fermilab in the USA.

antimony silver-white, brittle, semimetallic element (a metalloid), symbol Sb (from Latin *stibium*), atomic number 51, relative atomic mass 121.75. It occurs chiefly as the ore stibnite, and is used to make alloys harder; it is also used in photosensitive substances in colour photography, optical electronics, fireproofing, pigment, and medicine. It was employed by the ancient Egyptians in a mixture to protect the eyes from flies.

antinode in physics, the position in a ◊standing wave pattern at which the amplitude of vibration is greatest (compare ◊node). The standing wave of a stretched string vibrating in the fundamental mode has one antinode at its midpoint. A vibrating air column in a pipe has an antinode at the pipe's open end and at the place where the vibration is produced.

antioxidant any substance that prevents deterioration by oxidation in fats, oils, paints, plastics, and rubbers. When used as food ◊additives, antioxidants prevent fats and oils from becoming rancid when exposed to air, and thus extend their shelf life. They have ◊E numbers between 300 and 400.

Vegetable oils contain natural antioxidants, such as vitamin E, which prevent spoilage, but antioxidants are nevertheless added to most oils. They are not always listed on food labels because if a food manufacturer buys an oil to make a food product, and the oil has antioxidant already added, it does not have to be listed on the label of the product.

antiparticle in nuclear physics, a particle corresponding in mass and properties to a given ◊elementary particle but with the opposite electrical charge, magnetic properties, or coupling to other fundamental forces. For example, an electron carries a negative charge whereas its antiparticle, the positron, carries a positive one. When a particle and its antiparticle collide, they destroy each other, in the process called 'annihilation', their total energy being converted to lighter particles and/or photons. A substance consisting entirely of antiparticles is known as ◊antimatter.

Other antiparticles include the negatively charged antiproton and the antineutron.

antiseptic any substance that kills or inhibits the growth of microorganisms. The use of antiseptics was pioneered by English surgeon Joseph Lister. He used carbolic acid (◊phenol), which is a weak antiseptic; substances such as TCP are derived from this.

anus opening at the end of the alimentary canal that allows undigested food and associated materials to pass out of an animal. It is found in all types of multicellular animal except the coelenterates (sponges) and the platyhelminthes (flatworms), which have a mouth only.

aorta the chief ◊artery, the dorsal blood vessel carrying oxygenated blood from the left ventricle of the heart in birds and mammals. It branches to form smaller arteries, which in turn supply all body organs except the lungs. Loss of elasticity in the aorta provides evidence of atherosclerosis (thickening of the walls due to deposits of fatty material), which may lead to heart disease.

In fish a ventral aorta carries deoxygenated blood from the heart to the ◊gills, and the dorsal aorta carries oxygenated blood from the gills to other parts of the body.

a.p. in physics, abbreviation for *atmospheric pressure*.

Apache Point Observatory US observatory in the Sacramento Mountains of New Mexico containing a 3.5–m/138–in reflector, opened 1991, and operated by the Astrophysical Research Consortium (the universities of Washington, Chicago, Princeton, New Mexico, and Washington State).

apastron the point at which an object travelling in an elliptical orbit around a star is at its furthest from the star. The term is usually applied to the position of the minor component of a ◊binary star in relation to the primary. Its opposite is ◊periastron.

apatite common calcium phosphate mineral, $Ca_5(PO_4)_3(F,OH,Cl)$. Apatite has a hexagonal structure and occurs widely in igneous rocks, such as pegmatite, and in contact metamorphic rocks, such as marbles. It is used in the manufacture of fertilizer and as a source of phosphorus. Apatite is the chief constituent of tooth enamel while hydroxyapatite, $Ca_5(PO_4)_3(OH)_2$, is the chief inorganic constituent of bone marrow. Apatite ranks 5 on the ◊Mohs' scale of hardness.

apatosaurus large plant-eating dinosaur, formerly called *brontosaurus*, which flourished about 145 million years ago. Up to 21 m/69 ft long and 30 tonnes in weight, it stood on four elephantlike legs and had a long tail, long neck, and small head. It probably snipped off low-growing vegetation with peglike front teeth, and swallowed it whole to be ground by pebbles in the stomach.

Ape City Yerkes Regional Primate Center, Atlanta, Georgia, where large numbers of primates are kept for physiological and psychological experiment. A major area of research at Ape City is language.

aperture in photography, an opening in the camera that allows light to pass through the lens to strike the film. Controlled by shutter speed and the iris diaphragm, it can be set mechanically or electronically at various diameters.

aphelion the point at which an object, travelling in an elliptical orbit around the Sun, is at its furthest from the Sun.

apogee the point at which an object, travelling in an elliptical orbit around the Earth, is at its furthest from the Earth.

Apollo asteroid member of a group of ◊asteroids whose orbits cross that of the Earth. They are named after the first of their kind, Apollo, discovered 1932 and then lost until 1973. Apollo asteroids are so small and faint that they are difficult to see except when close to Earth (Apollo is about 2 km/1.2 mi across).

Apollo asteroids can collide with the Earth from time to time. In Jan 1991 the Apollo asteroid 1991 BA passed 170,000 km/105,000 mi from Earth, the closest observed approach of any asteroid. A collision with an Apollo asteroid 65 million years ago has been postulated as one of the causes of the extinction of the dinosaurs. A closely related group, the Amor asteroids, come close to Earth but do not cross its orbit.

Apollo project US space project to land a person on the Moon, achieved 20 July 1969, when Neil Armstrong was the first to set foot there. He was accompanied on the Moon's surface by Edwin E Aldrin Jr; Michael Collins remained in the orbiting command module.

The programme was announced 1961 by President Kennedy. The world's most powerful rocket, *Saturn V*, was built to launch the Apollo spacecraft, which carried three astronauts. When the spacecraft was in orbit around the Moon, two astronauts would descend to the surface in a lunar module to take samples of rock and set up experiments that would send data back to Earth. The first Apollo mission carrying a crew, *Apollo 7*, Oct 1968, was a test flight in orbit around the Earth. After three other preparatory flights, *Apollo 11* made the first lunar landing. Five more crewed landings followed, the last 1972. The total cost of the programme was over $24 billion.

Apollo–Soyuz test project joint US–Soviet space mission in which an Apollo and a Soyuz craft docked while in orbit around the Earth on 17 July 1975. The craft remained attached for two days and crew members were able to move from one craft to the other through an airlock attached to the nose of the Apollo. The mission was designed to test rescue procedures as well as having political significance.

APOLLO PROJECT: TRAVELLERS TO THE MOON

24 Americans have travelled to the Moon; 22 Americans have orbited the Moon; 12 Americans have walked on the Moon: Neil Armstrong, 'Buzz' Aldrin, Charles Conrad Jr, Alan L Bean, Alan Shepard, Edgar D Mitchell, David R Scott, James B Irwin, John W Young, Charles M Duke Jr, Eugene A Cernan, and Harrison S Schmitt; 3 Americans have been to the Moon twice: James A Lovell, Young, and Cernan; 2 Americans have orbited the Moon twice: Young and Cernan.

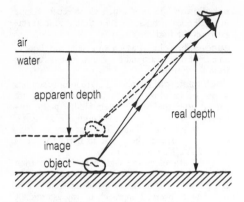

air

water

apparent depth

real depth

image

object

apparent depth

aposematic coloration in biology, the technical name for ◊warning coloration markings that make a dangerous, poisonous, or foul-tasting animal particularly conspicuous and recognizable to a predator. Examples include the yellow and black stripes of bees and wasps, and the bright red or yellow colours of many poisonous frogs. See also ◊mimicry.

apothecaries' weights obsolete units of mass, formerly used in pharmacy: 20 grains made one scruple; three scruples made one drachm; eight drachms made an apothecary's ounce (oz apoth.), and 12 such ounces made an apothecary's pound (lb apoth.). There are 7,000 grains in one pound avoirdupois (0.454 kg).

apparent depth depth that a transparent material such as water or glass appears to have when viewed from above. This is less than its real depth because of the ◊refraction that takes place when light passes into a less dense medium. The ratio of the real depth to the apparent depth of a transparent material is equal to its ◊refractive index.

appendix area of the mammalian gut, associated with the digestion of cellulose. In herbivores it may be large, containing millions of bacteria that secrete enzymes to digest grass. No vertebrate can produce the type of digestive enzyme that will digest cellulose, the main constituent of plant cell walls. Those herbivores that rely on cellulose for their energy have all evolved specialist mechanisms to make use of the correct type of bacteria.

Appleton layer band containing ionized gases in the Earth's upper atmosphere, above the ◊E layer (formerly the Kennelly–Heaviside layer). It can act as a reflector of radio signals, although its ionic composition varies with the sunspot cycle. It is named after the English physicist Edward Appleton.

applications program or *application* in computing, a program or job designed for the benefit of the end user, such as a payroll system or a ◊word processor. The term is used to distinguish such programs from those that control the computer or assist the programmer, such as a ◊compiler.

applications package in computing, the set of programs and related documentation (such as instruction manuals) used in a particular application. For example, a typical payroll applications package would consist of separate programs for the entry of data, updating the master files, and printing the pay slips, plus documentation in the form of program details and instructions for use.

appropriate technology simple or small-scale machinery and tools that, because they are cheap and easy to produce and maintain, may be of most use in the developing world; for example, hand ploughs and simple looms. This equipment may be used to supplement local crafts and traditional skills to encourage small-scale industrialization.

Many countries suffer from poor infrastructure and lack of capital but have the large supplies of labour needed for this level of technology. The use of appropriate technology was one of the recommendations of the Brandt Commission (1977–1983), established to examine the problems of developing countries and identify corrective measures that would command international support.

aqualung or *scuba* underwater breathing apparatus worn by divers, developed in the early 1940s by French diver Jacques Cousteau. Compressed-air cylinders strapped to the diver's back are regulated by a valve system and by a mouth tube to provide air to the diver at the same pressure as that of the surrounding water (which increases with the depth).

The vital component of an aqualung is the demand-regulator, a two-stage valve in the diver's mouthpiece. When the diver breathes in, air first passes from the compressed-air cylinders through a valve to the inner chamber of the mouthpiece. There, water that has entered the outer chamber pressurizes the air to the surrounding pressure before the diver takes in the air.

aquamarine blue variety of the mineral ◊beryl. A semiprecious gemstone, it is used in jewellery.

aquaplaning phenomenon in which the tyres of a road vehicle cease to make direct contact with the road surface, owing to the presence of a thin film of water. As a result, the vehicle can go out of control (particularly if the steered wheels are involved).

Aquaplaning can be prevented by fitting tyres with a good tread pattern at the correct pressure and by avoiding excessive speed when the roads are wet.

aqua regia (Latin 'royal water') mixture of three parts concentrated hydrochloric acid and one part concentrated nitric acid, which dissolves all metals except silver.

aquarium tank or similar container used for the study and display of living aquatic plants and animals. The same name is used for institutions that exhibit aquatic life. These have been common since Roman times, but the first modern public aquarium was opened in Regent's Park, London in 1853. A recent development is the oceanarium or seaquarium, a large display of marine life forms.

Aquarius zodiacal constellation a little south of the celestial equator near Pegasus. Aquarius is represented as a man pouring water from a jar. The Sun passes through Aquarius from late Feb to early March. In astrology, the dates for Aquarius are between about 20 Jan and 18 Feb (see ◊precession).

aquatic living in water. All life on Earth originated in the early oceans, because the aquatic environment has several advantages for organisms. Dehydration is almost impossible, temperatures usually remain stable, and the density of water provides physical support.

aqueduct any artificial channel or conduit for water, often an elevated structure of stone, wood, or iron built for conducting water across a valley. The Greeks built a tunnel 1,280 m/4,200 ft long near Athens, 2,500 years ago. Many Roman aqueducts are still standing, for example the one at Nîmes in S France, built about AD 18 (which is 48 m/160 ft high).

The largest Roman aqueduct is that at Carthage in Tunisia, which is 141 km/87 mi long and was built during the reign of Publius Aelius Hadrianus between AD 117 and 138. A recent aqueduct is the California State Water Project taking water from Lake Oroville in the north, through two power plants and across the Tehachapi mountains, more than 177 km/110 mi to S California.

The longest aqueduct in Britain is the Pont Cysylltau in Clwyd, Wales, opened 1805. It is 307 m/1,007 ft long, with 19 arches up to 36 m/121 ft high.

aqueous humour watery fluid found in the space between the cornea and lens of the vertebrate eye. Similar to blood serum in composition, it is renewed every four hours.

aqueous solution solution in which the solvent is water.

aquifer any rock formation containing water. The rock of an aquifer must be porous and permeable (full of interconnected holes) so that it can absorb water. Aquifers supply ◊artesian wells, and are actively sought in arid areas as sources of drinking and irrigation water.

An aquifer may be underlain, overlain, or sandwiched between impermeable layers, called *aquicludes*, which impede water movement. Sandstones and porous limestones make the best aquifers.

Aquila constellation on the celestial equator (see ◊celestial sphere). Its brightest star is first-magnitude ◊Altair, flanked by the stars Beta and Gamma Aquilae. It is represented by an eagle.

Arabic numerals or *Hindu-Arabic numerals* the symbols 0, 1, 2, 3, 4, 5, 6, 7, 8, 9, early forms of which were in use among the Arabs before being adopted by the peoples of Europe during the Middle Ages in place of ◊Roman numerals. The symbols appear to have originated in India and probably reached Europe by way of Spain.

Unlike Roman numerals, Arabic numerals include a symbol for zero which can be used as a place holder (indicating an empty space between digits in a positional, or place-value, number system).

arboretum collection of trees. An arboretum may contain a wide variety of species or just closely related species or varieties—for example, different types of pine tree.

Examples include ◊Kew Gardens, Bedgebury Pinetum, Kent, and Westonbirt Arboretum, Gloucestershire.

arc in geometry, a section of a curved line or circle. A circle has three types of arc: a *semicircle*, which is exactly half of the circle; *minor arcs*, which are less than the semicircle; and *major arcs*, which are greater than the semicircle.

An arc of a circle is measured in degrees, according to the angle formed by joining its two ends to the centre of that circle. A semicircle is therefore 180°, whereas a minor arc will always be less than 180° (acute or obtuse) and a major arc will always be greater than 180° but less than 360° (reflex).

Archaean or *Archaeozoic* the earliest eon of geological time; the first part of the Precambrian, from the formation of Earth up to about 2,500 million years ago. It was a time when no life existed, and with every new discovery of ancient life its upper boundary is being pushed further back.

archaebacteria three groups of bacteria whose DNA differs significantly from that of other bacteria (called the 'eubacteria'). All are strict anaerobes, that is, they are killed by oxygen. This is thought to be a primitive condition and to indicate that the archaebacteria are related to the earliest life forms, which appeared about 4 billion years ago, when there was little oxygen in the Earth's atmosphere. Archaebacteria are found in undersea vents, hot springs, the Dead Sea, and salt pans, and have even adapted to refuse tips.

archaeology study of history (primarily but not exclusively the prehistoric and ancient periods), based on the examination of physical remains. Principal activities include preliminary field (or site) surveys, excavation (where necessary), and the classification, dating, and interpretation of finds. Since 1958 radiocarbon dating has been used to establish the age of archaeological strata and associated materials.

history Interest in the physical remains of the past began in the Renaissance among dealers in and collectors of ancient art. It was further stimulated by discoveries made in Africa, the Americas, and Asia by Europeans during the period of imperialist colonization in the 16th–19th centuries, such as the antiquities discovered during Napoleon's Egyptian campaign in the 1790s. Towards the end of the 19th century archaeology became an academic study, making increasing use of scientific techniques and systematic methodologies.

Related disciplines that have been useful in archaeological reconstruction include stratigraphy (the study of geological strata), dendrochronology (the establishment of chronological sequences through the study of tree rings), palaeobotany (the study of ancient pollens, seeds, and grains), epigraphy (the study of inscriptions), and numismatics (the study of coins).

archegonium female sex organ found in bryophytes (mosses and liverworts), pteridophytes (ferns, club mosses, and horsetails), and some gymnosperms. It is a multicellular, flask-shaped structure consisting of two parts: the swollen base or venter containing the egg cell, and the long, narrow neck. When the egg cell is mature, the cells of the neck dissolve, allowing the passage of the male gametes, or ◊antherozoids.

archaeology: chronology

14th–16th centuries	The Renaissance revived interest in Classical Greek and Roman art and architecture, including ruins and buried art and artefacts.
1748	The buried Roman city of Pompeii was discovered under lava from Vesuvius.
1784	Thomas Jefferson excavated an Indian burial mound on the Rivanna River in Virginia and wrote a report on his finds.
1790	John Frere identified Old Stone Age (Palaeolithic) tools together with large extinct animals.
1822	Jean François Champollion deciphered Egyptian hieroglyphics.
1836	Christian Thomsen devised the Stone, Bronze, and Iron Age classification.
1840s	Austen Layard excavated the Assyrian capital of Nineveh.
1868	The Great Zimbabwe ruins in E Africa were first seen by Europeans.
1871	Heinrich Schliemann began excavations at Troy.
1879	Stone Age paintings were first discovered at Altamira, Spain.
1880s	Augustus Pitt-Rivers developed the concept of stratigraphy (identification of successive layers of soil within a site with successive archaeological stages; the most recent at the top).
1891	Flinders Petrie began excavating Akhetaton in Egypt.
1899–1935	Arthur Evans excavated Minoan Knossos in Crete.
1900–44	Max Uhle began the systematic study of the civilizations of Peru.
1911	The Inca city of Machu Picchu was discovered by Hiram Bingham in the Andes.
1911–12	The Piltdown skull was 'discovered'; it was proved to be a fake 1949.
1914–18	Osbert Crawford developed the technique of aerial survey of sites.
1917–27	John Eric Thompson (1898–1975) discovered the great Mayan sites in Yucatán, Mexico.
1922	Tutankhamen's tomb in Egypt was opened by Howard Carter.
1926	A kill site in Folsom, New Mexico, was found with human-made spearpoints in association with ancient bison.
1935	Dendrochronology (dating events in the distant past by counting tree rings) was developed by A E Douglas.
1939	An Anglo-Saxon ship-burial treasure was found at Sutton Hoo, England.
1947	The first of the Dead Sea Scrolls was discovered.
1948	The *Proconsul* prehistoric ape was discovered by Mary Leakey in Kenya.
1950s–1970s	Several early hominid fossils were found by Louis Leakey in Olduvai Gorge, Tanzania.
1953	Michael Ventris deciphered Minoan Linear B.
1960s	Radiocarbon and thermoluminescence measurement techniques were developed as aids for dating remains.
1961	The Swedish warship *Wasa* was raised at Stockholm.
1963	Walter Emery pioneered rescue archaeology at Abu Simbel before the site was flooded by the Aswan Dam.
1969	Human remains found at Lake Mungo, Australia, were dated at 26,000 years; earliest evidence of ritual cremation.
1974	The Tomb of Shi Huangdi was discovered in China. The footprints of a hominid called 'Lucy', 3 to 3.7 million years old, were found at Laetoli in Ethiopia.
1978	The tomb of Philip II of Macedon (Alexander the Great's father) was discovered in Greece.
1979	The Aztec capital Tenochtitlán was excavated beneath a zone of Mexico City.
1982	The English king Henry VIII's warship *Mary Rose* of 1545 was raised and studied with new techniques in underwater archaeology.
1985	The tomb of Maya, Tutankhamen's treasurer, was discovered at Saqqara, Egypt.
1988	The Turin Shroud was established as being of medieval origin by radiocarbon dating.
1989	The remains of the Globe and Rose Theatres, where many of Shakespeare's plays were originally performed, were discovered in London.
1991	Body of man from 5300 years ago, with clothing, bow, arrows, a copper axe, and other implements, found preserved in Italian Alps.
1992	The world's oldest surviving wooden structure, a well 15 m/49 ft deep made of huge oak timbers at Kückhoven, Germany, was dated by tree-rings to 5090 BC. The world's oldest sea-going vessel, dating from about 1400 BC, discovered Dover, southern England.

Archimedes' principle in physics, law stating that an object totally or partly submerged in a fluid displaces a volume of fluid that weighs the same as the apparent loss in weight of the object (which, in turn, equals the upwards force, or upthrust, experienced by that object). It was discovered by the Greek mathematician Archimedes.

If the weight of the object is less than the upthrust exerted by the fluid, it will float partly or completely above the surface; if its weight is equal to the upthrust, the object will come to equilibrium below the surface; if its weight is greater than the upthrust, it will sink.

Archimedes screw one of the earliest kinds of pump, thought to have been invented by Archimedes. It consists of a spiral screw revolving inside a close-fitting cylinder. It is used, for example, to raise water for irrigation.

archipelago group of islands, or an area of sea containing a group of islands. The islands of an archipelago are usually volcanic in origin, and they sometimes represent the tops of peaks in areas around continental margins flooded by the sea.

Volcanic islands are formed either when a hot spot within the Earth's mantle produces a chain of volcanoes on the surface, such as the Hawaiian Archipelago, or at a destructive plate margin (see ◊plate tectonics) where the subduction of one plate beneath another produces an arc-shaped island group, such as the Aleutian Archipelago. Novaya

Zemlya in the Arctic Ocean, the northern extension of the Ural Mountains, resulted from continental flooding.

arc lamp or *arc light* electric light that uses the illumination of an electric arc maintained between two electrodes. The British scientist Humphry Davy developed an arc lamp 1808, and its main use in recent years has been in cinema projectors. The lamp consists of two carbon electrodes, between which a very high voltage is maintained. Electric current arcs (jumps) between the two, creating a brilliant light.

The lamp incorporates a mechanism for automatically advancing the electrodes as they gradually burn away. Modern arc lamps (for example, searchlights) have the electrodes enclosed in an inert gas such as xenon.

arc minute, arc second units for measuring small angles, used in geometry, surveying, mapmaking, and astronomy. An arc minute (symbol ′) is one-sixtieth of a degree, and an arc second (symbol ″) is one-sixtieth of an arc minute. Small distances in the sky, as between two close stars or the apparent width of a planet's disc, are expressed in minutes and seconds of arc.

Arctic Circle imaginary line that encircles the North Pole at latitude 66° 32′ N. Within this line there is at least one day in the summer during which the Sun never sets, and at least one day in the winter during which the Sun never rises.

Arcturus or *Alpha Boötis* brightest star in the constellation Boötes and the fourth brightest star in the sky. Arcturus is a red giant about 28 times larger than the Sun and 70 times more luminous, 36 light years away from Earth.

ARCTURUS: A CANDLE IN THE SKY

The Earth receives as much heat from the orange star Arcturus, the brightest star in the Boötes constellation, as it would from a candle 8 km/5 mi away.

are metric unit of area, equal to 100 square metres (119.6 sq yd); 100 ares make one ◊hectare.

area the size of a surface. It is measured in square units, usually square centimetres (cm²), square metres (m²), or square kilometres (km²). Surface area is the area of the outer surface of a solid.

The areas of geometrical plane shapes with straight edges are determined using the area of a rectangle. Integration may be used to determine the area of shapes enclosed by curves.

Arecibo site in Puerto Rico of the world's largest single-dish ◊radio telescope, 305 m/1,000 ft in diameter. It is built in a natural hollow and uses the rotation of the Earth to scan the sky. It has been used both for radar work on the planets and for conventional radio astronomy, and is operated by Cornell University, USA.

arête (German *grat*; North American *comberidge*) sharp narrow ridge separating two ◊glacial troughs, or valleys. The typical U-shaped cross-sections of glacial troughs give arêtes very steep sides. Arêtes are common in glaciated mountain

regions such as the Rockies, the Himalayas, and the Alps.

ARECIBO: A MESSEGE TO THE STARS

In 1974 the giant radio telescope at Arecibo, Puerto Rico, beamed a message towards a cluster of stars. However, an answer cannot be expected until the year 50,000. It will take 25,000 years for the message to reach its target.

Argand diagram in mathematics, a method for representing ◊complex numbers by Cartesian coordinates (*x*, *y*). Along the *x*-axis (horizontal axis) are plotted the real numbers, and along the *y*-axis (vertical axis) the nonreal, or ◊imaginary, numbers.

argon (Greek *argos* 'idle') colourless, odourless, nonmetallic, gaseous element, symbol Ar, atomic number 18, relative atomic mass 39.948. It is grouped with the ◊inert gases, since it was long believed not to react with other substances, but observations now indicate that it can be made to combine with boron fluoride to form compounds. It constitutes almost 1% of the Earth's atmosphere, and was discovered 1894 by British chemists John Rayleigh (1842–1919) and William Ramsay after all oxygen and nitrogen had been removed chemically from a sample of air. It is used in electric discharge tubes and argon lasers.

argument in computing, the value on which a ◊function operates. For example, if the argument 16 is operated on by the function 'square root', the answer 4 is produced.

argument in mathematics, a specific value of the independent variable of a ◊function of *x*. It is also another name for ◊amplitude.

Ariane launch vehicle built in a series by the European Space Agency (first flight 1979). The launch site is at Kourou in French Guiana. Ariane is a three-stage rocket using liquid fuels. Small solid-fuel and liquid-fuel boosters can be attached to its first stage to increase carrying power.

Since 1984 it has been operated commercially by Arianespace, a private company financed by European banks and aerospace industries. A future version, *Ariane 5*, is intended to carry astronauts aboard the Hermes spaceplane.

arid region in earth science, a region that is very dry and has little vegetation. Aridity depends on temperature, rainfall, and evaporation, and so is difficult to quantify, but an arid area is usually defined as one that receives less than 250 mm/10 in of rainfall each year. (By comparison, New York City receives 1,120 mm/44 in per year.) There are arid regions in North Africa, Pakistan, Australia, the USA, and elsewhere. Very arid regions are ◊deserts.

The scarcity of fresh water is a problem for the inhabitants of arid zones, and constant research goes into discovering cheap methods of distilling sea water and artificially recharging natural groundwater reservoirs. Another problem is the eradication of salt in irrigation supplies from underground sources or where a surface deposit forms in poorly drained areas.

Ariel series of six UK satellites launched by the USA 1962–79, the most significant of which was *Ariel 5*, 1974, which made a pioneering survey of the sky at X-ray wavelengths.

Aries zodiacal constellation in the northern hemisphere between Pisces and Taurus, near Auriga, represented as the legendary ram whose golden fleece was sought by Jason and the Argonauts. Its most distinctive feature is a curve of three stars of decreasing brightness. The brightest of these is Hamal or Alpha Arietis, 65 light years from Earth.

The Sun passes through Aries from late April to mid-May. In astrology, the dates for Aries are between about 21 March and 19 April. The spring ♈equinox once lay in Aries, but has now moved into

areas of common plane shapes

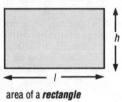

area of a ***rectangle***
= length × height
= *l* × *h*

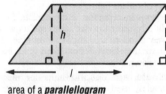

area of a ***parallellogram***
= base length × height
= *l* × *h*

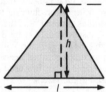
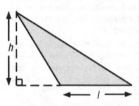

area of a ***triangle*** = $\frac{1}{2}$ base length × height = $\frac{1}{2}$ *l* × *h*

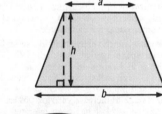

area of a ***trapezium***

= $\frac{1}{2}$ (sum of parallel sides) × height

= $\frac{1}{2}$ (*a* × *b*) × *h*

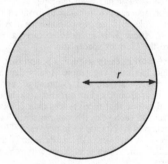

area of a ***circle***
= π × radius^2
= π*r*^{2}

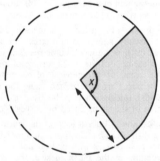

area of a ***sector***
= $\dfrac{\text{angle of sector}}{360}$ × π × radius^2

= $\dfrac{x}{360}$ × π*r*^{2}

Pisces through the effect of the Earth's ◊precession (wobble).

aril accessory seed cover other than a ◊fruit; it may be fleshy and sometimes brightly coloured, woody, or hairy. In flowering plants (◊angiosperms) it is often derived from the stalk that originally attached the ovule to the ovary wall. Examples of arils include the bright-red, fleshy layer surrounding the yew seed (yews are ◊gymnosperms so they lack true fruits), and the network of hard filaments that partially covers the nutmeg seed and yields the spice known as mace.

Another aril, the horny outgrowth found towards one end of the seed of the castor-oil plant *Ricinus communis*, is called a caruncle. It is formed from the integuments (protective layers enclosing the ovule) and develops after fertilization.

arithmetic branch of mathematics concerned with the study of numbers and their properties. The fundamental operations of arithmetic are addition, subtraction, multiplication, and division. Raising to powers (for example, squaring or cubing a number), the extraction of roots (for example, square roots), percentages, fractions, and ratios are developed from these operations.

Forms of simple arithmetic existed in prehistoric times. In China, Egypt, Babylon, and early civilizations generally, arithmetic was used for commercial purposes, records of taxation, and astronomy. During the Dark Ages in Europe, knowledge of arithmetic was preserved in India and later among the Arabs. European mathematics revived with the development of trade and overseas exploration. Hindu–Arabic numerals replaced Roman numerals, allowing calculations to be made on paper, instead of by the ◊abacus.

The essential feature of this number system was the introduction of zero, which allows us to have a *place–value* system. The decimal numeral system employs ten numerals (0,1,2,3,4,5,6,7,8,9) and is said to operate in 'base ten'. In a base-ten number, each position has a value ten times that of the position to its immediate right; for example, in the number 23 the numeral 3 represents three units (ones), and the number 2 represents two tens. The Babylonians, however, used a complex base-sixty system, residues of which are found today in the number of minutes in each hour and in angular measurement (6 × 60 degrees). The Mayans used a base-twenty system.

There have been many inventions and developments to make the manipulation of the arithmetic processes easier, such as the invention of ◊logarithms by Scottish mathematician John Napier 1614 and of the slide rule in the period 1620–30. Since then, many forms of ready reckoners, mechanical and electronic calculators, and computers have been invented.

Modern computers fundamentally operate in base two, using only two numerals (0,1), known as a binary system. In binary, each position has a value twice as great as the position to its immediate right, so that for example binary 111 is equal to 7 in the decimal system, and 1111 is equal to 15. Because the main operations of subtraction, multiplication, and division can be reduced mathematically to addition, digital computers carry out calculations by adding, usually in binary numbers in which the numerals 0 and 1 can be represented by off and on pulses of electric current.

Modular or modulo arithmetic, sometimes known as residue arithmetic, can take only a specific number of digits, whatever the value. For example, in modulo 4 (mod 4) the only values any number can take are 0, 1, 2, or 3. In this system, 7 is written as 3 mod 4, and 35 is also 3 mod 4. Notice 3 is the residue, or remainder, when 7 or 35 is divided by 4. This form of arithmetic is often illustrated on a circle. It deals with events recurring in regular cycles, and is used in describing the functioning of petrol engines, electrical generators, and so on. For example, in the mod 12, the answer to a question as to what time it will be in five hours if it is now ten o'clock can be expressed 10 + 5 = 3.

arithmetic and logic unit (ALU) in a computer, the part of the ◊central processing unit (CPU) that performs the basic arithmetic and logic operations on data.

arithmetic mean the average of a set of n numbers, obtained by adding the numbers and dividing by n. For example, the arithmetic mean of the set of 5 numbers 1, 3, 6, 8, and 12 is $(1 + 3 + 6 + 8 + 12)/5 = 30/5 = 6$

The term 'average' is often used to refer only to the arithmetic mean, even though the mean is in fact only one form of average (the others include ◊median and ◊mode).

arithmetic progression or *arithmetic sequence* sequence of numbers or terms that have a common difference between any one term and the next in the sequence. For example, 2, 7, 12, 17, 22, 27, ... is an arithmetic sequence with a common difference of 5.

The nth term in any arithmetic progression can be found using the formula:

$$n\text{th term} = a + (n - 1)\, d$$

where a is the first term and d is the common difference. An *arithmetic series* is the sum of the terms in an arithmetic sequence. The sum S of n terms is given by:

$$S = na + \tfrac{1}{2}n(n-1)d$$

armature in a motor or generator, the wire-wound coil that carries the current and rotates in a magnetic field. (In alternating-current machines, the armature is sometimes stationary.) The pole piece of a permanent magnet or electromagnet and the moving, iron part of a ◊solenoid, especially if the latter acts as a switch, may also be referred to as armatures.

armillary sphere earliest known astronomical device, in use from 3rd century BC. It showed the Earth at the centre of the universe, surrounded by a number of movable metal rings representing the Sun, Moon, and planets. The armillary sphere was originally used to observe the heavens and later for teaching navigators about the arrangements and movements of the heavenly bodies.

aromatic compound organic chemical compound in which some of the bonding electrons are delocalized (shared among several atoms within the molecule and not localized in the vicinity of the

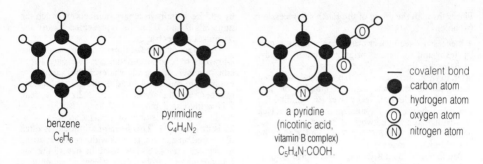

benzene
C_6H_6

pyrimidine
$C_4H_4N_2$

a pyridine
(nicotinic acid,
vitamin B complex)
$C_5H_4N \cdot COOH$.

— covalent bond
● carbon atom
○ hydrogen atom
⊙ oxygen atom
Ⓝ nitrogen atom

aromatic compound *Compounds whose molecules contain the benzene ring, or variations of it, are called aromatic. The term was originally used to distinguish sweet-smelling compounds from others.*

atoms involved in bonding). The commonest aromatic compounds have ring structures, the atoms comprising the ring being either all carbon or containing one or more different atoms (usually nitrogen, sulphur, or oxygen). Typical examples are benzene (C_6H_6) and pyridine (C_6H_5N).

arsenic brittle, greyish-white, semimetallic element (a metalloid), symbol As, atomic number 33, relative atomic mass 74.92. It occurs in many ores and occasionally in its elemental state, and is widely distributed, being present in minute quantities in the soil, the sea, and the human body. In larger quantities, it is poisonous. The chief source of arsenic compounds is as a by-product from metallurgical processes. It is used in making semiconductors, alloys, and solders.

As it is a cumulative poison, its presence in food and drugs is very dangerous. The symptoms of arsenic poisoning are vomiting, diarrhoea, tingling and possibly numbness in the limbs, and collapse. Its name derives from the Latin *arsenicum*.

artery vessel that carries blood from the heart to the rest of the body. It is built to withstand considerable pressure, having thick walls that are impregnated with muscle and elastic fibres. During contraction of the heart muscles, arteries expand in diameter to allow for the sudden increase in pressure that occurs; the resulting ◊pulse or pressure wave can be felt at the wrist. Not all arteries carry oxygenated (oxygen-rich) blood; the pulmonary arteries convey deoxygenated (oxygen-poor) blood from the heart to the lungs.

Arteries are flexible, elastic tubes, consisting of three layers, the middle of which is muscular; its rhythmic contraction aids the pumping of blood around the body. In middle and old age, the walls degenerate and are vulnerable to damage by the build-up of fatty deposits. These reduce elasticity, hardening the arteries and decreasing the internal bore. This condition, known as atherosclerosis, can lead to high blood pressure, loss of circulation, heart disease, and death. Research indicates that a typical Western diet, high in saturated fat, increases the chances of arterial disease developing.

artesian well well that is supplied with water rising from an underground water-saturated rock layer (◊aquifer). The water rises from the aquifer under its own pressure. Such a well may be drilled into an aquifer that is confined by impermeable rocks both above and below. If the water table (the top of the region of water saturation) in that aquifer is above the level of the well head, hydrostatic pressure will force the water to the surface.

Much use is made of artesian wells in E Australia, where aquifers filled by water in the Great Dividing Range run beneath the arid surface of the Simpson Desert. The artesian well is named after Artois, a French province, where the phenomenon was first observed.

arthropod member of the phylum Arthropoda; an invertebrate animal with jointed legs and a segmented body with a horny or chitinous casing (exoskeleton), which is shed periodically and replaced as the animal grows. Included are arachnids such as spiders and mites, as well as crustaceans, millipedes, centipedes, and insects.

artificial intelligence (AI) branch of science concerned with creating computer programs that can perform actions comparable with those of an intelligent human. Current AI research covers such areas as planning (for robot behaviour), language understanding, pattern recognition, and knowledge representation.

Early AI programs, developed in the 1960s, attempted simulations of human intelligence or were aimed at general problem-solving techniques. It is now thought that intelligent behaviour depends as much on the knowledge a system possesses as on its reasoning power. Present emphasis is on ◊knowledge-based systems, such as ◊expert systems. Britain's largest AI laboratory is at the Turing Institute, University of Strathclyde, Glasgow. In May 1990 the first International Robot Olympics was held there, including table-tennis matches between robots of the UK and the USA.

artificial radioactivity natural and spontaneous radioactivity arising from radioactive isotopes or elements that are formed when elements are bombarded with subatomic particles—protons, neutrons, or electrons—or small nuclei.

artificial selection in biology, selective breeding of individuals that exhibit the particular characteristics that a plant or animal breeder wishes to develop. In plants, desirable features might include resistance to disease, high yield (in crop plants), or attractive appearance. In animal breeding, selection has led to the development of particular breeds of cattle for improved meat production (such as the Aberdeen Angus) or milk production (such as Jerseys).

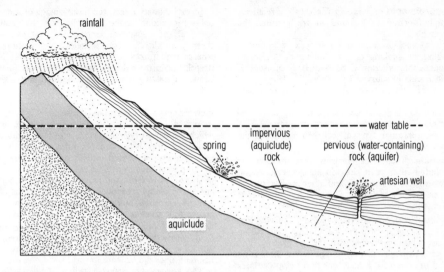

artesian well *In an artesian well, water rises from an underground water-containing rock layer under its own pressure. Rain falls at one end of the water-bearing layer, or aquifer, and percolates through the layer. The layer fills with water up to the level of the water table. Water will flow from a well under its own pressure if the well head is below the level of the water table.*

Artificial selection was practised by the Sumerians at least 5,500 years ago and carried on through the succeeding ages, with the result that all common vegetables, fruit and livestock are long modified by selective breeding. Artificial selection, particularly of pigeons, was studied by the English evolutionist Charles Darwin who saw a similarity between this phenomenon and the processes of natural selection.

ASA in photography, a numbering system for rating the speed of films, devised by the American Standards Association. It has now been superseded by ◊*ISO*, the International Standards Organization.

asbestos any of several related minerals of fibrous structure that offer great heat resistance because of their nonflammability and poor conductivity. Commercial asbestos is generally made from olivine, a ◊serpentine mineral, tremolite (a white ◊amphibole), and riebeckite (a blue amphibole, also known as crocidolite when in its fibrous form). Asbestos usage is now strictly controlled because exposure to its dust can cause cancer.

Asbestos has been used for brake linings, suits for fire fighters and astronauts, insulation of electric wires in furnaces, and fireproof materials for the building industry. Exposure to asbestos is a recognized cause of industrial cancer (mesothelioma), especially in the 'blue' form (from South Africa), rather than the more common 'white'. *Asbestosis* is a chronic lung inflammation caused by asbestos dust.

ASCII (acronym for *A*merican *s*tandard *c*ode for *i*nformation *i*nterchange) in computing, a coding system in which numbers are assigned to letters, digits, and punctuation symbols. Although computers work in ◊binary number code, ASCII numbers are usually quoted as decimal or ◊hexadecimal numbers. For example, the decimal number 45 (binary 0101101) represents a hyphen, and 65 (binary 1000001) a capital A. The first 32 codes are used for control functions, such as carriage return and backspace.

Strictly speaking, ASCII is a 7–bit binary code, allowing 128 different characters to be represented, but an eighth bit is often used to provide ◊parity or to allow for extra characters. The system is widely used for the storage of text and for the transmission of data between computers.

ASCII

character	binary code
A	1000001
B	1000010
C	1000011
D	1000100
E	1000101
F	1000110
G	1000111
H	1001000
I	1001001
J	1001010
K	1001011
L	1001100
M	1001101
N	1001110
O	1001111
P	1010000
Q	1010001
R	1010010
S	1010011
T	1010100
U	1010101
V	1010110
W	1010111
X	1011000
Y	1011001
Z	1011010

ascorbic acid or *vitamin C* $C_6H_8O_6$ a relatively simple organic acid found in fresh fruits and vegetables.

It is soluble in water and destroyed by prolonged boiling, so soaking or overcooking of vegetables reduces their vitamin C content. Lack of ascorbic acid results in scurvy.

In the human body, ascorbic acid is necessary for the correct synthesis of collagen. Lack of it causes skin sores or ulcers, tooth and gum problems, and burst capillaries (scurvy symptoms) owing to an abnormal type of collagen replacing the normal type in these tissues.

asepsis practice of ensuring that bacteria are excluded from open sites during surgery, wound dressing, blood sampling, and other medical procedures. Aseptic technique is a first line of defence against infection.

asexual reproduction in biology, reproduction that does not involve the manufacture and fusion of sex cells, nor the necessity for two parents. The process carries a clear advantage in that there is no need to search for a mate nor to develop complex pollinating mechanisms; every asexual organism can reproduce on its own. Asexual reproduction can therefore lead to a rapid population build-up.

In evolutionary terms, the disadvantage of asexual reproduction arises from the fact that only identical individuals, or clones, are produced—there is no variation. In the field of horticulture, where standardized production is needed, this is useful, but in the wild, an asexual population that cannot adapt to a changing environment or evolve defences against a new disease is at risk of extinction. Many asexually reproducing organisms are therefore capable of reproducing sexually as well.

Asexual processes include ◊binary fission, in which the parent organism splits into two or more 'daughter' organisms, and ◊budding, in which a new organism is formed initially as an outgrowth of the parent organism. The asexual reproduction of spores, as in ferns and mosses, is also common and many plants reproduce asexually by means of runners, rhizomes, bulbs, and corms; see also ◊vegetative reproduction.

aspartame noncarbohydrate sweetener used in foods under the tradename Nutrasweet. It is about 200 times as sweet as sugar and, unlike saccharine, has no aftertaste.

The aspartame molecule consists of two amino acids (aspartic acid and phenylalanine) linked by a methylene ($- CH_2-$) group. It breaks down slowly at room temperature and rapidly at higher temperatures. It is not suitable for people who suffer from phenylketonuria (a genetic condition in which the liver cannot control the level of phenylalanine, an amino acid derived from protein, in the bloodstream in the normal way, by excretion).

asphalt semisolid brown or black ◊bitumen, used in the construction industry. Asphalt is mixed with rock chips to form paving material, and the purer varieties are used for insulating material and for waterproofing masonry. It can be produced artificially by the distillation of ◊petroleum.

The availability of recycled coloured glass led in 1988 to the invention of *glassphalt*, asphalt that is 15% crushed glass. It is used to pave roads in New York.

Considerable natural deposits of asphalt occur around the Dead Sea and in the Philippines, Cuba, Venezuela, and Trinidad.

aspirin acetylsalicylic acid, a popular pain-relieving drug (◊analgesic) developed in the early 20th century for headaches and arthritis. It inhibits ◊prostaglandins, and is derived from the white willow tree *Salix alba*.

In the long term, even moderate use may cause stomach bleeding, kidney damage, and hearing defects, and aspirin is no longer considered suitable for children under 12, because of a suspected link with a rare disease, Reye's syndrome. However, recent medical research suggests that an aspirin a day may be of value in preventing heart attack and thrombosis.

assay in chemistry, the determination of the quantity of a given substance present in a sample. Usually it refers to determining the purity of precious metals.

The assay may be carried out by 'wet' methods, when the sample is wholly or partially dissolved in some reagent (often an acid), or by 'dry' or 'fire' methods, in which the compounds present in the sample are combined with other substances.

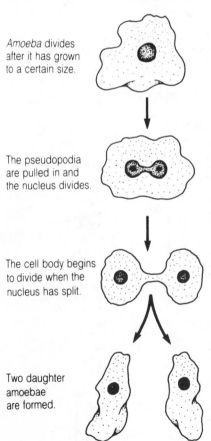

Amoeba divides after it has grown to a certain size.

The pseudopodia are pulled in and the nucleus divides.

The cell body begins to divide when the nucleus has split.

Two daughter amoebae are formed.

asexual reproduction

assembler in computing, a program that translates a program written in an assembly language into a complete ◊machine-code program that can be executed by a computer. Each instruction in the assembly language is translated into only one machine-code instruction.

assembly language low-level computer-programming language closely related to a computer's internal codes. It consists chiefly of a set of short sequences of letters (mnemonics), which are translated, by a program called an assembler, into ◊machine code for the computer's ◊central processing unit (CPU) to follow directly. In assembly language, for example, 'JMP' means 'jump' and 'LDA' means 'load accumulator'. Assembly language is used by programmers who need to write very fast or efficient programs.

Because they are much easier to use, high-level languages are normally used in preference to assembly languages. An assembly language may still be used in some cases, however, particularly when no suitable high-level language exists or where a very efficient machine-code program is required.

assimilation in animals, the process by which absorbed food molecules, circulating in the blood, pass into the cells and are used for growth, tissue repair, and other metabolic activities. The actual destiny of each food molecule depends not only on its type, but also on the body requirements at that time.

associative operation in mathematics, an operation in which the outcome is independent of the grouping of the numbers or symbols concerned. For example, multiplication is associative, as 4 × (3 × 2) = (4 × 3) × 2 = 24; however, division is not, as 12 ÷ (4 ÷ 2) = 6, but (12 ÷ 4) ÷ 2 = 1.5. Compare ◊commutative operation and ◊distributive operation.

assortative mating in ◊population genetics, selective mating in a population between individuals that are genetically related or have similar characteristics. If sufficiently consistent, assortative mating can theoretically result in the evolution of new species without geographical isolation (see ◊speciation).

astatine (Greek *astatos* 'unstable') nonmetallic, radioactive element, symbol At, atomic number 85, relative atomic mass 210. It is a member of the ◊halogen group, and is very rare in nature. Astatine is highly unstable, with at least 19 isotopes; the longest lived has a half-life of about eight hours.

asteroid or *minor planet* any of many thousands of small bodies, composed of rock and iron, that orbit the Sun. Most lie in a belt between the orbits of Mars and Jupiter, and are thought to be fragments left over from the formation of the ◊Solar System. About 100,000 may exist, but their total mass is only a few hundredths the mass of the Moon.

They include ◊Ceres (the largest asteroid, 940 km/584 mi in diameter), Vesta (which has a light-coloured surface, and is the brightest as seen from Earth), ◊Eros, and ◊Icarus. Some asteroids are on orbits that bring them close to Earth, and some, such as the ◊Apollo asteroids, even cross Earth's orbit; at least some of these may be former comets that have lost their gas. One group, the Trojans, moves along the same orbit as Jupiter, 60° ahead and behind the planet. One unusual asteroid, ◊Chiron, orbits beyond Saturn. The first asteroid was discovered by the Italian astronomer Guiseppe Piazzi at the Palermo Observatory, Sicily, 1 Jan 1801.

ASTEROID: A CLOSE ENCOUNTER

On 29 Oct 1991 the US spacecraft *Galileo* made the closest ever approach to an asteroid, Gaspra, flying within 1,600 km/ 1,000 mi. Photographs were radioed back to Earth using the craft's small antenna— the main antenna had failed to open.

---

asthenosphere division of the Earth's structure lying beneath the ◊lithosphere, at a depth of approximately 70 km/45 mi to 260 km/160 mi. It is thought to be the soft, partially molten layer of the ◊mantle on which the rigid plates of the Earth's surface move to produce the motions of ◊plate tectonics.

astigmatism aberration occurring in lenses, including that in the eye. It results when the curvature of the lens differs in two perpendicular planes, so that rays in one plane may be in focus while rays in the other are not. With astigmatic eyesight, the vertical and horizontal cannot be in focus at the same time; correction is by the use of a cylindrical lens that reduces the overall focal length of one plane so that both planes are seen in sharp focus.

astrolabe ancient navigational instrument, forerunner of the sextant. Astrolabes usually consisted of a flat disc with a sighting rod that could be

the six largest asteroids

name	diameter		average distance from Sun	orbital period
	(km)	(mi)	(Earth = 1)	(years)
Ceres	940	584	2.77	4.6
Pallas	588	365	2.77	4.6
Vesta	576	358	2.36	3.6
Hygeia	430	267	3.13	5.5
Interamnia	338	210	3.06	5.4
Davida	324	201	3.18	5.7

pivoted to point at the Sun or bright stars. From the altitude of the Sun or star above the horizon, the local time could be estimated.

astrometry measurement of the precise positions of stars, planets, and other bodies in space. Such information is needed for practical purposes including accurate timekeeping, surveying and navigation, and calculating orbits and measuring distances in space. Astrometry is not concerned with the surface features or the physical nature of the body under study.

Before telescopes, astronomical observations were simple astrometry. Precise astrometry has shown that stars are not fixed in position, but have a ◊proper motion caused as they and the Sun orbit the Milky Way galaxy. The nearest stars also show ◊parallax (apparent change in position), from which their distances can be calculated. Above the distorting effects of the atmosphere, satellites such as ◊*Hipparcos* can make even more precise measurements than ground telescopes, so refining the distance scale of space.

astronaut person making flights into space; the term *cosmonaut* is used in the West for any astronaut from the former Soviet Union.

astronautics science of space travel. See ◊rocket; ◊satellite; ◊space probe.

Astronomer Royal honorary post in British astronomy. Originally it was held by the director of the Royal Greenwich Observatory; since 1972 the title of Astronomer Royal has been awarded separately. The Astronomer Royal from 1991 is Arnold Wolfendale (1927–   ). A separate post of Astronomer Royal for Scotland is attached to the directorship of the Royal Observatory, Edinburgh.

astronomical unit unit (symbol AU) equal to the mean distance of the Earth from the Sun: 149,597,870 km/92,955,800 mi. It is used to describe planetary distances. Light travels this distance in approximately 8.3 minutes.

astronomy science of the celestial bodies: the Sun, the Moon, and the planets; the stars and galaxies; and all other objects in the universe. It is concerned with their positions, motions, distances, and physical conditions; and with their origins and evolution. Astronomy thus divides into fields such as astrophysics, celestial mechanics, and cosmology. See also ◊gamma-ray astronomy, ◊infrared astronomy, ◊radio astronomy, ◊ultraviolet astronomy, and ◊X-ray astronomy.

Astronomy is perhaps the oldest recorded science; there are observational records from ancient Babylonia, China, Egypt, and Mexico. The first true astronomers, however, were the Greeks, who deduced the Earth to be a sphere, and attempted to measure its size. Ancient Greek astronomers included Thales and Pythagoras. Eratosthenes of Cyrene measured the size of the Earth with considerable accuracy. Star catalogues were drawn up, the most celebrated being that of Hipparchus. The *Almagest*, by Ptolemy of Alexandria, summarized Greek astronomy, and survived in its Arab translation. However, the Greeks still regarded the Earth as the centre of the universe, although this was doubted by some philosophers, notably Aristarchus of Samos, who maintained that the Earth moves around the Sun.

Ptolemy, the last famous astronomer of the Greek school, died about AD 180, and little progress was made for some centuries. The Arabs revived the science, carrying out theoretical researches from the 8th and 9th centuries, and producing good star catalogues. Unfortunately, a general belief in the pseudoscience of astrology continued until the end of the Middle Ages (and has been revived from time to time).

The dawn of a new era came 1543, when a Polish canon, Copernicus, published a work entitled *De Revolutionibus Orbium Coelestium/About the Revolutions of the Heavenly Spheres*, in which he demonstrated that the Sun, not the Earth, is the centre of our planetary system. (Copernicus was wrong in many respects—for instance, he still believed that all celestial orbits must be perfectly circular.) Tycho Brahe, a Dane, increased the accuracy of observations by means of improved instruments allied to his own personal skill, and his observations were used by the German mathematician Johannes Kepler to prove the validity of the Copernican system. Considerable opposition existed, however, for removing the Earth from its central position in the universe; the Catholic Church was openly hostile to the idea, and, ironically, Brahe never accepted the idea that the Earth could move around the Sun. Yet before the end of the 17th century, the theoretical work of Isaac Newton had established celestial mechanics.

The refracting telescope was invented about 1608, by Hans Lippershey in Holland, and was first applied to astronomy by the Italian scientist Galileo in the winter of 1609–10. Immediately, Galileo made a series of spectacular discoveries. He found the four largest satellites (moons) of Jupiter, which gave strong support to the Copernican theory; he saw the craters of the Moon, the phases of Venus, and the myriad faint stars of our Galaxy, the Milky Way. Galileo's most powerful telescope magnified only 30 times, but before long, larger telescopes were built, and official observatories were established.

Galileo's telescope was a refractor; that is to say, it collected its light by means of a glass lens or object glass. Difficulties with his design led Newton, in 1671, to construct a reflector, in which the light is collected by means of a curved mirror.

Theoretical researches continued, and astronomy made rapid progress in many directions. New planets were discovered—Uranus 1781 by William Herschel, and Neptune 1846, following calculations by British astronomer John Couch Adams and French astronomer Urbain Jean Joseph Leverrier. Also significant was the first measurement of the distance of a star, when in 1838 the German astronomer Friedrich Bessel measured the ◊parallax of the star 61 Cygni, and calculated that it lies at a distance of about 6 light years (about half the correct value). Astronomical spectroscopy was developed, first by Fraunhofer in Germany and then by people such as Pietro Angelo Secchi and William Huggins, while Gustav Kirchhoff success-

The Comet of Doom and the search for ET

Comet sightings do not usually make headlines. At least two dozen are seen every year, about half of them reappearances of known objects. However, Comet Swift-Tuttle created a stir in 1992 when astronomers calculated that it might hit the Earth in August 2126. Newspapers christened it the Comet of Doom and announced the end of the world. Fortunately, the truth is not so alarming.

The comet was discovered in 1862 by US astronomers Lewis Swift and Horace Tuttle. Like all comets, it goes round the Sun on an elliptical orbit that regularly brings it back into view. It was expected to reappear around 1982 but, despite careful searches, nothing was seen. Was it lost or simply late?

Careful work by Dr Brian Marsden of the Harvard-Smithsonian Center for Astrophysics, Cambridge, Massachusetts, showed that the comet was probably the same as one seen in 1737. If so, its orbital period was nearer 130 years than the 120 previously calculated, so it would not return until late 1992.

The wanderer returns
Evidence of Swift-Tuttle's return came from observations of a shower of meteors (shooting stars) caused by dust in the comet's orbit. Visible in August every year, they are known as the Perseids, because they seem to radiate from the constellation Perseus. Amateur astronomers have recorded increasing numbers of Perseids over recent years, fuelling speculation that their parent comet was near.

In September 1992 Japanese comet hunter Tsuruhiko Kiuchi spotted a faint comet near the Big Dipper. Its direction of motion gave it away as Comet Swift-Tuttle.

Swift-Tuttle's orbit was difficult to calculate because jets of gas and dust, escaping from its central nucleus, push the comet off course like little rockets. New calculations by Marsden showed that if the comet were only two weeks later than expected next time round—not much, given the ten-year error this time—then the comet would strike in August 2126, during the Perseid meteor shower.

The consequences would be disastrous. The Earth would be shrouded in dust and water vapour, changing the climate and bringing famine to those parts that had escaped the earthquakes, tidal waves and fires from the initial strike. The death of the dinosaurs is believed to have resulted from a similar impact 65 million years ago.

So astronomers watched Comet Swift-Tuttle's passage round the Sun in 1992 with close interest. Analysis of these observations will help astronomers decide whether the comet could hit us on its next orbit or whether, as seems likely, the Comet of Doom is just another scare story.

NASA listens for extraterrestrials
In Steven Spielberg's film *ET*, the extraterrestrial landed on Earth and tried to phone home. Perhaps he should have phoned Earth before setting out; radio communication is easier than interstellar travel. That is the thinking behind NASA's search for extraterrestrial intelligence, which started on 12 October 1992.

NASA's survey uses two approaches. In an all-sky survey, radio telescopes will listen in all directions for signals at frequencies with least background noise, between 1,000 and 10,000 megahertz. The survey began with the 34–m/111–ft radio dish of NASA's Goldstone tracking station in California; the southern sky will be covered later by a similar dish at Canberra, Australia.

A targeted search will listen to a thousand stars like the Sun within 100 light-years' distance, using large radio telescopes worldwide. First observations were made with the world's largest radio dish, the 305–m/1,000–ft radio telescope at Arecibo, Puerto Rico.

The targeted search can pick up fainter signals, but might miss a transmission from some unexpected source, which the all-sky survey would pick up. The all-sky survey is less sensitive and might miss faint signals from a nearby star which the targeted search would hear. Sophisticated electronics have been developed to recognize a genuine alien message while filtering out terrestrial interference.

Is there anybody there?
No previous attempt to pick up alien radio messages has been as comprehensive. 'In the first few minutes, more searching will be accomplished than in all previous searches combined,' said John Billingham at NASA's Ames Research Center in California.

If a seemingly artificial signal is detected, the searchers will notify other radio observatories worldwide to check the discovery. If it is genuine rather than a false alarm, the world will be told. No attempt will be made to answer except by international agreement.

The survey will probably continue for ten years at a total cost of around $100 million. By early next century, we may know whether or not we are alone.

Ian Ridpath

fully interpreted the spectra of the Sun and stars. By the 1860s good photographs of the Moon had been obtained, and by the end of the century photographic methods had started to play a leading role in research.

William Herschel, probably the greatest observer in the history of astronomy, investigated the shape of our ◊Galaxy during the latter part of the 18th century, and concluded that its stars are arranged roughly in the form of a double-convex lens. Basically Herschel was correct, although he placed our Sun near the centre of the system; in fact, it is well out toward the edge, and lies 25,000 light years from the galactic nucleus. Herschel also studied the luminous 'clouds' or nebulae, and made the tentative suggestion that those nebulae capable of resolution into stars might be separate galaxies, far outside our own Galaxy. It was not until 1923 that US astronomer Edwin Hubble, using the 2.5 m/ 100 in reflector at the Mount Wilson Observatory, was able to verify this suggestion. It is now known that the 'starry nebulae' are galaxies in their own right, and that they lie at immense distances. The most distant galaxy visible to the naked eye, the Great Spiral in ◊Andromeda, is 2.2 million light years away; the most remote galaxy so far measured lies over 10 billion light years away. It was also found that galaxies tended to form groups, and that the groups were apparently receding from each other at speeds proportional to their distances. This concept of an expanding and evolving universe at first rested largely on ◊Hubble's law, relating the distance of objects to the amount their spectra shift towards red—the ◊red shift. Subsequent evidence derived from objects studied in other parts of the ◊electromagnetic spectrum, at radio and X-ray wavelengths, has provided confirmation. ◊Radio astronomy established its place in probing the structure of the universe by demonstrating in 1954 that an optically visible distant galaxy was identical with a powerful radio source known as Cygnus A. Later analysis of the comparative number, strength, and distance of radio sources suggested that in the distant past these, including the ◊quasars discovered 1963, had been much more powerful and numerous than today. This fact suggested that the universe has been evolving from an origin, and is not of infinite age as expected under a ◊steady-state theory. The discovery 1965 of microwave background radiation suggested that residue survived the tremendous thermal power of the giant explosion, or Big Bang, that brought the universe into existence.

Although the practical limit in size and efficiency of optical telescopes has apparently been reached, the siting of these and other types of telescope at new observatories in the previously neglected southern hemisphere has opened fresh areas of the sky to search. Australia has been in the forefront of these developments. The most remarkable recent extension of the powers of astronomy to explore the universe is in the use of rockets, satellites, space stations, and space probes. Even the range and accuracy of the conventional telescope may be greatly improved free from the Earth's atmosphere. The USA launched a large optical telescope, called the Hubble Space Telescope, permanently in space, in April 1990. It is the most powerful optical

telescope yet constructed, with a 2.4 m/94.5 in mirror. It detects celestial phenomena seven times more distant (up to 14 billion light-years) than any land telescope. See also ◊black hole, ◊cosmology, and ◊infrared radiation.

astrophotography use of photography in astronomical research. The first successful photograph of a celestial object was the daguerreotype plate of the Moon taken by John W Draper (1811–1882) of the USA in March 1840. The first photograph of a star, Vega, was taken by US astronomer William C Bond (1789–1859) in 1850. Modernday astrophotography uses techniques such as ◊charge-coupled devices (CCDs).

Before the development of photography, observations were gathered in the form of sketches made at the telescope. Several successful daguerreotypes were obtained prior to the introduction of wetplate collodion about 1850. The availability of this more convenient method allowed photography to be used on a more systematic basis, including the monitoring of sunspot activity. Dry plates were introduced in the 1870s, and in 1880 Henry Draper (1837–1882) obtained a photograph of the ◊Orion nebula. The first successful image of a comet was obtained 1882 by the Scottish astronomer David Gill (1843–1914), his plate displaying excellent star images. Following this, Gill and J C Kapteyn (1851–1922) compiled the first photographic atlas of the southern sky, cataloguing almost half a million stars.

Modern-day electronic innovations, notably charge-coupled devices (CCDs), provide a more efficient light-gathering capability than photographic film as well as enabling information to be transferred to a computer for analysis. However, CCD images are expensive and very small in size compared to photographic plates. Photographic plates are better suited to wide-field images, whereas CCDs are used for individual objects, which may be very faint, within a narrow field of sky.

astrophysics study of the physical nature of stars, galaxies, and the universe. It began with the development of spectroscopy in the 19th century, which allowed astronomers to analyse the composition of stars from their light. Astrophysicists view the universe as a vast natural laboratory in which they can study matter under conditions of temperature, pressure, and density that are unattainable on Earth.

asymptote in ◊coordinate geometry, a straight line that a curve approaches more and more closely but never reaches.

atavism (Latin *atavus* 'ancestor') in ◊genetics, the reappearance of a characteristic not apparent in the immediately preceding generations; in psychology, the manifestation of primitive forms of behaviour.

Atlas rocket US rocket, originally designed and built as an intercontinental missile, but subsequently adapted for space use. Atlas rockets launched astronauts in the Mercury series into orbit, as well as numerous other satellites and space probes.

atmosphere mixture of gases that surrounds the Earth, prevented from escaping by the pull of the

astronomy: chronology

2300 BC	Chinese astronomers made their earliest observations.
2000	Babylonian priests made their first observational records.
1900	Stonehenge was constructed: first phase.
365	The Chinese observed the moons of Jupiter with the naked eye.
3rd century AD	Aristarchus argued that the Sun is the centre of the Solar System.
2nd century	Ptolemy's complicated Earth-centred system was promulgated, which dominated the astronomy of the Middle Ages.
1543	Copernicus revived the ideas of Aristarchus in *De Revolutionibus*.
1608	Hans Lippershey invented the telescope, which was first used by Galileo 1609.
1609	Johannes Kepler's first two laws of planetary motion were published (the third appeared 1619).
1632	The world's first official observatory was established in Leiden in the Netherlands.
1633	Galileo's theories were condemned by the Inquisition.
1675	The Royal Greenwich Observatory was founded in England.
1687	Isaac Newton's *Principia* was published, including his 'law of universal gravitation'.
1705	Edmond Halley correctly predicted that the comet that had passed the Earth in 1682 would return in 1758; the comet was later to be known by his name.
1781	William Herschel discovered Uranus and recognized stellar systems beyond our Galaxy.
1796	Pierre Laplace elaborated his theory of the origin of the Solar System.
1801	Giuseppe Piazzi discovered the first asteroid, Ceres.
1814	Joseph von Fraunhofer first studied absorption lines in the solar spectrum.
1846	Neptune was identified by Johann Galle, following predictions by John Adams and Urbain Leverrier.
1859	Gustav Kirchhoff explained dark lines in the Sun's spectrum.
1887	The earliest photographic star charts were produced.
1889	Edward Barnard took the first photographs of the Milky Way.
1908	Fragment of comet fell at Tunguska, Siberia.
1920	Arthur Eddington began the study of interstellar matter.
1923	Edwin Hubble proved that the galaxies are systems independent of the Milky Way, and by 1930 had confirmed the concept of an expanding universe.
1930	The planet Pluto was discovered by Clyde Tombaugh at the Lowell Observatory, Arizona, USA.
1931	Karl Jansky founded radio astronomy.
1945	Radar contact with the Moon was established by Z Bay of Hungary and the US Army Signal Corps Laboratory.
1948	The 5–m/200–in Hale reflector telescope was installed at Mount Palomar Observatory, California, USA.
1957	The Jodrell Bank telescope dish in England was completed.
1957	The first Sputnik satellite (USSR) opened the age of space observation.
1962	The first X-ray source was discovered in Scorpius.
1963	The first quasar was discovered.
1967	The first pulsar was discovered by Jocelyn Bell and Antony Hewish.
1969	The first crewed Moon landing was made by US astronauts.
1976	A 6–m/240–in reflector telescope was installed at Mount Semirodniki, USSR.
1977	Uranus was discovered to have rings.
1977	The spacecrafts *Voyager 1* and *2* were launched, passing Jupiter and Saturn 1979–1981.
1978	The spacecrafts *Pioneer Venus 1* and *2* reached Venus.
1978	Pluto's moon, Charon, was discovered by James Christy of the US Naval Observatory.
1986	Halley's comet returned. *Voyager 2* flew past Uranus and discovered six new moons.
1987	Supernova SN1987A flared up, becoming the first supernova to be visible to the naked eye since 1604. The 4.2 m/165 in William Herschel Telescope on La Palma, Canary Islands, and the James Clerk Maxwell Telescope on Mauna Kea, Hawaii, began operation.
1988	The most distant individual star was recorded—a supernova, 5 billion light years away, in the AC118 cluster of galaxies.
1989	*Voyager 2* flew by Neptune and discovered eight moons and three rings.
1990	Hubble Space Telescope was launched into orbit by the US space shuttle.
1991	The space probe *Galileo* flew past the asteroid Gaspra.
1992	COBE satellite detected ripples from the Big Bang that mark the first stage in the formation of galaxies.

Earth's gravity. Atmospheric pressure decreases with height in the atmosphere. In its lowest layer, the atmosphere consists of nitrogen (78%) and oxygen (21%). The other 1% is largely argon, with very small quantities of other gases, including water vapour and carbon dioxide. The atmosphere plays a major part in the various cycles of nature (the ◊water cycle, ◊carbon cycle, and ◊nitrogen cycle). It is the principal industrial source of nitrogen, oxygen, and argon, which are obtained by fractional distillation of liquid air.

The lowest level of the atmosphere, the ◊troposphere, is heated by the Earth, which is warmed by infrared and visible radiation from the Sun. Warm air cools as it rises in the troposphere, causing rain and most other weather phenomena. However, infrared and visible radiations form only a part of the Sun's output of electromagnetic radiation. Almost all the shorter-wavelength ultraviolet radiation is filtered out by the upper layers of the atmosphere. The filtering process is an active one: at heights above about 50 km/31 mi ultraviolet photons collide with atoms, knocking out electrons to create a ◊plasma of electrons and positively charged ions. The resulting *ionosphere* acts as a reflector of radio waves, enabling radio trans-

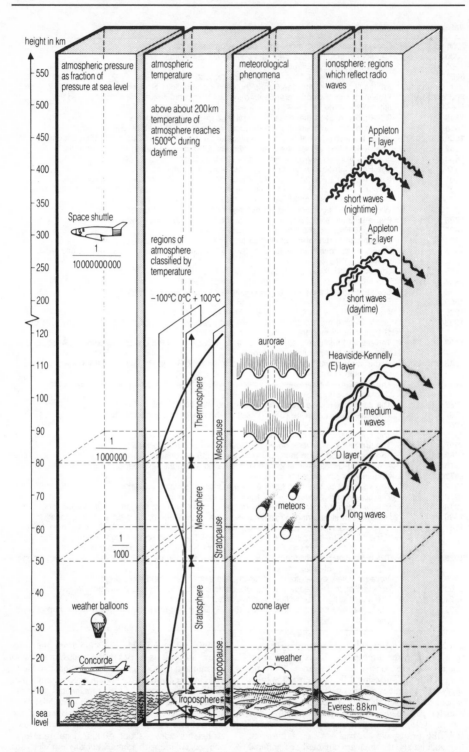

atmosphere *All but 1% of the Earth's atmosphere lies in a layer 30 km/19 mi above the ground. At a height of 5,500 m/18,000 ft, air pressure is half that at sea level. The temperature of the atmosphere varies greatly with height; this produces a series of layers, called the troposphere, stratosphere, mesosphere, and thermosphere.*

missions to 'hop' between widely separated points on the Earth's surface.

Waves of different wavelengths are reflected best at different heights. The collisions between ultraviolet photons and atoms lead to a heating of the upper atmosphere, although the temperature drops from top to bottom within the zone called the *thermosphere* as high-energy photons are progressively absorbed in collisions. Between the thermosphere and the tropopause (at which the warming effect of the Earth starts to be felt) there is a 'warm bulge' in the graph of temperature against height, at a level called the *stratopause*. This is due to longer-wavelength ultraviolet photons that have survived their journey through the upper layers; now they encounter molecules and split them apart into atoms. These atoms eventually bond together again, but often in different combinations. In particular, many ◊ozone molecules (oxygen atom triplets) are formed. Ozone is a better absorber of ultraviolet than ordinary (two-atom) oxygen, and it is the *ozone layer* that prevents lethal amounts of ultraviolet from reaching the Earth's surface.

Far above the atmosphere, as so far described, lie the *Van Allen radiation belts*. These are regions in which high-energy charged particles travelling outwards from the Sun (as the so-called solar wind) have been captured by the Earth's magnetic field. The outer belt (at about 1,600 km/1,000 mi) contains mainly protons, the inner belt (at about 2,000 km/1,250 mi) contains mainly electrons. Sometimes electrons spiral down towards the Earth, noticeably at polar latitudes, where the magnetic field is strongest. When such particles collide with atoms and ions in the thermosphere, light is emitted. This is the origin of the glows visible in the sky as the *aurora borealis* (northern lights) and the *aurora australis* (southern lights). A fainter, more widespread, *airglow* is caused by a similar mechanism.

During periods of intense solar activity, the atmosphere swells outwards; there is a 10–20% variation in atmosphere density. One result is to increase drag on satellites. This effect makes it impossible to predict exactly the time of re-entry of satellites.

atmosphere or *standard atmosphere* in physics, a unit (symbol atm) of pressure equal to 760 torr, 1013.25 millibars, or 1.01325×10^5 newtons per square metre. The actual pressure exerted by the atmosphere fluctuates around this value, which is assumed to be standard at sea level and 0°C, and is used when dealing with very high pressures.

ATMOSPHERE: JUST THIN AIR

If the Earth were the size of an apple, the atmosphere would be no thicker than the apple skin. Three-quarters of the atmosphere's mass lies below a height of 10 km/35,000 ft. The air at the top of Mount Everest is only one-third as thick as at sea level.

---

atmospheric pollution contamination of the atmosphere with the harmful by-products of human activity; see ◊air pollution.

atmospheric pressure the pressure at any point on the Earth's surface that is due to the weight of the column of air above it; it therefore decreases as altitude increases. At sea level the average pressure is 101 kilopascals (1,013 millibars; 760 mmHg, or 14.7 lb per sq in). Changes in atmospheric pressure, measured with a barometer, are used in weather forecasting. Areas of relatively high pressure are called ◊anticyclones; areas of low pressure are called ◊depressions.

atoll continuous or broken circle of ◊coral reef and low coral islands surrounding a lagoon.

atom smallest unit of matter that can take part in a chemical reaction, and which cannot be broken down chemically into anything simpler. An atom is made up of protons and neutrons in a central nucleus surrounded by electrons (see ◊atomic structure). The atoms of the various elements differ in atomic number, relative atomic mass, and chemical behaviour. There are 109 different types of atom, corresponding with the 109 known elements as listed in the ◊periodic table of the elements.

Atoms are much too small to be seen even by even the most powerful optical microscope (the largest, caesium, has a diameter of 0.0000005 mm/0.00000002 in), and they are in constant motion. (Modern electron microscopes, such as the

composition of the atmosphere

gas	symbol	volume (%)	role
nitrogen	N_2	78.08	cycled through human activities and through the action of microorganisms on animal and plant waste
oxygen	O_2	20.94	cycled mainly through the respiration of animals and plants and through the action of photosynthesis
carbon dioxide	CO_2	0.03	cycled through respiration and photosynthesis in exchange reactions with oxygen. It is also a product of burning fossil fuels
argon	Ar	0.093	chemically inert and with only a few industrial uses
neon	Ne	0.0018	as argon
helium	He	0.0005	as argon
krypton	Kr	trace	as argon
xenon	Xe	trace	as argon
ozone	O_3	0.00006	a product of oxygen molecules split into single atoms by the Sun's radiation and unaltered oxygen molecules
hydrogen	H_2	0.00005	unimportant

◊scanning tunnelling microscope (STM) and the ◊atomic force microscope (AFM), can produce images of individual atoms and molecules.) Belief in the existence of atoms dates back to the ancient Greek natural philosophers. The first scientist to gather evidence for the existence of atoms was British chemist John Dalton, in the 19th century, who believed that every atom was a complete unbreakable entity. New Zealand physicist Ernest Rutherford showed by experiment that an atom in fact consists of a nucleus surrounded by negatively charged particles called electrons.

. . . if your fist is as big as the nucleus of one atom then the atom is as big as St Paul's, and if it happens to be a hydrogen atom then it has a single electron flitting about like a moth in the empty cathedral, now by the dome, now by the altar . . . Every atom is a cathedral.

On the **atom** Tom Stoppard (1937–)
Hapgood 1988

atom, electronic structure of the arrangement of electrons around the nucleus of an atom, in distinct energy levels, also called orbitals or shells (see ◊orbital, atomic). These shells can be regarded as a series of concentric spheres, each of which can contain a certain maximum number of electrons; the noble gases have an arrangement in which every shell contains this number (see ◊noble gas structure). The energy levels are usually numbered beginning with the shell nearest to the nucleus. The outermost shell is known as the ◊valency shell as it contains the valence electrons.

The lowest energy level, or innermost shell, can contain no more than two electrons. Outer shells are considered to be stable when they contain eight electrons but additional electrons can sometimes be accommodated provided that the outermost shell has a stable configuration. Electrons in unfilled shells are available to take part in chemical bonding, giving rise to the concept of valency. In ions, the electron shells contain more or fewer electrons than are required for a neutral atom, generating negative or positive charges.

The atomic number of an element indicates the number of electrons in a neutral atom. From this it is possible to deduce its electronic structure. For example, sodium has atomic number 11 ($Z = 11$) and its electronic arrangement (configuration) is

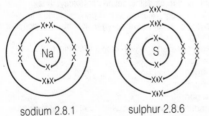

sodium 2.8.1   sulphur 2.8.6

atom, electronic structure of

two electrons in the first energy level, eight electrons in the second energy level and one electron in the third energy level—generally written as 2.8.1. Similarly for sulphur ($Z = 16$), the electron arrangement will be 2.8.6. The electronic structure dictates whether two elements will combine by ionic or covalent bonding (see ◊bond) or not at all.

atomic clock timekeeping device regulated by various periodic processes occurring in atoms and molecules, such as atomic vibration or the frequency of absorbed or emitted radiation.

The first atomic clock was the **ammonia clock**, invented at the US National Bureau of Standards 1948. It was regulated by measuring the speed at which the nitrogen atom in an ammonia molecule vibrated back and forth. The rate of molecular vibration is not affected by temperature, pressure, or other external influences, and can be used to regulate an electronic clock.

A more accurate atomic clock is the **caesium clock**. Because of its internal structure, a caesium atom produces or absorbs radiation of a very precise frequency (9,192,631,770 Hz) that varies by less than one part in 10 billion. This frequency has been used to define the second, and is the basis of atomic clocks used in international timekeeping.

Hydrogen maser clocks, based on the radiation from hydrogen atoms, are the most accurate. The hydrogen maser clock at the US Naval Research Laboratory, Washington DC, is estimated to lose one second in 1,700,000 years. Cooled hydrogen maser clocks could theoretically be accurate to within one second in 300 million years.

Atomic clocks are so accurate that minute adjustments must be made periodically to the length of the year to keep the calendar exactly synchronized with the Earth's rotation which has a tendency to speed up or slow down. There have been 17 adjustments made since 1972. In 1992 the northern hemisphere's summer was longer than usual—by one second. An extra second was added to the world's time at precisely 23 hours, 59 minutes and 60 seconds on June 30 1992. The adjustment was called for by the International Earth Rotation Service in Paris, which monitors the difference between Earth time and atomic time.

atomic energy another name for ◊nuclear energy.

atomic force microscope (AFM) microscope developed in the late 1980s that produces a magnified image using a diamond probe, with a tip so fine that it may consist of a single atom, dragged over the surface of a specimen to 'feel' the contours of the surface. In effect, the tip acts like the stylus of a phonograph or record player, reading the surface. The tiny up-and-down movements of the probe are converted to an image of the surface by computer, and displayed on a screen. The AFM is useful for examination of biological specimens since, unlike the ◊scanning tunnelling microscope, the specimen does not have to be electrically conducting.

atomicity number of atoms of an ◊element that combine together to form a molecule. A molecule of oxygen (O_2) has atomicity 2; sulphur (S_8) has atomicity 8.

atomic mass see ◊relative atomic mass.

atomic mass unit or *dalton unit* (symbol amu or u) unit of mass that is used to measure the relative mass of atoms and molecules. It is equal to one-twelfth of the mass of a carbon-12 atom, which is equivalent to the mass of a proton or 1.66 $\times$ 10^{-27} kg. The ◊relative atomic mass of an atom has no units; thus oxygen-16 has an atomic mass of 16 daltons, but a relative atomic mass of 16.

atomic number or *proton number* the number (symbol Z) of protons in the nucleus of an atom. It is equal to the positive charge on the nucleus. In a neutral atom, it is also equal to the number of electrons surrounding the nucleus. The 109 elements are arranged in the ◊periodic table of the elements according to their atomic number. See also ◊nuclear notation.

atomic radiation energy given out by disintegrating atoms during ◊radioactive decay, whether natural or synthesized. The energy may be in the form of fast-moving particles, known as ◊alpha particles and ◊beta particles, or in the form of high-energy electromagnetic waves known as ◊gamma radiation. Overlong exposure to atomic radiation can lead to ◊radiation sickness. Radiation biology studies the effect of radiation on living organisms.

atomic size or *atomic radius* size of an atom expressed as the radius in ◊angstroms or other units of length. The sodium atom has an atomic radius of 1.57 angstroms (1.57 $\times$ 10^{-8} cm). For metals, the size of the atom is always greater than the size of its ion. For non-metals the reverse is true.

atomic structure internal structure of an ◊atom. The core of the atom is the *nucleus*, a dense body only one ten-thousandth the diameter of the atom itself. The simplest nucleus, that of hydrogen, comprises a single stable positively charged particle, the *proton*. Nuclei of other elements contain more protons and additional particles, called *neutrons*, of about the same mass as the proton but with no electrical charge. Each element has its own characteristic nucleus with a unique number of protons, the atomic number. The number of neutrons may vary. Where atoms of a single element have different numbers of neutrons, they are called ◊isotopes. Although some isotopes tend to be unstable and exhibit ◊radioactivity, they all have identical chemical properties.

The nucleus is surrounded by a number of moving *electrons*, each of which has a negative charge equal to the positive charge on a proton, but which weighs only 1/1,839 times as much. In a neutral atom, the nucleus is surrounded by the same number of electrons as it contains protons. According to ◊quantum theory, the position of an electron is uncertain; it may be found at any point. However, it is more likely to be found in some places than others. The region of space in which an electron is most likely to be found is called an orbital (see ◊orbital, atomic). The chemical properties of an element are determined by the ease with which its atoms can gain or lose electrons from its outer orbitals.

High-energy physics research has discovered the existence of subatomic particles (see ◊particle physics) other than the proton, neutron, and electron. More than 300 kinds of particle are now known, and these are classified into several classes according to their mass, electric charge, spin, magnetic moment, and interaction. The ◊elementary particles, which include the electron, are indivisible and may be regarded as the fundamental units of matter; the *hadrons*, such as the proton and neutron, are composite particles made up of either two or three elementary particles called quarks.

Atoms are held together by the electrical forces of attraction between each negative electron and the positive protons within the nucleus. The latter repel one another with enormous forces; a nucleus holds together only because an even stronger force, called the *strong nuclear force*, attracts the protons and neutrons to one another. The strong force acts over a very short range – the protons and neutrons must be in virtual contact with one another (see ◊forces, fundamental). If, therefore, a fragment of a complex nucleus, containing some protons, becomes only slightly loosened from the main group of neutrons and protons, the natural repulsion between the protons will cause this fragment to fly apart from the rest of the nucleus at high speed. It is by such fragmentation of atomic nuclei (nuclear ◊fission) that nuclear energy is released.

atomic time time as given by ◊atomic clocks, which are regulated by natural resonance frequencies of particular atoms, and display a continuous count of seconds.

In 1967 a new definition of the second was adopted in the SI system of units: the duration of 9,192,631,770 periods of the radiation corresponding to the transition between two hyperfine levels of the ground state of the caesium-133 atom. The International Atomic Time Scale is based on clock data from a number of countries; it is a continuous scale in days, hours, minutes, and seconds from the origin on 1 Jan 1958, when the Atomic Time Scale was made 0 h 0 min 0 sec when Greenwich Mean Time was at 0 h 0 min 0 sec.

atomic weight another name for ◊relative atomic mass.

atomizer device that produces a spray of fine droplets of liquid. A vertical tube connected with a horizontal tube dips into a bottle of liquid, and at one end of the horizontal tube is a nozzle, at the other a rubber bulb. When the bulb is squeezed, air rushes over the top of the vertical tube and out through the nozzle. Following ◊Bernoulli's principle, the pressure at the top of the vertical tube is reduced, allowing the liquid to rise. The air stream picks up the liquid, breaks it up into tiny drops, and carries it out of the nozzle as a spray. Scent spray, paint spray guns and carburettors all use the principle of the atomizer.

ATP abbreviation for *adenosine triphosphate*, a nucleotide molecule found in all cells. It can yield large amounts of energy, and is used to drive the thousands of biological processes needed to sustain life, growth, movement, and reproduction. Green plants use light energy to manufacture ATP as part of the process of ◊photosynthesis. In animals, ATP is formed by the breakdown of glucose molecules, usually obtained from the carbohydrate component of a diet, in a series of reactions termed ◊respiration. It is the driving force behind muscle contraction and the synthesis of complex molecules needed by individual cells.

atrium one of the upper chambers of the heart, receiving blood under low pressure as it returns from the body. Atrium walls are thin and stretch easily to allow blood into the heart. On contraction, the atria force blood into the thick-walled ventricles, which then give a second, more powerful beat.

attrition in earth science, the process by which particles of rock being transported by river, wind, or sea are rounded and gradually reduced in size by being struck against one another.

The rounding of particles is a good indication of how far they have been transported. This is particularly true for particles carried by rivers, which become more rounded as the distance downstream increases.

auditory canal tube leading from the outer ◊ear opening to the eardrum. It is found only in animals whose eardrums are located inside the skull, principally mammals and birds.

Auriga constellation of the northern hemisphere, represented as a man driving a chariot. Its brightest star is the first-magnitude ◊Capella, about 45 light-years from Earth; Epsilon Aurigae is an ◊eclipsing binary star with a period of 27 years, the longest of its kind (last eclipse 1983).

The charioteer is usually identified as Erichthonius, legendary king of Athens, who invented the four-horse chariot.

aurora coloured light in the night sky near the Earth's magnetic poles, called *aurora borealis*, 'northern lights', in the northern hemisphere and *aurora australis* in the southern hemisphere. Although auroras are usually restricted to the polar skies, fluctuations in the ◊solar wind occasionally cause them to be visible at lower latitudes. An aurora is usually in the form of a luminous arch with its apex towards the magnetic pole followed by arcs, bands, rays, curtains, and coronas, usually green but often showing shades of blue and red, and sometimes yellow or white. Auroras are caused at heights of over 100 km/60 mi by a fast stream of charged particles from solar flares and low-density 'holes' in the Sun's corona. These are guided by the Earth's magnetic field towards the north and south magnetic poles, where they enter the upper atmosphere and bombard the gases in the atmosphere, causing them to emit visible light.

Australia Telescope giant radio telescope in New South Wales, Australia, operated by the Commonwealth Scientific and Industrial Research Organization (CSIRO). It consists of six 22–m/72–ft antennae at Culgoora, a similar antenna at Siding Spring Mountain, and the 64–m/210–ft ◊Parkes radio telescope—the whole simulating a dish 300 m/186 mi across.

autoclave pressurized vessel that uses superheated steam to sterilize materials and equipment such as surgical instruments. It is similar in principle to a pressure cooker.

autogiro or *autogyro* heavier-than-air craft that supports itself in the air with a rotary wing, or rotor. The Spanish aviator Juan de la Cierva designed the first successful autogiro 1923. The autogiro's rotor provides only lift and not propulsion; it has been superseded by the helicopter, in which the rotor provides both. The autogiro is propelled by an orthodox propeller.

The three-or four-bladed rotor on an autogiro spins in a horizontal plane on top of the craft, and is not driven by the engine. The blades have an aerofoil cross section, as a plane's wings. When the autogiro moves forward, the rotor starts to rotate by itself, a state known as autorotation. When travelling fast enough, the rotor develops enough lift from its aerofoil blades to support the craft.

autolysis in biology, the destruction of a ◊cell after its death by the action of its own ◊enzymes, which break down its structural molecules.

automatic pilot or *autopilot* or *gyropilot* control device that keeps an aeroplane flying automatically on a given course at a given height and speed. Devised by US business executive Lawrence Sperry 1912, the automatic pilot contains a set of ◊gyroscopes that provide references for the plane's course. Sensors detect when the plane deviates from this course and send signals to the control surfaces—the ailerons, elevators, and rudder—to take the appropriate action. Most airliners cruise on automatic pilot for much of the time. Autopilot is also used in missiles.

automaton mechanical figure imitating human or animal performance. Automatons are usually designed for aesthetic appeal as opposed to purely functional robots. The earliest recorded automaton is an Egyptian wooden pigeon of 400 BC.

autonomic nervous system in mammals, the part of the nervous system that controls the involuntary activities of the smooth muscles (of the digestive tract, blood vessels), the heart, and the glands. The *sympathetic* system responds to stress, when it speeds the heart rate, increases blood pressure and generally prepares the body for action. The *parasympathetic* system is more important when the body is at rest, since it slows the heart rate, decreases blood pressure, and stimulates the digestive system.

At all times, both types of autonomic nerves carry signals that bring about adjustments in visceral organs. The actual rate of heart beat is the net outcome of opposing signals. Today, it is known that the word 'autonomic' is not correct—the reflexes managed by this system are actually integrated by commands from the brain and spinal cord (the central nervous system).

autoradiography in biology, a technique for following the movement of molecules within an organism, especially a plant, by labelling with a radioactive isotope that can be traced on photographs. It is used to study ◊photosynthesis, where the pathway of radioactive carbon dioxide can be traced as it moves through the various chemical stages.

autosome any ◊chromosome in the cell other than a sex chromosome. Autosomes are of the same number and kind in both males and females of a given species.

autotroph any living organism that synthesizes organic substances from inorganic molecules by using light or chemical energy. Autotrophs are the *primary producers* in all food chains since the materials they synthesize and store are the energy sources of all other organisms. All green plants and many planktonic organisms are autotrophs, using sunlight to convert carbon dioxide and water into sugars by ◊photosynthesis.

The total ◊biomass of autotrophs is far greater than that of animals, reflecting the dependence of animals on plants, and the ultimate dependence of all life on energy from the Sun—green plants convert light energy into a form of chemical energy (food) that animals can exploit. Some bacteria use the chemical energy of sulphur compounds to synthesize organic substances. See also ◊heterotroph.

autumnal equinox see ◊equinox.

auxin plant ◊hormone that promotes stem and root growth in plants. Auxins influence many aspects of plant growth and development, including cell enlargement, inhibition of development of axillary buds, ◊tropisms, and the initiation of roots. *Synthetic auxins* are used in rooting powders for cuttings, and in some weedkillers, where high auxin concentrations cause such rapid growth that the plants die. They are also used to prevent premature fruitdrop in orchards. The most common naturally occurring auxin is known as indoleacetic acid, or IAA. It is produced in the shoot apex and transported to other parts of the plant.

avalanche (from French *avaler* 'to swallow') fall of a mass of snow and ice down a steep slope. Avalanches occur because of the unstable nature of snow masses in mountain areas.

Changes of temperature, sudden sound, or earth-borne vibrations may trigger an avalanche, particularly on slopes of more than 35°. The snow compacts into ice as it moves, and rocks may be carried along, adding to the damage caused.

average in statistics, a term used inexactly to indicate the typical member of a set of data. It usually refers to the ◊arithmetic mean. The term is also used to refer to the middle member of the set when it is sorted in ascending or descending order (the ◊median), and the most commonly occurring item of data (the ◊mode), as in 'the average family'.

aviation term used to describe both the science of powered ◊flight and also aerial navigation by means of an aeroplane.

Avogadro's hypothesis in chemistry, the law stating that equal volumes of all gases, when at the same temperature and pressure, have the same numbers of molecules. It was first propounded by Italian physicist Amedeo Avogadro.

Avogadro's number or *Avogadro's constant* the number of carbon atoms in 12 g of the carbon-12 isotope (6.022045×10^{23}). The relative atomic mass of any element, expressed in grams, contains this number of atoms.

avoirdupois system of units of mass based on the pound (0.45 kg), which consists of 16 ounces (each of 16 drams) or 7,000 grains (each equal to 65 mg).

axil upper angle between a leaf (or bract) and the stem from which it grows. Organs developing in the axil, such as shoots and buds, are termed axillary, or lateral.

axiom in mathematics, a statement that is assumed to be true and upon which theorems are proved by using logical deduction; for example, two straight lines cannot enclose a space. The Greek mathematician Euclid used a series of axioms that he considered could not be demonstrated in terms of simpler concepts to prove his geometrical theorems.

axis (plural *axes*) in geometry, one of the reference lines by which a point on a graph may be located. The horizontal axis is usually referred to as the x-axis, and the vertical axis as the y-axis. The term is also used to refer to the imaginary line about which an object may said to be symmetrical (*axis of symmetry*)— for example, the diagonal of a square—or the line about which an object may revolve (*axis of rotation*).

axon long threadlike extension of a ◊nerve cell that conducts electrochemical impulses away from the cell body towards other nerve cells, or towards an effector organ such as a muscle. Axons terminate in ◊synapses with other nerve cells, muscles, or glands.

azimuth in astronomy, the angular distance of an object eastwards along the horizon, measured from due north, between the astronomical ◊meridian (the vertical circle passing through the centre of the sky and the north and south points on the horizon) and the vertical circle containing the celestial body whose position is to be measured.

azo dye synthetic dye containing the azo group of two nitrogen atoms (N=N) connecting aromatic ring compounds. Azo dyes are usually red, brown, or yellow, and make up about half the dyes produced. They are manufactured from aromatic ◊amines.

AZT drug used in the treatment of AIDS; see ◊zidovudine.

B

Babbit metal soft, white metal, an ◊alloy of tin, lead, copper, and antimony, used to reduce friction in bearings, developed by the US inventor Isaac Babbit 1839.

bacillus member of a group of rodlike ◊bacteria that occur everywhere in the soil and air. Some are responsible for diseases such as anthrax or for causing food spoilage.

backcross a breeding technique used to determine the genetic makeup of an individual organism.

background radiation radiation that is always present in the environment. By far the greater proportion (87%) of it is emitted from natural sources. Alpha and beta particles, and gamma radiation are radiated by the traces of radioactive minerals that occur naturally in the environment and even in the human body, and by radioactive gases such as radon and thoron, which are found in soil and may seep upwards into buildings. Radiation from space (◊cosmic radiation) also contributes to the background level.

Because background radiation increases the count rate obtained during any experimental work on radioactive materials, a background count rate is usually subtracted from any data obtained.

backing storage in computing, memory outside the ◊central processing unit used to store programs and data that are not in current use. Backing storage must be nonvolatile—that is, its contents must not be lost when the power supply to the computer system is disconnected.

backup system in computing, a duplicate computer system that can take over the operation of a main computer system in the event of equipment failure. A large interactive system, such as an airline's ticket-booking system, cannot be out of action for even a few hours without causing considerable disruption. In such cases a complete duplicate computer system may be provided to take over and run the system should the main computer develop a fault or need maintenance. Backup systems include *incremental backup* and *full backup*.

bacteria (singular *bacterium*) microscopic unicellular organisms with prokaryotic cells (see ◊prokaryote). They usually reproduce by ◊binary fission (dividing into two equal parts), and since this may

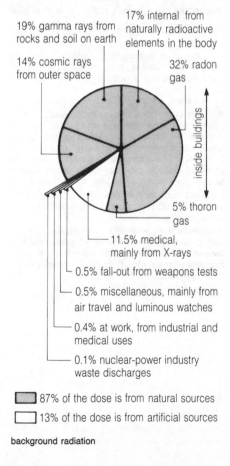

17% internal from naturally radioactive elements in the body

19% gamma rays from rocks and soil on earth

14% cosmic rays from outer space

32% radon gas

inside buildings

5% thoron gas

11.5% medical, mainly from X-rays

0.5% fall-out from weapons tests

0.5% miscellaneous, mainly from air travel and luminous watches

0.4% at work, from industrial and medical uses

0.1% nuclear-power industry waste discharges

87% of the dose is from natural sources

13% of the dose is from artificial sources

background radiation

occur approximately every 20 minutes, a single bacterium is potentially capable of producing 16 million copies of itself in a day. It is thought that 1–10% of the world's bacteria have been identified.

Bacteria are now classified biochemically, but their varying shapes provide a rough classification; for example, *cocci* are round or oval, *bacilli* are rodlike, *spirilla* are spiral, and *vibrios* are shaped like commas. Exceptionally, one bacterium has been found, *Gemmata obscuriglobus*, that does have a nucleus. Unlike ◊viruses, bacteria do not necessarily need contact with a live cell to become active.

Bacteria have a large loop of ◊DNA, sometimes called a bacterial chromosome. In addition there are often small, circular pieces of DNA known as ◊plasmids that carry spare genetic information. These plasmids can readily move from one bacterium to another, even though the bacteria may be of different species. In a sense they are parasites within the bacterial cell, but they survive by coding characteristics that promote the survival of their hosts. For example, some plasmids confer antibiotic resistance on the bacteria they inhabit. The rapid and problematic spread of antibiotic resistance among bacteria is due to plasmids, but they are also useful to humans in ◊genetic engineering.

Although generally considered harmful, certain types of bacteria are vital in many food and industrial processes, while others play an essential role in the ◊nitrogen cycle, which maintains soil fertility. For example, bacteria are used to break down waste products, such as sewage; make butter, cheese, and yoghurt; cure tobacco; tan leather; and (by virtue of the ability of certain bacteria to attack metal) clean ships' hulls and de-rust their tanks, and even extract minerals from mines.

Bacteria cannot normally survive temperatures above 100°C/212°F, such those produced in pasteurization; but those in deep sea hot vents in the eastern Pacific are believed to withstand temperatures of 350°C/662°F. *Thermus aquaticus*, or taq, grows freely in the boiling waters of hot springs, and an enzyme derived from it is used in genetic engineering to speed up the production of millions of copies of any DNA sequence, a reaction requiring very high temperatures.

Certain bacteria can influence the growth of others; for example, lactic acid bacteria will make conditions unfavourable for salmonella bacteria. Other strains produce nisin, which inhibits growth of listeria and botulism organisms. Plans in the food industry are underway to produce super strains of lactic acid bacteria to avoid food poisoning. In 1990, a British team of food scientists announced a new, rapid (five-minute) test for contamination

CLASSIFYING BACTERIA

Bacteria can be classified into two broad classes (called Gram positive and negative) according to their reactions to certain stains, or dyes, used in microscopy. The staining technique, called the Gram test after Danish bacteriologist Hans Gram, allows doctors to identify many bacteria quickly.

of food by listeria or salmonella bacteria. Fluorescent dyes, added to a liquidized sample of food, reveal the presence of bacteria under laser light.

bacteriology the study of ◊bacteria.

bacteriophage virus that attacks ◊bacteria. Such viruses are now of use in genetic engineering.

badlands barren landscape cut by erosion into a maze of ravines, pinnacles, gullies and sharp-edged ridges. Areas in South Dakota and Nebraska, USA, are examples.

Badlands, which can be created by overgrazing, are so called because of their total lack of value for agriculture and their inaccessibility.

Baikonur launch site for spacecraft, located at Tyuratam, Kazakhstan, near the Aral Sea: the first satellites and all Soviet space probes and crewed Soyuz missions were launched from here. It covers an area of 12,200 sq km/4,675 sq mi, much larger than its US equivalent, the ◊Kennedy Space Center in Florida.

Baily's beads bright spots of sunlight seen around the edge of the Moon for a few seconds immediately before and after a total ◊eclipse of the Sun, caused by sunlight shining between mountains at the Moon's edge. Sometimes one bead is much brighter than the others, producing the so-called *diamond ring* effect. The effect was described 1836 by the English astronomer Francis Baily (1774–1844), a wealthy stockbroker who retired in 1825 to devote himself to astronomy.

Bakelite first synthetic ◊plastic, created by US inventor Leo Baekeland 1909. Bakelite is hard, tough, and heatproof, and is used as an electrical insulator. It is made by the reaction of phenol with formaldehyde, producing a powdery resin that sets solid when heated. Objects are made by subjecting the resin to compression moulding (simultaneous heat and pressure in a mould).

It is one of the thermosetting plastics, which do not remelt when heated, and is often used for electrical fittings.

baking powder mixture of bicarbonate of soda (◊sodium hydrogencarbonate), an acidic compound, and a nonreactive filler (usually starch or calcium sulphate), used in baking as a raising agent. It gives a light open texture to cakes and scones, and is used as a substitute for yeast in making soda bread.

Several different acidic compounds (for example, tartaric acid, cream of tartar, sodium or calcium acid phosphates, and glucono-delta-lactone) may be used, any of which will react with the sodium hydrogencarbonate, in the presence of water and heat, to release the carbon dioxide that causes the cakemix or dough to rise.

balance apparatus for weighing or measuring mass. The various types include the *beam balance* consisting of a centrally pivoted lever with pans hanging from each end, and the *spring balance*, in which the object to be weighed stretches (or compresses) a vertical coil spring fitted with a pointer that indicates the weight on a scale. Kitchen and bathroom scales are balances.

balanced diet diet that includes carbohydrate, protein, fat, vitamins, water, minerals, and fibre (roughage). Although it is agreed that all these substances are needed if a person is to be healthy, argument occurs over how much of each type a person needs.

balance of nature in ecology, the idea that there is an inherent equilibrium in most ◊ecosystems, with plants and animals interacting so as to produce a stable, continuing system of life on Earth. Organisms in the ecosystem are adapted to each other — for example, waste products produced by one species are used by another and resources used by some are replenished by others; the oxygen needed by animals is produced by plants while the waste product of animal respiration, carbon dioxide, is used by plants as a raw material in photosynthesis. The nitrogen cycle, the water cycle, and the control of animal populations by natural predators are other examples. The activities of human beings can, and frequently do, disrupt the balance of nature.

baldness loss of hair from the upper scalp, common in older men. Its onset and extent are influenced by genetic make-up and the level of male sex ◊hormones. There is no cure, and expedients such as hair implants may have no lasting effect. Hair loss in both sexes may also occur as a result of ill health or radiation treatment, such as for cancer. *Alopecia*, a condition in which the hair falls out, is different from the 'male-pattern baldness' described above.

ball-and-socket joint a joint allowing considerable movement in three dimensions, for instance the joint between the pelvis and the femur. To facilitate movement, such joints are lubricated by cartilage and synovial fluid. The bones are kept in place by ligaments and moved by muscles.

ballistics study of the motion and impact of projectiles such as bullets, bombs, and missiles. For projectiles from a gun, relevant exterior factors include temperature, barometric pressure, and wind strength; and for nuclear missiles these extend to such factors as the speed at which the Earth turns.

balloon lighter-than-air craft that consists of a gasbag filled with gas lighter than the surrounding air and an attached basket, or gondola, for carrying passengers and/or instruments. In 1783, the first successful human ascent was in Paris, in a hot-air balloon designed by the Montgolfier brothers Joseph Michel and Etienne Jacques. In 1785, a hydrogen-filled balloon designed by French physicist Jacques Charles travelled across the English Channel.

During the French Revolution balloons were used for observation; in World War II they were used to defend London against low-flying aircraft. They are now used for recreation and as a means of meteorological, infrared, gamma-ray, and ultraviolet observation. The first transatlantic crossing by balloon was made 11–17 Aug 1978 by a US team.

ball valve valve that works by the action of external pressure raising a ball and thereby opening a hole.

An example is the valve used in lavatory cisterns to cut off the water supply when it reaches the correct level. It consists of a flat rubber washer at

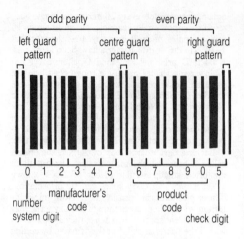

bar code *The bars of varying thicknesses and spacings represent two series of numbers, identifying the manufacturer and the product. Two longer, thinner bars mark the beginning and end of the manufacturer and product codes. The bar code is used on groceries, books, and most articles for sale in shops.*

one end of a pivoting arm and a hollow ball at the other. The ball floats on the water surface, rising as the cistern fills, and at the correct level the rubber washer is pushed against the water-inlet pipe, cutting off the flow.

bar unit of pressure equal to 10^5 pascals or 10^6 dyne cm^{-2}, approximately 750 mmHg or 0.987 atm. Its diminutive, the *millibar* (one-thousandth of a bar), is commonly used by meteorologists.

bar chart in statistics, a way of displaying data, using horizontal or vertical bars. The heights or lengths of the bars are proportional to the quantities they represent.

bar code pattern of bars and spaces that can be read by a computer. Bar codes are widely used in retailing, industrial distribution, and public libraries. The code is read by a scanning device; the computer determines the code from the widths of the bars and spaces.

The technique was patented 1949 but became popular only in 1973, when the food industry in North America adopted the Universal Product Code system.

barium (Greek *barytes* 'heavy') soft, silver-white, metallic element, symbol Ba, atomic number 56, relative atomic mass 137.33. It is one of the ◊alkaline-earth metals, found in nature as barium carbonate and barium sulphate. As the sulphate it is used in medicine: taken as a suspension (a 'barium meal'), its progress is followed by using X-rays to reveal abnormalities of the alimentary canal. Barium is also used in alloys, pigments, and safety matches and, with strontium, forms the emissive surface in cathode-ray tubes. It was first discovered in barytes or heavy spar.

barium chloride $BaCl_2$ white, crystalline solid, used in aqueous solution to test for the presence of the sulphate ion. When a solution of barium chloride is mixed with a solution of a sulphate, a white precipitate of barium sulphate is produced.

$$Ba^{2+}_{(aq)} + SO^{2-}_{4(aq)} \rightarrow BaSO_{4(s)}$$

barium nitrate $Ba(NO_3)_2$ compound that, like barium chloride, can be used to test for the presence of the sulphate ion as it readily dissolves in water. When sulphate is added, a white precipitate of barium sulphate is formed.

bark protective outer layer on the stems and roots of woody plants, composed mainly of dead cells. To allow for expansion of the stem, the bark is continually added to from within, and the outer surface often becomes cracked or is shed as scales.

Trees deposit a variety of chemicals in their bark, including poisons. Many of these chemical substances have economic value because they can be used in the manufacture of drugs. Quinine, derived from the bark of the *Cinchona* tree, is used to fight malarial infections; curare, an anaesthetic used in medicine, comes from the *Strychnus toxifera* tree in the Amazonian rainforest.

Bark technically includes all the tissues external to the vascular ◊cambium (the ◊phloem, cortex, and periderm), and its thickness may vary from 2.5 mm/0.1 in to 30 cm/12 in or more, as in the giant redwood *Sequoia* where it forms a thick, spongy layer.

Barnard's star second closest star to the Sun, six light years away in the constellation Ophiuchus. It is a faint red dwarf of 10th magnitude, visible only through a telescope. It is named after the US astronomer Edward Emerson Barnard (1857–1923), who discovered 1916 that it has the fastest proper motion of any star, crossing 1 degree of sky every 350 years. Some observations suggest that Barnard's star may be accompanied by planets.

barograph device for recording variations in atmospheric pressure. A pen, governed by the movements of an aneroid ◊barometer, makes a continuous line on a paper strip on a cylinder that rotates over a day or week to create a *barogram*, or permanent record of variations in atmospheric pressure.

barometer instrument that measures atmospheric pressure as an indication of weather. Most often used are the *mercury barometer* and the *aneroid barometer*.

In a mercury barometer a column of mercury in a glass tube, roughly 0.75 m/2.5 ft high (closed at one end, curved upwards at the other), is balanced by the pressure of the atmosphere on the open end; any change in the height of the column reflects a change in pressure. In an aneroid barometer, a shallow cylindrical metal box containing a partial vacuum expands or contracts in response to changes in pressure.

barrel unit of liquid capacity, the value of which depends on the liquid being measured. It is used for petroleum, a barrel of which contains 159 litres/35 imperial gallons; a barrel of alcohol contains 189 litres/41.5 imperial gallons.

barrier island long island of sand, lying offshore and parallel to the coast. Some are over 100 km/60 mi in length. Most barrier islands are derived from marine sands piled up by shallow longshore currents that sweep sand parallel to the seashore. Others are derived from former spits, connected

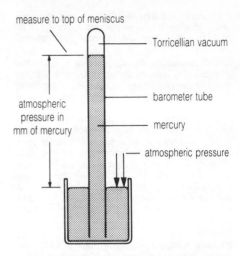

measure to top of meniscus

Torricellian vacuum

barometer tube

mercury

atmospheric pressure

atmospheric pressure in mm of mercury

barometer *In the mercury barometer, the weight of the column of mercury is balanced by the pressure of the atmosphere on the lower end. A change in height of the column indicates a change in atmospheric pressure. In an aneroid barometer, any change of atmospheric pressure causes a metal box which contains a vacuum to be squeezed or to expand slightly. The movements of the box sides are transferred to a pointer and scale via a chain of levers.*

to land and built up by drifted sand, that were later severed from the mainland.

Often several islands lie in a continuous row offshore. Coney Island and Jones Beach near New York City are well-known examples, as is Padre Island, Texas. The Frisian Islands are barrier islands along the coast of the Netherlands.

barrier reef ◊coral reef that lies offshore, separated from the mainland by a shallow lagoon.

baryon in nuclear physics, a heavy subatomic particle made up of three indivisible elementary particles called quarks. The baryons form a subclass of the ◊hadrons, and comprise the nucleons (protons and neutrons) and hyperons.

baryte barium sulphate, $BaSO_4$, the most common mineral of barium. It is white or light-coloured, and has a comparatively high density (specific gravity 4.6); the latter property makes it useful in the production of high-density drilling muds (muds used to cool and lubricate drilling equipment). Baryte occurs mainly in ore veins, where it is often found with calcite and with lead and zinc minerals. It crystallizes in the orthorhombic system and can form tabular crystals or radiating fibrous masses.

basal metabolic rate (BMR) amount of energy needed by an animal just to stay alive. It is measured when the animal is awake but resting, and includes the energy required to keep the heart beating, sustain breathing, repair tissues, and keep the brain and nerves functioning. Measuring the animal's consumption of oxygen gives an accurate value for BMR, because oxygen is needed to release energy from food.

A cruder measure of BMR estimates the amount of heat given off, some heat being released when food is used up. BMR varies from one species to another, and from males to females. In humans, it is highest in children and declines with age. Disease, including mental illness, can make it rise or fall. Hormones from the ◊thyroid gland control the BMR.

basalt commonest volcanic ◊igneous rock, and the principal rock type on the ocean floor; it is basic, that is, it contains relatively little silica: about 50%. It is usually dark grey, but can also be green, brown, or black.

The groundmass may be glassy or finely crystalline, sometimes with large ◊crystals embedded. Basaltic lava tends to be runny and flows for great distances before solidifying. Successive eruptions of basalt have formed the great plateaux of Colorado and the Indian Deccan. In some places, such as Fingal's Cave in the Inner Hebrides of Scotland and the Giant's Causeway in Antrim, Northern Ireland, shrinkage during the solidification of the molten lava caused the formation of hexagonal columns.

bascule bridge type of drawbridge in which one or two counterweighted deck members pivot upwards to allow shipping to pass underneath. One example is the double bascule Tower Bridge, London.

base in chemistry, a substance that accepts protons, such as the hydroxide ion (OH⁻) and ammonia (NH₃). Bases react with acids to give a salt. Those that dissolve in water are called ◊alkalis.

Inorganic bases are usually oxides or hydroxides of metals, which react with dilute acids to form a salt and water. A number of carbonates also react with dilute acids, additionally giving off carbon dioxide. Many organic compounds that contain nitrogen are bases.

base in mathematics, the number of different single-digit symbols used in a particular number system. In our usual (decimal) counting system of numbers (with symbols 0, 1, 2, 3, 4, 5, 6, 7, 8, 9) the base is 10. In the ◊binary number system, which has only the symbols 1 and 0, the base is two. A base is also a number that, when raised to a particular power (that is, when multiplied by itself a particular number of times as in $10^2 = 10 \times 10 = 100$), has a ◊logarithm equal to the power. For example, the logarithm of 100 to the base ten is 2.

In geometry, the term is used to denote the line or area on which a polygon or solid stands.

In general, any number system subscribing to a place-value system with base value b may be represented by ... $b^4, b^3, b^2, b^1, b^0, b^{-1}, b^{-2}, b^{-3}, ...$ Hence in base ten the columns represent ... $10^4, 10^3, 10^2, 10^1, 10^0, 10^{-1}, 10^{-2}, 10^{-3}$ ..., in base two ... $2^4, 2^3, 2^2, 2^1, 2^0, 2^{-1}, 2^{-2}, 2^{-3}$ ..., and in base eight ... $8^4, 8^3, 8^2, 8^1, 8^0, 8^{-1}, 8^{-2}, 8^{-3}$ ... For bases beyond 10, the denary numbers 10, 11, 12, and so on must be replaced by a single digit. Thus in base 16, all numbers up to 15 must be represented by single-digit 'numbers', since 10 in hexadecimal would mean 16 in decimal. Hence decimal 10, 11, 12, 13, 14, 15 are represented in hexadecimal by letters A, B, C, D, E, F.

base pair in biochemistry, the linkage of two base (purine or pyrimidine) molecules in ◊DNA. They are found in nucleotides, and form the basis of the genetic code.

One base lies on one strand of the DNA double helix, and one on the other, so that the base pairs link the two strands like the rungs of a ladder. In DNA, there are four bases: adenine and guanine (purines) and cytosine and thymine (pyrimidines). Adenine always pairs with thymine, and cytosine with guanine.

BASIC (acronym for *b*eginner's *a*ll-purpose *s*ymbolic *i*nstruction *c*ode) high-level computer-programming language, developed 1964, originally designed to take advantage of ◊multiuser systems (which can be used by many people at the same time). The language is relatively easy to learn and is popular among microcomputer users.

Most versions make use of an ◊interpreter, which translates BASIC into ◊machine code and allows programs to be entered and run with no intermediate translation. Some more recent versions of BASIC allow a ◊compiler to be used for this process.

basicity number of replaceable hydrogen atoms in an acid. Nitric acid (HNO₃) is monobasic, sulphuric acid (H₂SO₄) is dibasic, and phosphoric acid (H₃PO₄) is tribasic.

basic oxide compound formed by a metal and oxygen, containing the O²⁻ ion. If, like sodium oxide (Na₂O), it is soluble, it forms the metal hydroxide when added to water.

$$Na_2O_{(s)} + H_2O_{(l)} \rightarrow 2NaOH_{(aq)}$$

All basic oxides react with acids to form a salt and water.

$$CaO + 2HNO_3 \rightarrow Ca(NO_3)_2 + H_2O$$

basic-oxygen process most widely used method of steelmaking, involving the blasting of oxygen at supersonic speed into molten pig iron.

Pig iron from a blast furnace, together with steel scrap, is poured into a converter, and a jet of oxygen is then projected into the mixture. The excess carbon in the mix and other impurities quickly burn

base

binary (base 2)	octal (base 8)	decimal (base 10)	hexadecimal (base 16)
0	0	0	0
1	1	1	1
10	2	2	2
11	3	3	3
100	4	4	4
101	5	5	5
110	6	6	6
111	7	7	7
1000	10	8	8
1001	11	9	9
1010	12	10	A
1011	13	11	B
1100	14	12	C
1101	15	13	D
1110	16	14	E
1111	17	15	F
10000	20	16	10
11111111	377	255	FF
11111010001	3721	2001	7D1

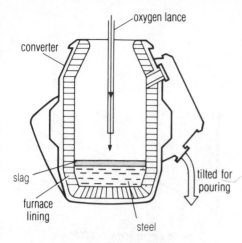

basic-oxygen process *The basic-oxygen process is the primary method used to produce steel. Oxygen is blown at high pressure through molten pig iron and scrap steel in a converter lined with basic refractory materials. The impurities, principally carbon, quickly burn out, producing steel.*

out or form a slag, and the converter is emptied by tilting. It takes only about 45 minutes to refine 350 tonnes/400 tons of steel. The basic-oxygen process was developed 1948 at a steelworks near the Austrian towns of Linz and Donawitz. It is a version of the ◊Bessemer process.

basidiocarp spore-bearing body, or 'fruiting body', of all basidiomycete fungi (see ◊fungus), except the rusts and smuts. A well known example is the edible mushroom *Agaricus brunnescens*. Other types include globular basidiocarps (puffballs) or flat ones that project from tree trunks (brackets). They are made up of a mass of tightly packed, intermeshed ◊hyphae.

The tips of these hyphae develop into the reproductive cells, or *basidia*, that form a fertile layer known as the hymenium, or **gills**, of the basidiocarp. Four spores are budded off from the surface of each basidium.

batch processing in computing, a system for processing data with little or no operator intervention. Batches of data are prepared in advance to be processed during regular 'runs' (for example, each night). This allows efficient use of the computer and is well suited to applications of a repetitive nature, such as a company payroll.

In ◊*interactive computing*, by contrast, data and instructions are entered while the processing program is running.

batholith large, irregular mass of igneous rock, usually granite, with an exposed surface of more than 100 sq km/40 sq mi. The mass forms by the intrusion or upswelling of magma (molten rock) through the surrounding rock. Batholiths form the core of all major mountain ranges.

According to plate tectonic theory, magma rises in subduction zones along continental margins where one plate sinks beneath another. The solidified magma becomes the central axis of a rising mountain range, resulting in the deformation (folding and overthrusting) of rocks on either side.

Gravity measurements indicate that the downward extent or thickness of many batholiths is some 10–15 km/6–9 mi.

In the UK, a batholith underlies SW England and is exposed in places to form areas of high ground such as Dartmoor and Land's End.

bathyal zone upper part of the ocean, which lies on the continental shelf at a depth of between 200 m/650 ft and 2,000 m/6,500 ft.

bathyscaph or *bathyscaphe* or *bathyscape* deep-sea diving apparatus used for exploration at great depths in the ocean. In 1960, Jacques Piccard and Don Walsh took the bathyscaph *Trieste* to a depth of 10,917 m/35,820 ft in the Challenger Deep in the ◊Mariana Trench off the island of Guam in the Pacific Ocean.

bathysphere watertight steel sphere used for observation in deep-sea diving. It is lowered into the ocean by cable to a maximum depth of 900 m/3,000 ft.

battery any energy-storage device allowing release of electricity on demand. It is made up of one or more electrical ◊cells. Primary-cell batteries are disposable; secondary-cell batteries, or ◊accumulators, are rechargeable. Primary-cell batteries are an extremely uneconomical form of energy, since they produce only 2% of the power used in their manufacture.

The common *dry cell* is a primary-cell battery based on the ◊Leclanché cell, and consists of a central carbon electrode immersed in a paste of manganese dioxide and ammonium chloride as the electrolyte. The zinc casing forms the other electrode. It is dangerous to try to recharge a primary-cell battery.

The lead–acid *car battery* is a secondary-cell battery. The car's generator continually recharges the battery. It consists of sets of lead (positive) and lead peroxide (negative) plates in an electrolyte of sulphuric acid (◊battery acid).

The introduction of rechargeable nickel-cadmium batteries has revolutionized portable electronic newsgathering (sound recording, video) and

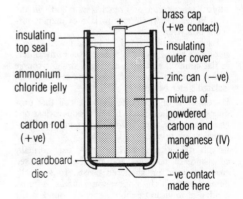

battery *The common dry cell relies on chemical changes occurring between the electrodes—the central carbon rod and the outer zinc casing—and the ammonium chloride electrolyte to produce electricity. The mixture of carbon and manganese dioxide is used to increase the life of the cell.*

information processing (computing). These batteries offer a stable, short-term source of power free of noise and other electrical hazards.

battery acid ◊sulphuric acid of approximately 70% concentration used in lead–acid cells (as found in car batteries).

The chemical reaction within the battery that is responsible for generating electricity also causes a change in the acid's composition. This can be detected as a change in its specific gravity: in a fully charged battery the acid's specific gravity is 1.270–1.290; in a half-charged battery it is 1.190–1.210; in a flat battery it is 1.110–1.130. See ◊battery.

baud in engineering, a unit of electrical signalling speed equal to one pulse per second, measuring the rate at which signals are sent between electronic devices such as telegraphs and computers; 300 baud is about 3,000 words a minute.

bauxite principal ore of ◊aluminium, consisting of a mixture of hydrated aluminium oxides and hydroxides, generally contaminated with compounds of iron, which give it a red colour. To produce aluminium the ore is processed into a white powder (alumina), which is then smelted by passing a large electric current through it. Chief producers of bauxite are Australia, Guinea, Jamaica, Russia, Kazakhstan, Surinam, and Brazil.

Bayesian statistics form of statistics that uses the knowledge of prior probability together with the probability of actual data to determine posterior probabilities, using Bayes' theorem.

Bayes' theorem in statistics, a theorem relating the ◊probability of particular events taking place to the probability that events conditional upon them have occurred.

For example, the probability of picking an ace at random out of a pack of cards is ⁴⁄₅₂. If two cards are picked out, the probability of the second card being an ace is conditional on the first card: if the first card was an ace the probability will be ³⁄₅₁; if not it will be ⁴⁄₅₁. Bayes' theorem gives the probability that given that the second card is an ace, the first card is also.

bayou (corruption of French *boyau* 'gut') in the Gulf States, USA, an ◊oxbow lake or marshy offshoot of a river.

Bayous may be formed, as in the lower Mississippi, by a river flowing in wide curves or meanders in flat country, and then cutting a straight course across them in times of flood, leaving loops of isolated water behind.

B cell or *B* ◊lymphocyte immune cell that produces ◊antibodies. Each B cell produces just one type of antibody, specific to a single ◊antigen. Lymphocytes are related to ◊T cells.

beach strip of land bordering the sea, normally consisting of boulders and pebbles on exposed coasts or sand on sheltered coasts. It is usually defined by the high-and low-water marks. A berm, a ridge of sand and pebbles, may be found at the farthest point that the water reaches.

The material of the beach consists of a rocky debris eroded from exposed rocks and headlands. The material is transported to the beach, and along the beach, by waves that hit the coastline at an

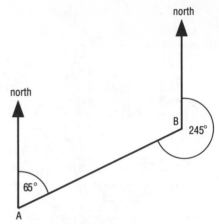

the bearing of B from A is 065°
the backbearing, or bearing of A from B, is 245°

bearing *Bearings are measured by compass and are given as three-digit numbers. The bearing for north is 000° and the numbers increase in a clockwise direction.*

angle, resulting in a net movement of the material in one particular direction. This movement is known as *longshore drift*. Attempts are often made to halt longshore drift by erecting barriers, or jetties, at right angles to the movement. Pebbles are worn into round shapes by being battered against one another by wave action and the result is called *shingle*. The finer material, the *sand*, may be subsequently moved about by the wind and form sand dunes. Apart from the natural process of longshore drift, a beach may be threatened by the commercial use of sand and aggregate, by the mineral industry—since particles of metal ore are often concentrated into workable deposits by the wave action—and by pollution (for example, by oil spilled or dumped at sea).

beam balance instrument for measuring mass (or weight). A simple form consists of a beam pivoted at its midpoint with a pan hanging at each end. The mass to be measured, in one pan, is compared with a variety of standard masses placed in the other. When the beam is balanced, the masses' turning effects or moments under gravity, and hence the masses themselves, are equal.

Bear, Great and Little common names (and translations of the Latin) for the constellations ◊Ursa Major and ◊Ursa Minor respectively.

bearing device used in a machine to allow free movement between two parts, typically the rotation of a shaft in a housing. *Ball bearings* consist of two rings, one fixed to a housing, one to the rotating shaft. Between them is a set, or race, of steel balls. They are widely used to support shafts, as in the spindle in the hub of a bicycle wheel.

The *sleeve*, or *journal bearing*, is the simplest bearing. It is a hollow cylinder, split into two halves. It is used for the big-end and main bearings on a car ◊crankshaft.

In some machinery the balls of ball bearings are

replaced by cylindrical rollers or thinner, **needle bearings**.

In precision equipment such as watches and aircraft instruments, bearings may be made from material such as ruby and are known as **jewel bearings**.

For some applications bearings made from nylon and other plastics are used. They need no lubrication because their surfaces are naturally waxy.

bearing the direction of a fixed point, or the path of a moving object, from a point of observation on the Earth's surface, expressed as an angle from the north. Bearings are taken by ◊compass and are measured in degrees (°), given as three-digit numbers increasing clockwise. For instance, north is 000°, northeast is 045°, south is 180°, and southwest is 225°.

True north differs slightly from magnetic north (the direction in which a compass needle points), hence NE may be denoted as 045M or 045T, depending on whether the reference line is magnetic (M) or true (T) north. True north also differs slightly from grid north since it is impossible to show a spherical Earth on a flat map.

beat regular variation in the loudness of the sound when two notes of nearly equal pitch or ◊frequency are heard together. Beats result from the ◊interference between the sound waves of the notes. The frequency of the beats equals the difference in frequency of the notes.

Musicians use the effect when tuning their instruments. A similar effect can occur in electrical circuits when two alternating currents are present, producing regular variations in the overall current.

Beaufort scale system of recording wind velocity, devised by British hydrographer Francis Beaufort 1806. It is a numerical scale ranging from 0 to 17, calm being indicated by 0 and a hurricane by 12; 13–17 indicate degrees of hurricane force.

In 1874, the scale received international recognition; it was modified 1926. Measurements are made at 10 m/33 ft above ground level.

becquerel SI unit (symbol Bq) of ◊radioactivity, equal to one radioactive disintegration (change in the nucleus of an atom when a particle or ray is given off) per second.

The becquerel is much smaller than the previous standard unit, the ◊curie, and so can be used for measuring smaller quantities of radioactivity. It is named after French physicist Antoine Henri Becquerel.

bed in geology, a single ◊sedimentary rock unit with a distinct set of physical characteristics or contained fossils, readily distinguishable from those of beds above and below. Well-defined partings called **bedding planes** separate successive beds or strata.

The depth of a bed can vary from a fraction of a centimetre to several metres or yards, and can extend over any area. The term is also used to indicate the floor beneath a body of water (lake bed) and a layer formed by a fall of particles (ash bed).

bel unit of sound measurement equal to ten ◊decibels. It is named after Scottish–US scientist Alexander Graham Bell.

benchmark in computing, a measure of the performance of a piece of equipment or software, usually consisting of a standard program or suite of programs. Benchmarks can indicate whether a computer is powerful enough to perform a particular task, and so enable machines to be compared. However, they provide only a very rough guide to practical performance, and may lead manufacturers to design systems that get high scores with the artificial benchmark programs but do not necessarily perform well with day-to-day programs or data.

Beaufort scale

number and description		features	air speed	
			mi per hr	m per sec
0	calm	smoke rises vertically; water smooth	less than 1	less than 0.3
1	light air	smoke shows wind direction; water ruffled	1–3	0.3–1.5
2	slight breeze	leaves rustle; wind felt on face	4–7	1.6–3.3
3	gentle breeze	loose paper blows around	8–12	3.4–5.4
4	moderate breeze	branches sway	13–18	5.5–7.9
5	fresh breeze	small trees sway, leaves blown off	19–24	8.0–10.7
6	strong breeze	whistling in telephone wires; sea spray from waves	25–31	10.8–13.8
7	moderate gale	large trees sway	32–38	13.9–17.1
8	fresh gale	twigs break from trees	39–46	17.2–20.7
9	strong gale	branches break from trees	47–54	20.8–24.4
10	whole gale	trees uprooted, weak buildings collapse	55–63	24.5–28.4
11	storm	widespread damage	64–72	28.5–32.6
12	hurricane	widespread structural damage	73–82	above 32.7
13			83–92	
14			93–103	
15			104–114	
16			115–125	
17			126–136	

Benchmark measures include *Whetstones*, *Dhrystones*, *SPECmarks*, and *TPC*. SPEC-marks are based on ten programs adopted by the Systems Performance Evaluation Cooperative for benchmarking workstations; the Transaction Processing Performance Council's TPC-B benchmark is used to test databases and on-line systems in banking (debit/credit) environments.

Benguela current cold ocean current in the S Atlantic Ocean, moving northwards along the west coast of Southern Africa and merging with the south equatorial current at a latitude of 15° S. Its rich plankton supports large, commercially exploited fish populations.

Benioff zone a seismically active zone dipping beneath a continent or continental margin behind a deep-sea trench. The zone is named after Hugo Benioff, a US seismologist who first described this feature. The zone is equivalent to what is called a subduction zone in plate tectonic theory, where one plate descends beneath another.

bentonite type of clay, consisting mainly of montmorillonite and resembling ◊fuller's earth, which swells when wet. It is used in papermaking, moulding sands, drilling muds for oil wells, and as a decolorant in food processing.

benzaldehyde C_6H_5CHO colourless liquid with the characteristic odour of almonds. It is used as a solvent and in the making of perfumes and dyes. It occurs in certain leaves, such as the cherry, laurel, and peach, and in a combined form in certain nuts and kernels. It can be extracted from such natural sources, but is usually made from ◊toluene.

benzene C_6H_6 clear liquid hydrocarbon of characteristic odour, occurring in coal tar. It is used as a solvent and in the synthesis of many chemicals.

The benzene molecule consists of a ring of six carbon atoms, all of which are in a single plane, and it is one of the simplest ◊cyclic compounds. Benzene is the simplest of a class of compounds collectively known as *aromatic compounds*. Some are considered carcinogenic (cancer-inducing).

benzoic acid C_6H_5COOH white crystalline solid, sparingly soluble in water, that is used as a preservative for certain foods and as an antiseptic. It is obtained chemically by the direct oxidation of benzaldehyde and occurs in certain natural resins, some essential oils, and as hippuric acid.

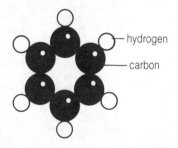

benzene *The molecule of benzene consists of six carbon atoms arranged in a ring, with six hydrogen atoms attached. The benzene ring structure is found in many naturally occurring organic compounds.*

benzpyrene one of a number of organic compounds associated with a particular polycyclic ring structure. Benzpyrenes are present in coal tar at low levels and are considered carcinogenic (cancer-inducing). Traces of benzpyrenes are present in wood smoke, and this has given rise to some concern about the safety of naturally smoked foods.

beriberi endemic polyneuritis, an inflammation of the nerve endings, mostly occurring in the tropics and resulting from a deficiency of vitamin B_1 (◊thiamine).

Beringia or *Bering Land Bridge* former land bridge 1,600 km/1,000 mi wide between Asia and North America; it existed during the ice ages that occurred before 35000 BC and during the period 9000–2400 BC. It is now covered by the Bering Strait and Chukchi Sea.

berkelium synthesized, radioactive, metallic element of the actinide series, symbol Bk, atomic number 97, relative atomic mass 247.

It was first produced 1949 by Glenn Seaborg and his team, at the University of California at Berkeley, USA, after which it is named.

berm on a beach, a ridge of sand or pebbles running parallel to the water's edge, formed by the action of the waves on beach material. Sand and pebbles are deposited at the farthest extent of swash (advance of water) on that particular beach. Berms can also be formed well up a beach following a storm, when they are known as *storm berms*.

Bernoulli's principle law stating that the speed of a fluid varies inversely with pressure, an increase in speed producing a decrease in pressure (such as a drop in hydraulic pressure as the fluid speeds up flowing through a constriction in a pipe) and vice versa. The principle also explains the pressure differences on each surface of an aerofoil, which gives lift to the wing of an aircraft. The principle was named after Swiss mathematician and physicist Daniel Bernoulli.

berry fleshy, many-seeded ◊fruit that does not split open to release the seeds. The outer layer of tissue, the exocarp, forms an outer skin that is often brightly coloured to attract birds to eat the fruit and thus disperse the seeds. Examples of berries are the tomato and the grape.

A *pepo* is a type of berry that has developed a hard exterior, such as the cucumber fruit. Another is the *hesperidium*, which has a thick, leathery outer layer, such as that found in citrus fruits, and fluid-containing vesicles within, which form the segments.

beryl a mineral, beryllium aluminium silicate, $Be_3Al_2Si_6O_{18}$, which forms crystals chiefly in granite. It is the chief ore of beryllium. Two of its gem forms are aquamarine (light-blue crystals) and emerald (dark-green crystals). Well-formed crystals of over 181 tons have been found.

beryllium hard, light-weight, silver-white, metallic element, symbol Be, atomic number 4, relative atomic mass 9.012. It is one of the ◊alkaline-earth metals, with chemical properties similar to those of magnesium; in nature it is found only in combination with other elements. It is used to make sturdy, light alloys and to control the speed of neutrons in nuclear reactors. Beryllium oxide was discovered in 1798 by French chemist Louis-Nicolas Vauquelin (1763–1829), but the element was not isolated until 1828, by German chemist Friedrich Wöhler and Antoine-Alexandre-Brutus Bussy independently. The name comes from Latin *beryllus*.

In 1992 large amounts of beryllium were unexpectedly discovered in six old stars in the Milky Way.

Bessemer process the first cheap method of making ◊steel, invented by Henry Bessemer in England 1856. It has since been superseded by more efficient steelmaking processes, such as the ◊basic-oxygen process. In the Bessemer process compressed air is blown into the bottom of a converter, a furnace shaped like a cement mixer, containing molten pig iron. The excess carbon in the iron burns out, other impurities form a slag, and the furnace is emptied by tilting.

beta decay the disintegration of the nucleus of an atom to produce a beta particle, or high-speed electron, and an electron-antineutrino. During beta decay, a proton in the nucleus changes into a neutron, thereby increasing the atomic number by one while the mass number stays the same. The mass lost in the change is converted into kinetic (movement) energy of the beta particle.

Beta decay is caused by the weak nuclear force, one of the fundamental ◊forces of nature operating inside the nucleus.

beta particle electron ejected with great velocity from a radioactive atom that is undergoing spontaneous disintegration. Beta particles do not exist in the nucleus but are created on disintegration, *beta decay*, when a neutron converts to a proton to emit an electron. The process transforms one element into another, increasing the atomic number by one while the mass number remains the same.

Beta particles are more penetrating than ◊alpha particles, but less so than ◊gamma radiation; they can travel several metres in air, but are stopped by 2–3 mm of aluminium. They are less strongly ionizing than alpha particles and, like cathode rays, are easily deflected by magnetic and electric fields.

Betelgeuse or *Alpha Orionis* red supergiant star in the constellation of Orion and the tenth brightest star in the sky, although its brightness varies. It is over 300 times the diameter of the Sun, about the same size as the orbit of Mars, is over 10,000 times as luminous as the Sun, and lies 650 light years from Earth.

BETELGEUSE: ORION'S GIANT STAR

The giant red star Betelgeuse—the red star in the shoulder of Orion—is 1,100 million km/700 million mi across, about 800 times larger than the Sun. Light takes 60 minutes to travel across the giant star.

Bezier curve curved line that connects a series of points (or 'nodes') in the smoothest possible way. The shape of the curve is governed by a series of complex mathematical formulae. They are used in ◊computer graphics and ◊CAD.

bhp abbreviation for *brake horsepower*.

bicarbonate common name for ◊hydrogencarbonate

bicarbonate indicator pH indicator sensitive enough to show a colour change as the concentration of the gas carbon dioxide increases. The indicator is used in photosynthesis and respiration experiments to find out whether carbon dioxide is being liberated. The initial red colour changes to yellow as the pH becomes more acidic.

Carbon dioxide, even in the concentrations found in exhaled air, will dissolve in the indicator to form a weak solution of carbonic acid, which will lower the pH and therefore give the characteristic colour change.

bicarbonate of soda or *baking soda* (technical name *sodium hydrogencarbonate*) $NaHCO_3$ white crystalline solid that neutralizes acids and is used in medicine to treat acid indigestion. It is also used in baking powders and effervescent drinks.

bicuspid valve in the left side of the ◊heart, a flap of tissue that prevents blood flowing back into the atrium when the ventricle contracts.

bicycle pedal-driven two-wheeled vehicle used in cycling. It consists of a metal frame mounted on two large wire-spoked wheels, with handlebars in front and a seat between the front and back wheels. The bicycle is an energy-efficient, nonpolluting form of transport, and it is estimated that 800 million bicycles are in use throughout the world—outnumbering cars three to one. China, India, Denmark, and the Netherlands are countries with a high use of bicycles. More than 10% of road spending in the Netherlands is on cycleways and bicycle parking.

The first bicycle was seen in Paris 1791 and was a form of hobby-horse. The first treadle-propelled cycle was designed by Scottish blacksmith Kirkpatrick Macmillan 1839. By the end of the 19th century wire wheels, metal frames (replacing wood), and pneumatic tyres (invented by Scottish veterinary surgeon John B Dunlop 1888) had been added. Among the bicycles of that time was the front-wheel-driven penny farthing with a large front wheel.

Recent technological developments have been related to reducing wind resistance caused by the frontal area and the turbulent drag of the bicycle. Most of an Olympic cyclist's energy is taken up in fighting wind resistance in a sprint. The first major innovation was the solid wheel, first used in competitive cycling in 1984, but originally patented as long ago as 1878. Further developments include handle bars that allow the cyclist to crouch and use the shape of the hands and forearms to divert air away from the chest. Modern racing bicycles now have a monocoque structure produced by laying carbon fibre around an internal mould and then baking them in an oven. Using all of these developments Chris Boardman set a speed record of 54.4 km/h (34 mph) on his way to winning a gold medal at the 1992 Barcelona Olympics.

biennial plant plant that completes its life cycle in two years. During the first year it grows vegetatively and the surplus food produced is stored in its ◊perennating organ, usually the root. In the following year these food reserves are used for the production of leaves, flowers, and seeds, after which the plant dies. Many root vegetables are biennials, including the carrot *Daucus carota* and parsnip *Pastinaca sativa*. Some garden plants that are grown as biennials are actually perennials, for example, the wallflower *Cheiranthus cheiri*.

Big Bang in astronomy, the hypothetical 'explosive' event that marked the origin of the universe as we know it. At the time of the Big Bang, the entire universe was squeezed into a hot, superdense state. The Big Bang explosion threw this compacted material outwards, producing the expanding universe (see ◊red shift). The cause of the Big Bang is unknown; observations of the current rate of expansion of the universe suggest that it took place about 10 to 20 billion years ago. The Big Bang theory began modern ◊cosmology.

According to a modified version of the Big Bang, called the *inflationary theory*, the universe underwent a rapid period of expansion shortly after the Big Bang, which accounts for its current large size and uniform nature. The inflationary theory is supported by the most recent observations of the ◊cosmic background radiation.

THE BIG BANG: THE BIGGEST EXPLOSION OF ALL TIME

Scientists have calculated that one million-million-million-million-million-millionth of a second after the Big Bang, the Universe was the size of a pea, and the temperature was 10,000 million million million million °C (18,000 million million million million °F).

One second after the Big Bang, the temperature was about 10,000 million °C (18,000 million °F).

Big Dipper North American name for the Plough, the seven brightest and most prominent stars in the constellation ◊Ursa Major.

bight coastal indentation, such as the Bight of Bonny in W Africa and the Great Australian Bight.

bile brownish fluid produced by the liver. In most vertebrates, it is stored in the gall bladder and emptied into the small intestine as food passes through. Bile consists of bile salts, bile pigments, cholesterol, and lecithin. *Bile salts* assist in the breakdown and absorption of fats; *bile pigments* are the breakdown products of old red blood cells that are passed into the gut to be eliminated with the faeces.

billion the cardinal number represented by a 1 followed by nine zeros (1,000,000,000 or 10^9), equivalent to a thousand million.

In Britain, this number was formerly known as a milliard, and a million million (1,000,000,000,000) as a billion, but the first definition is now internationally recognized.

bimetallic strip strip made from two metals each having a different coefficient of ◊thermal expansion; it therefore bends when subjected to a change in temperature. Such strips are used widely for temperature measurement and control.

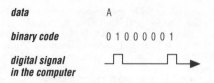

data	A
binary code	0 1 0 0 0 0 0 1
digital signal in the computer	

binary number system

bimodal in statistics, having two distinct peaks of ◊frequency distribution.

binary fission in biology, a form of ◊asexual reproduction, whereby a single-celled organism, such as the amoeba, divides into two smaller 'daughter' cells. It can also occur in a few simple multicellular organisms, such as sea anemones, producing two smaller sea anemones of equal size.

binary number system or *binary number code* system of numbers to ◊base two, using combinations of the digits 1 and 0. Codes based on binary numbers are used to represent instructions and data in all modern digital computers, the values of the binary digits (contracted to 'bits') being represented as the on/off states of switches and high/low voltages in circuits.

The value of any position in a binary number increases by powers of 2 (doubles) with each move from right to left (1, 2, 4, 8, 16, and so on). For example, 1011 in the binary number system means $(1 \times 8) + (0 \times 4) + (1 \times 2) + (1 \times 1)$, which adds up to 11 in the decimal system.

binary search in computing, a rapid technique used to find any particular record in a list of records held in sequential order. The computer is programmed to compare the record sought with the record in the middle of the ordered list. This being done, the computer discards the half of the list in which the record does not appear, thereby reducing the number of records left to search by half. This process of selecting the middle record and discarding the unwanted half of the list is repeated until the required record is found.

binary star pair of stars moving in orbit around their common centre of mass. Observations show that most stars are binary, or even multiple—for example, the nearest star system to the Sun, ◊Alpha Centauri.

A *spectroscopic binary* is a binary in which two stars are so close together that they cannot be seen separately, but their separate light spectra can be distinguished by a spectroscope. Another type is the ◊eclipsing binary.

BINARY STAR

The star called Epsilon Aurigae is really a binary system: two stars revolving around one another. One of the stars may be the largest known. Its diameter is 2,800 times that of the Sun. If it were put in place of the Sun, it would engulf all the planets except Uranus, Neptune, and Pluto.

---

binding energy in physics, the amount of energy needed to break the nucleus of an atom into the neutrons and protons of which it is made.

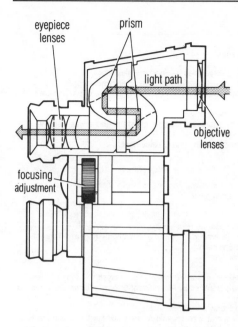

eyepiece lenses

prism

light path

objective lenses

focusing adjustment

binoculars *An optical instrument that allows the user to focus both eyes on the magnified image at the same time. The essential components of binoculars are objective lenses, eye pieces, and a system of prisms to invert and reverse the image. A focusing system provides a sharp image by adjusting the relative positions of these components.*

binoculars optical instrument for viewing an object in magnification with both eyes; for example, field glasses and opera glasses. Binoculars consist of two telescopes containing lenses and prisms, which produce a stereoscopic effect as well as magnifying the image. Use of prisms has the effect of 'folding' the light path, allowing for a compact design.

The first binocular telescope was constructed by the Dutch inventor Hans Lippershey (*c.* 1570–*c.* 1619), in 1608. Later development was largely due to the German Ernst Abbé (1840–1905) of Jena, who at the end of the 19th century designed prism binoculars that foreshadowed the instruments of today, in which not only magnification but also stereoscopic effect is obtained.

binocular vision vision in which both eyes can focus on an object at the same time. In humans, the eyes, which are about 7 cm apart, provide two slightly different images of the world, but these are coordinated to give a three-dimensional perception that allows the brain to judge accurately the position and speed of objects up to 60 m away.

binomial in mathematics, an expression consisting of two terms, such as $a + b$ or $a - b$.

binomial system of nomenclature in biology, the system in which all organisms are identified by a two-part Latinized name. Devised by the Swedish botanist Carolus Linnaeus, it is also known as the Linnaean system. The first name is capitalized and identifies the ◊genus; the second identifies the ◊species within that genus.

binomial theorem formula whereby any power of a binomial quantity may be found without performing the progressive multiplications. It was discovered by English physicist and mathematician Isaac Newton and first published 1676.

biochemistry science concerned with the chemistry of living organisms: the structure and reactions of proteins (such as enzymes), nucleic acids, carbohydrates, and lipids.

Its study has led to an increased understanding of life processes, such as those by which organisms synthesize essential chemicals from food materials, store and generate energy, and pass on their characteristics through their genetic material. A great deal of medical research is concerned with the ways in which these processes are disrupted. Biochemistry also has applications in agriculture and in the food industry (for instance, in the use of enzymes).

biodegradable capable of being broken down by living organisms, principally bacteria and fungi. In biodegradable substances, such as food and sewage, the natural processes of decay lead to compaction and liquefaction, and to the release of nutrients that are then recycled by the ecosystem. Nonbiodegradable substances, such as glass, heavy metals, and most types of plastic, present serious problems of disposal.

biodiversity (contraction of *biological diversity*) measure of the variety of the Earth's animal, plant, and microbial species; of genetic differences within species; and of the ecosystems that support those species. Its maintenance is important for ecological stability and as a resource for research into, for example, new drugs and crops. Research suggests that biodiversity is far greater than previously realized, especially among smaller organisms—for instance, it is thought that only 1–10% of the world's bacterial species have been identified. In the 20th century, however, the destruction of habitats is believed to have resulted in the most severe and rapid loss of diversity in the history of the planet.

The term came to public attention 1992 when an international convention for the preservation of biodiversity was signed by over 100 world leaders at the ◊Earth Summit in Brazil. The convention called on industrialized countries to give financial and technological help to developing countries in order to allow them to protect and manage their natural resources, and profit from growing commercial demand for genes and chemicals from wild species. However, the convention was weakened when the USA refused to sign because of fears that it would undermine the patents and licenses of biotechnology companies.

bioengineering the application of engineering to biology and medicine. Common applications include the design and use of artificial limbs, joints, and organs, including hip joints and heart valves.

biofeedback modification or control of a biological system by its results or effects. For example, a change in the position or ◊trophic level of one species affects all levels above it.

Many biological systems are controlled by negative feedback. When enough of the hormone thyroxine has been released into the blood, the

biochemistry: chronology

c. 1830	Johannes Müller discovered proteins.
1833	Anselme Payen and J F Persoz first isolated an enzyme.
1862	Haemoglobin was first crystallized.
1869	The genetic material DNA (deoxyribonucleic acid) was discovered by Friedrich Mieschler.
1899	Emil Fischer postulated the 'lock-and-key' hypothesis to explain the specificity of enzyme action.
1913	Leonor Michaelis and M L Menten developed a mathematical equation describing the rate of enzyme-catalysed reactions.
1915	The hormone thyroxine was first isolated from thyroid-gland tissue.
1920	The chromosome theory of heredity was postulated by Thomas H Morgan; growth hormone was discovered by Herbert McLean Evans and J A Long.
1921	Insulin was first isolated from the pancreas by Frederick Banting and Charles Best.
1926	Insulin was obtained in pure crystalline form.
1927	Thyroxine was first synthesized.
1928	Alexander Fleming discovered penicillin.
1931	Paul Karrer deduced the structure of retinol (vitamin A); vitamin D compounds were obtained in crystalline form by Adolf Windaus and Askew, independently of each other.
1932	Charles Glen King isolated ascorbic acid (vitamin C).
1933	Tadeus Reichstein synthesized ascorbic acid.
1935	Richard Kuhn and Karrer established the structure of riboflavin (vitamin B_2).
1936	Robert Williams established the structure of thiamine (vitamin B_1); biotin was isolated by Kogl and Tonnis.
1937	Niacin was isolated and identified by Conrad Arnold Elvehjem.
1938	Pyridoxine (vitamin B_6) was isolated in pure crystalline form.
1939	The structure of pyridoxine was determined by Kuhn.
1940	Hans Krebs proposed the citric acid (Krebs) cycle; Hickman isolated retinol in pure crystalline form; Williams established the structure of pantothenic acid; biotin was identified by Albert Szent-Györgyi, Vincent Du Vigneaud, and co-workers.
1941	Penicillin was isolated and characterized by Howard Florey and Ernst Chain.
1943	The role of DNA in genetic inheritance was first demonstrated by Oswald Avery, Colin MacLeod, and Maclyn McCarty.
1950	The basic components of DNA were established by Erwin Chargaff; the alpha-helical structure of proteins was established by Linus Pauling and R B Corey.
1953	James Watson and Francis Crick determined the molecular structure of DNA.
1956	Mahlon Hoagland and Paul Zamecnick discovered transfer RNA (ribonucleic acid); mechanisms for the biosynthesis of RNA and DNA were discovered by Arthur Kornberg and Severo Ochoa.
1957	Interferon was discovered by Alick Isaacs and Jean Lindemann.
1958	The structure of RNA was determined.
1960	Messenger RNA was discovered by Sydney Brenner and François Jacob.
1961	Marshall Nirenberg and Ochoa determined the chemical nature of the genetic code.
1965	Insulin was first synthesized.
1966	The immobilization of enzymes was achieved by Chibata.
1968	Brain hormones were discovered by Roger Guillemin and Andrew Schally.
1975	J Hughes and Hans Kosterlitz discovered encephalins.
1976	Guillemin discovered endorphins.
1977	J Baxter determined the genetic code for human growth hormone.
1978	Human insulin was first produced by genetic engineering.
1979	The biosynthetic production of human growth hormone was announced by Howard Goodman and J Baxter of the University of California, and by D V Goeddel and Seeburg of Genentech.
1982	Louis Chedid and Michael Sela developed the first synthesized vaccine.
1983	The first commercially available product of genetic engineering (Humulin) was launched.
1985	Alec Jeffreys devised genetic fingerprinting.
1993	UK researchers introduced a healthy version of the gene for cystic fibrosis into the lungs of mice with induced cystic fibrosis, restoring normal function.

hormone adjusts its own level by 'switching off' the gland that produces it. In ecology, as the numbers in a species rise, the food supply available to each individual is reduced. This acts to reduce the population to a sustainable level.

biofuel any solid, liquid, or gaseous fuel produced from organic (once living) matter, either directly from plants or indirectly from industrial, commercial, domestic, or agricultural wastes. There are three main ways for the development of biofuels: the burning of dry organic wastes (such as household refuse, industrial and agricultural wastes, straw, wood, and peat); the fermentation of wet wastes (such as animal dung) in the absence of oxygen to produce biogas (containing up to 60% methane); or the fermentation of sugar cane or corn to produce alcohol; and energy forestry (producing fast-growing wood for fuel).

biogenesis biological term coined 1870 by English scientist Thomas Henry Huxley to express the hypothesis that living matter always arises out of other similar forms of living matter. It superseded the opposite idea of ◊spontaneous generation or abiogenesis (that is, that living things may arise out of nonliving matter).

biogeography study of how and why plants and animals are distributed around the world, in the past as well as in the present; more specifically, a theory describing the geographical distribution of ◊species developed by Robert MacArthur and US zoologist Edward Osborne Wilson. The theory argues that for many species, ecological specializations mean that suitable habitats are patchy in their occurrence. Thus for a dragonfly, ponds in which to breed are separated by large tracts of land, and for edelweiss adapted to alpine peaks the deep valleys between cannot be colonized.

biological clock regular internal rhythm of activity, produced by unknown mechanisms, and not dependent on external time signals. Such clocks are known to exist in almost all animals, and also in many plants, fungi, and unicellular organisms. In higher organisms, there appears to be a series of clocks of graded importance. For example, although body temperature and activity cycles in human beings are normally 'set' to 24 hours, the two cycles may vary independently, showing that two clock mechanisms are involved.

biological computer proposed technology for computing devices based on growing complex organic molecules (biomolecules) as components. Its theoretical basis is that cells, the building blocks of all living things, have chemical systems that can store and exchange electrons and therefore function as electrical components. It is currently the subject of long-term research.

biological control control of pests such as insects and fungi through biological means, rather than the use of chemicals. This can include breeding resistant crop strains; inducing sterility in the pest; infecting the pest species with disease organisms; or introducing the pest's natural predator. Biological control tends to be naturally self-regulating, but as ecosystems are so complex, it is difficult to predict all the consequences of introducing a biological controlling agent.

The introduction of the cane toad to Australia 50 years ago to eradicate a beetle that was destroying sugar beet provides an example of the unpredictability of biological control. Since the cane toad is poisonous there are few Australian predators and it is now a pest, spreading throughout eastern and northern Australia at a rate of 35 km/22 mi a year.

biological oxygen demand (BOD) the amount of dissolved oxygen taken up by microorganisms in a sample of water. Since these microorganisms live by decomposing organic matter, and the amount of oxygen is proportional to their number and metabolic rate, BOD can be used as a measure of the extent to which the water is polluted with organic compounds.

biological shield shield around a nuclear reactor that is intended to protect personnel from the effects of ◊radiation. It usually consists of a thick wall of steel and concrete.

biological weathering form of ◊weathering caused by the activities of living organisms—for example, the growth of roots or the burrowing of animals. Tree roots are probably the most significant agents of biological weathering as they are capable of prising apart rocks by growing into cracks and joints.

biology science of life. Strictly speaking, biology includes all the life sciences—for example, anatomy and physiology, cytology, zoology and botany, ecology, genetics, biochemistry and biophysics, animal behaviour, embryology, and plant breeding. During the 1990s an important focus of biological research will be the international ◊Human Genome Project,

which will attempt to map the entire genetic code contained in the 23 pairs of human chromosomes.

bioluminescence production of light by living organisms. It is a feature of many deep-sea fishes, crustaceans, and other marine animals. On land, bioluminescence is seen in some nocturnal insects such as glow-worms and fireflies, and in certain bacteria and fungi. Light is usually produced by the oxidation of luciferin, a reaction catalysed by the ◊enzyme luciferase. This reaction is unique, being the only known biological oxidation that does not produce heat. Animal luminescence is involved in communication, camouflage, or the luring of prey, but its function in other organisms is unclear.

biomass the total mass of living organisms present in a given area. It may be specified for a particular species (such as earthworm biomass) or for a general category (such as herbivore biomass). Estimates also exist for the entire global plant biomass. Measurements of biomass can be used to study interactions between organisms, the stability of those interactions, and variations in population numbers.

The burning of biomass (defined either as natural areas of the ecosystem or as forest, grasslands and fuelwoods) produces 3.5 million tonnes of carbon in the form of carbon dioxide each year, accounting for up to 40% of the world's annual carbon dioxide production.

biome broad natural assemblage of plants and animals shaped by common patterns of vegetation and climate. Examples include the tundra biome and the desert biome.

biomechanics study of natural structures to improve those produced by humans. For example, mother-of-pearl is structurally superior to glass fibre, and deer antlers have outstanding durability because they are composed of microscopic fibres. Such natural structures may form the basis of high-tech composites.

biometry literally, the measurement of living things, but generally used to mean the application of mathematics to biology. The term is now largely obsolete, since mathematical or statistical work is an integral part of most biological disciplines.

bionics (from 'biological electronics') design and development of electronic or mechanical artificial systems that imitate those of living things. The bionic arm, for example, is an artificial limb (◊prosthesis) that uses electronics to amplify minute electrical signals generated in body muscles to work electric motors, which operate the joints of the fingers and wrist.

The first person to receive two bionic ears was Peter Stewart, an Australian journalist, 1989. His left ear was fitted with an array of 22 electrodes, replacing the hairs that naturally convert sounds into electrical impulses. Five years previously he had been fitted with a similar device in his right ear.

biophysics application of physical laws to the properties of living organisms. Examples include using the principles of ◊mechanics to calculate the strength of bones and muscles, and ◊thermodynamics to study plant and animal energetics.

biorhythm rhythmic change, mediated by

biology: chronology

c. 500 BC	First studies of the structure and behaviour of animals, by the Greek Alcmaeon of Creton.
c. 450	Hippocrates of Cos undertook the first detailed studies of human anatomy.
c. 350	Aristotle laid down the basic philosophy of the biological sciences and outlined a theory of evolution.
c. 300	Theophrastus carried out the first detailed studies of plants.
c. AD 175	Galen established the basic principles of anatomy and physiology.
c. 1500	Leonardo da Vinci studied human anatomy to improve his drawing ability and produced detailed anatomical drawings.
1628	William Harvey described the circulation of the blood and the function of the heart as a pump.
1665	Robert Hooke used a microscope to describe the cellular structure of plants.
1672	Marcelle Malphigi undertook the first studies in embryology by describing the development of a chicken egg.
1677	Anthony van Leeuwenhoek greatly improved the microscope and used it to describe spermatozoa as well as many microorganisms.
1682	Nehemiah Grew published the first textbook in botany.
1736	Carolus Linnaeus (Carl von Linné) published his systematic classification of plants, so establishing taxonomy.
1768–79	James Cook's voyages of discovery in the Pacific revealed an undreamed-of diversity of living species, prompting the development of theories to explain their origin.
1796	Edward Jenner established the practice of vaccination against smallpox, laying the foundations for theories of antibodies and immune reactions.
1809	Jean-Baptiste Lamarck advocated a theory of evolution through inheritance of acquired characters.
1839	Theodor Schwann proposed that all living matter is made up of cells.
1857	Louis Pasteur established that microorganisms are responsible for fermentation, creating the discipline of microbiology.
1859	Charles Darwin published *On the Origin of Species*, expounding his theory of the evolution of species by natural selection.
1866	Gregor Mendel pioneered the study of inheritance with his experiments on peas, but achieved little recognition.
1883	August Weismann proposed his theory of the continuity of the germ plasm.
1900	Mendel's work was rediscovered and the science of genetics founded.
1935	Konrad Lorenz published the first of many major studies of animal behaviour, which founded the discipline of ethology.
1953	James Watson and Francis Crick described the molecular structure of the genetic material, DNA.
1964	William Hamilton recognized the importance of inclusive fitness, so paving the way for the development of sociobiology.
1975	Discovery of endogenous opiates (the brain's own painkillers) opened up a new phase in the study of brain chemistry.
1976	Har Gobind Khorana and his colleagues constructed the first artificial gene to function naturally when inserted into a bacterial cell, a major step in genetic engineering.
1982	Gene databases were established at Heidelberg, Germany, for the European Molecular Biology Laboratory, and at Los Alamos, USA, for the US National Laboratories.
1985	The first human cancer gene, retinoblastoma, was isolated by researchers at the Massachusetts Eye and Ear Infirmary and the Whitehead Institute, Massachusetts.
1988	The Human Genome Organization (HUGO) was established in Washington, DC with the aim of mapping the complete sequence of DNA.
1991	Biosphere 2, an experiment that attempts to reproduce the world's biosphere in miniature within a sealed glass dome, was launched in Arizona, USA.
1992	Researchers at the University of California, USA, stimulated the multiplication of isolated brain cells of mice, overturning the axiom that mammalian brains cannot produce replacement cells once birth has taken place. The world's largest organism, a honey fungus with underground hyphae (filaments) spreading across 600 hectares/1,480 acres, was discovered in Washington State, USA.

◊hormones, in the physical state and activity patterns of certain plants and animals that have seasonal activities. Examples include winter hibernation, spring flowering or breeding, and periodic migration. The hormonal changes themselves are often a response to changes in day length (◊photoperiodism); they signal the time of year to the animal or plant. Other biorhythms are innate and continue even if external stimuli such as day length are removed. These include a 24–hour or ◊circadian rhythm, a 28–day or circalunar rhythm (corresponding to the phases of the Moon), and even a year-long rhythm in some organisms.

Such innate biorhythms are linked to an internal or ◊biological clock, whose mechanism is still poorly understood. Often both types of rhythm operate; thus many birds have a circalunar rhythm that prepares them for the breeding season, and a photoperiodic response. There is also a nonscientific and unproven theory that human activity is governed by three biorhythms: the *intellectual* (33 days), the *emotional* (28 days), and the *physical* (23 days). Certain days in each cycle are regarded as 'critical', even more so if one such day coincides with that of another cycle.

biosensor device based on microelectronic circuits that can directly measure medically significant variables for the purpose of diagnosis or monitoring treatment. One such device measures the blood sugar level of diabetics using a single drop of blood, and shows the result on a liquid crystal display within a few minutes.

biosphere the narrow zone that supports life on our planet. It is limited to the waters of the Earth, a fraction of its crust, and the lower regions of the atmosphere.

BioSphere 2 (BS2) ecological test project, a 'planet in a bottle', in Arizona, USA. Under a glass dome, several different habitats are recreated, with representatives of nearly 4,000 species, including eight humans, sealed in the biosphere for two years from summer 1991 to see how effectively the recycling of air, water, and waste can work in an enclosed environment and whether a stable ecosystem can be created; ultimately BS2 is a prototype space colony.

The sealed area covers a total of 3.5 acres and contains tropical rainforest, salt marsh, desert, coral reef, and savanna habitats, as well as a section for intensive agriculture. The people within are meant to be entirely self-sufficient, except for electricity, which is supplied by a 3.7 megawatt power station on the outside (solar panels were considered too expensive). Experiments with biospheres that contain relatively simple life forms have been carried out for decades, and a 21–day trial period 1989 that included humans preceded the construction of BS2. However, BS2 is not in fact the second in a series: the Earth is considered to be BioSphere 1.

In Jan 1993, the oxygen level in the biosphere declined mysteriously and oxygen had to be pumped in. In Feb 1993, several types of predator insects had to be introduced into the biosphere to destroy pests that were eating the sweet potato crop, causing a severe food shortage.

biosynthesis synthesis of organic chemicals from simple inorganic ones by living cells—for example, the conversion of carbon dioxide and water to glucose by plants during ◊photosynthesis. Other biosynthetic reactions produce cell constituents including proteins and fats.

Biosynthesis requires energy; in the initial stages of photosynthesis this is obtained from sunlight, but more often it is supplied by the ◊ATP molecule. The term is also used in connection with biotechnology processes.

Biosynthesis requires energy; in the initial or light-dependent stages of photosynthesis this is obtained from sunlight, but in all other instances, it is supplied chemically by ◊ATP and NADPH. The term is also used in connection with the products achieved through biotechnology processes.

biotechnology industrial use of living organisms to manufacture food, drugs, or other products. The brewing and baking industries have long relied on the yeast microorganism for ◊fermentation purposes, while the dairy industry employs a range of bacteria and fungi to convert milk into cheeses and yoghurts. ◊Enzymes, whether extracted from cells or produced artificially, are central to most biotechnological applications.

Recent advances include ◊genetic engineering, in which single-celled organisms with modified ◊DNA are used to produce insulin and other drugs.

biotic factor organic variable affecting an ecosystem—for example. the changing population of elephants and its effect on the African savanna.

biotin or *vitamin H* vitamin of the B complex, found in many different kinds of food; egg yolk, liver, legumes, and yeast contain large amounts. Its absence from the diet may lead to dermatitis.

biotite dark mica, $K(Mg, Fe)_3Al\,Si_3O_{10}(OH, F)_2$, a common silicate mineral It is brown to black with shiny surfaces, and like all micas, it splits into very thin flakes along its one perfect cleavage. Biotite is a mineral of igneous rocks such as granites, and metamorphic rocks such as schists and gneisses.

bird backboned animal of the class Aves, the biggest group of land vertebrates, characterized by warm blood, feathers, wings, breathing through lungs, and egg-laying by the female. There are nearly 8,500 species of birds.

Birds are bipedal, with the front limb modified to form a wing and retaining only three digits. The heart has four chambers, and the body is maintained at a high temperature (about 41°C/ 106°F). Most birds fly, but some groups (such as ostriches) are flightless, and others include flightless members. Many communicate by sounds, or by visual displays, in connection with which many species are brightly coloured, usually the males. Birds have highly developed patterns of instinctive behaviour. Hearing and eyesight are well developed, but the sense of smell is usually poor. Typically the eggs are brooded in a nest and, on hatching, the young receive a period of parental care. The study of birds is called ◊ornithology.

birth act of producing live young from within the body of female animals. Both viviparous and ovoviviparous animals give birth to young. In viviparous animals, embryos obtain nourishment from the mother via a ◊placenta or other means. In ovoviviparous animals, fertilized eggs develop and hatch in the oviduct of the mother and gain little or no nourishment from maternal tissues. See also ◊pregnancy.

bise cold dry northerly wind experienced in southern France and Switzerland.

bisect to divide a line or angle into two equal parts.

bismuth hard, brittle, pinkish-white, metallic element, symbol Bi, atomic number 83, relative atomic mass 208.98. It has the highest atomic number of all the stable elements (the elements from atomic number 84 up are radioactive). Bismuth occurs in ores and occasionally as a free metal (◊native metal). It is a poor conductor of heat and electricity, and is used in alloys of low melting point and in medical compounds to soothe gastric ulcers. The name comes from the Latin *besemutum*, from the earlier German *Wismut* (of unknown origin).

bistable circuit or *flip-flop* simple electronic circuit that remains in one of two stable states until it receives a pulse (logic 1 signal) through one of its inputs, upon which it switches, or 'flips', over to the other state. Because it is a two-state device, it can be used to store binary digits and is widely used in the ◊integrated circuits used to build computers.

bit (contraction of *binary digit*) in computing, a single binary digit, either 0 or 1. A bit is the smallest unit of data stored in a computer; all other data must be coded into a pattern of individual bits. A ◊byte represents sufficient computer memory to store a single character of data, and usually contains eight bits. For example, in the ◊ASCII code system used by most microcomputers the capital letter A would be stored in a single byte of memory as the bit pattern 01000001.

The maximum number of bits that a computer can normally process at one is called a *word*. Microcomputers are often described according to how many bits of information they can handle at

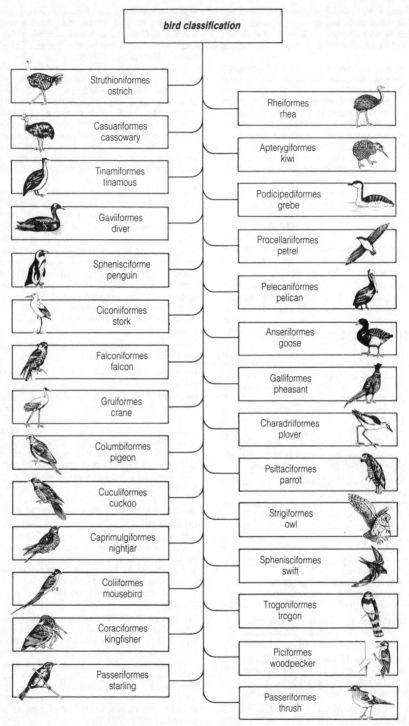

bird classification

Struthioniformes
ostrich

Casuariformes
cassowary

Tinamiformes
tinamous

Gaviiformes
diver

Sphenisciforme
penguin

Ciconiiformes
stork

Falconiformes
falcon

Gruiformes
crane

Columbiformes
pigeon

Cuculiformes
cuckoo

Caprimulgiformes
nightjar

Coliiformes
mousebird

Coraciformes
kingfisher

Passeriformes
starling

Rheiformes
rhea

Apterygiformes
kiwi

Podicipediformes
grebe

Procellariiformes
petrel

Pelecaniformes
pelican

Anseriformes
goose

Galliformes
pheasant

Charadriiformes
plover

Psittaciformes
parrot

Strigiformes
owl

Sphenisciformes
swift

Trogoniformes
trogon

Piciformes
woodpecker

Passeriformes
thrush

bird *This diagram shows a representative species from each of the 27 orders. There are nearly 8,500 species of birds, of which the largest is the N African ostrich, reaching a height of 2.74 m/9 ft and weighing 156 kg/345 lb. The smallest bird, the bee hummingbird of Cuba and the Isle of Pines, measures 57 mm/2.24 in in length and weighs a mere 1.6 g/0.056 oz. The Passeriformes (perching birds) is the largest order, containing about 5,000 species.*

once. For instance, the first microprocessor, the Intel 4004 (launched 1971), was a 4–bit device. In the 1970s several different 8–bit computers, many based on the Zilog Z80 or Rockwell 6502 processors, came into common use. During the early 1980s, the IBM personal computer was introduced, using the Intel 8088 processor, which combined a 16–bit processor with an 8–bit ◊data bus. Business micros of the later 1980s began to use 32–bit processors such as the Intel 80386 and Motorola 68030. Machines based on the first 64-bit processor, the Intel Pentium, appeared in 1993.

bit pad computer input device; see ◊graphics tablet.

bitumen impure mixture of hydrocarbons, including such deposits as petroleum, asphalt, and natural gas, although sometimes the term is restricted to a soft kind of pitch resembling asphalt.

Solid bitumen may have arisen as a residue from the evaporation of petroleum. If evaporation took place from a pool or lake of petroleum, the residue might form a pitch or asphalt lake, such as Pitch Lake in Trinidad. Bitumen was used in ancient times as a mortar, and by the Egyptians for embalming.

bivalent in biology, a name given to the pair of homologous chromosomes during reduction division (◊meiosis). In chemistry, the term is sometimes used to describe an element or group with a ◊valency of two, although the term 'divalent' is more common.

black body in physics, a hypothetical object that completely absorbs all thermal (heat) radiation striking it. It is also a perfect emitter of thermal radiation.

Although a black body is hypothetical, a practical approximation can be made by using a small hole in the wall of a constant-temperature enclosure. The radiation emitted by a black body is of all wavelengths, but with maximum radiation at a particular wavelength that depends on the body's temperature. As the temperature increases, the wavelength of maximum intensity becomes shorter (see ◊Wien's law). The total energy emitted at all wavelengths is proportional to the fourth power of the temperature (see ◊Stefan–Boltzmann law). Attempts to explain these facts failed until the development of ◊quantum theory 1900.

black box popular name for the unit containing an aeroplane's flight and voice recorders. These monitor the plane's behaviour and the crew's conversation, thus providing valuable clues to the cause of a disaster. The box is nearly indestructible and usually painted orange for easy recovery. The name also refers to any compact electronic device that can be quickly connected or disconnected as a unit.

The maritime equivalent is the *voyage recorder*, installed in ships from 1989. It has 350 sensors to record the performance of engines, pumps, navigation lights, alarms, radar, and hull stress.

black earth exceedingly fertile soil that covers a belt of land in NE North America, Europe, and Asia.

In Europe and Asia it extends from Bohemia through Hungary, Romania, S Russia, and Siberia,

as far as Manchuria, having been deposited when the great inland ice sheets melted at the close of the last ◊ice age. In North America, it extends from the Great Lakes E through New York State, having been deposited when the last glaciers melted from the terminal moraine.

black hole object in space whose gravity is so great that nothing can escape from it, not even light. Thought to form when massive stars shrink at the ends of their lives, a black hole sucks in more matter, including other stars, from the space around it. Matter that falls into a black hole is squeezed to infinite density at the centre of the hole. Black holes can be detected because gas falling towards them becomes so hot that it emits X-rays.

Satellites above the Earth's atmosphere have detected X-rays from a number of objects in our Galaxy that might be black holes. Massive black holes containing the mass of millions of stars are thought to lie at the centres of ◊quasars. Microscopic black holes may have been formed in the chaotic conditions of the ◊Big Bang. The English physicist Stephen Hawking has shown that such tiny black holes could 'evaporate' and explode in a flash of energy.

BLACK HOLES IN OUR GALAXY

Only four possible black holes have been identified in our galaxy: Cygnus X-1, discovered in 1971, an X-ray source in the constellation of Cygnus; A0620–00 in the constellation of Monoceros, one of the best black hole candidates in the galaxy, discovered by Jeffrey McClintock of the Harvard-Smithsonian Center for Astrophysics and Ronald Remillard of the Massachusetts Institute of Technology in the 1980s; V404 Cygni, close to Cygnus X1, discovered in 1992; Nova Muscae, identified as a black hole in 1992 by McClintock, Remillard, and Charles Bailyn of Yale University, lies roughly 18,000 light years from Earth.

---

bladder hollow elastic-walled organ in the ◊urinary systems of some fishes, most amphibians, some reptiles, and all mammals. Urine enters the bladder through two ureters, one leading from each kidney, and leaves it through the urethra.

blast freezing industrial method of freezing substances such as foods by blowing very cold air over them. See ◊deep freezing.

blast furnace smelting furnace in which temperature is raised by the injection of an air blast. It is used to extract metals from their ores, chiefly pig iron from iron ore.

The principle has been known for thousands of years, but the present blast furnace is a heavy engineering development combining a number of special techniques.

blastocyst in mammals, a stage in the development of the ◊embryo that is roughly equivalent to the ◊blastula of other animal groups.

blastomere in biology, a cell formed in the first

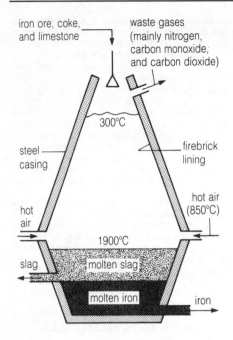

iron ore, coke, and limestone

waste gases (mainly nitrogen, carbon monoxide, and carbon dioxide)

300°C

steel casing

firebrick lining

hot air

hot air (850°C)

1900°C

slag   molten slag

molten iron   iron

blast furnace

stages of embryonic development, after the splitting of the fertilized ovum, but before the formation of the ◊blastula or blastocyst.

blastula early stage in the development of a fertilized egg, when the egg changes from a solid mass of cells (the morula) to a hollow ball of cells (the blastula), containing a fluid-filled cavity (the blastocoel). See also ◊embryology.

bleaching decolorization of coloured materials. The two main types of bleaching agent are the *oxidizing bleaches*, which bring about the ◊oxidation of pigments, and include the ultraviolet rays in sunshine, hydrogen peroxide, and chlorine in household bleaches; and the *reducing bleaches*, which bring about ◊reduction, and include sulphur dioxide.

Bleaching processes have been known from antiquity, mainly those acting through sunlight. Both natural and synthetic pigments usually possess highly complex molecules, the colour property often being due only to a part of the molecule. Bleaches usually attack only that small part, yielding another substance similar in chemical structure but colourless.

blight any of a number of plant diseases caused mainly by parasitic species of ◊fungus, which produce a whitish appearance on leaf and stem surfaces—for instance *potato blight Phytophthora infestans*. General damage caused by aphids or pollution is sometimes known as blight.

blind spot area where the optic nerve and blood vessels pass through the retina of the ◊eye. No visual image can be formed as there are no light-sensitive cells in this part of the retina. Thus the organism is blind to objects that fall in this part of the visual field.

BL Lacertae object starlike object that forms the centre of a distant galaxy, with a prodigious energy output. BL Lac objects, as they are called, seem to be related to ◊quasars and are thought to be the brilliant nuclei of elliptical galaxies. They are so named because the first to be discovered lies in the small constellation Lacerta.

block in computing, a group of records treated as a complete unit for transfer to or from ◊backing storage. For example, many disc drives transfer data in 512–byte blocks.

block and tackle type of ◊pulley.

blood liquid circulating in the arteries, veins, and capillaries of vertebrate animals; the term also refers to the corresponding fluid in those invertebrates that possess a closed ◊circulatory system. Blood carries nutrients and oxygen to individual cells and removes waste products, such as carbon dioxide. It is also important in the immune response and, in many animals, in the distribution of heat throughout the body.

In humans it makes up 5% of the body weight, occupying a volume of 5.5 l/10 pt in the average adult. It consists of a colourless, transparent liquid called *plasma*, containing microscopic cells of three main varieties. *Red cells* (erythrocytes) form nearly half the volume of the blood, with 5 billion cells per litre. Their red colour is caused by ◊haemoglobin. *White cells* (leucocytes) are of various kinds. Some (phagocytes) ingest invading bacteria

red cells

white cells

platelets

blood allowed to stand

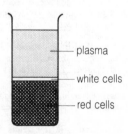

plasma

white cells

red cells

blood

The Body's Pump

After completing a preliminary medical course at the University of Cambridge, where would an ambitious young man in the 17th century go to get a really good medical training? To the University of Padua in Italy, where the great Italian anatomist Hieronymous Fabricius (1537–1619) taught. So this is where English physician William Harvey (1578–1657) naturally went.

The sluice gates

William Harvey had a consuming interest in the movement of the blood in the body. In 1579, Fabricius had publicly demonstrated the valves, which he termed sluice gates, in the veins: his principal anatomical work was an accurate and detailed description of them.

Galen, a Greek physician (c. 130–c. 200), had 1,500 years previously written a monumental treatise covering every aspect of medicine. In this work, he asserted that food turned to blood in the liver, ebbed and flowed in vessels and, on reaching the heart, flowed through pores in the septum (the dividing wall) from the right to left side, and was sent on its way by heart spasms. The blood did not circulate. This doctrine was still accepted and taught well into the 16th century.

Harvey was unconvinced. He had done a simple calculation. He worked out that for each human heart beat, about 60 cm³ of blood left the heart, which meant that the heart pumped out 259 litres every hour. This is more than three times the weight of the average man.

Harvey examined the heart and blood vessels of 128 mammals and found that the valve which separated the left side of the heart from the right ventricle is a one-way structure, as were the valves in the veins discovered by his tutor Fabricius.

For this reason he decided that the blood in the veins must flow only towards the heart.

One-way flow

Harvey was now in a position to do his famous experiment. He tied a tourniquet round the upper part of his arm. It was just tight enough to prevent the blood from flowing through the veins back into heart— but not so tight that arterial blood could not enter the arms. Below the tourniquet, the veins swelled up; above it, they remained empty. This showed that the blood could be entering the arm only through the arteries. Further, by carefully stroking the blood out of a short length of vein, Harvey showed that it could fill up only when blood was allowed to enter it from the end that was furthest away from the heart. He had proved that blood in the veins must flow only towards the heart.

Galen's pores in the septum of the heart had never been found. Belgian physician Andreas Versalius (1514–1564) was another alumnus of Padua University. Although brought up in the Galen tradition, he had carried out secret dissections to discover the pores, and had failed. He did, however, show that men and women had the same number of ribs!

Harvey clinched his researches into the movement of the blood when he demonstrated that no blood seeps through the septum of the heart. He reasoned that blood must pass from the right side of the heart to the left through the lungs. He had discovered the circulation of the blood, and thus, some 20 years after he left Padua, became the father of modern physiology. In 1628 Harvey published his proof of the circulation of the blood in his classic book *On the Motion of the Heart and Blood in Animals*. A new age in medicine and biology had begun.

---

and so protect the body from disease; these also help to repair injured tissues. Others (lymphocytes) produce antibodies, which help provide immunity. Blood *platelets* (thrombocytes) assist in the clotting of blood.

Blood cells constantly wear out and die, and are

MAKING BLOOD

There are about six million red cells in every millilitre of an adult's blood—a millilitre is about the size of an o. The body produces red blood cells at an average rate of 9,000 million per hour. Every day 200,000 million red blood cells die and break up.

replaced from the bone marrow. Dissolved in the plasma are salts, proteins, sugars, fats, hormones, and fibrinogen, which are transported around the body, the last having a role in clotting.

blood clotting complex series of events (known as the blood clotting cascade) that prevents excessive bleeding after injury. It is triggered by ◊vitamin K. The result is the formation of a meshwork of protein fibres (fibrin) and trapped blood cells over the cut blood vessels.

When platelets (cell fragments) in the bloodstream come into contact with a damaged blood vessel, they and the vessel wall itself release the enzyme *thrombokinase*, which brings about the conversion of the inactive enzyme *prothrombin* into the active *thrombin*. Thrombin in turn catalyses the conversion of the soluble protein *fibrinogen*, present in blood plasma, to the insoluble *fibrin*. This fibrous protein forms a net over the wound that traps red blood cells and seals the wound; the resulting jellylike clot hardens on exposure to air to form a hard scab. Calcium, vitamin K, and a variety of enzymes called factors are also necessary for efficient blood clotting.

blood glucose regulation the control of sugar concentration in the blood, so that it always remains within certain limits. Blood sugar levels are likely to vary, because an animal may have just eaten or have been physically active. However, the effects on the nervous system of an erratic blood sugar level are disastrous, leading in humans to coma or fits.

The maintenance of a steady state relies on the ability of the liver to store or release glucose as appropriate, topping up blood sugar levels when needed (for example, during or after physical exercise) and removing sugar from the blood after a heavy meal. This important homeostatic function (see ◊homeostasis) is in turn controlled by two hormones, insulin and glucagon, both of which are produced by the pancreas and have a powerful effect on the liver. ◊Diabetes occurs when insulin production by the pancreas becomes insufficient, leading to dangerously high blood sugar levels.

blood group any of the blood groups into which blood is classified according to antigenic activity. Red blood cells of one individual may carry molecules on their surface that act as ◊antigens in another individual whose red blood cells lack these molecules. The two main antigens are designated A and B. These give rise to four blood groups: having A only (A), having B only (B), having both (AB), and having neither (O). Each of these groups may or may not contain the ◊rhesus factor. Correct typing of blood groups is vital in transfusion, since incompatible types of donor and recipient blood will result in blood clotting, with possible death of the recipient.

These ABO blood groups were first described by Austrian scientist Karl Landsteiner 1902. Subsequent research revealed at least 14 main types of blood groupings, 11 of which are involved with induced ◊antibody production. Blood typing is also of importance in forensic medicine, cases of disputed paternity, and in anthropological studies.

blood vessel specialized tube that carries blood around the body of multicellular animals. Blood vessels are highly evolved in vertebrates, where the three main types—the arteries, veins, and capillaries—are all adapted for their particular role within the body.

bloom whitish powdery or waxlike coating over the surface of certain fruits that easily rubs off when handled. It often contains ◊yeasts that live on the sugars in the fruit. The term bloom is also used to describe a rapid increase in number of certain species of algae found in lakes, ponds, and oceans.

Such blooms may be natural but are often the result of nitrate pollution, in which artificial fertilizers, applied to surrounding fields, leach out into the waterways. This type of bloom can lead to the death of almost every other organism in the water; because light cannot penetrate the algal growth, the plants beneath can no longer photosynthesize and therefore do not release oxygen into the water. Only those organisms that are adapted to very low levels of oxygen survive.

blubber thick layer of ◊fat under the skin of marine mammals, which provides an energy store and an effective insulating layer, preventing the loss of body heat to the surrounding water. Blubber has been used (when boiled down) in engineering, food processing, cosmetics, and printing, but all of these products can now be produced synthetically, thus saving the lives of animals.

blue-green algae or *cyanobacteria* single-celled, primitive organisms that resemble bacteria in their internal cell organization, sometimes joined together in colonies or filaments. Blue-green algae are among the oldest known living organisms and, with bacteria, belong to the kingdom Monera; remains have been found in rocks up to 3.5 billion years old. They are widely distributed in aquatic habitats, on the damp surfaces of rocks and trees, and in the soil.

Blue-green algae and bacteria are prokaryotic organisms. Some can fix nitrogen and thus are necessary to the nitrogen cycle, while others follow a symbiotic existence—for example, living in association with fungi to form lichens. Fresh water can become polluted by nitrates and phosphates from fertilizers and detergents. This eutrophication, or over-enrichment, of the water causes multiplication of the algae in the form of algae blooms. The algae multiply and cover the water's surface, remaining harmless until they give off toxins as they decay. These toxins kill fish and other wildlife and can be harmful to domestic animals, cattle, and people.

blueprint photographic process used for copying engineering drawings and architectural plans, so called because it produces a white copy of the original against a blue background.

The plan to be copied is made on transparent tracing paper, which is placed in contact with paper sensitized with a mixture of iron ammonium citrate and potassium hexacyanoferrate. The paper is exposed to ◊ultraviolet radiation and then washed in water. Where the light reaches the paper, it turns blue (Prussian blue). The paper underneath the lines of the drawing is unaffected, so remains white.

Blueprint for Survival environmental manifesto published 1972 in the UK by the editors of the *Ecologist* magazine. The statement of support it attracted from a wide range of scientists helped draw attention to the magnitude of environmental problems.

blue shift in astronomy, a manifestation of the ◊Doppler effect in which an object appears bluer when it is moving towards the observer or the observer is moving towards it (blue light is of a higher frequency than other colours in the spectrum). The blue shift is the opposite of the ◊red shift.

BMR abbreviation for ◊*basal metabolic rate*.

Bode's law numerical sequence that gives the approximate distances, in astronomical units (distance between Earth and Sun = one astronomical unit), of the planets from the Sun by adding 4 to each term of the series 0, 3, 6, 12, 24, . . . and then dividing by 10. Bode's law predicted the existence of a planet between ◊Mars and ◊Jupiter, which led to the discovery of the asteroids.

The 'law' breaks down for ◊Neptune and ◊Pluto. The relationship was first noted 1772 by the German mathematician Johann Titius (1729–1796) 1772 (it is also known as the Titius–Bode law).

bog type of wetland where decomposition is slowed down and dead plant matter accumulates as ◊peat. Bogs develop under conditions of low temperature, high acidity, low nutrient supply, stagnant water, and oxygen deficiency. Typical bog plants are sphagnum moss, rushes and cotton grass; insectivorous plants such as sundews and bladderworts are common in bogs (insect prey make up for the lack of nutrients).

Large bogs are found in Ireland and northern Scotland. They are dominated by heather, cotton grass, and over 30 species of sphagnum mosses which make up the general bog matrix. The rolling blanket bogs of Scotland are a southern outcrop of the arctic tundra ecosystem, and have taken thousands of years to develop.

The eventual goal of science is to provide a single theory that describes the whole universe.

Stephen Hawking *A Brief History of Time* 1988

Bohr model model of the atom conceived by Danish physicist Neils Bohr 1913. It assumes that the following rules govern the behaviour of electrons: (1) electrons revolve in orbits of specific radius around the nucleus without emitting radiation; (2) within each orbit, each electron has a fixed amount of energy; electrons in orbits farther away from the nucleus have greater energies; (3) an electron may 'jump' from one orbit of high energy to another of lower energy causing the energy difference to be emitted as a ◊photon of electromagnetic radiation such as light. The Bohr model has been superseded by wave mechanics (see ◊quantum theory).

boiler any vessel that converts water into steam. Boilers are used in conventional power stations to generate steam to feed steam ◊turbines, which drive the electricity generators. They are also used in steamships, which are propelled by steam turbines, and in steam locomotives. Every boiler has a furnace in which fuel (coal, oil, or gas) is burned to produce hot gases, and a system of tubes in which heat is transferred from the gases to the water.

The common kind of boiler used in ships and power stations is the *water-tube* type, in which the water circulates in tubes surrounded by the hot furnace gases. The water-tube boilers at power stations produce steam at a pressure of up to 300 atmospheres and at a temperature of up to 600°C/1,100°F to feed to the steam turbines. It is more efficient than the *fire-tube* type that is used in steam locomotives. In this boiler the hot furnace gases are drawn through tubes surrounded by water.

boiling process of changing a liquid into its vapour, by heating it at the maximum possible temperature for that liquid (see ◊boiling point) at atmospheric pressure.

boiling point for any given liquid, the temperature at which the application of heat raises the temperature of the liquid no further, but converts it into vapour.

The boiling point of water under normal pressure is 100°C/212°F. The lower the pressure, the lower the boiling point and vice versa.

See also ◊elevation of boiling point.

bolometer sensitive ◊thermometer that measures the energy of radiation by registering the change in electrical resistance of a fine wire when it is exposed to heat or light. The US astronomer Samuel Langley devised it 1880 for measuring radiation from stars.

bolson basin without an outlet, found in desert regions. Bolsons often contain temporary lakes, called playa lakes, and become filled with sediment from inflowing intermittent streams.

Boltzmann constant in physics, the constant (symbol k) that relates the kinetic energy (energy of motion) of a gas atom or molecule to temperature. Its value is 1.380662×10^{-23} joules per kelvin. It is equal to the gas constant R, divided by ◊Avogadro's number.

bolus mouthful of chewed food mixed with saliva, ready for swallowing. Most vertebrates swallow food immediately, but grazing mammals chew their food a great deal, allowing a mechanical and chemical breakdown to begin.

bond in chemistry, the result of the forces of attraction that hold together atoms of an element or elements to form a molecule. The principal types of bonding are ◊ionic, ◊covalent, ◊metallic, and ◊intermolecular (such as hydrogen bonding).

The type of bond formed depends on the elements concerned and their electronic structure. In an ionic or electrovalent bond, common in inorganic compounds, the combining atoms gain or lose electrons to become ions; for example, sodium (Na) loses an electron to form a sodium ion (Na^+) while chlorine (Cl) gains an electron to form a chloride ion (Cl^-) in the ionic bond of sodium chloride (NaCl).

In a covalent bond, the atomic orbitals of two atoms overlap to form a molecular orbital containing two electrons, which are thus effectively shared between the two atoms. Covalent bonds are common in organic compounds, such as the four carbon-hydrogen bonds in methane (CH_4). In a dative covalent or coordinate bond, one of the combining atoms supplies both of the valence electrons in the bond.

A metallic bond joins metals in a crystal lattice; the atoms occupy lattice positions as positive ions, and valence electrons are shared between all the ions in an 'electron gas'.

In a hydrogen bond, a hydrogen atom joined to an electronegative atom, such as nitrogen or oxygen, becomes partially positively charged, and is weakly attracted to another electronegative atom on a neighbouring molecule.

bone hard connective tissue comprising the

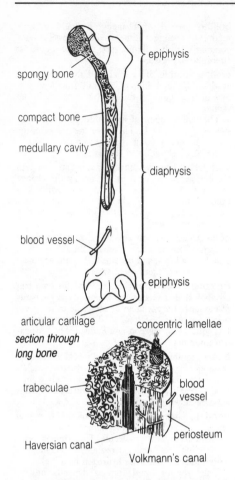

spongy bone

epiphysis

compact bone

medullary cavity

diaphysis

blood vessel

epiphysis

articular cartilage

*section through
long bone*

concentric lamellae

trabeculae

blood
vessel

Haversian canal

periosteum

Volkmann's canal

bone *Bone is a network of fibrous material impregnated with mineral salts and as strong as reinforced concrete. The upper end of the thighbone or femur is made up of spongy bone, which has a fine lacework structure designed to transmit the weight of the body. The shaft of the femur consists of hard compact bone designed to resist bending. Fine channels carrying blood vessels, nerves, and lymphatics maintain even the densest bone as living tissue.*

♢skeleton of most vertebrate animals. It consists of a network of collagen fibres impregnated with mineral salts (largely calcium phosphate and calcium carbonate), a combination that gives the bone great strength, comparable in some cases with that of reinforced concrete. Enclosed within this solid matrix are bone cells, blood vessels, and nerves. The interior of the long bones of the limbs consists of a spongy matrix filled with a soft marrow that produces blood cells.

There are two types of bone: those that develop by replacing ♢cartilage and those that form directly from connective tissue. The latter, which includes the bones of the cranium, are usually platelike in shape, and form in the skin of the developing embryo.

Humans have about 206 distinct bones in the skeleton, of which the smallest are the three ossicles in the middle ear.

bone marrow in vertebrates, soft tissue in the centre of some large bones that manufactures red and white blood cells.

booster first-stage rocket of a space-launching vehicle, or an additional rocket strapped to the main rocket to assist takeoff.

The US Delta rocket, for example, has a cluster of nine strap-on boosters that fire on lift off. Europe's Ariane 3 rocket uses twin strap-on boosters, as does the US space shuttle.

boot or **bootstrap** in computing, the process of starting up a computer. Most computers have a small, built-in boot program that starts automatically when the computer is switched on—its only task is to load a slightly larger program, usually from a disc, which in turn loads the main ♢operating system. In microcomputers the operating system is often held in the permanent ♢ROM memory and the boot program simply triggers its operation.

Some boot programs can be customized so that, for example, the computer, when switched on, always loads and runs a program from a particular backing store or always adopts a particular mode of screen display.

Boötes constellation of the northern hemisphere represented by a herdsman driving a bear (Ursa Major) around the pole. Its brightest star is ♢Arcturus (or Alpha Boötis) about 36 light-years from Earth.

borax hydrous sodium borate, $Na_2B_4O_7.10H_2O$, found as soft, whitish crystals or encrustations on the shores of hot springs and in the dry beds of salt lakes in arid regions, where it occurs with other borates, halite, and ♢gypsum. It is used in bleaches and washing powders.

A large industrial source is Borax Lake, California. Borax is also used in glazing pottery, in soldering, as a mild antiseptic, and as a metallurgical flux.

bore surge of tidal water up an estuary or a river, caused by the funnelling of the rising tide by a narrowing river mouth. A very high tide, possibly fanned by wind, may build up when it is held back by a river current in the river mouth. The result is a broken wave, a metre or a few feet high, that rushes upstream.

Famous bores are found in the rivers Severn (England), Seine (France), Hooghly (India), and Chiang Jiang (China), where bores of over 4 m/ 13 ft have been reported.

boric acid or **boracic acid** H_3BO_3 acid formed by the combination of hydrogen and oxygen with nonmetallic boron. It is a weak antiseptic and is used in the manufacture of glass and enamels. It is also an efficient insecticide against ants and cockroaches.

boron nonmetallic element, symbol B, atomic number 5, relative atomic mass 10.811. In nature it is found only in compounds, as with sodium and oxygen in borax. It exists in two allotropic forms (see ♢allotropy): brown amorphous powder and very hard, brilliant crystals. Its compounds are used in the preparation of boric acid, water softeners, soaps, enamels, glass, and pottery glazes. In alloys it is used to harden steel. Because it absorbs slow neutrons, it is used to make boron carbide control rods for nuclear reactors. It is a necessary trace element in the human diet. The element was named by English chemist Humphry Davy, who isolated it 1808, from ◊borax.

boson in physics, an elementary particle whose spin can only take values that are whole numbers or zero. Bosons may be classified as ◊gauge bosons (carriers of the four fundamental forces) or ◊mesons. All elementary particles are either bosons or ◊fermions.

Unlike fermions, more than one boson in a system (such as an atom) can possess the same energy state. When developed mathematically, this statement is known as the Bose–Einstein law, after its discoverers Indian physicist Satyendra Bose and Albert Einstein.

botanical garden place where a wide range of plants is grown, providing the opportunity to see a botanical diversity not likely to be encountered naturally. Among the earliest forms of botanical garden was the *physic garden*, devoted to the study and growth of medicinal plants; an example is the Chelsea Physic Garden in London, established 1673 and still in existence. Following increased botanical exploration, botanical gardens were used to test the commercial potential of new plants being sent back from all parts of the world.

Today a botanical garden serves many purposes: education, science, and conservation. Many are associated with universities and also maintain large collections of preserved specimens (see ◊herbarium), libraries, research laboratories, and gene banks.

botany the study of plants. It is subdivided into a number of specialized studies, such as the identification and classification of plants (taxonomy), their external formation (plant morphology), their internal arrangement (plant anatomy), their microscopic examination (plant histology), their functioning and life history (plant physiology), and their distribution over the Earth's surface in relation to their surroundings (plant ecology). Palaeobotany concerns the study of fossil plants, while economic botany deals with the utility of plants. Horticulture, agriculture, and forestry are branches of botany.

history The most ancient botanical record is carved on the walls of the temple at Karnak, Egypt, about 1500 BC. The Greeks in the 5th and 4th centuries BC used many plants for medicinal purposes, the first Greek *Herbal* being drawn up about 350 BC by Diocles of Carystus. Botanical information was collected into the works of Theophrastus of Eresus (380–287 BC), a pupil of Aristotle, who founded technical plant nomenclature. Cesalpino in the 16th century sketched out a system of classification based on flowers, fruits, and seeds, while Joachim Jungius (1587–1657) used flowers only as his criterion. The English botanist John Ray (1627–1705) arranged plants systematically, based on his findings on fruit, leaf, and flower, and described about 18,600 plants.

The Swedish botanist Carolus Linnaeus, who founded systematics in the 18th century, included in his classification all known plants and animals, giving each a ◊binomial descriptive label. His work greatly aided the future study of plants, as botanists found that all plants could be fitted into a systematic classification based on Linnaeus' work. Linnaeus was also the first to recognize the sexual nature of flowers. This was followed up by the English naturalist Charles Darwin and others.

Later work revealed the detailed cellular structure of plant tissues and the exact nature of ◊photosynthesis. Julius von Sachs (1832–1897) defined the function of ◊chlorophyll and the significance of plant ◊stomata. In the second half of the 20th century, much has been learned about cell function, repair, and growth by the hybridization of plant cells (the combination of the nucleus of one cell with the cytoplasm of another).

Bouguer anomaly in geophysics, an increase in the Earth's gravity observed near a mountain or dense rock mass. This is due to the gravitational force exerted by the rock mass. It is named after its discoverer, the French mathematician Pierre Bouguer (1698–1758), who first observed it 1735.

boulder clay another name for ◊till, a type of glacial deposit.

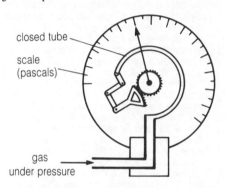

Bourdon gauge

Bourdon gauge instrument for measuring pressure, patented by French watchmaker Eugène Bourdon 1849. The gauge contains a C-shaped tube, closed at one end. When the pressure inside the tube increases, the tube uncurls slightly causing a small movement at its closed end. A system of levers and gears magnifies this movement and turns a pointer, which indicates the pressure on a circular scale. Bourdon gauges are often fitted to cylinders of compressed gas used in industry and hospitals.

Bowman's capsule in the vertebrate kidney, a microscopic filtering device used in the initial stages of waste-removal and urine formation.

There are approximately a million of these capsules in a human kidney, each made up of a tight knot of capillaries and each leading into a kidney tubule or nephron. Blood at high pressure passes into the capillaries where water, dissolved nutrients, and urea move through the capillary wall into the tubule.

Boyle's law law stating that the volume of a given mass of gas at a constant temperature is inversely proportional to its pressure. For example, if the pressure of a gas doubles, its volume will be reduced by a half, and vice versa. The law was discovered in 1662 by Irish physicist and chemist Robert Boyle. See also ◊gas laws.

Measuring the spring of the air

In the 1600s, orthodox science held that 'Nature abhorred a vacuum'. In 1643 Italian physicist Evangelista Torricelli (1608–1647) invented the mercury barometer, and suggested that it contained a vacuum. But in general, a vacuum was very rare. Scientists who believed that a vacuum could exist had to explain their scarcity. Irish-born Robert Boyle (1627–1691), who settled in Oxford in 1656, felt that air had an in-built expansive power, or 'spring', which made it expand to fill any vacuum.

The active spring
In 1660, Boyle set out to demonstrate and measure this expansive power. Later, describing his experiments, he wrote: 'Divers ways have been proposed to show both the Pressure of the Air, as the Atmosphere is a heavy Body, and the Air, especially when compressed by outward force, has a Spring that enables it to sustain or resist equal to that as much of the atmosphere, as can come to bear against it, and also to show, that such Air as we live in, and is not condensed by any human or Adventitious force, has not only a resisting Spring, but an active Spring (if I may so speak) in some measure, as when it distends a flaccid or breaks a full-blown bladder.'

Boyle began by demonstrating the 'active' spring of the air, helped by his assistant Robert Hooke (1635–1703), who had made an improved air pump for use in Boyle's experiments. Together they devised an apparatus comprising of a container from which air could be extracted, holding a small inner tube containing air trapped and compressed by mercury. When the air was pumped from the outer container, this compressed air expanded, pushing the mercury from the small tube, amply demonstrating the active spring of the trapped air.

The passive spring
Next, they studied the 'passive' or resisting spring of air when compressed by external pressure. They made a long glass tube 'crooked at the bottom . . . The orifice of the shorter leg . . . being hermetically sealed'. They pasted strips of paper, carefully marked with a scale in inches, along each arm of the apparatus, poured mercury in the long, open end, and tilted the tube to one side, so that 'the air in the enclosed tube should be of the same laxity (pressure) as the rest of the air about it'.

Then they added more mercury to increase the pressure on the trapped gas, until its volume decreased by half. The additional 'head' of mercury measured 29 in/73.7 cm. Earlier, they had used a Torricellian barometer to measure the atmospheric pressure: it was equivalent to 29 in of mercury. Hence, Boyle concluded, 'this observation does both very well agree with and confirm our hypothesis. . . that the greater the weight is, that leans upon the air, the more forcible is its endeavour of dilation and consequently its power of resistance (as other springs are stronger when bent by greater weights).'

At this point, the glass tube broke, scattering mercury around the laboratory. They constructed a new stronger tube, with a 'pretty bigness', placed it in a wooden box as a precaution against another breakage, and made a series of measurements of the relationship between the volume of the air and the weight of mercury needed to compress it.

They considered the effect of temperature on the results, putting a wet cloth around the tube to cool it; 'it sometimes seemed a little to shrink, but not so manifestly that we dare build anything upon it'. When they heated the closed end with a candle, 'the head had a more sensible operation' but, once again, no conclusion could be drawn. Boyle noted that 'a want of exactness . . . in such experiments is scarce avoidable'.

Boyle's Law
The next step was to investigate the effect of reduced pressure on trapped air, noting the expansion of the trapped gas. Boyle describes the apparatus: 'We provided a slender glass-pipe of about the bigness of a swan's quill'. The tube, with a paper scale marked in inches along its length and its top end sealed with wax, was inserted into a wide, mercury-filled tube so that about one inch extended above the mercury.

The procedure was to raise the slender tube gradually, reducing the pressure of the air inside. First, the mercury level inside rose until the weight of mercury and the reduced air pressure in the tube balanced the external air pressure. Then the volume of the trapped air gradually increased to double its original volume, when the mercury in the tube was about 14.8 in/ 37.6 cm above its original level. Atmospheric pressure on that day was 29.5 in/74.9 cm: the trapped air had doubled its volume when the pressure was halved. According to Boyle, this accorded well with 'the hypothesis that supposes the pressures and expansions to be in reciprocal proportions'—a proposition now known as Boyle's Law.

brachiopod any member of the phylum Brachiopoda, marine invertebrates with two shells, resembling but totally unrelated to bivalves. There are about 300 living species; they were much more numerous in past geological ages. They are suspension feeders, ingesting minute food particles from water. A single internal organ, the iophophore, handles feeding, aspiration, and excretion.

bract leaflike structure in whose ◊axil a flower or inflorescence develops. Bracts are generally green and smaller than the true leaves. However, in some plants they may be brightly coloured and conspicuous, taking over the role of attracting pollinating insects to the flowers, whose own petals are small; examples include poinsettia *Euphorbia pulcherrima* and bougainvillea.

A whorl of bracts surrounding an ◊inflorescence is termed an *involucre*. A *bracteole* is a leaflike organ that arises on an individual flower stalk, between the true bract and the ◊calyx.

brain in higher animals, a mass of interconnected ◊nerve cells, forming the anterior part of the ◊central nervous system, whose activities it coordinates and controls. In ◊vertebrates, the brain is contained by the skull. An enlarged portion of the upper spinal cord, the *medulla oblongata*, contains centres for the control of respiration, heartbeat rate and strength, and blood pressure. Overlying this is the *cerebellum*, which is concerned with coordinating complex muscular processes such as maintaining posture and moving limbs. The cerebral hemispheres (*cerebrum*) are paired outgrowths of the front end of the forebrain, in early vertebrates mainly concerned with the senses, but in higher vertebrates greatly developed and involved in the integration of all sensory input and motor output, and in intelligent behaviour.

In vertebrates, many of the nerve fibres from the two sides of the body cross over as they enter the brain, so that the left cerebral hemisphere is associated with the right side of the body and vice versa. In humans, a certain asymmetry develops in the two halves of the cerebrum. In right-handed people, the left hemisphere seems to play a greater role in controlling verbal and some mathematical skills, whereas the right hemisphere is more involved in spatial perception. In general, however, skills and abilities are not closely localized. In the brain, nerve impulses are passed across ◊synapses by neurotransmitters, in the same way as in other parts of the nervous system.

In mammals the cerebrum is the largest part of the brain, carrying the *cerebral cortex*. This consists of a thick surface layer of cell bodies (grey matter), below which fibre tracts (white matter) connect various parts of the cortex to each other and to other points in the central nervous system. As cerebral complexity grows, the surface of the brain becomes convoluted into deep folds. In higher mammals, there are large unassigned areas of the brain that seem to be connected with intelligence, personality, and higher mental faculties. Language is controlled in two special regions usually in the left side of the brain: *Broca's area* governs the ability to talk, and *Wernicke's area* is responsible for the comprehension of spoken and written words. In 1990, scientists at Johns Hopkins University, Baltimore, succeeded in culturing human brain cells.

If the cells and fibre in one human brain were all stretched out end to end, they would certainly reach to the moon and back. Yet the fact that they are not arranged end to end enabled man to go there himself. The astonishing tangle within our heads makes us what we are.
On the **brain** Colin Blakemore (1944–) BBC Reith Lecture 1976

brainstem region where the top of the spinal cord merges with the undersurface of the brain, consisting largely of the medulla oblongata and midbrain. The oldest part of the brain in evolutionary terms, the brainstem is the body's life-support centre, containing regulatory mechanisms for vital functions such as breathing, heart rate, and blood pressure. It is also involved in controlling the level of consciousness by acting as a relay station for nerve connections to and from the higher centres of the brain.

In many countries, death of the brainstem is now formally recognized as death of the person as a whole. Such cases are the principal donors of organs for transplantation. So-called 'beating-heart donors' can be maintained for a limited period by life-support equipment.

brake device used to slow down or stop the movement of a moving body or vehicle. The mechanically applied calliper brake used on bicycles uses a scissor action to press hard rubber blocks against the wheel rim. The main braking system of a car works hydraulically: when the driver depresses the brake pedal, liquid pressure forces pistons to apply brakes on each wheel.

Two types of car brakes are used.

Disc brakes are used on the front wheels of some cars and on all wheels of sports and performance cars, since they are the more efficient and less prone to fading (losing their braking power) when they get hot. Braking pressure forces brake pads against both sides of a steel disc that rotates with the wheel.

Drum brakes are fitted on the rear wheels of some cars and on all wheels of some passenger cars. Braking pressure forces brake shoes to expand outwards into contact with a drum rotating with the wheels. The brake pads and shoes have a tough ◊friction lining that grips well and withstands wear.

Many trucks and trains have *air brakes*, which work by compressed air. On landing, jet planes reverse the thrust of their engines to reduce their speed quickly. Space vehicles use retrorockets for braking in space, and use the air resistance, or drag of the atmosphere, to slow down when they return to Earth.

brass metal ◊alloy of copper and zinc, with not more than 5% or 6% of other metals. The zinc content ranges from 20% to 45%, and the colour of brass varies accordingly from coppery to whitish yellow. Brasses are characterized by the ease with which they may be shaped and machined; they are

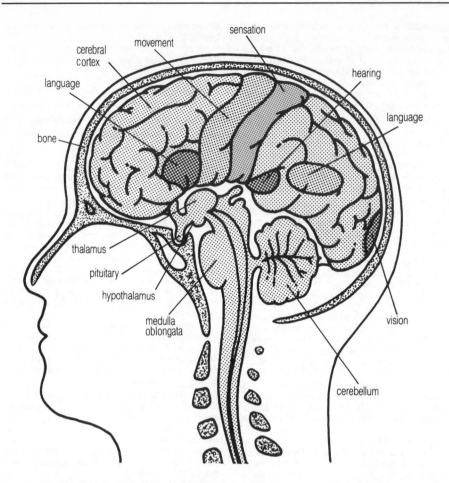

language
cerebral cortex
movement
sensation
hearing
language
bone
thalamus
pituitary
hypothalamus
medulla oblongata
vision
cerebellum

brain *The structure of the human brain. At the back of the skull lies the cerebellum, which coordinates reflex actions that control muscular activity. The medulla oblongata controls respiration, heartbeat, and blood pressure. The hypothalamus is concerned with instinctive drives and emotions. The thalamus relays signals to and from various parts of the brain. The pituitary gland controls the body's hormones. Distinct areas of the large convoluted cerebral hemispheres that fill most of the skull are linked to sensations, such as hearing and sight, and voluntary activities, such as movement.*

strong and ductile, resist many forms of corrosion, and are used for electrical fittings, ammunition cases, screws, household fittings, and ornaments.

Brasses are usually classed into those that can be worked cold (up to 25% zinc) and those that are better worked hot (about 40% zinc).

brazing method of joining two metals by melting an ◊alloy into the joint. It is similar to soldering but takes place at a much higher temperature. Copper and silver alloys are widely used for brazing, at temperatures up to about 900°C/1,650°F.

breast one of a pair of organs on the upper front of the human female, also known as a ◊mammary gland. Each of the two breasts contains milk-producing cells, and a network of tubes or ducts that lead to an opening in the nipple.

Milk-producing cells in the breast do not become active until a woman has given birth to a baby. Breast milk is made from substances extracted from the mother's blood as it passes through the breasts. It contains all the nourishment a baby needs, including antibodies to help fight infection.

Breathalyzer trademark for an instrument for on-the-spot checking by police of the amount of alcohol consumed by a suspect driver. The driver breathes into a plastic bag connected to a tube containing a chemical (such as a diluted solution of potassium dichromate in 50% sulphuric acid) that changes colour in the presence of alcohol. Another method is to use a gas chromatograph, again from a breath sample.

Breath testing was introduced in the UK in 1967. The approved device is now the Lion Intoximeter 3000, which is used by police to indicate the proportion of alcohol in the blood.

breathing in terrestrial animals, the muscular movements whereby air is taken into the lungs and then expelled, a form of ◊gas exchange. Breathing is sometimes referred to as external respiration, for true respiration is a cellular (internal) process.

Lungs are specialized for gas exchange but are

disc brake

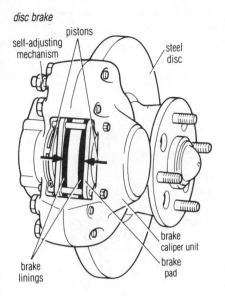

pistons
self-adjusting mechanism
steel disc
brake caliper unit
brake pad
brake linings

drum brake

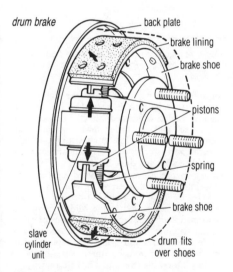

back plate
brake lining
brake shoe
pistons
spring
brake shoe
slave cylinder unit
drum fits over shoes

brake *Two common braking systems: the disc brake (top) and the drum brake (bottom). In the disc brake, increased hydraulic pressure of the brake fluid in the pistons forces the brake pads against the steel disc attached to the wheel. A self-adjusting mechanism balances the force on each pad. In the drum brake, increased pressure of the brake fluid within the slave cylinder forces the brake pad against the brake drum attached to the wheel.*

not themselves muscular, consisting of spongy material. In order for oxygen to be passed to the blood and carbon dioxide removed, air is forced in and out of the chest region by the ribs and accompanying intercostal muscles, the rate of breathing being controlled by the brain. High levels of activity lead to a greater demand for oxygen and a subsequent higher rate of breathing.

breathing rate the number of times a minute the lungs inhale and exhale. The rate increases during exercise because the muscles require an increased supply of oxygen and nutrients. At the same time very active muscles produce a greater volume of carbon dioxide, a waste gas that must be removed by the lungs via the blood.

The regulation of the breathing rate is under both voluntary and involuntary control, although a person can only forcibly stop breathing for a limited time. The regulatory system includes the use of chemoreceptors, which can detect levels of carbon dioxide in the blood. High concentrations of carbon dioxide, occurring for example during exercise, stimulate a fast breathing rate.

breccia coarse clastic ◊sedimentary rock, made up of broken fragments (clasts) of pre-existing rocks. It is similar to ◊conglomerate but the fragments in breccia are jagged in shape.

breed recognizable group of domestic animals, within a species, with distinctive characteristics that have been produced by ◊artificial selection.

breeder reactor or *fast breeder* or *fast reactor* in nuclear physics, a reactor in which more fissionable material is produced (breeding) than is consumed in running it. Breeder reactors are typically operated by and for the military to produce plutonium-239, the fissile material used in the nuclear weapons industry. Although uranium-235 is fissile, it comprises less than 1% of uranium ore, an amount too small to supply the weapons industry. Therefore, nonfissile uranium-238, which is plentiful, comprising 99% of uranium ore, is used instead, in breeder reactors, to manufacture fissile plutonium-239. There are no fast breeder reactors in Britain.

This is done by wrapping a blanket of U-238 around the core of a nuclear reactor. The fission process of the reactor-core material (usually U-235) produces excess neutrons that are captured by the U-238, which transmutes by beta decay into the highly unstable isotope U-239, which in turn through beta decay transmutes into the highly unstable isotope neptunium-239, which in turn undergoes beta decay to transmute into fissile Pu-239. See ◊breeding.

breeding in biology, the crossing and selection of animals and plants to change the characteristics of an existing ◊breed or ◊cultivar (variety), or to produce a new one.

Cattle may be bred for increased meat or milk yield, sheep for thicker or finer wool, and horses for speed or stamina. Plants, such as wheat or maize, may be bred for disease resistance, heavier and more rapid cropping, and hardiness to adverse weather.

breeding in nuclear physics, a process in a reactor in which more fissionable material is produced than is consumed in running the reactor.

For example, plutonium-239 can be made from the relatively plentiful (but nonfissile) uranium-238, or uranium-233 can be produced from thorium. The Pu-239 or U-233 can then be used to fuel other reactors.

brewster unit (symbol B) for measuring the reaction of optical materials to stress, defined in terms of the slowing down of light passing through the material when it is stretched or compressed.

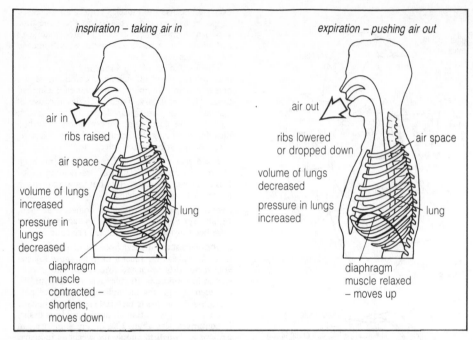

breathing *Breathing. To breathe in, the diaphragm muscle contracts. This allows the lungs to increase in volume, and air flows into the lungs. To breathe out, the diaphragm muscle is relaxed and moves up, forcing air from the lungs. Deep inspiration increases the volume of air breathed in, but there is always some air left in the lungs, the residual volume.*

brick common building material, rectangular in shape, made of clay that has been fired in a kiln. Bricks are made by kneading a mixture of crushed clay and other materials into a stiff mud and extruding it into a ribbon. The ribbon is cut into individual bricks, which are fired at a temperature of up to about 1,000°C/1,800°F. Bricks may alternatively be pressed into shape in moulds.

Refractory bricks used to line furnaces are made from heat-resistant materials such as silica and dolomite. They must withstand operating temperatures of 1,500°C/2,700°F or more. Sun-dried bricks of mud reinforced with straw were first used in Mesopotamia some 8,000 years ago. Similar mud bricks, called adobe, are still used today in Mexico and other areas where the climate is warm and dry.

Established in England by the Romans, brickmaking was later reintroduced in the 13th century, becoming widespread in domestic building only in the 19th century. Brick sizes were first regulated 1729.

bridge structure that provides a continuous path or road over water, valleys, ravines, or above other roads. The basic designs and composites of these are based on the way they bear the weight of the structure and its load. *Beam*, or *girder*, bridges are supported at each end by the ground with the weight thrusting downwards. *Cantilever* bridges are a complex form of girder. *Arch* bridges thrust outwards but downwards at their ends; they are in compression. *Suspension* bridges use cables under tension to pull inwards against anchorages on either side of the span, so that the roadway

hangs from the main cables by the network of vertical cables. The *cable-stayed* bridge relies on diagonal cables connected directly between the bridge deck and supporting towers at each end. Some bridges are too low to allow traffic to pass beneath easily, so they are designed with movable parts, like swing and draw bridges.

history In prehistory, people used logs or wove vines into ropes that were thrown across the obstacle. By 4000 BC arched structures of stone and/or brick were used in the Middle East, and the Romans built long arched spans, many of which are still standing. Wooden bridges proved vulnerable to fire and rot and many were replaced with cast and wrought iron, but these were disadvantaged by low tensile strength. The ◊Bessemer process produced steel that made it possible to build long-lived framed structures that support great weight over long spans.

Examples of the main types of bridges are: *arch* for example, Sydney Harbour Bridge, Australia, a steel arch with a span of 503 m/1,650 ft; *beam or girder* for example, Rio-Niteroi, Guanabara Bay, Brazil, centre span 300 m/984 ft; length 13,900 m/8 mi 3,380 ft; *cantilever* for example, Forth Rail Bridge, Scotland, 1,658 m/5,440 ft long with two main spans, two cantilevers each, one from each tower; *suspension* for example, Humber Bridge, England, with a centre span of 1,410 m/4,628 ft, which is the world's longest bridge span.

The Dartford bridge, 30 km downstream of central London, is the largest cable-stayed bridge in Europe with a main span of 450 m/1,476 ft. The longest cable-stayed bridge is the 490 m/1,608 ft

Ikuchi bridge between Honshu and Shikoku in Japan, completed 1991. Several larger cable-stayed bridges are under construction.

The single-span bridge under construction across the Messina Straits between Sicily and the mainland of Italy will be 3,320 m/10,892 ft long, the world's largest by far. Steel is pre-eminent in the construction of long-span bridges because of its high strength-to-weight ratio, but in other circumstances reinforced concrete has the advantage of lower maintenance costs. The Newport Transporter Bridge (built 1906 in Wales) is a high-level suspension bridge which carries a car suspended a few feet above the water. It was used in preference to a conventional bridge where expensive high approach roads would have to be built.

brine common name for a solution of sodium chloride (NaCl) in water. Brines are used extensively in the food-manufacturing industry for canning vegetables, pickling vegetables (sauerkraut manufacture), and curing meat. Industrially, brine is the source from which chlorine, caustic soda (sodium hydroxide), and sodium carbonate are made.

Brinell hardness test test of the hardness of a substance according to the area of indentation made by a 10 mm/0.4 in hardened steel or sintered tungsten carbide ball under standard loading conditions in a test machine. The resulting Brinell number is equal to the load (kg) divided by the surface area (mm^2) and is named after its inventor Scottish metallurgist Johann Brinell.

British Standards Institution (BSI) UK national standards body. Although government funded, the institute is independent. The BSI interprets international technical standards for the UK, and also sets its own.

For consumer goods, it sets standards which products should reach (the BS standard), as well as testing products to see that they conform to that standard (as a result of which the product may be given the BSI 'kite' mark).

British thermal unit imperial unit (symbol Btu) of heat, now replaced in the SI system by the ◊joule (one British thermal unit is approximately 1,055 joules). Burning one cubic foot of natural gas releases about 1,000 Btu of heat.

One British thermal unit is defined as the amount of heat required to raise the temperature of 0.45 kg/1 lb of water by 1°F. The exact value depends on the original temperature of the water.

brittle material material that breaks suddenly under stress at a point just beyond its elastic limit (see ◊elasticity). Brittle materials may also break suddenly when given a sharp knock. Pottery, glass, and cast iron are examples of brittle materials. Compare ◊ductile material.

broadcasting the transmission of sound and vision programmes by ◊radio and ◊television. Broadcasting may be organized under private enterprise, as in the USA, or may operate under a compromise system, as in Britain, where there is a television and radio service controlled by the state-regulated British Broadcasting Corporation (BBC) and also the commercial Independent Television

Commission (known as the Independent Broadcasting Authority before 1991).

In the USA, broadcasting is only limited by the issue of licences from the Federal Communications Commission to competing commercial companies; in Britain, the BBC is a centralized body appointed by the state and responsible to Parliament, but with policy and programme content not controlled by the state; in Japan, which ranks next to the USA in the number of television sets owned, there is a semigovernmental radio and television broadcasting corporation (NHK) and numerous private television companies. Television broadcasting entered a new era with the introduction of high-powered communications satellites in the 1980s. The signals broadcast by these satellites are sufficiently strong to be picked up by a small dish aerial located, for example, on the roof of a house. Direct broadcast by satellite thus became a feasible alternative to land-based television services. See also ◊cable television.

broad-leaved tree another name for a tree belonging to the ◊angiosperms, such as ash, beech, oak, maple, or birch. The leaves are generally broad and flat, in contrast to the needlelike leaves of most conifers. See also ◊deciduous tree.

bromide salt of the halide series containing the Br⁻ ion, which is formed when a bromine atom gains an electron.

The term 'bromide' is sometimes used to describe an organic compound containing a bromine atom, even though it is not ionic. Modern naming uses the term 'bromo-' in such cases. For example, the compound C_2H_5Br is now called bromoethane; its traditional name, still used sometimes, is ethyl bromide.

bromine (Greek *bromos* 'stench') dark, reddish-brown, nonmetallic element, a volatile liquid at room temperature, symbol Br, atomic number 35, relative atomic mass 79.904. It is a member of the ◊halogen group, has an unpleasant odour, and is very irritating to mucous membranes. Its salts are known as bromides.

Bromine was formerly extracted from salt beds but is now mostly obtained from sea water, where it occurs in small quantities. Its compounds are used in photography and in the chemical and pharmaceutical industries.

bromine water red-brown solution of bromine in water. It is used to test for unsaturation in organic compounds, as such compounds decolorize the solution as they react with the bromine.

$$CH_2=CH_2 + Br_2 \rightarrow CH_2BrCH_2Br$$

bronchiole small-bore air tube found in the vertebrate lung responsible for delivering air to the main respiratory surfaces. Bronchioles lead off from the larger bronchus and branch extensively before terminating in the many thousand alveoli that form the bulk of lung tissue.

bronchus one of a pair of large tubes (bronchii) branching off from the windpipe and passing into the vertebrate lung. Apart from their size, bronchii differ from the bronchioles in possessing cartilaginous rings, which give rigidity and prevent collapse during breathing movements.

Numerous glands secrete a slimy mucus, which

traps dust and other particles; the mucus is constantly being propelled upwards to the mouth by thousands of tiny hairs or cilia. The bronchus is adversely effected by several respiratory diseases and by smoking, which damages the cilia and therefore the lung-cleaning mechanism.

brontosaurus former name of a type of large, plant-eating dinosaur, now better known as ◊apatosaurus.

bronze alloy of copper and tin, yellow or brown in colour. It is harder than pure copper, more suitable for ◊casting, and also resists ◊corrosion. Bronze may contain as much as 25% tin, together with small amounts of other metals, mainly lead.

Bronze is one of the first metallic alloys known and used widely by early peoples during the period of history known as the Bronze Age (about 5000–1200 BC in the Middle East and about 2000–500 BC in Europe).

Bell metal, the bronze used for casting bells, contains 15% or more tin.

Phosphor bronze is hardened by the addition of a small percentage of phosphorus.

Silicon bronze (for telegraph wires) and *aluminium bronze* are similar alloys of copper with silicon or aluminium and small amounts of iron, nickel, or manganese, but usually no tin.

brown dwarf hypothetical object less massive than a star, but heavier than a planet. Brown dwarfs would not have enough mass to ignite nuclear reactions at their centres, but would shine by heat released during their contraction from a gas cloud. Because of the difficulty of detection, no brown dwarfs have been spotted with certainty, but some astronomers believe that vast numbers of them may exist throughout the Galaxy.

Brownian motion the continuous random motion of particles in a fluid medium (gas or liquid) as they are subjected to impact from the molecules of the medium. The phenomenon was explained by German physicist Albert Einstein in 1905 but was probably observed as long ago as 1827 by the Scottish botanist Robert Brown. It provides evidence for the ◊kinetic theory of matter.

brown ring test in analytical chemistry, a test for the presence of ◊nitrates.

To an aqueous solution containing the test substance is added iron(II) sulphate. Concentrated sulphuric acid is then carefully poured down the inside wall of the test tube so that it forms a distinct layer at the bottom. The formation of a brown colour at the boundary between the two layers indicates the presence of nitrate.

brush in certain electric motors, one of a pair of contacts that pass electric current into and out of the rotating coils by means of a device known as a ◊commutator. The brushes, which are often replaceable, are usually made of a soft, carbon material to reduce wear of the copper contacts on the rotating commutator.

bryophyte member of the Bryophyta, a division of the plant kingdom containing three classes: the Hepaticae (◊liverworts), Musci (◊mosses), and Anthocerotae (◊hornworts). Bryophytes are generally small, low-growing, terrestrial plants with no vascular (water-conducting) system as in higher

plants. Their life cycle shows a marked ◊alternation of generations. Bryophytes chiefly occur in damp habitats and require water for the dispersal of the male gametes (◊antherozoids).

In bryophytes, the ◊sporophyte, consisting only of a spore-bearing capsule on a slender stalk, is wholly or partially dependent on the ◊gametophyte for water and nutrients. In some liverworts the plant body is a simple ◊thallus, but in the majority of bryophytes it is differentiated into stem, leaves, and ◊rhizoids.

BSI abbreviation for ◊*British Standards Institution*.

BT abbreviation for *British Telecom*.

Btu symbol for ◊*British thermal unit*.

bubble chamber in physics, a device for observing the nature and movement of atomic particles, and their interaction with radiations. It is a vessel filled with a superheated liquid through which ionizing particles move and collide. The paths of these particles are shown by strings of bubbles, which can be photographed and studied. By using a pressurized liquid medium instead of a gas, it overcomes drawbacks inherent in the earlier ◊cloud chamber. It was invented by US physicist Donald Glaser 1952. See ◊particle detector.

bubble memory in computing, a memory device based on the creation of small 'bubbles' on a magnetic surface. Bubble memories typically store up to 4 megabits (4 million ◊bits) of information. They are not sensitive to shock and vibration, unlike other memory devices such as disc drives, yet, like magnetic discs, they are nonvolatile and do not lose their information when the computer is switched off.

bubble sort in computing, a technique for ◊sorting data. Adjacent items are continually exchanged until the data are in sequence.

buckminsterfullerene form of carbon, made up of molecules (buckyballs) consisting of 60 carbon atoms arranged in 12 pentagons and 20 hexagons to form a perfect sphere. It was named after the US architect and engineer Richard Buckminster Fuller because of its structural similarity to the geodesic dome that he designed. See ◊fullerene.

Buckminsterfullerene can be made into stable polymers using palladium.

buckyball popular name for a molecule of ◊buckminsterfullerene.

bud undeveloped shoot usually enclosed by protective scales; inside is a very short stem and numerous undeveloped leaves, or flower parts, or both. Terminal buds are found at the tips of shoots, while axillary buds develop in the ◊axils of the leaves, often remaining dormant unless the terminal bud is removed or damaged. Adventitious buds may be produced anywhere on the plant, their formation sometimes stimulated by an injury, such as that caused by pruning.

budding type of ◊asexual reproduction in which an outgrowth develops from a cell to form a new individual. Most yeasts reproduce in this way.

In a suitable environment, yeasts grow rapidly, forming long chains of cells as the buds themselves produce further buds before being separated from the parent. Simple invertebrates, such as ◊hydra, can also reproduce by budding.

In horticulture, the term is used for a technique of plant propagation whereby a bud (or scion) and a sliver of bark from one plant are transferred to an incision made in the bark of another plant (the stock). This method of ◊grafting is often used for roses.

buffer in chemistry, a mixture of compounds chosen to maintain a steady ◊pH. The commonest buffers consist of a mixture of a weak organic acid and one of its salts or a mixture of acid salts of phosphoric acid. The addition of either an acid or a base causes a shift in the ◊chemical equilibrium, thus keeping the pH constant.

buffer in computing, a part of the ◊memory used to store data temporarily while it is waiting to be used. For example, a program might store data in a printer buffer until the printer is ready to print it.

bug in computing, an ◊error in a program. It can be an error in the logical structure of a program or a syntax error, such as a spelling mistake. Some bugs cause a program to fail immediately; others remain dormant, causing problems only when a particular combination of events occurs. The process of finding and removing errors from a program is called *debugging*.

bulb underground bud with fleshy leaves containing a reserve food supply and with roots growing from its base. Bulbs function in vegetative reproduction and are characteristic of many monocotyledonous plants such as the daffodil, snowdrop, and onion. Bulbs are grown on a commercial scale in temperate countries, such as England and the Netherlands.

bulbil small bulb that develops above ground from a bud. Bulbils may be formed on the stem from axillary buds, as in members of the saxifrage family, or in the place of flowers, as seen in many species of onion *Allium*. They drop off the parent plant and develop into new individuals, providing a means of ◊vegetative reproduction and dispersal.

bulldozer earth-moving machine widely used in construction work for clearing rocks and tree stumps and levelling a site. The bulldozer is a kind of ◊tractor with a powerful engine and a curved, shovel-like blade at the front, which can be lifted and forced down by hydraulic rams. It usually has caterpillar tracks (a continuous flexible belt of metal plates instead of ordinary tyred wheels) so that it can move easily over rough ground.

bunsen burner gas burner used in laboratories, consisting of a vertical metal tube through which a fine jet of fuel gas is directed. Air is drawn in through airholes near the base of the tube and the mixture is ignited and burns at the tube's upper opening.

The invention of the burner is attributed to German chemist Robert von Bunsen 1855 but English chemist and physicist Michael Faraday is known to have produced a similar device at an earlier date. A later refinement was the metal collar that can be turned to close or partially close the airholes, thereby regulating the amount of air sucked in and hence the heat of the burner's flame.

buoy floating object used to mark channels for shipping or warn of hazards to navigation. Buoys come in different shapes, such as a pole (spar buoy), cylinder (car buoy), and cone (nun buoy). Light buoys carry a small tower surmounted by a flashing lantern, and bell buoys house a bell, which rings as the buoy moves up and down with the waves. Mooring buoys are heavy and have a ring on top to which a ship can be tied.

buoyancy lifting effect of a fluid on a body wholly or partly immersed in it. This was studied by Greek mathematician and physicist Archimedes in the 3rd century BC.

bur or *burr* in botany, a type of 'false fruit' or ◊pseudocarp, surrounded by numerous hooks; for instance, that of burdock *Arctium*, where the hooks are formed from bracts surrounding the flowerhead. Burs catch in the feathers or fur of passing animals, and thus may be dispersed over considerable distances.

The term is also used to include any type of fruit or seed-bearing hooks, such as that of goosegrass *Galium aparine* and wood avens *Geum urbanum*.

burette in chemistry, a piece of apparatus, used in ◊titration, for the controlled delivery of measured variable quantities of a liquid.

It consists of a long, narrow, calibrated glass tube, with a tap at the bottom, leading to a narrow-bore exit.

burning common name for ◊combustion.

bus in computing, the electrical pathway through which a computer processor communicates with some of its parts and/or peripherals. Physically, a bus is a set of parallel tracks that can carry digital signals; it may take the form of copper tracks laid down on the computer's ◊printed circuit boards (PCBs), or of an external cable or connection.

A computer typically has three internal buses laid down on its main circuit board: a *data bus*, which carries data between the components of the computer; an *address bus*, which selects the route to be followed by any particular data item travelling along the data bus; and a *control bus*, which is used to decide whether data is written to or read from the data bus. An external *expansion bus* is used for linking the computer processor to peripheral devices, such as modems and printers.

bushel dry or liquid measure equal to eight gallons or four pecks (2,219.36 cu in/36.37 litres) in the UK; some US states have different standards according to the goods measured.

butadiene or *buta-1,3–diene* $CH_2{:}CHCH{:}CH_2$ inflammable gas derived from petroleum, used in making synthetic rubber and resins.

butane C_4H_{10} one of two gaseous alkanes (paraffin hydrocarbons) having the same formula but differing in structure. Normal butane is derived from natural gas; isobutane is a by-product of petroleum manufacture. Liquefied under pressure, it is used as a fuel for industrial and domestic purposes (for example, in portable cookers).

butene C_4H_8 fourth member of the ◊alkene series of hydrocarbons. It is an unsaturated compound, containing one double bond.

butte steep-sided flat-topped hill, formed in horizontally layered sedimentary rocks, largely in arid areas. A large butte with a pronounced tablelike profile is a ◊mesa.

Buttes and mesas are characteristic of semi-arid areas where remnants of resistant rock layers protect softer rock underneath, as in the plateau regions of Colorado, Utah, and Arizona, USA.

by-product substance or product which arises out of the production of another product. For example, sulphur dioxide gas is a by-product of producing electricity through the burning of coal. Sheepskins are a by-product of the production of mutton for eating.

byte sufficient computer memory to store a single character of data. The character is stored in the byte of memory as a pattern of ◊bits (binary digits), using a code such as ◊ASCII. A byte usually contains eight bits—for example, the capital letter F can be stored as the bit pattern 01000110.

A single byte can specify 256 values, such as the decimal numbers from 0 to 255; in the case of a single-byte ◊pixel (picture element), it can specify 256 different colours. Three bytes (24 bits) can specify 16,777,216 values. Computer memory size is measured in *kilobytes* (1,024 bytes) or *megabytes* (1,024 kilobytes).

C high-level general-purpose computer-programming language popular on minicomputers and microcomputers. Developed in the early 1970s from an earlier language called BCPL, C was first used as the language of the operating system ◊Unix, though it has since become widespread beyond Unix. It is useful for writing fast and efficient systems programs, such as operating systems (which control the operations of the computer).

°C symbol for degrees ◊Celsius, sometimes called centigrade.

cable unit of length, used on ships, originally the length of a ship's anchor cable or 120 fathoms (219 m/720 ft), but now taken as one-tenth of a ◊nautical mile (185.3 m/608 ft).

cable car method of transporting passengers up steep slopes by cable. In the *cable railway*, passenger cars are hauled along rails by a cable wound by a powerful winch. A pair of cars usually operates together on the funicular principle, one going up as the other goes down. The other main type is the *aerial cable car*, where the passenger car is suspended from a trolley that runs along an aerial cableway.

A unique form of cable-car system has operated in San Francisco since 1873. The streetcars travel along rails and are hauled by moving cables under the ground.

cable television distribution of broadcast signals through cable relay systems.

Narrow-band systems were originally used to deliver services to areas with poor regular reception; systems with wider bands, using coaxial and fibreoptic cable, are increasingly used for distribution and development of home-based interactive services.

cache memory in computing, a reserved area of the ◊immediate-access memory used to increase the running speed of a computer program.

The cache memory may be constructed from ◊SRAM, which is faster but more expensive than the normal ◊DRAM. Most programs access the same instructions or data repeatedly. If these frequently used instructions and data are stored in a fast-access SRAM memory cache, the program will run more quickly. In other cases, the memory cache is normal DRAM, but is used to store frequently used instructions and data that would normally be accessed from ◊backing storage. Access to DRAM is faster than access to backing storage so, again, the program runs more quickly. This type of cache memory is often called a *disc cache*.

CAD (acronym for computer-aided design) the use of computers in creating and editing design drawings. CAD also allows such things as automatic testing of designs and multiple or animated three-dimensional views of designs. CAD systems are widely used in architecture, electronics, and engineering, for example in the motor-vehicle industry, where cars designed with the assistance of computers are now commonplace. A related development is ◊CAM (computer-assisted manufacturing).

cadmium soft, silver-white, ductile, and malleable metallic element, symbol Cd, atomic number 48, relative atomic mass 112.40. Cadmium occurs in nature as a sulphide or carbonate in zinc ores. It is a toxic metal that, because of industrial dumping, has become an environmental pollutant. It is used in batteries, electroplating, and as a constituent of alloys used for bearings with low coefficients of friction; it is also a constituent of an alloy with a very low melting point.

Cadmium is also used in the control rods of nuclear reactors, because of its high absorption of neutrons. It was named in 1817 by the German chemist Friedrich Strohmeyer (1776–1835) after Greek mythological character Cadmus.

caecum in the ◊digestive system of animals, a blind-ending tube branching off from the first part of the large intestine, terminating in the appendix. It has no function in humans but is used for the digestion of cellulose by some grass-eating mammals.

The rabbit caecum and appendix contains millions of bacteria that produce cellulase, the enzyme necessary for the breakdown of cellulose to glucose.

In order to be able to absorb nutrients released by the breakdown of cellulose, rabbits pass food twice down the intestine. They egest soft pellets which are then re-eaten. This is known as coprophagy.

caesium (Latin *caesius* 'bluish-grey') soft, silvery-white, ductile metallic element, symbol Cs, atomic number 55, relative atomic mass 132.905. It is one of the ◊alkali metals, and is the most electropositive of all the elements. In air it ignites spontaneously, and it reacts vigorously with water. It is used in the manufacture of photoelectric cells. The name comes from the blueness of its spectral line.

The rate of vibration of caesium atoms is used as the standard of measuring time. Its radioactive isotope Cs-137 (half-life 30.17 years) is a product of fission in nuclear explosions and in nuclear reactors; it is one of the most dangerous waste products of the nuclear industry, being a highly radioactive biological analogue for potassium.

caffeine ◊alkaloid organic substance found in tea, coffee, and kola nuts; it stimulates the heart and central nervous system. When isolated, it is a bitter crystalline compound, $C_8H_{10}N_4O_2$. Too much caffeine (more than six average cups of tea or coffee a day) can be detrimental to health.

caisson hollow cylindrical or boxlike structure, usually of reinforced ◊concrete, sunk into a riverbed to form the foundations of a bridge.

An *open caisson* is open at the top and at the bottom, where there is a wedge-shaped cutting edge. Material is excavated from inside, allowing the caisson to sink. A *pneumatic caisson* has a pressurized chamber at the bottom, in which workers carry out the excavation. The air pressure prevents the surrounding water entering; the workers enter and leave the chamber through an airlock, allowing for a suitable decompression period to prevent decompression sickness (the so-called bends).

cal symbol for ◊*calorie*.

CAL (acronym for computer-assisted *l*earning) the use of computers in education and training: the computer displays instructional material to a student and asks questions about the information given; the student's answers determine the sequence of the lessons.

calamine $ZnCO_3$ zinc carbonate, an ore of zinc. The term also refers to a pink powder made of a mixture of zinc oxide and iron(II) oxide used in lotions and ointments as an astringent for treating, for example, sunburn, eczema, measles rash, and insect bites and stings.

In the USA the term refers to zinc silicate $Zn_4Si_2O_7(OH)_2.H_2O$.

calcination ◊oxidation of metals by burning in air.

calcite colourless, white, or light-coloured common rock-forming mineral, calcium carbonate, $CaCO_3$. It is the main constituent of ◊limestone and marble, and forms many types of invertebrate shell.

Calcite often forms ◊stalactites and stalagmites in caves and is also found deposited in veins through many rocks because of the ease with which it is dissolved and transported by groundwater; ◊oolite is a rock consisting of spheroidal calcite grains. It rates 3 on the ◊Mohs' scale of hardness.

Large crystals up to 1 m/3 ft have been found in Oklahoma and Missouri, USA. ◊Iceland spar is a transparent form of calcite used in the optical industry; as limestone it is used in the building industry.

calcium (Latin *calcis* 'lime') soft, silvery-white metallic element, symbol Ca, atomic number 20, relative atomic mass 40.08. It is one of the ◊alkaline-earth metals. It is the fifth most abundant element (the third most abundant metal) in the Earth's crust. It is found mainly as its carbonate $CaCO_3$, which occurs in a fairly pure condition as chalk and limestone (see ◊calcite). Calcium is an essential component of bones, teeth, shells, milk, and leaves, and it forms 1.5% of the human body by mass.

Calcium ions in animal cells are involved in regulating muscle contraction, hormone secretion, digestion, and glycogen metabolism in the liver.

The element was discovered and named by the English chemist Humphry Davy in 1808. Its compounds include slaked lime (calcium hydroxide, $Ca(OH)_2$); plaster of Paris (calcium sulphate, $CaSO_4.H_2O$); calcium phosphate ($Ca_3(PO_4)_2$), the main constituent of animal bones; calcium hypochlorite ($CaOCl_2$), a bleaching agent; calcium nitrate ($Ca(NO_3)_2.H_2O$), a nitrogenous fertilizer; calcium carbide (CaC_2), which reacts with water to give ethyne (acetylene); calcium cyanamide ($CaCN_2$), the basis of many pharmaceuticals, fertilizers, and plastics, including melamine; calcium cyanide ($Ca(CN)_2$), used in the extraction of gold and silver and in electroplating; and others used in baking powders and fillers for paints.

calcium carbonate $CaCO_3$ white solid, found in nature as limestone, marble, and chalk. It is a valuable resource, used in the making of iron, steel, cement, glass, slaked lime, bleaching powder, sodium carbonate and bicarbonate, and many other industrially useful substances.

calcium hydrogencarbonate $Ca(HCO_3)_2$ substance found in ◊hard water, formed when rainwater passes over limestone rock.

$$CaCO_3 \text{ (s)} + CO_2 \text{ (g)} + H_2O \text{ (l)} \rightarrow Ca(HCO_3)_2 \text{ (aq)}$$

When this water is boiled it reforms calcium carbonate, removing the hardness; this type of hardness is therefore known as temporary hardness.

calcium hydrogenphosphate $Ca(H_2PO_4)_2$ substance made by heating calcium phosphate with 70% sulphuric acid. It is more soluble in water than calcium phosphate, and is used as a fertilizer.

calcium hydroxide or *slaked lime* $Ca(OH)_2$ white solid, slightly soluble in water. A solution of calcium hydroxide is called ◊limewater and is used in the laboratory to test for the presence of carbon dioxide. It is manufactured industrially by adding water to calcium oxide (quicklime) in a strongly exothermic reaction.

$$CaO + H_2O \rightarrow Ca(OH)_2$$

It is used to reduce soil acidity and as a cheap alkali in many industrial processes.

calcium oxide or *quicklime* CaO white solid compound, formed by heating ◊calcium carbonate.

$$CaCO_3 \rightarrow CaO + CO_2$$

When water is added it forms calcium hydroxide (slaked lime) in an ◊exothermic reaction.

$$CaO + H_2O \rightarrow Ca(OH)_2$$

It is a typical basic oxide, turning litmus blue.

calcium phosphate or *calcium orthophosphate* $Ca_3(PO_4)_2$ white solid, the main constituent of animal bones. It occurs naturally as the mineral apatite and in rock phosphate, and is used in the preparation of phosphate fertilizers.

calcium sulphate $CaSO_4$ white, solid compound, found in nature as gypsum and anhydrite. It dissolves slightly in water to form ◊hard water; this hardness is not removed by boiling, and hence is sometimes called permanent hardness.

calcium superphosphate common name for ◊calcium hydrogenphosphate.

calculator pocket-sized electronic computing device for performing numerical calculations. It can add, subtract, multiply, and divide; many calculators also compute squares and roots, and have advanced trigonometric and statistical functions. Input is by a small keyboard and results are shown on a one-line computer screen, typically a ◊liquid crystal display (LCD) or a light-emitting diode (LED). The first electronic calculator was manufactured by the Bell Punch Company in the USA 1963.

calculus (Latin 'pebble') branch of mathematics that permits the manipulation of continuously varying quantities, used in practical problems involving such matters as changing speeds, problems of flight, varying stresses in the framework of a bridge, and alternating current theory. *Integral calculus* deals with the method of summation or adding together the effects of continuously varying quantities. *Differential calculus* deals in a similar way with rates of change. Many of its applications arose from the study of the gradients of the tangents to curves.

There are several other branches of calculus, including calculus of errors and calculus of variation. Differential and integral calculus, each of which deals with small quantities which during manipulation are made smaller and smaller, compose the *infinitesimal calculus*. *Differential equations* relate to the derivatives of a set of variables and may include the variables. Many give the mathematical models for physical phenomena such as ◊simple harmonic motion. Differential equations are solved generally through integrative means, depending on their degrees. If no known mathematical processes are available, integration can be performed graphically or by computers.

history Calculus originated with Archimedes in the 3rd century BC as a method for finding the areas of curved shapes and for drawing tangents to curves. These ideas were not developed until the 17th century, when the French philosopher René Descartes introduced ◊coordinate geometry, showing how geometrical curves can be described and manipulated by means of algebraic expressions. Then the French mathematician Pierre de Fermat used these algebraic forms in the early stages of the development of differentiation. Later the German

philosopher Gottfried Leibniz and the English scientist Isaac Newton advanced the study.

caldera in geology, a very large basin-shaped ◊crater. Calderas are found at the tops of volcanoes, where the original peak has collapsed into an empty chamber beneath. The basin, many times larger than the original volcanic vent, may be flooded, producing a crater lake, or the flat floor may contain a number of small volcanic cones, produced by volcanic activity after the collapse.

Typical calderas are Kilauea, Hawaii; Crater Lake, Oregon, USA; and the summit of Olympus Mons, on Mars. Some calderas are wrongly referred to as craters, such as Ngorongoro, Tanzania.

calendar division of the ◊year into months, weeks, and days and the method of ordering the years. From year one, an assumed date of the birth of Jesus, dates are calculated backwards (BC 'before Christ' or BCE 'before common era') and forwards (AD, Latin *anno Domini* 'in the year of the Lord', or CE 'common era'). The *lunar month* (period between one new moon and the next) naturally averages 29.5 days, but the Western calendar uses for convenience a *calendar month* with a complete number of days, 30 or 31 (Feb has 28). For adjustments, since there are slightly fewer than six extra hours a year left over, they are added to Feb as a 29th day every fourth year (*leap year*), century years being excepted unless they are divisible by 400. For example, 1896 was a leap year; 1900 was not. 1996 is the next leap year.

The *month names* in most European languages were probably derived as follows: January from Janus, Roman god; February from *Februar*, Roman festival of purification; March from Mars, Roman god; April from Latin *aperire*, 'to open'; May from Maia, Roman goddess; June from Juno, Roman goddess; July from Julius Caesar, Roman general; August from Augustus, Roman emperor; September, October, November, December (originally the seventh–tenth months) from the Latin words meaning seventh, eighth, ninth, and tenth, respectively.

The *days of the week* are Monday named after the Moon; Tuesday from Tiu or Tyr, Anglo-Saxon and Norse god; Wednesday from Woden or Odin, Norse god; Thursday from Thor, Norse god; Friday from Freya, Norse goddess; Saturday from Saturn, Roman god; and Sunday named after the Sun.

All early calendars except the ancient Egyptian were lunar. The word calendar comes from the Latin *Kalendae* or *calendae*, the first day of each month on which, in ancient Rome, solemn proclamation was made of the appearance of the new moon.

The *Western* or *Gregorian calendar* derives from the *Julian calendar* instituted by Julius Caesar 46 BC. It was adjusted by Pope Gregory XIII 1582, who eliminated the accumulated error caused by a faulty calculation of the length of a year and avoided its recurrence by restricting century leap years to those divisible by 400. Other states only gradually changed from Old Style to New Style; Britain and its colonies adopted the Gregorian calendar 1752, when the error amounted to 11 days, and 3 Sept 1752 became 14 Sept (at the same time the beginning of the year was put back from 25 March to 1 Jan). Russia did not adopt it until the October Revolution of 1917, so that the event (then 25 Oct) is currently celebrated 7 Nov. The *Jewish calendar* is a complex combination of lunar and solar cycles, varied by considerations of religious observance. A year may have 12 or 13 months, each of which normally alternates between 29 and 30 days; the New Year (Rosh Hashanah) falls between 5 Sept and 5 Oct. The calendar dates from the hypothetical creation of the world (taken as 7 Oct 3761 BC). The *Chinese calendar* is lunar, with a cycle of 60 years. Both the traditional and, from 1911, the Western calendar are in use in China. The *Muslim calendar*, also lunar, has 12 months of alternately 30 and 29 days, and a year of 354 days. This results in the calendar rotating around the seasons in a 30–year cycle. The era is counted as beginning on the day Muhammad fled from Mecca AD 622.

calibration the preparation of a usable scale on a measuring instrument. A mercury ◊thermometer, for example, can be calibrated with a Celsius scale by noting the heights of the mercury column at two standard temperatures—the freezing point (0°C) and boiling point (100°C) of water—and dividing the distance between them into 100 equal parts and continuing these divisions above and below.

California current cold ocean ◊current in the E Pacific Ocean flowing southwards down the west coast of North America. It is part of the North Pacific ◊gyre (a vast, circular movement of ocean water).

californium synthesized, radioactive, metallic element of the actinide series, symbol Cf, atomic number 98, relative atomic mass 251. It is produced in very small quantities and used in nuclear reactors as a neutron source. The longest-lived isotope, Cf-251, has a half-life of 800 years.

It is named after the state of California, where it was first synthesized in 1950 by Glenn Seaborg and his team at the University of California at Berkeley.

calima (Spanish 'haze') dust cloud in Europe, coming from the Sahara Desert, which sometimes causes heatwaves and eye irritation.

callipers measuring instrument used, for example, to measure the internal and external diameters of pipes. Some callipers are made like a pair of compasses, having two legs, often curved, pivoting about a screw at one end. The ends of the legs are placed in contact with the object to be measured, and the gap between the ends is then measured against a rule. The slide calliper looks like an adjustable spanner, and carries a scale for direct measuring, usually with a ◊vernier scale for accuracy.

Callisto second largest moon of Jupiter, 4,800 km/3,000 mi in diameter, orbiting every 16.7 days at a distance of 1.9 million km/1.2 million mi from the planet. Its surface is covered with large craters.

callus in botany, a tissue that forms at a damaged plant surface. Composed of large, thin-walled ◊parenchyma cells, it grows over and around the wound, eventually covering the exposed area. In animals, a

calorie: average daily energy requirements (kcal/day)

	light activity	moderate activity	heavy activity
men	2,496	2,624	2,912
women	2,015	2,314	2,730

callus is a thickened pad of skin, formed where there is repeated rubbing against a hard surface. In humans, calluses often develop on the hands and feet of those involved in heavy manual work.

calomel (technical name *mercury(I) chloride*) Hg_2Cl_2 white, heavy powder formerly used as a laxative, now used as a pesticide and fungicide.

calorie cgs unit of heat, now replaced by the ◊joule (one calorie is approximately 4.2 joules). It is the heat required to raise the temperature of one gram of water by 1°C. In dietetics, the Calorie or kilocalorie is equal to 1,000 calories.

The kilocalorie measures the energy value of food in terms of its heat output: 28 g/1 oz of protein yields 120 kilocalories, of carbohydrate 110, of fat 270, and of alcohol 200.

calorific value the amount of heat generated by a given mass of fuel when it is completely burned. It is measured in joules per kilogram. Calorific values are measured experimentally with a bomb calorimeter.

calorimeter instrument used in physics to measure heat. A simple calorimeter consists of a heavy copper vessel that is polished (to reduce heat losses by radiation) and covered with insulating material (to reduce losses by convection and conduction).

In a typical experiment, such as to measure the heat capacity of a piece of metal, the calorimeter is filled with water, whose temperature rise is measured using a thermometer when a known mass of the heated metal is immersed in it. Chemists use a bomb calorimeter to measure the heat produced by burning a fuel completely in oxygen.

calotype paper-based photograph using a wax paper negative, the first example of the ◊negative/positive process invented by the English photographer William Henry Fox Talbot around 1834.

calyptra in mosses and liverworts, a layer of cells that encloses and protects the young ◊sporophyte (spore capsule), forming a sheathlike hood around the capsule. The term is also used to describe the root cap, a layer of ◊parenchyma cells covering the end of a root that gives protection to the root tip as it grows through the soil. This is constantly being worn away and replaced by new cells from a special ◊meristem, the calyptrogen.

calyx collective term for the ◊sepals of a flower, forming the outermost whorl of the ◊perianth. It surrounds the other flower parts and protects them while in bud. In some flowers, for example, the campions *Silene*, the sepals are fused along their sides, forming a tubular calyx.

cam part of a machine that converts circular motion to linear motion or vice versa. The *edge cam* in a car engine is in the form of a rounded projection on a shaft, the camshaft. When the camshaft turns, the cams press against linkages (plungers or followers) that open the valves in the cylinders.

A *face cam* is a disc with a groove in its face, in which the follower travels. A *cylindrical cam* carries angled parallel grooves, which impart a to-and-fro motion to the follower when it rotates.

CAM (acronym for computer-aided manufacturing) the use of computers to control production processes; in particular, the control of machine tools and ◊robots in factories. In some factories, the whole design and production system has been automated by linking ◊CAD (computer-aided design) to CAM.

Linking flexible CAD/CAM manufacturing to computer-based sales and distribution methods makes it possible to produce semicustomized goods cheaply and in large numbers.

cambium in botany, a layer of actively dividing cells (lateral ◊meristem), found within stems and roots, that gives rise to ◊secondary growth in perennial plants, causing an increase in girth. There are two main types of cambium: *vascular cambium*, which gives rise to secondary ◊xylem and ◊phloem tissues, and *cork cambium* (or phellogen), which gives rise to secondary cortex and cork tissues (see ◊bark).

Cambrian period of geological time 570–510 million years ago; the first period of the Palaeozoic era. All invertebrate animal life appeared, and marine algae were widespread. The earliest fossils with hard shells, such as trilobites, date from this period.

The name comes from Cambria, the medieval Latin name for Wales, where Cambrian rocks are typically exposed and were first described.

camcorder another name for a ◊video camera.

camera apparatus used in ◊photography, consisting of a lens system set in a light-proof box inside of which a sensitized film or plate can be placed. The lens collects rays of light reflected from the subject and brings them together as a sharp image on the film; it has marked numbers known as ◊apertures, or f-stops, that reduce or increase the amount of light. Apertures also control depth of field. A shutter controls the amount of time light has to affect the film. There are small-, medium-, and large-format cameras; the format refers to the size of recorded image and the dimensions of the print obtained.

A simple camera has a fixed shutter speed and aperture, chosen so that on a sunny day the correct amount of light is admitted. More complex cameras allow the shutter speed and aperture to be adjusted; most have a built-in exposure meter to help choose the correct combination of shutter speed and aperture for the ambient conditions and subject matter. The most versatile camera is the single lens reflex (◊SLR) which allows the lens to be removed and special lenses attached. A pin-hole camera has a small (pin-sized) hole instead of a lens. It must be left on a firm support during exposures, which are up to ten seconds with slow film, two seconds with fast film and five minutes for paper negatives in daylight. The pin-hole camera gives sharp images from close-up to infinity.

camera obscura darkened box with a tiny hole for projecting the inverted image of the scene outside on to a screen inside. For its development as a device for producing photographs, see ◊photography.

camouflage colours or structures that allow an animal to blend with its surroundings to avoid detection by other animals. Camouflage can take the form of matching the background colour, of countershading (darker on top, lighter below, to counteract natural shadows), or of irregular patterns that break up the outline of the animal's body. More elaborate camouflage involves closely resembling a feature of the natural environment, as with the stick insect; this is closely akin to ◊mimicry.

camphor $C_{10}H_{16}O$ volatile, aromatic ◊ketone substance obtained from the camphor tree *Cinnamomum camphora*. It is distilled from chips of the wood, and is used in insect repellents and medicinal inhalants and liniments, and in the manufacture of celluloid.

The camphor tree, a member of the family Lauraceae, is native to China, Taiwan, and Japan.

Camphylobacter genus of bacteria that cause serious outbreaks of gastroenteritis. They grow best at 43°C, and so are well suited to the digestive tract of birds. Poultry is therefore the most likely source of a *Camphylobacter* outbreak, although the bacteria can also be transmitted via beef or milk. *Camphylobacter* can survive in water for up to 15 days, so may be present in drinking water if supplies are contaminated by sewage or reservoirs are polluted by seagulls.

In 1990 the incidence of *Camphylobacter* poisoning equalled salmonella incidence in the UK.

canal artificial waterway constructed for drainage, irrigation, or navigation. *Irrigation canals* carry water for irrigation from rivers, reservoirs, or wells, and are designed to maintain an even flow of water over the whole length. *Navigation and ship canals* are constructed at one level between ◊locks, and frequently link with rivers or sea inlets to form a waterway system. The Suez Canal 1869 and the Panama Canal 1914 eliminated long trips around continents and dramatically shortened shipping routes.

Canaries current cold ocean current in the North Atlantic Ocean flowing SW from Spain along the NW coast of Africa. It meets the northern equatorial current at a latitude of 20°N.

cancer group of diseases characterized by abnormal proliferation of cells. Cancer (malignant) cells are usually degenerate, capable only of reproducing themselves (tumour formation). Malignant cells tend to spread from their site of origin by travelling through the bloodstream or lymphatic system.

There are more than 100 types of cancer. Some, like lung or bowel cancer, are common; others are rare. The likely cause remains unexplained. Triggering agents (◊carcinogens) include chemicals such as those found in cigarette smoke, other forms of smoke, asbestos dust, exhaust fumes, and many industrial chemicals. Some viruses can also trigger the cancerous growth of cells (see ◊oncogenes), as can X-rays and radioactivity. Dietary

Coding for cancer

Whatever its cause, a cancer takes several years to develop. The event or change in the living cell responsible may be triggered by ionizing radiation, a carcinogenic chemical or a virus. But between this event and the appearance of the cancer cells forming a tumour, the tissue involved will have renewed itself several times over. So the question that struck cancer researchers was: how did the cancer cell perpetuate itself?

Looking within the genes

It seemed likely that there must be some clue in the genes—within the DNA (deoxyribonucleic acid) of a cancer cell. The chromosomes, which are the gene carriers of the cell, should show characteristic abnormalities in some cancers. Towards the end of the 1970s, three separate teams of US researchers undertook a series of experiments to verify this hypothesis: Robert Weinberg of the Center for Cancer Research, a branch of the Massachusetts Institute of Technology; Geoffrey Cooper of the Sydney Forbes Cancer Institute of Boston; and Michael Wigler of the Cold Springs Laboratory on Long Island. They took the DNA from human cancer cells and, using cultured mouse cells (tissue culture),

grafted the DNA onto them. Some of the cultured cells became cancerous and these were isolated.

The DNA from these cells was extracted, and reduced selectively into fragments with restriction enzymes. The scientists again found that some of the cultured cells became cancerous when the DNA fragments were grafted to them. By repeating these procedures, it became clear that a particular segment of human DNA could be linked to cancer. But how and why?

One cancer-causing gene (oncogene) was isolated and analysed nucleotide by nucleotide, and an unusual sequence of bases (the molecules that define the genetic code) discovered. The oncogene concerned caused cancer of the bladder. It was found to comprise 930 bases and had only one small difference from the normal gene: at one point a single base was out of position. This one difference in the gene was enough to cause cancer. For the first time, a genetic change leading to cancer had been identified.

The discovery of oncogenes cannot be presented as the only key to cancer, but full understanding of how oncogene activity leads to abnormal cell growth may well lead to new therapies.

---

factors are important in some cancers; for example, lack of fibre in the diet may predispose people to bowel cancer and a diet high in animal fats and low in fresh vegetables and fruit increases the risk of breast cancer. In some families there is a genetic tendency towards a particular type of cancer.

Cancer is one of the leading causes of death in the industrialized world, yet it is by no means incurable, particularly in the case of certain tumours, including Hodgkin's disease, acute leukaemia, and testicular cancer. Cures are sometimes achieved with specialized treatments, such as surgery, chemotherapy with cytotoxic drugs (drugs that kill cancerous cells), and irradiation, or a combination of all three. ◊Monoclonal antibodies have been used therapeutically against some cancers, with limited success. There is also hope of combining a monoclonal antibody with a drug that will kill the cancer cell to produce a highly specific 'magic bullet' drug. In 1990 it was discovered that the presence in some patients of a particular protein, p-glycoprotein, actively protects the cancer cells from drugs intended to destroy them. If this action can be blocked, the cancer should become far easier to treat. However, at present public health programmes are more concerned with prevention and early detection.

Cancer faintest of the zodiacal constellations (its brightest stars are fourth magnitude). It lies in the northern hemisphere, between Leo and Gemini, and is represented as a crab. Cancer's most distinctive feature is the star cluster Praesepe, popularly

known as the Beehive. The Sun passes through the constellation during late July and early Aug. In astrology, the dates for Cancer are between about 22 June and 22 July (see ◊precession).

candela SI unit (symbol cd) of luminous intensity, which replaced the old units of candle and standard candle. It measures the brightness of a light itself rather than the amount of light falling on an object, which is called *illuminance* and measured in ◊lux.

One candela is defined as the luminous intensity in a given direction of a source that emits monochromatic radiation of frequency 540×10^{-12} Hz and whose radiant energy in that direction is $1/683$ watt per steradian.

candle vertical cylinder of wax (such as tallow or paraffin wax) with a central wick of string. A flame applied to the end of the wick melts the wax, thereby producing a luminous flame. The wick is treated with a substance such as alum so that it carbonizes but does not rapidly burn out.

Candles and oil lamps were an early form of artificial lighting. Accurately made candles—which burned at a steady rate—were calibrated along their lengths and used as a type of clock. The candle was also the name of a unit of luminous intensity, replaced 1940 by the ◊candela (cd), equal to 1/60 of the luminance of 1 sq cm of a black body radiator at a temperature of 2,042K (the temperature of solidification of platinum).

Canes Venatici constellation of the northern hemisphere near Ursa Major, identified with the hunting dogs of ◊Boötes, the herder. Its stars are faint, and it contains the Whirlpool galaxy (M51), the first spiral galaxy to be recognized.

canine in mammalian carnivores, any of the long, often pointed teeth found at the front of the mouth between the incisors and premolars. Canine teeth are used for catching prey, for killing, and for tearing flesh. They are absent in herbivores such as rabbits and sheep, and are much reduced in humans.

Canis Major brilliant constellation of the southern hemisphere, identified with one of the two dogs following at the heel of Orion. Its main star, ◊Sirius, is the brightest star in the sky.

Canis Minor small constellation along the celestial equator, identified with the second of the two dogs of Orion (the other dog is Canis Major). Its brightest star is Procyon.

Canopus or *Alpha Carinae* second brightest star in the sky (after Sirius), lying in the constellation Carina. It is a yellow-white supergiant about 120 light years from Earth, and thousands of times more luminous than the Sun.

cantilever beam or structure that is fixed at one end only, though it may be supported at some point along its length; for example, a diving board. The cantilever principle, widely used in construction engineering, eliminates the need for a second main support at the free end of the beam, allowing for more elegant structures and reducing the amount of materials required. Many large-span bridges have been built on the cantilever principle.

A typical cantilever bridge consists of two beams cantilevered out from either bank, each supported part way along, with their free ends meeting in the middle. The multiple-cantilever Forth Rail Bridge (completed 1890) across the Firth of Forth in Scotland has twin main spans of 521 m/1,710 ft.

canyon (Spanish *cañon* 'tube') deep, narrow valley or gorge running through mountains. Canyons are formed by stream down-cutting, usually in arid areas, where the stream or river receives water from outside the area.

There are many canyons in the western USA and in Mexico, for example the Grand Canyon of the Colorado River in Arizona, the canyon in Yellowstone National Park, and the Black Canyon in Colorado.

capacitance, electrical property of a capacitor that determines how much charge can be stored in it for a given potential difference between its terminals. It is equal to the ratio of the electrical charge stored to the potential difference. It is measured in ◊farads.

capacitor or *condenser* device for storing electric charge, used in electronic circuits; it consists of two or more metal plates separated by an insulating layer called a dielectric.

Its *capacitance* is the ratio of the charge stored on either plate to the potential difference between the plates. The SI unit of capacitance is the farad, but most capacitors have much smaller capacitances, and the microfarad (a millionth of a farad) is the commonly used practical unit.

Cape Canaveral promontory on the Atlantic coast of Florida, USA, 367 km/228 mi N of Miami, used as a rocket launch site by ◊NASA.

It was known as Cape Kennedy 1963–73. The ◊Kennedy Space Center is nearby.

Capella or *Alpha Aurigae* brightest star in the constellation Auriga and the sixth brightest star in the sky. It consists of a pair of yellow giant stars 41 light years from Earth, orbiting each other every 104 days.

capillarity spontaneous movement of liquids up or down narrow tubes, or capillaries. The movement is due to unbalanced molecular attraction at the boundary between the liquid and the tube. If liquid molecules near the boundary are more strongly attracted to molecules in the material of the tube than to other nearby liquid molecules, the liquid will rise in the tube. If liquid molecules are less attracted to the material of the tube than to other liquid molecules, the liquid will fall.

CAPILLARITY PUZZLE

A drop of water on a surface may be removed by touching it with a fine-bore tube, whereupon it is drawn up into the tube by capillary action. The fibre tip of modern pens acts in just this way, with the fine spaces between the fibres acting as many fine tubes. Clearly this is a useful idea, because it not only allows ink to be drawn from a reservoir in the body of the pen right down to the tip, but also stops the ink from dripping onto the paper. The problem is, how does the ink ever leave the pen when the action of the tip is to draw liquids up rather than let them go?
See page 651 for the answer.

---

capillary in biology, narrowest blood vessel in vertebrates, 0.008–0.02 mm in diameter, barely wider than a red blood cell. Capillaries are distributed as *beds*, complex networks connecting arteries and veins. Capillary walls are extremely thin, consisting of a single layer of cells, and so nutrients, dissolved gases, and waste products can easily pass through them. This makes the capillaries the main area of exchange between the fluid (◊lymph) bathing body tissues and the blood.

capillary in physics, a very narrow, thick-walled tube, usually made of glass, such as in a thermometer. Properties of fluids, such as surface tension and viscosity, can be studied using capillary tubes.

capitulum in botany, a flattened or rounded head (inflorescence) of numerous, small, stalkless flowers. The capitulum is surrounded by a circlet of petal-like bracts and has the appearance of a large, single flower. It is characteristic of plants belonging to the daisy family (Compositae) such as the daisy *Bellis perennis* and the garden marigold *Calendula officinalis*; but is also seen in parts of other families, such as scabious *Knautia* and teasels *Dipsacus*. The individual flowers are known as ◊florets.

Capricornus zodiacal constellation in the southern hemisphere next to Sagittarius. It is represented as a fish-tailed goat, and its brightest stars are third magnitude. The Sun passes through it late Jan to mid-Feb. In astrology, the dates for Capricornus (popularly known as Capricorn) are between about 22 Dec and 19 Jan (see ◊precession).

capsule in botany, a dry, usually many-seeded fruit formed from an ovary composed of two or more fused ◊carpels, which splits open to release the seeds. The same term is used for the spore-containing structure of mosses and liverworts; this is borne at the top of a long stalk or seta.

Capsules burst open (dehisce) in various ways, including lengthwise, by a transverse lid—for example, scarlet pimpernel *Anagallis arvensis*—or by a number of pores, either towards the top of the capsule, as in the poppy *Papaver*, or near the base, as in certain species of bellflower *Campanula*.

car small, driver-guided, passenger-carrying motor vehicle; originally the automated version of the horse-drawn carriage, meant to convey people and their goods over streets and roads. Over 50 million motor cars are produced each year worldwide. Most are four-wheeled and have water-cooled, piston-type internal-combustion engines fuelled by petrol or diesel. Variations have existed for decades that use ingenious and often nonpolluting power plants, but the motor industry long ago settled on this general formula for the consumer market. Experimental and sports models are streamlined, energy-efficient, and hand-built.

Although it is recorded that in 1479 Gilles de Dom was paid 25 livres (the equivalent of 25 pounds of silver) by the treasurer of Antwerp in the Low Countries for supplying a self-propelled vehicle, the ancestor of the automobile is generally agreed to be the cumbersome steam carriage made by Nicolas-Joseph Cugnot 1769, still preserved in Paris. Steam was an attractive form of power to the English pioneers, and in 1808 Richard Trevithick built a working steam carriage. Later in the 19th century, practical steam coaches were used for public transport until stifled out of existence by punitive road tolls and legislation.

Although a Frenchman, Jean Etienne Lenoir, patented the first internal combustion engine (gas-driven) 1860, and an Austrian, Siegfried Marcus, built a vehicle which was shown at the Vienna Exhibition (1873), two Germans, Gottleib Daimler and Karl Benz are generally regarded as the creators of the motorcar. In 1885 Daimler and Benz built and ran the first petrol-driven motorcar. The pattern for the modern motorcar was set by Panhard 1890 (front radiator, engine under bonnet, sliding-pinion gearbox, wooden ladder-chassis) and Mercedes 1901 (honeycomb radiator, in-line four-cylinder engine, gate-change gearbox, pressed-steel chassis) set the pattern for the modern car. Emerging with Haynes and Duryea in the early 1890s, US demand was so fervent that 300 makers existed by 1895; only 109 were left by 1900.

In England, cars were still considered to be light locomotives in the eyes of the law and, since the Red Flag Act 1865, had theoretically required someone to walk in front with a red flag (by night, a lantern). Despite these obstacles, which put UK

development another ten years behind all others, in 1896 Frederick Lanchester produced an advanced and reliable vehicle, later much copied. The period 1905–06 inaugurated a world motorcar boom continuing to the present day.

Among the legendary cars of the early 20th century are: De Dion Bouton, with the first practical high-speed engines; Mors, notable first for racing and later as a silent tourer ; Napier, the 24–hour record-holder at Brooklands 1907, unbeaten for 17 years; the incomparable Silver Ghost Rolls-Royce; the enduring Model T Ford; and the many types of Bugatti and Delage, from record-breakers to luxury tourers.

After World War I popular motoring began with the era of cheap, light (baby) cars made by Citroën, Peugeot, and Renault (France); Austin, Morris, Clyno, and Swift (England); Fiat (Italy); Volkswagen (Germany); and the cheap though bigger Ford, Chevrolet, and Dodge in the USA. During the interwar years a great deal of racing took place, and the experience gained benefited the everyday motorist in improved efficiency, reliability, and safety. There was a divergence between the lighter, economical European car, with its good handling, and the heavier US car, cheap, rugged, and well adapted to long distances on straight roads at speed. By this time motoring had become a universal pursuit. After World War II small European cars tended to fall into three categories, in about equal numbers: front engine and rear drive, the classic arrangement; front engine and front-wheel drive; rear engine and rear-wheel drive. Racing cars have the engine situated in the middle for balance. From the 1950s a creative resurgence produced in practical form automatic transmission for small cars, rubber suspension, transverse engine mounting, self-levelling ride, disc brakes, and safer wet-weather tyres. The drive against pollution (cars are responsible for almost a quarter of the world's carbon dioxide emissions) from the 1960s and the fuel crisis from the 1970s led to experiments with steam cars (cumbersome), diesel engines (slow and heavy, though economical), solar-powered cars, and hybrid cars using both electricity (in town centres) and petrol (on the open road). The industry brought on the market the stratified-charge petrol engine, using a fuel injector to achieve 20% improvement in petrol consumption (the average US car in 1991 did only 27 mi/gal); weight reduction in the body by the use of aluminium and plastics; and 'slippery' body designs with low air resistance, or drag. Microprocessors were also developed to measure temperature, engine speed, pressure, and oxygen/CO_2 content of exhaust gases, and readjust the engine accordingly.

Many of these developments have been introduced first by the large and vigorous Japanese motor industry. By the mid-1980s, Japan was building 8 million cars a year, on par with the US. The largest Japanese manufacturer, Toyota, was producing 2.5 million cars per year.

A typical present-day medium-sized saloon car has a semi-monocoque construction in which the body panels, suitably reinforced, support the road loads through independent front and rear sprung suspension, with seats located within the wheelbase for comfort. It is usually powered by a ◊petrol

car: chronology

1769	Nicholas-Joseph Cugnot in France built a steam tractor.
1801	Richard Trevithick built a steam coach.
1860	Jean Etienne Lenoir built a gas-fuelled internal-combustion engine.
1865	The British government passed the Red Flag Act, requiring a person to precede a 'horseless carriage' with a red flag.
1876	Nikolaus August Otto improved the gas engine, making it a practical power source.
1885	Gottlieb Daimler developed a successful lightweight petrol engine and fitted it to a bicycle to create he prototype of the present-day motorcycle; Karl Benz fitted his lightweight petrol engine to a three-wheeled carriage to pioneer the motorcar.
1886	Gottlieb Daimler fitted his engine to a four-wheeled carriage to produce a four-wheeled motorcar.
1891	René Panhard and Emile Levassor established the present design of cars by putting the engine in front.
1896	Frederick Lanchester introduced epicyclic gearing, which foreshadowed automatic transmission.
1899	Caamille Jenatzy broke the 100–kph barrier in an electric car *La Jamais Contente* at Achères, France, reaching 105.85 kph/65.60 mph.
1901	The first Mercedes took to the roads; it was the direct ancestor of the present car. Ransome Olds in the USA introduced mass production on an assembly line.
1904	Louis Rigolly broke the 100 mph barrier, reaching 166.61 kph/103.55 mph in a Gobron-Brillé at Nice, France.
1906	Rolls-Royce introduced the Silver Ghost, which established the company's reputation for superlatively engineered cars.
1908	Henry Ford also used assembly-line production to manufacture his celebrated Model T, nicknamed the Tin Lizzie because it used lightweight steel sheet for the body, which looked tinny.
1911	Cadillac introduced the electric starter and dynamo lighting.
1913	Ford introduced the moving conveyor belt to the assembly line, further accelerating production of the Model T.
1920	Duesenberg began fitting four-wheel hydraulic brakes.
1922	The Lancia Lambda featured unitary (all-in-one) construction and independent front suspension.
1927	Henry Segrave broke the 200 mph barrier in a Sunbeam, reaching 327.89 kph/203.79 mph.
1928	Cadillac introduced the synchromesh gearbox, greatly facilitating gear changing.
1934	Citroën pioneered front-wheel drive in their 7CV model.
1936	Fiat introduced their baby car, the Topolino, 500 cc.
1938	Germany produced its 'people's car', the Volkswagen Beetle.
1948	Jaguar launched the XK120 sports car; Michelin introduced the radial-ply tyre; Goodrich produced the tubeless tyre.
1950	Dunlop announced the disc brake.
1951	Buick and Chrysler introduced power steering.
1952	Rover's gas-turbine car set a speed record of 243 kph/152 mph.
1954	Carl Bosch introduced fuel injection for cars.
1955	Citroën produced the advanced DS-19 'shark-front' car with hydropneumatic suspension.
1957	Felix Wankel built his first rotary petrol engine.
1959	BMC (now Rover) introduced the Issigonis-designed Mini, with front-wheel drive, transverse engine, and independent rubber suspension.
1965	US car manufacturers were forced to add safety features after the publication of Ralph Nader's *Unsafe at Any Speed*.
1966	California introduced legislation regarding air pollution by cars.
1970	American Gary Gabelich drove a rocket-powered car, *Blue Flame*, to a new record speed of 1,001.473 kph/622.287 mph.
1972	Dunlop introduced safety tyres, which seal themselves after a puncture.
1979	American Sam Barrett exceeded the speed of sound in the rocket-engined *Budweiser Rocket*, reaching 1,190.377 kph/739.666 mph, a speed not officially recognized as a record because of timing difficulties.
1980	The first mass-produced car with four-wheel drive, the Audi Quattro, was introduced; Japanese car production overtook that of the USA.
1981	BMW introduced the on-board computer, which monitored engine performance and indicated to the driver when a service was required.
1983	British driver Richard Noble set an official speed record in the jet-engined *Thrust 2* of 1,019.4 kph/633.5 mph; Austin Rover introduced the Maestro, the first car with a 'talking dashboard' that alerted the driver to problems.
1987	The solar-powered *Sunraycer* travelled 3,000 km/1,864 mi from Darwin to Adelaide, Australia, in six days. Toyota Corona production tops 6 million in 29 years
1988	California introduces stringent controls on car emissions, aiming for widespread use of zero emission vehicles by 1998
1989	The first mass-produced car with four-wheel steering, the Mitsubishi Galant, was launched.
1990	Fiat of Italy and Peugeot of France launched electric passenger cars on the market.
1991	Satellite-based car navigation systems were launched in Japan. European Parliament voted to adopt stringent control of car emissions.
1992	Mazda and NEC of Japan developed an image-processing system for cars, which views the road ahead through a video camera, identifies road signs and markings, and helps the driver to avoid obstacles.

engine using a carburettor to mix petrol and air for feeding to the engine cylinders (typically four or six), and the engine is usually water cooled. In the 1980s high-performance diesel engines were being developed for use in private cars, and it is anticipated that this trend will continue for reasons of economy. From the engine, power is transmitted through a clutch to a four-or five-speed gearbox and from there, in a front-engine rear-drive car, through a drive (propeller) shaft to a ◊differential gear, which drives the rear wheels. In a front-engine, front-wheel drive car, clutch, gearbox, and

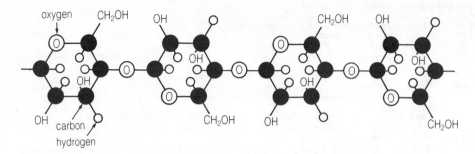

carbohydrate *A molecule of the polysaccharide glycogen (animal starch) is formed from linked glucose ($C_6H_{12}O_6$) molecules. A typical glycogen molecule has 100–1,000 glucose units.*

final drive are incorporated with the engine unit. An increasing number of high-performance cars are being offered with four-wheel drive. This gives superior roadholding in wet and icy conditions.

The UK, the Ministry of Transport was established 1919, roads were improved, and various laws and safety precautions imposed to govern the use of cars. A driver must possess a licence, and a vehicle must be registered with the local licensing authority, displaying the number assigned to it. A road tax is imposed, and the law also insists on insurance for third-party risks. Motoring organizations include the Automobile Association (AA) and the Royal Automobile Club (RAC). From 1951 to 1991 the number of cars on British roads increased from 2 million to 20 million.

caramel complex mixture of substances produced by heating sugars, without charring, until they turn brown. Caramel is used as colouring and flavouring in foods. Its production in the manufacture of sugar confection gives rise to a toffeelike sweet of the same name.

The intricate chemical reactions involved in the production of caramel (caramelization) are not fully understood, but are known to result in the formation of a number of compounds. Two compounds in particular (acetylformoin and 4–hydroxy-2,5–dimethyl-3–furanone) are thought to contribute to caramel's characteristic flavour. Commercially, the caramelization process is speeded up by the addition of selected ◊amino acids.

carat (Arabic *quirrat* 'seed') unit for measuring the mass of precious stones; it is equal to 0.2 g/0.00705 oz, and is part of the troy system of weights. It is also the unit of purity in gold (US karat). Pure gold is 24–carat; 22–carat (the purest used in jewellery) is 22 parts gold and two parts alloy (to give greater strength).

Originally, one carat was the weight of a carob seed.

carbide compound of carbon and one other chemical element, usually a metal, silicon, or boron.

Calcium carbide (CaC_2) can be used as the starting material for many basic organic chemical syntheses, by the addition of water and generation of ethyne (acetylene). Some metallic carbides are used in engineering because of their extreme hardness and strength. Tungsten carbide is an essential ingredient of carbide tools and high-speed tools. The 'carbide process' was used during World

War II to make organic chemicals from coal rather than from oil.

carbohydrate chemical compound composed of carbon, hydrogen, and oxygen, with the basic formula $C_m(H_2O)_n$, and related compounds with the same basic structure but modified ◊functional groups. As sugar and starch, carbohydrates form a major energy-providing part of the human diet.

The simplest carbohydrates are sugars (*monosaccharides*, such as glucose and fructose, and *disaccharides*, such as sucrose), which are soluble compounds, some with a sweet taste. When these basic sugar units are joined together in long chains or branching structures they form *polysaccharides*, such as starch and glycogen, which often serve as food stores in living organisms. Even more complex carbohydrates are known, including ◊chitin, which is found in the cell walls of fungi and the hard outer skeletons of insects, and ◊cellulose, which makes up the cell walls of plants. Carbohydrates form the chief foodstuffs of herbivorous animals.

carbolic acid common name for the aromatic compound ◊phenol.

carbon (Latin *carbo, carbonaris* 'coal') nonmetallic element, symbol C, atomic number 6, relative atomic mass 12.011. It occurs on its own as diamond, graphite, and as fullerenes, as compounds in carbonaceous rocks such as chalk and limestone, as carbon dioxide in the atmosphere, as hydrocarbons in petroleum, coal, and natural gas, and as a constituent of all organic substances.

In its amorphous form, it is familiar as coal, charcoal, and soot. Of the inorganic carbon compounds, the chief ones are **carbon dioxide**, a colourless gas formed when carbon is burned in an adequate supply of air; and **carbon monoxide** (CO), formed when carbon is oxidized in a limited supply of air. **Carbon disulphide** (CS_2) is a dense liquid with a sweetish odour.

Another group of compounds is the **carbon halides**, including ◊carbon tetrachloride (tetrachloromethane, CCl_4).

When added to steel, carbon forms a wide range of alloys with useful properties. In pure form, it is used as a moderator in nuclear reactors; as colloidal graphite it is a good lubricant and, when deposited on a surface in a vacuum, obviates photoelectric and secondary emission of electrons. Carbon is used as a fuel in the form of coal or coke. The

Great balls of carbon

Chemists regard a diamond as just another form of carbon, an element familiar to everyone as soot, which is practically pure carbon. Equally everyday is the graphite in a 'lead' pencil, another form of carbon. The chemical difference between graphite and diamond is that the carbon atoms in each substance are arranged differently. The carbon atoms of graphite are arranged in flat, hexagonal patterns, rather like the cells of a honeycomb. Because graphite molecules are flat, they slide over one another easily. In contrast, the carbon atoms in a diamond are interlinked three-dimensionally, giving the substance its extraordinary hardness. And there until recently the matter rested: carbon was an element that came in two forms—diamond and graphite. Now chemists are excited about a third form of carbon, in which the atoms are linked together in a molecule that looks very like a soccer ball. The new form of carbon is a cagelike molecule consisting of 60 carbon atoms that make a perfect sphere. It has been named 'buckminsterfullerene' in honour of the US architect Buckminster Fuller's work, which included spherical geodesic domes.

Soot, space, and lasers

The story of the discovery of these exotic new molecules involves soot, outer space, and lasers. It starts when two scientists, Donald Huffman and Wolfgang Kratschmer, were working at the Max Planck Institute for Nuclear Physics in Heidelberg, Germany. They were heating graphite rods under special conditions and examining the soot made in the process: they speculated that a similar process might take place in outer space, contributing to clouds of interstellar dust.

Meanwhile the team of Harold Kroto and David Walton at the University of Sussex had been on the trail of interstellar molecules made up of long chains of carbon atoms that might have originated in the atmosphere surrounding red giant stars. Enlisting the help of researchers at Rice University in Houston, Texas, who were using a giant laser to blast atoms from the surface of different substances, Kroto and his team soon found the long-chained carbon molecules. But they were struck by the discovery of a very stable molecule that contained exactly 60 carbon atoms.

Now Richard Smalley of the Rice team set out to make a model of the new molecule, using scissors, sticky tape, and paper. He soon found that a hexagonal arrangement of carbon atoms was impossible, but that a perfect sphere could be formed from 20 hexagons and 12 pentagons. Such a sphere has 60 vertices.

The buckyball's fingerprint

Ordinary soot absorbs ultraviolet light in a characteristic way, and the new molecule also showed a characteristic ultraviolet fingerprint. The problem was to make enough of the new carbon so that exact measurements could be made. If it could be crystallized, then an X-ray analysis would enable the precise distances between the carbon atoms to be determined. The Heidelberg team forged ahead, and produced milligrams of red-brown crystals by evaporating a solution of their product in benzene. The X-ray results confirmed that the molecules were indeed spherical, and that the paper model was correct.

This result was clinched when Kroto and his colleagues, using nuclear magnetic resonance spectroscopy, not only confirmed the new 60–atom structure but also provided evidence for a family of fullerenes, as the new forms of carbon are now called. Structures containing 28, 32, 50, 60, and 70 carbon atoms are known. Chemists affectionately term such molecules 'buckyballs'.

Promising future

In many ways, these new discoveries in carbon chemistry are as important as the key discovery more than a century ago of the structure of benzene. When in 1865 German chemist Friederich Kekulé proposed a ring structure for this important organic molecule, the whole field of aromatic chemistry opened up, leading to dyestuffs in the first instance, and millions of new substances since. The fullerene family holds similar promise.

Chemists at Exxon's laboratories in New Jersey have already played a part in the fullerene story, and are interested in the lubricating properties of the new materials. Sumio Iijima, a Japanese scientist, has synthesized tubelike structures based on the fullerene idea, which are naturally called 'buckytubes'.

Other teams have now done work that suggests that such molecules may have semiconducting abilities. Cagelike molecules can contain other atoms, such as metals: a group of researchers from the University of California at Los Angeles have produced a 'doped' fullerene that behaves as a superconductor. No evidence has been found that buckminsterfullerene exists in space, but some is almost certainly produced every time you light a candle.

radioactive isotope carbon-14 (half-life 5,730 years) is used as a tracer in biological research.

The element has the following characteristic reactions.

with air or oxygen It burns on heating to form carbon dioxide in excess air, or carbon monoxide in a limited supply of air.

$$C + O_2 \rightarrow CO_2$$
$$\Delta H = -394 \text{ kJ mol}^{-1}$$

$$2C + O_2 \rightarrow 2CO$$

with metal oxides It reduces many metal oxides at high temperatures.

$$Fe_2O_3 + 3C \rightarrow 2Fe + 3CO$$

with steam It forms water gas (a cheap, useful, industrial fuel) when steam is passed over white-hot coke.

$$C + H_2O \rightarrow CO + H_2$$

with concentrated acids With hot, concentrated sulphuric or nitric acids it forms carbon dioxide.

Life exists in the universe only because the carbon atom possesses certain exceptional properties.

On **carbon** James Jeans (1877–1946) *The Mysterious Universe*

---

carbonate CO_3^{2-} ion formed when carbon dioxide dissolves in water; any salt formed by this ion and another chemical element, usually a metal.

Carbon dioxide (CO_2) dissolves sparingly in water (for example, when rain falls through the air) to form carbonic acid (H_2CO_3), which unites with various basic substances to form carbonates. Calcium carbonate ($CaCO_3$) (chalk, limestone, and marble) is one of the most abundant carbonates known, being a constituent of mollusc shells and the hard outer skeletons of crustaceans.

carbonated water water in which carbon dioxide is dissolved under pressure. It forms the basis of many fizzy soft drinks such as soda water and lemonade.

carbonation in earth science, a form of chemical ◊weathering caused by rainwater that has absorbed carbon dioxide from the atmosphere and formed a weak carbonic acid. The slightly acidic rainwater is then capable of dissolving certain minerals in rocks. ◊Limestone is particularly vulnerable to this form of weathering.

carbon cycle sequence by which ◊carbon circulates and is recycled through the natural world. The carbon element from carbon dioxide, released into the atmosphere by living things as a result of ◊respiration, is taken up by plants during ◊photosynthesis and converted into carbohydrates; the oxygen component is released back into the atmosphere. The simplest link in the carbon cycle occurs when an animal eats a plant and carbon is transferred from, say, a leaf cell to the animal body.

Today, the carbon cycle is in danger of being disrupted by the increased consumption and burning of fossil fuels, and the burning of large tracts of tropical forests, as a result of which levels of carbon dioxide are building up in the atmosphere and probably contributing to the ◊greenhouse effect.

The European Community has pledged to stabilise CO_2 emissions at 1990 levels by the year 2000. They would otherwise rise by 11% per year over this period. Maintaining such levels will be achieved by burning less fossil fuel.

carbon dating alternative name for ◊radiocarbon dating.

carbon dioxide CO_2 colourless, odourless gas, slightly soluble in water and denser than air. It is formed by the complete oxidation of carbon.

It is produced by living things during the processes of respiration and the decay of organic matter, and plays a vital role in the carbon cycle. It is used as a coolant in its solid form (known as 'dry ice'), and in the chemical industry. Its increasing density contributes to the ◊greenhouse effect and ◊global warming.

Britain has 1% of the world's population, yet it produces 3% of CO_2 emissions; the USA has 5% of the world's population and produces 25% of CO_2 emissions.

carbon fibre fine, black, silky filament of pure carbon produced by heat treatment from a special grade of Courtelle acrylic fibre, used for reinforcing plastics. The resulting composite is very stiff and, weight for weight, has four times the strength of high-tensile steel. It is used in the aerospace industry, cars, and electrical and sports equipment.

carbonic acid H_2CO_3 weak, dibasic acid formed by dissolving carbon dioxide in water.

$$H_2O + CO_2 \rightleftharpoons H_2CO_3$$

It forms two series of salts: ◊carbonates and ◊hydrogencarbonates. Fizzy drinks are made by dissolving carbon dioxide in water under pressure; soda water is a solution of carbonic acid.

Carboniferous period of geological time 363–290 million years ago, the fifth period of the Palaeozoic era. In the USA it is divided into two periods: the Mississippian (lower) and the Pennsylvanian (upper). Typical of the lower-Carboniferous rocks are shallow-water ◊limestones, while upper-Carboniferous rocks have ◊delta deposits with ◊coal (hence the name). Amphibians were abundant, and reptiles evolved during this period.

carbon monoxide CO colourless, odourless gas formed when carbon is oxidized in a limited supply of air. It is a poisonous constituent of car exhaust fumes, forming a stable compound with haemoglobin in the blood, thus preventing the haemoglobin from transporting oxygen to the body tissues.

In industry, carbon monoxide is used as a reducing agent in metallurgical processes—for example, in the extraction of iron in ◊blast furnaces—and is a constituent of cheap fuels such as water gas. It burns in air with a luminous blue flame to form carbon dioxide.

carbon tetrachloride former name for ◊tetrachloromethane.

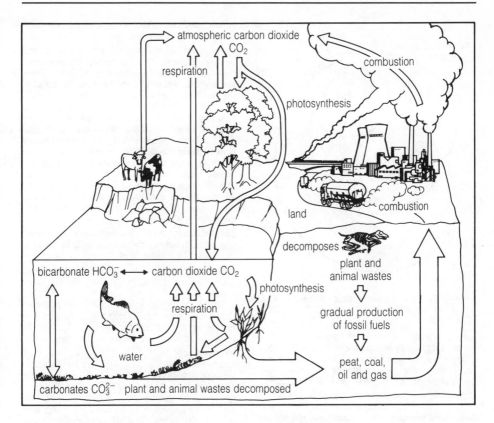

carbon cycle *The carbon cycle is necessary for the continuation of life. Since there is only a limited amount of carbon in the Earth and its atmosphere, carbon must be continuously recycled if life is to continue. Other chemicals necessary for life—nitrogen, sulphur, and phosphorus, for example—also circulate in natural cycles.*

Carborundum trademark for a very hard, black abrasive, consisting of silicon carbide (SiC), an artificial compound of carbon and silicon. It is harder than ◊corundum but not as hard as ◊diamond.

It was first produced 1891 by US chemist Edward Acheson (185–1931).

carboxyl group –COOH in organic chemistry, the acidic functional group that determines the properties of fatty acids (carboxylic acids) and amino acids.

carboxylic acid organic acid containing the carboxyl group (–COOH) attached to another group (R), which can be hydrogen (giving methanoic acid, HCOOH) or a larger molecule (up to 24 carbon atoms). When R is a straight-chain alkyl group (such as CH_3 or CH_3CH_2), the acid is known as a ◊fatty acid.

carburation mixing of a gas, such as air, with a volatile hydrocarbon fuel, such as petrol, kerosene, or fuel oil, in order to form an explosive mixture. The process, which increases the amount of potential heat energy released during combustion, is used in internal-combustion engines. In most petrol engines the liquid fuel is atomized and mixed with air by means of a device called a *carburettor.*

carcinogen any agent that increases the chance of a cell becoming cancerous (see ◊cancer), including various chemical compounds, some viruses, X-rays, and other forms of ionizing radiation. The term is often used more narrowly to mean chemical carcinogens only.

carcinoma malignant ◊tumour arising from the skin, the glandular tissues, or the mucous membranes that line the gut and lungs.

cardinal number in mathematics, one of the series of numbers 0, 1, 2, 3, 4, . . . Cardinal numbers relate to quantity, whereas ordinal numbers (first, second, third, fourth, . . .) relate to order.

cardioid heart-shaped curve traced out by a point on the circumference of a circle, resulting from the circle rolling around the edge of another circle of the same diameter.

The polar equation of the cardioid is of the form $r = a(1 + \cos \theta)$.

Carina constellation of the southern hemisphere, represented as a ship's keel. Its brightest star is Canopus; it also contains Eta Carinae, a massive and highly luminous star embedded in a gas cloud, perhaps 8,000 light years away. It has varied unpredictably in the past; some astronomers think it is likely to explode as a supernova within 10,000 years.

carnassial tooth powerful scissorlike pair of teeth, found in all mammalian carnivores except

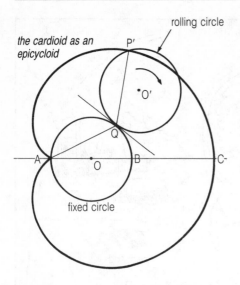

the cardioid as an epicycloid

rolling circle

P'

O'

Q

A

O

B

C

fixed circle

cardioid *The cardioid is the curve formed when one circle rolls around the edge of another circle of the same size. It is named after its heart shape.*

seals. Carnassials are formed from an upper premolar and lower molar, and are shaped to produce a sharp cutting surface. Carnivores such as dogs transfer meat to the back of the mouth, where the carnassials slice up the food ready for swallowing.

carnelian semiprecious gemstone variety of ◊chalcedony consisting of quartz (silica) with iron impurities, which give it a translucent red colour. It is found mainly in Brazil, India, and Japan.

carnivore animal that eats other animals. Although the term is sometimes confined to those that eat the flesh of ◊vertebrate prey, it is often used more broadly to include any animal that eats other animals, even microscopic ones. Carrion-eaters may or may not be included.

The mammalian order Carnivora includes cats, dogs, bears, badgers, and weasels.

carnivore in a food chain, a type of consumer that eats other animals. Carnivores therefore include not only the large vertebrates, such as sharks and tigers but also many invertebrates, including microscopic ones.

Carnot cycle series of changes in the physical condition of a gas in a reversible heat engine, necessarily in the following order: (1) isothermal expansion (without change of temperature), (2) adiabatic expansion (without change of heat content), (3) isothermal compression, and (4) adiabatic compression.

The principles derived from a study of this cycle are important in the fundamentals of heat and ◊thermodynamics.

carnotite potassium uranium vanadate, $K_2(UO_2)_2.(VO_4)_2.3H_2O$, a radioactive ore of vanadium and uranium with traces of radium. A yellow powdery mineral, it is mined chiefly in the Colorado Plateau, USA; Radium Hill, Australia; and Shaba, Zaire.

carotene naturally occurring pigment of the ◊carotenoid group. Carotenes produce the orange, yellow, and red colours of carrots, tomatoes, oranges, and crustaceans.

carotenoid any of a group of yellow, orange, red, or brown pigments found in many living organisms, particularly in the ◊chloroplasts of plants. There are two main types, the *carotenes* and the *xanthophylls*. Both types are long-chain lipids (◊fats).

Some carotenoids act as accessory pigments in ◊photosynthesis, and in certain algae they are the principal light-absorbing pigments functioning more efficiently than ◊chlorophyll in low-intensity light. Carotenoids can also occur in organs such as petals, roots, and fruits, giving them their characteristic colour, as in the yellow and orange petals of wallflowers *Cheiranthus*. They are also responsible for the autumn colours of leaves, persisting longer than the green chlorophyll, which masks them during the summer.

carotid artery one of a pair of major blood vessels, one on each side of the neck, supplying blood to the head.

carpel female reproductive unit in flowering plants (◊angiosperms). It usually comprises an ◊ovary containing one or more ovules, the stalk or style, and a ◊stigma at its top which receives the pollen. A flower may have one or more carpels, and they may be separate or fused together. Collectively the carpels of a flower are known as the ◊gynoecium.

carrying capacity in ecology, the maximum number of animals of a given species that a particular area can support. When the carrying capacity is exceeded, there is insufficient food (or other resources) for the members of the population. The population may then be reduced by emigration, reproductive failure, or death through starvation.

Cartesian coordinates in ◊coordinate geometry, components used to define the position of a point by its perpendicular distance from a set of two or more axes, or reference lines. For a two-dimensional area defined by two axes at right angles (a horizontal x-axis and a vertical y-axis), the coordinates of a point are given by its perpendicular distances from the y-axis and x-axis, written in the form (x,y). For example, a point P that lies three units from the y-axis and four units from the x-axis has Cartesian coordinates (3,4) (see ◊abscissa and ◊ordinate). In three-dimensional coordinate geometry, points are located with reference to a third, z-axis, mutually at right angles to the x and y axes.

The Cartesian coordinate system can be extended to any finite number of dimensions (axes), and is used thus in theoretical mathematics. It is named after the French mathematician, René Descartes.

The system is useful in creating technical drawings of machines or buildings, and in computer-aided design (◊CAD).

cartilage flexible bluish-white connective ◊tissue made up of the protein collagen. In cartilaginous fish it forms the skeleton; in other vertebrates it forms the greater part of the embryonic skeleton, and is replaced by ◊bone in the course of development, except in areas of wear such as bone endings, and the discs between the backbones. It also forms structural tissue in the larynx, nose, and external ear of mammals.

caryopsis dry, one-seeded ◊fruit in which the wall of the seed becomes fused to the carpel wall during its development. It is a type of ◊achene, and therefore develops from one ovary and does not split open to release the seed. Caryopses are typical of members of the grass family (Gramineae), including the cereals.

casein main protein of milk, from which it can be separated by the action of acid, the enzyme rennin, or bacteria (souring); it is also the main component of cheese. Casein is used commercially in cosmetics, glues, and as a sizing for coating paper.

Cassini joint space probe of the US agency NASA and the European Space Agency to the planet Saturn. *Cassini* is scheduled to be launched 1997 and to go into orbit around Saturn 2004, dropping off a sub-probe, *Huygens*, to land on Saturn's largest moon, Titan.

Cassiopeia prominent constellation of the northern hemisphere, named after the mother of Andromeda. It has a distinctive W-shape, and contains one of the most powerful radio sources in the sky, Cassiopeia A, the remains of a ◊supernova (star explosion).

cassiterite or *tinstone* chief ore of tin, consisting of reddish-brown to black stannic oxide (SnO_2), usually found in granite rocks. When fresh it has a bright ('adamantine') lustre. It was formerly extensively mined in Cornwall, England; today Malaysia is the world's main supplier. Other sources of cassiterite are Africa, Indonesia, and South America.

casting process of producing solid objects by pouring molten material into a shaped mould and allowing it to cool. Casting is used to shape such materials as glass and plastics, as well as metals and alloys.

The casting of metals has been practised for more than 6,000 years, using first copper and bronze, then iron. The traditional method of casting metal is *sand casting*.

Using a model of the object to be produced, a hollow mould is made in a damp sand and clay mix. Molten metal is then poured into the mould, taking its shape when it cools and solidifies.

The sand mould is broken up to release the casting. Permanent metal moulds called *dies* are also used for casting, in particular, small items in mass-production processes where molten metal is injected under pressure into cooled dies. *Continuous casting* is a method of shaping bars and slabs that involves pouring molten metal into a hollow, water-cooled mould of the desired cross section.

cast iron cheap but invaluable constructional material, most commonly used for car engine blocks. Cast iron is partly refined pig (crude) ◊iron, which is very fluid when molten and highly suitable for shaping by casting; it contains too many impurities (for example, carbon) to be readily shaped in any other way. Solid cast iron is heavy and can absorb great shock but is very brittle.

Castor or *Alpha Geminorum* second brightest star in the constellation Gemini and the 23rd brightest star in the sky. Along with ◊Pollux, it forms a prominent pair at the eastern end of Gemini. Castor is 45 light years from Earth, and is one of the finest binary stars in the sky for small telescopes. The two main components orbit each other over a period of 467 years. A third, much fainter, star orbits the main pair over a period probably exceeding 10,000 years. Each of the three visible components is a spectroscopic binary (◊binary star), making Castor a sextuple star system.

catabolism in biology, the destructive part of ◊metabolism where living tissue is changed into energy and waste products. It is the opposite of ◊anabolism. It occurs continuously in the body, but is accelerated during many disease processes, such as fever, and in starvation.

catalyst substance that alters the speed of, or makes possible, a chemical or biochemical reaction but remains unchanged at the end of the reaction. ◊Enzymes are natural biochemical catalysts. In practice most catalysts are used to speed up reactions.

catalytic converter device fitted to the exhaust system of a motor vehicle in order to reduce toxic emissions from the engine. It converts harmful exhaust products to relatively harmless ones by passing the exhaust gases over a mixture of catalysts coated on a metal or ceramic honeycomb (a structure that increases the surface area and therefore the amount of active catalyst with which the exhaust gases will come into contact). *Oxidation catalysts* (small amounts of precious palladium and platinum metals) convert hydrocarbons (unburnt fuel) and carbon monoxide into carbon dioxide and water, while *three-way catalysts* (platinum and rhodium metals) convert nitrogen oxide gases into nitrogen and oxygen.

Over the lifetime of a vehicle, a catalytic converter can reduce hydrocarbon emissions by 87%, carbon monoxide emissions by 85%, and nitrogen oxide emissions by 62%, but will cause a slight increase in the amount of carbon dioxide emitted. Catalytic converters are standard in the USA, where a 90% reduction in pollution from cars was achieved without loss of engine performance or fuel economy.

catastrophe theory mathematical theory developed by René Thom in 1972, in which he showed that the growth of an organism proceeds by a series of gradual changes that are triggered by, and in turn trigger, large-scale changes or 'catastrophic' jumps. It also has applications in engineering—for example, the gradual strain on the structure of a bridge that can eventually result in a sudden collapse—and has been extended to economic and psychological events.

catastrophism theory that the geological features of the Earth were formed by a series of sudden, violent 'catastrophes' beyond the ordinary workings of nature. The theory was largely the work of French comparative anatomist Georges Cuvier. It was later replaced by the concepts of ◊uniformitarianism and ◊evolution.

catchment area area from which water is collected by a river and its tributaries. In the social sciences the term may be used to denote the area from which people travel to obtain a particular service or product, such as the area from which a school draws its pupils.

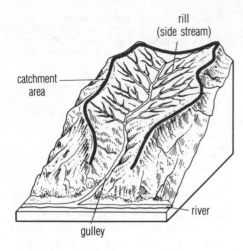

rill
(side stream)

catchment
area

river

gulley

catchment area *The catchment area of a river is the area from which it derives its waters.*

catecholamine chemical that functions as a ◊neurotransmitter or a ◊hormone. Dopamine, epinephrine (adrenaline), and norepinephrine (noradrenaline) are catecholamines.

catenary curve taken up by a flexible cable suspended between two points, under gravity; for example, the curve of overhead suspension cables that hold the conductor wire of an electric railway or tramway.

cathode in chemistry, the negative electrode of an electrolytic ◊cell, towards which positive particles (cations), usually in solution, are attracted. See ◊electrolysis.

A cathode is given its negative charge by connecting it to the negative side of an external electrical supply. This is in contrast to the negative electrode of an electrical (battery) cell, which acquires its charge in the course of a spontaneous chemical reaction taking place within the cell.

cathode in electronics, the part of an electronic device in which electrons are generated. In a thermionic valve, electrons are produced by the heating effect of an applied current; in a photoelectric cell, they are produced by the interaction of light and a semiconducting material. The cathode is kept at a negative potential relative to the device's other electrodes (anodes) in order to ensure that the liberated electrons stream away from the cathode and towards the anodes.

A suspension bridge takes up a catenary curve

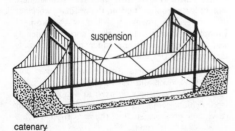

suspension

catenary

cathode-ray oscilloscope (CRO) instrument used to measure electrical potentials or voltages that vary over time and to display the waveforms of electrical oscillations or signals. Readings are displayed graphically on the screen of a ◊cathode-ray tube.

cathode-ray tube vacuum tube in which a beam of electrons is produced and focused onto a fluorescent screen. It is an essential component of television receivers, computer visual display units, and oscilloscopes.

cation ◊ion carrying a positive charge. During electrolysis, cations in the electrolyte move to the cathode (negative electrode).

catkin in flowering plants (◊angiosperms), a pendulous inflorescence, bearing numerous small, usually unisexual flowers. The tiny flowers are stalkless and the petals and sepals are usually absent or much reduced in size. Many types of trees bear catkins, including willows, poplars, and birches. Most plants with catkins are wind-pollinated, so the male catkins produce large quantities of pollen. Some ◊gymnosperms also have catkin-like structures that produce pollen, for example, the swamp cypress *Taxodium*.

CAT scan or *CT scan* (acronym for computerized axial *tomography*) sophisticated method of X-ray imaging. Quick and noninvasive, CAT scanning is used in medicine as an aid to diagnosis, helping to pinpoint problem areas without the need for exploratory surgery. It is also used in archaeology to examine mummies.

The CAT scanner passes a narrow fan of X-rays through successive slices of the suspect body part. These slices are picked up by crystal detectors in a scintillator and converted electronically into cross-sectional images displayed on a viewing screen. Gradually, using views taken from various angles, a three-dimensional picture of the organ or tissue can be built up and suspect irregularities analysed.

cat's eyes reflective studs used to mark the limits of traffic lanes, invented by Percy Shaw (1890–1976) in England, as a road safety device 1934.

A cat's eye stud has two pairs of reflective prisms (the eyes) set in a rubber pad, which reflect the light of a vehicle's headlamps back to the driver. When a vehicle goes over a stud, it moves down inside an outer rubber case; the surfaces of the prisms brush against the rubber and are thereby cleaned.

cauda tail, or tail-like appendage; part of the *cauda equina*, a bundle of nerves at the bottom of the spinal cord in vertebrates.

caustic soda former name for ◊sodium hydroxide (NaOH).

cave roofed-over cavity in the Earth's crust usually produced by the action of underground water or by waves on a seacoast. Caves of the former type commonly occur in areas underlain by limestone, such as Kentucky and many Balkan regions, where the rocks are soluble in water. A *pothole* is a vertical hole in rock caused by water descending a crack; it is thus open to the sky.

Cave animals often show loss of pigmentation or

sight, and under isolation, specialized species may develop. The scientific study of caves is called *speleology*.

During the ◊Ice Age, humans began living in caves leaving many layers of debris that archaeologists have unearthed and dated in the Old World and the New. They also left cave art, paintings of extinct animals often with hunters on their trail. Celebrated caves include the Mammoth Cave in Kentucky, USA, 6.4 km/4 mi long and 38 m/125 ft high; the Caverns of Adelsberg (Postumia) near Trieste, Italy, which extend for many miles; Carlsbad Cave, New Mexico, the largest in the USA; the Cheddar Caves, England; Fingal's Cave, Scotland, which has a range of basalt columns; and Peak Cavern, England.

cavitation the formation of partial vacuums in fluids at high velocities, produced by propellers or other machine parts in hydraulic engines, in accordance with ◊Bernoulli's principle. When these vacuums collapse, pitting, vibration, and noise can occur in the metal parts in contact with the fluids.

In earth science, cavitation, brought about by the forcing of air into rock cracks, is a major cause of ◊erosion. For example, the pounding of waves on the coast and the swirling of turbulent river currents exert great pressures, which eventually cause rocks to break apart.

cc symbol for *cubic centimetre*; abbreviation for *carbon copy/copies*.

CD abbreviation for ◊compact disc.

CD-I abbreviation for *Compact Disc-Interactive*; see ◊compact disc.

CD-ROM (abbreviation for *compact-disc read-only memory*) computer storage device developed from the technology of the audio ◊compact disc. It consists of a plastic-coated metal disc, on which binary digital information is etched in the form of microscopic pits. This can then be read optically by passing a light beam over the disc. CD-ROMs typically hold about 550 ◊megabytes of data, and are used in distributing large amounts of text and graphics, such as encyclopedias, catalogues, and technical manuals.

Standard CD-ROMs cannot have information written onto them by computer, but must be manufactured from a master. Although recordable CDs, called CD-R discs, have been developed for use as computer discs, they are as yet too expensive for widespread use. A compact disc that can be overwritten repeatedly by a computer has also been developed; see ◊optical disc. The compact disc, with its enormous storage capability, may eventually replace the magnetic disc as the most common form of backing store for computers.

Ceefax ('see facts') one of Britain's two ◊teletext systems (the other is Teletext), or 'magazines of the air', developed by the BBC and first broadcast 1973.

celestial mechanics the branch of astronomy that deals with the calculation of the orbits of celestial bodies, their gravitational attractions (such as those that produce the Earth's tides), and also the orbits of artificial satellites and space probes. It is based on the laws of motion and gravity laid down by English physicist and mathematician Isaac Newton.

Celestial Police group of astronomers in Germany 1800–15, who set out to discover a supposed missing planet thought to be orbiting the Sun between Mars and Jupiter, a region now known to be occupied by types of ◊asteroid. Although they did not discover the first asteroid (found 1801), they discovered the second, Pallas (1802), third, Juno (1804), and fourth, Vesta (1807).

celestial sphere imaginary sphere surrounding the Earth, on which the celestial bodies seem to lie. The positions of bodies such as stars, planets, and galaxies are specified by their coordinates on the celestial sphere. The equivalents of latitude and longitude on the celestial sphere are called ◊declination and ◊right ascension (which is measured in hours from 0 to 24). The *celestial poles* lie directly above the Earth's poles, and the *celestial equator* lies over the Earth's equator. The celestial sphere appears to rotate once around the Earth each day, actually a result of the rotation of the Earth on its axis.

celestine or *celestite* mineral consisting of strontium sulphate, $SrSO_4$, occurring as white or light blue crystals. It is the principal source of strontium.

Celestine is found in small quantities in Germany, Italy, and the USA.

cell in biology, a discrete, membrane-bound portion of living matter, the smallest unit capable of an independent existence. Each is formed by the cell division (◊mitosis or ◊meiosis) of another cell. All living organisms consist of one or more cells, with the exception of ◊viruses. Bacteria, protozoa, and many other microorganisms consist of single cells, whereas a human is made up of billions of cells.

Essential features of a cell are the membrane, which encloses it and restricts the flow of substances in and out; the jellylike material, or ◊cytoplasm, within; the ◊ribosomes, which carry out protein synthesis: and the ◊DNA, which forms the hereditary material. The cells of plants, bacteria, and fungi possess a rigid outer cell wall that protects the cell and maintains its shape.

In the ◊eukaryotes (protozoa, fungi, and higher animals and plants), DNA is organized into ◊chromosomes and contained within a ◊nucleus. The only cells of the human body that have no nucleus are the red blood cells. The nuclei of some cells contain a denser spot called the ◊nucleolus. The eukaryotic cell also possesses organelles such as ◊mitochondria, ◊chloroplasts, ◊endoplasmic reticulum, ◊Golgi apparatus and ◊centrioles, which perform specialized tasks.

In ◊prokaryotes (bacteria and cyanobacteria), the DNA forms a simple loop and there is no nucleus.

cell differentiation in developing embryos, the process by which cells acquire their specialization, such as heart cells, muscle cells, skin cells, and brain cells. The seven-day-old human pre-embryo consists of thousands of individual cells, each of which is destined to take up a particular role in the body.

Research has shown that the eventual function of a cell, in for example, a chicken embryo, is determined by the cell's position. The embryo can

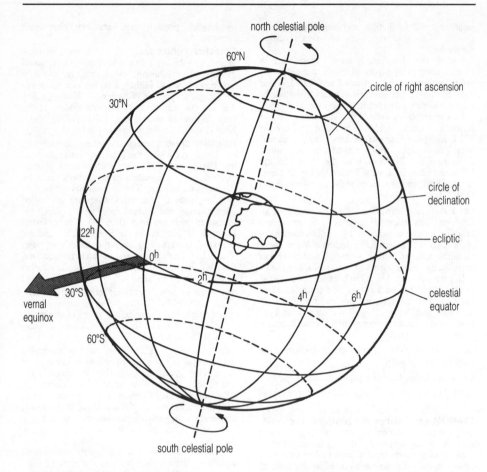

north celestial pole

60°N

circle of right ascension

30°N

circle of
declination

ecliptic

22ʰ

0ʰ

2ʰ

30°S

4ʰ

6ʰ

celestial
equator

vernal
equinox

60°S

south celestial pole

celestial sphere *The main features of the celestial sphere. The equivalents of latitude and longitude on the celestial sphere are declination and right ascension, which is measured eastward from the vernal equinox. Right ascension is measured in hours, one hour corresponding to 15° of longitude.*

be mapped into areas corresponding with the spinal cord, the wings, the legs, and many other tissues. If the embryo is relatively young, a cell transplanted from one area to another will develop according to its new position. As the embryo develops the cells lose their flexibility and become unable to change their destiny.

cell, electrical or *voltaic cell* or *galvanic cell* device in which chemical energy is converted into electrical energy; the popular name is ◊'battery', but this actually refers to a collection of cells in one unit. The reactive chemicals of a *primary cell* cannot be replenished, whereas *secondary cells*— such as storage batteries—are rechargeable: their chemical reactions can be reversed and the original condition restored by applying an electric current. It is dangerous to attempt to recharge a primary cell.

Each cell contains two conducting ◊electrodes immersed in an ◊electrolyte, in a container. A spontaneous chemical reaction within the cell generates a negative charge (excess of electrons) on one electrode, and a positive charge (deficiency of electrons) on the other. The accumulation of these

equal but opposite charges prevents the reaction from continuing unless an outer connection (external circuit) is made between the electrodes allowing the charges to dissipate. When this occurs, electrons escape from the cell's negative terminal and are replaced at the positive, causing a current to flow. After prolonged use, the cell will become flat (cease to supply current).

The first cell was made by Italian physicist Alessandro Volta in 1800. Types of primary cells include the Daniell, Lalande, Leclanché, and so-called 'dry' cells; secondary cells include the Planté, Faure, and Edison. Newer types include the Mallory (mercury depolarizer), which has a very stable discharge curve and can be made in very small units (for example, for hearing aids), and the Venner accumulator, which can be made substantially solid for some purposes.

cell, electrolytic device to which electrical energy is applied in order to bring about a chemical reaction; see ◊electrolysis.

cell membrane or *plasma membrane* thin layer of protein and fat surrounding cells that controls

plant cell

animal cell

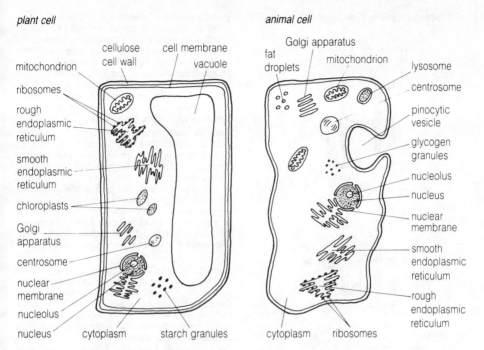

cell *Cells are the building units of living things. Most of them are tiny, having a diameter of less than 0.03 mm/0.001 in. Animal and plant cells have similar internal structures and composition, but plant cells because of their rigid cellulose walls are more rigid in structure. Animal cells lack chloroplasts, the photosynthesizing structures of plant cells.*

substances passing between the cytoplasm and the intercellular space. The cell membrane is semipermeable, allowing some substances to pass through and some not.

Generally, small molecules such as water, glucose, and amino acids can penetrate the membrane, while large molecules such as starch cannot. Membranes also play a part in ◊active transport, hormonal response, and cell metabolism.

cellophane transparent wrapping film made from wood ◊cellulose, widely used for packaging, first produced by Swiss chemist Jacques Edwin Brandenberger 1908.

Cellophane is made from wood pulp, in much the same way that the artificial fibre ◊rayon is made: the pulp is dissolved in chemicals to form a viscose solution, which is then pumped through a long narrow slit into an acid bath where the emergent viscose stream turns into a film of pure cellulose.

cell sap dilute fluid found in the large central vacuole of many plant cells. It is made up of water, amino acids, glucose, and salts. The sap has many functions, including storage of useful materials, and provides mechanical support for non-woody plants.

cellular phone or *cellphone* mobile radio telephone, one of a network connected to the telephone system by a computer-controlled communication system. Service areas are divided into small 'cells', about 5 km/3 mi across, each with a separate low-power transmitter.

The cellular system allows the use of the same set of frequencies with the minimum risk of interference. Nevertheless, in crowded city areas, cells can become overloaded. This has led to a move away from analogue transmissions to digital methods that allow more calls to be made within a limited frequency range.

celluloid transparent or translucent, highly flammable, plastic material (a ◊thermoplastic) made from cellulose nitrate and camphor. It was once used for toilet articles, novelties, and photographic film, but has now been replaced by the nonflammable substance ◊cellulose acetate.

cellulose complex ◊carbohydrate composed of long chains of glucose units. It is the principal constituent of the cell wall of higher plants, and a vital ingredient in the diet of many ◊herbivores. Molecules of cellulose are organized into long, unbranched microfibrils that give support to the cell wall. No mammal produces the enzyme (cellulase) necessary for digesting cellulose; mammals such as rabbits and cows are only able to digest grass because the bacteria present in their gut manufacture the appropriate enzyme.

Cellulose is the most abundant substance found in the plant kingdom. It has numerous uses in industry: in rope-making; as a source of textiles (linen, cotton, viscose, and acetate) and plastics (cellophane and celluloid); in the manufacture of nondrip paint; and in such foods as whipped dessert toppings.

cellulose acetate or *cellulose ethanoate* chemical (an ◊ester) made by the action of acetic acid (ethanoic acid) on cellulose. It is used in making transparent film, especially photographic film;

the basic principle of a electrical cell

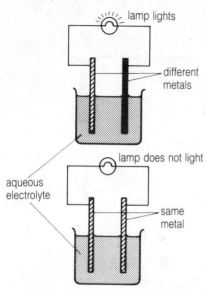

lamp lights

different metals

aqueous electrolyte

lamp does not light

same metal

a simple cell

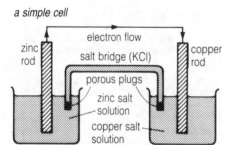

electron flow

zinc rod

salt bridge (KCl)

copper rod

porous plugs

zinc salt solution

copper salt solution

cell, electrical

unlike its predecessor, celluloid, it is not flammable.

cellulose nitrate or *nitrocellulose* series of esters of cellulose with up to three nitrate (NO_3) groups per monosaccharide unit. It is made by the action of concentrated nitric acid on cellulose (for example, cotton waste) in the presence of concentrated sulphuric acid. Fully nitrated cellulose (gun cotton) is explosive, but esters with fewer nitrate groups were once used in making lacquers, rayon, and plastics, such as coloured and photographic film, until replaced by the nonflammable cellulose acetate. ◊Celluloid is a form of cellulose nitrate.

cell wall in plants, the tough outer surface of the cell. It is constructed from a mesh of ◊cellulose and is very strong and relatively inelastic. Most living cells are turgid (swollen with water; see ◊turgor) and develop an internal hydrostatic pressure (wall pressure) that acts against the cellulose wall. The result of this turgor pressure is to give the cell, and therefore the plant, rigidity. Plants that are not woody are particularly reliant on this form of support.

The cellulose in cell walls plays a vital role in global nutrition. No vertebrate is able to produce cellulase, the enzyme necessary for the breakdown of cellulose into sugar. Yet most mammalian herbivores rely on cellulose, using secretions from microorganisms living in the gut to break it down. Humans cannot digest the cellulose of the cell walls; they possess neither the correct gut microorganisms nor the necessary grinding teeth. However, cellulose still forms a necessary part of the human diet as ◊fibre (roughage).

Celsius scale scale of temperature, previously called centigrade, in which the range from freezing to boiling of water is divided into 100 degrees, freezing point being 0 degrees and boiling point 100 degrees.

The degree centigrade (°) was officially renamed Celsius 1948 to avoid confusion with the angular measure known as the centigrade (one hundredth of a grade). The Celsius scale is named after the Swedish astronomer Anders Celsius (1701–1744), who devised it 1742 but in reverse (freezing point was 100°; boiling point 0°).

cement any bonding agent used to unite particles in a single mass or to cause one surface to adhere to another. *Portland cement* is a powder obtained from burning together a mixture of lime (or chalk) and clay, and when mixed with water and sand or gravel, turns into mortar or concrete. In geology, a chemically precipitated material such as carbonate that occupies the interstices of clastic rocks is called cement.

The term 'cement' covers a variety of materials, such as fluxes and pastes, and also bituminous products obtained from tar. In 1824 English bricklayer Joseph Aspdin (1779–1855) created and patented the first Portland cement, so named because its colour in the hardened state resembled that of Portland stone, a limestone used in building.

Cenozoic or *Caenozoic* era of geological time that began 65 million years ago and is still in process. It is divided into the Tertiary and Quaternary periods. The Cenozoic marks the emergence of mammals as a dominant group, including humans, and the formation of the mountain chains of the Himalayas and the Alps.

Centaurus large bright constellation of the southern hemisphere, represented as a centaur. Its brightest star, ◊Alpha Centauri, is a triple star and contains the closest star to the Sun, Proxima Centauri. Omega Centauri, the largest and brightest globular cluster of stars in the sky, is 16,000 light years away.

Centaurus A, a peculiar galaxy 15 million light years away, is a strong source of radio waves and X-rays.

centigrade former name for the ◊Celsius temperature scale.

central dogma in genetics and evolution, the fundamental belief that ◊genes can affect the nature of the physical body, but that changes in the body (for example, through use or accident) cannot be translated into changes in the genes.

central heating system of heating from a central source, typically of a house, larger building, or group of buildings, as opposed to heating each room individually. Steam heat and hot-water heat are the most common systems in use. Water is heated in a furnace burning oil, gas or solid fuel, and, as steam or hot water, is then pumped through radiators in each room. The level of temperature can be selected by adjusting a ◊thermostat on the burner or in a room.

Central heating has its origins in the hypocaust heating system introduced by the Romans nearly 2,000 years ago. From the 18th century, steam central heating, usually by pipe, was available in the West and installed in individual houses on an ad hoc basis. The Scottish engineer James Watt heated his study with a steam pipe connected to a boiler, and British factory owner Matthew Boulton installed steam heating in a friend's Birmingham house. Not until the latter half of the 20th century was central heating in general use.

Central heating systems are usually switched on and off by a time switch. Another kind of central heating system uses hot air, which is pumped through ducts (called risers) to grills in the rooms. Underfloor heating (called radiant heat) is used in some houses, the heat coming from electric elements buried in the floor.

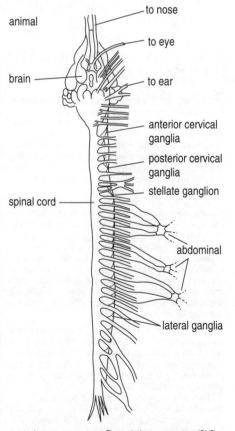

central nervous system *The central nervous system (CNS) with its associated nerves. The CNS controls and integrates body functions. In humans and other vertebrates it consists of a brain and a spinal cord, which are linked to the body's muscles and organs by means of the peripheral nervous system.*

central nervous system (CNS) the part of the nervous system that coordinates various body functions. It has a high concentration of nerve-cell bodies and synapses (junctions between ◊nerve cells). In ◊vertebrates, the CNS consists of a brain and a dorsal nerve cord (the spinal cord) enclosed and protected within the spinal column. In worms, insects, and crustaceans, it consists of a paired ventral nerve cord with concentrations of nerve-cell bodies, known as ◊ganglia in each segment, and a small brain in the head.

Some simple invertebrates, such as sponges and jellyfishes, have no CNS but a simple network of nerve cells called a **nerve net.**

central processing unit (CPU) main component of a computer, the part that executes individual program instructions and controls the operation of other parts. It is sometimes called the central processor or, when contained on a single integrated circuit, a microprocessor.

The CPU has three main components: the **arithmetic and logic unit** (ALU), where all calculations and logical operations are carried out; a **control unit**, which decodes, synchronizes, and executes program instruction; and the **immediate access memory**, which stores the data and programs on which the computer is currently working. All these components contain ◊registers, which are memory locations reserved for specific purposes.

centre of gravity the point in an object about which its weight is evenly balanced. In a uniform gravitational field, this is the same as the centre of mass.

centre of mass or **centre of gravity** point in or near an object from which its total weight appears to originate and can be assumed to act. A symmetrical homogeneous object such as a sphere or cube has its centre of mass at its physical centre; a hollow shape (such as a cup) may have its centre of mass in space inside the hollow.

For an object to be in stable equilibrium, a perpendicular line down through its centre of mass must run within the boundaries of its base; if tilted until this line falls outside the base, the object becomes unstable and topples over.

centrifugal force useful concept in physics, based on an apparent (but not real) force. It may be regarded as a force that acts radially outwards from a spinning or orbiting object, thus balancing the ◊centripetal force (which is real). For an object of mass m moving with a velocity v in a circle of radius r, the centrifugal force F equals mv^2/r (outwards).

centrifuge apparatus that rotates at high speeds, causing substances inside it to be thrown outwards. One use is for separating mixtures of substances of different densities.

The mixtures are usually spun horizontally in balanced containers ('buckets'), and the rotation sets up centrifugal forces, causing their components to separate according to their densities. A common example is the separation of the lighter plasma from the heavier blood corpuscles in certain blood tests. The **ultracentrifuge** is a very high-speed centrifuge, used in biochemistry for separating ◊colloids and organic substances; it may operate at several million revolutions per minute. The centrifuges used in the industrial separation of cream

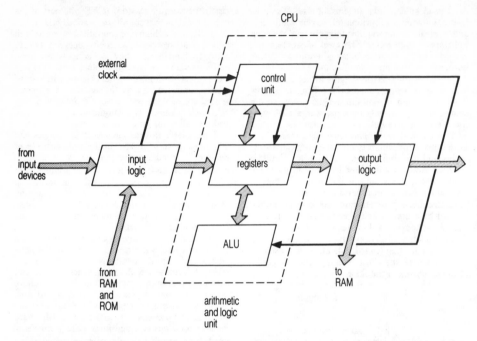

central processing unit *The central processing unit is the 'brain' of a computer; it is here that all the computer's work is done. The arithmetic and logic unit (ALU) does the arithmetic, using the registers to store intermediate results, supervised by the control unit. Input and output circuits connect the ALU to external memory, input, and output devices.*

from milk, and yeast from fermented wort (infused malt), operate by having mixtures pumped through a continually rotating chamber, the components being tapped off at different points. Large centrifuges are used for physiological research—for example, in astronaut training where bodily responses to gravitational forces many times the normal level are tested.

centriole structure found in the ◊cells of animals that plays a role in the processes of ◊meiosis and ◊mitosis (cell division).

centripetal force force that acts radially inwards on an object moving in a curved path. For example, with a weight whirled in a circle at the end of a length of string, the centripetal force is the tension in the string. For an object of mass m moving with a velocity v in a circle of radius r, the centripetal force F equals mv^2/r (inwards). The reaction to this force is the ◊centrifugal force.

centromere part of the ◊chromosome where there are no ◊genes. Under the microscope, it usually appears as a constriction in the strand of the chromosome, and is the point at which the spindle fibres are attached during ◊meiosis and ◊mitosis (cell division).

Centronics interface standard type of computer ◊interface, used to connect computers to ◊parallel devices, usually printers. (Centronics was an important printer manufacturer in the early days of microcomputing.)

cephalopod any predatory marine mollusc of the class Cephalopoda, with the mouth and head surrounded by tentacles. Cephalopods are the most

intelligent, the fastest-moving, and the largest of all animals without backbones, and there are remarkable luminescent forms which swim or drift at great depths. They have the most highly developed nervous and sensory systems of all invertebrates, the eye in some closely paralleling that found in vertebrates. Examples include octopus, squid, and cuttlefish. Shells are rudimentary or absent in most cephalopods.

Typically, they move by swimming with the mantle (fold of outer skin) aided by the arms, but can squirt water out of the siphon (funnel) to propel themselves backwards by jet propulsion. They grow very rapidly and may be mature in a year. The female common octopus lays 150,000 eggs after copulation, and stays to brood them for as long as six weeks. After they hatch the female dies, and, although reproductive habits of many cephalopods are not known, it is thought that dying after spawning may be typical.

Cepheid variable yellow supergiant star that varies regularly in brightness every few days or weeks as a result of pulsations. The time that a Cepheid variable takes to pulsate is directly related to its average brightness; the longer the pulsation period, the brighter the star.

This relationship, the *period luminosity law* (discovered by US astronomer Henrietta Leavitt), allows astronomers to use Cepheid variables as 'standard candles' to measure distances in our Galaxy and to nearby galaxies. They are named after their prototype, Delta Cephei, whose light variations were observed 1784 by English astronomer John Goodricke (1764–1786).

Cepheus constellation of the north polar region, named after King Cepheus of Greek mythology, husband of Cassiopeia and father of Andromeda. It contains the Garnet Star (Mu Cephei), a red supergiant of variable brightness that is one of the reddest-coloured stars known, and Delta Cephei, prototype of the ◊Cepheid variables.

ceramic nonmetallic mineral (clay) used to form articles that are then fired at high temperatures. Ceramics are divided into heavy clay products (bricks, roof tiles, drainpipes, sanitary ware), refractories or high-temperature materials (linings for furnaces used to manufacture steel, fuel elements in nuclear reactors), and pottery, which uses china clay, ball clay, china stone, and flint. Super-ceramics, such as silicon carbide, are lighter, stronger, and more heat-resistant than steel for use in motor and aircraft engines and have to be cast to shape since they are too hard to machine.

cereal grass grown for its edible, nutrient-rich, starchy seeds. The term refers primarily to wheat, oats, rye, and barley, but may also refer to corn, millet, and rice. Cereals contain about 75% complex carbohydrates and 10% protein, plus fats and fibre (roughage). They store well. If all the world's cereal crop were consumed as whole-grain products directly by humans, everyone could obtain adequate protein and carbohydrate; however, a large proportion of cereal production in affluent nations is used as animal feed to boost the production of meat, dairy products, and eggs.

The term also refers to breakfast foods prepared from the seeds of cereal crops. Some cereals require cooking (porridge oats), but most are ready to eat. Mass-marketed cereals include refined and sweetened varieties as well as whole cereals such as muesli. Whole cereals are more nutritious and provide more fibre than the refined cereals, which often have vitamins and flavourings added to replace those lost in the refining process.

cerebellum part of the brain of ◊vertebrate animals which controls muscular movements, balance, and coordination. It is relatively small in lower animals such as newts and lizards, but large in birds since flight demands precise coordination. The human cerebellum is also well developed, because of the need for balance when walking or running, and for coordinated hand movements.

cerebral hemisphere one of the two halves of the ◊cerebrum.

cerebrum part of the vertebrate ◊brain, formed from the two paired cerebral hemispheres. In birds and mammals it is the largest part of the brain. It is covered with an infolded layer of grey matter, the cerebral cortex, which integrates brain functions.

The cerebrum coordinates the senses, and is responsible for learning and other higher mental faculties.

Ceres the largest asteroid, 940 km/584 mi in diameter, and the first to be discovered (by Italian astronomer Giuseppe Piazzi 1801). Ceres orbits the Sun every 4.6 years at an average distance of 414 million km/257 million mi. Its mass is about one-seventieth of that of the Moon.

cerium malleable and ductile, grey, metallic element, symbol Ce, atomic number 58, relative atomic mass 140.12. It is the most abundant member of the lanthanide series, and is used in alloys, electronic components, nuclear fuels, and lighter flints. It was discovered 1804 by the Swedish chemists Jöns Berzelius and Wilhelm Hisinger (1766–1852), and, independently, by Martin Klaproth. The element was named after the then recently discovered asteroid Ceres.

cermet bonded material containing ceramics and metal, widely used in jet engines and nuclear reactors. Cermets behave much like metals but have the great heat resistance of ceramics. Tungsten carbide, molybdenum boride, and aluminium oxide are among the ceramics used; iron, cobalt, nickel, and chromium are among the metals.

CERN nuclear research organization founded 1954 as a cooperative enterprise among European governments. It has laboratories at Meyrin, near Geneva, Switzerland. It was originally known as the *Conseil Européen pour la Recherche Nucléaire* but subsequently renamed *Organisation Européenne pour la Recherche Nucléaire*, although still familiarly known as CERN. It houses the world's largest particle ◊accelerator, the ◊Large Electron Positron Collider (LEP), with which notable advances have been made in ◊particle physics.

In 1965 the original laboratory was doubled in size by extension across the border from Switzerland into France.

Cerro Tololo Inter-American Observatory observatory on Cerro Tololo mountain in the Chilean Andes operated by AURA (the Association of Universities for Research into Astronomy). Its main instrument is a 4–m/158–in reflector, opened 1974, a twin of that at Kitt Peak, Arizona.

Cetus (Latin 'whale') constellation straddling the celestial equator (see ◊celestial sphere), represented as a sea monster. Cetus contains the long-period variable star ◊Mira, and Tau Ceti, one of the nearest stars visible with the naked eye.

CFC abbreviation for ◊*chlorofluorocarbon*.

c.g.s. system system of units based on the centimetre, gram, and second, as units of length, mass, and time respectively. It has been replaced for scientific work by the ◊SI units to avoid inconsistencies in definition of the thermal calorie and electrical quantities.

chain reaction in chemistry, a succession of reactions, usually involving ◊free radicals, where the products of one stage are the reactants of the next. A chain reaction is characterized by the continual generation of reactive substances.

A chain reaction comprises three separate stages: *initiation*—the initial generation of reactive species; *propagation*—reactions that involve reactive species and generate similar or different reactive species; and *termination*—reactions that involve the reactive species but produce only stable, non-reactive substances. Chain reactions may occur slowly (for example, the oxidation of edible oils) or accelerate as the number of reactive species increases, ultimately resulting in explosion.

chain reaction in nuclear physics, a fission reaction that is maintained because neutrons released by the splitting of some atomic nuclei themselves go on to split others, releasing even more neutrons. Such a reaction can be controlled (as in a nuclear reactor) by using moderators to absorb excess neutrons. Uncontrolled, a chain reaction produces a nuclear explosion (as in an atom bomb).

chalaza glutinous mass of transparent albumen supporting the yolk inside birds' eggs. The chalaza is formed as the egg slowly passes down the oviduct, when it also acquires its coiled structure.

chalcedony form of quartz, SiO_2, in which the crystals are so fine-grained that they are impossible to distinguish with a microscope (cryptocrystalline). Agate, onyx, and carnelian are ◊gem varieties of chalcedony.

chalcopyrite copper iron sulphide, $CuFeS_2$, the most common ore of copper. It is brassy yellow in colour and may have an iridescent surface tarnish. It occurs in many different types of mineral vein, in rocks ranging from basalt to limestone.

chalk soft, fine-grained, whitish rock composed of calcium carbonate, $CaCO_3$, extensively quarried for use in cement, lime, and mortar, and in the manufacture of cosmetics and toothpaste. *Blackboard chalk* in fact consists of ◊gypsum (calcium sulphate, $CaSO_4.2H_2$).

Chalk was once thought to derive from the remains of microscopic animals or foraminifera. In 1953, however, it was seen under the electron microscope to be composed chiefly of ◊coccolithophores, unicellular lime-secreting algae, and hence primarily of plant origin. It is formed from deposits of deep-sea sediments called oozes.

Chalk was laid down in the later ◊Cretaceous period and covers a wide area in Europe. In England it stretches in a belt from Wiltshire and Dorset continuously across Buckinghamshire and Cambridgeshire to Lincolnshire and Yorkshire, and also forms the North and South Downs, and the cliffs of S and SE England.

chance likelihood, or ◊probability, of an event taking place, expressed as a fraction or percentage. For example, the chance that a tossed coin will land heads up is 50%.

As a science, it originated when the Chevalier de Méré consulted French mathematician Blaise Pascal about how to reduce his gambling losses. In 1664, in correspondence with another French mathematician, Pierre de Fermat, Pascal worked out the foundations of the theory of chance. This underlies the science of statistics.

change of state in science, a change in the physical state (solid, liquid, or gas) of a material; see ◊state change.

chaos theory or *chaology* branch of mathematics used to deal with chaotic systems—for example, an engineered structure, such as an oil platform, that is subjected to irregular, unpredictable wave stress.

chaparral thick scrub country of the southwestern USA. Thorny bushes have replaced what was largely evergreen oak trees.

characteristic in mathematics, the integral (whole-number) part of a ◊logarithm. The fractional part is the ◊mantissa.

For example, in base ten, $10^0 = 1$, $10^1 = 10$, $10^2 = 100$, and so on; the powers to which 10 is raised are the characteristics. To determine the power to which 10 must be raised to obtain a number between 10 and 100, say 20 (2×10, or log 2 + log 10), the logarithm for 2 is found (0.3010), and the characteristic 1 added to make 1.3010.

character set in computing, the complete set of symbols that can be used in a program or recognized by a computer. It may include letters, digits, spaces, punctuation marks, and special symbols.

character type check in computing, a ◊validation check to ensure that an input data item does not contain invalid characters. For example, an input name may be checked to ensure that it contains only letters of the alphabet or an input six-figure date may be checked to ensure it contains only numbers.

charcoal black, porous form of ◊carbon, produced by heating wood or other organic materials in the absence of air. It is used as a fuel in the smelting of metals such as copper and zinc, and by artists for making black line drawings. *Activated charcoal* has been powdered and dried so that it presents a much increased surface area for adsorption; it is used for filtering and purifying liquids and gases—for example, in drinking-water filters and gas masks.

Charcoal was traditionally produced by burning dried wood in a kiln, a process lasting several days. The kiln was either a simple hole in the ground, or an earth-covered mound. Today kilns are of brick or iron, both of which allow the waste gases to be collected and used.

Charcoal had many uses in earlier centuries. Because of the high temperature at which it burns (1,100°C), it was used in furnaces and blast furnaces before the development of ◊coke. It was also used in an industrial process for obtaining ethanoic acid (acetic acid), in producing wood tar and ◊wood pitch, and (when produced from alder or willow trees) as a component of gunpowder.

charge see ◊electric charge.

CHARGE-COUPLED DEVICE: SENSITIVE DETECTORS

Charge-coupled devices (CCDs) can detect 70% of the light entering a telescope. The earlier method of detecting and recording light, using a photographic plate, only detects 1/10 of 1% of the light. CCDs are 30 times more sensitive to light than a photographic plate and can reveal objects millions of times fainter than can be seen with the naked eye.

---

charge-coupled device (CCD) device for forming images electronically, using a layer of silicon that releases electrons when struck by incoming light. The electrons are stored in ◊pixels and read off into a computer at the end of the exposure. CCDs have now almost entirely replaced photographic film for applications such as astrophotography where extreme sensitivity to light is paramount.

charged particle beam high-energy beam of

RECENT PROGRESS IN MATHEMATICS

The mathematics of chaos

Why are tides predictable years ahead, whereas weather forecasts often go wrong within a few days?

Both tides and weather are governed by natural laws. Tides are caused by the gravitational attraction of the Sun and Moon; the weather by the motion of the atmosphere under the influence of heat from the Sun. The law of gravitation is not noticeably simpler than the laws of fluid dynamics; yet for weather the resulting behaviour seems to be far more complicated.

The reason for this is *chaos*, which lies at the heart of one of the most exciting and most rapidly expanding areas of mathematical research, the theory of nonlinear dynamic systems.

It has been known for a long time that dynamic systems—systems that change with time according to fixed laws—can exhibit regular patterns, such as repetitive cycles. Thanks to new mathematical techniques, emphasizing shape rather than number, and to fast and sophisticated computer graphics, we now know that dynamic systems can also behave randomly. The difference lies not in the complexity of the formulae that define their mathematics, but in the geometrical features of the dynamics. This is a remarkable discovery: random behaviour in a system whose mathematical description contains no hint whatsoever of randomness.

Simple geometric structure produces simple dynamics. For example, if the geometry shrinks everything towards a fixed point, then the motion tends towards a steady state. But if the dynamics keep stretching things apart and then folding them together again, the motion tends to be chaotic—like food being mixed in a bowl. The motion of the Sun and Moon, on the kind of timescale that matters when we want to predict the tides, is a series of regular cycles, so prediction is easy. The changing patterns of the weather involve a great deal of stretching and folding, so here chaos reigns.

Rubber sheets and computer graphics
The geometry of chaos can be explored using theoretical mathematical techniques such as topology—'rubber-sheet geometry'—but the most vivid pictures are obtained using computer graphics. The geometric structures of chaos are *fractals*: they have detailed form on all scales of magnification. Order and chaos, traditionally seen as opposites, are now viewed as two aspects of the same basic process, the evolution of a system in time.

Indeed, there are now examples where both order and chaos occur naturally within a single geometrical form.

The butterfly effect
Does chaos make randomness predictable? Sometimes. If what looks like random behaviour is actually governed by a dynamic system, then short-term prediction becomes possible. Long-term prediction is not as easy, however. In chaotic systems any initial error of measurement, however small, will grow rapidly and eventually ruin the prediction. This is known as the butterfly effect: if a butterfly flaps its wings in the Amazon rainforest, a month later the air disturbance created may cause a hurricane thousands of miles away in Texas.

Chaos can be applied to many areas of science, such as chemistry, engineering, computer sicence, biology, electronics, and astronomy. For example, although the short-term motions of the Sun and Moon are not chaotic, the longterm motion of the Solar System *is* chaotic. It is impossible to predict on which side of the Sun Pluto will lie in 200 million years' time. Saturn's satellite Hyperion tumbles chaotically. Chaos caused by Jupiter's gravitational field can fling asteroids out of orbit, towards the Earth. Disease epidemics, locust plagues, and irregular heartbeats are more down-to-earth examples of chaos, on a more human timescale.

Limitation or liberation?
Chaos places limits on science: it implies that even when we know the equations that govern a system's behaviour, we may not in practice be able to make effective predictions. On the other hand, it opens up new avenues for discovery, because it implies that apparently random phenomena may have simple, non-random explanations. So chaos is changing the way scientists think about what they do: the relation between determinism and chance, the role of experiment, the computability of the world, the prospects for prediction, and the interaction between mathematics, science, and nature.

Chaos cuts right across traditional subject boundaries, and distinctions between pure and applied mathematicians, between mathematicians and physicists, between physicists and biologists, become meaningless when compared to the unity revealed by their joint efforts.

Ian Stewart

electrons or protons. Such beams are being developed as weapons.

Charles's law law stating that the volume of a given mass of gas at constant pressure is directly proportional to its absolute temperature (temperature in kelvin). It was discovered by French physicist Jacques Charles 1787, and independently by French chemist Joseph Gay-Lussac 1802. See also ◊gas laws.

The gas increases by 1/273 of its volume at 0°C for each °C rise of temperature. This means that the coefficient of expansion of all gases is the same. The law is only approximately true and the coefficient of expansion is generally taken as 0.003663 per °C.

charm in physics, a property possessed by one type of ◊quark (very small particles found inside protons and neutrons), called the charm quark. The effects of charm are only seen in experiments with particle ◊accelerators. See ◊elementary particle.

check digit in computing, a digit attached to an important code number as a ◊validation check.

chelate chemical compound whose molecules consist of one or more metal atoms or charged ions joined to chains of organic residues by coordinate (or dative covalent) chemical ◊bonds.

The parent organic compound is known as a *chelating agent*—for example, EDTA (ethylenediaminetetraacetic acid), used in chemical analysis. Chelates are used in analytical chemistry, in agriculture and horticulture as carriers of essential trace metals, in water softening, and in the treatment of thalassaemia by removing excess iron, which may build up to toxic levels in the body. Metalloproteins (natural chelates) may influence the performance of enzymes or provide a mechanism for the storage of iron in the spleen and plasma of the human body.

chemical change change that occurs when two or more substances (reactants) interact with each other, resulting in the production of different substances (products) with different chemical compositions. A simple example of chemical change is the burning of carbon in oxygen to produce carbon dioxide.

chemical element alternative name for ◊element.

chemical equilibrium condition in which the products of a reversible chemical reaction are formed at the same rate at which they decompose back into the reactants, so that the concentration of each reactant and product remains constant.

The amounts of reactant and product present at equilibrium are defined by the *equilibrium constant* for that reaction and specific temperature.

chemical equation method of indicating the reactants and products of a chemical reaction by using chemical symbols and formulae. A chemical equation gives two basic pieces of information: (1) the reactants (on the left-hand side) and products (right-hand side); and (2) the reacting proportions (stoichiometry)—that is, how many units of each reactant and product are involved. The equation must balance; that is, the total number of atoms of a particular element on the left-hand side must be

the same as the number of atoms of that element on the right-hand side.

$$Na_2CO_3 + 2HCl \rightarrow 2NaCl + CO_2 + H_2O$$
reactants   products

This equation states that one molecule of sodium carbonate combines with two molecules of hydrochloric acid to form two molecules of sodium chloride, one of carbon dioxide, and one of water. Double arrows indicate that the reaction is reversible—in the formation of ammonia from hydrogen and nitrogen, the direction depends on the temperature and pressure of the reactants.

$$3H_2 + N_2 \rightleftharpoons 2NH_3$$

chemical oxygen demand (COD) measure of water and effluent quality, expressed as the amount of oxygen (in parts per million) required to oxidize the reducing substances present.

Under controlled conditions of time and temperature, a chemical oxidizing agent (potassium permanganate or dichromate) is added to the sample of water or effluent under consideration, and the amount needed to oxidize the reducing materials present is measured. From this the chemically equivalent amount of oxygen can be calculated. Since the reducing substances typically include remains of living organisms, COD may be regarded as reflecting the extent to which a sample is polluted. Compare ◊biological oxygen demand.

chemical weathering form of ◊weathering brought about by a chemical change in the rocks affected. Chemical weathering involves the 'rotting', or breakdown, of the minerals within a rock, and usually produces a claylike residue (such as china clay and bauxite). Some chemicals are dissolved and carried away from the weathering source.

A number of processes bring about chemical weathering, such as ◊carbonation (breakdown by weakly acidic rainwater), ◊hydrolysis (breakdown by water), ◊hydration (breakdown by the absorption of water), and ◊oxidation (breakdown by the oxygen in water).

chemiluminescence the emission of light from a substance as a result of a chemical reaction (rather than raising its temperature). See ◊luminescence.

chemisorption the attachment, by chemical means, of a single layer of molecules, atoms, or ions of gas to the surface of a solid or, less frequently, a liquid. It is the basis of catalysis (see ◊catalyst) and is of great industrial importance.

chemistry science concerned with the composition of matter (gas, liquid, or solid) and of the changes that take place in it under certain conditions.

All matter can exist in three states: gas, liquid, or solid. It is composed of minute particles termed *molecules*, which are constantly moving, and may be further divided into ◊atoms.

Molecules that contain atoms of one kind only are known as *elements*; those that contain atoms of different kinds are called *compounds*.

Chemical compounds are produced by a chemical action that alters the arrangement of the atoms in the reacting molecules. Heat, light, vibration, catalytic action, radiation, or pressure, as well as moisture (for ionization), may be necessary to produce a chemical change. Examination and possible breakdown of compounds to determine their components is *analysis*, and the building up of compounds from their components is *synthesis*. When substances are brought together without changing their molecular structures they are said to be *mixtures*.

Organic chemistry is the branch of chemistry that deals with carbon compounds. *Inorganic chemistry* deals with the description, properties, reactions, and preparation of all the elements and their compounds, with the exception of carbon compounds. *Physical chemistry* is concerned with the quantitative explanation of chemical phenomena and reactions, and the measurement of data required for such explanations. This branch studies in particular the movement of molecules and the effects of temperature and pressure, often with regard to gases and liquids. Symbols are used to denote the elements. The symbol is usually the first letter or letters of the English or Latin name of the element—for example, C for carbon; Ca for calcium; Fe for iron (*ferrum*). These symbols represent one atom of the element; molecules containing more than one atom of an element are denoted by a subscript figure—for example, water is H_2O. In some substances a group of atoms acts as a single entity, and these are enclosed in parentheses in the symbol—for example $(NH_4)_2SO_4$ denotes ammonium sulphate. The symbolic representation of a molecule is known as a *formula*. A figure placed before a formula represents the number of molecules of a substance taking part in, or being produced by, a chemical reaction—for example, $2H_2O$ indicates two molecules of water. Chemical reactions are expressed by means of *equations* as in:

$$NaCl + H_2SO_4 \rightarrow NaHSO_4 + HCl.$$

This equation states the fact that sodium chloride (NaCl) on being treated with sulphuric acid (H_2SO_4) is converted into sodium bisulphate (sodium hydrogensulphate, $NaHSO_4$) and hydrogen chloride (HCl). Elements are divided into *metals*, which have lustre and conduct heat and electricity, and *nonmetals*, which usually lack these properties. The *periodic system*, developed by John Newlands in 1863 and established by Dmitri Mendeleyev in 1869, classified elements according to their relative atomic masses. Those elements that resemble each other in general properties were found to bear a relation to one another by weight, and these were placed in groups or families. Certain anomalies in this system were later removed by classifying the elements according to their atomic numbers. The latter is equivalent to the positive charge on the nucleus of the atom.

history Ancient civilizations were familiar with certain chemical processes—for example, extracting metals from their ores, and making alloys. The alchemists endeavoured to turn base (nonprecious) metals into gold, and chemistry evolved towards the end of the 17th century from the techniques and insights developed during alchemical experiments. Robert Boyle defined elements as the simplest substances into which matter could be resolved. The alchemical doctrine of the four elements (earth, air, fire, and water) gradually lost its hold, and the theory that all combustible bodies contain a substance called phlogiston (a weightless 'fire element' generated during combustion) was discredited in the 18th century by the experimental work of Joseph Black, Antoine Lavoisier, and Joseph Priestley (who discovered the presence of oxygen in air). Henry Cavendish discovered the composition of water, and John Dalton put forward the atomic theory, which ascribed a precise relative weight to the 'simple atom' characteristic of each element. Much research then took place leading to the development of ◊biochemistry, ◊chemotherapy, and ◊plastics.

Chemistry . . . allows us to calculate the interactions of atoms, without knowing the internal structure of an atom's nucleus.

On **chemistry** Stephen Hawking *A Brief History of Time* 1988

chemosynthesis method of making ◊protoplasm (contents of a cell) using the energy from chemical reactions, in contrast to the use of light energy employed for the same purpose in ◊photosynthesis. The process is used by certain bacteria, which can synthesize organic compounds from carbon dioxide and water using the energy from special methods of ◊respiration.

Nitrifying bacteria are a group of chemosynthetic organisms which change free nitrogen into a form that can be taken up by plants; nitrobacteria, for example, oxidize nitrites to nitrates. This is a vital part of the ◊nitrogen cycle. As chemosynthetic bacteria can survive without light energy, they can live in dark and inhospitable regions, including the hydrothermal vents of the Pacific ocean. Around these vents, where temperatures reach up to 350°C/662°F, the chemosynthetic bacteria are the basis of a food web supporting fishes and other marine life.

chemotherapy any medical treatment with chemicals. It usually refers to treatment of cancer with cytotoxic and other drugs. The term was coined by the German bacteriologist Paul Ehrlich for the use of synthetic chemicals against infectious diseases.

chemotropism movement by part of a plant in response to a chemical stimulus. The response by the plant is termed 'positive' if the growth is towards the stimulus or 'negative' if the growth is away from the stimulus.

Fertilization of flowers by pollen is achieved because the ovary releases chemicals that produce a positive chemotropic response from the developing pollen tube.

Chernobyl town in central Ukraine; site of a nuclear power station. In April 1986 a leak, caused by overheating, occurred in a nonpressurized boiling-water nuclear reactor. The resulting clouds of radioactive isotopes were traced as far away as the UK; thousands of square miles were contaminated.

Current attempts to make the land around Chernobyl safe to farm again include scraping off the

chemistry: chronology

c. 3000 BC	Egyptians were producing bronze – an alloy of copper and tin.
c. 450 BC	Greek philosopher Empedocles proposed that all substances are made up of a combination of four elements – earth, air, fire, and water.
c. 400 BC	Greek philosopher Democritus theorized that matter consists of tiny, indivisible particles, *atomos*.
AD 1	Gold, silver, copper, lead, iron, tin, and mercury were known.
200	The techniques of solution, filtration, and distillation were known.
7th–17th centuries	Chemistry was dominated by alchemy, the attempt to transform nonprecious metals such as lead and copper into gold. Though misguided, it led to the discovery of many new chemicals and techniques, such as sublimation and distillation.
12th century	Alcohol was first distilled in Europe.
1242	Gunpowder introduced to Europe from the Far East.
1620	Scientific method of reasoning expounded by Francis Bacon in his *Novum Organum*
1661	Robert Boyle defined an element as any substance that cannot be broken down into still simpler substances.
1662	Boyle described the inverse relationship between the volume and pressure of a fixed mass of gas.
1697	Georg Stahl proposed the erroneous theory that substances burn because they are rich in a certain substance, called phlogiston.
1755	Joseph Black discovered carbon dioxide.
1774	Joseph Priestley discovered oxygen, which he called 'dephlogisticated air'. Antoine Lavoisier demonstrated his law of conservation of mass.
1777	Lavoisier showed air to be made up of a mixture of gases, and showed that one of these – oxygen – is the substance necessary for combustion (burning) and rusting to take place.
1792	Alessandra Volta demonstrated the electrochemical series.
1807	Humphry Davy passed electric current through molten compounds (the process of electrolysis) in order to isolate elements, such as potassium, that had never been separated by chemical means.
1808	John Dalton published his atomic theory, which states that every element consists of similar indivisible particles – called atoms – which differ from the atoms of other elements in their mass.
1811	Publication of Amedeo Avogadro's hypothesis on the relation between the volume and number of molecules of a gas, and its temperature and pressure.
1813–14	Berzelius devised the chemical symbols and formulae still used elements and compounds.
1828	Franz Wöhler converted ammonium cyanate into urea – the first synthesis of an organic compound from an inorganic substance.
1832–33	Michael Faraday expounded the laws of electrolysis.
1846	Thomas Graham expounded his law of diffusion.
1861	Organic chemistry was defined by German chemist Friedrich Kekulé as the chemistry of carbon compounds.
1869	Dmitri Mendeleyev expounded his periodic table of the elements (based on atomic weight).
1884	Swedish chemist Svante Arrhenius suggested that electrolytes (solutions or molten compounds that conduct electricity) dissociate into ions, atoms or groups of atoms that carry a positive or negative charge.
1894	William Ramsey and Lord Rayleigh discovered the first inert gases, argon.
1897	The electron was discovered by J J Thomson.
1909	Sören Sörensen devised the pH scale of acidity.
1912	Max von Laue showed crystals to be composed of regular, repeating arrays of atoms by studying the patterns in which they diffract X-rays.
1913–14	Henry Moseley equated the atomic number of an element with the positive charge on its nuclei, and drew up the periodic table, based on atomic number, that is used today.
1930	Electrophoresis suspension was invented by Arne Tiselius.
1940	Edwin McMillan and Philip Abelson showed that new elements with a higher atomic number than uranium can be formed by bombarding uranium with neutrons, and synthesized the first transuranic element, neptunium.
1942	Plutonium was first synthesized by Glenn T Seaborg and Edwin McMillan.
1954	Einsteinium and fermium were synthesized.
1955	Ilya Prigogine described the thermodynamics of irreversible processes.
1962	Neil Bartlett prepared the first compound of an inert gas, xenon hexafluoroplatinate.
1965	Robert B Woodward synthesized complex organic compounds.
1981	Quantum mechanics applied to predict course of chemical reactions by US chemist Roald Hoffmann and Kenichi Fukui of Japan.
1982	Element 109, unnilennium, synthesized.
1985	Fullerenes, a new class of carbon solids made up of closed cages of carbon atoms, were discovered by Harold Kroto and David Walton at the University of Sussex, England.
1990	Jean-Marie Lehn, Ulrich Koert, and Margaret Harding reported the synthesis of a new class of compounds, called nucleohelicates, that mimic the structure of DNA, turned inside out.

Industrial chemical processes: chronology

c. AD 1100	Alcohol was first distilled.
1746	John Roebuck invented the lead-chamber process for the manufacture of sulphuric acid.
1790	Nicolas Leblanc developed a process for making sodium carbonate from sodium chloride (common salt).
1827	John Walker invented phosphorus matches.
1831	Peregrine Phillips developed the contact process for the production of sulphuric acid; it was first used on an industrial scale 1875.
1834	Justus von Liebig developed melamine.
1835	Tetrachloroethene (vinyl chloride) was first prepared.
1850	Ammonia was first produced from coal gas.
1855	A technique was patented for the production of cellulose nitrate (nitrocellulose) fibres, the first artificial fibres.
1856	Henry Bessemer developed the Bessemer converter for the production of steel.
1857	William Henry Perkin set up the first synthetic-dye factory, for the production of mauveine.
1861	Ernest Solvay patented a method for the production of sodium carbonate from sodium chloride and ammonia; the first production plant was established 1863.
1862	Alexander Parkes produced the first known synthetic plastic (Parkesine, or xylonite) from cellulose nitrate, vegetable oils, and camphor; it was the forerunner of celluloid.
1864	William Siemens and Pierre Emile Martin developed the Siemens–Martin process (open-hearth method) for the production of steel.
1868	Henry Deacon invented the Deacon process for the production of chlorine by the catalytic oxidation of hydrogen chloride.
1869	Celluloid was first produced from cellulose nitrate and camphor.
1880	The first laboratory preparation of polyacrylic substances.
1886	Charles M Hall and Paul-Louis-Toussaint Héroult developed, independently of each other, a method for the production of aluminium by the electrolysis of aluminium oxide.
1891	Rayon was invented. Herman Frasch patented the Frasch process for the recovery of sulphur from underground deposits. Lindemann produced the first epoxy resins.
1894	Carl Kellner and Hamilton Castner developed, independently of each other, a method for the production of sodium hydroxide by the electrolysis of brine; collaboration gave rise to the Castner–Kellner process.
1895	The Thermit reaction for the reduction of metallic oxides to their molten metals was developed by Johann Goldschmidt.
1902	Friedrich Ostwald patented a process for the production of nitric acid by the catalytic oxidation of ammonia.
1908	Fritz Haber invented the Haber process for the production of ammonia from nitrogen and hydrogen. Heike Kamerlingh-Onnes prepared liquid helium.
1909	The first totally synthetic plastic (Bakelite) was produced by Leo Baekeland.
1912	I Ostromislensky patented the use of plasticizers, which rendered plastics mouldable.
1913	The thermal cracking of petroleum was established.
1919	Elwood Haynes patented non-rusting stainless steel.
1927	The commercial production of polyacrylic polymers began.
1930	Freons were first prepared and used in refrigeration plants. William Chalmers produced the polymer of methyl methacrylate (later marketed as Perspex).
1933	E W Fawcett and R O Gibson first produced polyethylene (polyethene) by the high-pressure polymerization of ethene.
1935	The catalytic cracking of petroleum was introduced. Triacetate film (used as base for photographic film) was developed.
1937	Wallace Carothers invented nylon. Polyurethanes were first produced .
1938	Roy Plunkett first produced polytetrafluoroethene (PTFE, marketed as Teflon).
1943	The industrial production of silicones was initiated. J R Whinfield invented Terylene.
1955	Artificial diamonds were first produced.
1959	The Du Pont company developed Lycra.
1963	Leslie Phillips and co-workers at the Royal Aircraft Establishment, Farnborough, England, invented carbon fibre.
1980	Nippon Oil patented the use of methyl-tert-butyl ether (MTBE) as a lead-free antiknock additive to petrol.
1984	About 2,500 people died in Bhopal, central India, when poisonous methyl isocyanate gas escaped from a chemical plant owned by US company Union Carbide.
1991	ICI began production of the hydrofluorocarbon HFA-134a, a substitute for CFCs in refrigerators and air-conditioning systems. Superconducting salts of buckminsterfullerene were discovered by researchers at AT & T Bell Laboratories, New Jersey, USA.
1992	Blue-light-emitting diodes based on the organic chemical poly(*p*-phenylene) were reported by Australian researchers.

top 3–4 cm/1–1.5 in of topsoil and then burying this 45 cm/18 in below without disturbing the intervening layer.

childbirth the expulsion of a baby from its mother's body following ◊pregnancy. In a broader sense, it is the period of time involving labour and delivery of the baby, plus the effort and pain involved.

chimera in biology, an organism composed of tissues that are genetically different. Chimeras can develop naturally if a ◊mutation occurs in a cell of a developing embryo, but are more commonly produced artificially by implanting cells from one organism into the embryo of another.

china clay clay mineral formed by the decomposition of ◊feldspars. The alteration of aluminium silicates results in the formation of *kaolinite*, $Al_2Si_2O_5(OH)_4$, from which *kaolin*, or white china clay, is derived.

Kaolinite is economically important in the

Tree bark and molecular machines

Chemistry is all around us; everything, including ourselves, is made of chemicals. Chemistry's effects underlie change throughout the universe. As we approach a new century our understanding of chemistry is increasing faster than ever.

Chemical synthesis

Chemists have hundreds of techniques for manipulating chemicals, from the simplest gases to complex anticancer drugs and supermolecules. In the 1960s, Elias J Corey made a breakthrough in organic (carbon-based) chemistry when he developed retrosynthesis, a powerful tool for building complex molecules from smaller, cheaper and more readily available ones. Retrosynthesis can be used to picture a molecule like a jigsaw, working backwards to find reactive components to complete the puzzle. Modern chemists use retrosynthesis to design everything from insect antifeedants for 'greener' farming to better drugs with fewer side-effects.

Recently, one such chemical puzzle, Taxol, has received worldwide attention. Extracted from Pacific yew tree bark, it is an effective treatment for advanced forms of ovarian, breast and other cancers, which resist traditional drugs. Retrosynthesis will help chemists design a laboratory production method for Taxol. Such a 'total' synthesis might need 25 individual chemical steps, making industrial scale-up difficult, but the payback could easily make it commercially viable.

Enzymes in chemistry

Traditional synthesis is powerful, but often requires numerous steps, and many reactions need high temperatures and pressures. Man-made catalysts help speed up some reactions, such as the conversion of methane (natural gas) into useful products. But some are expensive, and can pose disposal problems.

Chemists are turning to nature to solve such problems. Enzymes—nature's catalysts—have many advantages. They work at low temperatures, using less energy. Each type only catalyses certain reactions, so by-products are reduced. This does not limit their use because over 10,000 enzymes are known, and protein engineering could lead to even more.

Enzymes can distinguish between the right-and left-hand forms of a molecule. Some chemicals exist in two forms—enantiomers—with the chemical groups arranged as mirror images, like a pair of hands. The two enantiomers often interact very differently with other molecules. For example, one form of the morning sickness drug thalidomide is an effective tranquillizer; the other severely disturbs foetal development. Enzymes allow the chemist to determine the handedness of such reaction products.

Molecular recognition

Some researchers are working towards building 'molecular machines': tiny switches and transistors that respond to light; molecular wires and diodes that carry signals; and self-replicating systems and artificial enzymes. To do this they are exploiting 'molecular recognition'. Certain large molecules can 'recognize' and trap smaller molecules in their cavities. A simple example is the crown ethers—simple rings of alternating carbon and oxygen atoms. The number of atoms in the ring determines its size, and this determines which chemicals the molecule will recognize. The smallest crown ether can recognize and trap lithium metal ions, but nothing larger, and could be incorporated in a sensor to detect lithium in the presence of other metal ions.

A second important property is 'self-assembly'. If the component parts of the crown ether ring are mixed in a solution containing lithium ions the parts will spontaneously assemble around the ion to complete the ring. The lithium acts as a 'template' for the self-assembly of the ring.

Fraser Stoddart's research team at Birmingham University, UK, have recently built some long-chain compounds with cyclic molecules threaded on ring-shaped molecules, like beads on a string. They can make the 'beads' shuttle backwards and forwards between chemical groups incorporated along the chain. Once they can control this movement using an external input such as light they will have a switch for use in molecular-scale optoelectronic computers.

Julius Rebek's group at MIT, USA have found a molecule that could help explain the early replication processes at life's origins. It acts as a template, bringing molecules together and speeding up their self-assembly. The first template catalyses the formation of a second; the two templates then separate and each catalyses the next round of assembly. Rebek's self-replicating system is analogous to DNA replication in dividing cells. A single helical strand acts as a template for the self-assembly of the DNA base units forming a double helix. These two entwined templates then uncoil, resulting in two single-stranded templates, and so on.

David Bradley

ceramic and paper industries. It is mined in the UK, the USA, France, and the Czech Republic.

chinook (American Indian 'snow-eater') warm dry wind that blows downhill on the eastern side of the Rocky Mountains of North America. It often occurs in winter and spring when it produces a rapid thaw, and so is important to the agriculture of the area.

The chinook is similar to the ◊föhn in the valleys of the European Alps.

chip or *silicon chip* another name for an ◊integrated circuit, a complete electronic circuit on a slice of silicon (or other semiconductor) crystal only a few millimetres square.

Chiron unusual Solar-System object orbiting between Saturn and Uranus, discovered 1977 by US astronomer Charles T Kowal (1940–). Initially classified as an asteroid, it is now believed to be a giant cometary nucleus about 200 km/ 120 mi across, composed of ice with a dark crust of carbon dust.

chitin complex long-chain compound, or ◊polymer; a nitrogenous derivative of glucose. Chitin is widely found in invertebrates. It forms the ◊exoskeleton of insects and other arthropods. It combines with protein to form a covering that can be hard and tough, as in beetles, or soft and flexible, as in caterpillars and other insect larvae. It is insoluble in water and resistant to acids, alkalis, and many organic solvents. In crustaceans such as crabs, it is impregnated with calcium carbonate for extra strength.

Chitin also occurs in some ◊protozoans and coelenterates (such as certain jellyfishes), in the jaws of annelid worms, and as the cell-wall polymer of fungi.

In 1991 scientists discovered that chitin can be converted into carbomethylchitosan, a water-soluble, biodegradable material which is also nontoxic. Its uses include coating apples (still fresh after six months), coating seeds, clearing water of heavy metals, and dressing wounds.

chlamydia single-celled bacterium that can live only parasitically in animal cells. Chlamydiae are thought to be descendants of bacteria that have lost certain metabolic processes. In humans, they cause trachoma, a disease found mainly in the tropics (a leading cause of blindness); venereally transmitted chlamydiae cause genital and urinary infections.

chloral or *trichloroethanal* CCl_3CHO oily, colourless liquid with a characteristic pungent smell, produced by the action of chlorine on ethanol. It is soluble in water and its compound chloral hydrate is a powerful sleep-inducing agent.

chlorate any salt derived from an acid containing both chlorine and oxygen and possessing the negative ion ClO^-, ClO_2^-, ClO_3^-, or ClO_4^-. Common chlorates are those of sodium, potassium, and barium. Certain chlorates are used in weedkillers.

chloride Cl^- negative ion formed when hydrogen chloride dissolves in water, and any salt containing this ion, commonly formed by the action of hydrochloric acid (HCl) on various metals or by direct combination of a metal and chlorine. Sodium chloride (NaCl) is common table salt.

chlorinated solvent any liquid organic compound that contains chlorine atoms, often two or more. These compounds are very effective solvents for fats and greases, but many have toxic properties. They include trichloromethane (chloroform, $CHCl_3$), tetrachloromethane (carbon tetrachloride, CCl_4), and trichloroethene ($CH_2ClCHCl_2$).

chlorine (Greek *chloros* 'green') greenish-yellow, gaseous, nonmetallic element with a pungent odour, symbol Cl, atomic number 17, relative atomic mass 35.453. It is a member of the ◊halogen group and is widely distributed, in combination with the ◊alkali metals, as chlorates or chlorides.

In nature it is always found in the combined form, as in hydrochloric acid, produced in the mammalian stomach for digestion. Chlorine is obtained commercially by the electrolysis of concentrated brine and is an important bleaching agent and germicide, used for both drinking and swimming-pool water. As an oxidizing agent it finds many applications in organic chemistry. The pure gas (Cl_2) is a poison and was used in gas warfare in World War I, where its release seared the membranes of the nose, throat, and lungs,producing pneumonia. Chlorine is a component of chlorofluorocarbons (CFCs) and is partially responsible for the depletion of the ◊ozone layer; it is released from the CFC molecule by the action ofultraviolet radiation in the upper atmosphere, making it available to react with and destroy the ozone.

Chlorine was discovered 1774 by the German chemist Karl Scheele, but English chemist Humphry Davy first proved it to be an element 1810 and named it after its colour.

chlorofluorocarbon (CFC) synthetic chemical that is odourless, nontoxic, nonflammable, and chemically inert. CFCs have been used as propellants in ◊aerosol cans, as refrigerants in refrigerators and air conditioners, and in the manufacture of foam packaging. They are partly responsible for the destruction of the ◊ozone layer. In June 1990 representatives of 93 nations, including the UK and the USA, agreed to phase out production of CFCs and various other ozone-depleting chemicals by the end of the 20th century.

When CFCs are released into the atmosphere, they drift up slowly into the stratosphere, where, under the influence of ultraviolet radiation from the Sun, they break down into chlorine atoms which destroy the ozone layer and allow harmful radiation from the Sun to reach the Earth's surface. CFCs can remain in the atmosphere for more than 100 years.

Replacements for CFCs are being developed, and research into safe methods of destroying existing CFCs is being carried out. The European Community has agreed to ban by the end of 1995 the five 'full hydrogenated' CFCs that are restricted under the ◊Montréal Protocol and a range of CFCs used as industrial solvents, refrigerants, and in fire extinguishers.

chloroform (technical name *trichloromethane*) CCl_3 clear, colourless, toxic, carcinogenic liquid with a characteristic pungent, sickly sweet smell and taste, formerly used as an anaesthetic (now superseded by less harmful substances). It is used as a solvent and in the synthesis of organic chemical compounds.

chlorophyll green pigment present in most plants; it is responsible for the absorption of light energy during ◊photosynthesis. The pigment absorbs the red and blue-violet parts of sunlight but reflects the green, thus giving plants their characteristic colour.

Chlorophyll is found within chloroplasts, present in large numbers in leaves. Cyanobacteria (blue-green algae) and other photosynthetic bacteria also have chlorophyll, though of a slightly different type. Chlorophyll is similar in structure to ◊haemoglobin, but with magnesium instead of iron as the reactive part of the molecule.

chloroplast structure (◊organelle) within a plant cell containing the green pigment chlorophyll. Chloroplasts occur in most cells of the green plant that are exposed to light, often in large numbers. Typically, they are flattened and disclike, with a double membrane enclosing the stroma, a gel-like matrix. Within the stroma are stacks of fluid-containing cavities, or vesicles, where ◊photosynthesis occurs.

It is thought that the chloroplasts were originally free-living cyanobacteria (blue-green algae) which invaded larger, non-photosynthetic cells and developed a symbiotic relationship with them. Like ◊mitochondria, they contain a small amount of DNA and divide by fission. Chloroplasts are a type of ◊plastid.

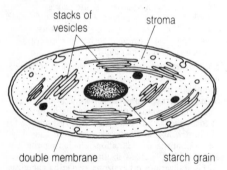

chloroplast *Green chlorophyll molecules on the membranes of the vesicle stacks capture light energy to produce food by photosynthesis.*

chlorosis abnormal condition of green plants in which the stems and leaves turn pale green or yellow. The yellowing is due to a reduction in the levels of the green chlorophyll pigments. It may be caused by a deficiency in essential elements (such as magnesium, iron, or manganese), a lack of light, genetic factors, or viral infection.

choke coil in physics, a coil employed to limit or suppress alternating current without stopping direct current, particularly the type used as a 'starter' in the circuit of fluorescent lighting.

cholecalciferol or *vitamin D* fat-soluble chemical found in fatty fish and margarine. It can be produced in the skin, provided that the skin is adequately exposed to sunlight. Lack of vitamin D leads to rickets and other bone diseases.

cholesterol white, crystalline ◊sterol found throughout the body, especially in fats, blood, nerve tissue, and bile; it is also provided in the diet by foods such as eggs, meat, and butter. A high level of cholesterol in the blood is thought to contribute to atherosclerosis (hardening of the arteries).

Cholesterol is an integral part of all cell membranes and the starting point for steroid hormones, including the sex hormones. It is broken down by the liver into bile salts, which are involved in fat absorption in the digestive system, and it is an essential component of *lipoproteins*, which transport fats and fatty acids in the blood. *Low-density lipoprotein cholesterol* (LDL-cholesterol), when present in excess, can enter the tissues and become deposited on the surface of the arteries, causing atherosclerosis. *High-density lipoprotein cholesterol* (HDL-cholesterol) acts as a scavenger, transporting fat and cholesterol from the tissues to the liver to be broken down. Blood cholesterol levels can be altered by reducing the amount of alcohol and fat in the diet and by substituting some of the saturated fat for polyunsaturated fat, which gives a reduction in LDL-cholesterol. HDL-cholesterol can be increased by exercise.

chord in geometry, a straight line joining any two points on a curve. The chord that passes through the centre of a circle (its longest chord) is the diameter. The longest and shortest chords of an ellipse (a regular oval) are called the major and minor axes respectively.

chordate animal belonging to the phylum Chordata, which includes vertebrates, sea squirts, amphioxi, and others. All these animals, at some stage of their lives, have a supporting rod of tissue (notochord or backbone) running down their bodies.

chorion outermost of the three membranes enclosing the embryo of reptiles, birds, and mammals; the ◊amnion is the innermost membrane.

choroid black layer found at the rear of the ◊eye beneath the retina. By absorbing light that has already passed through the retina, it stops back-reflection and so aids vision.

chromatography (Greek *chromos* 'colour') technique for separating or analysing a mixture of gases, liquids, or dissolved substances. This is brought about by means of two immiscible substances, one of which (*the mobile phase*) transports the sample mixture through the other (*the stationary phase*). The mobile phase may be a gas or a liquid; the stationary phase may be a liquid or a solid, and may be in a column, on paper, or in a thin layer on a glass or plastic support. The components of the mixture are absorbed or impeded by the stationary phase to different extents and therefore become separated. The technique is used for both qualitative and quantitative analyses in biology and chemistry.

In *thin-layer chromatography*, a wafer-thin layer of adsorbent medium on a glass plate replaces the filter paper. The mixture separates because of the differing solubilities of the components in the solvent flowing up the solid layer, and their differing tendencies to stick to the solid (adsorption). The same principles apply in *column chromatography*.

In *gas–liquid chromatography*, a gaseous mixture is passed into a long, coiled tube (enclosed in an oven) filled with an inert powder coated in a

paper chromatography

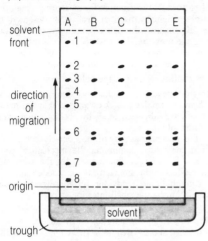

solvent front

direction of migration

origin

trough

solvent

column chromatography

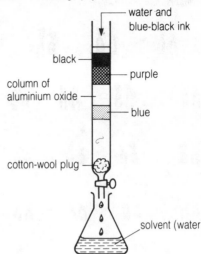

water and blue-black ink

black

purple

column of aluminium oxide

blue

cotton-wool plug

solvent (water

thin-layer chromatography

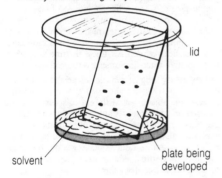

lid

solvent

plate being developed

gas–liquid chromatography; simplified diagram of apparatus

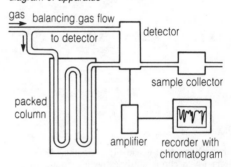

gas balancing gas flow

to detector

detector

packed column

sample collector

amplifier

recorder with chromatogram

chromatography

liquid. A carrier gas flows through the tube. As the mixture proceeds along the tube it separates as the components dissolve in the liquid to differing extents or stay as a gas. A detector locates the different components as they emerge from the tube. The technique is very powerful, allowing tiny quantities of substances (fractions of parts per million) to be separated and analysed.

Preparative chromatography is carried out on a large scale for the purification and collection of one or more of a mixture's constituents; for example, in the recovery of protein from abattoir wastes.

Analytical chromatography is carried out on far smaller quantities, often as little as one microgram (one-millionth of a gram), in order to identify and quantify the component parts of a mixture. It is used to determine the identities and amounts of amino acids in a protein, and the alcohol content of blood and urine samples.

The technique was first used in the separation of coloured mixtures into their component pigments.

chromite $FeCr_2O_3$, iron chromium oxide, the main chromium ore. It is one of the ◊spinel group of minerals, and crystallizes in dark-coloured octahedra of the cubic system. Chromite is usually found in association with ultrabasic and basic rocks; in Cyprus, for example, it occurs with ◊serpentine, and in South Africa it forms continuous layers in a layered ◊intrusion.

chromium (Greek *chromos* 'colour') hard, brittle, grey-white, metallic element, symbol Cr, atomic number 24, relative atomic mass 51.996. It takes a high polish, has a high melting point, and is very resistant to corrosion. It is used in chromium electroplating, in the manufacture of stainless steel and other alloys, and as a catalyst. Its compounds are used for tanning leather and for ◊alums. In human nutrition it is a vital trace element. In nature, it occurs chiefly as chrome iron ore or chromite $(FeCr_2O_4)$. Kazakhstan, Zimbabwe, and Brazil are sources.

The element was named 1797 by the French chemist Louis Vauquelin (1763–1829) after its brightly coloured compounds.

the 23 pairs of chromosomes of a normal human male (XY)

XY

chromosome

chromium ore essentially chromite, $FeCr_2O_3$, from which chromium is extracted. South Africa and Zimbabwe are major producers.

chromosome structure in a cell nucleus that carries the ◊genes. Each chromosome consists of one very long strand of DNA, coiled and folded to produce a compact body. The point on a chromosome where a particular gene occurs is known as its locus. Most higher organisms have two copies of each chromosome (they are ◊diploid) but some have only one (they are ◊haploid). There are 46 chromosomes in a normal human cell. See also ◊mitosis and ◊meiosis.

HUMAN CHROMOSOMES, AND OTHERS

There are 46 chromosomes in the living cells of a human being. Other living things also have chromosomes but the number varies widely from species to species. For example, a garden pea has 14 chromosomes, a potato 48, and a crayfish 200.

---

chromosphere (Greek 'colour' and 'sphere') layer of mostly hydrogen gas about 10,000 km/

6,000 mi deep above the visible surface of the Sun (the photosphere). It appears pinkish red during ◊eclipses of the Sun.

chronometer instrument for measuring time precisely, originally used at sea. It is designed to remain accurate through all conditions of temperature and pressure. The first accurate marine chronometer, capable of an accuracy of half a minute a year, was made 1761 by John Harrison in England.

chrysotile mineral in the ◊olivine group.

chyme general term for the stomach contents. Chyme resembles a thick creamy fluid and is made up of partly digested food, hydrochloric acid, and a range of enzymes.

The muscular activity of the stomach churns this fluid constantly, continuing the mechanical processes initiated by the mouth. By the time the chyme leaves the stomach for the duodenum, it is a smooth liquid ready for further digestion and absorption by the small intestine.

Cibachrome in photography, a process of printing directly from transparencies. It can be home-processed and the rich, saturated colours are highly resistant to fading. It was introduced 1963.

cilia (singular *cilium*) small threadlike organs on the surface of some cells, composed of contractile fibres that produce rhythmic waving movements. Some single-celled organisms move by means of cilia. In multicellular animals, they keep lubricated surfaces clear of debris. They also move food in the digestive tracts of some invertebrates.

ciliary body ring of muscle inside the vertebrate eye that controls the shape of the lens, allowing images of objects at different distances to be focused on the retina. When the muscle is relaxed the lens has its longest ◊focal length and focuses rays from distant objects. Contraction of the muscle reduces tension in the lens making it more curved; the lens therefore has a shorter focal length and focuses images of near objects.

ciliary muscle ring of muscle surrounding and controlling the lens inside the vertebrate eye, used in ◊accommodation (focusing). Suspensory ligaments, resembling spokes of a wheel, connect the lens to the ciliary muscle and pull the lens into a flatter shape when the muscle relaxes. On contraction, the lens returns to its normal spherical state.

cine camera camera that takes a rapid sequence of still photographs—24 frames (pictures) each second. When the pictures are projected one after the other at the same speed on to a screen, they appear to show movement, because our eyes hold on to the image of one picture before the next one appears.

The cine camera differs from an ordinary still camera in having a motor that winds the film on. The film is held still by a claw mechanism while each frame is exposed. When the film is moved between frames, a semicircular disc slides between the lens and the film and prevents exposure.

cinnabar mercuric sulphide, HgS, the only commercially useful ore of mercury. It is deposited in veins and impregnations near recent volcanic rocks and hot springs. The mineral itself is used as a red pigment, commonly known as **vermilion**. Cinnabar is found in the USA (California), Spain (Almadén), Peru, Italy, and Slovenia.

circadian rhythm metabolic rhythm found in most organisms, which generally coincides with the 24–hour day. Its most obvious manifestation is the regular cycle of sleeping and waking, but body temperature and the concentration of ◊hormones that influence mood and behaviour also vary over the day. In humans, alteration of habits (such as rapid air travel round the world) may result in the circadian rhythm being out of phase with actual activity patterns, causing malaise until it has had time to adjust.

circle perfectly round shape, the path of a point that moves so as to keep a constant distance from a fixed point (the centre). Each circle has a **radius** (the distance from any point on the circle to the centre), a **circumference** (the boundary of the circle), **diameters** (straight lines crossing the circle through the centre), **chords** (lines joining two points on the circumference), **tangents** (lines that touch the circumference at one point only), **sectors** (regions inside the circle between two radii), and **segments** (regions between a chord and the circumference).

The ratio of the distance all around the circle (the circumference) to the diameter is an ◊irrational number called π (*pi*), roughly equal to 3.1416. A circle of radius r and diameter d has a circumference $C = \pi d$, or $C = 2\pi r$, and an area $A = \pi r^2$.

The area of a circle can be shown by dividing it into very thin sectors and reassembling them to

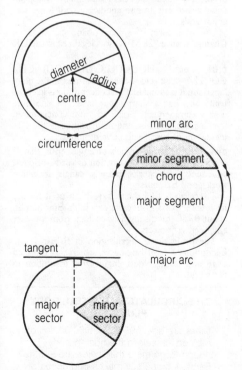

circle *Technical terms used in the geometry of the circle*

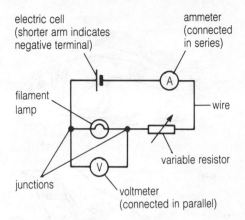

circuit diagram

make an approximate rectangle. The proof of $A = \pi r^2$ can be done only by using ◊integral calculus.

circuit in physics or electrical engineering, an arrangement of electrical components, through which a current can flow. There are two basic circuits, series and parallel. In a ◊series circuit, the components are connected end to end so that the current flows through all components one after the other. In a parallel circuit, components are connected side by side so that part of the current passes through each component. A circuit diagram shows in graphical form how components are connected together, using standard symbols for the components.

circuit breaker switching device designed to protect an electric circuit from excessive current. It has the same action as a ◊fuse, and many houses now have a circuit breaker between the incoming mains supply and the domestic circuits. Circuit breakers usually work by means of ◊solenoids. Those at electricity-generating stations have to be specially designed to prevent dangerous arcing (the release of luminous discharge) when the high-voltage supply is switched off. They may use an air blast or oil immersion to quench the arc.

circuit diagram simplified drawing of an electric circuit. The circuit's components are represented by internationally recognized symbols, and the connecting wires by straight lines. A dot indicates where wires join.

circulatory system system of vessels in an animal's body that transports essential substances (blood or other circulatory fluid) to and from the different parts of the body. Except for simple animals such as sponges and coelenterates (jellyfishes, sea anemones, corals), all animals have a circulatory system.

In fishes, blood passes once around the body before returning to a two-chambered heart (single circulation). In birds and mammals, blood passes to the lungs and back to the heart before circulating around the remainder of the body (double circulation). In all vertebrates, blood flows in one direction. Valves in the heart, large arteries, and veins prevent backflow, and the muscular walls of the arteries assist in pushing the blood around the body.

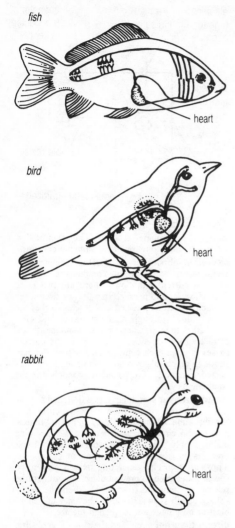

fish

heart

bird

heart

rabbit

heart

circulatory system *The circulatory systems of the fish, the bird, and the rabbit. The blood of a fish passes once around the body before returning to the heart. In birds and mammals, the blood passes to the lungs and returns to the heart before circulating around the body.*

Although most animals have a heart or hearts to pump the blood, normal body movements circulate the fluid in some small invertebrates. In the *open system*, found in snails and other molluscs, the blood (more correctly called ◊haemolymph) passes from the arteries into a body cavity (haemocoel), and from here is gradually returned to the heart, via the gills, by other blood vessels. Insects and other arthropods have an open system with a heart. In the *closed system* of earthworms, blood flows directly from the main artery to the main vein, via smaller lateral vessels in each body segment. Vertebrates, too, have a closed system with a network of tiny ◊capillaries carrying the blood from arteries to veins.

circumference in geometry, the curved line that encloses a curved plane figure, for example a ◊circle or an ellipse. Its length varies according to the nature of the curve, and may be ascertained by the appropriate formula. The circumference of a circle is πd or 2πr, where d is the diameter of the circle, r is its radius and π is the constant pi, approximately equal to 3.1416.

cirque French name for a ◊corrie, a steep-sided hollow in a mountainside.

CISC (acronym for *c*omplex *i*nstruction *s*et *c*omputer) in computing, a microprocessor (processor on a single chip) that can carry out a large number of ◊machine-code instructions—for example, the Intel 80386. The term was introduced to distinguish them from the more rapid ◊RISC (reduced instruction set computer) processors, which handle only a smaller set of instructions.

cistron in genetics, the segment of ◊DNA that is required to synthesize a complete polypeptide chain. It is the molecular equivalent of a ◊gene.

CITES (abbreviation for *Convention on International Trade in Endangered Species*) international agreement under the auspices of the ◊IUCN with the aim of regulating trade in ◊endangered species of animals and plants. The agreement came into force 1975 and by 1991 had been signed by 110 states. It prohibits any trade in a category of 8,000 highly endangered species and controls trade in a further 30,000 species.

citizens' band (CB) short-range radio communication facility (around 27 MHz) used by members of the public in the USA and many European countries to talk to one another or call for emergency assistance.

Use of a form of citizens' band called Open Channel (above 928 MHz) was legalized in the UK 1980.

citric acid $HOOCCH_2C(OH)(COOH)CH_2$. COOH organic acid widely distributed in the plant kingdom; it is found in high concentrations in citrus fruits and has a sharp, sour taste. At one time it was commercially prepared from concentrated lemon juice, but now the main source is the fermentation of sugar with certain moulds.

civil engineering branch of engineering that is concerned with the construction of roads, bridges, aqueducts, waterworks, tunnels, canals, irrigation works, and harbours.

The term is thought to have been used for the first time by British engineer John Smeaton in about 1750 to distinguish civilian from military engineering projects.

The professional organization in Britain is the Institution of Civil Engineers, which was founded 1818 and is the oldest engineering institution in the world.

THE CIRCULATORY SYSTEM: BODY PLUMBING

There are about 96,500 km/60,000 mi of blood vessels in the human body—enough to go more than twice around the Earth. A blood cell makes one circuit of the blood system in about 60 seconds.

---

cladistics method of biological ◊classification (taxonomy) that uses a formal step-by-step procedure for objectively assessing the extent to which organisms share particular characters, and for assigning them to taxonomic groups. Taxonomic groups (for example, ◊species, ◊genus, family) are termed *clades*.

cladode in botany, a flattened stem that is leaflike in appearance and function. It is an adaptation to dry conditions because a stem contains fewer ◊stomata than a leaf, and water loss is thus minimized. The true leaves in such plants are usually reduced to spines or small scales. Examples of plants with cladodes are butcher's-broom *Ruscus aculeatus*, asparagus, and certain cacti. Cladodes may bear flowers or fruit on their surface, and this distinguishes them from leaves.

Clarke orbit alternative name for ◊*geostationary orbit*, an orbit 35,900 km/22,300 mi high, in which satellites circle at the same speed as the Earth turns. This orbit was first suggested by space writer Arthur C Clarke 1945.

class in biological classification, a group of related ◊orders. For example, all mammals belong to the class Mammalia and all birds to the class Aves. Among plants, all class names end in 'idae' (such as Asteridae) and among fungi in 'mycetes'; there are no equivalent conventions among animals. Related classes are grouped together in a ◊phylum.

classification in biology, the arrangement of organisms into a hierarchy of groups on the basis of their similarities in biochemical, anatomical, or physiological characters. The basic grouping is a ◊species, several of which may constitute a ◊genus, which in turn are grouped into families, and so on up through orders, classes, phyla (in plants, sometimes called divisions), to kingdoms.

class interval in statistics, the range of each class of data, used when dealing with large amounts of data. To obtain an idea of the distribution, the data are broken down into convenient classes, which must be mutually exclusive and are usually equal. The class interval defines the range of each class; for example if the class interval is five and the data begin at zero, the classes are 0–4, 5–9, 10–14, and so on.

clathrate compound formed when the small molecules of one substance fill in the holes in the structural lattice of another, solid, substance—for example, sulphur dioxide molecules in ice crystals. Clathrates are therefore intermediate between mixtures and true compounds (which are held together by ◊ionic or covalent chemical bonds).

clathration in chemistry, a method of removing water from an aqueous solution, and therefore increasing the solution's concentration, by trapping it in a matrix with inert gases such as freons.

clausius in engineering, a unit of ◊entropy (the state of disorder in a system). It is defined as the ratio of energy to temperature above absolute zero.

clavicle the collar bone of many vertebrates. In humans it is vulnerable to fracture; falls involving a sudden force on the arm may result in very high stresses passing into the chest region by way of the clavicle and other bones.

claw hard, hooked, pointed outgrowth of the digits

of mammals, birds, and most reptiles. Claws are composed of the protein keratin, and grow continuously from a bundle of cells in the lower skin layer. Hooves and nails are modified structures with the same origin as claws.

clay very fine-grained sedimentary deposit that has undergone a greater or lesser degree of consolidation. When moistened it is plastic, and it hardens on heating, which renders it impermeable. It may be white, grey, red, yellow, blue, or black, depending on its composition. Clay minerals consist largely of hydrous silicates of aluminium and magnesium together with iron, potassium, sodium, and organic substances. The crystals of clay minerals have a layered structure, capable of holding water, and are responsible for its plastic properties. According to international classification, in mechanical analysis of soil, clay has a grain size of less than 0.002 mm/0.00008 in.

Types of clay include adobe, alluvial clay, building clay, brick, cement, china clay (or kaolinite), ferruginous clay, fireclay, fusible clay, puddle clay, refractory clay, and vitrifiable clay. Clays have a variety of uses, some of which, such as pottery and bricks, date back to prehistoric times.

clay mineral one of a group of hydrous silicate minerals that form most of the fine-grained particles in clays. Clay minerals are normally formed by weathering or alteration of other silicates. Virtually all have sheet silicate structures similar to the ◊micas. They exhibit the following useful properties: loss of water on heating, swelling and shrinking in different conditions, cation exchange with other media, and plasticity when wet. Examples are kaolinite, illite, and montmorillonite.

Kaolinite ($Al_2Si_2O_5(OH)_4$) is a common white clay mineral derived from alteration of aluminium silicates, especially feldspars. Illite contains the same constituents as kaolinite, plus potassium, and is the main mineral of clay sediments, mudstones, and shales; it is a weathering product of feldspars and other silicates. Montmorillonite contains the constituents of kaolinite plus sodium and magnesium; along with related magnesium-and iron-bearing clay minerals, it is derived from alteration and weathering of basic igneous rocks.

Kaolinite (the mineral name for kaolin or china clay) is economically important in the ceramic and paper industries. Illite, along with other clay minerals, may also be used in ceramics. Montmorillonite is the chief constituent of fuller's earth, and is also used in drilling muds. Vermiculite (similar to montmorillonite) will expand on heating to produce a material used in insulation.

cleanliness unit unit for measuring air pollution: the number of particles greater than 0.5 micrometres in diameter per cubic foot of air. A more usual measure is the weight of contaminants per cubic metre of air.

cleavage in mineralogy, the tendency of a mineral to split along defined, parallel planes related to its internal structure. It is a useful distinguishing feature in mineral identification. Cleavage occurs where bonding between atoms is weakest, and cleavages may be perfect, good, or poor, depending on the bond strengths; a given mineral may possess one, two, three, or more orientations along which it will cleave.

Some minerals have no cleavage, for example, quartz will fracture to give curved surfaces similar to those of broken glass. Some other minerals, such as apatite, have very poor cleavage that is sometimes known as a parting. Micas have one perfect cleavage and therefore split easily into very thin flakes. Pyroxenes have two good cleavages and break (less perfectly) into long prisms. Galena has three perfect cleavages parallel to the cube edges, and readily breaks into smaller and smaller cubes. Baryte has one perfect cleavage plus good cleavages in other orientations.

cleistogamy production of flowers that never fully open and that are automatically self-fertilized. Cleistogamous flowers are often formed late in the year, after the production of normal flowers, or during a period of cold weather, as seen in several species of violet *Viola*.

client-server architecture in computing, a system in which the mechanics of looking after data are separated from the programs that use the data. For example, the 'server' might be a central database, typically located on a large computer that is reserved for this purpose. The 'client' would be an ordinary program that requests data from the server as needed.

climate weather conditions at a particular place over a period of time. Climate encompasses all the meteorological elements and the factors that influence them. The primary factors that determine the variations of climate over the surface of the Earth are: (a) the effect of latitude and the tilt of the Earth's axis to the plane of the orbit about the Sun (66.5°); (b) the large-scale movements of different wind belts over the Earth's surface; (c) the temperature difference between land and sea; (d) contours of the ground; and (e) location of the area in relation to ocean currents. Catastrophic variations to climate may be caused by the impact of another planetary body, or by clouds resulting from volcanic activity. The most important local or global meteorological changes brought about by human activity are those linked with ◊ozone depleters and the ◊greenhouse effect.

How much heat the Earth receives from the Sun varies in different latitudes and at different times of the year. In the equatorial region the mean daily temperature of the air near the ground has no large seasonal variation. In the polar regions the temperature in the long winter, when there is no

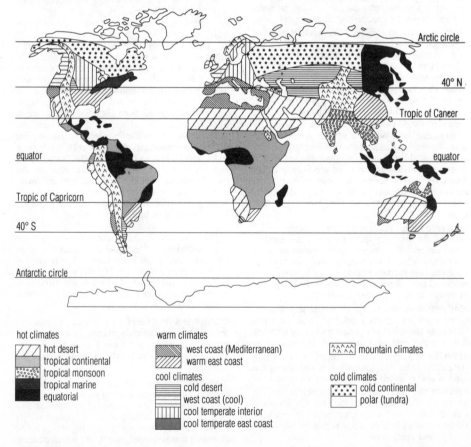

hot climates
- ⧄ hot desert
- ▦ tropical continental
- ▨ tropical monsoon
- ■ tropical marine
- ■ equatorial

warm climates
- ▩ west coast (Mediterranean)
- ▨ warm east coast

cool climates
- ▤ cold desert
- ▦ west coast (cool)
- ▥ cool temperate interior
- ■ cool temperate east coast

- ⋀⋀ mountain climates

cold climates
- ⋯ cold continental
- ▢ polar (tundra)

climate *The world's climatic zones.*

incoming solar radiation, falls far below the summer value. Climate types were first classified by Vladimir Köppen (1846–1940) in 1884.

The temperature of the sea, and of the air above it, varies little in the course of day or night, whereas the surface of the land is rapidly cooled by lack of solar radiation. In the same way the annual change of temperature is relatively small over the sea and great over the land. Continental areas are thus colder than the sea in winter and warmer in summer. Winds that blow from the sea are warm in winter and cool in summer, while winds from the central parts of continents are hot in summer and cold in winter.

On average, air temperature drops with increasing land height at a rate of 1°C/1.8°F per 90 m/ 300 ft. Thus places situated above mean sea level usually have lower temperatures than places at or near sea level. Even in equatorial regions, high mountains are snow-covered during the whole year.

Rainfall is produced by the condensation of water vapour in air. When winds blow against a range of mountains so that the air is forced to ascend, rain results, the amount depending on the height of the ground and the dampness of the air.

The complexity of the distribution of land and sea, and the consequent complexity of the general circulation of the atmosphere, have a direct effect on the distribution of the climate. Centred on the equator is a belt of tropical rainforest, which may be either constantly wet or monsoonal (seasonal with wet and dry seasons in each year). On each side of this is a belt of savannah, with lighter seasonal rainfall and less dense vegetation, largely in the form of grasses. Usually there is then a transition through steppe (semi-arid) to desert (arid), with a further transition through steppe to ◊Mediterranean climate with dry summer, followed by the moist temperate climate of middle latitudes. Next comes a zone of cold climate with moist winter. Where the desert extends into middle latitudes, however, the zones of Mediterranean and moist temperate climates are missing, and the transition is from desert to a cold climate with moist winter. In the extreme east of Asia a cold climate with dry winters extends from about 70°N to 35°N. The polar caps have ◊tundra and glacial climates, with little or no ◊precipitation (rain or snow).

climate model computer simulation, based on physical and mathematical data, of the entire climatic system of the Earth. It is used by researchers to study such topics as the possible long-term disruptive effects of the greenhouse gases, or of variations in the amount of radiation given off by the Sun.

climatic change change in the climate of an area or of the whole world over an appreciable period of time. The geological record shows that climatic changes have taken place regularly, most notably during the ◊Ice Age. Modern climatic changes are being brought on by pollution changing the composition of the atmosphere and producing a ◊greenhouse effect.

climatology study of climate, its global variations and causes.

climax community assemblage of plants and animals that is relatively stable in its environment (for example, oak woods in Britain). It is brought about by ecological ◊succession, and represents the point at which succession ceases to occur.

climax vegetation the state of equilibrium that is reached after a series of changes have occurred in the vegetation of a particular habitat. It is the final stage in a ◊succession, where the structure and species of a habitat do not develop further, provided conditions remain unaltered.

clinometer hand-held surveying instrument for measuring angles of slope.

clint one of a number of flat-topped limestone blocks that make up a ◊limestone pavement. Clints are separated from each other by enlarged joints called grykes.

clo unit of thermal insulation of clothing. Standard clothes have an insulation of about 1 clo; the warmest clothing is about 4 clo per 2.5 cm/1 in of thickness. See also ◊tog.

cloaca the common posterior chamber of most vertebrates into which the digestive, urinary, and reproductive tracts all enter; a cloaca is found in most reptiles, birds, and amphibians; many fishes; and, to a reduced degree, marsupial mammals. Placental mammals, however, have a separate digestive opening (the anus) and urinogenital opening. The cloaca forms a chamber in which products can be stored before being voided from the body via a muscular opening, the cloacal aperture.

clock any device that measures the passage of time, usually shown by means of pointers moving over a dial or by a digital display. Traditionally a timepiece consists of a train of wheels driven by a spring or weight controlled by a balance wheel or pendulum. The watch is a portable clock.

history In ancient Egypt the time during the day was measured by a shadow clock, a primitive form of sundial, and at night the water clock was used. Up to the late 16th century the only clock available for use at sea was the sand clock, of which the most familiar form is the hourglass. During the Middle Ages various types of sundial were widely used, and portable sundials were in use from the 16th to the 18th century. Watches were invented in the 16th century—the first were made in Nuremberg, Germany, shortly after 1500—but it was not until the 19th century that they became cheap enough to be widely available.

The first known public clock was set up in Milan, Italy, in 1353. The timekeeping of both clocks and watches was revolutionized in the 17th century by the application of pendulums to clocks and of balance springs to watches.

types of clock The **marine chronometer** is a precision timepiece of special design, used at sea for giving Greenwich mean time (GMT). Electric timepieces were made possible by the discovery early in the 19th century of the magnetic effects of electric currents. One of the earliest and most satisfactory methods of electrical control of a clock was invented by Matthaeus Hipp in 1842. In one kind of electric clock, the place of the pendulum or spring-controlled balance wheel is taken by a small synchronous electric motor, which counts up the alternations (frequency) of the incoming electric supply and, by a suitable train of wheels, records the time by means of hands on a dial. The *quartz crystal clock* (made possible by the ◊piezoelectric effect of certain crystals) has great precision, with a short-term variation in accuracy of about one-thousandth of a second per day. More accurate still is the ◊*atomic clock*. This utilizes the natural resonance of certain atoms (for example, caesium) as a regulator controlling the frequency of a quartz crystal ◊oscillator. It is accurate to within one-millionth of a second per day.

The first public clock in England was the Salisbury cathedral clock of 1386, which is still working. The British Royal Navy kept time by half-hour sand glasses until 1820.

clock interrupt in computing, an interrupt signal (see ◊interrupt) generated by the computer's internal electronic clock.

clock rate the frequency of a computer's internal electronic clock. Every computer contains an electronic clock, which produces a sequence of regular electrical pulses used by the control unit to synchronize the components of the computer and regulate the ◊fetch-execute cycle by which program instructions are processed. A fixed number of time pulses is required in order to execute each particular instruction. The speed at which a computer can process instructions therefore depends on the clock rate: increasing the clock rate will decrease the time required to complete each particular instruction.

Clock rates are measured in *megahertz* (MHz), or millions of pulses a second. Microcomputers commonly have a clock rate of 8–50 MHz.

clone group of cells or organisms arising by asexual reproduction from a single 'parent' individual. Clones therefore have exactly the same genetic make-up. The term has been adopted by computer technology, in which it describes a (nonexistent) device that mimics an actual one to enable certain software programs to run correctly.

closed set in mathematics, a set of data for which an operation (such as addition or multiplication) done on any members of the set gives a result that is also a member of the set.

For example, the set of even numbers is closed with respect to addition, because two even numbers added to each other always give another even number.

closed-circuit television (CCTV) localized television system in which programmes are sent over relatively short distances, the camera, receiver, and controls being linked by cable. Closed-circuit TV systems are used in department stores and large offices as a means of internal security, monitoring people's movements.

cloud water vapour condensed into minute water particles that float in masses in the atmosphere. Clouds, like fogs or mists, which occur at lower levels, are formed by the cooling of air containing water vapour, which generally condenses around tiny dust particles.

Clouds are classified according to the height at which they occur and their shape. *Cirrus* and *cirrostratus* clouds occur at around 10 km/33,000 ft. The former, sometimes called mares'-tails, consist of minute specks of ice and appear as feathery

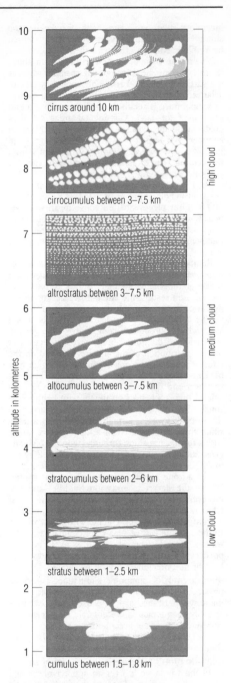

cirrus around 10 km

cirrocumulus between 3–7.5 km

altrostratus between 3–7.5 km

altocumulus between 3–7.5 km

stratocumulus between 2–6 km

stratus between 1–2.5 km

cumulus between 1.5–1.8 km

high cloud

medium cloud

low cloud

altitude in kolometres

cloud *Standard types of cloud. The height and nature of a cloud can be deduced from its name. Cirrus clouds are at high levels and have a wispy appearance. Stratus clouds form at low level and are layered. Middle-level clouds usually have names beginning with 'alto'. Cumulus clouds, ball or cottonwool clouds, occur over a range of heights.*

white wisps, while cirrostratus clouds stretch across the sky as a thin white sheet. Three types of cloud are found at 3–7 km/10,000–23,000 ft: cirrocumulus, altocumulus, and altostratus. *Cirrocumulus* clouds occur in small or large rounded tufts, sometimes arranged in the pattern called mackerel sky. *Altocumulus* clouds are similar, but larger, white clouds, also arranged in lines. *Altostratus* clouds are like heavy cirrostratus clouds and may stretch across the sky as a grey sheet.

Stratocumulus clouds are generally lower, occurring at 2–6 km/6,500–20,000 ft. They are dull grey clouds that give rise to a leaden sky that may not yield rain.

Two types of clouds, *cumulus* and *cumulonimbus*, are placed in a special category because they are produced by daily ascending air currents, which take moisture into the cooler regions of the atmosphere. Cumulus clouds have a flat base generally at 1.4 km/4,500 ft where condensation begins, while the upper part is dome-shaped and extends to about 1.8 km/6,000 ft. Cumulonimbus clouds have their base at much the same level, but extend much higher, often up to over 6 km/20,000 ft. Short heavy showers and sometimes thunder may accompany them. *Stratus* clouds, occurring below 1–2.5 km/3,000–8,000 ft, have the appearance of sheets parallel to the horizon and are like high fogs.

cloud chamber apparatus for tracking ionized particles. It consists of a vessel fitted with a piston and filled with air or other gas, supersaturated with water vapour. When the volume of the vessel is suddenly expanded by moving the piston outwards, the vapour cools and a cloud of tiny droplets forms on any nuclei, dust, or ions present. As fast-moving ionizing particles collide with the air or gas molecules, they show as visible tracks.

Much information about interactions between such particles and radiations has been obtained from photographs of these tracks. This system has been improved upon in recent years by the use of liquid hydrogen or helium instead of air or gas (see ◊particle detector).

The cloud chamber was devised in 1897 by Charles Thomson Rees Wilson (1869–1959) at Cambridge University.

clubroot disease affecting cabbages, turnips, and allied plants of the Cruciferae family. It is caused by a slime mould, *Plasmodiophora brassicae*. This attacks the roots of the plant, which send out knotty outgrowths. Eventually the whole plant decays.

Clubroot is popularly known as finger-and-toe disease after the characteristic outgrowths it produces.

clusec unit for measuring the power of a vacuum pump.

clutch any device for disconnecting rotating shafts, used especially in a car's transmission system. In a car with a manual gearbox, the driver depresses the clutch when changing gear, thus disconnecting the engine from the gearbox.

The clutch consists of two main plates, a pressure plate and a driven plate, which is mounted on a shaft leading to the gearbox. When the clutch is engaged, the pressure plate presses the driven plate against the engine ◊flywheel, and drive goes to the gearbox. Depressing the clutch springs the pressure plate away, freeing the driven plate. Cars with

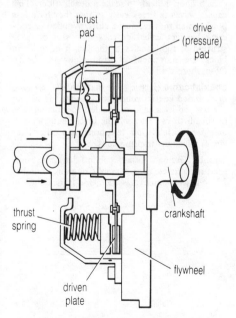

disengaged (pedal pressed down)

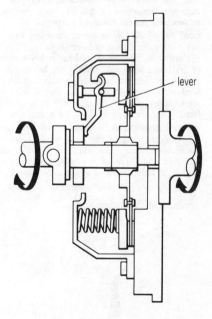

engaged (pedal up)

clutch The clutch consists of two main plates: a drive plate connected to the engine crankshaft and a driven plate connected to the wheels. When the clutch is disengaged, the drive plate does not press against the driven plate. When the clutch is engaged, the two plates are pressed into contact and the rotation of the crankshaft is transmitted to the wheels.

automatic transmission have no clutch. Drive is transmitted from the flywheel to the automatic gearbox by a liquid coupling or ◊torque converter.

cm symbol for **centimetre**.

CMOS abbreviation for ◊*complementary metal-oxide semiconductor* family of integrated circuits (chips) widely used in building electronic systems.

CMYK abbreviation for *c*yan, *m*agenta, *y*ellow, blac*k*—the four colour separations used in most colour-printing processes.

coal black or blackish mineral substance formed from the compaction of ancient plant matter in tropical swamp conditions. It is used as a fuel and in the chemical industry. Coal is classified according to the proportion of carbon it contains. The main types are ◊*anthracite* (shiny, with more than 90% carbon), *bituminous coal* (shiny and dull patches, more than 80% carbon), and *lignite* (woody, grading into peat, 70% carbon). Coal burning is one of the main causes of ◊acid rain.

In Britain, coal was mined on a small scale from Roman times until the Industrial Revolution. From about 1800, coal was carbonized commercially to produce ◊coal gas for gas lighting and ◊coke for smelting iron ore. By the second half of the 19th century, study of the byproducts (coaltar, pitch, and ammonia) formed the basis of organic chemistry, which eventually led to the development of the plastics industry in the 20th century. The York, Derby, and Notts coalfield is Britain's chief reserve, extending north of Selby.

coal gas gas produced when coal is destructively distilled or heated out of contact with the air. Its main constituents are methane, hydrogen, and carbon monoxide. Coal gas has been superseded by ◊natural gas for domestic purposes.

coal tar black oily material resulting from the destructive distillation of bituminous coal.

Further distillation of coal tar yields a number of fractions: light oil, middle oil, heavy oil, and anthracene oil; the residue is called pitch. On further fractionation a large number of substances are obtained, about 200 of which have been isolated. They are used as dyes and in medicines.

coastal erosion the erosion of the land by the constant battering of the sea's waves. This produces two effects. The first is a hydraulic effect, in which the force of the wave compresses air pockets in coastal rocks and cliffs, and the air then

expands explosively. The second is the effect of corrasion, in which rocks and pebbles are flung against the cliffs, wearing them away. Frost shattering (or freeze-thaw), caused by the expansion of frozen seawater in cavities, and ◊biological weathering, caused by the burrowing of rock-boring molluscs, also result in the breakdown of the rock.

Where resistant rocks form headlands, the sea erodes the coast in successive stages. First it creates cracks in cave openings and then gradually wears away the interior of the caves until their roofs are pierced through to form blowholes. In time, caves at either side of a headland may unite to form a natural arch. When the roof of the arch collapses, an isolated pillar, or stack, is formed. This may be worn down further to produce a stump and then a gently sloping surface (wave-cut platform).

In areas where there are beaches, the waves cause longshore drift, in which sand and stone fragments are carried parallel to the shore, causing buildups (sandspits) in some areas and beach erosion in others.

coastal protection measures taken to prevent ◊coastal erosion. Many stretches of coastline are so severely affected by erosion that beaches are swept away, threatening the livelihood of seaside resorts, and buildings become unsafe.

To reduce erosion, several different forms of coastal protection may be employed. Structures such as sea walls attempt to prevent waves reaching the cliffs by deflecting them back to sea. Such structures are expensive and of limited success. A currently preferred option is to add sediment (beach nourishment) to make a beach wider. This causes waves to break early so that they have less power when they reach the cliffs. Wooden or concrete barriers called groynes may also be constructed at right angles to the beach in order to block the movement of sand along the beach (◊longshore drift).

coaxial cable electric cable that consists of a solid or stranded central conductor insulated from and surrounded by a solid or braided conducting tube or sheath. It can transmit the high-frequency signals used in television, telephone, and other telecommunications transmissions.

cobalt (German *Kobalt* 'goblin') hard, lustrous, grey, metallic element, symbol Co, atomic number 27, relative atomic mass 58.933. It is found in

eroded headland at low tide

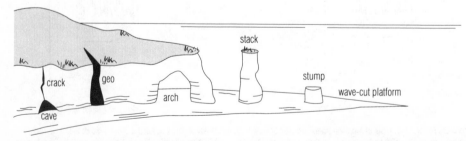

coastal erosion

various ores and occasionally as a free metal, sometimes in metallic meteorite fragments. It is used in the preparation of magnetic, wear-resistant, and high-strength alloys; its compounds are used in inks, paints, and varnishes.

The isotope Co-60 is radioactive (half-life 5.3 years) and is produced in large amounts for use as a source of gamma rays in industrial radiography, research, and cancer therapy. Cobalt was named in 1730 by Swedish chemist Georg Brandt (1694–1768); the name derives from the fact that miners considered its ore worthless because of its arsenic content.

cobalt ore cobalt is extracted from a number of minerals, the main ones being *smaltite*, (CoNi)As$_3$; *linnaeite*, Co$_3$S$_4$; *cobaltite*, CoAsS; and *glaucodot*, (CoFe)AsS.

All commercial cobalt is obtained as a byproduct of other metals. Zaire is the largest producer of cobalt, and it is obtained there as a byproduct of the copper industry. Other producers include Canada and Morocco. Cobalt is also found in the manganese nodules that occur on the ocean floor, and was successfully refined in 1988 from the Pacific Ocean nodules, although this process has yet to prove economic.

COBOL (acronym for *common business-oriented language*) high-level computer-programming language, designed in the late 1950s for commercial data-processing problems; it has become the major language in this field. COBOL features powerful facilities for file handling and business arithmetic. Program instructions written in this language make extensive use of words and look very much like English sentences. This makes COBOL one of the easiest languages to learn and understand.

cocaine alkaloid C$_{17}$H$_{21}$NO$_4$ extracted from the leaves of the coca tree. It has limited medical application, mainly as a local anaesthetic agent that is readily absorbed by mucous membranes (lining tissues) of the nose and throat. It is both toxic and addictive. Its use as a stimulant is illegal. ◊Crack is a derivative of cocaine.

Cocaine was first extracted from the coca plant in Germany in the 19th century. Most of the world's cocaine is produced from coca grown in Peru, Bolivia, Colombia, and Ecuador. Estimated annual production totals 215,000 tonnes, with most of the processing done in Colombia. Long-term use may cause mental and physical deterioration.

coccolithophore microscopic marine alga of a type that secretes a calcite shell. Coccolithophores were particularly abundant during the late ◊Cretaceous period and their calcite remains (coccoliths) form the chalk deposits characteristic of S England, N France, and Kansas, USA.

coccus (plural *cocci*) member of a group of globular bacteria, some of which are harmful to humans. The cocci contain the subgroups *streptococci*, where the bacteria associate in straight chains, and *staphylococci*, where the bacteria associate in branched chains.

cochlea part of the inner ◊ear. It is equipped with approximately 10,000 hair cells, which move in response to sound waves and thus stimulate nerve cells to send messages to the brain. In this way they turn vibrations of the air into electrical signals.

cocktail effect the effect of two toxic, or potentially toxic, chemicals when taken together rather than separately. Such effects are known to occur with some mixtures of chemicals, with one ingredient making the body more sensitive to another ingredient. This sometimes occurs because both chemicals require the same ◊enzyme to break them down. Chemicals such as pesticides and food additives are only ever tested singly, not in combination with other chemicals that may be consumed at the same time, so no allowance is made for cocktail effects.

COD abbreviation for ◊*chemical oxygen demand*, a measure of water and effluent quality.

codominance in genetics, the failure of a pair of alleles, controlling a particular characteristic, to show the classic recessive-dominant relationship. Instead, aspects of both alleles may show in the phenotype.

The snapdragon shows codominance in respect to colour. Two alleles, one for red petals and the other for white, will produce a pink colour if the alleles occur together as a heterozygous form.

codon in genetics, a triplet of bases (see ◊base pair) in a molecule of DNA or RNA that directs the placement of a particular amino acid during the process of protein (polypeptide) synthesis. There are 64 codons in the ◊genetic code.

coefficient the number part in front of an algebraic term, signifying multiplication. For example, in the expression $4x^2 + 2xy -x$, the coefficient of x^2 is 4 (because $4x^2$ means $4 \times x^2$), that of xy is 2, and that of x is -1 (because $-1 \times x = -x$).

In general algebraic expressions, coefficients are represented by letters that may stand for numbers; for example, in the equation $ax^2 + bx + c = 0$, a, b, and c are coefficients, which can take any number.

coefficient of relationship the probability that any two individuals share a given gene by virtue of being descended from a common ancestor. In sexual reproduction of diploid species, an individual shares half its genes with each parent, with its offspring, and (on average) with each sibling; but only a quarter (on average) with its grandchildren or its siblings' offspring; an eighth with its great-grandchildren, and so on.

In certain species of insects (for example honey bees), females have only one set of chromosomes (inherited from the mother), so that sisters are identical in genetic make-up; this produces a different set of coefficients. These coefficients are important in calculations of ◊inclusive fitness.

coelacanth lobe-finned fish *Latimeria chalumnae* up to 2 m/6 ft long. It has bone and muscle at the base of the fins, and is distantly related to the freshwater lobefins, which were the ancestors of all land animals with backbones. Coelacanths live in deep water (200 m/650 ft) around the Comoros Islands, off the coast of Madagascar. They were believed to be extinct until one was caught in 1938.

Coelacanths are now under threat; a belief that

fluid from the spine has a life-extending effect has made them much sought after.

coelenterate any freshwater or marine organism of the phylum Coelenterata, having a body wall composed of two layers of cells. They also possess stinging cells. Examples are jellyfish, hydra, and coral.

coelom in all but the simplest animals, the fluid-filled cavity that separates the body wall from the gut and associated organs, and allows the gut muscles to contract independently of the rest of the body.

coevolution evolution of those structures and behaviours within a species that can best be understood in relation to another species. For example, insects and flowering plants have evolved together: insects have produced mouthparts suitable for collecting pollen or drinking nectar, and plants have developed chemicals and flowers that will attract insects to them.

Coevolution occurs because both groups of organisms, over millions of years, benefit from a continuing association, and will evolve structures and behaviours that maintain this association.

coherence in physics, property of two or more waves of a beam of light or other electromagnetic radiation having the same frequency and the same ◊phase, or a constant phase difference.

cohesion in physics, a phenomenon in which interaction between two surfaces of the same material in contact makes them cling together (with two different materials the similar phenomenon is called adhesion). According to kinetic theory, cohesion is caused by attraction between particles at the atomic or molecular level. ◊Surface tension, which causes liquids to form spherical droplets, is caused by cohesion.

coil in medicine, another name for an ◊intrauterine device.

coke clean, light fuel produced when coal is strongly heated in an airtight oven. Coke contains 90% carbon and makes a useful domestic and industrial fuel (used, for example in the iron and steel industries and in the production of town gas).

The process was patented in England 1622, but it was only in 1709 that Abraham Darby devised a commercial method of production.

cold-blooded of animals, dependent on the surrounding temperature; see ◊*poikilothermy*.

cold fusion in nuclear physics, the fusion of atomic nuclei at room temperature. Were cold fusion to become possible it would provide a limitless, cheap, and pollution-free source of energy, and it has therefore been the subject of research around the world. In 1989, Martin Fleischmann (1927–) and Stanley Pons (1943–) of the University of Utah, USA, claimed that they had achieved cold fusion in the laboratory, but their results could not be substantiated.

Fleischmann and Pons reported that they had achieved the fusion of deuterium (heavy hydrogen) nuclei at room temperature by passing an electric current between palladium oxide electrodes suspended in heavy water (water containing deuterium). They claimed that the heat produced in the process could not be explained by any known electrical or chemical effect, and suggested that the palladium had absorbed deuterium atoms from the water, cramming their nuclei together so that they were close enough to fuse. In the following weeks a number of scientists claimed also to have observed the phenomenon, but researchers at the major research laboratories were unable to duplicate Fleischmann and Pons' results. A number of scientists subsequently withdrew their claims and gradually the evidence for cold fusion was undermined. Most scientists now believe that cold fusion is impossible; however, research has continued in some laboratories.

cold-working method of shaping metal at or near atmospheric temperature.

coleoptile the protective sheath that surrounds the young shoot tip of a grass during its passage through the soil to the surface. Although of relatively simple structure, most coleoptiles are very sensitive to light, ensuring that seedlings grow upwards.

collagen strong, rubbery ◊protein that plays a major structural role in the bodies of ◊vertebrates. Collagen supports the ear flaps and the tip of the nose in humans, as well as being the main constituent of tendons and ligaments. Bones are made up of collagen, with the mineral calcium phosphate providing increased rigidity.

collenchyma plant tissue composed of relatively elongated cells with thickened cell walls, in particular at the corners where adjacent cells meet. It is a supporting and strengthening tissue found in nonwoody plants, mainly in the stems and leaves.

colligative property property that depends on the concentration of particles in a solution. Such properties include osmotic pressure (see ◊osmosis), elevation of ◊boiling point, depression of ◊freezing point, and lowering of ◊vapour pressure.

collimator (1) small telescope attached to a larger optical instrument to fix its line of sight; (2) optical device for producing a nondivergent beam of light; (3) any device for limiting the size and angle of spread of a beam of radiation or particles.

collinear in mathematics, lying on the same straight line.

collision theory theory that explains how chemical reactions take place and why rates of reaction alter. For a reaction to occur the reactant particles must collide. Only a certain fraction of the total collisions cause chemical change; these are called *fruitful collisions*. The fruitful collisions have sufficient energy (activation energy) at the moment of impact to break the existing bonds and form new bonds, resulting in the products of the reaction. Increasing the concentration of the reactants and raising the temperature bring about more collisions and therefore more fruitful collisions, increasing the rate of reaction.

When a ◊catalyst undergoes collision with the reactant molecules, less energy is required for the chemical change to take place, and hence more collisions have sufficient energy for reaction to occur. The reaction rate therefore increases.

colloid substance composed of extremely small particles of one material (the dispersed phase) evenly and stably distributed in another material

a fruitful collision

unstable
activated
complex

an unfruitful collision

collision theory

(the continuous phase). The size of the dispersed particles (1–1,000 nanometres across) is less than that of particles in suspension but greater than that of molecules in true solution. Colloids involving gases include **aerosols** (dispersions of liquid or solid particles in a gas, as in fog or smoke) and **foams** (dispersions of gases in liquids). Those involving liquids include **emulsions** (in which both the dispersed and the continuous phases are liquids) and **sols** (solid particles dispersed in a liquid). Sols in which both phases contribute to a molecular three-dimensional network have a jelly-like form and are known as **gels**; gelatin, starch 'solution', and silica gel are common examples.

Milk is a natural emulsion of liquid fat in a watery liquid; synthetic emulsions such as some paints and cosmetic lotions have chemical emulsifying agents to stabilize the colloid and stop the two phases from separating out.

Colloids were first studied thoroughly by the British chemist Thomas Graham, who defined them as substances that will not diffuse through a semipermeable membrane (as opposed to what he termed crystalloids, solutions of inorganic salts, which will diffuse through).

colon in anatomy, the part of the large intestine between the caecum and rectum, where water and mineral salts are absorbed from digested food, and the residue formed into faeces or faecal pellets.

colonization in ecology, the spread of species into a new habitat, such as a freshly cleared field, a new motorway verge, or a recently flooded valley. The first species to move in are called **pioneers**, and may establish conditions that allow other animals and plants to move in (for example, by improving the condition of the soil or by providing shade). Over time a range of species arrives and the habitat matures; early colonizers will probably be replaced, so that the variety of animal and plant life present changes. This is known as ◊succession.

colour quality or wavelength of light emitted or reflected from an object. Visible white light consists of electromagnetic radiation of various wavelengths, and if a beam is refracted through a prism, it can be spread out into a spectrum, in which the various colours correspond to different wavelengths. From long to short wavelengths (from about 700 to 400 nanometres) the colours are red, orange, yellow, green, blue, indigo, and violet.

The light entering our eyes is either reflected from the objects we see, or emitted by hot or luminous objects. Sources of light have a characteristic ◊spectrum or range of wavelengths. Hot solid objects emit light with a broad range of wavelengths, the maximum intensity being at a wavelength which depends on the temperature. The hotter the object, the shorter the wavelengths emitted, as described by ◊Wien's law. Hot gasses , such as the vapour of sodium street lights, emit light at discrete wavelengths. The pattern of wavelengths emitted is unique to each gas and can be used to identify the gas (see ◊spectroscopy).

When an object is illuminated by white light, some of the wavelengths are absorbed and some are reflected to the eye of an observer. The object appears coloured because of the mixture of wavelengths in the reflected light. For instance, a red object absorbs all wavelengths falling on it except those in the red end of the spectrum. This process of subtraction also explains why certain mixtures of paints produce different colours. Blue and yellow paints when mixed together produce green because between them the yellow and blue pigments absorb all wavelengths except those around green. A suitable combination of three pigments—cyan (blue-green), magenta (blue-red), and yellow—can produce any colour when mixed. This fact is used in colour printing, although additional black pigment is also added in printing.

In the light sensitive lining of our eyeball (the ◊retina) cells called cones are responsible for colour vision. There are three kinds of cones. Each type is sensitive to one colour only, either red, green, or blue.The brain combines the signals sent from the set of cones to produce a sensation of colour. When all cones are stimulated equally the sensation is of white light. The three colours to which the cones respond are called the **primary colours**. By mixing lights of these three colours it is possible to produce any colour. This process is called colour mixing by addition, and is used to produce the colour on a television screen where glowing phosphor dots of red, green, and blue combine. Pairs of colours that produce white light, such as orange and blue, are called *complementary colours*.

Many schemes have been proposed for classifying colours. The most widely used is the Munsell scheme, which classifies colours according to their hue (dominant wavelength), saturation (the degree of whiteness), and brightness (intensity).

colouring food ◊additive used to alter or improve the colour of processed foods. Colourings include artificial colours, such as tartrazine and amaranth, which are made from petrochemicals, and the 'natural' colours such as chlorophyll, caramel, and carotene. Some of the natural colours are actually synthetic copies of the naturally occurring substances, and some of these, notably the synthetically produced caramels, may be injurious to health.

colour vision the ability of the eye to recognise different frequencies in the visible spectrum as colours. In most vertebrates, including humans, colour vision is due to the presence on the ◊retina of three types of light-sensitive cone cell, each of which responds to a different primary colour (red, green, or blue).

columbium (Cb) former name for the chemical element ◊niobium. The name is still used occasionally in metallurgy.

coma in astronomy, the hazy cloud of gas and dust that surrounds the nucleus of a ◊comet.

coma in optics, one of the geometrical aberrations of a lens, whereby skew rays from a point make a comet-shaped spot on the image plane instead of meeting at a point.

combination in mathematics, a selection of a number of objects from some larger number of objects when no account is taken of order within any one arrangement. For example, 123, 213, and 312 are regarded as the same combination of three digits from 1234. Combinatorial analysis is used in the study of ◊probability.

The number of ways of selecting r objects from a group of n is given by the formula $n!/[r!(n-r)!]$ (see ◊factorial). This is usually denoted by ^nC_r.

combined cycle generation system of electricity generation that makes use of both a gas turbine and a steam turbine. Combined cycle plants are more efficient than conventional generating plants, with an efficiency of energy conversion of around 40% (compared with under 38% for conventional plants).

In combined cycle generation, the gas turbine is powered by burning gas fuel, and turns an electric generator. The exhaust gases are then used to heat water to produce steam. The steam powers a steam turbine attached to an electric generator, producing additional electricity.

combined heat and power generation (CHP generation) simultaneous production of electricity and useful heat in a power station. The heat is often in the form of hot water or steam, which can be used for local district heating or in industry. The electricity output from a CHP plant is lower

than from a conventional station, but the overall efficiency of energy conversion is higher.

combustion burning, defined in chemical terms as the rapid combination of a substance with oxygen, accompanied by the evolution of heat and usually light. A slow-burning candle flame and the explosion of a mixture of petrol vapour and air are extreme examples of combustion.

comet small, icy body orbiting the Sun, usually on a highly elliptical path. A comet consists of a central nucleus a few kilometres across, and has been likened to a dirty snowball because it consists mostly of ice mixed with dust. As the comet approaches the Sun the nucleus heats up, releasing gas and dust which form a tenuous coma, up to 100,000 km/60,000 mi wide, around the nucleus. Gas and dust stream away from the coma to form one or more tails, which may extend for millions of kilometres.

Comets are believed to have been formed at the birth of the Solar System. Billions of them may reside in a halo (the ◊*Oort cloud*) beyond Pluto. The gravitational effect of passing stars pushes some towards the Sun, when they eventually become visible from Earth. Most comets swing around the Sun and return to distant space, never to be seen again for thousands or millions of years, although some, called *periodic comets*, have their orbits altered by the gravitational pull of the planets so that they reappear every 200 years or less. Of the 800 or so comets whose orbits have been calculated, about 160 are periodic. The brightest is ◊Halley's comet. The one with the shortest known period is Encke's comet, which orbits the Sun every 3.3 years. A dozen or more comets are discovered every year, some by amateur astronomers.

LIGHT COMET

Comets contain very little solid material. A comet's tail filling 41,500 cu km/10,000 cu mi of space might only contain 16.5 cu cm/1 cu in of solid matter. The mass of all the comets in the Solar System is only a few times that of the Earth.

---

comfort index estimate of how tolerable conditions are for humans in hot climates. It is calculated as the temperature in degrees Fahrenheit plus a quarter of the relative ◊humidity, expressed as a percentage. If the sum is less than 95, conditions are tolerable for those unacclimatized to the tropics.

command language in computing, a set of commands and the rules governing their use, by which users control a program. For example, an ◊operating system may have commands such as SAVE and DELETE, or a payroll program may have commands for adding and amending staff records.

commensalism in biology, a relationship between two ◊species whereby one (the commensal) benefits from the association, whereas the other neither benefits nor suffers. For example, certain species of millipede and silverfish inhabit the nests of army ants and live by scavenging on the refuse of their hosts, but without affecting the ants.

major comets

name	first recorded sighting	orbital period (years)	interesting facts
Halley's comet	240 BC	76	parent of Eta Aquarid and Orionid meteor showers
Comet Tempel-Tuttle	1366 AD	33	parent of Leonid meteors
Biela's comet	1772	6.6	broke in half 1846; not seen since 1852
Encke's comet	1786	3.3	parent of Taurid meteors
Comet Swift-Tuttle	1862	130	parent of Perseid meteors; reappeared 1992
Comet Ikeya-Seki	1965	880	so-called 'Sun-grazing' comet, passed 500,000 km/300,000 mi above surface of Sun on 21 Oct 1965
Comet Kohoutek	1973		observed from space by Skylab astronauts; period too long to calculate accurately
Comet West	1975	500,000	nucleus broke into four parts
Comet Bowell	1980		ejected from Solar System after close encounter with Jupiter
Comet IRAS–Araki–Alcock	1983		passed only 4.5 million km/2.8 million mi from Earth on 11 May 1983; period too long to calculate accurately
Comet Austin	1989		passed 20 million mi from Earth 1990

common logarithm another name for a ◊logarithm to the base ten.

communication in biology, the signalling of information by one organism to another, usually with the intention of altering the recipient's behaviour. Signals used in communication may be *visual* (such as the human smile or the display of colourful plumage in birds), *auditory* (for example, the whines or barks of a dog), *olfactory* (such as the odours released by the scent glands of a deer), *electrical* (as in the pulses emitted by electric fish), or *tactile* (for example, the nuzzling of male and female elephants).

communications satellite relay station in space for sending telephone, television, telex, and other messages around the world. Messages are sent to and from the satellites via ground stations. Most communications satellites are in ◊geostationary orbit, appearing to hang fixed over one point on the Earth's surface.

The first satellite to carry TV signals across the Atlantic Ocean was *Telstar* in July 1962. The world is now linked by a system of communications satellites called Intelsat. Other satellites are used by individual countries for internal communications, or for business or military use. A new generation of satellites, called **direct broadcast satellites**, are powerful enough to transmit direct to small domestic aerials. The power for such satellites is produced by solar cells (see ◊solar energy). The total energy requirement of a satellite is small; a typical communications satellite needs about 2 kW of power, the same as an electric heater.

community in ecology, an assemblage of plants, animals, and other organisms living within a circumscribed area. Communities are usually named by reference to a dominant feature such as characteristic plant species (for example, beech-wood community), or a prominent physical feature (for example, a freshwater-pond community).

commutative operation in mathematics, an operation that is independent of the order of the numbers or symbols concerned. For example, addition is commutative: the result of adding 4 + 2 is the same as that of adding 2 + 4; subtraction is not as 4 − 2 = 2, but 2 − 4 = −2. Compare ◊associative operation and ◊distributive operation.

commutator device in a DC (direct-current) electric motor that reverses the current flowing in the armature coils as the armature rotates. A DC generator, or ◊dynamo, uses a commutator to convert the AC (alternating current) generated in the armature coils into DC. A commutator consists of opposite pairs of conductors insulated from one another, and contact to an external circuit is provided by carbon or metal brushes.

compact disc or *CD* disc for storing digital information, about 12 cm/4.5 in across, mainly used for music, when it can have over an hour's playing time. Entirely different from a conventional LP (long-playing) gramophone record, the compact disc is made of aluminium with a transparent plastic coating; the metal disc underneath is etched by a ◊laser beam with microscopic pits that carry a digital code representing the sounds. During playback, a laser beam reads the code and produces signals that are changed into near-exact replicas of the original sounds.

CD-ROM, or *compact-disc read-only memory*, is used to store written text or pictures rather than music. The discs are ideal for large works, such as catalogues and encyclopedias. CD-I, or *compact-disc interactive*, is a form of CD-ROM used with a computerized reader, which responds intelligently to the user's instructions. These discs are used—for example, with audiovisual material—for training. Recordable CDs, called WORMs ('write once, read many times'), are used as computer discs, but are as yet too expensive for home use. Erasable CDs, which can be erased and recorded many times, are also used by the computer industry. These are coated with a compound

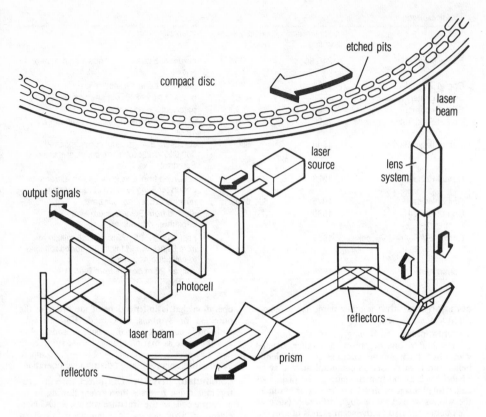

compact disc *The compact disc is a digital storage device; music is recorded as a series of etched pits representing numbers in digital code. During playing, a laser scans the pits and the pattern of reflected light reveals the numbers representing the sound recorded. The optical signal is converted to electrical form by a photocell and sent to the amplifiers and loudspeakers.*

of cobalt and gadolinium, which alters the polarization of light falling on it. In the reader, the light reflected from the disc is passed through polarizing filters and the changes in polarization are converted into electrical signals.

Compact discs were launched 1983.

COMPACT DISCS: LONG-DISTANCE MUSIC

The music track on a compact disc (CD) can be over 20 km/12 mi long. The tracks are so close together that 60 of them would fit into the width of a single groove on an ordinary LP record.

---

compass any instrument for finding direction. The most commonly used is a magnetic compass, consisting of a thin piece of magnetic material with the north-seeking pole indicated, free to rotate on a pivot and mounted on a compass card on which the points of the compass are marked. When the compass is properly adjusted and used, the north-seeking pole will point to the magnetic north, from which true north can be found from tables of magnetic corrections.

Compasses not dependent on the magnet are gyrocompasses, dependent on the ◊gyroscope, and radiocompasses, dependent on the use of radio. These are unaffected by the presence of iron and by magnetic anomalies of the Earth's magnetic field, and are widely used in ships and aircraft. See ◊navigation.

A compass (or pair of compasses) is also an instrument used for drawing circles or taking measurements, consisting of a pair of pointed legs connected by a central pivot.

compensation point in biology, the point at which there is just enough light for a plant to survive. At this point all the food produced by ◊photosynthesis is used up by ◊respiration. For aquatic plants, the compensation point is the depth of water at which there is just enough light to sustain life (deeper water = less light = less photosynthesis).

competition in ecology, the interaction between two or more organisms, or groups of organisms (for example, species), that use a common resource which is in short supply. Competition invariably results in a reduction in the numbers of one or both competitors, and in ◊evolution contributes both to the decline of certain species and to the evolution of ◊adaptations.

Thus plants may compete with each other for sunlight, or nutrients from the soil, while animals may compete amongst themselves for food, water, or refuge.

compiler computer program that translates programs written in a ◊high-level language into machine code (the form in which they can be run by the computer). The compiler translates each high-level instruction into several machine-code instructions—in a process called *compilation*—and produces a complete independent program that can be run by the computer as often as required, without the original source program being present.

Different compilers are needed for different high-level languages and for different computers. In contrast to using an ◊interpreter, using a compiler adds slightly to the time needed to develop a new program because the machine-code program must be recompiled after each change or correction. Once compiled, however, the machine-code program will run much faster than an interpreted program.

complement the set of the elements within the universal set that are not contained in the designated set. For example, if the universal set is the set of all positive whole numbers and the designated set S is the set of all even numbers, then the complement of S (denoted S') is the set of all odd numbers.

complementary angles two angles that add up to 90°.

complementary metal-oxide semiconductor (CMOS) in computing, a particular way of manufacturing integrated circuits (chips). The main advantage of CMOS chips is their low power requirement and heat dissipation, which enables them to be used in electronic watches and portable microcomputers. However, CMOS circuits are expensive to manufacture and have lower operating speeds than have circuits of the ◊transistor-transistor logic (TTL) family.

complementary number in number theory, the number obtained by subtracting a number from its base. For example, the complement of 7 in numbers to base 10 is 3. Complementary numbers are necessary in computing, as the only mathematical operation of which digital computers (including pocket calculators) are directly capable is addition. Two numbers can be subtracted by adding one number to the complement of the other; two numbers can be divided by using successive subtraction (which, using complements, becomes successive addition); and multiplication can be performed by using successive addition.

The four main operations of arithmetic can thus all be reduced to various types of addition and made with the capability of a digital computer, using the binary number system.

complementation in genetics, the interaction that can occur between two different mutant alleles of a gene in a ◊diploid organism, to make up for each other's deficiencies and allow the organism to function normally.

complex number in mathematics, a number written in the form $a + ib$, where a and b are ◊real numbers and i is the square root of -1 (that is, $i^2 = -1$); i used to be known as the 'imaginary' part of the complex number. Some equations in algebra,

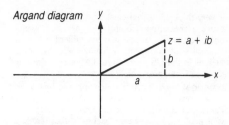

Argand diagram

complex number *A complex number can be represented graphically as a line whose end-point coordinates equal the real and imaginary parts of the complex number. This type of diagram is called an Argand diagram after the French mathematician Jean Robert Argand (1768–1822) who devised it.*

such as those of the form $x^2 + 5 = 0$, cannot be solved without recourse to complex numbers, because the real numbers do not include square roots of negative numbers.

The sum of two or more complex numbers is obtained by adding separately their real and imaginary parts, for example, $(a + bi) + (c + di) = (a + c) + (b + d)i$.

Complex numbers can be represented graphically on an Argand diagram, which uses rectangular ◊Cartesian coordinates in which the x-axis represents the real part of the number and the y-axis the imaginary part. Thus the number $z = a + bi$ is plotted as the point (a, b). Complex numbers have applications in various areas of science, such as the theory of alternating currents in electricity.

component in mathematics, one of the vectors produced when a single vector is resolved into two or more parts. The vector sum of the components gives the original vector.

Compositae daisy family, comprising dicotyledonous flowering plants characterized by flowers borne in composite heads (see ◊capitulum). It is the largest family of flowering plants, the majority being herbaceous. Birds seem to favour the family for use in nest 'decoration', possibly because many species either repel or kill insects. Species include the daisy and dandelion; food plants such as the artichoke, lettuce, and safflower; and the garden varieties of chrysanthemum, dahlia, and zinnia.

composite in industry, any purpose-designed engineering material created by combining single materials with complementary properties into a composite form. Most composites have a structure in which one component consists of discrete elements, such as fibres, dispersed in a continuous matrix. For example, lengths of asbestos, glass, or carbon steel, or 'whiskers' (specially grown crystals a few millimetres long) of substances such as silicon carbide may be dispersed in plastics, concrete, or steel.

composite function in mathematics, a function made up of two or more other functions carried out in sequence, usually denoted by ◊ or ○, as in the relation $(f ◊ g) x = f[g(x)]$.

Usually, composition is not commutative: $(f ◊ g)$ is not necessarily the same as $(g ◊ f)$.

composite volcano steep-sided conical ◊volcano formed at a ◊destructive margin. It is made up of alternate layers of ash and lava. The magma (molten rock) associated with composite volcanoes is very thick and often clogs up the vent. This can cause a tremendous buildup of pressure, which, once released, causes a very violent eruption. Examples of composite volcanoes are Mount St Helens in the USA and Mount Mayon in the Philippines. Compare ◊shield volcano.

compost organic material decomposed by bacteria under controlled conditions to make a nutrient-rich natural fertilizer for use in gardening or farming. A well-made compost heap reaches a high temperature (up to 66°C/150°F) during the composting process, killing most weed seeds that might be present.

compound chemical substance made up of two or more ◊elements bonded together, so that they cannot be separated by physical means. Compounds are held together by ionic or covalent bonds.

COMPOUND INTEREST

An estimated 10 million chemical compounds have been discovered; 400,000 new compounds are discovered each year.

---

compressor machine that compresses a gas, usually air, commonly used to power pneumatic tools, such as road drills, paint sprayers, and dentist's drills.
Reciprocating compressors use pistons moving in cylinders to compress the air. Rotary compressors use a varied rotor moving eccentrically inside a casing. The air compressor in jet and ◊gas turbine engines consists of a many-varied rotor rotating at high speed within a fixed casing, where the rotor blades slot between fixed, or stator, blades on the casing.

computer programmable electronic device that processes data and performs calculations and other symbol-manipulation tasks. There are three types: the ◊*digital computer*, which manipulates information coded as binary numbers (see ◊binary number system); the ◊*analogue computer*, which works with continuously varying quantities; and the *hybrid computer*, which has characteristics of both analogue and digital computers.
There are four types of digital computer, corresponding roughly to their size and intended use. *Microcomputers* are the smallest and most common, used in small businesses, at home, and in schools. They are usually single-user machines. *Minicomputers* are found in medium-sized businesses and university departments. They may support from 10 to 200 or so users at once. *Mainframes*, which can often service several hundred users simultaneously, are found in large organizations, such as national companies and government departments. *Supercomputers* are mostly used for highly complex scientific tasks, such as analysing the results of nuclear physics experiments and weather forecasting.
Microcomputers now come in a range of sizes from battery-powered pocket PCs and electronic organizers, notebook and laptop PCs to floor-standing tower systems that may serve local area ◊networks or work as minicomputers. Indeed, most

minicomputers are now built using low-cost microprocessors, and large-scale computers built out of multiple microprocessors are starting to challenge traditional mainframe and supercomputer designs.
history Computers are only one of the many kinds of ◊computing device. The first mechanical computer was conceived by English mathematician Charles Babbage 1835, but it never went beyond the design stage. In 1943, more than a century later, Thomas Flowers built Colossus, the first electronic computer. Working with him at the time was Alan Turing, a mathematician who seven years earlier had published a paper on the theory of computing machines that had a major impact on subsequent developments. John von Neumann's computer, EDVAC, built 1949, was the first to use binary arithmetic and to store its operating instructions internally. This design still forms the basis of today's computers.
basic components At the heart of a computer is the ◊central processing unit (CPU), which performs all the computations. This is supported by memory, which holds the current program and data, and 'logic arrays', which help move information around the system. A main power supply is needed and, for a mainframe or minicomputer, a cooling system. The computer's 'device driver' circuits control the ◊peripheral devices that can be attached. These will normally be keyboards and VDUs (◊visual display units) for user input and output, disc drive units for mass memory storage, and printers for printed output.

The real danger is not that computers will begin to think like men, but that men will begin to think like computers.

On **computers** S J Harris

---

computer-aided design use of computers to create and modify design drawings; see ◊CAD.

computer-aided manufacturing use of computers to regulate production processes in industry; see ◊CAM.

computer-assisted learning use of computers in education and training; see ◊CAL.

computer game or *video game* any computer-controlled game in which the computer (sometimes) opposes the human player. Computer games typically employ fast, animated graphics on a VDU (◊visual display unit), and synthesized sound.
Commercial computer games became possible with the advent of the ◊microprocessor in the mid-1970s and rapidly became popular as amusement-arcade games, using dedicated chips. Available games range from chess to fighter-plane simulations.

computer generation any of the five broad groups into which computers may be classified:
first generation the earliest computers, developed in the 1940s and 1950s, made from valves and wire circuits; *second generation* from the early 1960s, based on transistors and printed circuits; *third generation* from the late 1960s, using integrated circuits and often sold as families of computers, such as the IBM 360 series; *fourth generation* using ◊microprocessors, large-scale integration

typical mainframe computer system

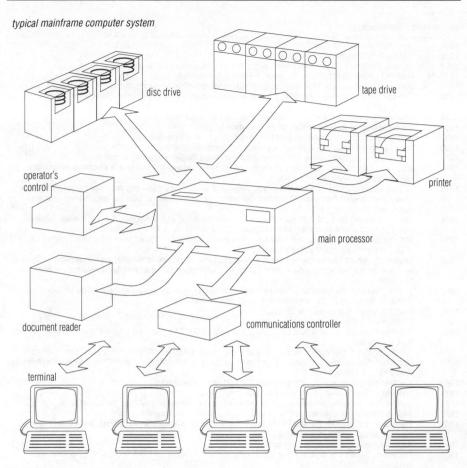

disc drive

tape drive

operator's control

printer

main processor

document reader

communications controller

terminal

computer

(LSI), and sophisticated programming languages, still in use in the 1990s; and **fifth generation** based on parallel processing and very large-scale integration, currently under development.

computer graphics use of computers to display and manipulate information in pictorial form. The output may be as simple as a pie chart, or as complex as an animated sequence in a science-fiction film, or a seemingly three-dimensional engineering blueprint. Input may be achieved by scanning an image, by drawing with a mouse or stylus on a graphics tablet, or by drawing directly on the screen with a light pen. The drawing is stored in the computer as ◊raster graphics or ◊vector graphics. Computer graphics are increasingly used in computer-aided design (◊CAD), and to generate models and simulations in engineering, meteorology, medicine and surgery, and other fields of science.

Recent developments in software mean that designers on opposite sides of the world will soon be able to work on complex three-dimensional computer models using ordinary PCs linked by telephone lines rather then powerful graphics workstations.

computerized axial tomography medical technique, usually known as ◊CAT scan, for looking inside bodies without disturbing them.

computer program coded instructions for a computer; see ◊program.

computer simulation representation of a real-life situation in a computer program. For example, the program might simulate the flow of customers arriving at a bank. The user can alter variables, such as the number of cashiers on duty, and see the effect.

More complex simulations can model the behaviour of chemical reactions or even nuclear explosions. Computers also control the actions of machines—for example, a ◊flight simulator models the behaviour of real aircraft and allows training to take place in safety. Computer simulations are very useful when it is too dangerous, time consuming, or simply impossible to carry out a real experiment or test.

computer terminal the device whereby the operator communicates with the computer; see ◊terminal.

computing device any device built to perform or help perform computations, such as the ◊abacus, ◊slide rule, or ◊computer.

computing: chronology

1614	John Napier invented logarithms.
1615	William Oughtred invented the slide rule.
1623	Wilhelm Schickard (1592–1635) invented the mechanical calculating machine.
1672–74	Gottfried Leibniz built his first calculator, the Stepped Reckoner.
1801	Joseph-Marie Jacquard developed an automatic loom controlled by punch cards.
1820	The first mass-produced calculator, the Arithometer, was developed by Charles Thomas de Colmar (1785–1870).
1822	Charles Babbage completed his first model for the difference engine.
1830s	Babbage created the first design for the analytical engine.
1890	Herman Hollerith developed the punched-card ruler for the US census.
1936	Alan Turing published the mathematical theory of computing.
1938	Konrad Zuse constructed the first binary calculator, using Boolean algebra.
1939	US mathematician and physicist J V Atanasoff (1903–　) became the first to use electronic means for mechanizing arithmetical operations.
1943	The Colossus electronic code-breaker was developed at Bletchley Park, England. The Harvard University Mark I or Automatic Sequence Controlled Calculator (partly financed by IBM) became the first program-controlled calculator.
1946	ENIAC (acronym for electronic numerator, integrator, analyser, and computer), the first general purpose, fully electronic digital computer, was completed at the University of Pennsylvania, USA.
1948	Manchester University (England) Mark I, the first stored-program computer, was completed. William Shockley of Bell Laboratories invented the transistor.
1951	Launch of Ferranti Mark I, the first commercially produced computer. Whirlwind, the first real-time computer, was built for the US air-defence system. Grace Murray Hopper of Remington Rand invented the compiler computer program.
1952	EDVAC (acronym for electronic discrete variable computer) was completed at the Institute for Advanced Study, Princeton, USA (by John Von Neumann and others).
1953	Magnetic core memory was developed.
1958	The first integrated circuit was constructed.
1963	The first minicomputer was built by Digital Equipment (DEC). The first electronic calculator was built by Bell Punch Company.
1964	Launch of IBM System/360, the first compatible family of computers. John Kemeny and Thomas Kurtz of Dartmouth College invented BASIC (Beginner's All-purpose Symbolic Instruction Code), a computer language similar to FORTRAN.
1965	The first supercomputer, the Control Data CD6600, was developed.
1971	The first microprocessor, the Intel 4004, was announced.
1974	CLIP–4, the first computer with a parallel architecture, was developed by John Backus at IBM.
1975	Altair 8800, the first personal computer (PC), or microcomputer, was launched.
1981	The Xerox Star system, the first WIMP system (acronym for windows, icons, menus, and pointing devices), was developed. IBM launched the IBM PC.
1984	Apple launched the Macintosh computer.
1985	The Inmos T414 transputer, the first 'off-the-shelf' microprocessor for building parallel computers, was announced.
1988	The first optical microprocessor, which uses light instead of electricity, was developed.
1989	Wafer-scale silicon memory chips, able to store 200 million characters, were launched.
1990	Microsoft released Windows 3, a popular windowing environment for PCs.
1992	Philips launched the CD-I (compact disc-interactive) player, based on CD audio technology, to provide interactive multimedia programs for the home user.
1993	Computers based on the first 64-bit processor, the Intel Pentium, went on sale.

The earliest known example is the abacus. Mechanical devices with sliding scales (similar to the slide rule) date from ancient Greece. In 1642, French mathematician Blaise Pascal built a mechanical adding machine and, in 1671, German mathematician Gottfried Leibniz produced a machine to carry out multiplication. The first mechanical computer, the ◊analytical engine, was designed by British mathematician Charles Babbage 1835. For the subsequent history of computing, see ◊computer.

concave of a surface, curving inwards, or away from the eye. For example, a bowl appears concave when viewed from above. In geometry, a concave polygon is one that has an interior angle greater than 180°. Concave is the opposite of ◊convex.

concave lens converging ◊lens—a parallel beam of light gets wider as it passes through such a lens. A concave lens is thinner at its centre than at the edges.

Common forms include *biconcave* (with both surfaces curved inwards) and *plano-concave* (with one flat surface and one concave). The whole lens may be further curved overall (making a *convexo-concave* or diverging meniscus lens, as in some lenses used for corrective purposes).

concentration in chemistry, the amount of a substance (◊solute) present in a specified amount of a solution. Either amount may be specified as a mass or a volume (liquids only). Common units used are ◊moles per cubic decimetre, grams per cubic decimetre, grams per 100 cubic centimetres, and grams per 100 grams. The term also refers to the process of increasing the concentration of a solution by removing some of the substance (◊solvent) in which the solute is dissolved. In a *concentrated solution*, the solute is present in large quantities. Concentrated brine is around 30% sodium chloride in water; concentrated caustic soda (caustic

RECENT PROGRESS IN COMPUTING

Multimedia: the video game grows up

Buying a computer has never been simpler. Almost every customer, large or small, wants an IBM PC-compatible with Intel 80486SX processor, four megabytes of memory, EISA expansion bus, Super VGA colour screen and Microsoft Windows 3.1 software. There are variations—faster processors, bigger hard discs and screens, more memory—but the market has a working standard: one that has made Microsoft and Intel very rich; one that is being extended, and challenged.

Beyond the desktop

Microsoft, Intel and numerous PC manufacturers want to expand beyond the desktop. Firms like Grid and NCR are selling battery-powered, portable 'pen-driven' computers for people who don't need a keyboard—van drivers making deliveries, or field service engineers. Home users can have interactive CD players with infrared controllers and Microsoft's Modular Windows. Corporate users can have multiprocessor 'servers' using Windows NT (New Technology) to replace minicomputers and small mainframes. RAID (redundant arrays of inexpensive discs) storage systems use multiple PC hard drives instead of large, expensive, mainframe discs.

Modular systems comprising PC processors and drives appeal to a market where large computers have been expensive and incompatible. Since PCs sell in large volumes, intense competition and economies of scale have driven down prices, and propagated a huge range of software.

Many firms have produced reduced instruction set computer (RISC) processors—simpler and faster than traditional designs like Intel's x86 line. Examples include AT&T's Hobbit, DEC's Alpha, the ARM (Acorn Risc Machine), Hewlett-Packard's Precision Architecture, Mips' R4000, and Sun's Sparc (scalable processor architecture). The Sparc chip is the most popular, but the market is confusing and confused. There is a plethora of alternative operating systems exploiting these various chips. One is Go's PenPoint, which AT&T, IBM and others are adopting for pen-driven computers and personal communicators. On minis and workstations, Unix is becoming dominant but, instead of backing the industry-standard Unix System V Release 4, several firms have developed variants, confusing the market and introducing unnecessary incompatibilities. The resulting 'Unix wars' over the last five years have provided Microsoft's NT with its opportunity.

CD-ROM

The biggest development is CD-ROM (compact disc, read-only memory). One disc can store more than 600 megabytes of data, equivalent to 450 standard floppies or 150 million words of text. Prices are low, because the industry benefits from the R&D investment and high-volume production facilities for audio CD, with which CD-ROM is compatible.

A large operating system or suite of programs, occupying from 25 to 80 or more floppy discs, can be packed onto a single CD. 'Multimedia' programs can be produced, with digital data, sound, still and moving images on the same disc. Typical discs include dictionaries that have pictures and can say the words you look up, atlases including national anthems, and encyclopedias with moving diagrams. Computer games are also appearing on CD-ROM.

Suppliers have started 'format wars' that make the VHS–Betamax video battle look trivial. Most CD-ROMs are designed for a specific system: there are products for PCs and Apple Macintoshes, Sun workstations, Commodore Amigas (the CDTV system), Fujitsu's FM-Towns, and Sega Megadrive and NEC games consoles. There are other consumer offerings, including CD-I (compact disc–interactive) from Philips and Sony, Photo CD from Philips and Kodak, the Video Information System from Tandy and Zenith, and Sony's Electronic Book. There is also a recordable format, CD-R.

Millions of CD-ROM drives will be sold for Sega, Nintendo and NEC games consoles, but the emerging standards are CD-I (incorporating Photo CD) and MPC, the Microsoft-backed Multimedia Personal Computer specification based on Windows 3.

The original PC was expensive, and its handling of colour graphics and sound primitive, limiting its appeal to the business market. The MPC is becoming affordable and suitable for use in homes and schools, ushering in a new age of electronic books in which people can speak, cows can moo, cartoons can be animated, and you can display the text any size you like. CDs of plays, operas and musicals can have moving video clips plus the libretto and score. Viewers will become interactive participants, making 'director's cuts' of films, steering plot-lines or choosing their own endings. Offered such power, will anyone want to resist?

Jack Schofield

vector graphics display before and after transformation

before   after

raster graphics display before and after transformation

before   after

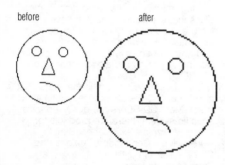

computer graphics

liquor) is around 40% sodium hydroxide; and concentrated sulphuric acid is 98% acid.

concentration gradient change in the concentration of a substance from one area to another.

Particles such as sugar molecules in a fluid appear move over time so that they become evenly distributed throughout the fluid. In particular, they seem to move from an area of high concentration to an area of low concentration; that is, they diffuse along the concentration gradient (see ◊diffusion).

This explains why oxygen in the lungs will diffuse into the blood supply. The oxygen molecules are more concentrated in the lungs than they are in the capillaries surrounding the ◊alveoli (air sacs). As it diffuses along the concentration gradient, oxygen will tend to pass into the blood. Gas exchange therefore depends on the maintenance of a concentration gradient, so that oxygen will continue to diffuse across the respiratory surface.

concentric circles two or more circles that share the same centre.

conceptacle flask-shaped cavities found in the swollen tips of certain brown seaweeds, notably the wracks, *Fucus*. The gametes are formed within them and released into the water via a small pore in the conceptacle, known as an ostiole.

Concorde the only supersonic airliner, which cruises at Mach 2, or twice the speed of sound, about 2,170 kph/1,350 mph. Concorde, the result of Anglo-French cooperation, made its first flight 1969 and entered commercial service seven years later. It is 62 m/202 ft long and has a wing span of nearly 26 m/84 ft.

concrete building material composed of cement, stone, sand, and water. It has been used since Egyptian and Roman times. During the 20th century, it has been increasingly employed as an economical alternative to materials such as brick and wood.

history
c. 5600 BC Earliest discovered use of concrete at Lepenski Vir, Yugoslavia (hut floors in Stone Age village).
2500 BC Concrete used in Great Pyramid at Giza by Egyptians.
2nd century BC Romans accidentally discovered the use of lime and silicon/alumina to produce 'pozzolanic' cement.
AD 127 Lightweight concrete (using crushed pumice as aggregate) used for walls of Pantheon, Rome.
Medieval times Concrete used for castles (infill in walls) and cathedrals (largely foundation work).
1756 John Smeaton produced the first high-quality cement since Roman times (for rebuilding of Eddystone lighthouse, England).
1824 Joseph Aspdin patented Portland cement in Britain.
1854 William Wilkinson patented reinforced concrete in Britain–first successful use in a building.
1880s First continuous-process rotary cement kiln installed (reducing costs of manufacturing cement).
1898 François Hennébique: first multistorey reinforced concrete building in Britain (factory in Swansea).
1926 Eugène Freysinnet began experiments on prestressed concrete in France.
1930s USA substituted concrete for limestone during federal building projects of the Great Depression; also used extensively for pavements, roadbeds, bridge approaches, dams, and sports facilities (stadiums, swimming pools, tennis courts, playgrounds).
1940s–1950s Much use of poured concrete to rebuild war-torn cities of Europe and the Middle East.
1960s Widespread use of concrete in industrialized countries as an economical house and office building material instead of traditional materials.

concurrent lines two or more lines passing through a single point; for example, the diameters of a circle are all concurrent at the centre of the circle.

condensation in organic chemistry, a reaction in which two organic compounds combine to form a larger molecule, accompanied by the removal of a smaller molecule (usually water). This is also known as an addition–elimination reaction. Polyamides (such as nylon) and polyesters (such as Terylene) are made by condensation ◊polymerization.

condensation number in physics, the ratio of the number of molecules condensing on a surface to the total number of molecules touching that surface.

condensation polymerization ◊polymerization reaction in which one or more monomers, with more than one reactive functional group, combine to form a polymer with the elimination of water or another small molecule.

condenser in optics, a ◊lens or combination of lenses with a short focal length used for concentrating a light source onto a small area, as used in a slide projector or microscope substage lighting unit. A condenser can also be made using a concave mirror.

condenser in electronic circuits, another name for a ◊capacitor.

conditioning in psychology, two major principles of behaviour modification. In *classical conditioning*, described by Ivan Pavlov, a new stimulus can evoke an automatic response by being repeatedly associated with a stimulus that naturally provokes a response. For example, the sound of a bell repeatedly associated with food will eventually trigger salivation, even if sounded without food being presented. In *operant conditioning*, described by US psychologists Edward Lee Thorndike (1874–1949) and Burrhus Frederic Skinner, the frequency of a voluntary response can be increased by following it with a reinforcer or reward.

condom or *sheath* or *prophylactic* barrier contraceptive, made of rubber, which fits over an erect penis and holds in the sperm produced by ejaculation. It is an effective means of preventing pregnancy if used carefully, preferably with a ◊spermicide. A condom with spermicide is 97% effective; one without spermicide is 85% effective. Condoms also give protection against sexually transmitted diseases, including AIDS.

conductance ability of a material to carry an electrical current, usually given the symbol G. For a direct current, it is the reciprocal of ◊resistance: a conductor of resistance R has a conductance of $1/R$. For an alternating current, conductance is the resistance R divided by the ◊impedance Z: $G = R/Z$. Conductance was formerly expressed in reciprocal ohms (or mhos); the SI unit is the siemens (S).

conduction, electrical flow of charged particles through a material that gives rise to electric current. Conduction in metals involves the flow of negatively charged free ◊electrons. Conduction in gases and some liquids involves the flow of ◊ions that carry positive charges in one direction and negative charges in the other. Conduction in a semiconductor such as silicon involves the flow of electrons and positive holes.

conduction, heat flow of heat energy through a material without the movement of any part of the material itself (compare ◊conduction, electrical). Heat energy is present in all materials in the form of the kinetic energy of their vibrating molecules, and may be conducted from one molecule to the next in the form of this mechanical vibration. In the case of metals, which are particularly good conductors of heat, the free electrons within the material carry heat around very quickly.

conductor any material that conducts heat or electricity (as opposed to an insulator, or nonconductor). A good conductor has a high electrical or heat conductivity, and is generally a substance rich in free electrons such as a metal. A poor conductor (such as the nonmetals glass and porcelain) has few free electrons. Carbon is exceptional in being nonmetallic and yet (in some of its forms) a relatively good conductor of heat and electricity. Substances such as silicon and germanium, with intermediate conductivities that are improved by heat, light, or voltage, are known as ◊semiconductors.

cone in geometry, a solid or surface consisting of the set of all straight lines passing through a fixed point (the vertex) and the points of a circle or ellipse whose plane does not contain the vertex.

A circular cone of perpendicular height, with its apex above the centre of the circle, is known as a *right circular cone*; it is generated by rotating an isosceles triangle or framework about its line of symmetry. A right circular cone of perpendicular height h and base of radius r has a volume $V = \frac{1}{3}\pi r^2 h$. The distance from the edge of the base of a cone to the vertex is called the slant height. In a right circular cone of slant height l, the curved surface area is $\pi r l$, and the area of the base is πr^2. Therefore the total surface area $A = \pi r l + \pi r^2 = \pi r(l + r)$.

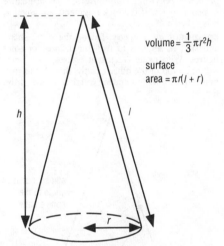

$$\text{volume} = \frac{1}{3}\pi r^2 h$$

$$\text{surface area} = \pi r(l + r)$$

cone *The volume and surface area of a cone are given by formulae involving a few simple dimensions.*

cone in botany, the reproductive structure of the conifers and cycads; also known as a ◊strobilus. It consists of a central axis surrounded by numerous, overlapping, scalelike, modified leaves (sporophylls) that bear the reproductive organs. Usually there are separate male and female cones, the former bearing pollen sacs containing pollen grains, and the larger female cones bearing the ovules that contain the ova or egg cells. The pollen is carried from male to female cones by the wind (◊anemophily). The seeds develop within the female cone and are released as the scales open in dry atmospheric conditions, which favour seed dispersal.

In some groups (for example, the pines) the cones take two or even three years to reach maturity. The cones of junipers have fleshy cone scales that fuse to form a berrylike structure. One group of ◊angiosperms, the alders, also bear conelike structures; these are the woody remains of the short female catkins, and they contain the alder ◊fruits.

In zoology, cones are a type of light-sensitive cell found in the retina of the ◊eye.

configuration in chemistry, the arrangement of atoms in a molecule or of electrons in atomic orbitals.

congenital disease in medicine, a disease that is present at birth. It is not necessarily genetic in origin; for example, congenital herpes may be acquired by the baby as it passes through the mother's birth canal.

conglomerate in mineralogy, coarse clastic ◊sedimentary rock, composed of rounded fragments (clasts) of pre-existing rocks cemented in a finer matrix, usually sand.

The fragments in conglomerates are pebble-to boulder-sized, and the rock can be regarded as the lithified equivalent of gravel. A ◊bed of conglomerate is often associated with a break in a sequence of rock beds (an unconformity), where it marks the advance of the sea over an old eroded landscape. An *oligomict conglomerate* contains one type of pebble; a *polymict conglomerate* has a mixture of pebble types. If the rock fragments are angular, it is called a ◊breccia.

congruent in geometry, having the same shape and size, as applied to two-dimensional or solid figures. With plane congruent figures, one figure will fit on top of the other exactly, though this may first require rotation and/or rotation of one of the figures.

conic section curve obtained when a conical surface is intersected by a plane. If the intersecting plane cuts both extensions of the cone, it yields a ◊hyperbola; if it is parallel to the side of the cone, it produces a ◊parabola. Other intersecting planes produce ◊circles or ◊ellipses.

The Greek mathematician Apollonius wrote eight books with the title *Conic Sections*, which superseded previous work on the subject by Aristarchus and Euclid.

conidium (plural *conidia*) asexual spore formed by some fungi at the tip of a specialized ◊hypha or conidiophore. The conidiophores grow erect, and cells from their ends round off and separate into conidia, often forming long chains. Conidia easily become detached and are dispersed by air movements.

conjugate in mathematics, a term indicating that two elements are connected in some way; for example, $(a + ib)$ and $(a − ib)$ are conjugate complex numbers.

conjugate angles two angles that add up to 360°.

conjugation in biology, the bacterial equivalent of sexual reproduction. A fragment of the ◊DNA from one bacterium is passed along a thin tube, the pilus, into the cell of another bacterium.

conjugation in organic chemistry, the alternation of double (or triple) and single carbon–carbon bonds in a molecule—for example,

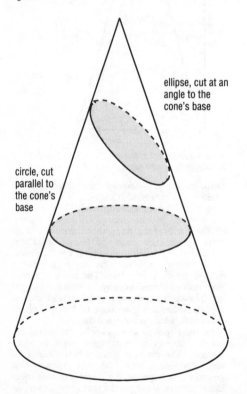

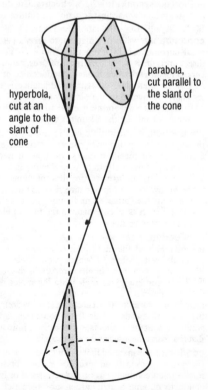

circle, cut parallel to the cone's base

ellipse, cut at an angle to the cone's base

hyperbola, cut at an angle to the slant of cone

parabola, cut parallel to the slant of the cone

conic section *The four types of curve that may be obtained by cutting a single or double right-circular cone with a plane (two-dimensional surface).*

in penta-1,3-diene, $H_2C=CH-CH=CH-CH_3$. Conjugation imparts additional stability as the double bonds are less reactive than isolated double bonds.

conjunction in astronomy, the alignment of two celestial bodies as seen from Earth. A ◊superior planet (or other object) is in conjunction when it lies behind the Sun. An ◊inferior planet (or other object) comes to *inferior conjunction* when it passes between the Earth and the Sun; it is at *superior conjunction* when it passes behind the Sun. *Planetary conjunction* takes place when a planet is closely aligned with another celestial object, such as the Moon, a star, or another planet.

Because the orbital planes of the inferior planets are tilted with respect to that of the Earth, they usually pass either above or below the Sun at inferior conjunction. If they line up exactly, a ◊transit will occur.

conjunctiva membrane covering the vertebrate ◊eye. It is continuous with the epidermis of the eyelids, and lies on the surface of the cornea.

connective tissue in animals, tissue made up of a noncellular substance, the ◊extracellular matrix, in which some cells are embedded. Skin, bones, tendons, cartilage, and adipose tissue (fat) are the main connective tissues. There are also small amounts of connective tissue in organs such as the brain and liver, where they maintain shape and structure.

conservation in the life sciences, action taken to protect and preserve the natural world, usually from pollution, overexploitation, and other harmful features of human activity. The late 1980s saw a great increase in public concern for the environment, with membership of conservation groups, such as ◊Friends of the Earth, rising sharply. Globally the most important issues include the depletion of atmospheric ozone by the action of chlorofluorocarbons (CFCs), the build-up of carbon dioxide in the atmosphere (thought to contribute to an intensification of the ◊greenhouse effect), and the destruction of the tropical rainforests (see ◊deforestation).

In the UK the conservation debate has centred on water quality, road-building schemes, the safety of nuclear power, and animal rights.

conservation of energy in chemistry, the principle that states that in a chemical reaction, the total amount of energy in the system remains unchanged.

For each component there may be changes in energy due to change of physical state, changes in the nature of chemical bonds, and either an input or output of energy. However, there is no net gain or loss of energy.

conservation of mass in chemistry, the principle that states that in a chemical reaction the sum of all the masses of the substances involved in the reaction (reactants) is equal to the sum of all of the masses of the substances produced by the reaction (products) — that is, no matter is gained or lost.

conservative margin in plate tectonics, a region on the Earth's surface in which one plate slides past another. An example is the San Andreas Fault, California, where the movement of the plates is

irregular and sometimes takes the form of sudden jerks, which cause the ◊earthquakes common in the San Francisco–Los Angeles area.

constant in mathematics, a fixed quantity or one that does not change its value in relation to ◊variables. For example, in the algebraic expression $y^2 = 5x - 3$, the numbers 3 and 5 are constants. In physics, certain quantities are regarded as universal constants, such as the speed of light in a vacuum.

constantan or *eureka* high-resistance alloy of approximately 40% nickel and 60% copper with a very low coefficient of ◊thermal expansion (measure of expansion on heating). It is used in electrical resistors.

constant composition, law of in chemistry, the law that states that the proportions of the amounts of the elements in a pure compound are always the same and are independent of the method by which the compound was produced.

constellation one of the 88 areas into which the sky is divided for the purposes of identifying and naming celestial objects. The first constellations were simple, arbitrary patterns of stars in which early civilizations visualized gods, sacred beasts, and mythical heroes.

The constellations in use today are derived from a list of 48 known to the ancient Greeks, who inherited some from the Babylonians. The current list of 88 constellations was adopted by the International Astronomical Union, astronomy's governing body, in 1930.

constructive margin in plate tectonics, a region in which two plates are moving away from each other. Magma, or molten rock, escapes to the surface along this margin to form new crust, usually in the form of a ridge. Over time, as more and more magma reaches the surface, the sea floor spreads — for example, the upwelling of magma at the Mid-Atlantic Ridge causes the floor of the Atlantic Ocean to grow at a rate of about 5 cm/2 in a year.

◊Volcanoes can form along the ridge and islands may result (for example, Iceland was formed in this way). Eruptions at constructive plate margins tend to be relatively gentle; the lava produced cools to form ◊basalt.

contact process the main industrial method of manufacturing the chemical ◊sulphuric acid. Sulphur dioxide (produced by burning sulphur) and air are passed over a hot (450°C) ◊catalyst of vanadium(V) oxide. The sulphur trioxide produced is absorbed in concentrated sulphuric acid to make fuming sulphuric acid (oleum), which is then diluted with water to give concentrated sulphuric acid (98%). Unreacted gases are recycled.

continent any one of the seven large land masses of the Earth, as distinct from the oceans. They are Asia, Africa, North America, South America, Europe, Australia, and Antarctica. Continents are constantly moving and evolving (see ◊plate tectonics). A continent does not end at the coastline; its boundary is the edge of the shallow continental shelf, which may extend several hundred kilometres or miles out to sea.

At the centre of each continental mass lies a shield or ◊craton, a deformed mass of old ◊meta-

The race to prevent mass extinction

About a million different living species have been identified so far. Recent studies in tropical forests—where biodiversity is greatest—suggest the true figure is nearer 30 million. Most are animals, and most of those are insects. Because the tropical forests are threatened, at least half the animal species could become extinct within the next century.

There is conflict within the conservation movement. Some believe that *habitats* (the places where animals and plants live) should be conserved; others prefer to concentrate upon individual *species*. Both approaches have strengths and weaknesses, and they must operate in harmony.

Habitat protection

Habitat protection has obvious advantages. Many species benefit if land is preserved. Animals need somewhere to live; unless the habitat is preserved it may not be worth saving the individual animal. Habitat protection *seems* cheap; for example, tropical forest can often be purchased for only a few dollars per hectare. But there are difficulties. Even when a protected area is designated a 'national park', its animals may not be safe. All five remaining species of rhinoceros are heavily protected in the wild, but are threatened by poaching. Early in 1991 Zimbabwe had 1,500 black rhinos—the world's largest population. Patrols of game wardens shoot poachers on sight. Yet by late 1992, 1,000 of the 1,500 had been poached. In many national parks worldwide, the habitat is threatened by the local farmers' need to graze their cattle.

Computer models and field studies show that wild populations need several hundred individuals to be viable. Smaller populations will eventually go extinct in the wild, because of accidents to key breeding individuals, or epidemics. The big predators need vast areas. One tiger may command hundreds of square kilometres; a viable population needs an area as big as Wales or Holland. Only one of the world's five remaining subspecies of tiger—a population of Bengals in India—occupies an area large enough to be viable. All the rest seem bound to die out.

Mosaic

Ecologists now emphasize the concept of *mosaic*. All animals need different things from their habitat; a failure of any one is disastrous. Giant pandas feed mainly on bamboo, but give birth in old hollow trees—of which there is a shortage. Birds commonly roost in one place, but feed in special areas far way. Nature reserves must either contain all essentials for an animal's life, or else allow access to such areas elsewhere. For many animals in a reserve, these conditions are not fulfilled. Hence year by year, after reserves are created, species go extinct.

Interest is increasing in *captive breeding*, carried out mainly by the world's 800 zoos. Their task is formidable; each captive species should include several hundred individuals. Zoos maintain such numbers through *cooperative breeding*, organized regionally and coordinated by the Captive Breeding Specialist Group or the World Conservation Union, based in Minneapolis, Minnesota. Each programme is underpinned by a studbook, showing which individuals are related to which.

Genetic diversity

Breeding for conservation is different from breeding for livestock improvement. Livestock breeders breed *uniform* creatures by selecting animals conforming to some prescribed ideal. Conservation breeders maintain *maximum genetic diversity* by encouraging every individual to breed, including those reluctant to breed in captivity; by equalizing family size, so one generation's genes are all represented in the next; and by swapping individuals between zoos, to prevent inbreeding.

Cooperative breeding programmes are rapidly diversifying; by the year 2000 there should be several hundred. They can only make a small impression on the 15 million endangered species, but they can contribute greatly to particular groups of animals, especially the land vertebrates—mammals, birds, reptiles and amphibians. There are 24,000 species of land vertebrate, of which 2,000 probably require captive breeding to survive. Zoos could save all 2,000; that would be a great contribution.

Captive breeding is not intended to establish 'museum' populations, but to provide a temporary 'lifeboat'. Things are hard for wild animals, but over the next few decades, it should be possible to establish more, safe national parks. Arabian oryx, California condor, black-footed ferret, red wolf, and Mauritius kestrel are among the creatures saved from extinction by captive breeding and returned to the wild. In the future, we can expect to see many more.

Colin Tudge

the constellations

constellation	popular name	constellation	popular name
Andromeda	—	Lacerta	Lizard
Antilla	Airpump	Leo	Lion
Apus	Bird of Paradise	Leo Minor	Little Lion
Aquarius	Water-bearer	Lepus	Hare
Aquila	Eagle	Libra	Balance
Ara	Altar	Lupus	Wolf
Aries	Ram	Lynx	—
Auriga	Charioteer	Lyra	Lyre
Boötes	Herdsman	Mensa	Table
Caelum	Chisel	Microscopium	Microscope
Camelopadalis	Giraffe	Monoceros	Unicorn
Cancer	Crab	Musca	Southern Fly
Canes Venatici	Hunting Dogs	Norma	Rule
Canis Major	Great Dog	Octans	Octant
Canis Minor	Little Dog	Ophiuchus	Serpent-bearer
Capricornus	Sea-goat	Orion	—
Carina	Keel	Pavo	Peacock
Cassiopeia	—	Pegasus	Flying Horse
Centaurus	Centaur	Perseus	—
Cepheus	—	Phoenox	Phoenix
Cetus	Whale	Pictor	Painter
Chamaeleon	Chameleon	Pisces	Fishes
Circinus	Compasses	Piscis Austrinus	Southern Fish
Columba	Dove	Puppis	Poop
Coma Berenices	Bernice's Hair	Pyxis	Compass
Corona Australis	Southern Crown	Reticulum	Net
Corona Borealis	Northern Crown	Sagitta	Arrow
Corvus	Crow	Sagittarius	Archer
Crater	Cup	Scorpius	Scorpion
Crux	Southern Cross	Sculptor	—
Cygnus	Swan	Scutum	Shield
Delphinus	Dolphin	Serpens	Serpent
Dorado	Goldfish	Sextans	Sextant
Draco	Dragon	Taurus	Bull
Equuleus	Foal	Telescopium	Telescope
Eridanus	River	Triangulum	Triangle
Fornax	Furnace	Triangulum Australe	Southern Triangle
Gemini	Twins	Tuscana	Toucan
Grus	Crane	Ursa Major	Great Bear
Hercules	—	Ursa Minor	Little Bear
Horologium	Clock	Vela	Sails
Hydra	Watersnake	Virgo	Virgin
Hydrus	Little Snake	Volans	Flying Fish
Indus	Indian	Vulpecula	Fox

morphic rocks dating from Precambrian times. The shield is thick, compact, and solid (the Canadian Shield is an example), having undergone all the mountain-building activity it is ever likely to, and is usually worn flat. Around the shield is a concentric pattern of fold mountains, with older ranges, such as the Rockies, closest to the shield, and younger ranges, such as the coastal ranges of North America, farther away. This general concentric pattern is modified when two continental masses have drifted together and they become welded with a great mountain range along the join, the way Europe and N Asia are joined along the Urals. If a continent is torn apart, the new continental edges have no fold mountains; for instance, South America has fold mountains (the Andes) along its western flank, but none along the east where it tore away from Africa 200 million years ago.

continental drift in geology, theory that, about 250–200 million years ago, the Earth consisted of a single large continent (◊Pangaea), which subsequently broke apart to form the continents known today. The theory was proposed 1912 by German meteorologist Alfred Wegener, but such vast continental movements could not be satisfactorily explained until the study of ◊plate tectonics in the 1960s.

The term 'continental drift' is not strictly correct, since land masses do not drift through the oceans.

ocean ridge formed where two plates move away from each other

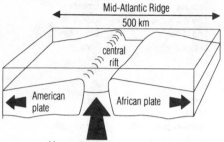

Magma escapes though the
ocean ridge, creating new crust
and causing the ocean floor to spread

constructive margin

The continents form part of a plate, and the amount of crust created at divergent plate margins must equal the amount of crust destroyed at subduction zones.

continental rise the portion of the ocean floor rising gently from the abyssal plain toward the steeper continental slope. The continental rise is a depositional feature formed from sediments transported down the slope mainly by turbidity currents. Much of the continental rise consists of coalescing submarine alluvial fans bordering the continental slope.

continental shelf the submerged edge of a continent, a gently sloping plain that extends into the ocean. It typically has a gradient of less than 1°. When the angle of the sea bed increases to 1°–5° (usually several hundred kilometres away from land), it becomes known as the *continental slope*.
The continental shelf around the UK contains large reserves of oil and gas.

continental slope sloping, submarine portion of a continent. It extends downward from the edge of the continental shelf. In some places, such as south of the Aleutian Islands of Alaska, continental slopes extend directly to the ocean deeps or abyssal plain. In others, such as the east coast of North America, they grade into the gentler continental rises that in turn grade into the abyssal plains.

continuity in mathematics, property of functions of a real variable that have an absence of 'breaks'. A function f is said to be continuous at a point a if $\lim f(x) = f(a)$.

continuous data data that can take any of an infinite number of values between whole numbers and so may not be measured completely accurately. This type of data contrasts with ◊discrete data, in which the variable can only take one of a finite set of values. For example, the sizes of apples on a tree form continuous data, whereas the numbers of apples form discrete data.

continuous variation the slight difference of an individual character, such as height, across a sample of the population. Although there are very tall and very short humans, there are also many people with an intermediate height. The same applies to weight. Continuous variation can result from the genetic make-up of a population, or from environmental influences, or from a combination of the two.

continuum in mathematics, a ◊set that is infinite and everywhere continuous, such as the set of points on a line.

contraceptive any drug, device, or technique that prevents pregnancy. The contraceptive pill (the ◊Pill) contains female hormones that interfere with egg production or the first stage of pregnancy. The 'morning-after' pill can be taken up to 72 hours after unprotected intercourse. Barrier contraceptives include ◊condoms (sheaths) and ◊diaphragms, also called caps or Dutch caps; they prevent the sperm entering the cervix (neck of the womb). ◊Intrauterine devices, also known as IUDs or coils, cause a slight inflammation of the lining of the womb; this prevents the fertilized egg from becoming implanted.

Other contraceptive methods include ◊sterilization (women) and ◊vasectomy (men); these are usually nonreversible. 'Natural' methods include withdrawal of the penis before ejaculation (coitus

Upper Carboniferous period

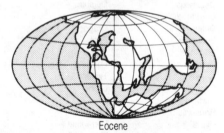

Eocene

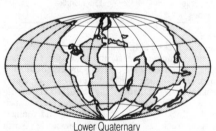

Lower Quaternary

continental drift *The drifting continents. The continents are slowly shifting their positions, driven by fluid motion beneath the Earth's crust. Over 200 million years ago, there was a single large continent called Pangaea. By 200 million years ago, the continents had started to move apart. By 50 million years ago, the continents were approaching their present positions.*

interruptus), and avoidance of intercourse at the time of ovulation (◊rhythm method). These methods are unreliable and normally only used on religious grounds. A new development is a sponge impregnated with spermicide that is inserted into the vagina. The use of any contraceptive (birth control) is part of family planning.

The effectiveness of a contraceptive method is often given as a percentage. To say that a method has 95% effectiveness means that, on average, out of 100 healthy couples using that method for a year, 95 will not conceive.

contractile root in botany, a thickened root at the base of a corm, bulb, or other organ that helps position it at an appropriate level in the ground. Contractile roots are found, for example, on the corms of plants of the genus *Crocus*. After they have become anchored in the soil, the upper portion contracts, pulling the plant deeper into the ground.

contractile vacuole tiny organelle found in many single-celled fresh-water organisms. It slowly fills with water, and then contracts, expelling the water from the cell.

Fresh-water protozoa such as *Amoeba* absorb water by the process of ◊osmosis, and this excess must be eliminated. The rate of vacuole contraction slows as the external salinity is increased, because the osmotic effect weakens; marine protozoa do not have a contractile vacuole.

control the process by which a tissue, an organism, a population, or an ecosystem maintains itself in a balanced, stable state. Blood sugar must be kept at a stable level if the brain is to function properly, and this steady-state is maintained by an interaction between the liver, the hormone insulin, and a detector system in the pancreas.

In the ecosystem, the activities of the human race are endangering the balancing mechanisms associated with the atmosphere in general and the ◊greenhouse effect in particular.

control bus in computing, the electrical pathway, or ◊bus, used to communicate control signals.

control experiment essential part of a scientifically valid experiment, designed to show that the factor being tested is actually responsible for the effect observed. In the control experiment all factors, apart from the one under test, are exactly the same as in the test experiments, and all the same measurements are carried out. In drug trials, a placebo (a harmless substance) is given alongside the substance being tested in order to compare effects.

control total in computing, a ◊validation check in which an arithmetic total of a specific field from a group of records is calculated. This total is input together with the data to which it refers. The program recalculates the control total and compares it with the one entered to ensure that no entry errors have been made.

control unit the component of the ◊central processing unit that decodes, synchronizes, and executes program instructions.

convection heat energy transfer that involves the movement of a fluid (gas or liquid). According to ◊kinetic theory, molecules of fluid in contact with the source of heat expand and tend to rise within the bulk of the fluid. Less energetic, cooler molecules sink to take their place, setting up convection currents. This is the principle of natural convection in many domestic hot-water systems and space heaters.

conventional current direction in which an electric current is considered to flow in a circuit. By convention, the direction is that in which positive-charge carriers would flow—from the positive terminal of a cell to its negative terminal. In circuit diagrams, the arrows shown on symbols for components such as diodes and transistors point in the direction of conventional current flow.

convectional rainfall rainfall associated with hot climates, resulting from the uprising of convection currents of warm air. Air that has been warmed by the extreme heat of the ground surface rises to great heights and is abruptly cooled. The water vapour carried by the air condenses and rain falls heavily. Convectional rainfall is usually accompanied by a ◊thunderstorm.

Convectional rainfall occurs occasionally in the UK during hot summers.

convergence in mathematics, the property of a series of numbers in which the difference between consecutive terms gradually decreases. The sum of a converging series approaches a limit as the number of terms tends to ◊infinity.

convergent evolution in biology, the independent evolution of similar structures in species (or other taxonomic groups) that are not closely related, as a result of living in a similar way. Thus, birds and bats have wings, not because they are descended from a common winged ancestor, but because their respective ancestors independently evolved flight.

converse in mathematics, the reversed order of a conditional statement; the converse of the statement 'if a, then b' is 'if b, then a'. The converse does not always hold true; for example, the converse of 'if $x = 3$, then $x^2 = 9$' is 'if $x^2 = 9$, then $x = 3$', which is not true, as x could also be -3.

convertiplane vertical takeoff and landing craft (VTOL) with rotors on its wings that spin horizontally for takeoff, but tilt to spin in a vertical plane for forward flight.

At takeoff it looks like a two-rotor helicopter, with both rotors facing skywards. As forward speed is gained, the rotors tilt slowly forward until they are facing directly ahead. There are several different forms of convertiplane. The LTV-Hillier-Ryan XC-142, designed in the USA, had wings, carrying the four engines and propellers, that rotated. The German VC-400 had two rotors on each of its wingtips. Neither of these designs went into production. A Boeing and Bell design under test in 1992, the V-22, uses a pair of tilting engines, with propellers 11.5 m/38 ft across, mounted at the end of the wings. It is intended eventually to carry about 50 passengers direct to city centres. Doubt was cast over the future of the V-22 in 1992 when a prototype crashed in the Potomac River in Virginia. Another prototype had crashed the previous year.

convex of a surface, curving outwards, or towards the eye. For example, the outer surface of a ball appears convex. In geometry, the term is used to

Lemons, crocodiles and pigs

Birth control, in one form or another, has a long history. That well-known lover Casanova (1725–1798) is reputed to have used a lemon as a spermicide, with less than complete success. The ancient Egyptians used crocodile faeces. The sheath was probably introduced by Gabriele Fallopia, the famous Italian anatomist (1523–1562), but mainly as a prophylactic against venereal disease. Knowledge of human genital systems and their physiology, particularly the female system, remained rudimentary until the beginning of the 20th century. The most reliable form of birth control up to this point was simple abstinence.

The population of the world in 1830 was 1 billion and by 1929 had increased to 2 billion. However, in the period following the carnage of World War I, people were more concerned to repopulate than to advance new methods of birth control. There were legal constraints on birth control as well. US social reformer Margaret Sanger (1883–1966) opened the first birth control clinic in 1916 in Brooklyn, New York, but it was closed by the police, and Sanger imprisoned. So, in addition to the necessary scientific advances, before successful birth control could be practised there had to be changes both in the law and in social awareness.

Preventing ovulation
The key step forward in scientific understanding came in the 1950s when US physiologist Gregory Pincus (1903–) discovered that the steroid hormone progesterone, found in greater concentrations during pregnancy, is responsible for the prevention of ovulation at that time.

The way had been paved in the 1920s when German physiologist Ludwig Haberlandt (1885–1932) conducted a crucial experiment. Starting from the idea that pregnancy somehow triggered the secretion of anti-ovulatory substances, he demonstrated that, by transplanting the ovaries from pregnant rabbits and rats to non-pregnant females, he could induce a temporary sterility. He also found that simple ovarian extracts could achieve the same result. He suggested that specific hormones secreted during pregnancy were responsible.

The modern study of sex hormones began with US endocrinologist Edgar Allen (1892–1943) who discovered oestrogen and showed experimentally that it induced the physiological changes normally found in the oestrous cycle. Allen and his colleagues extracted and crystallized a few milligrams of a hormone from the corpus luteum (part of the ovary) of the sow. To do this, they had to process the carcasses of several hundred sows. They suggested the name progesterone for the hormone.

A commercial source of progesterone
The story now moves to the Mexican jungle. In 1941, the US Bureau of Plant Industry received reports of the contraceptive properties of plant infusions used by the Indian women of Nevada. A US scientist, Russell Marker, found it difficult to pursue contraceptive studies in the prevailing climate of public opinion, and decided he would be better off in Mexico. Benefiting from local folklore, he investigated the wild yam as a possible source of progesterone.

After collecting considerable quantities of this tuber in the jungle, Marker succeeded in a remarkably short space of time in synthesizing 2 kg/4.4 lb of progesterone. It is related that he appeared one day in the offices of a small Mexican pharmaceutical firm and, placing four full 1 lb jars of progesterone on the manager's desk, enquired if he was interested in its industrial manufacture!

The pill arrives
There the matter rested, until the work of Gregory Pincus made it possible to conduct a series of trials, one among the women of Port-au-Prince in Haiti, and two in Puerto Rico. The trials gave conclusive results, and the firm of Searle, in Chicago, USA, was then able to produce the first contraceptive pill, Enovid.

Progesterone is now known to play a vital role in maintaining the normal course of pregnancy. In high doses, however, it acts to prevent ovulation. By taking a contraceptive pill, in which the amounts of progesterone and oestrogen activity are carefully balanced, the female body behaves as if conception has occurred. Other types of pill operate only through progesterone activity.

The population of the world in 1975 was 4 billion and in the year 2000 it will top 6 billion. Most of this staggering increase will happen in countries other than North America and Europe. The population is relatively stable in the developed countries where, in some instances, the birth rate has been cut by up to 40%. This is attributable both to advances in the scientific knowledge of birth and its control made during this century and, equally, to an increase in overall wealth.

bird wing

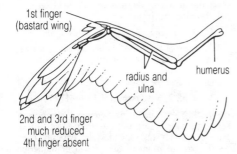

1st finger
(bastard wing)

radius and
ulna

humerus

2nd and 3rd finger
much reduced
4th finger absent

bat wing

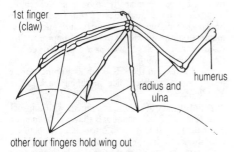

1st finger
(claw)

radius and
ulna

humerus

other four fingers hold wing out

convergent evolution *Convergent evolution produced the
superficially similar wings of the bird and the bat. However,
closer examination shows that in the bird's wing the 'fingers' of
the limb have been lost. In the bat's wing, the fingers are strongly
developed. In other words, similar results have evolved from very
different starting points.*

describe any polygon possessing no interior angle
greater than 180°. Convex is the opposite of
◊concave.

convex lens converging ◊lens—that is, a parallel
beam of light passing through it converges and is
eventually brought to a focus; it can therefore produce a real image on a screen. Such a lens is wider
at its centre than at the edges.

Common forms include *biconvex* (with both
surfaces curved outwards) and *plano-convex*
(with one flat surface and one convex). The whole
lens may be further curved overall, making a
concavo-convex or converging meniscus lens, as
in some lenses used in corrective eyewear.

coordinate in geometry, a number that defines
the position of a point relative to a point or axis
(reference line). ◊Cartesian coordinates define a
point by its perpendicular distances from two or
more axes drawn through a fixed point mutually at
right angles to each other; ◊polar coordinates
define a point in a plane by its distance from a
fixed point and direction from a fixed line.

coordinate geometry or *analytical geometry*
system of geometry in which points, lines, shapes,
and surfaces are represented by algebraic
expressions. In plane (two-dimensional) coordinate
geometry, the plane is usually defined by two axes
at right angles to each other, the horizontal *x*-axis

and the vertical *y*-axis, meeting at O, the origin. A
point on the plane can be represented by a pair of
◊Cartesian coordinates, which define its position
in terms of its distance along the *x*-axis and along
the *y*-axis from O. These distances are respectively
the *x* and *y* coordinates of the point.

Lines are represented as equations; for example,
$y = 2x + 1$ gives a straight line, and $y = 3x^2 + 2x$
gives a ◊parabola (a curve). The graphs of varying
equations can be drawn by plotting the coordinates
of points that satisfy their equations, and joining
up the points. One of the advantages of coordinate
geometry is that geometrical solutions can be
obtained without drawing but by manipulating
algebraic expressions. For example, the coordinates
of the point of intersection of two straight lines can
be determined by finding the unique values of *x*
and *y* that satisfy both of the equations for the lines,
that is, by solving them as a pair of ◊simultaneous
equations. The curves studied in simple coordinate
geometry are the ◊conic sections (circle, ellipse,
parabola, and hyperbola), each of which has a
characteristic equation.

copepod ◊crustacean of the subclass Copepoda,
mainly microscopic and found in plankton.

Cartesian coordinates

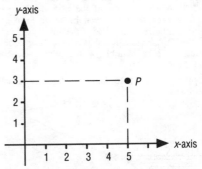

y-axis

x-axis

the Cartesian coordinates of *P* are (5,3)

polar coordinates

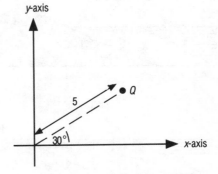

y-axis

5

30°

x-axis

the polar coordinates of *Q* are (5,30°)

coordinate

coplanar in geometry, describing lines or points that all lie in the same plane.

copper orange-pink, very malleable and ductile, metallic element, symbol Cu (from Latin *cuprum*), atomic number 29, relative atomic mass 63.546. It is used for its durability, pliability, high thermal and electrical conductivity, and resistance to corrosion.

It was the first metal used systematically for tools by humans; when mined and worked into utensils it formed the technological basis for the Copper Age in prehistory. When alloyed with tin it forms bronze, which strengthens the copper, allowing it to hold a sharp edge; the systematic production and use of this was the basis for the prehistoric Bronze Age. Brass, another hard copper alloy, includes zinc. The element's name comes from the Greek for Cyprus (*Kyprios*), where copper was mined.

copper ore any mineral from which copper is extracted, including native copper, Cu; chalcocite, Cu_2S; chalcopyrite, $CuFeS_2$; bornite, Cu_5FeS_4; azurite, $Cu_3(CO_3)_2(OH)_2$; malachite, $Cu_2CO_3(OH)_2$; and chrysocolla, $CuSiO_3.2H_2O$.

Native copper and the copper sulphides are usually found in veins associated with igneous intrusions. Chrysocolla and the carbonates are products of the weathering of copper-bearing rocks. Copper was one of the first metals to be worked, because it occurred in native form and needed little refining. Today the main producers are the USA, Russia, Kazakhstan, Georgia, Uzbekistan, Armenia, Zambia, Chile, Peru, Canada, and Zaire.

copper(II) sulphate $CuSO_4$ substance usually found as a blue, crystalline, hydrated salt $CuSO_4.5H_2O$ (also called blue vitriol). It is made from the action of dilute sulphuric acid on copper(II) oxide, hydroxide, or carbonate.

$$CuO + H_2SO_4 + 4H_2O \rightarrow CuSO_4.5H_2O$$

When the hydrated salt is heated gently it loses its water of crystallization and the blue crystals turn to a white powder. The reverse reaction is used as a chemical test for water.

$$CuSO_4.5H_2O \rightleftharpoons CuSO_4 + 5H_2O$$

coppicing woodland management practice of severe pruning where trees are cut down to near ground level at regular intervals, typically every 3–20 years, to promote the growth of numerous shoots from the base.

This form of ◊forestry was once commonly practised in Europe, principally on hazel and chestnut, to produce large quantities of thin branches for firewood, fencing, and so on; alder, eucalyptus, maple, poplar, and willow were also coppiced. The resulting thicket was known as a coppice or copse. See also ◊pollarding.

Some forests in the UK, such as Epping Forest near London, have coppice stretching back to the Middle Ages.

copulation act of mating in animals with internal ◊fertilization. Male mammals have a ◊penis or other organ that is used to introduce spermatozoa into the reproductive tract of the female. Most birds transfer sperm by pressing their cloacas (the openings of their reproductive tracts) together.

coral marine invertebrate of the class Anthozoa in the phylum Cnidaria, which also includes sea

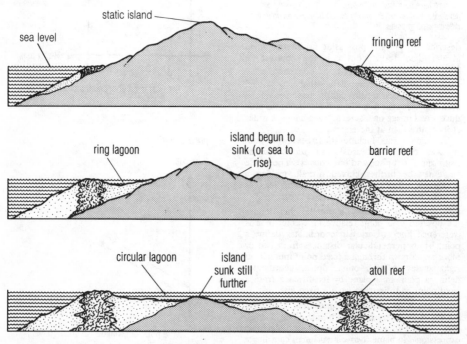

coral *The formation of a coral atoll by the gradual sinking of a volcanic island. The reefs fringing the island build up as the island sinks, eventually producing a ring of coral around the spot where the island sank.*

anemones and jellyfish. It has a skeleton of lime (calcium carbonate) extracted from the surrounding water. Corals exist in warm seas, at moderate depths with sufficient light. Some coral is valued for decoration or jewellery, for example, Mediterranean red coral *Corallum rubrum*.

Corals live in a symbiotic relationship with microscopic ◊algae (zooxanthellae), which are incorporated into the soft tissue. The algae obtain carbon dioxide from the coral polyps, and the polyps receive nutrients from the algae. Corals also have a relationship to the fish that rest or take refuge within their branches, and which excrete nutrients that make the corals grow faster. The majority of corals form large colonies although there are species that live singly. Their accumulated skeletons make up large coral reefs and atolls. The Great Barrier Reef, to the NE of Australia, is about 1,600 km/1,000 mi long, has a total area of 20,000 sq km/7,700 sq mi, and adds 50 million tonnes of calcium to the reef each year. The world's reefs cover an estimated 620,000 sq km/240,000 sq mi.

Fringing reefs are so called because they build up on the shores of continents or islands, the living animals mainly occupying the outer edges of the reef. *Barrier reefs* are separated from the shore by a saltwater lagoon, which may be as much as 30 km/20 mi wide; there are usually navigable passes through the barrier into the lagoon. *Atolls* resemble a ring surrounding a lagoon, and do not enclose an island. They are usually formed by the gradual subsidence of an extinct volcano, the coral growing up from where the edge of the island once lay.

cord unit for measuring the volume of wood cut for fuel. One cord equals 128 cubic feet (3.456 cubic metres), or a stack 8 feet (2.4 m) long, 4 feet (1.2 m) wide, and 4 feet high.

cordierite silicate mineral, $(Mg,Fe)_2Al_4Si_5O_{18}$, blue to purplish in colour. It is characteristic of metamorphic rocks formed from clay sediments under conditions of low pressure but moderate temperature; it is the mineral that forms the spots in spotted slate and spotted hornfels.

cordillera group of mountain ranges and their valleys, all running in a specific direction, formed by the continued convergence of two tectonic plates (see ◊plate tectonics) along a line.

core in earth science, the innermost part of the Earth. It is divided into an inner core, the upper boundary of which is 1,700 km/1,060 mi from the centre, and an outer core, 1,820 km/1,130 mi thick. Both parts are thought to consist of iron-nickel alloy, with the inner core being solid and the outer core being semisolid. The temperature may be 3,000°C/5,400°F.

Evidence for the nature of the core comes from seismology (observation of the paths of earthquake waves through the Earth), and calculations of the Earth's density.

Coriolis effect the effect of the Earth's rotation on the atmosphere and on all objects on the Earth's surface. In the northern hemisphere it causes moving objects and currents to be deflected to the right; in the southern hemisphere it causes deflection to the left. The effect is named after its discoverer, French mathematician Gaspard Coriolis (1792–1843).

The Coriolis effect can be easily observed by watching water go down a plughole—it does not flow directly downwards but spins to the right (clockwise) or to the left (anticlockwise), depending on whether the observer is in the northern or southern hemisphere.

THE CORIOLIS EFFECT AND LOPSIDED RIVERS

Rivers in the northern hemisphere scour their banks more severely on the right (viewed along the direction of flow) and the effect is more evident at high latitudes. This is a consequence of the Coriolis effect. For the same reason, long-range guns must allow for the deflection of their shells by the Earth's rotation.

cork light, waterproof outer layers of the bark of the stems and roots of almost all trees and shrubs. The cork oak *Quercus suber*, a native of S Europe and N Africa, is cultivated in Spain and Portugal; the exceptionally thick outer layers of its bark provide the cork that is used commercially.

corm short, swollen, underground plant stem, surrounded by protective scale leaves, as seen in the genus *Crocus*. It stores food, provides a means of ◊vegetative reproduction, and acts as a ◊perennating organ.

During the year, the corm gradually withers as the food reserves are used for the production of leafy, flowering shoots formed from axillary buds. Several new corms are formed at the base of these shoots, above the old corm.

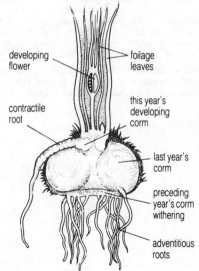

developing flower

foilage leaves

contractile root

this year's developing corm

last year's corm

preceding year's corm withering

adventitious roots

corm *Corms, found in plants such as the gladiolus and crocus, are underground storage organs. They provide the food for growth during adverse conditions such as winter cold or drought.*

cornea transparent front section of the vertebrate ◊eye. The cornea is curved and behaves as a fixed lens, so that light entering the eye is partly focused before it reaches the lens.

There are no blood vessels in the cornea and it relies on the fluid in the front chamber of the eye for nourishment. Further protection for the eye is provided by an outer membrane, called the conjunctiva. In humans, diseased or opaque parts may be replaced by grafts of corneal tissue from a donor.

cornified layer the upper layer of the skin where the cells have died, their cytoplasm being replaced by keratin, a fibrous protein also found in nails and hair. Cornification gives the skin its protective water-proof quality.

corolla collective name for the petals of a flower. In some plants the petal margins are partly or completely fused to form a *corolla tube*, for example in bindweed *Convolvulus arvensis*.

corona faint halo of hot (about 2,000,000°C/ 3,600,000°F) and tenuous gas around the Sun, which boils from the surface. It is visible at solar ◊eclipses or through a *coronagraph*, an instrument that blocks light from the Sun's brilliant disc. Gas flows away from the corona to form the ◊solar wind.

Corona Australis or *Southern Crown* constellation of the southern hemisphere, located near the constellation Sagittarius.

Corona Borealis or *Northern Crown* constellation of the northern hemisphere, between Hercules and Boötes, traditionally identified with the jewelled crown of Ariadne that was cast into the sky by Bacchus (in Greek mythology). Its brightest star is Alphecca (or Gemma), which is 78 light years from Earth.

corpuscular theory hypothesis about the nature of light championed by Isaac Newton, who postulated that it consists of a stream of particles or corpuscles. The theory was superseded at the beginning of the 19th century by English physicist Thomas Young's wave theory. ◊Quantum theory and wave mechanics embody both concepts.

corpus luteum temporary endocrine gland found in the mammalian ◊ovary. It is formed after ovulation from the Graafian follicle, a group of cells

associated with bringing the egg to maturity, and secretes the hormone progesterone.

After the release of an egg the follicle enlarges under the action of luteinizing hormone, released from the pituitary. The corpus luteum secretes the hormone progesterone, which maintains the uterus wall ready for pregnancy. If pregnancy does not occur the corpus luteum breaks down.

corrasion the grinding away of solid rock surfaces by particles carried by water, ice and wind. It is generally held to be the most significant form of ◊erosion. As the eroding particles are carried along they become eroded themselves due to the process of ◊attrition.

correlation the degree of relationship between two sets of information. If one set of data increases at the same time as the other, the relationship is said to be positive or direct. If one set of data increases as the other decreases, the relationship is negative or inverse. Correlation can be shown by plotting a best-fit line on a ◊scatter diagram.

In statistics, such relations are measured by the calculation of ◊coefficients of correlation. These generally measure correlation on a scale with 1 indicating perfect positive correlation, 0 no correlation at all, and –1 perfect inverse correlation.

Correlation coefficients for assumed linear relations include the Pearson product moment correlation coefficient (known simply as the correlation coefficient), Kendall's tau correlation coefficient, or Spearman's rho correlation coefficient, which is used in nonparametric statistics (where the data are measured on ordinal rather than interval scales). A high correlation does not always indicate dependence between two variables; it may be that there is a third (unstated) variable upon which both depend.

correspondence in mathematics, the relation between two sets where an operation on the members of one set maps some or all of them onto one or more members of the other. For example, if A is the set of members of a family and B is the set of months in the year, A and B are in correspondence if the operation is: '. . .has a birthday in the month of. . . '.

corrie (Welsh *cwm*; French, US *cirque*) Scottish term for a steep-sided hollow in the mountainside of a glaciated area. The weight and movement of

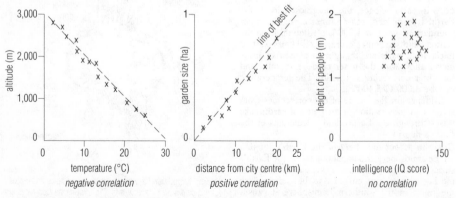

correlation *Scattergraphs showing different kinds of correlation.*

the ice has ground out the bottom and worn back the sides. A corrie is open at the front, and its sides and back are formed of ◊arêtes. There may be a lake in the bottom, called a tarn.

A corrie is formed as follows: (1) snow accumulates in a hillside hollow and turns to ice (enlarging the hollow by ◊nivation); (2) the hollow is deepened by ◊abrasion and ◊plucking; (3) the ice in the corrie rotates under the influence of gravity, deepening the hollow still further; (4) since the ice is thinner and moves more slowly at the foot of the hollow, a rock lip forms; (5) when the ice melts, a lake or tarn may be formed in the corrie. The steep back wall results from severe weathering by freeze-thaw, which provides material for further abrasion.

corrosion the eating away and eventual destruction of metals and alloys by chemical attack. The rusting of ordinary iron and steel is the most common form of corrosion. Rusting takes place in moist air, when the iron combines with oxygen and water to form a brown-orange deposit of ◊rust (hydrated iron oxide). The rate of corrosion is increased where the atmosphere is polluted with sulphur dioxide. Salty road and air conditions accelerate the rusting of car bodies.

Corrosion is largely an electrochemical process, and acidic and salty conditions favour the establishment of electrolytic cells on the metal, which cause it to be eaten away. Other examples of corrosion include the green deposit that forms on copper and bronze, called verdigris, a basic copper carbonate. The tarnish on silver is a corrosion product, a film of silver sulphide.

corrosion in earth science, an alternative name for ◊solution, the process by which water dissolves rocks such as limestone.

corruption of data introduction or presence of errors in data. Most computers use a range of ◊verification and ◊validation routines to prevent corrupt data from entering the computer system or detect corrupt data that are already present.

cortex in biology, the outer layer of a structure such as the brain, kidney, or adrenal gland. In botany the cortex includes non-specialized cells lying just beneath the surface cells of the root and stem.

corundum native aluminium oxide, Al_2O_3, the hardest naturally occurring mineral known apart

plant cortex cells

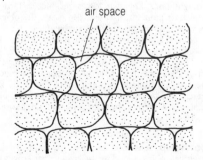

air space

cortex

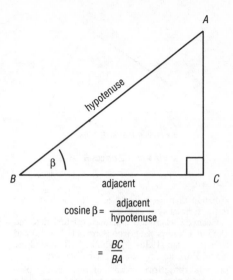

$$\cos\beta = \frac{adjacent}{hypotenuse}$$

$$= \frac{BC}{BA}$$

cosine *The cotangent of angle β is equal to the ratio of the length of the adjacent side to the length of the hypotenuse (the longest side, opposite to the right angle).*

from diamond (corundum rates 9 on the Mohs' scale of hardness); lack of ◊cleavage also increases its durability. Its crystals are barrel-shaped prisms of the trigonal system. Varieties of gem-quality corundum are **ruby** (red) and **sapphire** (any colour other than red, usually blue). Poorer-quality and synthetic corundum is used in industry, for example as an ◊abrasive.

Corundum forms in silica-poor igneous and metamorphic rocks. It is a constituent of emery, which is metamorphosed bauxite.

cosecant in trigonometry, a ◊function of an angle in a right-angled triangle found by dividing the length of the hypotenuse (the longest side) by the length of the side opposite the angle. Thus the cosecant of an angle A, usually shortened to cosec A, is always greater than (or equal to) 1. It is the reciprocal of the sine of the angle, that is, cosec A = $1/\sin A$.

cosine in trigonometry, a ◊function of an angle in a right-angled triangle found by dividing the length of the side adjacent to the angle by the length of the hypotenuse (the longest side). It is usually shortened to **cos**.

cosine rule in trigonometry, a rule that relates the sides and angles of triangles. The rule has the formula

$$a^2 = b^2 + c^2 - 2bc \cos A$$

where a, b, and c are the sides of the triangle, and A is the angle opposite a.

cosmic background radiation or *3° radiation* electromagnetic radiation left over from the original formation of the universe in the Big Bang around 15 billion years ago. It corresponds to an overall background temperature of 3K (–270°C/–454°F), or 3°C above absolute zero. In 1992 the Cosmic Background Explorer satellite, COBE, detected slight 'ripples' in the strength of the background

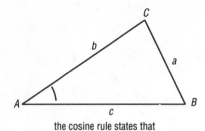

the cosine rule states that

$$a^2 = b^2 + c^2 - 2bc\cos A$$

cosine rule

radiation that are believed to mark the first stage in the formation of galaxies.

Cosmic background radiation was first detected 1965 by US physicists Arno Penzias (1933–) and Robert Wilson (1936–), who in 1978 shared the Nobel Prize for Physics for their discovery.

cosmic radiation streams of high-energy particles from outer space, consisting of protons, alpha particles, and light nuclei, which collide with atomic nuclei in the Earth's atmosphere, and produce secondary nuclear particles (chiefly ◊mesons, such as pions and muons) that shower the Earth.

Those of low energy seem to be galactic in origin, and those of high energy of extragalactic origin. The galactic particles may come from ◊supernova explosions or ◊pulsars. At higher energies, other sources are necessary, possibly the giants jets of gas whcih are emitted from some galaxies.

cosmid fragment of ◊DNA from the human genome inserted into a bacterial cell. The bacterium replicates the fragment along with its own DNA. In this way the fragments are copied for a gene library. Cosmids are characteristically 40,000 base pairs in length. The most commonly used bacterium is *Escherichia coli*. A ◊yeast artificial chromosome works in the same way.

cosmogony (Greek 'universe' and 'creation') study of the origin and evolution of cosmic objects, especially the Solar System.

cosmological principle in astronomy, a hypothesis that any observer anywhere in the ◊universe has the same view that we have; that is, that the universe is not expanding from any centre but all galaxies are moving away from one another.

cosmology study of the structure of the universe. Modern cosmology began in the 1920s with the discovery that the universe is expanding, which suggested that it began in an explosion, the ◊Big Bang. An alternative—now discarded—view, the ◊steady-state theory, claimed that the universe has no origin, but is expanding because new matter is being continually created.

cosmonaut term used in the West for any astronaut from the former Soviet Union.

Cosmos name used from the early 1960s for nearly all Soviet artificial satellites. Over 2,200 Cosmos satellites had been launched by the end of 1992.

cotangent in trigonometry, a ◊function of an angle in a right-angled triangle found by dividing the length of the side adjacent to the angle by the length of the side opposite it. It is usually written as cotan, or cot and it is the reciprocal of the tangent of the angle, so that cot A = 1/tan A, where A is the angle in question.

cotton gin machine that separates cotton fibres from the seed boll. Production of the gin (then called an en*gin*e) by US inventor Eli Whitney 1793 was a milestone in textile history.

The modern gin consists of a roller carrying a set of circular saws. These project through a metal grill in a hopper containing the seed bolls. As the roller rotates, the saws pick up the cotton fibres, leaving the seeds behind.

cotyledon structure in the embryo of a seed plant that may form a 'leaf' after germination and is commonly known as a seed leaf. The number of cotyledons present in an embryo is an important character in the classification of flowering plants (◊angiosperms).

Monocotyledons (such as grasses, palms, and lilies) have a single cotyledon, whereas dicotyledons (the majority of plant species) have two. In seeds that also contain ◊endosperm (nutritive tissue), the cotyledons are thin, but where they are the primary food-storing tissue, as in peas and beans, they may be quite large. After germination the cotyledons either remain below ground (hypogeal) or, more commonly, spread out above soil level (epigeal) and become the first green leaves. In gymnosperms there may be up to a dozen cotyledons within each seed.

coulomb SI unit (symbol C) of electrical charge. One coulomb is the quantity of electricity conveyed by a current of one ◊ampere in one second.

count rate number of particles emitted per unit time by a radioactive source. It is measured by a counter, such as a ◊Geiger counter, or ◊ratemeter.

country park pleasure ground or park, often located near an urban area, providing facilities for the public enjoyment of the countryside. Country parks were introduced in the UK following the 1968 Countryside Act and are the responsibility of local authorities with assistance from the Countryside Commission. They cater for a range of recreational activities such as walking, boating, and horse-riding.

couple in mechanics, a pair of forces acting on a rigid object that are equal in magnitude and opposite in direction, but do not act along the same straight line. The two forces produce a turning effect, or moment, that tends to rotate the object; however, no single resultant (unbalanced) force is produced and so the object is not moved from one position to another.

The moment of a couple is the product of the magnitude of either of the two forces and the perpendicular distance between those forces. If the magnitude of the force is F newtons and the distance is d metres then the moment, in newton-metres, is given by:

$$\text{moment} = Fd$$

courtship behaviour exhibited by animals as a prelude to mating. The behaviour patterns vary considerably from one species to another, but are

RECENT PROGRESS IN COSMOLOGY

Cosmology: have we seen the creation?

Lumpiness is the overriding characteristic of the universe. Most of its visible material is collected into stars, separated by several light years. The stars in turn are collected in vast aggregations called galaxies, millions of light years apart. On the grandest scale, even the galaxies are clustered.

In April 1992, US astronomers announced that they had detected the first steps in the formation of the overall structure of the universe, from the results of a satellite called *COBE*, the *Cosmic Background Explorer*. It was hailed variously as the discovery of the century and, more realistically, the discovery of the decade. But it also pointed out the unsettling fact that the visible stars and galaxies account for only a small fraction of all the matter in the universe.

'We have found the primordial, the oldest things you can imagine coming from the universe,' said an ecstatic Dr George Smoot of the University of California, Berkeley, leader of the team that made the the discovery. 'This begins the golden age of cosmology. It is going to change our view of the universe and our place within it.' Dr Michael Turner of the University of Chicago said: 'These astronomers have found the Holy Grail of cosmology.'

All modern cosmology starts from the basic fact, discovered in 1929 by Edwin Hubble at Mount Wilson Observatory in California, that the universe is expanding like a balloon. The inevitable conclusion, arrived at by imagining the expansion run in reverse, is that the universe must once have been compressed into a super-dense blob that exploded for some reason. That explosion is known as the Big Bang and, from the current rate of expansion of the universe, it is estimated to have occurred about 15 billion years ago.

Vestiges of the Big Bang

Direct evidence for the Big Bang turned up in 1965 when two physicists at the Bell Telephone Laboratories in New Jersey, Arno Penzias and Robert Wilson, detected a faint hiss from the universe at short radio wavelengths. This is known as the cosmic background radiation, and is interpreted as being the heat from the Big Bang explosion that has been cooled by the subsequent expansion of the universe. Penzias and Wilson won a Nobel prize for this discovery.

But there was a problem. The background radiation was too smooth to account for the lumpiness of the currently observed universe. Slight differences in temperature would be expected from denser areas of gas where the galaxies formed but, try as they might, astronomers could not find them. The Big Bang theory looked as though it might be in trouble.

Detecting ripples at last

COBE, launched by the US space agency NASA in Nov 1989, has changed all that. It has measured the temperature of the background radiation with new precision, 2.735°C above absolute zero. And, as Dr Smoot announced at the American Physical Society meeting in Washington, DC on 23 April 1992, it has detected the sought-after temperature variations, which appear like ripples on the background radiation. They amount to a mere 30 millionths of a degree, and it is a tribute to the precision of *COBE* that they could be detected at all.

The findings are in line with the predictions made by a modern version of the Big Bang, called the inflationary theory. According to this, the universe expanded very rapidly during the first fraction of a second of its existence, before settling down to the more leisurely rate of expansion observed today.

There is a catch, though. The density fluctuations detected by *COBE* would not, on their own, be enough to explain the formation of the galaxies. There must also be vast amounts of unseen dark matter in the universe to supply additional gravitational attraction.

'We need such invisible matter to explain how galaxies formed in the early universe and gathered themselves together into huge clusters. Ordinary matter would be attracted into regions of concentrated dark matter, and the universe as we know it today could develop, eventually leading to the formation of galaxies, stars, and planets,' said Dr Edward Wright of the University of California, Los Angeles.

If the astronomers are right, over 90% of the universe consists of unseen matter. At present, no one knows what this dark matter is, or how to detect it. The most popular view is that it consists of unknown atomic particles which have been given fanciful names such as axions, gravitinos, and photinos.

Other possibilities are immense black holes, or countless numbers of small, faint stars. Astronomers are hoping to get more clues to the nature of the dark matter from *COBE*.

Ian Ridpath

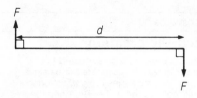

couple *Two equal but opposite forces (F) will produce a turning effect on a rigid body, provided that they do not act through the same straight line.*

often ritualized forms of behaviour not obviously related to courtship or mating (for example, courtship feeding in birds).

Courtship ensures that copulation occurs with a member of the opposite sex of the right species. It also synchronizes the partners' readiness to mate and allows each partner to assess the suitability of the other.

covalent bond chemical ◊bond produced when two atoms share one or more pairs of electrons (usually each atom contributes an electron). The bond is often represented by a single line drawn between the two atoms. Covalently bonded substances include hydrogen (H_2), water (H_2O), and most organic substances.

Double bonds, seen, for example, in the ◊alkenes, are formed when two atoms share two pairs of electrons (the atoms usually contribute a pair each); triple bonds, seen in the ◊alkynes, are formed when atoms share three pairs of electrons. Such bonds are represented by a double or triple line, respectively, between the atoms concerned.

Covalent compounds have the following general properties: they have low melting and boiling points; never conduct electricity; and are usually insoluble in water and soluble in organic solvents. Compare ◊ionic compound.

CP/M (abbreviation for *control program/monitor* or *control program for microcomputers*) one of the earliest ◊operating systems for microcomputers. It was written by Gary Kildall, who founded Digital Research, and became a standard for microcomputers based on the Intel 8080 and Zilog Z80 8-bit microprocessors. In the 1980s it was superseded by Microsoft's ◊MS-DOS, written for 16-bit microprocessors.

CPU in computing, abbreviation for ◊central processing unit.

Crab nebula cloud of gas 6,000 light years from Earth, in the constellation Taurus. It is the remains of a star that exploded as a ◊supernova (observed as a brilliant point of light on Earth 1054). At its centre is a ◊pulsar that flashes 30 times a second. The name comes from its crablike shape.

crack street name for a chemical derivative (bicarbonate) of ◊cocaine in hard, crystalline lumps; it is heated and inhaled (smoked) as a stimulant. Crack was first used in San Francisco in the early 1980s, and is highly addictive.

cracking reaction in which a large ◊alkane molecule is broken down by heat into a smaller alkane and a small ◊alkene molecule. The reaction is carried out at a high temperature (600°C or higher) and often in the presence of a catalyst. Cracking is a commonly used process in the ◊petrochemical industry.

courtship *The courtship display of gulls, with neck arching and sky-pointing. These displays help bring together opposite sexes of the same species at the right time and place. The pattern of signals displayed during courtship is unique to each species. This prevents individuals of different species from trying to mate.*

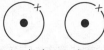

two hydrogen atoms

or H $\overset{x}{.}$ H, H–H
a molecule of hydrogen
sharing an electron pair

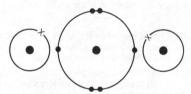

two hydrogen atoms and one
oxygen atom

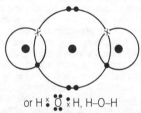

or H $\overset{x}{.}$ O $\overset{x}{.}$ H, H–O–H
a molecule of water
showing the two covalent bonds

covalent bond *The formation of a covalent bond between two hydrogen atoms to form a hydrogen molecule (H_2), and between two hydrogen atoms and an oxygen atom to form a molecule of water (H_2O). The sharing means that each atom has a more stable arrangement of electrons (its outer electron shells are full).*

It is the main method of preparation of alkenes and is also used to manufacture petrol from the higher-boiling-point ◊fractions that are obtained by fractional ◊distillation (fractionation) of crude oil.

crag in previously glaciated areas, a large lump of rock that a glacier has been unable to wear away. As the glacier passed up and over the crag, weaker rock on the far side was largely protected from erosion and formed a tapering ridge, or *tail*, of debris.

An example of a crag-and-tail feature is found in Edinburgh in Scotland; Edinburgh Castle was built on the crag (Castle Rock), which dominates the city beneath.

crane in engineering, a machine for raising, lowering, or placing in position heavy loads. The three main types are the jib crane, the overhead travelling crane, and the tower crane. Most cranes have the machinery mounted on a revolving turntable. This may be mounted on trucks or be self-propelled, often being fitted with caterpillar tracks (an endless flexible belt of metal plates which takes the place of ordinary tyred wheels).

The main features of a *jib crane* are a power winch, a rope or cable, and a movable arm or jib. The cable, which carries a pulley block, hangs from the end of the jib and is wound up and down by the winch. The *overhead travelling crane*, chiefly used in workshops, consists of a fixed horizontal arm, along which runs a trolley carrying the pulley block. *Tower cranes*, seen on large building sites, have a long horizontal arm able to revolve on top of a tall tower. The arm carries the trolley.

cranium the dome-shaped area of the vertebrate skull, consisting of several fused plates, that protects the brain. Fossil remains of the human cranium have aided the development of theories concerning human evolution.

The cranium has been studied as a possible indicator of intelligence or even of personality. The Victorian argument that a large cranium implies a large brain, which in turn implies a more profound intelligence, has been rejected.

crank handle bent at right angles and connected to the shaft of a machine; it is used to transmit motion or convert reciprocating (back-and-forwards or up-and-down) movement into rotary movement, or vice versa.

Although similar devices may have been employed in antiquity and as early as the 1st century in China and the 8th century in Europe, the earliest recorded use of a crank in a water-raising machine is by Arab mathematician al-Jazari in the 12th century. Not until the 15th century, however, did the crank become fully assimilated into developing European technology.

crankshaft essential component of piston engines that converts the up-and-down (reciprocating) motion of the pistons into useful rotary motion. The car crankshaft carries a number of cranks. The pistons are connected to the cranks by connecting rods and ◊bearings; when the pistons move up and down, the connecting rods force the offset crank pins to describe a circle, thereby rotating the crankshaft.

crater bowl-shaped depression, usually round and with steep sides. Craters are formed by explosive events such as the eruption of a volcano or by the impact of a meteorite. A ◊caldera is a much larger feature.

The Moon has more than 300,000 craters over 1 km/6 mi in diameter, formed by meteorite bombardment; similar craters on Earth have mostly been worn away by erosion. Craters are found on many other bodies in the Solar System.

CRATERS OF HEAVEN AND HELL

Hell is on the Moon. It is a crater 32 km/ 20 mi across and was named after 19th-century Hungarian astronomer Maximilian Hell. It has nothing to do with the flaming inferno. Utopia, on the other hand, is a smooth, low-lying region on Mars.

craton or *shield* core of a continent, a vast tract of highly deformed ◊metamorphic rock around which the continent has been built. Intense mountain-building periods shook these shield areas in Precambrian times before stable conditions set in.

Cratons exist in the hearts of all the continents, a typical example being the Canadian Shield.

creationism theory concerned with the origins of matter and life, claiming, as does the Bible in Genesis, that the world and humanity were created by a supernatural Creator, not more than 6,000 years ago. It was developed in response to Darwin's theory of ◊evolution; it is not recognized by most scientists as having a factual basis.

After a trial 1981–82 a US judge ruled unconstitutional an attempt in Arkansas schools to enforce equal treatment of creationism and evolutionary theory.

creep in civil and mechanical engineering, the property of a solid, typically a metal, under continuous stress that causes it to deform below its ◊yield point (the point at which any elastic solid normally stretches without any increase in load or stress). Lead, tin, and zinc, for example, exhibit creep at ordinary temperatures, as seen in the movement of the lead sheeting on the roofs of old buildings.

Copper, iron, nickel, and their alloys also show creep at high temperatures.

creosote black, oily liquid derived from coal tar, used as a wood preservative. Medicinal creosote, which is transparent and oily, is derived from wood tar.

crescent curved shape of the Moon when it appears less than half-illuminated. It also refers to any object or symbol resembling the crescent Moon. Often associated with Islam, it was first used by the Turks on their standards after the capture of Constantinople 1453, and appears on the flags of many Muslim countries. The *Red Crescent* is the Muslim equivalent of the Red Cross.

Cretaceous (Latin *creta* 'chalk') period of geological time 146–65 million years ago. It is the last period of the Mesozoic era, during which angiosperm (seed-bearing) plants evolved, and dinosaurs reached a peak before their almost complete extinction at the end of the period. Chalk is a typical rock type of the second half of the period.

crevasse deep crack in the surface of a glacier; it can reach several metres in depth. Crevasses often occur where a glacier flows over the break of slope, because the upper layers of ice are unable to stretch and cracks result. Crevasses may also form at the edges of glaciers owing to friction with the bedrock.

crith unit of mass used for weighing gases. One crith is the mass of one litre of hydrogen gas (H_2) at standard temperature and pressure.

critical angle in optics, for a ray of light passing from a denser to a less dense medium (such as from glass to air), the smallest angle of incidence at which the emergent ray grazes the surface of the denser medium—at an angle of refraction of 90°.

When the angle of incidence is less than the critical angle, the ray does not pass out into the

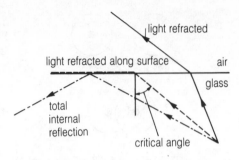

critical angle *The critical angle is the angle at which light from within a transparent medium just grazes the surface of the medium. In the diagram, the red beam is at the critical angle. Blue beams escape from the medium, at least partially. Green beams are totally reflected from the surface.*

less dense medium; when the angle of incidence is greater than the critical angle, the ray is not reflected back into the denser medium.

critical mass in nuclear physics, the minimum mass of fissile material that can undergo a continuous ◊chain reaction. Below this mass, too many ◊neutrons escape from the surface for a chain reaction to carry on; above the critical mass, the reaction may accelerate into a nuclear explosion.

critical path analysis procedure used in the management of complex projects to minimize the amount of time taken. The analysis shows which subprojects can run in parallel with each other, and which have to be completed before other subprojects can follow on. By identifying the time required for each separate subproject and the relationship between the subprojects, it is possible to produce a planning schedule showing when each subproject should be started and finished in order to complete the whole project most efficiently. Complex projects may involve hundreds of subprojects, and computer ◊applications packages for critical path analysis are widely used to help reduce the time and effort involved in their analysis.

critical temperature temperature above which a particular gas cannot be converted into a liquid by pressure alone. It is also the temperature at which a magnetic material loses its magnetism (the Curie temperature or point).

CRO abbreviation for ◊*cathode-ray oscilloscope.*

crop in birds, the thin-walled enlargement of the digestive tract between the oesophagus and stomach. It is an effective storage organ especially in seed-eating birds; a pigeon's crop can hold about 500 cereal grains. Digestion begins in the crop, by the moisturizing of food. A crop also occurs in insects and annelid worms.

crop circle circular area of flattened grain found in fields especially in SE England, with increasing frequency every summer since 1980. More than 1,000 such formations were reported in the UK 1991. The cause is unknown.

Most of the research into crop circles has been conducted by dedicated amateur investigators rather than scientists. Physicists who have studied the phenomenon have suggested that an electromagnetic whirlwind, or 'plasma vortex', can explain both the crop circles and some UFO sightings, but this does not account for the increasing geometrical complexity of crop circles, nor for the fact that until 1990 they were unknown outside the UK. Crop circles began to appear in the USA only after a US magazine published an article about them. A few people have confessed publicly to having made crop circles that were accepted as genuine by investigators.

crop rotation system of regularly changing the crops grown on a piece of land. The crops are grown in a particular order to utilize and add to the nutrients in the soil and to prevent the build-up of insect and fungal pests. Including a legume crop, such as peas or beans, in the rotation helps build up nitrate in the soil because the roots contain bacteria capable of fixing nitrogen from the air.

A simple seven-year rotation, for example, might include a three-year ley (temporary grassland sown to produce grazing and hay or silage) followed by two years of wheat and then two years of barley, before returning the land to temporary grass once more. In this way, the cereal crops can take advantage of the build-up of soil fertility that occurs during the period under grass.

In the 18th century, a four-year rotation was widely adopted with autumn-sown cereal, followed by a root crop, then spring cereal, and ending with a leguminous crop. Since then, more elaborate rotations have been devised with two, three, or four successive cereal crops, and with the root crop replaced by a cash crop such as sugar beet or potatoes, or by a legume crop such as peas or beans.

crossing over in biology, a process that occurs during ◊meiosis. While the chromosomes are lying alongside each other in pairs, each partner may twist around the other and exchange corresponding chromosomal segments. It is a form of genetic ◊recombination, which increases variation and thus provides the raw material of evolution.

cross linking in chemistry, lateral linking between two or more long-chain molecules in a ◊polymer. Cross linking gives the polymer a higher melting point and makes it harder. Examples of cross-linked polymers include Bakelite and vulcanized rubber.

CRT abbreviation for ◊cathode-ray tube.

crude oil the unrefined form of ◊petroleum.

crumple zone region at the front and rear of a motor vehicle that is designed to crumple gradually during a collision, so reducing the risk of serious injury to passengers. The progressive crumpling absorbs the kinetic energy of the vehicle more gradually than a rigid structure would, thereby diminishing the forces of deceleration acting on the vehicle and on the people inside.

crust the outermost part of the structure of Earth, consisting of two distinct parts, the oceanic crust and the continental crust. The *oceanic* crust is on average about 10 km/6.2 mi thick and consists mostly of basaltic types of rock. By contrast, the *continental* crust is largely made of granite and is more complex in its structure. Because of the

movements of ◊plate tectonics, the oceanic crust is in no place older than about 200 million years. However, parts of the continental crust are over 3 billion years old.

Beneath a layer of surface sediment, the oceanic crust is made up of a layer of basalt, followed by a layer of gabbro. The composition of the oceanic crust overall shows a high proportion of *s*ilicon and *m*agnesium oxides, hence named *sima* by geologists. The continental crust varies in thickness from about 40 km/25 mi to 70 km/45 mi, being deeper beneath mountain ranges. The surface layer consists of many kinds of sedimentary and igneous rocks. Beneath lies a zone of metamorphic rocks built on a thick layer of granodiorite. *S*ilicon and *al*uminium oxides dominate the composition and the name *sial* is given to continental crustal material.

crustacean one of the class of arthropods that includes crabs, lobsters, shrimps, woodlice, and barnacles. The external skeleton is made of protein and chitin hardened with lime. Each segment bears a pair of appendages that may be modified as sensory feelers (antennae), as mouthparts, or as swimming, walking, or grasping structures.

Crux constellation of the southern hemisphere, popularly known as the Southern Cross, the smallest of the 88 constellations. Its brightest stars are Alpha Crucis (or Acrux), a ◊double star about 400 light years from Earth, and Beta Crucis (or Mimosa). Near Beta Crucis lies a glittering star cluster known as the Jewel Box. The constellation also contains the Coalsack, a dark cloud of dust silhouetted against the bright starry background of the Milky Way.

cryogenics science of very low temperatures (approaching ◊absolute zero), including the production of very low temperatures and the exploitation of special properties associated with them, such as the disappearance of electrical resistance (◊superconductivity).

Low temperatures can be produced by the Joule-Thomson effect (cooling a gas by making it do work as it expands). Gases such as oxygen, hydrogen, and helium can be liquefied in this way, and temperatures of 0.3K can be reached. Further cooling requires magnetic methods; a magnetic material, in contact with the substance to be cooled and with liquid helium, is magnetized by a strong magnetic field. The heat generated by the process is carried away by the helium. When the material is then demagnetized, its temperature falls; temperatures of around 10^{-3}K have been achieved in this way. A similar process, called *nuclear adiabatic expansion*, was used to produce the lowest temperature recorded: 2×10^{-9}K, produced in 1989 by a team of Finnish scientists.

At temperatures near absolute zero, materials can display unusual properties. Some metals, such as mercury and lead, exhibit superconductivity. Liquid helium loses its viscosity and becomes a 'superfluid' when cooled to below 2K; in this state it flows up the sides of its container.

Cryogenics has several practical applications. *Cryotherapy* is a process used in eye surgery, in which a freezing probe is briefly applied to the outside of the eye to repair a break in the retina.

Electronic components called ◊Josephson junctions, which could be used in very fast computers, need low temperatures to function. Magnetic levitation (◊maglev) systems must be maintained at low temperatures. Food can be frozen for years, and it has been suggested that space travellers could be frozen for long journeys. Freezing people with terminal illnesses, to be revived when a cure has been developed, has also been suggested.

cryolite rare granular crystalline mineral (sodium aluminium fluoride), Na_3AlF_6, used in the electrolytic reduction of ◊bauxite to aluminium. It is chiefly found in Greenland.

cryptogam obsolete name applied to the lower plants. It included the algae, liverworts, mosses, and ferns (plus the fungi and bacteria in very early schemes of classification). In such classifications seed plants were known as ◊phanerogams.

cryptography science of creating and reading codes; for example, those produced by the Enigma coding machine used by the Germans in World War II and those used in commerce by banks encoding electronic fund-transfer messages, business firms sending computer-conveyed memos between headquarters, and in the growing field of electronic mail. No method of encrypting is completely unbreakable, but decoding can be made extremely complex and time consuming.

cryptosporidium parasite that has been found in drinking water in the UK and USA. It causes cramps, fever, and diarrhoea, and may lead to death. It is washed into rivers with slurry from farms.

Conventional filtration and chlorine disinfection are ineffective at removing the parasite. Slow-sand filtration is the best method of removal, but the existing systems were dismantled in the 1970s because of their slowness.

crystal substance with an orderly three-dimensional arrangement of its atoms or molecules, thereby creating an external surface of clearly defined smooth faces having characteristic angles between them. Examples are table salt and quartz.

Each geometrical form, many of which may be combined in one crystal, consists of two or more faces—for example, dome, prism, and pyramid. A mineral can often be identified by the shape of its crystals and the system of crystallization determined. A single crystal can vary in size from a submicroscopic particle to a mass some 30 m/100 ft in length.

Crystals fall into seven crystal systems or groups, classified on the basis of the relationship of three or four imaginary axes that intersect at the centre of any perfect, undistorted crystal.

crystal system all known crystalline substances crystallize in one of seven crystal systems, defined by symmetry. The elements of symmetry used for this purpose are: (1) planes of *mirror symmetry*, across which a mirror image is seen, and (2) axes of *rotational symmetry*, about which, in a 360° rotation of the crystal, equivalent faces are seen twice, three, four, or six times.

To be assigned to a particular crystal system, a mineral must possess a certain minimum symmetry, but it may also possess additional symmetry elements. Since crystal symmetry is related to

sodium chloride

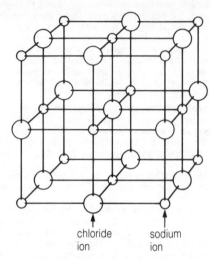

↑ chloride ion   ↑ sodium ion

crystal *The sodium chloride, or common salt, crystal is a regular cubic array of charged atoms (ions)—positive sodium atoms and negative chlorine atoms. Repetition of this structure builds up into cubic salt crystals.*

internal structure, a given mineral will always crystallize in the same system, although the crystals may not always grow into precisely the same shape. In cases where two minerals have the same chemical composition (for example graphite and diamond, or quartz and cristobalite), they will generally have different crystal systems because their internal structures are different.

Let us examine a crystal . . . the equality of the sides pleases us; that of the angles doubles the pleasure. On bringing to view a second face in all respects similar to the first, this pleasure appears to be squared; and bringing into view a third, it appears to be cubed, and so on.

On **crystals** Edgar Allen Poe (1809–1849) *The Rationale of Verse* 1843

---

crystallography the scientific study of crystals. In 1912 it was found that the shape and size of the repeating atomic patterns (unit cells) in a crystal could be determined by passing X-rays through a sample. This method, known as ◊X-ray diffraction, opened up an entirely new way of 'seeing' atoms. It has been found that many substances have a unit cell that exhibits all the symmetry of the whole crystal; in table salt (sodium chloride, NaCl), for instance, the unit cell is an exact cube.

Many materials were not even suspected of being crystals until they were examined by X-ray crystallography. It has been shown that purified biomolecules, such as proteins and DNA, can form crystals, and such compounds may now be studied by this method. Other applications include the study of metals and their alloys, and of rocks and soils.

CT scanner or *CAT scanner* medical device

crystal system

crystal system	minimum symmetry	possible shape	mineral examples
cubic	4 threefold axes	cube, octahedron, dodecahedron	diamond, garnet, pyrite
tetragonal	1 fourfold axis	square-based prism	zircon
orthorhombic	3 twofold axes or mirror planes	matchbox shape	baryte
monoclinic	1 twofold axis or mirror plane	matchbox distorted in one plane	gypsum
triclinic	no axes or mirror planes	matchbox distorted in three planes	plagioclase feldspar
trigonal	1 threefold axis	triangular prism, rhombohedron	calcite, quartz
hexagonal	1 sixfold axis	hexagonal prism	beryl

used to obtain detailed X-ray pictures of the inside of a patient's body; see ◊CAT scan.

cu abbreviation for *cubic* (measure).

cube in geometry, a regular solid figure whose faces are all squares. It has six equal-area faces and 12 equal-length edges. If the length of one edge is l, the volume V of the cube is given by:

$$V = l^3$$

and its surface area A by:

$$A = 6l^2.$$

cubic decimetre metric measure (symbol dm^3) of volume corresponding to the volume of a cube whose edges are all 1 dm (10 cm) long; it is equivalent to a capacity of one litre.

cubic equation any equation in which the largest power of x is x^3. For example, $x^3 + 3x^2y + 4y^2 = 0$ is a cubic equation.

cubit earliest known unit of length, which originated between 2800 and 2300 BC. It is approximately 50.5 cm/20.6 in long, which is about the length of the human forearm measured from the tip of the middle finger to the elbow.

cuboid six-sided three-dimensional prism whose faces are all rectangles. A brick is a cuboid.

cultivar variety of a plant developed by horticultural or agricultural techniques. The term derives from '*culti*vated *va*riety'.

culture in biology, the growing of living cells and tissues in laboratory conditions.

cumulative frequency in statistics, the total frequency of a given value up to and including a certain point in a set of data. It is used to draw the cumulative frequency curve, the ogive.

cuprite Cu_2O ore (copper(I) oxide), found in crystalline form or in earthy masses. It is red to black in colour, and is often called ruby copper.

cupronickel copper alloy (75% copper and 25% nickel), used in hardware products and for coinage.

In the UK in 1946, it was substituted for the 'silver' (50% silver, 40% copper, 5% nickel and 5% zinc) previously used in coins. US coins made with cupronickel include the dime, quarter, half-dollar, and dollar.

curie former unit (symbol Ci) of radioactivity, equal to 37×10^9 ◊becquerels. One gram of radium has a radioactivity of about one curie. It was named after French physicist Pierre Curie.

Curie temperature the temperature above which a magnetic material cannot be strongly magnetized. Above the Curie temperature, the energy of the atoms is too great for them to join together to form the small areas of magnetized material, or ◊domains, which combine to produce the strength of the overall magnetization.

curium synthesized, radioactive, metallic element of the *actinide* series, symbol Cm, atomic number 96, relative atomic mass 247. It is produced by bombarding plutonium or americium with neutrons. Its longest-lived isotope has a half-life of 1.7 $\times 10^7$ years.

Curium is used to generate heat and power in satellites or in remote places. It was first synthesized 1944 at the University of California at Berkeley and in 1946 was named after French scientists Pierre and Marie Curie by Glenn Seaborg, its synthesizer.

current the flow of a body of water or air, or of heat, moving in a definite direction. Ocean currents are fast-flowing currents of seawater generated by the wind or by variations in water density between two areas. They are partly responsible for transferring heat from the equator to the poles and thereby evening out the global heat imbalance.

The ◊Gulf Stream is an example of a warm ocean current; it flows from the Gulf of Mexico to NW Europe and is responsible for keeping the coast of Scandinavia free of ice in winter. The Peru current is a cold ocean current, transferring cold water from the Antarctic to the coasts of Chile and Peru. Small fish called anchovies thrive in these fertile waters, providing an important source of income for Peru. At approximate five-to-eight-year intervals, the phenomenon of ◊El Niño causes the Peru current to become warm and has disastrous results (as in 1982–83) for Peruvian wildlife and for the anchovy industry.

current, electric see ◊electric current.

cursor on a computer screen, the symbol that indicates the current entry position (position where the next character will appear). It usually consists of a solid rectangle or underline character, flashing on and off.

curve in geometry, the ◊locus of a point moving according to specified conditions. The circle is the locus of all points equidistant from a given point (the centre). Other common geometrical curves are the ◊ellipse, ◊parabola, and ◊hyperbola, which are

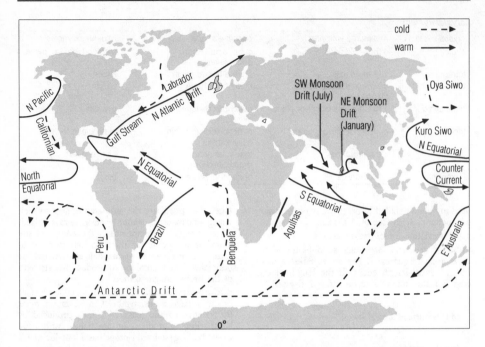

current *The main cold and warm ocean currents.*

also produced when a cone is cut by a plane at different angles.

Many curves have been invented for the solution of special problems in geometry and mechanics — for example, the cissoid (the inverse of a parabola) and the ◊cycloid.

cusp point where two branches of a curve meet and the tangents to each branch coincide.

cuticle the horny noncellular surface layer of many invertebrates such as insects; in botany, the waxy surface layer on those parts of plants that are exposed to the air, continuous except for ◊stomata and ◊lenticels. All types are secreted by the cells of the ◊epidermis. A cuticle reduces water loss and, in arthropods, acts as an ◊exoskeleton.

cutting technique of vegetative propagation involving taking a section of root, stem, or leaf and treating it so that it develops into a new plant.

cwt symbol for ◊*hundredweight*, a unit of weight equal to 112 pounds (50.802 kg); 100 lb (45.36 kg) in the USA.

cyanamide process process used in the manufacture of calcium cyanamide ($CaCN_2$), a colourless crystalline powder used as a fertilizer under the tradename Nitrolime. Calcium carbide is reacted with nitrogen in an electric furnace.

$$CaC_2 + N_2 = CaCN_2 + C$$

The calcium cyanamide reacts with water in the soil to form the ammonium ion and calcium carbonate. The ammonium is then oxidized to nitrate, which is taken up by plants. Calcium cyanamide can also be converted commercially into ammonia.

cyanide CN^- ion derived from hydrogen cyanide (HCN), and any salt containing this ion (produced when hydrogen cyanide is neutralized by alkalis), such as potassium cyanide (KCN). The principal cyanides are potassium, sodium, calcium, mercury, gold, and copper. Certain cyanides are poisons.

cyanobacteria (singular *cyanobacterium*) alternative name for ◊blue-green algae.

cyanocobalamin chemical name for ◊vitamin B_{12}, which is normally produced by microorganisms in the gut. The richest natural source is raw liver. The deficiency disease, pernicious anaemia, is the poor development of red blood cells with possible degeneration of the spinal chord. Sufferers develop extensive bruising and recover slowly from even minor injuries.

cybernetics (Greek *kubernan* 'to steer') science concerned with how systems organize, regulate, and reproduce themselves, and also how they evolve and learn. In the laboratory, inanimate objects are created that behave like living systems. Applications range from the creation of electronic artificial limbs to the running of the fully automated factory where decision-making machines operate up to managerial level.

Cybernetics was founded and named in 1947 by US mathematician Norbert Wiener. Originally, it was the study of control systems using feedback to produce automatic processes.

cyclamate derivative of cyclohexylsulphamic acid, formerly used as an artificial ◊sweetener, 30 times sweeter than sugar. It was first synthesized 1937.

Its use in foods was banned in the USA and the UK from 1970, when studies showed that massive doses caused cancer in rats.

cycle in physics, a sequence of changes that moves a system away from, and then back to, its original state. An example is a vibration that moves a particle first in one direction and then in the opposite direction, with the particle returning to its original position at the end of the vibration.

cyclic compound any of a group of organic chemicals that have rings of atoms in their molecules, giving them a closed-chain structure.
They may be alicyclic (cyclopentane), aromatic (benzene), or heterocyclic (pyridine). *Alicyclic compounds* (*aliphatic cyclic*) have localized bonding: all the electrons are confined to their own particular bonds, in contrast to *aromatic compounds*, where certain electrons have free movement between different bonds in the ring. Alicyclic compounds have chemical properties similar to their straight-chain counterparts; aromatic compounds, because of their special structure, undergo entirely different chemical reactions. *Heterocyclic compounds* have a ring of carbon atoms with one or more carbons replaced by another element, usually nitrogen, oxygen, or sulphur. They may be aliphatic or aromatic in nature.

The whole history of science has been the gradual realization that events do not happen in an arbitrary manner, but that they reflect a certain underlying order, which may or may not be divinely inspired.

Stephen Hawking *A Brief History of Time* 1988

---

cyclic polygon in geometry, a polygon in which each vertex (corner) lies on the circumference of a circle.

cycloid in geometry, a curve resembling a series of arches traced out by a point on the circumference of a circle that rolls along a straight line. Its applications include the study of the motion of wheeled vehicles along roads and tracks.

cyclone alternative name for a ◊depression, an area of low atmospheric pressure. A severe cyclone that forms in the tropics is called a tropical cyclone or ◊hurricane.

cyclotron circular type of particle ◊accelerator.

Cygnus large prominent constellation of the northern hemisphere, named after its shape (Latin 'swan'). Its brightest star is first-magnitude ◊Deneb.

Beta Cygni (Albireo) is a yellow and blue ◊double star, visible by small telescopes. The constellation contains the North America nebula (named after its shape), the Veil nebula (the remains of a supernova that exploded about 50,000

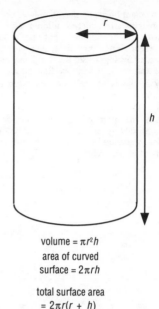

volume = $\pi r^2 h$
area of curved
surface = $2\pi r h$

total surface area
= $2\pi r(r + h)$

cylinder *The volume and area of a cylinder are given by simple formulae relating the dimensions of the cylinder.*

years ago), Cygnus A (apparently a double galaxy, and a powerful radio source), and the X-ray source Cygnus X-1, thought to mark the position of a black hole.

cylinder in geometry, a tubular solid figure with a circular base. In everyday use, the term applies to a *right cylinder*, the curved surface of which is at right angles to the base.

The volume V of a cylinder is given by the formula $V = \pi r^2 h$, where r is the radius of the base and h is the height of the cylinder. Its total surface area A has the formula $A = 2\pi r(h + r)$, where $2\pi rh$ is the curved surface area, and $2\pi r^2$ is the area of both circular ends.

cytochrome protein responsible for part of the process of ◊respiration by which food molecules are broken down in ◊aerobic organisms. Cytochromes are part of the electron transport chain, which uses energized electrons to reduce molecular oxygen (O_2) to oxygen ions (O^{2-}). These combine with hydrogen ions (H^+) to form water (H_2O), the end product of aerobic respiration. As electrons are passed from one cytochrome to another, energy is released and used to make ◊ATP.

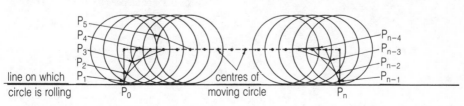

line on which P_1 centres of   circle is rolling P_0 moving circle P_n

cycloid *The cycloid is the curve traced out by a point on a circle as it rolls along a straight line. The teeth of gears are often cut with faces that are arcs of cycloids so that there is rolling contact when the gears are in use.*

cytokinin ◊plant hormone that stimulates cell division. Cytokinins affect several different aspects of plant growth and development, but only if ◊auxin is also present. They may delay the process of senescence, or ageing, break the dormancy of certain seeds and buds, and induce flowering.

cytology the study of ◊cells and their functions. Major advances have been made possible in this field by the development of ◊electron microscopes.

cytoplasm the part of the cell outside the ◊nucleus. Strictly speaking, this includes all the ◊organelles (mitochondria, chloroplasts, and so on), but often cytoplasm refers to the jellylike matter in which the organelles are embedded (correctly termed the cytosol).

In many cells, the cytoplasm is made up of two parts: the *ectoplasm* (or plasmagel), a dense gelatinous outer layer concerned with cell movement, and the *endoplasm* (or plasmasol), a more fluid inner part where most of the organelles are found.

cytoskeleton in a living cell, a matrix of protein filaments and tubules that occurs within the cytosol (the liquid part of the cytoplasm). It gives the cell a definite shape, transports vital substances around the cell, and may also be involved in cell movement.

DAC abbreviation for ◊*digital-to-analogue converter.*

daguerreotype in photography, a single-image process using mercury vapour and an iodine-sensitized silvered plate; it was invented by French-man Louis Daguerre 1838.

daisywheel printing head in a computer printer or typewriter that consists of a small plastic or metal disc made up of many spokes (like the petals of a daisy). At the end of each spoke is a character in relief. The daisywheel is rotated until the spoke bearing the required character is facing an inked ribbon, then a hammer strikes the spoke against the ribbon, leaving the impression of the character on the paper beneath.

The daisywheel can be changed to provide different typefaces; however, daisywheel printers cannot print graphics nor can they print more than one typeface in the same document. For these reasons, they are rapidly becoming obsolete.

dam structure built to hold back water in order to prevent flooding, provide water for irrigation and storage, and to provide hydroelectric power. The biggest dams are of the earth- and rock-fill type, also called *embankment dams.* Early dams in Britain, built before about 1800, had a core made from puddled clay (clay which has been mixed with water to make it impermeable). Such dams are generally built on broad valley sites. Deep, narrow gorges dictate a *concrete dam*, where the strength of reinforced concrete can withstand the water pressures involved. The first major all-concrete dam in Britain was built at Woodhead in 1876. The first dam in which concrete was used to seal the joints in the rocks below was built at Tunstall 1879.

A valuable development in arid regions, as in parts of Brazil, is the **underground dam**, where water is stored on a solid rock base, with a wall to ground level, so avoiding rapid evaporation.

Many concrete dams are triangular in cross section, with their vertical face pointing upstream. Their sheer weight holds them in position, and they are called *gravity dams.* Other concrete dams are more slightly built in the shape of an arch, with the curve facing upstream: the **arch dam** derives its strength from the arch shape, just as an arch bridge does.

Major dams include: Rogun (Tajikistan), the world's highest at 335 m/1,099 ft; New Cornelia Tailings (USA), the world's biggest in volume, 209 million cu m/7.4 billion cu ft; Owen Falls (Uganda), the world's largest reservoir capacity, 204.8 billion cu m/7.2 trillion cu ft; and Itaipu (Brazil/Paraguay), the world's most powerful, producing 12,700 megawatts of electricity.

daminozide (trade name *Alar*) chemical formerly used by fruit growers to make apples redder and crisper. In 1989 a report published in the USA found the consumption of daminozide to be linked with cancer, and the US Environment Protection Agency (EPA) called for an end to its use. The makers have now withdrawn it worldwide.

damper any device that deadens or lessens vibrations or oscillations; for example, one used to check vibrations in the strings of a piano. The term is also used for the movable plate in the flue of a stove or furnace for controlling the draught.

Daniell cell disposable, or primary, electrical cell invented 1836 by British chemist and meteorologist John Frederic Daniell (1790–1845). It consists of a central zinc ◊cathode dipping into a porous pot containing zinc sulphate solution. The porous pot is, in turn, immersed in a solution of copper sulphate contained in a copper can, which acts as the cell's ◊anode. The use of a porous barrier prevents polarization (the covering of the anode with small bubbles of hydrogen gas) and allows the cell to generate a continuous current of electricity.

darcy cgs unit (symbol D) of permeability, used mainly in geology to describe the permeability of rock (for example, to oil, gas, or water).

dark matter hypothetical matter that, according to current theories of ◊cosmology, makes up 90–99% of the mass of the universe but so far remains undetected. Astronomers are unsure if it

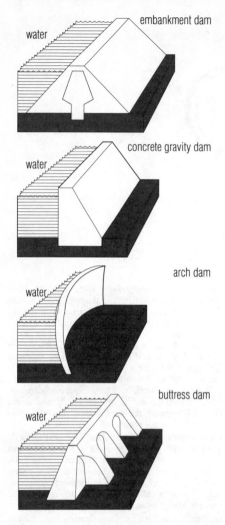

embankment dam

concrete gravity dam

arch dam

buttress dam

dam *There are two basic types of dam: the gravity dam and the arch dam. The gravity dam relies upon the weight of its material to resist the forces imposed upon it; the arch dam uses an arch shape to take the forces in a horizontal direction into the sides of the river valley. Buttress dams are used to hold back very wide rivers or lakes.*

consists of unknown atomic particles (cold dark matter) or fast-moving neutrinos (hot dark matter) or a combination of both.

DAT abbreviation for ◊*digital audio tape*.

data facts, figures, and symbols, especially as stored in computers. The term is often used to mean raw, unprocessed facts, as distinct from information, to which a meaning or interpretation has been applied.

Errors using inadequate data are much less than those using no data at all.

On **data** Charles Babbage (1792–1871)

---

database structured collection of data. The database makes data available to the various programs that need it, without the need for those programs to be aware of how the data are stored. There are three main types (or 'models'): hierarchical, network, and ◊relational, of which relational is the most widely used. A *free-text database* is one that holds the unstructured text of articles or books in a form that permits rapid searching.

A collection of databases is known as a *databank*. A database-management system (DBMS) program ensures that the integrity of the data is maintained by controlling the degree of access of the ◊application programs using the data. Databases are normally used by large organizations with mainframes or minicomputers.

A telephone directory stored as a database might allow all the people whose names start with the letter B to be selected by one program, and all those living in Chicago by another.

data bus in computing, the electrical pathway, or ◊bus, used to carry data between the components of the computer.

data capture collecting information for computer processing and analysis. Data may be captured automatically—for example, by a ◊sensor that continuously monitors physical conditions such as temperature—or manually; for example, by reading electricity meters.

data communications sending and receiving data via any communications medium, such as a telephone line. The term usually implies that the data are digital (such as computer data) rather than analogue (such as voice messages). However, in the ISDN (◊Integrated Services Digital Network) system, all data—including voices and video images—are transmitted digitally. See also ◊telecommunications.

data compression in computing, techniques for reducing the amount of storage needed for a given amount of data. They include word tokenization (in which frequently used words are stored as shorter codes), variable bit lengths (in which common characters are represented by fewer ◊bits than less common ones), and run-length encoding (in which a repeated value is stored once along with a count).

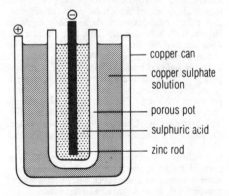

copper can

copper sulphate solution

porous pot

sulphuric acid

zinc rod

Daniell cell *The Daniell cell was the first reliable battery, supplying a steady current for a long time. It quickly became the standard form of battery.*

data flow chart diagram illustrating the possible routes that data can take through a system or program; see ◊flow chart.

data logging in computing, the process, usually automatic, of capturing and recording a sequence of values for later processing and analysis by computer. For example, the level in a water-storage tank might be automatically logged every hour over a seven-day period, so that a computer could produce an analysis of water use.

data processing (DP) use of computers for performing clerical tasks such as stock control, payroll, and dealing with orders. DP systems are typically ◊batch systems, running on mainframe computers. DP is sometimes called EDP (electronic data processing).

A large organization usually has a special department to support its DP activities, which might include the writing and maintenance of software (programs), control and operation of the computers, and an analysis of the organization's information requirements.

data protection safeguarding of information about individuals stored on computers, to protect privacy. The Council of Europe adopted, in 1981, a Data Protection Convention, which led in the UK to the Data Protection Act 1984. This requires computer databases containing personal information to be registered, and users to process only accurate information and to retain the information only for a necessary period and for specified purposes. Subject to certain exemptions, individuals have a right of access to their personal data and to have any errors corrected.

data terminator or *rogue value* in computing, a special value used to mark the end of a list of input data items. The computer must be able to detect that the data terminator is different from the input data in some way—for instance, a negative number might be used to signal the end of a list of positive numbers, or 'XXX' might be used to terminate the entry of a list of names.

dating science of determining the age of geological structures, rocks, and fossils, and placing them in the context of geological time. The techniques are of two types: relative dating and absolute dating.

Relative dating can be carried out by identifying fossils of creatures that lived only at certain times (marker fossils), and by looking at the physical relationships of rocks to other rocks of a known age.

Absolute dating is achieved by measuring how much of a rock's radioactive elements have changed since the rock was formed, using the process of ◊radiometric dating.

David Dunlap Observatory Canadian observatory at Richmond Hill, Ontario, operated by the University of Toronto, with a 1.88–m/74–in reflector, the largest optical telescope in Canada, opened 1935.

day time taken for the Earth to rotate once on its axis. The *solar day* is the time that the Earth takes to rotate once relative to the Sun. It is divided into 24 hours, and is the basis of our civil day. The *sidereal day* is the time that the Earth takes to rotate once relative to the stars. It is 3 minutes 56 seconds shorter than the solar day, because the Sun's position against the background of stars as seen from Earth changes as the Earth orbits it.

dBASE family of microcomputer programs used for manipulating large quantities of data; also, a related ◊fourth-generation language. The first version, dBASE II, appeared in 1981; it has since become the basis for a recognized standard for database applications, known as Xbase.

DCC abbreviation for ◊*digital compact cassette*.

DDT abbreviation for *dichloro-diphenyl-trichloroethane* ($ClC_6H_5)_2CHCHCl_2$, an insecticide discovered 1939 by Swiss chemist Paul Müller. It is useful in the control of insects that spread malaria, but resistant strains develop.

DDT is highly toxic and, because it is chemically stable, persists in the environment and in living tissue for several years. Since all living things are linked by ◊food chains, dangerous concentrations can build up at the top of the chains, with birds of prey and other predators receiving heavy doses. Its use is now banned in most countries, but it continues to be used on food plants in Latin America.

deamination removal of the amino group (-NH_2) from an unwanted ◊amino acid. This is the nitrogen-containing part, and it is converted into ammonia, uric acid, or urea (depending on the type of animal) to be excreted in the urine. In vertebrates, deamination occurs in the ◊liver.

death the cessation of all life functions, so that the molecules and structures associated with living things become disorganized and indistinguishable from similar molecules found in non-living things.

Large molecules such as proteins break down into simpler, soluble components. Eventually the organism partly dissoves into the soil or evaporates into the air, leaving behind skeletal material. Living organisms expend large amounts of energy preventing their complex molecules from breaking up; cellular repair and replacement are vital processes in multicellular organisms. At death this energy is no longer available, and the processes of disorganization become inevitable.

Biologists have a problem in explaining the phenomenon of death. If proteins, other complex molecules, and whole cells can be repaired or replaced, why cannot a multicellular organixm be immortal? The most favoured explanation is an evolutionary one. Organisms must die in order to make way for new ones, which, by virtue of ◊sexual reproduction, may vary slightly in relation to the previous generation. Most environments change constantly, if slowly; without this variation organisms would be unable to adapt to the changes.

debt-for-nature swap agreement under which a proportion of a country's debts are written off in exchange for a commitment by the debtor country to undertake projects for environmental protection. Debt-for-nature swaps were set up by environment groups in the 1980s in an attempt to reduce the debt problem of poor countries, while simultaneously promoting conservation.

To date, most debt-for-nature swaps have concentrated on setting aside areas of land, especially tropical rainforest, for protection and have involved private conservation foundations. The first swap took place 1987, when a US conservation group bought $650,000 of Bolivia's national debt from a bank for $100,000, and persuaded the Bolivian government to set aside a large area of rainforest as a nature reserve in exchange for never having to pay back the money owed. Other countries participating in debt-for-nature swaps are the Philippines, Costa Rica, Ecuador, and Poland. However, the debtor country is expected to ensure that the area of land remains adequately protected, and in practice this does not always happen. The practice has also produced complaints of neocolonialism.

debugging finding and removing errors, or ◊bugs, from a computer program or system.

decagon in geometry, a ten-sided ◊polygon.

decay, radioactive see ◊radioactive decay.

Decca navigation system radio-based aid to marine navigation, available in many parts of the world. The system consists of a master radio transmitter and two or three secondary transmitters situated within 100–200 km/60–120 mi from the master. The signals from the transmitters are detected by the ship's navigation receiver, and slight differences (phase differences) between the master and secondary transmitter signals indicate the position of the receiver.

The system can be used up to 550 km/344 mi from the master transmitter, and is accurate to within 50 m/160 ft at a range of 180 km/112 mi.

decibel unit (symbol dB) of measure used originally to compare sound intensities and subsequently electrical or electronic power outputs; now also used to compare voltages. An increase of 10 dB is equivalent to a 10-fold increase in intensity or power, and a 20-fold increase in voltage. A whisper has an intensity of 20 dB; 140 dB (a jet aircraft taking off nearby) is the threshold of pain.

The difference in decibels between two levels of intensity (or power) L_1 and L_2 is $10 \log_{10}(L_1/L_2)$; a difference of 1 dB thus corresponds to a change of about 25%. For two voltages V_1 and V_2, the difference in decibels is $20 \log_{10}(V_1/V_2)$; 1 dB corresponding in this case to a change of about 12%. Commonly such differences are given now not as ratios but as absolute values; for example, 10 dBV corresponds to the voltage level V_1 with V_2 set equal to 1 volt. In acoustics, the absolute reference used in the power ratio is 10^{-16} watt per sq cm.

deciduous of trees and shrubs, that shed their leaves at the end of the growing season or during a dry season to reduce ◊transpiration, the loss of water by evaporation.

Most deciduous trees belong to the ◊angiosperms, plants in which the seeds are enclosed within an ovary, and the term 'deciduous tree' is sometimes used to mean 'angiosperm tree', despite the fact that many angiosperms are evergreen, especially in the tropics, and a few ◊gymnosperms, plants in which the seeds are exposed, are deciduous (for example, larches). The term *broad-leaved* is now preferred to 'deciduous' for this reason.

Examples of deciduous trees are oak and beech.

decimal fraction a ◊fraction in which the denominator is any higher power of 10. Thus ³/₁₀, 51/100, and ²³/₁,₀₀₀ are decimal fractions and are normally expressed as 0.3, 0.51, 0.023. The use of decimals greatly simplifies addition and multiplication of fractions, though not all fractions can be expressed exactly as decimal fractions.

The regular use of the decimal point appears to have been introduced about 1585, but the occasional use of decimal fractions can be traced back as far as the 12th century.

decimal number system or *denary number system* the most commonly used number system, to the base ten. Decimal numbers do not necessarily contain a decimal point; 563, 5.63, and –563 are all decimal numbers. Other systems are mainly used in computing and include the ◊binary number system, ◊octal number system, and ◊hexadecimal number system.

Decimal numbers may be thought of as written under column headings based on the number ten. For example, the number 2,567 stands for:

1,000s	100s	10s	1s
(10^3)	(10^2)	(10^1)	(10^0)
2	5	6	7

Large decimal numbers may also be expressed in ◊floating-point notation.

decision table in computing, a method of describing a procedure for a program to follow, based on comparing possible decisions and their consequences. It is often used as an aid in systems design.

The top part of the table contains the conditions for making decisions (for example, if a number is negative rather than positive and is less than 1), and the bottom part describes the outcomes when those conditions are met. The program either ends or repeats the operation.

declination in astronomy, the coordinate on the ◊celestial sphere (imaginary sphere surrounding the Earth) that corresponds to latitude on the Earth's surface. Declination runs from 0° at the celestial equator to 90° at the north and south celestial poles.

decoder in computing, an electronic circuit used to select one of several possible data pathways. Decoders are, for example, used to direct data to individual memory locations within a computer's immediate access memory.

decomposer in biology, any organism that breaks down dead matter. Decomposers play a vital role in the ◊ecosystem by freeing important chemical substances, such as nitrogen compounds, locked up in dead organisms or excrement. They feed on some of the released organic matter, but leave the rest to filter back into the soil or pass in gas form into the atmosphere. The principal decomposers are bacteria and fungi, but earthworms and many other invertebrates are often included in this group. The ◊nitrogen cycle relies on the actions of decomposers.

decomposition process whereby a chemical compound is reduced to its component substances. In biology, it is the destruction of dead organisms either by chemical reduction or by the action of decomposers.

decontamination factor in radiological protection, a measure of the effectiveness of a decontamination process. It is the ratio of the original contamination to the remaining radiation after decontamination: 1,000 and above is excellent; 10 and below is poor.

decrepitation unusual features that accompany the thermal decomposition of some crystals, such as lead(II) nitrate. When these are heated, they spit and crackle and may jump out of the test tube before they decompose.

dedicated computer computer built into another device for the purpose of controlling or supplying information to it. Its use has increased dramatically since the advent of the ◊microprocessor: washing machines, digital watches, cars, and video recorders all now have their own processors.

A dedicated system is a general-purpose comuter system confined to performing only one function for reasons of efficiency or convenience. A word processor is an example.

deep freezing method of preserving food by rapid freezing and storage at –18°C/–1°F; see ◊food technology.

Commercial freezing is usually achieved by blasting with air at –40°C/–40°F; contact with hollow shelves in which a refrigerant is circulated; or spraying with liquid nitrogen. ◊Freeze-drying is another method of preserving food.

Deep-Sea Drilling Project research project initiated by the USA 1968 to sample the rocks of the ocean ◊crust. In 1985 it became known as the ◊Ocean Drilling Program (ODP).

deep-sea trench another term for ◊ocean trench.

deforestation destruction of forest for timber, fuel, charcoal burning, and clearing for agriculture and extractive industries, such as mining, without planting new trees to replace those lost (reafforestation) or working on a cycle that allows the natural forest to regenerate. Deforestation causes fertile soil to be blown away or washed into rivers, leading to ◊soil erosion, drought, flooding, and loss of wildlife.

Deforestation is taking place in both tropical rainforests and temperate forests.

degaussing neutralization of the magnetic field around a body by encircling it with a conductor through which a current is maintained. Ships were degaussed in World War II to prevent them from detonating magnetic mines.

degree in mathematics, a unit (symbol °) of measurement of an angle or arc. A circle or complete rotation is divided into 360°. A degree may be subdivided into 60 minutes (symbol ′), and each minute may be subdivided in turn into 60 seconds (symbol ″). *Temperature* is also measured in degrees, which are divided on a decimal scale. See also ◊Celsius, and ◊Fahrenheit scale.

A degree of latitude is the length along a meridian such that the difference between its north and south ends subtend an angle of 1° at the centre of the Earth. A degree of longitude is the length between two meridians making an angle of 1° at the centre of the Earth.

dehydration removal of water from a substance to give a product with a new chemical formula; it is not the same as ◊drying.

There are two types of dehydration. For substances such as hydrated copper sulphate ($CuSO_4.5H_2O$) that contain ◊water of crystallization, dehydration means removing this water to leave the anhydrous substance. This may be achieved by heating, and is reversible.

$$CuSO_4.5H_2O \rightarrow CuSO_4 + 5H_2O$$

Some substances, such as ethanol, contain the elements of water (hydrogen and oxygen) joined in a different form. *Dehydrating agents* such as concentrated sulphuric acid will remove these elements in the ratio 2:1.

$$C_2H_5OH \rightarrow CH_2=CH_2 + H_2O.$$

Deimos one of the two moons of Mars. It is irregularly shaped, 15 × 12 × 11 km/9 × 7.5 × 7 mi, orbits at a height of 24,000 km/15,000 mi every 1.26 days, and is not as heavily cratered as the other moon, Phobos. Deimos was discovered 1877 by US astronomer Asaph Hall (1829–1907), and is thought to be an asteroid captured by Mars's gravity.

deliquescence phenomenon of a substance absorbing so much moisture from the air that it ultimately dissolves in it to form a solution.

Deliquescent substances make very good drying agents and are used in the bottom chambers of ◊desiccators. Calcium chloride ($CaCl_2$) is one of the commonest.

delta tract of land at a river's mouth, composed of silt deposited as the water slows on entering the sea. Familiar examples of large deltas are those of the Mississippi, Ganges and Brahmaputra, Rhône, Po, Danube, and Nile; the shape of the Nile delta is like the Greek letter *delta* Δ, and thus gave rise to the name.

The *arcuate delta* of the Nile is only one form. Others are *birdfoot deltas*, like that of the Mississippi which is a seaward extension of the river's ◊levee system; and *tidal deltas*, like that of the

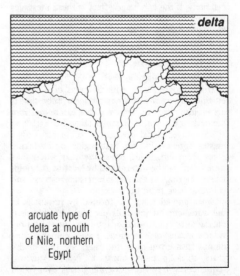

delta

arcuate type of delta at mouth of Nile, northern Egypt

delta *A delta containing a fan of minor channels with one edge bulging beyond the shoreline called 'arcuate'.*

Mekong, in which most of the material is swept to one side by sea currents.

Delta rocket US rocket used to launch many scientific and communications satellites since 1960, based on the Thor ballistic missile. Several increasingly powerful versions were produced as satellites and became larger and heavier. Solid-fuel boosters were attached to the first stage to increase lifting power.

delta wing aircraft wing shaped like the Greek letter *delta* Δ. Its design enables an aircraft to pass through the ◊sound barrier with little effect. The supersonic airliner ◊Concorde and the US ◊space shuttle have delta wings.

demodulation in radio, the technique of separating a transmitted audio frequency signal from its modulated radio carrier wave. At the transmitter the audio frequency signal (representing speech or music, for example) may be made to modulate the amplitude (AM broadcasting) or frequency (FM broadcasting) of a continuously transmitted radio-frequency carrier wave. At the receiver, the signal from the aerial is demodulated to extract the required speech or sound component. In early radio systems, this process was called detection. See ◊modulation.

denaturation irreversible changes occurring in the structure of proteins such as enzymes, usually caused by changes in pH or temperature. An example is the heating of egg albumen resulting in solid egg white.

The enzymes associated with digestion and metabolism become inactive if given abnormal conditions. Heat will damage their complex structure so that the usual interactions between enzyme and substrate can no longer occur.

dendrite part of a ◊nerve cell or neuron. The dendrites are slender filaments projecting from the cell body. They receive incoming messages from many other nerve cells and pass them on to the cell body. If the combined effect of these messages is strong enough, the cell body will send an electrical impulse along the axon (the threadlike extension of a nerve cell). The tip of the axon passes its message to the dendrites of other nerve cells. See ◊central nervous system.

dendrochronology analysis of the ◊annual rings of trees to date past events. Samples of wood are obtained by means of a narrow metal tube that is driven into a tree to remove a core extending from the bark to the centre. Samples taken from timbers at an archaeological site can be compared with a master core on file for that region or by taking cores from old, living trees; the year when they were felled can be determined by locating the point where the rings of the two samples correspond and counting back from the present.

Since annual rings are formed by variations in the water-conducting cells produced by the plant during different seasons of the year, they also provide a means of determining past climatic conditions in a given area (the rings are thin in dry years, thick in moist ones). In North America, sequences of tree rings extending back over 8,000 years have been obtained for the SW and N Mexico by using cores from the bristle-cone pine *Pinus*

aristata, which can live for over 4,000 years in that region. Also, the dryness of the area has preserved wood in SW archaeological sites; in wet temperate regions the soil acidity usually absorbs wood, so this dating technique cannot be used.

Deneb or *Alpha Cygni* brightest star in the constellation Cygnus, and the 19th brightest star in the sky. It is one of the greatest supergiant stars known, with a true luminosity of about 60,000 times that of the Sun. Deneb is about 1,800 light years from Earth.

denier unit used in measuring the fineness of yarns, equal to the mass in grams of 9,000 metres of yarn. Thus 9,000 metres of 15 denier nylon, used in nylon stockings, weighs 15 g/0.5 oz, and in this case the thickness of thread would be 0.00425 mm/0.0017 in. The term is derived from the French silk industry; the *denier* was an old French silver coin.

denitrification process occurring naturally in soil, where bacteria break down ◊nitrates to give nitrogen gas, which returns to the atmosphere. See ◊nitrogen cycle.

density measure of the compactness of a substance; it is equal to its mass per unit volume and is measured in kg per cubic metre/lb per cubic foot. Density is a ◊scalar quantity. The density D

densities of some common substances

	density (in kg m^{-3})
solids	
balsa wood	200
oak	700
butter	900
ice	920
ebony	120
sand (dry)	1,600
concrete	2,400
aluminium	2,700
steel	7,800
copper	8,900
lead	11,000
uranium	19,000
liquids	
water	1,000
petrol, paraffin	800
olive oil	900
milk	1,030
sea water	1,030
glycerine	1,300
Dead Sea brine	1,800
gases at standard temperature and pressure (0˚C and 1 atm)	
air	1.3
hydrogen	0.09
helium	0.18
methane	0.72
nitrogen	1.25
oxygen	1.43
carbon dioxide	1.98
propane	2.02
butane	2.65

of a mass m occupying a volume V is given by the formula:

$$D = m/V$$

◊Relative density is the ratio of the density of a substance to that of water at 4°C.

In photography, density refers to the degree of opacity of a negative; in population studies, it is the quantity or number per unit of area; in electricity, current density is the amount of current passing through a cross-sectional area of a conductor in a given amount of time (usually given in amperes per sq in or per sq cm).

DENSITY: THE LEAST-DENSE SUBSTANCE

The least-dense solid material is made from seaweed and is called Seagel. It was first made in 1992 at the US Government Lawrence Livermore National Laboratory in California. Seagel is lighter than air and would float away if it were not weighed down by the air trapped in its pores.

dental formula way of showing what an animal's teeth are like. The dental formula consists of eight numbers separated by a line into two rows. The four above the line represent the teeth in one side of the upper jaw, starting at the front. If this reads 2 1 2 3 (as for humans) it means two incisors, one canine, two premolars, and three molars (see ◊tooth). The numbers below the line represent the lower jaw. The total number of teeth can be calculated by adding up all the numbers and multiplying by two.

dentition type and number of teeth in a species. Different kinds of teeth have different functions; a grass-eating animal will have large molars for grinding its food, whereas a meat-eater will need powerful canines for catching and killing its prey and ◊carnassial teeth for slicing flesh. The teeth that are less useful may be reduced in size or missing altogether. An animal's dentition is represented diagramatically by a ◊dental formula.

denudation natural loss of soil and rock debris, blown away by wind or washed away by running water, that lays bare the rock below. Over millions of years, denudation causes a general lowering of the landscape.

deoxyribonucleic acid full name of ◊DNA.

depolarizer oxidizing agent used in dry-cell batteries that converts hydrogen released at the negative electrode into water. This prevents the build-up of gas, which would otherwise impair the efficiency of the battery. ◊Manganese(IV) oxide is used for this purpose.

depression or *cyclone* or *low* in meteorology, a region of low atmospheric pressure. A depression forms as warm, moist air from the tropics mixes with cold, dry polar air, producing warm and cold boundaries (◊fronts) and unstable weather—low cloud and drizzle, showers, or fierce storms. The warm air, being less dense, rises above the cold air to produce the area of low pressure on the ground. Air spirals in towards the centre of the depression in an anticlockwise direction in the northern hemi-

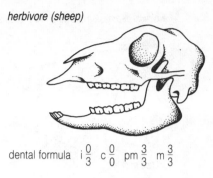

herbivore (sheep)

dental formula $i\,\dfrac{0}{3}$ $c\,\dfrac{0}{0}$ $pm\,\dfrac{3}{3}$ $m\,\dfrac{3}{3}$

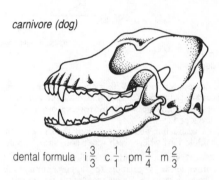

carnivore (dog)

dental formula $i\,\dfrac{3}{3}$ $c\,\dfrac{1}{1}$ $pm\,\dfrac{4}{4}$ $m\,\dfrac{2}{3}$

dentition *The dentition and dental formulae of a typical herbivore (sheep) and carnivore (dog). The dog has long pointed canines for puncturing and gripping its prey (they are also bared in threat), and has modified premolars and molars (carnassials) for shearing flesh. In the sheep, by contrast, there is a wide gap, or diastema, between the incisors and premolars; no canines exist.*

sphere, clockwise in the southern hemisphere, generating winds up to gale force. Depressions tend to travel eastwards and can remain active for several days.

A deep depression is one in which the pressure at the centre is very much lower than that round about; it produces very strong winds, as opposed to a shallow depression in which the winds are comparatively light. A severe depression in the tropics is called a ◊hurricane, tropical cyclone, or typhoon, and is a great danger to shipping; a ◊tornado is a very intense, rapidly swirling depression, with a diameter of only a few hundred metres or so.

derivative or *differential coefficient* in mathematics, the limit of the gradient of a chord linking two points on a curve as the distance between the points tends to zero; for a function with a single variable, $y = f(x)$, it is denoted by $f'(x)$, D$f(x)$, or dy/dx, and is equal to the gradient of the curve.

derrick simple lifting machine consisting of a pole carrying a block and tackle. Derricks are commonly used on ships that carry freight. In the oil industry the tower used for hoisting the drill pipes is known as a derrick.

desalination removal of salt, usually from sea water, to produce fresh water for irrigation or drinking. Distillation has usually been the method

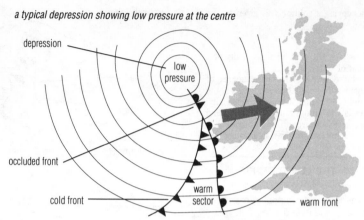

a typical depression showing low pressure at the centre

depression

low pressure

occluded front

cold front

warm sector

warm front

the fronts are associated with belts of rain (frontal rainfall)

depression

adopted, but in the 1970s a cheaper process, using certain polymer materials that filter the molecules of salt from the water by reverse osmosis, was developed.

Desalination plants occur along the shores of the Middle East where fresh water is in short supply.

desert arid area without sufficient rainfall and, consequently, vegetation to support human life. The term includes the ice areas of the polar regions (known as cold deserts). Almost 33% of Earth's land surface is desert, and this proportion is increasing.

The *tropical desert* belts of latitudes from 5° to 30° are caused by the descent of air that is heated over the warm land and therefore has lost its moisture. Other natural desert types are the *continental deserts*, such as the Gobi, that are too far from the sea to receive any moisture; *rain-shadow deserts*, such as California's Death Valley, that lie in the lee of mountain ranges, where the ascending air drops its rain only on the windward slopes; and *coastal deserts*, such as the Namib, where cold ocean currents cause local dry air masses to descend. Desert surfaces are usually rocky or gravelly, with only a small proportion being covered with sand. Deserts can be created by changes in climate, or by the human-aided process of desertification.

desertification creation of deserts by changes in climate, or by human-aided processes. Desertification can sometimes be reversed by special planting (marram grass, trees) and by the use of water-absorbent plastic grains, which, added to the soil, enable crops to be grown.

The processes leading to desertification include overgrazing, destruction of forest belts, and exhaustion of the soil by intensive cultivation without restoration of fertility—all of which may be prompted by the pressures of an expanding population or by concentration in land ownership. About 135 million people are directly affected by desertification, mainly in Africa, the Indian subcontinent, and South America.

desiccator airtight vessel, traditionally made of

glass, in which materials may be stored either to dry them or to prevent them, once dried, from reabsorbing moisture.

The base of the desiccator is a chamber in which is placed a substance with a strong affinity for water (such as calcium chloride or silica gel), which removes water vapour from the desiccator atmosphere and from substances placed in it.

desktop publishing (DTP) use of microcomputers for small-scale typesetting and page make-up. DTP systems are capable of producing camera-ready pages (pages ready for photographing and printing), made up of text and graphics, with text set in different typefaces and sizes. The page can be previewed on the screen before final printing on a laser printer.

destructive margin in ◊plate tectonics, a region on the Earth's crust in which two plates are moving towards one another. Usually one plate (the denser of the two) is forced to dive below the other into what is called the *subduction zone*. The descending plate melts to form a body of magma, which may then rise to the surface through cracks and faults to form volcanoes. If the two plates consist of more buoyant continental crust, subduction does not occur. Instead, the crust crumples gradually to form fold mountains, such as the Himalayas. See ◊continental drift.

detergent surface-active cleansing agent. The common detergents are made from ◊fats (hydrocarbons) and sulphuric acid, and their long-chain molecules have a type of structure similar to that of ◊soap molecules: a salt group at one end attached to a long hydrocarbon 'tail'. They have the advantage over soap in that they do not produce scum by forming insoluble salts with the calcium and magnesium ions present in hard water.

To remove dirt, which is generally attached to materials by means of oil or grease, the hydrocarbon 'tails' (soluble in oil or grease) penetrate the oil or grease drops, while the 'heads' (soluble in water but insoluble in grease) remain in the water and, being salts, become ionized. Consequently the oil drops become negatively charged

ocean trench and island arc formed by two plates moving towards one another

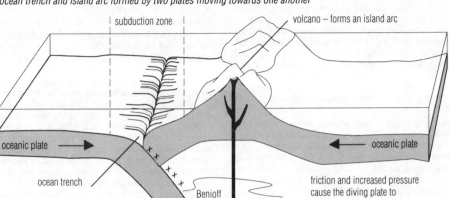

subduction zone

volcano – forms an island arc

oceanic plate →

← oceanic plate

ocean trench

Benioff zone magma

x earthquake foci

friction and increased pressure cause the diving plate to melt—magma escapes to the surface to form a volcano

destructive margin

and tend to repel one another; thus they remain in suspension and are washed away with the dirt.

Detergents were first developed from coal tar in Germany during World War I, and synthetic organic detergents came into increasing use after World War II. Domestic powder detergents for use in hot water have alkyl benzene as their main base, and may also include bleaches and fluorescers as whiteners, perborates to free stain-removing oxygen, and water softeners. Environment-friendly detergents contain no phosphates or bleaches. Liquid detergents for washing dishes are based on epoxyethane (ethylene oxide). Cold-water detergents consist of a mixture of various alcohols, plus an ingredient for breaking down the surface tension of the water, so enabling the liquid to penetrate fibres and remove the dirt. When these surface-active agents (surfactants) escape the normal processing of sewage, they cause troublesome foam in rivers; phosphates in some detergents can also cause the excessive enrichment (◊eutrophication) of rivers and lakes.

determinant in mathematics, an array of elements written as a square, and denoted by two vertical lines enclosing the array. For a 2 × 2 matrix, the determinant is given by the difference between the products of the diagonal terms. For example, the determinant of the matrix:

$$\begin{pmatrix} a & b \\ c & d \end{pmatrix} = \begin{vmatrix} a & b \\ c & d \end{vmatrix} = ad - bc$$

Determinants are used to solve sets of ◊simultaneous equations by matrix methods.

When applied to transformational geometry, the determinant of a 2 × 2 matrix signifies the ratio of the area of the transformed shape to the original and its sign (plus or minus) denotes whether the image is direct (the same way round) or indirect (a mirror image).

detonator or *blasting cap* or *percussion cap* small explosive charge used to trigger off a main charge of high explosive. The relatively unstable compounds mercury fulminate and lead acid are often used in detonators, being set off by a lighted fuse or, more commonly, an electric current.

detritus in biology, the organic debris produced during the ◊decomposition of animals and plants.

deuterium naturally occurring heavy isotope of hydrogen, mass number 2 (one proton and one neutron), discovered by Harold Urey 1932. It is sometimes given the symbol D. In nature, about one in every 6,500 hydrogen atoms is deuterium. Combined with oxygen, it produces 'heavy water' (D_2O), used in the nuclear industry.

DEUTERIUM: CONCENTRATED ENERGY

Deuterium is the most concentrated energy source known. When 1 kg of deuterium is converted to helium, it releases 3 million times more energy than burning the same weight of coal.

deuteron nucleus of an atom of deuterium (heavy hydrogen). It consists of one proton and one neutron, and is used in the bombardment of chemical elements to synthesize other elements.

Devarda's alloy mixture of aluminium and zinc in powder form used in the test for nitrate (NO_3^-). Sodium hydroxide solution is added to the solid or solution, followed by the Devarda's alloy, and the mixture is heated. The release of ammonia gas indicates the presence of a nitrate.

developing in photography, the process that produces a visible image on exposed photographic ◊film, involving the treatment of the exposed film with a chemical developer.

The developing liquid consists of a reducing agent that changes the light-altered silver salts in the film into darker metallic silver. The developed image is made permanent with a fixer, which dissolves away any silver salts which were not affected by light. The developed image is a negative, or reverse image: darkest where the strongest light hit the film, lightest where the least light fell. To produce a positive image, the negative is itself photographed, and the development process reverses the shading, producing the final print. Colour and black-and-white film can be obtained as direct reversal, slide, or transparency material. Slides and transparencies are used for projection or printing with a positive-to-positive process such as ♢Cibachrome.

development in biology, the process whereby a living thing transforms itself from a single cell into a vastly complicated multicellular organism, with structures, such as limbs, and functions, such as respiration, all able to work correctly in relation to each other. Most of the details of this process remain unknown, although some of the central features are becoming understood.

Apart from the sex cells (♢gametes), each cell within an organism contains exactly the same genetic code. Whether a cell develops into a liver cell or a brain cell depends therefore not on which ♢genes it contains, but on which genes are allowed to be expressed. The development of forms and patterns within an organism, and the production of different, highly specialized cells, is a problem of control, with genes being turned on and off according to the stage of development reached by the organism.

devil wind minor form of ♢tornado, usually occurring in fine weather; formed from rising thermals of warm air (as is a ♢cyclone). A fire creates a similar updraught.

A **fire devil** or **firestorm** may occur in oil-refinery fires, or in the firebombings of cities, for example Dresden, Germany, in World War II.

Devonian period of geological time 408–360 million years ago, the fourth period of the Palaeozoic era. Many desert sandstones from North America and Europe date from this time. The first land plants flourished in the Devonian period, corals were abundant in the seas, amphibians evolved from air-breathing fish, and insects developed on land.

The name comes from the county of Devon in SW England, where Devonian rocks were first studied.

dew precipitation in the form of moisture that collects on the ground. It forms after the temperature of the ground has fallen below the ♢dew point of the air in contact with it. As the temperature falls during the night, the air and its water vapour become chilled, and condensation takes place on the cooled surfaces.

dew point temperature at which the air becomes saturated with water vapour. At temperatures below the dew point, the water vapour condenses out of the air as droplets. If the droplets are large they become deposited on the ground as dew; if small they remain in suspension in the air and form mist or fog.

dewpond drinking pond for farm animals on arid hilltops such as chalk downs. In the UK, dewponds were excavated in the 19th century and lined with mud and clay. It is uncertain where the water

comes from but it may be partly rain, partly sea mist, and only a small part dew.

diabase alternative name for ♢dolerite (a form of basalt that contains very little silica), especially dolerite that has metamorphosed.

diabetes disease *diabetes mellitus* in which a disorder of the islets of Langerhans in the ♢pancreas prevents the body producing the hormone ♢insulin, so that sugars cannot be used properly. Treatment is by strict dietary control and oral or injected insulin, depending on the type of diabetes.

Sugar accumulates first in the blood, then in the urine. The patient experiences thirst, weight loss, and copious voiding, along with degenerative changes in the capillary system. Without treatment, the patient may go blind, ulcerate, lapse into diabetic coma, and die. Before the discovery of insulin by Frederick Banting and Charles Best, severe diabetics did not survive.

In 1989, it was estimated that 4% of the world's population had diabetes, and that there were 12 million sufferers in Canada and the USA. In the UK, about 2% of the population is affected by diabetes.

diagenesis or *lithification* in geology, the physical and chemical changes by which a sediment becomes a ♢sedimentary rock. The main processes involved include compaction of the grains, and the cementing of the grains together by the growth of new minerals deposited by percolating groundwater.

dialysis in medicine, the process used to mimic the effects of the kidneys. It may be life-saving in some types of poisoning. Dialysis is usually performed to compensate for failing kidneys; there are

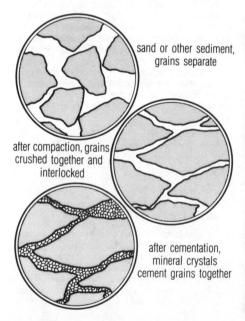

sand or other sediment, grains separate

after compaction, grains crushed together and interlocked

after cementation, mineral crystals cement grains together

diagenesis *The formation of sedimentary rock by diagenesis. Sand and other sediment grains are compacted and cemented together.*

two main methods, haemodialysis and peritoneal dialysis.

In *haemodialysis*, the patient's blood is passed through a pump, where it is separated from sterile dialysis fluid by a semipermeable membrane. This allows any toxic substances which have built up in the bloodstream, and which would normally be secreted by the kidneys, to diffuse out of the blood into the dialysis fluid. The red and white blood cells, however, are maintained in the circulation by the membrane. Haemodialysis is very expensive and requires the patient to attend a specialized unit.

Peritoneal dialysis uses one of the body's natural semipermeable membranes for the same purpose. About two litres of dialysis fluid is slowly instilled into the peritoneal cavity of the abdomen, and drained out again, over about two hours. During that time toxins from the blood diffuse into the peritoneal cavity across the peritoneal membrane. The advantage of peritoneal dialysis is that the patient can walk around while the dialysis is proceeding—this is known as continuous ambulatory peritoneal dialysis (CAPD).

In the long term dialysis is expensive and debilitating, and ◊transplants are now the treatment of choice for patients in chronic kidney failure.

diamagnetic material material which is weakly repelled by a magnet. All substances are diamagnetic but the behaviour is often masked by ◊paramagnetic or ◊ferromagnetic behaviour. Diamagnetic materials have a small negative magnetic ◊susceptibility.

diameter straight line joining two points on the circumference of a circle that passes through the centre of that circle. It divides a circle into two equal halves.

diamond generally colourless, transparent mineral, the hard crystalline form of carbon. It is regarded as a precious gemstone, and is the hardest substance known (10 on the ◊Mohs' scale). Industrial diamonds, which may be natural or synthetic, are used for cutting, grinding, and polishing.

Diamond crystallizes in the cubic system as octahedral crystals, some with curved faces and striations. The high refractive index of 2.42 and the high dispersion of light, or 'fire', account for the spectral displays seen in polished diamonds.

Diamonds were known before 3000 BC and until their discovery in Brazil 1725, India was the principal source of supply. Present sources are Australia, Zaire, Botswana, Russia (Yakut), South Africa, Namibia, and Angola; the first two produce large volumes of industrial diamonds. Today, about 80% of the world's rough gem diamonds are sold through the De Beers Central Selling Organization in London.

Diamonds may be found as alluvial diamonds on or close to the Earth's surface in riverbeds or dried watercourses; on the sea bottom (off SW Africa); or, more commonly, in diamond-bearing volcanic pipes composed of 'blue ground', ◊kimberlite or lamproite, where the original matrix has penetrated the Earth's crust from great depths. They are sorted from the residue of crushed ground by X-ray and other recovery methods.

There are four chief varieties of diamond: well-crystallized transparent stones, colourless or only slightly tinted, valued as gems; *boart*, poorly crystallized or inferior diamonds; *balas*, an industrial variety, extremely hard and tough; and *carbonado*, or industrial diamond, also called black diamond or carbon, which is opaque, black or grey, and very tough. Industrial diamonds are also produced synthetically from graphite. Some synthetic diamonds conduct heat 50% more efficiently than natural diamonds and are five times greater in strength. This is a great advantage in their use to disperse heat in electronic and telecommunication devices and in the production of laser components.

Because diamonds act as perfectly transparent windows and do not absorb infrared radiation, they were used aboard NASA space probes to Venus 1978. The tungsten-carbide tools used in steel mills are cut with industrial diamond tools.

Rough diamonds are often dull or greasy before being polished; around 50% are considered 'cuttable' (all or part of the diamond may be set into jewellery). Gem diamonds are valued by weight (◊carat), cut (highlighting the stone's optical properties), colour, and clarity (on a scale from internally flawless to having a large inclusion clearly visible to the naked eye). They are sawn and polished using a mixture of oil and diamond powder. The two most popular cuts are the brilliant, for thicker stones, and the marquise, for shallower ones. India is the world's chief cutting centre.

Noted rough diamonds include the Cullinan, or Star of Africa (3,106 carats, over 500 g/17.5 oz before cutting, South Africa, 1905); Excelsior (995.2 carats, South Africa, 1893); and Star of Sierra Leone (968.9 carats, Yengema, 1972).

DIANE (acronym from *D*irect *I*nformation *A*ccess *N*etwork for *E*urope) collection of information suppliers, or 'hosts', for the European computer network.

diapause period of suspended development that occurs in some species of insects, characterized by greatly reduced metabolism. Periods of diapause are often timed to coincide with the winter months, and improve the insect's chances of surviving adverse conditions.

diaphragm muscular sheet separating the thorax from the abdomen in mammals. Its rhythmical movements affect the size of the thorax and cause the pressure changes within the lungs that result in breathing.

The diaphragm muscle is under both voluntary and involuntary (automatic) control. Skilled divers can voluntarily hold their breath for up to five minutes, overriding the normal breathing cycles that we barely notice in day-to-day living.

diaphragm or *cap* or *Dutch cap* barrier ◊contraceptive that is pushed into the vagina and fits over the cervix (neck of the uterus), preventing sperm from entering the uterus. For a cap to be effective, a ◊spermicide must be used and the diaphragm left in place for 6–8 hours after intercourse. This method is 97% effective if practised correctly.

diapirism geological process in which a particularly light rock, such as rock salt, punches upwards through the heavier layers above. The resulting structure is called a salt dome, and oil is often trapped in the curled-up rocks at each side.

diatom microscopic alga of the division Bacillariophyta found in all parts of the world. Diatoms consist of single cells, sometimes grouped in colonies.

The cell wall is made up of two overlapping valves known as **frustules**, which are usually impregnated with silica, and which fit together like the lid and body of a pillbox. Diatomaceous earths (diatomite) are made up of the valves of fossil diatoms, and are used in the manufacture of dynamite and in the rubber and plastics industries.

DIATOM AT SEA

Diatoms are very plentiful in the seas: a litre of sea water may contain as many as 15,000 diatoms.

---

diatomic molecule molecule composed of two atoms joined together. In the case of an element such as oxygen (O_2), the atoms are identical.

dibasic acid acid containing two replaceable hydrogen atoms, such as sulphuric acid (H_2SO_4). The acid can form two series of salts, the normal salt (sulphate, SO_4^{2-}) and the acid salt (hydrogen-sulphate HSO_4^{-}).

dichloro-diphenyl-trichloroethane full name of the insecticide ◊DDT.

dichotomous key a method of identifying an organism. The investigator is presented with pairs of statements, for example 'flower has less than five stamens' and 'flower has five or more stamens'. By successively eliminating statements the field naturalist moves closer to a positive identification. Dichotomous keys assume a good knowledge of the subject under investigation.

dicotyledon major subdivision of the ◊angiosperms, containing the great majority of flowering plants. Dicotyledons are characterized by the presence of two seed leaves, or ◊cotyledons, in the embryo, which is usually surrounded by an ◊endosperm. They generally have broad leaves with net-like veins.

Dicotyledons may be small plants such as daisies and buttercups, shrubs, or trees such as oak and beech. The other subdivision of the angiosperms is the ◊monocotyledons.

diecasting form of ◊casting in which molten metal is injected into permanent metal moulds or dies.

dielectric substance (an insulator such as ceramic, rubber, or glass) capable of supporting electric stress. The dielectric constant, or relative permittivity, of a substance is the ratio of the capacitance of a capacitor with the substance as dielectric to that of a similar capacitor in which the dielectric is replaced by a vacuum.

diesel engine ◊internal-combustion engine that burns a lightweight fuel oil. The diesel engine operates by compressing air until it becomes sufficiently hot to ignite the fuel. It is a piston-in-cylinder engine, like the ◊petrol engine, but only air (rather than an air-and-fuel mixture) is taken into the cylinder on the first piston stroke (down). The piston moves up and compresses the air until it is at a very high temperature. The fuel oil is then injected into the hot air, where it burns, driving the piston

down on its power stroke. For this reason the engine is called a compression-ignition engine.

The principle of the diesel engine was first explained in England by Herbert Akroyd (1864–1937) in 1890, and was applied practically by Rudolf Diesel in Germany in 1892.

diesel oil lightweight fuel oil used in diesel engines. Like petrol, it is a petroleum product. When used in vehicle engines, it is also known as **derv**—*diesel-engine road vehicle*.

diet the range of foods eaten by an animal. The basic components of a diet are a group of chemicals: proteins, carbohydrates, fats, vitamins, minerals, and water. Different animals require these substances in different proportions, but the necessity of finding and processing an appropriate diet is a very basic drive in animal evolution. For instance, all guts are adapted for digesting and absorbing food, but different guts have adapted to cope with particular diets.

The diet an animal needs may vary over its lifespan, according to whether it is growing, reproducing, highly active, or approaching death. For instance, an animal may need increased carbohydrate for additional energy, or increased minerals during periods of growth.

difference in mathematics, the result obtained when subtracting one number from another. Also, those elements of one ◊set that are not elements of another.

difference engine mechanical calculating machine designed (and partly built 1822) by the British mathematician, Charles Babbage, to produce reliable tables of life expectancy. A precursor of the ◊analytical engine, it was to calculate mathematical functions by solving the differences between values given to ◊variables within equations. Babbage designed the calculator so that once the initial values for the variables were set it would produce the next few thousand values without error. In 1991, 200 years after Babbage's birth, the difference engine was finally constructed, at the Science Museum, London.

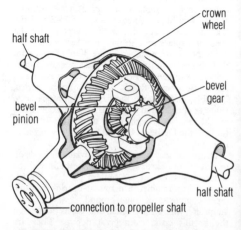

differential *The differential lies midway between the driving wheels of a motorcar. When the car is turning, the bevel pinions spin, allowing the outer wheel to turn faster than the inner wheel.*

differential arrangement of gears in the final drive of a vehicle's transmission system that allows the driving wheels to turn at different speeds when cornering. The differential consists of sets of bevel gears and pinions within a cage attached to the crown wheel. When cornering, the bevel pinions rotate to allow the outer wheel to turn faster than the inner.

differential calculus branch of ◊calculus involving applications such as the determination of maximum and minimum points and rates of change.

differentiation in mathematics, a procedure for determining the ◊derivative or gradient of the tangent to a curve f(x) at any point x. Part of a branch of mathematics called differential calculus, it is used to solve problems involving continuously varying quantities (such as the change in velocity or altitude of a rocket), to find the rates at which these variations occur and to obtain maximum and minimum values for the quantities.

The derivative may be regarded as the limit of the expression $[f(x + \delta x) - f(x)]/\delta x$ as δx tends to zero. Graphically, this is equivalent to the gradient (slope) of the curve represented by $y = f(x)$ at any point x.

differentiation in embryology, the process by which cells become increasingly different and specialized, giving rise to more complex structures that have particular functions in the adult organism. For instance, embryonic cells may develop into nerve, muscle, or bone cells.

diffraction the slight spreading of a light beam into a pattern of light and dark bands when it passes through a narrow slit or past the edge of an obstruction. A *diffraction grating* is a plate of glass or metal ruled with close, equidistant parallel lines used for separating a wave train such as a beam of incident light into its component frequencies (white light results in a spectrum).

The regular spacing of atoms in crystals are used to diffract X-rays, and in this way the structure of many substances has been elucidated, including

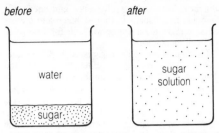

sugar and water molecules become evenly mixed

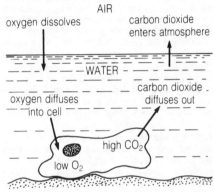

gas exchange in Amoeba

diffusion

recently that of proteins (see ◊X-ray diffraction). Sound waves can also be diffracted by a suitable array of solid objects.

diffusion spontaneous and random movement of molecules or particles in a fluid (gas or liquid) from a region in which they are at a high concentration to a region in which they are at a low concentration, until a uniform concentration is achieved throughout. No mechanical mixing or stirring is involved. For instance, if a drop of ink is added to water, its molecules will diffuse until their colour becomes evenly distributed throughout.

In biological systems, diffusion plays an essential role in the transport, over short distances, of molecules such as nutrients, respiratory gases, and neurotransmitters. It provides the means by which small molecules pass into and out of individual cells and microorganisms, such as amoebae.

One application of diffusion is the separation of isotopes, particularly those of uranium. When uranium hexafluoride diffuses through a porous plate, the ratio of the 235 and 238 isotopes is changed slightly. With sufficient number of passages, the separation is nearly complete. There are large plants in the USA and UK for obtaining enriched fuel for fast nuclear reactors and the fissile uranium-235, originally required for the first atom bombs. Another application is the diffusion pump, used extensively in vacuum work, in which the gas to be evacuated diffuses into a chamber from which it is carried away by the vapour of a suitable medium, usually oil or mercury.

digestion process whereby food eaten by an

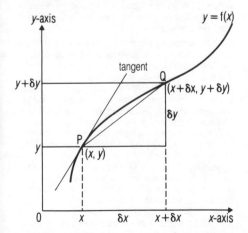

differentiation *A mathematical procedure for determining the gradient, or slope, of the tangent to any curve f(x) at any point x.*

animal is broken down physically, and chemically by ◊enzymes, usually in the ◊stomach and ◊intestines, to make the nutrients available for absorption and cell metabolism.

In some single-celled organisms, such as amoebae, a food particle is engulfed by the cell and digested in a ◊vacuole within the cell.

digestive system the mouth, stomach, intestine, and associated glands of animals, which are responsible for digesting food. The food is broken down by physical and chemical means in the stomach; digestion is completed, and most nutrients are absorbed in the small intestine; what remains is stored and concentrated into faeces in the large intestine. See ◊gut.

In smaller, simpler animals such as jellyfishes, the digestive system is simply a cavity (coelenteron or enteric cavity) with a 'mouth' into which food is taken; the digestible portion is dissolved and absorbed in this cavity, and the remains are ejected back through the mouth.

digit in mathematics, any of the numbers from 0 to 9 in the decimal system. Different bases have different ranges of digits. For example, the ◊hexadecimal system has digits 0 to 9 and A to F, whereas the binary system has two digits (or ◊bits), 0 and 1.

digital in electronics and computing, a term meaning 'coded as numbers'. A digital system uses two-state, either on/off or high/low voltage pulses, to encode, receive, and transmit information. A *digital display* shows discrete values as numbers (as opposed to an analogue signal, such as the continuous sweep of a pointer on a dial).

Digital electronics is the technology that underlies digital techniques. Low-power, miniature, integrated circuits (chips) provide the means for the coding, storage, transmission, processing, and reconstruction of information of all kinds.

digital audio tape (DAT) digitally recorded audio tape produced in cassettes that can carry two hours of sound on each side and are about half the size of standard cassettes. DAT players/recorders were developed 1987 but not marketed in the UK until 1989. Prerecorded cassettes are copy-protected. The first DAT for computer data was introduced 1988.

DAT machines are constructed like video cassette recorders (though they use metal audio tape), with a movable playback head, the tape winding in a spiral around a rotating drum. The tape can also carry additional information; for example, it can be programmed to skip a particular track and repeat another. The music industry delayed releasing prerecorded DAT cassettes because of fears of bootlegging, but a system has now been internationally agreed whereby it is not possible to make more than one copy of any prerecorded compact disc or DAT. DAT is mainly used in recording studios for making master tapes. The system was developed by Sony.

By 1990, DATs for computer data had been developed to a capacity of around 2.5 gigabytes per tape, achieved by using helical scan recording (in which the tape covers some 90% of the total head area of the rotating drum). This enables data from the tape to be read over 200 times faster than it can be written. Any file can be located within 60 seconds.

digital compact cassette (DCC) digitally recorded audio cassette that is roughly the same size as a standard cassette. It cannot be played on a normal tape recorder, though standard tapes can be played on a DCC machine; this is known as 'backwards compatibility'. The playing time is 90 minutes.

A DCC player has a stationary playback and recording head similar to that in ordinary tape decks, though the tape used is chrome video tape. The cassettes are copy-protected and can be individually programmed for playing order. Some DCC decks have a liquid-crystal digital-display screen, which can show track titles and other information encoded on the tape. DCC machines were launched in the UK in 1992, with some 500 prerecorded tapes to go with them. The system was developed by Philips.

digital computer computing device that operates on a two-state system, using symbols that are internally coded as binary numbers (numbers made up of combinations of the digits 0 and 1); see ◊computer.

digital data transmission in computing, a way of sending data by converting all signals (whether pictures, sounds, or words) into numeric (normally binary) codes before transmission, then reconverting them on receipt. This virtually eliminates any distortion or degradation of the signal during transmission, storage, or processing.

digital recording technique whereby the pressure of sound waves is sampled more than 30,000 times a second and the values converted by computer into precise numerical values. These are recorded and, during playback, are reconverted to sound waves.

This technique gives very high-quality reproduction. The numerical values converted by computer represent the original sound-wave form exactly and are recorded on compact disc. When this is played back by ◊laser, the exact values are retrieved. When the signal is fed via an amplifier to a loudspeaker, sound waves exactly like the original ones are reproduced.

digital sampling electronic process used in ◊telecommunications for transforming a constantly varying (analogue) signal into one composed of discrete units, a digital signal. In the creation of recorded music, sampling enables the composer, producer, or remix engineer to borrow discrete vocal or instrumental parts from other recorded work (it is also possible to sample live sound).

A telephone microphone changes sound waves into an analogue signal that fluctuates up and down like a wave. In the digitizing process the waveform is sampled thousands of times a second and each part of the sampled wave is given a binary code number (made up of combinations of the digits 0 and 1) related to the height of the wave at that point, which is transmitted along the telephone line. Using digital signals, messages can be transmitted quickly, accurately, and economically.

digital-to-analogue converter electronic circuit that converts a digital signal into an ◊analogue

An army marches on its stomach

On 6 June 1822 at Fort Mackinac, Michigan, USA, an 18-year-old French Canadian was accidentally wounded in the abdomen by the discharge of musket. He was brought to the army surgeon, US physician William Beaumont (1785–1853), who noted several serious wounds and, in particular, a hole in the abdominal wall and stomach. The surgeon observed that through this hole in the patient 'was pouring out the food he had taken for breakfast'.

The patient, Alexis St Martin, a trapper by profession, was serving with the army as a porter and general servant. Not surprisingly, St Martin was at first unable to keep food in his stomach. As the wound gradually healed, firm dressings were needed to retain the stomach contents. Beaumont tended his patient assiduously and tried during the ensuing months to close the hole in his stomach, without success. After 18 months, a small, protruding fleshy fold had grown to fill the aperture (fistula). This 'valve' could be opened simply by pressing it with a finger.

Digestion . . . inside and outside
At this point, it occurred to Beaumont that here was an ideal opportunity to study the process of digestion. His patient must have been an extremely tough character to have survived the accident at all. For the next nine years he was the subject of a remarkable series of pioneering experiments, in which Beaumont was able to vary systematically the conditions under which digestion took place and discover the chemical principles involved.

Beaumont attacked the problem of digestion in two ways. He studied how substances were actually digested in the stomach, and also how they were 'digested' outside the stomach in the digestive juices he extracted from St Martin. He found it was easy enough to drain out the juices from his patient 'by placing the subject on his left side, depressing the valve within the aperture, introducing a gum elastic tube and then turning him . . . on introducing the tube the fluid soon began to run.'

Beaumont was basically interested in the rate and temperature of digestion, and also the chemical conditions that favoured different stages of the process of digestion. He describes a typical experiment (he performed hundreds), where (a) digestion in the stomach is contrasted (b) with artificial digestion in glass containers kept at suitable temperatures, like this:
(a) 'At 9 o'clock he breakfasted on bread,

sausage and coffee, and kept exercising. 11 o'clock, 30 minutes, stomach two-thirds empty, aspects of weather similar, thermometer 298 [F], temperature of stomach 101½8 and 100¾48. The appearance of contraction and dilation and alternative piston motions were distinctly observed at this examination. 12 o'clock, 20 minutes, stomach empty.
(b) 'February 7. At 8 o'clock, 30 minutes A.M. I put twenty grains of boiled codfish into three drachms of gastric juice and placed them on the bath. At 1 o'clock, 30 minutes, P.M., the fish in the gastric juice on the bath was almost dissolved, four grains only remaining: fluid opaque, white, nearly the colour of milk. 2 o'clock, the fish in the vial all completely dissolved.'

All a matter of chemistry
Beaumont's research showed clearly for the first time just what happened during digestion and that digestion, as a process, could take place independently outside the body. He wrote that gastric juice: 'so far from being inert as water as some authors assert, is the most general solvent in nature of alimentary matter—even the hardest bone cannot withstand its action. It is capable, *even out of the stomach*, of effecting perfect digestion, with the aid of due and uniform degree of heat (100°Fahrenheit) and gentle agitation . . . I am impelled by the weight of evidence . . . to conclude that the change effected by it on the aliment, *is purely chemical*.'

Our modern understanding of the physiology of digestion as a process whereby foods are gradually broken down into their basic components follows logically from his work. An explanation of how the digestive juices flowed in the first place came in 1889, when Russian physiologist Ivan Pavlov (1849–1936) showed that their secretion in the stomach was controlled by the nervous system. By preventing the food eaten by a dog from actually entering the stomach, he found that the secretions of gastric juices began the moment the dog started eating, and continued as long as it did so. Since no food had entered the stomach, the secretions must be mediated by the nervous system.

Later it was found that the further digestion that takes place beyond the stomach was hormonally controlled. But it was Beaumont's careful scientific work, which was published in 1833 with the title *Experiments and Observations on the Gastric Juice and Physiology of Digestion*, that triggered subsequent research in this field.

(continuously varying) signal. Such a circuit is used to convert the digital output from a computer into the analogue voltage required to produce sound from a conventional loudspeaker.

digitizer in computing, a device that converts an analogue video signal into a digital format so that video images can be input, stored, displayed, and manipulated by a computer. The term is sometimes used to refer to a ◊graphics tablet.

dihybrid inheritance in genetics, a pattern of inheritance observed when two characteristics are studied in succeeding generations. The first experiments of this type, as well as in ◊monohybrid inheritance, were carried out by Austrian biologist Gregor Mendel using pea plants.

dilution process of reducing the concentration of a solution by the addition of a solvent.

The extent of a dilution normally indicates the final volume of solution required. A fivefold dilution would mean the addition of sufficient solvent to make the final volume five times the original.

dimension in science, any directly measurable physical quantity such as mass (M), length (L), and time (T), and the derived units obtainable by multiplication or division from such quantities. For example, acceleration (the rate of change of velocity) has dimensions (LT^{-2}), and is expressed in such units as km s^{-2}. A quantity that is a ratio, such as relative density or humidity, is dimensionless.

In geometry, the dimensions of a figure are the number of measures needed to specify its size. A point is considered to have zero dimension, a line to have one dimension, a plane figure to have two, and a solid body to have three.

dimethyl sulphoxide (DMSO) $(CH_3)_2SO_4$ colourless liquid used as an industrial solvent and an antifreeze. It is obtained as a by-product of the processing of wood to paper.

DIN (abbreviation for *D*eutsches *I*nstitut für *Nor*mung) German national standards body, which has set internationally accepted standards for (among other things) paper sizes and electrical connectors.

dinitrogen oxide alternative name for ◊nitrous oxide, or 'laughing gas', one of the nitrogen oxides.

dinosaur (Greek *deinos* 'terrible', *sauros* 'lizard') any of a group (sometimes considered as two separate orders) of extinct reptiles living between 230 million and 65 million years ago. Their closest living relations are crocodiles and birds. Many species of dinosaur evolved during the millions of years they were the dominant large land animals. Most were large (up to 27 m/90 ft), but some were as small as chickens. They disappeared 65 million years ago for reasons not fully understood, although many theories exist.

A currently popular theory of dinosaur extinction suggests that the Earth was struck by a giant meteorite or a swarm of comets 65 million years ago and this sent up such a cloud of debris and dust that climates were changed and the dinosaurs could not adapt quickly enough. The evidence for this includes a bed of rock rich in ◊iridium—an element rare on Earth but common in extraterrestrial bodies—dating from the time. An alternative theory suggests that changes in geography

brought about by the movements of continents and variations in sea level led to climate changes and the mixing of populations between previously isolated regions. This resulted in increased competition and the spread of disease.

Brachiosaurus, a long-necked plant-eater of the sauropod group, was about 12.6 m/40 ft to the top of its head, and weighed 80 tonnes. Compsognathus, a meat-eater, was only the size of a chicken, and ran on its hind legs. Stegosaurus, an armoured plant-eater 6 m/20 ft long, had a brain only about 3 cm/1.25 in long. Not all dinosaurs had small brains. At the other extreme, the hunting dinosaur Stenonychosaurus, 2 m/6 ft long, had a brain size comparable to that of a mammal or bird of today, stereoscopic vision, and grasping hands. Many dinosaurs appear to have been equipped for a high level of activity.

An almost complete fossil of a dinosaur skeleton was found in 1969 in the Andean foothills, South America; it had been a two-legged carnivore 6 ft tall and weighing more than 100 kg/220 lb. More than 230 million years old, it is the oldest known dinosaur. Eggs are known of some species. In 1982 a number of nests and eggs were found in 'colonies' in Montana, suggesting that some bred together like modern seabirds. In 1987 finds were made in China that may add much to the traditional knowledge of dinosaurs, chiefly gleaned from North American specimens. In 1989 and 1990 an articulated *Tyrannosaurus rex* was unearthed by a palaeontological team in Montana, with a full skull, one of only six known. Tyrannosaurus were huge, two-footed, meat-eating theropod dinosaurs of the Upper Cretaceous in North America and Asia.

NAMING DINOSAURS

Dinosaurs are usually named in Greek or Latin after the place where they were found, or for their special features: for example, Alamosaurus was found in Alamo, Texas, USA; Pentaceratops means five (*penta*) horns (*cerat*) around the eyes (*ops*).

diode combination of a cold anode and a heated cathode (or the semiconductor equivalent, which incorporates a *p–n* junction). Either device allows the passage of direct current in one direction only, and so is commonly used in a ◊rectifier to convert alternating current (AC) to direct current (DC).

dioecious of plants with male and female flowers borne on separate individuals of the same species. Dioecism occurs, for example, in the willows *Salix*. It is a way of avoiding self-fertilization.

dioptre optical unit in which the power of a ◊lens is expressed as the reciprocal of its focal length in metres. The usual convention is that convergent lenses are positive and divergent lenses negative. Short-sighted people need lenses of power about −0.66 dioptre; a typical value for long sight is about +1.5 dioptre.

diorite igneous rock intermediate in composition; the coarse-grained plutonic equivalent of ◊andesite.

dioxin any of a family of over 200 **organic**

dip circle

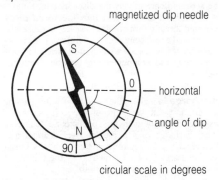

magnetized dip needle

horizontal

angle of dip

circular scale in degrees

dip, magnetic *A dip circle is used to measure the angle between the direction of the Earth's magnetic field and the horizontal at any point on the Earth's surface.*

chemicals, all of which are heterocyclic hydrocarbons (see ◊cyclic compounds). The term is commonly applied, however, to only one member of the family, 2,3,7,8-tetrachlorodibenzo-*p*-dioxin (2,3,7,8–TCDD), a highly toxic chemical that occurs, for example, as an impurity in the defoliant Agent Orange, used in the Vietnam War, and sometimes in the weedkiller 2,4,5–T. It has been associated with a disfiguring skin complaint (chloracne), birth defects, miscarriages, and cancer.

Disasters involving accidental release of large amounts of dioxin into the environment have occurred at Seveso, Italy, and Times Beach, Missouri, USA. Small amounts of dioxins are released by the burning of a wide range of chlorinated materials (treated wood, exhaust fumes from fuels treated with chlorinated additives, and plastics). The possibility of food becoming contaminated by dioxins in the environment has led the EC to significantly decrease the allowed levels of dioxin emissions from incinerators. Dioxin may be produced as a by-product in the manufacture of the bactericide ◊hexachlorophene.

UK government figures released in 1989 showed dioxin levels 100 times higher than guidelines set for environmental dioxin in breast milk, suggesting dioxin contamination is more widespread than previously thought.

diploblastic in biology, having a body wall composed of two layers. The outer layer is the *ectoderm*, the inner layer is the *endoderm*. This pattern of development is shown by ◊coelenterates.

diploid having two sets of ◊chromosomes in each cell. In sexually reproducing species, one set is derived from each parent, the ◊gametes, or sex cells, of each parent being ◊haploid (having only one set of chromosomes) due to ◊meiosis (reduction cell division).

dip, magnetic angle at a particular point on the Earth's surface between the direction of the Earth's magnetic field and the horizontal. It is measured using a *dip circle*, which has a magnetized needle suspended so that it can turn freely in the vertical plane of the magnetic field. In the northern hemisphere the needle dips below the horizontal, pointing along the line of the magnetic field towards its

north pole. At the magnetic north and south poles, the needle dips vertically and the angle of dip is 90°. See also ◊angle of dips.

dipole the uneven distribution of magnetic or electrical characteristics within a molecule or substance so that it behaves as though it possesses two equal but opposite poles or charges, a finite distance apart.

The uneven distribution of electrons within a molecule composed of atoms of different ◊electronegativities may result in an apparent concentration of electrons towards one end of the molecule and a deficiency towards the other, so that it forms a dipole consisting of apparently separated but equal positive and negative charges. The product of one charge and the distance between them is the *dipole moment*. A bar magnet behaves as though its magnetism were concentrated in separate north and south magnetic poles because of the uneven distribution of its magnetic field.

dipole in radio, a rod aerial, usually one half-wavelength or a whole wavelength long.

dipole, magnetic see ◊magnetic dipole.

direct access or *random access* type of ◊file access. A direct-access file contains records that can be accessed directly by the computer because each record has its own address on the storage disc.

direct combination method of making a simple salt by heating its two constituent elements together. For example, iron(II) sulphide can be made by heating iron and sulphur, and aluminium chloride by passing chlorine over hot aluminium.

$$Fe + S \rightarrow FeS$$

$$2Al + 3Cl_2 \rightarrow 2AlCl_3$$

direct current (DC) electric current that flows in one direction, and does not reverse its flow as ◊alternating current does. The electricity produced by a battery is direct current.

directed number an ◊integer with a positive (+) or negative (−) sign attached, for example +5 or −5. On a graph, a positive sign shows a movement to the right or upwards; a negative sign indicates movement downwards or to the left.

directory in computing, a list of file names, together with information that enables a computer to retrieve those files from ◊backing storage. The computer operating system will usually store and update a directory on the backing storage to which it refers. So, for example, on each ◊disc used by a computer a directory file will be created listing the disc's contents.

dirigible another name for ◊airship.

disaccharide sugar made up of two monosaccharides or simple sugars, such as glucose or fructose. Sucrose, $C_{12}H_{22}O_{11}$, or table sugar, is a disaccharide.

disc or *disk* in computing, a common medium for storing large volumes of data (an alternative is ◊magnetic tape). A *magnetic disc* is rotated at high speed in a disc-drive unit as a read/write (playback or record) head passes over its surfaces to record or read the magnetic variations that encode the data. Recently, *optical discs*, such as

cut-away view of a removable pack of hard discs in a drive unit

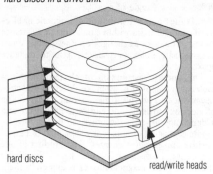

hard discs

read/write heads

disc

◊CD-ROM (compact-disc read-only memory) and ◊WORM (write once, read many times), have been used to store computer data. Data are recorded on the disc surface as etched microscopic pits and are read by a laser-scanning device. Optical discs have an enormous capacity—about 550 megabytes (million ◊bytes) on a compact disc, and thousands of megabytes on a full-size optical disc.

Magnetic discs come in several forms:

Fixed hard discs are built into the disc-drive unit, occasionally stacked on top of one another. A fixed disc cannot be removed: once it is full, data must be deleted in order to free space or a complete new disc drive must be added to the computer system in order to increase storage capacity. Large fixed discs, used with mainframe and minicomputers, provide up to 3,000 megabytes. Small fixed discs for use with microcomputers were introduced in the 1980s and typically hold 40–400 megabytes.

Removable hard discs are common in minicomputer systems. The discs are contained, individually or as stacks (disc packs), in a protective plastic case, and can be taken out of the drive unit and kept for later use. By swapping such discs around, a single hard-disc drive can be made to provide a potentially infinite storage capacity. However, access speeds and capacities tend to be lower than those associated with large fixed hard discs.

A *floppy disc* (or diskette) is the most common form of backing store for microcomputers. It is much smaller in size and capacity than a hard disc, normally holding 0.5–2 megabytes of data. The floppy disc is so called because it is manufactured from thin flexible plastic coated with a magnetic material. The earliest form of floppy disc was packaged in a card case and was easily damaged; more recent versions are contained in a smaller, rigid plastic case and are much more robust. All floppy discs can be removed from the drive unit.

disc drive mechanical device that reads data from and writes data to a magnetic ◊disc.

disc formatting in computing, preparing a blank magnetic disc so that data can be stored on it. Data are recorded on a disc's surface on circular tracks, each of which is divided into a number of sectors. In formatting a disc, the computer's operating system adds control information such as track and

sector numbers, which enables the data stored to be accessed correctly by the disc-drive unit.

Some hard discs, called **hard-sectored discs**, are sold already formatted. However, because different makes of computer use different disc formats, most new discs are sold unformatted, or **soft-sectored**, and computers are provided with the necessary ◊utility program to format these discs correctly before they are used.

discharge in a river, the volume of water passing a certain point per unit of time. It is usually expressed in cubic metres per second (cumecs). The discharge of a particular river channel may be calculated by multiplying the channel's cross-sectional area (in square metres) by the velocity of the water (in metres per second).

In the UK most small rivers have an average discharge of less than 1 cumec. The highest discharge usually occurs in early spring when melted snow combines with rainfall to increase the amount of water entering a river. Discharges are studied carefully by scientists as they can signal when a river is in danger of ◊flooding. Channel efficiency is an important factor in determining the likelihood of flooding at times of high discharge.

discharge tube device in which a gas conducting an electric current emits visible light. It is usually a glass tube from which virtually all the air has been removed (so that it 'contains' a near vacuum), with electrodes at each end. When a high-voltage current is passed between the electrodes, the few remaining gas atoms in the tube (or some deliberately introduced ones) ionize and emit coloured light as they conduct the current along the tube. The light originates as electrons change energy levels in the ionized atoms.

By coating the inside of the tube with a phosphor, invisible emitted radiation (such as ultraviolet light) can produce visible light; this is the principle of the fluorescent lamp.

Discman Sony trademark for a portable compact-disc player; it also comes in a model with a liquid crystal display for data discs.

surface of a magnetic disc

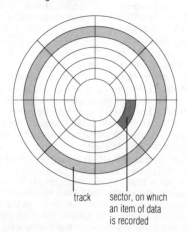

track   sector, on which
         an item of data
         is recorded

disc formatting

discrete data data that can take only whole-number or fractional values. The opposite is ◊continuous data, which can take all in-between values. Examples of discrete data include frequency and population data. However, measurements of time and other dimensions can give rise to continuous data.

disease any condition that impairs the normal state of an organism, and usually alters the functioning of one or more of its organs or systems. A disease is usually characterized by a set of specific symptoms and signs, although these may not always be apparent to the sufferer. Diseases may be inborn (congenital) or acquired through infection, injury, or other cause. Many diseases have unknown causes.

Not even medicine can master incurable diseases.

On **disease** Seneca (4 BC–AD 65) *Epistulae ad Lucilium* XCIV

disinfectant agent that kills, or prevents the growth of, bacteria and other microorganisms. Chemical disinfectants include carbolic acid (phenol, used by English surgeon Joseph Lister in surgery in the 1870s), ethanal, methanal, chlorine, and iodine.

dispersal phase of reproduction during which gametes, eggs, seeds, or offspring move away from the parents into other areas. The result is that overcrowding is avoided and parents do not find themselves in competition with their own offspring. The mechanisms are various, including a reliance on wind or water currents and, in the case of animals, locomotion. The ability of a species to spread widely through an area and to colonize new habitats has survival value in evolution.

dispersion in chemistry, the distribution of the microscopic particles of a ◊colloid. In colloidal sulphur the dispersion is the tiny particles of sulphur in the aqueous system.

dispersion in optics, the splitting of white light into a spectrum; for example, when it passes through a prism or a diffraction grating. It occurs because the prism (or grating) bends each component wavelength to a slightly different extent. The natural dispersion of light through raindrops creates a rainbow.

displacement activity in animal behaviour, an action that is performed out of its normal context, while the animal is in a state of stress, frustration, or uncertainty. Birds, for example, often peck at grass when uncertain whether to attack or flee from an opponent; similarly, humans scratch their heads when nervous.

displacement reaction chemical reaction in which a less reactive element is replaced in a compound by a more reactive one.

For example, the addition of powdered zinc to a solution of copper(II) sulphate displaces copper metal, which can be detected by its characteristic colour. See also ◊electrochemical series.

dissociation in chemistry, the process whereby a single compound splits into two or more smaller products, which may be capable of recombining to form the reactant.

Where dissociation is incomplete (not all the compound's molecules dissociate), a ◊chemical equilibrium exists between the compound and its dissociation products. The extent of incomplete dissociation is defined by a numerical value (dissociation constant).

distance ratio in a machine, the distance moved by the input force, or effort, divided by the distance moved by the output force, or load. The ratio indicates the movement magnification achieved, and is equivalent to the machine's ◊velocity ratio.

distillation technique used to purify liquids or to separate mixtures of liquids possessing different boiling points.

Simple distillation is used in the purification of liquids (or the separation of substances in solution from their solvents)—for example, in the production of pure water from a salt solution.

The solution is boiled and the vapours of the solvent rise into a separate piece of apparatus (the condenser) where they are cooled and condensed. The liquid produced (the distillate) is the pure solvent; the non-volatile solutes (now in solid form) remain in the distillation vessel to be discarded as impurities or recovered as required.

Mixtures of liquids (such as ◊petroleum or aqueous ethanol) are separated by *fractional distillation*, or fractionation. When the mixture is boiled, the vapours of its most volatile component rise into a vertical ◊fractionating column where they condense to liquid form. However, as this liquid runs back down the column it is reheated to boiling point by the hot rising vapours of the next-most-volatile component and so its vapours ascend the column once more. This boiling-condensing process occurs repeatedly inside the column, eventually bringing about a temperature gradient along its length. The vapours of the more volatile components therefore reach the top of the column and enter the condenser for collection before those of the less volatile components. In the fractional distillation of petroleum, groups of compounds

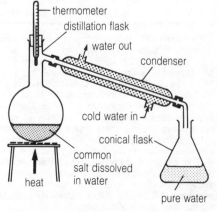

laboratory apparatus for simple distillation

thermometer
distillation flask
water out
condenser
cold water in
conical flask
common
salt dissolved
in water
heat
pure water

distillation

(fractions) possessing similar relative molecular masses and boiling points are tapped off at different points on the column.

The earliest-known reference to the process is to the distillation of wine in the 12th century by Adelard of Bath. The chemical retort used for distillation was invented by Muslims, and was first seen in the West about 1570.

distributive operation in mathematics, an operation, such as multiplication, that bears a relationship to another operation, such as addition, such that $a \times (b + c) = (a \times b) + (a \times c)$. For example, $3 \times (2 + 4) = (3 \times 2) + (3 \times 4) = 18$. Multiplication may be said to be distributive over addition. Addition is not, however, distributive over multiplication because $3 + (2 \times 4) = (3 + 2) \times (3 + 4)$

distributor device in the ignition system of a piston engine that distributes pulses of high-voltage electricity to the ◊spark plugs in the cylinders. The electricity is passed to the plug leads by the tip of a rotor arm, driven by the engine camshaft, and current is fed to the rotor arm from the ignition coil. The distributor also houses the contact point or breaker, which opens and closes to interrupt the battery current to the coil, thus triggering the high-voltage pulses. With electronic ignition it is absent.

DNA (*deoxyribonucleic acid*) complex giant molecule that contains, in chemically coded form, all the information needed to build, control, and maintain a living organism. DNA is a ladderlike double-stranded ◊nucleic acid that forms the basis of genetic inheritance in all organisms, except for a few viruses that have only ◊RNA. In organisms other than bacteria it is organized into ◊chromosomes and contained in the cell nucleus.

DNA is made up of two chains of ◊nucleotide sub-units, with each nucleotide containing either a purine (adenine or guanine) or pyrimidine (cytosine or thymine) base. The bases link up with each other (adenine linking with thymine, and cytosine with guanine) to form ◊base pairs that connect the two strands of the DNA molecule like the rungs of a twisted ladder. The specific way in which the pairs form means that the base sequence is preserved from generation to generation.

We have discovered the secret of life!

On **DNA** Francis Crick 1953

Hereditary information is stored as a specific sequence of bases. A set of three bases—known as a *codon*—acts as a blueprint for the manufacture of a particular ◊amino acid, the sub-unit of a protein molecule. Geneticists identify the codons by the initial letters of the constituent bases—for example, the base sequence of codon CAG is cytosine–adenine–guanine. The meaning of each of the codons in the ◊genetic code has been worked out by molecular geneticists. There are four different bases, which means that there must be $4 \times 4 \times 4 = 64$ different codons. Proteins are usually made up of only 20 different amino acids, so many amino acids have more than one codon (for example, GGT, GGC, GGA and GGG all code for the same amino acid, glycine.)

The information encoded by the codons is

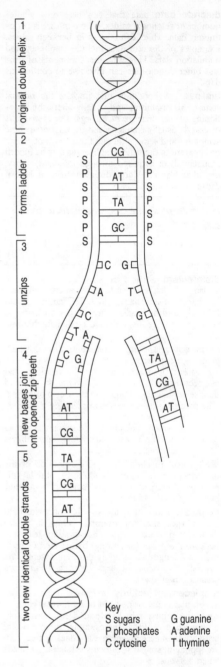

1 original double helix

2 forms ladder

S P S P S P S

CG
AT
TA
GC

S P S P S P S

3 unzips

C G
A T
C G
T A

4 new bases join onto opened zip teeth

C G

AT
CG
TA
CG
AT

TA
CG
AT

5 two new identical double strands

Key
S sugars   G guanine
P phosphates   A adenine
C cytosine   T thymine

DNA *How the DNA molecule divides. The DNA molecule consists of two strands wrapped around each other in a spiral or helix. The main strands consist of alternate sugar (S) and phosphate (P) groups, and attached to each sugar is a nitrogenous base-adenine (A), cytosine (C), guanine (G), or thymine (T). The sequence of bases carries the genetic code which specifies the characteristics of offspring. The strands are held together by weak bonds between the bases, cytosine to guanine, and adenine to thymine. The weak bonds allow the strands to split apart, allowing new bases to attach, forming another double strand.*

transcribed (see ◊transcription) by messenger RNA and is then translated into amino acids in the ribosomes and cytoplasm. The sequence of codons determines the precise order in which amino acids are linked up during manufacture and, therefore, the kind of protein that is to be produced. Because proteins are the chief structural molecules of living matter and, as enzymes, regulate all aspects of metabolism, it may be seen that the genetic code is effectively responsible for building and controlling the whole organism.

The sequence of bases along the length of DNA can be determined by cutting the molecule into small portions, using restriction enzymes. This technique can also be used for transferring specific sequences of DNA from one organism to another.

UNRAVELLING DNA

The DNA molecules in a cell nucleus contain up to 10 million atoms in two intertwined spirals. When a nucleus divides, the DNA molecule has to spin round at great speed—several hundred turns a second—in order to unravel. A car engine turning at this rate would break apart.

DNA fingerprinting or *DNA profiling* another name for ◊genetic fingerprinting.

document in computing, data associated with a particular application. For example, a **text document** might be produced by a ◊word processor and a **graphics document** might be produced with a ◊CAD package. An ◊OMR or ◊OCR document is a paper document containing data that can be directly input to the computer using a ◊document reader.

documentation in computing, the written information associated with a computer program or ◊applications package. Documentation is usually divided into two categories: program documentation and user documentation.

Program documentation is the complete technical description of a program, drawn up as the software is written and intended to support any later maintenance or development of that program. It typically includes details of when, where, and by whom the software was written; a general description of the purpose of the software, including recommended input, output, and storage methods; a detailed description of the way the software functions, including full program listings and ◊flow charts; and details of software testing, including sets of test data with expected results.

User documentation explains how to operate the software. It typically includes a nontechnical explanation of the purpose of the software; instructions for loading, running, and using the software; instructions for preparing any necessary input data; instructions for requesting and interpreting output data; and explanations of any error messages that the program may produce.

document reader in computing, an input device that reads marks or characters, usually on prepared forms and documents. Such devices are used to capture data by ◊optical mark recognition (OMR), ◊optical character recognition (OCR), and ◊mark sensing.

dodecahedron (plural *dodecahedra*) regular solid with 12 pentagonal faces and 12 vertices (corners). It is one of the five regular ◊polyhedra, or Platonic solids.

Dolby system electronic circuit that reduces the background noise, or hiss, during replay of magnetic tape recordings. The system was developed by US engineer Raymond Dolby (1933–   ) in 1966.

doldrums area of low atmospheric pressure along the equator, in the ◊intertropical convergence zone where the NE and SE trade winds converge. The doldrums are characterized by calm or very light winds, during which there may be sudden squalls and stormy weather. For this reason the areas are avoided as far as possible by sailing ships.

dolerite igneous rock formed below the Earth's surface, a form of basalt, containing relatively little silica (basic in composition).

Dolerite is a medium-grained (hypabyssal) basalt and forms in minor intrusions, such as dykes, which cut across the rock strata, and sills, which push between beds of sedimentary rock. When exposed at the surface, dolerite weathers into spherical lumps.

dolomite white mineral with a rhombohedral structure, calcium magnesium carbonate (CaMg $(CO_3)_2$). The term also applies to a type of limestone rock where the calcite content is replaced by the mineral dolomite. Dolomite rock may be white, grey, brown, or reddish in colour, commonly crystalline. It is used as a building material. The region of the Alps known as the Dolomites is a fine example of dolomite formation.

domain small area in a magnetic material that behaves like a tiny magnet. The magnetism of the material is due to the movement of electrons in the atoms of the domain. In an unmagnetized sample of material, the domains point in random directions, or form closed loops, so that there is no overall magnetization of the sample. In a magnetized sample, the domains are aligned so that their magnetic effects combine to produce a strong overall magnetism.

dome a geologic feature that is the reverse of a basin. It consists of anticlinally folded rocks that dip in all directions from a central high point, like an inverted but usually irregular cup.

Such structural domes are the result of pressure acting upward from below to produce an uplifted portion of the crust. Domes are often formed by the upwelling of plastic materials such as salt or magma. The salt domes along the North American Gulf Coast were produced by upwelling ancient sea salt deposits, while the Black Hills of South Dakota are the result of structural domes pushed up by intruding igneous masses.

dominance in genetics, the masking of one allele (an alternative form of a gene) by another allele. For example, if a ◊heterozygous person has one allele for blue eyes and one for brown eyes, his or her eye colour will be brown. The allele for blue eyes is described as recessive and the allele for brown eyes as dominant.

Too pretty not to be true

'We wish to suggest a structure for the salt of deoxyribose nucleic acid (DNA). This structure has novel features which are of considerable biological interest.'

So began a 900–word article that was published in the journal *Nature* in April 1953. Its authors were British molecular biologist Francis Crick (1916–) and US biochemist James Watson (1928–). The article described the correct structure of DNA, a discovery that many scientists have called the most important since Austrian botanist and monk Gregor Mendel (1822–1884) laid the foundations of the science of genetics. DNA is the molecule of heredity, and by knowing its structure, scientists can see exactly how forms of life are transmitted from one generation to the next.

The problem of inheritance

The story of DNA really begins with British naturalist Charles Darwin (1809–1882). When, in November 1859, he published an abstract entitled *On the Origin of Species by Means of Natural Selection* outlining his theory of evolution, he was unable to explain exactly how inheritance came about. For at that time it was believed that offspring inherited an average of the features of their parents. If this were so, as Darwin's critics pointed out, any remarkable features produced in a living organism by evolutionary processes would, in the natural course of events, soon disappear.

The work of Gregor Mendel, only rediscovered 20 years after Darwin's death, provided a clear demonstration that inheritance was not a 'blending' process at all. His description of the mathematical basis to genetics followed years of careful plant-breeding experiments. He concluded that each of the features he studied, such as colour or stem length, was determined by two 'factors' of inheritance, one coming from each parent. Each egg or sperm cell contained only one 'factor' of each pair. In this way a particular factor, say for the colour red, would be preserved through subsequent generations.

Today, we call Mendel's factors *genes*. Through the work of many scientists, it came to be realized that genes are part of the chromosomes located in the nucleus of living cells and that DNA, rather than protein as was first thought, was a hereditary material.

The double helix

In the early 1960s, scientists realized that X-ray crystallography, a method of using X-rays to obtain an exact picture of the atoms in a molecule, could be successfully applied to the large and complex molecules found in living cells.

It had been known since 1946 that genes consist of DNA. At King's College, London, New Zealand–British biophysicist Maurice Wilkins (1916–) had been using X-ray crystallography to examine the structure of DNA, together with his colleague, British X-ray crystallographer Rosalind Franklin (1920–1958) and had made considerable progress.

While in Copenhagen, James Watson had realized that one of the major unresolved problems of biology was the precise structure of DNA. In 1952, he came as a young postdoctoral student to join the Medical Research Council Unit at the Cavendish Laboratory, Cambridge, where Francis Crick was already working. Convinced that a gene must be some kind of molecule, the two scientists set to work on DNA.

Helped by the work of Wilkins, they were able in 1953 to build an accurate model of DNA. They showed that DNA had a double helical structure, rather like a spiral staircase. Because the molecule of DNA was made from two strands, they envisaged that as a cell divides, the strands unravel, and each could serve as a template as new DNA was formed in the resulting daughter cells. Their model also explained how genetic information might be coded in the sequence of the simpler molecules of which DNA is comprised. Here for the first time was a complete insight into the basis of her edity. James Watson truly commented that this result was 'too pretty not to be true!'

Cracking the code

Later, working with South African–British molecular biologist Sidney Brenner (1927–), Crick went on to work out the genetic code, and so ascribe a precise function to each specific region of the molecule of DNA. These triumphant results created a tremendous flurry of scientific activity around the world. The pioneering work of Crick, Wilkins and Watson was recognized in the award of the Nobel Prize for physiology or medicine in 1962.

The unravelling of the structure of DNA lead to a new scientific discipline, molecular biology, and laid the foundation stones for genetic engineering—a powerful and new technique that will revolutionize biology, medicine and food production through the purposeful adaptation of living organisms.

net of a dodecahedron

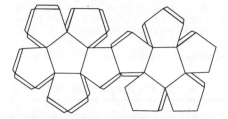

dodecahedron *A regular dodecahedron has 12 identical faces, each of which is a regular pentagon (five-sided figure with equal internal angles and all sides the same length).*

Dominion Astrophysical Observatory Canadian observatory near Victoria, British Columbia, the site of a 1.85–m/73–in reflector opened 1918, operated by the National Research Council of Canada. The associated Dominion Radio Astrophysical Observatory at Penticton, British Columbia, operates a 26–m/84–ft radio dish and an aperture synthesis radio telescope.

dopamine neurotransmitter, hydroxytyramine $C_8H_{11}NO_2$, an intermediate in the formation of adrenaline. There are special nerve cells (neurones) in the brain that use dopamine for the transmission of nervous impulses. One such area of dopamine neurones lies in the basal ganglia, a region that controls movement. Patients suffering from the tremors of Parkinson's disease show nerve degeneration in this region. Another dopamine brain area lies in the limbic system, a region closely involved with emotional responses. It has been found that schizophrenic patients respond well to drugs that act on limbic dopamine receptors in the brain.

Doppler effect change in the observed frequency (or wavelength) of waves due to relative motion between the wave source and the observer. The Doppler effect is responsible for the perceived change in pitch of a siren as it approaches and then recedes, and for the ◊red shift of light from distant stars. It is named after the Austrian physicist Christian Doppler (1803–1853).

Dorado constellation of the southern hemisphere, represented as a gold fish. It is easy to locate, since the Large ◊Magellanic Cloud marks its southern border. Its brightest star is Alpha Doradus, just under 200 light years from Earth.

dormancy in botany, a phase of reduced physiological activity exhibited by certain buds, seeds, and spores. Dormancy can help a plant to survive unfavourable conditions, as in annual plants that pass the cold winter season as dormant seeds, and plants that form dormant buds.

For various reasons many seeds exhibit a period of dormancy even when conditions are favourable for growth. Sometimes this dormancy can be broken by artificial methods, such as penetrating the seed coat to facilitate the uptake of water (chitting) or exposing the seed to light. Other seeds require a period of ◊after-ripening.

dorsal in vertebrates, the surface of the animal closest to the backbone. For most vertebrates and invertebrates this is the upper surface, or the surface furthest from the ground. For bipedal primates such as humans, where the dorsal surface faces backwards, then the word is 'back'.

Not all animals can be said to have a dorsal surface, just as many animals cannot be said to have a front. The lower invertebrates, for example jellyfish, anemones, and sponges, do not have a head or anterior section.

DOS (acronym for *d*isc *o*perating *s*ystem) computer ◊operating system specifically designed for use with disc storage; also used as an alternative name for a particular operating system, ◊MS-DOS.

dot matrix printer computer printer that produces each character individually by printing a pattern, or matrix, of very small dots. The printing head consists of a vertical line or block of either 9 or 24 printing pins. As the printing head is moved from side to side across the paper, the pins are pushed forwards selectively to strike an inked ribbon and build up the dot pattern for each character on the paper beneath.

A dot matrix printer is more flexible than a ◊daisywheel printer because it can print graphics and text in many different typefaces. It is cheaper to buy and maintain than a ◊laser printer or ◊inkjet printer, and, because its pins physically strike the paper, is capable of producing carbon copies. However, it is noisy in operation and cannot produce the high-quality printing associated with the nonimpact printers.

double bond two covalent bonds between adjacent atoms, as in the ◊alkenes (–C=C–) and ◊ketones (–C=O–).

double decomposition reaction between two chemical substances (usually ◊salts in solution) that results in the exchange of a constituent from each compound to create two different compounds.

For example, if silver nitrate solution is added to a solution of sodium chloride, there is an exchange of ions yielding sodium nitrate and silver chloride.

double star two stars that appear close together. Most double stars attract each other due to gravity, and orbit each other, forming a genuine ◊binary star, but other double stars are at different distances from Earth, and lie in the same line of sight

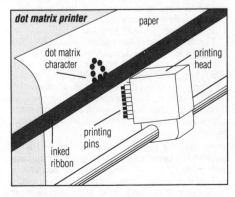

dot matrix printer

only by chance. Through a telescope both types of double star look the same.

dough mixture consisting primarily of flour, water, and yeast, which is used in the manufacture of bread.

The preparation of dough involves thorough mixing (kneading) and standing in a warm place to 'prove' (increase in volume) so that the ◊enzymes in the dough can break down the starch from the flour into smaller sugar molecules, which are then fermented by the yeast. This releases carbon dioxide, which causes the dough to rise.

downhole measurement in geology, any one of a number of experiments performed by instruments lowered down a borehole. Such instruments may be detectors that study vibrations passing through the rock from a generator, or may measure the electrical resistivity of the rock or the natural radiation.

dpi abbreviation for *dots per inch*, a measure of the ◊resolution of images produced by computer screens and printers.

Draco in astronomy, a large but faint constellation, representing a dragon coiled around the north celestial pole. The star Alpha Draconis (Thuban) was the pole star 4,800 years ago.

drag resistance to motion a body experiences when passing through a fluid—gas or liquid. The aerodynamic drag aircraft experience when travelling through the air represents a great waste of power, so they must be carefully shaped, or streamlined, to reduce drag to a minimum. Cars benefit from ◊streamlining, and aerodynamic drag is used to slow down spacecraft returning from space. Boats travelling through water experience hydrodynamic drag on their hulls, and the fastest vessels are ◊hydrofoils, whose hulls lift out of the water while cruising.

DRAM (acronym for *d*ynamic *r*andom-*a*ccess *m*emory) computer memory device in the form of a silicon chip commonly used to provide the ◊immediate access memory of microcomputers. DRAM loses its contents unless they are read and rewritten every 2 milliseconds or so. This process is called **refreshing** the memory. DRAM is slower but cheaper than ◊SRAM, an alternative form of silicon-chip memory.

driver in computing, a program that controls a peripheral device. Every device connected to the computer needs a driver program.

The driver ensures that communication between the computer and the device is successful.

For example, it is often possible to connect many different types of printer, each with its own special operating codes, to the same type of computer. This is because driver programs are supplied to translate the computer's standard printing commands into the special commands needed for each printer.

drug any of a range of chemicals voluntarily or involuntarily introduced into the bodies of humans and animals in order to enhance or suppress a biological function. Most drugs in use are medicines (pharmaceuticals), used to prevent or treat diseases, or to relieve their symptoms; they include

antibiotics, cytotoxic drugs, immunosuppressives, sedatives, and pain-relievers (analgesics).

The most popular drugs in use today, nicotine (in tobacco) and alcohol, operate on the nervous system and are potentially harmful.

A miracle drug is any drug that will do what the label says it will do.

On **drugs** Eric Hodgins (1899–1971) *Episode*

drumlin long streamlined hill created in formerly glaciated areas. Rocky debris (moraine) is gathered up by the glacial icesheet and moulded to form an egg-shaped mound, 8–60 m/25–200 ft in height and 0.5–1 km/0.3–0.6 mi in length. Drumlins commonly occur in groups on the floors of glacial troughs, producing a 'basket-of-eggs' landscape.

They are important indicators of the direction of ice flow, as their blunt ends point upstream, and their gentler slopes trail off downstream.

drupe fleshy ◊fruit containing one or more seeds which are surrounded by a hard, protective layer—for example cherry, almond, and plum. The wall of the fruit (◊pericarp) is differentiated into the outer skin (exocarp), the fleshy layer of tissues (mesocarp), and the hard layer surrounding the seed (endocarp).

The coconut is a drupe, but here the pericarp becomes dry and fibrous at maturity. Blackberries are an aggregate fruit composed of a cluster of small drupes.

dry-cleaning method of cleaning textiles based on the use of volatile solvents, such as trichloroethene (trichloroethylene), that dissolve grease. No water is used. Dry-cleaning was first developed in France 1849.

Some solvents are known to damage the ozone layer and one, tetrachloroethene (perchloroethylene), is toxic in water and gives off toxic fumes when heated.

dry ice solid carbon dioxide (CO_2), used as a refrigerant. At temperatures above –79°C/ –110.2°F, it sublimes (turns into vapour without passing through a liquid stage) to gaseous carbon dioxide.

drying removal of liquid water from a substance without altering its chemical composition (unlike ◊dehydration). Drying agents include deliquescent substances such as calcium chloride (see ◊deliquescence), concentrated acids such as sulphuric and nitric acids, and silica gel.

DTP abbreviation for ◊*desktop publishing*.

ductile material material that can sustain large deformations beyond its elastic limit (see ◊elasticity) without fracture. Metals are very ductile, and may be pulled out into wires, or hammered or rolled into thin sheets without breaking. Compare ◊brittle material.

ductless gland alternative name for an ◊endocrine gland.

dump in computing, the process of rapidly transferring data to external memory or to a printer. It is usually done to help with debugging (see ◊bug) or as part of an error-recovery procedure designed to provide data security. A ◊screen dump makes a printed copy of the current screen display.

dune mound or ridge of wind-drifted sand. Loose sand is blown and bounced along by the wind, up the windward side of a dune. The sand particles then fall to rest on the lee side, while more are blown up from the windward side. In this way a dune moves gradually downwind.

Dunes are features of sandy deserts and beach fronts. The typical crescent-shaped dune is called a *barchan*. *Seif dunes* are longitudinal and lie parallel to the wind direction, and *star-shaped dunes* are formed by irregular winds.

duodecimal system system of arithmetic notation using twelve as a base, at one time considered superior to the decimal number system in that 12 has more factors (2, 3, 4, 6) than 10 (2, 5).

It is now superseded by the universally accepted decimal system.

duodenum in vertebrates, a short length of alimentary canal found between the stomach and the small intestine. Its role is in digesting carbohydrates, fats, and proteins. The smaller molecules formed are then absorbed, either by the duodenum or the ileum.

Entry to the duodenum is controlled by the pyloric sphincter, a muscular ring at the base of the stomach. Once food has passed into the duodenum it is mixed with bile from the liver and with a range of enzymes secreted from the pancreas, a digestive gland near the top of the intestine. The bile neutralizes the acidity of the gastric juices passing out of the stomach and aids fat digestion.

duralumin lightweight aluminium ◊alloy widely used in aircraft construction, containing copper, magnesium, and manganese.

dust bowl area in the Great Plains region of North America (Texas to Kansas) that suffered extensive wind erosion as the result of drought and poor farming practice in once fertile soil. Much of the topsoil was blown away in the droughts of the 1930s and the 1980s.

Similar dust bowls are being formed in many areas today, noticeably across Africa, because of overcropping and overgrazing.

dust devil small dust storm caused by intense local heating of the ground in desert areas, particularly the Sahara. The air swirls upwards, carrying fine particles of dust with it.

Dutch elm disease disease of elm trees *Ulmus*, principally Dutch, English, and American elm, caused by the fungus *Certocystis ulmi*. The fungus is usually spread from tree to tree by the elm-bark beetle, which lays its eggs beneath the bark. The disease has no cure and control methods involve injecting insecticide into the trees annually to prevent infection, or the destruction of all elms in a broad band around an infected area, to keep the beetles out.

The disease was first described in the Netherlands and by the early 1930s had spread across Britain and continental Europe, as well as North America.

In the 1970s a new epidemic was caused by a much more virulent form of the fungus, probably brought to Britain from Canada.

dye substance that, applied in solution to fabrics, imparts a colour resistant to washing. *Direct dyes* combine with the material of the fabric, yielding a coloured compound; *indirect dyes* require the presence of another substance (a mordant), with which the fabric must first be treated; *vat dyes* are colourless soluble substances that on exposure to air yield an insoluble coloured compound.

Naturally occurring dyes include indigo, madder (alizarin), logwood, and cochineal, but industrial dyes (introduced in the 19th century) are usually synthetic: acid green was developed 1835 and bright purple 1856. Industrial dyes include ◊azo dyestuffs, ◊acridine, ◊anthracene, and ◊aniline.

common dyes and their properties

dye type	characteristics
natural dyes	
cochineal, lac	red dyes made from crushed insects; cochineal is used in food colouring
woad, indigo	blue or purple dye extracted from plants; also made synthetically
madder	red dye extracted from plants; also synthesized
saffron, safflower, annatto	yellow dyes extracted from plants; annatto is used in food colouring
logwood	black dye extracted from wood
Tyrian purple	purple dye extracted from molluscs; used for Roman emperors' togas
synthethic dyes	
basic dyes (for example, mauveine)	low fastness on most materials (that is, they tend to run and fade), but used for acrylics
acid dyes (for example, napthol green)	used on protein fibres such as wool and silk
direct dyes (for example, Congo red)	cheap and moderately fast
dispersed dyes	used for synthetic fibres that are difficult to dye with direct dyes
developed dyes (for example, azo dyes)	cheap and very fast; used for printed fabrics
fibre-reactive dyes	very fast to wash in and light, medium- priced, and easy to use; wide range of colours

dye-transfer print in photography, a print made by a relatively permanent colour process that uses red, yellow, and blue separation negatives printed together.

dyke in earth science, a sheet of ◊igneous rock created by the intrusion of magma (molten rock) across layers of pre-existing rock. (By contrast, a sill is intruded *between* layers of rock.) It may form a ridge when exposed on the surface. A dyke is also a human-made embankment built along a coastline (for example, in the Netherlands) to prevent the flooding of lowland coastal regions.

dynamics or *kinetics* in mechanics, the mathematical and physical study of the behaviour of bodies under the action of forces that produce changes of motion in them.

dynamite explosive consisting of a mixture of nitroglycerine and diatomaceous earth (diatomite, an absorbent, chalklike material). It was first devised by Alfred Nobel.

dynamo simple generator, or machine for transforming mechanical energy into electrical energy. A dynamo in basic form consists of a powerful field magnet between the poles of which a suitable conductor, usually in the form of a coil (armature), is rotated. The mechanical energy of rotation is thus converted into an electric current in the armature.

Present-day dynamos work on the principles described by English pysicist Michael Faraday 1830, that an ◊electromotive force is developed in a conductor when it is moved in a magnetic field.

dyne cgs unit (symbol dyn) of force. 10^5 dynes make one newton. The dyne is defined as the force that will accelerate a mass of one gram by one centimetre per second per second.

dysprosium (Greek *dusprositos* 'difficult to get near') silver-white, metallic element of the ◊lanthanide series, symbol Dy, atomic number 66, relative atomic mass 162.50. It is among the most magnetic of all known substances and has a great capacity to absorb neutrons.

It was discovered 1886 by French chemist Paul Lecoq de Boisbaudran (1838–1912).

ear organ of hearing in animals. It responds to the vibrations that constitute sound, and these are translated into nerve signals and passed to the brain. A mammal's ear consists of three parts: outer ear, middle ear, and inner ear. The **outer ear** is a funnel that collects sound, directing it down a tube to the **ear drum** (tympanic membrane), which separates the outer and **middle ear**. Sounds vibrate this membrane, the mechanical movement of which is transferred to a smaller membrane leading to the **inner ear** by three small bones, the auditory ossicles. Vibrations of the inner ear membrane move fluid contained in the snail-shaped cochlea, which vibrates hair cells that stimulate the auditory nerve connected to the brain. Three fluid-filled canals of the inner ear detect changes of position; this mechanism, with other sensory inputs, is responsible for the sense of balance.

When a loud noise occurs, muscles behind the eardrum contract automatically, suppressing the noise to enhance perception of sound and prevent injury.

earth electrical connection between an appliance

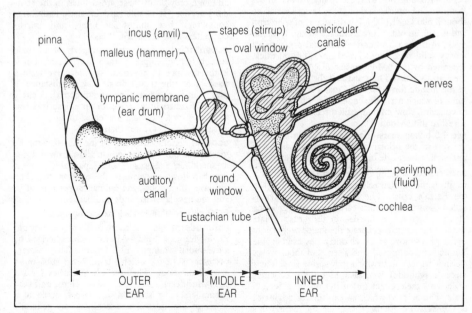

ear The structure of the ear. The three bones of the middle ear—hammer, anvil, and stirrup—vibrate in unison and magnify sounds about 20 times. The spiral-shaped cochlea is the organ of hearing. As sound waves pass down the spiral tube, they vibrate fine hairs lining the tube, which activate the auditory nerve connected to the brain. The semicircular canals are the organs of balance, detecting movements of the head.

and the ground. In the event of a fault in an electrical appliance, for example, involving connection between the live part of the circuit and the outer casing, the current flows to earth, causing no harm to the user.

In most domestic installations, earthing is achieved by a connection to a metal water-supply pipe buried in the ground before it enters the premises.

Earth third planet from the Sun. It is almost spherical, flattened slightly at the poles, and is composed of three concentric layers: the ◊core, the ◊mantle, and the ◊crust. 70% of the surface (including the north and south polar icecaps) is covered with water. The Earth is surrounded by a life-supporting atmosphere and is the only planet on which life is known to exist.

mean distance from the Sun 149,500,000 km/ 92,860,000 mi

equatorial diameter 12,756 km/7,923 mi

circumference 40,070 km/24,900 mi

rotation period 23 hr 56 min 4.1 sec

year (complete orbit, or sidereal period) 365 days 5 hr 48 min 46 sec. Earth's average speed around the Sun is 30 kps/18.5 mps; the plane of its orbit is inclined to its equatorial plane at an angle of 23.5°, the reason for the changing seasons.

atmosphere nitrogen 78.09%; oxygen 20.95%; argon 0.93%; carbon dioxide 0.03%; and less than 0.0001% neon, helium, krypton, hydrogen, xenon, ozone, radon

surface land surface 150,000,000 sq km/ 57,500,000 sq mi (greatest height above sea level 8,872 m/29,118 ft Mount Everest); water surface 361,000,000 sq km/139,400,000 sq mi (greatest depth 11,034 m/36,201 ft ◊Mariana Trench in the Pacific). The interior is thought to be an inner core about 2,600 km/1,600 mi in diameter, of solid iron and nickel; an outer core about 2,250 km/1,400 mi thick, of molten iron and nickel; and a mantle of mostly solid rock about 2,900 km/1,800 mi thick, separated by the ◊Mohorovičić discontinuity from the Earth's crust. The crust and the topmost layer of the mantle form about 12 major moving plates, some of which carry the continents. The plates are in constant, slow motion, called tectonic drift.

satellite the ◊Moon

age 4.6 billion years. The Earth was formed with the rest of the ◊Solar System by consolidation of interstellar dust. Life began 3.5–4 billion years ago.

earthquake shaking of the Earth's surface as a result of the sudden release of stresses built up in the Earth's crust. The study of earthquakes is called ◊seismology. Most earthquakes occur along ◊faults (fractures or breaks) in the crust. ◊Plate tectonic movements generate the major proportion: as two plates move past each other they can become jammed and deformed, and a series of shock waves (seismic waves) occur when they spring free. Their force (magnitude) is measured on the ◊Richter scale, and their effect (intensity) on the Mercalli scale. The point at which an earthquake originates is the *seismic focus*; the point on the Earth's surface directly above this is the *epicentre*.

In 1987 a California earthquake was successfully predicted by measurement of underground pressure waves; prediction attempts have also involved the study of such phenomena as the change in gases issuing from the ◊crust, the level of water in wells, and the behaviour of animals. The possibility of earthquake prevention is remote. However, rock slippage might be slowed at movement points or promoted at stoppage points by the extraction or injection of large quantities of water underground, since water serves as a lubricant. This would ease overall pressure.

earth science scientific study of the planet Earth as a whole, a synthesis of several traditional subjects such as ◊geology, ◊meteorology, oceanography, ◊geophysics, ◊geochemistry, and ◊palaeontology.

The mining and extraction of minerals and gems, the prediction of weather and earthquakes, the pollution of the atmosphere, and the forces that shape the physical world all fall within its scope of study. The emergence of the discipline reflects scientists' concern that an understanding of the global aspects of the Earth's structure and its past will hold the key to how humans affect its future, ensuring that its resources are used in a sustainable way.

Earth Summit (official name *United Nations Conference on Environment and Development*) international meeting in Rio de Janeiro, Brazil, June 1992 which drew up measures towards world environmental protection. Treaties were made to combat global warming and protect wildlife ('biodiversity') (the latter was not signed by the USA).

EBCDIC (abbreviation for *extended binary-coded decimal interchange code*) in computing, a code used for storing and communicating alphabetic and numeric characters. It is an 8–bit code, capable of holding 256 different characters, although only 85 of these are defined in the standard version. It is still used in many mainframe computers, but almost all mini- and microcomputers now use ◊ASCII code.

eccentricity in geometry, a property of a ◊conic section (circle, ellipse, parabola, or hyperbola). It is the distance of any point on the curve from a fixed point (the focus) divided by the distance of that point from a fixed line (the directrix). A circle has an eccentricity of zero; for an ellipse it is less than one; for a parabola it is equal to one; and for a hyperbola it is greater than one.

ecdysis periodic shedding of the ◊exoskeleton by insects and other arthropods to allow growth. Prior to shedding, a new soft and expandable layer is first laid down underneath the existing one. The old layer then splits, the animal moves free of it, and the new layer expands and hardens.

ECG abbreviation for ◊electrocardiogram.

echinoderm marine invertebrate of the phylum Echinodermata ('spiny-skinned'), characterized by a five-radial symmetry. Echinoderms have a water-vascular system which transports substances around the body. They include starfishes (or sea stars), brittlestars, sea-lilies, sea-urchins, and sea-cucumbers. The skeleton is external, made of a series of limy plates, and echinoderms generally move by using tube-feet, small water-filled sacs that can be protruded or pulled back to the body.

echo repetition of a sound wave, or of a ◊radar or ◊sonar signal, by reflection from a surface. By

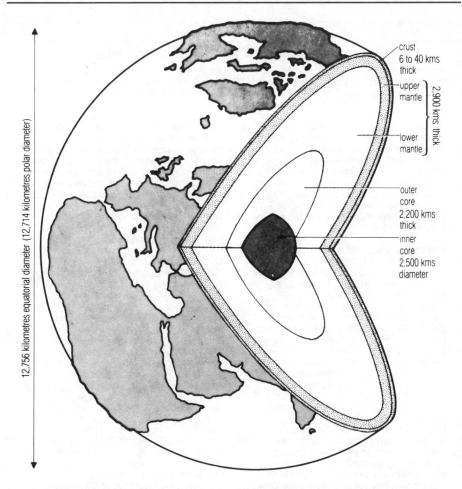

crust
6 to 40 kms
thick

upper
mantle

2,900 kms thick

lower
mantle

outer
core
2,200 kms
thick

inner
core
2,500 kms
diameter

12,756 kilometres equatorial diameter (12,714 kilometres polar diameter)

Earth *Inside the Earth. The surface of the Earth is a thin crust about 6 km/4 mi thick under the sea and 40 km/25 mi thick under the continents. Under the crust lies the mantle, about 2,900 km/1,800 mi thick and with a temperature of 1,500–3,000°C/2,700–5,400°F. The outer core is liquid iron and nickel at 4,000°C/7,200°F. The inner core is probably solid iron and nickel at about 5,000°C/9,000°F.*

accurately measuring the time taken for an echo to return to the transmitter, and by knowing the speed of a radar signal (the speed of light) or a sonar signal (the speed of sound in water), it is possible to calculate the range of the object causing the echo (◊echolocation).

A similar technique is used in echo sounders to estimate the depth of water under a ship's keel or the depth of a shoal of fish.

echolocation method used by certain animals, notably bats and dolphins, to detect the positions of objects by using sound. The animal emits a stream of high-pitched sounds, generally at ultrasonic frequencies (beyond the range of human hearing), and listens for the returning echoes reflected off objects to determine their exact location.

The location of an object can be established by the time difference between the emitted sound and its differential return as an echo to the two ears. Echolocation is of particular value under conditions when normal vision is poor (at night in the case of

bats, in murky water for dolphins). A few species of bird can also echolocate.

echo sounder or *sonar device* device that detects objects under water by means of ◊sonar— by using reflected sound waves. Most boats are equipped with echo sounders to measure the water depth beneath them. An echo sounder consists of a transmitter, which emits an ultrasonic pulse (see ◊ultrasound), and a receiver, which detects the pulse after reflection from the seabed. The time between transmission and receipt of the reflected signal is a measure of the depth of water. Fishing boats also use echo-sounders to detect shoals of fish.

eclipse passage of an astronomical body through the shadow of another. The term is usually employed for solar and lunar eclipses, which may be either partial or total, but also, for example, for eclipses by Jupiter of its satellites (moons). An eclipse of a star by a body in the Solar System is called an occultation.

A *solar eclipse* occurs when the Moon passes

Exploring the Earth from above and below

Recent years have seen basic geology merge with other sciences to produce the all-embracing discipline of earth science, and the advent of sophisticated data-gathering systems. There has been much debate amongst the scientists involved. Traditionalists maintain that the only way to study the Earth is go out and sample it, record it and interpret it directly. Progressives insist that the only way to a full picture of the Earth is through remote sensing and computerized assessment of the data. No matter how the science changes, there will always be room for both approaches.

Measuring the Earth
The greatest advances in remote sensing have been in distance measurement. The surface of the Earth can be measured extremely accurately, using global positioning geodesy (detecting signals from satellites by Earth-based receivers), satellite laser ranging (in which satellites reflect signals from ground transmitters back to ground receivers), and very-long-baseline interferometry, comparing signals received at ground-based receivers from distant extraterrestrial bodies. These techniques can measure distances of thousands of kilometres to accuracies of less than a centimetre. Movements of faults can be measured, as can the growth of the tectonic plates. Previously such speeds were calculated by averaging displacements measured over decades or centuries. The results show that in the oceanic crust plate growth is steady: from 12 mm/0.5 in per year across the Mid-Atlantic Ridge to 160 mm/6.5 in per year across the East Pacific Rise. The major continental faults seem to be very irregular in their movement; the Great Rift Valley of East Africa has remained stationary for 20 years, when long-term averages suggest that it would have opened up by about 100 mm/4 in in that time.

Offshore petroleum exploration
Petroleum is society's most important raw fuel. As the more accessible deposits become depleted, petroleum exploration is going into ever-deeper waters of the world's continental shelves. In 1991 Brazil, currently leading the way in offshore oil exploration, sank a production well in water 752 m/2,467 ft deep off the northeast coast. It is estimated that by the year 2000 more than 30% of the world's petroleum production will be from offshore wells.

However, the main oil reserves may lie somewhere else altogether. Accepted theory says that petroleum was formed from the remains of plant and animal matter, in the sedimentary rocks in which they were entombed. An alternative theory proposes inorganic origins for oil, from methane and hydrogen trapped when the Earth was formed in granites or metamorphic rocks, from which the oil seeps into the known traps in surrounding sedimentary basins. One team has been drilling into granite rocks in Sweden since 1987, with apparent success. Another will drill into granites in Canada in 1993, seeking vast new oil deposits in totally different geological sites from those traditionally investigated.

Climatic change
The burning of oil and other carbon-based fuels, pumping vast quantities of carbon dioxide into the atmosphere, raises the possibility of climatic change. The prospect is alarming, especially if such change is brought about by human interference with nature. However, the Earth's climate has never been stable, and human effects may be masked by natural variations. Beds of coal in Spitzbergen and glacial debris in the middle of the Australian desert attest to large-scale climate variations over hundreds of millions of years. Other studies show shorter-term variations: temperate tree fossils in Antarctica, 5° from the South Pole, and in the far north of Canada, show that climates were much warmer than now right up to the beginning of the Ice Age.

The international Pliocene Research Interpretation and Synoptic Mapping programme, begun in 1990, should produce fine details of climatic changes before the Ice Age. The Greenland Icecore project has shown that, during the Ice Age, temperatures over the ice caps varied between cold—with temperatures about 12°C lower than at present—and mild, with temperatures about 7° lower than at present. Each cold period lasted between 500 and 2,000 years, starting abruptly and ending gradually.

Several sudden, catastrophic climate changes may have been caused in the past by the impact of giant meteorites or comets. The crater of one, 214 million years old, has been found in Canada; this seems to relate to a mass extinction in the Triassic period. Scientists are still seeking remains of the meteorite believed to have wiped out the dinosaurs 65 million years ago.

Dougal Dixon

earth science: chronology

1735	English lawyer George Hadley described the circulation of the atmosphere as large-scale convection currents centred on the equator.
1746	A French expedition to Lapland proved the Earth to be flattened at the poles.
1743	Christopher Packe produced the first geological map, of S England.
1744	The first map produced on modern surveying principles was produced by César-François Cassini in France.
1745	In Russia, Mikhail Vasilievich Lomonosov published a catalogue of over 3,000 minerals.
1760	Lomonosov explained the formation of icebergs. John Mitchell proposed that earthquakes are produced when one layer of rock rubs against another.
1776	James Keir suggested that some rocks, such as those making up the Giant's Causeway in Ireland, may have formed as molten material that cooled and then crystallized.
1779	French naturalist Comte George de Buffon speculated that the Earth may be much older than the 6,000 years suggested by the Bible.
1785	Scottish geologist James Hutton proposed the theory of uniformitarianism: all geological features are the result of processes that are at work today, acting over long periods of time.
1786	German–Swiss Johann von Carpentier described the European ice age.
1793	Jean Baptiste Lamarck argued that fossils are the remains of once-living animals and plants.
1794	William Smith produced the first large-scale geological maps of England.
1804	French physicists Jean Biot and Joseph Gay-Lussac studied the atmosphere from a hot-air balloon.
1809	The first geological survey of the eastern USA was produced by William Maclure.
1815	In England, William Smith showed how rock strata (layers) can be identified by the fossils found in them.
1822	Mary Ann Mantell discovered on the English coast the first fossil to be recognized as that of a dinosaur (an iguanodon). In Germany, Friedrich Mohs introduced a scale for specifying mineral hardness.
1830	Scottish geologist Charles Lyell published the first volume of *The Principles of Geology*, which described the Earth as being several hundred million years old.
1839	In the USA, Louis Agassiz described the action of glaciers, confirming the reality of the ice ages.
1846	Irish physicist William Thomson (Lord Kelvin) estimated, using the temperature of the Earth, that the Earth is 100 million years old.
1850	US naval officer Matthew Fontaine Maury mapped the Atlantic Ocean, noting that it is deeper near its edges than at the centre.
1852	Edward Sabine in Ireland showed a link between sunspot activity and changes in the Earth's magnetic field.
1854	English astronomer George Airy calculated the mass of the Earth by measuring gravity in a coal mine.
1872	The beginning of the world's first major oceanographic expedition, the four-year voyage of the *Challenger*.
1882	Scottish physicist Balfour Stewart postulated the existence of the ionosphere (the ionized layer of the outer atmosphere) to account for differences in the Earth's magnetic field.
1884	German meteorologist Vladimir Köppen introduced a classification of the world's temperature zones.
1890	English geologist Arthur Holmes used radioactivity to date rocks, establishing the Earth to be 4.6 billion old.
1895	In the USA, Jeanette Picard launched the first balloon to be used for stratospheric research.
1896	Swedish chemist Svante Arrhenius discovered a link between the amount of carbon dioxide in the atmosphere and the global temperature.
1897	Norwegian-US meteorologist Jacob Bjerknes and his father Vilhelm developed the mathematical theory of weather forecasting.
1902	British physicist Oliver Heaviside and US engineer Arthur Edwin Kennelly predicted the existence of an electrified layer in the atmosphere that reflects radio waves. In France, Léon Teisserenc discovered layers of different temperatures in the atmosphere, which he called the troposphere and stratosphere.
1906	Richard Dixon Oldham proved the Earth to have a molten core by studying seismic waves.
1909	Yugoslav physicist Andrija Mohorovičić discovered a discontinuity in the Earth's crust, about 30 km/18 mi below the surface, that forms the boundary between the crust and the mantle.
1912	In Germany, Alfred Wegener proposed the theory of continental drift and the existence of a supercontinent, Pangaea, in the distant past.
1913	French physicist Charles Fabry discovered the ozone layer in the upper atmosphere.
1914	German-US geologist Beno Gutenberg discovered the discontinuity that marks the boundary between the Earth's mantle and the outer core.
1922	British meteorologist Lewis Fry Richardson developed a method of numerical weather forecasting.
1925	A German expedition discovered the Mid-Atlantic ridge by means of sonar. Edward Appleton discovered a layer of the atmosphere that reflects radio waves; it was later named after him.
1929	By studying the magnetism of rocks, Japanese geologist Motonori Matuyama showed that the Earth's magnetic field reverses direction from time to time.
1935	US seismologist Charles Francis Richter established a scale for measuring the magnitude of earthquakes.
1936	Danish seismologist Inge Lehmann postulated the existence of a solid inner core of the Earth from the study of seismic waves.
1939	In Germany, Walter Maurice Elsasser proposed that eddy currents in the molten iron core cause the Earth's magnetism.
1950	Hungarian-US mathematician John Von Neumann made the first 24–hour weather forecast by computer.
1956	US geologists Bruce Charles Heezen and Maurice Ewing discovered a global network of oceanic ridges and rifts that divide the Earth's surface into plates.
1958	Using rockets, US physicist James Van Allen discovered a belt of radiation around the Earth.
1960	The world's first weather satellite, *TIROS 1*, was launched.
1963	British geophysicists Fred Vine and Drummond Matthews analysed the magnetism of rocks in the Atlantic Ocean floor and found conclusive proof of seafloor spreading.
1985	A British expedition to the Antarctic discovered a hole in the ozone layer above the South Pole.
1991	A borehole in the Kola Peninsula in Arctic Russia, begun in the 1970s, reached a depth of 12,261 m/40,240 ft (where the temperature was 210°C/410°F). It is expected to reach a depth of 15,000 m/49,000 ft by 1995.

lunar eclipse

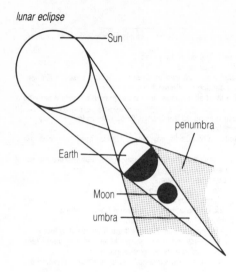

solar eclipse

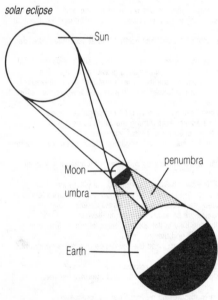

eclipse *The two types of eclipse: lunar and solar. A lunar eclipse occurs when the Moon passes through the shadow of the Earth. A solar eclipse occurs when the Moon passes between the Sun and the Earth, blocking out the Sun's light. During a total solar eclipse, when the Moon completely covers the Sun, the Moon's shadow sweeps across the Earth's surface from west to east at a speed of 3,200 kph/2,000 mph.*

in front of the Sun as seen from Earth, and can happen only at new Moon. During a total eclipse the Sun's ◊corona can be seen. A total solar eclipse can last up to 7.5 minutes. When the Moon is at its farthest from Earth it does not completely cover the face of the Sun, leaving a ring of sunlight visible. This is an **annular eclipse** (from the Latin word *annulus* 'ring'). Between two and five solar eclipses occur each year.

A *lunar eclipse* occurs when the Moon passes

into the shadow of the Earth, becoming dim until emerging from the shadow. Lunar eclipses may be partial or total, and they can happen only at full Moon. Total lunar eclipses last for up to 100 minutes; the maximum number each year is three.

ECLIPSES: LOOKING AHEAD

The first known prediction of an eclipse was made by the Greek philosopher Thales. He foresaw an eclipse which occurred on 25 May 585 BC. The eclipse suddenly darkened the sky at sunset, frightening the forces of King Alyattes of the Lydians and King Cyaxares of the Medes, who were engaged in battle at the time. The frightened soldiers quickly made peace and hurried home.

---

eclipsing binary binary (double) star in which the two stars periodically pass in front of each other as seen from Earth.

When one star crosses in front of the other the total light received on Earth from the two stars declines. The first eclipsing binary to be noticed was ◊Algol.

ecliptic path, against the background of stars, that the Sun appears to follow each year as the Earth orbits the Sun. It can be thought of as the plane of the Earth's orbit projected on to the ◊celestial sphere (imaginary sphere around the Earth).

The ecliptic is tilted at about 23.5° with respect to the celestial equator, a result of the tilt of the Earth's axis relative to the plane of its orbit around the Sun.

ecology (Greek *oikos* 'house') study of the relationship among organisms and the environments in which they live, including all living and nonliving components. The term was coined by the biologist Ernst Haeckel 1866.

Ecology may be concerned with individual organisms (for example, behavioural ecology, feeding strategies), with populations (for example, population dynamics), or with entire communities (for example, competition between species for access to resources in an ecosystem, or predator–prey relationships). Applied ecology is concerned with the management and conservation of habitats and the consequences and control of pollution.

ecosystem in ◊ecology, an integrated unit consisting of the ◊community of living organisms and the physical environment in a particular area. The relationships among species in an ecosystem are usually complex and finely balanced, and removal of any one species may be disastrous. The removal of a major predator, for example, can result in the destruction of the ecosystem through overgrazing by herbivores.

Energy and nutrients pass through organisms in an ecosystem in a particular sequence (see ◊food chain): energy is captured through ◊photosynthesis, and nutrients are taken up from the soil or water by plants; both are passed to herbivores that eat the plants and then to carnivores that feed on herbivores. These nutrients are returned to the soil through the ◊decomposition of excrement and dead organisms, thus completing a cycle that is crucial to the stability and survival of the ecosystem.

ectoparasite ◊parasite that lives on the outer surface of its host.

ectoplasm outer layer of a cell's ◊cytoplasm.

ectotherm 'cold-blooded' animal (see ◊poikilothermy), such as a lizard, that relies on external warmth (ultimately from the Sun) to raise its body temperature so that it can become active. To cool the body, ectotherms seek out a cooler environment.

eddy current electric current induced, in accordance with ◊Faraday's laws, in a conductor located in a changing magnetic field. Eddy currents can cause much wasted energy in the cores of transformers and other electrical machines.

edge connector in computing, an electrical connection formed by taking some of the metallic tracks on a ◊printed circuit board to the edge of the board and using them to plug directly into a matching socket.

Edge connectors are often used to connect the computer's main circuit board, or motherboard, to the expansion boards that provide the computer with extra memory or other facilities.

Edwards Air Force Base military USAF centre in California, situated on a dry lake bed, often used as a landing site by the space shuttle.

EEG abbreviation for ◊*electroencephalogram*.

EEPROM (acronym for *e*lectrically *e*rasable *p*rogrammable *r*ead-*o*nly *m*emory) computer memory that can record data and retain it indefinitely. The data can be erased with an electrical charge and new data recorded.

Some EEPROM must be removed from the computer and erased and reprogrammed using a special device. Other EEPROM, called *flash memory*, can be erased and reprogrammed without removal from the computer.

Effelsberg site, near Bonn, Germany, of the world's largest fully steerable radio telescope, the 100–m/328–ft radio dish of the Max Planck Institute for Radio Astronomy, opened 1971.

efficiency output of a machine (work done by the machine) divided by the input (work put into the machine), usually expressed as a percentage. Because of losses caused by friction, efficiency is always less than 100%, although it can approach this for electrical machines with no moving parts (such as a transformer).

Since the ◊mechanical advantage, or force ratio, is the ratio of the load (the output force) to the effort (the input force), and the ◊velocity ratio is the distance moved by the effort divided by the distance moved by the load, for certain machines the efficiency can also be defined as the mechanical advantage divided by the velocity ratio.

efflorescence loss of water of crystallization from crystals exposed to air, resulting in a dry powdery surface.

EFTPOS (acronym for *e*lectronic *f*unds *t*ransfer at *p*oint *o*f *s*ale) transfer of funds from one bank account to another by electronic means. For example, a customer inserts a plastic card into a point-of-sale computer terminal in a supermarket, and telephone lines are used to make an automatic debit from the customer's bank account to settle the bill.

egestion the removal of undigested food or faeces from the gut. In most animals egestion takes place via the anus, although the invertebrate flatworms must use the mouth because their gut has no exit. Egestion is the last part of a complex feeding process that starts with food capture and continues with digestion and assimilation.

egg in animals, the ovum, or female ◊gamete (reproductive cell).

section through a fertilized egg

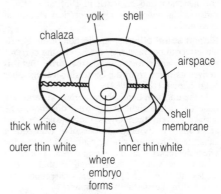

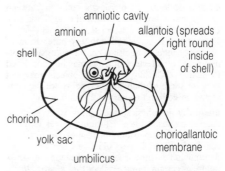

egg *Section through a fertilized bird egg. Inside a bird's egg is a complex structure of liquids and membranes designed to meet the needs of the growing embryo. The yolk, which is rich in fat, is gradually absorbed by the embryo. The white of the egg provides protein and water. The chalaza is a twisted band of protein which holds the yolk in place and acts as a shock absorber. The airspace allows gases to be exchanged through the shell. The allantois contains many blood vessels which carry gases between the embryo and the outside.*

After fertilization by a sperm cell, it begins to divide to form an embryo. Eggs may be deposited by the female (◊ovipary) or they may develop within her body (◊vivipary and ◊ovovivipary). In the oviparous reptiles and birds, the egg is protected by a shell, and well supplied with nutrients in the form of yolk.

eidophor television projection system that produces pictures up to 10 m/33 ft square at sports events, rock concerts, for example. The system uses three coloured beams of light, one of each primary colour (red, blue, and green), which scan the screen. The intensity of each beam is controlled by the strength of the corresponding colour in the television picture.

einsteinium synthesized, radioactive, metallic element of the actinide series, symbol Es, atomic number 99, relative atomic mass 254.

It was produced by the first thermonuclear explosion, in 1952, and discovered in fallout debris in the form of the isotope Es-253 (half-life 20 days). Its longest-lived isotope, Es-254, with a half-life of 276 days, allowed the element to be studied at length. It is now synthesized by bombarding lower-numbered ◊transuranic elements in particle accelerators. It was first identified by A Ghiorso and his team who named it in 1955 after Albert Einstein, in honour of his theoretical studies of mass and energy.

Ekman spiral effect an application of the ◊Coriolis effect to ocean currents, whereby the currents flow at an angle to the winds that drive them. It derives its name from the Swedish oceanographer Vagn Ekman (1874–1954).

In the northern hemisphere, surface currents are deflected to the right of the wind direction. The surface current then drives the subsurface layer at an angle to its original deflection. Consequent subsurface layers are similarly affected, so that the effect decreases with increasing depth. The result is that most water is transported at about right-angles to the wind direction. Directions are reversed in the southern hemisphere.

elastic collision in physics, a collision between two or more bodies in which the total ◊kinetic energy of the bodies is conserved (remains constant); none is converted into any other form of energy. ◊Momentum also is conserved in such collisions. The molecules of a gas may be considered to collide elastically, but large objects may not because some of their kinetic energy will be converted on collision to heat and sound.

elasticity in physics, the ability of a solid to recover its shape once deforming forces (stresses modifying its dimensions or shape) are removed. An elastic material obeys ◊Hooke's law: that is, its deformation is proportional to the applied stress up to a certain point, called the *elastic limit*, beyond which additional stress will deform it permanently. Elastic materials include metals and rubber; however, all materials have some degree of elasticity.

elastomer any material with rubbery properties, which stretches easily and then quickly returns to its original length when released. Natural and synthetic rubbers and such materials as poly-

chloroprene and butadiene copolymers are elastomers. The convoluted molecular chains making up theses materials are uncoiled by a stretching force, but return to their original position when released because there are relatively few crosslinks between the chains.

E layer (formerly called the Kennelly–Heaviside layer) the lower regions of the ◊ionosphere, which refract radio waves, allowing their reception around the surface of the Earth. The E layer approaches the Earth by day and recedes from it at night.

electrical cell device in which chemical energy is converted to electrical energy; see ◊cell, electrical.

electrical relay an electromagnetic switch; see ◊relay.

electric arc a continuous electric discharge of high current between two electrodes, giving out a brilliant light and heat. The phenomenon is exploited in the carbon-arc lamp, once widely used in film projectors. In the electric-arc furnace an arc struck between very large carbon rodes and the metal charge provides the heating. In arc ◊welding an electric arc provides the heat to fuse the metal. The discharges in low-pressure gases, as in neon and sodium lights, can also be broadly considered as electric arcs.

electric bell a bell that makes use of electromagnetism. At its heart is a wire-wound coil on an iron core (an electromagnet) which, when a direct current (from a battery) flows through it, attracts an iron ◊armature. The armature acts as a switch, whose movement causes contact with an adjustable contact point to be broken, so breaking the circuit. A spring rapidly returns the armature to the contact point, once again closing the circuit, and so bringing about the oscillation. The armature oscillates back and forth, and the clapper or hammer fixed to the armature strikes the bell.

electric charge property of some bodies that causes them to exert forces on each other. Two bodies both with positive or both with negative charges repel each other, whereas bodies with opposite or 'unlike' charges attract each other, since each is in the ◊electric field of the other. In atoms, ◊electrons possess a negative charge, and ◊protons an equal positive charge. The ◊SI unit of electric charge is the coulomb (symbol C).

Electric charge can be generated by friction induction or chemical change and shows itself as an accumulation of electrons (negative charge) or loss of electrons (positive charge) on an atom or body. Atoms have no charge but can sometimes gain electrons to become negative ions or lose them to become positive ions. So-called ◊static electricity, seen in such phenomena as the charging of nylon shirts when they are pulled on or off, or in brushing hair, is in fact the gain or loss of electrons from the surface atoms. A flow of charge (such as electrons through a copper wire) constitutes an electric current; the flow of current is measured in amperes (symbol A).

electric current the flow of electrically charged particles through a conducting circuit due to the presence of a ◊potential difference. The current at any point in a circuit is the amount of charge flowing per second; its SI unit is the ampere (coulomb per second).

Current carries electrical energy from a power supply, such as a battery of electrical cells, to the components of the circuit, where it is converted into other forms of energy, such as heat, light, or motion. It may be either ◊direct current or ◊alternating current.

heating effect When current flows in a component possessing resistance, electrical energy is converted into heat energy. If the resistance of the component is R ohms and the current through it is I amperes, then the heat energy W (in joules) generated in a time t seconds is given by the formula $W = I^2Rt$

magnetic effect A ◊magnetic field is created around all conductors that carry a current. When a current-bearing conductor is made into a coil it forms an ◊electromagnet with a magnetic field that is similar to that of a bar magnet, but which disappears as soon as the current is switched off. The strength of the magnetic field is directly proportional to the current in the conductor—a property that allows a small electromagnet to be used to produce a pattern of magnetism on recording tape that accurately represents the sound or data to be stored. The direction of the field created around a conducting wire may be predicted by using ◊Maxwell's screw rule.

motor effect A conductor carrying current in a magnetic field experiences a force, and is impelled to move in a direction perpendicular to both the direction of the current and the direction of the magnetic field. The direction of motion may be predicted by Fleming's left-hand rule (see ◊Fleming's rules). The magnitude of the force experienced depends on the length of the conductor and on the strengths of the current and the magnetic field, and is greatest when the conductor is at right angles to the field. A conductor wound into a coil that can rotate between the poles of a magnet forms the basis of an ◊electric motor.

electric field in physics, a region in which a particle possessing electric charge experiences a force owing to the presence of another electric charge. It is a type of electromagnetic field.

electricity all phenomena caused by ◊electric charge, whether static or in motion. Electric charge is caused by an excess or deficit of electrons in the charged substance, and an electric current by the movement of electrons around a circuit. Substances may be electrical conductors, such as metals, which allow the passage of electricity through them, or insulators, such as rubber, which are extremely poor conductors. Substances with relatively poor conductivities that can be improved by the addition of heat or light are known as ◊semiconductors.

Electricity generated on a commercial scale was available from the early 1880s and used for electric motors driving all kinds of machinery, and for lighting, first by carbon arc, but later by incandescent filaments (first of carbon and then of tungsten), enclosed in glass bulbs partially filled with inert gas under vacuum. Light is also produced by passing electricity through a gas or metal vapour or a fluorescent lamp. Other practical applications include telephone, radio, television, X-ray machines, and many other applications in ◊electronics.

The fact that amber has the power, after being rubbed, of attracting light objects, such as bits of straw and feathers, is said to have been known to Thales of Miletus and to the Roman naturalist Pliny. William Gilbert, Queen Elizabeth I's physician, found that many substances possessed this power, and he called it 'electric' after the Greek word meaning 'amber'.

In the early 1700s, it was recognized that there are two types of electricity and that unlike kinds attract each other and like kinds repel. The charge on glass rubbed with silk came to be known as positive electricity, and the charge on amber rubbed with wool as negative electricity. These two charges were found to cancel each other when brought together.

In 1800 Alessandro Volta found that a series of cells containing brine, in which were dipped plates of zinc and copper, gave an electric current, which later in the same year was shown to evolve hydrogen and oxygen when passed through water (◊electrolysis). Humphry Davy, in 1807, decomposed soda and potash (both thought to be elements) and isolated the metals sodium and potassium, a discovery that led the way to ◊electroplating. Other properties of electric currents discovered were the heating effect, now used in lighting and central heating, and the deflection of a magnetic needle, described by Hans Oersted 1820 and elaborated by André Ampère 1825. This work made possible the electric telegraph.

One day, Sir, you may tax it.

Michael Faraday (1791–1867) to Mr Gladstone, Chancellor of the Exchequer, when asked about the usefulness of **electricity**

For Michael Faraday, the fact that an electric current passing through a wire caused a magnet to move suggested that moving a wire or coil of wire rapidly between the poles of a magnet would induce an electric current. He demonstrated this 1831, producing the first ◊dynamo, which became the basis of electrical engineering. The characteristics of currents were crystallized about 1827 by Georg Ohm, who showed that the current passing along a wire was equal to the electromotive force (emf) across the wire multiplied by a constant, which was the conductivity of the wire. The unit of resistance (ohm) is named after Ohm, the unit of emf is named after Volta (volt), and the unit of current after Ampère (amp).

The work of the late 1800s indicated the wide interconnections of electricity (with magnetism, heat, and light), and about 1855 James Clerk Maxwell formulated a single electromagnetic theory. The universal importance of electricity was decisively proved by the discovery that the atom, up until then thought to be the ultimate particle of matter, is composed of a positively charged central core, the nucleus, about which negatively charged electrons rotate in various orbits.

Electricity is the most useful and most convenient form of energy, readily convertible into heat and light and used to power machines. Electricity can be generated in one place and distributed anywhere because it readily flows through wires. It is

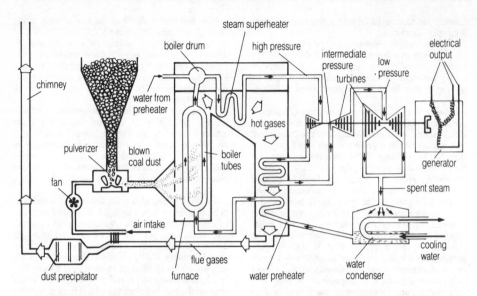

electricity *Coal-fired power station (highly simplified). Coal enters the system through the hopper on the left and enters the furnace after being pulverized. The coal burns inside the furnace, heating water in the boiler tube to steam. The hot gases are used to heat the steam further (superheat). The steam then passes to the turbines. There are usually three turbines—high, intermediate and low pressure—which extract all the energy of the steam and turn the electricity generator.*

generated at power stations where a suitable energy source is harnessed to drive ◊turbines that spin electricity generators. Current energy sources are coal, oil, water power (hydroelectricity), natural gas, and ◊nuclear energy. Research is under way to increase the contribution of wind, tidal, and geothermal power. Nuclear fuel has proved a more expensive source of electricity than initially anticipated and worldwide concern over radioactivity may limit its future development.

Electricity is generated at power stations at a voltage of about 25,000 volts, which is not a suitable voltage for long-distance transmission. For minimal power loss, transmission must take place at very high voltage (400,000 volts or more). The generated voltage is therefore increased ('stepped up') by a ◊transformer. The resulting high-voltage electricity is then fed into the main arteries of the ◊grid system, an interconnected network of power stations and distribution centres covering ·a large area. After transmission to a local substation, the line voltage is reduced by a step-down transformer and distributed to consumers.

Among specialized power units that convert energy directly to electrical energy without the intervention of any moving mechanisms, the most promising are thermionic converters. These use conventional fuels such as propane gas, as in portable military power packs, or, if refuelling is to be avoided, radioactive fuels, as in uncrewed navigational aids and spacecraft.

electric motor a machine that converts electrical energy into mechanical energy. There are various types, including direct-current and induction motors, most of which produce rotary motion. A linear induction motor produces linear (sideways) rather than rotary motion.

A simple *direct-current motor* consists of a horseshoe-shaped permanent ◊magnet with a wire-wound coil (◊armature) mounted so that it can rotate between the poles of the magnet. A ◊commutator reverses the current (from a battery) fed to the coil on each half-turn, which rotates because of the mechanical force exerted on a conductor carrying a current in a magnetic field.

An *induction motor* employs ◊alternating current. It comprises a stationary current-carrying coil (stator) surrounding another coil (rotor), which rotates because of the current induced in it by the magnetic field created by the stator; it thus requires no commutator.

electric power the rate at which an electrical machine uses electrical ◊energy or converts it into other forms of energy—for example, light, heat, mechanical energy. Usually measured in watts (equivalent to joules per second), it is equal to the product of the voltage and the current flowing.

An electric lamp that passes a current of 0.4 amps at 250 volts uses 100 watts of electrical power and converts it into light—in ordinary terms it is a 100—watt lamp. An electric motor that requires 6 amps at the same voltage consumes 1,500 watts (1.5 kilowatts), equivalent to delivering about 2 horsepower of mechanical energy.

electrocardiogram (ECG) graphic recording of the electrical changes in the heart muscle, as detected by electrodes placed on the chest. Electrocardiography is used in the diagnosis of heart disease.

electrochemical series list of chemical elements arranged in descending order of the ease with which they can lose electrons to form cations (positive ions). An element can be displaced (◊displacement reaction) from a compound by any element above it in the series.

simple direct-current motor

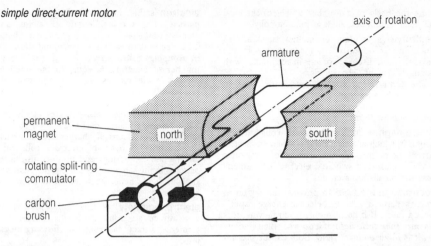

axis of rotation

armature

permanent magnet

north

south

rotating split-ring commutator

carbon brush

electric motor

electrochemistry the branch of science that studies chemical reactions involving electricity. The use of electricity to produce chemical effects, ◊electrolysis, is employed in many industrial processes, such as the manufacture of chlorine and the extraction of aluminium. The use of chemical reactions to produce electricity is the basis of electrical ◊cells, such as the dry cell and the ◊Leclanché cell.

Since all chemical reactions involve changes to the electronic structure of atoms, all reactions are now recognized as electrochemical in nature. Oxidation, for example, was once defined as a process in which oxygen was combined with a substance, or hydrogen was removed from a compound; it is now defined as a process in which electrons are lost.

electrode any terminal by which an electric current passes in or out of a conducting substance; for example, the anode or cathode in a battery or the carbons in an arc lamp. The terminals that emit and collect the flow of electrons in thermionic ◊valves (electron tubes) are also called electrodes: for example, cathodes, plates, and grids.

electrodynamics the branch of physics dealing with electric currents and associated magnetic forces. ◊Quantum electrodynamics (QED) studies the interaction between charged particles and their emission and absorption of electromagnetic radiation. This field combines quantum theory and relativity theory, making accurate predictions about subatomic processes involving charged particles such as electrons and protons.

electroencephalogram (EEG) graphic record of the electrical discharges of the brain, as detected by electrodes placed on the scalp. The pattern of electrical activity revealed by electroencephalography is helpful in the diagnosis of some brain disorders, such as epilepsy.

electrolysis in chemistry, the production of chemical changes by passing an electric current through a solution or molten salt (the electrolyte), resulting in the migration of ions to the electrodes: positive ions (cations) to the negative electrode (cathode) and negative ions (anions) to the positive electrode (anode).

During electrolysis, the ions react with the electrode, either receiving or giving up electrons. The resultant atoms may be liberated as a gas, or deposited as a solid on the electrode, in amounts that are proportional to the amount of current passed, as discovered by English chemist Michael Faraday. For instance, when acidified water is electrolysed, hydrogen ions (H^+) at the cathode receive electrons to form hydrogen gas; hydroxide ions (OH^-) at the anode give up electrons to form oxygen gas and water.

One application of electrolysis is *electroplating*, in which a solution of a salt, such as silver nitrate ($AgNO_3$), is used and the object to be plated acts as the negative electrode, thus attracting silver ions (Ag^+). Electrolysis is used in many industrial processes, such as coating metals for vehicles and ships, and refining bauxite into aluminium; it also

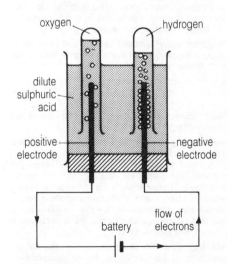

oxygen

hydrogen

dilute sulphuric acid

positive electrode

negative electrode

battery

flow of electrons

electrolysis *Passing an electric current through acidified water breaks down the water into its constituent elements—hydrogen and oxygen.*

forms the basis of a number of electrochemical analytical techniques, such as polarography.

electrolyte a solution or molten substance in which an electric current is made to flow by the movement and discharge of ions in accordance with Faraday's laws of ◊electrolysis.

The term 'electrolyte' is frequently used to denote a substance that, when dissolved in a specified solvent, usually water, produces an electrically conducting medium.

electromagnet an iron bar with coils of wire around it, which acts as a magnet when an electric current flows through the wire. Electromagnets have many uses: in switches, electric bells, solenoids, and metal-lifting cranes.

electromagnetic field in physics, the region in which a particle with an ◊electric charge experiences a force. If it does so only when moving, it is in a pure *magnetic field*; if it does so when stationary, it is in an *electric field*. Both can be present simultaneously.

electromagnetic force one of the four fundamental ◊forces of nature, the other three being gravity, the strong nuclear force, and the weak nuclear force. The ◊elementary particle that is the carrier for the electromagnetic (em) force is the photon.

electromagnetic induction in electronics, the production of an ◊electromotive force (emf) in a circuit by a change of magnetic flux through the circuit or by relative motion of the circuit and the magnetic flux. In a closed circuit an ◊induced current will be produced. All dynamos and generators make use of this effect. When magnetic tape is driven past the playback head (a small coil) of a tape-recorder, the moving magnetic field induces an emf in the head, which is then amplified to reproduce the recorded sounds.

If the change of magnetic flux is due to a variation in the current flowing in the same circuit, the phenomenon is known as self-induction; if it is due to a change of current flowing in another circuit it is known as mutual induction.

electromagnetic spectrum the complete range, over all wavelengths from the lowest to the highest, of ◊electromagnetic waves.

electromagnetic system of units former system of absolute electromagnetic units (emu) based on the ◊c.g.s. system and having, as its primary electrical unit, the unit magnetic pole. It was replaced by ◊SI units.

electromagnetic waves oscillating electric and magnetic fields travelling together through space at a speed of nearly 300,000 km/186,000 mi per second. The (limitless) range of possible wavelengths or ◊frequencies of electromagnetic waves, which can be thought of as making up the *electromagnetic spectrum*, includes radio waves, infrared radiation, visible light, ultraviolet radiation, X-rays, and gamma rays.

electromotive force (emf) loosely, the voltage produced by an electric battery or generator or, more precisely, the energy supplied by a source of electric power in driving a unit charge around an electrical circuit. The unit is the ◊volt.

electron stable, negatively charged ◊elementary particle; it is a constituent of all atoms, and a member of the class of particles known as ◊leptons. The electrons in each atom surround the nucleus in groupings called *shells*; in a neutral atom the number of electrons is equal to the number of protons in the nucleus. This electron structure is responsible for the chemical properties of the atom (see ◊atomic structure).

Electrons are the basic particles of electricity. Each carries a charge of 1.602192×10^{-19} coulomb, and all electrical charges are multiples of this quantity. A beam of electrons will undergo ◊diffraction (scattering) and produce interference patterns in the same way as ◊electromagnetic waves such as light; hence they may also be regarded as waves.

electronegativity the ease with which an atom can attract electrons to itself. Electronegative elements attract electrons, so forming negative ions.

Linus Pauling devised an electronegativity scale to indicate the relative power of attraction of elements for electrons. Fluorine, the most nonmetallic element, has a value of 4.0 on this scale; oxygen, the next most nonmetallic, has a value of 3.5.

In a covalent bond between two atoms of different electronegativities, the bonding electrons will be located close to the more electronegative atom, creating a ◊dipole.

The important thing in science is not so much to obtain new facts as to discover new ways of thinking about them.

William Lawrence Bragg *Beyond Reductionism* 1968

electron gun a part in many electronic devices consisting of a series of ◊electrodes, including a cathode for producing an electron beam. It plays an essential role in ◊cathode-ray tubes (television tubes) and ◊electron microscopes.

electronic flash discharge tube that produces a high-intensity flash of light, used for photography in dim conditions. The tube contains an inert gas such as krypton. The flash lasts only a few thousandths of a second.

electronic mail ◊telecommunications system that enables the users of a computer network to send messages to other users. Telephone wires are used to send the signals from terminal to terminal.

Subscribers to an electronic mail system type messages in ordinary letter form on a word processor, or microcomputer, and 'drop' the letters into a central computer's memory bank by means of a computer/telephone connector (a ◊modem). The recipient 'collects' the letter by calling up the central computer and feeding a unique password into the system.

electronic point of sale (EPOS) system used in retailing where the bar code on a product is scanned at the cash till and the information relayed to the store computer. The computer will then relay back the price of the item to the cash till. The

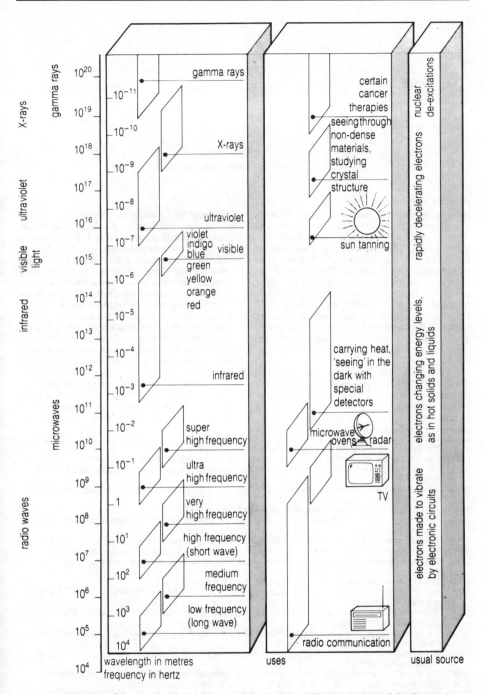

electromagnetic waves *Radio waves have the lowest frequency. Infrared radiation, visible light, ultraviolet radiation, X-rays, and gamma rays have progressively higher frequencies.*

customer can then be given an itemized receipt whist the computer can remove the item from stock figures.

In theory, EPOS could be used for computer stock control and reordering as well as giving a wealth of information about turnover, profitability on different lines, stock ratios, and other important financial indicators. In practice, retailers have been reluctant to use the system to its full potential.

electronics branch of science that deals with the

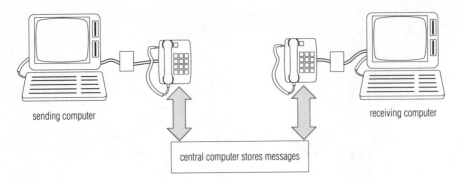

sending computer

receiving computer

central computer stores messages

electronic mail
emission of ◊electrons from conductors and ◊semi-conductors, with the subsequent manipulation of these electrons, and with the construction of electronic devices. The first electronic device was the thermionic ◊valve, or vacuum tube, in which electrons moved in a vacuum, and led to such inventions as ◊radio, ◊television, ◊radar, and the digital ◊computer. Replacement of valves with the comparatively tiny and reliable transistor in 1948 revolutionized electronic development. Modern electronic devices are based on minute ◊integrated circuits (silicon chips), which are wafer-thin crystal slices holding tens of thousands of electronic components.

By using solid-state devices such as integrated circuits, extremely complex electronic circuits can be constructed, leading to the development of ◊digital watches, pocket ◊calculators, powerful ◊microcomputers, and ◊word processors.

electronics: chronology

electron microscope instrument that produces a magnified image by using a beam of ◊electrons instead of light rays, as in an optical ◊microscope. An *electron lens* is an arrangement of electromagnetic coils that control and focus the beam. Electrons are not visible to the eye, so instead of an eyepiece there is a fluorescent screen or a photographic plate on which the electrons form an image. The wavelength of the electron beam is much shorter than that of light, so much greater magnification and resolution (ability to distinguish detail) can be achieved. The development of the electron microscope has made possible the observation of very minute organisms, viruses, and even large molecules.

A ◊transmission electron microscope passes the electron beam through a very thin slice of a specimen. A ◊scanning electron microscope looks at the exterior of a specimen. A ◊scanning transition

1897	The electron was discovered by English physicist John Joseph Thomson.
1904	Ambrose Fleming invented the diode valve, which allows flow of electricity in one direction only.
1906	The triode electron valve, the first device to control an electric current, was invented by US physicist Lee De Forest.
1947	John Bardeen, William Shockley, and Walter Brattain invented the junction germanium transistor at the Bell Laboratories, New Jersey, USA.
1952	British physicist G W A Dunner proposed the integrated circuit.
1953	Jay Forrester of the Massachusetts Institute of Technology, USA, built a magnetic memory smaller than existing vacuum-tube memories.
1954	The silicon transistor was developed by Gordon Teal of Texas Instruments, USA.
1958	The first integrated circuit, containing five components, was built by US electrical physicist Jack Kilby.
1959	The planar transistor, which is built up in layers, or planes, was designed by Robert Noyce of Fairchild Semiconductor Corporation, USA.
1961	Steven Hofstein designed the field-effect transistor used in integrated circuits.
1971	The first microprocessor, the Intel 4004, was designed by Ted Hoff in the USA; it contained 2,250 components and could add two 4-bit numbers in 11–millionths of a second.
1974	The Intel 8080 microprocessor was launched; it contained 4,500 components and could add two 8-bit numbers in 2.5–millionths of a second.
1979	The Motorola 68000 microprocessor was introduced; it contained 70,000 components and could multiply two 16–bit numbers in 3.2–millionths of a second.
1981	The Hewlett-Packard Superchip was introduced; it contained 450,000 components and could multiply two 32–bit numbers in 1.8–millionths of a second.
1985	The Inmos T414 transputer, the first microprocessor designed for use in parallel computers, was launched.
1988	The first optical microprocessor, which uses light instead of electricity, was developed.
1989	Wafer-scale silicon memory chips were introduced: the size of a beer mat, they are able to store 200 million characters.
1990	Memory chips capable of holding 4 million bits of information began to be mass-produced in Japan. The chips can store the equivalent of 520,000 characters, or the contents of a 16–page newspaper. Each chip contains 9 million components packed on a piece of silicon less than 15 mm long by 5 mm wide.
1992	Transistors made from high-temperature superconducting ceramics rather than semiconductors produced in Japan by Sanyo Electric. The new transistors are 10 times faster than semiconductor transistors.
1993	First 64-bit microprocessor, the Pentium, introduced by Intel.

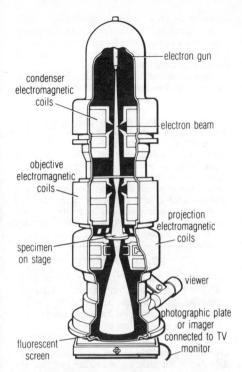

electron gun

condenser
electromagnetic
coils

electron beam

objective
electromagnetic
coils

projection
electromagnetic
coils

specimen
on stage

viewer

photographic plate
or imager
connected to TV
monitor

fluorescent
screen

electron microscope *The transmission electron microscope. The electromagnetic coils are the 'lenses' of the electron microscope, focusing the electron beam on the specimen. The beam travels through the specimen and is projected on to a screen or imagers. These microscopes can magnify by a million times.*

electron microscope (STEM) can produce a magnification of 90 million times. See also ◊atomic force microscope.

ELECTRON MICROSCOPES AND ELECTRON WAVES

In an electron microscope, electrons are accelerated to an energy at which their wavelength is about one-billionth of a centimetre (0.000000001 cm/ 0.0000000004 in). This is 10,000 times shorter than the wavelength of light (about 0.0001 cm/0.00004 in), so electron microscopes can reveal much smaller details than optical microscopes.

electron probe microanalyser modified ◊electron microscope in which the target emits X-rays when bombarded by electrons. Varying X-ray intensities indicate the presence of different chemical elements. The composition of a specimen can be mapped without the specimen being destroyed.

electrons, delocalized electrons that are not associated with individual atoms or identifiable chemical bonds, but are shared collectively by all the constituent atoms or ions of some chemical substances (such as metals, graphite, and ◊aromatic compounds).

A metallic solid consists of a three-dimensional arrangement of metal ions through which the delocalized electrons are free to travel. Aromatic compounds are characterized by the sharing of delocalized electrons by several atoms within the molecule.

electrons, localized a pair of electrons in a ◊covalent bond that are located in the vicinity of the nuclei of the two contributing atoms. Such electrons cannot move beyond this area.

electron spin resonance in archaeology, a nondestructive dating method applicable to teeth, bone, heat-treated flint, ceramics, sediments, and stalagmitic concretions. It enables electrons, displaced by natural radiation and then trapped in the structure, to be measured; their number indicates the age of the specimen.

electron volt unit (symbol eV) for measuring the energy of a charged particle (◊ion or ◊electron) in terms of the energy of motion an electron would gain from a potential difference of one volt. Because it is so small, more usual units are mega-(million) and giga-(billion) electron volts (MeV and GeV).

electrophoresis the ◊diffusion of charged particles through a fluid under the influence of an electric field. It can be used in the biological sciences to separate ◊molecules of different sizes, which diffuse at different rates. In industry, electrophoresis is used in paint-dipping operations to ensure that paint reaches awkward corners.

electroplating deposition of metals upon metallic surfaces by electrolysis for decorative and/or protective purposes. It is used in the preparation of printers' blocks, 'master' audio discs, and in many other processes.

A current is passed through a bath containing a solution of a salt of the plating metal, the object to be plated being the cathode (negative terminal); the anode (positive terminal) is either an inert substance or the plating metal. Among the metals most commonly used for plating are zinc, nickel, chromium, cadmium, copper, silver, and gold.

In *electropolishing*, the object to be polished is made the anode in an electrolytic solution and by carefully controlling conditions the high spots on the surface are dissolved away, leaving a high-quality stain-free surface. This technique is useful in polishing irregular stainless-steel articles.

electroporation in biotechnology, a technique of introducing foreign ◊DNA into pollen with a strong burst of electricity, used in creating genetically-engineered plants.

electropositivity in chemistry, a measure of the ability of elements (mainly metals) to donate electrons to form positive ions. The greater the metallic character, the more electropositive the element.

electrorheological fluid another name for ◊smart fluid, a liquid suspension that gels when an electric field is applied across it.

electroscope an apparatus for detecting ◊electric charge. The simple gold-leaf electroscope consists of a vertical conducting (metal) rod ending in a pair of rectangular pieces of gold foil, mounted inside and insulated from an earthed metal case. An electric charge applied to the end of the metal

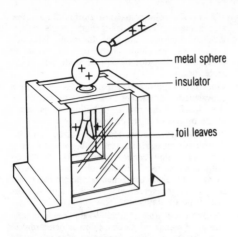

— metal sphere

— insulator

— foil leaves

electroscope *The electroscope, a simple means of detecting electric charge. The metal foil leaves diverge when a charge is applied to the metal sphere.*

rod makes the gold leaves diverge, because they each receive a similar charge (positive or negative) and so repel each other.

The polarity of the charge can be found by bringing up another charge of known polarity and applying it to the metal rod. A like charge has no effect on the gold leaves, whereas an opposite charge neutralizes the charge on the leaves and causes them to collapse.

electrostatic precipitator device that removes dust or other particles from air and other gases by electrostatic means. An electric discharge is passed through the gas, giving the impurities a negative electric charge. Positively charged plates are then used to attract the charged particles and remove them from the gas flow. Such devices are attached to the chimneys of coal-burning power stations to remove ash particles.

electrostatics the study of electric charges from stationary sources (not currents). See ◊static electricity.

electrovalent bond another name for an ◊ionic bond, a chemical bond in which the combining atoms lose or gain electrons to form ions.

electrum naturally occurring alloy of gold and silver used by early civilizations to make the first coins, about the 6th century BC.

element substance that cannot be split chemically into simpler substances. The atoms of a particular element all have the same number of protons in their nuclei (their atomic number). Elements are classified in the periodic table (see ◊periodic table of the elements). Of the 109 known elements, 95 are known to occur in nature (those with atomic numbers 1–95). Those from 96 to 109 do not occur in nature and are synthesized only, produced in particle accelerators. Eighty-one of the elements are stable; all the others, which include atomic numbers 43, 61, and from 84 up, are radioactive.

Elements are classified as metals, nonmetals, or metalloids (weakly metallic elements) depending on a combination of their physical and chemical properties; about 75% are metallic. Some elements occur abundantly (oxygen, aluminium); others occur moderately or rarely (chromium, neon); some, in particular the radioactive ones, are found in minute (neptunium, plutonium) or very minute (technetium) amounts.

Symbols (devised by Swedish chemist Jöns Berzelius) are used to denote the elements; the symbol is usually the first letter or letters of the English or Latin name (for example, C for carbon, Ca for calcium, Fe for iron, *ferrum*). The symbol represents one atom of the element.

According to current theories, hydrogen and helium were produced in the ◊Big Bang at the beginning of the universe. Of the other elements, those up to atomic number 26 (iron) are made by nuclear fusion within the stars. The more massive elements such as lead and uranium, are produced when an old star explodes; as its centre collapses, the gravitational energy squashes nuclei together to make new elements.

SOLID ELEMENTS

Of the 92 naturally occurring chemical elements, only two (bromine and mercury) are liquids at ordinary temperatures, and only 11 are gases. All the other elements are solids, mostly metals.

---

elementary particle in physics, a subatomic particle that is not made up of smaller particles, and so can be considered one of the fundamental units of matter. There are three groups of elementary particles: quarks, leptons, and gauge bosons.

Quarks, of which there are 12 types (up, down, charm, strange, top, and bottom, plus the antiparticles of each), combine in groups of three to produce heavy particles called baryons, and in groups of two to produce intermediate-mass particles called mesons. They and their composite particles are influenced by the strong nuclear force.

Leptons are light particles. Again, there are 12 types: the electron, muon, tau; their neutrinos, the electron neutrino, muon neutrino, and tau neutrino; and the antiparticles of each. These particles are influenced by the weak nuclear force.

Gauge bosons carry forces between other particles. There are four types: the gluons, photon, weakons, and graviton. The gluon carries the strong nuclear force, the photon the electromagnetic force, the weakons the weak nuclear force, and the graviton the force of gravity (see ◊forces, fundamental).

elevation of boiling point raising of the boiling point of a liquid above that of the pure solvent, caused by a substance being dissolved in it. The phenomenon is observed when salt is added to boiling water; the water ceases to boil because its boiling point has been elevated.

How much the boiling point is raised depends on the number of molecules of substance dissolved. For a single solvent, such as pure water, all substances in the same molecular concentration (expressed in ◊moles) produce the same elevation of boiling point. The elevation e produced by the presence of a solute of molar concentration C is given by the equation $e = KC$, where K is a constant

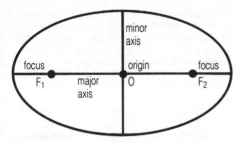

ellipse *The structure of an ellipse; for all points on the ellipse, the sum of the distances from the two foci, F₁ and F₂, is the same.*

(called the ebullioscopic constant) for the solvent concerned.

ellipse curve joining all points (loci) around two fixed points (foci) such that the sum of the distances from those points is always constant. The diameter passing through the foci is the major axis, and the diameter bisecting this at right angles is the minor axis. An ellipse is one of a series of curves known as ◊conic sections. A slice across a cone that is not made parallel to, and does not pass through, the base will produce an ellipse.

El Niño (Spanish 'the child') warm ocean surge of the ◊Peru Current. It is so called because it tends to occur at Christmas, recurring every 5–8 years or so in the E Pacific off South America. It involves a change in the direction of ocean currents, which prevents the upwelling of cold, nutrient-rich waters along the coast of Ecuador and Peru, killing fishes and plants. It is an important factor in global weather.

El Niño is believed to be caused by the failure of trade winds and, consequently, of the ocean currents normally driven by these winds. Warm surface waters then flow in from the east. The phenomenon can disrupt the climate of the area disastrously, and has played a part in causing famine in Indonesia, drought and bush fires in the Galápagos Islands, rainstorms in California and South America, and the destruction of Peru's anchovy harvest and wildlife 1982–1983; and algal blooms in Australia's drought-stricken rivers and an unprecedented number of typhoons in Japan 1991.

elongation in astronomy, the angular distance between the Sun and a planet or other solar-system object. This angle is 0° at ◊conjunction, 90° at ◊quadrature, and 180° at ◊opposition.

elution in chemistry, the washing of an adsorbed substance from the adsorbing material; it is used, for example, in the separation processes of chromatography and electrophoresis.

embryo early development stage of an animal or a plant following fertilization of an ovum (egg cell), or activation of an ovum by ◊parthenogenesis. In humans, the term embryo describes the fertilized egg during its first seven weeks of existence; from the eighth week onwards it is referred to as a fetus. In animals the embryo exists either within an

egg (where it is nourished by food contained in the yolk), or in mammals, in the ◊uterus of the mother. In mammals (except marsupials) the embryo is fed through the ◊placenta. The plant embryo is found within the seed in higher plants. It sometimes consists of only a few cells, but usually includes a root, a shoot (or primary bud), and one or two ◊cotyledons, which nourish the growing seedling.

embryology study of the changes undergone by an organism from its conception as a fertilized ovum (egg) to its emergence into the world at hatching or birth. It is mainly concerned with the changes in cell organization in the embryo and the way in which these lead to the structures and organs of the adult (the process of ◊differentiation).

Applications of embryology include embryo transplants, both commercial (for example, in building up a prize dairy-cow herd quickly at low cost) and in obstetric medicine (as a method for helping couples with fertility problems to have children). This usually involves the surgical removal of eggs from a female, their fertilization under laboratory conditions, and, once normal development is under way, their replacement in the womb.

embryo sac large cell within the ovule of flowering plants that represents the female ◊gametophyte when fully developed. It typically contains eight nuclei. Fertilization occurs when one of these nuclei, the egg nucleus, fuses with a male ◊gamete.

emerald a clear, green gemstone variety of the mineral ◊beryl. It occurs naturally in Colombia, the Ural Mountains, in Russia, Zimbabwe, and Australia.

emery greyish-black opaque metamorphic rock consisting of ◊corundum and magnetite, together with other minerals such as hematite. It is used as an ◊abrasive.

Emery occurs on the island of Naxos, Greece, and in Turkey.

emf in physics, abbreviation for ◊electromotive force.

emission spectroscopy in analytical chemistry, a technique for determining the identity or amount present of a chemical substance by measuring the amount of electromagnetic radiation it emits at specific wavelengths; see ◊spectroscopy.

empirical formula the simplest chemical formula for a compound. Quantitative analysis gives the proportion of each element present, and from this the empirical formula is calculated. It is related to the actual (molecular) formula by the relation:

$$(\text{empirical formula})_n = \text{molecular formula}$$

where n is a small whole number (1, 2,. . .)

emulator in computing, an item of software or firmware that allows one device to imitate the functioning of another. Emulator software is commonly used to allow one make of computer to run programs written for a different make of computer. This allows a user to select from a wider range of ◊applications programs, and perhaps to save money by running programs designed for an expensive computer on a cheaper model.

Many printers contain emulator firmware that enables them to imitate Hewlett Packard and

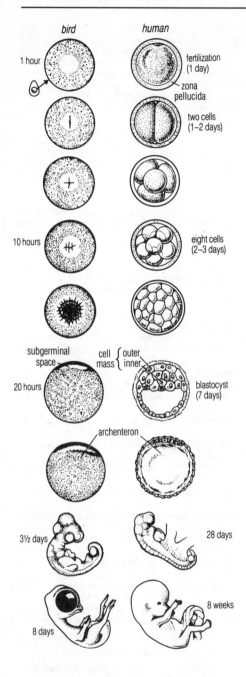

bird   human

1 hour

fertilization
(1 day)

zona
pellucida

two cells
(1-2 days)

10 hours

eight cells
(2-3 days)

subgerminal
space   cell { outer
mass { inner

20 hours

blastocyst
(7 days)

archenteron

3½ days

28 days

8 weeks

8 days

embryo *The development of a bird and a human embryo. In the human, division of the fertilized egg, or ovum, begins within hours of conception. Within a week, a hollow, fluid-containing ball—a blastocyst—with a mass of cells at one end has developed. After the third week, the embryo has changed from a mass of cells into a recognizable shape. At four weeks, the embryo is 3 mm/0.1 in long, with a large bulge for the heart and small pits for the ears. At six weeks, the embryo is 1.5 cm/0.6 in with a pulsating heart and ear flaps. At the eighth week, the embryo is 2.5 cm long and recognizably human, with eyelids, small fingers and toes.*

Epson printers, because so much software is written to work with these widely used machines.

emulsifier food ◊additive used to keep oils dispersed and in suspension, in products such as mayonnaise and peanut butter. Egg yolk is a naturally occurring emulsifier, but most of the emulsifiers in commercial use today are synthetic chemicals.

emulsion a stable dispersion of a liquid in another liquid—for example, oil and water in some cosmetic lotions.

encephalin a naturally occurring chemical produced by nerve cells in the brain that has the same effect as morphine or other derivatives of opium, acting as a natural painkiller. Unlike morphine, encephalins are quickly degraded by the body, so there is no buildup of tolerance to them, and hence no 'addiction'. Encephalins are a variety of ◊peptides, as are ◊endorphins, which have similar effects.

Encke's comet comet with the shortest known orbital period, 3.3 years. It is named after German mathematician and astronomer Johann Franz Encke (1791–1865), who calculated its orbit in 1819 from earlier sightings.

It was first seen in 1786 by the French astronomer Pierre Méchain (1744–1804). It is the parent body of the Taurid meteor shower and a fragment of it may have hit the Earth in the ◊Tunguska Event 1908.

In 1913, it became the first comet to be observed throughout its entire orbit when it was photographed near ◊aphelion (the point in its orbit furthest from the Sun) by astronomers at Mount Wilson Observatory in California, USA.

endangered species plant or animal species whose numbers are so few that it is at risk of becoming extinct. Officially designated endangered species are listed by the International Union for the Conservation of Nature (◊IUCN).

An example of an endangered species is the Javan rhinoceros. There are only about 50 alive today and, unless active steps are taken to promote this species' survival, it will probably be extinct within a few decades.

endocrine gland gland that secretes hormones into the bloodstream to regulate body processes. Endocrine glands are most highly developed in vertebrates, but are also found in other animals, notably insects. In humans the main endocrine glands are the pituitary, thyroid, parathyroid, adrenal, pancreas, ovary, and testis.

endolymph fluid found in the inner ◊ear, filling the central passage of the cochlea as well as the semicircular canals.

Sound waves travelling into the ear pass eventually through the three small bones of the middle ear and set up vibrations in the endolymph. These are detected by receptors in the cochlea, which send nerve impulses to the hearing centres of the brain.

endoparasite ◊parasite that lives inside the body of its host.

endoplasm inner, liquid part of a cell's ◊cytoplasm.

endoplasmic reticulum (ER) a membranous system of tubes, channels, and flattened sacs that form compartments within eukaryotic cells (see

The first embryologist

The essence of science is systematic observation of the natural world. This may seem obvious to us, but it is a hard-won insight. Some appreciation of the value of observation stirred in Greece about 2,000 years ago, when Aristotle began to use experiments and scientific methods in the study of biology.

The origins of animals and plants had been considered by Greek thinkers before Aristotle, beginning with Hippocrates in the 5th century BC. Did an organism form stage by stage or did it come into existence small but fully formed and then grow? The Hippocratic writings contain a clear suggestion for resolving the question: 'Take twenty eggs or more, and set them for brooding under two or more hens. Then on each day of incubation from the second to the last, that of hatching, remove one egg and open it for examination.'

There is no sign that the writer bothered to carry out the experiment; that was left to Aristotle, and he wrote a detailed description of what he saw. It seems that he extended the work to other birds, and that he understood something of the process which unfolded as the chick grew. He wrote:
'Generation from the egg proceeds in an identical manner with all birds, but the full periods from conception to birth differ. With the common hen after three days and three nights there is the first indication of the embryo; with larger birds the interval being longer, with smaller birds shorter. Meanwhile the yolk comes into being, rising towards the sharp end, where the primal element of the egg is situated, and where the egg gets hatched; and the heart appears, like a speck of blood, in the white of the egg. This point beats and moves as if endowed with life. . .'

On this Aristotle was wrong; the heart is not the first organ to develop. However, he correctly emphasizes that fetal development is closely linked to the development of blood vessels and membranes:
'. . . two vein-ducts with blood in them trend in a convoluted course . . . and a membrane carrying bloody fibres now envelops the yolk, leading off from the vein ducts. A little afterwards the body is differentiated, at first very small and white. The head is clearly distinguished, and in it the eyes, swollen out to a great extent'.

He understands correctly the role of the egg white and yolk:
'The life-element of the chick is in the white of the egg, and the nutrient comes through the navel-string out of the yolk'. And the functions of the membranes surrounding the chick: 'The disposition of the several constituent parts is as follows. First and outermost comes the membrane of the egg, not that of the shell, but underneath it. Inside this membrane is a white liquid; then comes the chick, and a membrane round about it, separating it off so as to keep the chick free from liquid; next after the chick comes the yolk, into which one of the two veins was described as leading, the other leading into the enveloping white substance'.

Day by day, Aristotle describes the development of the various organs:
'When the egg is now ten days old the chick and all its parts are distinctly visible. The head is still larger than the rest of its body, and the eyes are larger than the head, but still devoid of vision. . . . At this time also the larger internal organs are visible, as also the stomach and the arrangement of the viscera; and the veins seen to proceed from the heart are now close to the navel.

'About the twentieth day, if you open the egg and touch the chick, it moves inside and chirps; and it is already coming to be covered with down, when, after the twentieth day is past, the chick begins to break the shell. The head is situated over the right leg close to the flank, and the wing is placed over the head; and about this time is plain to be seen the membrane resembling an after-birth that comes next after the outermost membrane of the shell, into which membrane the one of the navel strings was described as leading . . .'

He details the way the various membranes change, shrivel and detach during birth, ending with the observation:
'By and by the yolk, diminishing gradually in size, at length becomes entirely used up and comprehended within the chick (so that, ten days after hatching, if you cut open the chick, a small remnant of the yolk is still left in connection with the gut) . . .'

Obviously, Aristotle believed in following up his experiments, seeking the unexpected. His description sounds like a modern scientist at work. For around 2,000 years, researchers could only refine the picture and correct a few small errors. Modern embryology is in the midst of an exciting revolution: we are beginning to understand the processes seen by Aristotle at molecular level—the level of genes. But the modern approach can still involve experiments on the embryo, and may mean opening chicken eggs and observing their development as Aristotle did.

endangered species

species	observation
plants	one-quarter of the world's plants are threatened with extinction by the year 2020
amphibians	worldwide decline in numbers; half of New Zealand's frog species are now extinct
birds	three-quarters of all bird species are declining or threatened with extinction
carnivores	almost all species of cats and bears are declining in numbers
fish	one-third of North American freshwater fish are threatened, or endangered; half the fish species in Lake Victoria, Africa's largest lake, are close to extinction due to predation by the Nile perch
invertebrates	about 100 species are lost each day due to deforestation; half the freshwater snails in the southeastern USA are now extinct or threatened; one-quarter of W German invertebrates are threatened
mammals	half of Australia's mammals are threatened; 40% of mammals in France, the Netherlands, Germany, and Portugal are threatened
primates	two-thirds of primate species are threatened
reptiles	over 40% of reptile species are threatened

◊eukaryote). It stores and transports proteins within cells and also carries various enzymes needed for the synthesis of ◊fats. The ◊ribosomes, or the organelles that carry out protein synthesis, are attached to parts of the ER.

Under the electron microscope, ER looks like a series of channels and vesicles, but it is in fact a large, sealed, baglike structure crumpled and folded into a convoluted mass. The interior of the 'bag', the ER lumen, stores various proteins needed elsewhere in the cell, then organizes them into

endocrine gland *The main human endocrine glands. These glands emit hormones—chemical messengers—which travel through the body to stimulate certain cells. The pituitary is the master gland which controls many of the other glands.*

transport vesicles formed by a small piece of ER membrane budding from the main membrane.

endorphin natural substance (a polypeptide) that modifies the action of nerve cells. Endorphins are produced by the pituitary gland and hypothalamus of vertebrates. They lower the perception of pain by reducing the transmission of signals between nerve cells.

Endorphins not only regulate pain and hunger, but are also involved in the release of sex hormones from the pituitary gland. Opiates act in a similar way to endorphins, but are not rapidly degraded by the body, as natural endorphins are, and thus have a long-lasting effect on pain perception and mood. Endorphin release is stimulated by exercise.

endoscopy examination of internal organs or tissues by an instrument allowing direct vision. An endoscope is equipped with an eyepiece, lenses, and its own light source to illuminate the field of vision. The endoscope that examines the alimentary canal is a flexible fibreoptic instrument swallowed by the patient.

There are various types of endoscope in use—some rigid, some flexible—with individual names prefixed by their site of application (for example, bronchoscope and laryngoscope). The value of endoscopy is in permitting diagnosis without the need for exploratory surgery. Biopsies (tissue samples) and photographs may be taken by way of the endoscope as an aid to diagnosis, or to monitor the effects of treatment. In some cases, treatment can be given during the course of an examination, using fine instruments introduced through the endoscope.

endoskeleton the internal supporting structure of vertebrates, made up of cartilage or bone. It provides support, and acts as a system of levers to which muscles are attached to provide movement. Certain parts of the skeleton (the skull and ribs) give protection to vital body organs.

Sponges are supported by a network of rigid, semirigid, spiky structures called spicules; a bath sponge is the proteinaceous skeleton of a sponge.

endosperm nutritive tissue in the seeds of most flowering plants. It surrounds the embryo and is produced by an unusual process that parallels the ◊fertilization of the ovum by a male gamete. A second male gamete from the pollen grain fuses with two female nuclei within the ◊embryo sac. Thus endosperm cells are triploid (having three sets of chromosomes); they contain food reserves such as starch, fat, and protein that are utilized by the developing seedling.

In 'non-endospermic' seeds, absorption of these food molecules by the embryo is completed early, so that the endosperm has disappeared by the time of germination.

endotherm 'warm-blooded', or homeothermic, animal. Endotherms have internal mechanisms for regulating their body temperatures to levels different from the environmental temperature. See ◊homeothermy.

endothermic reaction chemical reaction that requires an input of energy in the form of heat for it to proceed; the energy is absorbed from the surroundings by the reactants.

The dissolving of sodium chloride in water and the process of photosynthesis are both endothermic changes. See ◊energy of reaction.

end user the user of a computer program; in particular, someone who uses a program to perform a task (such as accounting or playing a computer game), rather than someone who writes programs (a programmer).

Energiya most powerful Soviet space rocket, first launched 15 May 1987. Used to launch the Soviet space shuttle, the Energiya ◊booster is capable, with the use of strap-on boosters, of launching payloads of up to 190 tonnes into Earth orbit.

energy capacity for doing ◊work. Potential energy (PE) is energy deriving from position; thus a stretched spring has elastic PE, and an object raised to a height above the Earth's surface, or the water in an elevated reservoir, has gravitational PE. A lump of coal and a tank of petrol, together with the oxygen needed for their combustion, have chemical energy. Other sorts of energy include electrical and nuclear energy, and light and sound. Moving bodies possess kinetic energy (KE). Energy can be converted from one form to another, but the total quantity stays the same (in accordance with the ◊conservation of energy principle). For example, as an apple falls, it loses gravitational PE but gains KE.

Although energy is never lost, after a number of conversions it tends to finish up as the kinetic energy of random motion of molecules (of the air, for example) at relatively low temperatures. This is 'degraded' energy in that it is difficult to convert it back to other forms.

So-called energy resources are stores of convertible energy. Nonrenewable resources include the fossil fuels (coal, oil, and gas) and nuclear-fission 'fuels'—for example, uranium-235. Renewable resources, such as wind, tidal, and geothermal power, have so far been less exploited. Hydroelectric projects are well established, and wind turbines and tidal systems are being developed.

Einstein's special theory of ◊relativity 1905 correlates any gain, E, in energy with a gain, m, in mass, by the equation $E = mc^2$, in which c is the speed of light. The equation applies universally, not just to nuclear reactions, although it is only for these that the percentage change in mass is large enough to detect.

the arrangement of the electrons in an atom of chlorine with three energy levels (shells)

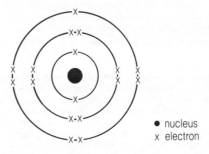

● nucleus
x electron

energy level

energy, alternative energy from sources that are renewable and ecologically safe, as opposed to sources that are nonrenewable with toxic by-products, such as coal, oil, or gas (fossil fuels), and uranium (for nuclear power). The most important alternative energy source is flowing water, harnessed as ◊hydroelectric power. Other sources include the oceans' tides and waves (see ◊tidal power station and ◊wave power), wind (harnessed by windmills and wind turbines), the Sun (◊solar energy), and the heat trapped in the Earth's crust (◊geothermal energy).

The Centre for Alternative Technology, near Machynlleth in mid-Wales, was established 1975 to research and demonstrate methods of harnessing wind, water, and solar energy.

energy conservation methods of reducing energy use through insulation, increasing energy efficiency, and changes in patterns of use. Profligate energy use by industrialized countries contributes greatly to air pollution and the ◊greenhouse effect when it draws on nonrenewable energy sources.

It has been calculated that increasing energy efficiency alone could reduce carbon dioxide emissions in several high-income countries by 1–2% a year. The average annual decrease in energy consumption in relation to gross national product 1973–87 was 1.2% in France, 2% in the UK, 2.1% in the USA, and 2.8% in Japan.

By applying existing conservation methods, UK electricity use could be reduced by 4 gigawatts by the year 2000—the equivalent of four Sizewell nuclear power stations—according to a study by the Open University. This would also be cheaper than building new generating plants.

energy level or **shell** or **orbital** the permitted energy that an electron can have in any particular system. Energy levels can be calculated using ◊quantum theory. The permitted energy levels depend mainly on the distance of the electron from the nucleus. See ◊orbital, atomic.

energy of reaction energy released or absorbed during a chemical reaction, also called **enthalpy of reaction** or **heat of reaction**. In a chemical reaction, the energy stored in the reacting molecules is rarely the same as that stored in the product molecules. Depending on which is the greater, energy is either released (an exothermic reaction) or absorbed (an endothermic reaction) from the surroundings (see ◊conservation of energy). The amount of energy released or absorbed by the quantities of substances represented by the chemical equation is the energy of reaction.

engine device for converting stored energy into useful work or movement. Most engines use a fuel as their energy store. The fuel is burnt to produce heat energy—hence the name 'heat engine'—which is then converted into movement. Heat engines can be classified according to the fuel they use (◊petrol engine or ◊diesel engine), or according to whether the fuel is burnt inside (◊internal-combustion engine) or outside (◊steam engine) the engine, or according to whether they produce a reciprocating or rotary motion (◊turbine or ◊Wankel engine).

engineering the application of science to the design, construction, and maintenance of works, machinery, roads, railways, bridges, harbour installations, engines, ships, aircraft and airports, spacecraft and space stations, and the generation, transmission, and use of electrical power. The main divisions of engineering are aerospace, chemical, civil, electrical, electronic, gas, marine, materials, mechanical, mining, production, radio, and structural.

To practise engineering professionally, university or college training in addition to practical experience is required, but technician engineers usually receive their training through apprenticeships or similar training schemes.

engineering drawing technical drawing that forms the plans for the design and construction of engineering components and structures. Engineering drawings show different projections, or views of objects, with the relevant dimensions, and show how all the separate parts fit together.

enthalpy in chemistry, alternative term for ◊energy of reaction, the heat energy associated with a chemical change.

entomology study of ◊insects.

entropy in ◊thermodynamics, a parameter representing the state of disorder of a system at the atomic, ionic, or molecular level; the greater the disorder, the higher the entropy. The fast-moving disordered molecules of water vapour have higher entropy than those of more ordered liquid water, which in turn have more entropy than the molecules in solid crystalline ice.

In a closed system undergoing change, entropy is a measure of the amount of energy unavailable for useful work. At ◊absolute zero (−273°C/−459.67°F/0K), when all molecular motion ceases and order is assumed to be complete, entropy is zero.

E number code number for additives that have been approved for use by the European Commission (EC). The E written before the number stands for European. E numbers do not have to be displayed on lists of ingredients, and the manufacturer may choose to list ◊additives by their name instead. E numbers cover all categories of additives apart from flavourings. Additives, other than flavourings, that are not approved by the EC, but

are still used in Britain, are represented by a code number without an E.

envelope in geometry, a curve that touches all the members of a family of lines or curves. For example, a family of three equal circles all touching each other and forming a triangular pattern (like a clover leaf) has two envelopes: a small circle that fits in the space in the middle, and a large circle that encompasses all three circles.

environment in ecology, the sum of conditions affecting a particular organism, including physical surroundings, climate, and influences of other living organisms. See also ◊biosphere and ◊habitat.

In common usage, 'the environment' often means the total global environment, without reference to any particular organism. In genetics, it is the external influences that affect an organism's development, and thus its ◊phenotype.

Population growth is the primary source of environmental damage.

On the **environment** Jacques Cousteau 1989

---

environmentalism theory emphasizing the primary influence of the environment on the development of groups or individuals. It stresses the importance of the physical, biological, psychological, or cultural environment as a factor influencing the structure or behaviour of animals, including humans.

In politics this has given rise in many countries to Green parties, which aim to 'preserve the planet and its people'.

Environmentally Sensitive Area (ESA) scheme introduced by the UK Ministry of Agriculture 1984, as a result of EC legislation, to protect some of the most beautiful areas of the British countryside from the loss and damage caused by agricultural change. The first areas to be designated ESAs are in the Pennine Dales, the North Peak District, the Norfolk Broads, the Breckland, the Suffolk River Valleys, the Test Valley, the South Downs, the Somerset Levels and Moors, West Penwith, Cornwall, the Shropshire Borders, the Cambrian Mountains, and the Lleyn Peninsula.

In these ESAs farmers are encouraged to use traditional methods to preserve the value of the land as a wildlife habitat. A farmer who joins the scheme agrees to manage the land in this way for at least five years. In return for this agreement, the Ministry of Agriculture pays the farmer a sum that reflects the financial losses incurred as a result of reconciling conservation with commercial farming. Further ESAs were under threat from budget cuts 1993.

Environmental Protection Agency (EPA) US agency set up 1970 to control water and air quality, industrial and commercial wastes, pesticides, noise, and radiation. In its own words, it aims to protect 'the country from being degraded, and its health threatened, by a multitude of human activities initiated without regard to long-ranging effects upon the life-supporting properties, the economic uses, and the recreational value of air, land, and water'.

E numbers: a selection of food additives

number	name	typical use	number	name	typical use
Colours					
E102	tartrazine	soft drinks	E215	sodium ethyl para-hydroxy-benzoate	
E104	quinoline yellow				
E110	sunset yellow	biscuits			
E120	cochineal	alcoholic drinks	E216	propyl para-hydroxy-benzoate	
E122	carmoisine	jams and preserves			
E123	amaranth		E218	methyl para-hydroxy-benzoate	
E124	ponceau 4R	dessert mixes			
E127	erythrosine	glacé cherries	E220	sulphur dioxide	
E131	patent blue V		E221	sodium sulphate	dried fruit, dehydrated vegetables, fruit juices and syrups, sausages, fruit-based dairy desserts, cider, beer, and wine; also used to prevent browning of peeled potatoes and to condition biscuit doughs
E132	indigo carmine		E222	sodium bisulphite	
E142	green S	pastilles	E223	sodium metabisulphite	
E150	caramel	beers, soft drinks, sauces, gravy browning	E224	potassium metabisulphite	
E151	black PN				
E160 (b)	annatto; bixin; norbixin	crisps			
E180	pigment rubine (lithol rubine BK)				
Antioxidants					
E310	propyl gallate	vegetable oils, chewing gum			
E311	octyl gallate		E226	calcium sulphite	
E312	dodecyl gallate		E227	calcium bisulphite	
E320	butylated hydroxynisole (BHA)	beef stock cubes, cheese spread	E249	potassium nitrite	
			E250	sodium nitrite	bacon, ham, cured meats, corned beef and some cheeses
E321	butylated hydroxytoluene (BHT)	chewing gum	E251	sodium nitrate	
Emulsifiers and stabilizers			E252	potassium nitrate	
E407	carageenan	quick-setting jelly mixes; milk shakes	**Others**		
			E450 (a)	disodium dihydrogen diphosphate trisodium diphosphate tetrasodium diphosphate tetrapotassium diphosphate	butters, sequestrants, emulsifying salts, stabilizers, texturizers
E413	tragacanth	salad dressings; processed cheese			
Preservatives					
E210	benzoic acid				
E211	sodium benzoate	beer, jam, salad cream, soft drinks, fruit pulp	E450 (b)	pentasodium triphosphate pentapotassium triphosphate	raising agents, used in whipping cream, fish and meat products, bread, processed cheese, canned vegetables
E212	potassium benzoate	fruit-based pie fillings, marinated herring and mackerel			
E213	calcium benzoate				
E214	ethyl para-hydroxy-benzoate				

environment–heredity controversy see ◊*nature–nurture controversy*.

enzyme biological ◊catalyst produced in cells, and capable of speeding up the chemical reactions necessary for life by converting one molecule (substrate) into another. Enzymes are not themselves destroyed by this process. They are large, complex ◊proteins, and are highly specific, each chemical reaction requiring its own particular enzyme. The enzyme fits into a 'slot' (active site) in the substrate molecule, forming an enzyme–substrate complex that lasts until the substrate is altered or split, after which the enzyme can fall away. The substrate may therefore be compared to a lock, and the enzyme to the key required to open it.

The activity and efficiency of enzymes are

RECENT PROGRESS ON ENVIRONMENTAL ISSUES

Losing the battle for a greener world?

In June 1992, Rio de Janeiro in Brazil hosted probably the largest-ever meeting of heads of state, to discuss the world's environment crisis at the United Nations Conference on Environment and Development (UNCED or Earth Summit). For many people it was the 'coming of age' of environmental issues: at last, world leaders were taking ecology and conservation seriously. But reactions to the conference have been mixed.

There were two concrete results. First, many nations committed themselves to protect biological diversity ('biodiversity') in terms of ecosystems, species and genetic variation within species. This should provide an important additional stimulus to conservation legislation in many countries. Secondly, a convention committed countries to combating the impact of global warming and resulting climate changes.

There were also notable failures: the lack of agreement on a global forest convention, and the richer nations' insistence on only modest additional funding to help developing countries enact some of the UNCED proposals. Many Southern politicians claimed the conference had given them new obligations without the funding to carry them out. Maurice Strong, secretary of UNCED, has suggested a global fund of $625 billion a year is needed to solve the world's environmental problems.

International trade

Despite the Earth Summit, some people believe international trade conditions may eventually be more important. The GATT (General Agreement on Tariffs and Trade) negotiations have concentrated on the risks of a trade war, but current discussions may drastically reduce individual countries' power to set individual standards on environmental safety if they interfere with 'free trade'.

For example, international determination of pesticide safety (by the Codex Alimentaris committee of the Food and Agricultural Organization) may mean the forcible reintroduction of previously banned pesticides into countries such as the USA, UK and Germany, with stricter standards than the FAO committee. Other trade agreements carry similar hidden environmental problems. In late 1992 it was suggested the European Commission might outlaw planting native wild flowers along British verges because it blocked the free trade of continental seed sources. Clearly, there is more controversy to come.

Pollution

Air pollution monitoring shows continuing deterioration in the global atmosphere. The 1991 UN survey of forest damage in Europe found 22% of trees suffering over 25% defoliation, largely because of the cocktail of air pollutants. The ozone layer in the stratosphere is still being degraded by pollutants such as chlorofluorocarbons (CFCs) and halons. World Meteorological Organization readings for Jan 1992 found average ozone levels over northern Europe 20% below normal, and those in Canada 16% below. The UN estimated that a sustained 10% ozone layer loss would increase skin cancer by 26%. In the UK, the Department of the Environment has published information on reducing risks from ozone pollution following evidence of an *increase* near ground level, where ozone—created as a side-effect of other air pollutants reacting in sunlight—can damage the health of plants and animals, including humans.

In Nov 1992, timetables for phasing out ozone-depleting chemicals were agreed by the 86 nations meeting in Copenhagen. They agreed to phase out CFCs by 1996, halons by 1994, and tetrachloromethane (carbon tetrachloride) and methyl chloroform by 1996.

Disappearing forests

Conservationists' fears about tropical deforestation are broadening to include temperate and boreal (Arctic) forests. In Oct 1992, the World Wide Fund for Nature (WWF) estimated that up to 90% of primary temperate forests, outside Russia, had been destroyed. In temperate regions, the area under trees is currently stable or even increasing, but with rapid loss of forest quality, as plantations and intensively managed forest areas replace natural or old-growth forests. They support less wildlife than natural forests and often create problems of soil erosion and loss of water quality.

The International Tropical Timber Agreement is up for renegotiation in 1993. The ITTA, and the associated International Tropical Timber Organization, have been criticized for insufficient action on tropical forest loss. Also, there are demands that the Agreement be broadened to include *all* timber, including that from temperate regions. This would accord with the wishes of conservation bodies promoting plans for the certification of timber from well-managed forests, which would have a special quality label.

Nigel Dudley

lock-and-key model of enzyme activity

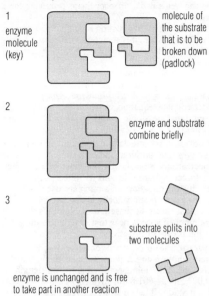

1
enzyme
molecule
(key)

molecule of
the substrate
that is to be
broken down
(padlock)

2

enzyme and substrate
combine briefly

3

substrate splits into
two molecules

enzyme is unchanged and is free
to take part in another reaction

enzyme

influenced by various factors, including temperature and pH conditions. Temperatures above 60°C/140°F damage (denature) the intricate structure of enzymes, causing reactions to cease. Each enzyme operates best within a specific pH range, and is denatured by excessive acidity or alkalinity.

Digestive enzymes include amylases (which digest starch), lipases (which digest fats), and proteases (which digest protein). Other enzymes play a part in the conversion of food energy into ◊ATP; the manufacture of all the molecular components of the body; the replication of ◊DNA when a cell divides; the production of hormones; and the control of movement of substances into and out of cells.

Enzymes have many medical and industrial uses, from washing powders to drug production, and as research tools in molecular biology. They can be extracted from bacteria and moulds, and ◊genetic engineering now makes it possible to tailor an enzyme for a specific purpose.

Eocene second epoch of the Tertiary period of geological time, 56.5–35.5 million years ago. Originally considered the earliest division of the Tertiary, the name means 'early recent', referring to the early forms of mammals evolving at the time, following the extinction of the dinosaurs.

eolith naturally shaped or fractured stone found in Lower Pleistocene deposits and once believed by some scholars to be the oldest known artefact type, dating to the pre-Palaeolithic era. They are now recognized as not having been made by humans.

eotvos unit unit (symbol E) for measuring small changes in the intensity of the Earth's ◊gravity with horizontal distance.

ephemeral plant plant with a very short life cycle, sometimes as little as six or eight weeks. It may complete several generations in one growing season.

A number of common weeds are ephemerals, for example groundsel *Senecio vulgaris*, as are many desert plants. The latter take advantage of short periods of rain to germinate and reproduce, passing the dry season as dormant seeds.

epicentre the point on the Earth's surface immediately above the seismic focus of an ◊earthquake. Most damage usually takes place at an earthquake's epicentre. The term sometimes refers to a point directly above or below a nuclear explosion ('at ground zero').

epicyclic gear or *sun-and-planet gear* gear system that consists of one or more gear wheels moving around another. Epicyclic gears are found in bicycle hub gears and in automatic gearboxes.

epicycloid in geometry, a curve resembling a series of arches traced out by a point on the circumference of a circle that rolls around another circle of a different diameter. If the two circles have the same diameter, the curve is a ◊cardioid.

epidermis outermost layer of ◊cells on an organism's body. In plants and many invertebrates such as insects, it consists of a single layer of cells. In vertebrates, it consists of several layers of cells.

The epidermis of plants and invertebrates often has an outer noncellular ◊cuticle that protects the organism from desiccation.

In vertebrates, such as reptiles, birds, and mammals, the outermost layer of cells is dead, forming a tough, waterproof layer, known as ◊skin.

epididymis in male vertebrates, a long coiled tube in the ◊testis, in which sperm produced in the seminiferous tube are stored. In men, it is a duct about 6 m/20 ft long, convoluted into a small space.

epigeal seed germination in which the ◊cotyledons (seed leaves) are borne above the soil.

epiglottis small flap found in the throats of

epicycloid *A seven-cusped epicycloid, formed by a point on the circumference of a circle (of diameter d) that rolls around another circle (of diameter 7d/3).*

mammals. It moves during swallowing to prevent food from passing into the windpipe and causing choking.

The action of the epiglottis is a highly complex reflex process involving two phases. During the first stage a mouthful of chewed food is lifted by the tongue towards the top and back of the mouth. This is accompanied by the cessation of breathing and by the blocking of the nasal areas from the mouth. The second phase involves the epiglottis moving over the larynx while the food passes down into the oesophagus.

epiphyte any plant that grows on another plant or object above the surface of the ground, and has no roots in the soil. An epiphyte does not parasitize the plant it grows on but merely uses it for support. Its nutrients are obtained from rainwater, organic debris such as leaf litter, or from the air.

The greatest diversity of epiphytes is found in tropical areas and includes many orchids.

EPNS abbreviation for *electroplated nickel silver*; see ◊electroplating.

epoch subdivision of a geological period in the geological time scale. Epochs are sometimes given their own names (such as the Palaeocene, Eocene, Oligocene, Miocene, and Pliocene epochs comprising the Tertiary period), or they are referred to as the late, early, or middle portions of a given period (as the Late Cretaceous or the Middle Triassic epoch).

epoxy resin synthetic ◊resin used as an ◊adhesive and as an ingredient in paints. Household epoxy resin adhesives come in component form as two separate tubes of chemical, one tube containing resin, the other a curing agent (hardener). The two chemicals are mixed just before application, and the mix soon sets hard.

EPROM (acronym for *erasable programmable read-only memory*) computer memory device in the form of an ◊integrated circuit (chip) that can record data and retain it indefinitely. The data can be erased by exposure to ultraviolet light, and new data recorded. Other kinds of computer memory chips are ◊ROM (read-only memory), ◊PROM (programmable read-only memory), and ◊RAM (random-access memory).

Epsilon Aurigae ◊eclipsing binary star in the constellation Auriga. One of the pair is an 'ordinary' star, but the other seems to be an enormous distended object whose exact nature remains unknown. The period (time between eclipses) is 27 years, the longest of its kind. The last eclipse was 1982–84.

Epsom salts $MgSO_4.7H_2O$ hydrated magnesium sulphate, used as a relaxant and laxative and added to baths to soothe the skin. The name is derived from a bitter saline spring at Epsom, Surrey, England, which contains the salt in solution.

equation in chemistry, representation of a chemical reaction by symbols and numbers; see ◊chemical equation.

equation in mathematics, expression that represents the equality of two expressions involving constants and/or variables, and thus usually includes an equals sign (=). For example, the equation $A = \pi r^2$ equates the area A of a circle of radius r to

the product πr^2. The algebraic equation $y = mx + c$ is the general one in coordinate geometry for a straight line.

If a mathematical equation is true for all variables in a given domain, it is sometimes called an identity and denoted by ≡. Thus $(x + y)^2 \equiv x^2 + 2xy + y^2$ for all $x, y \in R$.

An *indeterminate equation* is an equation for which there is an infinite set of solutions—for example, $2x = y$. A *diophantine equation* is an indeterminate equation in which the solution and terms must be whole numbers (after Diophantus of Alexandria, c. AD 250).

equations of motion mathematical equations that give the position and velocity of a moving object at any time. Given the mass of an object, the forces acting on it, and its initial position and velocity, the equations of motion are used to calculate its position and velocity at any later time. The equations must be based on ◊Newton's laws of motion or, if speeds near that of light are involved, on the theory of ◊relativity.

equator the *terrestrial equator* is the ◊great circle whose plane is perpendicular to the Earth's axis (the line joining the poles). Its length is 40,092 km/24,901.8 mi, divided into 360 degrees of longitude. The equator encircles the broadest part of the Earth, and represents 0° latitude. It divides the Earth into two halves, called the northern and the southern hemispheres.

The *celestial equator* is the circle in which the plane of the Earth's equator intersects the ◊celestial sphere.

equilateral of a geometrical figure, having all sides of equal length.

For example, a square and a rhombus are both equilateral four-sided figures. An equilateral triangle, to which the term is most often applied, has all three sides equal and all three angles equal (at 60°).

equilibrium in physics, an unchanging condition in which the forces acting on a particle or system of particles (a body) cancel out, or in which energy is distributed among the particles of a system in the most probable way; or the state in which a body is at rest or moving at constant velocity. A body is in *thermal equilibrium* with its surroundings if no heat enters or leaves it, so that all its parts are at the same temperature as the surroundings. See also ◊chemical equilibrium.

equinox the points in spring and autumn at which the Sun's path, the ◊ecliptic, crosses the celestial equator, so that the day and night are of approximately equal length. The *vernal equinox* occurs about 21 March and the *autumnal equinox*, 23 Sept.

era any of the major divisions of geological time, each including several periods, but smaller than an eon. The currently recognized eras all fall within the Phanerozoic eon—or the vast span of time, starting about 570 million years ago, when fossils are found to become abundant. The eras in ascending order are the Palaeozoic, Mesozoic, and Cenozoic. We are living in the Recent epoch of the Quaternary period of the Cenozoic era.

Eratosthenes' sieve a method for finding ◊prime

numbers. It involves writing in sequence all numbers from 2. Then, starting with 2, cross out every second number (but not 2 itself), thus eliminating numbers that can be divided by 2. Next, starting with 3, cross out every third number (but not 3 itself), and continue the process for 5, 7, 11, 13, and so on. The numbers that remain are primes.

erbium soft, lustrous, greyish, metallic element of the ◊lanthanide series, symbol Er, atomic number 68, relative atomic mass 167.26. It occurs with the element yttrium or as a minute part of various minerals. It was discovered 1843 by Carl Mosander (1797–1858), and named after the town of Ytterby, Sweden, near which the lanthanides (rare-earth elements) were first found.

erg cgs unit of work, replaced in the SI system by the ◊joule. One erg of work is done by a force of one ◊dyne moving through one centimetre.

ergonomics study of the relationship between people and the furniture, tools, and machinery they use at work. The object is to improve work performance by removing sources of muscular stress and general fatigue: for example, by presenting data and control panels in easy-to-view form, making office furniture comfortable, and creating a generally pleasant environment.

Eridanus in astronomy, the sixth largest constellation, which meanders from the celestial equator deep into the southern hemisphere of the sky. Its brightest star is ◊Achernar. Eridanus is represented as a river.

Eros in astronomy, an asteroid, discovered 1898, that can pass 22 million km/14 million mi from the Earth, as observed in 1975. Eros was the first asteroid to be discovered that has an orbit coming within that of Mars. It is elongated, measures about 36 × 12 km/22 × 7 mi, rotates around its shortest axis every 5.3 hours, and orbits the Sun every 1.8 years.

erosion wearing away of the Earth's surface, caused by the breakdown and transportation of particles of rock or soil (by contrast, ◊weathering does not involve transportation). Agents of erosion include the sea, rivers, glaciers, and wind. Water, consisting of sea waves and currents, rivers, and rain; ice, in the form of glaciers; and wind, hurling sand fragments against exposed rocks and moving dunes along, are the most potent forces of erosion. People also contribute to erosion by bad farming practices and the cutting down of forests, which can lead to the formation of dust bowls.

Agents of erosion include the sea, rivers, glaciers, and wind. There are several processes of erosion including ◊hydraulic action, ◊corrasion, ◊attrition, and ◊solution.

erratic in geology, a displaced rock that has been transported by a glacier or some other natural force to a site of different geological composition. For example, in East Anglia, England, erratics have been found that have been transported from as far away as Scotland and Scandinavia.

error in computing, a fault or mistake, either in the software or on the part of the user, that causes a program to stop running (crash) or produce unexpected results. Program errors, or bugs, are largely eliminated in the course of the programmer's initial testing procedure, but some will remain in most programs. All computer operating systems are designed to produce an **error message** (on the display screen, or in an error file or printout) whenever an error is detected, reporting that an error has taken place and, wherever possible, diagnosing its cause.

Errors can be categorized into several types: *syntax errors*, caused by the incorrect use of the programming language, and include spelling and keying mistakes. These errors are detected when the ◊compiler or ◊interpreter fails to translate the program into machine code (instructions that a computer can understand directly); *logical errors*, faults in the program design—for example, in the order of instructions. They may cause a program to respond incorrectly to the user's requests or to crash completely; *execution errors*, or *run-time errors*, caused by combinations of data that the programmer did not anticipate. A typical execution error is caused by attempting to divide a number by zero. This is impossible, and so the program stops running at this point. Execution errors occur only when a program is running, and cannot be detected by a compiler or interpreter.

Computers are designed to deal with a set range of numbers to a given range of accuracy. Many errors are caused by these limitations: *overflow error*, which occurs when a number is too large for the computer to deal with; *underflow error*, which occurs when a number is too small; *rounding* and *truncation errors*, caused by the need to round off decimal numbers, or to cut them off (truncate them) after the maximum number of decimal places allowed by the computer's level of accuracy.

error detection in computing, the techniques that enable a program to detect incorrect data. A common method is to add a check digit to important codes, such as account numbers and product codes. The digit is chosen so that the code conforms to a rule that the program can verify. Another technique involves calculating the sum (called the ◊hash total) of each instance of a particular item of data, and storing it at the end of the data.

erythrocyte another name for ◊red blood cell.

ESA abbreviation for ◊*European Space Agency*.

escape velocity in physics, minimum velocity with which an object must be projected for it to escape from the gravitational pull of a planetary body. In the case of the Earth, the escape velocity is 11.2 kps/6.9 mps; the Moon 2.4 kps/1.5 mps; Mars 5 kps/3.1 mps; and Jupiter 59.6 kps/37 mps.

escarpment or *cuesta* large ridge created by the erosion of dipping sedimentary rocks. It has one steep side (scarp) and one gently sloping side (dip). Escarpments are common features of chalk landscapes, such as the Chiltern Hills and the North Downs in England. Certain features are associated with chalk escarpments, including dry valleys (formed on the dip slope), combes (steep-sided valleys on the scarp slope), and springs.

esker geological feature of formerly glaciated areas consisting of a long, steep-walled narrow ridge, often sinuous and sometimes branching. Eskers consist of stratified glacial drift and are thought to

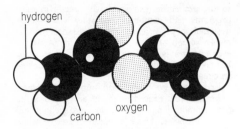

ester *Molecular model of the ester ethanoate,* $CH_3CH_2COOCH_3$.

form by the deposits of streams running through tunnels underneath melting stagnant ice. When the glacier finally disappeared, the old stream deposits were left standing as a high ridge. Eskers vary in height 3–30 m/10–100 ft and can run to about 160 km/100 mi in length.

ester organic compound formed by the reaction between an alcohol and an acid, with the elimination of water. Unlike ◊salts, esters are covalent compounds.

estradiol type of ◊oestrogen (female sex hormone).

estrogen alternative spelling of ◊oestrogen.

estuary river mouth widening into the sea, where fresh water mixes with salt water and tidal effects are felt. Estuaries are extremely rich in life forms and are breeding grounds for thousands of species. Water pollution threatens these ◊ecosystems.

Etesian wind north-northwesterly wind in the E Mediterranean and Aegean Sea, which blows June–Sept.

ethanal common name *acetaldehyde* CH_3CHO one of the chief members of the group of organic compounds known as ◊aldehydes. It is a colourless inflammable liquid boiling at 20.8°C/69.6°F. Ethanal is formed by the oxidation of ethanol or ethene and is used to make many other organic chemical compounds.

ethanal trimer common name *paraldehyde* $(CH_3CHO)_3$ colourless liquid formed from ethanal. It is soluble in water.

ethane CH_3CH_3 colourless, odourless gas, the second member of the ◊alkane series of hydrocarbons (paraffins).

ethane-1,2–diol technical name for ◊glycol.

ethanoate common name *acetate* $CH_3CO_2H^{2-}$ negative ion derived from ethanoic (acetic) acid; any salt containing this ion. In textiles, acetate rayon is a synthetic fabric made from modified cellulose (wood pulp) treated with ethanoic acid; in photography, acetate film is a non-flammable film made of cellulose ethanoate.

ethanoic acid common name *acetic acid* CH_3CO_2H one of the simplest fatty acids (a series of organic acids). In the pure state it is a colourless liquid with an unpleasant pungent odour; it solidifies to an icelike mass of crystals at 16.7°C/62.4°F, and hence is often called glacial ethanoic acid. Vinegar contains 5% or more ethanoic acid, produced by fermentation.

Cellulose (derived from wood or other sources)

may be treated with ethanoic acid to produce a cellulose ethanoate (acetate) solution, which can be used to make plastic items by injection moulding or extruded to form synthetic textile fibres.

ethanol common name *ethyl alcohol* C_2H_5OH alcohol found in beer, wine, cider, spirits, and other alcoholic drinks. When pure, it is a colourless liquid with a pleasant odour, miscible with water or ether; it burns in air with a pale blue flame. The vapour forms an explosive mixture with air and may be used in high-compression internal combustion engines. It is produced naturally by the fermentation of carbohydrates by yeast cells. Industrially, it can be made by absorption of ethene and subsequent reaction with water, or by the reduction of ethanal in the presence of a catalyst, and is widely used as a solvent.

Ethanol is used as a raw material in the manufacture of ether, chloral, and iodoform. It can also be added to petrol, where it improves the performance of the engine, or be used as a fuel in its own right (as in Brazil). Crops such as sugar cane may be grown to provide ethanol (by fermentation) for this purpose.

ethanolic solution solution produced when a solute is dissolved in ethanol; for example, ethanolic potassium hydroxide is a solution of potassium hydroxide in ethanol.

ethene common name *ethylene* C_2H_4 colourless, flammable gas, the first member of the ◊alkene series of hydrocarbons. It is the most widely used synthetic organic chemical and is used to produce the plastics polyethene (polyethylene), polychloroethene, and polyvinyl chloride (PVC). It is obtained from natural gas or coal gas, or by the dehydration of ethanol.

Ethene is produced during plant metabolism and is classified as a plant hormone. It is important in the ripening of fruit and in ◊abscission. Small amounts of ethene are often added to the air surrounding fruit to artificially promote ripening.

ether in chemistry, any of a series of organic chemical compounds having an oxygen atom linking the carbon atoms of two hydrocarbon radical groups (general formula R-O-R'); also the common name for ethoxyethane $C_2H_5OC_2H_5$ (also called diethyl ether). This is used as an anaesthetic and as an external cleansing agent before surgical operations. It is also used as a solvent, and in the extraction of oils, fats, waxes, resins, and alkaloids.

Ethoxyethane is a colourless, volatile, inflammable liquid, slightly soluble in water, and miscible with ethanol. It is prepared by treatment of ethanol with excess concentrated sulphuric acid at 140°C/284°F.

ether or *aether* in the history of science, a hypothetical medium permeating all of space. The concept originated with the Greeks, and has been revived on several occasions to explain the properties and propagation of light. It was supposed that light and other electromagnetic radiation—even in outer space—needed a medium, the ether, in which to travel. The idea was abandoned with the acceptance of ◊relativity.

ethnoarchaeology the study of human behaviour and the material culture of living societies, in order to see how materials enter the archaeological record, and that way to provide hypotheses explaining the production, use, and disposal patterns of ancient material culture.

ethnography study of living cultures, using anthropological techniques like participant observation (where the anthropologist lives in the society being studied) and a reliance on informants. Ethnography has provided many data of use to archaeologists as analogies.

ethnology study of contemporary peoples, concentrating on their geography and culture, as distinct from their social systems. Ethnologists make a comparative analysis of data from different cultures to understand how cultures work and why they change, with a view to deriving general principles about human society.

ethology comparative study of animal behaviour in its natural setting. Ethology is concerned with the causal mechanisms (both the stimuli that elicit behaviour and the physiological mechanisms controlling it), as well as the development of behaviour, its function, and its evolutionary history.

Ethology was pioneered during the 1930s by the Austrians Konrad Lorenz and Karl von Frisch who, with the Dutch zoologist Nikolaas Tinbergen, received the Nobel prize in 1973. Ethologists believe that the significance of an animal's behaviour can be understood only in its natural context, and emphasize the importance of field studies and an evolutionary perspective. A recent development within ethology is ◊sociobiology, the study of the evolutionary function of ◊social behaviour.

ethyl alcohol common name for ◊ethanol.

ethylene common name for ◊ethene.

ethylene glycol alternative name for ◊glycol.

ethyne common name *acetylene* CHCH colourless inflammable gas produced by mixing calcium carbide and water. It is the simplest member of the ◊alkyne series of hydrocarbons. It is used in the manufacture of the synthetic rubber neoprene, and in oxyacetylene welding and cutting.

Ethyne was discovered by Edmund Davy 1836 and was used in early gas lamps, where it was produced by the reaction between water and calcium carbide. Its combustion provides more heat, relatively, than almost any other fuel known (its calorific value is five times that of hydrogen). This means that the gas gives an intensely hot flame; hence its use in oxyacetylene torches.

etiolation in botany, a form of growth seen in plants receiving insufficient light. It is characterized by long, weak stems, small leaves, and a pale yellowish colour (◊chlorosis) owing to a lack of chlorophyll. The rapid increase in height enables a plant that is surrounded by others to quickly reach a source of light, after which a return to normal growth usually occurs.

eugenics (Greek 'well-born') study of ways in which the physical and mental quality of a people can be controlled and improved by selective breeding, and the belief that this should be done. The idea was abused by the Nazi Party in Germany during the 1930s to justify the attempted extermination of entire groups of people. Eugenics can try

to control the spread of inherited genetic abnormalities by counselling prospective parents.

The term was coined by English scientist Francis Galton (1822–1911) in 1883, and the concept was originally developed in the late 19th century with a view to improving human intelligence and behaviour.

In 1986 Singapore became the first democratic country to adopt an openly eugenic policy by guaranteeing pay increases to female university graduates when they give birth to a child, while offering grants towards house purchases for nongraduate married women on condition that they are sterilized after the first or second child.

eukaryote in biology, one of the two major groupings into which all organisms are divided. Included are all organisms, except bacteria and cyanobacteria (◊blue-green algae), which belong to the ◊prokaryote grouping.

The cells of eukaryotes possess a clearly defined nucleus, bounded by a membrane, within which DNA is formed into distinct chromosomes. Eukaryotic cells also contain mitochondria, chloroplasts, and other structures (organelles) that, together with a defined nucleus, are lacking in the cells of prokaryotes.

Euratom acronym for ◊*Euro*pean *Atom*ic Energy Commission, forming part of the European Community organization.

eureka in chemistry, alternative name for the copper–nickel alloy ◊constantan, which is used in electrical equipment.

Europa in astronomy, the fourth largest moon of the planet Jupiter, diameter 3,140 km/1,950 mi, orbiting 671,000 km/417,000 mi from the planet every 3.55 days. It is covered by ice and crisscrossed by thousands of thin cracks, each some 50,000 km/30,000 mi long.

European Atomic Energy Commission (Euratom) organization established by the second Treaty of Rome 1957, which seeks the cooperation of member states of the European Community in nuclear research and the rapid and large-scale development of nonmilitary nuclear energy.

European Southern Observatory observatory operated jointly by Belgium, Denmark, France, Germany, Italy, the Netherlands, Sweden, and Switzerland with head quarters near Munich. Its telescopes, located at La Silla, Chile, include a 3.6–m/142–in reflector opened 1976 and the 3.58–m/141–in New Technology Telescope opened 1990. By 1988 work began on the Very Large Telescope, consisting of four 8–m/315–in reflectors mounted independently but capable of working in combination.

European Space Agency (ESA) an organization of European countries (Austria, Belgium, Denmark, France, Germany, Ireland, Italy, the Netherlands, Norway, Spain, Sweden, Switzerland, and the UK) that engages in space research and technology. It was founded 1975, with headquarters in Paris.

ESA developed various scientific and communications satellites, the ◊Giotto space probe, and the ◊Ariane rockets. ESA built ◊Spacelab, plans to build its own space station, *Columbus*, for attachment to a US space station, and is working on its own shuttle project, Hermes.

europium soft, greyish, metallic element of the ◊lanthanide series, symbol Eu, atomic number 63, relative atomic mass 151.96. It is used in lasers and as the red phosphor in colour televisions; its compounds are used to make control rods for nuclear reactors.

It was named in 1901 by French chemist Eugène Demarçay (1852–1904) after the continent of Europe, where it was first found.

eusociality form of social life found in insects such as honey bees and termites, in which the colony is made up of special castes (for example, workers, drones, and reproductives) whose membership is biologically determined. The worker castes do not usually reproduce. Only one mammal, the naked mole rat, has a social organization of this type. See also ◊social behaviour.

Eustachian tube small air-filled canal connecting the middle ◊ear with the back of the throat. It is found in all land vertebrates and equalizes the pressure on both sides of the ear drum.

eustatic change worldwide rise or fall in sea level caused by a change in the amount of water in the oceans (by contrast, ◊isostasy involves a rising or sinking of the land). During the last ice age sea level fell because water became 'locked-up' in the form of ice and snow, and less water reached the oceans.

Eutelsat acronym for *Eu*ropean *Tel*ecommunications *Sat*ellite Organization.

eutrophication excessive enrichment of rivers, lakes, and shallow sea areas, primarily by nitrate fertilizers washed from the soil by rain, by phosphates from fertilizers and detergents in municipal sewage, and by sewage itself. These encourage the growth of algae and bacteria which use up the oxygen in the water, thereby making it uninhabitable for fishes and other animal life.

evaporation process in which a liquid turns to a vapour without its temperature reaching boiling point. A liquid left to stand in a saucer eventually evaporates because, at any time, a proportion of its molecules will be fast enough (have enough kinetic energy) to escape through the attractive intermolecular forces at the liquid surface into the atmosphere. The temperature of the liquid tends to fall because the evaporating molecules remove energy from the liquid. The rate of evaporation rises with increased temperature because as the mean kinetic energy of the liquid's molecules rises, so will the number possessing enough energy to escape.

A fall in the temperature of the liquid, known as the *cooling effect*, accompanies evaporation because as the faster proportion of the molecules escapes through the surface the mean energy of the remaining molecules falls. The effect may be noticed when wet clothes are worn, or as perspiration evaporates. ◊Refrigeration makes use of the cooling effect to extract heat from foodstuffs.

evaporite sedimentary deposit precipitated on evaporation of salt water. With a progressive evaporation of sea water, the most common salts are deposited in a definite sequence: calcite (calcium carbonate), gypsum (hydrous calcium sulphate),

halite (sodium chloride), and finally salts of potassium and magnesium.

Calcite precipitates when sea water is reduced to half its original volume, gypsum precipitates when the sea water body is reduced to one-fifth, and halite when the volume is reduced to one-tenth. Thus the natural occurrence of chemically precipitated calcium carbonate is common, of gypsum fairly common, and of halite less common. Because of the concentrations of different dissolved salts in sea water, halite accounts for about 95% of the chlorides precipitated if evaporation is complete. More unusual evaporite minerals include borates (for example borax, hydrous sodium borate) and sulphates (for example glauberite, a combined sulphate of sodium and calcium).

evergreen in botany, a plant such as pine, spruce, or holly, that bears its leaves all year round. Most conifers are evergreen. Plants that shed their leaves in autumn or during a dry season are described as ◊deciduous.

evolution slow process of change from one form to another, as in the evolution of the universe from its formation in the ◊Big Bang to its present state, or in the evolution of life on Earth. Some Christians and Muslims deny the theory of evolution as conflicting with the belief that God created all things (see ◊creationism).

Nature does not make jumps.

On **evolution** Carolus Linnaeus *Philosophia Botanica* 1751

The idea of continuous evolution can be traced as far back as the Roman philosopher Lucretius in the 1st century BC, but it did not gain wide acceptance until the 19th century following the work of Scottish geologist Charles Lyell, French naturalist Jean Baptiste Lamarck, English naturalist Charles Darwin, and English biologist Thomas Henry Huxley. Darwin assigned the major role in evolutionary change to ◊natural selection acting on randomly occurring variations (now known to be produced by spontaneous changes or ◊mutations in the genetic material of organisms). Natural selection occurs because those individuals better adapted to their particular environments reproduce more effectively, thus contributing their characteristics (in the form of genes) to future generations. The current theory of evolution, called ◊neo-Darwinism, combines Darwin's theory with Austrian biologist Gregor Mendel's theories on genetics. Although neither the general concept of evolution nor the importance of natural selection is doubted by biologists, there remains dispute over other possible processes involved in evolutionary change. Besides natural selection and ◊sexual selection, chance may play a large part in deciding which genes become characteristic of a population, a phenomenon called 'genetic drift'. It is now also clear that evolutionary change does not always occur at a constant rate, but that the process can have long periods of relative stability interspersed with periods of rapid change. This has led to new theories, such as ◊punctuated equilibrium model. See also ◊adaptive radiation.

evolutionary stable strategy (ESS) in ◊sociobiology, an assemblage of behavioural or physical characters (collectively termed a 'strategy') of a population that is resistant to replacement by any forms bearing new traits, because the new traits will not be capable of successful reproduction.

ESS analysis is based on ◊game theory and can be applied both to genetically determined physical characters (such as horn length), and to learned behavioural responses (for example, whether to fight or retreat from an opponent). An ESS may be conditional on the context, as in the rule 'fight if the opponent is smaller, but retreat if the opponent is larger'.

exclusion principle in physics, a principle of atomic structure originated by Austrian–US physicist Wolfgang Pauli. It states that no two electrons in a single atom may have the same set of ◊quantum numbers. Hence, it is impossible to pack together certain elementary particles, such as electrons, beyond a certain critical density, otherwise they would share the same location and quantum number. A white dwarf star is thus prevented from contracting further by the exclusion principle and never collapses.

excretion in biology, the removal of waste products from the cells of living organisms. In plants and simple animals, waste products are removed by diffusion, but in higher animals they are removed by specialized organs. In mammals, for example, carbon dioxide and water are removed via the lungs, and nitrogenous compounds and water via the liver, the kidneys, and the rest of the urinary system.

exobiology study of life forms that may possibly exist elsewhere in the universe, and of the effects of extraterrestrial environments on Earth organisms.

exocrine gland gland that discharges secretions, usually through a tube or a duct, onto a surface. Examples include sweat glands which release sweat onto the skin, and digestive glands which release digestive juices onto the walls of the intestine. Some animals also have ◊endocrine glands (ductless glands) that release hormones directly into the bloodstream.

exoskeleton the hardened external skeleton of insects, spiders, crabs, and other arthropods. It provides attachment for muscles and protection for the internal organs, as well as support. To permit growth it is periodically shed in a process called ◊ecdysis.

exosphere the uppermost layer of the ◊atmosphere. It is an ill-defined zone above the thermosphere, beginning at about 700 km/435 mi and fading off into the vacuum of space. The gases are extremely thin, with hydrogen as the main constituent.

exothermic reaction a chemical reaction during which heat is given out (see ◊energy of reaction).

expansion in physics, the increase in size of a constant mass of substance caused by, for example, increasing its temperature (◊thermal expansion) or its internal pressure. The *expansivity*, or coefficient of thermal expansion, of a material is its expansion (per unit volume, area, or length) per degree rise in temperature.

In mechanics, expansion refers to the increase in volume of steam in the cylinder of a steam engine after cutoff, or of gas in the cylinder of an internal-combustion engine after explosion. See also ◊Charles's law.

expansion board or *expansion card* printed circuit board that can be inserted into a computer in order to enhance its capabilities (for example, to increase its memory) or to add facilities (such as graphics).

experiment in science, a practical test designed with the intention that its results will be relevant to a particular theory or set of theories. Although some experiments may be used merely for gathering more information about a topic that is already well understood, others may be of crucial importance in confirming a new theory or in undermining long-held beliefs.

The manner in which experiments are performed, and the relation between the design of an experiment and its value, are therefore of central importance. In general an experiment is of most value when the factors that might affect the results (variables) are carefully controlled; for this reason most experiments take place in a well-managed environment such as a laboratory or clinic.

expert system computer program for giving advice (such as diagnosing an illness or interpreting the law) that incorporates knowledge derived from human expertise. It is a kind of ◊knowledge-based system containing rules that can be applied to find the solution to a problem. It is a form of ◊artificial intelligence.

Explorer series of US scientific satellites. *Explorer 1*, launched Jan 1958, was the first US satellite in orbit and discovered the Van Allen radiation belts around the Earth.

explosive any material capable of a sudden release of energy and the rapid formation of a large volume of gas, leading when compressed to the development of a high-pressure wave (blast).

Combustion and explosion differ essentially only in rate of reaction, and many explosives (called *low explosives*) are capable of undergoing relatively slow combustion under suitable conditions. *High explosives* produce uncontrollable blasts. The first low explosive was ◊gunpowder; the first high explosive was nitroglycerine. In 1867, Swedish chemist Alfred Nobel produced dynamite by mixing nitroglycerine with kieselguhr, a fine, chalk-like material. Other explosives now in use include trinitrotoluene (TNT); ANFO (a mixture of ammonium nitride and fuel oil), which is widely used in blasting; and pentaerythritol tetranitrate (PETN), a sensitive explosive with high power. Military explosives are often based on cyclonite (also called RDX), which is moderately sensitive but extremely powerful. Even more powerful explosives are made by mixing RDX with TNT and aluminium. *Plastic explosives*, such as Semtex, are based on RDX mixed with oils and waxes. The explosive force of *atomic and hydrogen bombs* arises from the conversion of matter to energy according to Einstein's mass–energy equation, $E = mc^2$.

exponent or *index* in mathematics, a number that

indicates the number of times a term is multiplied by itself; for example $x^2 = x \times x$, and $4^3 = 4 \times 4 \times 4$.

Exponents obey certain rules. Terms that contain them are multiplied together by adding the exponents; for example, $x^2 \times x^5 = x^7$. Division of such terms is done by subtracting the exponents; for example, $y^5 \div y^3 = y^2$. Any number with the exponent 0 is equal to 1; for example, $x^0 = 1$ and $99^0 = 1$.

exponential in mathematics, descriptive of a ◊function in which the variable quantity is an exponent (a number indicating the power to which another number or expression is raised).

Exponential functions and series involve the constant e = 2.71828.... Scottish mathematician John Napier devised natural ◊logarithms in 1614 with e as the base.

Exponential functions are basic mathematical functions, written as e^x or exp x. The expression e^x has five definitions, two of which are: (i) e^x is the solution of the differential equation $dx/dt = x$ ($x = 1$ if $t = 0$); (ii) e^x is the limiting sum of the infinite series $1 + x + (x^2/2!) + (x^3/3!) + \ldots + (x^n/n!)$.

Curves of the form $y = Ae^{-ax}$, $a < 0$ are known as decay functions; those of the form $y = Be^{bx}$, $b < 0$ are growth functions. *Exponential growth* is not constant. It applies, for example, to population growth, where the population doubles in a short time period. A graph of population number against time produces a curve that is characteristically rather flat at first but then shoots almost directly upwards.

export file in computing, a file stored by the computer in a standard format so that it can be accessed by other programs, possibly running on different makes of computer.

For example, a word-processing program running on an Apple ◊Macintosh computer may have a facility to save a file in a format that can be read by a word-processing program running on an IBM PC-compatible computer. When the file is being read by the second program or computer, it is often referred to as an *import file*.

exposure meter instrument used in photography for indicating the correct exposure—the length of time the camera shutter should be open under given light conditions. Meters use substances such as cadmium sulphide and selenium as light sensors. These materials change electrically when light strikes them, the change being proportional to the intensity of the incident light. Many cameras have a built-in exposure meter that sets the camera controls automatically as the light conditions change.

extensor a muscle that straightens a limb.

extinction in biology, the complete disappearance of a species. In the past, extinctions are believed to have occurred because species were unable to adapt quickly enough to a naturally changing environment. Today, most extinctions are due to human activity. Some species, such as the dodo of Mauritius, the moas of New Zealand, and the passenger pigeon of North America, were exterminated by hunting. Others became extinct when their habitat was destroyed. See also ◊endangered species.

Mass extinctions are episodes during which whole groups of species have become extinct, the

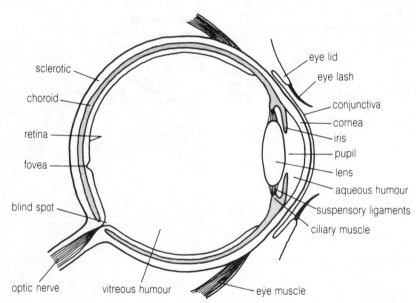

eye *The human eye. The retina of the eye contains about 137 million light sensitive cells in an area of about 650 sq mm/1 sq in. There are 130 million rod cells for black and white vision and 7 million cone cells for colour vision. The optic nerve contains about 1 million nerve fibres. The focusing muscles of the eye adjust about 100,000 times a day. To exercise the leg muscles to the same extent would need a 80 km/50 mi walk.*

best known being that of the dinosaurs, other large reptiles, and various marine invertebrates about 65 million years ago. Another mass extinction occurred about 10,000 years ago when many giant species of mammal died out. This is known as the 'Pleistocene overkill' because their disappearance was probably hastened by the hunting activities of prehistoric humans. The greatest mass extinction occurred about 250 million years ago marking the Permian-Triassic boundary (see ◊geological time) when up to 96% of all living species became extinct. It was proposed 1982 that mass extinctions occur periodically, at approximately 26-million-year intervals.

The current mass extinction is largely due to human destruction of habitats, as in the tropical forests and coral reefs; it is far more serious and damaging than mass extinctions of the past because of the speed at which it occurs. Man-made climatic changes and pollution also make it less likely that the biosphere can recover and evolve new species to suit a changed environment. The rate of extinction is difficult to estimate, since most losses occur in the rich environment of the tropical rainforest, where the total number of existent species is not known. Conservative estimates put the rate of loss due to deforestation alone at 4,000 to 6,000 species a year. Overall, the rate could be as high as one species an hour, with the loss of one species putting those dependent on it at risk. Australia has the worst record for extinction: 18 mammals have disappeared since Europeans settled there, and 40 more are threatened.

The last mouse-eared bat (*Myotis myotis*) in the UK died 1990. This is the first mammal to have become extinct in the UK for 250 years, since the last wolf was exterminated.

extracellular matrix strong material naturally occurring in animals and plants, made up of protein and long-chain sugars (polysaccharides) in which cells are embedded. It is often called a 'biological glue', and forms part of ◊connective tissues such as bone and skin.

The cell walls of plants and bacteria, and the ◊exoskeletons of insects and other arthropods, are also formed by types of extracellular matrix.

extrusion common method of shaping metals, plastics, and other materials. The materials, usually hot, are forced through the hole in a metal die and

take its cross-sectional shape. Rods, tubes, and sheets may be made in this way.

extrusive rock or *volcanic rock* ◊igneous rock formed on the surface of the Earth; for example, basalt. It is usually fine-grained (having cooled quickly), unlike the more coarse-grained ◊intrusive rocks (igneous rocks formed under the surface). The magma (molten rock) that cools to form extrusive rock may reach the surface through a crack, such as the ◊constructive margin at the Mid Atlantic Ridge, or through the vent of a ◊volcano.

Extrusive rock can be either *lava* (solidified from a flow) or a *pyroclastic deposit* (hot rocks or ash).

THE LARGEST EYE

The largest animal eyes are those of the giant squid. Each measures 40 cm/16 in across and contains more than 1,000 million light-sensitive cells. These cannot distinguish colours, however, so while a squid sees as well as we do, it is only in black and white.

---

eye the organ of vision. In the human eye, the light is focused by the combined action of the curved *cornea*, the internal fluids, and the *lens*. The insect eye is compound—made up of many separate facets—known as ommatidia, each of which collects light and directs it separately to a receptor to build up an image. Invertebrates have much simpler eyes, with no lens. Among molluscs, cephalopods have complex eyes similar to those of vertebrates. The mantis shrimp's eyes contain ten colour pigments with which to perceive colour; some flies and fishes have five, while the human eye has only three.

human eye This is a roughly spherical structure contained in a bony socket. Light enters it through the cornea, and passes through the circular opening (*pupil* in the iris (the coloured part of the eye). The ciliary muscles act on the lens (the rounded transparent structure behind the iris) to change its shape, so that images of objects at different distances can be focused on the ◊retina. (The ability to change the focus in this way is called ◊accommodation.) The retina is at the back of the eye, and is packed with light-sensitive cells (rods and cones), connected to the brain by the optic nerve.

°F symbol for degrees ◊*Fahrenheit*.

facies in geology, any assemblage of mineral, rock, or fossil features that reflect the environment in which rock was formed.

The set of characters that distinguish one facies from another in a given time stratigraphic unit is used to interpret local changes in simultaneously existing environments. Thus one facies in a body of rock might consist of porous limestone containing fossil reef-building organisms in their living positions. This facies might pass on the side into a reef-flank facies of steeply dipping deposits of rubble from the reef, which in turn might grade into an interreef basin composed of fine, clayey limestone.

Ancient floods and migrations of the seashore up or down can also be traced by changes in facies.

facsimile transmission full name for ◊*fax* or *telefax*.

factor a number that divides into another number exactly. For example, the factors of 64 are 1, 2, 4, 8, 16, 32, and 64. In algebra, certain kinds of polynomials (expressions consisting of several or many terms) can be factorized. For example, the factors of $x^2 + 3x + 2$ are $x + 1$ and $x + 2$, since $x^2 + 3x + 2 = (x + 1)(x + 2)$. This is called factorization. See also ◊prime number.

factorial of a positive number, the product of all the whole numbers (integers) inclusive between 1 and the number itself. A factorial is indicated by the symbol '!'. Thus $6! = 1 \times 2 \times 3 \times 4 \times 5 \times 6 = 720$. Factorial zero, 0!, is defined as 1.

faeces remains of food and other debris passed out of the digestive tract of animals. Faeces consist of quantities of fibrous material, bacteria and other microorganisms, rubbed-off lining of the digestive tract, bile fluids, undigested food, minerals, and water.

Fahrenheit scale temperature scale invented 1714 by Gabriel Fahrenheit which was commonly used in English-speaking countries up until the 1970s, after which the ◊Celsius scale was generally adopted, in line with the rest of the world. In the Fahrenheit scale, intervals are measured in degrees (°F); °F = (°C × ⅑) + 32.

Fahrenheit took as the zero point the lowest temperature he could achieve anywhere in the laboratory, and, as the other fixed point, body temperature, which he set at 96°F. On this scale, water freezes at 32°F and boils at 212°F.

Fallopian tube or *oviduct* in mammals, one of two tubes that carry eggs from the ovary to the uterus. An egg is fertilized by sperm in the Fallopian tubes, which are lined with cells whose ◊cilia move the egg towards the uterus.

fallout harmful radioactive material released into the atmosphere in the debris of a nuclear explosion (see ◊nuclear warfare) and descending to the surface of the Earth. Such material can enter the food chain, cause ◊radiation sickness, and last for hundreds of thousands of years (see ◊half-life).

false-colour imagery graphic technique that displays images in false (not true-to-life) colours so as to enhance certain features. It is widely used in displaying electronic images taken by spacecraft; for example, Earth-survey satellites such as *Landsat*. Any colours can be selected by a computer processing the received data.

falsificationism in philosophy of science, the belief that a scientific theory must be under constant scrutiny and that its merit lies only in how well it stands up to rigorous testing. It was first expounded by Austrian philosopher of science Karl Popper in his *Logic of Scientific Discovery* 1934.

Such thinking also implies that a theory can only be held to be scientific if it makes predictions that are clearly testable. Critics of this belief acknowledge the strict logic of this process, but doubt whether the whole of scientific method can be subsumed into so narrow a programme. Philosophers and historians such as the Americans Thomas Kuhn and Paul Feyerabend have attempted to use the history of science to show that scientific progress has resulted from a more complicated methodology than Popper suggests.

family in biological classification, a group of related genera (see ◊genus). Family names are not printed in italic (unlike genus and species names), and by convention they all have the ending –idae (animals) or –aceae (plants and fungi). For example, the genera of hummingbirds are grouped in the hummingbird family, Trochilidae. Related families are grouped together in an ◊order.

fanjet another name for ◊turbofan, the jet engine used by most airliners.

farad SI unit (symbol F) of electrical capacitance (how much electricity a ◊capacitor can store for a given voltage). One farad is a capacitance of one ◊coulomb per volt. For practical purposes the microfarad (one millionth of a farad) is more commonly used.
The farad is named after English scientist Michael Faraday.

faraday unit of electrical charge equal to the charge on one mole of electrons. Its value is 9.648 x 10⁴ coulombs.

Faraday's constant constant (symbol F) representing the electric charge carried on one mole of electrons. It is found by multiplying Avogadro's constant by the charge carried on a single electron, and is equal to 9.648×10^4 coulombs per mole. One *faraday* is this constant used as a unit. The constant is used to calculate the electric charge needed to discharge a particular quantity of ions during ◊electrolysis.

Faraday's laws three laws of electromagnetic induction, and two laws of electrolysis, all proposed originally by English scientist Michael Faraday:
induction (1) a changing magnetic field induces an electromagnetic force in a conductor; (2) the electromagnetic force is proportional to the rate of change of the field; (3) the direction of the induced electromagnetic force depends on the orientation of the field.
electrolysis (1) the amount of chemical change during electrolysis is proportional to the charge passing through the liquid; (2) the amount of chemical change produced in a substance by a given amount of electricity is proportional to the electrochemical equivalent of that substance.

fast breeder or *breeder reactor* alternative names for ◊fast reactor, a type of nuclear reactor.

fast reactor or *fast breeder reactor* ◊nuclear reactor that makes use of fast neutrons to bring about fission. Unlike other reactors used by the nuclear-power industry, it has little or no ◊moderator, to slow down neutrons. The reactor core is surrounded by a 'blanket' of uranium carbide. During operation, some of this uranium is converted into plutonium, which can be extracted and later used as fuel.
Fast breeder reactors can extract about 60 times the amount of energy from uranium that thermal reactors do. In the 1950s, when uranium stocks were thought to be dwindling, the fast breeder was considered to be the reactor of the future. Now, however, when new uranium reserves have been found and because of various technical difficulties in their construction, development of the fast breeder has slowed.

fat in the broadest sense, a mixture of ◊lipids—

chiefly triglycerides (lipids containing three ◊fatty acid molecules linked to a molecule of glycerol). More specifically, the term refers to a lipid mixture that is solid at room temperature (20°C); lipid mixtures that are liquid at room temperature are called *oils*. The higher the proportion of saturated fatty acids in a mixture, the harder the fat.
Boiling fats in strong alkali forms soaps (saponification). Fats are essential constituents of food for many animals, with a calorific value twice that of carbohydrates; however, eating too much fat, especially fat of animal origin, has been linked with heart disease in humans. In many animals and plants, excess carbohydrates and proteins are converted into fats for storage. Mammals and other vertebrates store fats in specialized connective tissues (◊adipose tissues), which not only act as energy reserves but also insulate the body and cushion its organs.
As a nutrient fat serves five purposes: it is a source of energy (9 kcal/g); makes the diet palatable; provides basic building blocks for cell structure; provides essential fatty acids (linoleic and linolenic); and acts as a carrier for fat-soluble vitamins (A, D, E, and K). Foods rich in fat are butter, lard, margarine, and cooking oils. Products high in monounsaturated or polyunsaturated fats are thought to be less likely to contribute to cardiovascular disease.

fathom (Anglo-Saxon *faethm* 'to embrace') in mining, seafaring, and handling timber, a unit of depth measurement (1.83 m/6 ft) used prior to metrication; it approximates to the distance between an adult man's hands when the arms are outstretched.

fatty acid or *carboxylic acid* organic compound consisting of a hydrocarbon chain, up to 24 carbon atoms long, with a carboxyl group (–COOH) at one end. The covalent bonds between the carbon atoms may be single or double; where a double bond occurs the carbon atoms concerned carry one instead of two hydrogen atoms. Chains with only single bonds have all the hydrogen they can carry, so they are said to be *saturated* with hydrogen. Chains with one or more double bonds are said to be *unsaturated* (see ◊polyunsaturate).
Saturated fatty acids include palmitic and stearic acids; unsaturated fatty acids include oleic (one double bond), linoleic (two double bonds), and linolenic (three double bonds). Linoleic acid accounts for more than one-third of some margerines. Supermarket brands that say they are high in polyunsaturates may contain as much as 39%. Fatty acids are generally found combined with glycerol in ◊lipids such as tryglycerides.

fault in geology, a fracture in the Earth's crust along which the two sides have moved as a result of differing strains in the adjacent rock bodies. Displacement of rock masses horizontally or vertically along a fault may be microscopic, or it may be massive, causing major ◊earthquakes.
If the movement has a major vertical component, the fault is termed a *normal fault*, where rocks on each side have moved apart, or a *reverse fault*, where one side has overridden the other (a low angle reverse fault is called a *thrust*). A *lateral fault*, or *tear fault*, occurs where the relative

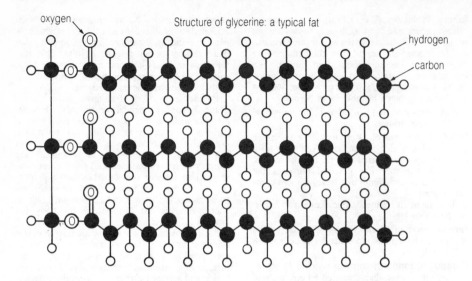

oxygen

Structure of glycerine: a typical fat

hydrogen

carbon

fat *The molecular structure of typical fat. The molecule consists of three fatty acid molecules linked to a molecule of glycerol.*

movement is sideways. A particular kind of fault found only in ocean ridges is the **transform fault** (a term coined by Canadian geophysicist John Tuzo Wilson 1965). On a map an ocean ridge has a stepped appearance. The ridge crest is broken into sections, each section offset from the next. Between each section of the ridge crest the newly generated plates are moving past one another, forming a transform fault.

Faults produce lines of weakness that are often exploited by processes of ◊weathering and ◊erosion. Coastal caves and geos (narrow inlets) often form along faults and, on a larger scale, rivers may follow the line of a fault.

fax (common name for *facsimile transmission* or *telefax*) the transmission of images over a ◊telecommunications link, usually the telephone network. When placed on a fax machine, the original image is scanned by a transmitting device and converted into coded signals, which travel via the telephone lines to the receiving fax machine, where an image is created that is a copy of the original. Photographs as well as printed text and drawings

can be sent. The standard transmission takes place at 4,800 or 9,600 bits of information per second.

The world's first fax machine, the pantélégraphe, was invented by Italian physicist Giovanni Caselli 1866, over a century before the first electronic model came on the market. Standing over 2 m/ 6.5 ft high, it transmitted by telegraph nearly 5,000 handwritten documents and drawings between Paris and Lyon in its first year.

feather rigid outgrowth of the outer layer of the skin of birds, made of the protein keratin. Feathers provide insulation and facilitate flight. There are several types, including long quill feathers on the wings and tail, fluffy down feathers for retaining body heat, and contour feathers covering the body. The colouring of feathers is often important in camouflage or in courtship and other displays. Feathers are replaced at least once a year.

fecundity the rate at which an organism reproduces, as distinct from its ability to reproduce (◊fertility). In vertebrates, it is usually measured as the number of offspring produced by a female each year.

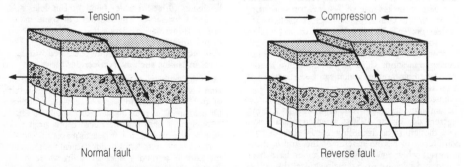

← Tension →   → Compression ←

Normal fault   Reverse fault

fault *Faults are caused by the movement of rock layers, producing such features as block mountains and rift valleys. A normal fault is caused by a tension or stretching force acting in the rock layers. A reverse fault is caused by compression forces. Faults can continue to move for thousands or millions of years.*

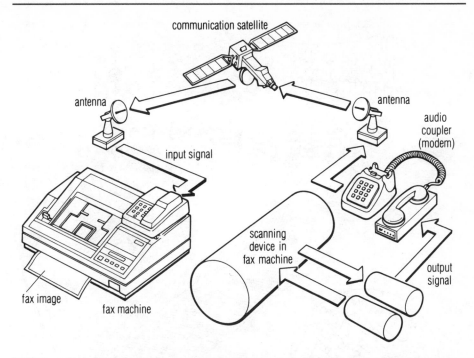

communication satellite

antenna

antenna

audio coupler (modem)

input signal

scanning device in fax machine

output signal

fax image

fax machine

fax *The fax machine scans a document to produce a digital (dot-by-dot) signal. The modem (which is usually built into the fax) converts the signal so that it can be sent over the telephone lines. At the receiving end, the process is reversed to reconstitute the original image.*

feedback general principle whereby the results produced in an ongoing reaction become factors in modifying or changing the reaction; it is the principle used in self-regulating control systems, from a simple ◊thermostat and steam-engine ◊governor to automatic computer-controlled machine tools. A fully computerized control system, in which there is no operator intervention, is called a *closed-loop feedback* system. A system that also responds to control signals from an operator is called an *open-loop feedback* system.

In self-regulating systems, information about what *is* happening in a system (such as level of temperature, engine speed, or size of workpiece) is fed back to a controlling device, which compares it with what *should* be happening. If the two are different, the device takes suitable action (such as switching on a heater, allowing more steam to the engine, or resetting the tools).

feedback in biology, another term for ◊biofeedback.

Fehling's test chemical test to determine whether an organic substance is a reducing agent (substance that donates electrons to other substances in a chemical reaction). It is usually used to detect reducing sugars (monosaccharides, such as glucose, and the disaccharides maltose and lactose) and aldehydes.

If the test substance is heated with a freshly prepared solution containing copper(II) sulphate, sodium hydroxide and sodium potassium tartrate, the production of a brick-red precipitate indicates the presence of a reducing agent.

feldspar one of a group of rock-forming minerals;

the chief constituents of ◊igneous rock. Feldspars all contain silicon, aluminium, and oxygen, linked together to form a framework; spaces within this structure are occupied by sodium, potassium, calcium, or occasionally barium, in various proportions. Feldspars form white, grey, or pink crystals and rank 6 on the ◊Mohs' scale of hardness.

The four extreme compositions of feldspar are *orthoclase*, $KAlSi_3O_8$; *albite*, $NaAlSi_3O_8$; *anorthite*, $CaAl_2Si_2O_8$; and *celsian*, $BaAl_2Si_2O_8$. These are grouped into *plagioclase feldspars*, which range from pure sodium feldspar (albite) through pure calcium feldspar (anorthite) with a negligible potassium content; and *alkali feldspars*, (including orthoclase), which have a high potassium content, less sodium, and little calcium.

The type known as ◊*moonstone* has a pearl-like effect and is used in jewellery. Approximately 4,000 tonnes of feldspar are used in the ◊ceramics industry annually.

feldspathoid any of a group of silicate minerals resembling feldspars but containing less silica. Examples are nepheline ($NaAlSiO_4$ with a little potassium) and leucite ($KAlSi_2O_6$). Feldspathoids occur in igneous rocks that have relatively high proportions of sodium and potassium. Such rocks may also contain alkali feldspar, but they do not generally contain quartz because any free silica would have combined with the feldspathoid to produce more feldspar instead.

femur the *thigh-bone*; also the upper bone in the hind limb of a four-limbed vertebrate.

Fermat's Principle in physics, the principle that

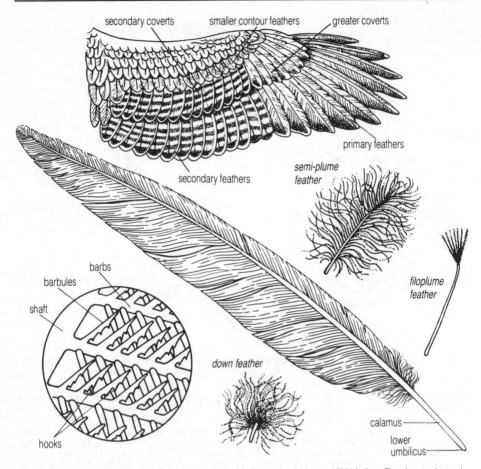

feather *Types of feather. A bird's wing is made up of two types of feather: contour feathers and flight feathers. The primary and secondary feathers are flight feathers. The coverts are contour feathers, used to streamline the bird. Semi-plume and down feathers insulate the bird's body and provide colour.*

a ray of light, or other radiation, moves between two points along the path that takes the minimum time. The principle is named after the French mathematician Pierre de Fermat who used it to deduce the laws of ◊reflection and ◊refraction.

fermentation the breakdown of sugars by bacteria and yeasts using a method of respiration without oxygen (◊anaerobic). Fermentation processes have long been utilized in baking bread, making beer and wine, and producing cheese, yoghurt, soy sauce, and many other foodstuffs.

In baking and brewing, yeasts ferment sugars to produce ◊ethanol and carbon dioxide; the latter makes bread rise and puts bubbles into beers and champagne. Many antibiotics are produced by fermentation; it is one of the processes that can cause food spoilage.

fermi unit of length equal to 10^{-15} m, used in atomic and nuclear physics. The unit is named after Italian–US physicist Enrico Fermi.

Fermilab (shortened form of *Fermi National Accelerator Laboratory*) US centre for ◊particle physics at Batavia, Illinois, near Chicago. It is named after Italian–US physicist Enrico Fermi. Fermilab was opened in 1972, and is the home of the Tevatron, the world's most powerful particle ◊accelerator. It is capable of boosting protons and antiprotons to speeds near that of light (to energies of 20 TeV).

fermion in physics, a subatomic particle whose spin can only take values that are half-integers, such as ½ or ³⁄₂. Fermions may be classified as leptons, such as the electron, and baryons, such as the proton and neutron. All elementary particles are either fermions or ◊bosons.

The exclusion principle, formulated by Austrian–US physicist Wolfgang Pauli 1925, asserts that no two fermions in the same system (such as an atom) can possess the same position, energy state, spin, or other quantized property.

fermium synthesized, radioactive, metallic element of the ◊actinide series, symbol Fm, atomic number 100, relative atomic mass 257. Ten isotopes are known, the longest-lived of which, Fm-257, has a half-life of 80 days. Fermium has been produced only in minute quantities in particle accelerators.

It was discovered in 1952 in the debris of the first thermonuclear explosion. The element was named 1955 in honour of Italian–US physicist Enrico Fermi.

fern plant of the class Filicales, related to horsetails and clubmosses. Ferns are spore-bearing, not flowering, plants, and most are perennial, spreading by low-growing roots. The leaves, known as fronds, vary widely in size and shape. Some taller types, such as tree-ferns, grow in the tropics. There are over 7,000 species.

Ferns found in Britain include the polypody *Polypodium vulgare*, shield fern *Polystichum*, male fern *Dryopteris filix-mas*, hart's-tongue *Phyllitis scolopend-*

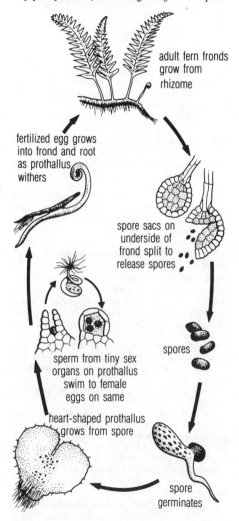

adult fern fronds grow from rhizome

fertilized egg grows into frond and root as prothallus withers

spore sacs on underside of frond split to release spores

spores

sperm from tiny sex organs on prothallus swim to female eggs on same

heart-shaped prothallus grows from spore

spore germinates

fern *The life cycle of a fern. Ferns have two distinct forms that alternate during their life cycle. For the main part of its life, a fern consists of a short stem (or rhizome) from which roots and leaves grow. The other part of its life is spent as a small heart-shaped plant called a prothallus.*

rium, maidenhair *Adiantum capillus-veneris*, and bracken *Pteridium aquilinum*, an agricultural weed.

ferric ion traditional name for the trivalent condition of iron, Fe^{3+}; the modern name is iron(III). Ferric salts are usually reddish or yellow in colour and form reddish-yellow solutions. $Fe_2(SO_4)_3$ is iron(III) sulphate (ferric sulphate).

ferrimagnetic material material in which adjacent molecular magnets are aligned anti-parallel, but have unequal strength, producing a strong overall magnetization. Ferrimagnetism is found in certain inorganic substances, such as ◊ferrites. See ◊magnetism.

ferrite ceramic ferrimagnetic material. Ferrites are iron oxides to which small quantities of ◊transition metal oxides (such as cobalt and nickel oxides) have been added. They are used in transformer cores, radio antenna, and, formerly, in computer memories.

ferro-alloy alloy of iron with a high proportion of elements such as manganese, silicon, chromium, and molybdenum. Ferro-alloys are used in the manufacture of alloy steels. Each alloy is generally named after the added metal—for example, ferro-chromium.

ferromagnetic material material that acquires very strong magnetism when placed in an external magnetic field. It may therefore be said to have a high magnetic ◊permeability and ◊susceptibility (which depends upon temperature). Examples are iron, cobalt, nickel, and their alloys. Some ferromagnetic materials retain their magnetism when the external magnetic field is removed and hence are used to make permanent magnets.

Ultimately, ferromagnetism is caused by spinning electrons in the atoms of the material, which act as tiny weak magnets. They align parallel to each other within small regions of the material to form ◊domains, or areas of stronger magnetism. In an unmagnetized material, the domains are aligned at random so there is no overall magnetic effect. If a magnetic field is applied to that material, the domains align to point in the same direction, producing a strong overall magnetic effect. Permanent magnetism arises if the domains remain aligned after the external field is removed. Ferromagnetic materials exhibit ◊hysteresis. See ◊magnetism.

ferrous ion traditional name for the divalent condition of iron, Fe^{2+}; the modern name is iron(II). Ferrous salts are usually green, and form yellow-green solutions. $FeSO_4$ is iron(II) sulphate (ferrous sulphate).

ferrous metal metal affected by magnetism. Iron, cobalt, and nickel are the three ferrous metals.

fertility an organism's ability to reproduce, as distinct from the rate at which it reproduces (see ◊fecundity). Individuals become infertile (unable to reproduce) when they cannot generate gametes (eggs or sperm) or when their gametes cannot yield a viable ◊embryo after fertilization.

fertilization in ◊sexual reproduction, the union of two ◊gametes (sex cells, often called egg and sperm) to produce a ◊zygote, which combines the genetic material contributed by each parent. In self-fertilization the male and female gametes come from the same plant; in cross-fertilization they

fertilization of a flowering plant

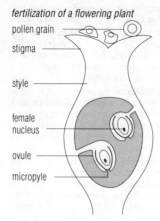
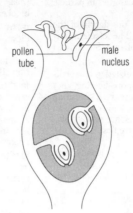
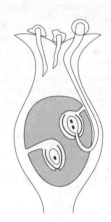

pollen grain
stigma
style
female nucleus
ovule
micropyle

pollen tube
male nucleus

fertilization

come from different plants. Self-fertilization rarely occurs in animals; usually even ◊hermaphrodite animals cross-fertilize each other.

In terrestrial insects, mammals, reptiles, and birds, fertilization occurs within the female's body; in the majority of fishes and amphibians, and most aquatic invertebrates, it occurs externally, when both sexes release their gametes into the water. In most fungi, gametes are not released, but the hyphae of the two parents grow towards each other and fuse to achieve fertilization. In higher plants, ◊pollination precedes fertilization.

fertilizer substance containing some or all of a range of about 20 chemical elements necessary for healthy plant growth, used to compensate for the deficiencies of poor or depleted soil. Fertilizers may be *organic*, for example farmyard manure, composts, bonemeal, blood, and fishmeal; or *inorganic*, in the form of compounds, mainly of nitrogen, phosphate, and potash, which have been used on a very much increased scale since 1945.

Because externally applied fertilizers tend to be in excess of plant requirements and drain away to affect lakes and rivers (see ◊eutrophication), attention has turned to the modification of crop plants themselves. Plants of the legume family, including the bean, clover, and lupin, live in symbiosis with bacteria located in root nodules, which fix nitrogen from the atmosphere. Research is now directed to producing a similar relationship between such bacteria and crops such as wheat.

fetch the distance of open water over which wind can blow to create ◊waves. The greater the fetch the more potential power waves have when they hit the coast. In the south and west of England the fetch stretches for several thousand kilometres, all the way to South America. This combines with the southwesterly ◊prevailing winds to cause powerful waves and serious ◊coastal erosion along south- and west-facing coastlines.

fetch-execute cycle or *processing cycle* in computing, the two-phase cycle used by the computer's central processing unit to process the instructions in a program. During the *fetch phase*, the next program instruction is transferred from

the computer's immediate-access memory to the instruction register (memory location used to hold the instruction while it is being executed). During the *execute phase*, the instruction is decoded and obeyed. The process is repeated in a continuous loop.

fetus or *foetus* stage in mammalian ◊embryo development. The human embryo is usually termed a fetus after the eighth week of development, when the limbs and external features of the head are recognizable.

In the UK, from 1989, the use of aborted fetuses for research and transplant purposes was approved provided that the mother's decision to seek an abortion is not influenced by consideration of this possible use. Each case has to be considered by an ethics committee, which may set conditions for the use of the fetal material.

fibre, dietary or *roughage* plant material that cannot be digested by human digestive enzymes; it consists largely of cellulose, a carbohydrate found in plant cell walls. Fibre adds bulk to the gut contents, assisting the muscular contractions that force food along the intestine. A diet low in fibre causes constipation and is believed to increase the risk of developing diverticulitis, diabetes, gall-bladder disease, and cancer of the large bowel—conditions that are rare in nonindustrialized countries, where the diet contains a high proportion of unrefined cereals.

Soluble fibre consists of indigestible plant carbohydrates (such as pectins, hemicelluloses, and gums) that dissolve in water. A high proportion of the fibre in such foods as oat bran, pulses, and vegetables is of this sort. Its presence in the diet has been found to reduce the amount of cholesterol in blood over the short term, although the mechanism for its effect is disputed.

fibreglass glass that has been formed into fine fibres, either as long continuous filaments or as a fluffy, short-fibred glass wool. Fibreglass is heat- and fire-resistant and a good electrical insulator. It has applications in the field of fibre optics and as a strengthener for plastics in ◊GRP (glass-reinforced plastics).

The long filament form is made by forcing molten glass through the holes of a spinneret, and is woven into textiles. Glass wool is made by blowing streams of molten glass in a jet of high-pressure steam, and is used for electrical, sound, and thermal insulation, especially for the roof space in houses.

fibre optics branch of physics dealing with the transmission of light and images through glass or plastic fibres known as ◊optical fibres.

fibrin an insoluble blood protein used by the body to stop bleeding. When an injury occurs fibrin is deposited around the wound in the form of a mesh, which dries and hardens, so that bleeding stops. Fibrin is developed in the blood from a soluble protein, fibrinogen.

The conversion of fibrinogen to fibrin is the final stage in blood clotting. Platelets, a type of cell found in blood, release the enzyme thrombin when they come into contact with damaged tissue, and the formation of fibrin then occurs. Calcium, vitamin K, and a variety of enzymes called factors are also necessary for efficient blood clotting.

fibula the rear lower bone in the hind leg of a vertebrate. It is paired and often fused with a smaller front bone, the tibia.

field in computing, a specific item of data. A field is usually part of a *record*, which in turn is part of a ◊file.

field in physics, a region of space in which an object exerts a force on another separate object because of certain properties they both possess. For example, there is a force of attraction between any two objects that have mass when one is in the gravitational field of the other.

Other fields of force include ◊electric fields (caused by electric charges) and ◊magnetic fields (caused by magnetic poles), either of which can involve attractive or repulsive forces.

field-length check ◊validation check in which the characters in an input field are counted to ensure that the correct number of characters have been entered. For example, a six-figure date field may be checked to ensure that it does contain exactly six digits.

field studies study of ecology, geography, geology, history, archaeology, and allied subjects, in the natural environment as opposed to the laboratory.

The Council for the Promotion of Field Studies was established in Britain 1943, in order to promote a wider knowledge and understanding of the natural environment among the public; Flatford Mill, Suffolk, was the first of its research centres to be opened.

fifth-generation computer anticipated new type of computer based on emerging microelectronic technologies with high computing speeds and ◊parallel processing. The development of very large-scale integration (◊VLSI) technology, which can put many more circuits on to an integrated circuit (chip) than is currently possible, and developments in computer hardware and software design may produce computers far more powerful than those in current use.

It has been predicted that such a computer will be able to communicate in natural spoken language with its user; store vast knowledge databases; search rapidly through these databases, making intelligent inferences and drawing logical conclusions; and process images and 'see' objects in the way that humans do.

In 1981 Japan's Ministry of International Trade and Industry launched a ten-year project to build the first fifth-generation computer, the 'parallel inference machine', consisting of over a thousand microprocessors operating in parallel with each other. By 1992, however, the project was behind schedule and had only produced 256–processor modules. It has since been suggested that research into other technologies, such as ◊neural networks, may present more promising approaches to artificial intelligence. See also ◊computer generations.

file in computing, a collection of data or a program stored in a computer's external memory (for example, on ◊disc). It might include anything from information on a company's employees to a program for an adventure game. *Serial files* hold information as a sequence of characters, so that, to read any particular item of data, the program must read all those that precede it. *Random-access files* allow the required data to be reached directly.

Files usually consist of a set of records, each having a number of *fields* for specific items of data. For example, the file for a class of schoolchildren might have a record for each child, with five fields of data in each record, storing: (1) family name; (2) first name; (3) house name or number; (4) street name; (5) town. To find out, for example, which children live in the same street, one would look in field 4.

file access in computing, the way in which the records in a file are stored, retrieved, or updated by computer. There are four main types of file organization, each of which allows a different form of access to the records.

Records in a *serial file* are not stored in any particular order, so a specific record can be accessed only by reading through all the previous records.

Records in a *sequential file* are sorted by reference to a key field (see ◊sorting) and the computer can use a searching technique, such as a binary search, to access a specific record. An *indexed sequential file* possesses an index, which records the position of each block of records and is created and updated with that file. By consulting the index, the computer can obtain the address of the block containing the required record, and search just that block rather than the whole file. A *direct-access* or *random-access file* contains records that can be accessed directly by the computer.

film, photographic strip of transparent material (usually cellulose acetate) coated with a light-sensitive emulsion, used in cameras to take pictures. The emulsion contains a mixture of light-sensitive silver halide salts (for example, bromide or iodide) in gelatin. When the emulsion is exposed to light, the silver salts are invisibly altered, giving a latent image, which is then made visible by the process of ◊developing. Films differ in their sensitivities to light, this being indicated by their speeds. Colour film consists of several layers of emulsion, each of

file structure

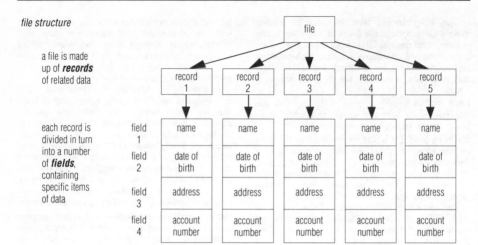

a file is made up of **records** of related data

each record is divided in turn into a number of **fields**, containing specific items of data

file *Structure of a computer file.*

which records a different colour in the light falling on it.

In *colour film* the front emulsion records blue light, then comes a yellow filter, followed by layers that record green and red light respectively. In the developing process the various images in the layers are dyed yellow, magenta (red), and cyan (blue), respectively. When they are viewed, either as a transparency or as a colour print, the colours merge to produce the true colour of the original scene photographed.

filter in chemistry, a porous substance, such as blotting paper, through which a mixture can be passed to separate out its solid constituents. In optics, a filter is a piece of glass or transparent material that passes light of one colour only.

filter in electronics, a circuit that transmits a signal of some frequencies better than others. A low-pass filter transmits signals of low frequency and direct current; a high-pass filter transmits high-frequency signals; a band-pass filter transmits signals in a band of frequencies.

filter in optics, a device that absorbs some parts of the visible ◊spectrum and transmits others. For example, a green filter will absorb or block all colours of the spectrum except green, which it allows to pass through. A yellow filter absorbs only light at the blue and violet end of the spectrum, transmitting red, orange, green, and yellow light.

filtrate liquid or solution that has passed through a filter.

filtration technique by which suspended solid particles in a fluid are removed by passing the mixture through a filter, usually porous paper, plastic, or cloth. The particles are retained by the filter to form a residue and the fluid passes through to make up the filtrate. For example, soot may be filtered from air, and suspended solids from water.

finsen unit unit (symbol FU) for measuring the intensity of ultraviolet (UV) light; for instance, UV light of 2 FUs causes sunburn in 15 minutes.

fiord alternative spelling of ◊fjord.

fire clay a ◊clay with refractory characteristics (resistant to high temperatures), and hence suitable for lining furnaces (firebrick). Its chemical composition consists of a high percentage of silicon and aluminium oxides, and a low percentage of the oxides of sodium, potassium, iron, and calcium. Fire clays underlie the coal seams in the UK.

firedamp gas that occurs in coal mines and is explosive when mixed with air in certain proportions. It consists chiefly of methane (CH_4, natural gas or marsh gas) but always contains small quantities of other gases, such as nitrogen, carbon dioxide, and hydrogen, and sometimes ethane and carbon monoxide.

fire-danger rating unit index used by the UK Forestry Commission to indicate the probability of a forest fire. 0 means a fire is improbable, 100 shows a serious fire hazard.

fire extinguisher device for putting out a fire. Fire extinguishers work by removing one of the three conditions necessary for fire to continue (heat, oxygen, and fuel), either by cooling the fire or by excluding oxygen.

The simplest fire extinguishers contain water, which when propelled onto the fire cools it down. Water extinguishers cannot be used on electrical fires, as there is a danger of electrocution, or on burning oil, as the oil will float on the water and spread the blaze.

Many domestic extinguishers contain liquid carbon dioxide under pressure. When the handle is pressed, carbon dioxide is released as a gas that blankets the burning material and prevents oxygen from reaching it. Dry extinguishers spray powder, which then releases carbon dioxide gas. Wet extinguishers are often of the soda-acid type; when activated, sulphuric acid mixes with sodium bicarbonate, producing carbon dioxide. The gas pressure forces the solution out of a nozzle, and a foaming agent may be added.

Some extinguishers contain ◊halons (hydrocarbons with one or more hydrogens substituted by a halogen such as chlorine, bromine, or fluorine).

These are very effective at smothering fires, but cause damage to the ◊ozone layer.

firewood the principal fuel for some 2 billion people, mainly in the Third World. In principle a renewable energy source, firewood is being cut far faster than the trees can regenerate in many areas of Africa and Asia, leading to ◊deforestation.

In Mali, for example, wood provides 97% of total energy consumption, and deforestation is running at an estimated 9,000 hectares a year. The heat efficiency of firewood can be increased by use of stoves, but many people cannot afford to buy these.

firmware computer program held permanently in a computer's ◊ROM (read-only memory) chips, as opposed to a program that is read in from external memory as it is needed.

fish aquatic vertebrate that uses gills for obtaining oxygen from fresh or sea water. There are three main groups, not closely related: the bony fishes or Osteichthyes (goldfish, cod, tuna); the cartilaginous fishes or Chondrichthyes (sharks, rays); and the jawless fishes or Agnatha (hagfishes, lampreys).

The *bony fishes* constitute the majority of living fishes (about 20,000 species). The skeleton is bone, movement is controlled by mobile fins, and the body is usually covered with scales. The gills are covered by a single flap. Many have a swim bladder with which the fish adjusts its buoyancy. Most lay eggs, sometimes in vast numbers; some cod can produce as many as 28 million. These are laid in the open sea, and probably no more than 28 of them will survive to become adults. Those species that produce small numbers of eggs very often protect them in nests, or brood them in their mouths. Some fishes are internally fertilized and retain eggs until hatching inside the body, then give birth to live young. Most bony fishes are ray-finned fishes, but a few, including lungfishes and coelacanths, are fleshy-finned.

The *cartilaginous fishes* are efficient hunters.

There are fewer than 600 known species of sharks and rays. The skeleton is cartilage, the mouth is generally beneath the head, the nose is large and sensitive, and there is a series of open gill slits along the neck region. They may lay eggs ('mermaid's purses') or bear live young. Some types of cartilaginous fishes, such as sharks, retain the shape they had millions of years ago.

Jawless fishes have a body plan like that of some of the earliest vertebrates that existed before true fishes with jaws evolved. There is no true backbone but a ◊notochord. The lamprey attaches itself to the fishes on which it feeds by a suckerlike rasping mouth. Hagfishes are entirely marine, very slimy, and feed on carrion and injured fishes.

fission in physics, the splitting of a heavy atomic nucleus into two or more major fragments. It is accompanied by the emission of two or three neutrons and the release of large amounts of energy (see ◊nuclear energy).

Fission occurs spontaneously in nuclei of uranium-235, the main fuel used in nuclear reactors. However, the process can also be induced by bombarding nuclei with neutrons because a nucleus that has absorbed a neutron becomes unstable and soon splits. The neutrons released spontaneously by the fission of uranium nuclei may therefore be used in turn to induce further fissions, setting up a ◊chain reaction that must be controlled if it is not to result in a nuclear explosion.

fission-track dating in geology, a dating method based on the natural and spontaneous nuclear fission of uranium-238 and its physical product, linear atomic displacements (tracks) created along the trajectory of released energized fission fragments. Knowing the rate of fission (a constant), the uranium content of the material and by counting the number of fission tracks, the age of the material can be determined. The method is most widely used to date volcanic deposits adjacent to archaeological material.

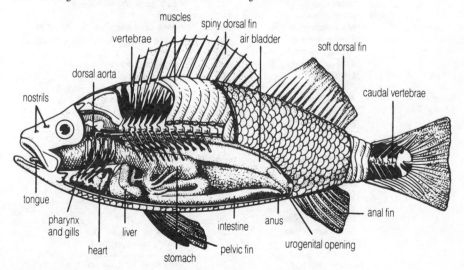

fish *The anatomy of a fish. All fishes move through water using their fins for propulsion. The bony fishes, like the specimen shown here, constitute the largest group of fishes with about 20,000 species.*

fish classification

superclass Agnatha (jawless fishes)

order	no of species	examples
Petromyzoniformes	30	lamprey
Myxiniformes	15	hagfish

superclass Gnathostomata (jawed fishes)

class Chondrichthyes (cartilaginous fishes)

subclass Elasmobranchii (sharks and rays)

Hexanchiformes	6	frilled shark, comb-toothed shark
Heterodontiformes	10	Port Jackson shark
Lamniformes	200	'typical' shark
Rajiformes	300	skate, ray

subclass Holocephali (rabbitfishes)

Chimaeriformes	20	chimaera, rabbitfish

class Osteichthyes (bony fishes)

subclass Sarcopterygii (fleshy-finned fishes)

Coelacanthiformes	1	coelacanth
Ceratodiformes	1	Australian lungfish
Lepidosireniformes	4	South American and African lungfish

subclass Actinopterygii (ray-finned fishes)

superorder Chondrostei

Polypteriformes	11	bichir, reedfish
Acipensiformes	25	paddlefish, sturgeon

superorder Holostei

Amiiformes	8	bowfin, garpike

superorder Teleostei

Elopiformes	12	tarpon, tenpounder
Anguilliformes	300	eel
Notacanthiformes	20	spiny eel
Clupeiformes	350	herring, anchovy
Osteoglossiformes	16	arapaima, African butterfly fish
Mormyriformes	150	elephant-trunk fish, featherback
Salmoniformes	500	salmon, trout, smelt, pike
Gonorhynchiformes	15	milkfish
Cypriniformes	350	carp, barb, characin, loach
Siluriformes	200	catfish
Myctophiformes	300	deep-sea lantern fish, Bombay duck
Percopsiformes	10	pirate perch, cave-dwelling amblyopsid
Batrachoidiformes	10	toadfish
Gobiesociformes	100	clingfish, dragonets
Lophiiformes	150	anglerfish
Gadiformes	450	cod, pollack, pearlfish, eelpout
Atheriniformes	600	flying fish, toothcarp, halfbeak
Lampridiformes	50	opah, ribbonfish
Beryciformes	150	squirrelfish
Zeiformes	60	John Dory, boarfish
Gasterosteiformes	150	stickleback, pipefish, seahorse
Channiformes	5	snakeshead
Synbranchiformes	7	cuchia
Scorpaeniformes	700	gurnard, miller's thumb, stonefish
Dactylopteriformes	6	flying gurnard
Pegasiformes	4	sea-moth
Pleuronectiformes	500	flatfish
Tetraodontiformes	250	puffer fish, triggerfish, sunfish
Perciformes	6500	perch, cichlid, damsel fish, gobie, wrasse, parrotfish, gourami, marlin, mackerel, tuna, swordfish, spiny eel, mullet, barracuda, sea bream, croaker, ice fish, butterfish

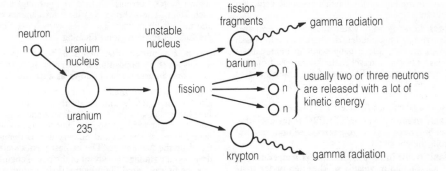

fission *Nuclear fission is harnessed in the reactors of nuclear power stations to generate energy. When a nucleus of uranium-235 is struck by a neutron (a subatomic particle) it quickly becomes unstable and splits into two parts—a barium nucleus and a krypton nucleus—liberating a large amount of energy and ejecting two or three neutrons. The emitted neutrons may then be used to trigger the fission of more neutrons, setting up a chain reaction.*

fitness in genetic theory, a measure of the success with which a genetically determined character can spread in future generations. By convention, the normal character is assigned a fitness of one, and variants (determined by other ◊alleles) are then assigned fitness values relative to this. Those with fitness greater than one will spread more rapidly and will ultimately replace the normal allele; those with fitness less than one will gradually die out. See also ◊inclusive fitness.

fixed point temperature that can be accurately reproduced and used as the basis of a temperature scale. In the Celsius scale, the fixed points are the temperature of melting ice, which is 0°C (32°F), and the temperature of boiling water (at standard atmospheric pressure), which is 100°C (212°F).

fixed-point notation system in which numbers are represented using a set of digits with the decimal point always in its correct position. For very large and very small numbers this requires a lot of digits. In computing, the size of the numbers that can be handled in this way is limited by the capacity of the computer, and so the slower ◊floating-point notation is often preferred.

fjord or *fiord* narrow sea inlet enclosed by high cliffs. Fjords are found in Norway, New Zealand, and western parts of Scotland. They are formed when an overdeepened U-shaped glacial valley is drowned by a rise in sea-level. At the mouth of the fjord there is a characteristic lip causing a shallowing of the water. This is due to reduced glacial erosion and the deposition of moraine at this point.

Fiordland is the deeply indented SW coast of South Island, New Zealand; one of the most beautiful inlets is Milford Sound.

flaccidity in botany, the loss of rigidity (turgor) in plant cells, caused by loss of water from the central vacuole so that the cytoplasm no longer pushes against the cellulose cell wall. If this condition occurs throughout the plant then wilting is seen.

Flaccidity can be induced in the laboratory by immersing the plant cell in a strong saline solution. Water leaves the cell by ◊osmosis causing the vacuole to shrink. In extreme cases the actual cytoplasm pulls away from the cell wall, a phenomenon known as plasmolysis.

flag in computing, an indicator that can be set or unset in order to signal whether a particular condition is true—for example, whether the end of a file has been reached, or whether an overflow error has occurred. The indicator usually takes the form of a single binary digit, or bit (either 0 or 1).

flagellum small hairlike organ on the surface of certain cells. Flagella are the motile organs of certain protozoa and single-celled algae, and of the sperm cells of higher animals. Unlike ◊cilia, flagella usually occur singly or in pairs; they are also longer and have a more complex whiplike action.

Each flagellum consists of contractile filaments producing snakelike movements that propel cells through fluids, or fluids past cells. Water movement inside sponges is also produced by flagella.

flame test in chemistry, the use of a flame to identify metal ◊cations present in a solid.

A nichrome or platinum wire is moistened with acid, dipped in a compound of the element, either powdered or in solution, and then held in a hot flame. The colour produced in the flame is characteristic of metals present; for example, sodium burns with an orange-yellow flame, and potassium with a lilac one.

flame test

element	colour of flame
sodium	orange-yellow
potassium	lilac
calcium	red or yellow-red
strontium, lithium	crimson
barium, manganese (manganese chloride)	pale green
copper, thallium, boron (boric acid)	bright green
lead, arsenic, antimony	livid blue
copper (copper(II) chloride)	bright blue

flare, solar brilliant eruption on the Sun above a ◊sunspot, thought to be caused by release of magnetic energy. Flares reach maximum brightness within a few minutes, then fade away over about an hour. They eject a burst of atomic particles into space at up to 1,000 kps/600 mps. When these particles reach Earth they can cause radio blackouts, disruptions of the Earth's magnetic field, and ◊auroras.

flash flood flood of water in a normally arid area brought on by a sudden downpour of rain. Flash floods are rare and usually occur in mountainous areas. They may travel many kilometres from the site of the rainfall. Because of the suddenness of flash floods, little warning can be given of their occurrence. In 1972 a flash flood at Rapid City, South Dakota, USA, killed 238 people.

flash memory type of ◊EEPROM memory that can be erased and reprogrammed without removal from the computer.

flash point in physics, the lowest temperature at which a liquid or volatile solid heated under standard conditions gives off sufficient vapour to ignite on the application of a small flame.

The *fire point* of a material is the temperature at which full combustion occurs. For safe storage of materials such as fuel or oil, conditions must be well below the flash and fire points to reduce fire risks to a minimum.

flatworm invertebrate of the phylum Platyhelminthes. Some are free-living, but many are parasitic (for example, tapeworms and flukes). The body is simple and bilaterally symmetrical, with one opening to the intestine. Many are hermaphroditic (with both male and female sex organs), and practise self-fertilization.

Fleming's rules memory aids used to recall the relative directions of the magnetic field, current, and motion in an electric generator or motor, using one's fingers. The three directions are represented by the thu*m*b (for *m*otion), *f*orefinger (for *f*ield) and

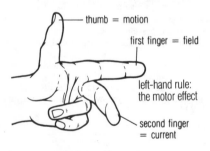

thumb = motion

first finger = field

left-hand rule:
the motor effect

second finger
= current

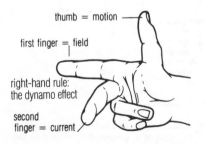

thumb = motion

first finger = field

right-hand rule:
the dynamo effect

second finger = current

Fleming's rules *Fleming's rules give the direction of the magnetic field, motion, and current in electrical machines. The left hand is used for motors, and the right hand for generators and dynamos.*

second finger (*c*urrent), all held at right angles to each other. The right hand is used for generators and the left for motors. The rules were devised by the English physicist John Fleming.

flexor any muscle that bends a limb. Flexors usually work in opposition to other muscles, the extensors, an arrangement known as antagonistic.

flight or *aviation* method of transport in which aircraft carry people and goods through the air. People first took to the air in ◊balloons and began powered flight 1852 in ◊airships, but the history of flying, both for civilian and military use, is dominated by the ◊aeroplane. The earliest planes were designed for gliding; the advent of the petrol engine saw the first powered flight by the Wright brothers 1903 in the USA. This inspired the development of aircraft throughout Europe. Biplanes were succeeded by monoplanes in the 1930s. The first jet plane (see ◊jet propulsion) was produced 1939, and after the end of World War II the development of jetliners brought about a continuous expansion in passenger air travel. In 1969 came the supersonic aircraft ◊Concorde.

history In Europe, at the beginning of the 20th century, France led in aeroplane design and Louis Blériot brought aviation much publicity by crossing the Channel 1909, as did the Reims air races of that year. The first powered flight in the UK was made by S F Cody 1908. In 1912 Sopwith and Bristol both built small biplanes. The first big twin-engined aeroplane was the Handley Page bomber 1917. The stimulus of World War I (1914–18) and rapid development of the petrol engine led to increased power, and speeds rose to 320 kph/ 200 mph. Streamlining the body of planes became imperative: the body, wings, and exposed parts were reshaped to reduce drag. Eventually the biplane was superseded by the internally braced monoplane structure, for example, the Hawker Hurricane and Supermarine Spitfire fighters and Avro Lancaster and Boeing Flying Fortress bombers of World War II (1939–45).

jet aircraft The German Heinkel 178, built 1939, was the first jet plane; it was driven, not by a ◊propeller as all planes before it, but by a jet of hot gases. The first British jet aircraft, the Gloster E.28/39, flew from Cranwell, Lincolnshire, on 15 May 1941, powered by a jet engine invented by British engineer Frank Whittle. Twin-jet Meteor fighters were in use by the end of the war. The rapid development of the jet plane led to enormous increases in power and speed until air-compressibility effects were felt near the speed of sound, which at first seemed to be a flight speed limit (the sound barrier). To attain ◊supersonic speed, streamlining the aircraft body became insufficient: wings were swept back, engines buried in wings and tail units, and bodies were even eliminated in all-wing delta designs. In the 1950s the first jet airliners, such as the Comet, were introduced into service. Today jet planes dominate both military and civilian aviation, although many light planes still use piston engines and propellers. The late 1960s saw the introduction of the ◊jumbo jet, and in 1976 the Anglo-French Concorde, which makes a transatlantic crossing in under three hours, came into commercial service.

flight: chronology

1783	First human flight, by Jean F Pilâtre de Rozier and the Marquis d'Arlandes, in Paris, using a hot-air balloon made by Joseph and Etienne Montgolfier; first ascent in a hydrogen-filled balloon by Jacques Charles and M N Robert in Paris.
1785	Jean-Pierre Blanchard and John J Jeffries made the first balloon crossing of the English Channel.
1852	Henri Giffard flew the first steam-powered airship over Paris.
1853	George Cayley flew the first true aeroplane, a model glider 1.5 m/5 ft long.
1891–96	Otto Lilienthal piloted a glider in flight.
1903	First powered and controlled flight of a heavier-than-air craft (aeroplane) by Orville Wright, at Kitty Hawk, North Carolina, USA.
1908	First powered flight in the UK by Samuel Cody.
1909	Louis Blériot flew across the English Channel in 36 minutes.
1914–18	World War I stimulated improvements in speed and power.
1919	First E–W flight across the Atlantic by Albert C Read, using a flying boat; first nonstop flight across the Atlantic E–W by John William Alcock and Arthur Whitten Brown in 16 hours 27 minutes; first complete flight from Britain to Australia by Ross Smith and Keith Smith.
1923	Juan de la Cierva flew the first autogiro with a rotating wing.
1927	Charles Lindbergh made the first W–E solo nonstop flight across the Atlantic.
1928	First transpacific flight, from San Francisco to Brisbane, by Charles Kinsford Smith and C T P Ulm.
1930	Frank Whittle patented the jet engine; Amy Johnson became the first woman to fly solo from England to Australia.
1937	The first fully pressurized aircraft, the Lockheed XC-35, came into service.
1939	Erich Warsitz flew the first Heinkel jet plane, in Germany; Igor Sikorsky designed the first helicopter, with a large main rotor and a smaller tail rotor.
1939–45	World War II—developments included the Hawker Hurricane and Supermarine Spitfire Fighters, and Avro Lancaster and Boeing Flying Fortress bombers.
1947	A rocket-powered plane, the Bell X-1, was the first aircraft to fly faster than the speed of sound.
1949	The de Havilland Comet, the first jet airliner, entered service; James Gallagher made the first nonstop round-the-world flight, in a Boeing Superfortress.
1953	The first vertical takeoff aircraft, the Rolls-Royce 'Flying Bedstead', was tested.
1968	The world's first supersonic airliner, the Russian TU-144, flew for the first time.
1970	The Boeing 747 jumbo jet entered service, carrying 500 passengers.
1976	Anglo-French Concorde, making a transatlantic crossing in under three hours, came into commercial service. A Lockheed SR-17A, piloted by Eldon W Joersz and George T Morgan, set the world air-speed record of 3,529.56 kmh/2,193.167 mph over Beale Air Force Base, California, USA.
1978	A US team made the first transatlantic crossing by balloon, in the helium-filled *Double Eagle II*.
1979	First crossing of the English Channel by a human-powered aircraft, *Gossamer Albatross*, piloted by Bryan Allen.
1981	The solar-powered *Solar Challenger* flew across the English Channel, from Paris to Kent, taking 5 hours for the 262 km/162.8 mi journey.
1986	Dick Rutan and Jeana Yeager made the first nonstop flight around the world without refuelling, piloting *Voyager*, which completed the flight in 9 days 3 minutes 44 seconds.
1987	Richard Branson and Per Lindstrand made the first transatlantic crossing by hot-air balloon, in *Virgin Atlantic Challenger*.
1988	*Daedalus*, a human-powered craft piloted by Kanellos Kanellopoulos, flew 118 km/74 mi across the Aegean Sea.
1991	Richard Branson and Per Lindstrand crossed the Pacific Ocean in the hot-air balloon *Virgin Otsouka Pacific Flyer* from the southern tip of Japan to NW Canada in 46 hours 15 minutes.
1992	US engineers demonstrated a model radio-controlled ornithopter, the first aircraft to be successfully propelled and manoeuvred by flapping wings.

other developments During the 1950s and 1960s research was done on V/STOL (vertical and/or short take-off) aircraft. The British Harrier jet fighter has been the only VTOL aircraft to achieve commercial success, but STOL technology has fed into subsequent generations of aircraft. The 1960s and 1970s also saw the development of variable geometry ('swing-wing') aircraft, the wings of which can be swept back in flight to achieve higher speeds. In the 1980s much progress was made in 'fly-by-wire' aircraft with computer-aided controls. International partnerships have developed both civilian and military aircraft. The Panavia Tornado is a joint project of British, German, and Italian companies. It is an advanced swing-wing craft of multiple roles—interception, strike, ground support, and reconnaissance. The airbus is a wide-bodied airliner built jointly by companies from France, Germany, the UK, the Netherlands, and Spain.

flight simulator computer-controlled pilot-training device, consisting of an artificial cockpit mounted on hydraulic legs, that simulates the experience of flying a real aircraft. Inside the cockpit, the trainee pilot views a screen showing a computer-controlled projection of the view from a real aircraft, and makes appropriate adjustments to the controls. The computer monitors these adjustments, changes both the alignment of the cockpit on its hydraulic legs, and the projected view seen by the pilot. In this way a trainee pilot can progress to quite an advanced stage of training without leaving the ground.

flint compact, hard, brittle mineral (a variety of chert), brown, black, or grey in colour, found in nodules in limestone or shale deposits. It consists of fine-grained silica, SiO_2 (usually ◊quartz), in cryptocrystalline form. Flint implements were widely used in prehistory.

When chipped, the flint nodules show a shell-like fracture and a sharp cutting edge. The earliest flint implements, belonging to Palaeolithic cultures and made by striking one flint against another, are simple, while those of the Neolithic are expertly chipped and formed, and are often ground or polished. The best flint, used for Neolithic tools, is *floorstone*, a shiny black flint that occurs deep within the chalk.

Because of their hardness (7 on the ◊Mohs' scale), flint splinters are used for abrasive purposes and, when ground into powder, added to clay during pottery manufacture. Flints have been used for making fire by striking the flint against steel, which produces a spark, and for discharging guns. Flints in cigarette lighters are made from cerium alloy.

Flints commonly occur in the ◊chalk downlands of southern England and were often used as building material.

flip-flop in computing, another name for a ◊bistable circuit.

floating state of equilibrium in which a body rests on or is suspended in the surface of a fluid (liquid or gas). According to ◊Archimedes' principle, a body wholly or partly immersed in a fluid will be subjected to an upward force, or upthrust, equal in magnitude to the weight of the fluid it has displaced.

If the ◊density of the body is greater than that of the fluid, then its weight will be greater than the upthrust and it will sink. However, if the body's density is less than that of the fluid, the upthrust will be the greater and the body will be pushed upwards towards the surface. As the body rises above the surface the amount of fluid that it displaces (and therefore the magnitude of the upthrust) decreases. Eventually the upthrust acting on the submerged part of the body will equal the body's weight, equilibrium will be reached, and the body will float.

(a)
weight greater
than upthrust –
stone sinks.

(b)
upthrust greater
than weight –
pingpong ball
rises

(c)
upthrust equals
weight –
pingpong ball
floats

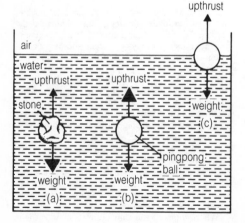

floating

floating-point notation system in which numbers are represented by means of a decimal fraction and an exponent. For example, in floating-point notation, 123,000,000,000 would be represented as 0.123 (the fraction, or mantissa) and 12 (the exponent). The exponent is the power of 10 by which the fraction must be multiplied in order to obtain the true value of the number. In other words, 123,000,000,000 can be represented by 0.123×10^{12}. In computing, floating-point notation enables programs to work with very large and very small numbers using only a few digits; however, it is slower than ◊fixed-point notation and suffers from small rounding errors.

In a computer, numbers expressed in floating-point notation are represented as pairs—for example, 97.8 as (.978, +2) or .978 E2. Decimal numbers are, of course, automatically converted by the computer into binary number code before storage and processing.

flooding the inundation of land that is not normally covered with water. Flooding from rivers commonly takes place after heavy rainfall or in the spring after winter snows have melted. The river's ◊discharge (volume of water carried in a given period) becomes too great, and water spills over the banks onto the surrounding flood plain. Small floods may happen once a year—these are called **annual floods** and are said to have a one-year return period. Much larger floods may occur on average only once every 50 years.

Flooding is least likely to occur in an efficient channel that is semicircular in shape. Flooding can also occur at the coast in stormy conditions (see ◊storm surge) or when there is an exceptionally high tide. The Thames Flood Barrier was constructed in 1982 to prevent the flooding of London from the sea.

flood plain area of periodic flooding along the course of river valleys. When river discharge exceeds the capacity of the channel, water rises over the channel banks and floods the adjacent low-lying lands. As water spills out of the channel some alluvium (silty material) will be deposited on the banks to form ◊levees (raised river banks). This water will slowly seep into the flood plain, depositing a new layer of rich fertile alluvium as it does so. Many important floodplains, such as the inner Niger delta in Mali, occur in arid areas where their exceptional productivity has great importance for the local economy.

A flood plain (often called inner ◊delta) can be regarded as part of a river's natural domain, statistically certain to be claimed by the river at repeated intervals. By plotting floods that have occurred and extrapolating from these data we can speak of ten-year floods, 100–year floods, 500–year floods, and so forth, based on the statistical probability of flooding across certain parts of the flood plain.

Even the most energetic flood-control plans (such as dams, dredging, and channel modification) will sometimes fail, and using flood plains as the site of towns and villages is always laden with risk. It is more judicious to use flood plains in ways compatible with flooding, such as for agriculture or parks.

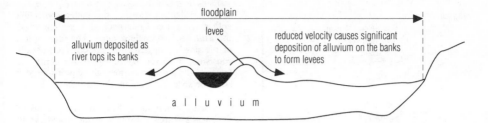

flood plain

Floodplain features include ◊meanders and ◊oxbow lakes.

floppy disc in computing, a storage device consisting of a light, flexible disc enclosed in a cardboard or plastic jacket. The disc is placed in a disc drive, where it rotates at high speed. Data are recorded magnetically on one or both surfaces.

Floppy discs were invented by IBM in 1971 as a means of loading programs into the computer. They were originally 20 cm/8 in in diameter and typically held about 240 ◊kilobytes of data. Present-day floppy discs, widely used on ◊microcomputers, are usually either 13.13 cm/5.25 in or 8.8 cm/3.5 in in diameter, and generally hold 0.5–2 ◊megabytes, depending on the disc size, recording method, and whether one or both sides are used.

Floppy discs are inexpensive, and light enough to send through the post, but have slower access speeds and are more fragile than hard discs.

floral diagram diagram showing the arrangement and number of parts in a flower, drawn in cross section. An ovary is drawn in the centre, surrounded by representations of the other floral parts, indicating the position of each at its base. If any parts such as the petals or sepals are fused, this is also indicated. Floral diagrams allow the structure of different flowers to be compared, and are usually shown with the floral formula.

floral formula symbolic representation of the structure of a flower. Each kind of floral part is represented by a letter (K for calyx, C for corolla, P for perianth, A for androecium, G for gynoecium) and a number to indicate the quantity of the part present, for example, C5 for a flower with five petals. The number is in brackets if the parts are fused. If the parts are arranged in distinct whorls within the flower, this is shown by two separate figures, such as A5 + 5, indicating two whorls of five stamens each.

A flower with radial symmetry is known as *actinomorphic*; a flower with bilateral symmetry as *zygomorphic*.

floret small flower, usually making up part of a larger, composite flower head. There are often two different types present on one flower head: disc florets in the central area, and ray florets around the edge which usually have a single petal known as the ligule. In the common daisy, for example, the disc florets are yellow, while the ligules are white.

flotation process common method of preparing mineral ores for subsequent processing by making use of the different wetting properties of various components. The ore is finely ground and then mixed with water and a specially selected wetting agent. Air is bubbled through the mixture, forming a froth; the desired ore particles attach themselves to the bubbles and are skimmed off, while unwanted dirt or other ores remain behind.

flow chart diagram, often used in computing, to show the possible paths that data can take through a system or program.

A *system flow chart*, or *data flow chart*, is used to describe the flow of data through a complete data-processing system. Different graphic symbols represent the clerical operations

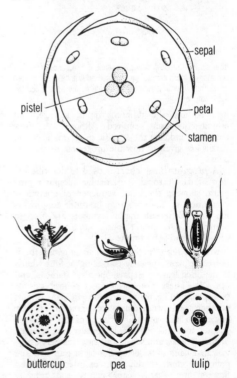

floral diagram *A floral diagram shows how the parts of a flower are arranged. The diagram is a ground plan showing the relative position and number of sepals, petals, stamens, and carpels of a flower. Both a floral diagram and a drawing of half a flower are needed to describe a flower fully.*

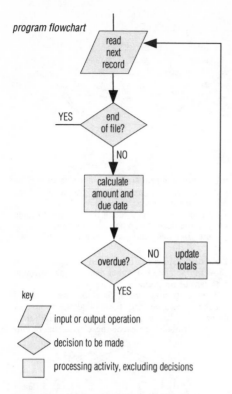

program flowchart

key

input or output operation

decision to be made

processing activity, excluding decisions

flow chart *A program flow chart shows the sequence of operations within a comuter program needed to achieve a task, such as reading customer accounts and calculating the amount due for each customer.*

involved and the different input, storage, and output equipment required. Although the flow chart may indicate the specific programs used, no details are given of how the programs process the data.

A *program flow chart* is used to describe the flow of data through a particular computer program, showing the exact sequence of operations performed by that program in order to process the data. Different graphic symbols are used to represent data input and output, decisions, branches, and ◊subroutines.

flower the reproductive unit of an ◊angiosperm or flowering plant, typically consisting of four whorls of modified leaves: ◊sepals, ◊petals, ◊stamens, and ◊carpels. These are borne on a central axis or ◊receptacle. The many variations in size, colour, number, and arrangement of parts are closely related to the method of pollination. Flowers adapted for wind pollination typically have reduced or absent petals and sepals and long, feathery ◊stigmas that hang outside the flower to trap airborne pollen. In contrast, the petals of insect-pollinated flowers are usually conspicuous and brightly coloured.

The sepals and petals form the *calyx* and *corolla* respectively and together comprise the perianth with the function of protecting the reproductive organs and attracting pollinators. The

stamens lie within the corolla, each having a slender stalk, or filament, bearing the pollen-containing anther at the top. Collectively they are known as the *androecium* (male organs). The inner whorl of the flower comprises the carpels, each usually consisting of an ◊ovary in which are borne the ◊ovules, and a stigma borne at the top of a slender stalk, or style. Collectively the carpels are known as the *gynoecium* (female organs).

In size, flowers range from the tiny blooms of duckweeds scarcely visible with the naked eye to the gigantic flowers of the Malaysian *Rafflesia*, which can reach over 1 m/3 ft across. Flowers may either be borne singly or grouped together in ◊inflorescences. The stalk of the whole inflorescence is termed a peduncle, and the stalk of an individual flower is termed a pedicel. A flower is termed hermaphrodite when it contains both male and female reproductive organs. When male and female organs are carried in separate flowers, they are termed monoecious; when male and female flowers are on separate plants, the term dioecious is used.

flowering plant term generally used for ◊angiosperms, which bear flowers with various parts, including sepals, petals, stamens, and carpels. Sometimes the term is used more broadly, to include both angiosperms and ◊gymnosperms, in which case the ◊cones of conifers and cycads are

system flowchart

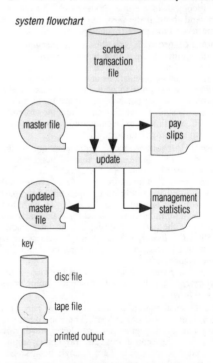

key

disc file

tape file

printed output

flow chart *A system flow chart describes a complete data-processing system, such as a payroll system. The symbol used for each data operation often resembles the medium used for that operation—for example, magnetic tape, hard discs (stacked in a disc drive), or paper.*

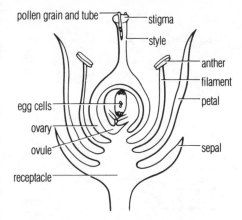

pollen grain and tube — stigma
— style
— anther
— filament
— petal
egg cells —
ovary —
ovule —
— sepal
receptacle —

flower *Cross-section of a typical flower showing its basic components: sepals, petals, stamens (anthers and filaments), and carpel (ovary and stigma). Flowers vary greatly in the size, shape, colour, and arrangement of these components.*

referred to as 'flowers'. Usually, however, the angiosperms and gymnosperms are referred to collectively as ◊seed plants, or spermatophytes.

flue-gas desulphurization process of removing harmful sulphur pollution from gases emerging from a boiler. Sulphur compounds such as sulphur dioxide are commonly produced by burning ◊fossil fuels, especially coal in power stations, and are the main cause of ◊acid rain.

fluid any substance, either liquid or gas, in which the molecules are relatively mobile and can 'flow'.

fluidization making a mass of solid particles act as a fluid by agitation or by gas passing through. Much earthquake damage is attributed to fluidization of surface soils during the earthquake shock.

fluid mechanics the study of the behaviour of fluids (liquids and gases) at rest and in motion. Fluid mechanics is important in the study of the weather, the design of aircraft and road vehicles, and in industries, such as the chemical industry, which deal with flowing liquids or gases.

fluid, supercritical fluid brought by a combination of heat and pressure to the point at which, as a near vapour, it combines the properties of a gas and a liquid. Supercritical fluids are used as solvents in chemical processes, such as the extraction of lubricating oil from refinery residues or the decaffeination of coffee, because they avoid the energy-expensive need for phase changes (from liquid to gas and back again) required in conventional distillation processes.

fluorescence in scientific usage, very short-lived ◊luminescence (a glow not caused by high temperature). Generally, the term is used for any luminescence regardless of the persistence. See ◊phosphorescence.

Fluorescence is used in strip and other lighting, and was developed rapidly during World War II because it was a more efficient means of illumination than the incandescent lamp. Recently, small bulb-size fluorescence lamps have reached the market. It is claimed that, if widely used, their greater efficiency could reduce demand for electricity. Other important applications are in fluorescent screens for television and cathode-ray tubes.

fluorescence microscopy technique for examining samples under a ◊microscope without slicing them into thin sections. Instead, fluorescent dyes are introduced into the tissue and used as a light source for imaging purposes. Fluorescent dyes can also be bonded to monoclonal antibodies and used to highlight areas where particular cell proteins occur.

fluoridation addition of small amounts of fluoride salts to drinking water by certain water authorities to help prevent tooth decay. Experiments in Britain, the USA, and elsewhere have indicated that a concentration of fluoride of 1 part per million in tap water retards the decay of teeth in children by more than 50%.

The recommended policy in Britain is to add sodium fluoride to the water to bring it up to the required amount, but implementation is up to each local authority.

fluoride negative ion (Fl⁻) formed when hydrogen fluoride dissolves in water; compound formed between fluorine and another element in which the fluorine is the more electronegative element (see ◊electronegativity, ◊halide).

In parts of India, the natural level of fluoride in water is 10 parts per million. This causes fluorosis, or chronic fluoride poisoning, mottling teeth and deforming bones.

fluorine pale yellow, gaseous, nonmetallic element, symbol F, atomic number 9, relative atomic mass 19. It is the first member of the halogen group of elements, and is pungent, poisonous, and highly reactive, uniting directly with nearly all the elements. It occurs naturally as the minerals fluorite (CaF_2) and cryolite (Na_3AlF_6). Hydrogen fluoride is used in etching glass, and the freons, which all contain fluorine, are widely used as refrigerants.

Fluorine was discovered by the Swedish chemist Karl Scheele in 1771 and isolated by the French chemist Henri Moissan in 1886. Combined with uranium as UF_6, it is used in the separation of uranium isotopes.

fluorite or *fluorspar* a glassy, brittle mineral, calcium fluoride CaF_2, forming cubes and octahedra; colourless when pure, otherwise violet or green.

Fluorite is used as a flux in iron and steel making; colourless fluorite is used in the manufacture of microscope lenses. It is also used for the glaze on pottery, and as a source of fluorine in the manufacture of hydrofluoric acid.

Deposits of fluorite occur in the N and S Pennines; the *blue john* from Derbyshire is a banded variety used as a decorative stone.

fluorocarbon compound formed by replacing the hydrogen atoms of a hydrocarbon with fluorine. Fluorocarbons are used as inert coatings, refrigerants, synthetic resins, and as propellants in aerosols.

There is concern that the release of fluorocarbons—particularly those containing chlorine (chlorofluorocarbons, CFCs)—depletes the ◊ozone layer, allowing more ultraviolet light from the Sun to penetrate the Earth's atmosphere, and increasing the incidence of skin cancer in humans.

fluvioglacial of a process or landform, associated with glacial meltwater. Meltwater, flowing beneath or ahead of a glacier, is capable of transporting rocky material and creating a variety of landscape features, including eskers, kames, and outwash plains.

flux in smelting, a substance that combines with the unwanted components of the ore to produce a fusible slag, which can be separated from the molten metal. For example, the mineral fluorite, CaF_2, is used as a flux in iron smelting; it has a low melting point and will form a fusible mixture with substances of higher melting point such as silicates and oxides.

flux in soldering, a substance that improves the bonding properties of solder by removing contamination from metal surfaces and preventing their oxidation, and by reducing the surface tension of the molten solder alloy. For example, with solder made of lead-tin alloys, the flux may be resin, borax, or zinc chloride.

flywheel heavy wheel in an engine that helps keep it running and smooths its motion. The ◊crankshaft in a petrol engine has a flywheel at one end, which keeps the crankshaft turning in between the intermittent power strokes of the pistons. It also comes into contact with the ◊clutch, serving as the connection between the engine and the car's transmission system.

FM in physics, abbreviation for ◊frequency modulation.

f-number measure of the relative aperture of a telescope or camera lens; it indicates the light-gathering power of the lens. In photography, each successive f-number represents a halving of exposure speed.

focal length or *focal distance* the distance from the centre of a lens or curved mirror to the focal point. For a concave mirror or convex lens, it is the distance at which parallel rays of light are brought to a focus to form a real image (for a mirror, this is half the radius of curvature). For a convex mirror or concave lens, it is the distance from the centre to the point at which a virtual image (an image produced by diverging rays of light) is formed.

With lenses, the greater the power (measured in dioptres) of the lens the shorter its focal length.

focus or *focal point* in optics, the point at which light rays converge, or from which they appear to diverge, to form a sharp image. Other electromagnetic rays, such as microwaves, and sound waves may also be brought together at a focus. Rays parallel to the principal axis of a lens or mirror are converged at, or appear to diverge from, the ◊principal focus.

focus in photography, the distance that a lens must be moved in order to focus a sharp image on the light-sensitive film at the back of the camera. The lens is moved away from the film to focus the image of closer objects. The focusing distance is often marked on a scale around the lens; however, some cameras now have an automatic focusing (auto-

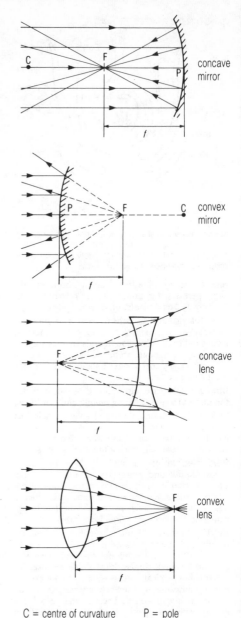

C = centre of curvature   P = pole
F = focus   f = focal length

focal length *The distance from the pole (P), or optical centre, of a lens or spherical mirror to its principal focus (F). The focal length of a spherical mirror is equal to half the radius of curvature (f = CP/2). The focal length of a lens is inversely proportional to the power of that lens (the greater the power the shorter the focal length).*

focus) mechanism that uses an electric motor to move the lens.

fog cloud that collects at the surface of the Earth, composed of water vapour that has condensed on particles of dust in the atmosphere. Cloud and fog

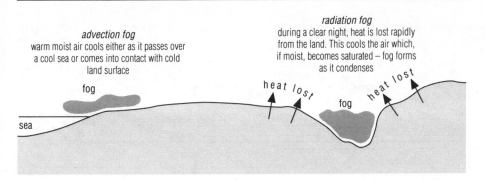

advection fog
warm moist air cools either as it passes over a cool sea or comes into contact with cold land surface

radiation fog
during a clear night, heat is lost rapidly from the land. This cools the air which, if moist, becomes saturated – fog forms as it condenses

fog

fog

are both caused by the air temperature falling below ◊dew point. The thickness of fog depends on the number of water particles it contains.

There are two types of fog. An *advection fog* is formed by the meeting of two currents of air, one cooler than the other, or by warm air flowing over a cold surface. Sea fogs commonly occur where warm and cold currents meet and the air above them mixes. A *radiation fog* forms on clear, calm nights when the land surface loses heat rapidly (by radiation); the air above is cooled to below its dew point and condensation takes place.

In drought areas, for example, Baja California, Canary Islands, Cape Verde islands, Namib Desert, Peru, and Chile, coastal fogs enable plant and animal life to survive without rain and are a potential source of water for human use (by means of water collectors exploiting the effect of condensation).

Officially, fog refers to a condition when visibility is reduced to 1 km or less, and mist or haze to that giving a visibility of 1–2 km. A mist is produced by condensed water particles, and a haze by smoke or dust. Industrial areas uncontrolled by pollution laws have a continual haze of smoke over them, and if the temperature falls suddenly, a dense yellow smog forms. At some airports since 1975 it has been possible for certain aircraft to land and take off blind in fog, using radar navigation.

föhn or *foehn* warm dry wind that blows down the leeward slopes of mountains.

The air heats up as it descends because of the increase in pressure, and it is dry because all the moisture was dropped on the windward side of the mountain. In the valleys of Switzerland it is regarded as a health hazard, producing migraine and high blood pressure. A similar wind, the chinook, is found on the eastern slopes of the Rocky Mountains in North America.

fold in geology, a bend in ◊beds or layers of rock. If the bend is arched in the middle it is called an *anticline*; if it sags downwards in the middle it is called a *syncline*. The line along which a bed of rock folds is called its axis. The axial plane is the plane joining the axes of successive beds.

folic acid a ◊vitamin of the B complex. It is found in liver and green leafy vegetables, and is also synthesized by the intestinal bacteria. It is essential for growth, and plays many other roles in the body.

Lack of folic acid causes anaemia because it is necessary for the synthesis of nucleic acids and the formation of red blood cells.

follicle in botany, a dry, usually many-seeded fruit that splits along one side only to release the seeds within. It is derived from a single ◊carpel, examples include the fruits of the larkspurs *Delphinium* and columbine *Aquilegia*. It differs from a pod, which always splits open (dehisces) along both sides.

follicle in zoology, a small group of cells that surround and nourish a structure such as a hair (hair follicle) or a cell such as an egg (Graafian follicle; see ◊menstrual cycle).

follicle-stimulating hormone (FSH) a ◊hormone produced by the pituitary gland. It affects the ovaries in women, triggering off the production of an egg cell. Luteinizing hormone is needed to complete the process. In men, FSH stimulates the testes to produce sperm.

Fomalhaut or *Alpha Piscis Austrini* the brightest star in the southern constellation Piscis Austrinus and the 18th brightest star in the sky. It is 22 light years from Earth, with a true luminosity 13 times that of the Sun.

Fomalhaut is one of a number of stars around which ◊IRAS (the Infra-Red Astronomy Satellite) detected excess infrared radiation, presumably from a region of solid particles around the star. This material may be a planetary system in the process of formation.

font or *fount* complete set of printed or display characters of the same typeface, size, and style (bold, italic, underlined, and so on). In the UK, font sizes are measured in points, a point being approximately 0.3 mm.

Fonts used in computer setting are of two main types: bit-mapped and outline. *Bit-mapped fonts* are stored in the computer memory as the exact arrangement of ◊pixels or printed dots required to produce the characters in a particular size on a screen or printer. *Outline fonts* are stored in the computer memory as a set of instructions for drawing the circles, straight lines, and curves that make up the outline of each character. They require a powerful computer because each character is separately generated from a set of instructions and this requires considerable computation.

Bit-mapped fonts become very ragged in appearance if they are enlarged and so a separate set of

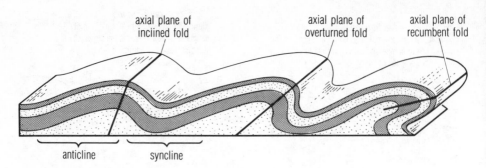

axial plane of inclined fold

axial plane of overturned fold

axial plane of recumbent fold

anticline   syncline

fold *The folding of rock strata occurs where compression causes them to buckle. Over time, folding can assume highly complicated forms, as can sometimes be seen in the rock layers of cliff faces or deep cuttings in the rock. Folding contributed to the formation of great mountain chains such as the Himalayas.*

bit maps is required for each font size. In contrast, outline fonts can be scaled to any size and still maintain exactly the same appearance.

food anything eaten by human beings and other animals to sustain life and health. The building blocks of food are nutrients, and humans can utilize the following nutrients: *carbohydrates*, as starches found in bread, potatoes, and pasta; as simple sugars in sucrose and honey; as fibres in cereals, fruit, and vegetables; *proteins* as from nuts, fish, meat, eggs, milk, and some vegetables; *fats* as found in most animal products (meat, lard, dairy products, fish), also in margarine, nuts and seeds, olives, and edible oils; *vitamins* are found in a wide variety of foods, except for vitamin B_{12}, which is mainly found in animal foods; *minerals* are found in a wide variety of foods; good sources of calcium are milk and broccoli, for example; iodine from seafood; iron from liver and green vegetables; *water* ubiquitous in nature; *alcohol* is found in fermented distilled beverages, from 40% in spirits to 0.01% in low-alcohol lagers and beers.

Food is needed both for energy, measured in calories or kilojoules, and nutrients, which are converted to body tissues. Some nutrients, such as fat, carbohydrate, and alcohol, provide mainly energy; other nutrients are important in other ways; for example, fibre is an aid to metabolism. Proteins provide energy and are necessary for building cell and tissue structure.

food chain in ecology, a sequence showing the feeding relationships between organisms in a particular ◊ecosystem. Each organism depends on the next lowest member of the chain for its food.

Energy in the form of food is shown to be transferred from ◊autotrophs, or producers, which are principally plants and photosynthetic microorganisms to a series of ◊heterotrophs, or consumers. The heterotrophs comprise the ◊herbivores, which feed on the producers; ◊carnivores, which feed on the herbivores; and ◊decomposers, which break down the dead bodies and waste products of all four groups (including their own), ready for recycling.

In reality, however, organisms have varied diets, relying on different kinds of foods, so that the food chain is an over-simplification. The more complex *food web* shows a greater variety of relationships,

but again emphasises that energy passes from plants to herbivores to carnivores.

Environmental groups have used the concept of the food chain to show how poisons and other forms of pollution can pass from one animal to another, eventually resulting in the death of rare animals such as the golden eagle *Aquila chrysaetos.*

food irradiation the exposure of food to low-level ◊irradiation to kill microorganisms; a technique used in ◊food technology. Irradiation is highly effective, and does not make the food any more radioactive than it is naturally. Irradiated food is used for astronauts and immunocompromised patients in hospitals. Some vitamins are partially destroyed, such as vitamin C, and it would be unwise to eat only irradiated fruit and vegetables.

Other damaging changes may take place in the food, such as the creation of ◊free radicals, but research so far suggests that the process is relatively safe.

food technology the application of science to the commercial processing of foodstuffs. Food is processed to make it more palatable or digestible, for which the traditional methods include boiling, frying, flour-milling, bread-making, yoghurt-and cheese-making, brewing, or to preserve it from spoilage caused by the action of ◊enzymes within the food that change its chemical composition, or the growth of bacteria, moulds, yeasts, and other microorganisms. Fatty or oily foods also suffer oxidation of the fats, which makes them rancid. Traditional forms of *food preservation*, include salting, smoking, pickling, drying, bottling, and preserving in sugar. Modern food technology also uses many novel processes and ◊additives, which allow a wider range of foodstuffs to be preserved.

Refrigeration below 5°C/41°F (or below 3°C/37°F for cooked foods) slows the processes of spoilage, but is less effective for foods with a high water content. This process cannot kill microorganisms, nor stop their growth completely, and a failure to realize its limitations causes many cases of food poisoning. Refrigerator temperatures should be checked as the efficiency of the machinery (see ◊refrigeration) can decline with age, and higher temperatures are dangerous.

Deep freezing (–18°C/–1°F or below) stops almost all spoilage processes, except residual

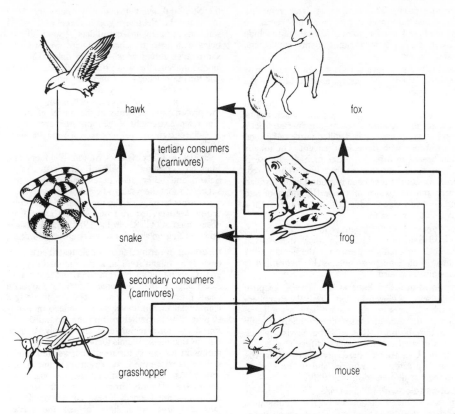

food chain *The complex interrelationships between animals and plants in a food chain. Food chains are normally only three or four links long. This is because most of the energy at each link is lost in respiration, and so cannot be passed on to the next link.*

enzyme activity in uncooked vegetables, which are blanched (dipped in hot water to destroy the enzymes) before freezing. Microorganisms cannot grow or divide while frozen, but most remain alive and can resume activity once defrosted. Some foods are damaged by freezing, notably soft fruits and salad vegetables, the cells of which are punctured by ice crystals, leading to loss of crispness. Fatty foods such as cow's milk and cream tend to separate. Various processes are used for ◊deep freezing foods commercially.

Pasteurization is used mainly for milk. By holding the milk at 72°C/161.6°F for 15 seconds, all disease-causing bacteria can be destroyed. Less harmful bacteria survive, so the milk will still go sour within a few days.

Ultra-heat treatment is used to produce UHT milk. This process uses higher temperatures than pasteurization, and kills all bacteria present, giving the milk a long shelf life but altering the flavour.

Drying is effective because both microorganisms and enzymes need water to be active. Products such as dried milk and instant coffee are made by spraying the liquid into a rising column of dry, heated air.

Freeze-drying is carried out under vacuum. It is less damaging to food than straight dehydration in the sense that foods reconstitute better, and is used for quality instant coffee and dried vegetables.

Canning relies on high temperatures to destroy microorganisms and enzymes. The food is sealed into a can to prevent recontamination. Drinks may be canned to preserve the carbon dioxide that makes them fizzy.

Pickling utilizes the effect of acetic acid, found in vinegar, in stopping the growth of moulds. In sauerkraut, lactic acid, produced by bacteria, has the same effect. Similar types of non-harmful, acid-generating bacteria are used to make yoghurt and cheese.

Curing of meat involves soaking in salt (sodium chloride) solution, with saltpetre (sodium nitrate) added to give the meat its pink colour and characteristic taste. Bacteria convert the nitrates in cured meats to nitrites and nitrosamines, which are potentially carcinogenic to humans.

Irradiation is a method of preserving food by subjecting it to low-level radiation (see ◊food irradiation).

Puffing is a method of processing cereal grains. They are subjected to high pressures, then suddenly ejected into a normal atmospheric pressure, causing the grain to expand sharply. This is used to make puffed wheat cereals and puffed rice cakes.

Chemical treatments are widely used, for example in margarine manufacture, in which hydrogen is bubbled through vegetable oils in the presence of a ◊catalyst to produce a more solid, spreadable fat. The catalyst is later removed. Chemicals introduced in processing that remain in the food are known as *food additives* and include flavourings, preservatives, antioxidants, emulsifiers, and colourings.

food test any of several types of simple test, easily performed in the laboratory, used to identify the main classes of food.

starch–iodine test Food is ground up in distilled water and iodine is added. A dense black colour indicates that starch is present. *sugar–Benedict's test* Food is ground up in distilled water and placed in a test tube with Benedict's reagent. The tube is then heated in a boiling water bath. If glucose is present the colour changes from blue to brick-red. *protein–Biuret test* Food is ground up in distilled water and a mixture of copper(II) sulphate and sodium hydroxide is added. If protein is present a mauve colour is seen.

foot imperial unit of length (symbol ft), equivalent to 0.3048 m, in use in Britain since Anglo-Saxon times. It originally represented the length of a human foot. One foot contains 12 inches and is one-third of a yard.

foot-and-mouth disease contagious eruptive viral disease of cloven-hoofed mammals, characterized by blisters in the mouth and around the hooves. In cattle it causes deterioration of milk yield and abortions. It is an airborne virus which makes its eradication extremely difficult.

In the UK, affected herds are destroyed; inoculation is practised in Europe, and in the USA, a vaccine was developed in the 1980s.

foot-candle unit of illuminance, now replaced by the lux. One foot-candle is the illumination received at a distance of one foot from an international candle. It is equal to 10.764 lux.

foot-pound imperial unit of energy (ft-lb), defined as the work done when a force of one pound moves through a distance of one foot. It has been superseded for scientific work by the joule: one foot-pound equals 1.356 joule.

force any influence that tends to change the state of rest or the uniform motion in a straight line of a body. The action of an unbalanced or resultant force results in the acceleration of a body in the direction of action of the force or it may, if the body is unable to move freely, result in its deformation (see ◊Hooke's law). Force is a vector quantity, possessing both magnitude and direction; its SI unit is the newton.

According to Newton's second law of motion the magnitude of a resultant force is equal to the rate of change of ◊momentum of the body on which it acts; the force F producing an acceleration a m s^{-2} on a body of mass m kilogrammes is therefore given by:

$$F = ma$$

See also ◊Newton's laws of motion.

force ratio the magnification of a force by a machine; see ◊mechanical advantage.

forces, fundamental in physics, the four fundamental interactions believed to be at work in the physical universe. There are two long-range forces: *gravity*, which keeps the planets in orbit around the Sun, and acts between all particles that have mass; and the *electromagnetic force*, which stops solids from falling apart, and acts between all particles with ◊electric charge. There are two very short-range forces which operate only inside the atomic nucleus: the *weak nuclear force*, responsible for the reactions that fuel the Sun and for the emission of ◊beta particles from certain nuclei; and the *strong nuclear force*, which binds together the protons and neutrons in the nuclei of atoms. The relative strengths of the four forces are: strong, 1; electromagnetic, 10^{-2}; weak, 10^{-6}; gravitational, 10^{-40}.

By 1971, US physicists Steven Weinberg and Sheldon Glashow, Pakistani physicist Abdus Salam, and others had developed a theory that suggested that the weak and electromagnetic forces were aspects of a single force called the *electroweak force*; experimental support came from observation at ◊CERN in the 1980s. Physicists are now working on theories to unify all four forces.

foredeep in earth science, an elongated structural basin lying inland from an active mountain system and receiving sediment from the rising mountains. According to plate tectonic theory, a mountain chain forming behind a subduction zone along a continental margin develops a foredeep or gently sloping trough parallel to it on the landward side. Foredeeps form rapidly and are usually so deep initially that the sea floods them through gaps in the mountain range. As the mountain system evolves, sediments choke the foredeep, pushing out marine water. As marine sedimentation stops, only nonmarine deposits from the rapidly eroding mountains are formed. These consist of alluvial fans and also rivers, flood plains, and related environments inland.

Before the advent of plate tectonic theory, such foredeep deposits and changes in sediments had been interpreted as sedimentary troughs, called geosynclines, that were supposed ultimately to build upward into mountains.

forensic science the use of scientific techniques to solve criminal cases. A multidisciplinary field embracing chemistry, physics, botany, zoology, and medicine, forensic science includes the identification of human bodies or traces. Traditional methods such as fingerprinting are still used, assisted by computers; in addition, blood analysis, forensic dentistry, voice and speech spectrograms, and ◊genetic fingerprinting are increasingly applied. Chemicals, such as poisons and drugs, are analysed by ◊chromatography. Ballistics (the study of projectiles, such as bullets), another traditional forensic field, makes use of tools such as the comparison microscope and the ◊electron microscope.

The first forensic laboratory was supposedly founded in Lyons, France in 1910 by Edmond Locard, although it is claimed that Locard's teacher, Alphonse Bertillon, had established one earlier, and that the laboratory in Lyons was founded by Jean Lacassagne. The science developed as a systematic discipline in the 1930s. In 1932 the US Federal Bureau of Investigation established a forensic science laboratory in Washington, DC, and in the UK the first such laboratory was founded in London in 1935.

How science is helping Poirot with his enquiries

The forensic scientist has to provide evidence that will stand scrutiny in a court of law. This is why crime laboratories need to keep up with the latest research and maintain the highest standards.

Every contact leaves a trace. The ordinary microscope is still a vital instrument for examining trace evidence—hairs, fibres, fragments of glass or paint—but the scanning electron microscope is also used. It provides high magnification with good resolution, and can also incorporate a microprobe that identifies the actual elements, particularly metallic ones, in the surface being examined. Surface elements absorb electrons and emit X-rays, which the microprobe converts into an X-ray emission spectrum whose characteristic pattern reveal the elements present. In this way, it is possible to detect and identify particles invisible to the optical microscope, such as those scattered from a firearm when discharged. These particles can indicate the type and make of ammunition used.

Anti-crime antibodies

Advances have also been made in analytical techniques called immunoassays, which use antibodies to detect and measure drugs, poisons, proteins and even explosives such as TNT and Semtex. When a foreign chemical, such as a disease organism, enters the human body, antibodies are produced which recognize and react with the foreign substance. The same process occurs in animals, which can be used to produce antibodies against a wide variety of chemicals. An animal is injected with a target substance—cocaine, for example—and the resulting antibodies can be separated and used to recognize the substance against a background of body fluids, or in a body swab.

When trying to detect a target compound, such as the presence of explosive residue on a person's hands, it is vital to take account of possible contamination, since the substance might have been picked up casually. Also, it is important that the method used detects the target compound and no other. The value of antibodies is that they are specific to the compound that triggered their production.

DNA fingerprinting

One impressive recent scientific advance is DNA profiling, or DNA fingerprinting. This has been used effectively in assault, rape and murder cases, and paternity disputes. It involves using enzymes to cut up a sample of DNA extracted from body cells. The resulting bits are separated by gel electrophoresis, blotted onto a special membrane, marked radioactively and then visualized as a sequence of bars. Only a very small amount of DNA—from just a few cells—is needed; the polymerase chain reaction (PCR) can amplify it into sufficient material for a profile.

DNA profiling must be treated with caution. If two profiles match, they may not necessarily come from the same person. If two bands on adjacent profiles correspond, but not exactly, statistical analysis may be needed to decide if there is a match. However, there can still be doubts. For example, the DNA profiles of people from small communities with significant inbreeding can be similar.

ESDA (electrostatic deposition analysis) is a recent technique used for revealing indentations on paper. Left on an underlying sheet, these indicate what has been written on the paper above. The method uses a high electrostatic voltage to transfer the indentations onto imaging film where they are visualized by photographic toner. If the resulting impressions are markedly uneven—perhaps one half is more heavily indented than the other—then the writing under examination (on the top sheet) may have been written at two different times, showing that the document has been tampered with.

Computers on the beat

The increasing power of computers has been responsible for great advances in forensic science. Computers have revolutionized the storage and retrieval of information. For example, all car registration numbers and owners' names are stored on computer for almost instant access. Information can be rapidly communicated to police officers in the field. This has increased the power of those who hold the information and, in addition to speeding the response to crime, has helped in the monitoring of 'undesirables' and in maintaining order on the streets.

Identification of a suspect fingerprint by comparison with thousands stored on file was once a time-consuming process. Nowadays, with a computer, the process takes minutes, even seconds. Even so, the final decision on fingerprint identification is still made visually by a trained expert. No matter how sophisticated the hardware, the human senses are still vital in forensic work.

John Broad

forest an area where trees have grown naturally for centuries, instead of being logged at maturity (about 150–200 years). A natural, or old-growth, forest has a nultistorey canopy and includes young and very old trees (this gives the canopy its range of heights). There are also fallen trees contributing to the very complex ecosystem, which may support over 150 species of mammals and many thousands of species of insects.

The Pacific forest of the west coast of North America is one of the few remaining old-growth forests in the temperate zone. It consists mainly of conifers and is threatened by logging—less than 10% of the original forest remains.

forestry the science of forest management. Recommended forestry practice aims at multipurpose crops, allowing the preservation of varied plant and animal species as well as human uses (lumbering, recreation). Forestry has often been confined to the planting of a single species, such as a rapid-growing conifer providing softwood for paper pulp and construction timber, for which world demand is greatest. In tropical countries, logging contributes to the destruction of ◊rainforests, causing global environmental problems. Small unplanned forests are ◊woodland.

The earliest planned forest dates from 1368 at Nuremberg, Germany; in Britain, planning of forests began in the 16th century. In the UK, Japan, and other countries, forestry practices have been criticized for concentration on softwood conifers to the neglect of native hardwoods.

forging one of the main methods of shaping metals, which involves hammering or a more gradual application of pressure. A blacksmith hammers red-hot metal into shape on an anvil, and the traditional place of work is called a forge. The blacksmith's mechanical equivalent is the drop forge. The metal is shaped by the blows from a falling hammer or ram, which is usually accelerated by steam or air pressure. Hydraulic presses forge by applying pressure gradually in a squeezing action.

formaldehyde common name for ◊methanal.

formalin aqueous solution of formaldehyde (methanal) used to preserve animal specimens.

formatting in computing, short for ◊disc formatting.

Formica trademark of the Formica Corporation for a heat-proof plastic laminate, widely used as a veneer on wipe-down kitchen surfaces and children's furniture. It is made from formaldehyde resins similar to ◊Bakelite. It was first put on the market 1913.

formic acid common name for ◊methanoic acid.

formula in chemistry, a representation of a molecule, radical, or ion, in which the component chemical elements are represented by their symbols. An *empirical formula* indicates the simplest ratio of the elements in a compound, without indicating how many of them there are or how they are combined. A *molecular formula* gives the number of each type of element present in one molecule. A *structural formula* shows the relative positions of the atoms and the bonds between them. For example, for ethanoic acid, the empirical formula is CH_2O, the molecular formula is $C_2H_4O_2$, and the structural formula is CH_3COOH.

formula in mathematics, a set of symbols and numbers that expresses a fact or rule. $A = \pi r^2$ is the formula for calculating the area of a circle. Einstein's famous formula relating energy and mass is $E = mc^2$.

FORTRAN (acronym for *fo*rmula *tran*slation) high-level computer-programming language suited to mathematical and scientific computations. Developed 1956, it is one of the earliest languages still in use. A recent version, Fortran 90, is now being used on advanced parallel computers. ◊BASIC was strongly influenced by FORTRAN and is similar in many ways.

fossil (Latin *fossilis* 'dug up') remains of an animal or plant preserved in rocks. Fossils may be formed by refrigeration (for example, Arctic ◊mammoths in ice); carbonization (leaves in coal); formation of a cast (dinosaur or human footprints in mud); or mineralization of bones, more generally teeth or shells. The study of fossils is called ◊palaeontology.

About 250,000 fossil species have been discovered—a figure that is believed to represents less than 1 in 20,000 of the species that ever lived.

THE COMMONEST FOSSILS

The most common fossils are brachiopods, or lamp-shells, which resemble bivalved (two-shelled) molluscs. At least 30,000 different species have existed during the past; there are about 300 species living today.

fossil fuel fuel, such as coal, oil, and natural gas, formed from the fossilized remains of plants that lived hundreds of millions of years ago. Fossil fuels are a ◊nonrenewable resource and will eventually run out. Extraction of coal and oil causes considerable environmental pollution, and burning coal contributes to problems of ◊acid rain and the ◊greenhouse effect.

four-colour process colour ◊printing using four printing plates, based on the principle that any colour is made up of differing proportions of the primary colours blue, red, and green. The first stage in preparing a colour picture for printing is to produce separate films, one each for the blue, red, and green respectively in the picture (colour separations). From these separations three printing plates are made, with a fourth plate for black (for shading or outlines). Ink colours complementary to those represented on the plates are used for printing—yellow for the blue plate, cyan for the red, and magenta for the green.

Fourdrinier machine papermaking machine patented by the Fourdrinier brothers Henry and Sealy in England 1803. On the machine, liquid pulp flows onto a moving wire-mesh belt, and water drains and is sucked away, leaving a damp paper web. This is passed first through a series of steam-heated rollers, which dry it, and then between heavy calendar rollers, which give it a smooth finish. Such machines can measure up to 90 m/300 ft in length, and are still in use.

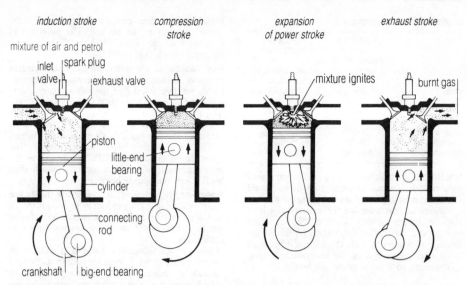

induction stroke compression stroke expansion of power stroke exhaust stroke

mixture of air and petrol
inlet | spark plug
valve | exhaust valve
piston
little-end bearing
cylinder
connecting rod
crankshaft | big-end bearing
mixture ignites
burnt gas

four-stroke cycle *The four-stroke cycle of a modern petrol engine. The cycle is also called the Otto cycle after German engineer Nikolaus Otto, who introduced it in 1876. It improved on earlier engine cycles by compressing the fuel mixture before it was ignited.*

four-stroke cycle the engine-operating cycle of most petrol and ◊diesel engines. The 'stroke' is an upward or downward movement of a piston in a cylinder. In a petrol engine the cycle begins with the induction of a fuel mixture as the piston goes down on its first stroke. On the second stroke (up) the piston compresses the mixture in the top of the cylinder. An electric spark then ignites the mixture, and the gases produced force the piston down on its third, power, stroke. On the fourth stroke (up) the piston expels the burned gases from the cylinder into the exhaust.

The four-stroke cycle is also called the *Otto cycle*. The diesel engine cycle works in a slightly different way to that of the petrol engine on the first two strokes.

fourth-generation language in computing, a type of programming language designed for the rapid programming of applications (see ◊applications program) but often lacking the ability to control the individual parts of the computer. Such a language typically provides easy ways of designing screens and reports, and of using databases. Other 'generations' (the term implies a class of language rather than a chronological sequence) are ◊machine code (first generation); ◊assembly languages, or low-level languages (second); and conventional high-level languages such as ◊BASIC and ◊PASCAL (third).

f.p.s. system system of units based on the foot, pound, and second as units of length, mass, and time, respectively. It has now been replaced for scientific work by the ◊SI system.

fractal (from Latin *fractus* 'broken') an irregular shape or surface produced by a procedure of repeated subdivision. Generated on a computer screen, fractals are used in creating models for geographical or biological processes (for example, the creation of a coastline by erosion or accretion, or the growth of plants).

Sets of curves with such discordant properties were developed in Germany by Georg Cantor (1845–1918) and Karl Weierstrass (1815–1897). The name was coined by the French mathematician Benoit Mandelbrod. Fractals are also used for computer art.

fraction (from Latin *fractus* 'broken') in mathematics, a number that indicates one or more equal parts of a whole. Usually, the number of equal parts into which the unit is divided (denominator) is written below a horizontal line, and the number of parts comprising the fraction (numerator) is written above; thus $\frac{2}{3}$ or $\frac{3}{4}$. Such fractions are called *vulgar* or *simple* fractions. The denominator can never be zero.

A *proper fraction* is one in which the numerator is less than the denominator. An *improper fraction* has a numerator that is larger than the denominator, for example $\frac{3}{2}$. It can therefore be expressed as a mixed number, for example, $1\frac{1}{2}$. A combination such as $\frac{5}{0}$ is not regarded as a fraction (an object cannot be divided into zero equal parts), and mathematically any number divided by 0 is equal to infinity. A *decimal fraction* has as its denominator a power of 10, and these are omitted by use of the decimal point and notation, for example 0.04, which is $\frac{4}{100}$. The digits to the right of the decimal point indicate the numerators of vulgar fractions whose denominators are 10, 100, 1,000, and so on. Most fractions can be expressed exactly as decimal fractions ($\frac{1}{3} = 0.333...$). Fractions are also known as the *rational numbers*, that is numbers formed by a ratio. *Integers* may be expressed as fractions with a denominator of 1.

fraction in chemistry, a group of similar compounds, the boiling points of which fall within a particular range and which are separated during fractional ◊distillation (fractionation).

crude petroleum fractions

fraction	approximate number of carbon atoms in hydrocarbon	approximate boiling range (°C) at atmospheric pressure
gases	1–4	below 25
petrol	4–12	40–100
naptha	7–14	90–150
kerosene	9–16	150–240
diesel oil	15–25	220–250
lubricating oil	20–70	250–350
bitumen residue	over 70	above 350

fractionating column device in which many separate ◊distillations can occur so that a liquid mixture can be separated into its components.

Various designs exist but the primary aim is to allow maximum contact between the hot rising vapours and the cooling descending liquid. As the mixture of vapours ascends the column it becomes progressively enriched in the lower-boiling-point components, so these separate out first.

laboratory apparatus for fractional distillation

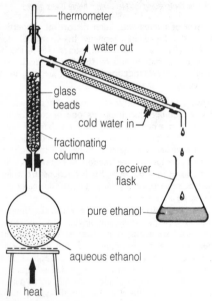

- thermometer
- water out
- glass beads
- cold water in
- fractionating column
- receiver flask
- pure ethanol
- aqueous ethanol
- heat

fractionating column

fractionation or *fractional distillation* process used to split complex mixtures (such as crude oil) into their components, usually by repeated heating, boiling, and condensation; see ◊distillation.

francium radioactive metallic element, symbol Fr, atomic number 87, relative atomic mass 223. It is one of the ◊alkali metals and occurs in nature in small amounts as a decay product of actinium. Its longest-lived isotope has a half-life of only 21 minutes. Francium was discovered and named in 1939 by Marguérite Perey to honour her country.

Frasch process process used to extract underground deposits of sulphur. Superheated steam is piped into the sulphur deposit and melts it. Com-pressed air is then pumped down to force the molten sulphur to the surface. The process was developed in the USA 1891 by German-born Herman Frasch (1851–1914).

free radical in chemistry, an atom or molecule that has an unpaired electron and is therefore highly reactive. Most free radicals are very short-lived. If free radicals are produced in living organisms they can be very damaging.

Free radicals are often produced by high temperatures and are found in flames and explosions. The action of ultraviolet radiation from the Sun splits chlorofluorocarbon (CFC) molecules in the upper atmosphere into free radicals, which then break down the ◊ozone layer.

freeze-drying method of preserving food; see ◊food technology. The product to be dried is frozen and then put in a vacuum chamber that forces out the ice as water vapour, a process known as sublimation.

Many of the substances that give products such as coffee their typical flavour are volatile, and would be lost in a normal drying process because they would evaporate along with the water. In the freeze-drying process these volatile compounds do not pass into the ice that is to be sublimed, and are therefore largely retained.

freeze-thaw form of physical ◊weathering, common in mountains and glacial environments, caused by the expansion of water as it freezes. Water in a crack freezes and expands in volume by 9% as it turns to ice. This expansion exerts great pressure on the rock causing the crack to enlarge. After many cycles of freeze-thaw, rock fragments may break off to form ◊scree slopes.

For freeze-thaw to operate effectively the temperature must fluctuate regularly above and below 0°C/32°F. It is therefore uncommon in areas of extreme and perpetual cold, such as the polar regions.

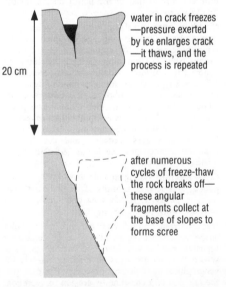

20 cm

water in crack freezes —pressure exerted by ice enlarges crack —it thaws, and the process is repeated

after numerous cycles of freeze-thaw the rock breaks off— these angular fragments collect at the base of slopes to forms scree

freeze-thaw

freezing change from liquid to solid state, as when water becomes ice. For a given substance, freezing occurs at a definite temperature, known as the *freezing point*, that is invariable under similar conditions of pressure, and the temperature remains at this point until all the liquid is frozen. The amount of heat per unit mass that has to be removed to freeze a substance is a constant for any given substance, and is known as the latent heat of fusion.

Ice is less dense than water since water expands just before its freezing point is reached. If pressure is applied, expansion is retarded and the freezing point will be lowered. The presence of dissolved substances in a liquid also lowers the freezing point (depression of ◊freezing point), the amount of lowering being proportional to the molecular concentration of the solution. Antifreeze mixtures for car radiators and the use of salt to melt ice on roads are common applications of this principle.

Animals in arctic conditions, for example insects or fish, cope with the extreme cold either by producing natural 'antifreeze' and staying active, or by allowing themselves to freeze in a controlled fashion, that is, they produce proteins to act as nuclei for the formation of ice crystals in areas that will not cause cellular damage, and so enable themselves to thaw back to life again.

freezing point, depression of lowering of a solution's freezing point below that of the pure solvent; it depends on the number of molecules of solute dissolved in it. For a single solvent, such as pure water, all solute substances in the same molar concentration produce the same lowering of freezing point. The depression d produced by the presence of a solute of molar concentration C is given by the equation $d = KC$, where K is a constant (called the cryoscopic constant) for the solvent.

Measurement of freezing-point depression is a useful method of determining the molecular weights of solutes. It is also used to detect the illicit addition of water to milk.

frequency in physics, the number of periodic oscillations, vibrations, or waves occurring per unit of time. The unit of frequency is the hertz (Hz), one hertz being equivalent to one cycle per second. Human beings can hear sounds from objects vibrating in the range 20–15,000 Hz. Ultrasonic frequencies well above 15,000 Hz can be detected by mammals such as bats. Infrasound (low frequency sound) can be detected by some animals and birds. Pigeons can detect sounds as low as 0.1 Hz; elephants communicate using sounds as low as 1 Hz.

One kilohertz (kHz) equals 1,000 hertz; one megahertz (MHz) equals 1,000,000 hertz.

frequency in statistics, the number of times an event occurs. For example, when two dice are thrown repeatedly and the two scores added together, each of the numbers 2 to 12 may have a frequency of occurrence. The set of data including the frequencies is called a *frequency distribution*, usually presented in a frequency table or shown diagramatically, by a frequency polygon.

frequency modulation (FM) method by which radio waves are altered for the transmission of

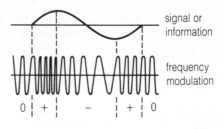

signal or information

frequency modulation

0 | + | − | + | 0

frequency modulation

broadcasting signals. FM is constant in amplitude and varies the frequency of the carrier wave in accordance with the signal being transmitted. Its advantage over AM (◊amplitude modulation) is its better signal-to-noise ratio.

friction in physics, the force that opposes the relative motion of two bodies in contact. The *coefficient of friction* is the ratio of the force required to achieve this relative motion to the force pressing the two bodies together.

Friction is greatly reduced by the use of lubricants such as oil, grease, and graphite. Air bearings are now used to minimize friction in high-speed rotational machinery. In other instances friction is deliberately increased by making the surfaces rough—for example, brake linings, driving belts, soles of shoes, and tyres.

FRICTION PUZZLE

As a car tyre becomes worn the tread disappears, and bald tyres are dangerous because of their lack of grip on the road surface. On the other hand, Formula One racing cars require exceptional grip but their tyres are normally smooth and treadless. Why is this? *See page 651 for the answer.*

Friends of the Earth (FoE or FOE) environmental pressure group, established in the UK 1971, that aims to protect the environment and to promote rational and sustainable use of the Earth's resources. It campaigns on issues such as acid rain; air, sea, river, and land pollution; recycling; disposal of toxic wastes; nuclear power and renewable energy; the destruction of rainforests; pesticides; and agriculture. FoE has branches in 30 countries.

fringing reef ◊coral reef that is attached to the coast without an intervening lagoon.

Frisch–Peierls memorandum a document revealing, for the first time, how small the critical mass (the minimum quantity of substance required for a nuclear chain reaction to begin) of uranium needed to be if the isotope uranium-235 was separated from naturally occurring uranium; the memo thus implied the feasibility of using this isotope to make an atom bomb. It was written by Otto Frisch and Rudolf Peierls (1907–) at the University of Birmingham 1940.

frond large leaf or leaflike structure; in ferns it is often pinnately divided. The term is also applied to the leaves of palms and less commonly to the

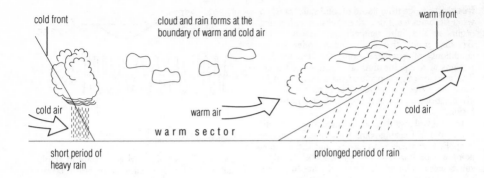

cold front

cloud and rain forms at the
boundary of warm and cold air

warm front

cold air

warm air

cold air

w a r m   s e c t o r

short period of
heavy rain

prolonged period of rain

front

plant bodies of certain seaweeds, liverworts, and lichens.

front in meteorology, the boundary between two air masses of different temperature or humidity. A *cold front* marks the line of advance of a cold air mass from below, as it displaces a warm air mass; a *warm front* marks the advance of a warm air mass as it rises up over a cold one. Frontal systems define the weather of the mid-latitudes, where warm tropical air is constantly meeting cold air from the poles.

Warm air, being lighter, tends to rise above the cold; its moisture is carried upwards and usually falls as rain or snow, hence the changeable weather conditions at fronts. Fronts are rarely stable and move with the air mass. An *occluded front* is a composite form, where a cold front overtakes a warm front, lifting warm air above the Earth's surface. An *inversion* occurs when the normal properties get reversed; this happens when a layer of air traps another near the surface, preventing the normal rising of surface air. Warm temperatures and pollution result from inversions.

front-end processor small computer used to coordinate and control the communications between a large mainframe computer and its input and output devices.

frost condition of the weather that occurs when the air temperature is below freezing, 0°C/32°F. Water in the atmosphere is deposited as ice crystals on the ground or exposed objects. As cold air is heavier than warm, ground frost is more common than hoar frost, which is formed by the condensation of water particles in the same way that ◊dew collects.

frost hollow depression or steep-sided valley in which cold air collects on calm, clear nights. Under clear skies, heat is lost rapidly from ground surfaces, causing the air above to cool and flow downhill (as ◊katabatic wind) to collect in valley bottoms. Fog may form under these conditions and, in winter, temperatures may be low enough to cause frost.

frost shattering alternative name for ◊freeze-thaw.

fructose $C_6H_{12}O_6$ a sugar that occurs naturally in honey, the nectar of flowers, and many sweet fruits; it is commercially prepared from glucose.

It is a monosaccharide, whereas the more familiar cane or beet sugar is a disaccharide, made up of two monosaccharide units: fructose and glucose. It is sweeter than cane sugar and can be used to sweeten foods for people with diabetes.

fruit (from Latin *frui* 'to enjoy') in botany, the ripened ovary in flowering plants that develops from one or more seeds or carpels and encloses one or more seeds. Its function is to protect the seeds during their development and to aid in their dispersal. Most fruits are borne by perennial plants. Fruits are often edible, sweet, juicy, and colourful; when they are eaten by animals, the seeds pass through the alimentary canal unharmed, and are passed out and dispersed in the faeces.

Fruits may be classified botanically as either *dry* (such as the ◊capsule, ◊follicle, ◊schizocarp, ◊nut, ◊caryopsis, pod or legume, ◊lomentum, and ◊achene) or *fleshy* (such as the ◊drupe and the ◊berry).

The fruit structure consists of the ◊pericarp or fruit wall, which is usually divided into a number of distinct layers. Sometimes parts other than the ovary are incorporated into the fruit structure, resulting in a false fruit or ◊pseudocarp, such as the apple and strawberry. Fruits may be dehiscent, which open to shed their seeds, or indehiscent, which remain unopened and are dispersed as a single unit.

Simple fruits (for example, peaches) are derived from a single ovary, whereas compositae or multiple fruits (for example, blackberries) are formed from the ovaries of a number of flowers.

frustule the cell wall of a ◊diatom (microscopic alga). Frustules are intricately patterned on the surface with spots, ridges, and furrows, each pattern being characteristic of a particular species.

frustum (from Latin for 'a piece cut off') in geometry, a 'slice' taken out of a solid figure by a pair of parallel planes. A conical frustum, for example, resembles a cone with the top cut off. The volume and area of a frustum are calculated by subtracting the volume or area of the 'missing' piece from those of the whole figure.

FSH abbreviation for ◊*follicle-stimulating hormone*.

f-stop in photography, another name for ◊*f-number*.

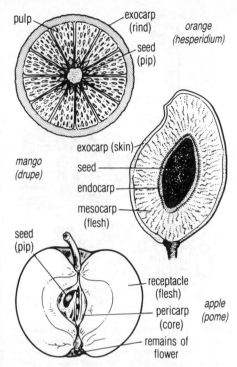

fruit *A fruit contains the seeds of a plant. Its outer wall is the exocarp, or epicarp; its inner layers are the mesocarp and endocarp. The orange is a hesperidium, a berry having a leathery rind and containing many seeds. The mango is a drupe, a fleshy fruit with a hard seed, or 'stone', at the centre. The apple is a pome, a fruit with a fleshy outer layer and a core containing the seeds.*

ft symbol for ◊*foot*, a measure of distance.

fuel any source of heat or energy, embracing the entire range of materials that burn (combustibles). A *nuclear fuel* is any material that produces energy by nuclear fission in a nuclear reactor.

fuel cell cell converting chemical energy directly to electrical energy. It works on the same principle as a battery but is continually fed with fuel, usually hydrogen. Fuel cells are silent and reliable (no moving parts) but expensive to produce.

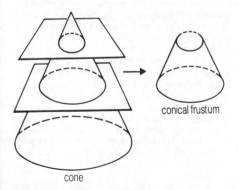

frustum *The frustum, a slice taken out of a cone.*

Hydrogen is passed over an ◊electrode (usually nickel or platinum) containing a ◊catalyst, which strips electrons off the atoms. These pass through an external circuit while hydrogen ions (charged atoms) pass through an ◊electrolyte to another electrode, over which oxygen is passed. Water is formed at this electrode (as a by-product) in a chemical reaction involving electrons, hydrogen ions, and oxygen atoms. If the spare heat also produced is used for hot water and space heating, 80% efficiency in fuel is achieved.

fuel injection injecting fuel directly into the cylinders of an internal combustion engine, instead of by way of a carburettor. It is the standard method used in ◊diesel engines, and is now becoming standard for petrol engines. In the diesel engine oil is injected into the hot compressed air at the top of the second piston stroke and explodes to drive the piston down on its power stroke. In the petrol engine, fuel is injected into the cylinder at the start of the first induction stroke of the ◊four-stroke cycle.

fullerene form of carbon, discovered 1985, based on closed cages of carbon atoms. The molecules of the most symmetrical of the fullerenes are called ◊buckminsterfullerenes. They are perfect spheres made up of 60 carbon atoms linked together in 12 pentagons and 20 hexagons fitted together like those of a spherical football. Other fullerenes with 28, 32, 50, 70, and 76 carbon atoms have also been identified.

Fullerenes can be made by arcing electricity between carbon rods. They may also occur in candle flames, and in clouds of interstellar gas. Fullerene chemistry may turn out to be as important as organic chemistry based on the benzene ring. Already, new molecules based on the ◊buckyball enclosing a metal atom, and 'buckytubes' (cylinders of carbon atoms arranged in hexagons), have been made. Applications envisaged include using the new molecules as lubricants, semiconductors, superconductors and as the starting point for making new drugs.

fuller's earth a soft, greenish-grey rock resembling clay, but without clay's plasticity. It is formed largely of clay minerals, rich in montmorillonite, but a great deal of silica is also present. Its absorbent properties make it suitable for removing oil and grease, and it was formerly used for cleaning fleeces ('fulling'). It is still used in the textile industry, but its chief application is in the purification of oils. Beds of fuller's earth are found in the southern USA, Germany, Japan, and the UK.

Beds of fuller's earth in the UK are linked to contemporary volcanic activity in Mesozoic times, when fine volcanic ash settled in water.

fulminate any salt of fulminic (cyanic) acid (HOCN), the chief ones being silver and mercury. The fulminates detonate (are exploded by a blow); see ◊detonator.

function in computing, a small part of a program that supplies a specific value—for example, the square root of a specified number, or the current date. Most programming languages incorporate a number of built-in functions; some allow programmers to write their own. A function may have one or more arguments (the values on which the function operates). A **function key** on a keyboard is one that, when pressed, performs a designated task, such as ending a program.

function in mathematics, a rule that maps each element in a given set (the domain) onto just one element (or image) in another set (the range). For example, the function:

$$f(x) = x + 3$$

maps every number onto one that is larger by three:

$$f(4) = 7$$

$$f(-6) = -3$$

An **inverse function** is responsible for mapping each image in the range back onto the original element in the domain. For example, the function:

$$f(x) = 2x + 1$$

has the inverse function:

$$f^{-1}(x) = (x - 1)/2$$

Functions are used in all branches of mathematics, physics, and science generally; for example, the formula

$$t = 2\pi\sqrt{(l/g)}$$

shows that for a simple pendulum the time of swing t is a function of its length l and of no other variable quantity (π and g, the acceleration due to gravity, are constants).

functional group in chemistry, a small number of atoms in an arrangement that determines the chemical properties of the group and of the molecule to which it is attached (for example, the carboxyl group COOH, or the amine group NH_2). Organic compounds can be considered as structural skeletons, with a high carbon content, with functional groups attached.

fundamental constant physical quantity that is constant in all circumstances throughout the whole universe. Examples are the electric charge of an electron, the speed of light, Planck's constant, and the gravitational constant.

fundamental forces see ◊forces, fundamental.

fundamental particle another term for ◊**elementary particle**.

fungicide any chemical ◊pesticide used to prevent fungus diseases in plants and animals. Inorganic and organic compounds containing sulphur are widely used.

fungus (plural **fungi**) any of a group of organisms in the kingdom Fungi. Fungi are not considered plants. They lack leaves and roots; they contain no chlorophyll and reproduce by spores. Moulds, yeasts, rusts, smuts, mildews, and mushrooms are all types of fungi.

Because fungi have no chlorophyll, they must get food from organic substances. They are either ◊parasites, existing on living plants or animals, or

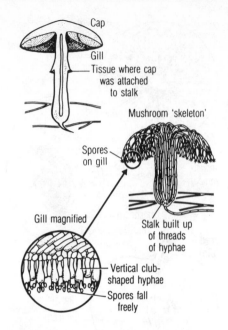

fungus Fungi grow from spores as fine threads, or hyphae. These have no distinct cellular structure. Mushrooms and toadstools are the fruiting bodies formed by the hyphae. Gills beneath the cap of these aerial structures produce masses of spores.

◊saprotrophs, living on dead matter. Some 50,000 different species have been identified. Some are edible, but many are highly poisonous.

Before the classification Fungi came into use, they were included within the division Thallophyta, along with algae and bacteria.

THE WORLD'S LARGEST ORGANISM IS A FUNGUS

In 1992 an individual honey fungus, *Armallaria ostoyae*, in Washington State, USA, was identified as the world's largest living thing—estimated to be between 500 and 1,000 years old, its underground network of hyphae covers 600 hectares/ 1,480 acres.

funicular railway railway with two cars connected by a wire cable wound around a drum at the top of a steep incline. Funicular railways of up to 1.5 km/1 mi exist in Switzerland. In Britain, the system is used only in seaside cliff railways.

fur the ◊hair of certain animals. Fur is an excellent insulating material and so has been used as clothing, although this is vociferously criticized by many groups on humane grounds. The methods of breeding or trapping animals are often cruel. Mink, chinchilla, and sable are among the most valuable, the wild furs being finer than the farmed. Fur such as mink is made up of a soft, thick, insulating layer called underfur and a top layer of longer, lustrous guard hairs.

Furs have been worn since prehistoric times and have long been associated with status and luxury (ermine traditionally worn by royalty, for example), except by certain ethnic groups like the Inuit. The fur trade had its origin in North America, exploited by the Hudson's Bay Company from the late 17th century. The chief centres of the fur trade are New York, London, St Petersburg, and Kastoria in Greece. It is illegal to import furs or skins of endangered species listed by ◊CITES, for example the leopard. Many synthetic fibres are widely used as substitutes.

furlong unit of measurement, originating in Anglo-Saxon England, equivalent to 220 yd (201.168 m).

A furlong consists of 40 rods, poles, or perches; 8 furlongs equal one statute ◊mile. Its literal meaning is 'furrow-long', and refers to the length of a furrow in the common field characteristic of medieval farming.

furnace structure in which fuel such as coal, coke, gas, or oil is burned to produce heat for various purposes. Furnaces are used in conjunction with ◊boilers for heating, to produce hot water, or steam for driving turbines—in ships for propulsion and in power stations for generating electricity. The largest furnaces are those used for smelting and refining metals, such as the ◊blast furnace, electric furnace, and ◊open-hearth furnace.

fuse in electricity, a wire or strip of metal designed to melt when excessive current passes through. It is a safety device to stop at that point in the circuit when surges of current would otherwise damage equipment and cause fires. In explosives, a fuse is a cord impregnated with chemicals so that it burns slowly at a predetermined rate. It is used to set off a main explosive charge, sufficient length of fuse being left to allow the person lighting it to get away to safety.

fusel oil liquid with a characteristic unpleasant smell, obtained as a by-product of the distillation of the product of any alcoholic fermentation, and used in paints, varnishes, essential oils, and plastics. It is a mixture of fatty acids, alcohols, and esters.

fusion in physics, the fusing of the nuclei of light elements, such as hydrogen, into those of a heavier element, such as helium. The resultant loss in their combined mass is converted into energy. Stars and thermonuclear weapons work on the principle of ◊nuclear fusion.

Very high temperatures and pressures are thought to be required in order for fusion to take place. Under these conditions the atomic nuclei can approach each other at high speeds and overcome the mutual repulsion of their positive charges. At very close range another force, the strong nuclear force, comes into play, fusing the particles together to form a larger nucleus. As fusion is accompanied by the release of large amounts of energy, the process might one day be harnessed to form the basis of commercial energy production. So far no successful fusion reactor—one able to produce the required conditions and contain the reaction—has been built. However, an important step along the road to fusion power was taken in November 1991. In an experiment that lasted 2 seconds, a 1.7 megawatt pulse of power was produced by the Joint European Torus (JET) at Culham, Oxfordshire, UK. This was the first time that a substantial amount of fusion power had been produced in a controlled experiment, as opposed to a bomb. See ◊nuclear energy and ◊cold fusion.

fuzzy logic in mathematics and computing, a form of knowledge representation suitable for notions (such as 'hot' or 'loud') that cannot be defined precisely but which depend on their context. For example, a jug of water may be described as too hot or too cold, depending on whether it is to be used to wash one's face or to make tea. The central idea of fuzzy logic is *probability of set membership*. For instance, referring to someone 5 ft 9 in tall, the statement 'this person is tall' (or 'this person is a member of the set of tall people') might be about 70% true if that person is a man, and about 85% true if that person is a woman. Fuzzy logic enables computerized devices to reason more like humans, responding effectively to complex messages from their control panels and sensors.

The term 'fuzzy logic' was coined in 1965 by Iranian computer scientist Lofti Zadeh of the University of California at Berkeley, although the core concepts go back to the work of Polish mathematician Jan Lukasiewicz in the 1920s. It has been largely ignored in Europe and the USA, but was taken up by Japanese manufacturers in the mid-1980s and has since been applied to hundreds of electronic goods and industrial machines. For example, a vacuum cleaner launched in 1992 by Matsushita uses fuzzy logic to adjust its sucking power in response to messages from its sensors about the type of dirt on the floor, its distribution, and its depth.

G

g symbol for ◊*gram*.

gabbro basic (low-silica) igneous rock formed deep in the Earth's crust. It contains pyroxene and calcium-rich feldspar, and may contain small amounts of olivine and amphibole. Its coarse crystals of dull minerals give it a speckled appearance.

Gabbro is the plutonic version of basalt (that is, derived from magma that has solidified below the Earth's surface), and forms in large, slow-cooling intrusions.

gadolinium silvery-white metallic element of the lanthanide series, symbol Gd, atomic number 64, relative atomic mass 157.25. It is found in the products of nuclear fission and used in electronic components, alloys, and products needing to withstand high temperatures.

Gaia hypothesis theory that the Earth's living and nonliving systems form an inseparable whole that is regulated and kept adapted for life by living organisms themselves. The planet therefore functions as a single organism, or a giant cell. Since life and environment are so closely linked, there is a need for humans to understand and maintain the physical environment and living things around them. The Gaia hypothesis was elaborated by British scientist James (Ephraim) Lovelock (1919–) in the 1970s.

When I first introduced Gaia, I had vague hopes that it might be denounced from the pulpit and thus made acceptable to my scientific colleagues. As it was, Gaia was embraced by theologians and by a wide range of New Age writers and thinkers but denounced by biologists.

On the **Gaia hypothesis** James Lovelock (1919–), article 'Gaia Takes Flight' in *Earthwatch* Sept/Oct 1992

---

gain in electronics, the ratio of the amplitude of the output signal produced by an amplifier to that of the input signal. In a ◊voltage amplifier the voltage gain is the ratio of the output voltage to the input voltage; in an inverting ◊operational amplifier (op-amp) it is equal to the ratio of the resistance of the feedback resistor to that of the input resistor.

gal symbol for ◊*gallon*, ◊*galileo*.

galaxy congregation of millions or billions of stars, held together by gravity. *Spiral galaxies*, such as the ◊Milky Way, are flattened in shape, with a central bulge of old stars surrounded by a disc of younger stars, arranged in spiral arms like a Catherine wheel. *Barred spirals* are spiral galaxies that have a straight bar of stars across their centre, from the ends of which the spiral arms emerge. The arms of spiral galaxies contain gas and dust from which new stars are still forming. *Elliptical galaxies* contain old stars and very little gas. They include the most massive galaxies known, containing a trillion stars. At least some elliptical galaxies are thought to be formed by mergers between spiral galaxies. There are also irregular galaxies. Most galaxies occur in clusters, containing anything from a few to thousands of members.

Our own galaxy, the Milky Way, is about 100,000 light years in diameter, and contains at least 100 billion stars. It is a member of a small cluster, the ◊Local Group. The Sun lies in one of its spiral arms, about 25,000 light years from the centre.

GALAXIES: COUNTING THE STARS

An average galaxy contains 100,000 million stars. If you counted the stars in a galaxy at a rate of one every second, it would take 3,000 years to count them all.

---

galena chief ore of lead, consisting of lead sulphide, PbS. It is lead-grey in colour, has a high metallic lustre and breaks into cubes because of its perfect cubic cleavage. It may contain up to 1% silver, and so the ore is sometimes mined for both metals. Galena occurs mainly among limestone

Hubble classification of galaxies

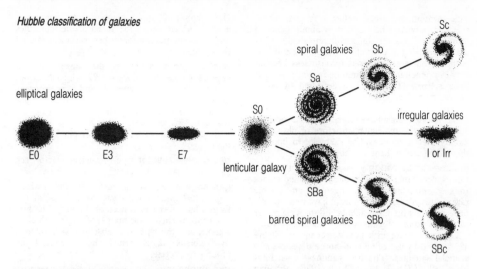

galaxy *Galaxies were classified by US astronomer Edwin Hubble in 1925. He placed the galaxies in a 'tuning-fork' pattern, in which the two prongs correspond to the barred and non-barred spiral galaxies.*

deposits in Australia, Mexico, Russia, Kazakhstan, the UK, and the USA.

galileo unit (symbol gal) of acceleration, used in geological surveying. One galileo is 10^{-2} metres per second per second. The Earth's gravitational field often differs by several milligals (thousandths of gals) in different places, because of the varying densities of the rocks beneath the surface.

Galileo spacecraft launched from the space shuttle *Atlantis* Oct 1989, on a six-year journey to Jupiter. It flew past Venus Feb 1990 and passed within 970 km/600 mi of Earth Dec 1990, using the gravitational fields of these two planets to increase its velocity. The craft flew past Earth again 8 December 1992 to receive its final boost towards Jupiter.

gall abnormal outgrowth on a plant that develops as a result of attack by insects or, less commonly, by bacteria, fungi, mites, or nematodes. The attack causes an increase in the number of cells or an enlargement of existing cells in the plant. Gall-forming insects generally pass the early stages of their life inside the gall. Gall wasps are responsible for the conspicuous bud galls forming on oak trees, 2.5 to 4 cm/1 to 1.5 in across, known as 'oak apples'.

gall bladder small muscular sac, part of the digestive system of most, but not all, vertebrates. In humans, it is situated on the underside of the liver and connected to the small intestine by the bile duct. It stores bile from the liver.

gallium grey metallic element, symbol Ga, atomic number 31, relative atomic mass 69.75. It is liquid at room temperature. Gallium arsenide (GaAs) crystals are used in microelectronics, since electrons travel a thousand times faster through them than through silicon. The element was discovered in 1875 by Lecoq de Boisbaudran (1838–1912).

gallon imperial liquid or dry measure, equal to 4.546 litres, and subdivided into four quarts or eight pints. The US gallon is equivalent to 3.785 litres. One imperial gallon equals 1.201 US gallons.

galvanizing process for rendering iron rust-proof, by plunging it into molten zinc (the dipping method), or by electroplating it with zinc.

galvanometer instrument for detecting small electric currents by their magnetic effect.

gamete cell that functions in sexual reproduction by merging with another gamete to form a ◊zygote. Examples of gametes include sperm and egg cells. In most organisms, the gametes are haploid (they contain half the number of chromosomes of the parent), owing to reduction division or ◊meiosis.

In higher organisms, gametes are of two distinct types: large immobile ones known as eggs or egg cells (see ◊ovum) and small ones known as ◊sperm. They come together at ◊fertilization. In some lower organisms the gametes are all the same, or they may belong to different mating strains but have no obvious differences in size or appearance.

game theory branch of mathematics that deals with strategic problems (such as those that arise in business, commerce, and warfare) by assuming that the people involved invariably try to win—that is, they are assumed to employ strategies that should give the greatest gain and the smallest loss. The theory was developed by Oscar Morgenstern (1902–1977) and John Von Neumann during World War II.

gametophyte the ◊haploid generation in the life cycle of a plant that produces gametes; see ◊alternation of generations.

gamma radiation very-high-frequency electromagnetic radiation, similar in nature to X-rays but of shorter wavelength, emitted by the nuclei of radioactive substances during decay or by the interactions of high-energy electrons with matter. Cosmic gamma rays have been identified as coming from pulsars, radio galaxies, and quasars, although they cannot penetrate the Earth's atmosphere.

Gamma rays are stopped only by direct collision

with an atom and are therefore very penetrating; they can, however, be stopped by about 4 cm/1.5 in of lead or by a very thick concrete shield. They are less ionizing in their effect than alpha and beta particles, but are dangerous nevertheless because they can penetrate deeply into body tissues such as bone marrow. They are not deflected by either magnetic or electric fields.

Gamma radiation is used to kill bacteria and other microorganisms, sterilize medical devices, and change the molecular structure of plastics to modify their properties (for example, to improve their resistance to heat and abrasion).

gamma-ray astronomy the study of gamma rays from space. Much of the radiation detected comes from collisions between hydrogen gas and cosmic rays in our galaxy. Some sources have been identified, including the Crab nebula and the Vela pulsar (the most powerful gamma-ray source detected).

Gamma rays are difficult to detect and are generally studied by use of balloon-borne detectors and artificial satellites. The first gamma-ray satellites were *SAS II* (1972) and *COS B* (1975), although gamma-ray detectors were carried on the *Apollo 15* and *16* missions. *SAS II* failed after only a few months, but *COS B* continued working until 1982, carrying out a complete survey of the galactic disc.

The US Gamma Ray Observatory was launched by US space shuttle *Atlantis* in April 1991 to study the gamma-ray sky for five years. The observatory cost $617 million and at 15 tonnes/17 tons was the heaviest payload ever carried by space shuttle.

ganglion (plural *ganglia*) solid cluster of nervous tissue containing many cell bodies and ◊synapses, usually enclosed in a tissue sheath; found in invertebrates and vertebrates.

In many invertebrates, the central nervous system consists mainly of ganglia connected by nerve cords. The ganglia in the head (cerebral ganglia) are usually well developed and are analogous to the brain in vertebrates. In vertebrates, most ganglia occur outside the central nervous system.

gangue the part of an ore deposit that is not itself economically valuable; for example, calcite may occur as a gangue mineral with galena.

Ganymede in astronomy, the largest moon of the planet Jupiter, and the largest moon in the Solar System, 5,260 km/3,270 mi in diameter (larger than the planet Mercury). It orbits Jupiter every 7.2 days at a distance of 1.1 million km/700,000 mi. Its surface is a mixture of cratered and grooved terrain.

garnet group of silicate minerals with the formula $X_3Y_2(SiO_4)_3$, when X is calcium, magnesium, iron, or manganese, and Y is iron, aluminium, or chromium. Garnets are used as semiprecious gems (usually pink to deep red) and as abrasives. They occur in metamorphic rocks such as gneiss and schist.

gas in physics, a form of matter, such as air, in which the molecules move randomly in otherwise empty space, filling any size or shape of container into which the gas is put.

A sugar-lump sized cube of air at room temperature contains 30 trillion molecules moving at an average speed of 500 metres per second

(1,800 kph/1,200 mph). Gases can be liquefied by cooling, which lowers the speed of the molecules and enables attractive forces between them to bind them together.

gas constant in physics, the constant R that appears in the equation $PV = nRT$, which describes how the pressure P, volume V, and temperature T of an ideal gas are related (n is the amount of gas in the specimen). This equation combines ◊Boyle's law and ◊Charles's law.

R has a value of 8.3143 joules per kelvin per mole.

gas-cooled reactor type of nuclear reactor; see ◊advanced gas-cooled reactor.

gas engine internal-combustion engine in which a gas (coal gas, producer gas, natural gas, or gas from a blast furnace) is used as the fuel. The first practical gas engine was built 1860 by Jean Etienne Lenoir, and the type was subsequently developed by Nikolaus August Otto, who introduced the ◊four-stroke cycle.

gas exchange movement of gases between an organism and the atmosphere, principally oxygen and carbon dioxide. All aerobic organisms (most animals and plants) take in oxygen in order to burn food and manufacture ◊ATP. The resultant oxidation reactions release carbon dioxide as a waste product to be passed out into the environment. Green plants also absorb carbon dioxide during ◊photosynthesis, and release oxygen as a waste product.

Specialized respiratory surfaces have evolved during evolution to make gas exchange more efficient. In humans and other tetrapods (four-limbed vertebrates), gas exchange occurs in the ◊lungs, aided by the breathing movements of the ribs. Many adult amphibia and terrestrial invertebrates can absorb oxygen directly through the skin. The bodies of insects and some spiders contain a system of air-filled tubes known as ◊tracheae. Fish have ◊gills as their main respiratory surface. In plants, gas exchange generally takes place via the ◊stomata and the air-filled spaces between the cells in the interior of the leaf.

gas laws physical laws concerning the behaviour of gases. They include ◊Boyle's law and ◊Charles's law, which are concerned with the relationships between the pressure, temperature, and volume of an ideal (hypothetical) gas. These two laws can be combined to give the **general** or **universal gas law**, which may be expressed as:

(pressure × volume)/temperature = constant

Van der Waals' law includes corrections for the nonideal behaviour of real gases.

gasohol motor fuel that is 90% petrol and 10% ethanol (alcohol). The ethanol is usually obtained by fermentation, followed by distillation, using maize, wheat, potatoes, or sugar cane. It was used in early cars before petrol became economical, and its use was revived during the 1940s war shortage and the energy shortage of the 1970s, for example in Brazil.

gasoline mixture of hydrocarbons derived from petroleum, whose main use is as a fuel for internal

in mammals

cells produce carbon dioxide from respiration, which is carried to the lungs in the blood

alveoli

air

carbon dioxide diffuses into alveoli

oxygen diffuses into blood through moist lining

blood vessel

oxygen is carried in the blood to cells, which use it for respiration

gas exchange

combustion engines. It is colourless and highly volatile.

In the UK, gasoline is called petrol.

gas syringe graduated piece of glass apparatus used to measure accurately the volumes of gases.

gastrolith stone that was once part of the digestive system of a dinosaur or other extinct animal. Rock fragments were swallowed to assist in the grinding process in the dinosaur digestive tract, much as some birds now swallow grit and pebbles to grind food in their crop. Once the animal has decayed, smooth round stones remain—often the only clue to their past use is the fact that they are geologically different from their surrounding strata.

gas turbine engine in which burning fuel supplies hot gas to spin a ◊turbine. The most widespread application of gas turbines has been in aviation. All jet engines (see under ◊jet propulsion) are modified gas turbines, and some locomotives and ships also use gas turbines as a power source.

They are also used in industry for generating and pumping purposes.

In a typical gas turbine a multivaned compressor draws in and compresses air. The compressed air enters a combustion chamber at high pressure, and fuel is sprayed in and ignited. The hot gases produced escape through the blades of (typically) two turbines and spin them around. One of the turbines drives the compressor; the other provides the external power that can be harnessed.

gate in electronics, short for ◊logic gate.

gauge any scientific measuring instrument—for example, a wire gauge or a pressure gauge. The term is also applied to the width of a railway or tramway track.

gauge boson or *field particle* any of the particles that carry the four fundamental forces of nature (see ◊forces, fundamental). Gauge bosons are ◊elementary particles that cannot be subdivided, and include the photon, the graviton, the gluons, and the weakons.

gauss cgs unit (symbol Gs) of magnetic flux density, replaced by the SI unit, the ◊tesla, but still commonly used. The Earth's magnetic field is about 0.5 Gs, and changes to it over time are measured in gammas (one gamma equals 10^{-5} gauss).

gear a toothed wheel that transmits the turning movement of one shaft to another shaft. Gear wheels may be used in pairs, or in threes if both shafts are to turn in the same direction. The gear ratio—the ratio of the number of teeth on the two wheels—determines the torque ratio, the turning force on the output shaft compared with the turning force on the input shaft. The ratio of the angular velocities of the shafts is the inverse of the gear ratio.

The common type of gear for parallel shafts is the *spur gear*, with straight teeth parallel to the shaft axis. The *helical gear* has teeth cut along sections of a helix or corkscrew shape; the double form of the helix gear is the most efficient for energy transfer. *Bevil gears*, with tapering teeth set on the base of a cone, are used to connect intersecting shafts.

Geiger counter any of a number of devices used for detecting nuclear radiation and/or measuring its intensity by counting the number of ionizing particles produced (see ◊radioactivity). It detects the momentary current that passes between ◊electrodes in a suitable gas when a nuclear particle or a radiation pulse causes the ionization of that gas. The electrodes are connected to electronic devices that enable the number of particles passing to be measured. The increased frequency of measured particles indicates the intensity of radiation. It is named after German physicist Hans Geiger.

Geiger–Müller, Geiger–Klemperer, and Rutherford–Geiger counters are all devices often referred to loosely as Geiger counters.

Geissler tube high-voltage ◊discharge tube in which traces of gas ionize and conduct electricity. Since the electrified gas takes on a luminous colour characteristic of the gas, the instrument is also used in ◊spectroscopy. It was developed 1858 by German physicist Heinrich Geissler (1814–1879).

gel solid produced by the formation of a three-dimensional cage structure, commonly of linked large-molecular-mass polymers, in which a liquid is trapped. It is a form of ◊colloid. A gel may be a jellylike mass (pectin, gelatin) or have a more rigid structure (silica gel).

gelatinous precipitate precipitate that is viscous and jellylike when formed; see ◊precipitation. ◊Aluminium hydroxide has this appearance.

gelignite type of ◊dynamite.

gem mineral valuable by virtue of its durability (hardness), rarity, and beauty, cut and polished for ornamental use, or engraved. Of 120 minerals known to have been used as gemstones, only about 25 are in common use in jewellery today; of these, the diamond, emerald, ruby, and sapphire are classified as precious, and all the others semi-precious, for example the topaz, amethyst, opal, and aquamarine.

Among the synthetic precious stones to have been successfully produced are rubies, sapphires, emeralds, and diamonds (first produced by General Electric in the USA 1955).

Gemini prominent zodiacal constellation in the northern hemisphere represented as the twins Castor and Pollux. Its brightest star is ◊Pollux; ◊Castor is a system of six stars. The Sun passes through Gemini from late June to late July. Each Dec, the Geminid meteors radiate from Gemini. In astrology, the dates for Gemini are between about 21 May and 21 June (see ◊precession).

Gemini project US space programme (1965–66) in which astronauts practised rendezvous and docking of spacecraft, and working outside their spacecraft, in preparation for the Apollo Moon landings (see ◊Apollo project).

Gemini spacecraft carried two astronauts and were launched by Titan rockets.

gemma (plural *gemmae*) unit of ◊vegetative reproduction, consisting of a small group of undifferentiated green cells. Gemmae are found in certain mosses and liverworts, forming on the surface of the plant, often in cup-shaped structures, or gemmae cups. Gemmae are dispersed by splashes of rain and can then develop into new plants. In many species, gemmation is more common than reproduction by ◊spores.

gene unit of inherited material, encoded by a strand of ◊DNA, and transcribed by ◊RNA. In higher organisms, genes are located on the ◊chromosomes. The term 'gene', coined 1909 by the Danish geneticist Wilhelm Johannsen (1857–1927), refers to the inherited factor that consistently affects a particular character in an individual—for example, the gene for eye colour. Also termed a Mendelian gene, after Austrian biologist Gregor Mendel, it occurs at a particular point or ◊locus on a particular chromosome and may have several variants or ◊alleles, each specifying a particular form of that character—for example, the alleles for blue or brown eyes. Some alleles show ◊dominance. These mask the effect of other alleles known as recessive (see ◊recessive gene).

In the 1940s, it was established that a gene could be identified with a particular length of DNA, which coded for a complete protein molecule, leading to the 'one-gene-one-enzyme' principle. Later it was realized that proteins can be made up of several ◊polypeptide chains, each with a separate gene, so this principle was modified to 'one-gene-one-polypeptide'. However, the fundamental idea remains the same, that genes produce their visible effects simply by coding for proteins; they control the structure of those proteins via the genetic code, as well as the amounts produced and the timing of production. In modern genetics, the gene is identified either with the ◊cistron (a set of ◊codons that determines a complete polypeptide) or with the unit of selection (a Mendelian gene that deter-

mines a particular character in the organism on which ◊natural selection can act). Genes undergo ◊mutation and ◊recombination to produce the variation on which natural selection operates. See also ◊nucleotide.

HUMAN GENE

Humans have approximately 100,000 genes ranging from 2,000 to 2,000,000 pairs of nucleotides; if the nucleotides were listed in a newspaper they would make a sequence of letters 8,000 km/5,000 mi long.

---

gene amplification technique by which selected DNA from a single cell can be repeatedly duplicated until there is a sufficient amount to analyse by conventional genetic techniques.

Gene amplification uses a procedure called the polymerase chain reaction. The sample of DNA is mixed with a solution of enzymes called polymerases, which enable it to replicate, and with a plentiful supply of nucleotides, the building blocks of DNA. The mixture is repeatedly heated and cooled. At each warming, the double-stranded DNA present separates into two single strands, and with each cooling the polymerase assembles a new paired strand for each single strand. Each cycle takes approximately 30 minutes to complete, so that after 10 hours there is one million times more DNA present than at the start.

The technique has been used to analyse DNA from a man who died 1959, showing the presence of sequences from the HIV virus in his cells. It can also be used to test for genetic defects in a single cell taken from an embryo, before the embryo is reimplanted in ◊in vitro fertilization.

gene bank collection of seeds or other forms of genetic material, such as tubers, spores, bacterial or yeast cultures, live animals and plants, frozen sperm and eggs, or frozen embryos. These are stored for possible future use in agriculture, plant and animal breeding, or in medicine, genetic engineering, or the restocking of wild habitats where species have become extinct. Gene banks will be increasingly used as the rate of extinction increases, depleting the Earth's genetic variety (biodiversity).

gene pool total sum of ◊alleles (variants of ◊genes) possessed by all the members of a given population or species alive at a particular time.

generation in computing, a stage of development in electronics (see ◊computer generation), or a class of programming language (see ◊fourth-generation language).

generator machine that produces electrical energy from mechanical energy, as opposed to an ◊electric motor, which does the opposite. A simple generator (dynamo) consists of a wire-wound coil (◊armature) that is rotated between the poles of a permanent magnet. The movement of the wire in the magnetic field induces a current in the coil by ◊electromagnetic induction, which can be fed by means of a ◊commutator as a continuous direct current into an external circuit. Slip rings instead of a commutator produce an alternating current, when the generator is called an alternator.

gene replacement therapy (GRT) hypothetical treatment for hereditary diseases in which affected cells from a sufferer would be removed from the body, the ◊DNA repaired in the laboratory (◊genetic engineering), and the functioning cells reintroduced. Successful experiments have been carried out on animals.

GRT could most readily be used to treat genetic blood diseases, because blood cells are easy to remove, grow well in tissue culture, and will repopulate the bone marrow successfully when injected. Gene therapy has been used for SCID (severe combined immune deficiency) with some success. Diseases of the nervous system, on the other hand, are unlikely to become treatable by GRT. Individuals cured by GRT might still pass on the disease to their children, because it is illegal to interfere with DNA contained in the germ cells (cells that will divide to form sperm or ova) . GRT is likely to be a very expensive therapy, which will strain the resources available for health care and widen the health divide between rich and poor nations.

gene shears new technique in ◊genetic engineering which may have practical applications in the future. The gene shears are pieces of messenger ◊RNA that can bind to other pieces of messenger RNA, recognizing specific sequences, and cut them at that point. If a piece of ◊DNA which codes for the shears can be inserted in the chromosomes of a plant or animal cell, that cell will then destroy all messenger RNA of a particular type. Genetic shears may be used to protect plants against viruses which infect them and cause disease. They might also be useful against ◊AIDS.

gene-splicing technique for inserting a foreign gene into laboratory cultures of bacteria to generate commercial biological products, such as synthetic insulin, hepatitis-B vaccine, and interferon. It was invented 1973 by the US scientists Stanley Cohen and Herbert Boyer, and patented in the USA 1984. See ◊genetic engineering.

gene therapy proposed medical technique for curing or alleviating inherited diseases or defects; see ◊gene replacement therapy. In 1990 a genetically engineered gene was used for the first time to treat a patient.

genetic code the way in which instructions for building proteins, the basic structural molecules of living matter, are 'written' in the genetic material ◊DNA. This relationship between the sequence of bases (the subunits in a DNA molecule) and the sequence of ◊amino acids (the subunits of a protein molecule) is the basis of heredity. The code employs ◊codons of three bases each; it is the same in almost all organisms, except for a few minor differences recently discovered in some protozoa.

genetic disease any disorder caused at least partly by defective genes or chromosomes. In humans there are some 3,000 genetic diseases, including cleft palate, cystic fibrosis, Down's syndrome, haemophilia, Huntington's chorea, some forms of anaemia, spina bifida, and Tay-Sachs disease.

genetic engineering deliberate manipulation of genetic material by biochemical techniques. It is often achieved by the introduction of new ◊DNA, usually by means of a virus or ◊plasmid. This can be for pure research or to breed functionally specific plants, animals, or bacteria. These organisms with a foreign gene added are said to be transgenic (see ◊transgenic organism).

In genetic engineering, the splicing and reconciliation of genes is used to increase knowledge of cell function and reproduction, but it can also achieve practical ends. For example, plants grown for food could be given the ability to fix nitrogen, found in some bacteria, and so reduce the need for expensive fertilizers, or simple bacteria may be modified to produce rare drugs. Developments in genetic engineering have led to the production of human insulin, human growth hormone, and a number of other bone-marrow stimulating hormones. New strains of animals have also been produced; a new strain of mouse was patented in the USA 1989 (the application was rejected in the European patent office). A ◊vaccine against a sheep parasite (a larval tapeworm) has been developed by genetic engineering; most existing vaccines protect against bacteria and viruses. There is a risk that when transplanting genes between different types of bacteria (*Escherichia coli*, which lives in the human intestine, is often used) new and harmful strains might be produced. For this reason strict safety precautions are observed, and the altered bacteria are disabled in some way so they are unable to exist outside the laboratory.

genetic fingerprinting technique used for determining the pattern of certain parts of the genetic material ◊DNA that is unique to each individual. Like skin fingerprinting, it can accurately distinguish humans from one another, with the exception of identical siblings from multiple births. It can be applied to as little material as a single cell.

Genetic fingerprinting involves isolating DNA from cells, then comparing and contrasting the sequences of component chemicals between individuals. The DNA pattern can be ascertained from a sample of skin, hair, or semen. Although differences are minimal (only 0.1% between unrelated people), certain regions of DNA, known as *hypervariable regions*, are unique to individuals.

Genetic fingerprinting was discovered by British scientist Alec Jeffreys (1950–), and is now allowed as a means of legal identification. It is used in paternity testing, forensic medicine, and inbreeding studies.

The results of DNA tests have been accepted as evidence in many thousands of US court cases, although several courts have challenged the validity of conventional genetic fingerprinting. A new method that makes it possible to express the individuals' information in digital code will mean that genetic fingerprinting will now be much more accurate than before.

genetics study of inheritance and of the units of inheritance (◊genes). The founder of genetics was Austrian biologist Gregor Mendel, whose experiments with plants, such as peas, showed that inheritance takes place by means of discrete 'particles', which later came to be called genes.

Before Mendel, it had been assumed that the characteristics of the two parents were blended

Genetic engineering's brave new world

The 1980s and 1990s have seen considerable advances in understanding of the structure of genes and how they work, thanks to the development of techniques for isolating and manipulating genes (recombinant DNA technology). We can determine the structure of the genetic material (DNA sequencing) and study minute quantities of DNA using PCR (polymerase chain reaction). We know more about embryogenesis (the process by which a single fertilized egg develops into a complex embryo), about the changes in the early stages of cancer, and about the basic defects in inherited diseases like cystic fibrosis, muscular dystrophy, and certain types of blindness.

Large-scale production of pure proteins
Each gene is responsible for making a protein. By purifying genes and transferring them to bacterial cells, scientists can harvest large quantities of proteins that would not normally be made by bacteria. Most of the insulin for diabetics is now made in bacteria. Several hormones and enzymes, whose deficiency results in inherited diseases, are manufactured in microorganisms. Other proteins are commercially produced in this way, including the bovine enzyme rennin used in making 'vegetarian' cheese.

Genetic engineering is being used to produce vaccines. These are normally made by growing pathogens (disease-causing organisms) in animals, or in organ or tissue culture, and then harvesting the live pathogen. The inactivated (attenuated) pathogens provoke our immune system, which reacts to the coat proteins covering the pathogen. The new method transfers the gene coding for the pathogen's coat protein into harmless bacteria or viruses which then make the protein themselves. The purified protein is used to stimulate the immune system. Vaccines against, for example, foot-and-mouth disease of cattle and hepatitis B in humans are proving much safer than attenuated vaccines.

Some animal vaccines now use live non-pathogenic viruses which have in addition to their own coat proteins also those of a pathogen. Such a modified 'friendly' virus has been successfully used to eradicate the rabies virus from populations of wild foxes in Belgium. The transfer of genes between viruses will play an important part in the development of new vaccines.

Crop improvement
The transfer of 'foreign' genes into plants has led to the development of crops with desirable characteristics. Species have been made resistant to weedkillers, permitting the removal of weeds without damage to the crop. A bacterial gene can be transferred conferring resistance to caterpillars and other insects that can devastate crops, reducing the need for insecticides and their consequent environmental damage. Transgenic plants resistant to the alfalfa mosaic virus and the tobacco ringspot virus are also now available.

Many Third World countries depend heavily on a single crop containing storage proteins deficient in essential amino acids. By modifying the genes for these proteins, the nutritional value of the food has been greatly improved.

Genetic fingerprinting
Mini-microsatellite DNA are types of DNA that differ so much between individuals that each person is effectively unique. Forensic scientists can determine the origin of tissue samples like blood and semen with a certainty that was previously impossible.

Medical applications
The fundamental genetic changes responsible for some inherited diseases have been identified, leading to deeper understanding of the causes of disease, and to diagnostic methods. Prenatal diagnosis allows parents the opportunity to abort an affected fetus and try again for a normal baby. Similar methods have been used on eggs fertilized in the test tube, permitting implantation of normal embryos only. In certain communities a particular inherited disease is so common that programmes of preventive medicine have been attempted. All pregnancies are screened; termination is offered to females shown to be carrying an affected fetus.

Two recent projects promise exciting developments. One is human gene therapy—transferring normal genes to people with inherited diseases. Animals have been successfully treated, and after much discussion, medical approval has been granted for human trials. The other is the Human Genome Project, a huge international research effort to identify and study all 100,000 human genes. The project, which will take about 10 years, should lead to deeper understanding of development and disease, and have unforeseen applications in many branches of biology.

Thomas Day

during inheritance, but Mendel showed that the genes remain intact, although their combinations change. Since Mendel, genetics has advanced greatly, first through ◊breeding experiments and light-microscope observations (classical genetics), later by means of biochemical and electron-microscope studies (molecular genetics). An advance was the elucidation of the structure of ◊DNA by US biologist James Watson and British molecular biologist Francis Crick, and the subsequent cracking of the ◊genetic code. These discoveries opened up the possibility of deliberately manipulating genes, or ◊genetic engineering. See also ◊genotype, ◊phenotype, and ◊monohybrid inheritance.

GENETICS PUZZLE

The nucleus of a human cell contains 23 pairs of chromosomes, each pair consisting of one chromosome from the father and one from the mother, the whole set being the information required to generate a particular individual.

In order to reproduce, the body splits the chromosome pairs back to single chromosomes, then makes special reproductive cells (gametes) containing 23 single chromosomes each. In the process, the chromosomes can be 'shuffled' so that the new gametes have some chromosomes from the individual's mother and some from the individual's father. When a male gamete (sperm) meets a female one (ovum), the single chromosomes pair up again to produce a new individual.

The problem is that there is only a limited number of ways of shuffling 23 pairs of chromosomes. Assuming that we all started from two individuals, the number of possibilities is 4^{23}. According to this argument, sooner or later a perfect copy of an earlier individual will emerge, perhaps Mozart or Einstein reborn. There should also be a large number of near copies, differing only by one or two chromosomes. But can this be true? *See page 651 for the answer.*

genitalia reproductive organs of sexually reproducing animals, particularly the external/visible organs of mammals: in males, the penis and the scrotum, which contains the testes, and in females, the clitoris and vulva.

genome the full complement of ◊genes carried by a single (haploid) set of ◊chromosomes. The term may be applied to the genetic information carried by an individual or to the range of genes found in a given species.

genotype the particular set of ◊alleles (variants of genes) possessed by a given organism. The term is usually used in conjunction with ◊phenotype, which is the product of the genotype and all environmental effects. See also ◊nature–nurture controversy.

genus (plural *genera*) group of ◊species with many characteristics in common. Thus all doglike species (including dogs, wolves, and jackals) belong to the genus *Canis* (Latin 'dog'). Species of the same genus are thought to be descended from a common ancestor species. Related genera are grouped into ◊families.

geochemical analysis archaeological technique that involves taking soil samples at regular intervals from the surface of a site and its surroundings to identify, through phosphorus concentrations in the soil, human settlements and activity and burial areas within sites. Excrement and bone are relatively high in phosphorus content, so human activity and remains tend to produce comparatively large concentrations of phosphates.

geochemistry science of chemistry as it applies to geology. It deals with the relative and absolute abundances of the chemical elements and their ◊isotopes in the Earth, and also with the chemical changes that accompany geologic processes.

geochronology the branch of geology that deals with the dating of the Earth by studying its rocks and contained fossils. The geologic time chart is a result of these studies, dividing Earth time into eons, eras, periods, and epochs determined on the basis of absolute and relative dating methods. Absolute dating methods involve the measurement of radioactive decay over time in certain chemical elements found in rocks, whereas relative dating methods establish the sequence of deposition of various rock layers by identifying and comparing their contained fossils.

geode in geology, a subspherical cavity into which crystals have grown from the outer wall into the centre. Geodes often contain very well-formed crystals of quartz (including amethyst), calcite, or other minerals.

geodesic dome spherical dome whose surface is formed by an arrangement of short rods arranged in triangles. The rods lie on geodesics (the shortest lines joining two points on a curved surface). This dome allows large spaces to be enclosed using the minimum of materials, and was patented by US engineer Buckminster Fuller 1954.

geodesy methods of surveying the Earth for making maps and correlating geological, gravitational, and magnetic measurements. Geodesic surveys, formerly carried out by means of various measuring techniques on the surface, are now commonly made by using radio signals and laser beams from orbiting satellites.

geography the study of the Earth's surface; its topography, climate, and physical conditions, and how these factors affect people and society. It is usually divided into *physical geography*, dealing with landforms and climates, and *human geography*, dealing with the distribution and activities of peoples on Earth.

geological time time scale embracing the history of the Earth from its physical origin to the present day. Geological time is traditionally divided into eons (Phanerozoic, Proterozoic, and Archaean), which in turn are divided into eras, periods, epochs, ages, and finally chrons.

geology science of the Earth, its origin, compo-

geological time chart

eon	era	period	epoch	millions of years ago	life forms
		Quaternary	Holocene	0.01	
			Pleistocene	1.64	humans appeared
	Cenozoic		Pliocene	5.2	
			Miocene	23.5	
		Terttiary	Oligocene	35.5	
			Eocene	56.6	
			Palaeocene	65	mammals flourished
Phanerozoic	Mesozoic	Cretaceous		146	heyday of dinosaurs
		Jurassic		208	first birds
		Triassic		245	first mammals and dinosaurs
	Palaeozoic	Permian		290	reptiles expanded
		Carboniferous		363	first reptiles
		Devonian		409	first amphibians
		Silurian		439	first land plants
		Ordovician		510	first fish
		Cambrian		570	first fossils
(Precambrian) Proterozoic				3,500	earliest living things
Archaean				4,600	

sition, structure, and history. It is divided into several branches: **mineralogy** (the minerals of Earth), **petrology** (rocks), **stratigraphy** (the deposition of successive beds of sedimentary rocks), **palaeontology** (fossils), and **tectonics** (the deformation and movement of the Earth's crust).

Geology is regarded as part of earth science, a more widely embracing subject that brings in meteorology, oceanography, geophysics, and geochemistry.

... Some drill and bore/The solid earth, and from the strata there/Extract a register, by which we learn/That he who made it, and reveal'd its date/To Moses, was mistaken in its age.

On **geology** William Cowper (1731–1800) *Task* Book iii *The Garden* 1895

---

geometric mean in mathematics, the *n*th root of the product of *n* positive numbers. The geometric mean *m* of two numbers *p* and *q* is such that $m = \sqrt{p} \times q$. For example, the mean of 2 and 8 is $\sqrt{2} \times 8 = \sqrt{16} = 4$

geometric progression or **geometric sequence** in mathematics, a sequence of terms (progression) in which each term is a constant multiple (called the **common ratio**) of the one preceding it. For example, 3, 12, 48, 192, 768, ... is a geometric progression with a common ratio 4, since each term is equal to the previous term multiplied by 4. Compare ◊arithmetic progression.

The sum of *n* terms of a *geometric series*

$$1 + r + r^2 + r^3 + \ldots + r^n - 1$$

is given by the formula

$$S_n = (1 - r^n)/(1 - r)$$

for all $r = 1$. For $r = 1$, the geometric series can be summed to infinity:

$$S\infty = 1/(1 - r)$$

In nature, many single-celled organisms reproduce by splitting in two so that one cell gives rise to 2, then 4, then 8 cells, and so on, forming a geometric sequence 1, 2, 4, 8, 16, 32, ..., in which the common ratio is 2.

geometry branch of mathematics concerned with the properties of space, usually in terms of plane (two-dimensional) and solid (three-dimensional) figures. The subject is usually divided into **pure geometry**, which embraces roughly the plane and solid geometry dealt with in Euclid's *Elements*, and **analytical** or ◊**coordinate geometry**, in which problems are solved using algebraic methods. A third, quite distinct, type includes the non-Euclidean geometries.

Where there is matter, there is geometry.

On **geometry** Johannes Kepler (1571–1630)

Geometry probably originated in ancient Egypt, in land measurements necessitated by the periodic inundations of the river Nile, and was soon extended into surveying and navigation. Early geometers were the Greek mathematicians Thales, Pythagoras, and Euclid. Analytical methods were introduced and developed by the French philosopher René Descartes in the 17th century. From the 19th century, various non-Euclidean geometries were devised by the Germans Karl Gauss and Georg Riemann, the Russian Nikolai Lobachevsky, and others. These proved significant in the development of the theory of relativity and in the formulation of atomic theory.

geomorphology branch of geology that deals

with the nature and origin of surface landforms such as mountains, valleys, plains, and plateaus.

geophysics branch of earth science using physics to study the Earth's surface, interior, and atmosphere. Studies also include winds, weather, tides, earthquakes, volcanoes, and their effects.

geostationary orbit circular path 35,900 km/ 22,300 mi above the Earth's equator on which a ◊satellite takes 24 hours, moving from west to east, to complete an orbit, thus appearing to hang stationary over one place on the Earth's surface. Geostationary orbits are particularly used for communications satellites and weather satellites. They were first thought of by the author Arthur C Clarke. A *geosynchronous orbit* lies at the same distance from Earth but is inclined to the equator.

geothermal energy energy extracted for heating and electricity generation from natural steam, hot water, or hot dry rocks in the Earth's crust. Water is pumped down through an injection well where it passes through joints in the hot rocks. It rises to the surface through a recovery well and may be converted to steam or run through a heat exchanger. Dry steam may be directed through turbines to produce electricity. It is an important source of energy in volcanically active areas such as Iceland and New Zealand.

germ colloquial term for a microorganism that causes disease, such as certain ◊bacteria and ◊viruses. Formerly, it was also used to mean something capable of developing into a complete organism (such as a fertilized egg, or the ◊embryo of a seed).

germanium brittle, grey-white, weakly metallic (◊metalloid) element, symbol Ge, atomic number 32, relative atomic mass 72.6. It belongs to the silicon group, and has chemical and physical properties between those of silicon and tin. Germanium is a semiconductor material and is used in the manufacture of transistors and integrated circuits. The oxide is transparent to infrared radiation, and is used in military applications. It was

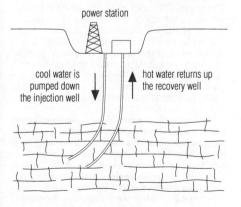

power station

cool water is pumped down the injection well

hot water returns up the recovery well

underground joint system forming underground reservoir where water is heated

geothermal energy

discovered 1886 by German chemist Clemens Winkler (1838–1904).

In parts of Asia, germanium and plants containing it are used to treat a variety of diseases, and it is sold in the West as a food supplement despite fears that it may cause kidney damage.

German silver or *nickel silver* silvery alloy of nickel, copper, and zinc. It is widely used for cheap jewellery and the base metal for silver plating. The letters EPNS on silverware stand for *electroplated nickel silver*.

germination in botany, the initial stages of growth in a seed, spore, or pollen grain. Seeds germinate when they are exposed to favourable external conditions of moisture, light, and temperature, and when any factors causing dormancy have been removed.

The process begins with the uptake of water by the seed. The embryonic root, or radicle, is normally the first organ to emerge, followed by the embryonic shoot, or plumule. Food reserves, either within the ◊endosperm or from the ◊cotyledons, are broken down to nourish the rapidly growing seedling. Germination is considered to have ended with the production of the first true leaves.

germ layer in ◊embryology, a layer of cells that can be distinguished during the development of a fertilized egg. Most animals have three such layers: the inner, middle, and outer.

The inner layer (*endoderm*) gives rise to the gut, the middle one (*mesoderm*) develops into most of the other organs, while the outer one (*ectoderm*) gives rise to the skin and nervous system. Simple animals, such as sponges, lack a mesoderm.

gestation in all mammals except the ◊monotremes (duck-billed platypus and spiny anteaters), the period from the time of implantation of the embryo in the uterus to birth. This period varies among species; in humans it is about 266 days, in elephants 18–22 months, in cats about 60 days, and in some species of marsupial (such as opossum) as short as 12 days.

geyser natural spring that intermittently discharges an explosive column of steam and hot water into the air due to the build-up of steam in underground chambers. One of the most remarkable geysers is Old Faithful, in Yellowstone National Park, Wyoming, USA. Geysers also occur in New Zealand and Iceland.

G-force force that pilots and astronauts experience when their craft accelerate or decelerate rapidly. One G is the ordinary pull of gravity. Early astronauts were subjected to launch and re-entry forces of up to six G or more; in the Space Shuttle, more than three G is experienced on liftoff. Pilots and astronauts wear G-suits that prevent their blood 'pooling' too much under severe G-forces, which can lead to unconsciousness.

gibberellin plant growth substance (see also ◊auxin) that promotes stem growth and may also affect the breaking of dormancy in certain buds and seeds, and the induction of flowering. Application of gibberellin can stimulate the stems of dwarf plants to additional growth, delay the ageing process in leaves, and promote the production of seedless fruit (◊parthenocarpy).

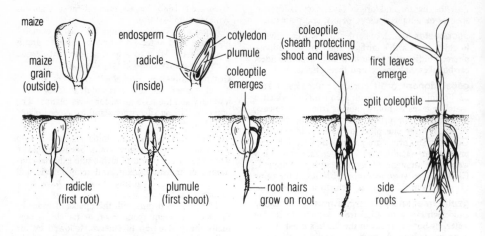

germination *The germination of a maize grain. The plumule and radicle emerge from the seed coat and begin to grow into a new plant. The coleoptile protects the emerging bud and the first leaves.*

Gibbs' function in ◊thermodynamics, an expression representing part of the energy content of a system that is available to do external work, also known as the free energy G. In an equilibrium system at constant temperature and pressure,
$$G = H - TS$$
where H is the enthalpy (heat constant), T the temperature, and S the ◊entropy (decrease in energy availability).

The function was named after US physicist Josiah Willard Gibbs.

giga- prefix signifying multiplication by 10^9 (1,000,000,000 or 1 billion), as in **gigahertz**, a unit of frequency equivalent to 1 billion hertz.

gigabyte in computing, a measure of ◊memory capacity, equal to 1,024 ◊megabytes. It is also used, less precisely, to mean 1,000 billion ◊bytes.

GIGO (acronym for *g*arbage *i*n, *g*arbage *o*ut) expression used in computing to emphasize that inaccurate input data will result in inaccurate output data.

gill in biology, the main respiratory organ of most fishes and immature amphibians, and of many aquatic invertebrates. In all types, water passes over the gills, and oxygen diffuses across the gill membranes into the circulatory system, while carbon dioxide passes from the system out into the water.

In aquatic insects, these gases diffuse into and out of air-filled canals called tracheae.

gill imperial unit of volume for liquid measure, equal to one-quarter of a pint or 5 fluid ounces (0.142 litre). It is used in selling alcoholic drinks.

In S England it is also called a noggin, but in N England the large noggin is used, which is two gills.

Giotto space probe built by the European Space Agency to study ◊Halley's comet. Launched by an Ariane rocket in July 1985, *Giotto* passed within 600 km/375 mi of the comet's nucleus on 13 March 1986. On 2 July 1990 it flew 23,000 km/14,000 mi from Earth, which diverted its path to encounter another comet, Grigg-Skjellerup, on 10 July 1992.

GIOTTO: COMMEMORATING A PAINTER

The *Giotto* spacecraft is named after Italian painter Giotto di Bondone, who in 1304 painted Halley's comet as the 'Star of Bethlehem' in a painting in a chapel at Padua, Italy.

---

gizzard muscular grinding organ of the digestive tract, below the ◊crop of birds, earthworms, and some insects, and forming part of the ◊stomach. The gizzard of birds is lined with a hardened horny layer of the protein keratin, preventing damage to the muscle layer during the grinding process. Most birds swallow sharp grit which aids maceration of food in the gizzard.

glacial deposition the laying-down of rocky material once carried by a glacier. When ice melts, it deposits the material that it has been carrying. The material dumped on the valley floor forms a deposit called till or boulder clay. It comprises angular particles of all sizes from boulders to clay. Till can be moulded by ice to form drumlins, egg-shaped hills. At the snout of the glacier, material piles up to form a ridge called a terminal moraine. Small depositional landforms may also result from glacial deposition, such as kames (small mounds) and kettle holes (small depressions, often filled with water).

Meltwater flowing away from a glacier will carry some of the till many kilometres away. This sediment will become rounded (by the water) and, when deposited, will form a gently sloping area called an outwash plain. Several landforms owe their existence to meltwater—these are called **fluvioglacial landforms** and include the long ridges called eskers. Meltwater may fill depressions eroded by the ice to form ribbon lakes.

glacial erosion the wearing-down and removal of rocks and soil by a glacier. Glacial erosion forms impressive landscape features, including the glacial trough (valley), arêtes (steep ridges), corries

features associated with glacial deposition

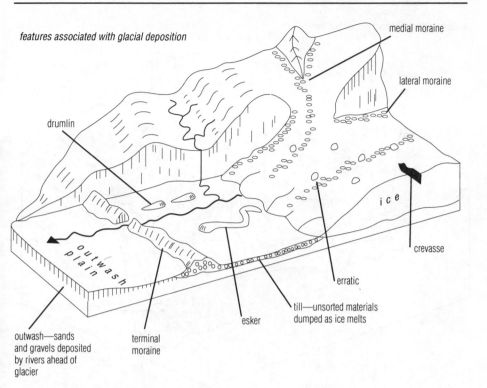

medial moraine

lateral moraine

drumlin

crevasse

i c e

erratic

till—unsorted materials
dumped as ice melts

esker

outwash—sands
and gravels deposited
by rivers ahead of
glacier

terminal
moraine

glacial deposition

(enlarged hollows), and pyramidal peaks (high mountain peaks with concave faces).

Ice is a powerful agent of erosion. It can bulldoze its way down a valley, eroding away spurs to form truncated spurs. Rock fragments below the ice will abrade the valley floor, leading to overdeepening. Loose fragments will be plucked away from the bedrock to form jagged surfaces such as those on a ◊roche moutonnée. Glacial erosion is made more effective by the process of frost shattering (a form of physical weathering), which weakens a rock surface and also provides material for abrasion.

In the UK, features of glacial erosion can be seen in Snowdonia, the Lake District, and the Highlands of Scotland.

glacial trough or *U-shaped valley* steep-sided, flat-bottomed valley formed by a glacier. The erosive action of the glacier and of the debris carried by it results in the formation not only of the trough itself but also of a number of associated features, such as truncated spurs (projections of rock that have been sheared off by the ice) and hanging valleys (smaller glacial valleys that enter the trough at a higher level than the trough floor). Features characteristic of glacial deposition, such as drumlins and eskers, are commonly found on the floor of the trough, together with linear lakes called ribbon lakes.

glacier tongue of ice, originating in mountains in snowfields above the snowline, which moves slowly downhill and is constantly replenished from its source. The scenery produced by the erosive action

of glaciers (◊glacial erosion) includes such features as ◊glacial troughs (U-shaped valleys), ◊corries, and ◊arêtes. In lowlands, the laying down of rocky debris once carried by glaciers (◊glacial deposition) produces a variety of landscape features.

Glaciers form where annual snowfall exceeds annual melting and drainage. The snow compacts to ice under the weight of the layers above. Under pressure the ice moves plastically (changing its shape permanently). When a glacier moves over an uneven surface, deep crevasses are formed in rigid upper layers of the ice mass; if it reaches the sea or a lake, it breaks up to form icebergs. A glacier that is formed by one or several valley glaciers at the base of a mountain is called a *piedmont* glacier. A body of ice that covers a large land surface or continent, for example Greenland or Antarctica, and flows outward in all directions is called an *ice sheet*.

THE LONGEST GLACIER

The longest glacier in the world is the Lambert Glacier in Australian Antarctic Territory. It is up to 64 km/40 mi wide and at least 400 km/250 mi long.

gland specialized organ of the body that manufactures and secretes enzymes, hormones, or other chemicals. In animals, glands vary in size from small (for example, tear glands) to large (for example, the pancreas), but in plants they are

features associated with glacial erosion

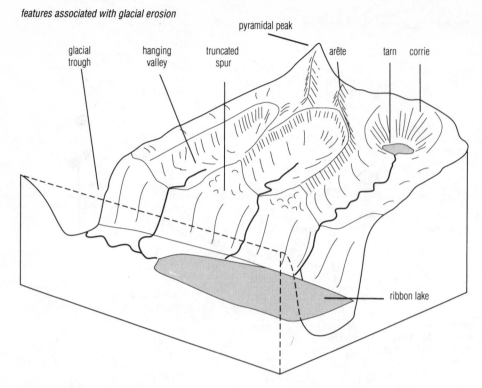

glacial erosion

always small, and may consist of a single cell. Some glands discharge their products internally, ◊endocrine glands, and others, ◊exocrine glands, externally. Lymph nodes are sometimes wrongly called glands.

glass transparent or translucent substance that is physically neither a solid nor a liquid. Although glass is easily shattered, it is one of the strongest substances known. It is made by fusing certain types of sand (silica); this fusion occurs naturally in volcanic glass (see ◊obsidian).

In the industrial production of common types of glass, the type of sand used, the particular chemicals added to it (for example, lead, potassium, barium), and refinements of technique determine the type of glass produced. Types of glass include: soda glass; flint glass, used in cut-crystal ware; optical glass; stained glass; heat-resistant glass; and glasses that exclude certain ranges of the light spectrum. Blown glass is either blown individually from molten glass (using a tube up to 1.5 m/4.5 ft long), as in the making of expensive crafted glass, or blown automatically into a mould—for example, in the manufacture of light bulbs and bottles; pressed glass is simply pressed into moulds, for jam jars, cheap vases, and light fittings; while sheet glass, for windows, is made by putting the molten glass through rollers to form a 'ribbon', or by floating molten glass on molten tin in the 'float glass' process; ◊fibreglass is made from fine glass fibres. Metallic glass is produced by treating alloys so that they take on the properties of glass while retaining the malleability and conductivity characteristic of metals.

Glauber's salt in chemistry, crystalline sodium sulphate decahydrate $Na_2SO_4.10H_2O$, which melts at 87.8°F/31°C; the latent heat stored as it solidifies makes it a convenient thermal energy store. It is used in medicine.

glaucophane in geology, a blue amphibole, $Na_2(Mg,Fe,Al)_5Si_8O_{22}(OH)_2$. It occurs in glaucophane schists (blue schists), formed from the ocean floor basalt under metamorphic conditions of high pressure and low temperature; these conditions are believed to exist in subduction systems associated with destructive plate boundaries (see ◊plate tectonics), and so the occurrence of glaucophane schists can indicate the location of such boundaries in geological history.

Global Positioning System (GPS) US satellite-based navigation system, a network of satellites in six orbits, each circling the Earth once every 24 hours. Each satellite sends out a continuous time signal, plus an identifying signal. To fix position, a user needs to be within range of four satellites, one to provide a reference signal and three to provide directional bearings. The user's receiver can then calculate the position from the difference in time between receiving the signals from each satellite.

The position of the receiver can be calculated to within 10 m/33 ft, although only the US military can tap the full potential of the system. Other users can obtain a position to within 100 m/330 ft. This is accurate enough to be of use to walkers and motorists, and suitable receivers are on the market.

global variable in computing, a ◊variable that can be accessed by any program instruction. See also ◊local variable.

global warming projected imminent climate change attributed to the ◊greenhouse effect.

globular cluster spherical or near-spherical ◊star cluster containing from approximately 10,000 to millions of stars. More than a hundred globular clusters are distributed in a spherical halo around our Galaxy. They consist of old stars, formed early in the Galaxy's history. Globular clusters are also found around other galaxies.

GLOBULAR CLUSTERS AND STARRY NIGHTS

The stars at the centre of a globular cluster are close together—sometimes only light days apart. If the Earth were at the centre of such a cluster, the night sky would be packed with bright stars. You would be able to see your shadow by the light of the stars.

---

glomerulus in the kidney, the blood capillaries responsible for forming the fluid that passes down the tubules and ultimately becomes urine. In the human kidney there are approximately one million tubules, each possessing its own glomerulus.

The structure of the glomerulus allows a wide range of substances including amino acids and sugar, as well as a large volume of water, to pass out of the blood. As the fluid moves through the tubules, most of the water and all of the sugars are reabsorbed, so that only waste remains, dissolved in a relatively small amount of water. This fluid collects in the bladder as urine.

glucose or **dextrose** or **grape-sugar** $C_6H_{12}O_6$ sugar present in the blood, and found also in honey and fruit juices. It is a source of energy for the body, being produced from other sugars and starches to form the 'energy currency' of many biochemical reactions also involving ◊ATP.

Glucose is prepared in syrup form by the hydrolysis of cane sugar or starch, and may be purified to a white crystalline powder. Glucose is a monosaccharide sugar (made up of a single sugar unit), unlike the more familiar sucrose (cane or beet sugar), which is a disaccharide (made up of two sugar units: glucose and fructose).

glue type of ◊adhesive.

gluon in physics, a ◊gauge boson that carries the ◊strong nuclear force, responsible for binding quarks together to form the strongly interacting subatomic particles known as ◊hadrons. There are eight kinds of gluon.

glyceride ◊ester formed between one or more acids and glycerol (propan-1,2,3–triol). A glyceride is termed a mono-, di-, or triglyceride, depending on the number of hydroxyl groups from the glycerol that have reacted with the acids.

Glycerides, chiefly triglycerides, occur naturally as esters of ◊fatty acids in plant oils and animal fats.

glycerine another name for ◊glycerol.

glycerol or **glycerine** or **propan-1,2,3–triol** $HOCH_2CH(OH)CH_2OH$ thick, colourless, odourless, sweetish liquid. It is obtained from vegetable and animal oils and fats (by treatment with acid, alkali, superheated steam, or an enzyme), or by fermentation of glucose, and is used in the manufacture of high explosives, in antifreeze solutions, to maintain moist conditions in fruits and tobacco, and in cosmetics.

glycine $CH_2(NH_2)COOH$ the simplest amino acid, and one of the main components of proteins. When purified, it is a sweet, colourless crystalline compound.

glycogen polymer (a polysaccharide) of the sugar ◊glucose made and retained in the liver as a carbohydrate store, for which reason it is sometimes called animal starch. It is a source of energy when needed by muscles, where it is converted back into glucose by the hormone ◊insulin and metabolized.

glycol or **ethylene glycol** or **ethane-1,2–diol** $(CH_2OH)_2$ thick, colourless, odourless, sweetish liquid. It is used in antifreeze solutions, in the preparation of ethers and esters (used for explosives), as a solvent, and as a substitute for glycerol.

GMT abbreviation for ◊Greenwich Mean Time.

gneiss coarse-grained ◊metamorphic rock, formed under conditions of increasing temperature and pressure, and often occurring in association with schists and granites. It has a foliated, laminated structure, consisting of thin bands of micas and/or amphiboles alternating with granular bands of quartz and feldspar. Gneisses are formed during regional metamorphism; *paragneisses* are derived from sedimentary rocks and *orthogneisses* from igneous rocks. Garnets are often found in gneiss.

Goddard Space Flight Center NASA installation at Greenbelt, Maryland, USA, responsible for the operation of NASA's unmanned scientific satellites, including the ◊Hubble Space Telescope. It is also home of the National Space Science Data centre, a repository of data collected by satellites.

SOFT GOLD

Gold is so soft that a single gram can be drawn into a wire 1 km/0.6 mi long. Sheets of gold can be produced that are less than 0.00013 mm/0.000005 in (400 atoms) thick.

---

gold heavy, precious, yellow, metallic element; symbol Au, atomic number 79, relative atomic mass 197.0. It is unaffected by oxygen and is highly resistant to acids. For manufacture, gold is alloyed with another strengthening metal (such as copper or silver), its purity being measured in ◊carats on a scale of 24.

In 1990 the three leading gold-producing countries were South Africa, 605.4 tonnes; USA, 295 tonnes; and Russia, 260 tonnes. In 1989 gold deposits were found in Greenland with an estimated yield of 12 tonnes per year.

Gold occurs naturally in veins, but following erosion it can be transported and redeposited. It has long been valued for its durability, malleability, and ductility, and its uses include dentistry, jewellery, and electronic devices.

golden section visually satisfying ratio, first constructed by the Greek mathematician Euclid and used in art and architecture. It is found by dividing a line AB at a point O such that the rectangle produced by the whole line and one of the segments is equal to the square drawn on the other segment. The ratio of the two segments is about 8:13 or 1:1.26, and a rectangle whose sides are in this ratio is called a *golden rectangle*.

In van Gogh's picture *Mother and Child*, for example, the Madonna's face fits perfectly into a golden rectangle.

Golgi apparatus or *Golgi body* stack of flattened membranous sacs found in the cells of ◊eukaryotes. Many molecules travel through the Golgi apparatus on their way to other organelles or to the endoplasmic reticulum. Some are modified or asssembled inside the sacs. The Golgi apparatus is named after the Italian physician Camillo Golgi (1844–1926). It produces the membranes that surround the cell vesicles or ◊lysosomes.

gonad the part of an animal's body that produces the sperm or egg cells (ova) required for sexual reproduction. The sperm-producing gonad is called a ◊testis, and the ovule-producing gonad is called an ◊ovary.

Gondwanaland or *Gondwana* southern land mass formed 200 million years ago by the splitting of the single world continent ◊Pangaea. (The northern land mass was ◊Laurasia.) It later fragmented into the continents of South America, Africa, Australia, and Antarctica, which then drifted slowly to their present positions. The baobab tree found in both Africa and Australia is a relic of this ancient land mass.

MAPPING GONDWANALAND

A database of the entire geology of Gondwanaland has been constructed by geologists in South Africa. The database, known as Gondwana Geoscientific Indexing Database (GO-GEOID), displays information as a map of Gondwana 155 million years ago, before the continents drifted apart.

---

gorge narrow steep-sided valley (or canyon) that may or may not have a river at the bottom. A gorge may be formed as a ◊waterfall retreats upstream, eroding away the rock at the base of a river valley; or it may be caused by ◊rejuvenation, when a river begins to cut downwards into its channel once again (for example, in response to a fall in sea level). Gorges are common in limestone country, where they are commonly formed by the collapse of the roofs of underground caverns.

Examples of gorges in the UK are Winnats Pass in Derbyshire and the Avon Gorge in Bristol.

Gossamer Albatross the first human-powered aircraft to fly across the English Channel, in June 1979. It was designed by Paul MacCready and piloted and pedalled by Bryan Allen. The Channel crossing took 2 hours 49 minutes. The same team was behind the first successful human-powered aircraft (*Gossamer Condor*) two years earlier.

governor in engineering, any device that controls the speed of a machine or engine, usually by regulating the intake of fuel or steam.

Scottish inventor James Watt invented the steam-engine governor in 1788. It works by means of heavy balls, which rotate on the end of linkages and move in or out because of ◊centrifugal force according to the speed of rotation. The movement of the balls closes or opens the steam valve to the engine. When the engine speed increases too much, the balls fly out, and cause the steam valve to close, so the engine slows down. The opposite happens when the engine speed drops too much.

Graafian follicle fluid-filled capsule that surrounds and protects the developing egg cell inside the ovary during the ◊menstrual cycle. After the egg cell has been released, the follicle remains and is known as a corpus luteum.

gradient on a graph, the slope of a straight or curved line. The slope of a curve at any given point is represented by the slope of the ◊tangent at that point.

Gradient is a measure of rate of change and can be used to represent such quantities as velocity (the gradient of a graph of distance moved against time) and acceleration (the gradient of a graph of a body's velocity against time).

grafting in medicine, the operation by which a piece of living tissue is removed from one organism and transplanted into the same or a different organism where it continues growing (see ◊transplant). In horticulture, it is a technique widely used for propagating plants, especially woody species. A bud or shoot on one plant, termed the *scion*, is inserted into another, the *stock*, so that they continue growing together, the tissues combining at the point of union. In this way some of the advantages of both plants are obtained.

Grafting is usually only successful between species that are closely related and is most commonly practised on roses and fruit trees. The grafting of non-woody species is more difficult but it is sometimes used to propagate tomatoes and cacti.

grain the smallest unit of mass in the three English systems (avoirdupois, troy, and apothecaries' weights) used in the UK and USA, equal to 0.0648 g. It was reputedly the weight of a grain of wheat. One pound avoirdupois equals 7,000 grains; one pound troy or apothecaries' weight equals 5,760 grains.

gram metric unit of mass; one-thousandth of a kilogram.

gramophone old-fashioned English name for a record player or stereo. US inventor Thomas Edison's original name for the machine, and the traditional US name, was *phonograph*.

grand unified theory (GUT) in physics, a sought-for theory that would combine the theory of the strong nuclear force (called ◊quantum chromodynamics) with the theory of the weak nuclear and electromagnetic forces. The search for the grand unified theory is part of a larger programme seeking a ◊unified field theory, which

gradient of a curve

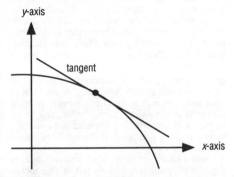

the gradient to a curve at any point is equal to the
gradient of the tangent drawn touching that point

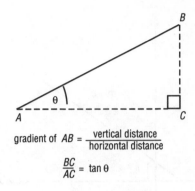

$$\text{gradient of } AB = \frac{\text{vertical distance}}{\text{horizontal distance}}$$

$$\frac{BC}{AC} = \tan\theta$$

gradient

would combine all the forces of nature (including
gravity) within one framework.

granite coarse-grained ◊igneous rock, typically
consisting of the minerals quartz, feldspar, and
mica. It may be pink or grey, depending on the
composition of the feldspar. Granites are chiefly
used as building materials.

Granites often form large intrusions in the core
of mountain ranges, and they are usually surrounded by zones of ◊metamorphic rock (rock that
has been altered by heat or pressure). Granite areas
have characteristic moorland scenery. In exposed
areas the bedrock may be weathered along joints
and cracks to produce a tor, consisting of rounded
blocks that appear to have been stacked upon one
another.

graph pictorial representation of numerical data,

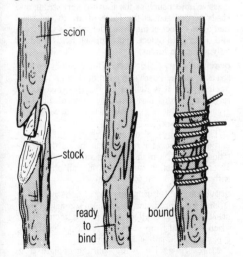

grafting *Grafting, a method of artificial propagation in plants, is
commonly used in the propagation of roses and fruit trees. A
relatively small part, the scion, of one plant is attached to another
plant so that growth continues. The plant receiving the transplanted
material is called the stock.*

such as statistical data, or a method of showing
the mathematical relationship between two or more
variables by drawing a diagram.

There are often two axes, or reference lines, at
right angles intersecting at the origin—the zero
point, from which values of the variables (for
example, distance and time for a moving object)
are assigned along the axes. Pairs of simultaneous
values (the distance moved after a particular time)
are plotted as points in the area between the axes,
and the points then joined by a smooth curve to
produce a graph.

graphical user interface (GUI) or *WIMP* in
computing, a type of ◊user interface in which programs and files appear as icons (small pictures),
user options are selected from pull-down menus,
and data are displayed in windows (rectangular
areas), which the operator can manipulate in various ways. The operator uses a pointing device,
typically a ◊mouse, to make selections and initiate
actions.

The concept of the graphical user interface was
developed by the Xerox Corporation in the 1970s,
was popularized with the Apple Macintosh computers in the 1980s, and is now available on many
types of computer—most notably as Windows, an
operating system for IBM PC-compatible microcomputers developed by the software company
Microsoft.

graphic equalizer control used in hi-fi systems
that allows the distortions introduced by unequal
amplification of different frequencies to be
corrected.

The frequency range of the signal is divided into
separate bands, usually third-octave bands. The
amplification applied to each band is adjusted by a
sliding contact; the position of the contact indicates
the strength of the amplification applied to each
frequency range.

graphics see ◊computer graphics.

graphics tablet or *bit pad* in computing, an input
device in which a stylus or cursor is moved, by
hand, over a flat surface. The computer can keep
track of the position of the stylus, so enabling the

graph of a straight line

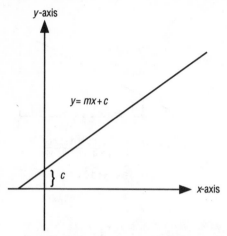

the equation of the straight-line graph takes the form $y = mx + c$, where m is the gradient (slope) of the line, and c is the y-intercept (the value of y where the line cuts the y-axis)

for example, a graph of the equation $y = -x + 4$ will have a gradient of -1 and will cut the y-axis at $y = 4$

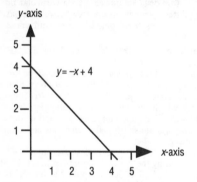

graph

operator to input drawings or diagrams into the computer.

A graphics tablet is often used with a form overlaid for users to mark boxes in positions that relate to specific registers in the computer, although recent developments in handwriting recognition may increase its future versatility.

graphite blackish-grey, laminar, crystalline form of ◊carbon. It is used as a lubricant and as the active component of pencil lead.

The carbon atoms are strongly bonded together in sheets, but the bonds between the sheets are weak so that the sheets are free to slide over one another. Graphite has a very high melting point (3,500°C/6,332°F), and is a good conductor of heat and electricity. In its pure form it is used as a moderator in nuclear reactors.

graph plotter alternative name for a ◊plotter.

grass plant of the large family Gramineae of monocotyledons, with about 9,000 species distributed worldwide except in the Arctic regions. The majority are perennial, with long, narrow leaves and jointed, hollow stems; hermaphroditic flowers are borne in spikelets; the fruits are grainlike. Included are bluegrass, wheat, rye, maize, sugarcane, and bamboo.

gravel coarse ◊sediment consisting of pebbles or small fragments of rock, originating in the beds of lakes and streams or on beaches. Gravel is quarried for use in road building, railway ballast, and for an aggregate in concrete. It is obtained from quarries known as gravel pits, where it is often found mixed with sand or clay.

Some gravel deposits also contain ◊placer deposits of metal ores (chiefly tin) or free metals (such as gold and silver).

gravimetric analysis in chemistry, a technique for determining, by weighing, the amount of a particular substance present in a sample. It usually involves the conversion of the test substance into a compound of known molecular weight that can be easily isolated and purified.

gravimetry study of the Earth's gravitational field. Small variations in the gravitational field (gravimetric anomalies) can be caused by varying densities of rocks and structure beneath the surface. Such variations are measured by a device called a gravimeter, which consists of a weighted spring that is pulled further downwards where the gravity is stronger (at a ◊Bouguer anomaly).

Gravimetry is used by geologists to map the subsurface features of the Earth's crust, such as underground masses of heavy rock like granite, or light rock like salt.

gravitational field the region around a body in which other bodies experience a force due to its gravitational attraction. The gravitational field of a massive object such as the Earth is very strong and easily recognized as the force of gravity, whereas that of an object of much smaller mass is very weak and difficult to detect. Gravitational fields produce only attractive forces.

gravitational field strength (symbol g) the strength of the Earth's gravitational field at a particular point. It is defined as the the gravitational force in newtons that acts on a mass of one kilogram. The value of g on the Earth's surface is taken to be 9.806 N kg^{-1}.

The symbol g is also used to represent the acceleration of a freely falling object in the Earth's gravitational field. Near the Earth's surface and in the absence of friction due to the air, all objects fall with an acceleration of 9.806 m s^{-2}.

gravitational lensing bending of light by a gravitational field, predicted by Einstein's general theory of relativity. The effect was first detected 1917 when the light from stars was found to be bent as it passed the totally eclipsed Sun.

More remarkable is the splitting of light from distant quasars into two or more images by intervening galaxies. In 1979 the first double image of a quasar produced by gravitational lensing was discovered and a quadruple image of another quasar was later found.

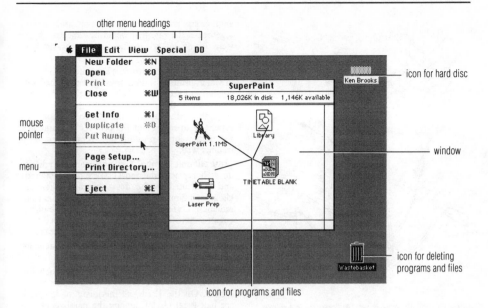

other menu headings

icon for hard disc

mouse pointer

menu

window

icon for deleting programs and files

icon for programs and files

graphical user interface

gravitational potential energy energy stored by an object when it is placed in a position from which it can fall under the influence of gravity. The gravitational potential energy E_p of an object of mass m kg placed at a height h m above the ground is given by the formula:

$$E_p = mgh$$

where ◊g is the gravitational field strength in N kg^{-1} of the Earth at the place.

In a ◊hydroelectric power station, gravitational potential energy stored in water held in a high-level reservoir is used to drive turbines to produce electricity.

graphics tablet

graviton in physics, the ◊gauge boson that is the postulated carrier of gravity.

gravity force of attraction that arises between objects by virtue of their masses. On Earth, gravity is the force of attraction between any object in the Earth's gravitational field and the Earth itself. It is regarded as one of the four ◊fundamental forces of nature, the other three being the ◊electromagnetic force, the ◊strong nuclear force, and the ◊weak nuclear force. The gravitational force is the weakest of the four forces, but it acts over great distances. The particle that is postulated as the carrier of the gravitational force is the ◊graviton.

According to Newton's law of gravitation, all objects fall to Earth with the same acceleration, regardless of mass. For an object of mass m_1 at a distance r from the centre of the Earth (mass m_2), the gravitational force of attraction F equals Gm_1m_2/r^2, where G is the gravitational constant. However, according to Newton's second law of motion, F also equals m_1g, where g is the acceleration due to gravity; therefore $g = Gm_2/r^2$ and is independent of the mass of the object; at the Earth's surface it equals 9.806 metres per second per second.

Were it not for gravity one man might hurl another by a puff of his breath into the depths of space, beyond recall for all eternity.

On **gravity** Ruggiero Boscovich (1711–1787)
Theoria

Einstein's general theory of relativity treats gravitation not as a force but as a curvature of space and time around a body. Relativity predicts the bending of light and the ◊red shift of light in a gravitational field; both have been observed. Another prediction of relativity is **gravitational**

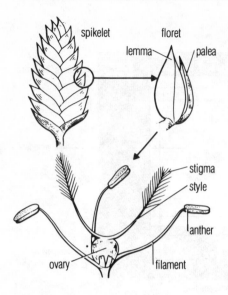

grass *The parts of a grass flower. The flowers are in a spikelet arranged on a central axis. The floret, on the right, contains three large anthers and feathery stigmas, catching the floating pollen grains like a net.*

waves, which should be produced when massive bodies are violently disturbed. These waves are so weak that they have not yet been detected with certainty, although observations of a ◊pulsar (which emits energy at regular intervals) in orbit around another star have shown that the stars are spiralling together at the rate that would be expected if they were losing energy in the form of gravitational waves.

gravure one of the three main ◊printing methods, in which printing is done from a plate etched with a pattern of recessed cells in which the ink is held. The greater the depth of a cell, the greater the strength of the printed ink. Gravure plates are expensive to make, but the process is economical for high-volume printing and reproduces illustrations well.

gray SI unit (symbol Gy) of absorbed radiation dose. It replaces the rad (1 Gy equals 100 rad), and is defined as the dose absorbed when one kilogram of matter absorbs one joule of ionizing radiation. Different types of radiation cause different amounts of damage for the same absorbed dose; the SI unit of *dose equivalent* is the ◊sievert.

Great Bear popular name for the constellation ◊Ursa Major.

great circle circle drawn on a sphere such that the diameter of the circle is a diameter of the sphere. On the Earth, all meridians of longitude are half great circles; among the parallels of latitude, only the equator is a great circle.

The shortest route between two points on the Earth's surface is along the arc of a great circle. These are used extensively as air routes although on maps, owing to the distortion brought about by ◊projection, they do not appear as straight lines.

Great Red Spot prominent oval feature, 14,000 km/8,500 mi wide and some 30,000 km/

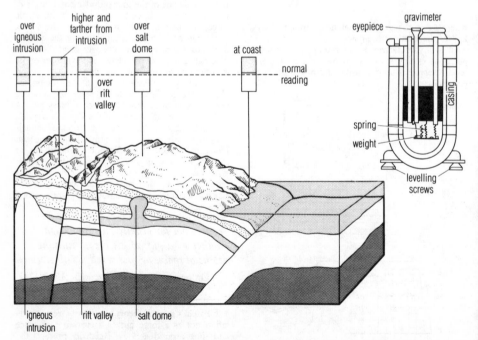

gravimetry *The gravimeter is an instrument for measuring the force of gravity at a particular location. Variations in the force of gravity acting on a weight suspended by a spring cause the spring to stretch. The gravimeter is flown in an aircraft over an area to be surveyed. Geological features such as intrusions and salt domes are revealed by the stretching of the spring.*

20,000 mi long, in the atmosphere of the planet ◊Jupiter, S of the equator. It was first observed in the 19th century. Space probes show it to be an anticlockwise vortex of cold clouds, coloured possibly by phosphorus.

Green Bank site in West Virginia, USA, of the National Radio Astronomy Observatory. Its main instruments are a 43–m/140–ft fully steerable dish, opened 1965, and three 26–m/85–ft dishes. A 90–m/300–ft partially steerable dish, opened 1962, collapsed 1988 because of metal fatigue; a replacement dish, 100 m/330 ft across, was under construction 1993.

green audit inspection of a company's accounts to assess the total environmental impact of its activities or of a particular product or process.

For example, a green audit of a manufactured product looks at the impact of production (including energy use and the extraction of raw materials used in manufacture), use (which may cause pollution and other hazards), and disposal (potential for recycling, and whether waste causes pollution). Companies are increasingly using green audits to find ways of reducing their environmental impact.

greenhouse effect phenomenon of the Earth's atmosphere by which solar radiation, trapped by the Earth and re-emitted from the surface, is prevented from escaping by various gases in the air. The result is a rise in the Earth's temperature. The main greenhouse gases are carbon dioxide, methane, and ◊chlorofluorocarbons (CFCs). Fossil-fuel consumption and forest fires are the main causes of carbon-dioxide buildup; methane is a byproduct of agriculture (rice, cattle, sheep). Water vapour is another greenhouse gas.

The United Nations Environment Programme estimates that by 2025 average world temperatures will have risen by 1.5°C with a consequent rise of 20 cm in sea level. Low-lying areas and entire countries would be threatened by flooding and crops would be affected by the change in climate. However, predictions about global warming and its possible climatic effects are tentative and often conflict with each other.

Dubbed the 'greenhouse effect' by Swedish scientist Svante Arrhenius, it was first predicted 1827 by French mathematician Joseph Fourier (1768–1830).

green movement collective term for the individuals and organizations involved in efforts to protect the environment. The movement encompasses political parties such as the Green Party and organizations like ◊Friends of the Earth and ◊Greenpeace.

Despite a rapid growth of public support, and membership of environmental organizations running into many millions worldwide, political green groups have failed to win significant levels of the vote in democratic societies.

Greenpeace international environmental pressure group, founded 1971, with a policy of nonviolent direct action backed by scientific research. During a protest against French atmospheric nuclear testing in the S Pacific 1985, its ship *Rainbow Warrior* was sunk by French intelligence agents, killing a crew member.

Greenwich Mean Time (GMT) local time on the zero line of longitude (the *Greenwich meridian*), which passes through the Old Royal Observatory at Greenwich, London. It was replaced 1986 by coordinated universal time (UTC); see ◊time.

grid network by which electricity is generated and distributed over a region or country. It contains many power stations and switching centres and allows, for example, high demand in one area to be met by surplus power generated in another.

The term is also used for any grating system, as in a cattle grid for controlling the movement of livestock across roads, and a conductor in a storage battery or electron gun. In trigonometry, a grid is a network of uniformly spaced vertical and horizontal lines as used to locate points on a map or to construct a graph.

grid reference on a map, numbers that are used to show location. The numbers at the bottom of the map (eastings) are given before those at the side (northings). A four-figure grid reference indicates a specific square, whereas a six-figure grid reference indicates a point within a square.

grooming in biology, the use by an animal of teeth, tongue, feet, or beak to clean fur or feathers. Grooming also helps to spread essential oils for waterproofing. In many social species, notably monkeys and apes, grooming of other individuals is used to reinforce social relationships.

ground water water collected underground in porous rock strata and soils; it emerges at the surface as springs and streams. The groundwater's upper level is called the *water table*. Sandy or other kinds of beds that are filled with groundwater are called *aquifers*. Recent estimates are that usable ground water amounts to more than 90% of all the fresh water on Earth; however, keeping such supplies free of pollutants entering the recharge areas is a critical environmental concern.

Most groundwater near the surface moves slowly through the ground while the water table stays in the same place. The depth of the water table reflects the balance between the rate of infiltration, called recharge, and the rate of discharge at springs or rivers or pumped water wells. The force of gravity makes underground water run 'downhill' underground just as it does above the surface. The greater the slope and the permeability, the greater the speed. Velocities vary from 40 in/100 cm per day to 0.2 in/0.5 cm.

The water table dropped in the UK during the drought of the early 1990s; however, under cities such as London and Liverpool the water table rose because the closure of industries meant that less water was being removed.

group in chemistry, a vertical column of elements in the ◊periodic table. Elements in a group have similar physical and chemical properties; for example, the group I elements (the alkali metals: lithium, sodium, potassium, rubidium, caesium, and francium) are all highly reactive metals that form univalent ions. There is a gradation of properties down any group: in group I, melting and boiling points decrease, and density and reactivity increase.

group in mathematics, a finite or infinite set of elements that can be combined by an operation;

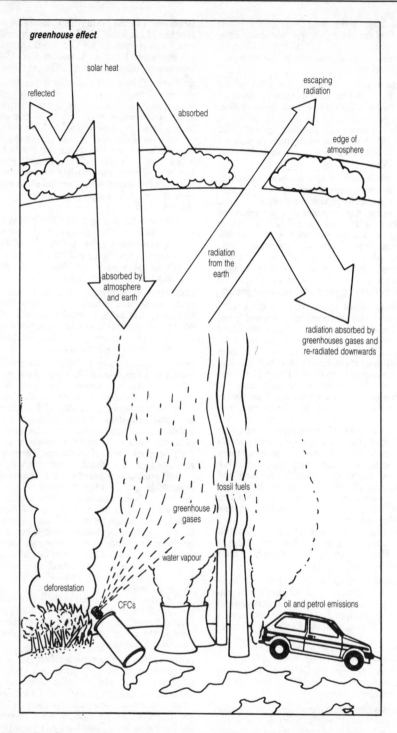

greenhouse effect *The warming effect of the Earth's atmosphere is called the greenhouse effect. Radiation from the Sun enters the atmosphere but is prevented from escaping back into space by gases such as carbon dioxide (produced, for example, by the burning of fossil fuels), nitrogen oxides (from car exhausts), and CFCs (from aerosols and refrigerators). As these gases build up in the atmosphere, the Earth's average temperature is expected to rise.*

formally, a group must satisfy certain conditions. For example, the set of all integers (positive or negative whole numbers) forms a group with regard to addition because: (1) addition is associative, that is, the sum of two or more integers is the same regardless of the order in which the integers are added; (2) adding two integers gives another integer; (3) the set includes an identity element 0, which has no effect on any integer to which it is added (for example, 0 + 3 = 3); and (4) each integer has an inverse (for instance, 7 has the inverse −7), such that the sum of an integer and its inverse is 0. *Group theory* is the study of the properties of groups.

growth in biology, the increase in size and weight during the development of an organism. Growth is an increase in biomass (mass of organic material, excluding water) and is associated with cell division.

All organisms grow, although the rate of growth varies over a lifetime. Typically, an organism shows an S-shaped curve, in which growth is at first slow, then fast, then, towards the end of life, nonexistent. Growth may even be negative during the period before death, with decay occurring faster than cellular replacement.

growth ring another name for ◊annual ring.

GRP abbreviation for *glass-reinforced plastic*, a plastic material strengthened by glass fibres, usually erroneously known as ◊fibreglass. GRP is a favoured material for boat hulls and for the bodies and some structural components of performance cars; it is also used in the manufacture of passenger cars.

Products are usually moulded, mats of glass fibre being sandwiched between layers of a polyester plastic, which sets hard when mixed with a curing agent.

gryke enlarged ◊joint that separates blocks of limestone (clints) in a ◊limestone pavement.

g scale scale for measuring force by comparing it with the force due to ◊gravity (*g*), often called ◊G-force.

guard cell in plants, a specialized cell on the undersurface of leaves for controlling gas exchange and water loss. Guard cells occur in pairs and are shaped so that a pore, or ◊stomata, exists between them. They can change shape with the result that the pore disappears. During warm weather, when a plant is in danger of losing excessive water, the guard cells close, cutting down evaporation from the interior of the leaf.

GUI in computing, abbreviation for ◊*graphical user interface*.

Gulf Stream warm ocean ◊current that flows north from the warm waters of Gulf of Mexico. Part of the current is diverted east across the Atlantic, where it is known as the *North Atlantic Drift*, and warms what would otherwise be a colder climate in the British Isles and NW Europe.

gum in botany, complex polysaccharides (carbohydrates) formed by many plants and trees, particularly by those from dry regions. They form four main groups: plant exudates (gum arabic); marine plant extracts (agar); seed extracts; and fruit and vegetable extracts. Some are made synthetically.

Gums are tasteless and odourless, insoluble in alcohol and ether but generally soluble in water.

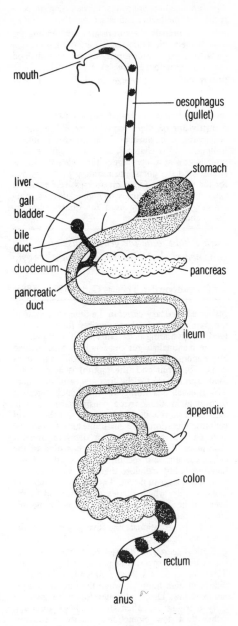

gut *The human gut, or alimentary canal. When food is swallowed, it is moved down the oesophagus by the action of muscles into the stomach. Digestion starts in the stomach as the food is mixed with enzymes and strong acid. After several hours, the food passes to the small intestine. Here more enzymes are added and digestion is completed. After all nutrients have been absorbed, the undigestible parts pass into the large intestine and thence to the rectum. The liver has many functions, such as storing minerals and vitamins and making bile, which is stored in the gall bladder until needed for the digestion of fats. The pancreas supplies enzymes. The appendix seems to have no function in human beings.*

They are used for adhesives, fabric sizing, in confectionery, medicine, and calico printing.

gum in mammals, the soft tissues surrounding the base of the teeth. Gums are liable to inflammation (gingivitis) or to infection by microbes from food deposits (periodontal disease).

gun metal type of ◊bronze, an alloy high in copper (88%), also containing tin and zinc, so-called because it was once used to cast cannons. It is tough, hard-wearing, and resists corrosion.

gunpowder or *black powder* the oldest known ◊explosive, a mixture of 75% potassium nitrate (saltpetre), 15% charcoal, and 10% sulphur. Sulphur ignites at a low temperature, charcoal burns readily, and the potassium nitrate provides oxygen for the explosion. Although progressively replaced since the late 19th century by high explosives, gunpowder is still widely used for quarry blasting, fuses, and fireworks.

Such I hold to be the genuine use of gunpowder; that it makes all men alike tall.

On **gunpowder** Thomas Carlyle (1795–1881)

---

gut or *alimentary canal* in the ◊digestive system, the part of an animal responsible for processing food and preparing it for entry into the blood.

The mammalian gut consists of a tube divided into regions capable of performing different functions. The front end (the mouth) is adapted for food capture and for the first stages of digestion. The stomach is a storage area, although digestion of protein by the enzyme pepsin starts here; in many herbivorous mammals this is also the site of cellulose digestion. The small intestine follows the stomach and is specialized for digestion and for absorption. The large intestine, consisting of the colon, caecum, and rectum, has a variety of functions, including cellulose digestion, water absorption, and storage of faeces. The gut has an excellent blood supply, which carries digested food to the liver via the hepatic portal vein, ready for assimilation by the cells.

In birds, additional digestive organs are the ◊crop and ◊gizzard.

guttation secretion of water on to the surface of leaves through specialized pores, or ◊hydathodes. The process occurs most frequently during conditions of high humidity when the rate of transpiration is low. Drops of water found on grass in early morning are often the result of guttation, rather than dew. Sometimes the water contains minerals in solution, such as calcium, which leaves a white crust on the leaf surface as it dries.

guyot truncated or flat-topped seamount. Such undersea mountains are found throughout the abyssal plains of major ocean basins, and most of them are covered by an appreciable depth of water. They are believed to have started as volcanic cones formed near mid-oceanic ridges, in relatively shallow water, and to have been truncated by wave action as their tops emerged above the surface. They were then transported to deeper waters as the ocean floor moved away on either side of the ridge.

gymnosperm (Greek 'naked seed') in botany, any plant whose seeds are exposed, as opposed to the structurally more advanced ◊angiosperms, where they are inside an ovary. The group includes conifers and related plants such as cycads and ginkgos, whose seeds develop in ◊cones. Fossil gymnosperms have been found in rocks about 350 million years old.

gynoecium or *gynaecium* collective term for the female reproductive organs of a flower, consisting of one or more ◊carpels, either free or fused together.

gypsum common ◊mineral, composed of hydrous calcium sulphate, $CaSO_4.2H_2O$. It ranks 2 on the Mohs' scale of hardness. Gypsum is used for making casts and moulds, and for blackboard chalk.

A fine-grained gypsum, called *alabaster*, is used for ornamental work. Burned gypsum is known as *plaster of Paris*, because for a long time it was obtained from the gypsum quarries of the Montmartre district of Paris.

gyre circular surface rotation of ocean water in each major sea (a type of ◊current). Gyres are large and permanent, and occupy the northern and southern halves of the three major oceans. Their movements are dictated by the prevailing winds and the ◊Coriolis effect. Gyres move clockwise in the northern hemisphere and anticlockwise in the southern hemisphere.

gyroscope mechanical instrument, used as a stabilizing device and consisting, in its simplest form, of a heavy wheel mounted on an axis fixed in a ring that can be rotated about another axis, which is also fixed in a ring capable of rotation about a third axis. Applications of the gyroscope principle include the gyrocompass, the gyropilot for automatic steering, and gyro-directed torpedoes.

The components of the gyroscope are arranged so that the three axes of rotation in any position pass through the wheel's centre of gravity. The wheel is thus capable of rotation about three mutually perpendicular axes, and its axis may take up any direction. If the axis of the spinning wheel is displaced, a restoring movement develops, returning it to its initial direction.

ha symbol for ◊*hectare*.

Haber process or *Haber–Bosch process* industrial process by which ammonia is manufactured by direct combination of its elements, nitrogen and hydrogen. The reaction is carried out at 400–500°C/752–932°F and at 200 atmospheres pressure. The two gases, in the proportions of 1:3 by volume, are passed over a ◊catalyst of finely divided iron. Around 10% of the reactants combine, and the unused gases are recycled. The ammonia is separated either by being dissolved in water or by being cooled to liquid form.

$$N_2 + 3H_2 \rightleftharpoons 2NH_3$$

habitat localized ◊environment in which an organism lives, and which provides for all (or almost all) of its needs. The diversity of habitats found within the Earth's ecosystem is enormous, and they are changing all the time. Many can be considered inorganic or physical, for example the Arctic ice cap, a cave, or a cliff face. Others are more complex, for instance a woodland or a forest floor. Some habitats are so precise that they are called *microhabitats*, such as the area under a stone where a particular type of insect lives. Most habitats provide a home for many species.

hacking unauthorized access to a computer, either for fun or for malicious or fraudulent purposes. Hackers generally use microcomputers and telephone lines to obtain access. In computing, the term is used in a wider sense to mean using software for enjoyment or self-education, not necessarily involving unauthorized access. See also computer ◊virus.

hadal zone the deepest level of the ocean, below the abyssal zone. It begins at depths of 6,000 m/ 19,500 ft and is the habitat of the ocean trenches.

Hadley cell in the atmosphere, a vertical circulation of air caused by convection. The typical Hadley cell (named after English meteorologist George Hadley) occurs in the tropics, where hot air over the equator in the ◊intertropical convergence zone (ITCZ) rises, giving the heavy rain associated with tropical rainforests. In the upper atmosphere this now-dry air then spreads north and south and, cooling, descends in the latitudes of the tropics, producing the northern and southern tropical desert belts. After that, the air is drawn back towards the equator, forming the northeast and southeast trade winds.

hadron in physics, a subatomic particle that experiences the strong nuclear force. Each is made up of two or three indivisible particles called ◊quarks. The hadrons are grouped into the ◊baryons (protons, neutrons, and hyperons) and the ◊mesons (particles with masses between those of electrons and protons).

haemoglobin protein used by all vertebrates and some invertebrates for oxygen transport because the two substances combine reversibly. In vertebrates it occurs in red blood cells (erythrocytes), giving them their colour.

In the lungs or gills where the concentration of oxygen is high, oxygen attaches to haemoglobin to form *oxyhaemoglobin*. This process effectively increases the amount of oxygen that can be carried in the bloodstream. The oxygen is later released in the body tissues where it is at a low concentration, and the deoxygenated blood returned to the lungs or gills. Haemoglobin will combine also with carbon monoxide to form carboxyhaemoglobin, but in this case the reaction is irreversible.

HAEMOGLOBIN: BLOOD MOLECULES

A healthy person has 6,000 million million million haemoglobin molecules, with 400 million million being destroyed every second and replaced by new ones.

---

haemolymph circulatory fluid of those molluscs and insects that have an 'open' circulatory system. Haemolymph contains water, amino acids, sugars, salts, and white cells like those of blood. Circulated by a pulsating heart, its main functions are to transport digestive and excretory products around the

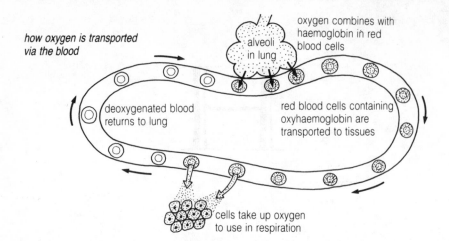

how oxygen is transported
via the blood

oxygen combines with
haemoglobin in red
blood cells

alveoli
in lung

deoxygenated blood
returns to lung

red blood cells containing
oxyhaemoglobin are
transported to tissues

cells take up oxygen
to use in respiration

haemoglobin

body. In molluscs, it also transports oxygen and carbon dioxide.

haemolysis destruction of red blood cells. Aged cells are constantly being lysed (broken down), but increased wastage of red cells is seen in some infections and blood disorders. It may result in jaundice (through the release of too much haemoglobin) and in ◊anaemia.

haemophilia any of several inherited diseases in which normal blood clotting is impaired. The sufferer experiences prolonged bleeding from the slightest wound, as well as painful internal bleeding without apparent cause.

Haemophilias are nearly always sex-linked, transmitted through the female line only to male infants; they have afflicted a number of European royal households. Males affected by the most common form are unable to synthesize Factor VIII, a protein involved in the clotting of blood. Treatment is primarily with Factor VIII (now mass-produced by recombinant techniques), but the haemophiliac remains at risk from the slightest incident of bleeding. The disease is a painful one that causes deformities of joints.

hafnium (Latin *Hafnia* 'Copenhagen') silvery, metallic element, symbol Hf, atomic number 72, relative atomic mass 178.49. It occurs in nature in ores of zirconium, the properties of which it resembles. Hafnium absorbs neutrons better than most metals, so it is used in the control rods of nuclear reactors; it is also used for light-bulb filaments.

It was named in 1923 by Dutch physicist Dirk Coster (1889–1950) and Hungarian chemist Georg von Hevesy after the city of Copenhagen, where the element was discovered.

hahnium name proposed by US scientists for the element also known as ◊unnilpentium (atomic number 105), in honour of German nuclear physicist Otto Hahn. The symbol is Ha.

hail precipitation in the form of pellets of ice (hailstones). It is caused by the circulation of moisture in strong convection currents, usually within cumulonimbus ◊clouds.

Water droplets freeze as they are carried upwards. As the circulation continues, layers of ice are deposited around the droplets until they become too heavy to be supported by the currents and they fall as a hailstorm.

hair threadlike structure growing from mammalian skin. Each hair grows from a pit-shaped follicle embedded in the second layer of the skin, the dermis. It consists of dead cells impregnated with the protein keratin.

There are about a million hairs on the average person's head. Each grows at the rate of 5–10 mm/0.2–0.4 in per month, lengthening for about three years before being replaced by a new one. A coat of hair helps to insulate land mammals by trapping air next to the body. It also aids camouflage and protection, and its colouring or erection may be used for communication. In 1990 scientists succeeded for the first time in growing human hair *in vitro*.

half-life during ◊radioactive decay, the time in which the strength of a radioactive source decays to half its original value. In theory, the decay process is never complete and there is always some residual radioactivity. For this reason, the half-life (the time taken for 50% of the isotope to decay) is measured, rather than the total decay time. It may vary from millionths of a second to billions of years.

Radioactive substances decay exponentially; thus the time taken for the first 50% of the isotope to decay will be the same as the time taken by the next 25%, and by the 12.5% after that, and so on. For example, carbon-14 takes about 5,730 years for half the material to decay; another 5,730 for half of the remaining half to decay; then 5,730 years for half of that remaining half to decay, and so on. Plutonium-239, one of the most toxic of all radioactive substances, has a half-life of about 24,000 years.

halftone process technique used in printing to reproduce the full range of tones in a photograph or other illustration. The intensity of the printed colour is varied from full strength to the lightest shades, even if one colour of ink is used. The

half-lives of some radioactive isotopes

isotope	half-life
(least stable)	
lithium-5	4.4×10^{-22} sec
polonium-213	4.2×10^{-6} sec
lead-211	36 min
lead-209	3.3 hours
uranium-238	4.551×10^9 years
thorium-232	1.39×10^{10} years
tellurium-128	1.5×10^{24} years
(most stable)	

picture to be reproduced is photographed through a screen ruled with a rectangular mesh of fine lines, which breaks up the tones of the original into areas of dots that vary in frequency according to the intensity of the tone. In the darker areas the dots run together; in the lighter areas they have more space between them.

Colour pictures are broken down into a pattern of dots in the same way, the original being photographed through a number of colour filters. The process is known as *colour separation*. Plates made from the separations are then printed in sequence, yellow, magenta (blue-red), cyan (blue-green), and black, which combine to give the full colour range (see ◊four-colour process).

halide any compound produced by the combination of a ◊halogen, such as chlorine or iodine, with a less electronegative element (see ◊electronegativity). Halides may be formed by ◊ionic bonds or by ◊covalent bonds.

halite mineral sodium chloride, NaCl, or common ◊salt. When pure it is colourless and transparent, but it is often pink, red, or yellow. It is soft and has a low density.

Halite occurs naturally in evaporite deposits that have precipitated on evaporation of bodies of salt water. As rock salt, it forms beds within a sedimentary sequence; it can also migrate upwards through surrounding rocks to form salt domes. It crystallizes in the cubic system.

Hall effect production of a voltage across a conductor or semiconductor carrying a current at a right angle to a surrounding magnetic field. It was discovered 1897 by the US physicist Edwin Hall (1855–1938). It is used in the *Hall probe* for measuring the strengths of magnetic fields and in magnetic switches.

Halley's comet comet that orbits the Sun about every 76 years, named after Edmond Halley who calculated its orbit. It is the brightest and most conspicuous of the periodic comets. Recorded sightings go back over 2,000 years. It travels around the Sun in the opposite direction to the planets. Its orbit is inclined at almost 20° to the main plane of the Solar System and ranges between the orbits of Venus and Neptune. It will next reappear 2061.

The comet was studied by space probes at its last appearance 1986. The European probe ◊*Giotto* showed that the nucleus of Halley's comet is a tiny and irregularly shaped chunk of ice, measuring some 15 km/10 m long by 8 km/5 m wide, coated by a layer of very dark material, thought to be composed of carbon-rich compounds. This surface coating has a very low ◊albedo, reflecting just 4% of the light it receives from the Sun. Although the comet is one of the darkest objects known, it has a glowing head and tail produced by jets of gas from fissures in the outer dust layer. These vents cover 10% of the total surface area and become active only when exposed to the Sun. The force of these jets affects the speed of the comet's travel in its orbit.

HALLEY'S COMET: THE LONG-HAIRED STAR

The *Anglo-Saxon Chronicle* records the 1066 visit of Halley's comet: 'Then there was seen all over England a sign such as no one had ever seen. Some said that the star was a comet, as some called the long-haired star. It had a tail streaming like smoke up to nearly half the sky.'

halogen any of a group of five nonmetallic elements with similar chemical bonding properties: fluorine, chlorine, bromine, iodine, and astatine. They form a linked group in the ◊periodic table of the elements, descending from fluorine, the most reactive, to astatine, the least reactive. They combine directly with most metals to form salts, such as common salt (NaCl). Each halogen has seven electrons in its valence shell, which accounts for the chemical similarities displayed by the group.

halon organic chemical compound containing one or two carbon atoms, together with ◊bromine and other ◊halogens. The most commonly used are halon 1211 (bromochlorodifluoromethane) and halon 1301 (bromotrifluoromethane). The halons are gases and are widely used in fire extinguishers. As destroyers of the ◊ozone layer, they are up to ten times more effective than ◊chlorofluorocarbons (CFCs), to which they are chemically related.

Levels in the atmosphere are rising by about 25% each year, mainly through the testing of fire-fighting equipment.

halophyte plant adapted to live where there is a high concentration of salt in the soil, for example, in salt marshes and mud flats.

Halophytes contain a high percentage of salts in their root cells, so that, despite the salt in the soil, water can still be taken up by the process of ◊osmosis. Some species also have fleshy leaves for storing water, such as sea blite *Suaeda maritima* and sea rocket *Cakile maritima*.

handshake in computing, an exchange of signals between two devices that establishes the communications channels and protocols necessary for the devices to send and receive data.

hanging valley valley that joins a larger glacial trough at a higher level than the trough floor. During glaciation the ice in the smaller valley was unable to erode as deeply as the ice in the trough, and so the valley was left perched high on the side of the trough when the ice retreated. A river or stream flowing along the hanging valley often forms a waterfall as it enters the trough.

haploid having a single set of ◊chromosomes in each cell. Most higher organisms are ◊diploid – that is, they have two sets—but their gametes (sex cells) are haploid. Some plants, such as mosses, liverworts, and many seaweeds, are haploid, and male honey bees are haploid because they develop from eggs that have not been fertilized. See also ◊meiosis.

hard disc in computing, a storage device usually consisting of a rigid metal ◊disc coated with a magnetic material. Data are read from and written to the disc by means of a disc drive. The hard disc may be permanently fixed into the drive or in the form of a disc pack that can be removed and exchanged with a different pack. Hard discs vary from large units with capacities of more than 3,000 megabytes, intended for use with mainframe computers, to small units with capacities as low as 20 megabytes, intended for use with microcomputers.

hardening of oils transformation of liquid oils to solid products by ◊hydrogenation.
Vegetable oils contain double covalent carbon-to-carbon bonds and are therefore examples of ◊unsaturated compounds. When hydrogen is added to these double bonds, the oils become saturated. The more saturated oils are waxlike solids.

hardness physical property of materials that governs their use. Methods of heat treatment can increase the hardness of metals. A scale of hardness was devised by German–Austrian mineralogist Friedrich Mohs in the 1800s, based upon the hardness of certain minerals from soft talc (Mohs hardness 1) to diamond (10), the hardest of all materials.
See also ◊Brinell hardness test, ◊hard water.

hard-sectored disc floppy disc that is sold already formatted, so that ◊disc formatting is not necessary. Usually sectors are marked by holes near the hub of the disc. This system is now obsolete.

hardware the mechanical, electrical, and electronic components of a computer system, as opposed to the various programs, which constitute ◊software.
Hardware associated with a microcomputer might include the power supply and housing of its processor unit, its circuit boards, VDU (screen), disc drive, keyboard, and printer.

hard water water that does not lather easily with soap, and produces 'fur' or 'scale' in kettles. It is caused by the presence of certain salts of calcium and magnesium.
Temporary hardness is caused by the presence of dissolved hydrogencarbonates (bicarbonates); when the water is boiled, they are converted to insoluble carbonates that precipitate as 'scale'. *Permanent hardness* is caused by sulphates and silicates, which are not affected by boiling. Water can be softened by ◊distillation, ◊ion exchange (the principle underlying commercial water softeners), addition of sodium carbonate or of large amounts of soap, or boiling (to remove temporary hardness).

Hardy–Weinberg equilibrium in population genetics, the theoretical relative frequency of different ◊alleles within a given population of a species, when the stable endpoint of evolution in an undisturbed environment is reached.

Hare's apparatus in physics, a specific kind of ◊hydrometer used to compare the relative densities of two liquids, or to find the density of one if the

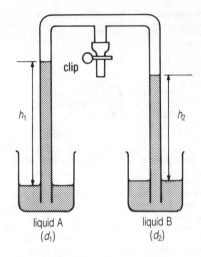

Hare's apparatus *Hare's apparatus, used to compare the density of two liquids. When air is removed from the top of the apparatus, the liquids rise in the tubes to heights which are inversely proportional to their densities.*

other is known. It was invented by US chemist Robert Hare (1781–1858).
It consists of a vertical E-shaped glass tube, with the long limbs dipping into the two liquids and a tap on the short limb. With the tap open, air is removed from the tops of the tubes and the liquids are pushed up the tubes by atmospheric pressure. When the tap is closed, the heights of the liquids are inversely proportional to their relative densities.
If a liquid of relative density d_1 rises to a height h_1, and liquid d_2 rises to h_2, then $d_1/d_2 = h_2/h_1$.

harmattan in meteorology, a dry and dusty NE wind that blows over W Africa.

Harrier the only truly successful vertical takeoff and landing fixed-wing aircraft, often called the *jump jet*. Built in Britain, it made its first flight 1966. It has a single jet engine and a set of swivelling nozzles. These deflect the jet exhaust vertically downwards for takeoff and landing, and to the rear for normal flight. Designed to fly from confined spaces with minimal ground support, it refuels in midair.

harrow agricultural implement used to break up the furrows left by the ◊plough and reduce the soil to a fine consistency or tilth, and to cover the seeds after sowing. The traditional harrow consists of spikes set in a frame; modern harrows use sets of discs.

hash total in computing, a ◊validation check in which an otherwise meaningless control total is calculated by adding together numbers (such as payroll or account numbers) associated with a set of records. The hash total is checked each time data are input, in order to ensure that no entry errors have been made.

haustorium (plural *haustoria*) specialized organ produced by a parasitic plant or fungus that penetrates the cells of its host to absorb nutrients. It may be either an outgrowth of hyphae (see ◊hypha), as in the case of parasitic fungi, or of the stems of

harmful/irritant

toxic

radioactive

explosive

flammable

corrosive

oxidizing/supports fire

hazard label

flowering parasitic plants, as in dodders (*Cuscuta*). The suckerlike haustoria of a dodder penetrate the vascular tissue of the host plant without killing the cells.

hazard label internationally recognized symbol indicating the potential dangers of handling certain substances.

hazardous substance waste substance, usually generated by industry, which represents a hazard to the environment or to people living or working nearby. Examples include radioactive wastes, acidic resins, arsenic residues, residual hardening salts, lead, mercury, nonferrous sludges, organic solvents, and pesticides. Their economic disposal or recycling is the subject of research.

The UK imported 41,000 tonnes of hazardous waste for disposal 1989, according to official estimates, the largest proportion of which came from Europe.

haze factor unit of visibility in mist or fog. It is the ratio of the brightness of the mist compared with that of the object.

HDTV abbreviation for ◊*high-definition television*.

headward erosion the backwards erosion of

material at the source of a river or stream. Broken rock and soil at the source is carried away by the river, causing erosion to take place in the opposite direction to the river's flow. The resulting lowering of the land behind the source will, over time, cause the river to flow backwards—possibly into a neighbouring valley to capture another river (see ◊river capture).

heart muscular organ that rhythmically contracts to force blood around the body of an animal with a circulatory system. Annelid worms and some other invertebrates have simple hearts consisting of thickened sections of main blood vessels that pulse regularly. An earthworm has ten such hearts. Vertebrates have one heart. A fish heart has two chambers—the thin-walled *atrium* (once called the auricle) that expands to receive blood, and the thick-walled *ventricle* that pumps it out. Amphibians and most reptiles have two atria and one ventricle; birds and mammals have two atria and two ventricles. The beating of the heart is controlled by the autonomic nervous system and an internal control centre or pacemaker, the sinoatrial node.

heartbeat the regular contraction and relaxation of the heart, and the accompanying sounds. As blood passes through the heart a double beat is heard. The first is produced by the sudden closure of the valves between the atria and the ventricles. The second, slightly delayed sound, is caused by the closure of the valves found at the entrance to the major arteries leaving the heart. Diseased valves may make unusual sounds, known as heart murmurs.

heart–lung machine apparatus used during heart surgery to take over the functions of the heart and the lungs temporarily. It has a pump to circulate the blood around the body and is able to add oxygen to the blood and remove carbon dioxide from it. A heart–lung machine was first used for open-heart surgery in the USA 1953.

heat form of internal energy possessed by a substance by virtue of the kinetic energy in the motion of its molecules or atoms. Heat energy is transferred by conduction, convection, and radiation. It always flows from a region of higher ◊temperature (heat intensity) to one of lower temperature. Its effect on a substance may be simply to raise its temperature, or to cause it to expand, melt (if a solid), vaporize (if a liquid), or increase its pressure (if a confined gas).

Quantities of heat are usually measured in units of energy, such as joules (J) or calories (cal). The *specific heat* of a substance is the ratio of the quantity of heat required to raise the temperature of a given mass of the substance through a given range of temperature to the heat required to raise the temperature of an equal mass of water through the same range. It is measured by a ◊calorimeter.

Conduction is the passing of heat along a medium to neighbouring parts with no visible motion accompanying the transfer of heat—for example, when the whole length of a metal rod is heated when one end is held in a fire.

Convection is the transmission of heat through a fluid (liquid or gas) in currents—for example, when the air in a room is warmed by a fire or radiator.

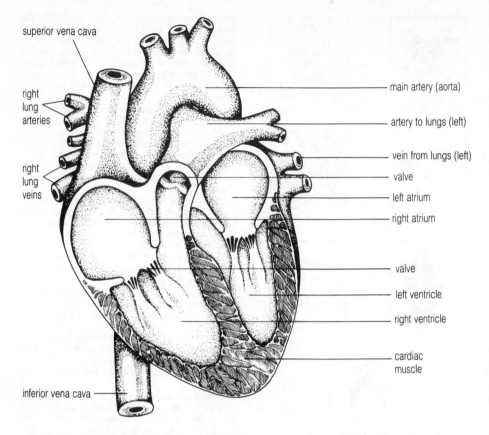

superior vena cava

right lung arteries

right lung veins

inferior vena cava

main artery (aorta)

artery to lungs (left)

vein from lungs (left)

valve

left atrium

right atrium

valve

left ventricle

right ventricle

cardiac muscle

heart *The structure of the human heart. During an average lifetime, the human heart beats more than 2,000 million times and pumps 500 million litres/110 million gallons of blood. The average pulse rate is 70–72 beats per minute at rest for adult males, and 78–82 beats per minute for adult females.*

Radiation is heat transfer by infrared rays. It can pass through a vacuum, travels at the same speed as light, can be reflected and refracted, and does not affect the medium through which it passes. For example, heat reaches the Earth from the Sun by radiation.

heat capacity in physics, the quantity of heat required to raise the temperature of an object by one degree. The *specific heat capacity* of a substance is the heat capacity per unit of mass, measured in joules per kilogram per kelvin (J kg^{-1} K^{-1}).

heat of reaction alternative term for ◊energy of reaction.

heat pump machine, run by electricity or another power source, that cools the interior of a building by removing heat from interior air and pumping it out or, conversely, heats the inside by extracting energy from the atmosphere or from a hot-water source and pumping it in.

heat shield any heat-protecting coating or system, especially the coating (for example, tiles) used in spacecraft to protect the astronauts and equipment inside from the heat of re-entry when returning to Earth. Air friction can generate temperatures of up to 1,500°C/2,700°F on re-entry into the atmosphere.

heat storage any means of storing heat for release later. It is usually achieved by using materials that undergo phase changes, for example, Glauber's salt and sodium pyrophosphate, which melts at 70°C/158°F. The latter is used to store off-peak heat in the home: the salt is liquefied by cheap heat during the night and then freezes to give off heat during the day.

Other developments include the use of plastic crystals, which change their structure rather than melting when they are heated. They could be incorporated in curtains or clothing.

heat treatment in industry, the subjection of metals and alloys to controlled heating and cooling after fabrication to relieve internal stresses and improve their physical properties. Methods include ◊annealing, ◊quenching, and ◊tempering.

heavy metal in chemistry, a metallic element of high relative atomic mass, such as platinum, gold, and lead. Many heavy metals are poisonous and tend to accumulate and persist in living systems—for example, high levels of mercury (from industrial waste and toxic dumping) accumulate in shellfish and fish, which are in turn eaten by humans.

Treatment of heavy-metal poisoning is difficult because available drugs are not able to distinguish between the heavy metals that are essential to living cells (zinc, copper) and those that are poisonous.

heavy water or **deuterium oxide** D_2O water containing the isotope deuterium instead of hydrogen. It has relative molecular mass 20 as opposed to 18 for ordinary water.

Its chemical properties are identical with those of ordinary water, but its physical properties differ slightly. It occurs in ordinary water in the ratio of about one part by mass of deuterium to 5,000 parts by mass of hydrogen, and can be concentrated by electrolysis, the ordinary water being more readily decomposed by this means than the heavy water. It has been used in the nuclear industry.

hectare metric unit of area equal to 100 ares or 10,000 square metres (2.47 acres), symbol ha.

Trafalgar Square, London's only metric square, was laid out as one hectare.

helicopter powered aircraft that achieves both lift and propulsion by means of a rotary wing, or rotor, on top of the fuselage. It can take off and land vertically, move in any direction, or remain stationary in the air. It can be powered by piston or jet engine. The ◊autogiro was a precursor.

The rotor of a helicopter has two or more blades of aerofoil cross-section like an aeroplane's wings. Lift and propulsion are achieved by angling the blades as they rotate. Experiments using the concept of helicopter flight date from the early 1900s, with the first successful lift-off and short flight 1907. Ukranian–US engineer Igor Sikorsky built the first practical single-rotor craft in the USA 1939. A single-rotor helicopter must also have a small tail rotor to counter the torque, or tendency of the body to spin in the opposite direction to the main rotor. Twin-rotor helicopters, like the Boeing Chinook, have their rotors turning in opposite directions to prevent the body from spinning. Helicopters are now widely used in passenger service, rescue missions on land and sea, police pursuits and traffic control, firefighting, and agriculture. In war they carry troops and equipment into difficult terrain, make aerial reconnaissance and attacks, and carry the wounded to aid stations.

Naval carriers are increasingly being built, with helicopters with depth charges and homing torpedoes being guided to submarine or surface targets beyond the carrier's attack range. The helicopter may also use dunking ◊sonar to find targets beyond the carrier's radar horizon. As many as 30 helicopters may be used on large carriers, in combination with V/STOL aircraft, such as the ◊Harrier. See also ◊autogiro, ◊convertiplane.

heliosphere region of space through which the ◊solar wind flows outwards from the Sun. The *heliopause* is the boundary of this region, believed to lie about 100 astronomical units from the Sun, where the flow of the solar wind merges with the interstellar gas.

helium (Greek *helios* 'Sun') colourless, odourless, gaseous, nonmetallic element, symbol He, atomic number 2, relative atomic mass 4.0026. It is grouped with the ◊inert gases, is nonreactive, and forms no compounds. It is the second most abundant element (after hydrogen) in the universe, and has the lowest boiling (−268.9°C/−452°F) and melting points (−272.2°C/−458°F) of all the elements. It is present in small quantities in the Earth's atmosphere from gases issuing from radioactive elements (from ◊alpha decay) in the Earth's crust; after hydrogen it is the second lightest element.

Helium is a component of most stars, including the Sun, where the nuclear-fusion process converts hydrogen into helium with the production of heat and light. It is obtained by compression and fractionation of naturally occurring gases. It is used for inflating balloons and as a dilutant for oxygen in deep-sea breathing systems. Liquid helium is used extensively in low-temperature physics (cryogenics).

helix in mathematics, a three-dimensional curve resembling a spring, corkscrew, or screw thread. It is generated by a line that encircles a cylinder or cone at a constant angle.

hematite principal ore of iron, consisting mainly of iron(III) oxide, Fe_2O_3. It occurs as *specular hematite* (dark, metallic lustre), *kidney ore* (reddish radiating fibres terminating in smooth, rounded surfaces), and as a red earthy deposit.

henry SI unit (symbol H) of ◊inductance (the reaction of an electric current against the magnetic field that surrounds it). One henry is the inductance of a circuit that produces an opposing voltage of one volt when the current changes at one ampere per second.

It is named after the US physicist Joseph Henry.

Henry Doubleday Research Association British gardening group founded 1954 by Lawrence Hills (1911–1990) to investigate organic growing techniques. It runs the *National Centre for Organic Gardening*, a 10–hectare/22–acre demonstration site, at Ryton-on-Dunsmore near Coventry, England. The association is named after the person who first imported Russian comfrey, a popular green-manuring crop.

herb any plant (usually a flowering plant) tasting sweet, bitter, aromatic, or pungent, used in cooking, medicine, or perfumery; technically, a herb is any plant in which the aerial parts do not remain above ground at the end of the growing season.

herbaceous plant plant with very little or no wood, dying back at the end of every summer. The herbaceous perennials survive winters as underground storage organs such as bulbs and tubers.

herbarium collection of dried, pressed plants used as an aid to identification of unknown plants and by taxonomists in the ◊classification of plants. The plant specimens are accompanied by information, such as the date and place of collection, by whom collected, details of habitat, flower colour, and local names.

Herbaria range from small collections containing plants of a limited region, to the large university and national herbaria (some at ◊botanical gardens) containing millions of specimens from all parts of the world.

The herbarium at the Royal Botanic Gardens, Kew, England, has over 5 million specimens.

herbicide any chemical used to destroy plants or check their growth; see ◊weedkiller.

herbivore animal that feeds on green plants (or photosynthetic single-celled organisms) or their products, including seeds, fruit, and nectar. The most numerous type of herbivore is thought to be the zooplankton, tiny invertebrates in the surface waters of the oceans that feed on small photosynthetic algae. Herbivores are more numerous than other animals because their food is the most abundant. They form a vital link in the food chain between plants and carnivores.

Mammalian herbivores that rely on cellulose as a major part of their diet, for instance cows and sheep, generally possess millions of specialized bacteria in their gut. These are capable of producing the enzyme cellulase, necessary for digesting cellulose; no mammal is able to manufacture cellulase on its own.

Hercules in astronomy, the fifth largest constellation, lying in the northern hemisphere. Despite its size it contains no prominent stars. Its most important feature is a ◊globular cluster of stars 22,500 light years from Earth, one of the best examples in the sky.

heredity in biology, the transmission of traits from parent to offspring. See also ◊genetics.

hermaphrodite organism that has both male and female sex organs. Hermaphroditism is the norm in species such as earthworms and snails, and is common in flowering plants. Cross-fertilization is the rule among hermaphrodites, with the parents functioning as male and female simultaneously, or as one or the other sex at different stages in their development.

Pseudo-hermaphrodites have the internal sex organs of one sex, but the external appearance of the other. The true sex of the latter becomes apparent at adolescence when the normal hormone activity appropriate to the internal organs begins to function.

hertz SI unit (symbol Hz) of frequency (the number of repetitions of a regular occurrence in one second). Radio waves are often measured in megahertz (MHz), millions of hertz, and the ◊clock rate of a computer is usually measured in megahertz. The unit is named after German physicist Heinrich Hertz.

Hertzsprung–Russell diagram in astronomy, a graph on which the surface temperatures of stars are plotted against their luminosities. Most stars, including the Sun, fall into a narrow band called the ◊*main sequence*. When a star grows old it moves from the main sequence to the upper right part of the graph, into the area of the giants and supergiants. At the end of its life, as the star shrinks to become a white dwarf, it moves again, to the bottom left area. It is named after the Dane Ejnar Hertzsprung (1873–1967) and the American Henry Norris Russell (1877–1957), who independently devised it in the years 1911–13.

heterogeneous reaction in chemistry, a reaction where there is an interface between the different components or reactants. Examples of heterogeneous reactions are those between a gas and a solid, a gas and a liquid, two immiscible liquids, or two different solids.

heterosis or *hybrid vigour* improvement in

physical capacities that sometimes occurs in the ◊hybrid produced by mating two genetically different parents.

The parents may be of different strains or varieties within a species, or of different species, as in the mule, which is stronger and has a longer life span than either of its parents (donkey and horse). Heterosis is also exploited in hybrid varieties of maize, tomatoes, and other crops.

heterostyly in botany, having ◊styles of different lengths. Certain flowers, such as primroses (*Primula vulgaris*), have different-sized ◊anthers and styles to ensure cross-fertilization (through ◊pollination) by visiting insects.

heterotroph any living organism that obtains its energy from organic substances produced by other organisms. All animals and fungi are heterotrophs, and they include herbivores, carnivores, and saprotrophs (those that feed on dead animal and plant material).

heterozygous in a living organism, having two different ◊alleles for a given trait. In ◊homozygous organisms, by contrast, both chromosomes carry the same allele. In an outbreeding population an individual organism will generally be heterozygous for some genes but homozygous for others.

For example, in humans, alleles for both blue- and brown-pigmented eyes exist, but the 'blue' allele is recessive to the dominant 'brown' allele. Only individuals with blue eyes are predictably homozygous for this trait; brown-eyed people can be either homozygous or heterozygous.

heuristics in computing, a process by which a program attempts to improve its performance by learning from its own experience.

hexachlorophene ($C_6HCl_3OH)_2CH_2$ white, odourless bactericide, used in minute quantities in soaps and surgical disinfectants.

Trichlorophenol is used in its preparation, and, without precise temperature control, the highly toxic TCDD (tetrachlorodibenzodioxin; see ◊dioxin) may form as a by-product.

hexadecimal number system number system to the base 16, used in computing. In hex (as it is commonly known) the decimal numbers 0–15 are represented by the characters 0, 1, 2, 3, 4, 5, 6, 7, 8, 9, A, B, C, D, E, F. Hexadecimal numbers are easy to convert to the computer's internal binary (base 2) code and are more compact than binary numbers.

Each place in a number increases in value by a power of 16 going from right to left; for instance, 8F is equal to $15 + (8 \times 16) = 143$ in decimal.

Hexadecimal numbers are often preferred by programmers writing in low-level languages because they are more easily converted to the computer's internal binary code than are decimal numbers, and because they are more compact than binary numbers and therefore more easily keyed, checked, and memorized.

HF in physics, the abbreviation for *high* ◊*frequency*.

HGV abbreviation for *heavy goods vehicle*.

hibernation state of ◊dormancy in which certain animals spend the winter. It is associated with a dramatic reduction in all metabolic processes,

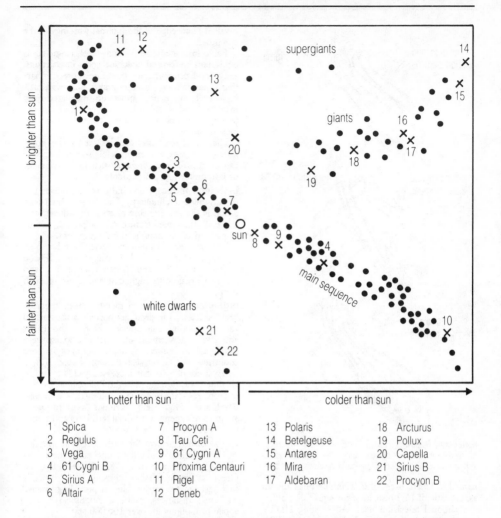

1	Spica	7	Procyon A	13	Polaris	18	Arcturus
2	Regulus	8	Tau Ceti	14	Betelgeuse	19	Pollux
3	Vega	9	61 Cygni A	15	Antares	20	Capella
4	61 Cygni B	10	Proxima Centauri	16	Mira	21	Sirius B
5	Sirius A	11	Rigel	17	Aldebaran	22	Procyon B
6	Altair	12	Deneb				

Hertzsprung–Russell diagram *Stars are plotted as dots against a vertical axis showing the brightness (or luminosity) of the star and a horizontal axis showing the star's temperature. Most stars fall within a narrow diagonal band called the main sequence. A star moves off the main sequence when it grows old. The Hertzsprung-Russell diagram is one of the most important diagrams in astrophysics.*

including body temperature, breathing, and heart rate. It is a fallacy that animals sleep throughout the winter.

The body temperature of the arctic ground squirrel falls to below 0°C/32°F during hibernation.

hide unit of measurement used in the 12th century to measure the extent of arable land; a hide was equal to 256 acres (104 hectares).

hi-fi (abbreviation of *high-fi*delity) faithful reproduction of sound from a machine that plays recorded music or speech. A typical hi-fi system includes a turntable for playing vinyl records, a cassette tape deck to play magnetic tape recordings, a tuner to pick up radio broadcasts, an amplifier to serve all the equipment, possibly a compact-disc player, and two or more loudspeakers.

Advances in mechanical equipment and electronics, such as digital recording techniques and

compact discs, have made it possible to eliminate many distortions in sound-reproduction processes.

Higgs particle or *Higgs boson* postulated ◊elementary particle whose existence would explain why particles, such as the ◊intermediate vector boson, have mass. The current theory of elementary particles, called the ◊standard model, cannot explain how mass arises, To overcome this difficulty, Peter Higgs (1929–   ) of the University of Edinburgh, Scotland, and Thomas Kibble (1932–   ) of Imperial College, London, England, proposed in 1964 a new particle that binds to other particles and gives them their mass. The Higgs particle has not yet been detected experimentally.

high-definition television (HDTV) ◊television system offering a significantly greater number of scanning lines, and therefore a clearer picture, than that provided by conventional systems. The Japanese HDTV system, or Vision as it is trade-

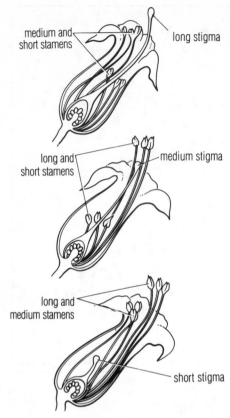

medium and short stamens

long stigma

long and short stamens

medium stigma

long and medium stamens

short stigma

heterostyly *Heterostyly, in which lengths of the stamens and stigma differ in flowers of different plants of the same species. This is a device to ensure cross-pollination by visiting insects.*

named in Japan, uses 1,125 scanning lines and an aspect ratio of 16:9 instead of the squarish 4:3 that conventional television uses. All existing HDTV systems are analogue-based. However, in 1993, the major electronics companies were developing digital systems, and the US government was considering a digital standard for HDTV.

high-level language in computing, a programming language designed to suit the requirements of the programmer; it is independent of the internal machine code of any particular computer. High-level languages are used to solve problems and are often described as *problem-oriented languages*—for example, ◊BASIC was designed to be easily learnt by first-time programmers; ◊COBOL is used to write programs solving business problems; and ◊FORTRAN is used for programs solving scientific and mathematical problems. In contrast, low-level languages, such as ◊assembly languages, closely reflect the machine codes of specific computers, and are therefore described as *machine-oriented languages*.

Unlike low-level languages, high-level languages are relatively easy to learn because the instructions bear a close resemblance to everyday language, and because the programmer does not require a

detailed knowledge of the internal workings of the computer.

Each instruction in a high-level language is equivalent to several machine-code instructions. High-level programs are therefore more compact than equivalent low-level programs. However, each high-level instruction must be translated into machine code—by either a ◊compiler or an ◊interpreter program—before it can be executed by a computer.

High-level languages are designed to be *portable*—programs written in a high-level language can be run on any computer that has a compiler or interpreter for that particular language.

high-tech industry any industry that makes use of advanced technology. The largest high-tech group is the fast-growing electronics industry and especially the manufacture of computers, microchips, and telecommunications equipment.

The products of these industries have low bulk but high value, as do their components. Silicon Valley in the USA and Silicon Glen in Scotland are two areas with high concentrations of such firms.

high-yield variety crop that has been specially bred or selected to produce more than the natural varieties of the same species. During the 1950s and 1960s, new strains of wheat and maize were developed to reduce the food shortages in poor countries (the Green Revolution). Later, IR8, a new variety of rice that increased yields by up to six times, was developed in the Philippines. Strains of crops resistant to drought and disease were also developed. High-yield varieties require large amounts of expensive artificial fertilizers and sometimes pesticides for best results.

Hipparcos (acronym for **hi**gh **p**recision **pa**rallax **co**llecting **s**atellite) satellite launched by the European Space Agency Aug 1989. Named after the Greek astronomer Hipparchus, it is the world's first ◊astrometry satellite and is providing precise positions, distances, colours, brightnesses, and apparent motions for over 100,000 stars.

histochemistry study of plant and animal tissue by visual examination, usually with a ◊microscope. ◊Stains are often used to highlight structural characteristics such as the presence of starch or distribution of fats.

histogram in statistics, a graph showing frequency of data, in which the horizontal axis is made up of discrete units or class boundaries, and the vertical axis represents the frequency. Blocks are drawn such that their areas (rather than their height as in a ◊bar chart) are proportional to the frequencies within a class or across several class boundaries. There are no spaces between blocks.

HIV abbreviation for *human immunodeficiency virus*, the infectious agent that causes ◊AIDS. It was first discovered 1983 by Luc Montagnier of the Pasteur Institute in Paris, who called it lymphocyte associated virus (LAV). Independently, US scientist Robert Gallo, of the National Cancer Institute in Bethesda, Maryland, claimed its discovery in 1984 and named it human T lymphocytotrophic virus 3 (HTLV-III).

Hoffman's voltameter in chemistry, an

apparatus for collecting gases produced by the ◊electrolysis of a liquid.

It consists of a vertical E-shaped glass tube with taps at the upper ends of the outer limbs and a reservoir at the top of the central limb. Platinum electrodes fused into the lower ends of the outer limbs are connected to a source of direct current. At the beginning of an experiment, the outer limbs are completely filled with electrolyte by opening the taps. The taps are then closed and the current switched on. Gases evolved at the electrodes bubble up the outer limbs and collect at the top, where they can be measured.

hogback geological formation consisting of a ridge with a short crest and abruptly sloping sides. Hogbacks are the result of differential erosion on steeply dipping rock strata composed of alternating resistant and soft beds. Exposed, almost vertical beds provide the sharp crests.

holdfast organ found at the base of many seaweeds, attaching them to the sea bed. It may be a flattened, suckerlike structure, or dissected and fingerlike, growing into rock crevices and firmly anchoring the plant.

holmium (Latin *Holmia* 'Stockholm') silvery, metallic element of the ◊lanthanide series, symbol Ho, atomic number 67, relative atomic mass 164.93. It occurs in combination with other rare-earth metals and in various minerals such as gadolinite. Its compounds are highly magnetic.

The element was discovered in 1878, spectroscopically, by the Swiss chemists L Soret and Delafontaine, and independently in 1879 by Swedish chemist Per Cleve (1840–1905), who named it after Stockholm, near which it was found.

Holocene epoch of geological time that began 10,000 years ago, the second and current epoch of the Quaternary period. The glaciers retreated, the climate became warmer, and humans developed significantly.

holography method of producing three-dimensional (3–D) images by means of ◊laser light. Holography uses a photographic technique (involving the splitting of a laser beam into two beams) to produce a picture, or hologram, that contains 3–D information about the object photographed. Some holograms show meaningless patterns in ordinary light and produce a 3–D image only when laser light is projected through them, but reflection holograms produce images when ordinary light is reflected from them (as found on credit cards).

Although the possibility of holography was suggested as early as 1947, it could not be demonstrated until a pure coherent light source, the laser, became available 1963.

The technique of holography is also applicable to sound, and bats may navigate by ultrasonic holography. Holographic techniques also have applications in storing dental records, detecting stresses and strains in construction and in retail goods, and detecting forged paintings and documents.

The technique of detecting strains is of widespread application. It involves making two different holograms of an object on one plate, the object being stressed between exposures. If the object has distorted during stressing, the hologram will be gretaly changed, and the distortion readily apparent.

homeostasis maintenance of a constant internal state in an organism, particularly with regard to pH, salt concentration, temperature, and blood sugar levels. Stable conditions are important for the efficient functioning of the ◊enzyme reactions within the cells, which affect the performance of the entire organism.

homeothermy maintenance of a constant body

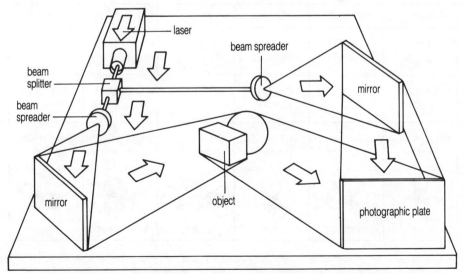

holography *Recording a transmission hologram. Light from a laser is divided into two beams. One beam goes directly to the photographic plate. The other beam reflects off the object before hitting the photographic plate. The two beams combine to produce a pattern on the plate which contains information about the 3–D shape of the object. If the exposed and developed plate is illuminated by laser light, the pattern can be seen as a 3–D picture of the object.*

temperature in endothermic (warm-blooded) animals, by the use of chemical body processes to compensate for heat loss or gain when external temperatures change. Such processes include generation of heat by the breakdown of food and the contraction of muscles, and loss of heat by sweating, panting, and other means.

Mammals and birds are homeotherms, whereas invertebrates, fish, amphibians, and reptiles are cold-blooded or poikilotherms. Homeotherms generally have a layer of insulating material to retain heat, such as fur, feathers, or fat (see ◊blubber). Their metabolism functions more efficiently due to homeothermy, enabling them to remain active under most climatic conditions.

homogeneous reaction in chemistry, a reaction where there is no interface between the components. The term applies to all reactions where only gases are involved or where all the components are in solution.

homologous in biology, a term describing an organ or structure possessed by members of different taxonomic groups (for example, species, genera, families, orders) that originally derived from the same structure in a common ancestor. The wing of a bat, the arm of a monkey, and the flipper of a seal are homologous because they all derive from the forelimb of an ancestral mammal.

homologous series any of a number of series of organic chemicals with similar chemical properties in which members differ by a constant relative molecular mass.

Alkanes (paraffins), alkenes (olefins), and alkynes (acetylenes) form such series in which members differ in mass by 14, 12, and 10 atomic mass units respectively. For example, the alkane homologous series begins with methane (CH_4), ethane (C_2H_6), propane (C_3H_8), butane (C_4H_{10}), and pentane (C_5H_{12}), each member differing from the previous one by a CH_2 group (or 14 atomic mass units).

homozygous in a living organism, having two identical ◊alleles for a given trait. Individuals homozygous for a trait always breed true; that is, they produce offspring that resemble them in appearance when bred with a genetically similar individual; inbred varieties or species are homozygous for almost all traits. Recessive alleles (see ◊recessive gene) are only expressed in the homozygous condition. See also ◊heterozygous.

honey sweet syrup produced by honey bees from the nectar of flowers. It is stored in honeycombs and made in excess of their needs as food for the winter. Honey comprises various sugars, mainly laevulose and dextrose, with enzymes, colouring matter, acids, and pollen grains. It has antibacterial properties and was widely used in ancient Egypt, Greece, and Rome as a wound salve. It is still popular for sore throats, in hot drinks or in lozenges.

honey guide in botany, line or spot on the petals of a flower that indicate to pollinating insects the position of the nectaries (see ◊nectar) within the flower. The orange dot on the lower lip of the toadflax flower (*Linaria vulgaris*) is an example. Sometimes the markings reflect only ultraviolet light, which can be seen by many insects although it is not visible to the human eye.

Hooke's law law stating that the deformation of a body is proportional to the magnitude of the deforming force, provided that the body's elastic limit (see ◊elasticity) is not exceeded. If the elastic limit is not reached, the body will return to its original size once the force is removed. It was

Alkane	Alcohol	Aldehyde	Ketone	Carboxylic acid	Alkene
CH_4 methane	$CH_3 OH$ methanol	HCHO methanal	–	HCOOH methanoic acid	–
$CH_3 CH_3$ ethane	$CH_3 CH_2 OH$ ethanol	$CH_3 CHO$ ethanal	–	$CH_3 COOH$ ethanoic acid	$CH_2 CH_2$ ethene
$CH_3 CH_2 CH_3$ propane	$CH_3 CH_2 CH_2 OH$ propanol	$CH_3 CH_2 CHO$ propanal	$CH_3 CO CH_3$ propanone	$CH_3 CH_2 COOH$ propanoic acid	$CH_2 CH CH_3$ propene
methane	methanol	methanal	propanone	methanoic acid	ethene

homologous series *The systematic naming of simple straight-chain organic molecules depends on two-part names. The first part of a name indicates the number of carbon atoms in the chain: one carbon, meth-; two carbons, eth-; three carbons, prop-; etc The second part of each name indicates the kind of bonding between the carbon atoms, or the atomic group attached to the chain. The name of a molecule containing only single bonds ends in –ane. Molecules with double bonds have names ending with –ene. Molecules containing the OH group have names ending in –anol; those containing –CO– groups have names ending in –anone; those containing the carboxyl group –COOH– have names ending in –anoic acid.*

discovered by English physicist Robert Hooke 1676.

For example, if a spring is stretched by 2 cm by a weight of 1 N, it will be stretched by 4 cm by a weight of 2 N, and so on; however, once the load exceeds the elastic limit for the spring, Hooke's law will no longer be obeyed and each successive increase in weight will result in a greater extension until finally the spring breaks.

Hope's apparatus in physics, an apparatus used to demonstrate the temperature at which water has its maximum density. It is named after Thomas Charles Hope (1766–1844).

It consists of a vertical cylindrical vessel fitted with horizontal thermometers through its sides near the top and bottom, and surrounded at the centre by a ledge that holds a freezing mixture (ice and salt). When the cylinder is filled with water, this gradually cools, the denser water sinking to the bottom; eventually the upper thermometer records 0°C/32°F (the freezing point of water) and the lower one has a constant reading of 4°C/39°F (temperature at which water is most dense).

horizon the limit to which one can see across the surface of the sea or a level plain, that is, about 5 km/3 mi at 1.5 m/5 ft above sea level, and about 65 km/40 mi at 300 m/1,000 ft.

hormone product of the ◊endocrine glands, concerned with control of body functions. The main glands are the thyroid, parathyroid, pituitary, adrenal, pancreas, uterus, ovary, and testis. Hormones bring about changes in the functions of various organs according to the body's requirements. The pituitary gland, at the base of the brain, is a centre for overall coordination of hormone secretion; the thyroid hormones determine the rate of general body chemistry; the adrenal hormones prepare the organism during stress for 'fight or flight'; and the sexual hormones such as oestrogen govern reproductive functions.

Many diseases due to hormone deficiency can be relieved with hormone preparations.

hormone-replacement therapy (HRT) use of oral ◊oestrogen and progestogen to help limit the effects of the menopause in women. The treatment was first used in the 1970s.

At the menopause, the ovaries cease to secrete natural oestrogen. This results in a number of symptoms, including hot flushes, anxiety, and a change in the pattern of menstrual bleeding. It is also associated with osteoporosis, or a thinning of bone, leading to an increased incidence of fractures, frequently of the hip, in older women. Oral oestrogens, taken to replace the decline in natural hormone levels, combined with regular exercise can help to maintain bone strength in women. In order to improve bone density, however, HRT must be taken for five years, during which time the woman will continue to menstruate. Many women do not find this acceptable.

hornblende green or black rock-forming mineral, one of the ◊amphiboles; it is a hydrous silicate of calcium, iron, magnesium, and aluminium. Hornblende is found in both igneous and metamorphic rocks.

hornfels ◊metamorphic rock formed by rocks heated by contact with a hot igneous body. It is fine-grained and brittle, without foliation.

Hornfels may contain minerals only formed under conditions of great heat, such as andalusite, Al_2SiO_5, and cordierite, $(Mg,Fe)_2Al_4Si_5O_{18}$. This rock, originating from sedimentary rock strata, is found in contact with large igneous ◊intrusions where it represents the heat-altered equivalent of the surrounding clays. Its hardness makes it suitable for road building and railway ballast.

hornwort plant of the class Anthocerotae; it is related to the ◊liverworts and ◊mosses, with which it forms the order Bryophyta. Hornworts are found in warm climates, growing on moist shaded soil.

Like the other bryophytes, hornworts exist in two alternating reproductive phases (see ◊alternation of generations)—a leafy gametophyte, which produces gametes, or sex cells, and a small horned sporophyte, which produces spores and grows upwards from the gametophyte. Unlike the sporophytes of mosses and liverworts, the hornwort sporophyte survives the death of the gametophyte.

horsepower imperial unit (abbreviation hp) of power, now replaced by the ◊watt. It was first used by the Scottish engineer James Watt, who employed it to compare the power of steam engines with that of horses.

Watt found a horse to be capable of 366 foot-pounds of work per second but, in order to enable him to use the term 'horsepower' to cover the additional work done by the more efficient steam engine, he exaggerated the pulling power of the horse by 50%. Hence, one horsepower is equal to 550 foot-pounds per second/745.7 watts, which is more than any real horse could produce. The metric horsepower is 735.5 watts; the standard US horsepower is 746 watts.

host organism that is parasitized by another. In ◊commensalism, the partner that does not benefit may also be called the host.

hot spot in geology, a hypothetical region of high thermal activity in the Earth's ◊mantle. It is believed to be the origin of many chains of ocean islands, such as Polynesia and the Galápagos.

A volcano forms on the ocean crust immediately above the hot spot, is carried away by ◊plate tectonic movement, and becomes extinct. A new volcano forms beside it, above the hot spot. The result is an active volcano and a chain of increasingly old and eroded volcanic stumps stretching away along the line of plate movement.

hour period of time comprising 60 minutes; 24 hours make one calendar day.

hovercraft vehicle that rides on a cushion of high-pressure air, free from all contact with the surface beneath, invented by British engineer Christopher Cockerell 1959. Hovercraft need a smooth terrain when operating overland and are best adapted to use on waterways. They are useful in places where harbours have not been established.

Large hovercraft (SR-N4) operate a swift car-ferry service across the English Channel, taking only about 35 minutes between Dover and Calais. They are fitted with a flexible 'skirt' that helps maintain the air cushion.

hp abbreviation for ◊*horsepower*.

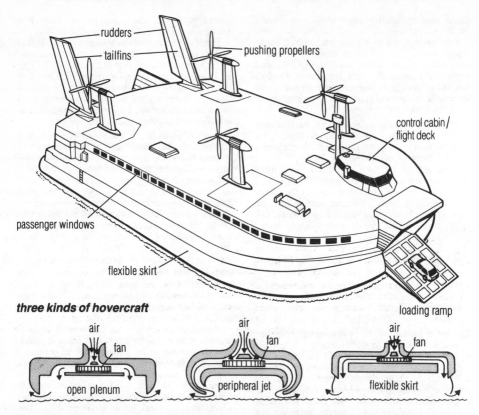

three kinds of hovercraft

hovercraft *There are several alternative ways of containing the cushion of air beneath the hull of a hovercraft. The passenger-carrying hovercraft that sails across the English Channel has a flexible skirt; other systems are the open plenum and the peripheral jet.*

Hubble's constant in astronomy, a measure of the rate at which the universe is expanding, named after Edwin Hubble. Observations suggest that galaxies are moving apart at a rate of 50–100 kps/ 30–60 mps for every million ◊parsecs of distance. This means that the universe, which began at one point according to the ◊Big Bang theory, is between 10 billion and 20 billion years old (probably closer to 20).

Hubble's law the law that relates a galaxy's distance from us to its speed of recession as the universe expands, announced in 1929 by US astronomer Edwin Hubble. He found that galaxies are moving apart at speeds that increase in direct proportion to their distance apart. The rate of expansion is known as Hubble's constant.

Hubble Space Telescope (HST) telescope placed into orbit around the Earth, at an altitude of 610 km/380 mi, by the space shuttle *Discovery* in April 1990. It has a main mirror 2.4 m/94 in wide, which suffers from spherical aberration and so cannot be focused properly. Yet, because it is above the atmosphere, the HST outperforms ground-based telescopes. Computer techniques are being used to improve the images from the telescope until the arrival of a maintenance mission to install corrective optics. The HST carries four scientific instruments.

The HST is the most expensive crewless space-craft so far made, costing $2.5 billion—five times the original estimate—and launched seven years late. An inquiry found both NASA and the contractors to blame for the manufacturing fault in the mirror, which was not recognized until it was in orbit.

hum, environmental disturbing sound of frequency about 40 Hz, heard by individuals sensitive to this range, but inaudible to the rest of the population. It may be caused by industrial noise pollution or have a more exotic origin, such as the jet stream, a fast-flowing high-altitude (about 15,000 m/50,000 ft) mass of air.

human body the physical structure of the human being. It develops from the single cell of the fertilized ovum, is born at 40 weeks, and usually reaches sexual maturity between 11 and 18 years of age. The bony framework (skeleton) consists of more than 200 bones, over half of which are in the hands and feet. Bones are held together by joints, some of which allow movement. The circulatory system supplies muscles and organs with blood, which provides oxygen and food and removes carbon dioxide and other waste products. Body functions are controlled by the nervous system and hormones. In the upper part of the trunk is the thorax, which contains the lungs and heart. Below this is the abdomen, containing the digestive system (stomach and intestines); the liver, spleen, and pancreas; the

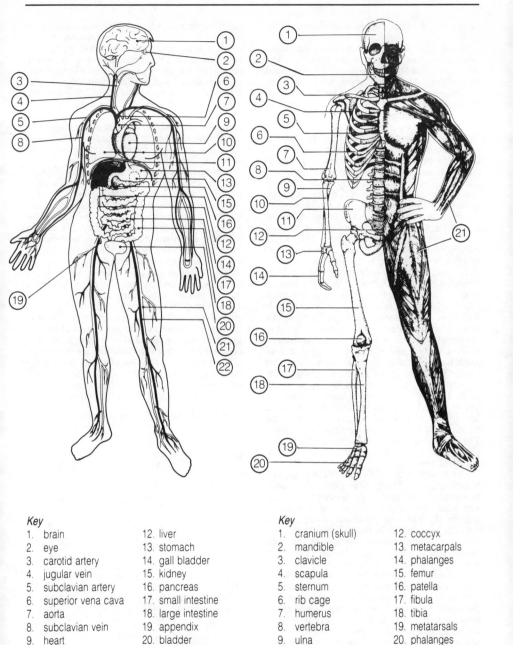

Key

1. brain	12. liver
2. eye	13. stomach
3. carotid artery	14. gall bladder
4. jugular vein	15. kidney
5. subclavian artery	16. pancreas
6. superior vena cava	17. small intestine
7. aorta	18. large intestine
8. subclavian vein	19. appendix
9. heart	20. bladder
10. lungs	21. femoral artery
11. diaphragm	22. femoral vein

Key

1. cranium (skull)	12. coccyx
2. mandible	13. metacarpals
3. clavicle	14. phalanges
4. scapula	15. femur
5. sternum	16. patella
6. rib cage	17. fibula
7. humerus	18. tibia
8. vertebra	19. metatarsals
9. ulna	20. phalanges
10. radius	21. superficial (upper)
11. pelvis	layer of muscles

human body *The adult human body has approximately 650 muscles, 100 joints, 100,000 km/60,000 mi of blood vessels and 13,000 nerve cells. There are 206 bones in the adult body, nearly half of them in the hands and feet.*

urinary system (kidneys, ureters, and bladder); and, in women, the reproductive organs (ovaries, uterus, and vagina). In men, the prostate gland and seminal vesicles only of the reproductive system are situated in the abdomen, the testes being in the scrotum, which, with the penis, is suspended in front of and below the abdomen. The bladder empties through a small channel (urethra); in the female this opens in the upper end of the vulval cleft, which also contains the opening of the vagina, or birth canal; in the male, the urethra is continued into the penis. In both sexes, the lower bowel terminates in the anus, a ring of strong muscle situated between the buttocks.

skeleton The skull is mounted on the spinal column, or spine, a chain of 24 vertebrae. The ribs, 12 on each side, are articulated to the spinal column behind, and the upper seven meet the breastbone (sternum) in front. The lower end of the spine rests on the pelvic girdle, composed of the triangular sacrum, to which are attached the hipbones (ilia), which are fused in front. Below the sacrum is the tailbone (coccyx). The shoulder blades (scapulae) are held in place behind the upper ribs by muscles, and connected in front to the breastbone by the two collarbones (clavicles). Each shoulder blade carries a cup (glenoid cavity) into which fits the upper end of the armbone (humerus). This articulates below with the two forearm bones (radius and ulna). These are articulated at the wrist (carpals) to the bones of the hand (metacarpals and phalanges). The upper end of each thighbone (femur) fits into a depression (acetabulum) in the hipbone; its lower end is articulated at the knee to the shinbone (tibia) and calf bone (fibula), which are articulated at the ankle (tarsals) to the bones of the foot (metatarsals and phalanges). At a moving joint, the end of each bone is formed of tough, smooth cartilage, lubricated by ◊synovial fluid. Points of special stress are reinforced by bands of fibrous tissue (ligaments).

Muscles are bundles of fibres wrapped in thin, tough layers of connective tissue (fascia); these are usually prolonged at the ends into strong, white cords (tendons, sinews) or sheets (aponeuroses), which connect the muscles to bones and organs, and by way of which the muscles do their work. Membranes of connective tissue also wrap organs and line the interior cavities of the body. The thorax has a stout muscular floor, the diaphragm, which expands and contracts the lungs in the act of breathing.

The blood vessels of the *circulatory system*, branching into multitudes of very fine tubes (capillaries), supply all parts of the muscles and organs with blood, which carries oxygen and food necessary for life. The food passes out of the blood to the cells in a clear fluid (lymph); this is returned with waste matter through a system of lymphatic vessels that converge into collecting ducts that drain into large veins in the region of the lower neck. Capillaries join together to form veins which return blood, depleted of oxygen, to the heart.

A finely branching *nervous system* regulates the function of the muscles and organs, and makes their needs known to the controlling centres in the central nervous system, which consists of the brain and spinal cord. The inner spaces of the brain and the cord contain cerebrospinal fluid. The body processes are regulated both by the nervous system and by hormones secreted by the endocrine glands. Cavities of the body that open onto the surface are coated with mucous membranes, which secrete a lubricating fluid (mucus). The exterior surface of the body is coated with *skin*. Within the skin are the sebaceous glands, which secrete sebum, an oily fluid that makes the skin soft and pliable, and the sweat glands, which secrete water and various salts. From the skin grow hairs, chiefly on the head, in the armpits, and around the sexual organs; and

nails shielding the tips of the fingers and toes; both hair and nail structures are modifications of skin tissue. The skin also contains nerves of touch, pain, heat, and cold. The human *digestive system* is nonspecialized and can break down a wide variety of foodstuffs. Food is mixed with saliva in the mouth by chewing and is swallowed. It enters the stomach, where it is gently churned for some time and mixed with acidic gastric juice. It then passes into the small intestine. In the first part of this, the duodenum, it is broken down further by the juice of the pancreas and duodenal glands, and mixed with bile from the liver, which splits up the fat. The jejunum and ileum continue the work of digestion and absorb most of the nutritive substances from the food. The large intestine completes the process, reabsorbing water into the body, and ejecting the useless residue as faeces. The body, to be healthy, must maintain water and various salts in the right proportions; the process is called *osmoregulation*. The blood is filtered in the two kidneys, which remove excess water, salts, and metabolic wastes. Together these form urine, which has a yellow pigment derived from bile, and passes down through two fine tubes (ureters) into the bladder, a reservoir from which the urine is emptied at intervals (micturition) through the urethra. Heat is constantly generated by the combustion of food in the muscles and glands, and by the activity of nerve cells and fibres. It is dissipated through the skin by conduction and evaporation of sweat, through the lungs in the expired air, and in other excreted substances. Average body temperature is about 38°C/100°F (37°C/98.4°F in the mouth).

composition of the human body by weight

class	chemical element or substance	body weight (%)
pure elements	oxygen	65
	carbon	18
	hydrogen	10
	nitrogen	3
	calcium	2
	phosphorus	1.1
	potassium	0.35
	sulphur	0.25
	sodium	0.15
	chlorine	0.15
	magnesium, iron, manganese, copper, iodine, cobalt, zinc	traces
water and solid matter	water	60–80
	total solid material	20–40
organic molecules	protein	15–20
	lipid	3–20
	carbohydrate	1–15
	small organic molecues	0–1

human–computer interaction exchange of information between a person and a computer, through the medium of a ◊user interface, studied as a branch of ergonomics.

Human Genome Project research scheme, begun 1988, to map the complete nucleotide (see ◊nucleic acid) sequence of human ◊DNA. There are approximately 80,000 different ◊genes in the human genome, and one gene may contain more than 2 million nucleotides. The knowledge gained is expected to help prevent or treat many crippling and lethal diseases, but there are potential ethical problems associated with knowledge of an individual's genetic make-up, and fears that it will lead to genetic engineering.

The Human Genome Organization (HUGO) coordinating the project expects to spend $1 billion over the first five years, making this the largest research project ever undertaken in the life sciences. Work is being carried out in more than 20 centres around the world. By the beginning of 1991, some 2,000 genes had been mapped. Concern that, for example, knowledge of an individual's genes may make that person an unacceptable insurance risk has led to planned legislation on genome privacy in the USA, and 3% of HUGO's funds have been set aside for researching and reporting on the ethical implications of the project. Each strand of DNA carries a sequence of chemical building blocks, the nucleotides. There are only four different types, but the number of possible combinations is immense. The different combinations of nucleotides produce different proteins in the cell, and thus determine the structure of the body and its individual variations. To establish the nucleotide sequence, DNA strands are broken into fragments, which are duplicated (by being introduced into cells of yeast or the bacterium *E. coli*) and distributed to the research centres. Genes account for only a small amount of the DNA sequence. Over 90% of DNA appears not to have any function, although it is perfectly replicated each time the cell divides, and handed on to the next generation. Many higher organisms have large amounts of redundant DNA and it may be that this is an advantage, in that there is a pool of DNA available to form new genes if an old one is lost by mutation.

human reproduction an example of ◊sexual reproduction, where the male produces sperm and the female eggs. These gametes contain only half the normal number of chromosomes, 23 instead of 46, so that on fertilization the resulting cell has the correct genetic complement. Fertilization is internal, which increases the chances of conception; unusually for mammals, copulation and pregnancy can occur at any time of the year. Human beings are also remarkable for the length of childhood and for the highly complex systems of parental care found in society. The use of contraception and the development of laboratory methods of insemination and fertilization are issues that make human reproduction more than a merely biological phenomenon.

human species, origins of evolution of humans from ancestral ◊primates. The African apes (gorilla and chimpanzee) are shown by anatomical and molecular comparisons to be the closest living relatives of humans. Humans are distinguished from apes by the size of their brain and jaw, their bipedalism, and their elaborate culture. Molecular studies put the date of the split between the human and African ape lines at 5–10 million years ago.

There are only fragmentary remains of ape and *hominid* (of the human group) fossils from this period; the oldest known hominids, found in Ethiopia and Tanzania, date from 3.5 to 4 million years

ago. These creatures are known as *Australopithecus afarensis*, and they walked upright. They were either direct ancestors or an offshoot of the line that led to modern humans. They may have been the ancestors of *Homo habilis* (considered by some to be a species of *Australopithecus*), who appeared about a million years later, had slightly larger bodies and brains, and were probably the first to use stone tools. *Australopithecus robustus* and *A. africanus* also lived in Africa at the same time, but these are not generally considered to be our ancestors.

Over 1.5 million years ago, *Homo erectus*, believed by some to be descended from *H. habilis*, appeared in Africa. The *erectus* people had much larger brains, and were probably the first to use fire and the first to move out of Africa. Their remains are found as far afield as China, western Asia, Spain, and S Britain. Modern humans, *H. sapiens sapiens*, and the Neanderthals, *H. sapiens neanderthalensis*, are probably descended from *H. erectus*.

Neanderthals were large-brained and heavily built, probably adapted to the cold conditions of the ice ages. They lived in Europe and the Middle East, and died out about 40,000 years ago, leaving *H. sapiens sapiens* as the only remaining species of the hominid group.

There are currently two major views of human evolution: the '**Out of Africa**' **model**, according to which *H. sapiens* emerged from *H. erectus* in Africa and then spread throughout the world; and the **multiregional model**, according to which selection pressures led to the emergence of similar advanced types of *H. sapiens* from *H. erectus* in different parts of the world at around the same time.

Analysis of DNA in recent human populations suggests that *H. sapiens* originated about 200,000 years ago in Africa from a single female ancestor, 'Eve'. The oldest known fossils of *H. sapiens* also come from Africa, dating from 150,000–100,000 years ago. Separation of human populations would have occurred later, with separation of Asian, European, and Australian populations taking place between 100,000 and 50,000 years ago.

Creationists believe that the origin of the human species is as written in the book of Genesis in the Old Testament of the Bible.

Descended from the apes? My dear, we will hope it is not true. But if it is, let us pray that it may not become generally known.

On the origins of the **human species**, remark by the wife of a canon of Worcester Cathedral

Humboldt Current former name of the ◊Peru Current.

humerus the upper bone of the forelimb of tetrapods. In humans, the humerus is the bone above the elbow.

humidity the quantity of water vapour in a given volume of the atmosphere (absolute humidity), or the ratio of the amount of water vapour in the atmosphere to the saturation value at the same temperature (relative humidity). At ◊dew point the relative humidity is 100% and the air is said to be saturated. Condensation (the conversion of vapour

Out of Africa and the Eve Hypothesis

Most palaeoanthropologists recognize the existence of two human species during the last million years—*Homo erectus*, now extinct, and *Homo sapiens* (including recent or 'modern' humans). In general, they believe that *Homo erectus* was the ancestor of *Homo sapiens*. How did the transition occur?

The multiregional model

There are two opposing views. The multiregional model says that *Homo erectus* gave rise to *Homo sapiens* across its whole range. About 700,000 years ago this included Africa, China, Java (Indonesia) and, probably, Europe. *Homo erectus*, following an African origin about 1.7 million years ago, dispersed around the Old World developing the regional variation that lies at the roots of modern 'racial' variation. Particular features in a given region persisted in the local descendant populations of today.

No definite representatives of *Homo erectus* have yet been discovered in Europe. The fossil record does not extend back as far as those of Africa and eastern Asia, although a possible *Homo erectus* jawbone more than one million years old was recently excavated in Georgia. Nevertheless, the multiregional model claims that European *Homo erectus* did exist, and evolved into a primitive form of *Homo sapiens*. Evolution in turn produced the Neanderthals: the ancestors of modern Europeans.

The multiregional model was first described in detail by Franz Weidenreich, the German palaeoanthropologist. It was developed further by American Carleton Coon, who tended to regard the regional lineages as genetically separate. Most recently, the model has become associated with researchers such as Milford Wolpoff (USA) and Alan Thorne (Australia), who have reemphasized the importance of gene flow between the regional lines. In fact, they regard the continuity in time and space between the various forms of *Homo erectus* and their descendants to be so complete that they should be regarded as representing only one species—*Homo sapiens*.

The Garden of Eden...

The opposing view is that *Homo sapiens* had a restricted origin in time and space. This is an old idea. Early in this century, workers such as Marcellin Boule (France) and Arthur Keith (United Kingdom) believed that the lineage of *Homo sapiens* was very ancient; it had developed in parallel with that of *Homo erectus* and the Neanderthals. However, much of the fossil evidence used to support their ideas has been reevaluated, and few workers now support the idea of a very ancient and separate origin for modern *Homo sapiens*.

The modern equivalent of such ideas focuses on a recent and restricted origin for modern *Homo sapiens*. This was dubbed the 'Garden of Eden' or 'Noah's Ark' model by the US anthropologist William Howells in 1976 because of the idea that all modern human variation had a localized origin from one centre. Howells did not specify the centre of origin, but research since 1976 points to Africa as especially important.

The consequent 'Out of Africa' model claims that *Homo erectus* evolved into *Homo sapiens* in Africa about 100,000–150,000 years ago. Part of the African stock of early modern humans spread into adjoining regions and eventually reached Australia, Europe and the Americas (probably by 45,000, 40,000 and 15,000 years ago respectively). Regional ('racial') variation only developed during and after the dispersal, so that there is no continuity of regional features between *Homo erectus* and present counterparts in the same regions.

Like the multiregional model, this view accepts that *Homo erectus* evolved into new forms of human in inhabited regions outside Africa, but argues that these non-African lineages became extinct without evolving into modern humans. Some, such as the Neanderthals, were displaced and then replaced by the spread of modern humans into their regions.

... and an African Eve?

In 1987, research on the genetic material called mitochondrial DNA in living humans led to the reconstruction of a hypothetical female ancestor for all present-day humanity. This 'Eve' was believed to have lived in Africa about 200,000 years ago. Recent reexamination has cast doubt on this hypothesis, but further support for an 'Out of Africa' model has come from genetic studies of nuclear DNA, which also point to a recent African origin for present-day *Homo sapiens*.

Studies of fossil material of the last 50,000 years also seem to indicate that many 'racial' features in the human skeleton have only developed over the last 30,000 years, in line with the 'Out of Africa' model, and at odds with the one-million-year timespan expected from the multiregional model.

Chris Stringer

to liquid) may then occur. Relative humidity is measured by various types of ◊hygrometer.

humus component of ◊soil consisting of decomposed or partly decomposed organic matter, dark in colour and usually richer towards the surface. It has a higher carbon content than the original material and a lower nitrogen content, and is an important source of minerals in soil fertility.

hundredweight imperial unit (abbreviation cwt) of mass, equal to 112 lb (50.8 kg). It is sometimes called the long hundredweight, to distinguish it from the short hundredweight or *cental*, equal to 100 lb (45.4 kg).

hurricane revolving storm in tropical regions, called *typhoon* in the N Pacific. It originates between 5° and 20° N or S of the equator, when the surface temperature of the ocean is above 27°C/80°F. A central calm area, called the eye, is surrounded by inwardly spiralling winds (anticlockwise in the northern hemisphere) of up to 320 kph/200 mph. A hurricane is accompanied by lightning and torrential rain, and can cause extensive damage. In meteorology, a hurricane is a wind of force 12 or more on the ◊Beaufort scale. The most intense hurricane recorded in the Caribbean/Atlantic sector was Hurricane Gilbert in 1988, with sustained winds of 280 kph/175 mph and gusts of over 320 kph/200 mph.

In Oct 1987 and Jan 1990, allegedly hurricane-force winds were experienced in S England. Although not technically hurricanes, they were the strongest winds there for three centuries.

Hyades V-shaped cluster of stars that forms the face of the bull in the constellation Taurus. It is 150 light years away and contains over 200 stars, although only about 12 are visible to the naked eye.

hybrid offspring from a cross between individuals of two different species, or two inbred lines within a species. In most cases, hybrids between species are infertile and unable to reproduce sexually. In plants, however, doubling of the chromosomes (see ◊polyploid) can restore the fertility of such hybrids.

Hybrids between different genera are extremely rare; an example is the *leylandii* cypress which, like many hybrids, shows exceptional vigour, or ◊heterosis. In the wild, a 'hybrid zone' may occur where the ranges of two related species meet.

hybridization the production of a ◊hybrid.

hydathode specialized pore, or less commonly, a hair, through which water is secreted by hydrostatic pressure from the interior of a plant leaf onto the surface. Hydathodes are found on many different plants and are usually situated around the leaf margin at vein endings. Each pore is surrounded by two crescent-shaped cells and resembles an open ◊stoma, but the size of the opening cannot be varied as in a stoma. The process of water secretion through hydathodes is known as ◊guttation.

Hydra in astronomy, the largest constellation, winding across more than a quarter of the sky between Cancer and Libra in the southern hemisphere. Hydra is named after the multiheaded monster slain by Heracles. Despite its size, it is not prominent; its brightest star is second-magnitude Alphard.

hydra in zoology, any member of the family Hydri-

dae, or freshwater polyps, of the phylum Cnidaria (coelenterates). The body is a double-layered tube (with six to ten hollow tentacles around the mouth), 1.25 cm/0.5 in long when extended, but capable of contracting to a small knob. Usually fixed to waterweed, hydras feed on minute animals that are caught and paralyzed by stinging cells on the tentacles.

Hydras reproduce asexually in the summer and sexually in the winter. They have no specialized organs except those of reproduction.

hydrate chemical compound that has discrete water molecules combined with it. The water is known as *water of crystallization* and the number of water molecules associated with one molecule of the compound is denoted in both its name and chemical formula: for example, $CuSO_4.5H_2O$ is copper(II) sulphate pentahydrate.

hydration in chemistry, the combination of water and another substance to produce a single product. It is the opposite of ◊dehydration. For example, anhydrous copper sulphate reacts with water to give copper sulphate pentahydrate.

$$CuSO_4 + 5H_2O = CuSO_4.5H_2O$$

hydraulic action erosive force exerted by water (as distinct from the forces exerted by rocky particles carried by water). It can wear away the banks of a river, particularly at the outer curve of a meander (bend in the river), where the current flows most strongly.

Hydraulic action occurs as a river tumbles over a waterfall to crash onto the rocks below. It will lead to the formation of a plunge pool below the waterfall. The hydraulic action of ocean waves and turbulent currents forces air into rock cracks, and therefore brings about erosion by ◊cavitation.

hydraulic radius measure of a river's channel efficiency (its ability to discharge water), used by water engineers to assess the likelihood of flooding. The hydraulic radius of a channel is defined as the ratio of its cross-sectional area to its wetted perimeter (the part of the cross-section that is in contact with the water).

The greater the hydraulic radius, the greater the efficiency of the channel and the less likely the river is to flood. The highest values occur when channels are deep, narrow, and semi-circular in shape.

hydraulics field of study concerned with utilizing the properties of water and other liquids, in particular the way they flow and transmit pressure, and with the application of these properties in engineering. It applies the principles of ◊hydrostatics and hydrodynamics. The oldest type of hydraulic machine is the *hydraulic press*, invented by Joseph Bramah in England 1795. The hydraulic principle of pressurized liquid increasing mechanical efficiency is commonly used on vehicle braking systems, the forging press, and the hydraulic systems of aircraft and excavators.

A hydraulic press consists of two liquid-connected pistons in cylinders, one of narrow bore, one of large bore. A force applied to the narrow piston applies a certain pressure (force per unit area) to the liquid, which is transmitted to the larger piston.

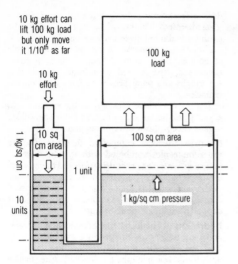

10 kg effort can lift 100 kg load but only move it 1/10ᵗʰ as far

10 kg effort ⇩

100 kg load

10 sq cm area ⇩ 1 unit

100 sq cm area

10 units

1 kg/sq cm pressure

1 kg/sq cm

hydraulics The hydraulic jack transmits the pressure on a small piston to a larger one. A larger total force is developed by the larger piston but it moves a smaller distance than the small piston.

Because the area of this piston is larger, the force exerted on it is larger. Thus the original force has been magnified, although the smaller piston must move a great distance to move the larger piston only a little, hence mechanical efficiency is gained in force but lost in movement.

hydride chemical compound containing hydrogen and one other element, and in which the hydrogen is the more electronegative element (see ◊electronegativity).

Hydrides of the more reactive metals may be ionic compounds containing a hydride anion (H⁻).

hydrocarbon any of a class of chemical compounds containing only hydrogen and carbon (for example, the alkanes and alkenes). Hydrocarbons are obtained industrially principally from petroleum and coal tar.

hydrochloric acid HCl solution of hydrogen chloride (a colourless, acidic gas) in water. The concentrated acid is about 35% hydrogen chloride and is corrosive. The acid is a typical strong, monobasic acid forming only one series of salts, the chlorides. It has many industrial uses, including recovery of zinc from galvanized scrap iron and the production of chlorine. It is also produced in the stomachs of animals for the purposes of digestion.

hydrocyanic acid or *prussic acid* solution of hydrogen cyanide gas (HCN) in water. It is a colourless, highly poisonous, volatile liquid, smelling of bitter almonds.

hydrodynamics science of nonviscous fluids (such as water, alcohol, and ether) in motion.

hydroelectric power (HEP) electricity generated by moving water. In a typical HEP scheme water stored in a reservoir, often created by damming a river, is piped into water ◊turbines, coupled to electricity generators. In ◊pumped storage plants, water flowing through the turbines is recycled. A ◊tidal power station exploits the rise and fall of the tides. About one-fifth of the world's electricity comes from HEP.

HEP plants have prodigious generating capacities. The Grand Coulee plant in Washington State, USA, has a power output of some 10,000 megawatts. The Itaipu power station on the Paraná River (Brazil/Paraguay) has a potential capacity of 12,000 megawatts.

hydrofoil wing that develops lift in the water in much the same way that an aeroplane wing develops lift in the air. A hydrofoil boat is one whose hull rises out of the water due to the lift, and the boat skims along on the hydrofoils. The first hydrofoil was fitted to a boat 1906. The first commercial hydrofoil went into operation 1956. One of the most advanced hydrofoil boats is the Boeing ◊jetfoil.

hydrogen (Greek *hydro* + *gen* 'water generator') colourless, odourless, gaseous, nonmetallic element, symbol H, atomic number 1, relative atomic mass 1.00797. It is the lightest of all the elements and occurs on Earth chiefly in combination with oxygen as water. Hydrogen is the most abundant element in the universe, where it accounts for 93% of the total number of atoms and 76% of the total mass. It is a component of most stars, including the Sun, whose heat and light are produced through the nuclear-fusion process that converts hydrogen into helium. When subjected to a pressure 500,000 times greater than that of the Earth's atmosphere, hydrogen becomes a solid with metallic properties, as in one of the inner zones of Jupiter. Hydrogen's common and industrial uses include the hardening of oils and fats by hydrogenation, the creation of high-temperature flames for welding, and as rocket fuel. It has been proposed as a fuel for road vehicles.

Its isotopes ◊deuterium and ◊tritium (half-life 12.5 years) are used in nuclear weapons, and deuterons (deuterium nuclei) are used in synthesizing elements. The element's name refers to the generation of water by the combustion of hydrogen, and was coined in 1787 by French chemist Louis Guyton de Morveau (1737–1816).

hydrogenation addition of hydrogen to an unsaturated organic molecule (one that contains ◊double bonds or triple bonds).

It is widely used in the manufacture of margarine and low-fat spreads by the addition of hydrogen to vegetable oils.

hydrogen bomb bomb that works on the principle of nuclear ◊fusion. Large-scale explosion results from the thermonuclear release of energy when hydrogen nuclei are fused to form helium nuclei. The first hydrogen bomb was exploded at Eniwetok Atoll in the Pacific Ocean by the USA 1952.

hydrogencarbonate or *bicarbonate* compound containing the ion HCO₃⁻, an acid salt of carbonic acid (solution of carbon dioxide in water). When heated or treated with dilute acids, it gives off carbon dioxide. The most important compounds are ◊sodium hydrogencarbonate (bicarbonate of soda), and ◊calcium hydrogencarbonate.

hydrogen cyanide HCN poisonous gas formed

by the reaction of sodium cyanide with dilute sulphuric acid; it is used for fumigation.

The salts formed from it are cyanides—for example sodium cyanide, used in hardening steel and extracting gold and silver from their ores. If dissolved in water, hydrogen cyanide gives hydrocyanic acid.

hydrogensulphate HSO_4^- compound containing the hydrogensulphate ion. Hydrogensulphates are ◊acid salts.

hydrogen sulphide H_2S poisonous gas with the smell of rotten eggs. It is found in certain types of crude oil where it is formed by decomposition of sulphur compounds. It is removed from the oil at the refinery and converted to elemental sulphur.

hydrograph graph showing how the discharge of a river varies with time. By studying hydrographs, water engineers can predict when flooding is likely and take action to prevent its taking place.

A hydrograph shows the time lag, or delay, between the occurrence of a rainstorm and the resultant rise in discharge, and the length of time taken for that discharge to peak. The shorter the time lag and the higher the peak the more likely it is that flooding will occur. Factors likely to give short time lags and high peaks include heavy rainstorms, steep slopes, deforestation, soil quality, and the covering of surfaces with impermeable substances such as tarmac and concrete. Actions taken by water engineers to increase time lags and lower peaks include planting trees in the drainage basin of a river.

I cannot believe that God plays dice with the cosmos.

Albert Einstein, quoted in the *Observer* 1953

God not only plays dice, he throws them where they can't be seen.

Stephen Hawking *A Brief History of Time* 1988

hydrography study and charting of Earth's surface waters in seas, lakes, and rivers.

hydrological cycle alternative name for the ◊water cycle, by which water is circulated between the Earth's surface and its atmosphere.

hydrology study of the location and movement of inland water, both frozen and liquid, above and below ground. It is applied to major civil engineering projects such as irrigation schemes, dams and hydroelectric power, and in planning water supply.

hydrolysis chemical reaction in which the action of water or its ions breaks down a substance into smaller molecules. Hydrolysis occurs in certain inorganic salts in solution, in nearly all nonmetallic chlorides, in esters, and in other organic substances. It is one of the mechanisms for the breakdown of food by the body, as in the conversion of starch to glucose.

hydrometer in physics, an instrument used to measure the density of liquids compared with that of water, usually expressed in grams per cubic centimetre. It consists of a thin glass tube ending in a sphere that leads into a smaller sphere, the latter being weighted so that the hydrometer floats upright, sinking deeper into less dense liquids than into denser liquids. It is used in brewing.

The hydrometer is based on ◊Archimedes' principle.

hydrophilic (Greek 'water-loving') in chemistry, a term describing ◊functional groups with a strong affinity for water, such as the carboxyl group (– COOH).

If a molecule contains both a hydrophilic and a ◊hydrophobic group (a group that repels water), it may have an affinity for both aqueous and nonaqueous molecules. Such compounds are used to stabilize ◊emulsions or as ◊detergents.

hydrophily ◊pollination in which the pollen is carried by water. Hydrophily is very rare but occurs in a few aquatic species. In Canadian pondweed *Elodea* and tape grass *Vallisneria*, the male flowers break off whole and rise to the water surface where they encounter the female flowers, which are borne on long stalks. In eel grasses *Zostera*, which are coastal plants growing totally submerged, the filamentous pollen grains are released into the water and carried by currents to the female flowers where they become wrapped around the stigmas.

hydrophobic (Greek 'water-hating') in chemistry, a term describing ◊functional groups that repel water (compare ◊hydrophilic).

hydrophone underwater ◊microphone and ancillary equipment capable of picking up waterborne sounds. It was originally developed to detect enemy submarines but is now also used, for example, for listening to the sounds made by whales.

hydrophyte plant adapted to live in water, or in waterlogged soil. Hydrophytes may have leaves with a very reduced or absent ◊cuticle and no ◊stomata (since there is no need to conserve water), a reduced root and water-conducting system, and less supporting tissue since water buoys plants up. There are often numerous spaces between the cells in their stems and roots to make ◊gas exchange with all parts of the plant body possible. Many have highly divided leaves, which lessens resistance to flowing water; an example is spiked water milfoil *Myriophyllum spicatum*.

hydroplane on a submarine, a moveable horizontal fin angled downwards or upwards when the vessel is descending or ascending. It is also a highly manoeuvrable motorboat with its bottom rising in steps to the stern, or a ◊hydrofoil boat that skims over the surface of the water when driven at high speed.

hydroponics cultivation of plants without soil, using specially prepared solutions of mineral salts. Beginning in the 1930s, large crops were grown by hydroponic methods, at first in California but since then in many other parts of the world.

Julius von Sachs (1832–1897) 1860 and W Knop 1865 developed a system of plant culture in water whereby the relation of mineral salts to plant growth could be determined, but it was not until about 1930 that large crops could be grown. The term was first coined by US scientist W F Gericke.

hydrosphere the water component of the Earth, usually encompassing the oceans, seas, rivers, streams, swamps, lakes, groundwater, and atmospheric water vapour.

hydrostatics in physics, the branch of ◊statics dealing with the mechanical problems of fluids in equilibrium—that is, in a static condition. Practical applications include shipbuilding and dam design.

Any solid lighter than a fluid will, if placed in the fluid, be so far immersed that the weight of the solid will be equal to the weight of the fluid displaced.

On **hydrostatics** Archimedes (*c.* 287–212 BC) *On Floating Bodies* I, proposition 5

hydrothermal in geology, pertaining to a fluid whose principal component is hot water, or to a mineral deposit believed to be precipitated from such a fluid.

hydrothermal vein a crack in rock filled with minerals precipitated through the action of circulating high-temperature fluids. Igneous activity often gives rise to the circulation of heated fluids that migrate outwards and move through the surrounding rock. When such solutions carry metallic ions, ore-mineral deposition occurs in the new surroundings on cooling.

hydrothermal vent hot fissure in the ocean floor, known as a ◊smoker.

hydroxide any inorganic chemical compound con-

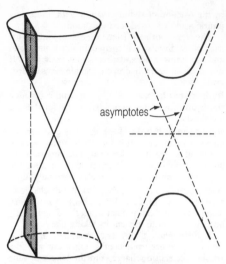

hyperbola *The hyperbola is produced when a cone is cut by a plane. It is one of a family of curves called conic sections: the circle, ellipse, and parabola. These curves are produced when the plane cuts the cone at different angles and positions.*

taining one or more hydroxyl (OH) groups and generally combined with a metal. Hydroxides include sodium hydroxide (caustic soda, NaOH), potassium hydroxide (caustic potash, KOH), and calcium hydroxide (slaked lime, Ca(OH)$_2$).

hydroxyl group an atom of hydrogen and an atom of oxygen bonded together and covalently bonded to an organic molecule. Common compounds containing hydroxyl groups are alcohols and phenols.

In chemical reactions, the hydroxyl group (–OH) frequently behaves as a single entity.

hydroxypropanoic acid technical name for ◊lactic acid.

hygrometer in physics, any instrument for measuring the humidity, or water vapour content, of a gas (usually air). A wet and dry bulb hygrometer consists of two vertical thermometers, with one of the bulbs covered in absorbent cloth dipped into water. As the water evaporates, the bulb cools producing a temperature difference between the two thermometers. The amount of evaporation, and hence cooling of the wet bulb, depends on the relative humidity of the air.

Other hygrometers work on the basis of a length of natural fibre, such as hair or a fine strand of gut, changing with variations in humidity. In a dew-point hygrometer, a polished metal mirror gradually cools until a fine mist of water (dew) forms on it. This gives a measure of the ◊dew point, from which the air's relative humidity can be calculated.

hygroscopic able to absorb moisture from the air without becoming wet.

hyperbola in geometry, a curve formed by cutting a right circular cone with a plane so that the angle between the plane and the base is greater than the angle between the base and the side of the cone. All hyperbolae are bounded by two asymptotes (straight lines which the hyperbola moves closer and closer to but never reaches). A hyperbola is a

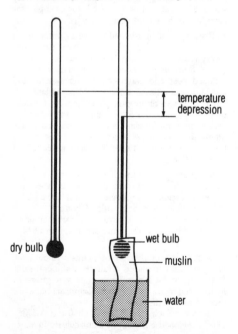

temperature depression

wet bulb

dry bulb

muslin

water

hygrometer *The most common hygrometer, or instrument for measuring the humidity of a gas, is the wet and dry bulb hygrometer. The wet bulb records a lower temperature because water evaporates from the muslin, taking heat from the wet bulb. The degree of evaporation and hence cooling depends upon the humidity of the surrounding air or other gas.*

member of the family of curves known as ◊conic sections.

A hyperbola can also be defined as a path traced by a point that moves such that the ratio of its distance from a fixed point (focus) and a fixed straight line (directrix) is a constant and greater than 1; that is, it has an ◊eccentricity greater than 1.

hypercharge in physics, a property of certain ◊elementary particles, analogous to electric charge, that accounts for the absence of some expected behaviour (such as decay) in terms of the short-range strong nuclear force, which holds atomic nuclei together.

◊Protons and ◊neutrons, for example, have a hypercharge of +1, whereas a π meson has a hypercharge of 0.

hyperon in physics, a ◊hadron; any of a group of highly unstable ◊elementary particles that includes all the ◊baryons with a mass greater than the ◊neutron. They are all composed of three quarks. The lambda, xi, sigma, and omega particles are hyperons.

Hypertext system for viewing information (both text and pictures) on a computer screen in such a way that related items of information can easily be reached. For example, the program might display a map of a country; if the user clicks (with a ◊mouse) on a particular city, the program will display some information about that city.

hypha (plural *hyphae*) delicate, usually branching filament, many of which collectively form the mycelium and fruiting bodies of a ◊fungus. Food molecules and other substances are transported along hyphae by the movement of the cytoplasm, known as 'cytoplasmic streaming'.

Typically hyphae grow by increasing in length from the tips and by the formation of side branches. Hyphae of the higher fungi (the ascomycetes and basidiomycetes) are divided by cross walls or septa at intervals, whereas those of lower fungi (for example, bread mould) are undivided. However, even the higher fungi are not truly cellular, as each septum is pierced by a central pore, through which cytoplasm, and even nuclei, can flow. The hyphal walls contain ◊chitin, a polysaccharide.

hypo in photography, a term for sodium thiosulphate, discovered 1819 by British astronomer and physicist John Herschel, and used as a fixative for photographic images since 1837.

hypocycloid in geometry, a cusped curve traced by a point on the circumference of a circle that rolls around the inside of another larger circle. (Compare ◊epicycloid.)

hypogeal term used to describe seed germination in which the ◊cotyledons remain below ground. It can refer to fruits that develop underground, such as peanuts *Arachis hypogea*.

hypotenuse the longest side of a right-angled triangle, opposite the right angle. It is of particular application in Pythagoras's theorem (the square on the hypotenuse equals the sum of the squares on the other two sides), and in trigonometry where the ratios ◊sine and ◊cosine are defined as the ratios opposite/hypotenuse and adjacent/hypotenuse respectively.

hypothalamus region of the brain below the ◊cerebrum which regulates rhythmic activity and physiological stability within the body, including water balance and temperature. It regulates the production of the pituitary gland's hormones and controls that part of the ◊nervous system regulating the involuntary muscles.

hypothesis in science, an idea concerning an event and its possible explanation. The term is one favoured by the followers of the Austrian philosopher of science Karl Popper, who argue that the merit of a scientific hypothesis lies in its ability to make testable predictions.

hypsometer (Greek *hypsos* 'height') instrument for testing the accuracy of a thermometer at the boiling point of water. It was originally used for determining altitude by comparing changes in the boiling point with changes in atmospheric pressure.

The name is also used for any of several instruments for measuring the heights of trees by ◊triangulation.

hysteresis phenomenon seen in the elastic and electromagnetic behaviour of materials, in which a lag occurs between the application or removal of a force or field and its effect.

If the magnetic field applied to a magnetic material is increased and then decreased back to its original value, the magnetic field inside the material does not return to its original value. The internal field 'lags' behind the external field. This behaviour results in a loss of energy, called the *hysteresis loss*, when a sample is repeatedly magnetized and demagnetized. Hence the materials used in transformer cores and electromagnets should have a low hysteresis loss.

Similar behaviour is seen in some materials when varying electric fields are applied (*electric hysteresis*). *Elastic hysteresis* occurs when a varying force repeatedly deforms an elastic material. The deformation produced does not completely disappear when the force is removed, and this results in energy loss on repeated deformations.

Hz in physics, the symbol for ◊hertz.

I

IAEA abbreviation for ◊International Atomic Energy Agency.

Iapetus Ocean or *Proto-Atlantic* sea that existed in early ◊Palaeozoic times between the continent that was to become Europe and that which was to become North America. The continents moved together in the late Palaeozoic, obliterating the ocean. When they moved apart once more, they formed the Atlantic.

IBM (abbreviation for *International Business Machines*) multinational company, the largest manufacturer of computers in the world. The company is a descendant of the Tabulating Machine Company, formed 1896 by US inventor Herman Hollerith to exploit his punched-card machines. It adopted its present name 1924. By 1991 it had an annual turnover of $64.8 billion and employed about 345,000 people. In 1992 it made a loss of $4.96 billion (about £3.5 billion).

IC abbreviation for ◊integrated circuit.

Icarus in astronomy, an ◊Apollo asteroid 1.5 km/ 1 mi in diameter, discovered 1949. It orbits the Sun every 409 days at a distance of 28–300 million km/18–186 million mi (0.19–2.0 astronomical units). It was the first asteroid known to approach the Sun closer than does the planet Mercury. In 1968 it passed 6 million km/4 million mi from the Earth.

ice solid formed by water when it freezes. It is colourless and its crystals are hexagonal. The water molecules are held together by hydrogen ◊bonds.

The freezing point of ice, used as a standard for measuring temperature, is 0° for the Celsius and Réaumur scales and 32° for the Fahrenheit. Ice expands in the act of freezing (hence burst pipes), becoming less dense than water (0.9175 at 5°C/ 41°F).

PERMANENT ICE

The world's largest permanent ice sheet lies over the continent of Antarctica. It covers an area equivalent to the USA and Europe together, or twice the size of Australia, and contains 95% of the world's permanent ice. The Greenland ice sheet, the largest in the northern hemisphere, contains about 10% of the world's total.

---

ice age any period of glaciation occurring in the Earth's history, but particularly that in the Pleistocene epoch (known as the Ice Age), immediately preceding historic times. On the North American continent, ◊glaciers reached as far south as the Great Lakes, and an ice sheet spread over N Europe, leaving its remains as far south as Switzerland. There were several glacial advances separated by interglacial stages during which the ice melted and temperatures were higher than today.

Formerly there were thought to have been only three or four glacial advances, but recent research has shown about 20 major incidences. For example, ocean-bed cores record the absence or presence in

H_2O

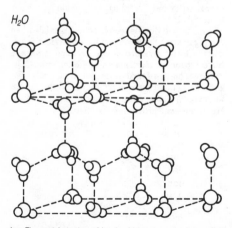

ice *The crystal structure of ice, in which water molecules are held together by hydrogen bonds.*

their various layers of such cold-loving small marine animals as radiolaria, which indicate a fall in ocean temperature at regular intervals. Other ice ages have occurred throughout geological time: there were four in the Precambrian era, one in the Ordovician, and one at the end of the the Carboniferous and beginning of the Permian.

The occurrence of an ice age is governed by a combination of factors (the *Milankovitch hypothesis*): (1) the Earth's change of attitude in relation to the Sun, that is, the way it tilts in a 41,000-year cycle and at the same time wobbles on its axis in a 22,000-year cycle, making the time of its closest approach to the Sun come at different seasons; and (2) the 92,000-year cycle of eccentricity in its orbit round the Sun, changing it from an elliptical to a near circular orbit, the severest period of an ice age coinciding with the approach to circularity. There is a possibility that the Pleistocene ice age is not yet over. It may reach another maximum in another 60,000 years.

The theory of ice ages was first proposed in the 19th century by, among others, Swiss civil engineer Ignace Venetz 1821 and Swiss naturalist Louis Agassiz 1837. (Before, most geologists had believed that the rocks and sediment they left behind were caused by the biblical flood.) The term 'ice age' was first used by botanist Karl Schimper 1837.

major ice ages

name	date (years ago)
Pleistocene	1.64 million–10,000
Permo-Carboniferous	330–250 million
Ordovician	440–430 million
Verangian	615–570 million
Sturtian	820–770 million
Gnejso	940–880 million
Huronian	2,700–1,800 million

iceberg floating mass of ice, about 80% of which is submerged, rising sometimes to 100 m/300 ft above sea level. Glaciers that reach the coast become extended into a broad foot; as this enters the sea, masses break off and drift towards temperate latitudes, becoming a danger to shipping.

Iceland spar form of ◊calcite, $CaCO_3$, originally found in Iceland. In its pure form Iceland spar is transparent and exhibits the peculiar phenomenon of producing two images of anything seen through it. It is used in optical instruments. The crystals cleave into perfect rhombohedra.

icon in computing, a small picture on the computer screen, or VDU (◊visual display unit), representing an object or function that the user may manipulate or otherwise use. It is a feature of ◊graphical user interface (GUI) systems. Icons make computers easier to use by allowing the user to point to and click with a ◊mouse on pictures, rather than type commands.

icosahedron (plural *icosahedra*) regular solid with 20 equilateral (equal-sided) triangular faces. It is one of the five regular ◊polyhedra or Platonic solids.

IEEE abbreviation for *Institute of Electrical and Electronic Engineers*, US institute which sets technical standards for electrical equipment and computer data exchange.

igneous rock rock formed from cooling magma or lava, and solidifying from a molten state. Igneous rocks are classified according to their crystal size, texture, chemical composition, or method of formation. They are largely composed of silica (SiO_2) and they are classified by their silica content into groups: acid (over 66% silica), intermediate (55%–66%), basic (45%–55%), and ultrabasic (under 45%). Igneous rocks that crystallize below the Earth's surface are called plutonic or intrusive, depending on the depth of formation. They have large crystals produced by slow cooling; examples include dolerite and granite. Those extruded at the surface are called extrusive or volcanic. Rapid cooling results in small crystals; basalt is an example.

ignition coil ◊transformer that is an essential part of a petrol engine's ignition system. It consists of two wire coils wound around an iron core. The primary coil, which is connected to the car battery, has only a few turns. The secondary coil, connected via the ◊distributor to the ◊spark plugs, has many turns.

The coil takes in a low voltage (usually 12 volts) from the battery and transforms it to a high voltage (about 20,000 volts) to ignite the engine.

When the engine is running, the battery current is periodically interrupted by means of the contact breaker in the distributor. The collapsing current in the primary coil induces a current in the secondary coil, a phenomenon known as ◊electromagnetic induction. The induced current in the secondary coil is at very high voltage, typically about 15,000–20,000 volts. This passes to the spark plugs to create sparks.

ignition temperature or *fire point* minimum temperature to which a substance must be heated before it will spontaneously burn independently of the source of heat; for example, ethanol has an ignition temperature of 425°C/798°F and a ◊flash point of 12°C/54°F.

iguanodon plant-eating ◊dinosaur of the order *Ornithiscia*, whose remains are found in deposits of the Lower Cretaceous age, together with the remains of other ornithiscians such as stegosaurus and triceratops. It was 5–10 m/16–32 ft long and, when standing upright, 4 m/13 ft tall. It walked on its hind legs, using its long tail to balance its body.

ileum part of the small intestine of the ◊digestive system, between the duodenum and the colon, that absorbs digested food.

Its wall is muscular so that waves of contraction (peristalsis) can mix the food and push it forward. Numerous fingerlike projections, or villi, point inward from the wall, increasing the surface area available for absorption. The ileum has an excellent blood supply, which receives the food molecules passing through the wall and transports them to the liver via the hepatic portal vein.

illumination the brightness or intensity of light falling on a surface. It depends upon the brightness, distance, and angle of any nearby light sources. The SI unit is the ◊lux.

ilmenite oxide of iron and titanium, iron titanate ($FeTiO_3$); an ore of titanium. The mineral is black,

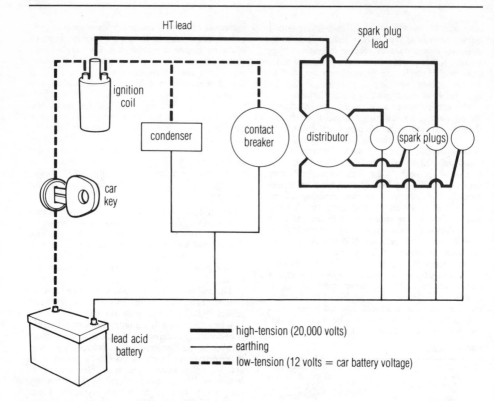

ignition coil *The ignition coil generates a high-tension (voltage) current which flows to the spark plugs through the distributor. The primary or low-tension current flows through the ignition coil from the battery. This current is continually interrupted by the contact breaker; the capacitor is needed to protect the breaker from burnout.*

with a metallic lustre. It is found in compact masses, grains, and sand.

ILS abbreviation for ◊*instrument landing system*, an automatic system for assisting aircraft landing at airports.

image picture or appearance of a real object, formed by light that passes through a lens or is reflected from a mirror. If rays of light actually pass through an image, it is called a *real image*. Real images, such as those produced by a camera or projector lens, can be projected onto a screen. An image that cannot be projected onto a screen, such as that seen in a flat mirror, is known as a *virtual image*.

image compression in computing, one of a number of methods used to reduce the amount of information required to represent an image, so that it takes up less computer memory and can be transmitted more rapidly and economically via telecommunications systems. It plays a major role in fax transmission and in videophone and multimedia systems.

image intensifier electronic device that brightens a dark image. Image intensifiers are used for seeing at night; for example, in military situations.

The intensifier first forms an image on a photo-cathode, which emits electrons in proportion to the intensity of the light falling on it. The electron flow is increased by one or more amplifiers. Finally, a fluorescent screen converts the electrons back into visible light, now bright enough to see.

imaginary number term often used to describe the non-real element of a ◊complex number. For the complex number $(a + ib)$, ib is the imaginary number where $i = \sqrt{-1}$, and b any real number.

The imaginary number is a fine and wonderful recourse of the divine spirit, almost an amphibian between being and not being.

On **imaginary numbers** Gottfried Liebniz (1646–1716)

imago sexually mature stage of an ◊insect.

immediate access memory in computing, ◊memory provided in the ◊central processing unit to store the programs and data in current use.

immiscible term describing liquids that will not mix with each other, such as oil and water. When two immiscible liquids are shaken together, a turbid mixture is produced. This normally forms separate layers on being left to stand.

immunity the protection that organisms have against foreign microorganisms, such as bacteria and viruses, and against cancerous cells (see ◊cancer). The cells that provide this protection are called white blood cells, or leucocytes, and make up the immune system. They include neutrophils and ◊macrophages, which can engulf invading organisms and other unwanted material, and natural killer cells that destroy cells infected by viruses and cancerous cells. Some of the most important immune cells are the ◊B cells and ◊T cells. Immune cells coordinate their activities by means of chemical messengers or ◊lymphokines, including the antiviral messenger ◊interferon. The lymph nodes play a major role in organizing the immune response.

Immunity is also provided by a range of physical barriers such as the skin, tear fluid, acid in the stomach, and mucus in the airways. ◊AIDS is one of many viral diseases in which the immune system is affected.

immunocompromised lacking a fully effective immune system. The term is most often used in connection with infections such as ◊AIDS where the virus interferes with the immune response (see ◊immunity).

Other factors that can impair the immune response are pregnancy, diabetes, old age, malnutrition and extreme stress, making someone susceptible to infections by microorganisms (such as listeria) that do not affect normal, healthy people.

Some people are ◊immunodeficient; others could be on ◊immunosuppressive drugs.

immunodeficient lacking one or more elements of a working immune system. Immune deficiency is the term generally used for patients who are born with such a defect, while those who acquire such a deficiency later in life are referred to as ◊immunocompromised or immunosuppressed.

A serious impairment of the immune system is sometimes known as SCID, or Severe Combined Immune Deficiency. At one time children born with this condition would have died in infancy. They can now be kept alive in a germ-free environment, then treated with a bone-marrow transplant from a relative, to replace the missing immune cells. At present, the success rate for this type of treatment is still fairly low.

immunoglobulin human globulin ◊protein that can be separated from blood and administered to confer immediate immunity on the recipient. It participates in the immune reaction as the antibody for a specific ◊antigen (disease-causing agent).

Normal immunoglobulin (gamma globulin) is the fraction of the blood serum that, in general, contains the most antibodies, and is obtained from plasma pooled from about a thousand donors. It is given for short-term (two to three months) protection when a person is at risk, mainly from hepatitis A (infectious hepatitis), or when a pregnant woman, not immunized against German measles, is exposed to the rubella virus.

Specific immunoglobulins are injected when a susceptible (nonimmunized) person is at risk of infection from a potentially fatal disease, such as hepatitis B (serum hepatitis), rabies, or tetanus. These immunoglobulins are prepared from blood pooled from donors convalescing from the disease.

impact printer computer printer that creates characters by striking an inked ribbon against the paper beneath. Examples of impact printers are

dot-matrix printers, daisywheel printers, and most types of line printer.

Impact printers are noisier and slower than nonimpact printers, such as ink-jet and laser printers, but can be used to produce carbon copies.

impedance the total opposition of a circuit to the passage of alternating electric current. It has the symbol Z. For an ◊alternating current (AC) it includes the reactance X (caused by ◊capacitance or ◊inductance); the impedance can then be found using the equation $Z^2 = R^2 + X^2$.

In acoustics, impedance refers to the ratio of the force per unit area to the volume displaced by the surface across which sound is being transmitted.

imperial system traditional system of units developed in the UK, based largely on the foot, pound, and second (f.p.s.) system.

In 1991 it was announced that the acre, pint, troy ounce, mile, yard, foot, and inch would remain in use indefinitely for beer, cider, and milk measures, and in road traffic signs and land registration. Other units, including the fathom and therm, would be phased out by 1994.

impermeable rock rock that does not allow water to pass through it—for example, clay, shale, and slate. Unlike ◊permeable rocks, which absorb water, impermeable rocks can support rivers. They therefore experience considerable erosion (unless, like slate, they are very hard) and commonly form lowland areas.

implantation in mammals, the process by which the developing ◊embryo attaches itself to the wall of the mother's uterus and stimulates the development of the ◊placenta.

In some species, such as seals and bats, implantation is delayed for several months, during which time the embryo does not grow; thus the interval between mating and birth may be a year, although the ◊gestation period is only seven or eight months.

import file in computing, a file that can be read by a program even though it was produced as an ◊export file by a different program or make of computer.

imprinting in ◊ethology, the process whereby a young animal learns to recognize both specific individuals (for example, its mother) and its own species.

Imprinting is characteristically an automatic response to specific stimuli at a time when the animal is especially sensitive to those stimuli (the *sensitive period*).

Thus, goslings learn to recognize their mother by following the first moving object they see after hatching; as a result, they can easily become imprinted on other species, or even inanimate objects, if these happen to move near them at this time. In chicks, imprinting occurs only between 10 and 20 hours after hatching. In mammals, the mother's attachment to her infant may be a form of imprinting made possible by a sensitive period; this period may be as short as the first hour after giving birth.

impulse in mechanics, the product of a force and the time over which it acts. An impulse applied to a body causes its ◊momentum to change and is equal to that change in momentum. It is measured in newton seconds.

For example, the impulse J given to a football when it is kicked is given by:

$$J = Ft$$

where F is the kick force in newtons and t is the time in seconds for which the boot is in contact with the ball.

in abbreviation for ◊*inch*, a measure of distance.

inbreeding in ◊genetics, the mating of closely related individuals. It is considered undesirable because it increases the risk that offspring will inherit copies of rare deleterious recessive alleles (see ◊recessive gene) from both parents and so suffer from disabilities.

incandescence emission of light from a substance in consequence of its high temperature. The colour of the emitted light from liquids or solids depends on their temperature, and for solids generally the higher the temperature the whiter the light. Gases may become incandescent through ◊ionizing radiation, as in the glowing vacuum ◊discharge tube.

The oxides of cerium and thorium are highly incandescent and for this reason are used in gas mantles. The light from an electric filament lamp is due to the incandescence of the filament, rendered white-hot when a current passes through it.

inch imperial unit of linear measure, a twelfth of a foot, equal to 2.54 centimetres. It was defined in statute by Edward II of England (1284–1327) as the length of three barley grains laid end to end.

incised meander in a river, a deep steep-sided meander (bend) formed by the severe downwards erosion of an existing meander. Such erosion is usually brought about by the ◊rejuvenation of a river (for example, in response to a fall in sea level).

There are several incised meanders along the middle course of the river Wye, near Chepstow, Gwent.

incisor sharp tooth at the front of the mammalian mouth. Incisors are used for biting or nibbling, as when a rabbit or a sheep eats grass. Rodents, such as rats and squirrels, have large continually-growing incisors, adapted for gnawing. The elephant tusk is a greatly enlarged incisor.

inclination angle between the ◊ecliptic and the plane of the orbit of a planet, asteroid, or comet. In the case of satellites (moons) orbiting a planet, it is the angle between the plane of orbit of the satellite and the equator of the planet.

inclusive fitness in ◊genetics, the success with which a given variant (or allele) of a ◊gene is passed on to future generations by a particular individual, after additional copies of the allele in the individual's relatives and their offspring have been taken into account.

The concept was formulated by W D Hamilton as a way of explaining the evolution of ◊altruism in terms of ◊natural selection. See also ◊fitness and ◊kin selection.

indeterminacy principle alternative name for ◊uncertainty principle.

index (plural *indices*) in mathematics, another term for ◊exponent, the number that indicates the power to which a term should be raised.

indexed sequential file in computing, a type of ◊file access in which an index is used to obtain the address of the ◊block containing the required record.

indicator in chemistry, a compound that changes its structure and colour in response to its environment. The commonest chemical indicators detect changes in ◊pH (for example, ◊litmus), or in the oxidation state of a system (redox indicators).

indicator species plant or animal whose presence or absence in an area indicates certain environmental conditions, such as soil type, high levels of pollution, or, in rivers, low levels of dissolved oxygen. Many plants show a preference for either alkaline or acid soil conditions, while certain trees require aluminium, and are found only in soils where it is present. Some lichens are sensitive to sulphur dioxide in the air, and absence of these species indicates atmospheric pollution.

indium (Latin *indicum* 'indigo') soft, ductile, silver-white, metallic element, symbol In, atomic number 49, relative atomic mass 114.82. It occurs in nature in some zinc ores, is resistant to abrasion, and is used as a coating on metal parts. It was discovered 1863 by German metallurgists Ferdinand Reich (1799–1882) and Hieronymus Richter (1824–1898), who named it after the two indigo lines of its spectrum.

induced current electric current that appears in a closed circuit when there is relative movement of its conductor in a magnetic field. The effect is known as the **dynamo effect**, and is used in all ◊dynamos and generators to produce electricity. See ◊electromagnetic induction.

There is no battery or other source of power in a circuit in which an induced current appears: the energy supply is provided by the relative motion of the conductor and the magnetic field. The magnitude of the induced current depends upon the rate at which the magnetic flux is cut by the conductor, and its direction is given by Fleming's right-hand rule (see ◊Fleming's rules).

moving a magnet into a coil

coil of many turns
of insulated wire

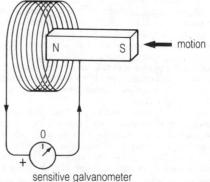

induced current

inductance in physics, a measure of the capability of an electronic circuit or circuit component to form a magnetic field or store magnetic energy when carrying a current. Its symbol is L, and its unit of measure is the ◊henry.

induction in obstetrics, deliberate intervention to initiate labour before it starts naturally; then it usually proceeds normally. Induction involves rupture of the fetal membranes (amniotomy) and the use of the hormone oxytocin to stimulate contractions of the womb. In biology, induction is a term used for various processes, including the production of an ◊enzyme in response to a particular chemical in the cell, and the ◊differentiation of cells in an ◊embryo in response to the presence of neighbouring tissues.

In obstetrics, induction is recommended as a medical necessity where there is risk to the mother or baby in waiting for labour to begin of its own accord.

induction in physics, an alteration in the physical properties of a body that is brought about by the influence of a field. See ◊electromagnetic induction and ◊magnetic induction.

induction coil type of electrical transformer, similar to an ◊ignition coil, that produces an intermittent high-voltage alternating current from a low-voltage direct current supply.

It has a primary coil consisting of a few turns of thick wire wound around an iron core and passing a low voltage (usually from a battery). Wound on top of this is a secondary coil made up of many turns of thin wire. An iron armature and make-and-break mechanism (similar to that in an ◊electric bell) repeatedly interrupts the current to the primary coil, producing a high, rapidly alternating current in the secondary circuit.

inductor device included in an electrical circuit because of its inductance.

inert gas or *noble gas* any of a group of six elements (helium, neon, argon, krypton, xenon, and radon), so named because they were originally thought not to enter into any chemical reactions. This is now known to be incorrect: in 1962, xenon was made to combine with fluorine, and since then, compounds of argon, krypton, and radon with fluorine and/or oxygen have been described.

The extreme unreactivity of the inert gases is due to the stability of their electronic structure. All the electron shells (see ◊atom, electronic structure) of inert gas atoms are full and, except for helium, they all have eight electrons in their outermost (◊valency) shell. The apparent stability of this electronic arrangement led to the formulation of the ◊octet rule to explain the different types of chemical bond found in simple compounds.

electronic structure of the inert gases

name	symbol	atomic number	electronic arrangement
helium	He	2	2.
neon	Ne	10	2.8.
argon	Ar	18	2.8.8.
krypton	Kr	36	2.8.18.8.
xenon	Xe	54	2.8.18.18.8.
radon	Rn	86	2.8.18.32.18.8.

inertia in physics, the tendency of an object to remain in a state of rest or uniform motion until an external force is applied, as stated by Isaac Newton's first law of motion (see ◊Newton's laws of motion).

inertial navigation navigation system that makes use of gyroscopes and accelerometers to monitor and measure a vehicle's movements. A computer calculates the vehicle's position relative to its starting position using the information supplied by the sensors. Inertial navigation is used in aircraft, submarines, spacecraft, and guided missiles.

inferior planet planet (Mercury or Venus) whose orbit lies within that of the Earth, best observed when at its greatest elongation from the Sun, either at eastern elongation in the evening (setting after the Sun) or at western elongation in the morning (rising before the Sun).

inferno in astrophysics, a unit for describing the temperature inside a star. One inferno is 1 billion K, or approximately 1 billion °C.

infinite series in mathematics, a series of numbers consisting of a denumerably infinite sequence of terms. The sequence n, n^2, n^3, ... gives the series $n + n^2 + n^3 + \ldots$. For example, $1 + 2 + 3 + \ldots$ is a divergent infinite arithmetic series, and $8 + 4 + 2 + 1 + \frac{1}{2} + \ldots$ is a convergent infinite geometric series that has a sum to infinity of 16.

infinity mathematical quantity that is larger than any fixed assignable quantity; symbol ∞. By convention, the result of dividing any number by zero is regarded as infinity.

inflorescence in plants, a branch, or system of branches, bearing two or more individual flowers. Inflorescences can be divided into two main types: cymose (or definite) and racemose (or indefinite). In a *cymose inflorescence*, the tip of the main axis produces a single flower and subsequent flowers arise on lower side branches, as in forget-me-not *Myosotis* and chickweed *Stellaria*; the oldest flowers are, therefore, found at the tip. A *racemose inflorescence* has an active growing region at the tip of its main axis, and bears flowers along its length, as in hyacinth *Hyacinthus*; the oldest flowers are found near the base or, in cases where the inflorescence is flattened, towards the outside.

The stalk of the inflorescence is called a peduncle; the stalk of each individual flower is called a pedicel.

Types of racemose inflorescence include the *raceme*, a spike of similar, stalked flowers, as seen in lupin *Lupinus*. A *corymb*, seen in candytuft *Iberis amara*, is rounded or flat-topped because the pedicels of the flowers vary in length, the outer pedicels being longer than the inner ones. A *panicle* is a branched inflorescence made up of a number of racemes; such inflorescences are seen in many grasses, for example, the oat *Avena*. The pedicels of an *umbel*, seen in members of the carrot family (Umbelliferae), all arise from the same point on the main axis, like the spokes of an umbrella. Other types of racemose inflorescence include the ◊catkin, a pendulous inflorescence, made up of many small stalkless flowers; the ◊*spadix*, in which tiny flowers are borne on a fleshy axis; and the ◊capitulum, in which the axis is flattened or rounded, bears many small flowers, and is surrounded by large petal-like bracts.

INFLORESCENCE: BLOOMS ON A LARGE SCALE

A rare Bolivian plant, *Puyu raimondii*, holds the record for the largest inflorescence, 2.4 m/8 ft in diameter and made up of 8,000 white flowers. It is also the world's slowest flowering plant—taking 150 years to produce a flower.

---

information technology collective term for the various technologies involved in processing and transmitting information. They include computing, telecommunications, and microelectronics.

infrared absorption spectrometry technique used to determine the mineral or chemical composition of artefacts and organic substances, particularly amber. A sample is bombarded by infrared radiation, which causes the atoms in it to vibrate at frequencies characteristic of the substance present, and absorb energy at those frequencies from the infrared spectrum, thus forming the basis for identification.

infrared astronomy study of infrared radiation produced by relatively cool gas and dust in space, as in the areas around forming stars. In 1983, the Infra-Red Astronomy Satellite (IRAS) surveyed the entire sky at infrared wavelengths. It found five new comets, thousands of galaxies undergoing bursts of star formation, and the possibility of planetary systems forming around several dozen stars.

infrared radiation invisible electromagnetic radiation of wavelength between about 0.75 micrometres and 1 millimetre—that is, between the limit of the red end of the visible spectrum and the shortest microwaves. All bodies above the ◊absolute zero of temperature absorb and radiate infrared radiation. Infrared radiation is used in medical photography and treatment, and in industry, astronomy, and criminology.

Infrared absorption spectra are used in chemical analysis, particularly for organic compounds. Objects that radiate infrared radiation can be photographed or made visible in the dark, or through mist or fog, on specially sensitized emulsions. This is important for military purposes and in detecting people buried under rubble. The strong absorption by many substances of infrared radiation is a useful method of applying heat, as in baking and toasting.

inhibition, neural in biology, the process in which activity in one ◊nerve cell suppresses activity in another. Neural inhibition in networks of nerve cells leading from sensory organs, or to muscles, plays an important role in allowing an animal to make fine sensory discriminations and to exercise fine control over movements.

inhibitor or *negative catalyst* ◊catalyst that reduces the rate of a reaction. Inhibitors are widely used in foods, medicines, and toiletries.

ink-jet printer computer printer that creates characters and graphics by spraying very fine jets of quick-drying ink onto paper. Ink-jet printers range in size from small machines designed to work with microcomputers to very large machines designed for high-volume commercial printing.

Because they produce very high-quality printing and are virtually silent, small ink-jet printers (along with ◊laser printers) are starting to replace impact printers, such as dot-matrix and daisywheel printers, for use with microcomputers.

inorganic chemistry branch of chemistry dealing with the chemical properties of the elements and their compounds, excluding the more complex covalent compounds of carbon, which are considered in ◊organic chemistry.

The origins of inorganic chemistry lay in observing the characteristics and experimenting with the uses of the substances (compounds and elements) that could be extracted from mineral ores. These could be classified according to their chemical properties: elements could be classified as metals or nonmetals; compounds as acids or bases, oxidizing or reducing agents, ionic compounds (such as salts), or covalent compounds (such as gases). The arrangement of elements into groups possessing similar properties led to Mendeleyev's ◊periodic table of the elements, which prompted chemists to predict the properties of undiscovered elements that might occupy gaps in the table. This, in turn, led to the discovery of new elements, including a number of highly radioactive elements that do not occur naturally.

inorganic compound compound found in organisms that is not typically biological. Water, sodium chloride, and potassium are inorganic compounds because they are widely found outside living cells. The term is also applied to those compounds that do not contain carbon and which are not manufactured by organisms. However, carbon dioxide is considered inorganic, contains carbon, and is manufactured by organisms during respiration. See ◊organic compound.

input device device for entering information into a computer. Input devices include keyboards, joysticks, mice, light pens, touch-sensitive screens, graphics tablets, speech-recognition devices, and vision systems. Compare ◊output device.

Input devices that are used commercially—for example, by banks, postal services, and supermarkets—must be able to read and capture large volumes of data very rapidly. Such devices include document readers for magnetic-ink character recognition (MICR), optical character recognition (OCR), and optical mark recognition (OMR); mark-sense readers; bar-code scanners; magnetic-strip readers; and point-of-sale (POS) terminals. Punched-card and paper-tape readers were used in earlier commercial applications but are now obsolete.

insect any member of the class Insecta among the ◊arthropods or jointed-legged animals. An insect's body is divided into head, thorax, and abdomen. The head bears a pair of feelers or antennae, and attached to the thorax are three pairs of legs and usually two pairs of wings. The scientific study of insects is termed entomology. More than 1 million species are known, and several thousand new ones are discovered every year. Insects vary in size from 0.02 cm/0.007 in to 35 cm/13.5 in in length.

anatomy The skeleton is external and is composed of ◊chitin. It is membranous at the joints, but elsewhere is hard. The head is the feeding and sensory centre. It bears the antennae, eyes, and mouthparts. By means of the antennae, the insect detects odours and experiences the sense of touch. The eyes include compound eyes and simple eyes (ocelli). Compound eyes are formed of a large number of individual facets or lenses; there are about 4,000 lenses to each compound eye in the housefly. The mouthparts include a labrum, or upper lip; a pair of principal jaws, or mandibles; a pair of accessory jaws, or maxillae; and a labium, or lower lip. These mouthparts are modified in the various insect groups, depending on the diet. The thorax is the locomotory centre, and is made up of three segments: the pro-, meso-, and metathorax. Each bears a pair of legs, and, in flying insects, the second and third of these segments also each bear a pair of wings.

Wings are composed of an upper and a lower membrane, and between these two layers they are strengthened by a framework of chitinous tubes known as veins. The abdomen is the metabolic and reproductive centre, where digestion, excretion, and the sexual functions take place. In the female, there is very commonly an egg-laying instrument,

or ovipositor, and many insects have a pair of tail feelers, or cerci. Most insects breathe by means of fine airtubes called tracheae, which open to the exterior by a pair of breathing pores, or spiracles. Reproduction is by diverse means. In most insects, mating occurs once only, and death soon follows.

growth and metamorphosis When ready to hatch from the egg, the young insect forces its way through the chorion, or eggshell, and growth takes place in cycles that are interrupted by successive moults. After moulting, the new cuticle is soft and pliable, and is able to adapt itself to increase in size and change of form.

Most of the lower orders of insects pass through a direct or incomplete metamorphosis. The young closely resemble the parents and are known as nymphs.

The higher groups of insects undergo indirect or complete metamorphosis. They hatch at an earlier stage of growth than nymphs and are termed larvae. The life of the insect is interrupted by a resting pupal stage when no food is taken.

During this stage, the larval organs and tissues are transformed into those of the imago, or adult. Before pupating, the insect protects itself by selecting a suitable hiding place, or making a cocoon of some material which will merge in with its surroundings. When an insect is about to emerge from

insect classification

class Insecta subclass	order	number of species	common names
Apterygota (wingless insects)			
	Thysanura	350	three-pronged bristletails, silverfish
	Diplura	400	two-pronged bristletails, campodeids, japygids
	Protura	50	minute insects living in soil
	Collembola	1500	springtails
Pterygota (winged insects or forms secondarily wingless)			
Exopterygota (young resemble adults but have externally developing wings)	Ephemeroptera	1,000	mayflies
	Odonata	5,000	dragonflies, damselflies
	Plecoptera	3,000	stoneflies
	Grylloblattodea	12	wingless soil-living insects of North America
	Orthoptera	20,000	crickets, grasshoppers, locusts, mantids, roaches
	Phasmida	2,000	stick insects, leaf insects
	Dermaptera	1,000	earwigs
	Embioptera	150	web-spinners
	Dictyoptera	5,000	cockroaches, praying mantises
	Isoptera	2,000	termites
	Zoraptera	16	tiny insects living in decaying plants
	Psocoptera	1,600	booklice, barklice, psocids
	Mallophaga	2,500	biting lice, mainly parasitic on birds
	Anoplura	250	sucking lice, mainly parasitic on mammals
	Hemiptera	55,000	true bugs, including aphids, shield- and bedbugs, froghoppers, pond skaters, water boatmen
	Thysanoptera	5,000	thrips
Endopterygota (young unlike adults, undergo sudden metamorphosis)	Neuroptera	4,500	lacewings, alder flies, snake flies, ant lions
	Mecoptera	300	scorpion flies
	Lepidoptera	165,000	butterflies, moths
	Trichoptera	3,000	caddis flies
	Diptera	70,000	true flies, including bluebottles, mosquitoes, leatherjackets, midges
	Siphonaptera	1,400	fleas
	Hymenoptera	100,000	bees, wasps, ants, sawflies
	Coloeoptera	350,000	beetles, including weevils, ladybirds, glow-worms, woodworms, chafers

the pupa, or protective sheath, it undergoes its final moult, which consists of shedding the pupal cuticle. Many insects are seen as *pests*. They may be controlled by chemical insecticides (these may also kill useful insects), by importation of natural predators (that may themselves become pests), or, more recently, by the use of artificially reared sterile insects, either the males only, or in 'population flushing' both sexes, so sharply reducing succeeding generations.

The *classification* of insects is largely based upon characters of the mouthparts, wings, and metamorphosis.

Insects are divided into two subclasses (one with two divisions) and 29 orders.

OUTNUMBERED BY INSECTS

It is estimated that there are around a billion billion individual insects alive at any one time. For every human being alive, there are about a million insects—and a million insects weigh about 12 times more than the average person.

insecticide any chemical pesticide used to kill insects. Among the most effective insecticides are synthetic organic chemicals such as ◊DDT and dieldrin, which are chlorinated hydrocarbons. These chemicals, however, have proved persistent in the environment and are also poisonous to all animal life, including humans, and are consequently banned in many countries. Other synthetic insecticides include organic phosphorus compounds such as malathion. Insecticides prepared from plants, such as derris and pyrethrum, are safer to use but need to be applied frequently and carefully.

insectivore any animal whose diet is made up largely or exclusively of insects. In particular, the name is applied to mammals of the order Insectivora, which includes the shrews, hedgehogs, moles, and tenrecs.

insectivorous plant plant that can capture and digest live prey (normally insects), to obtain nitrogen compounds that are lacking in its usual marshy habitat. Some are passive traps, for example, the pitcher plants *Nepenthes* and *Sarracenia*. One pitcher-plant species has container-traps holding 1.6 l/3.5 pt of the liquid that 'digests' its food, mostly insects but occasionally even rodents. Others, for example, sundews *Drosera*, butterworts *Pinguicula*, and Venus flytraps *Dionaea muscipula*, have an active trapping mechanism. Insectivorous plants have adapted to grow in poor soil conditions where the number of microorganisms recycling nitrogen compounds is very much reduced. In these circumstances other plants cannot gain enough nitrates to grow. See also ◊leaf.

inselberg or *kopje* prominent steep-sided hill of resistant solid rock, such as granite, rising out of a plain, usually in a tropical area. Its rounded appearance is caused by so-called onion-skin ◊weathering, in which the surface is eroded in successive layers.

The Sugar Loaf in Rio de Janeiro harbour in Brazil, and Ayers Rock in Northern Territory, Australia, are famous examples.

insolation the amount of solar radiation (heat energy from the Sun) that reaches the Earth's surface. Insolation varies with season and latitude, being greatest at the equator and least at the poles. At the equator the Sun is consistently high in the sky: its rays strike the equatorial region directly and are therefore more intense. At the poles the tilt of the Earth means that the Sun is low in the sky, and so its rays are slanted and spread out. Winds and ocean currents help to balance out the uneven spread of radiation.

instinct in ◊ethology, behaviour found in all equivalent members of a given species (for example, all the males, or all the females with young) that is presumed to be genetically determined.

Examples include a male robin's tendency to attack other male robins intruding on its territory and the tendency of many female mammals to care for their offspring. Instincts differ from ◊reflexes in that they involve very much more complex actions, and learning often plays an important part in their development.

instruction register in computing, a special memory location used to hold the instruction that the computer is currently processing. It is located in the control unit of the ◊central processing unit, and receives instructions individually from the immediate-access memory during the fetch phase of the ◊fetch-execute cycle.

instruction set in computing, the complete set of machine-code instructions that a computer's central processing unit can obey.

instrument landing system landing aid for aircraft that uses ◊radio beacons on the ground and instruments on the flight deck. One beacon (localizer) sends out a vertical radio beam along the centre line of the runway. Another beacon (glide slope) transmits a beam in the plane at right angles to the localizer beam at the ideal approach-path angle. The pilot can tell from the instruments how to manoeuvre to attain the correct approach path.

insulator any poor ◊conductor of heat, sound, or electricity. Most substances lacking free (mobile) ◊electrons, such as non-metals, are electrical or thermal insulators. Usually, devices of glass or porcelain, called insulators, are used for insulating and supporting overhead wires.

insulin protein ◊hormone, produced by specialized cells in the islets of Langerhans in the pancreas, that regulates the metabolism (rate of activity) of glucose, fats, and proteins. Insulin was discovered by Canadian physician Frederick Banting, who pioneered its use in treating ◊diabetes.

Normally, insulin is secreted in response to rising blood sugar levels (after a meal, for example), stimulating the body's cells to store the excess. Failure of this regulatory mechanism in *diabetes mellitus* requires treatment with insulin injections or capsules taken by mouth. Types vary from pig and beef insulins to synthetic and bioengineered ones. They may be combined with other substances to make them longer-or shorter-acting. Implanted, battery-powered insulin pumps deliver the hormone at a preset rate, to eliminate the unnatural rises and falls that result from conventional, subcutaneous (under the skin) delivery. Human insulin

the packaging of a silicon 'chip'

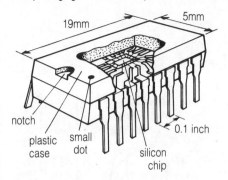

integrated circuit An integrated circuit (IC), or silicon chip. The IC is a piece of silicon about the size of a child's fingernail on which the components of an electronic circuit are etched. The IC is packaged in a plastic container with metal legs that connect it to the circuit board.

has now been produced from bacteria by ◊genetic engineering techniques, but may increase the chance of sudden, unpredictable hypoglycaemia, or low blood sugar. In 1990 the Medical College of Ohio developed gelatine capsules and an aspirinlike drug which helps the insulin pass into the bloodstream.

integer any whole number. Integers may be positive or negative; 0 is an integer, and is often considered positive. Formally, integers are members of the set $Z = \{\ldots -3, -2, -1, 0, 1, 2, 3, \ldots\}$ Fractions, such as ½ and 0.35, are known as nonintegral numbers ('not integers').

God made the integers, man made the rest.
On **integers** Leopold Kronecker (1823–1891)

integral calculus branch of mathematics using the process of ◊integration. It is concerned with finding volumes and areas and summing infinitesimally small quantities.

integrated circuit (IC), popularly called *silicon chip*, a miniaturized electronic circuit produced on a single crystal, or chip, of a semiconducting material—usually silicon. It may contain many thousands of components and yet measure only 5 mm/0.2 in square and 1 mm/0.04 in thick. The IC is encapsulated within a plastic or ceramic case, and linked via gold wires to metal pins with which it is connected to a ◊printed circuit board and the other components that make up such electronic devices as computers and calculators.

Integrated Services Digital Network (ISDN) internationally developed telecommunications system for sending signals in ◊digital format along optical fibres and coaxial cable. It involves converting the 'local loop'—the link between the user's telephone (or private automatic branch exchange) and the digital telephone exchange—from an ◊analogue system into a digital system, thereby greatly increasing the amount of information that can be carried. The first large-scale use of ISDN began in Japan 1988.

ISDN has advantages in higher voice quality, better-quality faxes, and the possibility of data transfer between computers faster than current modems. With ISDN's *Basic Rate Access*, a multiplexer divides one voice telephone line into three channels: two B bands and a D band. Each B band offers 64 kilobits per second and can carry one voice conversation or 50 simultaneous data calls at 1,200 bits per second. The D band is a data-signalling channel operating at 16 kilobits per second. With *Primary Rate Access*, ISDN provides 30 B channels.

British Telecom began offering ISDN to businesses 1991, with some 47,000 ISDN-equipped lines. Its adoption in the UK is expected to stimulate the use of data-communications services such as faxing, teleshopping, and home banking. New services may include computer conferencing, where both voice and computer communications take place simultaneously, and videophones.

integrated steelworks modern industrial complex where all the steelmaking processes—such as iron smelting and steel shaping—take place on the same site. In the UK, the Redcar/Lackenby works on Teesside is an example of integrated steelworks.

integration in mathematics, a method in ◊calculus of determining the solutions of definite or indefinite integrals. An example of a definite integral can be thought of as finding the area under a curve (as represented by an algebraic expression or function) between particular values of the function's variable.

In practice, integral calculus provides scientists with a powerful tool for doing calculations that involve a continually varying quantity (such as determining the position at any given instant of a space rocket that is accelerating away from Earth). Its basic principles were discovered in the late 1660s independently by the German philosopher Gottfried Leibniz and the British scientist Isaac Newton.

integument in seed-producing plants, the protective coat surrounding the ovule. In flowering plants there are two, in gymnosperms only one. A small hole at one end, the micropyle, allows a pollen tube to penetrate through to the egg during fertilization.

intelligent terminal in computing, a ◊terminal with its own processor which can take some of the processing load away from the main computer.

Intelsat International Telecommunications Satellite Organization, established 1964 to operate a worldwide system of communications satellites. More than 100 countries are members of Intelsat, with headquarters in Washington, DC. Intelsat satellites are stationed in geostationary orbit (maintaining their positions relative to the Earth) over the Atlantic, Pacific, and Indian Oceans. The first Intelsat satellite was *Early Bird*, launched 1965.

intensity in physics, the power (or energy per second) per unit area carried by a form of radiation or wave motion. It is an indication of the concentration of energy present and, if measured at varying distances from the source, of the effect of distance on this. For example, the intensity of light is a measure of its brightness, and may be shown to diminish with distance from its source in accordance with the ◊inverse square law (its intensity is

the back of a typical microcomputer showing range of interfaces provided

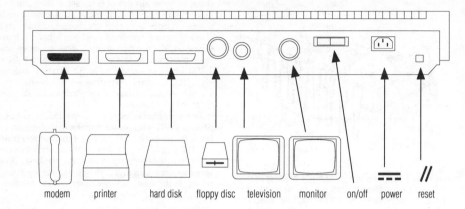

modem   printer   hard disk   floppy disc   television   monitor   on/off   power   reset

interface

inversely proportional to the square of the distance).

interactive computing in computing, a system for processing data in which the operator is in direct communication with the computer, receiving immediate responses to input data. In ◊batch processing, by contrast, the necessary data and instructions are prepared in advance and processed by the computer with little or no intervention from the operator.

interactive video (IV) computer-mediated system that enables the user to interact with and control information (including text, recorded speech, or moving images) stored on video disc. IV is most commonly used for training purposes, using analogue video discs, but has wider applications with digital video systems such as CD-I (Compact Disc Interactive, from Philips and Sony) which are based on the CD-ROM format derived from audio compact discs.

interface in computing, the point of contact between two programs or pieces of equipment. The term is most often used for the physical connection between the computer and a ◊peripheral device, which is used to compensate for differences in such operating characteristics as speed, data coding, voltage, and power consumption. For example, a *printer interface* is the cabling and circuitry used to transfer data from a computer to a printer, and to compensate for differences in speed and coding.

Common standard interfaces include the *Centronics interface*, used to connect parallel devices, and the *RS232 interface*, used to connect serial devices. For example, in many microcomputer systems, an RS232 interface is used to connect the microcomputer to a modem, and a Centronics device is used to connect it to a printer.

interference in physics, the phenomenon of two or more wave motions interacting and combining to produce a resultant wave of larger or smaller amplitude (depending on whether the combining waves are in or out of ◊phase with each other). Interference of white light (multiwavelength)

results in spectral coloured fringes; for example, the iridescent colours of oil films seen on water or soap bubbles (demonstrated by ◊Newton's rings). Interference of sound waves of similar frequency produces the phenomenon of beats, often used by musicians when tuning an instrument. With monochromatic light (of a single wavelength), interference produces patterns of light and dark bands. This is the basis of ◊holography, for example. Interferometry can also be applied to radio waves, and is a powerful tool in modern astronomy.

interferometer in physics, a device that splits a beam of light into two parts, the parts being recombined after travelling different paths to form an interference pattern of light and dark bands. Interferometers are used in many branches of science and industry where accurate measurements of distances and angles are needed.

In the Michelson interferometer, a light beam is split into two by a semisilvered mirror. The two beams are then reflected off fully silvered mirrors and recombined. The pattern of dark and light bands is sensitive to small alterations in the placing of the mirrors, so the interferometer can detect changes in their position to within one ten-millionth of a metre. Using lasers, compact devices of this kind can be built to measure distances, for example to check the accuracy of machine tools.

In radio astronomy, interferometers consist of separate radio telescopes, each observing the same distant object, such as a galaxy, in the sky. The signal received by each telescope is fed into a computer. Because the telescopes are in different places, the distance travelled by the signal to reach each differs and the overall signal is akin to the interference pattern in the Michelson interferometer. Computer analysis of the overall signal can build up a detailed picture of the source of the radio waves.

In space technology, interferometers are used in radio and radar systems. These include space-vehicle guidance systems, in which the position of the spacecraft is determined by combining the signals received by two precisely spaced antennae mounted on it.

interferon naturally occurring cellular protein that makes up part of the body's defences against viral disease. Three types (alpha, beta, and gamma) are produced by infected cells and enter the bloodstream and uninfected cells, making them immune to virus attack.

Interferon was discovered 1957 by Scottish virologist Alick Isaacs. At present, only alpha interferon has any proven therapeutic value, and may be used to treat a rare type of leukaemia.

interlocking spur one of a series of spurs (ridges of land) jutting out from alternate sides of a river valley. During glaciation its tip may be sheared off by erosion, creating a ◊truncated spur.

intermediate technology application of mechanics, electrical engineering, and other technologies, based on inventions and designs developed in scientifically sophisticated cultures, but utilizing materials, assembly, and maintenance methods found in technologically less advanced regions (known as the Third World).

Intermediate technologies aim to allow developing countries to benefit from new techniques and inventions of the 'First World', without the burdens of costly maintenance and supply of fuels and spare parts that in the Third World would represent an enormous and probably uneconomic overhead. See also ◊appropriate technology.

Intermediate Technology Development Group UK-based international aid organization established 1965 by German writer and economist Ernst Friedrich Schumacher to give advice and assistance on the appropriate choice of technologies for the rural poor of the Third World. It is an independent charity financed through donations.

Intermediate Technology concentrates mainly on renewable energy systems and small-scale manufacturing, and runs practical projects in many countries. The design and manufacture of a fuel-efficient stove and a bicycle trailer were among its 1980s successes.

intermediate vector boson alternative name for *weakon*, the elementary particle responsible for carrying the ◊weak nuclear force.

intermolecular force or *van der Waals' force* force of attraction between molecules. Intermolecular forces are relatively weak; hence simple molecular compounds are gases, liquids, or low-melting-point solids.

internal-combustion engine heat engine in which fuel is burned inside the engine, contrasting with an external combustion engine (such as the steam engine) in which fuel is burned in a separate unit. The ◊diesel and ◊petrol engine are both internal-combustion engines. Gas ◊turbines and ◊jet and ◊rocket engines are sometimes also considered to be internal-combustion engines because they burn their fuel inside their combustion chambers.

internal resistance or *source resistance* the resistance inside a power supply, such as a battery of cells, that limits the current that it can supply to a circuit.

International Atomic Energy Agency (IAEA) agency of the United Nations established 1957 to advise and assist member countries in the development and application of nuclear power, and to guard against its misuse. It has its headquarters in Vienna, and is responsible for research centres in Austria and Monaco, and the International Centre for Theoretical Physics, Trieste, Italy, established 1964.

international biological standards drugs (such as penicillin and insulin) of which the activity for a specific mass (called the international unit, or IU), prepared and stored under specific conditions, serves as a standard for measuring doses. For penicillin, one IU is the activity of 0.0006 mg of the sodium salt of penicillin, so a dose of a million units would be 0.6 g.

International Date Line (IDL) imaginary line that approximately follows the 180° line of longitude. The date is put forward a day when crossing the line going west, and back a day when going east. The IDL was chosen at the International Meridian Conference 1884.

International Standards Organization international organization founded 1947 to standardize technical terms, specifications, units, and so on. Its headquarters are in Geneva.

International Telecommunication Union body belonging to the Economic and Social Council of the United Nations. It aims to extend international cooperation by improving telecommunications of all kinds.

International Union for Conservation of Nature organization established by the United Nations to promote the conservation of wildlife and habitats as part of the national policies of member states.

It has formulated guidelines and established research programmes (for example, International Biological Programme, IBP) and set up advisory bodies (such as Survival Commissions, SSC). In 1980, it launched the *World Conservation Strategy* to highlight particular problems, designating a small number of areas as *World Heritage Sites* to ensure their survival as unspoiled habitats (for example, Yosemite National Park in the USA, and the Simen Mountains in Ethiopia).

interplanetary matter gas and dust thinly spread through the Solar System. The gas flows outwards from the Sun as the ◊solar wind. Fine dust lies in the plane of the Solar System, scattering sunlight to cause the ◊zodiacal light. Swarms of dust shed by comets enter the Earth's atmosphere to cause ◊meteor showers.

interpreter computer program that translates and executes a program written in a high-level language. Unlike a ◊compiler, which produces a complete machine-code translation of the high-level program in one operation, an interpreter translates the source program, instruction by instruction, each time that program is run.

Because each instruction must be translated each time the source program is run, interpreted programs run far more slowly than do compiled programs. However, unlike compiled programs, they can be executed immediately without waiting for an intermediate compilation stage.

interrupt in computing, a signal received by the computer's central processing unit that causes a temporary halt in the execution of a program while some other task is performed. Interrupts may be generated by the computer's internal electronic clock (clock interrupt), by an input or output device, or by a software routine. After the computer has completed the task to which it was diverted, control returns to the original program.

For example, many computers, while printing a long document, allow the user to carry on with other work. When the printer is ready for more data, it sends an interrupt signal that causes the computer to halt work on the user's program and transmit more data to the printer.

intersex individual that is intermediate between a normal male and a normal female in its appearance (for example, a genetic male that lacks external genitalia and so resembles a female).

Intersexes are usually the result of an abnormal hormone balance during development (especially during ◊gestation) or of a failure of the ◊genes controlling sex determination. The term ◊hermaphrodite is sometimes used for intersexes, but should be confined to animals that normally have both male and female organs.

interstellar molecules over 50 different types of molecule existing in gas clouds in our Galaxy. Most have been detected by their radio emissions, but some have been found by the absorption lines they produce in the spectra of starlight. The most complex molecules, many of them based on ◊carbon, are found in the dense clouds where stars are forming. They may be significant for the origin of life elsewhere in space.

intertropical convergence zone (ITCZ) area of heavy rainfall found in the tropics and formed as the trade winds converge and rise to form cloud and rain. It moves a few degrees northwards during the northern summer and a few degrees southwards during the southern summer, following the apparent movement of the Sun. The ITCZ is responsible for most of the rain that falls in Africa. The ◊doldrums are also associated with this zone.

intestine in vertebrates, the digestive tract from the stomach outlet to the anus. The human *small intestine* is 6 m/20 ft long, 4 cm/1.5 in in diameter, and consists of the duodenum, jejunum, and ileum; the *large intestine* is 1.5 m/5 ft long, 6 cm/2.5 in in diameter, and includes the caecum, colon, and rectum. Both are muscular tubes comprising an inner lining that secretes alkaline digestive juice, a submucous coat containing fine blood vessels and nerves, a muscular coat, and a serous coat covering all, supported by a strong peritoneum, which carries the blood and lymph vessels, and the nerves. The contents are passed along slowly by ◊peristalsis (waves of involuntary muscular action). The term intestine is also applied to the lower digestive tract of invertebrates.

intrauterine device IUD or coil, a contraceptive device that is inserted into the womb (uterus). It is a tiny plastic object, sometimes containing copper. By causing a mild inflammation of the lining of the uterus it prevents fertilized eggs from becoming implanted.

IUDs are not usually given to women who have

the position of the intertropical convergence zone in January and July in West Africa

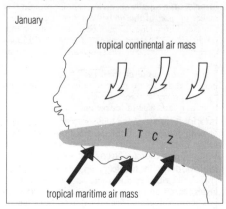

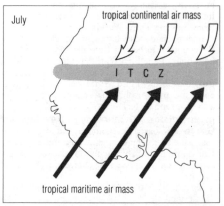

intertropical convergence zone

not had children. They are generally very reliable, as long as they stay in place, with a success rate of about 98%. Some women experience heavier and more painful periods, and there is a very small risk of a pelvic infection leading to infertility.

intrusion mass of ◊igneous rock that has formed by 'injection' of molten rock, or magma, into existing cracks beneath the surface of the Earth, as distinct from a volcanic rock mass which has erupted from the surface. Intrusion features include vertical cylindrical structures such as stocks, pipes, and necks; sheet structures such as dykes that cut across the strata and sills that push between them; laccoliths, which are blisters that push up the overlying rock; and batholiths, which represent chambers of solidified magma and contain vast volumes of rock.

intrusive rock ◊igneous rock formed within the Earth. Magma, or molten rock, cools slowly at these depths to form coarse-grained rocks, such as granite, with large crystals. (◊Extrusive rocks, which are formed on the surface, are usually fine-grained.) A mass of intrusive rock is called an intrusion.

intuitionism in mathematics, the theory that

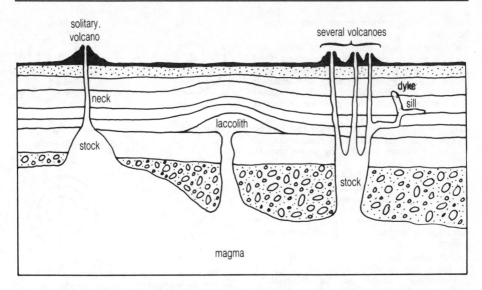

intrusion *Igneous intrusions can be a variety of shapes and sizes. Laccoliths are domed circular shapes, and can be many miles across. Sills are intrusions that flow between rock layers. Pipes or necks connect the underlying magma chamber to surface volcanoes.*

propositions can be built up only from intuitive concepts that we all recognize easily, such as unity or plurality. The concept of ◊infinity, of which we have no intuitive experience, is thus not allowed.

Invar trademark for an alloy of iron containing 36% nickel, which expands or contracts very little when the temperature changes. It is used to make precision instruments (such as pendulums and tuning forks) whose dimensions must not alter.

inverse square law in physics, the statement that the magnitude of an effect (usually a force) at a point is inversely proportional to the square of the distance between that point and the point location of its cause.

Light, sound, electrostatic force (Coulomb's law), gravitational force (Newton's law) and magnetic force (see ◊magnetism) all obey the inverse square law.

inverse video or *reverse video* in computing, a display mode in which images on a display screen are presented as a negative of their normal appearance. For example, if the computer screen normally displays dark images on a light background, inverse video will change all or part of the screen to a light image on a dark background.

Inverse video is commonly used to highlight parts of a display or to mark out text and pictures that the user wishes the computer to change in some way. For example, the user of a word-processing program might use a pointing device such as a ◊mouse to mark in inverse video a paragraph of text that is to be deleted from the document.

invertebrate animal without a backbone. The invertebrates comprise over 95% of the million or so existing animal species and include sponges, coelenterates, flatworms, annelids, arthropods, molluscs, echinoderms, and primitive aquatic chordates, such as sea squirts and lancelets.

inverted file in computing, a file that reorganizes the structure of an existing data file to enable a

rapid search to be made for all records having one field falling within set limits.

For example, a file used by an estate agent might store records on each house for sale, using a reference number as the key field for ◊sorting. One field in each record would be the asking price of the house. To speed up the process of drawing up lists of houses falling within certain price ranges, an inverted file might be created in which the records are rearranged according to price. Each record would consist of an asking price, followed by the reference numbers of all the houses offered for sale at this approximate price.

in vitro process biological experiment or technique carried out in a laboratory, outside the body of a living organism (literally 'in glass', for example in a test tube). By contrast, an *in vivo* process takes place within the body of an organism.

in vitro fertilization (IVF) ('fertilization in glass') allowing eggs and sperm to unite in a laboratory to form embryos. The embryos produced may then either be implanted into the womb of the otherwise infertile mother (an extension of artificial insemination), or used for research. The first baby to be produced by this method was born 1978 in the UK. In cases where the Fallopian tubes are blocked, fertilization may be carried out by **intra-vaginal culture**, in which egg and sperm are incubated (in a plastic tube) in the mother's vagina, then transferred surgically into the uterus.

Recent extensions of the *in vitro* technique have included the birth of a baby from a frozen embryo (Australia 1984) and from a frozen egg (Australia 1986). Pioneers in the field have been the British doctors Robert Edwards (1925–) and Patrick Steptoe. As yet the success rate is relatively low; only 15–20% of IVFs result in babies.

in vivo process biological experiment or technique carried out within a living organism; by contrast, an *in vitro* process takes place outside the organism, in an artificial environment such as a laboratory.

involuntary action behaviour not under conscious control, for example the contractions of the gut during peristalsis or the secretion of adrenaline by the adrenal glands. Breathing and urination reflexes are involuntary, although both can be controlled to some extent. These processes are part of the autonomic nervous system.

involute (Latin 'rolled in') ◊spiral that can be thought of as being traced by a point at the end of a taut nonelastic thread being wound onto or unwound from a spool.

Io in astronomy, the third largest moon of the planet Jupiter, 3,630 km/2,260 mi in diameter, orbiting in 1.77 days at a distance of 422,000 km/ 262,000 mi. It is the most volcanically active body in the Solar System, covered by hundreds of vents that erupt not lava but sulphur, giving Io an orange-coloured surface.

iodide compound formed between iodine and another element in which the iodine is the more electronegative element (see ◊electronegativity, ◊halide).

iodine (Greek *iodes* 'violet') greyish-black non-metallic element, symbol I, atomic number 53, relative atomic mass 126.9044. It is a member of the ◊halogen group. Its crystals give off, when heated, a violet vapour with an irritating odour resembling that of chlorine. It only occurs in combination with other elements. Its salts are known as iodides, which are found in sea water. As a mineral nutrient it is vital to the proper functioning of the thyroid gland, where it occurs in trace amounts as part of the hormone thyroxine. Iodine is used in photography, in medicine as an antiseptic, and in making dyes.

Its radioactive isotope ^{131}I (half-life of eight days) is a dangerous fission product from nuclear explosions and from the nuclear reactors in power plants, since, if ingested, it can be taken up by the thyroid and damage it. It was discovered 1811 by French chemist Bernard Courtois (1777–1838).

iodoform (chemical name *triiodomethane*) CHI_3, an antiseptic that crystallizes into yellow hexagonal plates. It is soluble in ether, alcohol, and chloroform, but not in water.

ion atom, or group of atoms, which is either positively charged (◊cation) or negatively charged (◊anion), as a result of the loss or gain of electrons during chemical reactions or exposure to certain forms of radiation.

ion engine rocket engine that uses ◊ions (charged particles) rather than hot gas for propulsion. Ion engines have been successfully tested in space, where they will eventually be used for gradual rather than sudden velocity changes. In an ion engine, atoms of mercury, for example, are ionized (given an electric charge by an electric field) and then accelerated at high speed by a more powerful electric field.

ion exchange process whereby an ion in one compound is replaced by a different ion, of the same charge, from another compound. It is the basis of a type of ◊chromatography in which the components of a mixture of ions in solution are

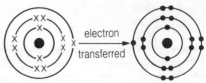

electronic arrangement, 2.8.1 of a sodium atom,

electronic arrangement, 2.8.7 of a chlorine atom,

becomes a sodium ion, Na^+, with an electron arrangement 2.8

becomes a chloride ion, Cl^-, with an electron arrangement 2.8.8

ionic bond *The formation of an ionic bond between a sodium atom and a chlorine atom to form a molecule of sodium chloride. The sodium atom transfers an electron from its outer electron shell (becoming the positive ion Na^+) to the chlorine atom (which becomes the negative chloride ion Cl^-). The opposite charges mean that the ions are strongly attracted to each other. The formation of the bond means that each atom becomes more stable, having a full quota of electrons in its outer shell.*

separated according to the ease with which they will replace the ions on the polymer matrix through which they flow. The exchange of positively charged ions is called cation exchange; that of negatively charged ions is called anion exchange.

Ion-exchange is used in commercial water softeners to exchange the dissolved ions responsible for the water's hardness with others that do not have this effect. For example, when hard water is passed over an ion-exchange resin, the dissolved calcium and magnesium ions are replaced by either sodium or hydrogen ions, so the hardness is removed.

ionic bond or *electrovalent bond* bond produced when atoms of one element donate electrons to atoms of another element, forming positively and negatively charged ◊ions respectively. The electrostatic attraction between the oppositely charged ions constitutes the bond. Sodium chloride (Na^+Cl^-) is a typical ionic compound.

Each ion has the electronic structure of an inert gas (see ◊noble gas structure). The maximum number of electrons that can be gained is usually two.

ionic compound substance composed of oppositely charged ions. All salts, most bases, and some acids are examples of ionic compounds. They possess the following general properties: they are crystalline solids with a high melting point; are soluble in water and insoluble in organic solvents; and always conduct electricity when molten or in aqueous solution. A typical ionic compound is sodium chloride (Na^+Cl^-).

ionization chamber any device for measuring ◊ionizing radiation. The radiation ionizes the gas in the chamber and the ions formed are collected and measured as an electric charge. Ionization chambers are used for determining the intensity of X-rays or the disintegration rate of radioactive materials.

ionization potential measure of the energy required to remove an ◊electron from an ◊atom. Elements with a low ionization potential readily lose electrons to form ◊cations.

ionizing radiation radiation that knocks electrons from atoms during its passage, thereby leaving ions in its path. Alpha and beta particles are far more ionizing in their effect than are neutrons or gamma radiation.

ionosphere ionized layer of Earth's outer ◊atmosphere (60–1,000 km/38–620 mi) that contains sufficient free electrons to modify the way in which radio waves are propagated, for instance by reflecting them back to Earth. The ionosphere is thought to be produced by absorption of the Sun's ultraviolet radiation.

ion plating method of applying corrosion-resistant metal coatings. The article is placed in argon gas, together with some coating metal, which vaporizes on heating and becomes ionized (acquires charged atoms) as it diffuses through the gas to form the coating. It has important applications in the aerospace industry.

ir in physics, abbreviation for **infrared**.

IRAS acronym for *I*nfra*r*ed *A*stronomy *S*atellite, a joint US–UK–Dutch satellite launched 1983 to survey the sky at infrared wavelengths, studying areas of star formation, distant galaxies, possible embryo planetary systems around other stars, and discovering five new comets in our own solar system.

iridium (Latin *iridis* 'rainbow') hard, brittle, silver-white, metallic element, symbol Ir, atomic number 77, relative atomic mass 192.2. It is twice as heavy as lead and is resistant to tarnish and corrosion. It is one of the so-called platinum group of metals; it occurs in platinum ores and as a free metal (◊native metal) with osmium in osmiridium, a natural alloy that includes platinum, ruthenium, and rhodium.

It is alloyed with platinum for jewellery and used for watch bearings and in scientific instruments. It was named in 1804 by English chemist Smithson Tennant (1761–1815) for its iridescence in solution.

iris in anatomy, the coloured muscular diaphragm that controls the size of the pupil in the vertebrate eye. It contains radial muscle that increases the pupil diameter and circular muscle that constricts the pupil diameter. Both types of muscle respond involuntarily to light intensity.

iron hard, malleable and ductile, silver-grey, metallic element, symbol Fe (from Latin *ferrum*), atomic number 26, relative atomic mass 55.847. It is the fourth most abundant element (the second most abundant metal, after aluminium) in the Earth's crust. Iron occurs in concentrated deposits as the ores hematite (Fe_2O_3), spathic ore ($FeCO_3$), and magnetite (Fe_3O_4). It sometimes occurs as a

free metal, occasionally as fragments of iron or iron–nickel meteorites.

Iron is the most common and most useful of all metals; it is strongly magnetic and is the basis for ◊steel, an alloy with carbon and other elements (see also ◊cast iron). In electrical equipment it is used in all permanent magnets and electromagnets, and forms the cores of transformers and magnetic amplifiers. In the human body, iron is an essential component of haemoglobin, the molecule in red blood cells that transports oxygen to all parts of the body. A deficiency in the diet causes a form of anaemia.

iron ore any mineral from which iron is extracted. The chief iron ores are ◊*magnetite*, a black oxide; ◊*hematite*, or kidney ore, a reddish oxide; ◊*limonite*, brown, impure oxyhydroxides of iron; and *siderite*, a brownish carbonate.

Iron ores are found in a number of different forms, including distinct layers in igneous intrusions, as components of contact metamorphic rocks, and as sedimentary beds. Much of the world's iron is extracted in Russia, Kazakhstan, and the Ukraine. Other important producers are the USA, Australia, France, Brazil, and Canada; over 40 countries produce significant quantities of ore.

iron pyrites or *pyrite* FeS_2 common iron ore. Brassy yellow, and occurring in cubic crystals, it is often called 'fool's gold', since only those who have never seen gold would mistake it.

irradiation in technology, subjecting anything to radiation, including cancer tumours. See also ◊food irradiation.

In optics, the term refers to the apparent enlargement of a brightly lit object when seen against a dark background.

irrational number a number that cannot be expressed as an exact ◊fraction. Irrational numbers include some square roots (for example, $\sqrt{2}$, $\sqrt{3}$, and $\sqrt{5}$ are irrational) and numbers such as π (the ratio of the circumference of a circle to its diameter, which is approximately equal to 3.14159) and e (the base of ◊natural logarithms, approximately 2.71828).

irrigation artificial water supply for dry agricultural areas by means of dams and channels. Drawbacks are that it tends to concentrate salts, ultimately causing infertility, and that rich river silt is retained at dams, to the impoverishment of the land and fisheries below them.

Irrigation has been practised for thousands of years, in Eurasia as well as the Americas. An example is the channelling of the annual Nile flood in Egypt, which has been done from earliest times to its present control by the Aswan High Dam.

ISDN abbreviation for ◊*Integrated Services Digital Network*, a telecommunications system.

island area of land surrounded entirely by water. Australia is classed as a continent rather than an island, because of its size.

Islands can be formed in many ways. *Continental islands* were once part of the mainland, but became isolated (by tectonic movement, erosion, or a rise in sea level, for example). *Volcanic islands*, such as Japan, were formed by the explosion of underwater volcanoes. *Coral islands* consist mainly of ◊coral, built up over many years. An *atoll* is a circular coral reef surrounding a lagoon; atolls were formed when a coral reef grew up around a volcanic island that subsequently sank or was submerged by a rise in sea level. *Barrier islands* are found by the shore in shallow water, and are formed by the deposition of sediment eroded from the shoreline.

island arc curved chain of islands produced by volcanic activity caused by rising magma behind the ocean trench formed by one tectonic plate sliding beneath another in a subduction zone. The arc shape is produced as the descending plate cuts into the curved surface of the Earth.

Such island arcs are often later incorporated into continental margins during mountain-building episodes. Island arcs are common in the W and N Pacific where they ring the ocean on both sides; the Aleutian Islands of Alaska are an example.

islets of Langerhans groups of cells within the pancreas responsible for the secretion of the hormone insulin. They are sensitive to the blood sugar, producing more hormone when glucose levels rise.

ISO in photography, a numbering system for rating the speed of films, devised by the International Standards Organization.

isobar line drawn on maps and weather charts linking all places with the same atmospheric pressure (usually measured in millibars). When used in weather forecasting, the distance between the isobars is an indication of the barometric gradient.

Where the isobars are close together, cyclonic weather is indicated, bringing strong winds and a depression, and where far apart anticyclonic, bringing calmer, settled conditions.

isoline on a map, a line that joins places of equal value. Examples are contour lines (joining places of equal height), ◊isobars (for equal pressure), isotherms (for equal temperature), and isohyets (for equal rainfall). Isolines are most effective when values change gradually and when there is plenty of data.

isomer chemical compound having the same

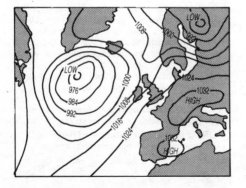

isobar *The isobars around a low-pressure area or depression. In the northern hemisphere, winds blow anticlockwise around lows, approximately parallel to the isobars, and clockwise around highs. In the southern hemisphere, the winds blow in the opposite directions.*

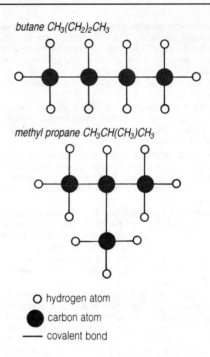

butane $CH_3(CH_2)_2CH_3$

methyl propane $CH_3CH(CH_3)CH_3$

○ hydrogen atom

● carbon atom

── covalent bond

isomer *The chemicals butane and methyl propane are isomers. Each has the molecular formula $CH_3CH(CH_3)CH_3$, but has a different spatial arrangements of the atoms in their molecules.*

molecular composition and mass as another, but with different physical or chemical properties owing to the different structural arrangement of its constituent atoms. For example, the organic compounds butane ($CH_3(CH_2)_2CH_3$) and methyl propane ($CH_3CH(CH_3)CH_3$) are isomers, each possessing four carbon atoms and ten hydrogen atoms but differing in the way that these are arranged with respect to each other.

Structural isomers have obviously different constructions, but *geometrical* and *optical isomers* must be drawn or modelled in order to appreciate the difference in their three-dimensional arrangement. Geometrical isomers have a plane of symmetry and arise because of the restricted rotation of atoms around a bond; optical isomers are mirror images of each other. For instance, 1,1–dichloroethene ($CH_2=CCl_2$) and 1,2–dichloroethene ($CHCl=CHCl$) are structural isomers, but there are two possible geometric isomers of the latter (depending on whether the chlorine atoms are on the same side or on opposite sides of the plane of the carbon-carbon double bond).

isomorphism the existence of substances of different chemical composition but with similar crystalline form.

isoprene $CH_2CHC(CH_3)CH_2$ (technical name *methylbutadiene*) colourless, volatile fluid obtained from petroleum and coal, used to make synthetic rubber.

isostasy the theoretical balance in buoyancy of

all parts of the Earth's ◊crust, as though they were floating on a denser layer beneath. High mountains, for example, have very deep roots, just as an iceberg floats with most of its mass submerged.

Similarly, during an ◊ice age the weight of the ice sheet pushes that continent into the earth's mantle; once the ice has melted, the continent rises again. This accounts for shoreline features being found some way inland in regions that were heavily glaciated during the Pleistocene period.

isotherm line on a map linking all places having the same temperature at a given time.

isotope one of two or more atoms that have the same atomic number (same number of protons), but which contain a different number of neutrons, thus differing in their atomic masses. They may be stable or radioactive, naturally occurring or synthesized. The term was coined by English chemist Frederick Soddy, pioneer researcher in atomic disintegration.

isthmus narrow strip of land joining two larger land masses. The Isthmus of Panama joins North and South America.

iteration in computing, a method of solving a problem by performing the same steps repeatedly until a certain condition is satisfied. For example, in one method of ◊sorting, adjacent items are repeatedly exchanged until the data are in the required sequence.

iteroparity in biology, the repeated production of offspring at intervals throughout the life cycle. It is usually contrasted with ◊semelparity, where each individual reproduces only once during its life. Most vertebrates are iteroparous.

IUCN (abbreviation for *International Union for the Conservation of Nature*) organization established by the United Nations to promote the conservation of wildlife and habitats as part of the national policies of member states.

It has formulated guidelines and established research programmes (for example, International Biological Programme, IBP) and set up advisory bodies (such as Survival Services Commissions, SSC).

In 1980, it launched the *World Conservation Strategy* to highlight particular problems, designating a small number of areas as *World Heritage Sites* to ensure their survival as unspoiled habitats (for example, Yosemite National Park in California, and the Simen Mountains in Ethiopia).

IUE (acronym for *International Ultraviolet Explorer*) joint NASA-ESA (US and European Space Agency) orbiting ultraviolet telescope with a 45-cm/18-in mirror, launched 1978.

IUPAC (abbreviation for *International Union of Pure and Applied Chemistry*) organization that recommends the nomenclature to be used for naming substances, the units to be used, and which conventions are to be adopted when describing particular changes.

ivory the hard white substance of which the teeth and tusks of certain mammals are composed. Among the most valuable are elephants' tusks, which are of unusual hardness and density. Ivory is used in carving and other decorative work, and is so valuable that poachers continue to destroy the remaining wild elephant herds in Africa to obtain it illegally.

Poaching for ivory has led to the decline of the African elephant population from 2 million to approximately 600,000, with the species virtually extinct in some countries. Trade in ivory was halted by Kenya 1989, but Zimbabwe continued its policy of controlled culling to enable the elephant population to thrive and to release ivory for export. China and Hong Kong have refused to obey an international ban on ivory trading. *Vegetable ivory* is used for buttons, toys, and cheap ivory goods. It consists of the hard albumen of the seeds of a tropical palm *Phytelephas macrocarpa*, and is imported from Colombia.

J in physics, the symbol for *joule*, the SI unit of energy.

jacinth or *hyacinth* red or yellowish-red gem, a variety of ◊zircon.

jack tool or machine for lifting, hoisting, or moving heavy weights, such as motor vehicles. A *screw jack* uses the principle of the screw to magnify an applied effort; in a car jack, for example, turning the handle many times causes the lifting screw to rise slightly, and the effort is magnified to lift heavy weights. A *hydraulic jack* uses a succession of piston strokes to increase pressure in a liquid and force up a lifting ram.

jade semiprecious stone consisting of either jadeite, $NaAlSi_2O_6$ (a pyroxene), or nephrite, $Ca_2(Mg,Fe)_5Si_8O_{22}(OH,F)_2$ (an amphibole), ranging from colourless through shades of green to black according to the iron content. Jade ranks 5.5–6.5 on the Mohs' scale of hardness.

The early Chinese civilization discovered jade, bringing it from E Turkestan, and carried the art of jade-carving to its peak. The Olmecs, Aztecs, Maya, and the Maoris have also used jade for ornaments, ceremony, and utensils.

jansky unit of radiation received from outer space, used in radio astronomy. It is equal to 10^{-26} watts per square metre per hertz, and is named after US engineer Karl Jansky.

Japan Current or *Kuroshio* warm ocean ◊current flowing from Japan to North America.

Jarvik 7 the first successful artificial heart intended for permanent implantation in a human being. Made from polyurethane plastic and aluminium, it is powered by compressed air. Barney Clark became the first person to receive a Jarvik 7, in Salt Lake City, Utah, USA, in Dec 1982; it kept him alive for 112 days.

In 1986 a similar heart was implanted temporarily in a British patient waiting for a human heart transplant. Recently the US Food and Drug Administration withdrew approval for artificial heart transplants owing to evidence of adverse reactions.

jasper hard, compact variety of ◊chalcedony SiO_2, usually coloured red, brown, or yellow. Jasper can be used as a gem.

jaw one of two bony structures that form the framework of the mouth in all vertebrates except lampreys and hagfishes (the agnathous or jawless vertebrates). They consist of the upper jawbone (maxilla), which is fused to the skull, and the lower jawbone (mandible), which is hinged at each side to the bones of the temple by ◊ligaments.

jellyfish marine invertebrate of the phylum Cnidaria (coelenterates) with an umbrella-shaped body composed of a semi-transparent gelatinous substance, with a fringe of stinging tentacles. Most adult jellyfishes move freely, but during parts of their life cycle many are polylike and attached. They feed on small animals that are paralyzed by stinging cells in the jellyfishes' tentacles.

jet hard, black variety of lignite, a type of coal. It is cut and polished for use in jewellery and ornaments. Articles made of jet have been found in Bronze Age tombs.

In Britain, jet occurs in quantity near Whitby and along the Yorkshire coast. It became popular as mourning jewellery after the death of Prince Albert 1861.

JET (abbreviation for *Joint European Torus*) ◊tokamak machine built in England to conduct experiments on nuclear fusion. It is the focus of the European effort to produce a practical fusion-power reactor. On 9 November 1991, the JET tokamak produced a 1.7 megawatt pulse of power in an experiment that lasted two seconds. This was the first time that a substantial amount of energy has been produced by nuclear power in a controlled experiment.

jetfoil advanced type of ◊hydrofoil boat built by Boeing, propelled by water jets. It features horizontal, fully submerged hydrofoils fore and aft and has a sophisticated computerized control system to maintain its stability in all waters.

Jetfoils have been in service worldwide since 1975. A jetfoil service operates across the English Channel between Dover and Ostend, Belgium, with a passage time of about 1.5 hours. Cruising speed of the jetfoil is about 80 kph/50 mph.

jet lag the effect of a sudden switch of time zones in air travel, resulting in tiredness and feeling 'out of step' with day and night. In 1989 it was suggested that use of the hormone melatonin helped to lessen the effect of jet lag by resetting the body clock. See also ◊circadian rhythm.

jet propulsion method of propulsion in which an object is propelled in one direction by a jet, or stream of gases, moving in the other. This follows from Isaac Newton's third law of motion: 'To every action, there is an equal and opposite reaction.' The most widespread application of the jet principle is in the jet engine, the most common kind of aircraft engine.

The *jet engine* is a kind of ◊gas turbine. Air, after passing through a forward-facing intake, is compressed by a compressor, or fan, and fed into a combustion chamber. Fuel (usually kerosene) is sprayed in and ignited. The hot gas produced expands rapidly rearwards, spinning a turbine that drives the compressor before being finally ejected from a rearward-facing tail pipe, or nozzle, at very high speed. Reaction to the jet of gases streaming backwards produces a propulsive thrust forwards, which acts on the aircraft through its engine-mountings, not from any pushing of the hot gas stream against the static air. Thrust is proportional to the mass of the gas ejected multiplied by the acceleration imparted to it, and is stated in units of pounds force (lbf) or kilograms force (kgf), both now being superseded by the international unit, the Newton (N). The ◊*turbojet* is the simplest form of gas turbine, used in aircraft well into the supersonic range. The ◊*turboprop* used for moderate speeds and altitudes (up to 725 kph/450 mph and 10,000 m/30,000 ft incorporates extra stages of turbine that absorb most of the energy from the gas stream to drive the propeller shaft via a speed reduction gear. The ◊*turbofan* is best suited to high subsonic speeds. It is fitted with an extra compressor or fan in front, and some of the airflow bypasses the core engine, and mixes with the jet exhaust stream, to give it lower temperature and

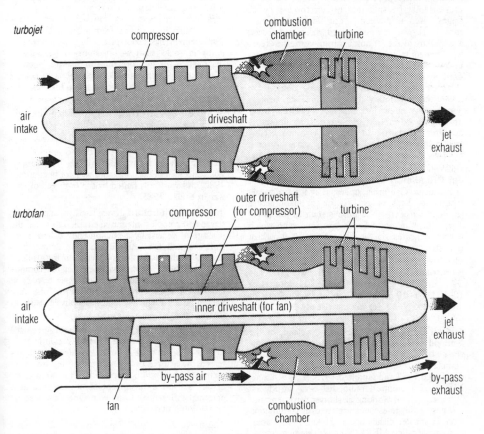

jet propulsion *Two forms of jet engine. In the turbojet, air passing into the air intake is compressed by the compressor and fed into the combustion chamber where fuel burns. The hot gases formed are expelled at high speed from the rear of the engine, driving the engine forwards and turning a turbine which drives the compressor. In the turbofan, some air flows around the combustion chamber and mixes with the exhaust gases. This arrangement is more efficient and quieter than the turbojet.*

velocity. This results in greater economy, efficiency and quietness compared with the turbojet, and a higher speed than the turboprop. The **turboshaft** is used to drive the main and tail rotors of ◊helicopters, and in ◊hovercraft, ships, trains, as well as in power stations and pumping equipment. It is effectively a turboprop without its propeller, power from an extra turbine being delivered to a reduction gearbox or directly to an output shaft. Most of the gas energy drives the compressors and provides shaft power, so residual thrust is low. Turboshaft power is normally quoted as shaft horsepower (shp) or kilowatts (kW). The **ramjet** is used for some types of missiles. At twice the speed of sound (Mach 2), pressure in the forward-facing intake of a jet engine is seven times that of the outside air, a compression ratio which rapidly mounts with increased speed (to Mach 8), with the result that no compressor or turbine is needed. The ramjet comprises merely an open-ended rather barrel-shaped tube, burning fuel in its widest section. It is cheap, light and easily made. However, fuel consumption is high and it needs rocket-boosting to its operational speed.

Variants and additional capabilities of jets include **multi-spool engines** in which the compressors may be split into two or three parts or stages driven by independent turbines, so that each runs at its own optimum speed; **vectored thrust** a swivelling of the jet nozzles from vertical to rearward horizontal to achieve vertical take-off followed by level flight (as in the ◊Harrier); **reverse thrust** used to slow down a jet plane on landing, and achieved by blocking off the jet pipe with special doors and redirecting the gases forward through temporarily opened cascades; **reheat** (afterburning), used in military aircraft to obtain short-duration thrust increase of up to 70% by the controlled burning of fuel in the gas stream after it has passed through the turbine, but increases fuel consumption.

Jet Propulsion Laboratory NASA installation at Pasadena, California, operated by the California Institute of Technology. It is the command centre for NASA's deep-space probes such as the ◊Voyager, ◊Magellan, and ◊Galileo missions, with which it communicates via the Deep Space Network of radio telescopes at Goldstone, California; Madrid, Spain; and Canberra, Australia.

jet stream narrow band of very fast wind (velocities of over 150 kph/95 mph) found at altitudes of 10–16 km/6–10 mi in the upper troposphere or lower stratosphere. Jet streams usually occur about the latitudes of the Westerlies (35°–60°).

Jodrell Bank site in Cheshire, England, of the Nuffield Radio Astronomy Laboratories of the University of Manchester. Its largest instrument is the 76 m/250 ft radio dish (the Lovell Telescope), completed 1957 and modified 1970. A 38 × 25 m/ 125 × 82 ft elliptical radio dish was introduced 1964, capable of working at shorter wave lengths.

These radio telescopes are used in conjunction with six smaller dishes up to 230 km/143 mi apart in an array called MERLIN (**m**ulti-**e**lement **r**adio-**l**inked **i**nterferometer **n**etwork) to produce detailed maps of radio sources.

Johnson Space Center NASA installation at Houston, Texas, home of mission control for crewed space missions. It is the main centre for the selection and training of astronauts.

joint in any animal with a skeleton, a point of movement or articulation. In vertebrates, it is the point where two bones meet. Some joints allow no motion (the sutures of the skull), others allow a very small motion (the sacroiliac joints in the lower back), but most allow a relatively free motion. Of these, some allow a gliding motion (one vertebra of the spine on another), some have a hinge action (elbow and knee), and others allow motion in all directions (hip and shoulder joints), by means of a ball-and-socket arrangement. The ends of the bones at a moving joint are covered with cartilage for greater elasticity and smoothness, and enclosed in an envelope (capsule) of tough white fibrous tissue lined with a membrane which secretes a lubricating and cushioning ◊synovial fluid. The joint is further strengthened by ligaments.

In invertebrates with an ◊exoskeleton, the joints are places where the exoskeleton is replaced by a more flexible outer covering, the arthrodial membrane, which allows the limb (or other body part) to bend at that point.

joint in earth science, a vertical crack in a rock, formed by compression; it is usually several metres in length. A joint differs from a ◊fault in that no displacement has taken place. The weathering of joints in rocks such as limestone and granite is responsible for the formation of features such as ◊limestone pavements and ◊tors. Joints in coastal rocks are often exploited by the sea to form erosion features such as caves and geos.

Joint European Torus experimental nuclear-fusion machine, known as ◊JET.

Jonglei Canal N African civil-engineering project to divert the White Nile in order to improve the water supply to Egypt from Sudan. The project was conceived in the 1940s but was not begun until 1982, and work was halted by the outbreak of civil war in Sudan 1983.

Josephson junction device used in 'superchips' (large and complex integrated circuits) to speed the passage of signals by a phenomenon called 'electron tunnelling'. Although these superchips respond a thousand times faster than the ◊silicon chip, they have the disadvantage that the components of the Josephson junctions operate only at temperatures close to ◊absolute zero. They are named after English theoretical physicist Brian Josephson.

joule SI unit (symbol J) of work and energy, replacing the ◊calorie (one joule equals 4.2 calories).

It is defined as the work done (energy transferred) by a force of one newton acting over one metre. It can also be expressed as the work done in one second by a current of one ampere at a potential difference of one volt. One watt is equal to one joule per second.

Joule–Kelvin effect in physics, the fall in temperature of a gas as it expands adiabatically (without loss or gain of heat to the system) through a narrow jet. It can be felt when, for example, compressed air escapes through the valve of an inflated bicycle

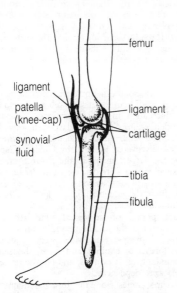

hinge joint (knee)

- femur
- ligament
- patella (knee-cap)
- synovial fluid
- ligament
- cartilage
- tibia
- fibula

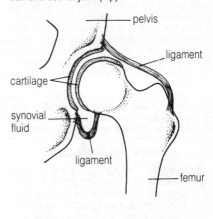

ball-and-socket joint (hip)

- pelvis
- ligament
- cartilage
- synovial fluid
- ligament
- femur

joint *The knee joint is the largest and most complex in the human body. It is similar to a hinge joint but with a slight rotation which allows the leg to lock into a rigid postion when extended. The ligaments surrounding the joint help prevent overextension. The ligament surrounding the patella or kneecap helps to protect the joint.*

tyre. Only hydrogen does not exhibit the effect. It is the basic principle of most refrigerators.

It was named after the British scientists James Prescott Joule and William Thomson (Lord Kelvin).

COUNTING THE JOULES

A bolt of lightning releases up to 3,000 million joules of energy. When a ton of TNT explodes, about 5,000 million joules of chemical energy are released. More than 100,000 million are needed to send a rocket into space. A tropical hurricane releases about 100,000 million million joules. In 1960 an earthquake in Chile released 10 million million million joules.

---

joystick in computing, an input device that signals to a computer the direction and extent of displacement of a hand-held lever. It is similar to the joystick used to control the flight of an aircraft.

Joysticks are sometimes used to control the movement of a cursor (marker) across a display screen, but are much more frequently used to provide fast and direct input for moving the characters and symbols that feature in computer games. Unlike a ◊mouse, which can move a pointer in any direction, simple games joysticks are often only capable of moving an object in one of eight different directions.

jugular vein one of two veins in the necks of vertebrates; they return blood from the head to the

superior (or anterior) vena cava and thence to the heart.

jumbo jet popular name for a generation of huge wide-bodied airliners including the *Boeing 747*, which is 71 m/232 ft long, has a wingspan of 60 m/ 196 ft, a maximum takeoff weight of nearly 400 tonnes, and can carry more than 400 passengers.

jump in computing, a programming instruction that causes the computer to branch to a different part of a program, rather than execute the next instruction in the program sequence. Unconditional jumps are always executed; conditional jumps are only executed if a particular condition is satisfied.

jungle popular name for ◊rainforest.

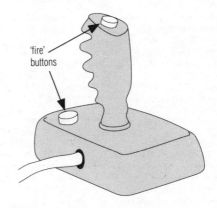

'fire' buttons

joystick

Jupiter the fifth planet from the Sun, and the largest in the Solar System (equatorial diameter 142,800 km/88,700 mi), with a mass more than twice that of all the other planets combined, 318 times that of the Earth's. It takes 11.86 years to orbit the Sun, at an average distance of 778 million km/484 million mi, and has at least 16 moons. It is largely composed of hydrogen and helium, liquefied by pressure in its interior, and probably with a rocky core larger than the Earth. Its main feature is the Great Red Spot, a cloud of rising gases, revolving anticlockwise, 14,000 km/8,500 mi wide and some 30,000 km/20,000 mi long.

JUPITER: RELATIVE SIZES

Jupiter is much smaller than the Sun. If the Sun were the size of a beach ball, 60 cm/ 2 ft across, Jupiter would be the size of a golf ball, 5 cm/2 in across. On this scale, the Earth would be less than 2.5 mm/0.1 in across.

Its visible surface consists of clouds of white ammonia crystals, drawn out into belts by the planet's high speed of rotation (9 hr 51 min at the equator, the fastest of any planet). Darker orange and brown clouds at lower levels may contain sulphur, as well as simple organic compounds.

Further down still, temperatures are warm, a result of heat left over from Jupiter's formation, and it is this heat that drives the turbulent weather patterns of the planet. The Great Red Spot was first observed 1664. Its top is higher than the surrounding clouds; its colour is thought to be due to red phosphorus. Jupiter's strong magnetic field gives rise to a large surrounding magnetic 'shell', or magnetosphere, from which bursts of radio waves are detected. The Southern Equatorial Belt in which the Great Red Spot occurs is subject to unexplained fluctuation. In 1989 it sustained a dramatic and sudden fading. The four largest moons, Io, Europa, Ganymede, and Callisto, are the *Galilean satellites*, discovered in 1610 by Galileo (Ganymede is the largest moon in the Solar System). Three small moons were discovered in 1979 by the Voyager space probes, as was a faint ring of dust around Jupiter's equator, 55,000 km/34,000 mi above the cloud tops.

Jurassic period of geological time 208–146 million years ago; the middle period of the Mesozoic era. Climates worldwide were equable, creating forests of conifers and ferns, dinosaurs were abundant, birds evolved, and limestones and iron ores were deposited.

The name comes from the Jura mountains in France and Switzerland, where the rocks formed during this period were first studied.

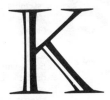

K symbol for *kelvin*, a scale of temperature.

k symbol for *kilo–*, as in kg (kilogram) and km (kilometre).

Kagoshima Space Centre headquarters of Japan's Institute of Space and Astronautical Science (ISAS), situated in S Kyushu Island.

ISAS is responsible for the development of satellites for scientific research; other aspects of the space programme fall under the National Space Development Agency which runs the ◊Tanegashima Space Centre. Japan's first satellite was launched from Kagoshima 1970. By 1988 ISAS had launched 17 satellites and space probes.

kame geological feature, usually in the form of a mound or ridge, formed by the deposition of rocky material carried by a stream of glacial meltwater. Kames are commonly laid down in front of or at the edge of a glacier (kame terrace), and are associated with the disintegration of glaciers at the end of an ice age.

Kames are made of well-sorted rocky material, usually sands and gravels. The rock particles tend to be rounded (by attrition) because they have been transported by water.

kaolinite white or greyish ◊clay mineral, hydrated

aluminium silicate, $Al_2Si_2O_5(OH)_4$, formed mainly by the decomposition of feldspar in granite. China clay (kaolin), a pure white clay used in the manufacture of porcelain and other ceramics, paper, rubber, paint, textiles, and medicines, is derived from it. It is mined in France, the UK, Germany, China, and the USA.

karst landscape characterized by remarkable surface and underground forms, created as a result of the action of water on permeable limestone. The feature takes its name from the Karst region on the Adriatic coast in Slovenia and Croatia, but the name is applied to landscapes throughout the world, the most dramatic of which is found near the city of Guilin in the Guangxi province of China.

Limestone is soluble in the weak acid of rainwater. Erosion takes place most swiftly along cracks and joints in the limestone and these open up into gullies called grikes. The rounded blocks left upstanding between them are called clints.

karyotype in biology, the set of ◊chromosomes characteristic of a given species. It is described as the number, shape, and size of the chromosomes in a single cell of an organism. In humans for example, the karyotype consists of 46 chromosomes, in mice 40, crayfish 200, and in fruit flies 8.

The diagrammatic representation of a complete chromosome set is called a *karyogram*.

katabatic wind cool wind that blows down a valley on calm clear nights. (By contrast, an ◊anabatic wind is warm and moves up a valley in the early morning.) When the sky is clear, heat escapes rapidly from ground surfaces, and the air above the ground becomes chilled. The cold dense air moves downhill, forming a wind that tends to blow most strongly just before dawn.

Cold air blown by a katabatic wind may collect in a depression or valley bottom to create a ◊frost hollow.

Katabatic winds are most likely to occur in the late spring and autumn because of the greater daily temperature differences.

formation of kame features

| kame delta— a stream flowing into a small lake deposits material to form a delta | kame— material washed into crevasses | kame terrace— a lake infilled with sediment, left on a hillside to form a terrace |

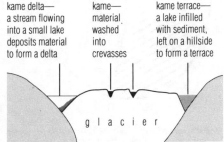

kame

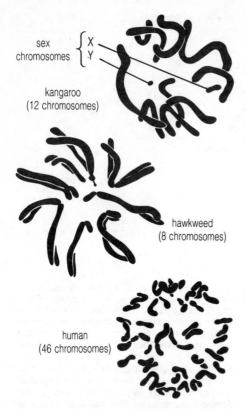

sex chromosomes { X Y

kangaroo (12 chromosomes)

hawkweed (8 chromosomes)

human (46 chromosomes)

karyotype *The characteristics, or karyotype, of the chromosomes vary according to species. The kangaroo has 12 chromosomes, the hawkweed has 8, and a human being has 46.*

kayser unit of wave number (number of waves in a unit length), used in spectroscopy. It is expressed as waves per centimetre, and is the reciprocal of the wavelength. A wavelength of 0.1 cm has a wave number of 10 kaysers.

kcal symbol for *kilocalorie* (see ◊calorie).

Keck Telescope world's largest optical telescope, situated on Mauna Kea, Hawaii. It has a primary mirror 10 m/33 ft in diameter, unique in that it consists of 36 hexagonal sections, each controlled and adjusted by a computer to generate single images of the objects observed. It received its first images Nov 1990.

An identical telescope, to be named Keck II, is under construction next to it and is due for completion 1996. Both telescopes are jointly owned by the California Institute of Technology and the University of California.

kelvin scale temperature scale used by scientists. It begins at ◊absolute zero (−273.15°C) and increases by the same degree intervals as the Celsius scale; that is, 0°C is the same as 273 K and 100°C is 373 K.

Kennedy Space Center ◊NASA launch site on Merritt Island, near Cape Canaveral, Florida, used for Apollo and space-shuttle launches. The first flight to land on the Moon (1969) and *Skylab*,

the first orbiting laboratory (1973), were launched here.

The Center is dominated by the Vehicle Assembly Building, 160 m/525 ft tall, used for assembly of ◊Saturn rockets and space shuttles.

Kennelly–Heaviside layer former term for the ◊E layer.

Kepler's laws three laws of planetary motion formulated 1609 and 1619 by German mathematician and astronomer Johannes Kepler: (1) the orbit of each planet is an ellipse with the Sun at one of the foci; (2) the radius vector of each planet sweeps out equal areas in equal times; (3) the squares of the periods of the planets are proportional to the cubes of their mean distances from the Sun.

keratin fibrous protein found in the ◊skin of vertebrates and also in hair, nails, claws, hooves, feathers, and the outer coating of horns in animals such as cows and sheep.

If pressure is put on some parts of the skin, more keratin is produced, forming thick calluses that protect the layers of skin beneath.

kernel the inner, softer part of a ◊nut, or of a seed within a hard shell.

kerosene thin oil obtained from the distillation of petroleum; a highly refined form is used in jet aircraft fuel. Kerosene is a mixture of hydrocarbons of the ◊paraffin series.

ketone member of the group of organic compounds containing the carbonyl group (C=O)

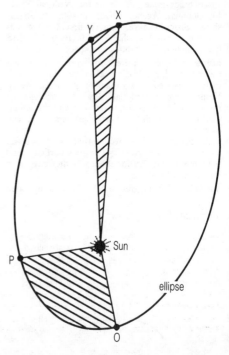

Kepler's laws *Kepler's second law states that the two shaded areas are equal if the planet moves from P to O in the same time that it moves from X to Y. The law says, in effect, that a planet moves fastest when it is closest to the Sun.*

bonded to two atoms of carbon (instead of one carbon and one hydrogen as in ◊aldehydes). Ketones are liquids or low-melting-point solids, slightly soluble in water.

An example is propanone (acetone, CH_3COCH_3), used as a solvent.

kettle hole pit or depression formed when a block of ice from a receding glacier becomes isolated and buried in glacial debris (till). As the block melts the till collapses to form a hollow, which may become filled with water to form a kettle lake or pond. Kettle holes range from 5 m/15 ft to 13 km/8 mi in diameter, and may exceed 33 m/100 ft in depth.

Kew Gardens popular name for the Royal Botanic Gardens, Kew, Surrey, England. They were founded 1759 by the mother of King George III as a small garden and passed to the nation by Queen Victoria 1840. By then they had expanded to almost their present size of 149 hectares/368 acres and since 1841 have been open daily to the public. They contain a collection of over 25,000 living plant species and many fine buildings. The gardens are also a centre for botanical research.

The ◊herbarium is the biggest in the world, with over 5 million dried plant specimens. Kew also has a vast botanical library, the Jodrell Laboratory, and three museums. The buildings include the majestic Palm House 1848, the Temperate House 1862, both designed by Decimus Burton, and the Chinese Pagoda, some 50 m/165 ft tall, designed by William Chambers 1761. More recently, two additions have been made to the glasshouses: the Alpine House 1981 and the Princess of Wales Conservatory, a futuristic building for plants from ten different climatic zones, 1987. Much of the collection of trees at Kew was destroyed by a gale 1987.

Since 1964 there have been additional grounds at Wakehurst Place, Ardingly, West Sussex (the seeds of 5,000 species are preserved there in the seed physiology department, 2% of those known to exist in the world).

keyboard in computing, an input device resembling a typewriter keyboard, used to enter instructions and data. There are many variations on the layout and labelling of keys. Extra numeric keys may be added, as may special-purpose function keys, whose effects can be defined by programs in the computer.

key field selected field, or portion, of a record that is used to identify that record uniquely; in a file of records it is the field used as the basis for ◊sorting the file. For example, in a file containing details of a bank's customers, the customer account number would probably be used as the key field.

kg symbol for ◊*kilogram*.

khamsin hot southeasterly wind that blows from the Sahara desert over Egypt and parts of the Middle East from late March to May or June. It is called *sharav* in Israel.

kidney in vertebrates, one of a pair of organs responsible for water regulation, excretion of waste products, and maintaining the ionic composition of the blood. The kidneys are situated on the rear wall of the abdomen. Each one consists of a number of long tubules; the outer parts filter the aqueous components of blood, and the inner parts selectively reabsorb vital salts, leaving waste products in the remaining fluid (urine), which is passed through the ureter to the bladder.

The action of the kidneys is vital, although if one is removed, the other enlarges to take over its function. A patient with two defective kidneys may continue near-normal life with the aid of a kidney machine or continuous ambulatory peritoneal ◊dialysis (CAPD).

kiln high-temperature furnace used commercially for drying timber, roasting metal ores, or for making cement, bricks, and pottery. Oil- or gas-fired kilns are used to bake ceramics at up to 1,760°C/3,200°F; electric kilns do not generally reach such high temperatures.

kilo- prefix denoting multiplication by 1,000, as in kilohertz, a unit of frequency equal to 1,000 hertz.

kilobyte (K or KB) in computing, a unit of memory equal to 1,024 ◊bytes. It is sometimes used, less precisely, to mean 1,000 bytes.

In the metric system, the prefix 'kilo-' denotes multiplication by 1,000 (as in kilometre, a unit

typical computer keyboard

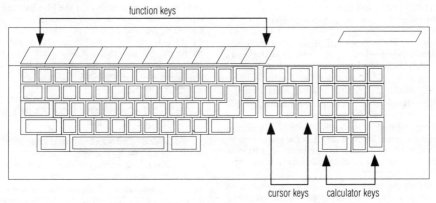

function keys

cursor keys   calculator keys

keyboard

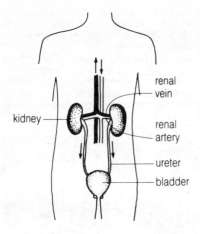

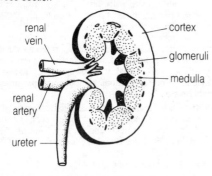

cross section

kidney *Blood enters the kidney through the renal artery. The blood is filtered through the glomeruli to extract the nitrogenous waste products and excess water that make up urine. The urine flows through the ureter to the bladder; the cleaned blood then leaves the kidney along the renal vein.*

equal to 1,000 metres). However, computer memory size is based on the ◊binary number system, and the most convenient binary equivalent of 1,000 is 2^{10}, or 1,024.

kilogram SI unit (symbol kg) of mass equal to 1,000 grams (2.24 lb). It is defined as a mass equal to that of the international prototype, a platinum-iridium cylinder held at the International Bureau of Weights and Measures at Sèvres, France.

kilometre unit (symbol km) of length equal to 1,000 metres (3,280.89 ft or about ⅝ of a mile).

kilowatt unit (symbol kW) of power equal to 1,000 watts or about 1.34 horsepower.

kilowatt-hour commercial unit of electrical energy (symbol kWh), defined as the work done by a power of 1,000 watts in one hour. It is used to calculate the cost of electrical energy taken from the domestic supply.

Kimball tag stock-control device commonly used in clothes shops, consisting of a small ◊punched card attached to each item offered for sale. The tag carries information about the item (such as its serial number, price, colour, and size), both in the form of printed details (which can be read by the customer) and as a pattern of small holes. When the item is sold, the tag (or a part of the tag) is removed and kept as a computer-readable record of sales.

kimberlite an igneous rock that is ultrabasic (containing very little silica); a type of alkaline ◊peridotite containing mica in addition to olivine and other minerals. Kimberlite represents the world's principal source of diamonds.

Kimberlite is found in carrot-shaped pipelike ◊intrusions called *diatremes*, where mobile material from very deep in the Earth's crust has forced itself upwards, expanding in its ascent. The material, brought upwards from near the boundary between crust and mantle, often altered and fragmented, includes diamonds. Diatremes are found principally near Kimberley, South Africa, from which the name of the rock is derived, and in the Yakut area of Siberia, Russia.

kinesis (plural *kineses*) in biology, a nondirectional movement in response to a stimulus; for example, woodlice move faster in drier surroundings. *Taxis* is a similar pattern of behaviour, but there the response is directional.

kinetic energy the energy of a body resulting from motion. It is contrasted with ◊potential energy.

kinetics branch of ◊dynamics dealing with the action of forces producing or changing the motion of a body; *kinematics* deals with motion without reference to force or mass.

kinetics the branch of chemistry that investigates the rates of chemical reactions.

kinetic theory theory describing the physical properties of matter in terms of the behaviour—principally movement—of its component atoms or molecules. The temperature of a substance is dependent on the velocity of movement of its constituent particles, increased temperature being accompanied by increased movement. A gas consists of rapidly moving atoms or molecules and, according to kinetic theory, it is their continual impact on the walls of the containing vessel that accounts for the pressure of the gas. The slowing of molecular motion as temperature falls, according to kinetic theory, accounts for the physical properties of liquids and solids, culminating in the concept of no molecular motion at ◊absolute zero (0K/ −273°C). By making various assumptions about the nature of gas molecules, it is possible to derive from the kinetic theory the various gas laws (such as ◊Avogadro's hypothesis, ◊Boyle's law, and ◊Charles's law).

KINETIC THEORY: MOLECULES IN THE AIR

At standard temperature and pressure (0°C, 760 mmHg) a litre of air contains 30,000 million million million molecules, moving at an average speed of 450 m/ 1,500 ft per second. Each molecule undergoes 5,000 million collisions every second.

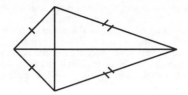

kite *A kite has two pairs of adjacent sides equal in length. The square and the rhombus are special types of kite, having four equal sides.*

kingdom the primary division in biological ◊classification. At one time, only two kingdoms were recognized: animals and plants. Today most biologists prefer a five-kingdom system, even though it still involves grouping together organisms that are probably unrelated. One widely accepted scheme is as follows: *Kingdom Animalia* (all multicellular animals); *Kingdom Plantae* (all plants, including seaweeds and other algae except blue-green); *Kingdom Fungi* (all fungi, including the unicellular yeasts, but not slime moulds); *Kingdom Protista* or *Protoctista* (protozoa, diatoms, dinoflagellates, slime moulds, and various other lower organisms with eukaryotic cells); and *Kingdom Monera* (all prokaryotes—the bacteria and cyanobacteria, or ◊blue-green algae). The first four of these kingdoms make up the eukaryotes.

When only two kingdoms were recognized, any organism with a rigid cell wall was a plant, and so bacteria and fungi were considered plants, despite their many differences. Other organisms, such as the photosynthetic flagellates (euglenoids), were claimed by both kingdoms. The unsatisfactory nature of the two-kingdom system became evident during the 19th century, and the biologist Ernst Haeckel was among the first to try to reform it. High-power microscopes have revealed more about the structure of cells; it has become clear that there is a fundamental difference between cells without a nucleus (◊prokaryotes) and those with a nucleus (◊eukaryotes). However, these differences are larger than those between animals and higher plants, and are unsuitable for use as kingdoms. At present there is no agreement on how many kingdoms there are in the natural world. Although the five-kingdom system is widely favoured, some schemes have as many as 20.

kin selection in biology, the idea that ◊altruism shown to genetic relatives can be worthwhile, because those relatives share some genes with the individual that is behaving altruistically, and may continue to reproduce. See ◊inclusive fitness.

Alarm-calling in response to predators is an example of a behaviour that may have evolved through kin selection: relatives that are warned of danger can escape and continue to breed, even if the alarm caller is caught.

Kirchhoff's first law law formulated by German physicist Gustav Kirchhoff 1845 that states that the total current entering a junction in a circuit must equal the total current leaving it. It is based on the law of conservation of electric charge: as no charge can be created or destroyed at a junction, the flow of charge (current), into and out of the junction must be the same.

kite quadrilateral (four-sided figure) with two pairs of adjacent equal sides.

Kitt Peak National Observatory observatory in the Quinlan Mountains near Tucson, Arizona, USA, operated by AURA (Association of Universities for Research into Astronomy). Its main telescopes are the 4–m/158–in Mayall reflector, opened 1973, and the McMath Solar Telescope, opened 1962, the world's largest of its type. Among numerous other telescopes on the site is a 2.3–m/90–in reflector owned by the Steward Observatory of the University of Arizona.

km symbol for ◊*kilometre.*

knocking in a spark-ignition petrol engine, a phenomenon that occurs when unburned fuel-air mixture explodes in the combustion chamber before being ignited by the spark. The resulting shock waves produce a metallic knocking sound. Loss of power occurs, which can be prevented by reducing the compression ratio, re-designing the geometry of the combustion chamber, or increasing the octane number of the petrol (usually by the use of lead tetraethyl antiknock additives).

knot in navigation, unit by which a ship's speed is measured, equivalent to one ◊nautical mile per hour (one knot equals about 1.15 miles per hour). It is also sometimes used in aviation.

knowledge-based system (KBS) computer program that uses an encoding of human knowledge to help solve problems. It was discovered during research into ◊artificial intelligence that adding heuristics (rules of thumb) enabled programs to tackle problems that were otherwise difficult to solve by the usual techniques of computer science.

Chess-playing programs have been strengthened by including knowledge of what makes a good position, or of overall strategies, rather than relying solely on the computer's ability to calculate variations.

Königsberg bridge problem long-standing puzzle that was solved by topology (the geometry of those properties of a figure which remain the same under distortion). In the city of Königsberg (now Kaliningrad in Russia), seven bridges connect

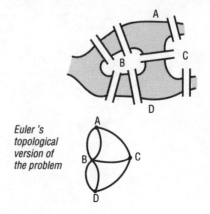

Euler's topological version of the problem

Königsberg bridge problem

the banks of the River Pregol'a and the island in the river. For many years, people were challenged to cross each of the bridges in a single tour and return to their starting point. In 1736 Swiss mathematician Leonhard Euler converted the puzzle into a topological network, in which the islands and river banks were represented as nodes (junctions), and the connecting bridges as lines. By analysing this network he was able to show that it is not traversible—that is, it is impossible to cross each of the bridges once only and return to the point at which one started.

Kourou river and second-largest town of French Guiana, NW of Cayenne, site of the Guiana Space Centre of the European Space Agency.

Situated near the equator, it is an ideal site for launches of satellites into ◊geostationary orbit.

kph or **km/h** symbol for *kilometres per hour.*

Krebs cycle or *citric acid cycle* or *tricarboxylic acid cycle* final part of the chain of biochemical reactions by which organisms break down food using oxygen to release energy (respiration). It takes place within structures called ◊mitochondria in the body's cells, and breaks down food molecules in a series of small steps, producing energy-rich molecules of ◊ATP.

krypton (Greek *kryptos* 'hidden') colourless, odourless, gaseous, nonmetallic element, symbol Kr, atomic number 36, relative atomic mass 83.80.

It is grouped with the inert gases and was long believed not to enter into reactions, but it is now known to combine with fluorine under certain conditions; it remains inert to all other reagents. It is present in very small quantities in the air (about 114 parts per million). It is used chiefly in fluorescent lamps, lasers, and gas-filled electronic valves.

Krypton was discovered 1898 in the residue from liquid air by British chemists William Ramsay and Morris Travers; the name refers to their difficulty in isolating it.

K-T boundary geologists' shorthand for the boundary between the rocks of the ◊Cretaceous and the ◊Tertiary periods 65 million years ago. It marks the extinction of the dinosaurs and in many places reveals a layer of iridium, possibly deposited by a meteorite that may have caused the extinction by its impact.

kurchatovium name proposed by Soviet scientists for the element currently known as ◊unnilquadium (atomic number 104), to honour Soviet nuclear physicist Igor Kurchatov (1903–1960).

kW symbol for ◊*kilowatt.*

kyanite aluminium silicate, Al_2SiO_5, a pale-blue mineral occurring as blade-shaped crystals. It is an indicator of high-pressure conditions in metamorphic rocks formed from clay sediments. Andalusite, kyanite, and sillimanite are all polymorphs.

l symbol for ◊*litre*, a measure of liquid volume.

labelled compound or *tagged compound* chemical compound in which a radioactive isotope is substituted for a stable one. The path taken by such a compound through a system can be followed, for example by measuring the radiation emitted.

This powerful and sensitive technique is used in medicine, chemistry, biochemistry, and industry.

labellum lower petal of an orchid flower; it is a different shape from the two lateral petals and gives the orchid its characteristic appearance. The labellum is more elaborate and usually larger than the other petals. It often has distinctive patterning to encourage ◊pollination by insects; sometimes it is extended backwards to form a hollow spur containing nectar.

laboratory room or building equipped for scientific experiments, research, and teaching.

In Europe, private research laboratories have been used since the 16th century, and by the 18th century laboratories were also used for teaching. The German chemist Justus van Liebig established a laboratory 1826 in Giessen, Germany, where he not only instructed his students in the art of experimental chemistry, but also undertook chemical research with them.

A first-rate laboratory is one in which mediocre scientists can produce outstanding work.

On **laboratories** Patrick Blackett (1897–1974)

---

laccolith intruded mass of igneous rock that forces apart two strata and forms a round lens-shaped mass many times wider than thick. The overlying layers are often pushed upward to form a dome. A classic development of laccoliths is illustrated in the Henry, La Sal, and Abajo mountains of SE Utah, USA, found on the Colorado plateau.

lactation secretion of milk from the mammary glands of mammals. In late pregnancy, the cells lining the lobules inside the mammary glands begin extracting substances from the blood to produce milk. The supply of milk starts shortly after birth with the production of colostrum, a clear fluid consisting largely of water, protein, antibodies, and vitamins. The production of milk continues practically as long as the infant continues to suck.

lacteal small vessel responsible for absorbing fat in the small intestine. Occurring in the fingerlike villi of the ◊ileum, lacteals have a milky appearance and drain into the lymphatic system.

Before fat can pass into the lacteal, bile from the liver causes its emulsification into droplets small enough for attack by the enzyme lipase. The products of this digestion form into even smaller droplets, which diffuse into the villi. Large droplets re-form before entering the lacteal and this causes the milky appearance.

lactic acid or *2–hydroxypropanoic acid* $CH_3CHOHCOOH$ organic acid, a colourless, almost odourless liquid, produced by certain bacteria during fermentation and by active muscle cells when they are exercised hard and are experiencing ◊oxygen debt. It occurs in yoghurt, buttermilk, sour cream, poor wine, and certain plant extracts, and is used in food preservation and in the preparation of pharmaceuticals.

lactose white sugar, found in solution in milk; it forms 5% of cow's milk. It is commercially prepared from the whey obtained in cheese-making. Like table sugar (sucrose), it is a disaccharide, consisting of two basic sugar units (monosaccharides), in this case, glucose and galactose. Unlike sucrose, it is tasteless.

lagoon coastal body of shallow salt water, usually with limited access to the sea. The term is normally used to describe the shallow sea area cut off by a ◊coral reef or barrier islands.

Lagrangian points five locations in space where the centrifugal and gravitational forces of two bodies neutralize each other; a third, less massive body located at any one of these points will be held in equilibrium with respect to the other two. Three of the points, L1–L3, lie on a line joining the two large bodies. The other two points, L4 and L5, which are the most stable, lie on either side of this line. Their existence was predicted in 1772 by French mathematician Joseph Louis Lagrange.

The *Trojan asteroids* lie at Lagrangian points L4 and L5 in Jupiter's orbit around the Sun. Clouds of dust and debris may lie at the Lagrangian points of the Moon's orbit around the Earth.

lahar mudflow formed of a fluid mixture of water and volcanic ash. During a volcanic eruption, melting ice may combine with ash to form a powerful flow capable of causing great destruction. The lahars created by the eruption of Nevado del Ruiz in Colombia, South America, in 1985 buried 22,000 people in 8 m/26 ft of mud.

lake body of still water lying in depressed ground without direct communication with the sea. Lakes are common in formerly glaciated regions, along the courses of slow rivers, and in low land near the sea. The main classifications are by origin: *glacial lakes*, formed by glacial scouring; *barrier lakes*, formed by landslides and glacial moraines; *crater lakes*, found in volcanoes; and *tectonic lakes*, occurring in natural fissures.

Most lakes are freshwater, such as the Great Lakes in North America, but in hot regions where evaporation is excessive they may contain many salts, the Dead Sea is an example. In the 20th century large artificial lakes have been created in connection with hydroelectric and other works. Some lakes have become polluted as a result of human activity. Sometimes ◊eutrophication (a state of overnourishment) occurs, when agricultural fertilizers leaching into lakes cause an explosion of aquatic life, which then depletes the lake's oxygen supply until it is no longer able to support life.

Lamarckism theory of evolution, now discredited, advocated during the early 19th century by French naturalist Jean Baptiste Lamarck. It differed from the Darwinian theory of evolution.

The theory was based on the idea that ◊acquired characteristics were inherited: he argued that particular use of an organ or limb strengthens it, and that this development may be 'preserved by reproduction'. For example, he suggested that giraffes have long necks because they are continually stretching them to reach high leaves; according to the theory, giraffes that have lengthened their necks by stretching will pass this characteristic on to their offspring.

lambert unit of luminance (the light shining from a surface), equal to one ◊lumen per square centimetre. In scientific work the ◊candela per square metre is preferred.

lamina in flowering plants (◊angiosperms), the blade of the ◊leaf on either side of the midrib. The lamina is generally thin and flattened, and is usually the primary organ of ◊photosynthesis. It has a network of veins through which water and nutrients are conducted. More generally, a lamina is any thin, flat plant structure, such as the ◊thallus of many seaweeds.

lamp, electric device designed to convert electrical energy into light energy.

In a *filament lamp* such as a light bulb an electric current causes heating of a long thin coil of fine high-resistance wire enclosed at low pressure inside a glass bulb. In order to give out a good light the wire must glow white-hot and therefore must be made of a metal, such as tungsten, that has a high melting point. The efficiency of filament lamps is low because most of the electrical energy is converted to heat.

A *fluorescent light* uses an electrical discharge or spark inside a gas-filled tube to produce light. The inner surface of the tube is coated with a fluorescent material that converts the ultraviolet light generated by the discharge into visible light. Although a high voltage is needed to start the discharge, these lamps are far more efficient than filament lamps at producing light.

land breeze gentle breeze blowing from the land towards the sea and affecting coastal areas. It forms at night in the summer or autumn, and tends to be cool. By contrast, a ◊sea breeze blows from the sea towards the land.

Landsat series of satellites used for monitoring Earth resources. The first was launched in 1972.

landslide sudden downward movement of a mass of soil or rocks from a cliff or steep slope. Landslides happen when a slope becomes unstable, usually because the base has been undercut or because materials within the mass have become wet and slippery.

A *mudflow* happens when soil or loose material is soaked so that it no longer adheres to the slope; it forms a tongue of mud that reaches downhill from a semicircular hollow. A *slump* occurs when the material stays together as a large mass, or several smaller masses, and these may form a tilted steplike structure as they slide. A *landslip* is formed when ◊beds of rock dipping towards a cliff slide along a lower bed. Earthquakes may precipitate landslides.

lanolin sticky, purified wax obtained from sheep's wool and used in cosmetics, soap, and leather preparation.

lanthanide any of a series of 15 metallic elements (also known as rare earths) with atomic numbers 57 (lanthanum) to 71 (lutetium). One of its members, promethium, is radioactive. All occur in nature. Lanthanides are grouped because of their chemical similarities (they are all bivalent), their properties differing only slightly with atomic number.

Lanthanides were called rare earths originally because they were not widespread and were difficult to identify and separate from their ores by their discoverers. The series is set out in a band in the periodic table of the elements, as are the ◊actinides.

lanthanum (Greek *lanthanein* 'to be hidden') soft, silvery, ductile and malleable, metallic element, symbol La, atomic number 57, relative atomic mass 138.91, the first of the lanthanide series. It is used in making alloys. It was named 1839 by Swedish chemist Carl Mosander (1797–1858).

lapis lazuli rock containing the blue mineral lazurite in a matrix of white calcite with small amounts of other minerals. It occurs in silica-poor igneous rocks and metamorphic limestones found in

mudflow landslide

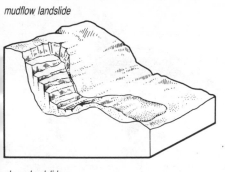

slump landslide

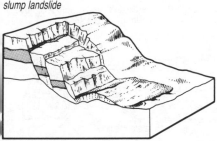

landslip landslide

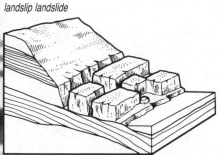

landslide *Types of landslide. A mudflow is a tongue of mud that slides downhill. A slump is a fall of a large mass that stays together after the fall. A landslip occurs when beds of rock move along a lower bed.*

Afghanistan, Siberia, Iran, and Chile. Lapis lazuli was a valuable pigment of the Middle Ages, also used as a gemstone and in inlaying and ornamental work. It was formerly used in the manufacture of ultramarine pigment.

laptop computer portable microcomputer, small enough to be used on the operator's lap. It consists of a single unit, incorporating a keyboard, ◊floppy disc or ◊hard disc drives, and a screen. The screen often forms a lid that folds back in use. It uses a liquid-crystal or gas-plasma display, rather than the bulkier and heavier cathode-ray tubes found in most display terminals. A typical laptop computer measures about 210 × 297 mm/8.3 × 11.7 in (A4), is 5 cm/2 in thick, and weighs less than 3 kg/6 lb 9 oz.

Large Electron Positron Collider (LEP) the world's largest particle ◊accelerator, in operation from 1989 at the CERN laboratories near Geneva in Switzerland. It occupies a tunnel 3.8 m/12.5 ft wide and 27 km/16.7 mi long, which is buried 180 m/590 ft underground and forms a ring consisting of eight curved and eight straight sections. In 1989 the LEP was used to measure the mass and lifetime of the Z particle, carrier of the weak nuclear force.

Electrons and positrons enter the ring after passing through the Super Proton Synchrotron accelerator. They travel in opposite directions around the ring, guided by 3,328 bending magnets and kept within tight beams by 1,272 focusing magnets. As they pass through the straight sections, the particles are accelerated by a pulse of radio energy. Once sufficient energy is accumulated, the beams are allowed to collide. Four giant detectors are used to study the resulting shower of particles.

larva stage between hatching and adulthood in those species in which the young have a different appearance and way of life from the adults. Examples include tadpoles (frogs) and caterpillars (butterflies and moths). Larvae are typical of the invertebrates, some of which (for example, shrimps) have two or more distinct larval stages. Among vertebrates, it is only the amphibians and some fishes that have a larval stage.

The process whereby the larva changes into another stage, such as a pupa (chrysalis) or adult, is known as ◊metamorphosis.

larynx in mammals, a cavity at the upper end of the trachea (windpipe), containing the vocal cords. It is stiffened with cartilage and lined with mucous membrane. Amphibians and reptiles have much simpler larynxes, with no vocal cords. Birds have a similar cavity, called the *syrinx*, found lower down the trachea, where it branches to form the bronchi. It is very complex, with well-developed vocal cords.

Las Campanas Observatory site in Chile of the 2.5-m/100-in Du Pont telescope of the Carnegie Institution of Washington, opened 1977.

laser (acronym for *light amplification by stimulated emission of radiation*) a device for producing a narrow beam of light, capable of travelling over vast distances without dispersion, and of being focused to give enormous power densities (10^8 watts per cm^2 for high-energy lasers). The laser operates on a principle similar to that of the ◊maser (a high-frequency microwave amplifier or oscillator). The uses of lasers include communications (a laser beam can carry much more information than can radio waves), cutting, drilling, welding, satellite tracking, medical and biological research, reading compact discs and bar codes, and surgery.

Any substance the majority of whose atoms or molecules can be put into an excited energy state can be used as laser material. Many solid, liquid, and gaseous substances have been used, including synthetic ruby crystal (used for the first extraction of laser light in 1960, and giving a high-power pulsed output) and a helium–neon gas mixture, capable of continuous operation, but at a lower power.

A blue shortwave laser was developed in Japan in 1988. Its expected application is in random access memory (◊RAM) and ◊compact disc recording, where its shorter wavelength will allow a greater concentration of digital information to be stored and read. A gallium arsenide chip, produced by

IBM in 1989, contains the world's smallest lasers in the form of cylinders of ◊semiconductor roughly one tenth of the thickness of a human hair; a million lasers can fit on a chip 1 cm/2.5 in square.

Sound wave vibrations from the window glass of a room can be picked up by a reflected laser beam. Lasers are also used as entertainment in theatres, concerts, and light shows.

LASER PUZZLE

Lasers have the power to burn holes in targets. Clearly the energy of the laser is absorbed by the target, which then melts or even evaporates. However, the same energy was originally absorbed into the laser by its 'pump'. Why then did the laser itself not melt? *See page 651 for the answer.*

laser printer computer printer in which the image to be printed is formed by the action of a laser on a light-sensitive drum, then transferred to paper by means of an electrostatic charge. Laser printers are page printers, printing a complete page at a time. The printed image, which can take the form of text or pictures, is made up of tiny dots, or ink particles. The quality of the image generated depends on the fineness of these dots—most laser printers can print up to 120 dots per cm/300 dots per in across the page.

A typical desktop laser printer can print about 4–20 pages per minute. The first low-cost laser printer suitable for office use appeared in 1984.

Laser printers range in size from small machines designed to work with microcomputers to very large machines designed for high-volume commercial printing. Because they produce very high-quality print and are virtually silent, small laser printers (along with ◊ink-jet printers) have replaced ◊dot matrix and ◊daisywheel printers as the most popular type of microcomputer printer.

lat. abbreviation for ◊*latitude.*

latent heat in physics, the heat absorbed or radiated by a substance as it changes state (for example, from solid to liquid) at constant temperature and pressure.

lateral line system system of sense organs in fishes and larval amphibians (tadpoles) that detects water movement. It usually consists of a row of interconnected pores on either side of the body that divide into a system of canals across the head.

lateral moraine linear ridge of rocky debris deposited near the edge of a ◊glacier. Much of the debris is material that has fallen from the valley side onto the glacier's edge, having been weathered by ◊freeze-thaw (the alternate freezing and thawing of ice in cracks); it will, therefore, tend to be angular in nature. Where two glaciers merge, two lateral moraines may join together to form a *medial moraine* running along the centre of the merged glacier.

laterite red residual soil characteristic of tropical rainforests. It is formed by the weathering of basalts, granites, and shales and contains a high percentage of aluminium and iron hydroxides. It

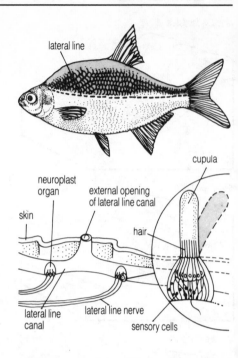

lateral line system *In fishes, the lateral line system detects water movement. Arranged along a line down the length of the body are two water-filled canals, just under the skin. The canals are open to the outside, and water movements cause water to move in the canals. Nerve endings detect the movements.*

may form an impermeable and infertile layer that hinders plant growth.

latex (Latin 'liquid') fluid of some plants (such as the rubber tree and poppy), an emulsion of resins, proteins, and other organic substances. It is used as the basis for making rubber. The name is also applied to a suspension in water of natural or synthetic rubber (or plastic) particles used in rubber goods, paints, and adhesives.

lathe machine tool, used for *turning.* The workpiece to be machined, usually wood or metal, is held and rotated while cutting tools are moved against it. Modern lathes are driven by electric motors, which can drive the spindle carrying the workpiece at various speeds.

latitude and longitude imaginary lines used to locate position on the globe. Lines of latitude are drawn parallel to the equator, with 0° at the equator and 90° at the north and south poles. Lines of longitude are drawn at right angles to these, with 0° (the Prime Meridian) passing through Greenwich, England.

laughing gas popular name for ◊nitrous oxide, an anaesthetic.

Laurasia northern land mass formed 200 million years ago by the splitting of the single world continent ◊Pangaea. (The southern land mass was ◊Gondwanaland.) It consisted of what was to become North America, Greenland, Europe, and Asia, and is believed to have broken up about 100

Point X lies on longitude 60°W

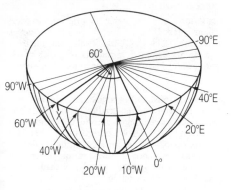

Point X lies on latitude 20°S

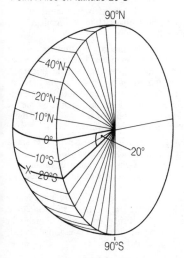

Together longitude 60°W latitude 20°S places point X on a precise position on the globe.

latitude and longitude *Locating a point on a globe using latitude and longitude. Longitude is the angle between the terrestrial meridian through a place and the standard meridian 0° passing through Greenwich, England. Latitude is the angular distance of a place from the equator.*

million years ago with the separation of North America from Europe.

lava molten rock that erupts from a ◊volcano and cools to form extrusive ◊igneous rock. It differs from magma in that it is molten rock on the surface; magma is molten rock below the surface. Lava that is high in silica is viscous and sticky and does not flow far; it forms a steep-sided conical volcano. Low-silica lava can flow for long distances and forms a broad flat volcano.

lawrencium synthesized, radioactive, metallic element, the last of the actinide series, symbol Lr, atomic number 103, relative atomic mass 262. Its only known isotope, Lr-257, has a half-life of 4.3

seconds and was originally synthesized at the University of California at Berkeley 1961 by bombarding californium with boron nuclei. The original symbol, Lw, was officially changed 1963.

The element was named after Ernest Lawrence (1901–1958), the US inventor of the cyclotron particle ◊accelerator.

lb (Latin 'libra') symbol for ◊*pound* (weight).

LCD abbreviation for ◊*liquid crystal display*.

LDR abbreviation for ◊*light-dependent resistor*.

leaching process by which substances are washed out of the soil. Fertilizers leached out of the soil drain into rivers, lakes, and ponds and cause water pollution. In tropical areas, leaching of the soil after the destruction of forests removes scarce nutrients and can lead to a dramatic loss of soil fertility. The leaching of soluble minerals in soils can lead to the formation of distinct soil horizons as different minerals are deposited at successively lower levels.

lead heavy, soft, malleable, grey, metallic element, symbol Pb (from Latin *plumbum*), atomic number 82, relative atomic mass 207.19. Usually found as an ore (most often in galena), it occasionally occurs as a free metal (◊native metal), and is the final stable product of the decay of uranium. Lead is the softest and weakest of the commonly used metals, with a low melting point; it is a poor conductor of electricity and resists acid corrosion.

As a cumulative poison, lead enters the body from lead water pipes, lead-based paints, and leaded petrol. (In humans, exposure to lead shortly after birth is associated with impaired mental health between the ages of two and four.) The metal is an effective shield against radiation and is used in batteries, glass, ceramics, and alloys such as pewter and solder.

lead–acid cell type of ◊accumulator (storage battery).

leaded petrol petrol containing the lead compound tetraethyl lead, $(C_2H_5)_4Pb$, and dibromoethane, a mixture that acts as an ◊antiknock agent.

It improves the combustion of petrol and the performance of a car engine. The lead from the exhaust fumes enters the atmosphere, mostly as simple lead compounds. Compare with ◊unleaded petrol.

lead(II) nitrate $Pb(NO_3)_2$ one of only two common water-soluble compounds of lead. When heated, it decrepitates (see ◊decrepitation) and decomposes readily into oxygen, brown nitrogen(IV) oxide gas and the red-yellow solid lead(II) oxide.

$$2Pb(NO_3)_2 \rightarrow 2PbO + 4NO_2 + O_2$$

lead ore any of several minerals from which lead is extracted. The main primary ore is galena or lead sulphite PbS. This is unstable, and on prolonged exposure to the atmosphere it oxidizes into the minerals cerussite $PbCO_3$ and anglesite $PbSO_4$. Lead ores are usually associated with other metals, particularly silver—which can be mined at the same time—and zinc, which can cause problems during smelting.

Most commercial deposits of lead ore are in the form of veins, where hot fluids have leached the ore from cooling igneous masses (see ◊igneous rock) and deposited it in cracks in the surrounding country rock, and in thermal metamorphic zones (see ◊metamorphic rock), where the heat of igneous intrusions has altered the minerals of surrounding rocks. Lead is mined in over 40 countries, but half of the world's output comes from the USA, Russia, Kazakhstan, Uzbekistan, Canada, and Australia.

leaf lateral outgrowth on the stem of a plant, and in most species the primary organ of ◊photosynthesis. The chief leaf types are cotyledons (seed leaves), scale leaves (on underground stems), foliage leaves, and bracts (in the axil of which a flower is produced).

Typically leaves are composed of three parts: the sheath or leaf base, the petiole or stalk, and the lamina or blade. The lamina has a network of veins through which water and nutrients are conducted. Structurally the leaf is made up of ◊mesophyll cells surrounded by the epidermis and usually, in addition, a waxy layer, termed the cuticle, which prevents excessive evaporation of water from the leaf tissues by transpiration. The epidermis is interrupted by small pores, or stomata, through which gas exchange between the plant and the atmosphere occurs.

A *simple leaf* is undivided, as in the beech or oak. A *compound leaf* is composed of several leaflets, as in the blackberry, horse-chestnut, or ash tree (the latter being a ◊pinnate leaf). Leaves that fall in the autumn are termed *deciduous*, while evergreen leaves are termed *persistent*.

least action principle in science, the principle that nature 'chooses' the easiest path for moving objects, rays of light, and so on; also known in biology as the *principle of parsimony*.

Le Chatelier's principle or *Le Chatelier-Braun principle* in science, the principle that if a change in conditions is imposed on a system in equilibrium, the system will react to counteract that change and restore the equilibrium.

First stated in 1884 by French chemist Henri le Chatelier (1850–1936), it has been found to apply widely outside the field of chemistry.

lecithin lipid (fat), containing nitrogen and phosphorus, that forms a vital part of the cell membranes of plant and animal cells. The name is from the Greek *lekithos* 'egg yolk', eggs being a major source of lecithin.

Leclanché cell disposable, or primary, electrical cell invented 1866 by French engineer Georges Leclanché (1839–1882), which still forms the basis of most dry batteries. It consists of a carbon rod (the ◊anode) inserted into a mixture of powdered carbon and manganese dioxide within a porous pot. The pot sits in a glass jar containing an ◊electrolyte (conducting medium) of ammonium chloride solution, into which a zinc ◊cathode is inserted. The cell produces a continuous current, the carbon mixture acting as a depolarizer; that is, it prevents hydrogen bubbles from forming on the anode and increasing resistance to the flow of current. In modern dry batteries, the electrolyte is made in the form of a paste with starch.

LED abbreviation for ◊*light-emitting diode*.

left-hand rule in physics, a memory aid used to recall the relative directions of motion, magnetic field, and current in an electric motor. It was devised by English physicist John Fleming. (See ◊Fleming's rules.)

legume plant of the family Leguminosae, which has a pod containing dry seeds. The family includes peas, beans, lentils, clover, and alfalfa (lucerne). Legumes are important in agriculture because of their specialized roots, which have nodules containing bacteria capable of fixing nitrogen from the air and increasing the fertility of the soil. The edible seeds of legumes are called *pulses*.

lek in biology, a closely spaced set of very small ◊territories each occupied by a single male during the mating season. Leks are found in the mating systems of several ground-dwelling birds (such as grouse) and a few antelopes.

The lek is a traditional site where both males and females congregate during the breeding season. The males display to passing females in the hope of attracting them to mate. Once mated, the females go elsewhere to lay their eggs or to complete gestation.

lens in optics, a piece of a transparent material, such as glass, with two polished surfaces—one

section through a leaf blade

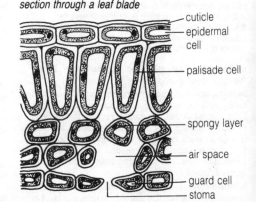

- cuticle
- epidermal cell
- palisade cell
- spongy layer
- air space
- guard cell
- stoma

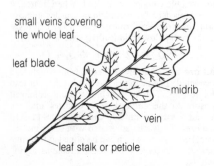

small veins covering the whole leaf

leaf blade

midrib

vein

leaf stalk or petiole

leaf *Leaf shapes and arrangements on the stem are many and varied; in cross-section, a leaf is a complex arrangement of cells surrounded by the epidermis. This is pierced by the stomata through which gases enter and leave.*

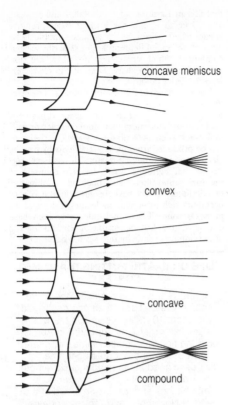

concave meniscus

convex

concave

compound

lens *The passage of light through lenses. The concave lens diverges a beam of light from a distant source. The convex and compound lenses focus light from a distant source to a point. The distance between the focus and the lens is called the focal length. The shorter the focus, the more powerful the lens.*

concave or convex, and the other plane, concave, or convex—that modifies rays of light. A convex lens brings rays of light together; a concave lens makes the rays diverge. Lenses are essential to spectacles, microscopes, telescopes, cameras, and almost all optical instruments.

The image formed by a single lens suffers from several defects or ◊aberrations, notably *spherical aberration* in which a straight line becomes a curved image, and *chromatic aberration* in which an image in white light tends to have coloured edges. Aberrations are corrected by the use of compound lenses, which are built up from two or more lenses of different refractive index.

lensing, gravitational see ◊gravitational lensing.

lenticel small pore on the stems of woody plants or the trunks of trees. Lenticels are means of gas exchange between the stem interior and the atmosphere. They consist of loosely packed cells with many air spaces in between, and are easily seen on smooth-barked trees such as cherries, where they form horizontal lines on the trunk.

Lenz's law in physics, a law stating that the direction of an electromagnetically induced current (generated by moving a magnet near a wire or a wire in a magnetic field) will oppose the motion producing it. It is named after the German physicist

Heinrich Friedrich Lenz (1804–1865), who announced it in 1833.

Leo zodiacal constellation in the northern hemisphere represented as a lion. The Sun passes through Leo from mid-Aug to mid-Sept. Its brightest star is first-magnitude ◊Regulus at the base of a pattern of stars called the Sickle. In astrology, the dates for Leo are between about 23 July and 22 Aug (see ◊precession).

lepidoptera order of insects, including butterflies and moths, which have overlapping scales on their wings; the order consists of some 165,000 species.

lepton any of a class of light ◊elementary particles that are not affected by the strong nuclear force; they do not interact strongly with other particles or nuclei. The leptons are comprised of the ◊electron, ◊muon, and ◊tau, and their ◊neutrinos (the electron neutrino, muon neutrino, and tau neutrino), plus their six ◊antiparticles.

leucite silicate mineral, $KAlSi_2O_6$, occurring frequently in some potassium-rich volcanic rocks. It is dull white to grey, and usually opaque. It is used as a source of potassium for fertilizer.

leucocyte another name for a ◊white blood cell.

levee naturally formed raised bank along the side of a river channel. When a river overflows its banks, the rate of flow in the flooded area is less than that in the channel, and silt is deposited. After the waters have withdrawn the silt is left as a bank that grows with successive floods. Eventually the river, contained by the levee, may be above the surface of the surrounding flood plain. Notable levees are found on the lower reaches of the Mississippi in the USA and the Po in Italy.

level or *spirit level* instrument for finding horizontal level, or adjusting a surface to an even level, used in surveying, building construction, and archaeology. It has a glass tube of coloured liquid, in which a bubble is trapped, mounted in an elongated frame. When the tube is horizontal, the bubble moves to the centre.

lever simple machine consisting of a rigid rod pivoted at a fixed point called the fulcrum, used for shifting or raising a heavy load or applying force in a similar way. Levers are classified into orders according to where the effort is applied, and the load-moving force developed, in relation to the position of the fulcrum.

A *first-order* lever has the load and the effort on opposite sides of the fulcrum—for example, a see-saw or pair of scissors. A *second-order* lever has the load and the effort on the same side of the fulcrum, with the load nearer the fulcrum—for example, nutcrackers or a wheelbarrow. A *third-order* lever has the effort nearer the fulcrum than the load, with both on the same side of it—for example, a pair of tweezers or tongs. The mechanical advantage of a lever is the ratio of load to effort, equal to the perpendicular distance of the effort's line of action from the fulcrum divided by the distance to the load's line of action. Thus tweezers, for instance, have a mechanical advantage of less than one.

LF in physics, abbreviation for *low* ◊frequency.

LH abbreviation for ◊*luteinizing hormone*.

first-order lever

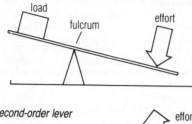

second-order lever

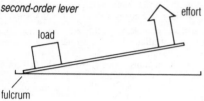

third-order lever

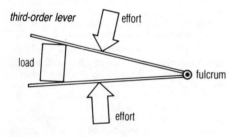

lever *Types of lever. Practical applications of the first-order lever include the crowbar, seesaw, and scissors. The wheelbarrow and nutcrackers are second-order levers. A pair of tweezers, or tongs, is a third-order lever.*

Libra faint zodiacal constellation in the southern hemisphere adjoining Scorpius, and represented as the scales of justice. The Sun passes through Libra during Nov. The constellation was once considered to be a part of Scorpius, seen as the scorpion's claws. In astrology, the dates for Libra are between about 23 Sept and 23 Oct (see ◊precession).

library program one of a collection, or library, of regularly used software routines, held in a computer backing store. For example, a programmer might store a routine for sorting a file into ◊key field order, and so could incorporate it easily into any new program being developed instead of having to rewrite it.

Lick Observatory observatory of the University of California and Mount Hamilton, California. Its main instruments are the 3.04-m/120-in Shane reflector, opened 1959, and a 91-cm/36-in refractor, opened 1888, the second-largest refractor in the world.

lie detector instrument that records graphically certain body activities, such as thoracic and abdominal respiration, blood pressure, pulse rate, and galvanic skin response (changes in electrical resistance of the skin). Marked changes in these activities when a person answers a question may indicate that the person is lying.

life the ability to grow, reproduce, and respond to such stimuli as light, heat, and sound. It is thought that life on Earth began about 4,000 million years ago. Over time, life has evolved from primitive single-celled organisms to complex multicellular ones. Problems of defining life exist partly because computers, which are clearly not alive, are now complex enough to share some of the characteristics of living things.

Life originated in the primitive oceans. The original atmosphere, 4,000 million years ago, consisted of carbon dioxide, nitrogen, and water. It has been shown in the laboratory that more complex organic molecules, such as ◊amino acids and ◊nucleotides, can be produced from these ingredients by passing electric sparks through a mixture. It has been suggested that lightning was extremely common in the early atmosphere, and that this combination of conditions could have resulted in the oceans becoming rich in organic molecules, the so-called primeval soup. These molecules may then have organized into clusters, capable of reproducing and of developing eventually into simple cells.

LIFE ON EARTH: PUTTING IT ALL INTO CONTEXT

If one year were to represent the entire span of life on Earth, then one day would represent 10 million years. On this scale, algae-like organisms appeared in August and the first soft-bodied animals appeared towards the end of September. The first fish developed in the middle of November, along with the first land plants. Amphibians and insects emerged at the end of November, and reptiles appeared in early December. The first dinosaur appeared in the middle of December. A week later, the first birds flew. By the beginning of the last week of the year the dinosaurs had became extinct. The earliest apes appeared halfway through that week. Humans did not come on the scene until the evening of 31 December.

---

life cycle in biology, the sequence of developmental stages through which members of a given species pass. Most vertebrates have a simple life cycle consisting of ◊fertilization of sex cells or ◊gametes, a period of development as an ◊embryo, a period of juvenile growth after hatching or birth, an adulthood including ◊sexual reproduction, and finally death. Invertebrate life cycles are generally more complex and may involve major reconstitution of the individual's appearance (◊metamorphosis) and completely different styles of life. Plants have a special type of life cycle with two distinct phases, known as ◊alternation of generations. Many insects such as cicadas, dragonflies, and mayflies have a long larvae or pupae phase and a short adult phase. Dragonflies live an aquatic life as larvae and an aerial life during the adult phase. In many invertebrates and protozoa there is a sequence of stages in the life cycle, and in parasites different stages often occur in different host organisms.

life expectancy average lifespan of a person at birth. It depends on nutrition, disease control, environmental contaminants, war, stress, and living standards in general.

There is a marked difference between industrialized countries, which generally have an ageing population, and the poorest countries, where life expectancy is much shorter. In Bangladesh, life expectancy is currently 48; in Nigeria 49. In famine-prone Ethiopia it is only 41.

In the UK, average life expectancy for both sexes currently stands at 75 and heart disease is the main cause of death.

life sciences scientific study of the living world as a whole, a new synthesis of several traditional scientific disciplines including ◊biology, ◊zoology, and ◊botany, and newer, more specialized areas of study such as ◊biophysics and ◊sociobiology. This approach has led to many new ideas and discoveries as well as to an emphasis on ◊ecology, the study of living organisms in their natural environments.

lifespan time between birth and death. Lifespan varies enormously between different species. In humans it is also known as ◊life expectancy and varies according to sex, country, and occupation.

life table way of summarizing the probability that an individual will give birth or die during successive periods of life. From this, the proportion of individuals who survive from birth to any given age (*survivorship*) and the mean number of offspring produced (*net reproductive rate*) can be determined.

ligament strong flexible connective tissue, made of the protein collagen, which joins bone to bone at moveable joints. Ligaments prevent bone dislocation (under normal circumstances) but permit joint flexion.

light electromagnetic waves in the visible range, having a wavelength from about 400 nanometres in the extreme violet to about 770 nanometres in the extreme red. Light is considered to exhibit particle and wave properties, and the fundamental particle, or quantum, of light is called the photon. The speed of light (and of all electromagnetic radiation) in a vacuum is approximately 300,000 km/186,000 mi per second, and is a universal constant denoted by *c*.

Issac Newton was the first to discover, 1666, that sunlight is composed of a mixture of light of different colours in certain proportions and that it could be separated into its components by dispersion. Before his time it was supposed that dispersion of light produced colour instead of separating already existing colours.

The ancients believed that light travelled at infinite speed; its speed was first measured accurately by Danish astronomer Olaus Roemer 1681.

light-dependent resistor (LDR) component of electronic circuits whose resistance varies with the level of illumination on its surface. Usually resistance decreases as illumination rises. LDRs are used in light-measuring or light-sensing instruments (for example, in the exposure-meter circuit of an automatic camera) and in switches (such as those that switch on street lights at dusk). LDRs are made from ◊semiconductors, such as cadmium sulphide.

light-emitting diode (LED) means of displaying symbols in electronic instruments and devices. An LED is made of ◊semiconductor material, such as gallium arsenide phosphide, that glows when electricity is passed through it. The first digital watches and calculators had LED displays, but many later models use ◊liquid crystal displays.

lighthouse structure carrying a powerful light to warn ships or aeroplanes that they are approaching a place dangerous or important to navigation. The light is magnified and directed out to the horizon or up to the zenith by a series of mirrors or prisms. Increasingly lighthouses are powered by electricity and automated rather than staffed; the more recent models also emit radio signals.

lightning high-voltage electrical discharge between two charged rainclouds or between a cloud and the Earth, caused by the build-up of electrical charges. Air in the path of lightning ionizes (becomes conducting), and expands; the accompanying noise is heard as thunder. Currents of 20,000 amperes and temperatures of 30,000°C/54,000°F are common.

LIGHTNING SPEED

The upward flash of a lightning stroke can travel at up to 140,000 km/87,000 mi per second, which is nearly half the speed of light. The downward stroke of a lightning flash is much slower at 1,600 km/1,000 mi per second.

---

light pen in computing, a device resembling an ordinary pen, used to indicate locations on a computer screen. With certain computer-aided design (◊CAD) programs, the light pen can be used to instruct the computer to change the shape, size, position, and colours of sections of a screen image.

The pen has a photoreceptor at its tip that emits signals when light from the screen passes beneath it. From the timing of this signal and a gridlike representation of the screen in the computer memory, a computer program can calculate the position of the light pen.

light second unit of length, equal to the distance travelled by light in one second. It is equal to 2.997925×10^8 m/9.835592×10^8 ft. See ◊light year.

light watt unit of radiant power (brightness of light). One light watt is the power required to produce a perceived brightness equal to that of light at a wavelength of 550 nanometres and 680 lumens.

light year in astronomy, the distance travelled by a beam of light in a vacuum in one year, approximately 9.46 trillion (million million) km/5.88 trillion miles.

lignin naturally occurring substance produced by plants to strengthen their tissues. It is difficult for ◊enzymes to attack lignin, so living organisms cannot digest wood, with the exception of a few specialized fungi and bacteria. Lignin is the essential ingredient of all wood and is, therefore, of great commercial importance.

Chemically, lignin is made up of thousands of

And All was Light

Alexander Pope, the poet, wrote of Isaac Newton's work: 'Nature, and Nature's Laws lay hid in Night: God said, *Let Newton be*! and All was *Light*.' Isaac Newton, the English physicist and mathematician (1642–1727), would have been remembered as a great scientist for any one of his many discoveries. He made outstanding contributions to mathematics, astronomy, mechanics, and to understanding gravitation and the nature of light.

The first microscopes revealed the world of the very small to scientists for the very first time. They saw in great detail what microorganisms were really like. In the same way, at the other end of the scale, the first telescopes revealed to astronomers vast numbers of stars and the beauty of the Earth's planets. But there was a problem. Any image formed by the combination of the lenses then available to make microscopes or telescopes was surrounded by a colour fringe. This blurred the outline—an effect which became worse at higher magnifications.

From prisms and rainbows . . .
People had long been familiar with the rainbow colours produced when light shone through a chandelier. Now, in order to improve their instruments, scientists needed to know why these colours were formed. Newton ground his own glass lenses and had, since he was an undergraduate student at Cambridge, been interested in the effect of a glass prism on sunlight.

Newton wrote: 'In the year 1666 (at which time I applied myself to the grinding of optick glass of other figures than spherical) I procured me a triangular glass prism, to try the celebrated phaenomena of colours.'

Newton made a small hole in the shutter covering the window in his room to let in a beam of sunlight. He placed the prism in front of this hole and viewed the vivid and intense spectrum of rainbow colours cast on the opposite wall.

Newton now examined these colours individually, using a second prism. He took two boards, each with a small hole in it. The first he placed behind the prism at the window. He positioned the second board so that only a single colour produced by the first prism fell on the hole in it. The second prism was placed so as to cast the light passing through the hole in the second board on the wall. Newton saw at once that the coloured beam of light passing through the second prism was unchanged. This proved to be true for all the rainbow colours produced by the first prism.

Newton had performed the crucial experiment, because it had been assumed previously that light was basically white, and that colours could be added to it. Now it was clear that white light was a mixture of the colours of the rainbow; the prism simply split them up or refracted the light. A second prism could not 'split' them up further.

. . . to modern telescopes
In his book *Opticks*, published in 1704, Newton described a further experiment in which he used the second prism to recombine the rainbow colours of the spectrum to produce white light. Here for the first time was a simple explanation of the nature of light. The book had a great impact on 18th-century writers. What Newton had to say about what happened when white light passed through a prism could be immediately applied to rainbows, and this captured the imagination of artists and poets.

Because Newton suspected that the colour fringes or chromatic aberration produced by lenses in telescopes could not be avoided, in 1668 he designed a telescope that depended instead on the use of a curved mirror. Light reflected from a surface produces no colour effects, unlike light passing through a lens. Nowadays all large astronomical telescopes are of the reflecting type.

As a result of his elegant experiments with prisms and light, Newton also speculated about the ultimate nature of light itself. He proposed that light consisted of small 'corpuscules' (small particles) which are shot out from the source of light, rather in the same way that pellets are ejected from a shot gun. He was able to explain many of the known properties of light with this theory, which was widely accepted. Light has proved remarkably difficult to understand; nowadays it is regarded as having both particle-like and wave-like properties.

---

rings of carbon atoms joined together in a long chain. The way in which they are linked up varies along the chain.

lignite type of ◊coal that is brown and fibrous, with a relatively low carbon content. In Scandinavia it is burned to generate power.

lime or *quicklime* CaO (technical name *calcium oxide*) white powdery substance used in making mortar and cement. It is made commercially by heating calcium carbonate ($CaCO_3$), obtained from limestone or chalk, in a ◊lime kiln. Quicklime readily absorbs water to become calcium hydroxide

light pen *The tiny photoreceptor at the tip of a light pen can recognize individual pixels (picture elements) and can therefore be used very precisely to indicate locations on a computer screen.*

(CaOH), known as slaked lime, which is used to reduce soil acidity.

lime kiln oven used to make quicklime (calcium oxide, CaO) by heating limestone (calcium carbonate, $CaCO_3$) in the absence of air. The carbon dioxide is carried away to heat other kilns and to ensure that the reversible reaction proceeds in the right direction.

$$CaCO_3 \rightleftharpoons CaO + CO_2$$

limestone sedimentary rock composed chiefly of calcium carbonate $CaCO_3$, either derived from the shells of marine organisms or precipitated from solution, mostly in the ocean. Various types of limestone are used as building stone.

◊Marble is metamorphosed limestone. Certain so-called marbles are not in fact marbles but fine-grained fossiliferous limestones that take an attractive polish. Caves commonly occur in limestone. ◊Karst is a type of limestone landscape.

limestone pavement bare rock surface resembling a block of chocolate, found on limestone plateaus. It is formed by the weathering of limestone into individual upstanding blocks, called clints, separated from each other by joints, called grikes. The weathering process is thought to entail a combination of freeze-thaw (the alternate freezing and thawing of ice in cracks) and carbonation (the dissolving of minerals in the limestone by weakly acidic rainwater). Malham Tarn in North Yorkshire is an example of a limestone pavement.

limewater common name for a dilute solution of slaked lime (calcium hydroxide, $Ca(OH)_2$). In chemistry, it is used to detect the presence of carbon dioxide.

If a gas containing carbon dioxide is bubbled through limewater, the solution turns milky owing to the formation of calcium carbonate ($CaCO_3$). Continued bubbling of the gas causes the limewater to clear again as the calcium carbonate is converted to the more soluble calcium hydrogencarbonate ($Ca(HCO_3)_2$).

limnology study of lakes and other bodies of open fresh water, in terms of their plant and animal biology, and their physical properties.

limonite iron ore, mostly poorly crystalline iron oxyhydroxide, but usually mixed with ◊hematite and other iron oxides. Also known as brown iron ore, it is often found in bog deposits.

linac contraction of ◊linear accelerator, a type of particle accelerator in which the particles are accelerated along a straight tube.

features of a limestone region

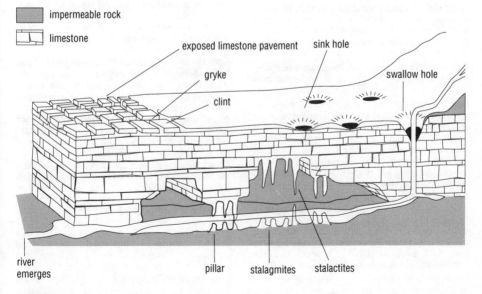

impermeable rock

limestone

exposed limestone pavement   sink hole

gryke   swallow hole

clint

river emerges   pillar   stalagmites   stalactites

limestone

linear accelerator or *linac* in physics, a type of particle ◊accelerator in which the particles move along a straight tube. Particles pass through a linear accelerator only once—unlike those in a cyclotron (a ring-shaped accelerator), which make many revolutions, gaining energy each time.

The world's longest linac is the Stanford Linear Collider, in which electrons and positrons are accelerated along a straight track 3.2 km/2 mi long and then steered into a head-on collision with other particles.

The first linear accelerator was built in 1928 by Norwegian engineer Ralph Wideröe to investigate the behaviour of heavy ions (large atoms with one or more electrons removed), but devices capable of accelerating smaller particles such as protons and electrons could not be built until after World War II and the development of high-power radio- and microwave-frequency generators.

linear equation in mathematics, a relationship between two variables that, when plotted on Cartesian axes produces a straight-line graph; the equation has the general form $y = mx + c$, where m is the slope of the line represented by the equation and c is the y-intercept, or the value of y where the line crosses the y-axis in the ◊Cartesian coordinate system. Sets of linear equations can be used to describe the behaviour of buildings, bridges, trusses, and other static structures.

linear motor type of electric motor, an induction motor in which the fixed stator and moving armature are straight and parallel to each other (rather than being circular and one inside the other as in an ordinary induction motor). Linear motors are used, for example, to power sliding doors. There is a magnetic force between the stator and armature; this force has been used to support a vehicle, as in the experimental ◊maglev linear motor train.

linear programming in mathematics and economics, a set of techniques for finding the maxima or minima of certain variables governed by linear equations or inequalities. These maxima and minima are used to represent 'best' solutions in terms of goals such as maximizing profit or minimizing cost.

line of force in physics, an imaginary line representing the direction of force at any point in a magnetic, gravitational, or electrical field.

line printer computer ◊printer that prints a complete line of characters at a time. Line printers can achieve very high printing speeds of up to 2,500 lines a minute, but can print in only one typeface, cannot print graphics, and are very noisy. Until the late 1980s they were the obvious choice for high-volume printing, but high-speed ◊page printers, such as laser printers, are now preferred.

linkage in genetics, the association between two or more genes that tend to be inherited together because they are on the same chromosome. The closer together they are on the chromosome, the less likely they are to be separated by crossing over (one of the processes of ◊recombination) and they are then described as being 'tightly linked'.

Linotype trademark for a typesetting machine once universally used for newspaper work, which sets complete lines (slugs) of hot-metal type as operators type the copy at a keyboard. It was invented in the USA in 1884 by German-born Ottmar Mergenthaler. It has been replaced by phototypesetting.

lipase the enzyme responsible for breaking down fats into fatty acids and glycerol. It is produced by the ◊pancreas and requires a slightly alkaline environment. The products of fat digestion are absorbed by the intestinal wall.

lipid any of a large number of esters of fatty acids, commonly formed by the reaction of a fatty acid with glycerol (see ◊glycerides). They are soluble in alcohol but not in water. Lipids are the chief constituents of plant and animal waxes, fats, and oils.

Phospholipids are lipids that also contain a phosphate group, usually linked to an organic base; they are major components of biological cell membranes.

lipophilic (Greek 'fat-loving') in chemistry, a term describing ◊functional groups with an affinity for fats and oils.

lipophobic (Greek 'fat-hating') in chemistry, a term describing ◊functional groups that tend to repel fats and oils.

liquefaction the process of converting a gas to a liquid, normally associated with low temperatures and high pressures (see ◊condensation).

liquefaction in earth science, the conversion of a soft deposit, such as clay, to a jellylike state by severe shaking. During an earthquake buildings and lines of communication built on materials prone to liquefaction will sink and topple. In the Alaskan earthquake of 1964 liquefaction led to the destruction of much of the city of Anchorage.

liquefied petroleum gas (LPG) liquid form of butane, propane, or pentane, produced by the distillation of petroleum during oil refining. At room temperature these substances are gases, although they can be easily liquefied and stored under pressure in metal containers. They are used for heating and cooking where other fuels are not available: camping stoves and cigarette lighters, for instance, often use liquefied butane as fuel.

liquid state of matter between a ◊solid and a ◊gas. A liquid forms a level surface and assumes the shape of its container. Its atoms do not occupy fixed positions as in a crystalline solid, nor do they have freedom of movement as in a gas. Unlike a gas, a liquid is difficult to compress since pressure applied at one point is equally transmitted throughout (Pascal's principle). ◊Hydraulics makes use of this property.

liquid air air that has been cooled so much that it has liquefied. This happens at temperatures below about −196°C/−321°F. The various constituent gases, including nitrogen, oxygen, argon, and neon, can be separated from liquid air by the technique of ◊fractionation.

Air is liquefied by the *Linde process*, in which air is alternately compressed, cooled, and expanded, the expansion resulting each time in a considerable reduction in temperature.

liquid crystal display (LCD) display of numbers (for example, in a calculator) or pictures (such as on a pocket television screen) produced by

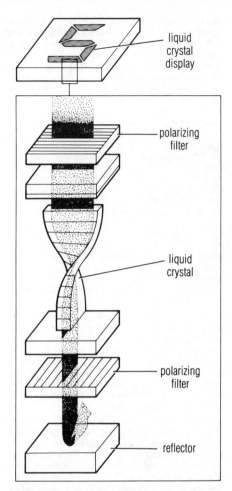

liquid
crystal
display

polarizing
filter

liquid
crystal

polarizing
filter

reflector

liquid crystal display *A liquid crystal display consists of a liquid crystal sandwiched between polarizing filters similar to polaroid sunglasses. When a segment of the seven-segment display is electrified, the liquid crystal twists the polarized light from the front filter, allowing the light to bounce off the rear reflector and illuminate the segment.*

molecules of a substance in a semiliquid state with some crystalline properties, so that clusters of molecules align in parallel formations. The display is a blank until the application of an electric field, which 'twists' the molecules so that they reflect or transmit light falling on them.

LISP (acronym for *list processing*) high-level computer-programming language designed for manipulating lists of data items. It is used primarily in research into ◊artificial intelligence (AI).

Developed in the 1960s, and until recently common only in university laboratories, LISP is used more in the USA than in Europe, where the language ◊PROLOG is often preferred for AI work.

lithification another term for ◊diagenesis.

lithium (Greek *lithos* 'stone') soft, ductile, silver-white, metallic element, symbol Li, atomic number

3, relative atomic mass 6.941. It is one of the ◊alkali metals, has a very low density (far less than most woods), and floats on water (specific gravity 0.57); it is the lightest of all metals. Lithium is used to harden alloys, and in batteries; its compounds are used in medicine to treat manic depression.

Lithium was named in 1818 by Swedish chemist Jöns Berzelius, having been discovered the previous year by his student Johan A Arfwedson (1792–1841). Berzelius named it after 'stone' because it is found in most igneous rocks and many mineral springs.

lithography printmaking technique originated in 1798 by Aloys Senefelder, based on the antipathy of grease and water. A drawing is made with greasy crayon on an absorbent stone, which is then wetted. The wet stone repels ink (which is greasy) applied to the surface and the crayon attracts it, so that the drawing can be printed. Lithographic printing is used in book production and has developed this basic principle into complex processes.

lithosphere topmost layer of the Earth's structure, forming the jigsaw of plates that take part in the movements of ◊plate tectonics. The lithosphere comprises the ◊crust and a portion of the upper ◊mantle. It is regarded as being rigid and moves about on the semi-molten ◊asthenosphere. The lithosphere is about 75 km/47 mi thick.

litmus dye obtained from various lichens and used in chemistry as an indicator to test the acidic or alkaline nature of aqueous solutions; it turns red in the presence of acid, and blue in the presence of alkali.

litre metric unit of volume (symbol l), equal to one cubic decimetre (1.76 pints). It was formerly defined as the volume occupied by one kilogram of pure water at 4°C at standard pressure, but this is slightly larger than one cubic decimetre.

Little Dipper another name for the most distinctive part of ◊Ursa Minor, the Little Bear.

The name has also been applied to the stars in the ◊Pleiades open star cluster, the overall shape of the cluster being similar to that of the ◊Big Dipper.

liver large organ of vertebrates, which has many regulatory and storage functions. The human liver is situated in the upper abdomen, and weighs about 2 kg/4.5 lb. It receives the products of digestion, converts glucose to glycogen (a long-chain carbohydrate used for storage), and breaks down fats. It removes excess amino acids from the blood, converting them to urea, which is excreted by the kidneys. The liver also synthesizes vitamins, produces bile and blood-clotting factors, and removes damaged red cells and toxins such as alcohol from the blood.

liverwort plant of the class Hepaticae, of the bryophyte division of nonvascular plants. Related to mosses and hornworts, it is found growing in damp places.

The main sexual generation consists of a ◊thallus, which may be flat, green, and lobed, like a small leaf, or leafy and mosslike. The spore-bearing generation is smaller, typically parasitic on the thallus, and throws up a capsule from which spores are spread.

loam type of fertile soil, a mixture of sand, silt, clay, and organic material. It is porous, which allows for good air circulation and retention of moisture.

local area network (LAN) in computing, a ◊network restricted to a single room or building. Local area networks enable around 500 devices to be connected together.

Local Group in astronomy, a cluster of about 30 galaxies that includes our own, the Milky Way. Like other groups of galaxies, the Local Group is held together by the gravitational attraction among its members, and does not expand with the expanding universe. Its two largest galaxies are the Milky Way and the Andromeda galaxy; most of the others are small and faint.

local variable in computing, a ◊variable that can be accessed only by the instructions within a particular ◊subroutine.

lock construction installed in waterways to allow boats or ships to travel from one level to another. The earliest form, the *flash lock*, was first seen in the East in 1st-century-AD China and in the West in 11th-century Holland. By this method barriers temporarily dammed a river and when removed allowed the flash flood to impel the waiting boat through any obstacle. This was followed in 12th-century China and 14th-century Holland by the *pound lock*. In this system the lock has gates at each end. Boats enter through one gate when the levels are the same both outside and inside. Water is then allowed in (or out of) the lock until the level rises (or falls) to the new level outside the other gate.

Locks are important to shipping where canals link oceans of differing levels, such as the Panama Canal, or where falls or rapids are replaced by these adjustable water 'steps'.

lock and key devices that provide security, usually fitted to a door of some kind. In 1778 English locksmith Robert Barron made the forerunner of the *mortise lock*, which contains levers that the key must raise to an exact height before the bolt can be moved. The *Yale lock*, a pin-tumbler cylinder design, was invented by US locksmith Linus Yale, Jr, in 1865. More secure locks include *combination locks*, with a dial mechanism that must be turned certain distances backwards and forwards to open, and *time locks*, which are set to be opened only at specific times.

Locks originated in the Far East over 4,000 years ago. The Romans developed the warded lock, which contains obstacles (wards) that the key must pass to turn.

locomotion the ability to move independently from one place to another, occurring in most animals but not in plants. The development of locomotion as a feature of animal life is closely linked to another vital animal feature, that of nutrition. Animals cannot make their food, as can plants, but must find it first; often the food must be captured and killed, which may require great powers of speed. Locomotion is also important in finding a mate, in avoiding predators, and in migrating to favourable areas.

locomotive engine for hauling railway trains. In

1804 Richard Trevithick built the first steam engine to run on rails. Locomotive design did not radically improve until British engineer George Stephenson built the *Rocket* 1829, which featured a multitube boiler and blastpipe, standard in all following *steam locomotives*. Today most locomotives are diesel or electric: *diesel locomotives* have a powerful diesel engine, and *electric locomotives* draw their power from either an overhead cable or a third rail alongside the ordinary track.

In a steam locomotive, fuel (usually coal, sometimes wood) is burned in a furnace. The hot gases and flames produced are drawn through tubes running through a huge water-filled boiler and heat up the water to steam. The steam is then fed to the cylinders, where it forces the pistons back and forth. Movement of the pistons is conveyed to the wheels by cranks and connecting rods. Diesel locomotives have a powerful diesel engine, burning oil. The engine may drive a generator to produce electricity to power electric motors that turn the wheels, or the engine drives the wheels mechanically or through a hydraulic link. A number of *gas-turbine locomotives* are in use, in which a turbine spun by hot gases provides the power to drive the wheels.

locus (Latin 'place') in mathematics, traditionally the path traced out by a moving point, but now defined as the set of all points on a curve satisfying given conditions. For example, the locus of a point that moves so that it is always at the same distance from another fixed point is a circle; the locus of a point that is always at the same distance from two fixed points is a straight line that perpendicularly bisects the line joining them.

lode geological deposit rich in certain minerals, generally consisting of a large vein or set of veins containing ore minerals. A system of veins that can be mined directly forms a lode, for example the mother lode of the California gold rush.

Lodes form because hot hydrothermal liquids and gases from magmas penetrate surrounding rocks, especially when these are limestones; on cooling, veins of ores formed from the magma then extend from the igneous mass into the local rock.

lodestar or *loadstar* a star used in navigation or astronomy, often ◊Polaris, the Pole Star.

loess yellow loam, derived from glacial meltwater deposits and accumulated by wind in periglacial regions during the ◊ice ages. Loess usually attains considerable depths, and the soil derived from it is very fertile. There are large deposits in central Europe (Hungary), China, and North America. It was first described in 1821 in the Rhine area, and takes its name from a village in Alsace.

log in mathematics, abbreviation for ◊logarithm.

logarithm or *log* the ◊exponent or index of a number to a specified base—usually 10. For example, the logarithm to the base 10 of 1,000 is 3 because $10^3 = 1,000$; the logarithm of 2 is 0.3010 because $2 = 10^{0.3010}$. Before the advent of cheap electronic calculators, multiplication and division could be simplified by being replaced with the addition and subtraction of logarithms.

For any two numbers x and y (where $x = b^a$ and $y = b^c$) $x \times y = b^a \times b^c = b^{a+c}$; hence we would add the logarithms of *x* and *y*, and look up this answer in antilogarithm tables. Tables of logarithms and antilogarithms are available that show conversions of numbers into logarithms, and vice versa. For example, to multiply 6,560 by 980, one looks up their logarithms (3.8169 and 2.9912), adds them together (6.8081), then looks up the antilogarithm of this to get the answer (6,428,800). *Natural* or *Napierian logarithms* are to the base *e*, an ◊irrational number equal to approximately 2.7183.

The principle of logarithms is also the basis of the slide rule. With the general availability of the electronic pocket calculator, the need for logarithms has been reduced. The first log tables (to base *e*) were published by the Scottish mathematician John Napier in 1614. Base-ten logs were introduced by the Englishman Henry Briggs (1561–1631) and Dutch mathematician Adriaen Vlacq (1600–1667).

logic gate or *logic circuit* in electronics, one of the basic components used in building ◊integrated circuits. The five basic types of gate make logical decisions based on the functions NOT, AND, OR, NAND (NOT AND), and NOR (NOT OR). With the exception of the NOT gate, each has two or more inputs.

Information is fed to a gate in the form of binary-coded input signals (logic value 0 stands for 'off' or 'low-voltage pulse', logic 1 for 'on' or 'high-voltage'), and each combination of input signals yields a specific output (logic 0 or 1). An *OR* gate will give a logic 1 output if one or more of its inputs receives a logic 1 signal; however, an *AND* gate will yield a logic 1 output only if it receives a logic 1 signal through both its inputs. The output of a *NOT* or *inverter* gate is the opposite of the signal received through its single input, and a *NOR* or *NAND* gate produces an output signal that is the opposite of the signal that would have been produced by an OR or AND gate respectively. The properties of a logic gate, or of a combination of gates, may be defined and presented in the form of a diagram called a *truth table*, which lists the output that will be triggered by each of the possible combinations of input signals. The process has close parallels in computer programming, where it forms the basis of binary logic.

LOGO (Greek *logos* 'word') high-level computer programming language designed to teach mathematical concepts. Developed about 1970 at the Massachusetts Institute of Technology, it became popular in schools and with home computer users because of its 'turtle graphics' feature. This allows the user to write programs that create line drawings on a computer screen, or drive a small mobile robot (a 'turtle' or 'buggy') around the floor.

LOGO encourages the use of languages in a logical and structured way, leading to 'microworlds', in which problems can be solved by using a few standard solutions.

log on or *log in* in computing, the process by which a user identifies himself or herself to a multiuser computer and enters the system. Logging on usually requires the user to enter a password before access is allowed.

lomentum fruit similar to a pod but constricted between the seeds. When ripe, it splits into one-seeded units, as seen, for example, in the fruit of sainfoin *Onobrychis viciifolia* and radish *Raphanus raphanistrum*. It is a type of ◊schizocarp.

lone pair in chemistry, a pair of electrons in the outermost shell of an atom that are not used in bonding. In certain circumstances, they will allow the atom to bond with atoms, ions, or molecules (such as boron trifluoride, BF_3) that are deficient in electrons, forming coordinate covalent (dative) bonds in which they provide both of the bonding electrons.

long. abbreviation for *longitude*; see ◊latitude and longitude.

longitude see ◊latitude and longitude.

longitudinal wave ◊wave in which the displacement of the medium's particles is in line with or parallel to the direction of travel of the wave motion.

It is characterized by its alternating compressions and rarefactions. In the *compressions* the

circuit symbols

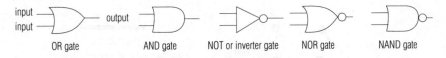

| OR gate | AND gate | NOT or inverter gate | NOR gate | NAND gate |

truth tables

inputs		output
0	0	0
0	1	1
1	0	1
1	1	1

OR gate

inputs		output
0	0	0
0	1	0
1	0	0
1	1	1

AND gate

inputs	output
0	1
1	0

NOT gate

inputs		output
0	0	1
0	1	0
1	0	0
1	1	0

NOR gate

inputs		output
0	0	1
0	1	1
1	0	1
1	1	0

NAND gate

logic gate

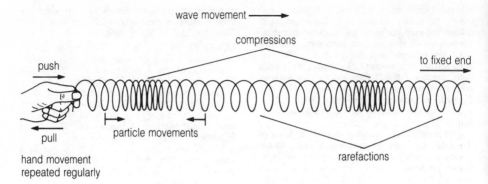

longitudinal wave

particles are pushed together (in a gas the pressure rises); in the *rarefactions* they are pulled apart (in a gas the pressure falls). Sound travels through air as a longitudinal wave.

longshore drift the movement of material along a ◊beach. When a wave breaks obliquely, pebbles are carried up the beach in the direction of the wave (swash). The wave draws back at right angles to the beach (backwash), carrying some pebbles with it. In this way, material moves in a zigzag fashion along a beach. Longshore drift is responsible for the erosion of beaches and the formation of spits (ridges of sand or shingle projecting into the water). Attempts are often made to halt longshore drift by erecting barriers, or groynes, at right angles to the shore.

loom any machine for weaving yarn or thread into cloth. The first looms were used to weave sheep's wool about 5000 BC. A loom is a frame on which a set of lengthwise threads (warp) is strung. A second set of threads (weft), carried in a shuttle, is inserted at right angles over and under the warp.

In most looms the warp threads are separated by a device called a treddle to create a gap, or shed, through which the shuttle can be passed in a straight line. A kind of comb called a reed presses each new line of weave tight against the previous ones. All looms have similar features, but on the power loom, weaving takes place automatically at great speed. Mechanization of weaving began in 1733 when British inventor John Kay invented the flying shuttle. In 1785 British inventor Edmund Cartwright introduced a steam-powered loom. Among recent developments are shuttleless looms, which work at very high speed, passing the weft through the warp by means of 'rapiers', and jets of air or water.

loop in computing, short for ◊program loop.

Loran navigation system long-range radio-based aid to navigation. The current system (Loran C) consists of a master radio transmitter and two to four secondary transmitters situated within 1,000–2,000 km/600–1200 mi. Signals from the transmitters are detected by the ship or aircraft's navigation receiver, and slight differences (phase differences) between the master and secondary transmitter signals indicate the position of the receiver. The system is accurate to within 500 m/1640 ft at a range of 2000 km/1080 mi, and covers the N Atlantic and N Pacific oceans, and adjacent areas such as the Mediterranean and Arabian Gulf.

The original Loran stem (Loran A) was used to

the action of longshore drift

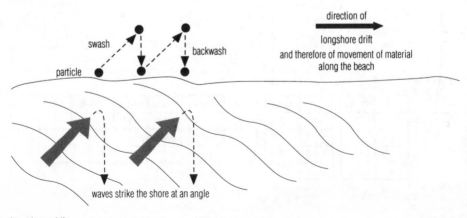

longshore drift

guide bombers on night raids in Europe and N Africa in 1944.

Lotus 1-2-3 ◊spreadsheet computer program, produced by Lotus Development Corporation. It first appeared in 1982 and its combination of spreadsheet, graphics display, and data management contributed to the rapid acceptance of the IBM Personal Computer in businesses.

loudness subjective judgement of the level or power of sound reaching the ear. The human ear cannot give an absolute value to the loudness of a single sound, but can only make comparisons between two different sounds. The precise measure of the power of a sound wave at a particular point is called its ◊intensity. Accurate comparisons of sound levels may be made using sound-level meters, which are calibrated in units called ◊decibels.

loudspeaker electromechanical device that converts electrical signals into sound waves, which are radiated into the air. The most common type of loudspeaker is the *moving-coil speaker*. Electrical signals from, for example, a radio are fed to a coil of fine wire wound around the top of a cone. The coil is surrounded by a magnet. When signals pass through it, the coil becomes an electromagnet, which by moving causes the cone to vibrate, setting up sound waves.

Lowell Observatory US astronomical observatory founded by Percival Lowell at Flagstaff, Arizona, with a 61–cm/24–in refractor opened in 1896. The observatory now operates other telescopes at a nearby site on Anderson Mesa including the 1.83–m/72–in Perkins reflector of Ohio State and Ohio Wesleyan universities.

low-level language in computing, a programming language designed for a particular computer and reflecting its internal ◊machine code; low-level languages are therefore often described as *machine-oriented* languages. They cannot easily be converted to run on a computer with a different central processing unit, and they are relatively difficult to learn because a detailed knowledge of the internal working of the computer is required. Since they must be translated into machine code by an ◊assembler program, low-level languages are also called ◊assembly languages.

A mnemonic-based low-level language replaces binary machine-code instructions, which are very hard to remember, write down, or correct, with short codes chosen to remind the programmer of the instructions they represent. For example, the binary-code instruction that means 'store the contents of the accumulator' may be represented with the mnemonic STA.

In contrast, ◊high-level languages are designed to solve particular problems and are therefore described as *problem-oriented languages*.

LSI (abbreviation for *large-scale integration*) the technology that enables whole electrical circuits to be etched into a piece of semiconducting material just a few millimetres square.

By the late 1960s a complete computer processor could be integrated on a single chip, or ◊integrated circuit, and in 1971 the US electronics company Intel produced the first commercially available ◊microprocessor. Very large-scale integration (◊VLSI) results in even smaller chips.

lubricant substance used between moving surfaces to reduce friction. Carbon-based (organic) lubricants, commonly called grease and oil, are recovered from petroleum distillation.

Extensive research has been carried out on chemical additives to lubricants, which can reduce corrosive wear, prevent the accumulation of 'cold sludge' (often the result of stop-start driving in city traffic jams), keep pace with the higher working temperatures of aviation gas turbines, or provide radiation-resistant greases for nuclear power plants. Silicon-based spray-on lubricants are also used; they tend to attract dust and dirt less than carbon-based ones.

A solid lubricant is graphite, an allotropic form of carbon, either flaked or emulsified (colloidal) in water or oil.

lumen SI unit (symbol lm) of luminous flux (the amount of light passing through an area per second).

The lumen is defined in terms of the light falling on a unit area at a unit distance from a light source of luminous intensity of one ◊candela. One lumen at a wavelength of 5,550 angstroms equals 0.0014706 watts.

luminescence emission of light from a body when its atoms are excited by means other than raising its temperature. Short-lived luminescence is called fluorescence; longer-lived luminescence is called phosphorescence.

When exposed to an external source of energy, the outer electrons in atoms of a luminescent substance absorb energy and 'jump' to a higher energy level. When these electrons 'jump' back to their former level they emit their excess energy as light. Many different exciting mechanisms are possible: visible light or other forms of **electromagnetic**

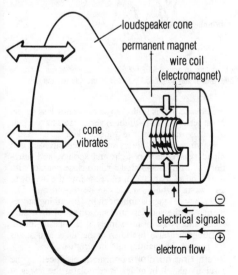

loudspeaker cone

permanent magnet

wire coil (electromagnet)

cone vibrates

electrical signals

electron flow

loudspeaker *A moving-coil loudspeaker. Electrical signals flowing through the wire coil turn it into an electromagnet, which moves as the signals vary. The attached cone vibrates, producing sound waves.*

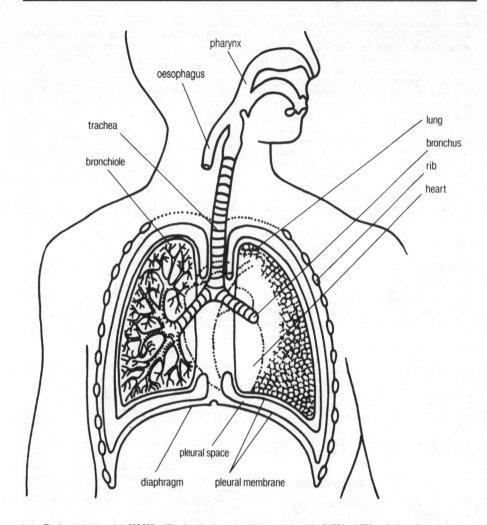

lung *The human lungs contain 300,000 million tiny blood vessels which would stretch for 2,400 km/1,500 mi if laid end to end. A healthy adult at rest breathes 12 times a minute; a baby breathes at twice this rate. Each breath brings 350 millilitres of fresh air into the lungs, and expels 150 millilitres of stale air from the nose and throat.*

radiation (ultraviolet rays or X-rays), electron bombardment, chemical reactions, friction, and ◊radioactivity. Certain living organisms produce ◊bioluminescence.

luminosity or *brightness* the amount of light emitted by a star, measured in ◊magnitudes. The apparent brightness of an object decreases in proportion to the square of its distance from the observer. The luminosity of a star or other body can be expressed in relation to that of the Sun.

luminous paint preparation containing a mixture of pigment, oil, and a phosphorescent sulphide, usually calcium or barium. After exposure to light it is luminous in the dark. The luminous paint used on watch faces contains radium, is radioactive and therefore does not require exposure to light.

lung large cavity of the body, used for ◊gas exchange. It is essentially a sheet of thin, moist membrane that is folded so as to occupy less space. Most tetrapod (four-limbed) vertebrates have a pair of lungs occupying the thorax. The lung tissue, consisting of multitudes of air sacs (alveoli) and blood vessels, is very light and spongy, and functions by bringing inhaled air into close contact with the blood so that oxygen can pass into the organism and waste carbon dioxide can be passed out. The efficiency of lungs is enhanced by ◊breathing movements, by the thinness and moistness of their surfaces, and by a constant supply of circulating blood.

Lungs are also found in some slugs and snails (as part of the mantle) and in lungfish.

luteinizing hormone ◊hormone produced by the pituitary gland. In males, it stimulates the testes to produce androgens (male sex hormones). In females, it works together with follicle-stimulating hormone to initiate production of egg cells by the ovary. If fertilization occurs, it plays a part in maintaining the pregnancy by controlling the levels of the hormones oestrogen and progesterone.

lutetium (Latin *Lutetia* 'Paris') silver-white, metallic element, the last of the ◊lanthanide series, symbol Lu, atomic number 71, relative atomic mass 174.97. It is used in the 'cracking', or breakdown, of petroleum and in other chemical processes. It was named by its discoverer, French chemist Georges Urbain, (1872–1938) after his native city.

lux SI unit (symbol lx) of illuminance or illumination (the light falling on an object). It is equivalent to one ◊lumen per square metre or to the illuminance of a surface one metre distant from a point source of one ◊candela.

The lux replaces the foot-candle (one foot-candle equals 10.76 lux).

LW abbreviation for *long wave*, a radio wave with a wavelength of over 1,000 m/3,300 ft; one of the main wavebands into which radio frequency transmissions are divided.

lymph fluid found in the lymphatic system of vertebrates.

Lymph is drained from the tissues by lymph capillaries, which empty into larger lymph vessels (lymphatics). These lead to lymph nodes (small, round bodies chiefly situated in the neck, armpit, groin, thorax, and abdomen), which process the ◊lymphocytes produced by the bone marrow, and filter out harmful substances and bacteria. From the lymph nodes, vessels carry the lymph to the thoracic duct and the right lymphatic duct, which lead into the large veins in the neck. Some vertebrates, such as amphibians, have a lymph heart, which pumps lymph through the lymph vessels.

It carries nutrients, oxygen, and white blood cells to the tissues, and waste matter away from them.

It exudes from capillaries into the tissue spaces between the cells and is made up of blood plasma, plus white cells.

lymph nodes small masses of lymphatic tissue in the body that occur at various points along the major lymphatic vessels. Tonsils and adenoids are large lymph nodes. As the lymph passes through them it is filtered, and bacteria and other microorganisms are engulfed by cells known as macrophages.

Lymph nodes are sometimes mistakenly called 'lymph glands', and the term 'swollen glands' refers to swelling of the lymph nodes caused by infection.

lymphocyte type of white blood cell with a large nucleus, produced in the bone marrow. Most occur in the ◊lymph and blood, and around sites of infection. *B-lymphocytes* or ◊B cells are responsible for producing ◊antibodies. *T-lymphocytes* or ◊T cells have several roles in the formation of ◊immunity.

lymphokine chemical messenger produced by a lymphocyte that carries messages between the cells of the immune system (see ◊immunity). Examples of lymphokines include interferon, which initiates defensive reactions to viruses, and the interleukins, which activate specific immune cells.

lyophilization technical term for the ◊freeze-drying process used for foods and drugs and in the preservation of organic archaeological remains.

Lyra small but prominent constellation of the northern hemisphere, representing the lyre of Orpheus. Its brightest star is ◊Vega.

Epsilon Lyrae, the 'double double', is a system of four gravitationally linked stars. Beta Lyrae is an eclipsing binary. The Ring nebula, M57, is a ◊planetary nebula.

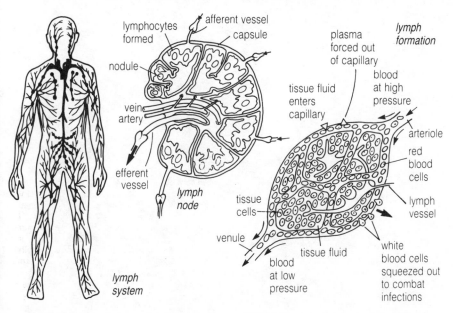

lymph *Lymph is the fluid that carries nutrients, oxygen, and white blood cells to the tissues. Lymph enters the tissue from the capillaries (right) and is drained from the tissues by lymph vessels. The lymph vessels form a network (left) called the lymphatic system. At various points in the lymphatic system, lymph nodes (centre) filter and clean the lymph.*

lysis in biology, any process that destroys a cell by rupturing its membrane or cell wall (see ◊lysosome).

lysosome membrane-enclosed structure, or organelle, inside a ◊cell, principally found in animal cells. Lysosomes contain enzymes that can break down proteins and other biological substances. They play a part in digestion, and in the white blood cells known as phagocytes the lysosome enzymes attack ingested bacteria.

M

m symbol for ◊*metre*.

McDonald Observatory observatory of the University of Texas on Mount Locke, Texas. It is the site of a 2.72–m/107–in reflector opened 1969 and a 2.08–m/82–in reflector opened 1939.

machine device that allows a small force (the effort) to overcome a larger one (the load). There are three basic machines: the inclined plane (ramp), the lever, and the wheel and axle. All other machines are combinations of these three basic types. Simple machines derived from the inclined plane include the wedge, the gear, and the screw; the spanner is derived from the lever; the pulley from the wheel.

The principal features of a machine are its ◊mechanical advantage, which is the ratio of load to effort, its ◊velocity ratio, and its ◊efficiency, which is the work done by the load divided by the work done by the effort; the latter is expressed as a percentage. In a perfect machine, with no friction, the efficiency would be 100%. All practical machines have efficiencies of less than 100%, otherwise perpetual motion would be possible.

machine code in computing, a set of instructions that a computer's central processing unit (CPU) can understand and obey directly, without any translation. Each type of CPU has its own machine code. Because machine-code programs consist entirely of binary digits (bits), most programmers write their programs in an easy-to-use ◊high-level language. A high-level program must be translated into machine code—by means of a ◊compiler or ◊interpreter program—before it can be executed by a computer.

Where no suitable high-level language exists or where very efficient machine code is required, programmers may choose to write programs in a low-level, or assembly, language, which is eventually translated into machine code by means of an ◊assembler program.

Microprocessors (CPUs based on a single integrated circuit) may be classified according to the number of machine-code instructions that they are capable of obeying: ◊CISC (complex instruction set computer) microprocessors support up to 200 instructions, whereas ◊RISC (reduced instruction set computer) microprocessors support far fewer instructions but execute programs more rapidly.

machine tool automatic or semi-automatic power-driven machine for cutting and shaping metals. Machine tools have powerful electric motors to force cutting tools into the metal: these are made from hardened steel containing heat-resistant metals such as tungsten and chromium. The use of precision machine tools in mass-production assembly methods ensures that all duplicate parts produced are virtually identical.

Many machine tools now work under computer control and are employed in factory automation. The most common machine tool is the ◊lathe, which shapes shafts and similar objects. A ◊milling machine cuts metal with a rotary toothed cutting wheel. Other machine tools cut, plane, grind, drill, and polish.

Mach number ratio of the speed of a body to the speed of sound in the undisturbed medium through which the body travels. Mach 1 is reached when a body (such as an aircraft) has a velocity greater than that of sound ('passes the sound barrier'), namely 331 m/1,087 ft per second at sea level. It is named after Austrian physicist Ernst Mach (1838–1916).

Macintosh range of microcomputers produced by Apple Computers. The Apple Macintosh, introduced in 1984, was the first popular microcomputer with a ◊graphical user interface.

The success of the Macintosh prompted other manufacturers and software companies to create their own graphical user interfaces. Most notable of these are Microsoft Windows, which runs on IBM PC-compatible microcomputers, and OSF/Motif, from the Open Software Foundation, which is used with many Unix systems.

macro in computer programming, a new command created by combining a number of existing ones. For example, if a programming language has separate commands for obtaining data from the keyboard and for displaying data on the screen, the programmer might create a macro that performs both these tasks with one command. A *macro key* on the keyboard combines the effects of pressing several individual keys.

macromolecule in chemistry, a very large molecule, generally a ◊polymer.

macrophage type of ◊white blood cell, or leucocyte, found in all vertebrate animals. Macrophages specialize in the removal of bacteria and other microorganisms, or of cell debris after injury. Like phagocytes, they engulf foreign matter, but they are larger than phagocytes and have a longer life span. They are found throughout the body, but mainly in the lymph and connective tissues, and especially the lungs, where they ingest dust, fibres, and other inhaled particles.

Magellan NASA space probe to ◊Venus, launched May 1989; it went into orbit around Venus Aug 1990 to make a detailed map of the planet by radar. It revealed volcanoes, meteorite craters, and fold mountains on the planet's surface.

Magellanic Clouds in astronomy, the two galaxies nearest to our own galaxy. They are irregularly shaped, and appear as detached parts of the ◊Milky Way, in the southern constellations Dorado and Tucana.

The Large Magellanic Cloud is 169,000 light years from Earth, and about a third the diameter of our galaxy; the Small Magellanic Cloud, 180,000 light years away, is about a fifth the diameter of our galaxy. They are named after the navigator Ferdinand Magellan, who first described them.

magic numbers in atomic physics certain numbers of ◊neutrons or ◊protons (2, 8, 20, 28, 50, 82, 126) in the nuclei of elements of outstanding stability, such as lead and helium. Such stability is the result of neutrons and protons being arranged in completed 'layers' or 'shells'.

magic square in mathematics, a square array of different numbers in which the rows, columns, and diagonals add up to the same total. A simple example employing the numbers 1 to 9, with a total of 15, is:

$$
\begin{array}{ccc}
6 & 7 & 2 \\
1 & 5 & 9 \\
8 & 3 & 4 \\
\end{array}
$$

maglev (acronym from *mag*netic *lev*itation) high-speed surface transport using the repellent force of superconductive magnets (see ◊superconductivity) to propel and support, for example, a train above a track.

Technical trials on a maglev train track began in Japan in the 1970s, and a speed of 500 kph/310 mph has been reached, with a cruising altitude of 10 cm/4 in. The train is levitated by electromagnets and forward thrust is provided by linear motors aboard the cars, propelling the train along a reaction plate. A small low-power maglev monorail system is in use at Birmingham Airport, England. A subway line using maglev carriages began operating in Osaka, Japan, 1990.

magma molten rock material beneath the Earth's surface from which ◊igneous rocks are formed.

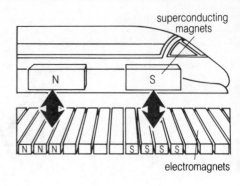

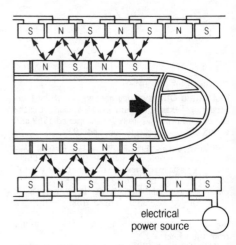

maglev *The repulsion of superconducting magnets and electromagnets in the track keeps a maglev train suspended above the track. By varying the strength and polarity of the track electromagnets, the train can be driven forward.*

◊Lava is magma that has reached the surface and solidified, losing some of its components on the way.

magnesia common name for ◊magnesium oxide.

magnesium lightweight, very ductile and malleable, silver-white, metallic element, symbol Mg, atomic number 12, relative atomic mass 24.305. It is one of the ◊alkaline-earth metals, and the lightest of the commonly used metals. Magnesium silicate, carbonate, and chloride are widely distributed in nature. The metal is used in alloys and flash photography. It is a necessary trace element in the human diet, and green plants cannot grow without it since it is an essential constituent of the photosynthetic pigment ◊chlorophyll ($C_{55}H_{72}MgN_4O_5$).

It was named after the ancient Greek city of Magnesia, near where it was first found. It was first recognized as an element by Scottish chemist Joseph Black 1755 and discovered in its oxide by English chemist Humphry Davy 1808. Pure magnesium was isolated 1828 by French chemist Antoine-Alexandre-Brutus Bussy.

magnesium carbonate $MgCO_3$ white solid that

occurs in nature as the mineral magnesite. It is a commercial antacid, and the anhydrous form is used as a drying agent in table salt. When rainwater containing dissolved carbon dioxide flows over magnesite rocks, the carbonate dissolves to form magnesium hydrogencarbonate, one of the causes of temporary hardness in water.

$$H_2O + CO_2 + MgCO_3 \rightarrow Mg(HCO_3)_2$$

magnesium oxide or *magnesia* MgO white powder or colourless crystals, formed when magnesium is burned in air or oxygen; a typical basic oxide. It is used to treat acidity of the stomach, and in some industrial processes; for example, as a lining brick in furnaces, because it is very stable when heated (refractory oxide).

magnesium sulphate $MgSO_4$ white solid that occurs in nature in several hydrated forms. The heptahydrate $MgSO_4.7H_2O$ is known as Epsom Salts and is used as a laxative. Industrially it is used in tanning leather and in the manufacture of fertilizers and explosives.

magnet any object that forms a magnetic field (displays ◊magnetism), either permanently or temporarily through induction, causing it to attract materials such as iron, cobalt, nickel, and alloys of these. It always has two ◊magnetic poles, called north and south.

magnetic declination see ◊angle of declination.

magnetic dip see ◊dip, magnetic and ◊angle of dip.

magnetic dipole the pair of north and south magnetic poles, separated by a short distance, that makes up all magnets. Individual magnets are often called 'magnetic dipoles'. Single magnetic poles, or monopoles, have never been observed despite being searched for. See also magnetic ◊domain.

magnetic field region around a permanent magnet, or around a conductor carrying an electric current, in which a force acts on a moving charge or on a magnet placed in the field. The field can

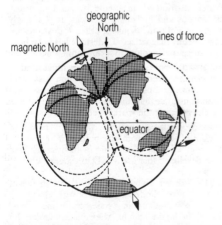

magnetic field *The Earth's magnetic field is similar to that of a bar magnet with poles near, but not exactly at, the geographic poles. Compass needles align themselves with the magnetic field, which is horizontal near the equator and vertical at the magnetic poles.*

be represented by lines of force, which by convention link north and south poles and are parallel to the directions of a small compass needle placed on them. Its magnitude and direction are given by the ◊magnetic flux density, expressed in ◊teslas.

magnetic flux measurement of the strength of the magnetic field around electric currents and magnets. Its SI unit is the ◊weber; one weber per square metre is equal to one tesla.

The amount of magnetic flux through an area equals the product of the area and the magnetic field strength at a point within that area. It is a measure of the number of magnetic field lines passing through the area.

magnetic induction the production of magnetic properties in unmagnetized iron or other ferromagnetic material when it is brought close to a magnet. The material is influenced by the magnet's magnetic field and the two are attracted. The induced magnetism may be temporary, disappearing as soon as the magnet is removed, or permanent depending on the nature of the iron and the strength of the magnet.

◊Electromagnets make use of temporary induced magnetism to lift sheets of steel: the magnetism induced in the steel by the approach of the electromagnet enables it to be picked up and transported. To release the sheet, the current supplying the electromagnet is temporarily switched off and the induced magnetism disappears.

magnetic-ink character recognition (MICR) in computing, a technique that enables special characters printed in magnetic ink to be read and input rapidly to a computer. MICR is used extensively in banking because magnetic-ink characters are difficult to forge and are therefore ideal for marking and identifying cheques.

magnetic material one of a number of substances that are strongly attracted by magnets and can be magnetised. These include iron, nickel, and cobalt, and all those ◊alloys that contain a proportion of these metals.

Soft magnetic materials can be magnetized very easily, but the magnetism induced in them (see ◊magnetic induction) is only temporary. They include Stalloy, an alloy of iron with 4% silicon used to make the cores of electromagnets and transformers, and the materials used to make 'iron' nails and paper clips.

Hard magnetic materials can be permanently magnetized by a strong magnetic field. Steel and special alloys such as Alcomax, Alnico, and Ticonal, which contain various amounts of aluminium, nickel, cobalt, and copper, are used to make permanent magnets. The strongest permanent magnets are ceramic, made under high pressure and at high temperature from powders of various metal oxides.

magnetic pole region of a magnet in which its magnetic properties are strongest. Every magnet has two poles, called north and south. The north (or north-seeking) pole is so named because a freely suspended magnet will turn so that this pole points towards the Earth's magnetic north pole. The north pole of one magnet will be attracted to the south pole of another, but will be repelled by its north pole. Like poles may therefore be said to attract, unlike poles to repel.

magnetic resonance imaging (MRI) diagnostic scanning system based on the principles of nuclear magnetic resonance. MRI yields finely detailed three-dimensional images of structures within the body without exposing the patient to harmful radiation. The technique is useful for imaging the soft tissues of the body, such as the brain and the spinal cord.

Claimed as the biggest breakthrough in diagnostic imaging since the discovery of X-rays, MRI is a noninvasive technique using the principle that atomic nuclei in a strong magnetic field can be made to give off electromagnetic radiation, the characteristics of which depend on the environment of the nuclei.

magnetic storm in meteorology, a sudden disturbance affecting the Earth's magnetic field, causing anomalies in radio transmissions and magnetic compasses. It is probably caused by ◊sunspot activity.

magnetic strip or *magnetic stripe* thin strip of magnetic material attached to a plastic card (such as a credit card) and used for recording data.

magnetic tape narrow plastic ribbon coated with an easily magnetizable material on which data can be recorded. It is used in sound recording, audiovisual systems (videotape), and computing. For mass storage on commercial mainframe computers, large reel-to-reel tapes are still used, but cartridges are coming in. Various types of cartridge are now standard on minis and PCs, while audio cassettes are sometimes used with home computers.

Magnetic tape was first used in **sound recording** 1947, and made overdubbing possible, unlike the direct-to-disc system it replaced. Two-track tape was introduced in the 1950s and four-track in the early 1960s; today, studios use 16-, 24-, or 32-track tape, from which the tracks are mixed down to a stereo master tape.

In computing, magnetic tape was first used to record data and programs in 1951 as part of the UNIVAC 1 system. It was very popular as a storage medium for external memory in the 1950s and 1960s. Since then it has been largely replaced by magnetic ◊discs as a working medium, although tape is still used to make backup copies of important data. Information is recorded on the tape in binary form, with two different strengths of signal representing 1 and 0.

magnetism phenomena associated with ◊magnetic fields. Magnetic fields are produced by moving charged particles: in electromagnets, electrons flow through a coil of wire connected to a battery; in permanent magnets, spinning electrons within the atoms generate the field.

Substances differ in the extent to which they can be magnetized by an external field (susceptibility). Materials that can be strongly magnetized, such as iron, cobalt, and nickel, are said to be **ferromagnetic**; this is due to the formation of areas called ◊domains in which atoms, weakly magnetic because of their spinning electrons, align to form areas of strong magnetism. Magnetic materials lose their magnetism if heated to the ◊Curie temperature. Most other materials are **paramagnetic**, being

only weakly pulled towards a strong magnet. This is because their atoms have a low level of magnetism and do not form domains. *Diamagnetic* materials are weakly repelled by a magnet since electrons within their atoms act as electromagnets and oppose the applied magnetic force. *Antiferromagnetic* materials have a very low susceptibility that increases with temperature; a similar phenomenon in materials such as ferrites is called *ferrimagnetism*.

Apart from its universal application in dynamos, electric motors, and switch gears, magnetism is of considerable importance in advanced technology— for example, in particle ◊accelerators for nuclear research, memory stores for computers, tape recorders, and ◊cryogenics.

MAGNETISM PUZZLE

The lines of force around a bar magnet are said to be the path that an isolated pole would follow, but magnetic poles of course always come in pairs, one north and one south. The concept of an isolated pole is therefore hypothetical—or is it? Consider the following: Take a steel sphere and cut it through three planes at right-angles, so that it forms eight equal pieces, each of which has a curved outer surface at one end and a point at the other, corresponding to the centre of the sphere. Arrange suitable holes so that the pieces can be bolted together again, and magnetize each piece so that the curved surface is a north pole and the point is a south pole. Now bolt the sphere together again so that the entire outer surface is a north pole. Will it now behave as an isolated north pole? *See page 651 for the answer.*

---

magnetite black iron ore, iron oxide (Fe_3O_4). Widely distributed, magnetite is found in nearly all igneous and metamorphic rocks. It is strongly magnetic and some deposits, called *lodestone*, are permanently magnetized. Lodestone has been used as a compass since the first millennium BC.

magneto simple electric generator, often used to provide the electricity for the ignition system of motorcycles and used in early cars. It consists of a rotating magnet that sets up an electric current in a coil, providing the spark.

magnetohydrodynamics (MHD) field of science concerned with the behaviour of ionized gases or liquid in a magnetic field. Systems have been developed that use MHD to generate electrical power.

MND driven ships have been tested in Japan. In 1991 two cylindrical thrusters with electrodes and niobium–titanium superconducting coils, soaked in liquid helium, were placed under the passenger boat *Yamato 1*. The boat, 30 m/100 ft long, was designed to travel at 8 knots. An electric current passed through the electrodes accelerates water through the thrusters, like air through a jet engine, propelling the boat forward.

magnetomotive force or *magnetic potential* the work done in carrying a unit magnetic pole around a magnetic circuit. The concept is analogous to ◊electromotive force in an electric circuit.

magnetosphere volume of space, surrounding a planet, controlled by the planet's magnetic field, and acting as a magnetic 'shell'. The Earth's magnetosphere extends 64,000 km/40,000 mi towards the Sun, but many times this distance on the side away from the Sun.

The extension away from the Sun is called the *magnetotail*. The outer edge of the magnetosphere is the *magnetopause*. Beyond this is a turbulent region, the *magnetosheath*, where the ◊solar wind is deflected around the magnetosphere. Inside the magnetosphere, atomic particles follow the Earth's lines of magnetic force. The magnetosphere contains the ◊Van Allen radiation belts. Other planets have magnetospheres, notably Jupiter.

magnetron thermionic ◊valve (electron tube) for generating very high-frequency oscillations, used in radar and to produce microwaves in a microwave oven. The flow of electrons from the tube's cathode to one or more anodes is controlled by an applied magnetic field.

magnification measure of the enlargement or reduction of an object in an imaging optical system. *Linear magnification* is the ratio of the size (height) of the image to that of the object. *Angular magnification* is the ratio of the angle subtended at the observer's eye by the image to the angle subtended by the object when viewed directly.

magnitude in astronomy, measure of the brightness of a star or other celestial object. The larger the number denoting the magnitude, the fainter the object. Zero or first magnitude indicates some of the brightest stars. Still brighter are those of negative magnitude, such as Sirius, whose magnitude is −1.46. *Apparent magnitude* is the brightness of an object as seen from Earth; *absolute magnitude* is the brightness at a standard distance of 10 parsecs (32.6 light years).

Each magnitude step is equal to a brightness difference of 2.512 times. Thus a star of magnitude 1 is $(2.512)^5$ or 100 times brighter than a sixth-magnitude star just visible to the naked eye. The apparent magnitude of the Sun is −26.8, its absolute magnitude +4.8.

Magnox early type of nuclear reactor used in the UK, for example in Calder Hall, the world's first commercial nuclear power station. This type of reactor uses uranium fuel encased in tubes of magnesium alloy called Magnox. Carbon dioxide gas is used as a coolant to extract heat from the reactor core. See also ◊nuclear energy.

mail merge in computing, a feature offered by some word-processing packages that enables a list of personal details, such as names and addresses, to be combined with a general document outline to produce individualized documents.

For example, a club secretary might create a file containing a mailing list of the names and addresses of the club members. Whenever a letter is to be sent to all club members, a general letter outline is prepared with indications as to where individual names and addresses need to be added. The mail-merge feature then combines the file of names and addresses with the letter outline to produce and print individual letters addressed to each club member.

mainframe large computer used for commercial data processing and other large-scale operations. Because of the general increase in computing power, the differences between the mainframe, ◊supercomputer, ◊minicomputer, and ◊microcomputer (personal computer) are becoming less marked.

Mainframe manufacturers include IBM, Amdahl, Fujitsu, and Hitachi. Typical mainframes have from 32 to 256 Mb of memory and tens of gigabytes of disc storage.

main sequence in astronomy, the part of the ◊Hertzsprung–Russell diagram that contains most of the stars, including the Sun. It runs diagonally from the top left of the diagram to the lower right. The most massive (and hence brightest) stars are at the top left, with the least massive (coolest) stars at the bottom right.

malachite common ◊copper ore, basic copper carbonate, $Cu_2CO_3(OH)_2$. It is a source of green pigment and is polished for use in jewellery, ornaments, and art objects.

malaria infectious parasitic disease of the tropics transmitted by mosquitoes, marked by periodic fever and an enlarged spleen. When a female mosquito of the *Anopheles* genus bites a human who has malaria, it takes in with the human blood one of four malaria protozoa of the genus *Plasmodium*. This matures within the insect and is then transferred when the mosquito bites a new victim. Malaria affects some 200 million people a year on a recurring basis.

Inside the human body the parasite settles first in the liver, then multiplies to attack the red blood cells. Within the red blood cells the parasites multiply, eventually causing the cells to rupture and other cells to become infected. The cell rupture tends to be synchronized, occurring every 2–3 days, when the symptoms of malaria become evident. ◊Quinine was the first drug used against malaria, now replaced by synthetics, such as atabrine to prevent the disease and chloroquine to treat it. Tests on a vaccine were begun 1986 in the USA. In Brazil a malaria epidemic broke out among new settlers in the Amazon region, with 287,000 cases 1983 and 500,000 cases 1988.

malic acid $COOHCH_2CH(OH)COOH$ organic crystalline acid that can be extracted from apples, plums, cherries, grapes, and other fruits, but occurs in all living cells in smaller amounts, being one of the intermediates of the ◊Krebs cycle.

maltase enzyme found in plants and animals that breaks down the disaccharide maltose into glucose.

maltose $C_{12}H_{22}O_{11}$ a ◊disaccharide sugar in which both monosaccharide units are glucose.

It is produced by the enzymic hydrolysis of starch and is a major constituent of malt, produced in the early stages of beer and whisky manufacture.

mammal an animal characterized by having mammary glands in the female; these are used for suckling the young. Other features of mammals are

◊hair (very reduced in some species, such as whales); a middle ear formed of three small bones (ossicles); a lower jaw consisting of two bones only; seven vertebrae in the neck; and no nucleus in the red blood cells.

Mammals are divided into three groups: *placental mammals*, where the young develop inside the ◊uterus, receiving nourishment from the blood of the mother via the ◊placenta; *marsupials*, where the young are born at an early stage of development and develop further in a pouch on the mother's body; *monotremes*, where the young hatch from an egg outside the mother's body and are then nourished with milk. The monotremes are the least evolved and have been largely displaced by more sophisticated marsupials and placentals, so that there are only a few types surviving (platypus and echidna). Placentals are considered the most sophisticated and have spread to all parts of the globe. Where placentals have competed with marsupials, the placentals have in general displaced marsupial types. However, marsupials occupy many specialized niches in South America and, especially, Australasia. (The theory that marsupials succeed only where they do not compete with placentals was shaken in 1992, when a tooth, 55 million years old and belonging to a placental mammal, was found in Murgon, Australia, indicating that placental animals appeared in Australia at the same time as the marsupials. The marsupials, however, still prevailed.) There are over 4,000 species of mammals, adapted to almost every way of life. The smallest shrew weighs only 2 g/0.07 oz, the largest whale up to 140 tonnes.

mammals: classification

order	typical species
Monotremata	echidna, platypus
Marsupiala	kangaroo, koala, opossum
Insectivora	shrew, hedgehog, mole
Chiroptera	bat
Primates	lemur, monkey, ape, human
Edentata	anteater, armadillo, sloth
Pholidota	pangolin
Dermoptera	flying lemur
Rodentia	rat, mouse, squirrel, porcupine
Lagomorpha	rabbit, hare, pika
Cetacea	whale, dolphin
Carnivora	cat, dog, weasel, bear
Pinnipedia	seal, walrus
Artiodactyla	pig, deer, cattle, camel, giraffe
Perissodactyla	horse, rhinocerous, tapir
Sirenia	dugong, manatee
Tubulidentata	aardvark
Hyracoidea	hyrax
Proboscidea	elephant

mammary gland in female mammals, a milk-producing gland derived from epithelial cells underlying the skin, active only after the production of young. In all but monotremes (egg-laying mammals), the mammary glands terminate in teats which aid infant suckling. The number of glands and their position vary between species. In humans there are two, in cows four, and in pigs between ten and fourteen.

The hatched young of monotremes simply lick milk from a specialized area of skin on the mother's abdomen.

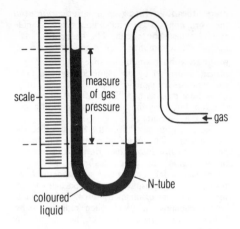

manometer *The manometer indicates gas pressure by the rise of liquid in the tube.*

mammoth extinct elephant of genus *Mammuthus*, whose remains are found worldwide. Some were 50% taller than modern elephants.

The woolly mammoth *Elephas primigenius* of northern zones, the size of an Indian elephant, had long fur and large inward-curving tusks. Various species of mammoth were abundant in both the Old World and the New World in Pleistocene times, and were hunted by humans for food.

manganese hard, brittle, grey-white metallic element, symbol Mn, atomic number 25, relative atomic mass 54.9380. It resembles iron (and rusts), but it is not magnetic and is softer. It is used chiefly in making steel alloys, also alloys with aluminium and copper. It is used in fertilizers, paints, and industrial chemicals. It is a necessary trace element in human nutrition. The name is old, deriving from the French and Italian forms of Latin for *magnesia* (MgO), the white tasteless powder used as an antacid from ancient times.

manganese ore any mineral from which manganese is produced. The main ores are the oxides, such as *pyrolusite*, MnO_2; *hausmannite*, Mn_3O_4; and *manganite*, MnO(OH).

Manganese ores may accumulate in metamorphic rocks or as sedimentary deposits, frequently forming nodules on the sea floor (since the 1970s many schemes have been put forward to harvest deep-sea manganese nodules). The world's main producers are Georgia, Ukraine, South Africa, Brazil, Gabon, and India.

manganese(IV) oxide or *manganese dioxide* MnO_2 brown solid that acts as a ◊depolarizer in dry batteries by oxidizing the hydrogen gas produced to water; without this process, the performance of the battery is impaired.

manometer instrument for measuring the pressure of liquids (including human blood pressure) or gases. In its basic form, it is a U-tube partly filled with coloured liquid; pressure of a gas entering at one side is measured by the level to which the liquid rises at the other.

mantissa in mathematics, the decimal part of a ◊logarithm. For example, the logarithm of 347.6 is

2.5411; in this case, the 0.5411 is the mantissa, and the integral (whole number) part of the logarithm, the 2, is the ◊characteristic.

mantle intermediate zone of the Earth between the crust and the core, accounting for 82% of the Earth's volume. It is thought to consist of silicate minerals such as olivine.

The mantle is separated from the crust by the ◊Mohorovičić discontinuity, and from the core by the Gutenberg discontinuity. The patterns of seismic waves passing through it show that its uppermost as well as its lower layers are solid. However, from 72 km/45 mi to 250 km/155 mi in depth is a zone through which seismic waves pass more slowly (the 'low-velocity zone'). The inference is that materials in this zone are close to their melting points and they are partly molten. The low-velocity zone is considered the ◊asthenosphere on which the solid lithosphere rides.

map diagrammatic representation of an area—for example, part of the Earth's surface or the distribution of the stars. Modern maps of the Earth are made using satellites in low orbit to take a series of overlapping stereoscopic photographs from which a three-dimensional image can be prepared. The earliest accurate large-scale maps appeared about 1580.

Conventional aerial photography, laser beams, microwaves, and infrared equipment are also used for land surveying. Many different kinds of ◊map projection (the means by which a three-dimensional body is shown in two dimensions) are used in map-making. Detailed maps requiring constant updating are kept in digital form on computer so that minor revisions can be made without redrafting.

The Ordnance Survey is the official body responsible for the mapping of Britain; it produces maps in a variety of scales, such as the Landranger series (scale 1:50,000). Large-scale maps—for example, 1:25,000—show greater detail at a local level than small-scale maps; for example, 1:100,000.

map projection ways of depicting the spherical surface of the Earth on a flat piece of paper. Traditional projections include the *conic*, *azimuthal*, and *cylindrical*. The most common cylindrical projection is the Mercator projection, which dates from 1569. The weakness of these systems is that countries in different latitudes are disproportionately large, and lines of longitude and latitude appear distorted. In 1973 German historian Arno Peters devised the *Peters projection* in which countries of the world retain their relative areas.

The theory behind traditional map projection is that, if a light were placed at the centre of a transparent Earth, the surface features could be thrown as shadows on a piece of paper close to the surface. This paper may be flat and placed on a pole (azimuthal or zenithal), or may be rolled around the equator (cylindrical), or may be in the form of a tall cone resting on the equator (conical). The resulting maps differ from one another, distorting either area or direction, and each is suitable for a particular purpose. For example, projections distorting area the least are used for distribution maps,

and those with least distortion of direction are used for navigation charts.

marble metamorphosed ◊limestone that takes and retains a good polish; it is used in building and sculpture. In its pure form it is white and consists almost entirely of calcite $CaCO_3$. Mineral impurities give it various colours and patterns. Carrara, Italy, is known for white marble.

mare (plural *maria*) dark lowland plain on the Moon. The name comes from Latin 'sea', because these areas were once wrongly thought to be water.

margarine butter substitute made from animal fats and/or vegetable oils. The French chemist Hippolyte Mège-Mouries invented margarine 1889. Today, margarines are usually made with vegetable oils, such as soy, corn, or sunflower oil, giving a product low in saturated fats (see ◊polyunsaturate) and fortified with vitamins A and D.

Mariana Trench lowest region on the Earth's surface; the deepest part of the sea floor. The trench is 2,400 km/1,500 mi long and is situated 300 km/200 mi E of the Mariana Islands, in the NW Pacific Ocean. Its deepest part is the gorge known as the Challenger Deep, which extends 11,034 m/36,210 ft below sea level.

MARIANA TRENCH: SLOW FALL

A heavy object dropped into the ocean over the 11–km/6.7 mi-deep Mariana Trench in the Pacific would take over one hour to sink to the bottom.

Mariner spacecraft series of US space probes that explored the planets Mercury, Venus, and Mars 1962–75.

Mariner 1 (to Venus) had a failed launch. *Mariner 2* 1962 made the first fly-by of Venus, at 34,000 km/21,000 mi, confirmed the existence of ◊solar wind, and measured Venusian temperature. *Mariner 3* did not achieve its intended trajectory to Mars. *Mariner 4* 1965 passed Mars at a distance of 9,800 km/6,100 mi, and took photographs, revealing a dry, cratered surface. *Mariner 5* 1967 passed Venus at 4,000 km/2,500 mi, and measured Venusian temperature, atmosphere, mass, and diameter. *Mariner 6* and *7* 1969 photographed Mars' equator and southern hemisphere respectively, and also measured temperature, atmospheric pressure and composition, and diameter. *Mariner 8* (to Mars) had a failed launch. *Mariner 9* 1971 mapped the entire Martian surface, and photographed Mars' moons. Its photographs revealed the changing of the polar caps, and the extent of volcanism, canyons, and features, which suggested that there might once have been water on Mars. *Mariner 10* 1974–75 took close-up photographs of Mercury and Venus, and measured temperature, radiation, and magnetic fields.

Mariner 11 and *12* were renamed *Voyager 1* and *2* (see ◊Voyager probes).

Markov chain in statistics, an ordered sequence of discrete states (random variables) $x_1, x_2, \ldots, x_i, \ldots, x_n$ such that the probability of x_i depends only on n and/or the state x_{i-1} which has preceded it. If independent of n, the chain is said to be homogeneous.

mark sensing in computing, a technique that enables pencil marks made in predetermined positions on specially prepared forms to be rapidly read and input to a computer. The technique makes use of the fact that pencil marks contain graphite and therefore conduct electricity. A *mark sense reader* scans the form by passing small metal brushes over the paper surface. Whenever a brush touches a pencil mark a circuit is completed and the mark is detected.

marl crumbling sedimentary rock, sometimes called *clayey limestone*, including various types of calcareous ◊clays and fine-grained ◊limestones. Marls are often laid down in freshwater lakes and are usually soft, earthy, and of a white, grey, or brownish colour. They are used in cement-making and as fertilizer.

Mars fourth planet from the Sun, average distance 227.9 million km/141.6 million mi. It revolves around the Sun in 687 Earth days, and has a rotation period of 24 hr 37 min. It is much smaller than Venus or Earth, with a diameter 6,780 km/ 4,210 mi, and mass 0.11 that of Earth. Mars is slightly pearshaped, with a low, level northern hemisphere, which is comparatively uncratered and geologically 'young', and a heavily cratered 'ancient' southern hemisphere.

The landscape is a dusty, red, eroded lava plain; red atmospheric dust whipped up by winds of up to 200 kph/125 mph accounts for the light pink sky. Mars has white polar caps (water ice and frozen carbon dioxide) that advance and retreat with the seasons. There are four enormous volcanoes near the equator, of which the largest is Olympus Mons 24 km/15 m high, with a base 600 km/375 mi across, and a crater 65 km/40 mi wide. The atmosphere is 95% carbon dioxide, 3% nitrogen, 1.5% argon, and 0.15% oxygen. Recorded temperatures vary from −100°C/−148°F to 0°C/32°F. The atmospheric pressure is 7 millibars, equivalent to the pressure 35 km/22 mi above Earth. No proof of life on Mars has been obtained. There are two small satellites (moons): ◊Phobos and Deimos.

Mars may approach Earth to within 54.7 million km/34 million mi. The first human-made object to orbit another planet was *Mariner 9*. *Viking 1* and *2*, which landed, also provided much information. Studies in 1985 showed that enough water might exist to sustain prolonged missions by space crews. To the east of the four volcanoes lies a high plateau cut by a system of valleys, Valles Marineris, some 4,000 km/2,500 mi long, up to 200 km/120 mi wide and 6 km/4 mi deep; these features are apparently caused by faulting and wind erosion.

marsh low-lying wetland. Freshwater marshes are common wherever groundwater, surface springs, streams, or run-off causes frequent flooding or more or less permanent shallow water. A marsh is alkaline whereas a ◊bog is acid. Marshes develop on inorganic silt or clay soils. Rushes are typical marsh plants. Large marshes dominated by papyrus, cattail, and reeds, with standing water throughout the year, are commonly called ◊swamps. Near the sea, ◊salt marshes may form.

Marshall Space Flight Center NASA installation at Huntsville, Alabama, where the series of ◊Saturn rockets and the space-shuttle engines were developed. It also manages various payloads for the space shuttle, including the ◊Spacelab space station.

marsh gas gas consisting mostly of ◊methane. It is produced in swamps and marshes by the action of bacteria on dead vegetation.

Mars Observer NASA space probe launched 1992 to orbit Mars and survey the planet, its atmosphere, and the polar caps. Due to go into orbit Aug 1993 and to have a survey period of two years, the probe is also scheduled to communicate information back to Earth from the landers delivered by the Commonwealth of Independent States' Mars-94 mission.

marsupial (Greek *marsupion* 'little purse') mammal in which the female has a pouch where she carries her young (born tiny and immature) for a considerable time after birth. Marsupials include omnivorous, herbivorous, and carnivorous species, among them the kangaroo, wombat, opossum, phalanger, bandicoot, dasyure, and wallaby. Exceptionally, the marsupial anteater *Myrmecobius* has no pouch.

maser (acronym for *m*icrowave *a*mplification by *s*timulated *e*mission of *r*adiation) in physics, a high-frequency microwave amplifier or oscillator in which the signal to be amplified is used to stimulate unstable atoms into emitting energy at the same frequency. Atoms or molecules are raised to a higher energy level and then allowed to lose this energy by radiation emitted at a precise frequency. The principle has been extended to other parts of the electromagnetic spectrum as, for example, in the ◊laser.

The two-level ammonia-gas maser was first suggested 1954 by US physicist Charles Townes at Columbia University, New York, and independently the same year by Nikolai Basov and Aleksandr Prokhorov in Russia. The solid-state three-level maser, the most sensitive amplifier known, was envisaged by Nicolaas Bloembergen (1920–) at Harvard 1956. The ammonia maser is used as a frequency standard oscillator (see ◊clock), and the three-level maser as a receiver for satellite communications and radio astronomy.

mass in physics, the quantity of matter in a body as measured by its inertia. Mass determines the acceleration produced in a body by a given force acting on it, the acceleration being inversely proportional to the mass of the body. The mass also determines the force exerted on a body by ◊gravity on Earth, although this attraction varies slightly from place to place. In the SI system, the base unit of mass is the kilogram.

At a given place, equal masses experience equal gravitational forces, which are known as the weights of the bodies. Masses may, therefore, be compared by comparing the weights of bodies at the same place. The standard unit of mass to which all other masses are compared is a platinum-iridium cylinder of 1 kg, which is kept at the International Bureau of Weights and Measures in Sèvres, France.

mass action, law of in chemistry, a law stating that at a given temperature the rate at which a chemical reaction takes place is proportional to the product of the active masses of the reactants. The active mass is taken to be the molar concentration of the each reactant.

mass–energy equation Albert Einstein's equation $E = mc^2$, denoting the equivalence of mass and energy, where E is the energy in joules, m is the mass in kilograms, and c is the speed of light, in a vacuum, in metres per second.

mass extinction an event that produced the extinction of many species at about the same time. One notable example is the boundary between the Cretaceous and Tertiary periods (known as the ◊K-T boundary) that saw the extinction of the dinosaurs and other big reptiles, and many of the marine invertebrates as well. Mass extinctions have taken place several times during Earth's history.

mass number or *nucleon number* sum (symbol A) of the numbers of protons and neutrons in the nucleus of an atom. It is used along with the ◊atomic number (the number of protons) in ◊nuclear notation: in symbols that represent nuclear isotopes, such as $^{14}_{6}C$, the lower number is the atomic number, and the upper number is the mass number.

mass spectrometer in physics, an apparatus for analysing chemical composition. Positive ions (charged particles) of a substance are separated by an electromagnetic system, which permits accurate measurement of the relative concentrations of the various ionic masses present, particularly isotopes.

mass storage system in computing, a backing-store system, such as a library of magnetic-tape cartridges, capable of storing very large amounts of data.

mastodon any of an extinct family (Mastodontidae) of mammals of the elephant order (Proboscidae). They differed from elephants and mammoths in the structure of their grinding teeth. There were numerous species, among which the American mastodon *Mastodon americanum*, about 3 m/10 ft high, of the Pleistocene era, is well known. They were hunted by humans for food.

mathematical induction formal method of proof in which the proposition $P(n + 1)$ is proved true on the hypothesis that the proposition $P(n)$ is true. The proposition is then shown to be true for a particular value of n, say k, and therefore by induction the proposition must be true for $n = k + 1$, $k + 2$, $k + 3$, ... In many cases $k = 1$, so then the proposition is true for all positive integers.

mathematics science of spatial and numerical relationships. The main divisions of *pure mathematics* include geometry, arithmetic, algebra, calculus, and trigonometry. Mechanics, statistics, numerical analysis, computing, the mathematical theories of astronomy, electricity, optics, thermodynamics, and atomic studies come under the heading of *applied mathematics*.

early history Prehistoric human beings probably learned to count at least up to ten on their fingers. The Chinese, Hindus, Babylonians, and Egyptians all devised methods of counting and measuring that were of practical importance in their everyday lives. The first theoretical mathematician is held to be Thales of Melitus (*c.* 580 BC) who is believed to have proposed the first theorems in plane geometry.

His disciple Pythagoras established geometry as a recognized science among the Greeks. The later school of Alexandrian geometers (4th and 3rd centuries BC) included Euclid and Archimedes. Our present decimal numerals are based on a Hindu-Arabic system that reached Europe about AD 100 from Arab mathematicians of the Middle East such as Khwārizmī.

Europe Western mathematics began to develop from the 15th century. Geometry was revitalized by the invention of coordinate geometry by René Descartes 1637; Blaise Pascal and Pierre de Fermat developed probability theory, John Napier invented logarithms, and Isaac Newton and Gottfried Leibniz developed calculus. In Russia, Nikolai Lobachevsky rejected Euclid's parallelism and developed non-Euclidean geometry, a more developed form of which (by Georg Riemann) was later utilized by Einstein in his relativity theory.

the present Higher mathematics has a powerful tool in the high-speed electronic computer, which can create and manipulate mathematical 'models' of various systems in science, technology, and commerce. Modern additions to school syllabuses such as sets, group theory, matrices, and graph theory are sometimes referred to as 'new' or 'modern' mathematics.

Mathematics is the door and the key to the sciences.

On **mathematics** Roger Bacon (1214–1292)
Opus Major 1266

matrix in mathematics, a square ($n \times n$) or rectangular ($m \times n$) array of elements (numbers or algebraic variables). They are a means of condensing information about mathematical systems and can be used for, among other things, solving simultaneous linear equations (see ◊linear equation) and transformations.

Much early matrix theory was developed by the British mathematician Arthur Cayley, although the term was coined by his contemporary James Sylvester (1814–1897).

matrix in biology, usually refers to the ◊extracellular matrix.

matter in physics, anything that has mass and can be detected and measured. All matter is made up of ◊atoms, which in turn are made up of ◊elementary particles; it exists ordinarily as a solid, liquid, or gas. The history of science and philosophy is largely taken up with accounts of theories of matter, ranging from the hard 'atoms' of Democritus to the 'waves' of modern quantum theory.

Mauna Kea astronomical observatory in Hawaii, USA, built on a dormant volcano at 4,200 m/ 13,784 ft above sea level. Because of its elevation high above clouds, atmospheric moisture, and artificial lighting, Mauna Kea is ideal for infrared astronomy. The first telescope on the site was installed 1970.

Telescopes include the 2.24–m/88–in University of Hawaii reflector 1970. In 1979 three telescopes were erected: the 3.8 m/150 in United Kingdom Infrared Telescope (UKIRT) (also used for optical observations); the 3–m/120–in NASA

mathematics: chronology

c. 2500 BC	The people of Mesopotamia (now Iraq) developed a positional numbering (place-value) system, in which the value of a digit depends in its position in a number.
c. 2000 BC	Mesopotamian mathematicians solved quadratic equations (algebraic equations in which the highest power of a variable is 2).
876 BC	A symbol for zero was used for the first time, in India.
c. 550 BC	Greek mathematician Pythagoras formulated a theorem relating the lengths of the sides of a right-angled triangle. The theorem was already known by earlier mathematicians in China, Mesopotamia, and Egypt.
c. 450 BC	Hipparcos of Metapontum discovered that some numbers are irrational (cannot be expressed as the ratio of two integers).
300 BC	Euclid laid out the laws of geometry in his book *Elements*, which was to remain a standard text for 2,000 years.
c. 230 BC	Eratosthenes developed a method for finding all prime numbers.
c. 100 BC	Chinese mathematicians began using negative numbers.
c. 190 BC	Chinese mathematicians used powers of 10 to express magnitudes.
c. AD 210	Diophantus of Alexandria wrote the first book on algebra.
c. 600	A decimal number system was developed in India.
829	Persian mathematician Muhammad ibn-Mūsā al-Khwārizmī published a work on algebra that made use of the decimal number system.
1202	Italian mathematician Leonardo Fibonacci studied the sequence of numbers (1, 1, 2, 3, 5, 8, 13, 21, . . .) in which each number is the sum of the two preceding ones.
1550	In Germany, Rheticus published trigonometrical tables that simplified calculations involving triangles.
1614	Scottish mathematician John Napier invented logarithms, which enable lengthy calculations involving multiplication and division to be carried out by addition and subtraction.
1623	Wilhelm Schickard invented the mechanical calculating machine.
1637	French mathematician and philosopher René Descartes introduced coordinate geometry.
1654	In France, Blaise Pascal and Pierre de Fermat developed probability theory.
1666	Isaac Newton developed differential calculus, a method of calculating rates of change.
1675	German mathematician Gottfried Wilhelm Leibniz introduced the modern notation for integral calculus, a method of calculating volumes.
1679	Leibniz introduced binary arithmetic, in which only two symbols are used to represent all numbers.
1684	Leibniz published the first account of differential calculus.
1718	Jakob Bernoulli in Switzerland published his work on the calculus of variations (the study of functions that are close to their minimum or maximum values.
1746	In France, Jean le Rond d'Alembert developed the theory of complex numbers.
1747	D'Alembert used partial differential equations in mathematical physics.
1798	Norwegian mathematician Caspar Wessel introduced the vector representation of complex numbers.
1799	Karl Friedrich Gauss of Germany proved the fundamental theorem of algebra: the number of solutions of an algebraic equation is the same as the exponent of the highest term.
1810	In France, Jean Baptiste Joseph Fourier published his method of representing functions by a series of trigonometric functions.
1812	French mathematician Pierre Simon Laplace published the first complete account of probability theory.
1822	In the UK, Charles Babbage began construction of the first mechanical computer, the difference machine.
1827	Gauss introduced differential geometry, in which small features of curves are described by analytical methods.
1829	In Russia, Nikolai Ivanovich Lobachevsky developed hyperbolic geometry, in which a plane is regarded as part of a hyperbolic surface, shaped like a saddle. In France, Evariste Galois introduced the theory of groups (collections whose members obey certain simple rules of addition and multiplication).
1844	French mathematician Joseph Liouville found the first transcendental number, which cannot be expressed as an algebraic equation with rational coefficients. In Germany, Hermann Grassmann studied vectors with more than three dimensions.
1854	George Boole in the UK published his system of symbolic logic, now called Boolean algebra.
1858	English mathematician Arthur Cayley developed calculations using ordered tables called matrices.
1865	August Ferdinand Möbius in Germany described how a strip of paper can have only one side and one edge.
1892	German mathematician Georg Cantor showed that there are different kinds of infinity and studied transfinite numbers.
1895	Jules Henri Poincaré published the first paper on topology, often called 'the geometry of rubber sheets'.
1931	In the USA, Austrian-born mathematician Kurt Gödel proved that any formal system strong enough to include the laws of arithmetic is either incomplete or inconsistent.
1937	English mathematician Alan Turing published the mathematical theory of computing.
1944	John Von Neumann and Oscar Morgenstern developed game theory in the USA.
1945	The first general purpose, fully electronic digital computer, ENIAC (electronic numerator, integrator, analyser, and computer), was built at the University of Pennsylvania, USA.
1961	Meteorologist Edward Lorenz at the Massachusetts Institute of Technology, USA, discovered a mathematical system with chaotic behaviour, leading to a new branch of mathematics—chaos theory.
1962	Benit Mandelbrot in the USA invented fractal images, using a computer that repeats the same mathematical pattern over and over again.
1975	US mathematician Mitchell Feigenbaum discovered a new fundamental constant (approximately 4.669201609103), which plays an important role in chaos theory.
1980	Mathematicians worldwide completed the classification of all finite and simple groups, a task that took over a hundred mathematicians more than 35 years to complete and whose results took up more than 14,000 pages in mathematical journals.
1989	A team of US computer mathematicians at Amdahl Corporation, California, discovered the highest known prime number (it contains 65,087 digits).

The electronic mathematician

Mathematicians have always dreamed of possessing machines that would remove the drudgery from their work. The inventor of logarithms, John Napier, also invented a system of carved ivory rods for doing multiplication, known as Napier's bones. Blaise Pascal built the first mechanical calculator in 1642. In 1835 Charles Babbage designed a calculating machine that could modify its own instructions, a forerunner of today's computers. Two of the true parents of the computer, John Von Neumann and Alan Turing, were mathematicians.

Not just number crunchers
Until the 1970s, computers were used as glorified calculators, for 'number crunching'—performing what were essentially just long and complicated calculations in arithmetic. Many mathematical problems, however, require understanding, not just a numerical answer. More and more, computers are being used by mathematicians as 'experimental' tools: to investigate aspects of mathematical problems, test predictions, and prove the correctness of theories. Computer scientists have also responded to mathematicians' needs by devising symbolic computation systems. These manipulate algebraic expressions in the same way that a human mathematician would—only faster and more accurately. The result might be called 'computer-assisted mathematics': the computer does not make mathematicians obsolete—on the contrary, it adds enormously to their power, bringing within their range problems that had hitherto seemed impossible.

A good example is the proof in 1972 of the Four Colour Theorem by Kenneth Appel and Wolfgang Haken. In 1850 Francis Guthrie conjectured that no more than four colours need be used in colouring a map in order to ensure that no two adjacent countries share the same colour. Mathematicians quickly proved that five colours would suffice, but had no success whatsoever in reducing that number to four. A direct attack by computer would not be possible, for how could a computer consider all possible maps? But Appel and Haken came up with a list of 1,936 particular maps, and showed that if each had a rather complicated property, then the conjecture must be true. They then checked this property, case by case, on a computer, taking about 1,200 hours.

Symbolic computation systems
There are now many different symbolic computation systems, such as Macsyma, Reduce, Maple, and Mathematica. Their use is becoming almost routine among research mathematicians. Their power is immense. In 1847 the French mathematician Charles Delaunay spent 20 years calculating a formula for the position of the Moon, and the end result occupies an entire book. In 1970 three researchers at the Boeing Laboratories in Seattle checked his calculation by symbolic computation, taking only 20 *hours*. They found that Delaunay had made three errors, none serious.

Symbolic computation has been used to answer long-standing questions about dynamic systems, to help prove new results in number theory, to investigate questions in algebraic geometry, to devise new minimal surfaces, and even in the mathematics of games. In March 1991 Uri Zwick and Mike Patterson of Warwick University, England, used Mathematica to work out a winning strategy for the memory game pelmanism, in which pairs of identical cards are laid face down and players take turns to turn two of them over. If they match, the player removes them and takes an extra turn. The winner is whoever removes the most cards. The strategy is quite simple, but the proof that it works—and the experiments needed to discover it—would not have been possible without the computer.

New perspectives
Computation and mathematics have always been closely related, and as the century draws to a close they are becoming intimately intertwined. This strong interaction between computing and mathematics will let the mathematician of the future spend more time thinking about concepts, and less time performing routine calculations. Computers, moreover, open up a whole new range of problems that mathematicians would not otherwise have thought of, and offer new perspectives from which to find answers.

The technology is not yet perfect. Symbolic computation programs seldom work well without a lot of careful structuring by their human user. Left to their own devices, the programs tend to suffer 'memory explosions', leaving copies of intermediate steps all over the computer's memory, filling it up, and grinding to an ignominious halt. But the technology is improving so fast that by the turn of the century there will be few mathematicians who do not 'collaborate' with a computer on a regular basis.

Ian Stewart

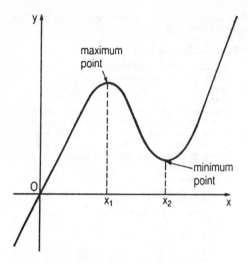

maximum and minimum *A maximum point on a curve is higher than the points immediately on either side of it; it is not necessarily the highest point on the curve. Similarly, a minimum point is lower than the points immediately on either side. At the maximum and minimum points of a curve the gradient (slope) of the tangent is zero.*

Infrared Telescope Facility (IRTF); and the 3.6–m/142–in Canada–France–Hawaii Telescope (CFHT), designed for optical and infrared work. The 15–m/50–ft diameter UK/Netherlands James Clerk Maxwell Telescope (JCMT) is the world's largest telescope specifically designed to observe millimetre wave radiation from nebulae, stars, and galaxies. The JCMT is operated via satellite links by astronomers in Europe.

The world's largest optical telescope, the ◊Keck Telescope, is also situated on Mauna Kea.

maximum and minimum in ◊coordinate geometry, points at which the slope of a curve representing a ◊function changes from positive to negative (maximum), or from negative to positive (minimum). A tangent to the curve at a maximum or minimum has zero gradient.

Maxima and minima can be found by differentiating the function for the curve and setting the differential to zero (the value of the slope at the turning point). For example, differentiating the function for the ◊parabola $y = 2x^2 - 8x$ gives $dy/dx = 4x - 8$. Setting this equal to zero gives $x = 2$, so that $y = -8$ (found by substituting $x = 2$ into the parabola equation). Thus the function has a minimum at the point $(2, -8)$.

maxwell cgs unit (symbol Mx) of magnetic flux (the strength of a ◊magnetic field in an area multiplied by the area). It is now replaced by the SI unit, the ◊weber (one maxwell equals 10^{-8} weber). The ◊maxwell is a very small unit, representing a single line of magnetic flux. It is named after the Scottish physicist James Clerk Maxwell.

Maxwell–Boltzmann distribution in physics, a statistical equation describing the distribution of velocities among the molecules of a gas. It is named after James Maxwell and Ludwig Boltzmann, who

derived the equation, independently of each other, in the 1860s.

One form of the distribution is $n = Ne^{-E/RT}$, where 'N' is the total number of molecules, n is the number of molecules with energy in excess of E, T is the absolute temperature (temperature in kelvin), R is the ◊gas constant, and 'e' is the exponential constant.

Maxwell's screw rule in physics, a rule formulated by Scottish physicist James Maxwell that predicts the direction of the magnetic field produced around a wire carrying electric current. It states that if a right-handed screw is turned so that it moves forwards in the same direction as the current, its direction of rotation will give the direction of the magnetic field.

mean in mathematics, a measure of the average of a number of terms or quantities. The simple *arithmetic mean* is the average value of the quantities, that is, the sum of the quantities divided by their number. The *weighted mean* takes into account the frequency of the terms that are summed; it is calculated by multiplying each term by the number of times it occurs, summing the results and dividing this total by the total number of occurrences. The *geometric mean* of n quantities is the nth root of their product. In statistics, it is a measure of central tendency of a set of data.

meander loop-shaped curve in a river flowing across flat country. As a river flows, any curve in its course is accentuated by the current. The current is fastest on the outside of the curve where it cuts into the bank; on the curve's inside the current is slow and deposits any transported material. In this way the river changes its course across the flood plain.

A loop in a river's flow may become so accentuated that it becomes cut off from the normal course and forms an ◊oxbow lake. The word comes from the river Menderes in Turkey.

mean deviation in statistics, a measure of the spread of a population from the ◊mean.

Thus if there are n observations with a mean of

magnetic field direction

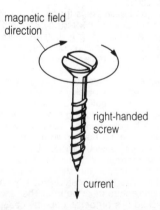

right-handed screw

current

Maxwell's screw rule *Maxwell's screw rule gives the direction of the magnetic field produced when electric current flows through a wire. If the tip of a right-handed screw is imagined to point in the same direction as the current, the direction of the magnetic field will be the same as the rotation of the screw.*

m, the mean deviation is the sum of the moduli (absolute values) of the differences of the observation values from *m*, divided by *n*.

mean free path in physics, the average distance travelled by a particle, atom, or molecule between successive collisions. It is of importance in the ◊kinetic theory of gases.

mechanical advantage (MA) in physics, the number of times the load moved by a machine is greater than the effort applied to that machine. In equation terms: MA = load/effort.

The exact value of a working machine's MA is always less than its predicted value because there will always be some frictional resistance that increases the effort necessary to do the work.

mechanical equivalent of heat in physics, a constant factor relating the calorie (the c.g.s. unit of heat) to the joule (the unit of mechanical energy), equal to 4.1868 joules per calorie. It is redundant in the SI system of units, which measures heat and all forms of energy in ◊joules (so that the mechanical equivalent of heat is 1).

mechanical weathering in earth science, an alternative name for ◊physical weathering.

mechanics branch of physics dealing with the motions of bodies and the forces causing these motions, and also with the forces acting on bodies in ◊equilibrium. It is usually divided into ◊dynamics and ◊statics.

Quantum mechanics is the system based on the ◊quantum theory that has superseded Newtonian mechanics in the interpretation of physical phenomena on the atomic scale.

media (singular *medium*) in computing, the collective name for materials on which data can be recorded. For example, paper is a medium that can be used to record printed data; a floppy disc is a medium for recording magnetic data.

medial moraine linear ridge of rocky debris running along the centre of a glacier. Medial moraines are commonly formed by the joining of two ◊lateral moraines when two glaciers merge.

median in mathematics and statistics, the middle number of an ordered group of numbers. If there is no middle number (because there is an even number of terms), the median is the ◊mean (average) of the two middle numbers. For example, the median of the group 2, 3, 7, 11, 12 is 7; that of 3, 4, 7, 9, 11, 13 is 8 (the average of 7 and 9).

In geometry, the term refers to a line drawn from

the vertex of a triangle to the midpoint of the opposite side.

medicine science of preventing, diagnosing, alleviating, or curing disease, both physical and mental; also any substance used in the treatment of disease. The basis of medicine is anatomy (the structure and form of the body) and physiology (the study of the body's functions).

In the West, medicine increasingly relies on new drugs and sophisticated surgical techniques, while diagnosis of disease is more and more by noninvasive procedures. The time and cost of Western-type medical training makes it inaccessible to many parts of the Third World; where health care of this kind is provided it is often by auxiliary medical helpers trained in hygiene and the administration of a limited number of standard drugs for the prevalent diseases of a particular region.

The prime goal is to alleviate suffering, and not to prolong life. And if your treatment does not alleviate suffering, but only prolongs life, that treatment should be stopped.

On **medicine** Christiaan Barnard (1922–)

Mediterranean climate climate characterized by hot dry summers and warm wet winters. Mediterranean zones are situated in either hemisphere on the western side of continents, between latitudes of 30° and 60°.

During the winter rain is brought by the ◊Westerlies; in summer Mediterranean zones are under the influence of the ◊trade winds. The regions bordering the Mediterranean Sea, California, central Chile, the Cape of Good Hope, and parts of S Australia have such climates.

medulla central part of an organ. In the mammalian kidney, the medulla lies beneath the outer cortex and is responsible for the reabsorption of water from the filtrate. In plants, it is a region of packing tissue in the centre of the stem. In the vertebrate brain, the medulla is the posterior region responsible for the coordination of basic activities, such as breathing and temperature control.

medusa the free-swimming phase in the life cycle of a coelenterate, such as a jellyfish or ◊coral. The other phase is the sedentary *polyp*.

meerschaum aggregate of minerals, usually the soft white mineral *sepiolite*, hydrous magnesium silicate. It floats on water and is used for making pipe bowls.

mega- prefix denoting multiplication by a million. For example, a megawatt (MW) is equivalent to a million watts.

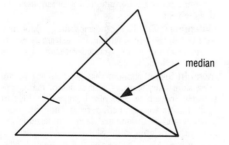

median

MEGA NUMBERS

If you were to count at the rate of one number each second and keep counting, it would take just 11 days to count to one million, but 32 years to reach one billion (1,000 million).

Western medicine: chronology

c. 400 BC	Hippocrates recognized that disease had natural causes.
c. AD 200	Galen consolidated the work of the Alexandrian doctors.
1543	Andreas Vesalius gave the first accurate account of the human body.
1628	William Harvey discovered the circulation of the blood.
1768	John Hunter began the foundation of experimental and surgical pathology.
1785	Digitalis was used to treat heart disease; the active ingredient was isolated 1904.
1798	Edward Jenner published his work on vaccination.
1877	Patrick Manson studied animal carriers of infectious diseases.
1882	Robert Koch isolated the bacillus responsible for tuberculosis.
1884	Edwin Klebs isolated the diphtheria bacillus.
1885	Louis Pasteur produced a vaccine against rabies.
1890	Joseph Lister demonstrated antiseptic surgery.
1895	Wilhelm Röntgen discovered X-rays.
1897	Martinus Beijerinck discovered viruses.
1899	Felix Hoffman developed aspirin; Sigmund Freud founded psychiatry.
1900	Karl Landsteiner identified the first three blood groups, later designated A, B, and O.
1910	Paul Ehrlich developed the first specific antibacterial agent, Salvarsan, a cure for syphilis.
1922	Insulin was first used to treat diabetes.
1928	Alexander Fleming discovered penicillin.
1932	Gerhard Domagk discovered the first antibacterial sulphonamide drug, Prontosil.
1937	Electro-convulsive therapy (ECT) was developed.
1940s	Lithium treatment for manic-depressive illness was developed.
1950s	Antidepressant drugs and beta-blockers for heart disease were developed. Manipulation of the molecules of synthetic chemicals became the main source of new drugs. Peter Medawar studied the body's tolerance of transplanted organs and skin grafts.
1950	Proof of a link between cigarette smoking and lung cancer was established.
1953	Francis Crick and James Watson announced the structure of DNA. Jonas Salk developed a vaccine against polio.
1958	Ian Donald pioneered diagnostic ultrasound.
1960s	A new generation of minor tranquillizers called benzodiazepines was developed.
1967	Christiaan Barnard performed the first human heart-transplant operation.
1971	Viroids, disease-causing organisms even smaller than viruses, were isolated outside the living body.
1972	The CAT scan, pioneered by Godfrey Hounsfield, was first used to image the human brain.
1975	César Milstein developed monoclonal antibodies.
1978	World's first 'test-tube baby' was born in the UK.
1980s	AIDS (acquired immune-deficiency syndrome) was first recognized in the USA. Barbara McClintock's discovery of the transposable gene was recognized.
1980	The World Health Organization reported the eradication of smallpox.
1983	The virus responsible for AIDS, now known as human immunodeficiency virus (HIV), was identified by Luc Montagnier at the Institut Pasteur, Paris; Robert Gallo at the National Cancer Institute, Maryland, USA discovered the virus independently 1984.
1984	The first vaccine against leprosy was developed.
1987	The world's longest-surviving heart-transplant patient died in France, 18 years after his operation.
1989	Grafts of fetal brain tissue were first used to treat Parkinson's disease.
1990	Gene for maleness discovered by UK researchers.
1991	First successful use of gene therapy (to treat severe combined immune deficiency) was reported in the USA.

megabyte (Mb) in computing, a unit of memory equal to 1,024 ◊kilobytes. It is sometimes used, less precisely, to mean 1 million bytes.

megatherium genus of extinct giant ground sloth of North and South America. Various species lived from about 7 million years ago until geologically recent times. They were plant-eaters, and some grew to 6 m/20 ft long.

megaton one million (10^6) tons. Used with reference to the explosive power of a nuclear weapon, it is equivalent to the explosive force of one million tons of trinitrotoluene (TNT).

meiosis in biology, a process of cell division in which the number of ◊chromosomes in the cell is halved. It only occurs in eukaryotic cells (see ◊eukaryote), and is part of a life cycle that involves sexual reproduction because it allows the genes of two parents to be combined without the total number of chromosomes increasing.

In sexually reproducing ◊diploid animals (having two sets of chromosomes per cell), meiosis occurs during formation of the ◊gametes (sex cells, sperm and egg), so that the gametes are ◊haploid (having only one set of chromosomes). When the gametes unite during ◊fertilization, the diploid condition is restored. In plants, meiosis occurs just before spore formation. Thus the spores are haploid and in lower plants such as mosses they develop into a haploid plant called a gametophyte which produces the gametes (see ◊alternation of generations). See also ◊mitosis.

melamine $C_3N_6H_6$ ◊thermosetting ◊polymer based on urea– formaldehyde. It is extremely resistant to heat and is also scratch-resistant. Its uses include synthetic resins.

melanin brown pigment that gives colour to the eyes, skin, hair, feathers, and scales of many vertebrates. In humans, melanins help protect the skin against ultraviolet radiation from sunlight. Both genetics and environmental factors determine the amount of melanin in the skin.

melanism black coloration of animal bodies caused by large amounts of the pigment melanin.

Medicine: the pace of progress slows down

Medicine has become synonymous with progress. In the four decades since World War II virtually every year seems to have brought some revolutionary new method of curing or modifying human illness. The rate of change was most marked in the 1950s and 1960s when most of the drugs currently used in medical practice were discovered, when infectious illness succumbed to the twin onslaughts of antibiotics and vaccination, and when the most dramatic of operations that are now considered routine were introduced—particularly open-heart surgery and transplants.

Medical innovation has slowed down in recent years for two main reasons. First, there is a limit to technological development. Once the ultimate operation of replacing a diseased heart with a healthy one has been achieved, as happened when the first transplant was performed in 1968, there is a limit to the extent that cardiac surgery can progress further.

Secondly, medicine still remains remarkably ignorant about the causes and pathogenesis of most diseases, and this places an intellectual block to further development. Thus although doctors can treat diabetes with insulin, without understanding why diabetes occurs in the first place it is difficult to know how to prevent it or cure it completely. The same is true for the common diseases of middle life—rheumatoid arthritis, multiple sclerosis, hypertension—and for the chronic degenerative diseases associated with ageing such as cancer, heart disease and strokes.

Medicine has continued to progress in the last few years but with fewer dramatic breakthroughs and more fine tuning and improvement of treatments and techniques that are already established. There are, however, exceptions to this rule.

Surgery

Surgical techniques have been transformed in the last three years with the introduction of 'minimally invasive surgery'. This is best illustrated by the operation for removal of the gall bladder—a cholecystectomy—usually necessitated by the accumulation of gallstones, causing pain and recurrent infection. For decades the standard cholecystectomy entailed making a large incision in the abdominal wall, identifying the gall bladder nestling under the liver, tying off the bile duct and arteries, and removing it. Now the sameoperation can be performed by making four minute holes in the abdomen, one for the laparoscope through which the

gallbladder is visualized, two through which the surgeon manipulates his delicate instruments, and one to remove any fluid and debris.

The advantages to the patient are immeasurable; they are out of hospital in a couple of days and back to work within a fortnight. Minimally invasive surgery will in the next decade transform such a practice out of all recognition. It has already been applied to conditions as diverse as resecting cancer of the colon and repairing hernias.

Obstetrics

The most significant progress in obstetrics has paradoxically been a retreat from the high-tech management of labour, with its emphasis on induction and constant fetal monitoring, to a more 'natural' approach. The rate of Caesarean section remains high, as this is the best way of managing difficult births when there is a risk of damage to the fetus.

The other main development has been the improvements in ultrasound and amniocentesis to detect many more congenital abnormalities of the fetus early in pregnancy. With ultrasound it is now possible to diagnose not just spina bifida and hydrocephalus, but subtler abnormalities of the heart and kidneys. Amniocentesis—where cells are removed from the amniotic sac and examined for chromosomal defects such as Down's syndrome—can now be done much earlier in pregnancy, or the same information can be obtained by removing a few cells from the placenta (chorionic villus sampling). Early diagnosis obviates the trauma for the mother of a late abortion of her fetus if it is found to be abnormal. The same procedure can be used to diagnose genetic disease when the specific genetic abnormality is known, as in sickle cell disease and cystic fibrosis.

The third technological development in obstetrics has been the use of Doppler ultrasound to assess the blood flow through the placenta which compared results in the retardation of growth or even death of the fetus.

Infectious disease

There is a tantalising theory that many chronic diseases, such as rheumatoid arthritis and multiple sclerosis, may be caused by some as yet unidentified infectious agent. This is certainly so for peptic ulcers of the stomach or duodenum, which used to be thought to arise from excess acid production and so were amenable to treatment with drugsthat reduce acid secretions, such as

Zantac and Tagamet. However, 80% of treated ulcers are active again within a year. It now appears that in many the underlying causes are bacteria— *Helicobacter pylori*—which inhabit the lining of the gut while making it more sensitive to acid damage. Patients with *Helicobacter* can be effectively cured of their recurring ulcers if treated with a course of antibiotics and the anti-infective agent bismuth.

Treatment of viral infections has been the other main development in this field, though antiviral drugs remain much less effective than antibiotics. The drug Acyclovir, originally introduced for the treatment of genital herpes, has now been shown to modify the course of chickenpox, and the related condition shingles, by reducing the duration of the illness. There is still no sign of a definitive cure for AIDS, though the antiviral drug Zidovudine (or AZT) if given to those with asymptomatic HIV infection may delay the development of the full-blown AIDS syndrome.

Heart disease
Heart disease is still the commonest cause of premature death in middle-aged males. Its incidence has fallen steadily in the USA and several European countries, including the UK. The reason is not clear; it cannot be explained in terms of conventional beliefs about the role of fat in the diet as there have been only trivial changes in the pattern of food consumption. Major progress has been made in the treatment of acute heart attack with the use of clot-busting drugs such as Strepokinaise, which dissolves the thrombus that forms in the coronary arteries. Given early enough, it reduces the mortality rate by 50%.

The other main form of heart disease, angina pectoris—caused by narrowing of the coronary arteries—is now increasingly treated by passing a wire with a balloon on the tip directly into the coronary arteries, dilating them 'like a footprint in the snow'. The enormous advantage of this technique of coronary angioplasty is that it is much less traumatic than open-heart bypass surgery.

Genetics
Recent progress in genetics has far outstripped that of any other medical discipline. It has been known for many years that all biological functions are determined by genes made up of discrete segments of DNA arranged on chromosomes. The revolution in genetics has been made possible by the ability to identify specific genes, find out exactly what they do, discover the abnormalities in gene sequences that cause disease, and use genes to make hormones and—

more speculatively—cure diseases. The key to this complex technology lies in the ability to make 'gene probes', to cut up DNA into manageable segments by enzymes known as restriction endonucleases, and introduce genes into bacterial viruses with a view to altering their function in a specific way. The practical applications of these techniques are many.

Antenatal diagnosis
The great uncertainty for families with a history of genetic disease such as sickle cell or cystic fibrosis is whether their children are affected. Now that the genes for several of these disorders have been identified it is possible, by removing a few cells from the placenta early in pregnancy, to determine whether they contain the abnormal genes responsible for the defect. If present, the pregnancy can be aborted and the couple can then try again with the ultimate intention of producing a normal child.

As virtually every disease has a genetic component—whether it is alcoholism, schizophrenia or heart disease—it should be possible in future, when the full genetic code has been worked out, to screen fetuses to ensure they are as perfect as can be. The logistics of this are likely to be very complex, and this type of diagnosis is likely to be limited for the foreseeable future to highly inherited conditions caused by defects in a single gene.

Genetic engineering
It is now possible to make hormones like insulin and growth hormone, and the factors involved in blood clotting by inserting the human gene that controls for them into bacteria such as *Escherichia coli* so the bacterium manufactures them in a pure form.

Gene therapy
In genetic disease the abnormality is present in every cell. So, for example, in all the lung cells in patients with the inherited lung condition emphysema the amount of the enzyme alpha 1 antitrypsin will be either reduced or absent. The challenge for gene therapy is to get a normal alpha 1 antitrypsin gene into as many of these cells as possible so that they can start manufacturing the normal enzymes. The tentative answer seems to be to insert the normal gene into a virus to which the patient is then exposed, in the hope that it will affect sufficient numbers of the abnormal cells and correct the defect by replacing the abnormal gene with the normal.

James Le Fanu

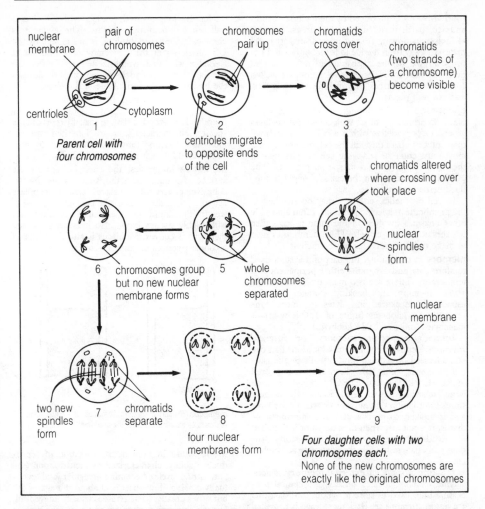

nuclear membrane / pair of chromosomes / chromosomes pair up / chromatids cross over / chromatids (two strands of a chromosome) become visible

centrioles / cytoplasm

1
Parent cell with four chromosomes

2
centrioles migrate to opposite ends of the cell

3

chromatids altered where crossing over took place

nuclear spindles form

6 chromosomes group but no new nuclear membrane forms

5 whole chromosomes separated

4

nuclear membrane

two new spindles form

7 chromatids separate

8
four nuclear membranes form

9
Four daughter cells with two chromosomes each.
None of the new chromosomes are exactly like the original chromosomes

meiosis *Meiosis is a type of cell division that produces gametes (sex cells, sperm and egg). This sequence shows an animal cell but only four chromosomes are present in the parent cell (1). There are two stages in the division process. In the first stage (2–6), the chromosomes come together in pairs and exchange genetic material. This is called crossing over. In the second stage (7–9), the cell divides to produce four gamete cells, each with only one copy of each chromosome from the parent cell.*

Melanin is of significance in insects, because melanic ones warm more rapidly in sunshine than do pale ones, and can be more active in cool weather. A fall in temperature may stimulate such insects to produce more melanin. In industrial areas, dark insects and pigeons match sooty backgrounds and escape predation, but they are at a disadvantage in rural areas where they do not match their backgrounds. This is known as *industrial melanism.*

meltdown the melting of the core of a nuclear reactor, due to overheating. To prevent such accidents all reactors have equipment intended to flood the core with water in an emergency. The reactor is housed in a strong containment vessel, designed to prevent radiation escaping into the atmosphere. The result of a meltdown is an area radioactively contaminated for 25,000 years or more.

At Three Mile Island, Pennsylvania, USA, in March 1979, a partial meltdown occurred caused by a combination of equipment failure and operator error, and some radiation was released into the air. In April 1986, a reactor at Chernobyl, near Kiev, Ukraine, exploded, causing a partial meltdown of the core. Radioactive ◊fallout was detected as far away as Canada and Japan.

melting change of state from a solid to a liquid, associated with an intake of energy (for example, if the temperature rises).

melting point temperature at which a substance melts, or changes from a solid to liquid form. A pure substance under standard conditions of pressure (usually one atmosphere) has a definite melting point. If heat is supplied to a solid at its melting point, the temperature does not change until the melting process is complete. The melting point of ice is 0°C or 32°F.

meltwater water produced by the melting of snow and ice, particularly in glaciated areas. Streams of meltwater flowing from glaciers transport rocky materials away from the ice to form ◊outwash. Features formed by the deposition of debris carried by meltwater or by its erosive action are called *fluvioglacial features*; they include eskers, kames, and outwash plains.

membrane in living things, a continuous layer, made up principally of fat molecules, that encloses a ◊cell or ◊organelles within a cell. Certain small molecules can pass through the cell membrane, but most must enter or leave the cell via channels in the membrane made up of special proteins. The ◊Golgi apparatus within the cell is thought to produce certain membranes.

In cell organelles, enzymes may be attached to the membrane at specific positions, often alongside other enzymes involved in the same process, like workers at a conveyor belt. Thus membranes help to make cellular processes more efficient.

memory in computing, the part of a system used to store data and programs either permanently or temporarily. There are two main types: immediate access memory and backing storage. Memory capacity is measured in ◊bytes or, more conveniently, in kilobytes (units of 1,024 bytes) or megabytes (units of 1,024 kilobytes).

Immediate access memory, or *internal memory*, describes the memory locations that can be addressed directly and individually by the central processing unit. It is either read-only (stored in ROM, PROM, and EPROM chips) or read/write (stored in RAM chips). Read-only memory stores information that must be constantly available and is unlikely to be changed. It is nonvolatile—that is, it is not lost when the computer is switched off. Read/write memory is volatile—it stores programs and data only while the computer is switched on.

Backing storage, or *external memory*, is nonvolatile memory, located outside the central processing unit, used to store programs and data that are not in current use. Backing storage is provided by such devices as magnetic ◊discs (floppy and hard discs), ◊magnetic tape (tape streamers and cassettes), optical discs (such as ◊CD-ROM), and ◊bubble memory. By rapidly switching blocks of information between the backing storage and the immediate-access memory, the limited size of the immediate-access memory may be increased artificially. When this technique is used to give the appearance of a larger internal memory than physically exists, the additional capacity is referred to as ◊virtual memory.

A memory is what is left when something happens and does not completely unhappen.'

On **memory** Edward de Bono *The Mechanism of Mind* 1969

---

mendelevium synthesized, radioactive metallic element of the ◊actinide series, symbol Md, atomic number 101, relative atomic mass 258. It was first produced by bombardment of Es-253 with helium nuclei. Its longest-lived isotope, Md-258, has a half-life of about two months. The element is chemically similar to thulium. It was named by the US physicists at the University of California at Berkeley who first synthesized it 1955 after the Russian chemist Mendeleyev, who in 1869 devised the basis for the periodic table of the elements.

Mendelism in genetics, the theory of inheritance originally outlined by Austrian biologist Gregor Mendel. He suggested that, in sexually reproducing species, all characteristics are inherited through indivisible 'factors' (now identified with ◊genes) contributed by each parent to its offspring.

meniscus in physics, the curved shape of the surface of a liquid in a thin tube, caused by the cohesive effects of ◊surface tension (capillary action). When the walls of the container are made wet by the liquid, the meniscus is concave, but with highly viscous liquids (such as mercury) the meniscus is convex. Meniscus is also the name of a concavo-convex or convexo-concave ◊lens.

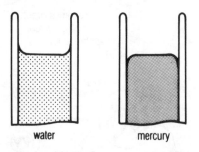

water   **mercury**

meniscus *The curved shape, or meniscus, of a liquid surface is caused by the attraction or repulsion between liquid and container molecules.*

menopause in women, the cessation of reproductive ability, characterized by menstruation (see ◊menstrual cycle) becoming irregular and eventually ceasing. The onset is at about the age of 50, but varies greatly. Menopause is usually uneventful, but some women suffer from complications such as flushing, excessive bleeding, and nervous disorders. Since the 1950s, ◊hormone replacement therapy (HRT), using ◊oestrogen alone or with ◊progesterone, has been developed to counteract such effects.

Long-term use of HRT was previously associated with an increased risk of cancer of the uterus, and of clot formation in the blood vessels, but newer formulations using natural oestrogens are not associated with these risks. Without HRT there is increased risk of osteoporosis (thinning of the bones) leading to broken bones, which may be indirectly fatal, particularly in the elderly.

The menopause is also known as the 'change of life'.

menstrual cycle cycle that occurs in female mammals of reproductive age, in which the body is prepared for pregnancy. At the beginning of the cycle, a Graafian (egg) follicle develops in the ovary, and the inner wall of the uterus forms a soft spongy lining. The egg is released from the ovary, and the lining of the uterus becomes vascularized (filled with blood vessels). If fertilization does not occur, the corpus luteum (remains of the Graafian follicle)

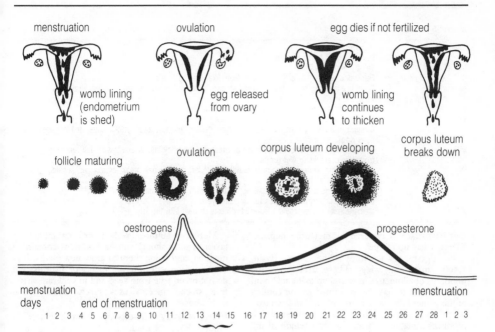

menstruation ovulation egg dies if not fertilized

womb lining (endometrium is shed) egg released from ovary womb lining continues to thicken

follicle maturing ovulation corpus luteum developing corpus luteum breaks down

oestrogens progesterone

menstruation menstruation
days end of menstruation
1 2 3 4 5 6 7 8 9 10 11 12 13 14 15 16 17 18 19 20 21 22 23 24 25 26 27 28 1 2 3

start of menstruation copulation could result in fertilization

menstrual cycle *From puberty to the menopause, most women produce a regular rhythm of hormones that stimulate the various stages of the menstrual cycle. The change in hormone level may cause premenstrual tension.*

degenerates, and the uterine lining breaks down, and is shed. This is what causes the loss of blood that marks menstruation. The cycle then begins again. Human menstruation takes place from puberty to menopause, occurring about every 28 days.

The cycle is controlled by a number of ◊hormones, including ◊oestrogen and ◊progesterone. If fertilization occurs, the corpus luteum persists and goes on producing progesterone.

menu in computing, a list of options, displayed on screen, from which the user may make a choice—for example, the choice of services offered to the customer by a bank cash dispenser: withdrawal, deposit, balance, or statement. Menus are used extensively in ◊graphical user interface (GUI) systems, where the menu options are often selected using a pointing device called a ◊mouse.

Mercalli scale scale used to measure the intensity of an ◊earthquake. It differs from the ◊Richter scale, which measures **magnitude**. It is named after the Italian seismologist Giuseppe Mercalli (1850–1914).

Intensity is a subjective value, based on observed phenomena, and varies from place to place with the same earthquake.

mercury or **quicksilver** heavy, silver-grey, metallic element, symbol Hg (from Latin *hydrargyrum*), atomic number 80, relative atomic mass 200.59. It is a dense, mobile liquid with a low melting point (−38.87°C/−37.96°F). Its chief source is the mineral cinnabar, HgS, but it sometimes occurs in nature as a free metal.

Its alloys with other metals are called amalgams (a silver-mercury amalgam is used in dentistry for filling cavities in teeth). Industrial uses include drugs and chemicals, mercury-vapour lamps, arc rectifiers, power-control switches, barometers, and thermometers.

Mercury is a cumulative poison that can contaminate the food chain, and cause intestinal disturbance, kidney and brain damage, and birth defects in humans. (The World Health Organization's 'safe' limit for mercury is 0.5 milligrams of mercury per kilogram of muscle tissue). The discharge into the sea by industry of organic mercury compounds such as dimethylmercury is the chief cause of mercury poisoning in the latter half of the 20th century. Between 1953 and 1975, 684 people in the Japanese fishing village of Minamata were poisoned (115 fatally) by organic mercury wastes that had been dumped into the bay and had accumulated in the bodies of fish and shellfish.

The element was known to the ancient Chinese and Hindus, and is found in Egyptian tombs of about 1500 BC. It was named by the alchemists after the fast-moving god, for its fluidity.

Mercury in astronomy, the closest planet to the Sun, at an average distance of 58 million km/36 million mi. Its diameter is 4,880 km/3,030 mi, its mass 0.056 that of Earth. Mercury orbits the Sun every 88 days, and spins on its axis every 59 days. On its sunward side the surface temperature reaches over 400°C/752°F, but on the 'night' side it falls to −170°C/−274°F. Mercury has an atmosphere with minute traces of argon and helium. In 1974 the US space probe *Mariner 10* discovered

Mercalli scale

intensity value	description
I	Only detected by instrument.
II	Felt by people resting.
III	Felt indoors; hanging objects swing; feels like passing traffic.
IV	Feels like passing heavy traffic; standing cars rock; windows, dishes, and doors rattle; wooden frames creak.
V	Felt outdoors; sleepers are woken; liquids spill; doors swing open.
VI	Felt by everybody; people stagger; windows break; trees and bushes rustle; weak plaster cracks.
VII	Difficult to stand upright; noticed by vehicle drivers; plaster, loose bricks, tiles, and chimneys fall; bells ring.
VIII	Car steering affected; some collapse of masonry; chimney stacks and towers fall; branches break from trees; cracks in wet ground.
IX	General panic; serious damage to buildings; underground pipes break; cracks and subsidence in ground.
X	Most buildings destroyed; landslides; water thrown out of canals.
XI	Rails bent; underground pipes totally destroyed.
XII	Damage nearly total; rocks displaced; objects thrown into the air.

that its surface is cratered by meteorite impacts. Mercury has no moons.

Its largest known feature is the Caloris Basin, 1,400 km/870 mi wide. There are also cliffs hundreds of kilometres long and up to 4 km/2.5 mi high, thought to have been formed by the cooling of the planet billions of years ago. Inside is an iron core three-quarters of the planet's diameter, which produces a magnetic field 1% the strength of the Earth's.

THE IRON CORE OF MERCURY

The planet Mercury has a core of iron slightly larger than the Moon. If this could be mined, it would supply enough iron to last thousands of millions of years.

mercury fulminate highly explosive compound used in detonators and percussion caps. It is a grey, sandy powder and extremely poisonous.

Mercury project US project to put a human in space in the one-seat Mercury spacecraft 1961–63.

The first two Mercury flights, on Redstone rockets, were short flights to the edge of space and back. The orbital flights, beginning with the third in the series (made by John Glenn), were launched by Atlas rockets.

meridian half a ◊great circle drawn on the Earth's surface passing through both poles and thus through all places with the same longitude. Terrestrial longitudes are usually measured from the Greenwich Meridian.

An astronomical meridian is a great circle passing through the celestial pole and the zenith (the point immediately overhead).

meristem region of plant tissue containing cells that are actively dividing to produce new tissues (or have the potential to do so). Meristems found in the tip of roots and stems, the apical meristems, are responsible for the growth in length of these organs.

The ◊cambium is a lateral meristem that is responsible for increase in girth in perennial plants. Some plants also have intercalary meristems, as in the stems of grasses, for example. These are responsible for their continued growth after cutting or grazing has removed the apical meristems of the shoots.

Meristem culture involves growing meristems taken from shoots on a nutrient-containing medium, and using them to grow new plants. It is used to propagate infertile plants or hybrids that do not breed true from seed and to generate virus-free stock, since viruses rarely infect apical meristems.

MERLIN array radiotelescope network centred on ◊Jodrell Bank, N England.

mesa (Spanish 'table') flat-topped steep-sided plateau, consisting of horizontal weak layers of rock topped by a resistant formation; in particular, those found in the desert areas of the USA and Mexico. A small mesa is called a butte.

mesoglea noncellular tissue that separates the endoderm and ectoderm in ◊coelenterates.

meson in physics, an unstable subatomic particle made up of two indivisible elementary particles called ◊quarks. It has a mass intermediate between that of the electron and that of the proton, is found in cosmic radiation, and is emitted by nuclei under bombardment by very high-energy particles.

The mesons form a subclass of the hadrons and include the kaons and pions. Their existence was predicted in 1935 by Japanese physicist Hideki Yukawa.

mesophyll the tissue between the upper and lower epidermis of a leaf blade (◊lamina), consisting of parenchyma-like cells containing numerous ◊chloroplasts.

In many plants, mesophyll is divided into two distinct layers. The *palisade mesophyll* is usually just below the upper epidermis and is composed of regular layers of elongated cells. Lying below them is the *spongy mesophyll*, composed of loosely arranged cells of irregular shape. This layer contains fewer chloroplasts and has many intercellular spaces for the diffusion of gases (required for ◊respiration and ◊photosynthesis), linked to the outside by means of ◊stomata.

mesosphere layer in the Earth's ◊atmosphere above the stratosphere and below the thermosphere. It lies between about 50 km/31 mi and 80 km/50 mi above the ground.

Mesozoic era of geological time 245–65 million years ago, consisting of the Triassic, Jurassic, and Cretaceous periods. At the beginning of the era, the continents were joined together as Pangaea;

dinosaurs and other giant reptiles dominated the sea and air; and ferns, horsetails, and cycads thrived in a warm climate worldwide. By the end of the Mesozoic era, the continents had begun to assume their present positions, flowering plants were dominant, and many of the large reptiles and marine fauna were becoming extinct.

metabolism the chemical processes of living organisms: a constant alternation of building up (*anabolism*) and breaking down (*catabolism*). For example, green plants build up complex organic substances from water, carbon dioxide, and mineral salts (photosynthesis); by digestion animals partially break down complex organic substances, ingested as food, and subsequently resynthesize them in their own bodies.

metal any of a class of chemical elements with certain chemical characteristics (◊metallic character) and physical properties: they are good conductors of heat and electricity; opaque but reflect light well; malleable, which enables them to be cold-worked and rolled into sheets; and ductile, which permits them to be drawn into thin wires.

Metallic elements compose about 75% of the 109 elements shown in the ◊periodic table of the elements. They form alloys with each other, ◊bases with the hydroxyl radical (OH), and replace the hydrogen in an ◊acid to form a salt. The majority are found in nature in the combined form only, as compounds or mineral ores; about 16 of them also occur in the elemental form, as ◊native metals. Their chemical properties are largely determined by the extent to which their atoms can lose one or more electrons and form positive ions (cations). They have been put to many uses, both structural and decorative, since prehistoric times, and the Copper Age, Bronze Age, and Iron Age are named for the metal that formed the technological base for that stage of human evolution.

The following are widely used in commerce: *precious metals*: gold, silver, mercury, and platinum, used principally in jewellery; *heavy metals*: iron, copper, zinc, tin, and lead, the common metals of engineering; *rarer heavy metals*: nickel, cadmium, chromium, tungsten, molybdenum, manganese, cobalt, vanadium, antimony, and bismuth, used principally for alloying with the heavy metals; *light metals*: aluminium and magnesium; *alkali metals*: sodium, potassium, and lithium; and *alkaline-earth metals*: calcium, barium, and strontium, used principally for chemical purposes. Other metals have come to the fore because of special nuclear requirements—for example, technetium, produced in nuclear reactors, is corrosion-inhibiting; zirconium may replace aluminium and magnesium alloy in canning uranium in reactors.

metal detector electronic device for detecting metal, usually below ground, developed from the wartime mine detector. In the head of the metal detector is a coil, which is part of an electronic circuit. The presence of metal causes the frequency of the signal in the circuit to change, setting up an audible note in the headphones worn by the user.

They are used to survey areas for buried metallic objects, occasionally by archaeologists. However, their indiscriminate use by treasure hunters has led to their being banned on recognized archaeological sites in some countries.

In Britain the law forbids the use of metal detectors on 'scheduled' (that is, nationally important) sites.

metal fatigue condition in which metals fail or fracture under relatively light loads, when these loads are applied repeatedly. Structures that are subject to flexing, such as the airframes of aircraft, are prone to metal fatigue.

metallic bond the force of attraction operating in a metal that holds the atoms together. In the metal the ◊valency electrons are able to move within the crystal and these electrons are said to be delocalized (see ◊electrons, delocalized). Their movement creates short-lived, positively charged ions. The electrostatic attraction between the delocalized electrons and the ceaselessly forming ions constitutes the metallic bond.

metallic character chemical properties associated with those elements classed as metals. These properties, which arise from the element's ability to lose electrons, are: the displacement of hydrogen from dilute acids; the formation of ◊basic oxides; the formation of ionic chlorides; and their reducing reaction, as in the ◊thermite process (see ◊reduction).

In the ◊periodic table of the elements, metallic character increases down any group and across a period from right to left.

metallic glass substance produced from metallic materials (non-corrosive alloys rather than simple metals) in a liquid state which, by very rapid cooling, are prevented from reverting to their regular metallic structure. Instead they take on the properties of glass, while retaining the metallic properties of malleability and relatively good electrical conductivity.

metalloid or *semimetal* any chemical element having some of but not all the properties of metals; metalloids are thus usually electrically semiconducting. They comprise the elements germanium, arsenic, antimony, and tellurium.

metallurgy the science and technology of producing metals, which includes extraction, alloying, and hardening. Extractive, or *process, metallurgy* is concerned with the extraction of metals from their ◊ores and refining and adapting them for use. *Physical metallurgy* is concerned with their properties and application. *Metallography* establishes the microscopic structures that contribute to hardness, ductility, and strength.

Metals can be extracted from their ores in three main ways: *dry processes*, such as smelting, volatilization, or amalgamation (treatment with mercury); *wet processes*, involving chemical reactions; and *electrolytic processes*, which work on the principle of ◊electrolysis.

The foundations of metallurgical science were laid about 3500 BC in Egypt, Mesopotamia, China, and India, where the art of ◊smelting metals from ores was discovered, starting with the natural alloy bronze. Later, gold, silver, copper, lead, and tin were worked in various ways, although they had been cold-hammered as native metals for thousands of years. The smelting of iron was discovered

metamorphic rocks

typical depth and temperature formation	main primary material (before metamorphism)		
	shale with several minerals	sandstone with only quartz	limestone with only calcite
15 km/300°C	slate	quartzite	marble
20 km/400°C	schist		
25 km/500°C	gneiss		
30 km/600°C	hornfels	quartzite	marble

about 1500 BC. The Romans hardened and tempered iron into steel, using ◊heat treatment. From then until about AD 1400, advances in metallurgy came into Europe by way of Arabian chemists. ◊Cast iron began to be made in the 14th century in a crude blast furnace. The demands of the Industrial Revolution led to an enormous increase in ◊wrought iron production. The invention by British civil engineer Henry Bessemer of the ◊Bessemer process in 1856 made cheap steel available for the first time, leading to its present widespread use and the industrial development of many specialized steel alloys.

metamorphic rock rock altered in structure and composition by pressure, heat, or chemically active fluids after original formation. (If heat is sufficient to melt the original rock, technically it becomes an igneous rock upon cooling.) The term was coined in 1833 by Scottish geologist Charles Lyell (1797–1875).

The mineral assemblage present in a metamorphic rock depends on the composition of the starting material (which may be sedimentary or igneous) and the temperature and pressure conditions to which it is subjected. There are two main types of metamorphism. *Thermal metamorphism* is brought about by the baking of solid rocks in the vicinity of an igneous intrusion (molten rock, or magma, in a crack in the Earth's crust). It is responsible, for example, for the conversion of limestone to marble. *Regional metamorphism* results from the heat and intense pressures associated with the movements and collision of tectonic plates (see ◊plate tectonics). It brings about the conversion of shale to slate, for example.

metamorphism geological term referring to the changes in rocks of the Earth's crust caused by increasing pressure and temperature. The resulting rocks are metamorphic rocks. All metamorphic changes take place in solid rocks. If the rocks melt and then harden, they become ◊igneous rocks.

metamorphosis period during the life cycle of many invertebrates, most amphibians, and some fish, during which the individual's body changes from one form to another through a major reconstitution of its tissues. For example, adult frogs are produced by metamorphosis from tadpoles, and butterflies are produced from caterpillars following metamorphosis within a pupa.

metazoa another name for animals. It reflects an earlier system of classification, in which there were two main divisions within the animal kingdom, the multicellular animals, or metazoa, and the single-celled 'animals' or protozoa. The ◊protozoa are no longer included in the animal kingdom, so only the metazoa remain.

meteor flash of light in the sky, popularly known as a *shooting* or *falling star*, caused by a particle of dust, a *meteoroid*, entering the atmosphere at speeds up to 70 kps/45 mps and burning up by friction at a height of around 100 km/60 mi. On any clear night, several *sporadic meteors* can be seen each hour.

Several times each year the Earth encounters swarms of dust shed by comets, which give rise to a *meteor shower*. This appears to radiate from one particular point in the sky, after which the shower is named; the Perseid meteor shower in Aug appears in the constellation Perseus. A brilliant meteor is termed a *fireball*. Most meteoroids are smaller than grains of sand. The Earth sweeps up an estimated 16,000 tonnes of meteoric material every year.

METEORS: ARRIVALS FROM SPACE

Every day, over one million meteors visible only with telescopes and over 500,000 meteors visible to the naked eye burn up in the Earth's atmosphere. On average, three or four meteors weighing about 4.5 kg/10 lb each arrive each day (only a fraction of this weight reaches the ground). Every month a meteor weighing 5 tonnes arrives (not more than 450 kg/1,000 lb reaches the ground). A 50–tonne meteor arrives every 30 years; a 250–tonne meteor arrives every 150 years; and a 50,000–tonne meteor arrives every 100,000 years. A small asteroid with a diameter of a few kilometres arrives every 50 million years.

meteor-burst communications technique for sending messages by bouncing radio waves off the fiery tails of ◊meteors. High-speed computer-controlled equipment is used to sense the presence of a meteor and to broadcast a signal during the short time that the meteor races across the sky.

The system, first suggested in the late 1920s, remained impracticable until data-compression techniques were developed, enabling messages to be sent in automatic high-speed bursts each time a meteor trail appeared. There are usually enough meteor trails in the sky at any time to permit continuous transmission of a message. The technique offers a communications link that is difficult to jam, undisturbed by storms on the Sun, and would not be affected by nuclear war.

meteorite piece of rock or metal from space that reaches the surface of the Earth, Moon, or other body. Most meteorites are thought to be fragments from asteroids, although some may be pieces from the heads of comets. Most are stony, although some are made of iron and a few have a mixed rock-iron composition. Meteorites provide evidence for the nature of the solar system and may be similar to the Earth's core and mantle, neither of which can be observed directly.

Thousands of meteorites hit the Earth each year, but most fall in the sea or in remote areas and are never recovered. The largest known meteorite is one composed of iron, weighing 60 tonnes, which lies where it fell in prehistoric times at Grootfontein, Namibia. Meteorites are slowed down by the Earth's atmosphere, but if they are moving fast enough they can form a ◊crater on impact. Meteor Crater in Arizona, about 1,200 m/4,000 ft in diameter and 200 m/650 ft deep, is the site of a meteorite impact about 50,000 years ago.

meteorite: the 11 largest meteorites

name and location	weight (tonnes)
Hoba West, Grootfontein, Namibia	60
Ahnighito, Greenland	30
Bacuberito, Mexico	27
Mbosi, Tanzania	26
Agpalik, Greenland	20
Armanty, Mongolia	20
Chupaderos, Mexico	14
Willamette, Oregon, USA	14
Campo del Cielo, Argentina	13
Mundrabilla, W Australia	12
Morito, Mexico	11

meteoroid chunk of rock in interplanetary space. There is no official distinction between meteoroids and asteroids, except that the term asteroid is generally reserved for objects larger than 1.6 km/1 mi in diameter, whereas meteoroids can range anywhere from pebble-size up.

Meteoroids are believed to result from the fragmentation of asteroids after collisions. Some meteoroids strike the Earth's atmosphere, and their fiery trails are called meteors. If they fall to Earth, they are named meteorites.

meteorology scientific observation and study of the ◊atmosphere, so that weather can be accurately forecast. Data from meteorological stations and weather satellites are collated by computer at central agencies, and forecast and ◊weather maps based on current readings are issued at regular intervals. Modern analysis can give useful forecasts for up to six days ahead.

At meteorological stations readings are taken of the factors determining weather conditions: atmospheric pressure, temperature, humidity, wind (using the ◊Beaufort scale), cloud cover (measuring both type of cloud and coverage), and precipitation such as rain, snow, and hail (measured at 12–hour intervals). Satellites are used either to relay information transmitted from the Earth-based stations, or to send pictures of cloud development, indicating wind patterns, and snow and ice cover.

As well as supplying reports for the media, the Meteorological Office in Bracknell, near London, does specialist work for industry, agriculture, and transport.

meter any instrument used for measurement. The term is often compounded with a prefix to denote a specific type of meter: for example, ammeter, voltmeter, flowmeter, or pedometer.

methanal (common name *formaldehyde*) HCHO gas at ordinary temperatures, condensing to a liquid at $-21°C/-5.8°F$. It has a powerful, penetrating smell. Dissolved in water, it is used as a biological preservative. It is used in the manufacture of plastics, dyes, foam (for example urea-formaldehyde foam, used in insulation), and in medicine.

methane CH_4 the simplest hydrocarbon of the paraffin series. Colourless, odourless, and lighter than air, it burns with a bluish flame and explodes when mixed with air or oxygen. It is the chief constituent of natural gas and also occurs in the explosive firedamp of coal mines. Methane emitted by rotting vegetation forms marsh gas, which may ignite by spontaneous combustion to produce the pale flame seen over marshland and known as will-o'-the-wisp.

Methane causes about 38% of the warming of the globe through the ◊greenhouse effect; amount of methane in the air is predicted to double over the next 60 years. An estimated 15% of all methane gas into the atmosphere is produced by cows and other cud-chewing animals.

methanogenic bacteria one of a group of primitive bacteria (◊archaebacteria). They give off methane gas as a by-product of their metabolism, and are common in sewage treatment plants and hot springs, where the temperature is high and oxygen is absent.

methanoic acid (common name *formic acid*) HCOOH, a colourless, slightly fuming liquid that freezes at $8°C/46.4°F$ and boils at $101°C/213.8°F$. It occurs in stinging ants, nettles, sweat, and pine needles, and is used in dyeing, tanning, and electroplating.

methanol (common name *methyl alcohol*) CH_3OH the simplest of the alcohols. It can be made by the dry distillation of wood (hence it is also known as wood alcohol), but is usually made from coal or natural gas. When pure, it is a colourless, flammable liquid with a pleasant odour, and is highly poisonous.

Methanol is used to produce formaldehyde (from which resins and plastics can be made), methyl-ter-butyl ether (MTB, a replacement for lead as an octane-booster in petrol), vinyl acetate (largely used in paint manufacture), and petrol.

methyl alcohol common name for ◊methanol.

methylated spirit alcohol that has been rendered undrinkable, and is used for industrial purposes, as a fuel for spirit burners or a solvent. It is nevertheless drunk by some individuals, resulting eventually in death. One of the poisonous substances in it is ◊methanol, or methyl alcohol, and this gives it its name. (The 'alcohol' of alcoholic drinks is ethanol.)

methyl benzene alternative name for ◊toluene.

methyl orange $C_{14}H_{14}N_3NaO_3S$ orange-yellow powder used as an acid–base indicator in chemical tests, and as a stain in the preparation of slides of biological material. Its colour changes with pH; below pH 3.1 it is red, above pH 4.4 it is yellow.

metre SI unit (symbol m) of length, equivalent to

Weather satellites: forecasting from around the world

Forecasting the weather has always posed problems, especially in medium-to-high latitudes, because the weather here is modified daily by moving air masses— cyclonic depressions. These bring rain and wind and have variable speeds, making forecasting difficult. The daily weather is also influenced by complex changes in the upper atmosphere, best observed from space.

Today a huge amount of information is available, especially from remote-sensing techniques. This is constantly relayed to sophisticated computers which can assess the data and produce a forecast almost immediately. The computer is given current weather information and compares this with the situation over, for example, the previous few hours. It can map features such as rain belts as they would be expected to develop in the future, based on past information.

Weather satellites

Today we expect to see images from space on our TV weather forecasts, showing the cloud systems swirling above us, and computer-generated models predicting where rain belts will be in a few hours' time. Such information has only become available through the use of many satellites hundreds of miles in the sky, monitoring and relaying the progress of air masses back to powerful computers in the earthbound weather stations.

There are two types of meteorological satellite. In one system, satellites orbit at the equator, in the same direction and at the same speed as the Earth, over 37,000 km/23,000 mi above the surface. They are stationary with relation to the ground and can follow the development of weather patterns. Alternatively, satellites orbit the Earth from one pole to the other at a lower altitude, about 800 km/500 mi. They sweep along a different longitude with each orbit, and produce a series of strip-like scans of developing weather systems.

In 1961, when President Kennedy called for the landing of a man on the Moon, he also asked Congress to finance a network of weather-monitoring satellites. TIROS 1 was the first; it produced images of cloud systems. These satellites were primitive by modern standards, taking available-light television photographs of clouds, which were then fed back to Earth stations. Newer generations of the TIROS satellites use infrared scanners to detect moisture and temperature at different levels in the sky. Radar is extensively used to show cloud images and the intensity of rainfall (radar maps are frequently broadcast on TV forecasts). Microwave radar gives high-quality images, seeing through any cloud cover.

By the mid 1980s the USA had five GEOS (geostationary operational satellites); together with the European Meteosats, India's Insat-B and Japan's GMS2, these formed the World Weather Watch network. International cooperation allows rapid access of information and the monitoring of changing weather systems on a global scale.

Weather patterns and ozone concentrations

The USA has had many meteorological satellites in orbit, including the NOAA machines, launched into near-polar or polar orbits. These cover the planet from about 840 km/520 mi altitude, twice every 24 hours. Some are equipped with high-resolution radiometers recording in the infrared spectrum. They can collect accurate data on sea temperature and can map areas of ice and snow cover and ozone concentrations. Infrared imaging also allows night-time photography of clouds.

The US weather satellites were developed from military machines, and some can be used to monitor military targets. The former USSR also developed many satellites for defence and these were adapted for meteorological use. The USSR put its original Meteor 1 into orbit in 1969. Many derivatives have since been employed, often in near-polar orbits, since Russian forecasting requires the monitoring of severe weather in the northern polar seas. These more recent Meteor satellites are similar to the USA's NOAA machines, and use infrared cameras as well as available-light TV scanners. These satellites can be used to monitor other atmospheric phenomena such as the depletion of the ozone shield around the Earth, pollution of the atmosphere and the movement of potentially dangerous hurricanes and ocean currents like El Niño.

Europe in space

Recent NOAAs have carried instruments provided by the UK Meteorological Office. The European Space Agency made its first steps into space in the early 1980s with a series of Meteosats, hanging in the sky off the African coast. The original satellite reads in the infrared band and, with later Meteosats, collects data on sea temperature, atmospheric radiation, atmospheric moisture and wind speeds.

Chris Pellant

1.093 yards. It is defined by scientists as the length of the path travelled by light in a vacuum during a time interval of 1/299,792,458 of a second.

metric system system of weights and measures developed in France in the 18th century and recognized by other countries in the 19th century. In 1960 an international conference on weights and measures recommended the universal adoption of a revised International System (Système International d'Unités, or SI), with seven prescribed 'base units': the metre (m) for length, kilogram (kg) for mass, second (s) for time, ampere (A) for electric current, kelvin (K) for thermodynamic temperature, candela (cd) for luminous intensity, and mole (mol) for quantity of matter.

Two supplementary units are included in the SI system—the radian (rad) and steradian (sr)—used to measure plane and solid angles. In addition, there are recognized derived units that can be expressed as simple products or divisions of powers of the basic units, with no other integers appearing in the expression; for example, the watt.

Some non-SI units, well established and internationally recognized, remain in use in conjunction with SI: minute, hour, and day in measuring time; multiples or submultiples of base or derived units which have long-established names, such as tonne for mass, the litre for volume; and specialist measures such as the metric carat for gemstones.

Prefixes used with metric units are tera (T) million million times; giga (G) billion (thousand million) times; mega (M) million times; kilo (k) thousand times; hecto (h) hundred times; deca (da) ten times; deci (d) tenth part; centi (c) hundredth part; milli (m) thousandth part; micro (μ) millionth part; nano (n) billionth part; pico (p) trillionth part; femto (f) quadrillionth part; atto (a) quintillionth part.

The metric system was made legal for most purposes in the UK and USA in the 19th century. The UK government agreed to the adoption of SI as the primary system of weights and measures in 1965, but compulsion was abandoned in 1978, although Britain will have to conform to European Community regulations. A Metric Act was passed in the USA in 1975.

mg symbol for *milligram.*

MHD abbreviation for ◊*magnetohydrodynamics.*

mho SI unit of electrical conductance, now called the ◊siemens; equivalent to a reciprocal ohm.

mi symbol for ◊*mile.*

mica group of silicate minerals that split easily into thin flakes along lines of weakness in their crystal structure (perfect basal cleavage). They are glossy, have a pearly lustre, and are found in many igneous and metamorphic rocks. Their good thermal and electrical insulation qualities make them valuable in industry.

Their chemical composition is complicated, but they are silicates with silicon-oxygen tetrahedra arranged in continuous sheets, with weak bonding between the layers, resulting in perfect cleavage. A common example of mica is muscovite (white mica), $KAl_2Si_3AlO_{10}(OH,F)_2$.

MICR abbreviation for ◊magnetic-ink character recognition.

micro- prefix (symbol μ) denoting a one-millionth part (10^{-6}). For example, a micrometre, μm, is one-millionth of a metre.

microbe another name for ◊microorganism.

microbiology the study of organisms that can only be seen under the microscope, mostly viruses and single-celled organisms such as bacteria, protozoa, and yeasts. The practical applications of microbiology are in medicine (since many microorganisms cause disease); in brewing, baking, and other food and beverage processes, where the microorganisms carry out fermentation; and in genetic engineering, which is creating increasing interest in the field of microbiology.

microchip popular name for the silicon chip, or ◊integrated circuit.

microclimate the climate of a small area, such as a woodland, lake, or even a hedgerow. Significant differences can exist between the climates of two neighbouring areas—for example, a town is usually warmer than the surrounding countryside (forming a heat island), and a woodland cooler, darker, and less windy than an area of open land.

Microclimates play a significant role in agriculture and horticulture, as different crops require different growing conditions.

microcomputer or *micro* or *personal computer* small desktop or portable computer, typically designed to be used by one person at a time, although individual computers can be linked in a network so that users can share data and programs. Its central processing unit is a ◊microprocessor, contained on a single integrated circuit.

Microcomputers are the smallest of the four classes of computer (the others are ◊supercomputer, ◊mainframe, and ◊minicomputer). Since the appearance in 1975 of the first commercially available microcomputer, the Altair 8800, micros have become widely accepted in commerce, industry, and education.

microfiche sheet of film on which printed text is photographically reduced. See ◊microform.

microform generic name for media on which text or images are photographically reduced. The main examples are *microfilm* (similar to the film in an ordinary camera) and *microfiche* (flat sheets of film, generally 105 mm/4 in × 148 mm/6 in, holding the equivalent of 420 A4 sheets). Microform has the advantage of low reproduction and storage costs, but it requires special devices for reading the text. It is widely used for archiving and for storing large volumes of text, such as library catalogues.

Computer data may be output directly and quickly in microform by means of COM (computer output on microfilm/microfiche) techniques.

microlight aircraft very light aircraft with a small engine, rather like a powered hang-glider.

micrometer instrument for measuring minute lengths or angles with great accuracy; different types of micrometer are used in astronomical and engineering work.

The type of micrometer used in astronomy consists of two fine wires, one fixed and the other movable, placed in the focal plane of a telescope; the movable wire is fixed on a sliding plate and can be positioned parallel to the other until the object

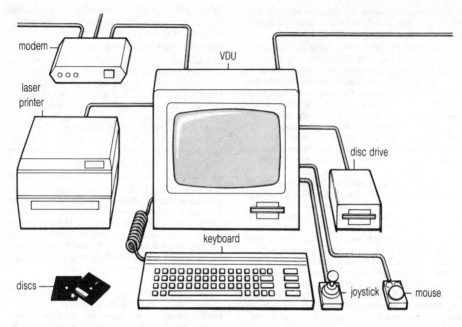

microcomputer *A microcomputer and associated input and output devices. The keyboard, joystick, and mouse are used to input data. The VDU (visual display unit) displays words, numbers, or graphics. It often houses the electronic heart (the CPU, or central processing unit) of the computer, and sometimes has a built-in disc drive. A disc drive reads data and program instructions stored on discs. The laser printer produces a written output. The modem allows the computer to be connected to other computers using telephone lines.*

appears between the wires. The movement is then indicated by a scale on the adjusting screw.

The **micrometer gauge**, of great value in engineering, has its adjustment effected by an extremely accurate fine-pitch screw (◊vernier).

micrometre one-millionth of a ◊metre (symbol μm).

microminiaturization reduction in size and weight of electronic components. The first size reduction in electronics was brought about by the introduction of the ◊transistor. Further reductions were achieved with ◊integrated circuits and the ◊silicon chip.

micron obsolete name for the micrometre, one millionth of a metre.

microorganism or **microbe** living organism invisible to the naked eye but visible under a microscope. Microorganisms include viruses and single-celled organisms such as bacteria, protozoa, yeasts, and some algae. The term has no taxonomic significance in biology. The study of microorganisms is known as microbiology.

MICROORGANISMS: CREATURES THAT ARE ALWAYS WITH US

There are more living organisms on the skin of a single human than there are human beings on the surface of the Earth.

---

microphone primary component in a sound-reproducing system, whereby the mechanical energy of sound waves is converted into electrical signals by means of a ◊transducer. One of the simplest is the telephone receiver mouthpiece, invented by Scottish–US inventor Alexander Graham Bell 1876; other types of microphone are used with broadcasting and sound-film apparatus.

Telephones have a **carbon microphone**, which reproduces only a narrow range of frequencies. For live music, a **moving-coil microphone** is often used. In it, a diaphragm that vibrates with sound waves moves a coil through a magnetic field, thus generating an electric current. The **ribbon microphone** combines the diaphragm and coil. The **condenser microphone** is most commonly used in recording and works by a ◊capacitor.

microprocessor complete computer ◊central processing unit contained on a single ◊integrated circuit, or chip. The appearance of the first microprocessor 1971 designed by Intel for a pocket calculator manufacturer heralded the introduction of the microcomputer. The microprocessor has led to a dramatic fall in the size and cost of computers, and ◊dedicated computers can now be found in washing machines, cars, and so on. Examples of microprocessors are the Intel 8086 family and the Motorola 68000 family.

micropropagation the mass production of plants by placing tiny pieces of plant tissue in sterile glass containers along with nutrients. Perfect clones of superplants are produced in sterile cabinets, with filtered air and carefully controlled light, temperature, and humidity. The system is used for the house-plant industry and for forestry—micropropagation gives immediate results, whereas obtaining genetically homogenous tree seed by traditional means would take over 100 years.

micropyle in flowering plants, a small hole towards one end of the ovule. At pollination the pollen tube growing down from the ◊stigma eventually passes through this pore. The male gamete is contained within the tube and is able to travel to the egg in the interior of the ovule. Fertilization can then take place, with subsequent seed formation and dispersal.

microscope instrument for magnification with high resolution for detail. Optical and electron microscopes are the ones chiefly in use; other types include acoustic, ◊scanning tunnelling, and ◊atomic force microscopes. In 1988 a scanning tunnelling microscope was used to photograph a single protein molecule for the first time.

The *optical microscope* usually has two sets of glass lenses and an eyepiece. It was invented 1609 in the Netherlands by Zacharias Janssen (1580–c. 1638). *Fluorescence microscopy* makes use of fluorescent dyes to illuminate samples, or to highlight the presence of particular substances within a sample. Various illumination systems are also used to highlight details.

The ◊*transmission electron microscope*, developed from 1932, passes a beam of electrons, instead of a beam of light, through a specimen. Since electrons are not visible, the eyepiece is replaced with a fluorescent screen or photographic plate; far higher magnification and resolution are possible than with the optical microscope.

The ◊*scanning electron microscope* (SEM), developed in the mid-1960s, moves a fine beam of electrons over the surface of a specimen, the reflected electrons being collected to form the image. The specimen has to be in a vacuum chamber.

The *acoustic microscope* passes an ultrasonic (ultrahigh-frequency sound) wave through the specimen, the transmitted sound being used to form an image on a computer screen.

The *scanned-probe microscope*, developed in the late 1980s, runs a probe, with a tip so fine that it may consist only of a single atom, across the surface of the specimen, which requires no special preparation. In the *scanning tunnelling microscope*, an electric current that flows through the probe is used to construct an image of the specimen. In the *atomic force microscope*, the force felt by the probe is measured and used to form the image. These instruments can magnify a million times and give images of single atoms.

microtubule tiny tube found in almost all cells with a nucleus. Microtubules help to define the shape of a cell by forming scaffolding for cilia and form fibres of mitotic spindle.

microwave ◊electromagnetic wave with a wavelength in the range 0.3 to 30 cm/0.1 in to 12 in, or 300–300,000 megahertz (between radio waves and ◊infrared radiation). They are used in radar, as carrier waves in radio broadcasting, and in microwave heating and cooking.

microwave heating heating by means of microwaves. Microwave ovens use this form of heating for the rapid cooking or reheating of foods, where heat is generated throughout the food simultaneously. If food is not heated completely, there is a danger of bacterial growth that may lead to food poisoning. Industrially, microwave heating is used for destroying insects in grain and enzymes in processed food, pasteurizing and sterilizing liquids, and drying timber and paper.

Mid-Atlantic Ridge ◊ocean ridge, formed by the movement of plates described by ◊plate tectonics, that runs along the centre of the Atlantic Ocean, parallel to its edges, for some 14,000 km/8,800 mi – almost from the Arctic to the Antarctic.

The Mid-Atlantic Ridge is central because the ocean crust beneath the Atlantic Ocean has continually grown outwards from the ridge at a steady rate during the past 200 million years. Iceland straddles the ridge and was formed by volcanic outpourings.

MIDI (acronym for *m*usical *i*nstruments *d*igital *i*nterface) standard ◊interface that enables electronic musical instruments to be connected to a computer. A computer with a MIDI interface can input and store the sounds produced by the connected instruments, and can then manipulate these sounds in many different ways. For example, a single keystroke may change the key of an entire composition. Even a full written score for the composition may be automatically produced.

migration the movement, either seasonal or as

eyepiece lens
light paths
coarse focusing adjustment
fine focusing adjustment
barrel
alternative objective lenses
objective lens
slide
moves slide
stage light source
condenser
stage
mirror   stand   condenser focus adjuster

microscope *Terms used to describe an optical microscope. In essence, the optical microscope consists of an eyepiece lens and an objective lens, which are used to produce an enlarged image of a small object by focusing light from a light source. Optical microscopes can achieve magnifications of up to 1,500–2,000. Higher magnifications and resolutions are obtained by electron microscopes.*

Seeing the unseeable

In Jan 1991, the Japanese electronics firm Hitachi produced the smallest writing of all time. The message read 'Peace 91', followed by the initials of the Hitachi Central Research Laboratory where the work was done. The letters were less than 1.5 nanometres (1.5 thousand-millionths of a metre) high. They were written by removing individual sulphur atoms from the surface of a crystal of molybdenum disulphide. This was not the first time that a message had been written with individual atoms: in late 1990, researchers at IBM wrote their company name by placing 35 xenon atoms on the surface of a nickel crystal.

Scanning tunnelling microscopes
Both teams used scanning tunnelling microscopes (STMs) to write their messages. These belong to a family of remarkable new microscopes, called scanned probe microscopes, developed in the late 1980s. These microscopes run a metal probe, with a tip so fine that it may consist only of a single atom, across the surface of the specimen. In the scanning tunnelling microscope, an electric current that flows through the probe is used to construct an image of the specimen on a computer screen. A related instrument, the atomic force microscope, uses the force felt by the probe to form the image. These microscopes can magnify a million times and can give images of single atoms and molecules.

The STM can also be used to manipulate individual atoms, picking them up and moving them to new positions. The IBM message was written by bringing the tip of the STM probe very close to a xenon atom, and then increasing the current through the probe. This pins the atom down, allowing it to be dragged to a new position. The Japanese technique was to bring the probe tip very close to an atom in the crystal surface and subject the atom to a pulse of electricity. This blasts the atom away, leaving a space in the regular atomic arrangement of the crystal.

Using atoms to store data?
The ability to manipulate individual atoms could be used to produce miniature data storage devices. The presence or absence of an atom at a particular site in a regular crystal arrangement could represent the 1s or 0s used in binary computer code. Hitachi claim that one thousand million bits of data could be held on a square 10 millionths of a metre across. Currently a computer disc 10 cm/4 in across would be needed to store this much data.

The photon STM
Probe technology has also produced an optical microscope that is capable of seeing details finer than the wavelength of visible light. Called the photon STM, the new microscope can see details at least ten times smaller than the best conventional optical microscopes. The photon STM has an advantage over the electron STM in that the specimen does not need to be electrically conducting, hence it can be used to examine biological samples. Biological samples examined by electron STMs must be coated with a thin film of conducting material before an image can be obtained. The photon STM, however, can produce images of both conducting and non-conducting materials, as long as they are transparent. Scientists at the Oak Ridge National Laboratory, where the new microscope was developed, have produced images of bacteria and viruses.

In the photon STM, the specimen is placed on the upper surface of a transparent prism. Light is shone upwards so that it reflects from the inside of the prism surface. A fine fibre optic probe is brought close to the specimen, and particles of light—photons—are collected by the probe as it travels across the specimen. The photons are collected by a photomultiplier and converted into an electrical signal. A computer forms an image from these signals.

Researchers expect to be able to increase the power of the photon STM in a number of ways. Firstly, they plan to reduce the wavelength of light used. At present they are using visible light, but plan to change to shorter-wavelength ultraviolet light, and then to X-rays. Another improvement would be to use a probe with a finer tip. This could be achieved by chemically etching the quartz probe until the tip is just 150 nanometres across. Finally, resolution could be increased by improving control of the scanning mechanism. As in all scanned probe microscopes, the distance between the probe and the specimen must be controlled and measured very accurately. These improvements should enable images of individual atoms to be produced, although the resolution of the photon STM will never match that of the electron STM.

Peter Lafferty

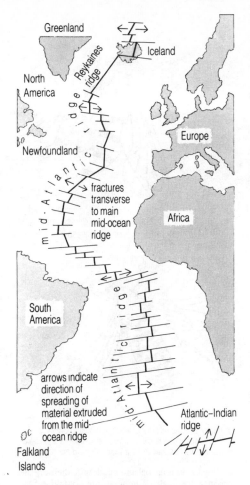

Mid-Atlantic Ridge *The Mid-Atlantic Ridge is the boundary between the crustal plates that form America, and Europe and Africa. An oceanic ridge cannot be curved since the material welling up to form the ridge flows at a right angle to the ridge. The ridge takes the shape of small straight sections offset by fractures transverse to the main ridge.*

part of a single life cycle, of certain animals, chiefly birds and fish, to distant breeding or feeding grounds.

The precise methods by which animals navigate and know where to go are still obscure. Birds have much sharper eyesight and better visual memory of ground clues than humans, but in long-distance flights appear to navigate by the Sun and stars, possibly in combination with a 'reading' of the Earth's magnetic field through an inbuilt 'magnetic compass', which is a tiny mass of tissue between the eye and brain in birds. Similar cells occur in 'homing' honeybees and in certain bacteria that use it to determine which way is 'down'. Most striking, however, is the migration of young birds that have never flown a route before and are unaccompanied by adults. It is postulated that they may inherit as part of their genetic code an overall 'sky chart' of their journey that is triggered into use when they become aware of how the local sky pattern, above the place in which they hatch, fits into it. Similar theories have been advanced in the case of fish, such as eels and salmon, with whom vision obviously plays a lesser role, but for whom currents and changes in the composition and temperature of the sea in particular locations may play a part— for example, in enabling salmon to return to the precise river in which they were spawned. Migration also occurs with land animals; for example, lemmings and antelope.

Species migration is the spread of the home range of a species over many years, for example the spread of the collared dove (*Streptopelia decaocto*) from Turkey to Britain over the period 1920–52. Any journey that takes an animal outside its normal home range is called *individual migration*; when the animal does not return to its home range it is called *removal migration*. An example of *return migration* is the movement of birds that fly south for the winter and return to their home ranges in the spring. Many types of whale also make return migrations. In *remigration*, the return leg of the migration is completed by a subsequent generation; for example, locust swarms migrate, but each part of the circuit is completed by a different generation.

Milankovitch hypothesis the combination of factors governing the occurrence of ◊ice ages proposed in 1930 by the Yugoslav geophysicist Milutin Milankovitch (1879–1958). These include the variation in the angle of the Earth's axis, and the geometry of the Earth's orbit around the Sun.

mile imperial unit of linear measure. A statute mile is equal to 1,760 yards (1.60934 km), and an international nautical mile is equal to 2,026 yards (1,852 m).

milk secretion of the ◊mammary glands of female mammals, with which they suckle their young (during ◊lactation). Over 85% is water, the remainder comprising protein, fat, lactose (a sugar), calcium, phosphorus, iron, and vitamins. The milk of cows, goats, and sheep is often consumed by humans, but regular drinking of milk after infancy is principally a Western practice; for people in most of the world, milk causes flatulence and diarrhoea.

Milk composition varies among species, depending on the nutritional requirements of the young; human milk contains less protein and more lactose than that of cows.

Skimmed milk is what remains when the cream has been separated from milk. It is readily dried and is available in large quantities at low prices, so it is often sent as food aid to Third World countries. *Evaporated milk* is milk reduced by heat until it reaches about half its volume. *Condensed milk* is concentrated to about a third of its original volume with added sugar.

The average consumption of milks in the UK is about 2.5–3 1/4–5 pt per week.

milk of magnesia common name for a suspension of magnesium hydroxide in water. It is commonly used as an antacid.

Milky Way faint band of light crossing the night sky, consisting of stars in the plane of our Galaxy. The name Milky Way is often used for the Galaxy itself. It is a spiral ◊galaxy, about 100,000 light years in diameter, containing at least 100 billion stars. The Sun is in one of its spiral arms, about 25,000 light years from the centre.

The densest parts of the Milky Way, towards the Galaxy's centre, lie in the constellation Sagittarius. In places, the Milky Way is interrupted by lanes of dark dust that obscure light from the stars beyond, such as the Coalsack nebula in Crux (the Southern Cross).

THE MILKY WAY: OUR GALAXY

Most of the galaxy is empty space, with stars spread as thinly as a handful of sand grains scattered across an area the size of the British Isles. If stars were the size of sand grains, our galaxy would be 40,000 km/24,000 mi across, about three times bigger than the Earth.

---

milli- prefix (symbol m) denoting a one-thousandth part (10^{-3}). For example, a millimetre, mm, is one thousandth of a metre.

millibar unit of pressure, equal to one-thousandth of a ◊bar.

millilitre one-thousandth of a litre (ml), equivalent to one cubic centimetre (cc).

millimetre of mercury unit (symbol mmHg) of pressure, used in medicine for measuring blood pressure defined as the pressure exerted by a column of mercury one millimetre high, under the action of gravity.

milling metal machining method that uses a rotating toothed cutting wheel to shape a surface. The term also applies to grinding grain, cacao, coffee, pepper, and other spices.

Mills Cross type of ◊radio telescope consisting of two rows of aerials at right angles to each other, invented 1953 by the Australian radio astronomer Bernard Mills (1920–   ). The cross-shape produces a narrow beam useful for pinpointing the positions of radio sources.

mimicry imitation of one species (or group of species) by another. The most common form is *Batesian mimicry* (named after English naturalist H W Bates), where the mimic resembles a model that is poisonous or unpleasant to eat, and has aposematic, or warning, coloration; the mimic thus benefits from the fact that predators have learned to avoid the model. Hoverflies that resemble bees or wasps are an example. Appearance is usually the basis for mimicry, but calls, songs, scents, and other signals can also be mimicked.

In *Mullerian mimicry*, two or more equally poisonous or distasteful species have a similar colour pattern, thereby reinforcing the warning each gives to predators. In some cases, mimicry is not for protection, but allows the mimic to prey on, or parasitize, the model.

mineral naturally formed inorganic substance with a particular chemical composition and a regularly repeating internal structure. Either in their perfect crystalline form or otherwise, minerals are the constituents of ◊rocks. In more general usage, a mineral is any substance economically valuable for

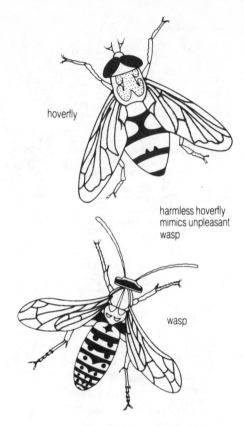

hoverfly

harmless hoverfly mimics unpleasant wasp

wasp

mimicry *Batesian mimicry in which a harmless hoverfly is coloured like an unpleasant wasp in order to confuse a predator. A predator that has tried to eat a wasp will avoid the hoverfly.*

mining (including coal and oil, despite their organic origins). Ice is also a mineral, the crystalline form of water, H_2O.

mineral dressing preparing a mineral ore for processing. Ore is seldom ready to be processed when it is mined; it often contains unwanted rock and dirt. Therefore it is usually crushed into uniform size and then separated from the dirt, or gangue. This may be done magnetically (some iron ores), by washing (gold), by treatment with chemicals (copper ores), or by flotation.

mineral extraction recovery of valuable ores from the Earth's crust. The processes used include open-cast mining, shaft mining, and quarrying, as well as more specialized processes such as those used for oil and sulphur (see, for example, ◊Frasch process).

mineralogy study of minerals. The classification of minerals is based chiefly on their chemical composition and the kind of chemical bonding that holds these atoms together. The mineralogist also studies their crystallographic and physical characters, occurrence, and mode of formation.

The systematic study of minerals began in the 18th century, with the division of minerals into four classes: earths, metals, salts, and bituminous substances, distinguished by their reactions to heat and water.

mineral oil oil obtained from mineral sources, for example coal or petroleum, as distinct from oil obtained from vegetable or animal sources.

mineral salt in nutrition, a simple inorganic chemical that is required by living organisms. Plants usually obtain their mineral salts from the soil, while animals get theirs from their food. Important mineral salts include iron salts (needed by both plants and animals), magnesium salts (needed mainly by plants, to make chlorophyll), and calcium salts (needed by animals to make bone or shell). A ◊trace element is required only in tiny amounts.

minicomputer multiuser computer with a size and processing power between those of a ◊mainframe and a ◊microcomputer. Nowadays almost all minicomputers are based on microprocessors.

Minicomputers are often used in medium-sized businesses and in university departments handling ◊database or other commercial programs and running scientific or graphical applications.

Mini Disc digital audio disc that resembles a computer floppy disc in a 5 cm/2 in square case, with up to an hour's playing time. The system was developed by Sony for release 1993.

Their small size makes Mini Discs particularly suitable for personal and car stereos, and the players are equipped with memory chips that store up to three seconds of music in order to compensate for any skips and jumps in playback.

minor planet another name for an ◊asteroid.

minute unit of time consisting of 60 seconds; also a unit of angle equal to one sixtieth of a degree.

Miocene ('middle recent') fourth epoch of the Tertiary period of geological time, 23.5–5.2 million years ago. At this time grasslands spread over the interior of continents, and hoofed mammals rapidly evolved.

mips (acronym for *m*illion *i*nstructions *p*er *s*econd) in computing, a measure of the speed of a processor. It does not equal the computer power in all cases.

The original IBM PC had a speed of one-quarter mips, but now 50 mips PCs and 100 mips workstations are available.

Mir (Russian 'peace' or 'world') Soviet space station, the core of which was launched 20 Feb 1986. *Mir* is intended to be a permanently occupied space station.

Mir weighs almost 21 tonnes, is approximately 13.5 m/44 ft long, and has a maximum diameter of 4.15 m/13.6 ft. It carries a number of improvements over the earlier ◊Salyut series of space stations, including six docking ports; four of these can have scientific and technical modules attached to them. The first of these was the *Kvant* (quantum) astrophysics module, launched 1987. This had two main sections: a main experimental module, and a service module that would be separated in orbit. The experimental module was 5.8 m/19 ft long and had a maximum diameter matching that of *Mir*. When attached to the *Mir* core, *Kvant* added a further 40 cu m/1,413 cu ft of working space to that already there. Among the equipment carried

by *Kvant* were several X-ray telescopes and an ultraviolet telescope.

Mira or **Omicron Ceti** brightest long-period pulsating ◊variable star, located in the constellation ◊Cetus. Mira was the first star discovered to vary periodically in brightness.

In 1596 Dutch astronomer David Fabricus noticed Mira as a third-magnitude object. Because it did not appear on any of the star charts available at the time, he mistook it for a ◊nova. The German astronomer Johann Bayer included it on his star atlas 1603 and designated it Omicron Ceti. The star vanished from view again, only to reappear within a year. Continued observation revealed a periodic variation between third or fourth magnitude and ninth magnitude over an average period of 331 days. Mira can sometimes reach second magnitude and once, in 1779, it almost attained first magnitude.

mirage illusion seen in hot climates of water on the horizon, or of distant objects being enlarged. The effect is caused by the ◊refraction, or bending, of light.

Light rays from the sky bend as they pass through the hot layers of air near the ground, so that they appear to come from the horizon. Because the light is from a blue sky, the horizon appears blue and watery. If, during the night, cold air collects near the ground, light can be bent in the opposite direction, so that objects below the horizon appear to float above it. In the same way, objects such as trees or rocks near the horizon can appear enlarged.

mirror any polished surface that reflects light; often made from 'silvered' glass (in practice, a mercury-alloy coating of glass). A plane (flat) mirror produces a same-size, erect 'virtual' image located behind the mirror at the same distance from it as the object is in front of it. A spherical concave mirror produces a reduced, inverted real image in front or an enlarged, erect virtual image behind it (as in a shaving mirror), depending on how close the object is to the mirror. A spherical convex mirror produces a reduced, erect virtual image behind it (as in a car's rear-view mirror).

In a plane mirror the light rays appear to come from behind the mirror but do not actually do so. The inverted real image from a spherical concave mirror is an image in which the rays of light pass through it. The ◊focal length f of a spherical mirror is half the radius of curvature; it is related to the image distance v and object distance u by the equation $1/v + 1/u = 1/f$.

miscarriage spontaneous expulsion of a fetus from the womb before it is capable of independent survival. Often, miscarriages are due to an abnormality in the developing fetus.

missile rocket-propelled weapon, which may be nuclear-armed (see ◊nuclear warfare). Modern missiles are often classified as surface-to-surface missiles (SSM), air-to-air missiles (AAM), surface-to-air missiles (SAM), or air-to-surface missiles (ASM). A *cruise missile* is in effect a pilotless, computer-guided aircraft; it can be sea-launched from submarines or surface ships, or launched from the air or the ground.

Rocket-propelled weapons were first used by the

Chinese about AD 1100, and were encountered in the 18th century by the British forces. The rocket missile was then re-invented by William Congreve in England around 1805, and remained in use with various armies in the 19th century. The first wartime use of a long-range missile was against England in World War II, by the jet-powered German V1 (*Vergeltungswaffe*, 'revenge weapon' or Flying Bomb), a monoplane (wingspan about 6 m/ 18 ft, length 8.5 m/26 ft); the first rocket-propelled missile with a preset guidance system was the German V2, also launched by Germany against Britain in World War II.

Modern missiles are also classified as strategic or tactical: strategic missiles are the large, long-range *intercontinental ballistic missiles* (ICBMs, capable of reaching targets over 5,500 km/3,400 mi), and tactical missiles are the short-range weapons intended for use in limited warfare (with a range under 1,100 km/680 mi).

Not all missiles are large. There are many missiles that are small enough to be carried by one person. The Stinger, for example, is an anti-aircraft missile fired by a single soldier from a shoulder-held tube. Most fighter aircraft are equipped with missiles to use against enemy aircraft or against ground targets. Other small missiles are launched from a type of truck, called a MLRS (multiple-launch rocket system), that can move around a battlefield. Ship-to-ship missiles like the Exocet have proved very effective in naval battles.

The vast majority of missiles have systems that guide them to their target. The guidance system may consist of radar and computers, either in the missile or on the ground. These devices track the missile and determine the correct direction and distance required for it to hit its target. In the radio-guidance system, the computer is on the ground, and guidance signals are radio-transmitted to the missile. In the inertial guidance system, the computer is on board the missile. Some small missiles have heat-seeking devices fitted to their noses to seek out the engines of enemy aircraft, or are guided by laser light reflected from the target. Others (called TOW missiles) are guided by signals sent along wires that trail behind the missile in flight.

Outside the industrialized countries, 22 states had active ballistic-missile programmes by 1989, and 17 had deployed these weapons: Afghanistan, Argentina, Brazil, Cuba, Egypt, India, Iran, Iraq, Israel, North Korea, South Korea, Libya, Pakistan, Saudi Arabia, South Africa, Syria, and Taiwan. Non-nuclear short-range missiles were used during the Iran–Iraq War 1980–88 against Iraqi cities.

Battlefield missiles used in the 1991 Gulf War include antitank missiles and short-range attack missiles. NATO announced in 1990 that it was phasing out ground-launched nuclear battlefield missiles, and these are being replaced by types of tactical air-to-surface missile (TASM), also with nuclear warheads.

In the Falklands conflict 1982, small, conventionally armed sea-skimming missiles were used (the French Exocet) against British ships by the Argentine forces, and similar small missiles have been used against aircraft and ships elsewhere.

Mississippian US term for the Lower or Early ◊Carboniferous period of geological time, 363–323 million years ago. It is named after the state of Mississippi.

mistral cold, dry, northerly wind that occasionally blows during the winter on the Mediterranean coast of France. It has been known to reach a velocity of 145 kph/90 mph.

mitochondria (singular *mitochondrion*) membrane-enclosed organelles within eukaryotic cells (see ◊eukaryote), containing enzymes responsible for energy production during ◊aerobic respiration. These rodlike or spherical bodies are thought to be derived from free-living bacteria that, at a very early stage in the history of life, invaded larger cells and took up a symbiotic way of life inside. Each still contains its own small loop of DNA called mitochondrial DNA, and new mitochondria arise by division of existing ones.

mitosis in biology, the process of cell division. The genetic material of eukaryotic cells (see ◊eukaryote) is carried on a number of ◊chromosomes. To control their movements during cell division so that both new cells get a full complement, a system of protein tubules, known as the spindle, organizes the chromosomes into position in the middle of the cell before they replicate. The spindle then controls the movement of chromosomes as the cell goes through the stages of division: *interphase, prophase, metaphase, anaphase*, and *telophase*. See also ◊meiosis.

mixture in chemistry, a substance containing two or more compounds that still retain their separate physical and chemical properties. There is no chemical bonding between them and they can be separated from each other by physical means (compare ◊compound).

m.k.s. system system of units in which the base units metre, kilogram, and second replace the centimetre, gram, and second of the ◊c.g.s. system. From it developed the SI system (see ◊SI units).

It simplifies the incorporation of electrical units into the metric system, and was incorporated in SI. For application to electrical and magnetic phenomena, the ampere was added, creating what is called the m.k.s.a. system.

ml symbol for *millilitre*.

mm symbol for *millimetre*.

mmHg symbol for ◊*millimetre of mercury*.

mnemonic in computing, a short sequence of letters used in low-level programming languages (see ◊low-level language) to represent a ◊machine-code instruction.

mobile electrons another term for delocalized electrons (see ◊electrons, delocalized). Mobile electrons are found in metals, graphite, and unsaturated molecules.

mobile ion in chemistry, an ion that is free to move; such ions are found only in the aqueous solutions or melts (molten masses) of an ◊electrolyte. The mobility of the ions in an electrolyte is what allows it to conduct electricity.

Möbius strip structure made by giving a half twist to a flat strip of paper and joining the ends together. It has certain remarkable properties, arising from

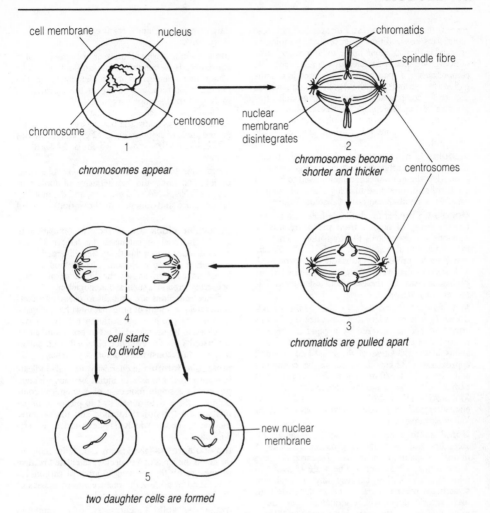

mitosis *The stages of mitosis, the process of cell division that takes place when a plant or animal cell divides for growth or repair. The two daughter cells each receive the same number of chromosomes as were in the original cell.*

the fact that it has only one edge and one side. If cut down the centre of the strip, instead of two new strips of paper, only one long strip is produced. It was invented by the German mathematician August Möbius.

mode in mathematics, the element that appears most frequently in a given set of data. For example, the mode for the data 0, 0, 9, 9, 9, 12, 87, 87 is 9.

modem (acronym for *mo*dulator/*dem*odulator) device for transmitting computer data over telephone lines. Such a device is necessary because the ◊digital signals produced by computers cannot, at present, be transmitted directly over the telephone network, which uses ◊analogue signals. The modem converts the digital signals to analogue, and back again. Modems are used for linking remote terminals to central computers and enable computers to communicate with each other anywhere in the world.

moderator in a ◊nuclear reactor, a material such as graphite or heavy water used to reduce the speed of high-energy neutrons. Neutrons produced by nuclear fission are fast-moving and must be slowed to initiate further fission so that nuclear energy continues to be released at a controlled rate.

Slow neutrons are much more likely to cause ◊fission in a uranium-235 nucleus than to be captured in a U-238 (nonfissile uranium) nucleus. By using a moderator, a reactor can thus be made to work with fuel containing only a small proportion of U-235.

modulation in radio transmission, the intermittent change of frequency, or amplitude, of a radio carrier wave, in accordance with the audio characteristics of the speaking voice, music, or other signal being transmitted. See ◊pulse-code modulation, ◊AM (amplitude modulation), and ◊FM (frequency modulation).

module in construction, a standard or unit that

governs the form of the rest. For example, Japanese room sizes are traditionally governed by multiples of standard tatami floor mats; today prefabricated buildings are mass-produced in a similar way. The components of a spacecraft are designed in coordination; for example, for the Apollo Moon landings the craft comprised a command module (for working, eating, sleeping), service module (electricity generators, oxygen supplies, manoeuvring rocket), and lunar module (to land and return the astronauts).

modulus in mathematics, a number that divides exactly into the difference between two given numbers. Also, the multiplication factor used to convert a logarithm of one base to a logarithm of another base. Also, another name for ◊absolute value.

Mohole US project for drilling a hole through the Earth's crust, so named from the ◊Mohorovičić discontinuity that marks the transition from crust to mantle. Initial tests were made in the Pacific 1961, but the project was subsequently abandoned.

The cores that were brought up illuminated the geological history of the Earth and aided the development of geophysics.

Mohorovičić discontinuity also *Moho* or *M-discontinuity* boundary that separates the Earth's crust and mantle, marked by a rapid increase in the speed of earthquake waves. It follows the variations in the thickness of the crust and is found approximately 32 km/20 mi below the continents and about 10 km/6 mi below the oceans. It is named after the Yugoslav geophysicist Andrija Mohorovičić (1857–1936) who suspected its presence after analysing seismic waves from the Kulpa Valley earthquake 1909.

Mohs' scale scale of hardness for minerals (in ascending order): 1 talc; 2 gypsum; 3 calcite; 4 fluorite; 5 apatite; 6 orthoclase; 7 quartz; 8 topaz; 9 corundum; 10 diamond. The scale is useful in mineral identification because any mineral will scratch any other mineral lower on the scale than itself, and similarly it will be scratched by any other mineral higher on the scale. It is named after German–Austrian mineralogist Friedrich Mohs.

Mohs' table

number	defining mineral	other substances compared
1	talc	
2	gypsum	2½ fingernail
3	calcite	3½ copper coin
4	fluorite	
5	apatite	5½ steel blade
6	orthoclase	5¾ glass
7	quartz	7 steel file
8	topaz	
9	corundum	
10	diamond	

Note: The scale is not regular; diamond, at number 10 the hardest natural substance, is 90 times harder in absolute terms than corundum, number 9.

molarity in chemistry, ◊concentration of a solution expressed as the number of ◊moles in grams of solute per cubic decimetre of solution.

molar solution in chemistry, solution that contains one ◊mole of a substance per litre of solvent.

molar volume volume occupied by one ◊mole (the molecular mass in grams) of any gas at standard temperature and pressure, equal to 2.24136 $\times 10^{-2}$ m³.

mole SI unit (symbol mol) of the amount of a substance. It is defined as the amount of a substance that contains as many elementary entities (atoms, molecules, and so on) as there are atoms in 12 g of the ◊isotope carbon-12.

One mole of an element that exists as single atoms weighs as many grams as its ◊atomic number (so one mole of carbon weighs 12 g), and it contains 6.022045×10^{23} atoms, which is ◊Avogadro's number.

molecular biology study of the molecular basis of life, including the biochemistry of molecules such as DNA, RNA, and proteins, and the molecular structure and function of the various parts of living cells.

molecular clock use of rates of ◊mutation in genetic material to calculate the length of time elapsed since two related species diverged from each other during evolution. The method can be based on comparisons of the DNA or of widely occurring proteins, such as haemoglobin.

Since mutations are thought to occur at a constant rate, the length of time that must have elapsed in order to produce the difference between two species can be estimated. This information can be compared with the evidence obtained from palaeontology to reconstruct evolutionary events.

molecular formula in chemistry, formula indicating the actual number of atoms of each element present in a single molecule of a chemical compound. This is determined by two pieces of information: the ◊empirical formula and the ◊relative molecular mass, which is determined experimentally.

molecular mass (also known as ◊relative molecular mass) the mass of a molecule, calculated relative to one-twelfth the mass of an atom of carbon-12. It is found by adding the relative atomic masses of the atoms that make up the molecule.

molecular solid in chemistry, solid composed of molecules that are held together by relatively weak ◊intermolecular forces. Such solids are low-melting and tend to dissolve in organic solvents. Examples of molecular solids are sulphur, ice, sucrose, and solid carbon dioxide.

molecule group of two or more ◊atoms bonded together. A molecule of an element consists of one or more like ◊atoms; a molecule of a compound consists of two or more different atoms bonded together. Molecules vary in size and complexity from the hydrogen molecule (H_2) to the large ◊macromolecules of proteins. They are held together by ionic bonds, in which the atoms gain or lose electrons to form ◊ions, or by covalent bonds, where electrons from each atom are shared in a new molecular orbital.

According to the molecular or ◊kinetic theory of matter, molecules are in a state of constant motion, the extent of which depends on their temperature, and exert forces on one another.

The shape of a molecule profoundly affects its chemical, physical, and biological properties. Optical ◊isomers (molecules that are mirror images of each other) rotate plane ◊polarized light in opposite directions; isomers of drug molecules may have different biological effects; and ◊enzyme reactions are crucially dependent on the shape of the enzyme and the substrate on which it acts.

The symbolic representation of a molecule is known as its formula. The presence of more than one atom is denoted by a subscript figure—for example, one molecule of the compound water, having two atoms of hydrogen and one atom of oxygen, is shown as H_2O.

The existence of molecules was inferable from Italian physicist Amedeo ◊Avogadro's hypothesis 1811, but only became generally accepted 1860 when proposed by Italian chemist Stanislao Cannizzaro.

LARGE MOLECULES

Most molecules are made up of small numbers of atoms but some contain rather more. For example, a molecule of aspirin contains 21 atoms. A molecule of haemoglobin, a substance found in blood, contains 758 carbon atoms, 1,203 hydrogen atoms, 195 oxygen atoms, 218 nitrogen atoms, 1 iron atom, and 3 sulphur atoms. Rubber molecules may have up to 65,000 atoms.

---

mollusc any invertebrate of the phylum Mollusca with a body divided into three parts, a head, a foot, and a visceral mass. The majority of molluscs are marine animals, but some inhabit fresh water, and a few are terrestrial. They include bivalves, mussels, octopuses, oysters, snails, slugs, and squids. The body is soft, limbless, and cold-blooded. There is no internal skeleton, but many species have a hard shell covering the body.

Molluscs vary in diet, the carnivorous species feeding chiefly upon other members of the phylum. Some are vegetarian. Reproduction is by means of eggs and is sexual; many species are hermaphrodite. The shells of molluscs take a variety of forms: univalve (snail), bivalve (clam), chambered (nautilus), and many other variations. In some cases (for example cuttlefish and squid), the shell is internal. Every mollusc has a fold of skin, the mantle, which covers the whole body or the back only, and secretes the calcareous substance forming the shell. The lower ventral surface forms the locomotory organ, or foot.

Shellfish (oysters, mussels, clams) are commercially valuable, especially when artificially bred and 'farmed'. The Romans, and in the 17th century the Japanese, experimented with advanced methods of farming shellfish, and raft culture of oysters is now widely practised. The cultivation of pearls, pioneered by Kokichi Mikimoto, began in the 1890s and became an important export industry after World War I.

molten state of a solid that has been heated until it melts. Hot molten solids are sometimes called *melts*.

molluscs: classification

phylum	*Mollusca*
class *Monoplacophora*	primitive marine forms, including Neopilina (2 species)
class *Amphineura*	(1,150 species)
1 Aplacophora	wormlike marine forms
2 Polyplacophora	chitons, coat-of-mail shells
class *Gastropoda*	snail-like molluscs, with single or no shell (9,000 species)
1 Prosobranchia	limpets, winkles, whelks
2 Opisthobranchia	seaslugs
3 Pulmonata	land and freshwater snails, slugs
class *Scaphopoda*	tusk shells, marine burrowers (350 species)
class *Bivalvia*	molluscs with a double (two-valved) shell (15,000 species): mussels, oysters, clams, cockles, scallops, tellins, razor shells, shipworms
class *Cephalopoda*	molluscs with shell generally reduced, arms to capture prey, and beaklike mouth; body bilaterally symmetrical and nervous system well developed (750 species): squids, cuttlefish, octopuses, pearly nautilus, argonaut

---

molybdenite molybdenum sulphide, MoS_2, the chief ore mineral of molybdenum. It possesses a hexagonal crystal structure similar to graphite, has a blue metallic lustre, and is very soft (1–1.5 on Mohs' scale). The largest source of molybdenite is a deposit in Colorado, USA.

molybdenum (Greek *malybdos* 'lead') heavy, hard, lustrous, silver-white, metallic element, symbol Mo, atomic number 42, relative atomic mass 95.94. The chief ore is the mineral molybdenite. The element is highly resistant to heat and conducts electricity easily. It is used in alloys, often to harden steels. It is a necessary trace element in human nutrition. It was named 1781 by Swedish chemist Karl Scheele, after its isolation by P J Hjelm (1746–1813), for its resemblance to lead ore.

It has a melting point of 2,620°C, and is not found in the free state. As an aid to lubrication, molybdenum disulphide (MoS_2) greatly reduces surface friction between ferrous metals. Producing countries include Canada, the USA, and Norway.

moment of a force in physics, measure of the turning effect, or torque, produced by a force acting on a body. It is equal to the product of the force and the perpendicular distance from its line of action to the point, or pivot, about which the body will turn. Its unit is the newton metre.

If the magnitude of the force is F newtons and the perpendicular distance is d metres then the moment is given by:

$$moment = Fd$$

See also ◊couple.

moment of inertia in physics, the sum of all the point masses of a rotating object multiplied by the squares of their respective distances from the axis of rotation. It is analogous to the ◊mass of a stationary object or one moving in a straight line.

In linear dynamics, Newton's second law of motion states that the force F on a moving object equals the products of its mass m and acceleration *a* (*F* = *ma*); the analogous equation in rotational dynamics is *T* = *I*α, where *T* is the torque (the turning effect of a force) that causes an angular acceleration α and *I* is the moment of inertia. For a given object, *I* depends on its shape and the position of its axis of rotation.

momentum in physics, the product of the mass of a body and its linear velocity. The *angular momentum* of an orbiting or rotating body is the product of its moment of inertia and its angular velocity. The momentum of a body does not change unless it is acted on by an external force; angular momentum does not change unless it is acted upon by a turning force, or torque.

The law of conservation of momentum is one of the fundamental concepts of classical physics. It states that the total momentum of all bodies in a closed system is constant and unaffected by processes occurring within the system. Angular momentum is similarly conserved in a closed system.

monazite mineral, $(Ce,La,Th)PO_4$, yellow to red, valued as a source of ◊lanthanides or rare earths, including cerium and europium; generally found in placer deposit (alluvial) sands.

monocarpic or *hapaxanthic* describing plants that flower and produce fruit only once during their life cycle, after which they die. Most ◊annual plants and ◊biennial plants are monocarpic, but there are also a small number of monocarpic ◊perennial plants that flower just once, sometimes after as long as 90 years, dying shortly afterwards, for example, century plant *Agave* and some species of bamboo *Bambusa*. The general biological term related to organisms that reproduce only once during their lifetime is ◊semelparity.

monoclinic sulphur allotropic form of sulphur, formed from sulphur that crystallizes above 96°C. Once formed, it very slowly changes into the rhombic allotrope.

monoclonal antibody (MAB) antibody produced by fusing an antibody-producing lymphocyte with a cancerous myeloma (bone-marrow) cell. The resulting fused cell, called a hybridoma, is immortal and can be used to produce large quantities of a single, specific antibody. By choosing antibodies that are directed against antigens found on cancer cells, and combining them with cytotoxic drugs, it is hoped to make so-called magic bullets that will be able to pick out and kill cancers.

It is the antigens on the outer cell walls of germs entering the body that provoke the production of antibodies as a first line of defence against disease. Antibodies 'recognize' these foreign antigens, and, in locking on to them, cause the release of chemical signals in the bloodstream to alert the immune system for further action. MABs are copies of these natural antibodies, with the same ability to recognize specific antigens. Introduced into the body, they can be targeted at disease sites.

The full potential of these biological missiles, developed by César Milstein and others at Cambridge University, England, 1975, is still under investigation. However, they are already in use in blood-grouping, in pinpointing viruses and other sources of disease, in tracing cancer sites, and in developing vaccines.

monocotyledon angiosperm (flowering plant) having an embryo with a single cotyledon, or seed leaf (as opposed to ◊dicotyledons, which have two). Monocotyledons usually have narrow leaves with parallel veins and smooth edges, and hollow or soft stems. Their flower parts are arranged in threes. Most are small plants such as orchids, grasses, and lilies, but some are trees such as palms.

monoecious having separate male and female flowers on the same plant. Maize (*Zea mays*), for example, has a tassel of male flowers at the top of the stalk and a group of female flowers (on the ear, or cob) lower down. Monoecism is a way of avoiding self-fertilization. ◊Dioecious plants have male and female flowers on separate plants.

monohybrid inheritance pattern of inheritance seen in simple ◊genetics experiments, where the two animals (or two plants) being crossed are genetically identical except for one gene.

This gene may code for some obvious external features such as seed colour, with one parent having green seeds and the other having yellow seeds. The offspring are monohybrids, that is, hybrids for one gene only, having received one copy of the gene from each parent. Known as the F1 generation, they are all identical, and usually resemble one parent, whose version of the gene (the dominant ◊allele) masks the effect of the other version (the recessive allele). Although the characteristic coded for by the recessive allele (for example, green seeds) completely disappears in this generation, it can reappear in offspring of the next generation if they have two recessive alleles. On average, this will occur in one out of four offspring from a cross between two of the monohybrids. The next generation (called F2) show a 3:1 ratio for the characteristic in question, 75% being like the original parent with the recessive allele. Austrian biologist Gregor Mendel first carried out experiments of this type (crossing varieties of artificially bred plants, such as peas) and they revealed the principles of genetics. The same basic mechanism underlies all inheritance, but in most plants and animals there are so many genetic differences interacting to produce the external appearance (phenotype) that such simple, clear-cut patterns of inheritance are not evident.

monomer chemical compound composed of simple molecules from which ◊polymers can be made. Under certain conditions the simple molecules (of the monomer) join together (polymerize) to form a very long chain molecule (macromolecule) called a polymer. For example, the polymerization of ethene (ethylene) monomers produces the polymer polyethene (polyethylene).

$$2nCH_2=CH_2 \rightarrow (CH_2-CH_2-CH_2-CH_2)_n$$

monorail railway that runs on a single rail; the cars can be balanced on it or suspended from it. It was invented 1882 to carry light loads, and when run by electricity was called a *telpher*.

The Wuppertal Schwebebahn, which has been running in Germany since 1901, is a suspension monorail, where the passenger cars hang from an arm fixed to a trolley that runs along the rail. Today most monorails are of the straddle type, where the passenger cars run on top of the rail. They are used to transport passengers between terminals at some airports; Japan has a monorail (1964) running from the city centre to Hameda airport.

monosaccharide or *simple sugar* ◊carbohydrate that cannot be hydrolyzed (split) into smaller carbohydrate units. Examples are glucose and fructose, both of which have the molecular formula $C_6H_{12}O_6$.

monosodium glutamate (MSG) $NaC_5H_8NO_4$ a white, crystalline powder, the sodium salt of glutamic acid (an ◊amino acid found in proteins that plays a role in the metabolism of plants and animals). It is used to enhance the flavour of many packaged and 'fast foods', and in Chinese cooking. Ill effects may arise from its overconsumption, and some people are very sensitive to it, even in small amounts. It is commercially derived from vegetable protein.

monotreme any member of the order Monotremata, the only living egg-laying mammals, found in Australasia. They include the echidnas and the platypus.

monsoon wind pattern that brings seasonally heavy rain to S Asia; it blows towards the sea in winter and towards the land in summer. The monsoon may cause destructive flooding all over India and SE Asia from April to Sept. Thousands of people are rendered homeless each year.

The monsoon cycle is believed to have started about 12 million years ago with the uplift of the Himalayas.

Montréal Protocol international agreement, signed 1987, to stop the production of chemicals that are ◊ozone depleters by the year 2000.

Originally the agreement was to reduce the production of ozone depleters by 35% by 1999. The green movement criticized the agreement as inadequate, arguing that an 85% reduction in ozone depleters would be necessary just to stabilize the ozone layer at 1987 levels. The protocol (under the Vienna Convention for the Protection of the Ozone Layer) was reviewed 1992. Amendments added another 11 chemicals to the original list of eight chemicals suspected of harming the ozone layer. A controversial amendment concerns a fund established to pay for the transfer of ozone-safe technology to poor countries.

moon in astronomy, any natural ◊satellite that orbits a planet. Mercury and Venus are the only planets in the Solar System that do not have moons.

Moon natural satellite of Earth, 3,476 km/2,160 mi in diameter, with a mass 0.012 (approximately one-eightieth) that of Earth. Its surface gravity is only 0.16 (one-sixth) that of Earth. Its average distance from Earth is 384,400 km/238,855 mi, and it orbits in a west-to-east direction every 27.32 days (the *sidereal month*). It spins on its axis with one side permanently turned towards Earth. The Moon has no atmosphere or water.

The Moon is illuminated by sunlight, and goes through a cycle of phases of shadow, waxing from *new* (dark) via *first quarter* (half Moon) to *full*, and waning back again to new every 29.53 days (the *synodic month*, also known as a *lunation*). On its sunlit side, temperatures reach 110°C/

230°F, but during the two-week lunar night the surface temperature drops to −170°C/−274°F.

The origin of the Moon is still open to debate. Scientists suggest the following theories: that it split from the Earth; that it was a separate body captured by Earth's gravity; that it formed in orbit around Earth; or that it was formed from debris thrown off when a body the size of Mars struck Earth. Future exploration of the Moon may detect water permafrost, which could be located at the permanently shadowed lunar poles.

Much of our information about the Moon has been derived from photographs and measurements taken by US and Soviet Moon probes; from geological samples brought back by US Apollo astronauts and by Soviet Luna probes; and from experiments set up by the US astronauts 1969–72. The Moon's composition is rocky, with a surface heavily scarred by ◊meteorite impacts that have formed craters up to 240 km/150 mi across. Rocks brought back by astronauts show the Moon is 4.6 billion years old, the same age as Earth. It differs from Earth in that most of the Moon's surface features were formed within the first billion years of its history when it was hit repeatedly by meteorites. The youngest craters are surrounded by bright rays of ejected rock. The largest scars have been filled by dark lava to produce the lowland plains called seas, or *maria* (plural of ◊mare). These dark patches form the so-called 'man-in-the-Moon' pattern.

One small step for a man, one big step for mankind.

On stepping onto the **Moon** Neil Armstrong (1930–) 21 July 1969

Moon probe crewless spacecraft used to investigate the Moon. Early probes flew past the Moon or crash-landed on it, but later ones achieved soft landings or went into orbit. Soviet probes included the Luna/Lunik series. US probes (Ranger, Surveyor, Lunar Orbiter) prepared the way for the Apollo crewed flights.

The first space probe to hit the Moon was the Soviet *Luna 2*, on 13 Sept 1959 (*Luna 1* had missed the Moon eight months earlier). In Oct 1959, *Luna 3* sent back the first photographs of the Moon's far side. *Luna 9* was the first probe to soft-land on the Moon, on 3 Feb 1966, transmitting photographs of the surface to Earth. *Luna 16* was the first probe to return automatically to Earth carrying Moon samples, in Sept 1970, although by then Moon rocks had already been brought back by US Apollo astronauts. *Luna 17* landed in Nov 1970 carrying a lunar rover, Lunokhod, which was driven over the Moon's surface by remote control from Earth.

The first successful US Moon probe was *Ranger 7*, which took close-up photographs before it hit the Moon on 31 July 1964. *Surveyor 1*, on 2 June 1966, was the first US probe to soft-land on the lunar surface. It took photographs, and later Surveyors analysed the surface rocks. Between 1966 and 1967 a series of five Lunar Orbiters photographed the entire Moon in detail, in preparation for the Apollo landings 1969–72.

In March 1990 Japan put a satellite, *Hagoromo*, into orbit around the Moon. *Hagoromo* weighs only 11 kg/26 lb and was released from a larger Japanese space probe.

moonstone translucent, pearly variety of potassium sodium ◊feldspar, found in Sri Lanka and Myanmar, and distinguished by a blue, silvery, or red opalescent tint. It is valued as a gem.

moraine rocky debris or ◊till carried along and deposited by a ◊glacier. Material eroded from the side of a glaciated valley and carried along the glacier's edge is called lateral moraine; that worn from the valley floor and carried along the base of the glacier is called ground moraine. Rubble dropped at the foot of a melting glacier is called terminal moraine.

When two glaciers converge their lateral moraines unite to form a medial moraine. Debris that has fallen down crevasses and become embedded in the ice is termed englacial moraine; when this is exposed at the surface due to partial melting it becomes ablation moraine.

morphology in biology, the study of the physical structure and form of organisms, in particular their soft tissues.

Morse code international code for transmitting messages by wire or radio using signals of short (dots) and long (dashes) duration, originated by US inventor Samuel Morse for use on his invention, the telegraph (see ◊telegraphy).

The letters SOS (3 short, 3 long, 3 short) form the international distress signal, being distinctive and easily transmitted (popularly but erroneously *save our souls*). By radio telephone the distress call is 'Mayday', for similar reasons (popularly alleged to derive from French *m'aidez*, help me).

moss small nonflowering plant of the class Musci (10,000 species), forming with the ◊liverworts and the ◊hornworts the order Bryophyta. The stem of each plant bears ◊rhizoids that anchor it; there are no true roots. Leaves spirally arranged on its lower portion have sexual organs at their tips. Most mosses flourish best in damp conditions where other vegetation is thin. The peat or bog moss *Sphagnum* was formerly used for surgical dressings.

Mössbauer effect the recoil-free emission of gamma rays from atomic nuclei under certain conditions. The effect was discovered 1958 by German physicist Rudolf Mössbauer (1929–), and used 1960 to provide the first laboratory test of Einstein's general theory of relativity.

The absorption and subsequent re-emission of a gamma ray by an atomic nucleus usually causes it to recoil, so affecting the wavelength of the emitted ray. Mössbauer found that at low temperatures, crystals will absorb gamma rays of a specific wavelength and resonate so that the crystal as a whole recoils while individual nuclei do not. The wavelength of the re-emitted gamma rays is therefore virtually unaltered by recoil and may be measured to a high degree of accuracy. Changes in the wavelength may therefore be studied as evidence of the effect of, say, neighbouring electrons or gravity. For example, the effect provided the first verification of the general theory of relativity by showing that gamma-ray wavelengths become longer in a gravitational field, as predicted by Einstein.

motherboard ◊printed circuit board that contains the main components of a microcomputer. The power, memory capacity, and capability of the microcomputer may be enhanced by adding expansion boards to the motherboard.

motility the ability to move. The term is often restricted to those cells that are capable of independent locomotion, such as spermatozoa. Many single-celled organisms are motile, for example the amoeba. Research has shown that cells capable of movement, including vertebrate muscle cells, have certain biochemical features in common. Filaments of the proteins actin and myosin are associated with motility, as are the metabolic processes needed for breaking down the energy-rich compound ATP (adenosine triphosphate).

motor anything that produces or imparts motion; a machine that provides mechanical power—for example, an ◊electric motor. Machines that burn fuel (petrol, diesel) are usually called engines, but the internal-combustion engine that propels vehicles has long been called a motor, hence 'motoring' and 'motorcar'. Actually the motor is a part of the car engine.

motorcar another term for ◊car.

motorcycle or *motorbike* two-wheeled vehicle propelled by a ◊petrol engine. The first successful motorized bicycle was built in France 1901, and British and US manufacturers first produced motorbikes 1903.

In 1868 Ernest and Pierre Michaux in France experimented with a steam-powered bicycle, but the steam power unit was too heavy and cumbersome. Gottlieb Daimler, a German engineer, created the first motorcycle when he installed his lightweight petrol engine in a bicycle frame 1885. Daimler soon lost interest in two wheels in favour of four and went on to pioneer the ◊car.

The first really successful two-wheel design was devised by Michael and Eugene Werner in France 1901. They adopted the classic motorcycle layout with the engine low down between the wheels. Harley Davidson in the USA and Triumph in the UK began manufacture 1903. Road races like the Isle of Man TT (Tourist Trophy), established 1907, helped improve motorcycle design and it soon evolved into more or less its present form. Until the 1970s British manufacturers predominated but today Japanese motorcycles, such as Honda, Kawasaki, Suzuki, and Yamaha, dominate the world market. They make a wide variety of machines, from mopeds (lightweights with pedal assistance) to streamlined superbikes capable of speeds up to 250 kph/160 mph. There is still a smaller but thriving Italian motorcycle industry, making more specialist bikes. Laverda, Moto Guzzi, and Ducati continue to manufacture in Italy.

The lightweight bikes are generally powered by a two-stroke petrol engine (see ◊two-stroke cycle), while bikes with an engine capacity of 250 cc or more are generally four-strokes (see ◊four-stroke cycle). However, many special-use larger bikes (such as those developed for off-road riding and racing) are two-stroke. Most motorcycles are air-cooled—their engines are surrounded by metal fins to offer a large surface area—although some have a water-cooling system similar to that of a car. Most small bikes have single-cylinder engines, but larger machines can have as many as six. The single-cylinder engine is economical and was popular in British manufacture, then the Japanese developed multiple-cylinder models, but there has recently been some return to single-cylinder engines. In the majority of bikes a chain carries the drive from the engine to the rear wheel, though some machines are now fitted with shaft drive.

motor effect tendency of a wire carrying an electric current in a magnetic field to move. The direction of the movement is given by the left-hand rule (see ◊Fleming's rules). This effect is used in the ◊electric motor. It also explains why streams of electrons produced, for instance, in a television tube can be directed by electromagnets.

motor nerve in anatomy, any nerve that transmits impulses from the central nervous system to muscles or body organs. Motor nerves cause voluntary and involuntary muscle contractions, and stimulate glands to secrete hormones.

mould mainly saprophytic fungi (see ◊fungus) living on foodstuffs and other organic matter, a few being parasitic on plants, animals, or each other. Many moulds are of medical or industrial importance; for example, penicillin.

moulding use of a pattern, hollow form, or matrix to give a specific shape to something in a plastic or molten state. It is commonly used for shaping plastics, clays, and glass. When metals are used, the process is called ◊casting.

In *injection moulding*, molten plastic, for example, is injected into a water-cooled mould and takes the shape of the mould when it solidifies. In *blow moulding*, air is blown into a blob of molten plastic inside a hollow mould. In *compression moulding*, synthetic resin powder is simultaneously heated and pressed into a mould.

moulting periodic shedding of the hair or fur of mammals, feathers of birds, or skin of reptiles. In mammals and birds, moulting is usually seasonal and is triggered by changes of day length.

The term is also often applied to the shedding of the ◊exoskeleton of arthropods, but this is more correctly called ◊ecdysis.

mountain natural upward projection of the Earth's surface, higher and steeper than a hill. The process of mountain building (orogeny) consists of volcanism, folding, faulting, and thrusting, resulting from the collision and welding together of two tectonic plates.

This process deforms the rock and compresses the sediment between the two plates into mountain chains.

Mount Palomar astronomical observatory, 80 km/ 50 mi NE of San Diego, California, USA. It has a 5–m/200–in diameter reflector called the Hale. Completed 1948, it was the world's premier observatory during the 1950s.

Mount Stromlo Observatory astronomical observatory established in Canberra, Australia, in 1923. Important observations have been made there on the Magellanic Clouds, which can be seen clearly from southern Australia.

Mount Wilson site near Los Angeles, California, of the 2.5–m/100–in Hooker telescope, opened 1917, with which Edwin Hubble discovered the expansion of the universe. It was closed in 1985 when the Carnegie Institution withdrew its support. Two solar telescopes in towers 18.3 m/60 ft and 45.7 m/150 ft tall, and a 1.5–m/60–in reflector opened 1908, still operate there.

mouse in computing, an input device used to control a pointer on a computer screen. It is a feature of ◊graphical user interface (GUI) systems. The mouse is about the size of a pack of playing cards, is connected to the computer by a wire, and incorporates one or more buttons that can be pressed. Moving the mouse across a flat surface causes a corresponding movement of the pointer. In this way, the operator can manipulate objects on the screen and make menu selections.

The mouse was invented 1963 at the Stanford Research Institute, USA, by Douglas Engelbart, and developed by the Xerox Corporation in the 1970s. The first was made of wood; the Microsoft mouse was introduced 1983, and the Apple Macintosh mouse 1984. Mice work either mechanically (with electrical contacts to sense the movement in two planes of a ball on a level surface), or optically (photoelectric cells detecting movement by recording light reflected from a grid on which the mouse is moved).

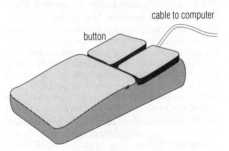
cable to computer

button

mouse *The mouse, which controls the movement of a pointer on a computer screen, is a major development in computer ergonomics, freeing the user, to a large extent, from the more complicated layout of the keyboard.*

mouth cavity forming the entrance to the digestive tract. In land vertebrates, air from the nostrils enters the mouth cavity to pass down the trachea. The mouth in mammals is enclosed by the jaws, cheeks, and palate.

moving-coil meter instrument used to detect and measure electrical current. A coil of wire pivoted between the poles of a permanent magnet is turned by the motor effect of an electric current (by which a force acts on a wire carrying a current in a magnetic field). The extent to which the coil turns can then be related to the magnitude of the current.

The sensitivity of the instrument depends directly upon the strength of the permanent magnet used, the number of turns making up the moving coil, and the coil's area. It depends inversely upon the strength of the controlling springs used to restrain the rotation of the coil. By the addition of a suitable resistor, a moving-coil meter can be adapted to read potential difference in volts.

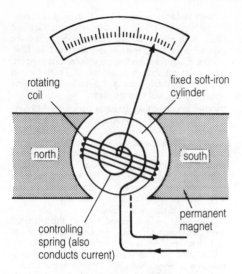

rotating coil

fixed soft-iron cylinder

north

south

permanent magnet

controlling spring (also conducts current)

moving-coil meter *A simple moving-coil meter. Direct electric current (DC) flowing through the wire coil combined with the presence of a magnetic field causes the coil to rotate; this in turn moves a pointer across a calibrated scale so that the degree of rotation can be related to the magnitude of the current.*

mp in chemistry, abbreviation for *melting point*.

MS-DOS (abbreviation for *Microsoft Disc Operating System*) computer ◊operating system produced by Microsoft Corporation, widely used on ◊microcomputers with Intel × 86 family microprocessors. A version called PC-DOS is sold by IBM specifically for its personal computers. MS-DOS and PC-DOS are usually referred to as DOS. MS-DOS first appeared 1981, and was similar to an earlier system from Digital Research called CP/M.

MTBF abbreviation for *mean time between failures*, the statistically average time a component can be used before it goes wrong. The MTBF of a computer hard disc, for example, is around 150,000 hours.

mucous membrane thin skin lining all animal body cavities and canals that come into contact with the air (for example, eyelids, breathing and digestive passages, genital tract). It secretes mucus, a moistening, lubricating, and protective fluid.

mucus lubricating and protective fluid, secreted by mucous membranes in many different parts of the body. In the gut, mucus smooths the passage of food and keeps potentially damaging digestive enzymes away from the gut lining. In the lungs, it traps airborne particles so that they can be expelled.

mudstone fine-grained sedimentary rock made up of clay-to silt-sized particles (up to 0.0625 mm/ 0.0025 in).

Mullard Radio Astronomy Observatory radio observatory of the University of Cambridge, England. Its main instrument is the Ryle Telescope, eight dishes 12.8 m/42 ft wide in a line of 5 km/3 mi long, opened 1972.

multimedia computer system that combines audio and video components to create an interactive application that uses text, sound, and graphics (still, animated, and video sequences). For example, a multimedia database of musical instruments may allow a user not only to search and retrieve text about a particular instrument but also to see pictures of it and hear it play a piece of music.

Multimedia PCs are frequently fitted with CD drives because of the high storage capacity of CD-ROM discs.

multiple birth in humans, the production of more than two babies from one pregnancy. Multiple births can be caused by more than two eggs being produced and fertilized (often as the result of hormone therapy to assist pregnancy), or by a single fertilized egg dividing more than once before implantation.

Multiple Mirror Telescope telescope on Mount Hopkins, Arizona, USA, opened 1979, consisting of six 1.83-m/72-in mirrors mounted in a hexagon, the light-collecting area of which equals that of a single mirror of 4.5 m/176 in diameter. It is planned to replace the six mirrors with a single mirror 6.5 m/256 in wide.

multiple proportions, law of in chemistry, the principle that if two elements combine with each other to form more than one compound, then the ratio of the masses of one of them that combine with a particular mass of the other is a small whole number.

multiplexer in telecommunications, a device that allows a transmission medium to carry a number of separate signals at the same time—enabling, for example, several telephone conversations to be carried by one telephone line, and radio signals to be transmitted in stereo.

multiplexing in telephony and telegraphy, sending several messages simultaneously along a similar wire. In *frequency-division multiplexing*, signals of different frequency, each carrying a different message, are transmitted. Electrical frequency filters separate the message at the receiving station. In *time-division multiplexing*, the messages are broken into sections and the sections of several messages interleaved during transmission. ◊Pulse-code modulation allows hundreds of messages to be sent simultaneously over a single link.

multistage rocket rocket launch vehicle made up of several rocket stages (often three) joined end to end. The bottom stage fires first, boosting the vehicle to high speed, then it falls away. The next stage fires, thrusting the now lighter vehicle even faster. The remaining stages fire and fall away in turn, boosting the vehicle's payload (cargo) to an orbital speed that can reach 28,000 kph/ 17,500 mph.

multitasking or *multiprogramming* in computing, a system in which one processor appears to run several different programs (or different parts of the same program) at the same time. All the programs are held in memory together and each is allowed to run for a certain period.

For example, one program may run while other programs are waiting for a ◊peripheral device to work or for input from an operator. The ability to multitask depends on the ◊operating system rather than the type of computer. Unix is one of the commonest.

multiuser system or *multiaccess system* in computing, an operating system that enables several users to access the same computer at the same time. Each user has a terminal, which may be local (connected directly to the computer) or remote (connected to the computer via a modem and a telephone line). Multiaccess is usually achieved by *time-sharing*: the computer switches very rapidly between terminals and programs so that each user has sole use of the computer for only a fraction of a second but can work as if she or he had continuous access.

muon an ◊elementary particle similar to the electron except for its mass which is 207 times greater than that of the electron. It has a half-life of 2 millionths of a second, decaying into electrons and ◊neutrinos. The muon was originally thought to be a ◊meson and is thus sometimes called a mu meson, although current opinion is that it is a ◊lepton.

muscle contractile animal tissue that produces locomotion and maintains the movement of body substances. Muscle is made of long cells that can contract to between one-half and one-third of their relaxed length.

Striped muscles are activated by ◊motor nerves under voluntary control; their ends are usually attached via tendons to bones. *Involuntary* or *smooth* muscles are controlled by motor nerves of the ◊autonomic nervous system, and are located in the gut, blood vessels, iris, and various ducts. *Cardiac* muscle occurs only in the heart, and is also controlled by the autonomic nervous system.

An artificial muscle fibre was developed in the USA 1990. Besides replacing muscle fibre, it can be used for substitute ligaments and blood vessels and to prevent tissues sticking together after surgery.

muscovite white mica, $KAl_2Si_3AlO_{10}(OH,F)_2$, a common silicate mineral. It is colourless to silvery white with shiny surfaces, and like all micas it splits into thin flakes along its one perfect cleavage. Muscovite is a metamorphic mineral occurring mainly in schists; it is also found in some granites, and appears as shiny flakes on bedding planes of some sandstones.

mushroom fruiting body of certain fungi, consisting of an upright stem and a spore-producing cap with radiating gills on the undersurface. There are many edible species belonging to the genus *Agaricus*. See also ◊fungus and ◊toadstool.

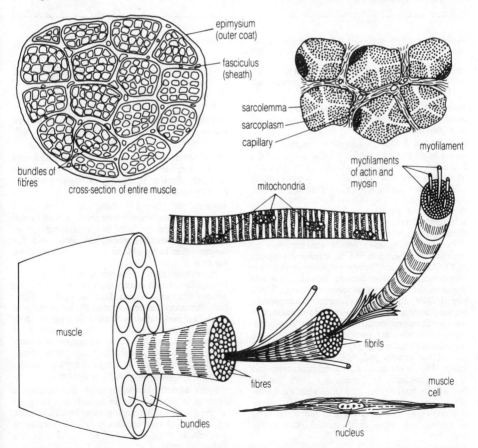

epimysium (outer coat)

fasciculus (sheath)

bundles of fibres

cross-section of entire muscle

sarcolemma

sarcoplasm

capillary

myofilament

mitochondria

myofilaments of actin and myosin

muscle

fibrils

fibres

muscle cell

bundles

nucleus

muscle *The movements of the arm depend on two muscles, the biceps and the triceps. To lift the arm, the biceps shortens and the triceps lengthens. To lower the arm, the opposite occurs: the biceps lengthens and the triceps shortens.*

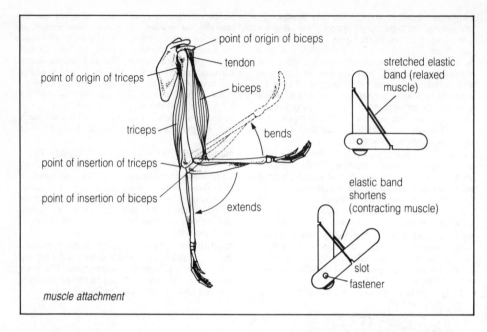

point of origin of biceps

tendon

point of origin of triceps

biceps

stretched elastic band (relaxed muscle)

triceps

bends

point of insertion of triceps

point of insertion of biceps

extends

elastic band shortens (contracting muscle)

slot

fastener

muscle attachment

muscle *Muscles make up 35–45% of the body weight; there are over 650 skeletal muscles. Muscle cells may be up to 20 cm/0.8 in long; they are arranged in bundles, fibres, fibrils, and myofilaments.*

musical instrument digital interface manufacturer's standard for digital music equipment; see ◊MIDI.

mutagen any substance that makes ◊mutation of genes more likely. A mutagen is likely to also act as a ◊carcinogen.

mutation in biology, a change in the genes produced by a change in the ◊DNA that makes up the hereditary material of all living organisms. Mutations, the raw material of evolution, result from mistakes during replication (copying) of DNA molecules. Only a few improve the organism's performance and are therefore favoured by ◊natural selection. Mutation rates are increased by certain chemicals and by radiation.

Common mutations include the omission or insertion of a base (one of the chemical subunits of DNA); these are known as *point mutations*. Larger-scale mutations include removal of a whole segment of DNA or its inversion within the DNA strand. Not all mutations affect the organism, because there is a certain amount of redundancy in the genetic information. If a mutation is 'translated' from DNA into the protein that makes up the organism's structure, it may be in a nonfunctional part of the protein and thus have no detectable effect. This is known as a *neutral mutation*, and is of importance in ◊molecular clock studies because such mutations tend to accumulate gradually as time passes. Some mutations do affect genes that control protein production or functional parts of protein, and most of these are lethal to the organism.

mutual induction in physics, the production of an electromotive force (emf) or voltage in an electric circuit caused by a changing ◊magnetic flux in a neighbouring circuit. The two circuits are often coils of wire, as in a ◊transformer, and the size of the induced emf depends largely on the numbers of turns of wire in each of the coils.

mutualism or ◊*symbiosis* an association between two organisms of different species whereby both profit from the relationship.

mycelium interwoven mass of threadlike filaments or ◊hyphae, forming the main body of most fungi. The reproductive structures, or 'fruiting bodies', grow from the mycelium.

mycorrhiza mutually beneficial (mutualistic) association occurring between plant roots and a soil fungus. Mycorrhizal roots take up nutrients more efficiently than non-mycorrhizal roots, and the fungus benefits by obtaining carbohydrates from the tree.

An *ectotrophic mycorrhiza* occurs on many tree species, which usually grow much better, most noticeably in the seeding stage, as a result. Typically the roots become repeatedly branched and coral-like, penetrated by hyphae of a surrounding fungal ◊mycelium. In an *endotrophic mycorrhiza*, the growth of the fungus is mainly inside the root, as in orchids. Such plants do not usually grow properly, and may not even germinate, unless the appropriate fungus is present.

myelin sheath insulating layer that surrounds nerve cells in vertebrate animals. It acts to speed up the passage of nerve impulses. Myelin is made up of fats and proteins and is formed from up to a hundred layers, laid down by special cells, the *Schwann cells*.

myoglobin globular protein, closely related to ◊haemoglobin and located in vertebrate muscle.

Oxygen binds to myoglobin and is released only when the haemoglobin can no longer supply adequate oxygen to muscle cells.

myrmecophyte plant that lives in association with a colony of ants and possesses specialized organs in which the ants live. For example, *Myrmecodia*, an epiphytic plant from Malaysia, develops root tubers containing a network of cavities inhabited by ants.

Several species of *Acacia* from tropical America have specialized hollow thorns for the same purpose. This is probably a mutualistic (mutually beneficial) relationship, with the ants helping to protect the plant from other insect pests and in return receiving shelter.

nadir the point on the celestial sphere vertically below the observer and hence diametrically opposite the *zenith*. The term is used metaphorically to mean the low point of a person's fortunes.

nail in biology, a hard, flat, flexible outgrowth of the digits of primates (humans, monkeys, and apes). Nails are derived from the ◊claws of ancestral primates.

NAND gate a type of ◊logic gate.

nano- prefix used in ◊SI units of measurement, equivalent to a one-billionth part (10^{-9}). For example, a nanosecond is one-billionth of a second.

nanotechnology the building of devices on a molecular scale. Micromachines, such as gears smaller in diameter than a human hair, have been made at the AT&T Bell laboratories in New Jersey, USA. Building large molecules with useful shapes has been accomplished by research groups in the USA. A robot small enough to travel through the bloodstream and into organs of the body, inspecting or removing diseased tissue, was under development in Japan 1990.

The ◊scanning electron microscope can be used to see and position single atoms and molecules, and to drill holes a nanometre (billionth of a metre) across in a variety of materials. The instrument can be used for ultrafine etching; the entire 28 volumes of the *Encyclopedia Britannica* could be engraved on the head of a pin. A complete electric motor has been built in the USA; it is less than 0.1 mm across with a top speed of 600,000 rpm. It is etched out of silicon, using the ordinary methods of chip manufacturers.

The idea of manipulating material on a nanometre scale – atom by atom—was first discussed by US physicist Richard Feynman 1959. Nanotechnology enthusiasts say that it will eventually be possible to build computers on the molecular scale, produce ultrastrong materials, and allow the molecular correction of most diseases, even the repair of ageing cells.

naphtha the mixtures of hydrocarbons obtained by destructive distillation of petroleum, coal tar, and shale oil. It is raw material for the petrochemical and plastics industries. The term was originally applied to naturally occurring liquid hydrocarbons.

naphthalene $C_{10}H_8$ solid, white, shiny, aromatic hydrocarbon obtained from coal tar. The smell of moth-balls is due to their napthalene content. It is used in making indigo and certain azo dyes, as a mild disinfectant, and as an insecticide.

narcotic pain-relieving and sleep-inducing drug. The chief narcotics induce dependency, and include opium, its derivatives and synthetic modifications (such as morphine and heroin); alcohols (such as ethanol); and barbiturates.

NASA (acronym for *N*ational *A*eronautics and *S*pace *A*dministration) US government agency, founded 1958, for spaceflight and aeronautical research. Its headquarters are in Washington, DC and its main installation is at the ◊Kennedy Space Center in Florida. NASA's early planetary and lunar programs included Pioneer spacecraft from 1958, which gathered data for the later crewed missions, the most famous of which took the first people to the Moon in *Apollo 11* on 16–24 July 1969.

nastic movement plant movement that is caused by an external stimulus, such as light or temperature, but is directionally independent of its source, unlike ◊tropisms. Nastic movements occur as a result of changes in water pressure within specialized cells or differing rates of growth in parts of the plant. Examples include the opening and closing of crocus flowers following an increase or decrease in temperature (*thermonasty*), and the opening and closing of evening-primrose *Oenothera* flowers on exposure to dark and light (*photonasty*).

The leaf movements of the Venus flytrap *Dionaea muscipula* following a tactile stimulus, and the rapid collapse of the leaflets of the sensitive plant *Mimosa pudica* are examples of **haptonasty**. Sleep movements, where the leaves or flowers of some plants adopt a different position at night, are described as **nyctinasty**. Other movement types include **hydronasty**, in response to a change in the atmospheric humidity, and *chemonasty*, in response to a chemical stimulus.

national park land set aside and conserved for public enjoyment. The first was Yellowstone National Park, USA, established 1872. National parks include not only the most scenic places, but also places distinguished for their historic, prehistoric, or scientific interest, or for their superior recreational assets. They range from areas the size of small countries to pockets of just a few hectares.

In England and Wales under the National Park Act 1949 the Peak District, Lake District, Snowdonia, and other areas of natural beauty were designated national parks. Some countries have ◊wilderness areas, with no motorized traffic, no overflying aircraft, no hotels, hostels, shops, or cafés, no industry, and the minimum of management. In the UK national parks are not wholly wilderness or conservation areas, but merely places where planning controls on development are stricter than elsewhere.

National Physical Laboratory (NPL) research establishment, set up 1900 at Teddington, England, under the control of the Department of Industry; the chair of the visiting committee is the president of the Royal Society. In 1944 it began work on a project to construct a digital computer, called the ACE (Automatic Computing Engine), one of the first ever built. It was completed in Nov 1950, embodying many of the ideas of British mathematician Alan Turing.

National Rivers Authority (NRA) UK government agency launched 1989. It is responsible for managing water resources, investigating and regulating pollution, and taking over flood controls and land drainage from the former ten regional water authorities of England and Wales.

native element any nongaseous element that occurs naturally, uncombined with any other element(s). Examples of native nonmetals are carbon and sulphur.

native metal or *free metal* any of the metallic elements that occur in nature in the chemically uncombined or elemental form (in addition to any combined form). They include bismuth, cobalt, copper, gold, iridium, iron, lead, mercury, nickel, osmium, palladium, platinum, ruthenium, rhodium, tin, and silver. Some are commonly found in the free state, such as gold; others occur almost exclusively in the combined state, but under unusual conditions do occur as native metals, such as mercury.

Natural Environment Research Council (NERC) UK organization established by royal charter 1965 to undertake and support research in the earth sciences, to give advice both on exploiting natural resources and on protecting the environment, and to support education and training of scientists in these fields of study. Research areas include geothermal energy, industrial pollution, waste disposal, satellite surveying, acid rain, biotechnology, atmospheric circulation, and climate. Research is carried out principally within the UK but also in Antarctica and in many Third World countries. It comprises 13 research bodies.

Within the NERC, the research bodies are:

Freshwater Biological Association, British Geological Survey, Institute of Hydrology, Plymouth Marine Laboratory, Institute of Oceanographic Sciences, Deacon Laboratory, Proudman Oceanographic Laboratory, Institute of Virology, Institute of Terrestrial Ecology, Scottish Marine Biological Association, Sea Mammal Research Unit, Unit of Comparative Plant Ecology, and the NERC Unit for Thematic Information Systems.

natural frequency the frequency at which a mechanical system will vibrate freely. A pendulum, for example, always oscillates at the same frequency when set in motion. More complicated systems, such as bridges, also vibrate with a fixed natural frequency. If a varying force with a frequency equal to the natural frequency is applied to such an object the vibrations can become violent, a phenomenon known as ◊resonance.

natural gas mixture of flammable gases found in the Earth's crust (often in association with petroleum), now one of the world's three main fossil fuels (with coal and oil). Natural gas is a mixture of ◊hydrocarbons, chiefly methane, with ethane, butane, and propane.

Before the gas is piped to storage tanks and on to consumers, butane and propane are removed and liquefied to form 'bottled gas'. Natural gas is liquefied for transport and storage, and is therefore often used where other fuels are scarce and expensive.

Test flights of the first aircraft powered by liquefied natural gas began 1989. The Soviet-produced craft will save 9 tonnes/8.9 tons of kerosene on a journey of 2,000 km/1,250 mi.

In the UK from the 1970s, natural gas from the North Sea has superseded coal gas, or town gas, both as a domestic fuel and as an energy source for power stations.

natural logarithm in mathematics, the ◊exponent of a number expressed to base *e*, where *e* represents the ◊irrational number 2.71828... Natural ◊logarithms are also called Napierian logarithms, after their inventor, the Scottish mathematician John Napier.

natural radioactivity radioactivity generated by those radioactive elements that exist in the Earth's crust. All the elements from polonium (atomic number 84) to uranium (atomic number 92) are radioactive. ◊Radioisotopes of some lighter elements are also found in nature (for example potassium-40).

natural selection the process whereby gene frequencies in a population change through certain individuals producing more descendants than others because they are better able to survive and reproduce in their environment. The accumulated effect of natural selection is to produce ◊adaptations such as the insulating coat of a polar bear or the spadelike forelimbs of a mole. The process is slow, relying firstly on random variation in the genes of an organism being produced by ◊mutation and secondly on the genetic ◊recombination of sexual reproduction. It was recognized by Charles Darwin and English naturalist Alfred Russel Wallace as the main process driving ◊evolution.

I have called this principle, by which each slight variation, if useful, is preserved, by the term of Natural Selection.

On **natural selection** Charles Darwin *On the Origin of Species* 1859

nature the living world, including plants, animals, fungi, and all microorganisms, and naturally formed features of the landscape, such as mountains and rivers.

Nature Conservancy Council (NCC) former name of UK government agency divided 1991 into English Nature, Scottish Natural Heritage, and the Countryside Council for Wales.

The NCC was established by act of Parliament 1973 (Nature Conservancy created by royal charter 1949) with the aims of designating and managing national nature reserves and other conservation areas; identifying Sites of Special Scientific Interest; advising government ministers on policies; providing advice and information; and commissioning or undertaking relevant scientific research. In 1991 the Nature Conservancy Council was dissolved and its three regional bodies became autonomous agencies.

nature–nurture controversy or *environment- -heredity controversy* long-standing dispute among philosophers and psychologists over the relative importance of environment, that is upbringing, experience and learning ('nurture'), and heredity, that is genetic inheritance ('nature'), in determining the make-up of an organism, as related to human personality and intelligence.

One area of contention is the reason for differences between individuals, for example, in performing intelligence tests. The environmentalist position assumes that individuals do not differ significantly in their inherited mental abilities and that subsequent differences are due to learning, or to differences in early experiences. Opponents insist that certain differences in the capacities of individuals (and hence their behaviour) can be attributed to inherited differences in their genetic make-up.

nature reserve area set aside to protect a habitat and the wildlife that lives within it, with only restricted admission for the public. A nature reserve often provides a sanctuary for rare species. The world's largest is Etosha Reserve, Namibia; area 99,520 sq km/38,415 sq mi.

nautical mile unit of distance used in navigation, an internationally agreed-on standard (since 1959) equalling the average length of one minute of arc on a great circle of the Earth, or 1,852 m/ 6,076.12 ft. The term formerly applied to various units of distance used in navigation. In the UK the nautical mile was formerly defined as 6,082 ft.

navel small indentation in the centre of the abdomen of mammals, the remains of the site of attachment of the ◊umbilical cord, which connects the fetus to the ◊placenta.

navigation the science and technology of finding the position, course, and distance travelled by a ship, plane, or other craft. Traditional methods include the magnetic ◊compass and ◊sextant. Today the gyrocompass is usually used, together with highly sophisticated electronic methods, employing beacons of radio signals, such as ◊Decca, ◊Loran, and ◊Omega. Satellite navigation uses satellites that broadcast time and position signals.

The US ◊Global Positioning System (GPS), when complete, will feature 24 Navstar satellites that will enable users (including eventually motorists and walkers) to triangulate their position (from any three satellites) to within 15 m/50 ft.

In 1992, 85 nations agreed to take part in trials of a new navigation system which makes use of surplus military space technology left over from the Cold War. The new system, known as FANS or Future Navigation System, will make use of the 24 Russian Glonass satellites and the 24 US GPS satellites. Small computers will gradually be fitted to civil aircraft to process the signals from the satellite, allowing aircraft to navigate with pinpoint accuracy anywhere in the world. The signals from at least three satellites will guide the craft to within a few metres of accuracy. FANS will be used in conjunction with four Inmarsat satellites to provide worldwide communications between pilots and air-traffic controllers.

The winds and waves are always on the side of the ablest of navigators.

On **navigation** Edward Gibbon (1737–1794) *The Decline and Fall of the Roman Empire* 1776–88

navigation, biological the ability of animals or insects to navigate. Although many animals navigate by following established routes or known landmarks, many animals can navigate without such aids; for example, birds can fly several thousand miles back to their nest site, over unknown terrain. Such feats may be based on compass information derived from the position of the Sun, Moon, or stars, or on the characteristic patterns of Earth's magnetic field.

Biological navigation refers to the ability to navigate both in long-distance ◊migrations and over shorter distances when foraging (for example, the honey bee finding its way from the hive to a nectar site and back). Where reliant on known landmarks, birds may home on features that can be seen from very great distances (such as the cloud caps that often form above isolated mid-ocean islands). Even smells can act as a landmark. Aquatic species like salmon are believed to learn the characteristic taste of the river where they hatch and return to it, often many years later. Brain cells in some birds have been found to contain ◊magnetite and may therefore be sensitive to the Earth's magnetic field.

NBS abbreviation for *National Bureau of Standards*, the US federal standards organization, on whose technical standards all US weights and measures are based.

near point the closest position to the eye to which an object may be brought and still be seen clearly. For a normal human eye the near point is about 25 cm; however, it gradually moves further away with age, particularly after the age of 40.

nebula cloud of gas and dust in space. Nebulae are the birthplaces of stars, but some nebulae are produced by gas thrown off from dying stars (see ◊planetary nebula; ◊supernova). Nebulae are classified depending on whether they emit, reflect, or absorb light.

An *emission nebula*, such as the ◊Orion nebula, glows brightly because its gas is energized by stars that have formed within it. In a *reflection nebula*, starlight reflects off grains of dust in the nebula, such as surround the stars of the ◊Pleiades cluster. A *dark nebula* is a dense cloud, composed of molecular hydrogen, which partially or completely absorbs light behind it. Examples include the Coalsack nebula in ◊Crux and the Horsehead nebula in Orion.

THE LARGEST NEBULA

The Tarantula nebula is the largest nebula known: if it lay as close to us as the Orion nebula, it would cast shadows. It is thought to contain a huge star over 1,000 times the mass of the Sun, ten times more massive than any star in the Milky Way.

---

neck the structure between the head and the trunk in animals. In the back of the neck are the upper seven vertebrae, and there are many powerful muscles that support and move the head. In front, the neck region contains the pharynx and trachea, and behind these the oesophagus. The large arteries (carotid, temporal, maxillary) and veins (jugular) that supply the brain and head are also located in the neck.

nectar sugary liquid secreted by some plants from a nectary, a specialized gland usually situated near the base of the flower. Nectar often accumulates in special pouches or spurs, not always in the same location as the nectary. Nectar attracts insects, birds, bats, and other animals to the flower for ◊pollination and is the raw material used by bees in the production of honey.

Néel temperature the temperature at which the ◊susceptibility of an antiferromagnetic material has a maximum value; see also ◊magnetism.

negative/positive in photography, a reverse image, which when printed is again reversed, restoring the original scene. It was invented by English pioneer of photography William Henry Fox Talbot about 1834.

Today the most common snapshots and photos are printed positives (on paper) made from film negatives.

nematode unsegmented worm of the phylum Nematoda. Nematodes are pointed at both ends, with a tough, smooth outer skin. They include many free-living species found in soil and water, including the sea, but a large number are parasites, such as the roundworms and pinworms that live in humans, or the eelworms that attack plant roots. They differ from ◊flatworms in that they have two openings to the gut (a mouth and an anus).

nemesis theory theory of animal extinction, suggesting that a sister star to the Sun caused the extinction of groups of animals such as dinosaurs.

The theory holds that the movement of this as yet undiscovered star disrupts the ◊Oort cloud of comets every 26 million years, resulting in the Earth suffering an increased bombardment from comets at these times. The theory was proposed in 1984 to explain the newly discovered layer of iridium—an element found in comets and meteorites—in rocks dating from the end of dinosaur times. However, many palaeontologists deny any evidence for a 26–million-year cycle of extinctions.

neo-Darwinism the modern theory of ◊evolution, built up since the 1930s by integrating the 19th-century English scientist Charles Darwin's theory of evolution through natural selection with the theory of genetic inheritance founded on the work of the Austrian biologist Gregor Mendel.

neodymium yellowish metallic element of the ◊lanthanide series, symbol Nd, atomic number 60, relative atomic mass 144.24. Its rose-coloured salts are used in colouring glass, and neodymium is used in lasers.

It was named in 1885 by Austrian chemist Carl von Welsbach (1858–1929), who fractionated it away from didymium (originally thought to be an element but actually a mixture of rare-earth metals consisting largely of neodymium, praseodymium, and cerium).

neon (Greek *neon* 'new') colourless, odourless, nonmetallic, gaseous element, symbol Ne, atomic number 10, relative atomic mass 20.183. It is grouped with the ◊inert gases, is non-reactive, and forms no compounds. It occurs in small quantities in the Earth's atmosphere.

Tubes containing neon are used in electric advertising signs, giving off a fiery red glow; it is also used in lasers. Neon was discovered by Scottish chemist William Ramsay and the Englishman Morris Travers.

neoprene synthetic rubber, developed in the USA 1931 from the polymerization of chloroprene. It is much more resistant to heat, light, oxidation, and petroleum than is ordinary rubber.

neoteny in biology, the retention of some juvenile characteristics in an animal that seems otherwise mature. An example is provided by the axolotl, a salamander that can reproduce sexually although still in its larval form.

It has been suggested that new species could arise in this way, and that our own species evolved from its apelike ancestors by neoteny, on the grounds that facially we resemble a young ape.

neper unit used in telecommunications to express a ratio of powers and currents. It gives the attenuation of amplitudes as the natural logarithm of the ratio.

nephron microscopic unit in vertebrate kidneys that forms **urine**. A human kidney is composed of over a million nephrons. Each nephron consists of a filter cup surrounding a knot of blood capillaries and a long narrow collecting tubule in close association with yet more capillaries. Waste materials and water pass from the bloodstream into the filter cup, and essential minerals and some water are reabsorbed from the tubule back into the blood. The urine that is left eventually passes out from the body.

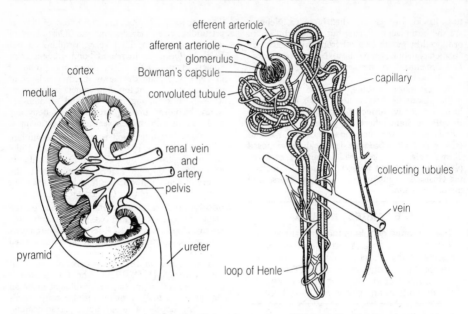

nephron *The kidney (left) contains more than a million filtering units, or nephrons (right), consisting of the glomerulus, Bowman's capsule, and the loop of Henle. Blood flows through the glomerulus—a tight knot of fine blood vessels from which water and chemicals filter into Bowman's capsule. The filtered fluid flows through the convoluted tubule and loop of Henle where most of the water and useful chemicals are reabsorbed into the blood via the capillaries. The waste materials flow to the collecting tubule as urine.*

Neptune in astronomy, the eighth planet in average distance from the Sun. Neptune orbits the Sun every 164.8 years at an average distance of 4.497 billion km/2.794 billion mi. It is a giant gas (hydrogen, helium, methane) planet, with a diameter of 48,600 km/30,200 mi and a mass 17.2 times that of Earth. Its rotation period is 16 hours 7 minutes. The methane in its atmosphere absorbs red light and gives the planet a blue colouring. It is believed to have a central rocky core covered by a layer of ice. Neptune has eight known moons.

Neptune was located 1846 by German astronomers Johann Gottfried Galle and Heinrich d'Arrest (1822–1875) after calculations by English astronomer John Couch Adams and French mathematician Urbain Leverrier had predicted its existence from disturbances in the movement of Uranus. *Voyager 2*, which passed Neptune in Aug 1989, revealed various cloud features, notably an Earth-sized oval storm cloud, the Great Dark Spot, similar to the Great Red Spot on Jupiter.

Neptune has three faint rings and of its eight moons, two (◊Triton and Nereid) are visible from Earth. Six were discovered by the *Voyager 2* probe in 1989, of which Proteus (diameter 415 km/260 mi) is larger than Nereid (300 km/200 mi).

neptunium silvery, radioactive metallic element of the ◊actinide series, symbol Np, atomic number 93, relative atomic mass 237.048. It occurs in nature in minute amounts in ◊pitchblende and other uranium ores, where it is produced from the decay of neutron-bombarded uranium in these ores. The longest-lived isotope, Np-237, has a half-life of 2.2 million years. The element can be produced by bombardment of U-238 with neutrons and is chemically highly reactive.

It was first synthesized in 1940 by US physicists E McMillan (1907–) and P Abelson (1913–), who named it for the planet Neptune (since it comes after uranium as the planet Neptune comes after Uranus). Neptunium was the first ◊transuranic element to be synthesized.

NERC abbreviation for ◊*Natural Environment Research Council*.

nerve strand of nerve cells enclosed in a sheath of connective tissue joining the ◊central and the ◊autonomic nervous systems with receptor and effector organs. A single nerve may contain both ◊motor and sensory nerve cells, but they act independently.

nerve cell or *neuron* elongated cell, part of the ◊nervous system, that transmits information rapidly between different parts of the body. When nerve cells are collected together in significant numbers (as in the ◊brain), they not only transfer information but also process it. The unit of information is the *nerve impulse*, a travelling wave of chemical and electrical changes affecting the membrane of the nerve cell.

The impulse involves the passage of sodium and potassium ions across the nerve-cell membrane. Sequential changes in the permeability of the membrane to positive sodium (Na^+) ions and potassium (K^+) ions, produce electrical signals called action potentials. Impulses are received by the cell body and passed, as a pulse of electric charge, along a long extension, called the ◊axon. The axon terminates at the ◊synapse, a specialized area closely linked to the next cell (which may be another nerve cell or a specialized effector cell such as a muscle). On reaching the synapse, the impulse releases a chemical ◊neurotransmitter, which diffuses across

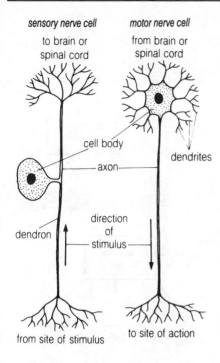

sensory nerve cell   motor nerve cell

to brain or spinal cord   from brain or spinal cord

cell body

dendrites

axon

dendron

direction of stimulus

from site of stimulus   to site of action

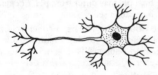

intermediate nerve cell

nerve cell *The anatomy and action of a nerve cell. The nerve cell, or neuron, consists of a cell body with the nucleus and projections called dendrites which pick up messages. An extension of the cell, the axon, connects one cell to the dendrites of the next. When a nerve cell is stimulated, waves of sodium (Na+) and potassium (K+) ions carry an electrical impulse down the axon.*

to the neighbouring cell and there stimulates another impulse or the action of the effector cell.

Nerve impulses travel quickly—in humans, they may be reach speeds of 160 m/525 ft per second along a nerve cell.

nervous system the system of interconnected ◊nerve cells of most invertebrates and all vertebrates. It is composed of the ◊central and ◊autonomic nervous systems. It may be as simple as the nerve net of coelenterates (for example, jellyfishes) or as complex as the mammalian nervous system, with a central nervous system comprising brain and spinal cord, and a peripheral nervous system connecting up with sensory organs, muscles, and glands.

network in computing, a method of connecting computers so that they can share data and ◊peripheral devices, such as printers. The main types are classified by the pattern of the connections—star or ring network, for example—or by the degree of geographical spread allowed; for example, local area networks (LANs) for communication within a

room or building, and wide area networks (WANs) for more remote systems.

One of the most common networking systems is Ethernet, developed in the 1970s (released 1980) at Xerox's Palo Alto Research Center, California, by Rich Seifert, Bob Printis, and Dave Redell.

neural network artificial network of processors that attempts to mimic the structure of nerve cells (neurons) in the human brain. Neural networks may be electronic, optical, or simulated by computer software.

A basic network has three layers of processors: an input layer, an output layer, and a 'hidden' layer in between. Each processor is connected to every other in the network by a system of 'synapses'; every processor in the top layer connects to every one in the hidden layer, and each of these connects to every processor in the output layer. This means that each nerve cell in the middle and bottom layers receives input from several different sources; only when the amount of input exceeds a critical level does the cell fire an output signal.

The chief characteristic of neural networks is

local area network

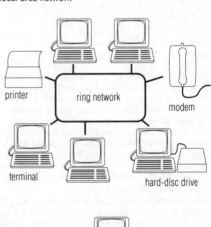

printer   ring network   modem

terminal   hard-disc drive

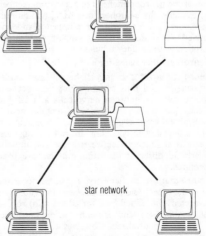

star network

network

wide area network

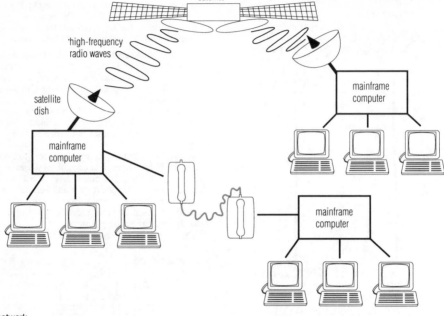

network

their ability to sum up large amounts of imprecise data and decide whether they match a pattern or not. Networks of this type may be used in developing robot vision, matching fingerprints, and analysing fluctuations in stock-market prices. However, it is thought unlikely by scientists that such networks will ever be able accurately to imitate the human brain, which is very much more complicated; it contains around 10 billion nerve cells, whereas current artificial networks contain only a few hundred processors.

neurohormone chemical secreted by nerve cells and carried by the blood to target cells. The function of the neurohormone is to act as a messenger; for example, the neurohormone ADH (antidiuretic hormone) is secreted in the pituitary gland and carried to the kidney, where it promotes water reabsorption in the kidney tubules.

neuron another name for a ◊nerve cell.

neurotransmitter chemical that diffuses across a ◊synapse, and thus transmits impulses between ◊nerve cells, or between nerve cells and effector organs (for example, muscles). Common neurotransmitters are norepinephrine (which also acts as a hormone) and acetylcholine, the latter being most frequent at junctions between nerve and muscle. Nearly 50 different neurotransmitters have been identified.

neutralization in chemistry, a process occurring when the excess acid (or excess base) in a substance is reacted with added base (or added acid) so that the resulting substance is neither acidic nor basic.

In theory neutralization involves adding acid or base as required to achieve ◊pH 7. When the colour of an ◊indicator is used to test for neutraliz-

ation, the final pH may differ from pH 7 depending upon the indicator used.

neutral oxide oxide that has neither acidic nor basic properties (see ◊oxide). Neutral oxides are only formed by ◊nonmetals. Examples are carbon monoxide, water, and nitrogen(I) oxide.

neutral solution solution of pH 7, in which the concentrations of $H^{+(aq)}$ and $OH^{-(aq)}$ ions are equal.

neutrino in physics, any of three uncharged ◊elementary particles (and their antiparticles) of the ◊lepton class, having a mass too close to zero to be measured. The most familiar type, the antiparticle of the electron neutrino, is emitted in the beta decay of a nucleus. The other two are the muon neutrino and the tau neutrino.

Supernova 1987A was the first object outside the Solar System to be observed by neutrino emission. The Sun emits neutrinos, but in smaller numbers than theoretically expected. The shortage of neutrinos is one of the biggest mysteries in modern astrophysics.

THE ELUSIVE NEUTRINO

A neutrino could pass through a solid piece of lead 965 million million km/600 million million mi thick without being absorbed. An estimated 10 million neutrinos from the Sun will pass through your body while you read this sentence. Neutrinos travel at the speed of light: by the time you have read this paragraph, the neutrinos that passed through your body will be further away than the Moon.

Discovering the neutrino

Most fundamental particles are discovered experimentally. Some are predicted by theory, but are often not taken seriously until detected. The neutrino is an exception. Postulated in 1930, its existence was taken as an act of faith for 25 years.

There were good reasons for predicting the neutrino's existence. Unexplained energy losses in certain radioactive decays (called beta decay) suggested the laws of conservation of energy and angular momentum were invalid at subatomic level. The Danish physicist Niels Bohr maintained these laws held only on average, not for each individual process. Austrian physicist Wolfgang Pauli explained it more simply: in beta decay, there was a previously undiscovered neutral particle carrying away the missing energy and momentum. Italian physicist Enrico Fermi introduced the term 'neutrino': Italian for 'little neutral one'. He constructed a successful theory of beta decay based on the neutrino's existence.

An explosive suggestion
Experimental observation of neutrinos proved difficult. They pass easily through solid material, penetrating lead 965 million million km/600 million million mi thick without being absorbed. This makes them extremely hard to detect. In the 1950s, US physicists Frederick Reines and Clyde Cowan put forward a far-fetched scheme using a nuclear weapon as neutrino source. Cowan described the proposed scheme several years later:

'We would dig a shaft near the centre of the explosion about 10 ft in diameter and about 150 feet deep. We would put a tank, 10 ft in diameter and 75 ft long on end at the bottom of the shaft. We would then suspend a detector from the top of the tank, along with its recording apparatus, and back-fill the shaft about the tank.

'As the time for the explosion approached, we would start vacuum pumps and evacuate the tank as highly as possible. Then, when the countdown reached zero, we would break the suspension with a small explosive, allowing the detector to fall freely in the vacuum. For about two seconds, the falling detector would be seeing neutrinos and recording the pulses from them while the earth shock from the blast passed harmlessly by, rattling the tank mightily but not disturbing our falling detector. When all was relatively quiet, the detector would reach the bottom of the tank, landing on a thick pile of foam rubber and feathers.

'We would return to the site of the shaft in a few days (when surface radioactivity had died away sufficiently), dig down to the tank, recover the detector and discover the truth about neutrinos.'

The crucial experiment
Before the spectacular experiment could be set up, Reines and Cowan realized that, with suitable modifications, they could utilize the much smaller neutrino flux from a nuclear reactor. They set up their experiment next to the Savannah River nuclear plant's reactor. The detector comprised a tank containing cadmium chloride dissolved in water. A neutrino passing through the tank reacts with protons in the water, producing a positron and a neutron. The positron combines with electrons, producing gamma rays. The neutron also produces gamma rays by reacting with the cadmium. When a gamma ray falls on a scintillation detector it produces a flash of light, detected using a photomultiplier, and shows that a neutrino has passed through the tank.

At first, there were about 1.63 'events' (or signals) per hour with the reactor operating and 0.4 with it shut down. Subsequent experiments with various adjustments raised the rate to nearly three events per hour. The next task was to confirm that these signals were produced by neutrinos. This involved eliminating all other possible sources: principally neutrons, gamma rays, electrons and protons from the reactor.

The tanks were surrounded by lead walls and floor 10 cm/4 in thick and roof and doors 20 cm/8 in thick to eliminate radiation from the reactor. To prove the lead's effectiveness, tests were done with additional shields. One experiment involved packing wet sawdust in bags round the detectors to reduce spurious neutron flow. There was no appreciable change in signal, and so it was concluded that it was not due to radiation produced outside the detectors. The neutrino had indeed been detected.

Nowadays, neutrinos from space are routinely detected by underground 'telescopes' similar to the Reines and Cowan set-up. Some modern neutrino telescopes comprise large tanks of ultrapure water, weighing up to 3,000 tonnes. The tank, in a mine, has walls lined with photoelectric detectors. An alternative type uses carbon tetrachloride; when it captures a neutrino, it produces radioactive argon whose presence may be detected by analysis of the liquid. There are large-scale neutrino detectors in Japan, Russia, Italy and the USA.

neutron one of the three main subatomic particles, the others being the proton and the electron. The neutron is a composite particle, being made up of three ◊quarks, and therefore belongs to the ◊baryon group of the ◊hadrons. Neutrons have about the same mass as protons but no electric charge, and occur in the nuclei of all atoms except hydrogen. They contribute to the mass of atoms but do not affect their chemistry.

For instance, the ◊isotopes of a single element differ only in the number of neutrons in their nuclei but have identical chemical properties. Outside a nucleus, a free neutron is radioactive, decaying with a half-life of 11.6 minutes into a proton, an electron, and an antineutrino. The neutron was discovered by the British chemist James Chadwick 1932.

neutron activation analysis chemical analysis used to determine the composition of a wide range of materials found in archaeological contexts. A specimen is bombarded with neutrons, which interact with nuclei in the sample to form radioactive isotopes that emit gamma rays as they decay. The energy spectrum of the emitted rays is detected with a counter and constituent elements and concentrations are identified by the characteristic energy spectrum of emitted rays and its intensity.

neutron beam machine nuclear reactor or accelerator producing a stream of neutrons, which can 'see' through metals. It is used in industry to check molecular changes in metal under stress.

neutron bomb small hydrogen bomb for battlefield use that kills by radiation without destroying buildings and other structures. See ◊nuclear warfare.

neutron star very small, 'superdense' star composed mostly of ◊neutrons. They are thought to form when massive stars explode as ◊supernovae, during which the protons and electrons of the star's atoms merge, owing to intense gravitational collapse, to make neutrons. A neutron star may have the mass of up to three Suns, compressed into a globe only 20 km/12 mi in diameter.

If its mass is any greater, its gravity will be so strong that it will shrink even further to become a ◊black hole. Being so small, neutron stars can spin very quickly. The rapidly flashing radio stars called ◊pulsars are believed to be neutron stars. The flashing is caused by a rotating beam of radio energy similar in behaviour to a lighthouse beam of light.

NEUTRON STAR: DENSE STAR

The density of a neutron star is about 100 million million times that of water. A teaspoonful of neutron star material would weigh more than 100 million tonnes. Your weight on a neutron star would be 10,000 million times greater than it is on Earth.

New General Catalogue catalogue of star clusters and nebulae compiled by the Danish astronomer John Louis Emil Dreyer (1852–1926) and published 1888. Its main aim was to revise, correct, and expand upon the *General Catalogue* compiled by English astronomer John Herschel, which appeared 1864.

New Madrid seismic fault zone the largest system of geological faults in the eastern USA, centred on New Madrid, Missouri. Geologists estimate that there is a 50% chance of a magnitude 6 earthquake in the area by the year 2000. This would cause much damage because the solid continental rocks would transmit the vibrations over a wide area, and buildings in the region have not been designed with earthquakes in mind.

The zone covers a rift in the continental plate running 200 km/125 mi along the Mississippi River from northern Arkansas to southern Illinois. There are several hundred earthquakes along the fault every year, most of them too small to be felt. A series of very severe earthquakes in 1811–12 created a lake 22 km/14 mi long and were felt as far away as Washington, DC and Canada.

new technology collective term applied to technological advances made in such areas as ◊telecommunications, ◊nuclear energy, space ◊satellites and ◊computers.

New Technology Telescope optical telescope that forms part of the ◊European Southern Observatory, La Silla, Chile; it came into operation 1991. It has a thin, lightweight mirror, 3.38 m/141 in across, which is kept in shape by computer-adjustable supports to produce a sharper image than is possible with conventional mirrors. Such a system is termed *active optics*.

newton SI unit (symbol N) of ◊force. One newton is the force needed to accelerate an object with mass of one kilogram by one metre per second per second. The weight of a medium size (100 g/3 oz) apple is one newton.

Newtonian physics ◊physics based on the concepts of English physicist and mathematician Isaac Newton, before the formulation of ◊quantum theory or ◊relativity theory.

Newton's laws of motion three laws that form the basis of Newtonian mechanics.

(1) Unless acted upon by an external resultant, or unbalanced, force, an object at rest stays at rest, and a moving object continues moving at the same speed in the same straight line. Put more simply, the law says that, if left alone, stationary objects will not move and moving objects will keep on moving at a constant speed in a straight line. (2) A resultant or unbalanced force applied to an object produces a rate of change of ◊momentum that is directly proportional to the force and is in the direction of the force. For an object of constant mass m, this law may be rephrased as: a resultant or unbalanced force F applied to an object gives it an acceleration a that is directly proportional to, and in the direction of, the force applied and inversely proportional to the mass. This relationship is represented by the equation:

$$a = F/m,$$

which is usually rearranged in the form:

$$F = ma.$$

(3) When an object A applies a force to an object B, B applies an equal and opposite force to A; that

is, to every action there is an equal and opposite reaction.

Newton's rings in optics, an ◊interference phenomenon seen (using white light) as concentric rings of spectral colours where light passes through a thin film of transparent medium, such as the wedge of air between a large-radius convex lens and a flat glass plate. With monochromatic light (light of a single wavelength), the rings take the form of alternate light and dark bands. They are caused by interference (interaction) between light rays reflected from the plate and those reflected from the curved surface of the lens.

niacin one of the 'B group' ◊vitamins, deficiency of which gives rise to ◊pellagra.

Niacin is the collective name for compounds that satisfy the dietary need for this function. Nicotinic acid ($C_5H_5N.COOH$) and nicotinamide ($C_5H_5N.-CONH_2$) are both used by the body. Common natural sources are yeast, wheat, and meat.

niche in ecology, the 'place' occupied by a species in its habitat, including all chemical, physical, and biological components, such as what it eats, the time of day at which the species feeds, temperature, moisture, the parts of the habitat that it uses (for example, trees or open grassland), the way it reproduces, and how it behaves.

It is believed that no two species can occupy exactly the same niche, because they would be in direct competition for the same resources at every stage of their life cycle.

Nichrome trade name for a series of alloys containing mainly nickel and chromium, with small amounts of other substances such as iron, magnesium, silicon, and carbon. Nichrome has a high melting point and is resistant to corrosion. It is therefore used in electrical heating elements and as a substitute for platinum in the ◊flame test.

nickel hard, malleable and ductile, silver-white metallic element, symbol Ni, atomic number 28, relative atomic mass 58.71. It occurs in igneous rocks and as a free metal (◊native metal), occasionally occurring in fragments of iron-nickel meteorites. It is a component of the Earth's core, which is held to consist principally of iron with some nickel. It has a high melting point, low electrical and thermal conductivity, and can be magnetized. It does not tarnish and therefore is much used for alloys, electroplating, and for coinage.

It was discovered in 1751 by Swedish mineralogist Axel Cronstedt (1722–1765) and the name given as an abbreviated form of *kopparnickel*, Swedish 'false copper', since the ore in which it is found resembles copper but yields none.

nickel ore any mineral ore from which nickel is obtained. The main minerals are arsenides such as chloanthite ($NiAs_2$), and the sulphides millerite (NiS) and pentlandite ($(Ni,Fe)_9S_8$), the commonest ore. The chief nickel-producing countries are Canada, Russia, Kazakhstan, Cuba, and Australia.

nicotine $C_{10}H_{14}N_2$ an ◊alkaloid (nitrogenous compound) obtained from the dried leaves of the tobacco plant *Nicotiana tabacum* and used as an insecticide. A colourless oil, soluble in water, it turns brown on exposure to the air.

Nicotine in its pure form is one of the most

powerful poisons known. It is the component of cigarette smoke that causes physical addiction. It is named after a 16th-century French diplomat, Jacques Nicot, who introduced tobacco to France.

nicotinic acid water-soluble ◊vitamin ($C_5H_5N.-COOH$) of the B complex, found in meat, fish, and cereals. Absence of nicotinic acid from the diet leads to the disease pellagra. See ◊niacin.

nielsbohrium name proposed by Soviet scientists for the element currently known as ◊unnilpentium (atomic number 105), to honour Danish physicist Niels Bohr.

niobium soft, grey-white, somewhat ductile and malleable, metallic element, symbol Nb, atomic number 41, relative atomic mass 92.906. It occurs in nature with tantalum, which it resembles in chemical properties. It is used in making stainless steel and other alloys for jet engines and rockets and for making superconductor magnets.

Niobium was discovered in 1801 by English chemist Charles Hatchett (1765–1847), who named it columbium (symbol Cb), a name that is still used in metallurgy. In 1844 it was renamed after Niobe by German chemist Heinrich Rose (1795–1864) because of its similarity to tantalum (Niobe is the daughter of Tantalus in Greek mythology).

nitrate any salt of nitric acid, containing the NO_3^- ion. Nitrates of various kinds are used in explosives, in the chemical industry, in curing meat (see ◊nitre), and as inorganic fertilizers. They are the most water-soluble salts known.

Nitrates in the soil, whether naturally occurring or from inorganic or organic fertilizers, can be used by plants to make proteins and nucleic acids. Being soluble in water, nitrates are leached out by rain into streams and reservoirs. High levels are now found in drinking water in arable areas. These may be harmful to newborn babies, and it is possible that they contribute to stomach cancer, although the evidence for this is unproven.

The UK current standard is 100 milligrams per litre, double the EC limits.

nitrate pollution the contamination of water by nitrates. Nitrates in the soil, whether naturally occurring or from agricultural fertilizers, are used by plants to make proteins. However, increased use of artifical fertilizers and land cultivation means that higher levels of nitrates are being washed from the soil into rivers, lakes, and aquifers. There they cause an excessive enrichment of the water (◊eutrophication), leading to a rapid growth of algae, which in turn darkens the water and reduces its oxygen content. The water is expensive to purify and many plants and animals die.

nitre or *saltpetre* potassium nitrate, KNO_3, a mineral found on and just under the ground in desert regions; used in explosives. Nitre occurs in Bihar, India, Iran, and Cape Province, South Africa. The salt was formerly used for the manufacture of gunpowder, but the supply of nitre for explosives is today largely met by making the salt from nitratine (also called Chile saltpetre, $NaNO_3$). Saltpetre is a ◊preservative and is widely used for curing meats.

nitric acid or *aqua fortis* HNO_3 fuming acid obtained by the oxidation of ammonia or the action of sulphuric acid on potassium nitrate. It is a highly corrosive acid, dissolving most metals, and a strong oxidizing agent. It is used in the nitration and esterification of organic substances, and in the making of sulphuric acid, nitrates, explosives, plastics, and dyes.

nitrification process that takes place in soil when bacteria oxidize ammonia, turning it into nitrates. Nitrates can be absorbed by the roots of plants, so this is a vital stage in the ◊nitrogen cycle.

nitrite salt or ester of nitrous acid, containing the nitrite ion (NO_2^-). Nitrites are used as preservatives (for example, to prevent the growth of botulism spores) and as colouring agents in cured meats such as bacon and sausages.

nitrocellulose alternative name for ◊cellulose nitrate.

nitrogen (Greek *nitron* 'native soda', sodium or potassium nitrate) colourless, odourless, tasteless, gaseous, nonmetallic element, symbol N, atomic number 7, relative atomic mass 14.0067. It forms almost 80% of the Earth's atmosphere by volume and is a constituent of all plant and animal tissues (in proteins and nucleic acids). Nitrogen is obtained for industrial use by the liquefaction and fractional distillation of air. Its compounds are used in the manufacture of foods, drugs, fertilizers, dyes, and explosives.

Nitrogen has been recognized as a plant nutrient, found in manures and other organic matter, from early times, long before the complex cycle of ◊nitrogen fixation was understood. It was isolated in 1772 by English chemist Daniel Rutherford (1749–1819) and named in 1790 by French chemist Jean Chaptal (1756–1832).

Nitrogen is used in the Haber process to make ammonia, NH_3, and to provide an inert atmosphere for certain chemical reactions.

nitrogen cycle the process of nitrogen passing through the ecosystem. Nitrogen, in the form of inorganic compounds (such as nitrates) in the soil, is absorbed by plants and turned into organic compounds (such as proteins) in plant tissue. A proportion of this nitrogen is eaten by ◊herbivores, with some of this in turn being passed on to the carnivores, which feed on the herbivores. The nitrogen is ultimately returned to the soil as excrement and when organisms die and are converted back to inorganic form by ◊decomposers.

Although about 78% of the atmosphere is nitrogen, this cannot be used directly by most organisms. However, certain bacteria and cyanobacteria (see ◊blue-green algae) are capable of nitrogen fixation. Some nitrogen-fixing bacteria live mutually with leguminous plants (peas and beans) or other plants (for example, alder), where they form characteristic nodules on the roots. The presence of such plants increases the nitrate content, and hence the fertility, of the soil.

nitrogen fixation the process by which nitrogen in the atmosphere is converted into nitrogenous compounds by the action of microorganisms, such as cyanobacteria (see ◊blue-green algae) and bacteria, in conjunction with certain ◊legumes. Several chemical processes duplicate nitrogen fixation to produce fertilizers; see ◊nitrogen cycle.

nitrogen oxide any chemical compound that contains only nitrogen and oxygen. All nitrogen oxides are gases. Nitrogen monoxide and nitrogen dioxide contribute to air pollution. See also ◊nitrous oxide.

Nitrogen monoxide (NO) is a colourless gas released when metallic copper reacts with concentrated ◊nitric acid. It is also produced when nitrogen and oxygen combine at high temperature. On contact with air it is oxidized to nitrogen dioxide.

Nitrogen dioxide (nitrogen(IV) oxide, NO_2) is a brown, acidic, pungent gas that is harmful if inhaled and contributes to the formation of ◊acid rain, as it dissolves in water to form nitric acid. It is the most common of the nitrogen oxides and is obtained by heating most nitrate salts (for example ◊lead(II) nitrate, $Pb(NO_3)_2$). If liquefied, it gives a colourless solution (N_2O_4). It has been used in rocket fuels.

In high-temperature combustion some nitrogen and oxygen from the air combine together to form nitrogen(II) oxide, $N_2 + O_2 \rightarrow 2NO$. When this oxide cools in the presence of air it is further oxidized to nitrogen(IV) oxide, $2NO + O_2 \rightarrow 2NO_2$. Consequently, in notation, NO_x or NOX are used when discussing oxides of nitrogen and their emission, since both gases are present in the air. The NO_2 gas dissolves in water to give a solution of nitric acid.

nitroglycerine $C_3H_5(ONO_2)_3$ flammable, explosive oil produced by the action of nitric and sulphuric acids on glycerol. Although poisonous, it is used in cardiac medicine. It explodes with great violence if heated in a confined space and is used in the preparation of dynamite, cordite, and other high explosives.

nitrous acid HNO_2 weak acid that, in solution with water, decomposes quickly to form nitric acid and nitrogen dioxide.

nitrous oxide or *dinitrogen oxide* N_2O colourless, nonflammable gas that reduces sensitivity to pain. In higher doses it is an anaesthetic. Well tolerated, but less potent than some other anaesthetic gases, it is often combined with other drugs to allow lower doses to be used. It may be self-administered; for example, in childbirth. It is popularly known as 'laughing gas'.

nivation in earth science, a complex of physical processes, operating beneath or adjacent to snow, believed to be responsible for the enlargement of the hollows in which snow collects. It is also thought to play a role in the early formation of ◊corries. The processes involved include freeze-thaw (weathering by the alternate freezing and melting of ice), mass movement (the downhill movement of substances under gravity), and erosion by meltwater.

nobelium synthesized, radioactive, metallic element of the ◊actinide series, symbol No, atomic number 102, relative atomic mass 259. It is synthesized by bombarding curium with carbon nuclei.

It was named in 1957 for the Nobel Institute in Stockholm, Sweden, where it was claimed to have been first synthesized. Later evaluations determined that this was in fact not so, as the successful 1958 synthesis at the University of California at Berkeley produced a different set of data. The name was not, however, challenged.

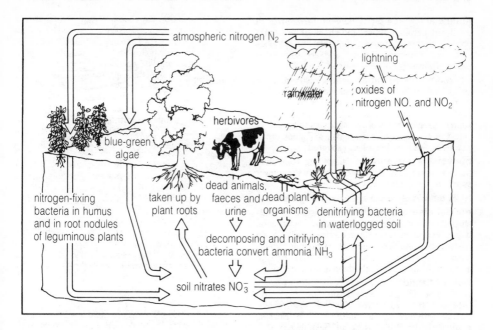

nitrogen cycle *The nitrogen cycle is one of a number of cycles during which the chemicals necessary for life are recycled. The carbon, sulphur, and phosphorus cycles are others. Since there is only a limited amount of these chemicals in the Earth and its atmosphere, the chemicals must be continuously recycled if life is to go on.*

Nobel prize annual international prize, first awarded 1901 under the will of Alfred Nobel, Swedish chemist, who invented dynamite. The interest on the Nobel endowment fund is divided annually among the persons who have made the greatest contributions in the fields of physics, chemistry, medicine, literature, and world peace. A sixth prize, for economics, financed by the Swedish National Bank, was first awarded 1969. The prizes have a large cash award and are given to organizations—such as the United Nations peacekeeping forces, which received the Nobel Peace Prize in 1988—as well as individuals.

noble gas alternative name for ◊inert gas.

noble gas structure the configuration of electrons in noble or ◊inert gases (helium, neon, argon, krypton, xenon, and radon).

This is characterized by full electron shells (see ◊atom, electronic structure) around the nucleus of an atom, which render the element stable. Any ion, produced by the gain or loss of electrons, that achieves an electronic configuration similar to one of the inert gases is said to have a noble gas structure.

node in physics, a position in a ◊standing wave pattern at which there is no vibration. Points at which there is maximum vibration are called *anti-nodes*. Stretched strings, for example, can show nodes when they vibrate. Guitarists can produce special effects (harmonics) by touching a sounding string lightly to produce a node.

nodule in geology, a lump of mineral or other matter found within rocks or formed on the seabed surface; mining technology is being developed to exploit them.

noise unwanted sound. Permanent, incurable loss of hearing can be caused by prolonged exposure to high noise levels (above 85 decibels). Over 55 decibels on a daily outdoor basis is regarded as an unacceptable level, to which an estimated 130 million people were exposed in 1991.

If the noise is in a narrow frequency band, temporary hearing loss can occur even though the level is below 85 decibels or exposure is only for short periods. Lower levels of noise are an irritant, but seem not to increase fatigue or affect efficiency to any great extent. Roadside meter tests, introduced by the Ministry of Transport in Britain in 1968, allowed 87 decibels as the permitted limit for cars and 92 for lorries. Loud noise is a major pollutant in towns and cities. In the UK the worst source of noise nuisance is noise from neighbours, suffered by 20% of the population.

electronic noise takes the form of unwanted signals generated in electronic circuits and in recording processes by stray electrical or magnetic fields, or by temperature variations. In electronic recording and communication systems, 'white noise' frequently appears in the form of high frequencies, or hiss. The main advantages of digital systemsare their relative freedom from such noise and their ability to recover and improve noise-affected signals.

LOUDEST NOISE

The low-frequency call of the humpback whale is the loudest noise made by a living creature. At 190 decibels, it is louder than Concorde taking off, and can be detected up to 800 km/500 mi away.

---

nonmetal one of a set of elements (around 20 in total) with certain physical and chemical properties opposite to those of metals. Nonmetals accept electrons (see ◊electronegativity) and are sometimes called electronegative elements.

nonrenewable resource natural resource, such as coal or oil, that takes thousands or millions of years to form naturally and can therefore not be replaced once it is consumed. The main energy sources used by humans are nonrenewable; renewable sources, such as solar, tidal, and geothermal power, have so far been less exploited.

nonvolatile memory in computing, ◊memory that does not lose its contents when the power supply to the computer is disconnected.

NOR gate in electronics, a type of ◊logic gate.

normal distribution curve the distinctive bell-shaped curve obtained when ◊continuous variation within a population is expressed graphically. When a statistician studies height or intelligence, most people have an intermediate or 'normal' score, with a few individuals scoring either high or low.

North Atlantic Drift warm ocean ◊current in the N Atlantic Ocean; an extension of the ◊Gulf Stream. It flows east across the Atlantic and has a mellowing effect on the climate of NW Europe, particularly the British Isles and Scandinavia.

northern lights common name for the aurora borealis (see ◊aurora).

nose in humans, the upper entrance of the respiratory tract; the organ of the sense of smell. The external part is divided down the middle by a septum of ◊cartilage. The nostrils contain plates of cartilage that can be moved by muscles and have a growth of stiff hairs at the margin to prevent foreign objects from entering. The whole nasal cavity is lined with a ◊mucous membrane that warms and moistens the air and ejects dirt. In the upper parts of the cavity the membrane contains 50 million olfactory receptor cells (cells sensitive to smell).

nostril in vertebrates, the opening of the nasal cavity, in which cells sensitive to smell are located. (In fish, these cells detect water-borne chemicals, so they are effectively organs of taste.) In vertebrates with lungs (lungfish and tetrapod vertebrates), the nostrils also take in air. In humans, and most other mammals, the nostrils are located on a ◊nose.

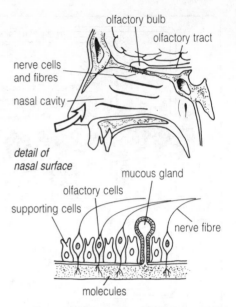

detail of nasal surface

nose *The structure of the nose. The organs of smell are confined to a small area in the roof of the nasal cavity. The olfactory cells are stimulated when certain molecules reach them. Smell is one of our most subtle senses: tens of thousands of smells can be distinguished. By comparison, taste, although closely related to smell, is a crude sensation. All the subtleties of taste depend upon smell.*

notebook computer small ◊laptop computer. Notebook computers became available in the early 1990s and, even complete with screen and hard-disc drive, are no larger than a standard A4 notebook.

NOT gate or *inverter gate* in electronics, a type of ◊logic gate.

notochord the stiff but flexible rod that lies between the gut and the nerve cord of all embryonic and larval chordates, including the vertebrates. It forms the supporting structure of the adult lancelet, but in vertebrates it is replaced by the vertebral column, or spine.

ENERGETIC NOVA

A typical nova explosion releases about as much energy as the Sun emits in ten thousand years, or as much as in a thousand million million nuclear bombs.

---

nova (plural *novae*) faint star that suddenly erupts in brightness by 10,000 times or more. Novae are believed to occur in close ◊binary star systems, where gas from one star flows to a companion ◊white dwarf. The gas ignites and is thrown off in an explosion at speeds of 1,500 kps/930 mps or more. Unlike a ◊supernova, the star is not completely disrupted by the outburst. After a few weeks or months it subsides to its previous state; it may erupt many more times.

The name comes from the Latin 'new', although novae are not new stars at all.

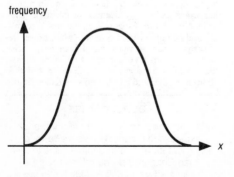

frequency

x

normal distribution curve

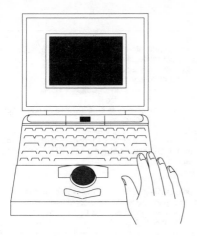

notebook computer

NTP abbreviation for *n*ormal *t*emperature and *p*ressure, former name for *STP* (◊*s*tandard *t*emperature and *p*ressure).

NTSC abbreviation for *National Television System Committee*, the US body responsible for setting up the world's first successful coding system for colour-television broadcasting 1953. See ◊television.

nuclear energy energy from the inner core or ◊nucleus of the atom, as opposed to energy released in chemical processes, which is derived from the electrons surrounding the nucleus.

Nuclear fission, as in an atom bomb, is achieved by allowing a ◊neutron to strike the nucleus of an atom of fissile material (such as uranium-235 or plutonium-239), which then splits apart to release two or three other neutrons. If the uranium-235 is pure, a ◊chain reaction is set up when these neutrons in turn strike other nuclei. This happens very quickly, resulting in the tremendous release of energy seen in nuclear weapons. The process is controlled inside the reactor of a nuclear power plant by absorbing excess neutrons in control rods and slowing down their speed.

Nuclear fusion is the process whereby hydrogen nuclei fuse to helium nuclei with an accompanying release of energy. It is a continuing reaction in the Sun and other stars. Nuclear fusion is the principle behind thermonuclear weapons (the ◊hydrogen bomb). Attempts to harness fusion for commercial power production have so far been unsuccessful, although the Joint European Torus (or ◊JET) laboratory at Culham, Oxfordshire, England, achieved fusion in 1991.

Nuclear energy is incomparably greater than the molecular energy which we use today ... What is lacking is the match to set the bonfire alight ... The Scientists are looking for this.

On **nuclear energy** Winston Churchill (1874–1965) *Thoughts and Adventures* 1932

nuclear fusion process whereby two atomic nuclei are fused, with the release of a large amount of energy. Very high temperatures and pressures are thought to be required in order for the process to happen. Under these conditions the atoms involved are stripped of all their electrons so that the remaining particles, which together make up ◊plasma, can come close together at very high speeds and overcome the mutual repulsion of the positive charges on the atomic nuclei. At very close range another nuclear force will come into play, fusing the particles together to form a larger nucleus. As fusion is accompanied by the release of large amounts of energy, the process might one day be harnessed to form the basis of commercial energy production. Methods of achieving controlled fusion are therefore the subject of research around the world.

Fusion is the process by which the Sun and the other stars produce their energy.

nuclear notation method used for labelling an atom according to the composition of its nucleus. The atoms or isotopes of a particular element are represented by the symbol $A_Z X$ where A is the mass number of their nuclei, Z is their atomic number, and X is the chemical symbol for that element.

nuclear physics the study of the properties of the nucleus of the ◊atom, including the structure of nuclei; nuclear forces; the interactions between particles and nuclei; and the study of radioactive decay. The study of elementary particles is ◊particle physics.

nuclear reaction reaction involving the nuclei of atoms. Atomic nuclei can undergo changes either as a result of radioactive decay, as in the decay of radium to radon (with the emission of an alpha particle) or as a result of particle bombardment in a machine or device, as in the production of cobalt-60 by the bombardment of cobalt-59 with neutrons.

$$^{226}_{88}\text{Ra} \rightarrow {}^{222}_{86}\text{Rn} + {}^{4}_{2}\text{He}$$

$$^{59}_{37}\text{Co} + {}^{1}_{0}\text{n} \rightarrow {}^{60}_{27}\text{Co} + \gamma$$

Nuclear ◊fission and nuclear ◊fusion are examples of nuclear reactions. The enormous amounts of energy released arise from the mass–energy relation put forward by Einstein, stating that $E = mc^2$ (where E is energy, m is mass, and c is the velocity of light).

In nuclear reactions the sum of the masses of all the products (on the atomic mass unit scale) is less than the sum of the masses of the reacting particles. This lost mass is converted to energy according to Einstein's equation.

nuclear reactor device for producing ◊nuclear energy in a controlled manner. There are various types of reactor in use, all using nuclear fission.

In a *gas-cooled reactor*, a circulating gas under pressure (such as carbon dioxide) removes heat from the core of the reactor, which usually contains natural uranium. The efficiency of the fission process is increased by slowing neutrons in the core by using a ◊moderator such as carbon. The reaction is controlled with neutron-absorbing rods made of boron. An *advanced gas-cooled reactor* (AGR) generally has enriched uranium as its fuel.

A *water-cooled reactor*, such as the steam-

nuclear energy: chronology

1896	French physicist Henri Becquerel discovered radioactivity.
1905	In Switzerland, Albert Einstein showed that mass can be converted into energy.
1911	New Zealand physicist Ernest Rutherford proposed the nuclear model of the atom.
1919	Rutherford split the atom, by bombarding a nitrogen nucleus with alpha particles.
1939	Otto Hahn, Fritz Strassman, and Lise Meitner announced the discovery of nuclear fission.
1942	Enrico Fermi built the first nuclear reactor, in a squash court at the University of Chicago, USA.
1946	The first fast reactor, called Clementine, was built at Los Alamos, New Mexico.
1951	The Experimental Breeder Reator, Idaho, USA, produced the first electricity to be generated by nuclear energy.
1954	The first reator for generating electricity was built in the USSR, at Obninsk.
1956	The world's first commercial nuclear power station, Calder Hall, came into operation in the UK.
1957	The release of radiation from Windscale (now Sellafield) nuclear power station, Cumbria, England, caused 39 deaths to 1977. In Kyshym, USSR, the venting of plutonium waste caused high but undisclosed casualties (30 small communities were deleted from maps produced in 1958).
1959	Experimental fast reactor built in Dounreay, N Scotland.
1979	Nuclear-reactor accident at Three Mile Island, Pennsylvania, USA.
1986	An explosive leak from a reactor at Chernobyl, the Ukraine, resulted in clouds of radioactive material spreading as far as Sweden; 31 people were killed and thousands of square kilometres were contaminated.
1991	The first controlled and substantial production of nuclear-fusion energy (a two-second pulse of 1.7 MW) was achieved at JET, the Joint European Torus, at Culham, Oxfordshire, England.

generating heavy water (deuterium oxide) reactor, has water circulating through the hot core. The water is converted to steam, which drives turbo-alternators for generating electricity.

The most widely used reactor is the *pressurized-water reactor* (PWR), which contains a sealed system of pressurized water that is heated to form steam in heat exchangers in an external circuit.

The *fast reactor* has no moderator and uses fast neutrons to bring about fission. It uses a mixture of plutonium and uranium oxide as fuel. When operating, uranium is converted to plutonium, which can be extracted and used later as fuel. It is also called the fast breeder because it produces more plutonium than it consumes. Heat is removed from the reactor by a coolant of liquid sodium.

Public concern over the safety of nuclear reactors has been intensified by explosions and accidental release of radioactive materials. The safest system allows for the emergency cooling of a reactor by automatically flooding an overheated core with water. Other concerns about nuclear power centre on the difficulties of reprocessing nuclear fuel and disposing safely of nuclear waste, and the cost of maintaining nuclear power stations and of decommissioning them at the end of their lives.

In 1989, the UK government decided to postpone the construction of new nuclear power stations; in the USA, no new stations have been commissioned in over a decade. Rancho Seco, near Sacramento, California, was the first nuclear power station to be closed, by popular vote, in 1989. Sweden is committed to decommissioning its reactors. Some countries, such as France, are pressing ahead with their nuclear programmes.

nuclear safety measures to avoid accidents in the operation of nuclear reactors and in the production and disposal of nuclear weapons and of ◊nuclear waste. There are no guarantees of the safety of any of the various methods of disposal.

nuclear accidents

Chernobyl, Ukraine. In April 1986 there was an explosive leak, caused by overheating, from a nonpressurized boiling-water reactor, one of the

NUCLEAR REACTORS IN CONTEXT

To generate the same amount of electricity as a modern nuclear power station, one would need: 150 sq km/60 sq mi of solar panels; 300 wind turbines, each one 600 m/660 yd in diameter, covering an area about the size of a large city; 100 km/62 mi of wave energy converters; a tidal barrier about 16 km/10 mi long; or a hydroelectric power station about 10 times the size of the largest existing UK station (130 megawatts).

largest in Europe. The resulting clouds of radioactive material spread as far as the UK; 31 people were killed in the explosion (many more are expected to die or become ill because of the long-term effects of radiation), and thousands of square kilometres of land were contaminated by fallout.

Three Mile Island, Harrisburg, Pennsylvania, USA. In 1979, a combination of mechanical and electrical failure, as well as operator error, caused a pressurized water reactor to leak radioactive matter.

Church Rock, New Mexico, USA. In July 1979, 380 million litres/100 million gallons of radioactive water containing uranium leaked from a pond into the Rio Purco, causing the water to become over 6,500 times as radioactive as safety standards allow for drinking water.

Ticonderoga, 130 km/80 mi off the coast of Japan. In 1965 a US Navy Skyhawk jet bomber fell off the deck of this ship, sinking in 4,900 m/16,000 ft of water. It carried a one-megaton hydrogen bomb. The accident was only revealed in 1989.

Windscale (now Sellafield), Cumbria, England. In 1957, fire destroyed the core of a reactor, releasing large quantities of radioactive fumes into the atmosphere.

In 1990 a scientific study revealed an increased risk of leukemia in children whose fathers had worked at Sellafield between 1950 and 1985. Sellafield (UK) is the world's greatest discharger of radioactive waste, followed by Hanford, Washington (USA).

Nuclear energy: where are we now?

In 1992 a dozen nations depended on nuclear energy for at least a quarter of their electricity. France, with 72.7%, and Belgium and Sweden with well over 50%, top the list of 30 countries operating a total of 420 nuclear reactors around the world. Hungary and Korea are close behind with 48.4% and 47.5% respectively.

Three new reactors started up during 1991, in Bulgaria, Canada, and Japan, while seven closed down. Four of the closures were in Eastern Germany, where the condition of old reactors had been causing concern, and two in the former Soviet Union.

However, 25 new units are under construction in the ex-USSR, a third of the worldwide total of 76. Ten are being build in Japan and seven in India. The UK has one building, Sizewell B, to add to its current total of 37 operating reactors. The United States has 111 operating reactors and three more under construction. The 612.6 billion units (terawatt-hours or TWh) of nuclear electricity currently produced in the USA is just under a third of the world total—and provides just over a fifth of the country's needs.

country	reactors in operation	reactors being built	nuclear electricity (% of total)
Argentina	2	1	19.1
Belgium	7		59.3
Brazil	1	1	0.6
Bulgaria	6		34.0
Canada	20	2	16.4
China	1	2	
Cuba		2	
CSFR	8	6	28.6
Finland	4		33.3
France	56	5	72.7
Germany	21		27.6
Hungary	4		48.4
India	7	7	1.8
Iran		2	
Japan	42	10	23.8
Korea	9	3	47.5
Mexico	1	1	3.6
Netherlands	2		4.9
Pakistan	1		0.8
Romania		5	
South Africa	2		5.9
Spain	9		35.9
Sweden	12		51.6
Switzerland	5		40.0
Taiwan	6		37.8
UK	37	1	20.6
USA	111	3	21.7
ex-USSR	45	25	12.6
ex-Yugoslavia	1		6.3
TOTAL	420	76	

nuclear warfare war involving the use of nuclear weapons. Nuclear-weapons research began in Britain 1940, but was transferred to the USA after it entered World War II. The research programme, known as the Manhattan Project, was directed by J Robert Oppenheimer.

atom bomb The original weapon relied on use of a chemical explosion to trigger a chain reaction. The first test explosion was at Alamogordo, New Mexico, 16 July 1945; the first use in war was by the USA against Japan 6 Aug 1945 over Hiroshima and three days later at Nagasaki.

hydrogen bomb A much more powerful weapon than the atom bomb, it relies on the release of thermonuclear energy by the condensation of hydrogen nuclei to helium nuclei (as happens in the Sun). The first detonation was at Eniwetok Atoll, Pacific Ocean, 1952 by the USA.

neutron bomb or enhanced radiation weapon (ERW) A very small hydrogen bomb that has relatively high radiation but relatively low blast, designed to kill (in up to six days) by a brief neutron radiation that leaves buildings and weaponry intact.

nuclear methods of attack now include aircraft bombs, missiles (long-or short-range, surface to surface, air to surface, and surface to air), depth charges, and high-powered landmines ('atomic

The first nuclear reactor

The world's first nuclear reactor was built in a squash court under Stagg Field football ground at the University of Chicago. On 2 December 1942 a team led by Italian physicist Enrico Fermi demonstrated a controlled self-sustaining nuclear chain reaction, opening the way to modern nuclear power and nuclear weapons.

The reactor—or 'atomic pile'—comprised 400 tonnes of graphite made into 45,000 bricks and built into a structure about 6 m/20 ft high. Sixty tonnes of uranium was pressed into 22,000 small cylinders fitted into slots in the graphite, which slowed the neutrons produced by the splitting of the uranium nuclei. Cadmium control rods in the pile soaked up excess neutrons to control the reaction.

The pile took less than a month to build, following intensive research starting with the discovery of nuclear fission in 1939. Fermi had worked on the design for six months, calculating the amount of uranium needed, and the best arrangement of the fuel and graphite. The work was dirty; black graphite dust got everywhere. The team was never free of it; even after showering, it continued to ooze out of their skin pores. They worked non-stop, knowing that the Nazis were also working to master nuclear power. Fermi, however, believed in regular routine. He always stopped at midday for lunch, and always went home at 5 pm. Team member Albert Wattenberg recalls: 'He made everything look very logical and straightforward and he never made mistakes. It was eight years before I remember him making a mathematical mistake.'

Time for lunch
During the morning of 2 December, Fermi began fine-tuning the pile, measuring the neutrons being produced. A single cadmium control rod regulated the neutron flow. The task was to determine how far to withdraw the rod for a chain reaction to proceed. Shortly before noon, there was a loud bang. Alarm followed—until the team realized that the safety control rod has operated automatically to shut the reactor down. The cutoff had been set too low. Fermi decided it was time for a break. 'I'm hungry; let's go to lunch,' he said. Lunch was quiet; no-one talked about the experiment. Fermi sat silent and preoccupied.

Back at work at 2 pm, they eased the control rod out further. The instruments showing the reaction rate went off the scale and had to be reset; but the reaction was still not self-sustaining. Finally at 3.25 pm, Fermi asked for the control rod to be withdrawn another foot. 'This is going to do it. Now it will become self-sustaining. The trace will climb and continue to climb—it will not level off.' The rod was withdrawn and the reaction rate climbed. Team member Harold Agnew remembers: 'Hearing the counters going faster and faster and seeing the pen recorders continue to rise was scary.' Fermi did rapid calculations on his sliderule. The reaction counters clicked increasingly rapidly until they became a continuous buzz. After a few minutes, Fermi looked up from his calculations with a broad smile. 'The reaction is self-sustaining,' he announced.

The reactor was allowed to operate for 28 minutes; at 3.53 pm Fermi told Wally Zinn to insert the control rod. The neutron rate dropped immediately. The first controlled nuclear chain reaction had stopped. Agnew recalls: 'No smoke, no smell. It was over and we were back where we were that morning. And the outside world was unaware of what had happened.'

The scientists applauded softly, aware of their work's implications. A bottle of chianti was produced and the team drank to their success from paper cups. The wicker basket around the bottle was signed by several of the experimenters. All 43 people present knew what the future held. They fell silent and turned towards Fermi. Hungarian scientist Leo Szilard, an important team member, said, 'This will go down as a black day in the history of mankind.' Fermi's reply is not recorded.

Enrico sinks the admiral
The breakthrough was not announced publicly. Wartime secrecy prevailed; even Fermi's wife Laura did not learn of the success until after the war. At a party held after the successful experiment, she asked Leona Woods, the only woman in the team, what was being celebrated. 'Enrico has sunk a Japanese admiral,' was the cryptic reply. The need for secrecy was paramount when Arthur Compton, in charge of the US nuclear weapons project, telephoned a colleague at Harvard with the news: 'The Italian navigator has landed in the New World,' he said. 'How were the natives?' came the reply. 'Very friendly,' Compton answered.

The reactor did not operate again. In February 1943 it was dismantled and some of its parts used to make another reactor. The graphite blocks were eventually ground down to make ink to print invitations to the 40th anniversary celebrations, and to make into pencils given away as souvenirs.

demolition munitions') to destroy bridges and roads.

Major subjects of disarmament negotiations are: *intercontinental ballistic missiles* (ICBMs), which have from 1968 been equipped with clusters of warheads (which can be directed to individual targets) and are known as multiple independently targetable re-entry vehicles (MIRVs). The 1980s US-designed MX (Peacekeeper) carries up to ten warheads in each missile. In 1989, the UK agreed to purchase submarine-launched Trident missiles from the USA. Each warhead has eight independently targetable re-entry vehicles (each nuclear-armed) with a range of about 6,400 km/4,000 mi to eight separate targets within about 240 km/150 mi of the central aiming point. The Trident system was scheduled to enter service within the Royal Navy in the mid-1990s.

nuclear methods of defence include: *antiballistic missile* (ABM) Earth-based systems with two types of missile, one short-range with high acceleration, and one comparatively long-range for interception above the atmosphere; *Strategic Defense Initiative* (announced by the USA 1983 to be operative from 2000 but cancelled 1993; popularly known as the 'Star Wars' programme) 'directed energy weapons' firing laser beams would be mounted on space stations, and by burning holes in incoming missiles would either collapse them or detonate their fuel tanks.

nuclear waste the radioactive and toxic by-products of the nuclear-energy and nuclear-weapons industries. Nuclear waste may have an active life of several thousand years. Reactor waste is of three types: *high-level* spent fuel, or the residue when nuclear fuel has been removed from a reactor and reprocessed; *intermediate*, which may be long-or short-lived; and *low-level*, but bulky, waste from reactors, which has only short-lived radioactivity. Disposal, by burial on land or at sea, has raised problems of safety, environmental pollution, and security. In absolute terms, nuclear waste cannot be safely relocated or disposed of.

The dumping of nuclear waste at sea officially stopped 1983, when a moratorium was agreed by the members of the London Dumping Convention (a United Nations body that controls disposal of wastes at sea). Covertly, the USSR continued dumping, and deposited thousands of tonnes of nuclear waste and three faulty reactors in the sea 1964–86.

Waste from a site where uranium is mined or milled may have an active life of several thousand years, and spent (irradiated) fuel is dangerous for tens of thousands of years. Sea disposal has occurred at many sites, for example 450 km/300 mi off Land's End, England, but there is no guarantee of the safety of this method of disposal, even for low-activity waste. There have been proposals to dispose of high-activity waste in old mines, granite formations, and specially constructed bunkers. The most promising proposed method is by vitrification into solid glass cylinders, which would be placed in titanium-cobalt alloy containers and deposited on dead planets in space. Beneath the sea the containers would start to corrode after 1,000 years,

and the cylinders themselves would dissolve within the next 1,000 years.

About one-third of the fuel from nuclear reactors becomes spent each year. It is removed to a *reprocessing* plant where radioactive waste products are chemically separated from remaining uranium and plutonium, in an expensive and dangerous process. This practice increases the volume of radioactive waste more than a hundred times.

nucleic acid complex organic acid made up of a long chain of nucleotides. The two types, known as DNA (deoxyribonucleic acid) and RNA (ribonucleic acid), form the basis of heredity. The nucleotides are made up of a sugar (deoxyribose or ribose), a phosphate group, and one of four purine or pyrimidine bases. The order of the bases along the nucleic acid strand contains the genetic code.

nucleolus in biology, a structure found in the nucleus of eukaryotic cells (see ◊eukaryote). It produces the RNA that makes up the ◊ribosomes, from instructions in the DNA.

nucleon in particle physics, either a ◊proton or a ◊neutron, both particles present in the atomic nucleus. *Nucleon number* is an alternative name for the ◊mass number of an atom.

nucleotide organic compound consisting of a purine (adenine or guanine) or a pyrimidine (thymine, uracil, or cytosine) base linked to a sugar (deoxyribose or ribose) and a phosphate group. ◊DNA and ◊RNA are made up of long chains of nucleotides.

nucleus in physics, the positively charged central part of an ◊atom, which constitutes almost all its mass. Except for hydrogen nuclei, which have only protons, nuclei are composed of both protons and neutrons. Surrounding the nuclei are electrons, which contain a negative charge equal to the protons, thus giving the atom a neutral charge.

The nucleus was discovered by New Zealand physicist Ernest Rutherford in 1911 as a result of experiments in passing alpha particles through very thin gold foil.

A few of the particles were deflected back, and Rutherford, astonished, reported: 'It was almost as incredible as if you fired a 15–inch shell at a piece of tissue paper and it came back and hit you!' The deflection, he deduced, was due to the positively charged alpha particles being repelled by approaching a small but dense positively charged nucleus.

nucleus in biology, the central, membrane-enclosed part of a eukaryotic cell (see ◊eukaryote, containing the chromosomes.

nuclide in physics, one of two or more atoms having the same atomic number (number of protons) and mass number (number of nucleons); compare ◊isotope.

nuée ardente glowing white-hot cloud of ash and gas emitted by a volcano during a violent eruption. In 1902 a nuée ardente produced by the eruption of Mount Pelee in Martinique swept down the volcano in a matter of seconds and killed 28,000 people in the nearby town of St Pierre.

number symbol used in counting or measuring. In mathematics, there are various kinds of numbers. The everyday number system is the decimal

---

The most incredible event

What is an atom? Until about 100 years ago, scientists were pretty confident in regarding atoms as the permanent bricks of which the whole universe was built. All the changes of the universe amounted to nothing more drastic than simple rearrangements of permanent, indestructible atoms. Atoms seemed like the bricks in a child's box of toys, which could be used to build many different buildings in turn.

Chipped bricks . . .
This comfortable picture was changed by the investigations of British physicist Joseph John Thomson (1856–1940) into cathode rays. Thomson showed conclusively that not only could atomic 'bricks' be chipped, but that the fragments produced were identical, no matter what atom they came from. They were of equal weight or mass, and carried the same negative electrical charge. The fragments were called electrons.

Thomson's experiments enabled him to calculate the ratio of the mass of the electron to its charge: it was about a thousand times smaller than the value that had already been calculated for a hydrogen ion in the electrolysis of liquids. This was the proof, announced in April 1897, that electrons were fundamental particles of matter, far smaller than any atom. But atoms could not consist of negatively charged electrons alone. Particles with like electrical charge repel one another and, anyway, atoms are electrically neutral.

Thomson, who built up the Cavendish Laboratory at Cambridge into a great research school, was succeeded as professor by New Zealand physicist Ernest Rutherford (1871–1937), his pupil and one of the greatest pioneers of subatomic physics. Rutherford, while professor of physics at Manchester University, had shown that α-particles were doubly ionized helium atoms, using a Geiger–Müller counter. His assistant, German physicist Hans Wilhelm Geiger (1882–1945), with a colleague Walther

Müller, invented this instrument, which is still used to measure ionizing radiation.

. . . and indestructible tissue paper?
Rutherford's idea for attacking the problem of the nature of an atom was to study the scattering of α-particles passing through thin metal foils. This painstaking work was carried out by Geiger in 1910 with his colleague Marsden, with astonishing results. They found that most of the particles were only slightly deflected as they passed through the foil. But a very small proportion, about 1 in 8,000, were widely deflected. Rutherford described this result as 'quite the most incredible event that has ever happened to me in my life . . . It was almost as if you fired a 15–inch shell at a piece of tissue paper and it came back and hit you.'

Rutherford concluded when he published the results in 1911 that almost all the mass of the atom was concentrated in a very small region and that most of the atom was 'empty space.' This crucial experiment established the idea of the nuclear atom. The obstacle that deflected the α-particles could only be the missing positive charges of the atom. These were carried by the minute central nucleus. The electrons, Rutherford supposed, must be in motion around the nucleus, otherwise they would be drawn to it.

The number of positive charges in the nucleus equalled the number of electrons, and so accounted for the electrical neutrality of the atom. The proton, later discovered as part of the nucleus, carried an equal but opposite charge to the electron. But on theoretical grounds, Rutherfords's planetary model of the atom was unstable. Since electrons are charged particles, an atom ought to radiate energy by virtue of their motion. Danish theoretical physicist Niels Bohr (1885–1962), who was Rutherford's pupil, provided the theory that accounted for the stability of the atom. He showed that electrons must reside in 'stationary' orbits, in which they did not radiate energy.

('proceeding by tens') system, using the base ten. ◊*Real numbers* include all rational numbers (integers, or whole numbers, and fractions) and irrational numbers (those not expressible as fractions). ◊*Complex numbers* include the real and unreal numbers (real-number multiples of the square root of –1). The ◊binary number system, used in computers, has two as its base.

The ordinary numerals, 0, 1, 2, 3, 4, 5, 6, 7, 8, and 9, give a counting system that, in the decimal system, continues 10, 11, 12, 13, and so on. These

are whole numbers (integers), with fractions represented as, for example, ¼, ½, ¾, or as decimal fractions (0.25, 0.5, 0.75). They are also *rational numbers*. *Irrational numbers* cannot be represented in this way and require symbols, such as √2, π, and e. They can be expressed numerically only as the (inexact) approximations 1.414, 3.142 and 2.718 (to three places of decimals) respectively. The symbols π and e are also examples of *transcendental numbers*, because they (unlike √2) cannot be derived by solving a ◊polynomial equation (an equation with one ◊variable quantity) with rational ◊coefficients (multiplying factors). Complex numbers, which include the real numbers as well as unreal numbers, take the general form $a + bi$, where $i = \sqrt{-1}$ (that is, $i^2 = -1$), and a is the real part and bi the unreal part. **history** The ancient Egyptians, Greeks, Romans, and Babylonians all evolved number systems, although none had a zero, which was introduced from India by way of Arab mathematicians in about the 6th century AD and allowed a place-value system to be devised on which the decimal system is based. Other number systems have since evolved and have found applications. For example, numbers to base two (binary numbers), using only 0 and 1, are commonly used in digital computers to represent the two-state 'on' or 'off' pulses of electricity. Binary numbers were first developed by German mathematician Gottfried Leibniz in the late 17th century.

Numbers constitute the only universal language.

On **numbers** Nathanael West (1903–1940)
Miss Lonelyhearts 1933

---

number theory in mathematics, the abstract study of the structure of number systems and the properties of positive integers (whole numbers). For example, the theories of factors and prime numbers fall within this area as do the work of mathematicians Giuseppe Peano (1858–1932), Pierre de Fermat, and Karl Gauss.

nunatak mountain peak protruding through an ice sheet. Such peaks are common in Antarctica.

nut any dry, single-seeded fruit that does not split open to release the seed, such as the chestnut. A nut is formed from more than one carpel, but only one seed becomes fully formed, the remainder aborting. The wall of the fruit, the pericarp, becomes hard and woody, forming the outer shell.

Examples of true nuts are the acorn and hazelnut. The term also describes various hard-shelled fruits and seeds, including almonds and walnuts, which are really the stones of ◊drupes, and brazil nuts and shelled peanuts, which are seeds.

The kernels of most nuts provide a concentrated, nutritious food, containing vitamins, minerals, and enzymes, about 50% fat, and 10–20% protein, although a few, such as chestnuts, are high in carbohydrates and have only a moderate protein content of 5%. Nuts also provide edible and indus-

trial oils. Most nuts are produced by perennial trees and shrubs.

Whereas the majority of nuts are obtained from plantations, considerable quantities of pecans and brazil nuts are still collected from the wild. World production in the mid-1980s was about 4 million tonnes per year.

nutation in astronomy, a slight 'nodding' of the Earth in space, caused by the varying gravitational pulls of the Sun and Moon. Nutation changes the angle of the Earth's axial tilt (average 23.5°) by about 9 seconds of arc to either side of its mean position, a complete cycle taking just over 18.5 years.

nutation in botany, the spiral movement exhibited by the tips of certain stems during growth; it enables a climbing plant to find a suitable support. Nutation sometimes also occurs in tendrils and flower stalks.

The direction of the movements, clockwise or anticlockwise, is usually characteristic for particular species.

nutrition the stragegy adopted by an organism to obtain the chemicals it needs to live, grow, and reproduce. There are two main types: autotrophic nutrition (plants) and heterotrophic nutrition (animals).

autotrophic nutrition The Sun's energy is used to drive the reactions of ◊photosynthesis, in which carbon dioxide and water are combined to make ◊carbohydrates, which are subsequently used as food. Plants therefore do not (in general) need to eat.

heterotrophic nutrition This involves the ingestion (taking in) of food and its subsequent digestion and absorption. This food directly or indirectly comes from plant matter. All animals ultimately depend on plants for their nutrition.

nylon synthetic long-chain polymer similar in chemical structure to protein. Nylon was the first all-synthesized fibre, made from petroleum, natural gas, air, and water by the Du Pont firm in 1938. It is used in the manufacture of moulded articles, textiles, and medical sutures. Nylon fibres are stronger and more elastic than silk and are relatively insensitive to moisture and mildew. Nylon is used for hosiery and woven goods, simulating other materials such as silks and furs; it is also used for carpets.

It was developed in the USA 1937 by the chemist W H Carothers and his associates.

nymph in entomology, the immature form of insects that do not have a pupal stage; for example, grasshoppers and dragonflies. Nymphs generally resemble the adult (unlike larvae), but do not have fully formed reproductive organs or wings.

oasis area of land made fertile by the presence of water near the surface in an otherwise arid region. The occurrence of oases affects the distribution of plants, animals, and people in the desert regions of the world.

object-oriented programming (OOP) computer programming based on 'objects', in which data are closely linked to the procedures that operate on them. For example, a circle on the screen might be an object: it has data, such as a centre point and a radius, as well as procedures for moving it, erasing it, changing its size, and so on.

The technique originated with the Simula and Smalltalk languages in the 1960s and early 1970s, but it has now been incorporated into many general-purpose programming languages.

object program in computing, the ◊machine-code translation of a program written in a ◊source language.

observatory site or facility for observing astronomical or meteorological phenomena. The earliest recorded observatory was in Alexandria, N Africa, built by Ptolemy Soter in about 300 BC. The modern observatory dates from the invention of the telescope. Observatories may be ground-based, carried on aircraft, or sent into orbit as satellites, in space stations, and on the space shuttle.

The erection of observatories was revived in W Asia about AD 1000, and extended to Europe. The observatory built on the island of Hven (now Ven) in Denmark 1576 for Tycho Brahe (1546–1601) was elaborate, but survived only to 1597. It was followed by those in Paris 1667, Greenwich (the ◊Royal Greenwich Observatory) 1675, and Kew, England. Most early observatories were near towns, but with the advent of big telescopes, clear skies with little background light, and hence high, remote sites, became essential.

The most powerful optical telescopes covering the sky are at ◊Mauna Kea, Hawaii; ◊Mount Palomar, California; ◊Kitt Peak National Observatory, Arizona; La Palma, Canary Islands; ◊Cerro Tololo Inter-American Observatory, and the ◊European Southern Observatory, Chile; ◊Siding Spring Mountain, Australia; and ◊Zelenchukskaya in the Caucasus.

Radio astronomy observatories include ◊Jodrell Bank, Cheshire, England; the ◊Mullard Radio Astronomy Observatory, Cambridge, England; ◊Arecibo, Puerto Rico; ◊Effelsberg, Germany; and ◊Parkes, Australia. The ◊Hubble Space Telescope was launched into orbit April 1990. The Very Large Telescope is under construction by the European Southern Observatory in the mountains of N Chile, for completion by 1997.

obsidian black or dark-coloured glassy volcanic rock, chemically similar to ◊granite, but formed by cooling rapidly on the Earth's surface at low pressure.

The glassy texture is the result of rapid cooling, which inhibits the growth of crystals. Obsidian was valued by the early civilizations of Mexico for making sharp-edged tools and ceremonial sculptures.

obtuse angle an angle greater than 90° but less than 180°.

occluded front weather ◊front formed when a cold front catches up with a warm front. It brings cloud and rain as air is forced to rise upwards along the front, cooling and condensing as it does so.

occultation in astronomy, the temporary obscuring of a star by a body in the Solar System. Occultations are used to provide information about changes in an orbit, and the structure of objects in space, such as radio sources.

The exact shapes and sizes of planets and asteroids can be found when they occult stars. The rings of Uranus were discovered when that planet occulted a star 1977.

ocean great mass of salt water. Strictly speaking three oceans exist—the Atlantic, Indian, and Pacific—to which the Arctic is often added. They cover approximately 70% or 363,000,000 sq km/ 140,000,000 sq mi of the total surface area of the Earth. Water levels recorded in the world's oceans have shown an increase of 10–15 cm/4–6 in over the past 100 years.

depth (average) 3,660 m/12,000 ft, but shallow ledges 180 m/600 ft run out from the continents, beyond which the continental slope reaches down to the ◊abyssal zone, the largest area, ranging from 2,000–6,000 m/6,500–19,500 ft. Only the ◊deep-sea trenches go deeper, the deepest recorded being 11,034 m/36,201 ft (by the *Vityaz*, USSR) in the Mariana Trench of the W Pacific 1957 **features** deep trenches (off E and SE Asia, and western South America), volcanic belts (in the W Pacific and E Indian Ocean), and ocean ridges (in the mid-Atlantic, E Pacific, and Indian Ocean) **temperature** varies on the surface with latitude (–2°C to +29°C); decreases rapidly to 370 m/1,200 ft, then more slowly to 2,200 m/7,200 ft; and hardly at all beyond that **water contents** salinity averages about 3%; minerals commercially extracted include bromine, magnesium, potassium, salt; those potentially recoverable include aluminium, calcium, copper, gold, manganese, silver.

How inappropriate to call this planet Earth when quite clearly it is Ocean.

On **oceans** Arthur C Clarke *Nature* 1990

---

oceanarium large display tank in which aquatic animals and plants live together much as they would in their natural environment. The first oceanarium was created by the explorer and naturalist W Douglas Burden 1938 in Florida, USA.

ocean current fast-flowing body of seawater; see ◊current.

Ocean Drilling Program (ODP, formerly the *Deep-Sea Drilling Project* 1968–85) research project initiated by the USA 1968 to sample the rocks of the ocean ◊crust. The operation became international 1975, when Britain, France, West Germany, Japan, and the USSR also became involved.

Boreholes were drilled in all the oceans using the ships *Glomar Challenger* and JOIDES *Resolution*, and knowledge of the nature and history of the ocean basins was increased dramatically. The technical difficulty of drilling the seabed to a depth of 2,000 m/6,500 ft was overcome by keeping the ship in position with side-thrusting propellers and satellite navigation, and by guiding the drill using a radiolocation system. The project is intended to continue until 2005.

oceanography study of the oceans, their origin, composition, structure, history, and wildlife (seabirds, fish, plankton, and other organisms). Much oceanography uses computer simulations to plot the possible movements of the waters, and many studies are carried out by remote sensing.

Oceanography involves the study of water movements – currents, waves, and tides – and the chemical and physical properties of the seawater. It deals with the origin and topography of the ocean floor – ocean trenches and ridges formed by ◊plate tectonics, and continental shelves from the submerged portions of the continents.

The World Ocean Circulation Experiment, a seven-year project begun 1990, involves

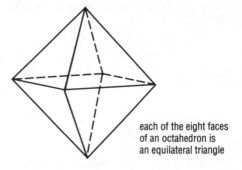

each of the eight faces of an octahedron is an equilateral triangle

octahedron, regular

researchers from 44 countries examining the physics of the ocean and its role in the climate of the Earth. It is based at Southampton University, England.

ocean ridge mountain range on the seabed indicating the presence of a constructive plate margin (where tectonic plates are moving apart and magma rises to the surface; see ◊plate tectonics). Ocean ridges, such as the ◊Mid-Atlantic Ridge, consist of many segments offset along ◊faults, and can rise thousands of metres above the surrounding seabed.

Ocean ridges usually have a ◊rift valley along their crests, indicating where the flanks are being pulled apart by the growth of the plates of the ◊lithosphere beneath. The crests are generally free of sediment; increasing depths of sediment are found with increasing distance down the flanks.

ocean trench deep trench in the seabed indicating the presence of a destructive margin (produced by the movements of ◊plate tectonics). The subduction or dragging downwards of one plate of the ◊lithosphere beneath another means that the ocean floor is pulled down. Ocean trenches are found around the edge of the Pacific Ocean and the NE Indian Ocean; minor ones occur in the Caribbean and near the Falkland Islands.

Ocean trenches represent the deepest parts of the ocean floor, the deepest being the ◊Mariana Trench which has a depth of 11,034 m/36,201 ft. At depths of below 6 km/3.6 mi there is no light and very high pressure; ocean trenches are inhabited by crustaceans, coelenterates (for example, sea anemones), polychaetes (a type of worm), molluscs, and echinoderms.

OCR abbreviation for ◊*optical character recognition*.

octahedron, regular regular solid comprised of eight faces, each of which is an equilateral triangle. It is one of the five regular polyhedra, or Platonic solids. The figure made by joining the midpoints of the faces is a perfect cube and the vertices of the octahedron are themselves the midpoints of the faces of a surrounding cube. For this reason, the cube and the octahedron are called dual solids.

octal number system number system to the base eight, used in computing. The highest digit that can appear in the octal system is seven. Normal decimal, or base-ten, numbers may be considered to be written under column headings based on the number ten. For example, the decimal number 567 stands for:

100s	10s	1s
(10^2)	(10^1)	(10^0)
5	6	7

Octal, or base-eight, numbers can be thought of as written under column headings based on the number eight. For example, the octal number 567 stands for:

64s	8s	1s
(8^2)	(8^1)	(8^0)
5	6	7

The octal number 567 is therefore equivalent to the decimal number 375, since $(5 \times 64) + (6 \times 8) + (7 \times 1) = 375$.

The octal number system is sometimes used by computer programmers as an alternative to the ◊hexadecimal number system.

octane rating numerical classification of petroleum fuels indicating their combustion characteristics.

The efficient running of an ◊internal-combustion engine depends on the ignition of a petrol–air mixture at the correct time during the cycle of the engine. Higher-rated petrol burns faster than lower-rated fuels. The use of the correct grade must be matched to the engine.

Octans constellation in the southern hemisphere containing the southern celestial pole. The closest naked-eye star to the south celestial pole is fifth-magnitude Sigma Octantis.

octet rule in chemistry, rule stating that elements combine in a way that gives them the electronic structure of the nearest ◊inert gas. All the inert gases except helium have eight electrons in their outermost shell, hence the term octet.

oersted cgs unit (symbol Oe) of ◊magnetic field strength, now replaced by the SI unit ampere per metre. The Earth's magnetic field is about 0.5 oersted; the field near the poles of a small bar magnet is several hundred oersteds; and a powerful ◊electromagnet can have a field strength of 30,000 oersteds.

oesophagus passage by which food travels from mouth to stomach. The human oesophagus is about 23 cm/9 in long. Its upper end is at the bottom of the ◊pharynx, immediately behind the windpipe.

oestrogen group of hormones produced by the ◊ovaries of vertebrates; the term is also used for various synthetic hormones that mimic their effects. The principal oestrogen in mammals is oestradiol. Oestrogens promote the development of female secondary sexual characteristics; stimulate egg production; and, in mammals, prepare the lining of the uterus for pregnancy.

oestrus in mammals, the period during a female's reproductive cycle (also known as the oestrus cycle or ◊menstrual cycle) when mating is most likely to occur. It usually coincides with ovulation.

office automation introduction of computers and other electronic equipment, such as fax machines, to support an office routine. Increasingly, computers are used to support administrative tasks such as document processing, filing, mail, and schedule management; project planning and management accounting have also been computerized.

offset printing the most common method of ◊printing, which uses smooth (often rubber) printing plates. It works on the principle of ◊lithography: that grease and water repel one another.

The printing plate is prepared using a photographic technique, resulting in a type image that attracts greasy printing ink. On the printing press the plate is wrapped around a cylinder and wetted and inked in turn. The ink adheres only to the type area, and this image is then transferred via an intermediate blanket cylinder to the paper.

ohm SI unit (symbol Ω) of electrical ◊resistance (the property of a substance that restricts the flow of electrons through it).

It was originally defined with reference to the resistance of a column of mercury, but is now taken as the resistance between two points when a potential difference of one volt between them produces a current of one ampere.

ohmic heating method of heating used in the food-processing industry, in which an electric current is passed through foodstuffs to sterilize them before packing. The heating effect is similar to that obtained by microwaves in that electrical energy is transformed into heat throughout the whole volume of the food, not just at the surface.

Ohmic heating is suitable for foods containing chunks of meat or fruit. It is an alternative to in-can sterilization and has been used to produce canned foods such as meat chunks, prawns, baked beans, fruit, and vegetables.

Ohm's law law that states that the current flowing in a metallic conductor maintained at constant temperature is directly proportional to the potential difference (voltage) between its ends. The law was discovered by German physicist Georg Ohm 1827.

If a current of I amperes flows between two points in a conductor across which the potential difference is V volts, then V/I is a constant called the ◊resistance R ohms between those two points. Hence:

$$V/I = R$$

or

$$V = IR$$

Not all conductors obey Ohm's law; those that do are called **ohmic conductors**.

oil flammable substance, usually insoluble in water, and composed chiefly of carbon and hydrogen. Oils may be solids (fats and waxes) or liquids. The three main types are: **essential oils**, obtained from plants; **fixed oils**, obtained from animals and plants; and **mineral oils**, obtained chiefly from the refining of ◊petroleum.

Essential oils are volatile liquids that have the odour of their plant source and are used in perfumes, flavouring essences, and in aromatherapy. Fixed oils are mixtures of ◊lipids, of varying consistency, found in both animals (for example, fish oils) and plants (in nuts and seeds); they are used as foods and as lubricants, and in the making of soaps, paints, and varnishes. Mineral oils are composed of a mixture of hydrocarbons, and are used as fuels and lubricants. Eight of the 14 top-earning companies in the USA in 1990 (led by Exxon with $7 billion in sales) are in the global petroleum industry.

oil spill oil released by damage to or discharge from a tanker or oil installation. An oil spill kills all shore life, clogging up the feathers of birds and suffocating other creatures. At sea toxic chemicals leach into the water below, poisoning sea life. Mixed with dust, the oil forms globules that sink to the seabed, poisoning sea life there as well.

In March 1989 the tanker *Exxon Valdez* spilled oil in Alaska's Prince William Sound, covering 12,400 sq km/4,800 sq mi and killing at least 34,400 sea birds, 10,000 sea otters, and up to 16 whales. The incident led to the US Oil Pollution Act of 1990, which requires tankers operating in US waters to have double hulls. The world's largest oil spill was in the Persian Gulf in Feb 1991 as a direct result of hostilities during the Gulf War. Around 6–8 million barrels of oil were spilled, polluting 675 km/420 mi of Saudi coastline. In some places, the oil was 30 cm/12 in deep in the sand.

The amount of oil entering oceans from shipping operations decreased by 60% 1981–1991.

major oil spills

year	place	source	quantity tonnes litres
1967	off Cornwall, England	Torrey Canyon	107,100
1968	off South Africa	World Glory	51,194,000
1972	Gulf of Oman	Sea Star	103,500
1977	North Sea	Ekofisk oilfield	31,040,000
1978	off France	Amoco Cadiz	200,000
1979	Gulf of Mexico	Ixtoc 1 oil well	535,000
1979	off Trinidad and Tobago	collision of Atlantic Empress and Aegean Captain	270,000
1983	Persian Gulf	Nowruz oilfield	540,000
1983	off South Africa	Castillo de Beliver	225,000
1989	off Alaska	Exxon Valdez	40,504,000
1991	Persian Gulf	oil wells in Kuwait and Iraq	
1993	Shetland Islands, Scotland	Braer	85,000

Olbers' paradox question put forward 1826 by Heinrich Olbers, who asked: If the universe is infinite in extent and filled with stars, why is the sky dark at night? The answer is that the stars do not live infinitely long, so there is not enough starlight to fill the universe. A wrong answer, frequently given, is that the expansion of the universe weakens the starlight.

olefin common name for ◊alkene.

Oligocene third epoch of the Tertiary period of geological time, 35.5–3.25 million years ago. The name, from Greek, means 'a little recent', referring to the presence of the remains of some modern types of animals existing at that time.

oligosaccharide ◊carbohydrate comprising a few ◊monosaccharide units linked together. It is a general term used to indicate that a carbohydrate is larger than a simple di-or trisaccharide but not as large as a polysaccharide.

olivenite basic copper arsenate, $Cu_2AsO_4(OH)$, occurring as a mineral in olive-green prisms.

olivine greenish mineral, magnesium iron silicate, $(Mg,Fe)_2SiO_4$. It is a rock-forming mineral, present in, for example, peridotite, gabbro, and basalt. Olivine is called *peridot* when pale green and transparent, and used in jewellery.

Omega navigation system long-range radio-based aid to navigation, giving worldwide coverage. There are eight Omega transmitting stations, located in Norway, Liberia, Hawaii, Réunion, Argentina, Australia, USA, and Japan. The very-low-frequency signals from the transmitters are detected by a ship's navigation receiver, and slight differences (phase differences) between the signals indicate the position of the receiver. The system is accurate to within 4 km/2.5 mi during the day and 7 km/4 mi at night.

omnivore animal that feeds on both plant and animal material. Omnivores have digestive adaptations intermediate between those of ◊herbivores and ◊carnivores, with relatively unspecialized digestive systems and gut microorganisms that can digest a variety of foodstuffs.

Examples include the chimpanzee, the cockroach, and the ant.

OMR abbreviation for ◊*optical mark recognition*.

oncogene a gene carried by a virus that induces a cell to divide abnormally forming a ◊tumour. Oncogenes arise from mutations in genes (proto-oncogenes) found in all normal cells. They are usually also found in viruses that are capable of transforming normal cells to tumour cells. Such viruses are able to insert their oncogenes into the host cell's DNA, causing it to divide uncontrollably. More than one oncogene may be necessary to transform a cell in this way.

on-line system in computing, originally a system that allows the computer to work interactively with its users, responding to each instruction as it is given and prompting users for information when necessary. Since almost all the computers people use now work this way, 'on-line system' is now used to refer to large database, electronic mail, and conferencing systems accessed via a dial-up modem. These often have tens or hundreds of users from different places—sometimes from different countries—'on line' at the same time.

ontogeny process of development of a living organism, including the part of development that takes place after hatching or birth. The idea that 'ontogeny recapitulates phylogeny' (the development of an organism goes through the same stages as its evolutionary history), proposed by the German scientist Ernst Heinrich Haeckel, is now discredited.

onyx semiprecious variety of chalcedonic ◊silica (SiO_2) in which the crystals are too fine to be detected under a microscope, a state known as cryptocrystalline. It has straight parallel bands of different colours: milk-white, black, and red.

Sardonyx, an onyx variety, has layers of brown or red carnelian alternating with lighter layers of onyx. It can be carved into cameos.

oolite limestone made up of tiny spherical carbonate particles called *ooliths*. Ooliths have a concentric structure with a diameter up to 2 mm/0.08 in.

They were formed by chemical precipitation and accumulation on ancient sea floors.

The surface texture of oolites is rather like that of fish roe. The late Jurassic limestones of the British Isles are mostly oolitic in nature.

Oort cloud spherical cloud of comets beyond Pluto, extending out to about 100,000 astronomical units (1.5 light years) from the Sun. The gravitational effect of passing stars and the rest of our Galaxy disturbs comets from the cloud so that they fall in towards the Sun on highly elongated orbits, becoming visible from Earth. As many as 10 trillion comets may reside in the Oort cloud, named after Dutch astronomer Jan Oort who postulated it 1950.

oosphere another name for the female gamete, or ◊ovum, of certain plants such as algae.

ooze sediment of fine texture consisting mainly of organic matter found on the ocean floor at depths greater than 2,000 m/6,600 ft. Several kinds of ooze exist, each named after its constituents.

Siliceous ooze is composed of the ◊silica shells of tiny marine plants (diatoms) and animals (radiolarians). *Calcareous ooze* is formed from the ◊calcite shells of microscopic animals (foraminifera) and floating algae (coccoliths).

opal form of ◊silica (SiO₂), often occurring as stalactites and found in many types of rock. The common opal is translucent, milk-white, yellow, red, blue, or green, and lustrous. Precious opal is opalescent, the characteristic play of colours being caused by close-packed silica spheres diffracting light rays within the stone.

Opal is cryptocrystalline, that is, the crystals are too fine to be detected under a microscope. Opals are found in Hungary; New South Wales, Australia (black opals were first discovered there 1905); and Mexico (red fire opals).

opencast mining or *open-pit mining* or *strip mining* mining at the surface rather than underground. Coal, iron ore, and phosphates are often extracted by opencast mining. Often the mineral deposit is covered by soil, which must first be stripped off, usually by large machines such as walking draglines and bucket-wheel excavators. The ore deposit is then broken up by explosives.

One of the largest excavations in the world has been made by opencast mining at the Bingham Canyon copper mine in Utah, USA, measuring 790 m/2,590 ft deep and 3.7 km/2.3 mi across.

open-hearth furnace method of steelmaking, now largely superseded by the ◊basic-oxygen process. It was developed in 1864 in England by German-born William and Friedrich Siemens, and improved by Pierre and Emile Martin in France in the same year. In the furnace, which has a wide, saucer-shaped hearth and a low roof, molten pig iron and scrap are packed into the shallow hearth and heated by overhead gas burners using preheated air.

operating system (OS) in computing, a program that controls the basic operation of a computer. A typical OS controls the ◊peripheral devices, organizes the filing system, provides a means of communicating with the operator, and runs other programs.

Some operating systems were written for specific computers, but some are accepted standards. These include CP/M (by Digital Research) and MS-DOS (by Microsoft) for microcomputers. Unix (developed at AT&T's Bell Laboratories) is the standard on workstations, minicomputers, and super computers; it is also used on desktop PCs and mainframes.

operational amplifier (op-amp) type of electronic circuit that is used to increase the size of an alternating voltage signal without distorting it.

Operational amplifiers are used in a wide range of electronic measuring instruments. The name arose because they were originally designed to carry out mathematical operations and solve equations.

operculum small cap covering the spore-producing body of many mosses. It is pushed aside when the spores are mature and ready to be ejected.

operon group of genes that are found next to each other on a chromosome, and are turned on and off as an integrated unit. They usually produce enzymes that control different steps in the same biochemical pathway. Operons were discovered 1961 (by the French biochemists François Jacob and Jacques Monod) in bacteria.

They are less common in higher organisms where the control of metabolism is a more complex process.

Ophiuchus large constellation along the celestial equator, known as the serpent bearer because the constellation Serpens is wrapped around it. The Sun passes through Ophiuchus each Dec, but the constellation is not part of the zodiac. Ophiuchus contains ◊Barnard's star.

opiate, endogenous naturally produced chemical in the body that has effects similar to morphine and other opiate drugs; a type of neurotransmitter. Examples include ◊endorphins and ◊encephalins.

opium drug extracted from the unripe seeds of the opium poppy *Papaver somniferum* of SW Asia. An addictive narcotic, it contains several alkaloids, including *morphine*, one of the most powerful natural painkillers and addictive narcotics known, and *codeine*, a milder painkiller.

Heroin is a synthetic derivative of morphine and even more powerful as a drug. Opium is still sometimes given as a tincture, dissolved in alcohol and known as *laudanum*. Opium also contains the highly poisonous alkaloid *thebaine*.

opposition in astronomy, the moment at which a body in the Solar System lies opposite the Sun in the sky as seen from the Earth and crosses the ◊meridian at about midnight.

Although the ◊inferior planets cannot come to opposition, it is the best time for observation of the superior planets as they can then be seen all night.

optical aberration see ◊aberration, optical.

optical activity in chemistry, the ability of certain crystals, liquids, and solutions to rotate the plane of ◊polarized light as it passes through them. The phenomenon is related to the three-dimensional arrangement of the atoms making up the molecules concerned. Only substances that lack any form of structural symmetry exhibit optical activity.

optical character recognition (OCR) in computing, a technique for inputting text to a computer by means of a document reader. First, a ◊scanner produces a digital image of the text; then character-recognition software makes use of stored knowledge about the shapes of individual characters to convert the digital image to a set of internal codes that can be stored and processed by computer.

OCR originally required specially designed characters but current devices can recognize most standard typefaces and even handwriting. OCR is used, for example, by gas and electricity companies to input data collected on meter-reading cards.

optical computer computer in which both light and electrical signals are used in the ◊central processing unit. The technology is still not fully developed, but such a computer promises to be faster and less vulnerable to outside electrical interference than one that relies solely on electricity.

optical disc in computing, a storage medium in which laser technology is used to record and read large volumes of digital data. Types include ◊CD-ROM, ◊WORM, and erasable optical disc.

optical emission spectrometry another term for emission ◊spectroscopy.

optical fibre very fine, optically pure glass fibre through which light can be reflected to transmit an image or information from one end to the other. Optical fibres are increasingly being used to replace copper wire in telephone cables, the messages being coded as pulses of light rather than a fluctuating electric current. Although expensive to produce and install, optical fibres can carry more data than traditional cables, and are less susceptible to interference.

Bundles of optical fibres are also used in endoscopes to inspect otherwise inaccessible parts of machines or of the living body (see ◊endoscopy).

optical illusion scene or picture that fools the eye. An example of a natural optical illusion is that the Moon appears bigger when it is on the horizon than when it is high in the sky, owing to the ◊refraction of light rays by the Earth's atmosphere.

optical instrument instrument that makes use of one or more lenses or mirrors, or of a combination of these, in order to change the path of light rays and produce an image. Optical instruments such as magnifying glasses, ◊microscopes, and ◊telescopes are used to provide a clear, magnified image of the very small or the very distant. Others, such as ◊cameras, photographic enlargers, and film ◊projectors, may be used to store or reproduce images.

optical mark recognition (OMR) in computing, a technique that enables marks made in predetermined positions on computer-input forms to be detected optically and input to a computer. An *optical mark reader* shines a light beam onto the input document and is able to detect the marks because less light is reflected back from them than from the paler, unmarked paper.

optic nerve large nerve passing from the eye to the brain, carrying visual information. In mammals, it may contain up to a million nerve fibres, connecting the sensory cells of the retina to the optical centres in the brain. Embryologically, the optic nerve develops as an outgrowth of the brain.

optics branch of physics that deals with the study of ◊light and vision—for example, shadows and

mirror images, lenses, microscopes, telescopes, and cameras. For all practical purposes light rays travel in straight lines, although US physicist Albert Einstein demonstrated that they may be 'bent' by a gravitational field. On striking a surface they are reflected or refracted with some absorption of energy, and the study of this is known as geometrical optics.

optoelectronics branch of electronics concerned with the development of devices (based on the ◊semiconductor gallium arsenide) that respond not only to the ◊electrons of electronic data transmission, but also to ◊photons.

In 1989, scientists at IBM in the USA built a gallium arsenide microprocessor ('chip') containing 8,000 transistors and four photodetectors. The densest optoelectronic chip yet produced, this can detect and process data at a speed of 1 billion bits per second.

orbit path of one body in space around another, such as the orbit of Earth around the Sun, or the Moon around Earth. When the two bodies are similar in mass, as in a ◊binary star, both bodies move around their common centre of mass. The movement of objects in orbit follows Johann Kepler's laws, which apply to artificial satellites as well as to natural bodies.

As stated by the laws, the orbit of one body around another is an ellipse. The ellipse can be highly elongated, as are comet orbits around the Sun, or it may be almost circular, as are those of some planets. The closest point of a planet's orbit to the Sun is called *perihelion*; the most distant point is called *aphelion*. (For a body orbiting the Earth, the closest and furthest points of the orbit are called *perigee* and *apogee*.)

orbital, atomic region around the nucleus of an atom (or, in a molecule, around several nuclei) in which an ◊electron is most likely to be found. According to ◊quantum theory, the position of an electron is uncertain; it may be found at any point. However, it is more likely to be found in some places than in others, and it is these that make up the orbital.

An atom or molecule has numerous orbitals, each of which has a fixed size and shape. An orbital is characterized by three numbers, called ◊quantum numbers, representing its energy (and hence size), its angular momentum (and hence shape), and its orientation. Each orbital can be occupied by one or (if their spins are aligned in opposite directions) two electrons.

order in biological classification, a group of related ◊families. For example, the horse, rhinoceros, and tapir families are grouped in the order Perissodactyla, the odd-toed ungulates, because they all have either one or three toes on each foot. The names of orders are not shown in italic (unlike genus and species names) and by convention they have the ending '-formes' in birds and fish; '-a' in mammals, amphibians, reptiles, and other animals; and '-ales' in fungi and plants. Related orders are grouped together in a ◊class.

ordinal number in mathematics, one of the series first, second, third, fourth, ... Ordinal numbers relate to order, whereas ◊cardinal numbers (1, 2, 3, 4, . . .) relate to quantity, or count.

orbital, atomic *The shapes of atomic orbitals. An atomic orbital is a picture of the 'electron cloud' that surrounds the nucleus of an atom. There are four basic shapes for atomic orbitals: spherical, dumbbell, clover-leaf, and complex (shown at bottom left).*

ordinate in ◊coordinate geometry, the y coordinate of a point; that is, the vertical distance of the point from the horizontal or x-axis. For example, a point with the coordinates (3,4) has an ordinate of 4. See ◊abscissa.

Ordovician period of geological time 510–439 million years ago; the second period of the ◊Palaeozoic era. Animal life was confined to the sea: reef-building algae and the first jawless fish are characteristic.

The period is named after the Ordovices, an ancient Welsh people, because the system of rocks formed in the Ordovician period was first studied in Wales.

ore body of rock, a vein within it, or a deposit of sediment, worth mining for the economically valuable mineral it contains. The term is usually applied to sources of metals. Occasionally metals are found uncombined (native metals), but more often they occur as compounds such as carbonates, sulphides, or oxides. The ores often contain unwanted impurities that must be removed when the metal is extracted.

Commercially valuable ores include bauxite (aluminium oxide, Al_2O_3) hematite (iron(III) oxide, Fe_2O_3), zinc blende (zinc sulphide, ZnS), and rutile (titanium dioxide, TiO_2).

Hydrothermal ore deposits are formed from fluids such as saline water passing through fissures in the host rock at an elevated temperature. Examples are the 'porphyry copper' deposits of Chile and Bolivia, the submarine copper–zinc–iron sulphide deposits recently discovered on the East Pacific Rise, and the limestone lead–zinc deposits that occur in the southern USA and in the Pennines of Britain.

Other ores are concentrated by igneous processes, causing the ore metals to become segregated from a magma—for example, the bands rich in chromite and platinum metal within the Bushveld, South Africa. Erosion and transportation in rivers of material from an existing rock source can lead to further concentration of heavy minerals in a deposit, for example, Malaysian tin deposits.

Weathering of rocks in situ can result in residual metal-rich soils, such as the nickel-bearing laterites of New Caledonia.

organ in biology, part of a living body, such as the liver or brain, that has a distinctive function or set of functions.

organelle discrete and specialized structure in a living cell; organelles include mitochondria, chloroplasts, lysosomes, ribosomes, and the nucleus.

organic chemistry branch of chemistry that deals with carbon compounds. Organic compounds form the chemical basis of life and are more abundant than inorganic compounds. In a typical organic compound, each carbon atom forms bonds covalently with each of its neighbouring carbon atoms in a chain or ring, and additionally with other atoms, commonly hydrogen, oxygen, nitrogen, or sulphur.

The basis of organic chemistry is the ability of carbon to form long chains of atoms, branching chains, rings, and other complex structures. Compounds containing only carbon and hydrogen are known as *hydrocarbons*.

Organic chemistry is largely the chemistry of a great variety of homologous series—those in which the molecular formulae, when arranged in ascending order, form an arithmetical progression. The physical properties undergo a gradual change from one member to the next.

The linking carbon atoms that form the backbone of an organic molecule may be built up from beginning to end without branching; or may throw off branches at one or more points. Sometimes, however, the ropes of carbon atoms curl round and form rings (*cyclic compounds*), usually of five, six, or seven atoms. Open-chain and cyclic compounds may be classified as ◊aliphatic or ◊aromatic depending on the nature of the bonds between their atoms. Compounds containing oxygen, sulphur, or nitrogen within a carbon ring are called *heterocyclic compounds*.

Many organic compounds (such as proteins and carbohydrates) are made only by living organisms,

common organic molecule groupings

formula	name	structural formula
CH_3	methyl	$H-\overset{\displaystyle H}{\underset{\displaystyle H}{C}}-$
CH_2CH_3	ethyl	$-\overset{\displaystyle H}{\underset{\displaystyle H}{C}}-\overset{\displaystyle H}{\underset{\displaystyle H}{C}}-H$
CC	double bond	$\overset{\diagdown}{\diagup}C=C\overset{\diagup}{\diagdown}$
CHO	aldehyde	$-C\overset{\displaystyle H}{\underset{\displaystyle O}{}}$
CH_2OH	alcohol	$-\overset{\displaystyle H}{\underset{\displaystyle H}{C}}-OH$
CO	ketone	$\overset{\diagdown}{\diagup}C=O$
$COOH$	acid	$-C\overset{\displaystyle O}{\underset{\displaystyle OH}{}}$
CH_2NH_2	amine	$-\overset{\displaystyle H}{\underset{\displaystyle H}{C}}-N\overset{\diagup H}{\diagdown H}$
C_6H_6	benzene ring	benzene ring structure

organic chemistry *Common organic-molecule groupings.*
Organic chemistry is the study of carbon compounds, which
make up over 90% of all chemical compounds. This diversity
arises because carbon atoms can combine in many different ways
with other atoms, forming a wide variety of loops and chains.

and it was once believed that organic compounds
could not be made by any other means. This was
disproved when Wöhler synthesized urea, but the
name 'organic' (that is 'living') chemistry has
remained in use. Many organic compounds are
derived from petroleum, which represents the

chemical remains of millions of microscopic marine
organisms.
   In inorganic chemistry, a specific formula usually
represents one substance only, but in organic
chemistry, it is exceptional for a molecular formula
to represent only one substance. Substances having
the same molecular formula are called *isomers*,
and the relationship is known as *isomerism*.
   Hydrocarbons form one of the most prolific of
the many organic types; fuel oils are largely made
up of hydrocarbons. Typical groups containing only
carbon, hydrogen, and oxygen are alcohols, aldehydes, ketones, ethers, esters, and carbohydrates.
Among groups containing nitrogen are amides,
amines, nitro-compounds, amino acids, proteins,
purines, alkaloids, and many others, both natural
and artificial. Other organic types contain sulphur,
phosphorus, or halogen elements.
   The most fundamental of all natural processes
are oxidation, reduction, hydrolysis, condensation,
polymerization, and molecular rearrangement. In
nature, such changes are often brought about
through the agency of promoters known as
enzymes, which act as catalytic agents in promoting specific reactions. The most fundamental of all
natural processes is *synthesis*, or building up. In
living plant and animal organisms, the energy
stored in carbohydrate molecules, derived originally from sunlight, is released by slow oxidation
and utilized by the organisms. The complex carbohydrates thereby revert to carbon dioxide and
water, from where they were built up with absorption of energy. Thus, a so-called carbon food cycle
exists in nature. In a corresponding nitrogen food
cycle, complex proteins are synthesized in nature
from carbon dioxide, water, soil nitrates, and
ammonium salts, and these proteins ultimately
revert to the elementary raw materials from which
they came, with the discharge of their energy of
chemical combination.

organic compound in biochemistry, one of the
class of compounds whose behaviour is influenced
by the chemistry of carbon. All organic compounds
contain carbon, which was once thought to be present only in living things. The original distinction
between organic and inorganic compounds was
based on the belief that the molecules of living
systems were unique, and could not be synthesized
in the laboratory. Today it is routine to manufacture thousands of organic chemicals both in
research and in the drug industry.

organic farming farming without the use of synthetic fertilizers (such as ◊nitrates and phosphates)
or ◊pesticides (herbicides, insecticides, and fungicides) or other agrochemicals (such as hormones,
growth stimulants, or fruit regulators).
   In place of artificial fertilizers, compost, manure,
seaweed, or other substances derived from living
things are used (hence the name 'organic'). Growing a crop of a nitrogen-fixing plant such as
lucerne, then ploughing it back into the soil, also
fertilizes the ground. Some organic farmers use
naturally occurring chemicals such as nicotine or
pyrethrum to kill pests, but control by non-chemi-
cal methods is preferred. Those methods include
removal by hand, intercropping (planting with companion plants which deter pests), mechanical bar-

riers to infestation, crop rotation, better cultivation methods, and ◊biological control. Weeds can be controlled by hoeing, mulching (covering with manure, straw, or black plastic), or burning off. Organic farming methods produce food without pesticide residues and greatly reduce pollution of the environment. They are more labour intensive, and therefore more expensive, but use less fossil fuel. Soil structure is greatly improved by organic methods, and recent studies show that a conventional farm can lose four times as much soil through erosion as an organic farm, although the loss may not be immediately obvious.

organizer in embryology, a part of the embryo that causes changes to occur in another part, through ◊induction, thus 'organizing' development and ◊differentiation.

OR gate in electronics, a type of ◊logic gate.

orimulsion fuel made by mixing ◊bitumen and water that can be burnt in the same way as heavy oil. It is cheap to make but the smoke produced has a high sulphur content.

Orion in astronomy, a very prominent constellation in the equatorial region of the sky, identified with the hunter of Greek mythology.

It contains the bright stars Betelgeuse and Rigel, as well as a distinctive row of three stars that make up Orion's belt. Beneath the belt, marking the sword of Orion, is the Orion nebula; nearby is one of the most distinctive dark nebulae, the Horsehead.

Orion nebula luminous cloud of gas and dust 1,500 light years away, in the constellation Orion, from which stars are forming. It is about 15 light years in diameter, and contains enough gas to make a cluster of thousands of stars.

At the nebula's centre is a group of hot young stars, called the **Trapezium**, which make the surrounding gas glow. The nebula is visible to the naked eye as a misty patch below the belt of Orion.

ORION NEBULA

If the distance between the Earth and the Sun were 30 cm/1 ft, the Orion nebula would be about 190 km/120 mi across.

---

ormolu (French *or moulu* 'ground gold') alloy of copper, zinc, and sometimes tin, used for furniture decoration.

ornithology study of birds. It covers scientific aspects relating to their structure and classification, and their habits, song, flight, and value to agriculture as destroyers of insect pests. Worldwide scientific banding (or the fitting of coded rings to captured specimens) has resulted in accurate information on bird movements and distribution. There is an International Council for Bird Preservation with its headquarters at the Natural History Museum, London.

Interest in birds has led to the formation of societies for their protection, of which the Society for the Protection of Birds 1889 in Britain was the first; it received a royal charter 1904. The Audubon Society 1905 in the USA has similar aims; other countries now have similar societies. The head-

quarters of the British Trust for Ornithology is at Beech Grove, Tring, Hertfordshire. Migration, age, and pollution effects on birds are monitored by ringing (trained government-licensed operators fit numbered metal rings to captured specimens with a return address). Legislation in various countries to protect wild birds followed from a British act of Parliament 1880.

ornithophily ◊pollination of flowers by birds. Ornithophilous flowers are typically brightly coloured, often red or orange. They produce large quantities of thin, watery nectar, and are scentless because most birds do not respond well to smell. They are found mostly in tropical areas, with hummingbirds being important pollinators in North and South America, and the sunbirds in Africa and Asia.

orogeny or *orogenesis* the formation of mountains. It is brought about by the movements of the rigid plates making up the Earth's crust (described by ◊plate tectonics). Where two plates collide at a destructive margin rocks become folded and lifted to form chains of fold mountains (such as the ◊young fold mountains of the Himalayas).

orographic rainfall rainfall that occurs when an airstream is forced to rise over a mountain range. The air becomes cooled and precipitation takes place. In the UK, the Pennine hills, which extend southwards from Northumbria to Derbyshire in N England, cause orographic rainfall and are partly responsible for the west of the UK being wetter than the east. The orographic effect can sometimes occur in large cities, when air rises over tall buildings.

orrery mechanical device for demonstrating the motions of the heavenly bodies. Invented about 1710 by George Graham, it was named after his patron, the 4th Earl of Orrery. It is the forerunner of the planetarium.

orthochromatic photographic film or paper of decreased sensitivity, which can be processed with a red safelight. Using it, blue objects appear lighter and red ones darker because of increased blue sensitivity.

OS/2 single-user computer ◊operating system produced jointly by Microsoft Corporation and IBM for use on large microcomputers. Its main features are ◊multitasking and the ability to access large amounts of internal ◊memory.

It was announced 1987. Microsoft abandoned it 1992 due to its lack of market acceptance.

oscillating universe in astronomy, a theory that states that the gravitational attraction of the mass within the universe will eventually slow down and stop the expansion of the universe. The outward motions of the galaxies will then be reversed, eventually resulting in a 'Big Crunch' where all the matter in the universe would be contracted into a small volume of high density. This could undergo a further ◊Big Bang, thereby creating another expansion phase. The theory suggests that the universe would alternately expand and collapse through alternate Big Bangs and Big Crunches.

oscillation one complete to-and-fro movement of a vibrating object or system. For any particular vibration, the time for one oscillation is called its

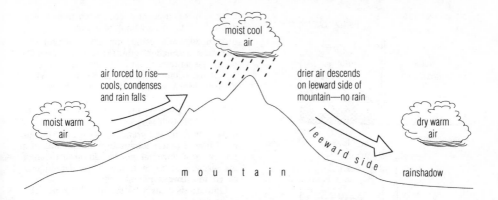

orographic rainfall

◊period and the number of oscillations in one second is called its ◊frequency. The maximum displacement of the vibrating object from its rest position is called the ◊amplitude of the oscillation.

oscillator any device producing a desired oscillation (vibration). There are many types of oscillator for different purposes, involving various arrangements of thermionic ◊valves or components such as ◊transistors, ◊inductors, ◊capacitors, and ◊resistors.

An oscillator is an essential part of a radio transmitter, generating the high-frequency carrier signal necessary for radio communication. The ◊frequency is often controlled by the vibrations set up in a crystal (such as quartz).

oscillograph instrument for displaying or recording the values of rapidly changing oscillations, electrical or mechanical.

oscilloscope or *cathode-ray oscilloscope* (CRO) instrument used to measure electrical voltages that vary over time and to display the waveforms of electrical oscillations or signals, by means of the deflection of a beam of ◊electrons. Readings are displayed graphically on the screen of a ◊cathode-ray tube.

osmium (Greek *osme* 'odour') hard, heavy, bluish-white, metallic element, symbol Os, atomic number 76, relative atomic mass 190.2. It is the densest of the elements, and is resistant to tarnish and corrosion. It occurs in platinum ores and as a free metal (see ◊native metal) with iridium in a natural alloy called osmiridium, containing traces of platinum, ruthenium, and rhodium. Its uses include pen points and light-bulb filaments; like platinum, it is a useful catalyst.

Osmium was discovered in 1803 and named in 1804 by English chemist Smithson Tennant (1761–1815) after the irritating smell of one of its oxides.

osmoregulation process whereby the water content of living organisms is maintained at a constant level. If the water balance is disrupted, the concentration of salts will be too high or too low, and vital functions, such as nerve conduction, will be adversely affected.

In mammals, loss of water by evaporation is counteracted by increased intake and by mechanisms in the kidneys that enhance the rate at which water is resorbed before urine production. Both

or

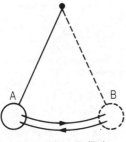

one complete oscillation or cycle is from A to B and back to A

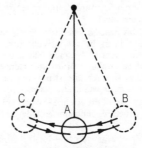

one complete oscillation or cycle is from A to B to C and back to A, moving in the same direction again

oscillation *The oscillation of a pendulum. The period of oscillation is the time taken for the pendulum bob to swing from and return to position A.*

before osmosis

semipermeable membrane

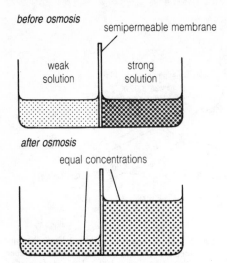

| weak solution | strong solution |

after osmosis

equal concentrations

osmosis *Osmosis, the movement of liquid through a semipermeable membrane separating solutions of different concentrations, is essential for life. Osmosis transports water from the soil into the roots of plants, for example. It is applied in kidney dialysis to remove impurities from the blood.*

these responses are mediated by hormones, primarily those of the adrenal cortex (see ◊adrenal gland).

osmosis movement of solvent (liquid) through a semipermeable membrane separating solutions of different concentrations. The solvent passes from a less concentrated solution to a more concentrated solution until the two concentrations are equal. Applying external pressure to the solution on the more concentrated side arrests osmosis, and is a measure of the osmotic pressure of the solution.

Many cell membranes behave as semipermeable membranes, and osmosis is a vital mechanism in the transport of fluids in living organisms—for example, in the transport of water from the roots up the stems of plants.

Fish have protective mechanisms to counteract osmosis, which would otherwise cause fluid transport between the body of the animal and the surrounding water (outwards in saltwater fish, inwards in freshwater ones).

ossification process whereby bone is formed in vertebrate animals by special cells (*osteoblasts*) that secrete layers of ◊extracellular matrix on the surface of the existing ◊cartilage. Conversion to bone occurs through the deposition of calcium phosphate crystals within the matrix.

osteology part of the science of ◊anatomy, dealing with the structure, function, and development of bones.

Otto cycle alternative name for the ◊four-stroke cycle, introduced by the German engineer Nikolaus Otto (1832–1891) in 1876. It improved on existing piston engines by compressing the fuel mixture in the cylinder before it was ignited.

ounce unit of mass, one-sixteenth of a pound ◊avoirdupois, equal to 437.5 grains (28.35 g); also one-twelfth of a pound troy, equal to 480 grains. The *fluid ounce* is a measure of capacity. In the UK it is equivalent to one-twentieth of a pint; in the USA to one-sixteenth of a pint.

output device in computing, any device for displaying, in a form intelligible to the user, the results of processing carried out by a computer.

The most common output devices are the VDU (◊visual display unit, or screen) and the printer. Other output devices include graph plotters, speech synthesizers, and COM (computer output on microfilm/microfiche).

outwash sands and gravels deposited by streams of meltwater (water produced by the melting of a glacier). Such material may be laid down ahead of the glacier's snout to form a large flat expanse called an *outwash plain*.

Outwash is usually well sorted, the particles being deposited by the meltwater according to their size—the largest are deposited next to the snout while finer particles are deposited further downstream.

ovary in female animals, the organ that generates the ◊ovum. In humans, the ovaries are two whitish rounded bodies about 25 mm/1 in by 35 mm/1.5 in, located in the abdomen near the ends of the ◊Fallopian tubes. Every month, from puberty to the onset of the menopause, an ovum is released from the ovary. This is called ovulation, and forms part of the ◊menstrual cycle. In botany, an ovary is the expanded basal portion of the ◊carpel of flowering plants, containing one or more ◊ovules. It is hollow with a thick wall to protect the ovules. Following fertilization of the ovum, it develops into the fruit wall or pericarp.

The ovaries of female animals secrete the hormones responsible for the secondary sexual characteristics of the female, such as smooth, hairless facial skin and enlarged breasts. An ovary in a half-grown human fetus contains 5 million eggs, and so the unborn baby already contains the female genetic information for the next generation.

In botany, the relative position of the ovary to the other floral parts is often a distinguishing character in classification; it may be either inferior or superior, depending on whether the petals and sepals are inserted above or below.

overfishing fishing at rates that exceed the ◊sustained-yield cropping of fish species, resulting in a net population decline. For example, in the North Atlantic, herring has been fished to the verge of extinction and the cod and haddock populations are severely depleted. In the Third World, use of huge factory ships, often by fisheries from industrialized countries, has depleted stocks for local people who cannot obtain protein in any other way.

overtone note that has a frequency or pitch that is a multiple of the fundamental frequency, the sounding body's ◊natural frequency. Each sound source produces a unique set of overtones, which gives the source its quality or timbre.

ovipary method of animal reproduction in which eggs are laid by the female and develop outside her body, in contrast to ovovivipary and vivipary. It is the most common form of reproduction.

ovovivipary method of animal reproduction in which fertilized eggs develop within the female (unlike ovipary), and the embryo gains no nutritional substances from the female (unlike vivipary). It occurs in some invertebrates, fishes, and reptiles.

ovulation in female animals, the process of releasing egg cells (ova) from the ◊ovary. In mammals it occurs as part of the ◊menstrual cycle.

ovule structure found in seed plants that develops into a seed after fertilization. It consists of an ◊embryo sac containing the female gamete (◊ovum or egg cell), surrounded by nutritive tissue, the nucellus. Outside this there are one or two coverings that provide protection, developing into the testa, or seed coat, following fertilization.

In ◊angiosperms (flowering plants) the ovule is within an ◊ovary, but in ◊gymnosperms (conifers and their allies) the ovules are borne on the surface of an ovuliferous (ovule-bearing) scale, usually within a ◊cone, and are not enclosed by an ovary.

ovum (plural *ova*) female gamete (sex cell) before fertilization. In animals it is called an egg, and is produced in the ovaries. In plants, where it is also known as an egg cell or oosphere, the ovum is produced in an ovule. The ovum is nonmotile. It must be fertilized by a male gamete before it can develop further, except in cases of ◊parthenogenesis.

oxalic acid $(COOH)_2.2H_2O$ white, poisonous solid, soluble in water, alcohol, and ether. Oxalic acid is found in rhubarb, and its salts (oxalates) occur in wood sorrel (genus *Oxalis*, family Oxalidaceae) and other plants. It is used in the leather and textile industries, in dyeing and bleaching, ink manufacture, metal polishes, and for removing rust and ink stains.

oxbow lake curved lake found on the flood plain of a river. Oxbows are caused by the loops of ◊meanders being cut off at times of flood and the river subsequently adopting a shorter course. In the USA, the term ◊bayou is often used.

oxidation in chemistry, the loss of ◊electrons, gain of oxygen, or loss of hydrogen by an atom, ion, or molecule during a chemical reaction.

Oxidation may be brought about by reaction with another compound (oxidizing agent), which simultaneously undergoes ◊reduction, or electrically at the anode (positive electrode) of an electrolytic cell.

oxidation in earth science, a form of ◊chemical weathering caused by the chemical reaction that takes place between certain iron-rich minerals in rock and the oxygen in water. It tends to result in the formation of a red-coloured soil or deposit. The inside walls of canal tunnels and bridges often have deposits formed in this way.

oxidation number Roman numeral often seen in a chemical name, indicating the ◊valency of the element immediately before the number. Examples are lead(II) nitrate, manganese(IV) oxide, and potassium manganate(VII).

oxide compound of oxygen and another element, frequently produced by burning the element or a compound of it in air or oxygen.

Oxides of metals are normally ◊bases and will react with an acid to produce a ◊salt in which the metal forms the cation (positive ion). Some of them will also react with a strong alkali to produce a salt in which the metal is part of a complex anion

the formation of an oxbow lake

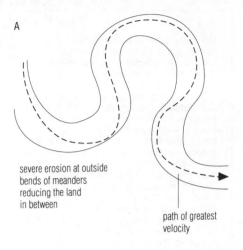

A

severe erosion at outside bends of meanders reducing the land in between

path of greatest velocity

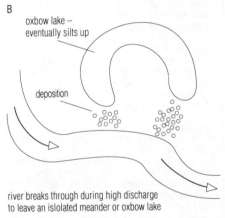

B

oxbow lake – eventually silts up

deposition

river breaks through during high discharge to leave an isolated meander or oxbow lake

oxbow lake

(negative ion; see ◊amphoteric). Most oxides of nonmetals are acidic (dissolve in water to form an ◊acid). Some oxides display no pronounced acidic or basic properties.

oxidizing agent substance that will oxidize another substance (see ◊oxidation).

In a redox reaction, the oxidizing agent is the substance that is itself reduced. Common oxidizing agents include oxygen, chlorine, nitric acid, and potassium manganate(VII).

oxyacetylene torch gas torch that burns ethene (acetylene) in pure oxygen, producing a high-temperature (3,000°C/5,400°F) flame. It is widely used in welding to fuse metals. In the cutting torch, a jet of oxygen burns through metal already melted by the flame.

oxygen (Greek *oxys* 'acid' *genes* 'forming') colourless, odourless, tasteless, nonmetallic, gaseous element, symbol O, atomic number 8, relative atomic mass 15.9994. It is the most abundant element in the Earth's crust (almost 50% by mass), forms about 21% by volume of the atmosphere,

The French Revolution

Not many scientists are credited with founding a major branch of science, but Antoine Lavoisier, the French aristocrat (1743–1794), is universally regarded as the founder of modern chemistry.

Before Lavoisier, chemistry had been dominated for 100 years by an influential theory of matter proposed by the German chemist Georg Stahl (1660–1743). This theory held that a combustible material burned because it contained a substance called *phlogiston*. Charcoal was a prime example of a phlogiston-rich material. The fact that metallurgists obtained some metals from their ores by heating them with charcoal seemed to lend support to the phlogiston theory of combustion. With hindsight, the reasons why such a theory persisted are obvious enough. Chemists had no clear idea of what a chemical element was, nor any understanding of the nature of gases.

Lavoisier restructured chemistry and gave it its modern form. He provided a firm foundation for the atomic theory proposed by British chemist and physicist John Dalton (1766–1844), and his chemical elements were later classified in the Periodic Table.

Fixed air?

The key experiments Lavoisier did seem simple enough. He burned phosphorus, lead and other solid elements in a sealed container. He noted that although the weight of the container and its contents did not increase, the weights of the solids in it did. He showed that the increase in weight of the solid was exactly compensated by the diminished weight of air present.

In 1772, he recorded the observation that when phosphorus and sulphur burn, their weight increases. He deduced that they combine with air. It followed that there should be a way of releasing this 'fixed' air. It should be explained that in the 18th century, because of the imperfect understanding scientists had of the nature of gases, they generally and confusingly referred to them as different sorts of 'air'.

Then in 1774, using a very large burning glass that belonged to the French Academy of Sciences, Lavoisier showed that when litharge (lead oxide, a compound of lead containing oxygen) is roasted with charcoal, an enormous volume of 'air' was indeed liberated. Although other scientists had already made similar experiments, it was Lavoisier who finally drew all the evidence together, and discovered the oxygen theory of combustion. He did not come to his conclusions immediately, but as the

result of several related steps.

In 1774, Lavoisier was visited in Paris by the English chemist, Joseph Priestley (1773–1804). Priestley had heated a calx (he used the oxide of mercury) in a closed apparatus, and collected the gas that was liberated in the process. Priestley had found that this gas supported combustion better than air: a candle flame burned with a marvellous brilliance when placed in a container full of it.

The acid former

Lavoisier repeated Priestley's experiments, and convinced himself of the presence in air of a gas which combined with substances when they burn, and that it was the same gas given off when the oxide of mercury was heated. He named this gas 'oxygine', or 'acid former' (from the Greek), because he believed all acids contained oxygen.

Lavoisier had in the meantime identified the other main component of air, nitrogen, which he named 'azote', from the Greek for 'no life'. He also demonstrated that when hydrogen, which chemists of the day called 'inflammable air', was burned with oxygen, water was formed.

Lavoisier could now put the final nail in the coffin of the phlogiston theory because he could now explain that when metals dissolved in an acid, the hydrogen evolved came from the water, and not from the metal, as the phlogistonists asserted.

From combustion to respiration

The oxygen theory of combustion was now complete: when a substance burned it combined with the oxygen in the air. To publish these results, Lavoisier and his colleagues started the journal *Annales de Chimie* in 1778. His textbook *An Elementary Treatise of Chemistry* was published the following year, and remained a model for modern chemical instruction for several decades.

Lavoisier also realized that respiration was like a slow form of combustion, in which the oxygen breathed in 'burned' the carbon in foodstuffs, which was then exhaled. As with the experiments on combustion, Lavoisier and his colleagues carried out exact measurements, using a guinea-pig. This is how the phrase 'to be a guinea-pig' originated.

Lavoisier was guillotined on 8 May 1794, executed on trumped-up charges during the Age of Terror. The mathematician Joseph Lagrange (1736–1813) said: 'It required only a moment to sever his head, and probably one hundred years will not suffice to produce another like it.'

and is present in combined form in water and many other substances. Life on Earth evolved using oxygen, which is a by-product of ◊photosynthesis and the basis for ◊respiration in plants and animals.

Oxygen is very reactive and combines with all other elements except the ◊inert gases and fluorine. It is present in carbon dioxide, silicon dioxide (quartz), iron ore, calcium carbonate (limestone). In nature it exists as a molecule composed of two atoms (O_2); single atoms of oxygen are very short-lived owing to their reactivity. They can be produced in electric sparks and by the Sun's ultraviolet radiation in space, where they rapidly combine with molecular oxygen to form ozone (an allotrope of oxygen).

Oxygen is obtained for industrial use by the fractional distillation of liquid air, by the electrolysis of water, or by heating manganese(IV) oxide with potassium chlorate. It is essential for combustion, and is used with ethyne (acetylene) in high-temperature oxyacetylene welding and cutting torches.

The element was first identified by English chemist Joseph Priestley 1774 and independently in the same year by Swedish chemist Karl Scheel. It was named by French chemist Antoine Lavoisier 1777.

oxygen debt physiological state produced by vigorous exercise, in which the lungs cannot supply all the oxygen that the muscles need.

Oxygen is required for the release of energy from food molecules (aerobic ◊respiration). Instead of breaking food molecules down fully, muscle cells switch to a form of partial breakdown that does not require oxygen (anaerobic respiration) so that they can continue to generate energy. This partial breakdown produces ◊lactic acid, which results in a sensation of fatigue when it reaches certain levels in the muscles and the blood. Once the vigorous muscle movements cease, the body breaks down the lactic acid, using up extra oxygen to do so. Panting after exercise is an automatic reaction to 'pay off' the oxygen debt.

oz abbreviation for ◊*ounce.*

Ozalid process trademarked copying process used to produce positive prints from drawn or printed materials or film, such as printing proofs from film images. The film is placed on top of chemically treated paper and then exposed to ultraviolet light. The image is developed dry using ammonia vapour.

ozone O_3 highly reactive pale-blue gas with a penetrating odour. Ozone is an allotrope of oxygen (see ◊allotropy), made up of three atoms of oxygen. It is formed when the molecule of the stable form of oxygen (O_2) is split by ultraviolet radiation or electrical discharge. It forms a thin layer in the upper atmosphere, which protects life on Earth from ultraviolet rays, a cause of skin cancer. At lower atmospheric levels it is an air pollutant and contributes to the ◊greenhouse effect.

At ground level, ozone can cause asthma attacks, stunted growth in plants, and corrosion of certain materials. It is produced by the action of sunlight on air pollutants, including car exhaust fumes, and is a major air pollutant in hot summers. Ozone is a powerful oxidizing agent and is used industrially in bleaching and air conditioning.

A continent-sized hole has formed over Antarctica as a result of damage to the ozone layer. This has been caused in part by ◊chlorofluorocarbons (CFCs), but many reactions destroy ozone in the stratosphere: nitric oxide, chlorine, and bromine atoms are implicated. In 1989 ozone depletion was 50% over the Antarctic compared with 3% over the Arctic. In April 1991 satellite data from NASA revealed that the ozone layer had depleted by 4–8% in the N hemisphere and by 6–10% in the S hemisphere between 1978 and 1990. It is believed that the ozone layer is depleting at a rate of about 5% every 10 years over N Europe, with depletion extending south to the Mediterranean and southern USA. However, ozone depletion over the polar regions is the most dramatic manifestation of a general global effect. At ground level, ozone is so dangerous that the US Environment Protection Agency recommends people should not be exposed for more than one hour a day to ozone levels of 120 parts per billion (ppb), while the World Health Organization recommends a lower 76–100 ppb. It is known that even at levels of 60 ppb ozone causes respiratory problems, and may cause the yields of some crops to fall. In the USA, the annual economic loss due to ozone has been estimated at $5.4 billion.

THE OZONE HOLE: GETTING LARGER

In 1992, the hole in the ozone layer over the Antarctic opened earlier than expected, according to information gathered by the *Nimbus-7* satellite. On 23 Sept 1992, the hole measured 23 million sq km/8.9 million sq mi (up 15% on 1991), an area only marginally less than that of the North American continent.

---

ozone depleter any chemical that destroys the ozone in the stratosphere. Most ozone depleters are chemically stable compounds containing chlorine or bromine, which remain unchanged for long enough to drift up to the upper atmosphere. The best known are ◊chlorofluorocarbons (CFCs), but many other ozone depleters are known, including halons, used in some fire extinguishers; methyl chloroform and carbon tetrachloride, both solvents; some CFC substitutes; and the pesticide methyl bromide.

pacemaker or *sinoatrial node* (SAN) in vertebrates, a group of muscle cells in the wall of the heart that contracts spontaneously and rhythmically, setting the pace for the contractions of the rest of the heart. The pacemaker's intrinsic rate of contraction is increased or decreased, according to the needs of the body, by stimulation from the ◊autonomic nervous system.

The term also refers to a medical device implanted under the skin of a patient whose heart beats irregularly. It delivers minute electric shocks to stimulate the heart muscles at regular intervals and restores normal heartbeat. The latest pacemakers are powered by radioactive isotopes for long life and weigh no more than 15 g/0.5 oz.

Pacific Ocean world's largest ocean, extending from Antarctica to the Bering Strait; area 166,242,500 sq km/64,170,000 sq mi; average depth 4,188 m/13,749 ft; greatest depth of any ocean 11,034 m/36,210 ft in the ◊Mariana Trench.

packet switching in computing, a method of transmitting data between computers connected in a ◊network. A complete packet consists of the data being transmitted and information about which computer is to receive the data. The packet travels around the network until it reaches the correct destination.

paedomorphosis in biology, an alternative term for ◊neoteny.

page-description language in computing, a control language used to describe the contents and layout of a complete printed page. Page-description languages are frequently used to control the operation of ◊laser printers. The most popular page-description languages are Adobe Postscript and Hewlett-Packard Printer Control Language.

page printer computer ◊printer that prints a complete page of text and graphics at a time. Page printers use electrostatic techniques, very similar to those used by photocopiers, to form images of pages, and range in size from small ◊laser printers designed to work with microcomputers to very large machines designed for high-volume commercial printing.

paging method of increasing a computer's apparent memory capacity. See ◊virtual memory.

paint any of various materials used to give a protective and decorative finish to surfaces or for making pictures. A paint consists of a pigment suspended in a vehicle, or binder, usually with added solvents. It is the vehicle that dries and hardens to form an adhesive film of paint. Among the most common kinds are cellulose paints (or lacquers), oil-based paints, emulsion paints, and special types such as enamels and primers.

Lacquers consist of a synthetic resin (such as an acrylic resin or cellulose acetate) dissolved in a volatile organic solvent, which evaporates rapidly to give a very quick-drying paint. A typical *oil-based paint* has a vehicle of a natural drying oil (such as linseed oil), containing a prime pigment of iron, lead, titanium, or zinc oxide, to which coloured pigments may be added. The finish—gloss, semimatte, or matte—depends on the amount of inert pigment (such as clay or silicates). Oil-based paints can be thinned, and brushes cleaned, in a solvent such as turpentine or white spirit (a petroleum product). *Emulsion paints*, sometimes called latex paints, consist of pigments dispersed in a water-based emulsion of a polymer (such as polyvinyl chloride [PVC] or acrylic resin). They can be thinned with water, which can also be used to wash the paint out of brushes and rollers. *Enamels* have little pigment, and they dry to an extremely hard, high-gloss film. *Primers* for the first coat on wood or metal, on the other hand, have a high pigment content (as do undercoat paints). Aluminium or bronze powder may be used for priming or finishing objects made of metal.

PAL abbreviation for *phase alternation by line*, the coding system for colour-television broadcasting adopted by the UK, West Germany, the Netherlands, and Switzerland 1967. See ◊television.

Palaeocene (Greek 'old' + 'recent') first epoch

of the Tertiary period of geological time, 65–56.5 million years ago. Many types of mammals spread rapidly after the disappearance of the great reptiles of the Mesozoic. Flying mammals replaced the flying reptiles, swimming mammals replaced the swimming reptiles, and all the ecological niches vacated by the reptiles were adopted by mammals.

palaeomagnetism science of the reconstruction of the Earth's ancient magnetic field and the former positions of the continents from the evidence of *remanent magnetization* in ancient rocks; that is, traces left by the Earth's magnetic field in ◊igneous rocks before they cool. Palaeomagnetism shows that the Earth's magnetic field has reversed itself—the magnetic north pole becoming the magnetic south pole, and vice versa—at approximate half-million-year intervals, with shorter reversal periods in between the major spans.

Starting in the 1960s, this known pattern of magnetic reversals was used to demonstrate seafloor spreading or the formation of new ocean crust on either side of mid-oceanic ridges. As new material hardened on either side of a ridge, it would retain the imprint of the magnetic field, furnishing datable proof that material was spreading steadily outward. Palaeomagnetism is also used to demonstrate ◊continental drift by determining the direction of the magnetic field of dated rocks from different continents.

palaeontology in geology, the study of ancient life that encompasses the structure of ancient organisms and their environment, evolution, and ecology, as revealed by their ◊fossils. The practical aspects of palaeontology are based on using the presence of different fossils to date particular rock strata and to identify rocks that were laid down under particular conditions, for instance giving rise to the formation of oil.

The use of fossils to trace the age of rocks was pioneered in Germany by Johann Friedrich Blumenbach (1752–1830) at Göttingen, followed by Georges Cuvier and Alexandre Brongniart (1770–1847) in France 1811.

The term palaeontology was first used in 1834, during the period when the first ◊dinosaur remains were discovered.

Palaeozoic era of geological time 570–245 million years ago. It comprises the Cambrian, Ordovician, Silurian, Devonian, Carboniferous, and Permian periods. The Cambrian, Ordovician, and Silurian constitute the Lower or Early Palaeozoic; the Devonian, Carboniferous, and Permian make up the Upper or Late Palaeozoic. The era includes the evolution of hard-shelled multicellular life forms in the sea; the invasion of land by plants and animals; and the evolution of fish, amphibians, and early reptiles. The earliest identifiable fossils date from this era.

The climate at this time was mostly warm with short ice ages. The continents were very different from the present ones but, towards the end of the era, all were joined together as a single world continent called ◊Pangaea.

palate in mammals, the ceiling of the mouth. The bony front part is the hard palate, the muscular rear part the soft palate. Incomplete fusion of the

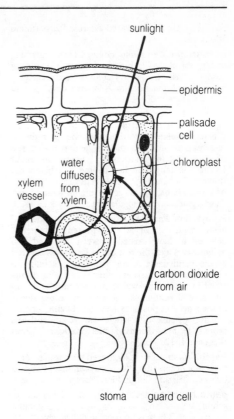

palisade cell *Palisade cells are closely packed, columnar cells, lying in the upper surfaces of leaves. They contain many chloroplasts (the structures responsible for photosynthesis) and are well adapted to receive and process the components necessary for photosynthesis—carbon dioxide, water, and sunlight. For instance, their vertical arrangement means that there are fewer cross-walls to interfere with the passage of sunlight.*

two lateral halves of the palate causes interference with speech.

Technically, the mammalian palate is a secondary structure and parallel to the original mouth roof of fishes, amphibians, reptiles, and birds.

palisade cell cylindrical cell lying immediately beneath the upper epidermis of a leaf. Palisade cells normally exist as one closely packed row and contain many chloroplasts. During the hours of daylight palisade cells are photosynthetic, using the energy of the sun to create carbohydrates from water and carbon dioxide.

palladium lightweight, ductile and malleable, silver-white, metallic element, symbol Pd, atomic number 46, relative atomic mass 106.4.

It is one of the so-called platinum group of metals, and is resistant to tarnish and corrosion. It often occurs in nature as a free metal (see ◊native metal) in a natural alloy with platinum. Palladium is used as a catalyst, in alloys of gold (to make white gold) and silver, in electroplating, and in dentistry.

It was discovered 1803 by British physicist William Wollaston (1766–1828), and named after the then recently discovered asteroid Pallas (found 1802).

panchromatic in photography, a term describing highly sensitive black-and-white film made to render all visible spectral colours in correct grey tones. Panchromatic film is always developed in total darkness.

pancreas in vertebrates, an accessory gland of the digestive system located close to the duodenum. When stimulated by the hormone secretin, it secretes enzymes into the duodenum that digest starches, proteins, and fats. In humans, it is about 18 cm/7 in long, and lies behind and below the stomach. It contains groups of cells called the *islets of Langerhans*, which secrete the hormones insulin and glucagon that regulate the blood sugar level.

Pangaea or *Pangea* (Greek 'all-land') single land mass, made up of all the present continents, believed to have existed between 250 and 200 million years ago; the rest of the Earth was covered by the Panthalassa ocean. Pangaea split into two land masses ◊Laurasia in the north and ◊Gondwanaland in the south—which subsequently broke up into several continents. These then drifted slowly to their present positions (see ◊continental drift).

The existence of a single 'supercontinent' was proposed by German meteorologist Alfred Wegener 1812.

Panthalassa ocean that covered the surface of the Earth not occupied by the world continent Pangaea between 250 and 200 million years ago.

pantothenic acid $C_9H_{17}NO_5$ one of the water-soluble B ◊vitamins, occurring widely throughout a normal diet. There is no specific deficiency disease associated with pantothenic acid but it is known to be involved in the breakdown of fats and carbohydrates. It was first isolated from liver 1933.

paper thin, flexible material made in sheets from vegetable fibres (such as wood pulp) or rags and used for writing, drawing, printing, packaging, and various household needs. The name comes from papyrus, a form of writing material made from water reed, used in ancient Egypt. The invention of true paper, originally made of pulped fishing nets and rags, is credited to Tsai Lun, Chinese minister of agriculture, AD 105.

Paper came to the West with Arabs who had learned the secret from Chinese prisoners of war in Samarkand in 768. It spread to Moorish Spain and to Byzantium in the 11th century, then to the rest of Europe. All early paper was handmade within frames.

With the spread of literacy there was a great increase in the demand for paper. Production by hand of single sheets could not keep pace with this demand, which led to the invention, by Louis Robert (1761–1828) in 1799, of a machine to produce a continuous reel of paper. The process was developed and patented in 1801 by Francois Didot, Robert's employer. Today most paper is made from ◊wood pulp on a Foudrinier machine, then cut to size. Paper products absorb 35% of the world's annual commercial wood harvest; recycling avoids some of the enormous waste of trees, and most papermakers plant and replant their own forests of fast-growing stock.

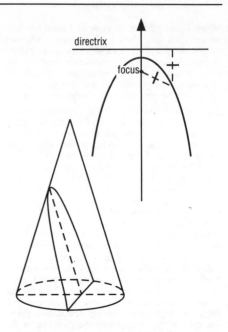

parabola *The parabola is a curve produced when a cone is cut by a plane. It is one of a family of curves called conic sections, which also includes the circle, ellipse, and hyperbole. These curves are produced when the plane cuts the cone at different angles and positions.*

The first English paper mill was established at Stevenage in the 15th century.

pappus (plural *pappi*) in botany, a modified ◊calyx comprising a ring of fine, silky hairs, or sometimes scales or small teeth, that persists after fertilization. Pappi are found in members of the daisy family (Compositae) such as the dandelions *Taraxacum*, where they form a parachutelike structure that aids dispersal of the fruit.

parabola in mathematics, a curve formed by cutting a right circular cone with a plane parallel to the sloping side of the cone. A parabola is one of the family of curves known as ◊conic sections. The graph of $y = x^2$ is a parabola.

It can also be defined as a path traced out by a point that moves in such a way that the distance from a fixed point (focus) is equal to its distance from a fixed straight line (directrix); it thus has an ◊eccentricity of 1.

The trajectories of missiles within the Earth's gravitational field approximate closely to parabolas (ignoring the effect of air resistance). The corresponding solid figure, the paraboloid, is formed by rotating a parabola about its axis. It is a common shape for headlight reflectors, dish-shaped microwave and radar aerials, and radiotelescopes, since a source of radiation placed at the focus of a paraboloidal reflector is propagated as a parallel beam.

parachute any canopied fabric device strapped to a person or a package, used to slow down descent from a high altitude, or returning spent missiles or parts to a safe speed for landing, or sometimes to aid (through braking) the landing of a plane or missile. Modern designs enable the parachutist to exercise considerable control of direction, as in skydiving.

Leonardo da Vinci sketched a parachute design, but the first descent, from a balloon at a height of 2,200 ft over Paris, was not made until 1797 by André-Jacques Garnerin (1769–1823). The first descent from an aircraft was made by Capt Albert Berry 1912 from a height of 1,500 ft over Missouri. A parachute is typically folded into a pack from which it is released by a rip cord or other device. It originally consisted of some two dozen panels of silk (later nylon) in a circular canopy with shroud lines to a harness. Modern parachutes are variously shaped, often small and rectangular.

In **parascending** the parachuting procedure is reversed, the canopy (parafoil) to which the person is attached being towed behind a vehicle to achieve an ascent.

paraffin common name for ◊alkane, any member of the series of hydrocarbons with the general formula C_nH_{2n+2}. The lower members are gases, such as methane (marsh or natural gas). The middle ones (mainly liquid) form the basis of petrol, kerosene, and lubricating oils, while the higher ones (paraffin waxes) are used in ointment and cosmetic bases.

The fuel commonly sold as paraffin in Britain is more correctly called kerosene.

paraldehyde common name for ◊ethanal trimer.

parallax the change in the apparent position of an object against its background when viewed from two different positions. In astronomy, nearby stars show a shift owing to parallax when viewed from different positions on the Earth's orbit around the Sun. A star's parallax is used to deduce its distance. Nearer bodies such as the Moon, Sun, and

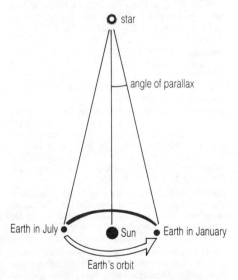

parallax *The parallax of a star, the apparent change of its position during the year, can be used to find the star's distance from the Earth. The star appears to change its position because it is viewed at a different angle in July and January. By measuring the angle of parallax, and knowing the diameter of the Earth's orbit, simple geometry can be used to calculate the distance to the star.*

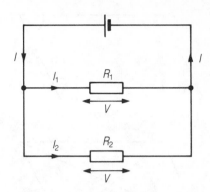

parallel circuit *Two resistors, R_1 and R_2, connected in a parallel circuit. The potential difference V across each resistor is the same, but the current passing through each is divided up in the ratio of their resistances ($I_1:I_2$ is the same as $R_1:R_2$). The advantage of such an arrangement is that if one of the components fails, the other parallel components will continue to receive current.*

planets also show a parallax caused by the motion of the Earth. **Diurnal parallax** is caused by the Earth's rotation.

parallel circuit electrical circuit in which current is split between two or more parallel paths or conductors. The division of the current across each conductor is in the ratio of their resistances. If the currents across two conductors of resistance R_1 and R_2, connected in parallel, are I_1 and I_2 respectively, then the ratio of those currents is given by the equation:

$$I_1/I_2 = R_2/R_1.$$

The total resistance R of those conductors is given by:

$$1/R = 1/R_1 + 1/R_2$$

Compare ◊series circuit.

parallel device in computing, a device that communicates binary data by sending the bits that represent each character simultaneously along a set of separate data lines, unlike a ◊serial device.

parallel lines and parallel planes in mathematics, straight lines or planes that always remain a constant distance from one another no matter how far they are extended. This is a principle of Euclidean geometry. Some non-Euclidean geometries, such as elliptical and hyperbolic geometry, however, reject Euclid's parallel axiom.

parallelogram in mathematics, a quadrilateral (four-sided plane figure) with opposite pairs of sides equal in length and parallel, and opposite angles equal. The diagonals of a parallelogram bisect each other. Its area is the product of the length of one side and the perpendicular distance between this and the opposite side. In the special case when all four sides are equal in length, the parallelogram is known as a rhombus, and when the internal angles are right angles, it is a rectangle or square.

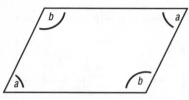

(i) opposite sides and anngles are equal

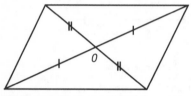

(ii) diagonals bisect each other at O

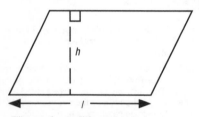

(iii) area of a parallelogram $l \times h$

parallelogram *Some properties of a parallelogram.*

parallelogram of forces in physics and applied mathematics, a method of calculating the resultant (combined effect) of two different forces acting together on an object. Because a force has both magnitude and direction it is a ◊vector quantity and can be represented by a straight line. A second force acting at the same point in a different

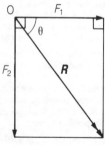

R is the resultant of F_1 and F_2

parallelogram of forces *The combined effect of two forces F_1 and F_2 acting on the same point may be found by drawing up a parallelogram of forces. The two forces are drawn as straight lines, with their lengths representing their magnitudes and their directions corresponding to the actual direction of each force. If these are considered to be the adjacent sides of a parallelogram, their combined effect R is given—both in length and direction—by the diagonal of the completed parallelogram.*

direction can be represented by another line drawn at an angle to the first. By completing the parallelogram (of which the two lines are sides) a diagonal may be drawn from the original angle to the opposite corner to represent the resultant force vector.

parallel processing emerging computer technology that allows more than one computation at the same time. Although in the 1980s this technology enabled only a small number of computer processor units to work in parallel, in theory thousands or millions of processors could be used at the same time.

Parallel processing, which involves breaking down computations into small parts and performing thousands of them simultaneously, rather than in a linear sequence, offers the prospect of a vast improvement in working speed for certain repetitive applications.

paramagnetic material material that is weakly pulled towards a strong magnet. The effect is caused by unpaired electrons in the atoms of the material, which cause the atoms to act as weak magnets. A paramagnetic material has a fairly low magnetic ◊susceptibility that is inversely proportional to temperature. See ◊magnetism.

parameter variable factor or characteristic. For example, length is one parameter of a rectangle; its height is another. In computing, it is frequently useful to describe a program or object with a set of variable parameters rather than fixed values.

For example, if a programmer writes a routine for drawing a rectangle using general parameters for the length, height, line thickness, and so on, any rectangle can be drawn by this routine by giving different values to the parameters.

Similarly, in a word-processing application that stores parameters for fount, page layout, type of justification, and so on, these can be changed by the user.

paraquat $CH_3(C_5H_4N)_2CH_3.2CH_3SO_4$ (technical name *1,1–dimethyl-4,4–dipyridylium*) nonselective herbicide (weedkiller). Although quickly degraded by soil microorganisms, it is deadly to human beings if ingested.

parasite organism that lives on or in another organism (called the 'host'), and depends on it for nutrition, often at the expense of the host's welfare. Parasites that live inside the host, such as liver flukes and tapeworms, are called *endoparasites*; those that live on the outside, such as fleas and lice, are called *ectoparasites*.

parathyroid one of a pair of small ◊endocrine glands. Most tetrapod vertebrates, including humans, possess two such pairs, located behind the ◊thyroid gland. They secrete parathyroid hormone, which regulates the amount of calcium in the blood.

parenchyma plant tissue composed of loosely packed, more or less spherical cells, with thin cellulose walls. Although parenchyma often has no specialized function, it is usually present in large amounts, forming a packing or ground tissue. It usually has many intercellular spaces.

parental care in biology, the time and energy spent by a parent in order to rear its offspring to maturity. Among animals, it ranges from the simple provision of a food supply for the hatching young at the time the eggs are laid (for example, many wasps) to feeding and protection of the young after hatching or birth, as in birds and mammals. In the more social species, parental care may include the teaching of skills—for example, female cats teach their kittens to hunt.

parental generation in genetic crosses, the set of individuals at the start of a test, providing the first set of gametes from which subsequent generations (known as the F1 and F2) will arise.

parity of a number, the state of being either even or odd. In computing, the term refers to the number of 1s in the binary codes used to represent data. A binary representation has *even parity* if it contains an even number of 1s and *odd parity* if it contains an odd number of 1s.

For example, the binary code 1000001, commonly used to represent the character 'A', has even parity because it contains two 1s and the binary code 1000011, commonly used to represent the character 'C', has odd parity because it contains three 1s. A *parity bit* is sometimes added to each binary representation to adjust its parity and enable a ◊validation check to be carried out each time data are transferred from one part of the computer to another. The parity bit is added as either a 1 or a 0 so that, after it has been added, every binary representation has the same parity. So, for example, the codes 1000001 and 1000011 could have parity bits added and become *0*1000001 and *1*1000011, both with even parity. If any bit in these codes should be altered in the course of processing the parity would change and the error would be quickly detected.

parity

character	binary code	parity	base-ten representation
A	1000001	even	65
B	1000010	even	66
C	1000011	odd	67
D	1000100	even	68

parity check a form of ◊validation of data.

Parkes site in New South Wales of the Australian National Radio Astronomy Observatory, featuring a radio telescope of 64 m/210 ft aperture, run by the Commonwealth Scientific and Industrial Research Organization.

parsec in astronomy, a unit (symbol pc) used for distances to stars and galaxies. One parsec is equal to 3.2616 ◊light years, 2.063×10^5 ◊astronomical units, and 3.086×10^{13} km.

It is the distance at which a star would have a ◊parallax (apparent shift in position) of one second of arc when viewed from two points the same distance apart as the Earth's distance from the Sun; or the distance at which one astronomical unit subtends an angle of one second of arc.

parthenocarpy in botany, the formation of fruits without seeds. This phenomenon, of no obvious benefit to the plant, occurs naturally in some plants, such as bananas. It can also be induced in some fruit crops, either by breeding or by applying certain plant hormones.

parthenogenesis development of an ovum (egg)

without any genetic contribution from a male. Parthenogenesis is the normal means of reproduction in a few plants (for example, dandelions) and animals (for example, certain fish). Some sexually reproducing species, such as aphids, show parthenogenesis at some stage in their life cycle.

In most cases, there is no fertilization at all, but in a few the stimulus of being fertilized by a sperm is needed to initiate development, although the male's chromosomes are not absorbed into the nucleus of the ovum. Parthenogenesis can be artificially induced in many animals (such as rabbits) by cooling, pricking, or applying acid to an egg.

particle detector one of a number of instruments designed to detect subatomic particles and track their paths.

The earliest particle detector was the ◊cloud chamber, which contains a super-saturated vapour in which particles leave a trail of droplets, in much the same way that a jet aircraft leaves a trail of vapour in the sky. A ◊bubble chamber contains a superheated liquid in which a particle leaves a trail of bubbles. A ◊spark chamber contains a series of closely-packed parallel metal plates, each at a high voltage. As particles pass through the chamber, they leave a visible spark between the plates. A modern multiwire chamber consists of an array of fine, closely-packed wires, each at a high voltage. As a particle passes through the chamber, it produces an electrical signal in the wires. A computer analyses the signal and reconstructs the path of the particles. Multiwire detectors can be used to detect X-ray and gamma rays, and are used as detectors in ◊positron electron tomography (PET).

particle physics study of the particles that make up all atoms, and of their interactions. More than 300 subatomic particles have now been identified by physicists, categorized into several classes according to their mass, electric charge, spin, magnetic moment, and interaction. Subatomic particles include the ◊elementary particles (◊quarks, ◊leptons, and ◊gauge bosons), which are believed to be indivisible and so may be considered the fundamental units of matter; and the ◊hadrons (baryons, such as the proton and neutron, and mesons), which are composite particles, made up of two or three quarks. The proton, electron, and neutrino are the only stable particles (the neutron being stable only when in the atomic nucleus). The unstable particles decay rapidly into other particles, and are known from experiments with particle accelerators and cosmic radiation. See ◊atomic structure.

Pioneering research took place at the Cavendish laboratory, Cambridge, England. In 1895 English physicist Joeseph John Thomson discovered that all atoms contain identical, negatively charged particles (◊electrons), which can easily be freed. By 1911 New Zealand physicist Ernest Rutherford had shown that the electrons surround a very small, positively-charged ◊nucleus. In the case of hydrogen, this was found to consist of a single positively charged particle, a ◊proton (identified by English physicist James Chadwick in 1932). The nuclei of other elements are made up of protons and uncharged particles called ◊neutrons.

1932 also saw the discovery of a particle (whose

existence had been predicted by British theoretical physicist Paul Dirac in 1928) with the mass of an electron, but an equal and opposite charge—the ◊positron. This was the first example of ◊antimatter; it is now believed that almost all particles have corresponding antiparticles. In 1934 Italian–US physicist Enrico Fermi argued that a hitherto unsuspected particle, the ◊neutrino, must accompany electrons in beta-emission.

particles and fundamental forces By the mid-1930s, four types of fundamental ◊force interacting between particles had been identified. The ◊electromagnetic force acts between all particles with electric charge, and is thought to be related to the exchange between these particles of ◊gauge bosons called ◊photons, packets of electromagnetic radiation. In 1935 Japanese physicist Hideki Yukawa suggested that the ◊strong nuclear force (binding protons and neutrons together in the nucleus) was transmitted by the exchange of particles with a mass about one-tenth of that of a proton; these particles, called ◊pions (originally pi mesons), were found by British physicist Cecil Powell in 1946. Yukawa's theory was largely superseded from 1973 by the theory of ◊quantum chromodynamics, which postulates that the strong nuclear force is transmitted by the exchange of gauge bosons called ◊gluons between the quarks and antiquarks making up protons and neutrons. Theoretical work on the ◊weak nuclear force began with Enrico Fermi in the 1930s. The existence of the gauge bosons that carry this force, the ◊weakons (W and Z particles), was confirmed in 1983 at CERN, the European nuclear research organization. The fourth fundamental force, ◊gravity, is experienced by all matter; the postulated carrier of this force has been named the ◊graviton.

leptons The electron, muon, tau, and their neutrinos comprise the ◊leptons—particles with half-integral spin that 'feel' the weak nuclear force but not the strong force. The muon (found by US physicist Carl Anderson in cosmic radiation in 1937) produces the muon neutrino when it decays; the tau, a surprise discovery of the 1970s, produces the tau neutrino when it decays.

mesons and baryons The hadrons (particles that 'feel' the strong nuclear force) were found in the 1950s and 1960s. They are classified into ◊mesons, with whole-number or zero spins, and ◊baryons (which include protons and neutrons), with half-integral spins. It was shown in the early 1960s that if hadrons of the same spin are represented as points on suitable charts, simple patterns are formed. This symmetry enabled a hitherto unknown baryon, the omega-minus, to be predicted from a gap in one of the patterns; it duly turned up in experiments.

quarks In 1964, US physicists Murray Gell-Mann and George Zweig suggested that all hadrons were built from three 'flavours' of a new particle with half-integral spin and a charge of magnitude either ⅓ or ⅔ that of an electron; Gell-Mann named the particle the *quark*. Mesons are quark-antiquark pairs (spins either add to one or cancel to zero), and baryons are quark triplets. To account for new mesons such as the psi (J) particle the number of quark flavours had risen to six by 1985.

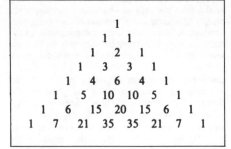

Pascal's triangle *In Pascal's triangle, each number is the sum of the numbers above it—for example, 2 is the sum of 1 and 1, and 6 is the sum of 3 and 3. Furthermore, the sum of each row equals the corresponding power of 2; for example, the sum of the third row $1 + 2 + 1$ equals $4 = 2^2$. Also, each row read across equals the corresponding power of 11; for example, $121 = 11^2$.*

particle, subatomic in physics, a particle that is smaller than an atom; see ◊particle physics.

PASCAL (French acronym for *program appliqué à la selection et la compilation automatique de la litterature*) a high-level computer-programming language. Designed by Niklaus Wirth (1934–   ) in the 1960s as an aid to teaching programming, it is still widely used as such in universities, but is also recognized as a good general-purpose programming language. It was named after 17th-century French mathematician Blaise Pascal.

pascal SI unit (symbol Pa) of pressure, equal to one newton per square metre. It replaces ◊bars and millibars (10^5 Pa equals one bar). It is named after the French scientist Blaise Pascal.

Pascal's triangle a triangular array of numbers (with 1 at the apex), in which each number is the sum of the pair of numbers above it. It is named after French mathematician Blaise Pascal, who used it in his study of probability. When plotted at equal distances along a horizontal axis, the numbers in the rows give the binomial probability distribution (with equal probability of success and failure) of an event, such as the result of tossing a coin.

pasteurization treatment of food to reduce the number of microorganisms it contains and so protect consumers from disease. Harmful bacteria are killed and the development of others is delayed. For milk, the method involves heating it to 72°C/161°F for 15 seconds followed by rapid cooling to 10°C/50°F or lower. The process also kills beneficial bacteria and reduces the nutritive property of milk.

The experiments of Louis Pasteur on wine and beer in the 1850s and 1860s showed how heat treatment slowed the multiplication of bacteria and thereby the process of souring. Pasteurization of milk made headway in the dairy industries of Scandinavia and the USA before 1900 because of the realization that it also killed off bacteria associated with the diseases of tuberculosis, typhoid, diphtheria, and dysentery.

In Britain, progress was slower but with encouragement from the 1922 Milk and Dairies Act the number of milk-processing plants gradually increased in the years before the World War II. In the 1990s nearly all liquid milk sold in the UK is heat treated and available in pasteurized, sterilized, or ultra-heat-treated (UHT) form. UHT milk is heated to at least 132°C/269°F for one second to give it a shelf life of several months.

patella or *knee cap* a flat bone embedded in the knee tendon of birds and mammals, which protects the joint from injury.

pathogen (Greek 'disease producing') in medicine, a bacterium or virus that causes disease. Most pathogens are ◊parasites, and the diseases they cause are incidental to their search for food or shelter inside the host. Nonparasitic organisms, such as soil bacteria or those living in the human gut and feeding on waste foodstuffs, can also become pathogenic to a person whose immune system or liver is damaged. The larger parasites that can cause disease, such as nematode worms, are not usually described as pathogens.

PC in computing, the abbreviation for ◊*personal computer*.

PCM abbreviation for ◊*pulse-code modulation*.

pd abbreviation for ◊*potential difference*.

pearl shiny, hard, rounded abnormal growth composed of nacre (or mother-of-pearl), a chalky substance. Nacre is secreted by many molluscs, and deposited in thin layers on the inside of the shell around a parasite, a grain of sand, or some other irritant body. After several years of the mantle (the layer of tissue between the shell and the body mass) secreting this nacre, a pearl is formed.

Although commercially valuable pearls are obtained from freshwater mussels and oysters, most precious pearls come from the various species of the family Pteriidae (the pearl oysters) found in tropical waters off N and W Australia, off the Californian coast, in the Persian Gulf, and in the Indian Ocean. Because of their rarity, large mussel pearls of perfect shape are worth more than those from oysters.

Artificial pearls were first cultivated in Japan in 1893. A tiny bead of shell from a clam, plus a small piece of membrane from another pearl oyster's mantle (to stimulate the secretion of nacre) is inserted in oysters kept in cages in the sea for three years, and then the pearls are harvested.

peat fibrous organic substance found in bogs and formed by the incomplete decomposition of plants such as sphagnum moss. N Asia, Canada, Finland, Ireland, and other places have large deposits, which have been dried and used as fuel from ancient times. Peat can also be used as a soil additive.

Peat bogs began to be formed when glaciers retreated, about 9,000 years ago. They grow at the rate of only a millimetre a year, and large-scale digging can result in destruction both of the bog and of specialized plants growing there.

A number of ancient corpses, thought to have been the result of ritual murders, have been found preserved in peat bogs, mainly in Scandinavia. In 1984, Lindow Man, dating from about 500 BC, was found in mainland Britain, near Wilmslow, Cheshire. In 1990 the third largest peat bog in Britain, on the borders of Shropshire and Clwyd,

was bought by the Nature Conservancy Council, the largest purchase ever made by the NCC.

peck obsolete unit of dry measure, equalling eight quarts or a quarter bushel (9.002 litres).

pectoral in vertebrates, the upper area of the thorax associated with the muscles and bones used in moving the arms or forelimbs. In birds, the *pectoralis major* is the very large muscle used to produce a powerful downbeat of the wing during flight.

pedicel the stalk of an individual flower, which attaches it to the main floral axis, often developing in the axil of a bract.

Pegasus in astronomy, a constellation of the northern hemisphere, near Cygnus, and represented as the winged horse of Greek mythology.

It is the seventh largest constellation in the sky, and its main feature is a square outlined by four stars, one of which (Alpherat) is actually part of the adjoining constellation Andromeda. Diagonally across is Markab (or Alpha Pegasi), about 100 light years distant.

pegmatite extremely coarse-grained ◊igneous rock of any composition found in veins; pegmatites are usually associated with large granite masses.

Peking man Chinese representative of an early species of human, found as fossils, 500,000–750,000 years old, in the cave of Choukoutien 1927 near Beijing (Peking). Peking man used chipped stone tools, hunted game, and used fire. Similar varieties of early human have been found in Java and E Africa. Their classification is disputed: some anthropologists classify them as *Homo erectus*, others as *Homo sapiens pithecanthropus*.

pellagra chronic disease of subtropical countries in which the staple food is maize, caused by deficiency of ◊nicotinic acid (one of the B vitamins), which is contained in protein foods, beans and peas, and yeast. Symptoms include digestive disorders, skin eruptions, and mental disturbances.

Peltier effect in physics, a change in temperature at the junction of two different metals produced when an electric current flows through them. The extent of the change depends on what the conducting metals are, and the nature of change (rise or fall in temperature) depends on the direction of current flow. It is the reverse of the ◊Seebeck effect. It is named after the French physicist Jean Charles Peltier (1785–1845) who discovered it 1834.

pelvis in vertebrates, the lower area of the abdomen featuring the bones and muscles used to move the legs or hindlimbs. The *pelvic girdle* is a set of bones that allows movement of the legs in relation to the rest of the body and provides sites for the attachment of relevant muscles.

pendulum weight (called a 'bob') swinging at the end of a rod or cord. The regularity of a pendulum's swing was used in making the first really accurate clocks in the 17th century. Pendulums can be used for measuring the acceleration due to gravity (an important constant in physics), and in prospecting for oils and minerals.

Specialized pendulums are used to measure velocities (ballistic pendulum) and to demonstrate the Earth's rotation (Foucault's pendulum).

penicillin any of a group of ◊antibiotic (bacteria killing) compounds obtained from filtrates of moulds of the genus *Penicillium* (especially *P. notatum*) or produced synthetically. Penicillin was the first antibiotic to be discovered (by Scottish bacteriologist Alexander Fleming); it kills a broad spectrum of bacteria, many of which cause disease in humans.

The use of the original type of penicillin is limited by the increasing resistance of ◊pathogens and by allergic reactions in patients. Since 1941, numerous other antibiotics of the penicillin family have been discovered which are more selective against, or resistant to, specific microorganisms.

peninsula land surrounded on three sides by water but still attached to a larger landmass. Florida, USA, is an example.

penis male reproductive organ, used for internal fertilization; it transfers sperm to the female reproductive tract. In mammals, the penis is made erect by vessels that fill with blood, and in most mammals (but not humans) is stiffened by a bone. It also contains the urethra, through which urine is passed.

Snakes and lizards have a paired structure that serves as a penis, other reptiles have a single organ. A few birds, mainly ducks and geese, also have a type of penis, as do snails, barnacles, and some other invertebrates. Many insects have a rigid, non-erectile male organ, usually referred to as an intromittent organ.

Pennsylvanian US term for the Upper or Late ◊Carboniferous period of geological time, 323–290 million years ago; it is named after the US state.

pentadactyl limb typical limb of the mammals, birds, reptiles, and amphibians. These vertebrates (animals with backbone) are all descended from primitive amphibians whose immediate ancestors were fleshy-finned fish. The limb which evolved in those amphibians had three parts: a 'hand/foot' with five digits (fingers/toes), a lower limb containing two bones, and an upper limb containing one bone.

This basic pattern has persisted in all the terrestrial vertebrates, and those aquatic vertebrates (such as seals) which are descended from them. Natural selection has modified the pattern to fit different ways of life. In flying animals (birds and bats) it is greatly altered and in some vertebrates, such as whales and snakes, the limbs are greatly reduced or lost. Pentadactyl limbs of different species are an example of ◊homologous organs.

penumbra the region of partial shade between the totally dark part (umbra) of a ◊shadow and the fully illuminated region outside. It occurs when a source of light is only partially obscured by a shadow-casting object. The darkness of a penumbra varies gradually from total darkness at one edge to full brightness at the other. In astronomy, a penumbra is a region of the Earth from which only a partial ◊eclipse of the Sun or Moon can be seen.

pepsin enzyme that breaks down proteins during digestion. It requires a strongly acidic environment and is found in the stomach.

peptide molecule comprising two or more ◊amino acid molecules (not necessarily different) joined by

peptide bonds, whereby the acid group of one acid is linked to the amino group of the other (–CO.NH). The number of amino acid molecules in the peptide is indicated by referring to it as a di-, tri-, or polypeptide (two, three, or many amino acids).

Proteins are built up of interacting polypeptide chains with various types of bonds occurring between the chains. Incomplete hydrolysis (splitting up) of a protein yields a mixture of peptides, examination of which helps to determine the sequence in which the amino acids occur within the protein.

peptide bond bond that joins two peptides together within a protein. The carboxyl (–COOH) group on one ◊amino acid reacts with the amino (–NH₂) group on another amino acid to form a peptide bond (–CO–NH–) with the elimination of water.

perborate any salt formed by the action of hydrogen peroxide on borates. Perborates contain the radical BO₃.

percentage way of representing a number as a ◊fraction of 100. Thus 45 percent (45%) equals ⁴⁵/₁₀₀, and 45% of 20 is 45/100 × 20 = 9.

In general, if a quantity x changes to y, the percentage change is calculated as $100(x − y)/x$. Thus, if the number of people in a room changes from 40 to 50, the percentage increase is $(100 × 10)/40 = 25\%$. To express a fraction as a percentage, its denominator must first be converted to 100, for example, $⅛ = ¹²·⁵/₁₀₀ = 12.5\%$. The use of percentages often makes it easier to compare fractions that do not have a common denominator.

The percentage sign is thought to have been derived as an economy measure when recording in the old counting houses; writing in the numeric symbol for ²⁵/₁₀₀ of a cargo would take two lines of parchment, and hence the '100' denominator was put alongside the 25 and rearranged to '%'.

perennating organ in plants, that part of a ◊biennial plant or herbaceous perennial that allows it to survive the winter; usually a root, tuber, rhizome, bulb, or corm.

perennial plant plant that lives for more than two years. Herbaceous perennials have aerial stems and leaves that die each autumn. They survive the winter by means of an underground storage (perennating) organ, such as a bulb or rhizome. Trees and shrubs or woody perennials have stems that persist above ground throughout the year, and may be either ◊deciduous or ◊evergreen. See also ◊annual plant, ◊biennial plant.

perianth in botany, a collective term for the outer whorls of the ◊flower, which protect the reproductive parts during development. In most ◊dicotyledons the perianth is composed of two distinct whorls, the calyx of ◊sepals and the corolla of ◊petals, whereas in many ◊monocotyledons the sepals and petals are indistinguishable and the segments of the perianth are then known individually as tepals.

periastron in astronomy, the point at which an object travelling in an elliptical orbit around a star is at its closest to the star; the point at which it is furthest is known as the apastron.

GREAT EXPERIMENTS AND DISCOVERIES

Chance favours the prepared mind

'I think the discovery and development of penicillin may be looked on as quite one of the luckiest accidents that have occurred in medicine.' With these words, Professor Howard Florey, an Australian pathologist, concluded a lecture he gave in 1943 at the Royal Institution in London. The topic was penicillin: the first, and still perhaps overall the best, of a range of natural chemotherapeutic agents called antibiotics.

Together with Ernst Chain, Howard Florey had made the practical exploitation of penicillin possible on a worldwide scale.

A chance discovery

The story of the discovery of penicillin started with a chance observation made by Scottish bacteriologist Alexander Fleming (1881–1955), while he was working in 1928 in his laboratory at St Mary's Hospital in London.

Fleming was doing research on staphylococcus, a species of bacterium that can cause disease in humans and animals. In the laboratory such bacteria are grown in dishes containing a culture medium—a jelly-like substance that contains their food. Fleming noticed that one of his dishes had been accidentally contaminated with a mould. The mould could have entered the laboratory through an open window. All around where the specks of green mould were growing, the staphylococcus bacteria had disappeared.

Intrigued, because he had correctly concluded that the mould must contain a substance that killed the bacterium, Fleming isolated the mould and grew more of it in a culture broth. He found that the broth acquired a high antibacterial activity. He tested the action of the broth with a wide variety of pathogenic bacteria, and found that many of them were quickly destroyed.

Fleming was also able to demonstrate that the white corpuscles in human blood were not sensitive to the broth. This suggested to him that other human cells in the body would not be affected by it.

The mould that Fleming's acute powers of observation had noted in a single culture dish, was found to be *Penicillium notatum*, a species related to the mould that commonly grows on stale bread. Fleming named the drug penicillin after this mould.

There the matter rested for 15 years. At the time that penicillin was discovered, it would have been extremely difficult to isolate and purify the drug using the chemical techniques then available. It would also have been quite easily destroyed by them. Without any of the pure substance, Fleming had no way of knowing its extraordinarily high antibacterial activity and its almost negligible toxicity. In fact, penicillin diluted 80,000,000 times will still inhibit the growth of staphylococcus. To translate this number of noughts into something tangible: if one drop of water were diluted 80,000,000 times it would fill over 6,000 whisky bottles.

The first practical application

Then in 1935, German–British biochemist Ernst Chain (1906–1979) joined Professor Florey at Oxford. There he began in 1939 a survey of antimicrobial substances. One of the first he looked at was Fleming's mould. It was chosen because it was already known to be active against staphylococcus, and because, since it was difficult to purify without destroying its antibacterial effect, it represented a biochemical challenge.

A method of purification using primitive apparatus was discovered, and an experiment using penicillin to treat infected mice was carried out with remarkable success. Eventually, enough penicillin was made to treat the first human patient in the terminal stage of a generalized infection. The patient showed an astonishing, although temporary, recovery (the stock of penicillin became exhausted before the infection could be completely cured). Five more seriously ill patients were successfully treated.

Because these patients had already failed to respond to sulphonamide drugs, the value of penicillin was now clearly apparent. For although suphonamides are much more toxic to bacteria than to leucocytes (white blood cells), they do have some poisonous action on the whole human organism.

The new lifesaver

England, then in the midst of World War II, had insufficient resources to manufacture the new wonder drug on a sufficiently large scale. Florey went to the USA to try to interest the large pharmaceutical companies in penicillin. As a result of an outstanding effort, enough penicillin was produced in time to treat all the battle casualties of the Normandy landing in 1944. The use of penicillin in wartime saved countless lives, as it has continued to do ever since.

pericarp wall of a ◊fruit. It encloses the seeds and is derived from the ◊ovary wall. In fruits such as the acorn, the pericarp becomes dry and hard, forming a shell around the seed. In fleshy fruits the pericarp is typically made up of three distinct layers. The *epicarp*, or *exocarp*, forms the tough outer skin of the fruit, while the *mesocarp* is often fleshy and forms the middle layers. The innermost layer or *endocarp*, which surrounds the seeds, may be membranous or thick and hard, as in the ◊drupe (stone) of cherries, plums, and apricots.

peridot pale-green, transparent gem variety of the mineral ◊olivine.

peridotite rock consisting largely of the mineral olivine; pyroxene and other minerals may also be present. Peridotite is an ultrabasic rock containing less than 45% silica by weight. It is believed to be one of the rock types making up the Earth's upper mantle, and is sometimes brought from the depths to the surface by major movements, or as inclusions in lavas.

perigee the point at which an object, travelling in an elliptical orbit around the Earth, is at its closest to the Earth. The point at which it is furthest from the Earth is the apogee.

periglacial bordering a glacial area but not actually covered by ice. Periglacial areas today include parts of Siberia, Greenland and North America. The soil in these areas is frozen to a depth of several metres (◊permafrost) with only the top few centimetres thawing during the brief summer. The vegetation is characteristic of ◊tundra.

During the last ice age all of southern England was periglacial. Weathering by ◊freeze-thaw (the alternate freezing and thawing of ice in rock cracks) would have been severe, and ◊solifluction would have taken place on a large scale, causing wet topsoil to slip from valley sides.

perihelion the point at which an object, travelling in an elliptical orbit around the Sun, is at its closest to the Sun. The point at which it is furthest from the Sun is the aphelion.

perimeter or *boundary* line drawn around the edge of an area or shape. For example, the perimeter of a rectangle is the sum of its four sides; the perimeter of a circle is known as its *circumference*.

period another name for menstruation; see ◊menstrual cycle.

period in physics, the time taken for one complete cycle of a repeated sequence of events. For example, the time taken for a pendulum to swing from side to side and back again is the period of the pendulum.

periodic table of the elements in chemistry, a table in which the elements are arranged in order of their atomic number. The table summarizes the major properties of the elements and enables predictions to be made about their behaviour.

There are striking similarities in the chemical properties of the elements in each of the vertical columns (called *groups*), which are numbered I–VII and 0 to reflect the number of electrons in the outermost unfilled shell and hence the maximum ◊valency. A gradation of properties may be traced along the horizontal rows (called *periods*). Metallic

character increases across a period from right to left, and down a group. A large block of elements, between groups II and III, contains the transition elements, characterized by displaying more than one valency state.

These features are a direct consequence of the electronic (and nuclear) structure of the atoms of the elements. The relationships established between the positions of elements in the periodic table and their major properties has enabled scientists to predict the properties of other elements— for example, technetium, atomic number 43, first synthesized 1937.

The first periodic table was devised by Russian chemist Dmitri Mendeleyev 1869; the elements being arranged by atomic mass (rather than atomic number) in accordance with Mendeleyev's statement 'the properties of elements are in periodic dependence upon their atomic weight'.

peripheral device in computing, any item of equipment attached to and controlled by a computer. Peripherals are typically for input from and output to the user (for example, a keyboard or printer), storing data (for example, a disc drive), communications (such as a modem), or for performing physical tasks (such as a robot).

periscope optical instrument designed for observation from a concealed position such as from a submerged submarine. In its basic form it consists of a tube with parallel mirrors at each end, inclined at 45° to its axis. The periscope attained prominence in naval and military operations of World War I.

peristalsis wavelike contractions, produced by the contraction of smooth muscle, that pass along tubular organs, such as the intestines. The same term describes the wavelike motion of earthworms and other invertebrates, in which part of the body contracts as another part elongates.

permafrost condition in which a deep layer of soil does not thaw out during the summer. Permafrost occurs under ◊periglacial conditions. It is claimed that 26% of the world's land surface is permafrost.

Permafrost gives rise to a poorly drained form of grassland typical of N Canada, Siberia, and Alaska known as ◊tundra.

permanent hardness hardness of water that cannot be removed by boiling (see ◊hard water).

permeability in physics, the degree to which the presence of a substance alters the magnetic field around it. Most substances have a small constant permeability. When the permeability is less than 1, the material is a ◊diamagnetic material; when it is greater than 1, it is a ◊paramagnetic material. ◊Ferrimagnetic materials have very large permeabilities. See also ◊magnetism.

permeable rock rock that allows water to pass through it. Rocks are permeable if they have cracks or joints running through them or if they are porous, containing many interconnected pores. Examples of permeable rocks include limestone (which is heavily jointed) and chalk (porous).

Unlike ◊impermeable rocks, which do not allow water to pass through, permeable rocks rarely support rivers and are therefore subject to less erosion. As a result they commonly form upland areas (such

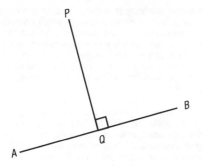

PQ is the perpendicular to AB

perpendicular

as the chalk downs of SE England, and the limestone Pennines of N England).

Permian period of geological time 290–245 million years ago, the last period of the Palaeozoic era. Its end was marked by a significant change in marine life, including the extinction of many corals and trilobites. Deserts were widespread, and terrestrial amphibians and mammal-like reptiles flourished. Cone-bearing plants (gymnosperms) came to prominence.

permutation in mathematics, a specified arrangement of a group of objects. It is the arrangement of *a* distinct objects taken *b* at a time in all possible orders. It is given by *a*!/(*a* −*b*)!, where '!' stands for ◊factorial. For example, the number of permutations of four letters taken from any group of six different letters is 6!/2! = (1 × 2 × 3 × 4 x 5 × 6)/(1 × 2) = 360. The theoretical number of four-letter 'words' that can be made from an alphabet of 26 letters is 26!/22! = 358,800.
    See also ◊combination.

perpendicular in mathematics, at a right angle; also, a line at right angles to another or to a plane. For a pair of skew lines (lines in three dimensions that do not meet), there is just one common perpendicular, which is at right angles to both lines; the nearest points on the two lines are the feet of this perpendicular.

perpetual motion the idea that a machine can be designed and constructed in such a way that, once started, it will continue in motion indefinitely without requiring any further input of energy (motive power). Such a device contradicts the two laws of thermodynamics that state that (1) energy can neither be created nor destroyed (the law of conservation of energy) and (2) heat cannot by itself flow from a cooler to a hotter object. As a result, all practical (real) machines require a continuous supply of energy, and no heat engine is able to convert all the heat into useful work.

Perseus in astronomy, a constellation of the northern hemisphere, near Cassiopeia, and represented as the mythological hero. The head of the decapitated Gorgon, Medusa, is marked by the variable star Algol. Perseus lies in the Milky Way and contains the Double Cluster, a twin cluster of stars. Every August the Perseid meteor shower radiates from its northern part.

personal computer (PC) another name for ◊microcomputer. The term is also used, more specifically, to mean the IBM Personal Computer and computers compatible with it.
    The first IBM PC was introduced in 1981; it had 64 kilobytes of random access memory (RAM) and one floppy-disc drive. It was followed in 1983 by the XT (with a hard-disc drive) and in 1984 by the AT (based on a more powerful ◊microprocessor). Many manufacturers have copied the basic design, which is now regarded as a standard for business microcomputers. Computers designed to function like an IBM PC are called *IBM-compatible computers*.

Perspex trade name for a clear, lightweight, tough plastic first produced 1930. It is widely used for watch glasses, advertising signs, domestic baths, motorboat windshields, aircraft canopies, and protective shields. Its chemical name is polymethylmethacrylate (PMMA). It is manufactured under other names: Plexiglas (in the USA), Oroglas (in Europe), and Lucite.

perspiration excretion of water and dissolved substances from the ◊sweat glands of the skin of mammals. Perspiration has two main functions: body cooling by the evaporation of water from the skin surface, and excretion of waste products such as salts.

Peru Current formerly known as *Humboldt Current* cold ocean ◊current flowing north from the Antarctica along the west coast of South America to S Ecuador, then west. It reduces the coastal temperature, making the west slopes of the Andes arid because winds are already chilled and dry when they meet the coast.

pervious rock another name for ◊permeable rock.

pest in biology, any insect, fungus, rodent, or other living organism that has a harmful effect on human beings, other than those that directly cause human diseases. Most pests damage crops or livestock, but the term also covers those that damage buildings, destroy food stores, and spread disease.

pesticide any chemical used in farming, gardening or indoors to combat pests. Pesticides are of three main types: *insecticides* (to kill insects), *fungicides* (to kill fungal diseases), and *herbicides* (to kill plants, mainly those considered weeds). Pesticides cause a number of pollution problems through spray drift onto surrounding areas, direct contamination of users or the public, and as residues on food. The safest pesticides are those made from plants, such as the insecticides pyrethrum and derris.
    More potent are synthetic products, such as chlorinated hydrocarbons. These products, including DDT and dieldrin, are highly toxic to wildlife and human beings, so their use is now restricted by law in some areas and is declining. Safer pesticides such as malathion are based on organic phosphorus compounds, but they still present hazards to health. The aid organization Oxfam estimates that pesticides cause about 10,000 deaths worldwide every year.
    Pesticides were used to deforest SE Asia during the Vietnam War, causing death and destruction to

the area's ecology and lasting health and agricultural problems.

In 1991 Central America was the world's highest consumer of pesticides per head of the population.

There are around 4,000 cases of acute pesticide poisoning a year in the UK. In 1985, 400 different chemicals were approved in the UK for use as pesticides.

petal part of a flower whose function is to attract pollinators such as insects or birds. Petals are frequently large and brightly coloured and may also be scented. Some have a nectary at the base and markings on the petal surface, known as honey guides, to direct pollinators to the source of the nectar. In wind-pollinated plants, however, the petals are usually small and insignificant, and sometimes absent altogether. Petals are derived from modified leaves, and are known collectively as a corolla.

Some insect-pollinated plants also have inconspicuous petals, with large colourful ◊bracts (leaflike structures) or ◊sepals taking over their role, or strong scents that attract pollinators such as flies.

petiole in botany, the stalk attaching the leaf blade, or ◊lamina, to the stem. Typically it is continuous with the midrib of the leaf and attached to the base of the lamina, but occasionally it is attached to the lower surface of the lamina, as in the nasturtium (a peltate leaf). Petioles that are flattened and leaflike are termed phyllodes. Leaves that lack a petiole are said to be sessile.

petrochemical chemical derived from the processing of petroleum (crude oil). *Petrochemical industries* are those that obtain their raw materials from the processing of petroleum and natural gas. Polymers, detergents, solvents, and nitrogen fertilizers are all major products of the petrochemical industries. Inorganic chemical products include carbon black, sulphur, ammonia, and hydrogen peroxide.

In the UK, ICI has a large petrochemical plant on Teesside.

petrol mixture of hydrocarbons derived from petroleum, mainly used as a fuel for internal combustion engines. It is colourless and highly volatile. *Leaded petrol* contains antiknock (a mixture of tetraethyl lead and dibromoethane), which improves the combustion of petrol and the performance of a car engine. The lead from the exhaust fumes enters the atmosphere, mostly as simple lead compounds. There is strong evidence that it can act as a nerve poison on young children and cause mental impairment. This has prompted a gradual switch to the use of *unleaded petrol* in the UK.

The changeover to leaded petrol gained momentum from 1989 owing to a change in the tax on petrol, making it cheaper to buy unleaded fuel. Unleaded petrol contains a different mixture of hydrocarbons, and has a lower ◊octane rating than leaded petrol.

In the USA, petrol is called gasoline.

petrol engine the most commonly used source of power for motor vehicles, introduced by the German engineers Gottlieb Daimler and Karl Benz 1885. The petrol engine is a complex piece of machinery made up of about 150 moving parts. It is a reciprocating piston engine, in which a number of pistons move up and down in cylinders. The motion of the pistons rotate a crankshaft, at the end of which is a heavy flywheel. From the flywheel the power is transferred to the car's driving wheels via the transmission system of clutch, gearbox, and final drive.

The parts of the petrol engine can be subdivided into a number of systems. The *fuel system* pumps fuel from the petrol tank into the carburettor. There it mixes with air and is sucked into the engine cylinders. (With electronic fuel injection, it goes directly from the tank into the cylinders by way of an electronic monitor.) The *ignition system* supplies the sparks to ignite the fuel mixture in the cylinders. By means of an ignition coil and contact breaker, it boosts the 12-volt battery voltage to pulses of 18,000 volts or more. These go via a distributor to the spark plugs in the cylinders, where they create the sparks. (Electronic ignitions replace these parts.) Ignition of the fuel in the cylinders produces temperatures of 700°C/1,300°F or more, and the engine must be cooled to prevent overheating. Most engines have a *water-cooling system*, in which water circulates through channels in the cylinder block, thus extracting the heat. It flows through pipes in a radiator, which are cooled by fan-blown air. A few cars and most motorbikes are air-cooled, the cylinders being surrounded by many fins to present a large surface area to the air. The *lubrication system* also reduces some heat, but its main job is to keep the moving parts coated with oil, which is pumped under pressure to the camshaft, crankshaft, and valve-operating gear.

petroleum or *crude oil* natural mineral oil, a thick greenish-brown flammable liquid found underground in permeable rocks. Petroleum consists of hydrocarbons mixed with oxygen, sulphur, nitrogen, and other elements in varying proportions. It is thought to be derived from ancient organic material that has been converted by, first, bacterial action, then heat and pressure (but its origin may be chemical also). From crude petroleum, various products are made by distillation and other processes; for example, fuel oil, petrol, kerosene, diesel, lubricating oil, paraffin wax, and petroleum jelly.

The organic material in petroleum was laid down millions of years ago (hence it is known as a fossil fuel). Petroleum is found trapped in porous reservoir rocks, such as sandstone or limestone, in anticlines and other traps below impervious rock layers. Oil may flow from wells under natural pressure, causing it to rise up the borehole, but many oil wells require pumping to bring the oil to the surface.

The occurrence of petroleum was known in ancient times, and it was used medicinally by American Indians, but the exploitation of oilfields began with the first commercial well in Pennsylvania 1859.

The USA led in production until the 1960s, when the Middle East outproduced other areas, their immense reserves leading to a worldwide dependence on cheap oil for transport and industry. In 1961 the Organization of the Petroleum Exporting Countries (OPEC) was established to avoid

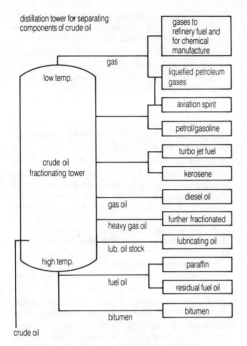

distillation tower for separating
components of crude oil

gas

low temp.

crude oil
fractionating tower

gas oil

heavy gas oil

lub. oil stock

high temp.

fuel oil

bitumen

crude oil

gases to
refinery fuel and
for chemical
manufacture

liquefied petroleum
gases

aviation spirit

petrol/gasoline

turbo jet fuel

kerosene

diesel oil

further fractionated

lubricating oil

paraffin

residual fuel oil

bitumen

petroleum *Refining petroleum using a distillation column. The crude petroleum is fed in at the bottom of the column where the temperature is high. The gases produced rise up the column, cooling as they travel. At different heights up the column, different gases condense to liquids called fractions, and are drawn off.*

exploitation of member countries; after OPEC's price rises in 1973, the International Energy Agency (IEA) was established 1974 to protect the interests of oil-consuming countries. New technologies were introduced to pump oil from offshore and from the Arctic (the Alaska pipeline) in an effort to avoid a monopoly by OPEC.

Petroleum products and chemicals are used in large quantities in the manufacture of detergents, artificial fibres, plastics, insecticides, fertilizers, pharmaceuticals, toiletries, and synthetic rubber. Aviation fuel is a volatile form of petrol.

The burning of petroleum fuel is one cause of air pollution. The transport of oil can lead to major catastrophes—for example, the *Torrey Canyon* tanker lost off SW England 1967, which led to an agreement by the international oil companies 1968 to pay compensation for massive shore pollution. The 1989 oil spill in Alaska from the *Exxon Valdez* damaged the area's fragile environment, despite clean-up efforts. Drilling for oil involves the risks of accidental spillage and drilling-rig accidents. The problems associated with oil have led to the various ◊alternative energy technologies.

A new kind of bacterium was developed during the 1970s in the USA, capable of 'eating' oil as a means of countering oil spills. Its creation gave rise to the so-called Frankenstein law (a ruling that allowed new forms of life created in laboratories to be patented).

petrology branch of geology that deals with the study of rocks, their mineral compositions, and their origins.

pewter any of various alloys of mostly tin with varying amounts of lead, copper, or antimony. Pewter has been known for centuries and was once widely used for domestic utensils but is now used mainly for ornamental ware.

pH scale from 0 to 14 for measuring acidity or alkalinity. A pH of 7.0 indicates neutrality, below 7 is acid, while above 7 is alkaline. Strong acids, such as those used in car batteries, have a pH of about 2; strong alkalis such as sodium hydroxide are pH 13.

Acidic fruits such as citrus fruits are about pH 4. Fertile soils have a pH of about 6.5 to 7.0, while weak alkalis such as soap are 9 to 10.

The pH of a solution can be measured by using a broad-range indicator, either in solution or as a

petroleum: chronology

1859	Edwin Drake drilled the world's first successful oil well in Titusville, Pennsylvania, USA, to a depth of 18 m/70 ft.
1865	The first oil pipeline, 9,750 m/32,000 ft long, was constructed at Oil Creek, Pennsylvania, USA, to carry oil from well to nearby coalfield.
1896	The first offshore wells were drilled from piers off the California coast.
1899	The first gravity meter was produced.
1914	Reginald Fessenden patented the seismograph.
1939	The aircraft-borne magnetometer was developed to measure magnetism of rocks.
1966	Oil was discovered beneath the North Sea.
1967	The worst oil spill in British waters from the *Torrey Canyon*, which struck rocks at Lands End. Over 108,000 tonnes of oil were spilled.
1974	The world's deepest oil well, 10,941 m/31,441 ft, was drilled in Oklahoma, USA.
1979	The oil rig *Ixtoc I* in the Gulf of Mexico accidentally released 545,000 tonnes of oil into the sea. The slick spread for 400 miles. Later the same year, the worst tanker spillage occurred. Two tankers, the *Atlantic Empress* and the *Aegean Captain*, collided off the island of Tobago, in the Caribbean Sea. More than 230,000 tonnes of oil was spilled.
1984	An exploratory well was drilled off the coast of New England, USA, in water of depth 2,116 m/6,942 ft—a world record.
1988	The Piper Alpha drilling rig in the North Sea caught fire in July, killing 167 people.
1989	The worst spill in American waters occurred when 55,000 tonnes of oil escaped from the *Exxon Valdez* off the Alaskan coast, near Prince William Sound.
1991	The worst spill to date was a consequence of the Gulf War when Iraqi forces opened the pipeline into the Persian Gulf and coalition forces caused further damage during Operation Desert Storm.

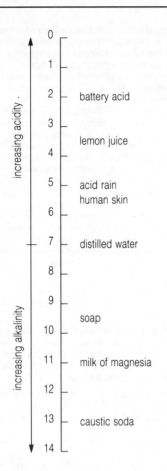

	0
	1
	2 — battery acid
	3
	— lemon juice
	4
	5 — acid rain
	— human skin
	6
	7 — distilled water
	8
	9
	— soap
	10
	11 — milk of magnesia
	12
	13 — caustic soda
	14

increasing acidity

increasing alkalinity

pH *The pHs of some common substances. The lower the pH the more acid the substance.*

paper strip. The colour produced by the indicator is compared with a colour code related to the pH value. An alternative method is to use a pH meter fitted with a glass electrode.

phage another name for a ◊bacteriophage, a virus that attacks bacteria.

phagocyte type of white blood cell, or ◊leucocyte, that can engulf a bacterium or other invading microorganism. Phagocytes are found in blood, lymph, and other body tissues, where they also ingest foreign matter and dead tissue. A ◊macrophage differs in size and life span.

Phagocytosis is the engulfing of foreign bodies or food by phagocytes.

phanerogam obsolete term for a plant that bears flowers or cones and reproduces by means of seeds, that is an ◊angiosperm and ◊gymnosperm, or a ◊seed plant. Plants such as mosses, fungi, and ferns were known as *cryptogams*.

Phanerozoic (Greek *phanero* 'visible') eon in Earth history, consisting of the most recent 570 million years. It comprises the Palaeozoic, Mesozoic, and Cenozoic eras. The vast majority of fossils

come from this eon, owing to the evolution of hard shells and internal skeletons. The name means 'interval of well-displayed life'.

pharmacology study of the origins, applications, and effects of chemical substances on living organisms. Products of the pharmaceutical industry range from aspirin to anticancer agents.

A wide range of resources has been investigated in the search for new and better drugs (human and plant molecular biology, newly discovered soil-grown moulds, genetically engineered compounds, and monoclonal antibiotics).

pharynx interior of the throat, the cavity at the back of the mouth. Its walls are made of muscle strengthened with a fibrous layer and lined with mucous membrane. The internal nostrils lead backwards into the pharynx, which continues downwards into the oesophagus and (through the epiglottis) into the windpipe. On each side, a Eustachian tube enters the pharynx from the middle ear cavity.

The upper part (nasopharynx) is an airway, but the remainder is a passage for food. Inflammation of the pharynx is named pharyngitis.

phase in astronomy, the apparent shape of the Moon or a planet when all or part of its illuminated hemisphere is facing the Earth. The Moon undergoes a full cycle of phases from new (when between the Earth and the Sun) through first quarter (when at 90° eastern elongation from the Sun), full (when opposite the Sun), and last quarter (when at 90° western elongation from the Sun).

The Moon is gibbous (more than half but less than fully illuminated) when between first quarter and full or full and last quarter. Mars can appear gibbous at quadrature (when it is at right angles to the Sun in the sky). The gibbous appearance of Jupiter is barely noticeable.

The planets whose orbits lie within that of the Earth can also undergo a full cycle of phases, as can an asteroid passing inside the Earth's orbit.

PHASE: NO FULL MOON

February is the only month during which there may be no full or new Moon. This is because there are 29 days between new moons, but only 28 days in February (29 in leap years).

phase in chemistry, a physical state of matter: for example, ice and liquid water are different phases of water; a mixture of the two is termed a two-phase system.

phase in physics, a stage in an oscillatory motion, such as a wave motion: two waves are in phase when their peaks and their troughs coincide. Otherwise, there is a *phase difference*, which has consequences in ◊interference phenomena and ◊alternating current electricity.

phenol member of a group of aromatic chemical compounds with weakly acidic properties, which are characterized by a hydroxyl (OH) group

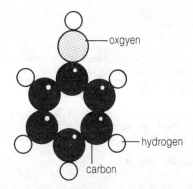

phenol *The phenol molecule with its ring of six carbon atoms and a hydroxyl (OH) group attached. Phenol was first extracted from coal tar in 1834. It is used to make phenolic and epoxy resins, explosives, pharmaceuticals, perfumes, and nylon.*

attached directly to an aromatic ring. The simplest of the phenols, derived from benzene, is also known as phenol and has the formula C_6H_5OH. It is sometimes called ***carbolic acid*** and can be extracted from coal tar.

Pure phenol consists of colourless, needle-shaped crystals, which take up moisture from the atmosphere. It has a strong and characteristic smell and was once used as an antiseptic. It is, however, toxic by absorption through the skin.

phenolphthalein acid–base indicator that is clear below pH 8 and red above pH 9.6. It is used in titrating weak acids against strong bases.

phenotype in genetics, visible traits, those actually displayed by an organism. The phenotype is not a direct reflection of the ◊genotype because some alleles are masked by the presence of other, dominant alleles (see ◊dominance). The phenotype is further modified by the effects of the environment (for example, poor nutrition stunts growth).

pheromone chemical signal (such as an odour) that is emitted by one animal and affects the behaviour of others. Pheromones are used by many animal species to attract mates.

PHEROMONE: SEEKING A MATE

A pheromone released by a female Atlas moth to advertise that she is ready to mate can be detected by a male moth more than 8 km/5 mi away. He picks up her scent in the air with his huge antennae.

---

phloem tissue found in vascular plants whose main function is to conduct sugars and other food materials from the leaves, where they are produced, to all other parts of the plant.

Phloem is composed of sieve elements and their associated companion cells, together with some ◊sclerenchyma and ◊parenchyma cell types. Sieve elements are long, thin-walled cells joined end to end, forming sieve tubes; large pores in the end walls allow the continuous passage of nutrients. Phloem is usually found in association with ◊xylem, the water-conducting tissue, but unlike the latter it is a living tissue.

phlogiston hypothetical substance formerly believed to have been produced during combustion. The term was invented by the German chemist Georg Stahl. The phlogiston theory was replaced by the theory of oxygen gain and loss, first enunciated by the French chemist Antoine Lavoisier.

Phobos one of the two moons of Mars, discovered 1877 by the US astronomer Asaph Hall (1829–1907). It is an irregularly shaped lump of rock, cratered by ◊meteorite impacts. Phobos is 27 × 22 × 19 km/17 × 13 x 12 mi across, and orbits Mars every 0.32 days at a distance of 9,400 km/ 5,840 mi from the planet's centre. It is thought to be an asteroid captured by Mars' gravity.

phon unit of loudness, equal to the value in decibels of an equally loud tone with frequency 1,000 Hz. The higher the frequency, the louder a noise sounds for the same decibel value; thus an 80–decibel tone with a frequency of 20 Hz sounds as loud as 20 decibels at 1,000 Hz, and the phon value of both tones is 20. An aircraft engine has a loudness of around 140 phons.

phonograph the name US inventor Thomas Edison gave to his sound-recording apparatus, which developed into the record player. The word 'phonograph' is still used in the USA.

phosphate salt or ester of ◊phosphoric acid. Incomplete neutralization of phosphoric acid gives rise to acid phosphates (see ◊acid salts and ◊buffer). Phosphates are used as fertilizers, and are required for the development of healthy root systems. They are involved in many biochemical processes, often as part of complex molecules, such as ◊ATP.

phospholipid any ◊lipid consisting of a glycerol backbone, a phosphate group, and two long chains. Phospholipids are found everywhere in living systems as the basis for biological membranes.

One of the long chains tends to be hydrophobic and the other hydrophilic (that is, they interrelate with water in opposite ways). This means that phospholipids will line up the same way round when in solution.

phosphor any substance that is phosphorescent, that is, gives out visible light when it is illuminated by a beam of electrons or ultraviolet light. The television screen is coated on the inside with phosphors that glow when beams of electrons strike them. Fluorescent lamp tubes are also phosphor-coated. Phosphors are also used in Day-Glo paints, and as optical brighteners in detergents.

phosphorescence in physics, the emission of light by certain substances after they have absorbed energy, whether from visible light, other electromagnetic radiation such as ultraviolet rays or X-rays, or cathode rays (a beam of electrons). When the stimulating energy is removed phosphorescence ceases, although it may persist for a short time after (unlike ◊fluorescence, which stops immediately).

phosphoric acid acid derived from phosphorus and oxygen. Its commonest form (H_3PO_4) is also known as orthophosphoric acid, and is produced by the action of phosphorus pentoxide (P_2O_5) on water. It is used in rust removers and for rust-proofing iron and steel.

phosphorus (Greek *phosphoros* 'bearer of light')

highly reactive, nonmetallic element, symbol P, atomic number 15, relative atomic mass 30.9738. It occurs in nature as phosphates (commonly in the form of the mineral ◊apatite), and is essential to plant and animal life. Compounds of phosphorus are used in fertilizers, various organic chemicals, for matches and fireworks, and in glass and steel.

Phosphorus was first identified 1674 by German alchemist Hennig Brand (*c.* 1630–?), who prepared it from urine. The element has three allotropic forms: a black powder; a white-yellow, waxy solid that ignites spontaneously in air to form the poisonous gas phosphorus pentoxide; and a red-brown powder that neither ignites spontaneously nor is poisonous.

photocell or *photoelectric cell* device for measuring or detecting light or other electromagnetic radiation, since its electrical state is altered by the effect of light. In a *photoemissive* cell, the radiation causes electrons to be emitted and a current to flow (◊photoelectric effect); a *photovoltaic* cell causes an ◊electromotive force to be generated in the presence of light across the boundary of two substances. A *photoconductive* cell, which contains a semiconductor, increases its conductivity when exposed to electromagnetic radiation.

Photocells are used for photographers' exposure meters, burglar and fire alarms, automatic doors, and in solar energy arrays.

photochemical reaction any chemical reaction in which light is produced or light initiates the reaction. Light can initiate reactions by exciting atoms or molecules and making them more reactive: the light energy becomes converted to chemical energy. Many photochemical reactions set up a ◊chain reaction and produce ◊free radicals.

This type of reaction is seen in the bleaching of dyes or the yellowing of paper by sunlight. It is harnessed by plants in ◊photosynthesis and by humans in ◊photography. Chemical reactions that produce light are most commonly seen when materials are burned. Light-emitting reactions are used by living organisms in ◊bioluminescence. One photochemical reaction is the action of sunlight on car exhaust fumes, which results in the production of ◊ozone. Some large cities, such as Los Angeles, and Santiago, Chile, now suffer serious pollution due to photochemical smog.

photocopier machine that uses some form of photographic process to reproduce copies of documents or illustrations. Most modern photocopiers, as pioneered by the Xerox Corporation, use electrostatic photocopying, or ◊xerography ('dry writing'). This employs a drum coated with a light-sensitive material such as selenium, which holds a pattern of static electricity charges corresponding to the dark areas of an image projected on to the drum by a lens. Finely divided pigment (toner) of opposite electric charge sticks to the charged areas of the drum and is transferred to a sheet of paper, which is heated briefly to melt the toner and stick it to the paper.

Additional functions of photocopiers include enlargement and reduction, copying on both sides of the sheet of paper, copying in colour, collating, and stapling.

photodiode semiconductor ◊*p–n* junction diode used to detect light or measure its intensity. The photodiode is encapsulated in a transparent plastic case that allows light to fall onto the junction. When this occurs, the reverse-bias resistance (high resistance in the opposite direction to normal current-flow) drops and allows a larger reverse-biased current to flow through the device. The increase in current can then be related to the amount of light falling on the junction.

Photodiodes that can detect small changes in light level are used in alarm systems, camera exposure-controls, and optical communication links.

photoelectric effect in physics, the emission of ◊electrons from a substance (usually a metallic surface) when it is struck by ◊photons (quanta of electromagnetic radiation), usually those of visible light or ultraviolet radiation.

photogram picture produced on photographic material by exposing it to light, but without using a camera.

photography process for reproducing images on sensitized materials by various forms of radiant energy, including visible light, ultraviolet, infrared, X-rays, atomic radiations, and electron beams. Photography was developed in the 19th century; among the pioneers were Louis Daguerre in France and William Henry Fox Talbot in the UK. Colour photography dates from the early 20th century.

The most familiar photographic process depends upon the fact that certain silver compounds (called ◊halides) are sensitive to light. A photographic film is coated with these compounds and, in a camera, is exposed to light. An image, or picture, of the scene before the camera is formed on the film because the silver halides become activated (light-altered) where light falls but not where light does not fall. The image is made visible by the process of ◊developing, made permanent by fixing, and, finally, is usually printed on paper. Motion-picture photography uses a camera that exposes a roll of film to a rapid succession of views that, when developed, are projected in equally rapid succession to provide a moving image.

photogravure ◊printing process that uses a plate prepared photographically, covered with a pattern of recessed cells in which the ink is held. See ◊gravure.

photometer instrument that measures luminous intensity, usually by comparing relative intensities from different sources. Bunsen's grease-spot photometer 1844 compares the intensity of a light source with a known source by each illuminating one half of a translucent area. Modern photometers use ◊photocells, as in a photographer's exposure meter. A photomultiplier can also be used as a photometer.

photomultiplier instrument that detects low levels of electromagnetic radiation (usually visible light or ◊infrared radiation) and amplifies it to produce a detectable signal.

One type resembles a ◊photocell with an additional series of coated ◊electrodes (dynodes) between the ◊cathode and ◊anode. Radiation striking the cathode releases electrons (primary

photography: chronology

1515   Leonardo da Vinci described the camera obscura.
1750   The painter Canaletto used a camera obscura as an aid to his painting in Venice.
1790   Thomas Wedgwood in England made photograms – placing objects on leather, sensitized using silver nitrate.
1826   Nicephore Niépce (1765–1833), a French doctor, produced the world's first photograph from nature on pewter plates with a camera obscura and an eight-hour exposure.
1835   Niépce and L J M Daguerre produced the first Daguerreotype camera photograph.
1839   Daguerre was awarded an annuity by the French government and his process given to the world.
1840   Invention of the Petzval lens, which reduced exposure time by 90%. Herschel discovered sodium thiosulphate as a fixer for silver halides.
1841   Fox Talbot's calotype process was patented—the first multicopy method of photography using a negative/positive process, sensitized with silver iodide.
1851   Fox Talbot used a one-thousandth of a second exposure to demonstrate high-speed photography. Invention of the wet-collodion-on-glass process and the waxed-paper negative.
1855   Roger Fenton made documentary photographs of the Crimean War from a specially constructed caravan with portable darkroom.
1859   Nadar in Paris made photographs underground using battery powered arc lights.
1861   The single-lens reflex plate camera was patented by Thomas Sutton. The principles of three-colour photography were demonstrated by Scottish physicist James Clerk Maxwell.
1862   Nadar took aerial photographs over Paris.
1871   Gelatin-silver bromide was developed.
1878   In the USA Eadweard Muybridge analysed the movements of animals through sequential photographs, using a series of cameras.
1879   The photogravure process was invented.
1880   A silver bromide emulsion was fixed with hypo. Photographs were first reproduced in newspapers in New York using the half-tone engraving process. The first twin-lens reflex camera was produced in London. Gelatin-silver chloride paper was introduced.
1884   George Eastman produced flexible negative film.
1889   The Eastman Company in the USA produced the Kodak No 1 camera and roll film, facilitating universal, hand-held snapshots.
1891   The first telephoto lens. The interference process of colour photography was developed by the French doctor Gabriel Lippmann.
1902   In Germany, Deckel invented a prototype leaf shutter and Zeiss introduced the Tessar lens.
1904   The autochrome colour process was patented by the Lumière brothers.
1907   The autochrome process began to be factory-produced.
1914   Oskar Barnack designed a prototype Leica camera for Leitz in Germany.
1924   Leitz launched the first 35mm camera, the Leica, delayed because of World War I.
1929   Rolleiflex produced a twin-lens reflex camera in Germany.
1935   In the USA, Mannes and Godowsky invented Kodachrome transparency film, which produced sharp images and rich colour quality. Electronic flash was invented in the USA.
1940   Multigrade enlarging paper by Ilford was made available in the UK.
1942   Kodacolour negative film was introduced.
1947   Polaroid black and white instant process film was invented by Dr Edwin Land, who set up the Polaroid corporation in Boston, Massachusetts. The principles of holography were demonstrated in England by Dennis Gabor.
1955   Kodak introduced Tri-X, a black and white 200 ASA film.
1959   The zoom lens was invented by the Austrian firm of Voigtlander.
1960   The laser was invented in the USA, making holography possible. Polacolor, a self-processing colour film, was introduced by Polaroid, using a 60–second colour film and dye diffusion technique.
1963   Cibachrome, paper and chemicals for printing directly from transparencies, was made available by Ciba-Geigy of Switzerland.
1969   Photographs were taken on the Moon by US astronauts.
1970   A charge-coupled device was invented at Bell Laboratories in New Jersey, USA, to record very faint images (for example in astronomy).
1972   The SX70 system, a single-lens reflex camera with instant prints, was produced by Polaroid.
1975   The Center for Creative Photography was established at the University of Arizona.
1980   *Voyager 1* sent photographs of Saturn back to Earth across space.
1985   The Minolta Corporation in Japan introduced the Minolta 7000 – the world's first body-integral autofocus single-lens reflex camera.
1988   The electronic camera, which stores pictures on magnetic disc instead of on film, was introduced in Japan.
1990   Kodak introduced PhotoCD which converts 35mm camera pictures (on film) into digital form and stores them on compact disc (CD) for viewing on TV.
1992   Japanese company Canon introduced a camera with autofocus controlled by the user's eye. The camera focuses on whatever the user is looking

emission) which hit the first dynode, producing yet more electrons (◊secondary emission), which strike the second dynode. Eventually this produces a measurable signal up to 100 million times larger than the original signal by the time it leaves the anode. Similar devices, called image intensifiers, are used in television camera tubes that 'see' in the dark.

photon in physics, the ◊elementary particle or 'package' (quantum) of energy in which light and other forms of electromagnetic radiation are emitted. The photon has both particle and wave properties; it has no charge, is considered massless but possesses momentum and energy. It is one of the ◊gauge bosons, a particle that cannot be subdivided, and is the carrier of the ◊electromagnetic force, one of the fundamental forces of nature.

According to ◊quantum theory the energy of a photon is given by the formula $E = hf$, where h is Planck's constant and f is the frequency of the radiation emitted.

SHOWERS OF PHOTONS

In sunlight, a million million photons fall on a pinhead every second. When you look at a faint star (such as the faintest star in the Little Dipper), your eye receives only about 500 photons per second.

photoperiodism biological mechanism that determines the timing of certain activities by responding to changes in day length. The flowering of many plants is initiated in this way. Photoperiodism in plants is regulated by a light-sensitive pigment, *phytochrome*. The breeding seasons of many temperate-zone animals are also triggered by increasing or declining day length, as part of their ◊biorhythms.

Autumn-flowering plants (for example, chrysanthemum and soya bean) and autumn-breeding mammals (such as goats and deer) require days that are shorter than a critical length; spring-flowering and spring-breeding ones (such as radish and lettuce, and birds) are triggered by longer days.

photosphere visible surface of the Sun, which emits light and heat. About 300 km/200 mi deep, it consists of incandescent gas at a temperature of 5,800K (5,530°C/9,980°F).

Rising cells of hot gas produce a mottling of the photosphere known as *granulation*, each granule being about 1,000 km/620 mi in diameter. The photosphere is often marked by large, dark patches called ◊sunspots.

photosynthesis process by which green plants trap light energy and use it to drive a series of chemical reactions, leading to the formation of carbohydrates. All animals ultimately depend on photosynthesis because it is the method by which the basic food (sugar) is created. For photosynthesis to occur, the plant must possess ◊chlorophyll and must have a supply of carbon dioxide and water. Actively photosynthesizing green plants store excess sugar as starch (this can be tested for in the laboratory using iodine).

The chemical reactions of photosynthesis occur in two stages. During the *light reaction* sunlight is used to split water (H_2O) into oxygen (O_2), protons (hydrogen ions, H^+), and electrons, and oxygen is given off as a by-product. In the second-stage *dark reaction*, for which sunlight is not required, the protons and electrons are used to convert carbon dioxide (CO_2) into carbohydrates ($C_m(H_2O)_n$). Photosynthesis depends on the ability of chlorophyll to capture the energy of sunlight and to use it to split water molecules. The initial charge separation occurs in less than a billionth of a second, a speed that compares with current computers.

Other pigments, such as ◊carotenoids, are also involved in capturing light energy and passing it on to chlorophyll. Photosynthesis by cyanobacteria was responsible for the appearance of oxygen in the Earth's atmosphere 2 billion years ago, and photosynthesis by plants maintains the oxygen level today.

phototropism movement of part of a plant toward or away from a source of light. Leaves are positively phototropic, detecting the source of light and orientating themselves to receive the maximum amount.

phyllite ◊metamorphic rock produced under increasing temperature and pressure, in which mica crystals are aligned so that the rock splits along their plane of orientation, the resulting break being shiny and smooth. It is intermediate between slate and schist.

phyllotaxis the arrangement of leaves on a plant stem. Leaves are nearly always arranged in a regular pattern and in the majority of plants they are inserted singly, either in a *spiral* arrangement up the stem, or on *alternate* sides. Other principal forms are opposite leaves, where two arise from the same node, and whorled, where three or more arise from the same node.

phylogeny historical sequence of changes that occurs in a given species during the course of its evolution. It was once erroneously associated with ontogeny (the process of development of a living organism).

phylum (plural *phyla*) major grouping in biological classification. Mammals, birds, reptiles, amphibians, fishes, and tunicates belong to the phylum Chordata; the phylum Mollusca consists of snails, slugs, mussels, clams, squid, and octopuses; the phylum Porifera contains sponges; and the phylum Echinodermata includes starfish, sea urchins, and sea cucumbers. In classifying plants (where the term 'division' often takes the place of 'phylum'), there are between four and nine phyla depending on the criteria used; all flowering plants belong to a single phylum, Angiospermata, and all conifers to another, Gymnospermata. Related phyla are grouped together in a ◊kingdom; phyla are subdivided into ◊classes.

physical change in chemistry, a type of change that does not produce a new chemical substance, does not involve large energy changes, and that can be easily reversed (the opposite of a ◊chemical change). Boiling and melting are examples of physical change.

physical chemistry branch of chemistry concerned with examining the relationships between the chemical compositions of substances and the physical properties that they display. Most chemical

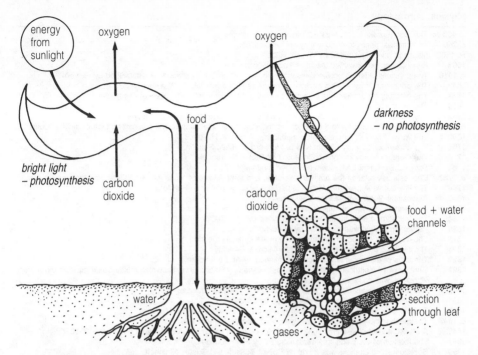

energy from sunlight

oxygen

oxygen

food

darkness – no photosynthesis

bright light – photosynthesis   carbon dioxide

carbon dioxide

food + water channels

water

section through leaf

gases

photosynthesis *Process by which green plants and some bacteria manufacture carbohydrates from water and atmospheric carbon dioxide, using the energy of sunlight. Photosynthesis depends on the ability of chlorophyll molecules within plant cells to trap the energy of light to split water molecules, giving off oxygen as a by-product. The hydrogen of the water molecules is then used to reduce carbon dioxide to simple carbohydrates.*

reactions exhibit some physical phenomenon (change of state, temperature, pressure, or volume, or the use or production of electricity), and the measurement and study of such phenomena has led to many chemical theories and laws.

physical weathering or *mechanical weathering* in earth science, the form of ◊weathering responsible for the mechanical breakdown of rocks but involving no chemical change. Processes involved include ◊freeze-thaw (the alternate freezing and melting of ice in rock cracks) and exfoliation (the alternate expansion and contraction of rocks in response to extreme changes in temperature).

physics branch of science concerned with the laws that govern the structure of the universe, and the forms of matter and energy and their interactions. For convenience, physics is often divided into branches such as nuclear physics, particle physics, solid-and liquid-state physics, electricity, electronics, magnetism, optics, acoustics, heat, and thermodynamics. Before this century, physics was known as *natural philosophy*.

Every statement in physics has to state relations between observable quantities.

On **physics** Ernst Mach (Mach's principle) 1863

physiology branch of biology that deals with the functioning of living animals, as opposed to anatomy, which studies their structures.

phytomenadione vitamin K, a fat-soluble chemi-

cal found in leafy vegetables and liver. Lack of this vitamin can lead to difficulties in the healing of injured blood vessels.

pi an ◊irrational number, symbol π, the ratio of the circumference of a circle to its diameter. The value of pi is 3.1415926, correct to seven decimal places. Common approximations to pi are $^{22}/_7$ and 3.14, although the value 3 can be used as a rough estimation.

The most exciting phrase to hear in science, the one that heralds new discoveries, is not "Eureka!" (I've found it!) but "That's funny..."

Isaac Asimov

pie chart method of displaying proportional information by dividing a circle up into different-sized sectors (slices of pie). The angle of each sector is proportional to the size, expressed as a percentage, of the group of data that it represents.

For example, data from a traffic survey could be presented in a pie chart in the following way:

(1) convert each item of data to a percentage figure;

(2) 100% will equal 360 degrees of the circle, therefore each 1% = 360/100 = 3.6 degrees;

(3) calculate the angle of the segment for each item of data, and plot this on the circle. The diagram may be made clearer by adding colours or shadings to each group, together with a key.

physics: chronology

c. 400 BC	The first 'atomic' theory was put forward by Democritus.
c. 250	Archimedes' principle of buoyancy was established.
AD 1600	Magnetism was described by William Gilbert.
1608	Hans Lippershey invented the refracting telescope.
c. 1610	The principle of falling bodies descending to earth at the same speed was established by Galileo.
1642	The principles of hydraulics were put forward by Blaise Pascal.
1643	The mercury barometer was invented by Evangelista Torricelli.
1656	The pendulum clock was invented by Christiaan Huygens.
1662	Boyle's law concerning the behaviour of gases was established by Robert Boyle.
c. 1665	Isaac Newton put forward the law of gravity, stating that the Earth exerts a constant force on falling bodies.
1690	The wave theory of light was propounded by Christiaan Huygens.
1704	The corpuscular theory of light was put forward by Isaac Newton.
1714	The mercury thermometer was invented by Daniel Fahrenheit.
1764	Specific and latent heats were described by Joseph Black.
c. 1787	Charles's law relating the pressure, volume, and temperature of a gas was established by Jacques Charles.
1798	The link between heat and friction was discovered by Benjamin Rumford.
1800	Alessandro Volta invented the Voltaic cell.
1801	Interference of light was discovered by Thomas Young.
1808	The 'modern' atomic theory was propounded by John Dalton.
1815	Refraction of light was explained by Augustin Fresnel.
1819	The discovery of electromagnetism was made by Hans Oersted.
1821	The dynamo principle was described by Michael Faraday.
1822	The laws of electrodynamics were established by André Ampère.
1827	Ohm's law of electrical resistance was established by Georg Ohm; Brownian motion resulting from molecular vibrations was observed by Robert Brown.
1829	The law of gaseous diffusion was established by Thomas Graham.
1831	Electromagnetic induction was discovered by Faraday.
1842	The principle of conservation of energy was observed by Julius von Mayer.
c. 1847	The mechanical equivalent of heat was described by James Joule.
1849	A measurement of speed of light was put forward by French physicist Armand Fizeau (1819–1896).
1851	The rotation of the Earth was demonstrated by Jean Foucault.
1859	Spectrographic analysis was made by Robert Bunsen and Gustav Kirchhoff.
1861	Osmosis was discovered.
1873	Light was conceived as electromagnetic radiation by James Maxwell.
1887	The existence of radio waves was predicted by Heinrich Hertz.
1895	X-rays were discovered by Wilhelm Röntgen.
1896	The discovery of radioactivity was made by Antoine Becquerel.
1897	Joseph Thomson discovered the electron.
1899	Ernest Rutherford discovered alpha and beta rays.
1900	Quantum theory was propounded by Max Planck; the discovery of gamma rays was made by French physicist Paul-Ulrich Villard (1860–1934).
1905	Albert Einstein propounded his special theory of relativity.
1908	The Geiger counter was invented by Hans Geiger and Rutherford.
1911	The discovery of the atomic nucleus was made by Rutherford.
1913	The orbiting electron atomic theory was propounded by Danish physicist Niels Bohr.
1915	X-ray crystallography was discovered by William and Lawrence Bragg.
1916	Einstein put forward his general theory of relativity; mass spectrography was discovered by William Aston.
1926	Wave mechanics was introduced by Erwin Schrödinger.
1931	The cyclotron was developed by Ernest Lawrence.
1932	The discovery of the neutron was made by James Chadwick.
1933	The positron, the antiparticle of the electron, was discovered by Carl Anderson.
1939	The discovery of nuclear fission was made by Otto Hahn and Fritz Strassmann.
1942	The first controlled nuclear chain reaction was achieved by Enrico Fermi.
1956	The neutrino, an elementary particle, was discovered by Clyde Cowan and Fred Reines.
1960	The Mössbauer effect of atom emissions was discovered by Rudolf Mössbauer; the first maser was developed by US physicist Theodore Maiman (1927–).
1963	Maiman developed the first laser.
1964	Murray Gell-Mann and George Zweig discovered the quark.
1967	Jocelyn Bell (now Burnell) and Antony Hewish discovered pulsars.
1971	The theory of superconductivity was announced, where electrical resistance in some metals vanishes above absolute zero.
1979	The asymmetry of elementary particles was discovered by US physicists James W Cronin and Val L Fitch.
1983	Evidence of the existence of weakons (W and Z particles) was confirmed at CERN, validating the link between the weak nuclear force and the electromagnetic force.
1986	The first high-temperature superconductor was discovered, able to conduct electricity without resistance at a temperature of −238°C/−396°F.
1989	CERN's Large Electron Positron Collider (LEP), a particle accelerator with a circumference of 27 km/16.8 mi, came into operation.
1991	LEP experiments demonstrated the existence of three generations of elementary particles, each with two quarks and two leptons.
1992	Japanese researchers developed a material that becomes superconducting at −103°C/−153°F (about 45°C/80°F warmer than the previous record).

The missing neutrinos, and optical molasses

Progress in physics can be expected or unexpected. Particle physicists fully expect to detect the top quark, with one of their existing accelerators. Yet the discovery of superconductivity in thin films of potassium-doped carbon-60 molecules was completely unexpected.

Solar neutrinos

The scarcity of solar neutrinos is a mystery. Neutrinos are particles without electric charge or mass, produced in processes such as radioactive beta decay. There are three types—electron, muon and tau. The Sun produces energy by combining four hydrogen nuclei into a helium nucleus— the process of thermonuclear fusion. It converts 600 tonnes of hydrogen every second in a web of reactions, several producing neutrinos. According to calculations, 65 billion solar electron neutrinos should penetrate every square centimetre of the Earth's surface every second. Detection of these neutrinos would confirm physicists' theories about the Sun. Neutrinos can travel through large objects, and are extremely difficult to detect.

The first experiment to detect solar neutrinos was set up in South Dakota, USA, in the 1960s. Despite running for 20 years, it only detected about a quarter of the predicted quantity; a second experiment in Japan detected less than half. Both experiments were sensitive to less than 10% of solar neutrinos. Neither could detect the neutrinos produced by protons fusing into heavy hydrogen (deuterium) at the start of the solar thermonuclear process. Physicists had hopes for two experiments (Gallex in Italy and SAGE in Russia) using detectors containing gallium, sensitive to most of these neutrinos. Both have now reported their first results: higher than the previous experiments, but not as high as predicted by theory.

Both experiments will be improved in the next few years; more sophisticated detectors are planned for 1996. If the shortage of neutrinos persists, perhaps neutrinos have mass after all, allowing electron neutrinos to change into muon or tau neutrinos on passing through the Sun. Proof of neutrino mass would be a significant development for particle physics.

Laser cooling

When an atom absorbs a photon of light its velocity increases. This increase occurs in the direction of the light. The atom loses this velocity again when it emits a photon. Since this loss can occur in any direction, on average an atom travelling in the opposite direction to a laser beam will lose speed. With two opposed lasers, the atoms can be confined in a plane at right angles to the beams. With three pairs of lasers at right angles, some atoms will be trapped where all six beams overlap. Because temperature is related to velocity the atoms—now almost stationary—are 'cooled'. By making the laser wavelength slightly longer than the atom's normal absorption and emission wavelength, physicists can use the Doppler effect to make the atoms even cooler, in devices called 'optical molasses'.

Optical molasses have been used to cool sodium atoms to a few millionths of a degree above absolute zero. They are one of a variety of devices being used to trap and cool atoms and ions, to test the fundamentals of quantum mechanics and provide new standards for time and frequency. The power to manipulate and control atoms and large particles heralds new devices with colourful names like atomic fountains and trampolines, and optical tweezers.

Multilayer structures

Molecular beam epitaxy (MBE) allows scientists to grow layers only a few atoms thick, producing materials with completely new electronic, optical and magnetic properties. The first semiconductor lasers to work at blue-green wavelengths were made with MBE, by growing layers of zinc selenide less than 10 nanometres (thousand millionths of a metre) thick, interspersed with layers containing cadmium or sulphur.

Most forms of information storage rely on thin magnetic films. Many recording and playback heads incorporate magnetoresistance (MR) sensors. MR is a property of most metals: their electrical resistance changes slightly in a magnetic field. In 1988 French researchers discovered that the resistance of MBE-grown thin films of alternate iron and chromium was halved in a magnetic field. They called them magnetic superlattices, and the effect 'giant magnetoresistance' (GMR). GMR could lead to the next generation of magnetic recording techniques.

Top quark

US physicists may have detected the elusive top quark at the Fermilab particle physics centre near Chicago. The particle, generated by the Tevatron particle collider on 29 October 1992, is only the second candidate for the sixth quark. Confirmation of the result is awaited.

Peter Rodgers

piezoelectric effect property of some crystals (for example, quartz) to develop an electromotive force or voltage across opposite faces when subjected to a mechanical strain, and, conversely, to expand or contract in size when subjected to an electromotive force. Piezoelectric crystal ◊oscillators are used as frequency standards (for example, replacing balance wheels in watches), and for producing ◊ultrasound.

The crystals are also used in gramophone pick-ups, transducers in ultrasonics, and certain gas lighters.

pig iron or *cast iron* the quality of iron produced in a ◊blast furnace. It contains around 4% carbon plus some other impurities.

Pill, the commonly used term for the contraceptive pill, based on female hormones. The combined pill, which contains synthetic hormones similar to oestrogen and progesterone, stops the production of eggs, and makes the mucus produced by the cervix hostile to sperm. It is the most effective form of contraception apart from sterilization, being more than 99% effective.

The *minipill* or progesterone-only pill prevents implantation of a fertilized egg into the wall of the uterus. The minipill has a slightly higher failure rate, especially if not taken at the same time each day, but has fewer side effects and is considered safer for long-term use. Possible side effects of the Pill include migraine or headache and high blood pressure. More seriously, oestrogen-containing pills can slightly increase the risk of a clot forming in the blood vessels. This risk is increased in women over 35 if they smoke. Controversy surrounds other possible health effects of taking the Pill. The evidence for a link with cancer is slight (and the Pill may protect women from some forms of cancer). Once a woman ceases to take it, there is an increase in the chance of conceiving identical twins.

Piltdown man fossil skull fragments 'discovered' by Charles Dawson at Piltdown, E Sussex, England, in 1913, and believed to be the earliest European human remains until proved a hoax in 1953 (the jaw was that of an orang-utan).

The most likely perpetrator was Samuel Woodhead, a lawyer friend of Dawson, who was an amateur palaeontologist.

PIN (acronym for *p*ersonal *i*dentification *n*umber) in banking, a unique number used as a password to establish the identity of a customer using an automatic cash dispenser. The PIN is normally encoded into the magnetic strip of the customer's bank card and is known only to the customer and to the bank's computer. Before a cash dispenser will issue money or information, the customer must insert the card into a slot in the machine (so that the PIN can be read from the magnetic strip) and enter the PIN correctly at a keyboard. This effectively prevents stolen cards from being used to obtain money from cash dispensers.

pineal body or *pineal gland* a cone-shaped outgrowth of the vertebrate brain. In some lower vertebrates, it develops a rudimentary lens and retina, which show it to be derived from an eye, or pair of eyes, situated on the top of the head in ancestral vertebrates. In fishes that can change colour to

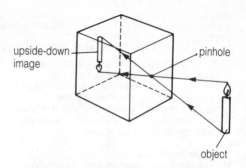

pinhole camera *A pinhole camera has no lens but can nevertheless produce a sharp inverted image because only one ray from a particular point can enter the tiny 'pinhole' aperture, and so no blurring takes place. However, the very low amount of light entering the camera also means that the film at the back must be exposed for a long time before a photographic image is produced; the camera is therefore only suitable for photographing stationary objects.*

match the background, the pineal perceives the light level and controls the colour change. In birds, the pineal detects changes in daylight and stimulates breeding behaviour as spring approaches. Mammals also have a pineal gland, but it is located deeper within the brain. It secretes a hormonelike substance, melatonin, thought to influence rhythms of activity. In humans, it is a small piece of tissue attached to the posterior wall of the third ventricle of the brain.

pingo landscape feature of tundra terrain consisting of a hemispherical mound about 30 m/100 ft high, covered with soil that is cracked at the top. The core consists of ice, probably formed from the water of a former lake. The lake that forms when such a feature melts after an ice age is also called a pingo.

pinhole camera the simplest type of camera, in which a pinhole rather than a lens is used to form an image. Light passes through the pinhole at one end of a box to form a sharp inverted image on the inside surface of the opposite end. The image is equally sharp for objects placed at different distances from the camera because only one ray can enter through a particular distance or direction can enter through the tiny pinhole, and so only one corresponding point of light will be produced on the image. A photographic film or plate fitted inside the box will, if exposed for a long time, record the image.

pinna in botany, the primary division of a ◊pinnate leaf. In mammals, the pinna is the external part of the ear.

pinnate leaf leaf that is divided up into many small leaflets, arranged in rows along either side of a midrib, as in ash trees (*Fraxinus*). It is a type of compound leaf. Each leaflet is known as a *pinna*, and where the pinnae are themselves divided, the secondary divisions are known as pinnules.

pint imperial dry or liquid measure of capacity equal to 20 fluid ounces, half a quart, one-eighth of a gallon, or 0.568 litre. In the USA, a liquid pint is equal to 0.473 litre, while a dry pint is equal to 0.550 litre.

pion or *pi meson* in physics, any of three ◊mesons (positive, negative, neutral) that play a role in binding together the neutrons and protons in the nucleus of an atom. They belong to the ◊hadron class of ◊elementary particles.

The mass of a positive or negative pion is 273 times that of an electron; the mass of a neutral pion is 264 times that of an electron.

Pioneer probe any of a series of US Solar-System space probes 1958–78. The probes *Pioneer 4–9* went into solar orbit to monitor the Sun's activity during the 1960s and early 1970s. *Pioneer 5*, launched 1960, was the first of a series to study the solar wind between the planets. *Pioneer 10*, launched March 1972, was the first probe to reach Jupiter (Dec 1973) and to leave the Solar System 1983. *Pioneer 11*, launched April 1973, passed Jupiter Dec 1974, and was the first probe to reach Saturn (Sept 1979), before also leaving the Solar System.

Pioneer 10 and *11* carry plaques containing messages from Earth in case they are found by other civilizations among the stars. Pioneer Venus probes were launched May and Aug 1978. One orbited Venus, and the other dropped three probes onto the surface. The orbiter finally burned up in the atmosphere of Venus 1992. In 1992 *Pioneer 10* was more than 8 billion km/4.4 billion mi from the Sun. Both it and *Pioneer 11* were still returning data measurements of starlight intensity to Earth.

Pioneer 1, 2, and *3,* launched 1958, were intended Moon probes, but *Pioneer 2*'s launch failed, and *1* and *3* failed to reach their target, although they did measure the ◊Van Allen radiation belts. *Pioneer 4* began to orbit the Sun after passing the Moon.

Pioneer 6 (1965), *7* (1966), *8* (1967), and *9* (1968) monitored solar activity.

pioneer species in ecology, those species that are the first to colonize and thrive in new areas. Coal tips, recently cleared woodland, and new roadsides are areas where pioneer species will quickly appear. As the habitat matures other species take over, a process known as *succession*.

Piper Alpha disaster accident aboard the North Sea oil platform Piper Alpha on 6 July 1988, in which 167 people died. The rig was devastated by a series of explosions, caused initially by a gas leakage. An official inquiry held into the disaster highlighted the vulnerability of offshore rigs.

pipette device for the accurate measurement of a known volume of liquid, usually for transfer from one container to another, used in chemistry and biology laboratories.

A pipette is a glass tube, often with an enlarged bulb, which is calibrated in one or more positions, or it may be a plastic device with an adjustable plunger, fitted with one or more disposable plastic tips.

Pisces zodiac constellation, mainly in the northern hemisphere between Aries and Aquarius, near Pegasus. It is represented by two fish tied together by their tails. The Circlet, a delicate ring of stars, marks the head of the western fish in Pisces. The constellation contains the **vernal equinox**, the point at which the Sun's path around the sky (the *ecliptic*) crosses the celestial equator. The Sun reaches this point around 21 March each year as it passes through Pisces from mid-March to late April. In astrology, the dates for Pisces are between about 19 Feb and 20 March (see ◊precession).

Piscis Austrinus or *Southern Fish* constellation of the southern hemisphere near Capricornus. Its brightest star is ◊Fomalhaut.

pistil general term for the female part of a flower, either referring to one single ◊carpel or a group of several fused carpels.

piston barrel-shaped device used in reciprocating engines (steam, petrol, diesel oil) to harness power. Pistons are driven up and down in cylinders by expanding steam or hot gases. They pass on their motion via a connecting rod and crank to a crankshaft, which turns the driving wheels. In a pump or compressor, the role of the piston is reversed, being used to move gases and liquids. See also ◊internal-combustion engine.

pitch in chemistry, a black, sticky substance, hard when cold, but liquid when hot, used for waterproofing, roofing, and paving. It is made by the destructive distillation of wood or coal tar, and has been used since antiquity for caulking wooden ships.

pitch in mechanics, the distance between the adjacent threads of a screw or bolt. When a screw is turned through one full turn it moves up or down a distance equal to the pitch of its thread. A screw thread is a simple type of machine, acting like a rolled-up inclined plane, or ramp (as may be illustrated by rolling a long paper triangle around a pencil). A screw has a ◊mechanical advantage greater than one.

pitchblende or *uraninite* brownish-black mineral, the major constituent of uranium ore, consisting mainly of uranium oxide (UO_2). It also contains some lead (the final, stable product of uranium decay) and variable amounts of most of the naturally occurring radioactive elements, which are products of either the decay or the fissioning of uranium isotopes. The uranium yield is 50–80%; it is also a source of radium, polonium, and actinium. Pitchblende was first studied by Pierre and Marie Curie, who found radium and polonium in its residues in 1898.

Pitot tube instrument that measures fluid (gas and liquid) flow. It is used to measure the speed of aircraft, and works by sensing pressure differences in different directions in the airstream. It was invented in the 1730s by the French scientist Henri Pitot (1695–1771).

The Pitot tube is a small, L-shaped tube that is inserted vertically into a flowing fluid with its open end facing upstream, thus measuring the total pressure of the fluid and indirectly the velocity of its flow.

pituitary gland major ◊endocrine gland of vertebrates, situated in the centre of the brain. The anterior lobe secretes hormones, some of which control the activities of other glands (thyroid, gonads, and adrenal cortex); others are direct-acting hormones affecting milk secretion and controlling growth. Secretions of the posterior lobe control body water balance and contraction of the uterus. The posterior lobe is regulated by nerves from the ◊hypothalamus, and thus forms a link between the nervous and hormonal systems.

pixel (acronym for *pic*ture *el*ement) single dot on a computer screen. All screen images are made up of a collection of pixels, with each pixel being either off (dark) or on (illuminated, possibly in colour). The number of pixels available determines the screen's resolution. Typical resolutions of micro-computer screens vary from 320 × 200 pixels to 640 × 480 pixels, but screens with over 1,000 × 1,000 pixels are now quite common for high-quality graphic (pictorial) displays.

The number of bits (binary digits) used to repre-sent each pixel determines how many colours it can display: a two-bit pixel can have four colours; an eight-bit (one-byte) pixel can have 256 colours. The higher the resolution of a screen and the more colours it is capable of displaying, the more memory will be needed in order to store that screen's contents.

placenta organ that attaches the developing ◊embryo or ◊fetus to the ◊uterus in placental mam-mals (mammals other than marsupials, platypuses, and echidnas). Composed of maternal and embry-onic tissue, it links the blood supply of the embryo to the blood supply of the mother, allowing the exchange of oxygen, nutrients, and waste products. The two blood systems are not in direct contact, but are separated by thin membranes, with materials diffusing across from one system to the other. The placenta also produces hormones that maintain and regulate pregnancy. It is shed as part of the afterbirth.

It is now understood that a variety of materials, including drugs and viruses, can pass across the placental membrane. HIV, the infection agent that causes ◊AIDS, can be transmitted in this way.

The tissue in plants that joins the ovary to the ovules is also called a placenta.

placer deposit detrital concentration of an econ-omically important mineral, such as gold, but also other minerals such as cassiterite, chromite, and platinum metals. The mineral grains become con-centrated during transport by water or wind because they are more dense than other detrital minerals such as quartz, and (like quartz) they are relatively resistant to chemical breakdown. Examples are the Witwatersrand gold deposits of South Africa, which are gold-and uranium-bearing conglomerates laid down by ancient rivers, and the placer tin deposits of the Malay Peninsula.

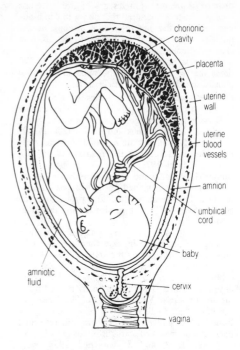

placenta *The placenta is a disc-shaped organ about 25 cm/10 in in diameter and 3 cm/1 in thick. It is connected to the fetus by the umbilical cord.*

plain or *grassland* land, usually flat, upon which grass predominates. The plains cover large areas of the Earth's surface, especially between the deserts of the tropics and the rainforests of the equator, and have rain in one season only. In such regions the climate belts move north and south during the year, bringing rainforest conditions at one time and desert conditions at another. Temper-ate plains include the North European Plain, the High Plains of the USA and Canada, and the Rus-sian Plain also known as the steppe.

Planck's constant in physics, a fundamental con-stant (symbol h) that is the energy of one quantum of electromagnetic radiation (the smallest possible 'packet' of energy; see ◊quantum theory) divided by the frequency of its radiation. Its value is 6.626196×10^{-34} joule seconds. It is named after German physicist Max Planck.

plane figure in geometry, a two-dimensional figure. All ◊polygons are plane figures.

planet large celestial body in orbit around a star, composed of rock, metal, or gas. There are nine planets in the Solar System: Mercury, Venus, Earth, Mars, Jupiter, Saturn, Neptune, Uranus, and Pluto. The inner four, called the *terrestrial planets*, are small and rocky, and include the planet Earth. The outer planets, with the exception of Pluto, are called the giant planets, large balls of rock, liquid, and gas; the largest is Jupiter, which contains more than twice as much mass as all the other planets combined. Planets do not produce

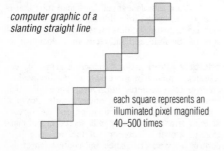

computer graphic of a slanting straight line

each square represents an illuminated pixel magnified 40–500 times

pixel

planets

planet	main constituents	atmosphere	average distance from Sun in millions of km	time for one orbit in Earth-years	diameter in thousands of km	average density if density of water is 1 unit
Mercury	rocky, ferous	–	58	0.241	4.88	5.4
Venus	rocky, ferrous	carbon dioxide	108	0.615	12.10	5.2
Earth	rocky, ferrous	nitrogen, oxygen	150	1.00	12.76	5.5
Mars	rocky	carbon dioxide	228	1.88	6.78	3.9
Jupiter	liquid hydrogen, helium	–	778	11.86	142.80	1.3
Saturn	hydrogen, helium	–	1,427	29.46	120.00	0.7
Uranus	icy, hydrogen, helium	hydrogen, helium	2,870	84.00	50.80	1.3
Neptune	icy, hydrogen, helium	hydrogen, helium	4,497	164.80	48.60	1.8
Pluto	icy, rocky	methane	5,900	248.50	2.25	about 2

light, but reflect the light of their parent star. See also ◊inferior planet, ◊superior planet.

planetarium optical projection device by means of which the motions of stars and planets are reproduced on a domed ceiling representing the sky.

The planetarium of the Heureka Science Centre, Finland, opened 1989, is the world's first to use fibre optics.

planetary nebula shell of gas thrown off by a star at the end of its life. Planetary nebulae have nothing to do with planets. They were named by British astronomer William Herschel, who thought their rounded shape resembled the disc of a planet. After a star such as the Sun has expanded to become a ◊red giant, its outer layers are ejected into space to form a planetary nebula, leaving the core as a ◊white dwarf at the centre.

planimeter simple integrating instrument for measuring the area of a regular or irregular plane surface. It consists of two hinged arms: one is kept fixed and the other is traced around the boundary of the area. This actuates a small graduated wheel; the area is calculated from the wheel's change in position.

plankton small, often microscopic, forms of plant and animal life that live in the upper layers of fresh and salt water, and are an important source of food for larger animals. Marine plankton is concentrated in areas where rising currents bring mineral salts to the surface.

plant organism that carries out ◊photosynthesis, has cellulose cell walls and complex cells, and is immobile. A few parasitic plants have lost the ability to photosynthesize but are still considered to be plants.

Plants are autotrophs, that is, they make carbohydrates from water and carbon dioxide, and are the primary producers in all food chains, so that all animal life is dependent on them. They play a vital part in the carbon cycle, removing carbon dioxide from the atmosphere and generating oxygen. The study of plants is known as botany.

Many of the lower plants (the algae and bryophytes) consist of a simple body, or thallus, on which the organs of reproduction are borne. Simplest of all are the threadlike algae, for example *Spirogyra*, which consist of a chain of cells. The seaweeds (algae) and mosses and liverworts (bryophytes) represent a further development, with simple, multicellular bodies that have specially modified areas in which the reproductive organs

are carried. Higher in the morphological scale are the ferns, club mosses, and horsetails (pteridophytes). Ferns produce leaflike fronds bearing sporangia on their undersurface in which the spores are carried. The spores are freed and germinate to produce small independent bodies carrying the sexual organs; thus the fern, like other pteridophytes and some seaweeds, has two quite separate generations in its life cycle (see ◊alternation of generations).

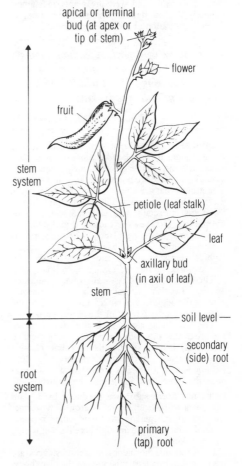

plant

The pteridophytes have supportive water-conducting tissues, which identify them as vascular plants, a group which includes all seed plants, that is the gymnosperms (conifers, yews, cycads, and ginkgo) and the angiosperms (flowering plants).

The seed plants are the largest group, and structurally the most complex. They are usually divided into three parts: root, stem, and leaves. Stems grow above or below ground. Their cellular structure is designed to carry water and salts from the roots to the leaves in the ◊xylem, and sugars from the leaves to the roots in the ◊phloem. The leaves manufacture the food of the plant by means of photosynthesis, which occurs in the ◊chloroplasts they contain. Flowers and cones are modified leaves arranged in groups, enclosing the reproductive organs from which the fruits and seeds result.

Cooksonia pertoni, a tiny fossil plant only a few centimetres high and 400 million years old, is considered to be the ancestor of all higher (vascular) plants. This relationship was only recently confirmed by fossil evidence found in Shropshire even though the first specimens were originally found 1937. The fossil demonstrates many features critical to higher plants, most notably small strengthened tubes which allowed the plants to support themselves, grow upright, and transport water to the cells at the top of the their branches.

The earth brought forth vegetation, plants yielding seed according to their own kinds, and trees bearing fruit in which is their seed, each according to its kind.

On **plants** The Bible, Genesis 1:12

---

plant classification taxonomy or classification of plants. Originally the plant kingdom included bacteria, diatoms, dinoflagellates, fungi, and slime moulds, but these are not now thought of as plants. The groups that are always classified as plants are the bryophytes (mosses and liverworts), pteridophytes (ferns, horsetails, and club mosses), gymnosperms (conifers, yews, cycads, and ginkgos), and angiosperms (flowering plants). The angiosperms are split into monocotyledons (for example, orchids, grasses, lilies) and dicotyledons (for example, oak, buttercup, geranium, and daisy).

The basis of plant classification was established by the Swedish naturalist, Carolus Linnaeus. Among the angiosperms, it is largely based on the number and arrangement of the flower parts.

The unicellular algae, such as *Chlamydomonas*, are often now put with the protists (single-celled organisms) instead of the plants. Some classification schemes even classify the multicellular algae (seaweeds and freshwater weeds) in a new kingdom, the Protoctista, along with the protists.

plant hormone substance produced by a plant that has a marked effect on its growth, flowering, leaf fall, fruit ripening, or some other process. Examples include ◊auxin, ◊gibberellin, ◊ethylene, and ◊cytokinin.

Unlike animal hormones, these substances are not produced by a particular area of the plant body, and they may be less specific in their effects. It has

therefore been suggested that they should not be described as hormones at all.

plant propagation production of plants. Botanists and horticulturalists can use a wide variety of means for propagating plants. There are the natural techniques of ◊vegetative reproduction, together with ◊cuttings, ◊grafting, and ◊micropropagation. The range is wide because most plant tissue, unlike animal tissue, can give rise to the complete range of tissue types within a particular species.

Plaskett's Star the most massive ◊binary star known, consisting of two supergiants of about 40 and 50 solar masses, orbiting each other every 14.4 days. Plaskett's star lies in the constellation Monoceros and is named after the Canadian astronomer John S Plaskett (1865–1941), who identified it as a binary star and discovered its massive nature in 1922.

plasma in biology, the liquid part of the ◊blood.

plasma in physics, an ionized gas produced at extremely high temperatures, as in the Sun and other stars, which contains positive and negative charges in approximately equal numbers. It is a good electrical conductor. In thermonuclear reactions the plasma produced is confined through the use of magnetic fields.

plasmid small, mobile piece of ◊DNA found in bacteria and used in ◊genetic engineering. Plasmids are separate from the bacterial chromosome but still multiply during cell growth. Their size ranges from 3% to 20% of the size of the chromosome. There is usually only one copy of a single plasmid per cell, but occasionally several are found. Some plasmids carry 'fertility genes' that enable them to move from one bacterium to another and transfer genetic information between strains. Plasmid genes determine a wide variety of bacterial properties including resistance to antibiotics and the ability to produce toxins.

plasmolysis the separation of the plant cell cytoplasm from the cell wall as a result of water loss. As moisture leaves the vacuole the total volume of the cytoplasm decreases while the cell itself, being rigid, hardly changes. Plasmolysis is induced in the laboratory by immersing a plant cell in a strongly saline or sugary solution, so that water is lost by osmosis. Plasmolysis is unlikely to occur in the wild except in severe conditions.

plaster of Paris form of calcium sulphate, obtained from gypsum; it is mixed with water for making casts and moulds.

plastic any of the stable synthetic materials that are fluid at some stage in their manufacture, when they can be shaped, and that later set to rigid or semi-rigid solids. Plastics today are chiefly derived from petroleum. Most are polymers, made up of long chains of identical molecules.

Processed by extrusion, injection-moulding, vacuum-forming and compression, plastics emerge in consistencies ranging from hard and inflexible to soft and rubbery. They replace an increasing number of natural substances, being lightweight, easy to clean, durable, and capable of being rendered very strong—for example, by the addition of carbon fibres—for building aircraft and other engineering projects.

Thermoplastics soften when warmed, then reharden as they cool. Examples of thermoplastics include polystyrene, a clear plastic used in kitchen utensils or (when expanded into a 'foam' by gas injection) in insulation and ceiling tiles; polyethylene (polyethene), used for containers and wrapping; and polyvinyl chloride (PVC), used for drainpipes, floor tiles, audio discs, shoes, and handbags.

Thermosets remain rigid once set, and do not soften when warmed. They include bakelite, used in electrical insulation and telephone receivers; epoxy resins, used in paints and varnishes, to laminate wood, and as adhesives; polyesters, used in synthetic textile fibres and, with fibreglass reinforcement, in car bodies and boat hulls; and polyurethane, prepared in liquid form as a paint or varnish, and in foam form for upholstery and in lining materials (where it may be a fire hazard). One group of plastics, the silicones, are chemically inert, have good electrical properties, and repel water. Silicones find use in silicone rubber, paints, electrical insulation materials, laminates, waterproofing for walls, stain-resistant textiles, and cosmetics.

Shape-memory polymers are plastics that can be crumpled or flattened and will resume their original shape when heated. They include transpolyisoprene and polynorbornene. The initial shape is determined by heating the polymer to over 35°C/95°F and pouring it into a metal mould. The shape can be altered with boiling water and the substance solidifies again when its temperature falls below 35°C/95°F.

Biodegradable plastics are increasingly in demand: Biopol was developed in 1990. Soil microorganisms are used to build the plastic in their cells from carbon dioxide and water (it constitutes 80% of their cell tissue). The unused parts of the microorganism are dissolved away by heating in water. The discarded plastic can be placed in landfill sites where it breaks back down into carbon dioxide and water. It costs three to five times as much as ordinary plastics to produce. Another plastic digested by soil microorganisms is polyhydroxybutyrate (PHB), which is made from sugar.

plastid general name for a cell ◊organelle of plants that is enclosed by a double membrane and contains a series of internal membranes and vesicles. Plastids contain ◊DNA and are produced by division of existing plastids. They can be classified into two main groups: the *chromoplasts*, which contain pigments such as carotenes and chlorophyll, and the *leucoplasts*, which are colourless; however, the distinction between the two is not always clear-cut.

◊Chloroplasts are the major type of chromoplast. They contain chlorophyll, are responsible for the green coloration of most plants, and perform ◊photosynthesis. Other chromoplasts give flower petals and fruits their distinctive colour. Leucoplasts are food-storage bodies and include amyloplasts, found in the roots of many plants, which store large amounts of starch.

plate according to plate tectonics, one of a number of slabs of solid rock, about a hundred kilometres thick and often several thousands of kilometres across, making up the Earth's surface. Together, the plates make up the ◊lithosphere.

Plates are made up of two types of crustal material: oceanic crust (sima) and continental crust (sial), both of which are underlain by a solid layer of the ◊mantle. *Oceanic crust* is heavy and consists largely of basalt. It is formed at constructive margins. *Continental crust* is less dense and is often rich in granite. It is made up of volcanic islands and folded sediments, and is usually associated with destructive margins.

plateau elevated area of fairly flat land, or a mountainous region in which the peaks are at the same height. An *intermontane plateau* is one surrounded by mountains. A *piedmont plateau* is one that lies between the mountains and low-lying land. A *continental plateau* rises abruptly from low-lying land or the sea. Examples are the Tibetan Plateau and the Massif Central in France.

platelet tiny 'cell' found in the blood, which helps it to clot. Platelets are not true cells, but membrane-bound cell fragments that bud off from large cells in the bone marrow.

plate tectonics theory formulated in the 1960s to explain the phenomena of ◊continental drift and seafloor spreading, and the formation of the major physical features of the Earth's surface. The Earth's outermost layer is regarded as a jigsaw of rigid major and minor plates up to 100 km/62 mi thick, which move relative to each other, probably under the influence of convection currents in the mantle beneath. Major landforms occur at the margins of the plates, where plates are colliding or moving apart—for example, volcanoes, fold mountains, ocean trenches, and ocean ridges.

constructive margin Where two plates are moving apart from each other, molten rock from the mantle wells up in the space left between the plates and hardens to form new crust, usually in the form of an ocean ridge (such as the ◊Mid-Atlantic Ridge). The newly formed crust accumulates on either side of the ocean ridge, causing the seafloor to spread—for example, the floor of the Atlantic Ocean is growing by 5 cm/2 in each year because of the upwelling of new material at the Mid-Atlantic Ridge.

destructive margin Where two plates are moving towards each other, the denser of the two plates may be forced under the other into a region called the subduction zone. The descending plate melts to form a body of magma, which may then rise to the surface through cracks and faults to form volcanoes. If the two plates consist of more buoyant continental crust, subduction does not occur. Instead, the crust crumples gradually to form ranges of young fold mountains, such as the

the plates of the Earth's lithosphere

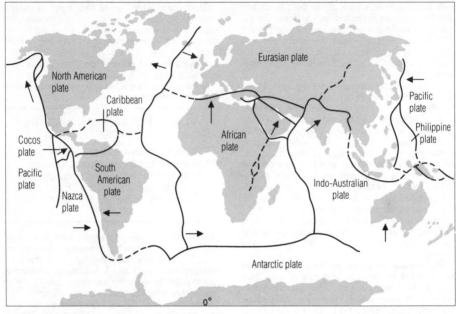

——— plate margins   – – – – uncertain plate margins

plate tectonics

Himalayas in Asia, the Andes in South America, and the Rockies in North America.

conservative margin Sometimes two plates will slide past each other—an example is the San Andreas Fault, California, where the movement of the plates sometimes takes the form of sudden jerks, causing the earthquakes common in the San Francisco–Los Angeles area. Most of the earthquake and volcano zones of the world are, in fact, found in regions where two plates meet or are moving apart.

The concept of continental drift was first put forward in 1915 by the German geophysicist Alfred Wegener; plate tectonics was formulated by Canadian geophysicist John Tuzo Wilson (1908–1993) 1965 and has gained widespread acceptance among earth scientists.

platinum (Spanish *platina* 'little silver') heavy, soft, silver-white, malleable and ductile, metallic element, symbol Pt, atomic number 78, relative atomic mass 195.09. It is the first of a group of six metallic elements (platinum, osmium, iridium, rhodium, ruthenium, and palladium) that possess similar traits, such as resistance to tarnish, corrosion, and attack by acid, and that often occur as free metals (◊native metals). They often occur in natural alloys with each other, the commonest of which is osmiridium. Both pure and as an alloy, platinum is used in dentistry, jewellery, and as a catalyst.

playa temporary lake in a region of interior drainage. Such lakes are common features in arid desert basins fed by intermittent streams. The streams bring dissolved salts to the lakes, and when the lakes shrink during dry spells, the salts precipitate as evaporite deposits.

Pleiades in astronomy, a star cluster about 400 light years away in the constellation Taurus, representing the Seven Sisters of Greek mythology. Its brightest stars (highly luminous, blue-white giants only a few million years old) are visible to the naked eye, but there are many fainter ones.

The stars of the Pleiades are still surrounded by traces of the reflection ◊nebula from which they formed, visible on long-exposure photographs.

THE PLEIADES: MANY MORE SISTERS

The Pleiades (or Seven Sisters) cluster of stars seems to contain seven stars, but with the aid of a telescope about 400 stars can be seen.

---

pleiotropy process whereby a given gene influences several different observed characteristics of an organism. For example, in the fruit fly *Drosophila* the vestigial gene reduces the size of wings, modifies the halteres, changes the number of egg strings in the ovaries, and changes the direction of certain bristles.

Many human syndromes are caused by pleiotropic genes, for example Marfan's syndrome where the slender physique, hypermobility of the joints, elongation of the limbs, dislocation of the lens, and susceptibility to heart disease are all caused by one gene.

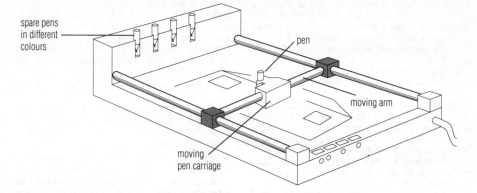

spare pens in different colours

pen

moving arm

moving pen carriage

plotter

Pleistocene first epoch of the Quaternary period of geological time, beginning 1.64 million years ago and ending 10,000 years ago. The polar ice caps were extensive and glaciers were abundant during the ice age of this period, and humans evolved into modern *Homo sapiens sapiens* about 100,000 years ago.

Plesetsk rocket-launching site 170 km/105 mi S of Archangel, Russia. From 1966 the USSR launched artificial satellites from here, mostly military.

plesiosaur prehistoric carnivorous marine reptile of the Jurassic and Cretaceous periods, which reached a length of 12 m/36 ft, and had a long neck and paddlelike limbs. The pliosaurs evolved from the plesiosaurs.

Plimsoll line loading mark painted on the hull of merchant ships, first suggested by English politician Samuel Plimsoll. It shows the depth to which a vessel may be safely (and legally) loaded.

Pliocene ('almost recent') fifth and last epoch of the Tertiary period of geological time, 5.2–1.64 million years ago. The earliest hominid, the humanlike ape 'australopithecines', evolved in Africa.

pliosaur prehistoric carnivorous marine reptile, descended from the plesiosaurs, but with a shorter neck, and longer head and jaws. It was approximately 5 m/15 ft long. In 1989 the skeleton of one of a previously unknown species was discovered in northern Queensland, Australia. A hundred million years ago, it lived in the sea which once covered the Great Artesian Basin.

plotter or *graph plotter* device that draws pictures or diagrams under computer control. Plotters are often used for producing business charts, architectural plans, and engineering drawings. *Flatbed plotters* move a pen up and down across a flat drawing surface, whereas *roller plotters* roll the drawing paper past the pen as it moves from side to side.

Plough, the in astronomy, a popular name for the most prominent part of the constellation ◊Ursa Major.

plough agricultural implement used for tilling the soil. The plough dates from about 3500 BC, when oxen were used to pull a simple wooden blade, or ard. In about 500 BC the iron ploughshare came into use. By about AD 1000 horses as well as oxen were being used to pull wheeled ploughs, equipped with a ploughshare for cutting a furrow, a blade for forming the walls of the furrow (called a coulter), and a mouldboard to turn the furrow. In the 18th century an innovation introduced by Robert Ransome (1753–1830), led to a reduction in the number of animals used to draw a plough: from 8–12 oxen, or 6 horses, to a 2-or 4-horse plough.

Steam ploughs came into use in some areas in the 1860s, superseded half a century later by tractor-drawn ploughs. The modern plough consists of many 'bottoms', each comprising a curved ploughshare and angled mouldboard. The bottom is designed so that it slices into the ground and turns the soil over.

plucking in earth science, a process of glacial erosion. Water beneath a glacier will freeze fragments of loose rock to the base of the ice. When the ice moves, the rock fragment is 'plucked' away from the underlying bedrock. Plucking is thought to be responsible for the formation of steep, jagged slopes such as the backwall of the corrie and the downslope-side of the roche moutonnee.

plumbago alternative name for the mineral ◊graphite.

plumule part of a seed embryo that develops into the shoot, bearing the first true leaves of the plant. In most seeds, for example the sunflower, the plumule is a small conical structure without any leaf structure. Growth of the plumule does not occur until the ◊cotyledons have grown above ground. This is ◊epigeal germination. However, in seeds such as the broad bean, a leaf structure is visible on the plumule in the seed. These seeds develop by the plumule growing up through the soil with the cotyledons remaining below the surface. This is known as ◊hypogeal germination.

Pluto in astronomy, the smallest and, usually, outermost planet of the Solar System. The existence of Pluto was predicted by calculation by Percival Lowell and the planet was located by Clyde Tombaugh 1930, both US astronomers. It orbits the Sun every 248.5 years at an average distance of 5.8 billion km/3.6 billion mi. Its highly elliptical orbit occasionally takes it within the orbit of Neptune, as in 1979–99. Pluto has a diameter of about 2,300 km/1,400 mi, and a mass about 0.002 of that of Earth. It is of low density, composed of rock and ice, with frozen methane on its surface and a thin atmosphere.

Charon, Pluto's moon, was discovered 1978 by James Walter Christy (1938–). It is about 1,200 km/750 mi in diameter, half the size of Pluto, making it the largest moon in relation to its parent planet in the Solar System. It orbits about 20,000 km/12,500 mi from the planet's centre every 6.39 days—the same time that Pluto takes to spin on its axis. Some astronomers have suggested that Pluto was a former moon of Neptune that escaped, but it is more likely that it was an independent body that was captured.

PLUTO: DISTANT PLANET

A supersonic aircraft such as Concorde would take 310 years to travel from the Earth to Pluto at its usual cruising speed of 2,170 kph/1,350 mph.

---

plutonic rock igneous rock derived from magma that has cooled and solidified deep in the crust of the Earth; granites and gabbros are examples of plutonic rocks.

plutonium silvery-white, radioactive, metallic element of the ◊actinide series, symbol Pu, atomic number 94, relative atomic mass 239.13. It occurs in nature in minute quantities in ◊pitchblende and other ores, but is produced in quantity only synthetically. It has six allotropic forms (see ◊allotropy) and is one of three fissile elements (elements capable of splitting into other elements—the others are thorium and uranium). The element has awkward physical properties and is the most toxic substance known.

Because Pu-239 is so easily synthesized from abundant uranium, it has been produced in large quantities by the weapons industry.

It has a long half-life (24,000 years) during which time it remains highly toxic. Plutonium is dangerous to handle, difficult to store, and impossible to dispose of. It was first synthesized in 1940 by Glenn Seaborg and his team at the University of California at Berkeley, by bombarding uranium with deuterons; this was the second transuranic element to be synthesized, the first being neptunium.

plywood manufactured panel of wood widely used in building. It consists of several thin sheets, or plies, of wood, glued together with the grain (direction of the wood fibres) of one sheet at right angles to the grain of the adjacent plies. This construction gives plywood equal strength in every direction.

pneumatic drill drill operated by compressed air, used in mining and tunnelling, for drilling shot holes (for explosives), and in road repairs for breaking up pavements. It contains an air-operated piston that delivers hammer blows to the drill bit many times a second. The French engineer Germain Sommeiller (1815–1871) developed the pneumatic drill 1861 for tunnelling in the Alps.

pneumatophore erect root that rises up above

the soil or water and promotes ◊gas exchange. Pneumatophores, or breathing roots, are formed by certain swamp-dwelling trees, such as mangroves, since there is little oxygen available to the roots in waterlogged conditions. They have numerous pores or ◊lenticels over their surface, allowing gas exchange.

pneumothorax the presence of air in the pleural cavity, between a lung and the chest wall. It may be due to a penetrating injury of the lung or to lung disease, or it may arise without apparent cause. Prevented from expanding normally, the lung is liable to collapse.

p–n junction diode in electronics, a two-terminal semiconductor device that allows electric current to flow in only one direction, the **forward-bias** direction. A very high resistance prevents current flow in the opposite, or **reverse-bias**, direction. It is used as a ◊rectifier, converting alternating current (AC) to direct current (DC).

The diode is cut from a single crystal of a semiconductor (such as silicon) to which special impurities have been added during manufacture such that the crystal is now composed of two distinct regions. One region contains semiconductor material of the *p*-type, which contains more *p*ositive charge carriers than negative; the other contains material of the *n*-type, which has more *n*egative charge carriers than positive. The region of contact between the two types is called the *p–n* junction, and it is this that acts as the barrier preventing current from flowing, in conventional current terms, from the *n*-type to the *p*-type (in the reverse-bias direction).

pod in botany, a type of fruit that is characteristic of legumes (plants belonging to the Leguminosae family), such as peas and beans. It develops from a single ◊carpel and splits down both sides when ripe to release the seeds.

In certain species the seeds may be ejected explosively due to uneven drying of the fruit wall, which sets up tensions within the fruit. In agriculture, the name 'legume' is used for the crops of the pea and bean family. 'Grain legume' refers to those that are grown mainly for their dried seeds, such as lentils, chick peas, and soya beans.

podzol or *podsol* type of light-coloured soil found predominantly under coniferous forests and on moorlands in cool regions where rainfall exceeds evaporation. The constant downward movement of water leaches nutrients from the upper layers, making podzols poor agricultural soils.

The leaching of minerals such as iron, lime, and alumina leads to the formation of a bleached zone, often also depleted of clay.

These minerals can accumulate lower down the soil profile to form a hard, impermeable layer which restricts the drainage of water through the soil.

poikilothermy the condition in which an animal's body temperature is largely dependent on the temperature of the air or water in which it lives. It is characteristic of all animals except birds and mammals, which maintain their body temperatures by ◊homeothermy (they are 'warm-blooded').

Poikilotherms have behavioural means of temperature control; they can warm themselves up by basking in the sun, or shivering, and can cool themselves down by sheltering from the Sun under a rock or by bathing in water.

Poikilotherms are often referred to as 'cold-blooded animals', but this is not really correct: their internal temperatures, regulated by behavioural means, are often as high as those of birds and mammals during the times they need to be active for feeding and reproductive purposes, and may be higher, for example in very hot climates. The main difference is that their body temperatures fluctuate more than those of homeotherms.

point of sale in business premises, the point where a sale is transacted, for example, a super-market checkout. In conjunction with electronic funds transfer, point of sale is part of the termin-ology of 'cashless shopping', enabling buyers to transfer funds directly from their bank accounts to the shop's (see ◊EFTPOS).

point-of-sale terminal (POS terminal) computer terminal used in shops to input and output data at the point where a sale is transacted; for example, at a supermarket checkout. The POS terminal inputs information about the identity of each item sold, retrieves the price and other details from a central computer, and prints out a fully itemized receipt for the customer. It may also input sales data for the shop's computerized stock-control system.

A POS terminal typically has all the facilities of a normal till, including a cash drawer and a sales register, plus facilities for the direct capture of sales information—commonly, a laser scanner for reading bar codes. It may also be equipped with a device to read customers' bank cards, so that pay-ment can be transferred electronically from the customers' bank accounts to the shop's (see ◊EFTPOS).

poise cgs unit (symbol P) of dynamic ◊viscosity (the property of liquids that determines how readily they flow). It is equal to one dyne-second per square centimetre. For most liquids the centipoise (one hundredth of a poise) is used. Water at 20°C/68°F has a viscosity of 1.002 centipoise.

Poiseuille's formula in physics, a relationship describing the rate of flow of a fluid through a narrow tube. For a capillary (very narrow) tube of length l and radius r with a pressure difference p between its ends, and a liquid of ◊viscosity η, the velocity of flow expressed as the volume per second is $\pi pr^4/8l\ \eta$. The formula was devised 1843 by French physicist Jean Louis Poiseuille (1799–1869).

polar coordinates in mathematics, a way of defining the position of a point in terms of its distance r from a fixed point (the origin) and its angle θ to a fixed line or axis. The coordinates of the point are (r,θ).

Often the angle is measured in ◊radians, rather than degrees. The system is useful for defining positions on a plane in programming the operations of, for example, computer-controlled cloth-and metal-cutting machines.

Polaris or *Pole Star* or *North Star* the bright star closest to the north celestial pole, and the brightest star in the constellation Ursa Minor. Its position is indicated by the 'pointers' in Ursa

Major. Polaris is a yellow ◊supergiant about 500 light years away.

It currently lies within 1° of the north celestial pole; ◊precession (Earth's axial wobble) will bring Polaris closest to the celestial pole (less than 0.5° away) about AD 2100. It is also known as *Alpha Ursae Minoris*.

polarized light light in which the electromagnetic vibrations take place in one particluar direction. In ordinary (unpolarized) light, the electric and magnetic fields vibrate in all directions perpendicu-lar to the direction of propagation. After reflection from a polished surface or transmission through certain materials (such as Polaroid), the electric and magnetic fields are confined to one direction, and the light is said to be *plane polarized*. In *circularly polarized* and *elliptically polarized* light, the magnetic and electric fields are confined to one direction, but the direction rotates as the light propagates. Polarized light is used to test the strength of sugar solutions, to measure stresses in transparent materials, and to prevent glare.

polarography electrochemical technique for the analysis of oxidizable and reducible compounds in solution. It involves the diffusion of the substance to be analysed onto the surface of a small electrode, usually a bead of mercury, where oxidation or reduction occurs at an electric potential character-istic of that substance.

Polaroid camera instant-picture camera, invented by Edwin Land in the USA 1947. The original camera produced black-and-white prints in about one minute. Modern cameras can produce black-and-white prints in a few seconds, and colour prints in less than a minute. An advanced model has automatic focusing and exposure. It ejects a piece of film on paper immediately after the picture has been taken.

The film consists of layers of emulsion and colour dyes together with a pod of chemical devel-oper. When the film is ejected the pod bursts and processing occurs in the light, producing a paper-backed print.

polar reversal changeover in polarity of the Earth's magnetic poles. Studies of the magnetism retained in rocks at the time of their formation have shown that in the past the Earth's north mag-netic pole repeatedly became the south magnetic pole, and vice versa.

Polar reversal seems to be relatively frequent, taking place three or four times every million years. The last occasion was 700,000 years ago. Distinc-tive sequences of magnetic reversals are used in dating rock formations. Movements of the Earth's molten core are thought to be responsible for both the Earth's magnetic field and its reversal.

It is calculated that in about 1,200 years' time the north magnetic pole will become the south magnetic pole.

polder area of flat reclaimed land that used to be covered by a river, lake, or the sea. Polders have been artificially drained and protected from flood-ing by building dykes. They are common in the Netherlands, where the total land area has been increased by nearly one-fifth since AD 1200. Such schemes as the Zuider Zee project have provided some of the best agricultural land in the country.

pole either of the geographic north and south points of the axis about which the Earth rotates. The geographic poles differ from the magnetic poles, which are the points towards which a freely suspended magnetic needle will point.

In 1985 the magnetic north pole was some 350 km/218 mi NW of Resolute Bay, Northwest Territories, Canada. It moves northwards about 10 km/6 mi each year, although it can vary in a day about 80 km/50 mi from its average position. It is relocated every decade in order to update navigational charts. It is thought that periodic changes in the Earth's core cause a reversal of the magnetic poles (see ◊polar reversal, ◊magnetic field). Many animals, including migrating birds and fish, are believed to orientate themselves partly using the Earth's magnetic field. A permanent scientific base collects data at the South Pole.

Pole Star ◊Polaris, the northern pole star. There is no bright star near the southern celestial pole.

pollarding type of pruning whereby the young branches of a tree are severely cut back, about 2–4 m/6–12 ft above the ground, to produce a stumplike trunk with a rounded, bushy head of thin new branches.

It is often practised on willows, where the new branches or 'poles' are cut at intervals of a year or more, and used for fencing and firewood. Pollarding is also used to restrict the height of many street trees. See also ◊coppicing.

pollen the grains of ◊seed plants that contain the male gametes. In ◊angiosperms (flowering plants) pollen is produced within ◊anthers; in most ◊gymnosperms (cone-bearing plants) it is produced in male cones. A pollen grain is typically yellow and, when mature, has a hard outer wall. Pollen of insect-pollinated plants (see ◊pollination) is often sticky and spiny and larger than the smooth, light grains produced by wind-pollinated species.

The outer wall of pollen grains from both insect-pollinated and wind-pollinated plants is often elaborately sculptured with ridges or spines so distinctive that individual species or genera of plants can be recognized from their pollen. Since pollen is extremely resistant to decay, useful information on the vegetation of earlier times can be gained from the study of fossil pollen. The study of pollen grains is known as palynology.

pollen tube outgrowth from a pollen grain that grows towards the ◊ovule, following germination of the grain on the ◊stigma. In ◊angiosperms (flowering plants) the pollen tube reaches the ovule by growing down through the ◊style, carrying the male gametes inside. The gametes are discharged into the ovule and one fertilizes the egg cell.

pollination the process by which pollen is transferred from one plant to another. The male ◊gametes are contained in pollen grains, which must be transferred from the anther to the stigma in ◊angiosperms (flowering plants), and from the male cone to the female cone in ◊gymnosperms (cone-bearing plants). Fertilization (not the same as pollination) occurs after the growth of the pollen tube to the ovary. Self-pollination occurs when pollen is transferred to a stigma of the same flower, or to another flower on the same plant; cross-pollination occurs when pollen is transferred to another plant. This involves external pollen-carrying agents, such as wind (see ◊anemophily), water, insects, birds (see ◊ornithophily), bats, and other small mammals.

Animal pollinators carry the pollen on their bodies and are attracted to the flower by scent, or by the sight of the petals. Most flowers are adapted for pollination by one particular agent only. Those that rely on animals generally produce nectar, a sugary liquid, or surplus pollen, or both, on which the pollinator feeds. Thus the relationship between pollinator and plant is an example of mutualism, in which both benefit. However, in some plants the pollinator receives no benefit (as in ◊pseudocopulation), while in others, nectar may be removed by animals that do not effect pollination.

polling in computing, a technique for transferring data from a terminal to the central computer of a ◊multiuser system. The computer automatically makes a connection with each terminal in turn, interrogates it to check whether it is holding data for transmission, and, if it is, collects the data.

pollinium group of pollen grains that is transported as a single unit during pollination. Pollinia are common in orchids.

polluter-pays principle the idea that whoever causes pollution is responsible for the cost of repairing any damage. The principle is accepted in British law but has in practice often been ignored; for example, farmers causing the death of fish through slurry pollution have not been fined the full costs of restocking the river.

pollution the harmful effect on the environment of by-products of human activity, principally industrial and agricultural processes—for example, noise, smoke, car emissions, chemical and radioactive effluents in air, seas, and rivers, pesticides, radiation, sewage (see ◊sewage disposal), and household waste. Pollution contributes to the ◊greenhouse effect.

Pollution control involves higher production costs for the industries concerned, but failure to implement adequate controls may result in irreversible environmental damage and an increase in the incidence of diseases such as cancer. See also ◊nuclear safety.

Natural disasters may also cause pollution; volcanic eruptions, for example, cause ash to be ejected into the atmosphere and deposited on land surfaces.

POLLUTION: DIRTY CITIES

A small town can produce over 5,000 kg/ 11,000 lb of air pollutants—dust, smoke, soot and harmful gases—in a day. After directing traffic in Tokyo, policemen need to have two hours' oxygen treatment to combat the effects of pollution. In London, over 3 billion l/0.6 billion gall of sewage is produced each day. This is about 365 l/ 80 gall for every man, woman, and child in the city.

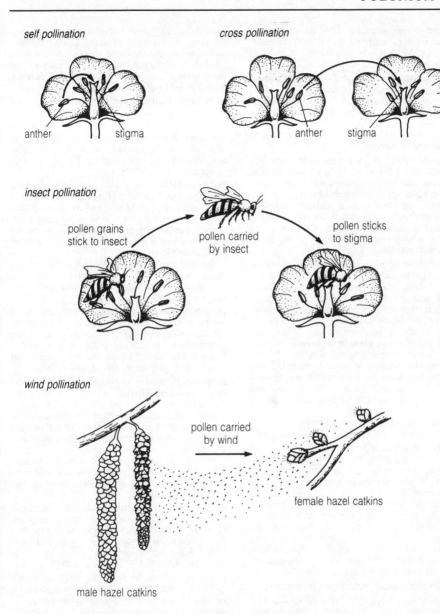

self pollination

cross pollination

anther   stigma   anther   stigma

insect pollination

pollen grains stick to insect

pollen carried by insect

pollen sticks to stigma

wind pollination

pollen carried by wind

female hazel catkins

male hazel catkins

pollen *Pollination, the process by which pollen grains transfer their male nuclei (gametes) to the ovary of a flower. (1) The pollen grains land on the stigma and (2) form a pollen tube that (3) grows down into the ovary. The male nuclei travel along the pollen tube.*

In the UK in 1987 air pollution caused by carbon monoxide emission from road transport was measured at 5.26 million tonnes. In Feb 1990 the UK had failed to apply 21 European Community Laws on air and water pollution and faced prosecution before the European Court of Justice on 31 of the 160 EC directives in force.

The existence of 1,300 toxic waste tips in the UK in 1990 posed a considerable threat for increased water pollution.

Pollux or *Beta Geminorum* the brightest star in the constellation Gemini (the twins), and the 17th brightest star in the sky. Pollux is a yellowish star with a true luminosity 45 times that of the Sun. It is 35 light years away.

polonium radioactive, metallic element, symbol Po, atomic number 84, relative atomic mass 210. Polonium occurs in nature in small amounts and was isolated from ◊pitchblende. It is the element having the largest number of isotopes (27) and is 5,000 times as radioactive as radium, liberating considerable amounts of heat. It was the first element to have its radioactive properties recognized and investigated.

Polonium was isolated in 1898 from the pitch-blende residues analysed by French scientists Pierre and Marie Curie, and named after Marie Curie's native Poland.

polychlorinated biphenyl (PCB) any of a group of chlorinated isomers of biphenyl $(C_6H_5)_2$. They are dangerous industrial chemicals, valuable for their fire-resisting qualities. They constitute an environmental hazard because of their persistent toxicity. Since 1973 their use has been limited by international agreement.

polyester synthetic resin formed by the ◊conden-sation of polyhydric alcohols (alcohols containing more than one hydroxyl group) with dibasic acids (acids containing two replaceable hydrogen atoms). Polyesters are thermosetting ◊plastics, used in making synthetic fibres, such as Dacron and Tery-lene, and constructional plastics. With glass fibre added as reinforcement, polyesters are used in car bodies and boat hulls.

polyethylene or *polyethene* polymer of the gas ethylene (technically called ethene, C_2H_4). It is a tough, white, translucent, waxy thermoplastic (which means it can be repeatedly softened by heat-ing). It is used for packaging, bottles, toys, wood preservation, electric cable, pipes and tubing.

Polyethylene is produced in two forms: low-density polyethylene, made by high-pressure poly-merization of ethylene gas, and high-density poly-ethylene, which is made at lower pressure by using catalysts. This form, first made by German chemist Karl Ziegler, is more rigid at low temperatures and softer at higher temperatures than the low-density type.

In the UK it is better known under the trade-mark Polythene.

polygon in geometry, a plane (two-dimensional) figure with three or more straight-line sides. Common polygons have names which define the number of sides (for example, triangle, quadrilat-eral, pentagon).

These are all convex polygons, having no interior angle greater than 180°. The sum of the internal angles of a polygon having n sides is given by the formula $(2n - 4) \times 90°$; therefore, the more sides a polygon has, the larger the sum of its internal angles and, in the case of a convex polygon, the more closely it approximates to a circle.

polygraph technical name for a ◊lie detector.

polyhedron in geometry, a solid figure with four or more plane faces. The more faces there are on a polyhedron, the more closely it approximates to a sphere. Knowledge of the properties of polyhedra is needed in crystallography and stereochemistry to determine the shapes of crystals and molecules.

There are only five types of regular polyhedron (with all faces the same size and shape), as was deduced by early Greek mathematicians; they are the tetrahedron (four equilateral triangular faces), cube (six square faces), octahedron (eight equi-lateral triangles), dodecahedron (12 regular penta-gons) and icosahedron (20 equilateral triangles).

polymer compound made up of a large long-chain or branching matrix composed of many repeated simple units (*monomers*). There are many poly-mers, both natural (cellulose, chitin, lignin) and synthetic (polyethylene and nylon, types of plastic). Synthetic polymers belong to two groups: thermo-softening and thermosetting (see ◊plastic).

The size of the polymer matrix is determined by the amount of monomer used; it therefore does not form a molecule of constant molecular size or mass.

polymerase chain reaction phenomenon used in ◊gene amplification.

polymerase chain reaction (PCR) a technique developed during the 1980s to clone short strands of DNA from the ◊genome of an organism. The aim is to produce enough of the DNA to be able to sequence and identify it.

polymerization chemical union of two or more (usually small) molecules of the same kind to form a new compound. *Addition polymerization* pro-duces simple multiples of the same compound. *Condensation polymerization* joins molecules together with the elimination of water or another small molecule.

Addition polymerization uses only a single mono-mer (basic molecule); condensation polymerization may involve two or more different monomers (*co-polymerization*).

polymorphism in genetics, the coexistence of several distinctly different types in a ◊population (groups of animals of one species). Examples include the different blood groups in humans, dif-ferent colour forms in some butterflies, and snail shell size, length, shape, colour, and stripiness.

polymorphism in mineralogy, the ability of a sub-stance to adopt different internal structures and external forms, in response to different conditions of temperature and/or pressure. For example, dia-mond and graphite are both forms of the element carbon, but they have very different properties and appearance.

Silica (SiO_2) also has several polymorphs, includ-ing quartz, tridymite, cristobalite, and stishovite (the latter a very high pressure form found in met-eoritic impact craters).

polynomial in mathematics, an algebraic expression that has one or more ◊variables (denoted by letters). A polynomial of degree one, that is, whose highest ◊power of x is 1, as in $2x + 1$, is called a linear polynomial; $3x^2 + 2x + 1$ is quadratic; $4x^3 + 3x^2 + 2x + 1$ is cubic.

polyp in zoology, the sedentary stage in the life cycle of a coelenterate (such as a ◊coral or jellyfish), the other being the free-swimming *medusa*.

polypeptide long-chain ◊peptide.

polyploid in genetics, possessing three or more sets of chromosomes in cases where the normal complement is two sets (◊diploid). Polyploidy arises spontaneously and is common in plants (mainly among flowering plants), but rare in animals. Many crop plants are natural polyploids, including wheat, which has four sets of chromosomes per cell (durum wheat) or six sets (common wheat). Plant breeders can induce the formation of polyploids by treatment with a chemical, colchicine.

Matings between polyploid individuals and normal diploid ones are invariably sterile. Hence, an individual that develops polyploidy through a genetic aberration can initially only reproduce veg-etatively, by parthenogenesis, or by self-fertilization

addition of ethene molecules to form polyethene

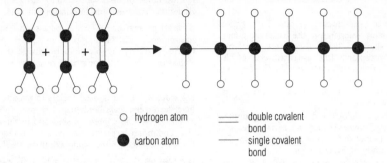

○ hydrogen atom   ═══ double covalent bond

● carbon atom   ─── single covalent bond

polymerization

(modes of reproduction that are common only among plants). Once a polyploid population is established, however, they can reproduce sexually. An example is cord-grass *Spartina anglica*, which is a polyploid of a European grass and a related North American grass, accidentally introduced. The resulting polyploid has spread dramatically.

polysaccharide long-chain ◊carbohydrate made up of hundreds or thousands of linked simple sugars (monosaccharides) such as glucose and closely related molecules.

The polysaccharides are natural polymers. They either act as energy-rich food stores in plants (starch) and animals (glycogen), or have structural roles in the plant cell wall (cellulose, pectin) or the tough outer skeleton of insects and similar creatures (chitin). See also ◊carbohydrate.

polystyrene type of ◊plastic used in kitchen utensils or, in an expanded form in insulation and ceiling tiles.

polytetrafluoroethene (PTFE) polymer made from the monomer tetrafluoroethene (CF_2CF_2). It is a thermosetting plastic with a high melting point that is used to produce 'non-stick' surfaces on pans and to coat bearings. Its trade name is Teflon.

Polythene trade name for a variety of ◊polyethylene.

polyunsaturate type of ◊fat or oil containing a high proportion of triglyceride molecules whose ◊fatty-acid chains contain several double bonds. By contrast, the fatty-acid chains of the triglycerides

in saturated fats (such as lard) contain only single bonds. Medical evidence suggests that polyunsaturated fats, used widely in margarines and cooking fats, are less likely to contribute to cardiovascular disease than saturated fats, but there is also some evidence that they may have adverse effects on health.

The more double bonds the fatty-acid chains contain, the lower the melting point of the fat. Unsaturated chains with several double bonds produce oils, such as vegetable and fish oils, which are liquids at room temperature. Saturated fats, with no double bonds, are solids at room temperature. The polyunsaturated fats used for margarines are produced by taking a vegetable or fish oil and turning some of the double bonds to single bonds, so that the product is semi-solid at room temperature. This is done by bubbling hydrogen through the oil in the presence of a catalyst, such as platinum. The catalyst is later removed.

Monounsaturated oils, such as olive oil, whose fatty-acid chains contain a single double bond, are probably healthier than either saturated or polyunsaturated fats. Butter contains both saturated and unsaturated fats, together with ◊cholesterol, which also plays a role in heart disease.

polyurethane polymer made from the monomer urethane. It is a thermoset ◊plastic, used in liquid form as a paint or varnish, and in foam form for upholstery and in lining materials (where it may be a fire hazard).

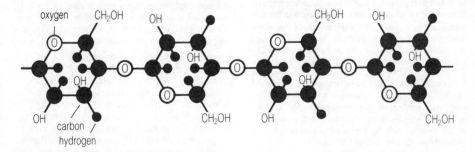

polysaccharide *A typical polysaccharide molecule, glycogen (animal starch), is formed from linked glucose ($C_6H_{12}O_6$) molecules. A glycogen molecule has 100–1,000 linked glucose units.*

polyvinyl chloride (PVC) a type of ◊plastic used for drainpipes, floor tiles, audio discs, shoes and handbags.

pome type of ◊pseudocarp, or false fruit, typical of certain plants belonging to the Rosaceae family. The outer skin and fleshy tissues are developed from the ◊receptacle (the enlarged end of the flower stalk) after fertilization, and the five ◊carpels (the true fruit) form the pome's core, which surrounds the seeds. Examples of pomes are apples, pears, and quinces.

population in biology and ecology, a group of animals of one species, living in a certain area and able to interbreed; the members of a given species in a ◊community of living things.

population cycle in biology, regular fluctuations in the size of a population, as seen in lemmings, for example. Such cycles are often caused by density-dependent mortality: high mortality due to overcrowding causes a sudden decline in the population, which then gradually builds up again. Population cycles may also result from an interaction between a predator and its prey.

population genetics the branch of genetics that studies the way in which the frequencies of different ◊alleles (alternative forms of a gene) in populations of organisms change, as a result of natural selection and other processes.

porphyry any ◊igneous rock containing large crystals in a finer matrix.

port in computing, a socket that enables a computer processor to communicate with an external device. It may be an **input port** (such as a joystick port), or an **output port** (such as a printer port), or both (an **i/o port**).

Microcomputers may provide ports for cartridges, televisions and/or monitors, printers, and modems, and sometimes for hard discs and musical instruments (MIDI, the musical-instrument digital interface). Ports may be serial or parallel.

portability in computing, the characteristic of certain programs that enables them to run on different types of computer with minimum modification. Programs written in a ◊high-level language can usually be run on any computer that has a compiler or interpreter for that particular language.

portable computer computer that can be carried from place to place. The term embraces a number of very different computers—from those that would be carried only with some reluctance to those, such as ◊laptop computers and ◊notebook computers, that can be comfortably carried and used in transit.

positron in physics, the antiparticle of the electron; an ◊elementary particle having the same magnitude of mass and charge as an electron but exhibiting a positive charge. The positron was discovered in 1932 by US physicist Carl Anderson at Caltech, USA, its existence having been predicted by the British physicist Paul Dirac 1928.

positron emission tomography (PET) technique that enables doctors to observe the operation of the human body by following the progress of a radioactive chemical that has been inhaled or injected. PET scanners pinpoint the location of the chemical by detecting ◊gamma radiation given out when ◊positrons emitted by the chemical are annihilated. The technique has been used to study a wide range of diseases including schizophrenia, Alzheimer's disease and Parkinson's disease.

potash general name for any potassium-containing mineral, most often applied to potassium carbonate (K_2CO_3) or potassium hydroxide (KOH). Potassium carbonate, originally made by roasting plants to ashes in earthenware pots, is commercially produced from the mineral sylvite (potassium chloride, KCl) and is used mainly in making artificial fertilizers, glass, and soap.

The potassium content of soils and fertilizers is also commonly expressed as potash, although in this case it usually refers to potassium oxide (K_2O).

potassium (Dutch *potassa* 'potash') soft, waxlike, silver-white, metallic element, symbol K (Latin *kalium*), atomic number 19, relative atomic mass 39.0983. It is one of the ◊alkali metals and has a very low density—it floats on water, and is the second lightest metal (after lithium). It oxidizes rapidly when exposed to air and reacts violently with water. Of great abundance in the Earth's crust, it is widely distributed with other elements and found in salt and mineral deposits in the form of potassium aluminium silicates.

Potassium, with sodium, plays a role in the transmission of impulses by nerve cells, and so is essential for animals; it is also required by plants for growth. The element was discovered and named in 1807 by English chemist Humphry Davy, who isolated it from potash in the first instance of a metal being isolated by electric current.

potassium dichromate $K_2Cr_2O_7$ orange, crystalline solid, soluble in water, that is a strong ◊oxidizing agent in the presence of dilute sulphuric acid. As it oxidizes other compounds it is itself reduced to potassium chromate (K_2CrO_4), which is green. Industrially it is used in the manufacture of dyes and glass and in tanning, photography, and ceramics.

potassium manganate(VII) $KMnO_4$ or *potassium permanganate* dark purple, crystalline solid, soluble in water, that is a strong ◊oxidizing agent in the presence of dilute sulphuric acid. In the process of oxidizing other compounds it is itself reduced to manganese(II) salts (containing the Mn^{2+} ion), which are colourless.

potential difference (pd) measure of the electrical potential energy converted to another form for every unit charge moving between two points in an electric circuit (see ◊potential, electric). The unit of potential difference is the volt.

potential divider or *voltage divider* two resistors connected in series in an electrical circuit in order to obtain a fraction of the potential difference, or voltage, across the battery or electrical source. The potential difference is divided across the two resistors in direct proportion to their resistances.

When one of the resistors is a variable resistor, or *potentiometer*, the potential difference across it can be varied continuously from zero to the value of the input voltage by sliding a contact along the resistor. Devices like this are used in electronic

standard potential divider

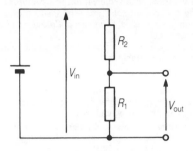

potentiometer used as a potential divider

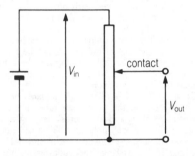

potential divider

equipment as the variable controls for functions such as volume, tone, and brightness control.

potential, electric in physics, the relative electrical state of an object. A charged conductor, for example, has a higher potential than the Earth, whose potential is taken by convention to be zero. An electric ◊cell (battery) has a potential in relation to emf (◊electromotive force), which can make current flow in an external circuit. The difference in potential between two points—the ◊potential difference—is expressed in ◊volts; that is, a 12 V battery has a potential difference of 12 volts between its negative and positive terminals.

potential energy ◊energy possessed by an object by virtue of its relative position or state (for example, as in a compressed spring). It is contrasted with kinetic energy, the form of energy possessed by moving bodies.

potentiometer in physics, an electrical ◊resistor that can be divided so as to compare, measure, or control voltages. In radio circuits, any rotary variable resistance (such as volume control) is referred to as a potentiometer.

A simple type of potentiometer consists of a length of uniform resistance wire (about 1 m/3 ft long) carrying a constant current provided by a battery connected across the ends of the wire. The source of potential difference (voltage) to be measured is connected (to oppose the cell) between one end of the wire, through a ◊galvanometer (instrument for measuring small currents), to a contact free to slide along the wire. The sliding contact is moved until the galvanometer shows no deflection. The ratio of the length of potentiometer wire in the galvanometer circuit to the total length of wire is then equal to the ratio of the unknown potential difference to that of the battery.

pothole small hollow in the rock bed of a river. Potholes are formed by the erosive action of rocky material carried by the river (corrasion), and are commonly found along the river's upper course, where it tends to flow directly over solid bedrock.

pound imperial unit (abbreviation lb) of mass. The commonly used avoirdupois pound, also called the *imperial standard pound* (7,000 grains/0.45 kg), differs from the *pound troy* (5,760 grains/0.37 kg), which is used for weighing precious metals. It derives from the Roman *libra*, which weighed 0.327 kg.

poundal imperial unit (abbreviation pdl) of force, now replaced in the SI system by the ◊newton. One poundal equals 0.1383 newtons.

It is defined as the force necessary to accelerate a mass of one pound by one foot per second per second.

powder metallurgy method of shaping heat-resistant metals such as tungsten. Metal is pressed into a mould in powdered form and then sintered (heated to very high temperatures).

power in mathematics, that which is represented by an ◊exponent or index, denoted by a superior small numeral. A number or symbol raised to the power of 2, that is, multiplied by itself, is said to be squared (for example, 3^2, x^2), and when raised to the power of 3, it is said to be cubed (for example, 2^3, y^3).

power in optics, a measure of the amount by which a lens will deviate light rays. A powerful converging lens will converge parallel rays steeply, bringing them to a focus at a short distance from the lens. The unit of power is the *dioptre*, which is equal to the reciprocal of focal length in metres. By convention, the power of a converging (or convex) lens is positive and that of a diverging (or concave) lens negative.

power in physics, the rate of doing work or consuming energy. It is measured in watts (joules per second) or other units of work per unit time.

If the work done or energy consumed is W joules and the time taken is t seconds, then the power P is given by the formula:

$$P = W/t$$

power station building where electrical energy is generated from a fuel or from another form of energy. Fuels used include fossil fuels such as coal, gas, and oil, and the nuclear fuel uranium. Renewable sources of energy include ◊gravitational potential energy, used to produce ◊hydroelectric power, and ◊wind power.

The energy supply is used to turn ◊turbines either directly by means of water or wind pressure, or indirectly by steam pressure, steam being generated by burning fossil fuels or from the heat released by the fission of uranium nuclei. The turbines in their turn spin alternators, which generate electricity at very high voltage.

ppm abbreviation for *parts per million*. An alternative (but numerically equivalent) unit used in chemistry is milligrams per litre (mg l^{-1}).

prairie the central North American plain, formerly grass-covered, extending over most of the region between the Rocky Mountains on the west and the Great Lakes and Ohio River on the east.

praseodymium (Greek *praseo* 'leek-green' + *dymium*) silver-white, malleable, metallic element of the ◊lanthanide series, symbol Pr, atomic number 59, relative atomic mass 140.907. It occurs in nature in the minerals monzanite and bastnasite, and its green salts are used to colour glass and ceramics. It was named in 1885 by Austrian chemist Carl von Welsbach (1858–1929).

He fractionated it from dydymium (originally thought to be an element but actually a mixture of rare-earth metals consisting largely of neodymium, praseodymium, and cerium), and named it for its green salts and spectroscopic line.

preadaptation in biology, the fortuitous possession of a character that allows an organism to exploit a new situation. In many cases, the character evolves to solve a particular problem that a species encounters in its preferred habitat, but once evolved may allow the organism to exploit an entirely different situation. The ability to extract oxygen directly from the air evolved in some early fishes, probably in response to life in stagnant, deoxygenated pools; this later made it possible for their descendants to spend time on land, so giving rise eventually to the air-breathing amphibians.

Precambrian in geology, the time from the formation of Earth (4.6 billion years ago) up to 570 million years ago. Its boundary with the succeeding Cambrian period marks the time when animals first developed hard outer parts (exoskeletons) and so left abundant fossil remains. It comprises about 85% of geological time and is divided into two periods: the Archaean, in which no life existed, and the Proterozoic, in which there was life in some form.

precession slow wobble of the Earth on its axis, like that of a spinning top. The gravitational pulls of the Sun and Moon on the Earth's equatorial bulge cause the Earth's axis to trace out a circle on the sky every 25,800 years. The position of the celestial poles (see ◊celestial sphere) is constantly changing owing to precession, as are the positions of the equinoxes (the points at which the celestial equator intersects the Sun's path around the sky). The *precession of the equinoxes* means that there is a gradual westward drift in the ecliptic— the path that the Sun appears to follow—and in the coordinates of objects on the celestial sphere.

This is why the dates of the astrological signs of the zodiac no longer correspond to the times of year when the Sun actually passes through the constellations. For example, the Sun passes through Leo from mid-Aug to mid-Sept, but the astrological dates for Leo are between about 23 July and 22 Aug.

All problems are finally scientific problems.

George Bernard Shaw *The Doctor's Dilemma*
Preface

precipitation in chemistry, the formation of an insoluble solid in a liquid as a result of a reaction within the liquid between two or more soluble substances. If the solid settles, it forms a *precipitate*; if the particles of solid are very small, they will remain in suspension, forming a *colloidal precipitate* (see ◊colloid).

In an equation the precipitate is often represented by the symbol (s).

$$AgNO_3{}_{(aq)} + NaCl \rightarrow AgCl_s + NaNO_3{}_{(aq)}$$

precipitation in meteorology, water that falls to the Earth from the atmosphere. It includes rain, snow, sleet, hail, dew, and frost.

pregnancy in humans, the period during which an embryo grows within the womb. It begins at conception and ends at birth, and the normal length is 40 weeks. Menstruation usually stops on conception. About one in five pregnancies fails, but most of these failures occur very early on, so the woman may notice only that her period is late. After the second month, the breasts become tense and tender, and the areas round the nipples become darker. Enlargement of the uterus can be felt at about the end of the third month, and thereafter the abdomen enlarges progressively. Pregnancy in animals is called ◊gestation.

Occasionally the fertilized egg implants not in the womb but in the ◊Fallopian tube (the tube between the ovary and the uterus), leading to an ectopic ('out of place') pregnancy. This will cause the woman severe abdominal pain and vaginal bleeding. If the growing fetus ruptures the tube, life-threatening shock may ensue.

prehistoric life the diverse organisms that inhabited Earth from the origin of life about 3.5 billion years ago to the time when humans began to keep written records, about 3500 BC. During the course of evolution, new forms of life developed and many other forms, such as the dinosaurs, became extinct. Prehistoric life evolved over this vast timespan from simple bacteria-like cells in the oceans to algae and protozoans and complex multicellar forms such as worms, molluscs, crustaceans, fishes, insects, land plants, amphibians, reptiles, birds, and mammals. On a geological timescale human beings evolved relatively recently, about 4 million years ago, although the exact dating is a matter of some debate. See also ◊geological time.

premolar in mammals, one of the large teeth toward the back of the mouth. In herbivores they are adapted for grinding. In carnivores they may be carnassials. Premolars are present in milk ◊dentition as well as permanent dentition.

preservative substance (◊additive) added to a food in order to inhibit the growth of bacteria, yeasts, mould, and other microorganisms, and therefore extend its shelf-life. The term sometimes refers to ◊antioxidants (substances added to oils and fats to prevent their becoming rancid) as well. All preservatives are potentially damaging to health if eaten in sufficient quantity. Both the amount used, and the foods in which they can be used, are restricted by law.

Alternatives to preservatives include faster turnover of food stocks, refrigeration, better hygiene in preparation, sterilization and pasteurization (see ◊food technology).

pressure in physics, the force acting normally (at

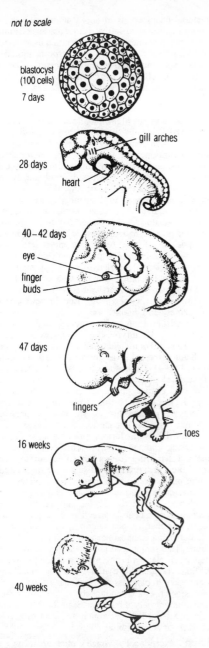

not to scale

blastocyst
(100 cells)
7 days

gill arches

28 days

heart

40–42 days

eye

finger
buds

47 days

fingers

toes

16 weeks

40 weeks

pregnancy *The development of a human embryo. Division of the fertilized egg, or ovum, begins within hours of conception. Within a week a ball of cells—a blastocyst—has developed. After the third week, the embryo has changed from a mass of cells into a recognizable shape. At four weeks, the embryo is 3 mm/0.1 in long, with a large bulge for the heart and small pits for the ears. At six weeks, the embryo is 1.5 cm/0.6 in with a pulsating heart and ear flaps. At the eighth week, the embryo is 2.5 cm/1 in long and recognizably human, with eyelids, small fingers, and toes. From the end of the second month, the embryo is almost fully formed and further development is mainly by growth. After this stage, the embryo is termed a fetus.*

right angles) to a body per unit surface area. The SI unit of pressure is the pascal (newton per square metre), equal to 0.01 millibars. In a fluid (liquid or gas), pressure increases with depth. At the edge of Earth's atmosphere, pressure is zero, whereas at sea level ◊atmospheric pressure due to the weight of the air above is about 100 kilopascals (1,013 millibars or 1 atmosphere). Pressure is commonly measured by means of a ◊barometer, ◊manometer, or ◊Bourdon gauge.

Pressure at a depth h in a fluid of density d is equal to hdg, where g is the acceleration due to gravity.

pressure cooker closed pot in which food is cooked in water under pressure, where water boils at a higher temperature than normal boiling point (100°C/212°F) and therefore cooks food quickly. The modern pressure cooker has a quick-sealing lid and a safety valve that can be adjusted to vary the steam pressure inside.

The French scientist Denis Papin invented the pressure cooker in England 1679.

pressurized water reactor (PWR) a ◊nuclear reactor design used in nuclear power stations in many countries, and in nuclear-powered submarines. In the PWR, water under pressure is the coolant and ◊moderator. It circulates through a steam generator, where its heat boils water to provide steam to drive power ◊turbines.

Prestel the ◊viewdata service provided by British Telecom (BT), which provides information on the television screen via the telephone network. BT pioneered the service 1975.

prevailing wind the direction from which the wind most commonly blows in a locality. In NW Europe, for example, the prevailing wind is southwesterly, blowing from the Atlantic Ocean in the southwest and bringing moist and warm conditions.

primary sexual characteristic in males, the primary sexual characteristic is the ◊testis; in females it is the ◊ovary. Both are endocrine glands that produce hormones responsible for secondary sexual characteristics, such as facial hair and a deep voice in males and smooth facial skin and breasts in females.

primate in zoology, any member of the order of mammals that includes monkeys, apes, and humans (together called **anthropoids**), as well as lemurs, bushbabies, lorises, and tarsiers (together called **prosimians**). Generally, they have forward-directed eyes, gripping hands and feet, opposable thumbs, and big toes. They tend to have nails rather than claws, with gripping pads on the ends of the digits, all adaptations to the arboreal, climbing mode of life.

There are one hundred and ninety-three living species of monkeys and apes. One hundred and ninety-two of them are covered with hair. The exception is a naked ape self-named Homo sapiens.

On human beings as **primates** Desmond Morris (1928–) *The Naked Ape* 1967

prime number a number that can be divided only by 1 or itself, that is, having no other factors. There is an infinite number of primes, the first ten of which are 2, 3, 5, 7, 11, 13, 17, 19, 23, and 29 (by definition, the number 1 is excluded from the set of prime numbers). The number 2 is the only even prime because all other even numbers have 2 as a factor.

Over the centuries mathematicians have sought general methods (algorithms) for calculating primes, from ◊Eratosthenes' sieve to programs on powerful computers.

In 1989 researchers at Amdahl Corporation, Sunnyvale, California, calculated the largest known prime number. It has 65,087 digits, and is more than a trillion trillion trillion times as large as the previous record holder. It took over a year of computation to locate the number and prove it was a prime.

UNUSUAL PRIME NUMBER

The number formed by writing 1,031 ones in a row is a prime.

---

principal focus in optics, the point at which incident rays parallel to the principal axis of a ◊lens converge, or appear to diverge, after refraction. The distance from the lens to its principal focus is its ◊focal length.

printed circuit board (PCB) electrical circuit created by laying (printing) 'tracks' of a conductor such as copper on one or both sides of an insulating board. The PCB was invented 1936 by Austrian scientist Paul Eisler, and was first used on a large scale 1948.

Components such as integrated circuits (chips), resistors and capacitors can be soldered to the surface of the board (surface-mounted) or, more commonly, attached by inserting their connecting pins or wires into holes drilled in the board. PCBs include ◊motherboards, ◊expansion boards, and adaptors.

printer in computing, an output device for producing printed copies of text or graphics. Types include the ◊*daisywheel printer*, which produces good-quality text but no graphics; the ◊*dot matrix printer*, which produces text and graphics by printing a pattern of small dots; the ◊*ink-jet printer*, which creates text and graphics by spraying a fine jet of quick-drying ink onto the paper; and the ◊*laser printer*, which uses electrostatic technology very similar to that used by a photocopier to produce high-quality text and graphics.

Printers may be classified as *impact printers* (such as daisywheel and dot-matrix printers), which form characters by striking an inked ribbon against the paper, and *nonimpact printers* (such as ink-jet and laser printers), which use a variety of techniques to produce characters without physical impact on the paper.

A further classification is based on the basic unit of printing, and categorizes printers as character printers, line printers, or page printers, according to whether they print one character, one line of characters, or a complete page at a time.

printing reproduction of text or illustrative material on paper, as in books or newspapers, or on an increasing variety of materials; for example, on plastic containers. The first printing used woodblocks, followed by carved wood type or moulded metal type and hand-operated presses. Modern printing is effected by electronically controlled machinery. Current printing processes include electronic phototypesetting with ◊offset printing, and ◊gravure print.

In China the art of printing from a single wooden block was known by the 6th century AD, and movable type was being used by the 11th century. In Europe printing was unknown for another three centuries, and it was only in the 15th century that movable type was reinvented, traditionally by Johannes Gutenberg in Germany. William Caxton introduced printing to England. There was no further substantial advance until, in the 19th century, steam power replaced hand-operation of printing presses, making possible long 'runs'; hand-composition of type (each tiny metal letter was taken from the case and placed individually in the narrow stick that carried one line of text) was replaced by machines operated by a keyboard.

Linotype, a hot-metal process (it produced a line of type in a solid slug) used in newspapers, magazines, and books, was invented by Ottmar Mergenthaler 1886 and commonly used until the 1980s. The *Monotype*, used in bookwork (it produced a series of individual characters, which could be hand-corrected), was invented by Tolbert Lanston (1844–1913) in the USA 1889. Important as these developments were, they represented no fundamental change but simply a faster method of carrying out the same basic typesetting operations. The actual printing process still involved pressing the inked type on to paper, by ◊*letterpress*.

In the 1960s this form of printing began to face increasing competition from ◊*offset printing*, a method that prints from an inked flat surface, and from the ◊*gravure* method (used for high-circulation magazines), which uses recessed plates.

The introduction of electronic phototypesetting machines, also in the 1960s, allowed the entire process of setting and correction to be done in the same way that a typist operates, thus eliminating the hot-metal composing room (with its hazardous fumes, lead scraps, and noise) and leaving only the making of plates and the running of the presses to be done traditionally. By the 1970s some final steps were taken to plateless printing, using various processes, such as a computer-controlled laser beam, or continuous jets of ink acoustically broken up into tiny equal-sized drops, which are electrostatically charged under computer control.

prion exceptionally small microorganism, a hundred times smaller than a virus. Composed of protein, and without any detectable amount of nucleic acid (genetic material), it is thought to cause diseases such as scrapie in sheep, and certain degenerative diseases of the nervous system in humans. How it can operate without nucleic acid is not yet known.

The prion was claimed to have been discovered at the University of California 1982.

prism in mathematics, a solid figure whose cross section is constant in planes drawn perpendicular

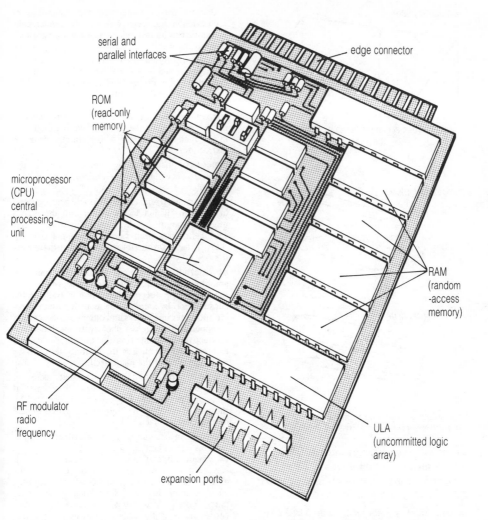

serial and
parallel interfaces

edge connector

ROM
(read-only
memory)

microprocessor
(CPU)
central
processing
unit

RAM
(random
-access
memory)

RF modulator
radio
frequency

ULA
(uncommitted logic
array)

expansion ports

printed circuit board *A typical microcomputer printed circuit board, or PCB. The PCB contains sockets for the integrated circuits, or chips, and the connecting tracks. The edge connectors allow the board to be connected to the power supply, printers, and display unit.*

to its axis. A cube, for example, is a rectangular prism with all faces (bases and sides) the same shape and size. A cylinder is a prism with a circular cross section.

prism in optics, a triangular block of transparent material (plastic, glass, silica) commonly used to 'bend' a ray of light or split a beam into its spectral colours. Prisms are used as mirrors to define the optical path in binoculars, camera viewfinders, and periscopes. The dispersive property of prisms is used in the spectroscope (an instrument for dispersing electromagnetic radiation).

probability likelihood, or chance, that an event will occur, often expressed as odds, or in mathematics, numerically as a fraction or decimal. In general, the probability that n particular events will happen out of a total of m possible events is n/m. A certainty has a probability of 1; an impossibility has a probability of 0. Empirical probability is

defined as the number of successful events divided by the total possible number of events.

In tossing a coin, the chance that it will land 'heads' is the same as the chance that it will land 'tails', that is, 1 to 1 or even; mathematically, this probability is expressed as ½ or 0.5. The odds against any chosen number coming up on the roll of a fair die are 5 to 1; the probability is ⅙ or 0.1666.... If two dice are rolled there are 6 x 6 = 36 different possible combinations. The probability of a double (two numbers the same) is ⁶⁄₃₆ or 1/6 since there are six doubles in the 36 events: (1,1), (2,2), (3,3), (4,4), (5,5), and (6,6).

Probability theory was developed by the French mathematicians Blaise Pascal and Pierre de Fermat in the 17th century, initially in response to a request to calculate the odds of being dealt various hands at cards. Today probability plays a major part in

triangular prism

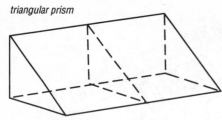

cross-section is the same throughout
the prism's length

trapezoidal prism

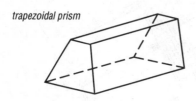

pentagonal prism

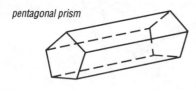

prism

the mathematics of atomic theory and finds application in insurance and statistical studies.

---

*There was once a brainy baboon,/Who
always breathed down a bassoon,/For he
said, 'It appears/That in millions of years/
I shall certainly hit on a tune.'*

On **probability** Arthur Eddington *New Pathways
in Science* 1935

---

procedure in computing, a small part of a computer program that performs a specific task, such as clearing the screen or sorting a file. A *procedural language*, such as BASIC, is one in which the programmer describes a task in terms of how it is to be done, as opposed to a *declarative language*, such as PROLOG, in which it is described in terms of the required result. See ◊programming.

Careful use of procedures is an element of ◊structured programming. In some programming languages there is an overlap between procedures, ◊functions, and ◊subroutines.

process control automatic computerized control of a manufacturing process, such as glassmaking. The computer receives ◊feedback information from sensors about the performance of the machines involved, and compares this with ideal performance data stored in its control program. It then outputs instructions to adjust automatically the machines' settings.

Because the computer can monitor and reset each machine hundreds of times each minute, performance can be maintained at levels that are very close to the ideal.

processing cycle in computing, the sequence of steps performed repeatedly by a computer in the execution of a program. The computer's CPU (central processing unit) continuously works through a loop, involving fetching a program instruction from memory, fetching any data it needs, operating on the data, and storing the result in the memory, before fetching another program instruction.

processor in computing, another name for the ◊central processing unit or ◊microprocessor of a computer.

Procyon or *Alpha Canis Minoris* brightest star in the constellation Canis Minor and the eighth brightest star in the sky. Procyon is a white star 11.4 light years from Earth, with a mass of 1.7 Suns. It has a ◊white dwarf companion that orbits it every 40 years.

productivity, biological in an ecosystem, the amount of material in the food chain produced by the primary producers (plants) that is available for consumption by animals. Plants turn carbon dioxide and water into sugars and other complex carbon compounds by means of photosynthesis. Their net productivity is defined as the quantity of carbon compounds formed, less the quantity used up by the respiration of the plant itself.

progesterone ◊steroid hormone that occurs in vertebrates. In mammals, it regulates the menstrual cycle and pregnancy. Progesterone is secreted by the corpus luteum (the ruptured Graafian follicle of a discharged ovum).

program in computing, a set of instructions that controls the operation of a computer. There are two main kinds: ◊applications programs, which carry out tasks for the benefit of the user—for example, word processing; and ◊systems programs, which control the internal workings of the computer. A ◊utility program is a systems program that carries out specific tasks for the user. Programs can be written in any of a number of ◊programming languages but are always translated into machine code before they can be executed by the computer.

program counter in computing, an alternative name for ◊sequence-control register.

program documentation ◊documentation that provides a complete technical description of a program, built up as the software is written, and is intended to support any later maintenance or development of the program.

program flow chart type of ◊flow chart used to describe the flow of data through a particular computer program.

program loop a part of a computer program that is repeated several times. The loop may be repeated a fixed number of times (*counter-controlled loop*) or until a certain condition is satisfied (*condition-controlled loop*). For example, a counter-controlled loop might be used to repeat an input routine until exactly ten numbers have been input; a condition-controlled loop might be used to repeat an input routine until the ◊data terminator 'XXX' is entered.

programming writing instructions in a programming language for the control of a computer. *Applications programming* is for end-user programs, such as accounts programs or word-processing packages. *Systems programming* is for operating systems and the like, which are concerned more with the internal workings of the computer.

There are several programming styles: *procedural programming*, in which programs are written as lists of instructions for the computer to obey in sequence, is by far the most popular. It is the 'natural' style, closely matching the computer's own sequential operation; *declarative programming*, as used in the programming language PROLOG, does not describe how to solve a problem, but rather describes the logical structure of the problem. Running such a program is more like proving an assertion than following a procedure; *functional programming* is a style based largely on the definition of functions. There are very few functional programming languages, HOPE and ML being the most widely used, though many more conventional languages (for example, C) make extensive use of functions; *object-oriented programming*, the most recently developed style, involves viewing a program as a collection of objects that behave in certain ways when they are passed certain 'messages'. For example, an object might be defined to represent a table of figures, which will be displayed on screen when a 'display' message is received.

programming language in computing, a special notation in which instructions for controlling a computer are written. Programming languages are designed to be easy for people to write and read, but must be capable of being mechanically translated (by a ◊compiler or an ◊interpreter) into the ◊machine code that the computer can execute. Programming languages may be classified as ◊high-level languages or ◊low-level languages. See also ◊source language.

program trading in finance, buying and selling a group of shares using a computer program to generate orders automatically whenever there is an appreciable movement in prices.

One form in use in the USA in 1989 was *index arbitrage*, in which a program traded automatically whenever there was a difference between New York and Chicago prices of an equivalent number of shares. Program trading comprised some 14% of daily trading on the New York Stock Exchange by volume in Sept 1989, but was widely criticized for lessening market stability. It has been blamed, among other factors, for the Stock Market crashes of 1987 and 1989.

progression sequence of numbers each formed by a specific relationship to its predecessor. An *arithmetic progression* has numbers that increase or decrease by a common sum or difference (for example, 2, 4, 6, 8); a *geometric progression* has numbers each bearing a fixed ratio to its predecessor (for example, 3, 6, 12, 24); and a *harmonic progression* has numbers whose ◊reciprocals are in arithmetical progression, for example 1, ½, ⅓, ¼.

projectile particle that travels with both horizontal and vertical motion in the Earth's gravitational field. The two components of its motion can generally be analysed separately: its vertical motion will be accelerated due to its weight in the gravitational field; its horizontal motion may be assumed to be at constant velocity if the frictional forces of air resistance are ignored. In a uniform gravitational field and in the absence of frictional forces the path of a projectile is a parabola.

projection see ◊map projection.

projector any apparatus that projects a picture on to a screen. In a *slide projector*, a lamp shines a light through the photographic slide or transparency, and a projection ◊lens throws an enlarged image of the slide onto the screen. A *film projector* has similar optics, but incorporates a mechanism that holds the film still while light is transmitted through each frame (picture). A shutter covers the film when it moves between frames.

prokaryote in biology, an organism whose cells lack organelles (specialized segregated structures such as nuclei, mitochondria, and chloroplasts). Prokaryote DNA is not arranged in chromosomes but forms a coiled structure called a *nucleoid*. The prokaryotes comprise only the *bacteria* and *cyanobacteria* (see ◊blue-green algae); all other organisms are eukaryotes.

PROLOG (acronym for *pro*gramming in *log*ic) high-level computer-programming language based on logic. Invented in 1971 at the University of Marseille, France, it did not achieve widespread use until more than ten years later. It is used mainly for ◊artificial-intelligence programming.

PROM (acronym for *p*rogrammable *r*ead-*o*nly *m*emory) in computing, a memory device in the form of an integrated circuit (chip) that can be programmed after manufacture to hold information permanently. PROM chips are empty of information when manufactured, unlike ROM (read-only memory) chips, which have information built into them. Other memory devices are ◊EPROM (erasable programmable read-only memory) and ◊RAM (random-access memory).

promethium radioactive, metallic element of the ◊lanthanide series, symbol Pm, atomic number 61, relative atomic mass 145. It occurs in nature only in minute amounts, produced as a fission product/by-product of uranium in ◊pitchblende and other uranium ores; for a long time it was considered not to occur in nature. The longest-lived isotope has a half-life of slightly more than 20 years.

Promethium is synthesized by neutron bombardment of neodymium, and is a product of the fission of uranium, thorium, or plutonium; it can be isolated in large amounts from the fission-product debris of uranium fuel in nuclear reactors. It is used in phosphorescent paints and as an X-ray source.

It was named 1949 after the Greek Titan Prometheus.

prominence bright cloud of gas projecting from the Sun into space 100,000 km/60,000 mi or more. *Quiescent prominences* last for months, and are

programming languages

language	main uses	description
Ada	defence applications	high level
assembler languages	jobs needing detailed control of the hardware, fast execution, and small program sizes	fast and efficient but require considerable effort and skill
BASIC (beginners'all-purpose symbolic instruction code)	mainly in education, business, and the home, and among nonprofessional programmers, such as engineers	easy to learn; early versions lacked the features of other languages
C	systems programming; general programming	fast and efficient; widely used as a general-purpose language; especially popular among professional programmers
COBOL (common business-oriented language)	business programming	strongly oriented towards commercial work; easy to learn but very verbose; widely used on mainframes
FORTH	control applications	reverse Polish notation language
FORTRAN (formula translation)	scientific and computational work	based on mathematical formulae; popular among engineers, scientists, and mathematicians
LISP (list processing)	artificial intelligence	symbolic language with a reputation for being hard to learn; popular in the academic and research communities
Modula-2	systems and real-time programming; general programming	highly structured; intended to replace Pascalfor 'real-world' applications
OBERON	general programming	small, compact language incorporating many of the features of PASCAL and Modula-2
PASCAL (program appliqué à la sélection et la compilation automatique de la littérature)	general-purpose language	highly structured; widely used for teaching programming in universities
PROLOG (programming in logic)	artificial intelligence	symbolic-logic programming system, originally intended for theorem solving but now used more generally in artificial intelligence

held in place by magnetic fields in the Sun's corona. *Surge prominences* shoot gas into space at speeds of 1,000 kps/600 mps. *Loop prominences* are gases falling back to the Sun's surface after a solar ◊flare.

proof spirit numerical scale used to indicate the alcohol content of an alcoholic drink. Proof spirit (or 100% proof spirit) acquired its name from a solution of alcohol in water which, when used to moisten gunpowder, contained just enough alcohol to permit it to burn.

In practice, the degrees proof of an alcoholic drink is based on the specific gravity of an aqueous solution containing the same amount of alcohol as the drink. Typical values are: whisky, gin, rum 70 degrees proof (40% alcohol); vodka 65 degrees proof; sherry 28 degrees proof; table wine 20 degrees proof; beer 4 degrees proof. The USA uses a different proof scale to the UK; a US whisky of 80 degrees proof on the US scale would be 70 degrees proof on the UK scale.

propane C_3H_8 gaseous hydrocarbon of the ◊alkane series, found in petroleum and used as fuel.

propanol or *propyl alcohol* third member of the homologous series of ◊alcohols. Propanol is usually a mixture of two isomeric compounds (see ◊isomer): propan-1-ol ($CH_3CH_2CH_2OH$) and propan-2-ol ($CH_3CHOHCH_3$). Both are colourless liquids that can be mixed with water and are used in perfumery.

propanone CH_3COCH_3 (common name *acetone*) colourless flammable liquid used extensively as a solvent, as in nail-varnish remover. It boils at

56.5°C/133.7°F, mixes with water in all proportions, and has a characteristic odour.

propellant substance burned in a rocket for propulsion. Two propellants are used: oxidizer and fuel are stored in separate tanks and pumped independently into the combustion chamber. Liquid oxygen (oxidizer) and liquid hydrogen (fuel) are common propellants, used, for example, in the space-shuttle main engines. The explosive charge that propels a projectile from a gun is also called a propellant.

propeller screwlike device used to propel some ships and aeroplanes. A propeller has a number of curved blades that describe a helical path as they rotate with the hub, and accelerate fluid (liquid or gas) backwards during rotation. Reaction to this backward movement of fluid sets up a propulsive thrust forwards. The marine screw propeller was developed by Francis Pettit Smith in the UK and Swedish-born John Ericson in the USA and was first used 1839.

propene $CH_3CH:CH_2$ (common name *propylene*) second member of the alkene series of hydrocarbons. A colourless, flammable gas, it is widely used by industry to make organic chemicals, including polypropylene plastics.

propenoic acid $H_2C:CHCOOH$ (common name *acrylic acid*) acid obtained from the aldehyde propenal (acrolein) derived from glycerol or fats. Glasslike thermoplastic resins are made by polymerizing ◊esters of propenoic acid or methyl propenoic acid and used for transparent components, lenses, and dentures. Other acrylic compounds are used for adhesives, artificial fibres, and artists' acrylic paint.

proper motion gradual change in the position of a star that results from its motion in orbit around our galaxy, the Milky Way. Proper motions are slight and undetectable to the naked eye, but can be accurately measured on telescopic photographs taken many years apart. Barnard's Star is the star with the largest proper motion, 10.3 arc seconds per year.

properties in chemistry, the characteristics a substance possesses by virtue of its composition.

Physical properties of a substance can be measured by physical means, for example boiling point, melting point, hardness, elasticity, colour, and physical state. *Chemical properties* are the way it reacts with other substances; whether it is acidic or basic, an oxidizing or a reducing agent, a salt, or stable to heat, for example.

proportion two variable quantities x and y are proportional if, for all values of x, $y = kx$, where k is a constant. This means that if x increases, y increases in a linear fashion.

A graph of x against y would be a straight line passing through the origin (the point $x = 0$, $y = 0$). y is inversely proportional to x if the graph of y against $1/x$ is a straight line through the origin. The corresponding equation is $y = k/x$. Many laws of science relate quantities that are proportional (for example, ◊Boyle's law).

prop root or *stilt root* modified root that grows from the lower part of a stem or trunk down to the ground, providing a plant with extra support. Prop roots are common on some woody plants, such as mangroves, and also occur on a few herbaceous plants, such as maize. *Buttress roots* are a type of prop root found at the base of tree trunks, extended and flattened along the upper edge to form massive triangular buttresses; they are common on tropical trees.

propyl alcohol common name for ◊propanol.

propylene common name for ◊propene.

prosimian or *primitive primate* in zoology, any animal belonging to the suborder Strepsirhin of ◊primates. Prosimians are characterized by a wet nose with slitlike nostrils, the tip of the nose having a prominent vertical groove. Examples are lemurs, pottos, tarsiers, and the aye-aye.

prostaglandin any of a group of complex fatty acids that act as messenger substances between cells. Effects include stimulating the contraction of smooth muscle (for example, of the womb during birth), regulating the production of stomach acid, and modifying hormonal activity. In excess, prostaglandins may produce inflammatory disorders such as arthritis. Synthetic prostaglandins are used to induce labour in humans and domestic animals.

The analgesic actions of substances such as aspirin are due to inhibition of prostaglandin synthesis.

prostate gland gland surrounding and opening into the ◊urethra at the base of the ◊bladder in male mammals. The prostate gland produces an alkaline fluid that is released during ejaculation; this fluid activates sperm, and prevents their clumping together. In humans, the prostate often enlarges

and obstructs the urethra; this is treated by prostatectomy.

prosthesis replacement of a body part with an artificial substitute. Prostheses include artificial limbs, hearing aids, false teeth and eyes, and for the heart, a ◊pacemaker and plastic heart valves and blood vessels.

Prostheses in the form of artificial limbs, such as wooden legs and metal hooks for hands, have been used for centuries, although artificial limbs are now more natural-looking and comfortable to wear. The comparatively new field of ◊bionics has developed myoelectric, or bionic, arms, which are electronically operated and worked by minute electrical impulses from body muscles.

protactinium (Latin *proto* 'before' + actinium) silver-grey, radioactive, metallic element of the ◊actinide series, symbol Pa, atomic number 91, relative atomic mass 231.036. It occurs in nature in very small quantities, in ◊pitchblende and other uranium ores. It has 14 known isotopes; the longest-lived, Pa-231, has a half-life of 32,480 years.

The element was discovered in 1913 (Pa-234, with a half-life of only 1.2 minutes) as a product of uranium decay. Other isotopes were later found and the name was officially adopted in 1949, although it had been in use since 1918.

protandry in a flower, the state where the male reproductive organs reach maturity before those of the female. This is a common method of avoiding self-fertilization. See also ◊protogyny.

protease general term for an enzyme capable of splitting proteins. Examples include pepsin, found in the stomach, and trypsin, found in the small intestine.

protein complex, biologically important substance composed of amino acids joined by ◊peptide bonds. Other types of bond, such as sulphur–sulphur bonds, hydrogen bonds, and cation bridges between acid sites, are responsible for creating the protein's characteristic three-dimensional structure, which may be fibrous, globular, or pleated.

Proteins are essential to all living organisms. As *enzymes* they regulate all aspects of metabolism. Structural proteins such as *keratin* and *collagen* make up the skin, claws, bones, tendons, and ligaments; *muscle* proteins produce movement; *haemoglobin* transports oxygen; and *membrane* proteins regulate the movement of substances into and out of cells.

For humans, protein is an essential part of the diet (60 g per day is required), and is found in greatest quantity in soya beans and other grain legumes, meat, eggs, and cheese. Protein provides 4 kcal of energy per gram.

protein engineering the creation of synthetic proteins designed to carry out specific tasks. For example, an enzyme may be designed to remove grease from soiled clothes and remain stable at the high temperatures in a washing machine.

protein synthesis manufacture, within the cytoplasm of the cell, of the ◊proteins an organism needs. The building blocks of proteins are ◊amino acids, of which there are 20 types. The pattern in which the amino acids are linked decides what kind of protein is produced. In turn it is the genetic

amino acids, where R is one of many possible side chains

peptide bond

protein *A protein molecule is a long chain of amino acids linked by peptide bonds. The properties of a protein are determined by the order, or sequence, of amino acids in its molecule, and by the three-dimensional structure of the molecular chain. The chain folds and twists, often forming a spiral shape.*

code, contained within ◊DNA, that determines the precise order in which the amino acids are linked up during protein manufacture. Interestingly, DNA is found only in the nucleus, yet protein synthesis only occurs in the cytoplasm. The information necessary for making the proteins is carried from the nucleus to the cytoplasm by another nucleic acid, ◊RNA.

Proterozoic eon of geological time, possible 3.5 billion to 570 million years ago, the second division of the Precambrian. It is defined as the time of simple life, since many rocks dating from this eon show traces of biological activity, and some contain the fossils of bacteria and algae.

prothallus short-lived gametophyte of many ferns and other ◊pteridophytes (such as horsetails or clubmosses). It bears either the male or female sex organs, or both. Typically it is a small, green, flattened structure that is anchored in the soil by several ◊rhizoids (slender, hairlike structures, acting as roots) and needs damp conditions to survive. The reproductive organs are borne on the lower surface close to the soil. See also ◊alternation of generations.

protist in biology, a single-celled organism which has a eukaryotic cell, but which is not member of the plant, fungal, or animal kingdoms. The main protists are ◊protozoa.

Single-celled photosynthetic organisms, such as diatoms and dinoflagellates, are classified as protists or algae. Recently the term has also been used for members of the kingdom Protista, which features in certain five-kingdom classifications of the living world (see also ◊plant classification). This kingdom may include slime moulds, all algae (seaweeds as well as unicellular forms), and protozoa.

protocol in computing, an agreed set of standards for the transfer of data between different devices. They cover transmission speed, format of data, and the signals required to synchronize the transfer. See also ◊interface.

protogyny in a flower, the state where the female

reproductive organs reach maturity before those of the male. Like protandry, in which the male organs reach maturity first, this is a method of avoiding self-fertilization, but it is much less common.

proton (Greek 'first') in physics, a positively charged subatomic particle, a constituent of the nucleus of all atoms. It belongs to the ◊baryon subclass of the ◊hadrons.

A proton is extremely long-lived, with a lifespan of at least 10^{32} years. It carries a unit positive charge equal to the negative charge of an ◊electron. Its mass is almost 1,836 times that of an electron, or 1.67×10^{-24} g. The number of protons in the atom of an element is equal to the atomic number of that element.

protonema young ◊gametophyte of a moss, which develops from a germinating spore (see ◊alternation of generations). Typically it is a green, branched, threadlike structure that grows over the soil surface bearing several buds that develop into the characteristic adult moss plants.

proton number alternative name for ◊atomic number.

Proton rocket Soviet space rocket introduced 1965, used to launch heavy satellites, space probes, and the Salyut and Mir space stations.

Proton consists of up to four stages as necessary. It has never been used to launch humans into space.

protoplasm contents of a living cell. Strictly speaking it includes all the discrete structures (organelles) in a cell, but it is often used simply to mean the jellylike material in which these float. The contents of a cell outside the nucleus are called ◊cytoplasm.

Prototheria sub-class of mammals made up of the egg-laying ◊monotremes. It contains only the echidna (spiny anteater) and the platypus.

prototype in technology, any of the first few machines of a new design. Prototypes are tested for performance, reliability, economy, and safety;

then the main design can be modified before full-scale production begins.

protozoa group of single-celled organisms without rigid cell walls. Some, such as amoeba, ingest other cells, but most are ◊saprotrophs or parasites. The group is polyphyletic (containing organisms which have different evolutionary origins).

protractor instrument used to measure a flat ◊angle.

PROXIMA CENTAURI: STAR TRAVEL

A car travelling at 100 kph/60 mph would take 48 million years to reach the nearest star, Proxima Centauri. This is about 660,000 average human lifetimes. If you did manage to reach Proxima Centauri, the Sun would appear to be a bright star in the constellation of Cassiopeia.

Proxima Centauri the closest star to the Sun, 4.2 light years away. It is a faint ◊red dwarf, visible only with a telescope, and is a member of the Alpha Centauri triple-star system.

It is called Proxima because it is about 0.1 light years closer to us than its two partners.

prussic acid former name for ◊hydrocyanic acid.

pseudocarp fruitlike structure that incorporates tissue that is not derived from the ovary wall. The additional tissues may be derived from floral parts such as the ◊receptacle and ◊calyx. For example, the coloured, fleshy part of a strawberry develops from the receptacle and the true fruits are small ◊achenes—the 'pips' embedded in its outer surface. Rose hips are a type of pseudocarp that consists of a hollow, fleshy receptacle containing a number of achenes within. Different types of pseudocarp include pineapples, figs, apples, and pears.

A *coenocarpium* is a fleshy, multiple pseudocarp derived from an ◊inflorescence rather than a single flower. The pineapple has a thickened central axis surrounded by fleshy tissues derived from the receptacles and floral parts of many flowers. A fig is a type of pseudocarp called a *syconium*, formed from a hollow receptacle with small flowers attached to the inner wall. After fertilization the ovaries of the female flowers develop into one-seeded achenes. Apples and pears are ◊pomes, another type of pseudocarp.

pseudocopulation attempted copulation by a male insect with a flower. It results in ◊pollination of the flower and is common in the orchid family, where the flowers of many species resemble a particular species of female bee. When a male bee attempts to mate with a flower, the pollinia (groups of pollen grains) stick to its body. They are transferred to the stigma of another flower when the insect attempts copulation again.

pseudomorph mineral that has replaced another *in situ* and has retained the external crystal shape of the original mineral.

psychology systematic study of human and animal behaviour. The first psychology laboratory was founded 1879 by Wilhelm Wundt at Leipzig, Germany. The subject includes diverse areas of study and application, among them the roles of instinct, heredity, environment, and culture; the processes of sensation, perception, learning and memory; the bases of motivation and emotion; and the functioning of thought, intelligence, and language. Significant psychologists have included Gustav Fechner (1801–1887) founder of psychophysics; Wolfgang Köhler (1887–1967), one of the gestalt or 'whole' psychologists; Sigmund Freud and his associates Carl Jung, Alfred Adler, and Hermann Rorschach (1884–1922); William James, Jean Piaget; Carl Rogers; Hans Eysenck; J B Watson, and B F Skinner.

Experimental psychology emphasizes the application of rigorous and objective scientific methods to the study of a wide range of mental processes and behaviour, whereas social psychology concerns the study of individuals within their social environment; for example, within groups and organizations. This has led to the development of related fields such as *occupational psychology*, which studies human behaviour at work, and *educational psychology*. *Clinical psychology* concerns the understanding and treatment of mental health disorders, such as anxiety, phobias, or depression; treatment may include behaviour

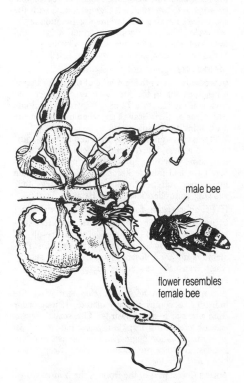

pseudocopulation *The male bee, attracted to the orchid because of its resemblance to a female bee, attempts to mate with the flower. The bee's efforts cover its body with pollen, which is carried to the next flower it visits.*

therapy, cognitive therapy, counselling, psycho-analysis, or some combination of these.

Modern studies have been diverse, for example the psychological causes of obesity; the nature of religious experience; and the underachievement of women seen as resulting from social pressures. Other related subjects are the nature of sleep and dreams, and the possible extensions of the senses, which leads to the more contentious ground of parapsychology (the study of phenomena beyond the range of established science, for example, extra-sensory perception).

Psychology has a long past, but only a short history.

On **psychology** Hermann Ebbinghaus (1850–1909) *Summary of Psychology* 1908

pt symbol for ◊*pint*.

pteridophyte simple type of ◊vascular plant. The pteridophytes comprise four classes: the Psilosida, including the most primitive vascular plants, found mainly in the tropics; the Lycopsida, including the club mosses; the Sphenopsida, including the horse-tails; and the Pteropsida, including the ferns. They do not produce seeds.

They are mainly terrestrial, non-flowering plants characterized by the presence of a vascular system; the possession of true stems, roots, and leaves; and by a marked ◊alternation of generations, with the sporophyte forming the dominant generation in the life cycle. The pteridophytes formed a large and dominant flora during the Carboniferous period, but many are now known only from fossils.

pterodactyl genus of ◊pterosaur.

pterosaur extinct flying reptile of the order Pterosauria, existing in the Mesozoic age. They ranged from starling size to a 12 m/40 ft wingspan. Some had horns on their heads that, when in flight, made a whistling to roaring sound.

Pterosaurs were formerly assumed to be smooth-skinned gliders, but recent discoveries show that at least some were furry, probably warm-blooded, and may have had muscle fibres and blood vessels on their wings, stiffened by moving the hind legs, thus allowing controlled and strong flapping flight.

PTFE abbreviation for ◊*polytetrafluoroethene*.

ptomaine any of a group of toxic chemical substances (alkaloids) produced as a result of decomposition by bacterial action on proteins. Ptomaine is not the relevant factor in food poisoning, which is usually caused by bacteria of the genus ◊*Salmonella*.

puberty stage in human development when the individual becomes sexually mature. It may occur from the age of ten upwards. The sexual organs take on their adult form and pubic hair grows. In girls, menstruation begins, and the breasts develop; in boys, the voice breaks and becomes deeper, and facial hair develops.

pubes lowest part of the front of the human trunk, the region where the external generative organs are situated. The underlying bony structure, the pubic arch, is formed by the union in the midline of the two pubic bones, which are the front portions of

the hip bones. In women this is more prominent than in men, to allow more room for the passage of the child's head at birth, and it carries a pad of fat and connective tissue, the *mons veneris* (mountain of Venus), for its protection.

puddle clay clay, with sand or gravel, that has had water added and mixed thoroughly so that it becomes watertight. The term was coined 1762 by the canal builder James Brindley, although the use of such clay in dams goes back to Roman times.

pull-down menu in computing, a list of options provided as part of a ◊graphical user interface. The presence of pull-down menus is normally indicated by a row of single words at the top of the screen. When the user points at a word with a ◊mouse, a full menu appears (is pulled down) and the user can then select the required option.

In some graphical user interfaces the menus appear from the bottom of the screen and in others they may appear at any point on the screen when a special menu button is pressed on the mouse.

pulley simple machine consisting of a fixed, grooved wheel, sometimes in a block, around which a rope or chain can be run. A simple pulley serves only to change the direction of the applied effort (as in a simple hoist for raising loads). The use of more than one pulley results in a mechanical advantage, so that a given effort can raise a heavier load.

The mechanical advantage depends on the arrangement of the pulleys. For instance, a block and tackle arrangement with three ropes supporting the load will lift it with one-third of the effort needed to lift it directly (if friction is ignored), giving a mechanical advantage of 3.

pulsar celestial source that emits pulses of energy at regular intervals, ranging from a few seconds to a few thousandths of a second. Pulsars are thought to be rapidly rotating ◊neutron stars, which flash at radio and other wavelengths as they spin. They were discovered in 1967 by Jocelyn Bell (now Burnell) and Antony Hewish at the Mullard Radio Astronomy Observatory, Cambridge, England. Over 500 radio pulsars are now known in our Galaxy, although a million or so may exist.

Pulsars gradually slow down as they get older, and eventually the flashes fade. Of the 500 known radio pulsars, 20 are millisecond pulsars (flashing a thousand times a second). Such pulsars are thought to be more than a billion years old. Two pulsars, one (estimated to be a thousand years old) in the Crab nebula and one (estimated to be 11,000 years old) in the constellation Vela, give out flashes of visible light.

Geminga is the closest pulsar to the Earth, about 120 light years away. First detected in the 1970s, it was finally dated in 1991 (from the rate of change of the pulsar) and found to be 370,000 years old.

pulse impulse transmitted by the heartbeat throughout the arterial systems of vertebrates. When the heart muscle contracts, it forces blood into the ◊aorta (the chief artery). Because the arteries are elastic, the sudden rise of pressure causes a throb or sudden swelling through them. The actual flow of the blood is about 60 cm/2 ft a second in humans. The pulse rate is generally about 70 per minute. The pulse can be felt where

The great cookie-cutter experiment

One of the basic questions examined by psychology is: 'How do humans perceive things?' Perception is the process of forming a coherent picture of the world. But how does perception work?

Until the 1960s, a rather simplistic view was taken of perception. The organs of perception—the eyes and ears, for example—were considered as passive receptors of data from the surroundings. It was thought that the act of perception simply consisted of constructing a world-picture from the external data. However, there were some early indications that this could not be the whole story.

Stimulus ... and no response

Early experiments on perception maintained a person in as passive a condition as possible, so that he or she became a simple receptor. The person was then subjected to stimuli of various sorts—sounds, flashing light, and so on—to see how the stimuli were perceived. These experiments produced alarming results: a person held in a completely passive condition did not perceive the world as made up of things and, furthermore, after a short time did not perceive anything at all! These experiments should have warned psychologists that something was wrong with their theories of perception. But psychologists are very conservative, and in the middle of the 20th century the experimental study of perception continued along lines established in the 19th century.

Another very simple experiment demonstrates that perception cannot be regarded as a completely passive process. When the head is moved, the images produced on the retina of the eye move; yet the world is perceived as stationary. Presumably, when a moving object passes in front of a stationary eye, the images on the retina move in much the same way. In this case, however, we perceive that it is the object which is moving. In other words, the same retinal stimulus can produce different perceptions.

The mechanism of perception was clarified by the work of US psychologist James Gibson in the early 1960s. Gibson was born in McConnelsville, Ohio, in 1904. He was educated at Princeton University. In 1932 he married his wife Eleanor, also a psychologist, and much of his later work was done in collaboration with her. Most of his work was done at Cornell University. He is credited with a discovery that revolutionized the teaching of pilots to land. He found that, as an aircraft descends, only one point appears not to change in relation to its surroundings; this is the point at which the aircraft will touch down. This discovery perhaps revealed to Gibson the importance of those things which appear unchanging in our environment. He put forward a new theory that perception was based on an active exploration of the world. According to Gibson, our world-picture is built up from observation of invariants or unchanging quantities.

Active exploration

To prove his ideas correct, Gibson undertook a series of experiments using cookie cutters. The cutters were of different shapes—some square, some star-shaped, and so on. In the first part of the experiment, the cutters were pressed onto the upturned palm of a hand with a standard pressure. The subject was held still, and he could not see the cutter. This corresponded to the passive condition so well loved by earlier experimenters. It was found that only 29% of the subjects could correctly identify the shape of the cutter. Next, the subject was allowed to use his fingers to feel the cutter, to turn it and actively explore the shape. Now 95% of the subjects identified the shapes correctly. Another experiment was done in which the cutters were pressed to the palm of the hand, and then rotated gently. In this case, 72% of the subjects correctly identified the shapes.

The best explanation of the results is that accurate perception of shape requires changing stimuli. The unchanging stimulus when the cutter is pressed onto the palm is least helpful in perceiving shape. The most accurate perception occurs when the fingers are used to actively explore a cutter; in this case, the subject receives changing stimuli not only from the hand but also from the arm and hand muscles. When the cutter is rotated on the palm, stimuli are received from the hand but not the arm—an intermediate case. From among the changing stimuli, the brain extracts the unchanging information and constructs a picture of the cutter from this.

In this simple experiment, Gibson demonstrated that perception is a process of active exploration, not a passive reception of information. The reason why we know whether an object is moving, or whether our head is moving, is now clear. We receive information from our head muscles and neck joints when the head moves. Put another way, we use information which does not arise from external stimuli when we perceive the world.

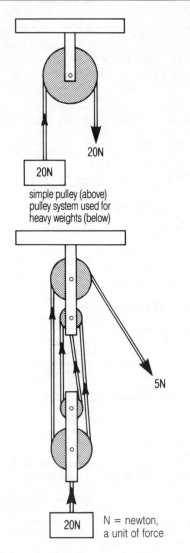

20N

20N

simple pulley (above)
pulley system used for
heavy weights (below)

5N

20N   N = newton,
       a unit of force

pulley *The mechanical advantage of a pulley increases with the number of rope strands. If a pulley system has four ropes supporting the load, the mechanical advantage is four, and a 5 N force will lift a 20 N load.*

an artery is near the surface, for example in the wrist or the neck.

pulse-code modulation (PCM) in physics, a form of digital ◊modulation in which microwaves or light waves (the carrier waves) are switched on and off in pulses of varying length according to a binary code. It is a relatively simple matter to transmit data that are already in binary code, such as those used by computer, by these means. However, if an analogue audio signal is to be transmitted, it must first be converted to a *pulse-amplitude modulated* signal (PAM) by regular sampling of its amplitude. The value of the amplitude is then converted into a binary code for transmission on the carrier wave.

pumice light volcanic rock produced by the frothing action of expanding gases during the solidification of lava. It has the texture of a hard sponge and is used as an abrasive.

pump any device for moving liquids and gases, or compressing gases. Some pumps, such as the traditional *lift pump* used to raise water from wells, work by a reciprocating (up-and-down) action. Movement of a piston in a cylinder with a one-way valve creates a partial vacuum in the cylinder, thereby sucking water into it.

Gear pumps, used to pump oil in a car's lubrication system, have two meshing gears that rotate inside a housing, and the teeth move the oil. *Rotary pumps* contain a rotor with vanes projecting from it inside a casing, sweeping the oil round as they move.

pumped storage hydroelectric plant that uses surplus electricity to pump water back into a high-level reservoir. In normal working conditions the water flows from this reservoir through the ◊turbines to generate power for feeding into the grid. At times of low power demand, electricity is taken from the grid to turn the turbines into pumps that then pump the water back again. This ensures that there is always a maximum 'head' of water in the reservoir to give the maximum output when required. An example of a pumped-storage plant is in Dinorwig, North Wales.

punched card in computing, an early form of data storage and input, now almost obsolete. The 80–column card widely used in the 1960s and 1970s was a thin card, measuring 190 mm/7.5 in × 84 mm/3.33 in, holding up to 80 characters of data encoded as small rectangular holes.

The punched card was invented by French textile manufacturer Joseph-Marie Jacquard (1752–1834) about 1801 to control weaving looms. The first data-processing machine using punched cards was developed by US inventor Herman Hollerith in the 1880s for the US census.

punctuated equilibrium model evolutionary theory developed by Niles Eldridge and US palaeontologist Stephen Jay Gould 1972 to explain discontinuities in the fossil record. It claims that periods of rapid change alternate with periods of relative stability (stasis), and that the appearance of new lineages is a separate process from the gradual evolution of adaptive changes within a species.

The pattern of stasis and more rapid change is now widely accepted, but the second part of the theory remains unsubstantiated.

punnett square graphic technique used in genetics for determining the likely outcome, in statistical terms, of a genetic cross. It resembles a game of noughts and crosses, in which the genotypes of the parental generation gametes are entered first, so that the subsequent combinations can then be calculated.

pupa nonfeeding, largely immobile stage of some insect life cycles, in which larval tissues are broken down, and adult tissues and structures are formed.

In many insects, the pupa is *exarate*, with the appendages (legs, antennae, wings) visible outside the pupal case; in butterflies and moths, it is called

Little green men?

Making a major scientific discovery is not as many people imagine; often, there is no 'eureka moment' or single instant of discovery. Discovery is more often a process of checking and rechecking, of gradually eliminating spurious effects, until the truth is apparent. The discovery of the first pulsar shows this process in action.

Luck played a part in the discovery. In 1967 Antony Hewish at the Mullard Radio Astronomy Laboratory, Cambridge, constructed a new type of radio telescope—a large array of 2048 aerials covering an area of 1.8 hectares/4.4 acres—to study the 'twinkling' or scintillation of radio galaxies. This is caused by clouds of ionized gas ejected from the Sun. It is most noticeable at metre wavelengths, so the new telescope had to be sensitive to radiation of this wavelength.

Most radio telescopes achieve high sensitivity by averaging incoming signals for several seconds, and so are unsuitable for studying rapidly varying signals. They collect radiation with wavelengths of around a centimetre. The new telescope's ability to detect rapidly varying signals of metre wavelength was just what was needed to detect a pulsar.

Twinkle, twinkle, little galaxy

The telescope began work in July 1967. Its first task was to locate all radio galaxies twinkling in the area of sky accessible to it. Each day, as the Earth rotated, the telescope swept its radio eye across a band of sky. A complete scan took four days. Initially a graduate student, Jocelyn Bell, ran the survey and analysed the results, output on about 30 m/100 ft of chart paper each day.

After a few weeks, Bell noticed an unusual signal: not a single blip, but an untidy bunch of squiggles on the chart. It was nothing like the signals she was looking for, so she marked a query on the chart and did not investigate further. Later, when she saw the same signal on another chart, she realized it merited closer attention. The signal came from a part of the sky where scintillations were normally weak. It occurred at night, and scintillations are strongest during the day. A faster chart recorder was installed for a more detailed look at the signal. It would stretch out the signal over a longer chart, like a photographic enlargement.

For a while, Bell and Hewish's efforts were frustrated—the signal weakened and vanished. For a month, there was no sign of it. The researchers feared that they had seen a one-off event: possibly a star flaring brightly for a short time. If so, they had missed the chance to study it in detail. However, on 28 November, it returned. This time, the new recorder revealed the true nature of the signal. It was a series of short pulses about 1.3 seconds apart. Timing the pulses more accurately showed that they were in step to within one-millionth of a second. Their short duration indicated that they were coming from a very small object. Something peculiar was going on, but what?

The task now was to rule out spurious effects. Did the signal result from a machine malfunction? Was it caused by a satellite signal? Or a radar echo from the Moon? The fact that the signal appeared at regular intervals hinted that it was not a machine malfunction.

Having calculated when the signal would next appear, at the appointed time, the research team stood around the recorder. Nothing! Hewish and the others began to wander away. But before they reached the door, they were called back. 'Here it is!' said a student. They had miscalculated when the signal would be picked up. They knew now that it came from outside the laboratory.

Is there anybody there?

Further study established that the signal rotated with the stars, and came from beyond the Solar System, but from within our galaxy. At one stage, Hewish thought that the signal might be a message from an extraterrestrial civilization. It was given the name 'LGM', for 'little green men'. However, this possibility was ruled out. Other beings must live on a planet circling round a star. The planetary motion would show up as a slight variation in the pulse rate. After carefully timing the pulses for several weeks, the idea of a planetary origin was given up.

On 21 December 1967 Bell discovered a second signal elsewhere in the sky. This clinched the reality of the phenomenon, and the name 'pulsar' was quickly coined to describe the object radiating the pulsation. The results were published in February 1968; within a year it was generally accepted that the signals came from rapidly spinning neutron stars. Since then more than 500 pulsars have been discovered. Antony Hewish shared the 1974 Nobel Prize for physics with British radio astronomer Martin Ryle for their work in radio astronomy and, in particular, the discovery of pulsars.

PCM microwave

an analogue signal

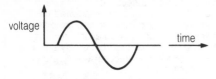

pulse-amplitude-modulated signal (PAM)

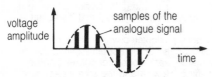

pulse-code modulation

a chrysalis, and is *obtect*, with the appendages developing inside the case.

putrefaction decomposition of organic matter by microorganisms.

PVC abbreviation for *polyvinylchloride*, a type of ◊plastic derived from vinyl chloride (CH_2:$CHCl$).

PWR abbreviation for ◊*pressurized water reactor*, a type of nuclear reactor.

pyramid in geometry, a three-dimensional figure with triangular side-faces meeting at a common vertex (point) and with a ◊polygon as its base. The volume V of a pyramid is given by $V = \frac{1}{3}Bh$, where B is the area of the base and h is the perpendicular height.

Pyramids are generally classified by their bases. For example, the Egyptian pyramids have square bases, and are therefore called square pyramids. Triangular pyramids are also known as tetrahedra ('four sides').

pyramidal peak angular mountain peak with concave faces found in glaciated areas; for example, the Matterhorn in Switzerland. It is formed when

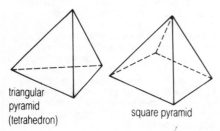

triangular
pyramid
(tetrahedron)

square pyramid

pyramid

three or four ◊corries (steep-sided hollows) are eroded, back-to-back, around the sides of a mountain, leaving an isolated peak in the middle.

pyramid of numbers in ecology, a diagram that shows how many plants and animals there are at different levels in a ◊food chain.

There are always far fewer individuals at the bottom of the chain than at the top because only about 10% of the food an animal eats is turned into flesh, so the amount of food flowing through the chain drops at each step. In a pyramid of numbers, the primary producers (usually plants) are represented at the bottom by a broad band, the plant-eaters are shown above by a narrower band, and the animals that prey on them by a narrower band still. At the top of the pyramid are the 'top carnivores' such as lions and sharks, which are present in the smallest number.

Progress in science depends on new techniques, new discoveries, and new ideas, probably in that order.

Sidney Brenner *Nature* May 1980

pyridine C_5H_5N a heterocyclic compound (see ◊cyclic compounds). It is a liquid with a sickly smell that occurs in coal tar. It is soluble in water, acts as a strong ◊base, and is used as a solvent, mainly in the manufacture of plastics.

pyridoxine or *vitamin B₆* $C_8H_{11}NO_3$ water-soluble ◊vitamin of the B complex. There is no clearly identifiable disease associated with deficiency but its absence from the diet can give rise to malfunction of the central nervous system and general skin disorders. Good sources are liver, meat, milk, and cereal grains. Related compounds may also show vitamin B_6 activity.

pyrite iron sulphide FeS_2; also called *fool's gold* because of its yellow metallic lustre. Pyrite has a hardness of 6–6.5 on the Mohs' scale. It is used in the production of sulphuric acid.

pyroclastic in geology, pertaining to fragments of solidified volcanic magma, ranging in size from fine ash to large boulders, that are extruded during an explosive volcanic eruption; also the rocks that are formed by consolidation of such material. Pyroclastic rocks include tuff (ash deposit) and agglomerate (volcanic breccia).

pyroclastic deposit deposit made up of fragments of rock, ranging in size from fine ash to large boulders, ejected during an explosive volcanic eruption.

pyrogallol $C_6H_3OH_3$ (technical name *trihydroxybenzene*) derivative of benzene, prepared from gallic acid. It is used in gas analysis for the measurement of oxygen because its alkaline solution turns black as it rapidly absorbs oxygen. It is also used as a developer in photography.

pyrolysis decomposition of a substance by heating it to a high temperature in the absence of air. The process is used to burn and dispose of old tyres, for example, without contaminating the atmosphere.

pyrometer instrument used for measuring high temperatures.

pyroxene any one of a group of minerals, silicates of calcium, iron, and magnesium with a general formula X,YSi_2O_6, found in igneous and metamorphic rocks. The internal structure is based on single chains of silicon and oxygen. Diopside (X = Ca, Y = Mg) and augite (X = Ca, Y = Mg,Fe,Al) are common pyroxenes.

Jadeite ($NaAlSi_2O_6$), which is considered the more valuable form of jade, is also a pyroxene.

Pythagoras' theorem in geometry, a theorem stating that in a right-angled triangle, the area of the square on the hypotenuse (the longest side) is equal to the sum of the areas of the squares drawn on the other two sides. If the hypotenuse is h units long and the lengths of the other sides are a and b, then $h^2 = a^2 + b^2$.

The theorem provides a way of calculating the length of any side of a right-angled triangle if the lengths of the other two sides are known. It is also used to determine certain trigonometrical relationships such as $\sin^2 \theta + \cos^2 \theta = 1$.

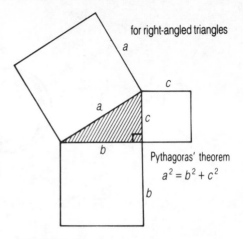

for right-angled triangles

Pythagoras' theorem
$$a^2 = b^2 + c^2$$

Pythagoras' theorem *Pythagoras' theorem for right-angled triangles is likely to have been known long before the time of Pythagoras. It was probably used by the ancient Egyptians to lay out the pyramids.*

quadrat in environmental studies, a square structure used to study the distribution of plants in a particular place, for instance a field, rocky shore, or mountainside. The size varies, but is usually 0.5 or 1 metre square, small enough to be carried easily. The quadrat is placed on the ground and the abundance of species estimated. By making such measurements a reliable understanding of species distribution is obtained.

quadratic equation in mathematics, a polynomial equation of second degree (that is, an equation containing as its highest power the square of a variable, such as x^2). The general formula of such equations is $ax^2 + bx + c = 0$, in which a, b, and c are real numbers, and only the coefficient a cannot equal 0. In ◊coordinate geometry, a quadratic function represents a ◊parabola.

Some quadratic equations can be solved by factorization, or the values of x can be found by using the formula for the general solution

$$x = [-b \pm \sqrt{(b^2 - 4ac)}]/2a$$

Depending on the value of the discriminant $b^2 - 4ac$, a quadratic equation has two real, two equal, or two complex roots (solutions). When $b^2 - 4ac < 0$, there are two distinct real roots. When $b^2 - 4ac = 0$, there are two equal real roots. When $b^2 - 4ac > 0$, there are two distinct complex roots.

quadrature position of the Moon or an outer planet where a line between it and Earth makes a right angle with a line joining Earth to the Sun.

quadrilateral a plane (two-dimensional) figure with four straight sides. The following are all quadrilaterals, each with distinguishing properties: *square* with four equal angles and sides, four axes of symmetry; *rectangle* with four equal angles, opposite sides equal, two axes of symmetry; *rhombus* with four equal sides, two axes of symmetry; *parallelogram* with two pairs of parallel sides, rotational symmetry; and *trapezium* one pair of parallel sides.

qualitative analysis in chemistry, a procedure for determining the identity of the component(s) of a single substance or mixture. A series of simple reactions and tests can be carried out on a compound to determine the elements present.

quantitative analysis in chemistry, a procedure for determining the precise amount of a known component present in a single substance or mixture. A known amount of the substance is subjected to particular procedures. *Gravimetric analysis* determines the mass of each constituent present; ◊*volumetric analysis* determines the concentration of a solution by ◊titration against a solution of known concentration.

quantum chromodynamics (QCD) in physics, a theory describing the interactions of ◊quarks, the elementary particles that make up all ◊hadrons (subatomic particles such as protons and neutrons). In quantum chromodynamics, quarks are considered to interact by exchanging particles called gluons, which carry the ◊strong nuclear force, and whose role is to 'glue' quarks together. The mathematics involved in the theory is complex, and although a number of successful predictions have been made, as yet the theory does not compare in accuracy with ◊quantum electrodynamics, upon which it is modelled. See ◊elementary particle and ◊forces, fundamental.

quantum electrodynamics (QED) in physics, a theory describing the interaction of charged subatomic particles within electric and magnetic fields. It combines ◊quantum theory and ◊relativity, and considers charged particles to interact by the exchange of photons. QED is remarkable for the accuracy of its predictions—for example, it has been used to calculate the value of some physical quantities to an accuracy of ten decimal places, a feat equivalent to calculating the distance between New York and Los Angeles to within the thickness of a hair. The theory was developed by US physicists Richard Feynman and Julian Schwinger, and by Japanese physicist Sin-Itiro Tomonaga 1948.

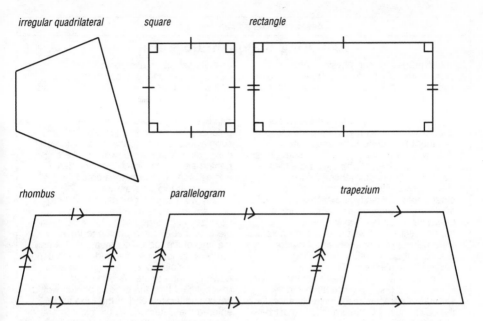

irregular quadrilateral   square   rectangle

rhombus   parallelogram   trapezium

quadrilateral

quantum number in physics, one of a set of four numbers that uniquely characterize an ◊electron and its state in an ◊atom. The *principal quantum number n* defines the electron's main energy level. The *orbital quantum number l* relates to its angular momentum. The *magnetic quantum number m* describes the energies of electrons in a magnetic field. The *spin quantum number m*ₛ gives the spin direction of the electron.

The principal quantum number, defining the electron's energy level, corresponds to shells (energy levels) also known by their spectroscopic designations K, L, M, and so on. The orbital quantum number gives rise to a series of subshells designated *s*, *p*, *d*, *f*, and so on, of slightly different energy levels. The magnetic quantum number allows further subdivision of the subshells (making three subdivisions p_x, p_y, and p_z in the *p* subshell, for example, of the same energy level). No two electrons in an atom can have the same set of quantum numbers (the Pauli exclusion principle).

quantum theory or *quantum mechanics* in physics, the theory that ◊energy does not have a continuous range of values, but is, instead, absorbed or radiated discontinuously, in multiples of definite, indivisible units called quanta. Just as earlier theory showed how light, generally seen as a wave motion, could also in some ways be seen as composed of discrete particles (◊photons), quantum theory shows how atomic particles such as electrons may also be seen as having wavelike properties. Quantum theory is the basis of particle physics, modern theoretical chemistry, and the solid-state physics that describes the behaviour of the silicon chips used in computers.

The theory began with the work of Max Planck 1900 on radiated energy, and was extended by Albert Einstein to electromagnetic radiation gener-

ally, including light. Danish physicist Niels Bohr used it to explain the ◊spectrum of light emitted by excited hydrogen atoms. Later work by Erwin Schrödinger, Werner Heisenberg, Paul Dirac, and others elaborated the theory to what is called quantum mechanics (or wave mechanics).

quark in physics, the ◊elementary particle that is the fundamental constituent of all ◊hadrons (baryons, such as neutrons and protons, and mesons). There are six types, or 'flavours': up, down, top, bottom, strange, and charm, each of which has three varieties, or 'colours': red, yellow, and blue (visual colour is not meant, although the analogy is useful in many ways). To each quark there is an antiparticle, called an antiquark. See ◊quantum chromodynamics.

Three quarks for Muster Mark!

Supposedly the origin of the name **quark**, James Joyce (1882–1941) *Finnegans Wake* 1939

---

quart imperial liquid or dry measure, equal to two pints or 1.136 litres. In the USA, a liquid quart is equal to 0.946 litre, while a dry quart is equal to 1.101 litres.

quartz crystalline form of ◊silica SiO_2, one of the most abundant minerals of the Earth's crust (12% by volume). Quartz occurs in many different kinds of rock, including sandstone and granite. It ranks 7 on the Mohs' scale of hardness and is resistant to chemical or mechanical breakdown. Quartzes vary according to the size and purity of their crystals. Crystals of pure quartz are coarse, colourless, and transparent, and this form is usually called rock crystal. Impure coloured varieties, often used as gemstones, include ◊agate, citrine quartz, and ◊amethyst. Quartz is used in ornamental work and

Exploring the quantum world

In the 1930s German physicist Albert Einstein and his colleagues Boris Podolsky and Nathan Rosen proposed an experiment which they thought would expose a flaw in quantum theory. The experiment was based on the behaviour of pairs of particles that were once close together but had become separated. Such pairs are formed when an atom gives out two particles of light, called photons. According to quantum theory, if we observe one photon of such a pair, the other photon instantly knows that we have made an observation and adjusts its behaviour in certain ways. This would happen even if we waited until the photons were billions of kilometres apart before making our observation. According to Albert Einstein's theory of relativity no information can travel faster than the speed of light, so how can such widely separated particles react instantly after the observation? Quantum theory must be wrong, said Einstein.

Einstein was never comfortable with quantum theory. He could not accept the idea that in the subatomic world events happen almost by chance, and particles do not have exact positions. Instead there is a range of positions where the particle might be found and all we can do is calculate the probability of it being at any particular point. There is a similar uncertainty about other physical quantities, such as energy and momentum. Einstein felt that behind the uncertainty of quantum theory there must be an exact reality. Paradoxically, when his suggested experiment was eventually performed, it proved him wrong.

Sunglasses for photons

It took a long time to convert Einstein's idea into a practicable experiment. However, in Paris in the early 1980s a series of experiments was carried out by a team of French scientists led by Alain Aspect, Jean Dalibard and Gerard Roger. The apparatus used consisted of a source of light—calcium atoms—halfway along a long tube. Pairs of photons simultaneously emitted by the calcium atoms split so that each photon travelled towards opposite ends of the tube. At each end of the tube there were devices which would detect photons. Just before the detectors were further devices, called a light switch and polarizer, that could change a certain property— the polarization—of a photon within 0.000,000,01 second of the photon reaching it.

The light switches were ingenious devices. Each consisted of a small cell of water in which two vibrating piezoelectric crystals set up an ultrasonic wave. Depending upon the exact state of the ultrasonic vibration when a photon arrived, the photon either passed through the cell, or was deflected at right angles. The photons then passed through the polarizer, which worked just like sunglasses to change the polarization of the photon. There were two polarizers behind each light switch, set at different angles, so that the photons could be polarized in two different ways.

Why were such high-speed switches needed? Because the speed of the polarization change had to be faster than the time that it took a photon to travel to the end of the tube. This meant that, when the polarization was changed, there must be no possibility of a 'message' passing down the tube to the second photon (because by then the second photon would have reached the end of the tube and entered the detector). As the experiment proceeded, the light switch operated to continually change the polarization of the photons passing through it. The scientists measured the polarization of the photons arriving at each end of the tube. Photons which arrived simultaneously at the tube ends were obviously emitted simultaneously by the same calcium atom and therefore, according to quantum theory, should be linked in some mysterious way.

There were two possible results of this experiment. If quantum theory was correct, the polarization of photons arriving simultaneously at the detectors would always be the same. If quantum theory was incorrect, then the polarization of the photons would not always be the same.

Faster than light?

The results went against Einstein. Aspect and his team found that if one of a pair of photons was polarized in a certain way, its twin at the other end of the apparatus would always be polarized in the same way. It is as if the photons know what is happening to each other, even though there can be no possible communication between them. So, there is a mysterious instantaneous faster-than-light 'action at a distance' between once-linked photons, and presumably between once-linked particles, too. Scientists and philosophers are still examining the implications of this result. According to the Big Bang theory, all particles now in existence originate from a common point at the birth of the universe. Does this mean that there is a hidden connection between all the particles in the universe? How does this web of connections manifest itself?

industry, where its reaction to electricity makes it valuable in electronic instruments (see ◊piezoelectric effect). Quartz can also be made synthetically.

Crystals that would take millions of years to form naturally can now be 'grown' in pressure vessels to a standard that allows them to be used in optical and scientific instruments and in electronics, such as quartz wristwatches.

quartzite ◊metamorphic rock consisting of pure quartz sandstone that has recrystallized under increasing heat and pressure.

quasar (from *quasi*-stell*ar* object or QSO) one of the most distant extragalactic objects known, discovered 1964–65. Quasars appear starlike, but each emits more energy than 100 giant galaxies. They are thought to be at the centre of galaxies, their brilliance emanating from the stars and gas falling towards an immense ◊black hole at their nucleus.

Quasar light shows a large ◊red shift, indicating that they are very distant. Some quasars emit radio waves (see ◊radio astronomy), which is how they were first identified 1963, but most are radio-quiet. The furthest are over 10 billion light years away.

quasi-atom particle assemblage resembling an atom, in which particles not normally found in atoms become bound together for a brief period. Quasi-atoms are generally unstable structures, either because they are subject to matter-antimatter annihilation (positronium), or because one or more of their constituents is unstable (muonium).

FAINT QUASARS

From the Earth, quasars appear about as bright as a candle on the Moon. Astronomers need to amplify the light from quasars by 10 million times in order to study them. If a radio telescope collected the energy from a quasar for 10,000 years, there would only be enough energy to light a small bulb for a fraction of a second.

Quaternary period of geological time that began 1.64 million years ago and is still in process. It is divided into the ◊Pleistocene and ◊Holocene epochs.

quenching ◊heat treatment used to harden metals. The metals are heated to a certain temperature and then quickly plunged into cold water or oil.

quicklime common name for ◊calcium oxide.

quicksilver former name for the element ◊mercury.

quinine antimalarial drug extracted from the bark of the cinchona tree. Peruvian Indians taught French missionaries how to use the bark in 1630, but quinine was not isolated until 1820. It is a bitter alkaloid $C_{20}H_{24}N_2O_2$.

Other drugs against malaria have since been developed with fewer side effects, but quinine derivatives are still valuable in the treatment of unusually resistant strains.

raceme in botany, a type of ◊inflorescence.

rad unit of absorbed radiation dose, now replaced in the SI system by the ◊gray (one rad equals 0.01 gray), but still commonly used. It is defined as the dose when one kilogram of matter absorbs 0.01 joule of radiation energy (formerly, as the dose when one gram absorbs 100 ergs).

radar (acronym for *radio direction and ranging*) device for locating objects in space, direction finding, and navigation by means of transmitted and reflected high-frequency radio waves.

The direction of an object is ascertained by transmitting a beam of short-wavelength (1–100 cm/½–40 in), short-pulse radio waves, and picking up the reflected beam. Distance is determined by timing the journey of the radio waves (travelling at the speed of light) to the object and back again. Radar is also used to detect objects underground, for example service pipes, and in archaeology. Contours of remains of ancient buildings can be detected down to 20 m/66 ft below ground.

Radar is essential to navigation in darkness, cloud, and fog, and is widely used in warfare to detect enemy aircraft and missiles. To avoid detection, various devices, such as modified shapes (to reduce their radar cross-section), radar-absorbent paints and electronic jamming are used. To pinpoint small targets ◊laser 'radar' instead of microwaves, has been developed. Chains of ground radar stations are used to warn of enemy attack—for example, North Warning System 1985, consisting of 52 stations across the Canadian Arctic and N Alaska. Radar is also used in ◊meteorology and ◊astronomy.

radar astronomy bouncing of radio waves off objects in the Solar System, with reception and analysis of the 'echoes'. Radar contact with the Moon was first made 1945 and with Venus 1961. The travel time for radio reflections allows the distances of objects to be determined accurately. Analysis of the reflected beam reveals the rotation period and allows the object's surface to be mapped. The rotation periods of Venus and Mercury were first determined by radar. Radar maps of Venus were obtained first by Earth-based radar and subsequently by orbiting space probes.

radial circuit circuit used in household electric wiring in which all electrical appliances are connected to cables that radiate out from the main supply point or fuse box. In more modern systems, the appliances are connected in a ring, or ◊ring circuit, with each end of the ring connected to the fuse box.

radian SI unit (symbol rad) of plane angles, an alternative unit to the ◊degree. It is the angle at the centre of a circle when the centre is joined to the two ends of an arc (part of the circumference) equal in length to the radius of the circle. There are 2π (approximately 6.284) radians in a full circle (360°).

One radian is approximately 57°, and 1° is $\pi/180$ or approximately 0.0175 radians. Radians are commonly used to specify angles in ◊polar coordinates.

radiant heat energy that is radiated by all warm or hot bodies. It belongs to the infrared part of the electromagnetic ◊spectrum and causes heating when absorbed. Radiant heat is invisible and should not be confused with the red glow associated with very hot objects, which belongs to the visible part of the spectrum.

Infrared radiation can travel through a vacuum and it is in this form that the radiant heat of the Sun travels through space. It is the trapping of this radiation by carbon dioxide and methane in the atmosphere that gives rise to the ◊greenhouse effect.

radiation in physics, emission of radiant ◊energy as particles or waves—for example, heat, light, alpha particles, and beta particles (see ◊electromagnetic waves and ◊radioactivity). See also ◊atomic radiation.

radiation biology study of how living things are affected by radioactive (ionizing) emissions (see ◊radioactivity) and by electromagnetic (nonionizing) radiation (◊electromagnetic waves). Both are potentially harmful and can cause mutations as well as leukaemia and other cancers; even low levels of radioactivity are very dangerous. Both are, however, used therapeutically, for example to treat cancer, when the radiation dose is very carefully controlled (◊radiotherapy or X-ray therapy).

Radioactive emissions are more harmful. Exposure to high levels produces radiation burns and radiation sickness, plus genetic damage (resulting in birth defects) and cancers in the longer term. Exposure to low-level ionizing radiation can also cause genetic damage and cancers, particularly leukaemia.

Electromagnetic radiation is usually harmful only if exposure is to high-energy emissions, for example close to powerful radio transmitters or near radar-wave sources. Such exposure can cause organ damage, cataracts, loss of hearing, leukaemia and other cancers, or premature ageing. It may also affect the nervous system and brain, distorting their electrical nerve signals and leading to depression, disorientation, headaches, and other symptoms. Individual sensitivity varies and some people are affected by electrical equipment such as televisions, computers, and refrigerators.

Background radiation is the natural radiation produced by cosmic rays and radioactive rocks such as granite. Readings of background radiation must be taken into account when calculating any effect produced by nuclear accidents or contamination from power stations.

radiation sickness sickness resulting from exposure to radiation, including X-rays, gamma rays, neutrons, and other nuclear radiation, as from weapons and fallout. Such radiation ionizes atoms in the body and causes nausea, vomiting, diarrhoea, and other symptoms. The body cells themselves may be damaged even by very small doses, causing leukaemia; genetic changes may be induced in the germ plasm, causing infants to be born damaged or mutated.

radiation units units of measurement for radioactivity and radiation doses. Continued use of the units introduced earlier this century (the curie, rad, rem, and roentgen) has been approved while the derived SI units (becquerel, gray, sievert, and coulomb) become familiar. One curie equals 3.7×10^{-10} becquerels (activity); one rad equals 10^{-2} gray (absorbed dose); one rem equals 10^{-2} sievert (dose equivalent); one roentgen equals 2.58×10^{-4} coulomb/kg (exposure to ionizing radiation).

The average radiation exposure per person per year in the USA is one millisievert (0.1 rem), of which 50% is derived from naturally occurring radon.

radical in chemistry, a group of atoms forming part of a molecule, which acts as a unit and takes part in chemical reactions without disintegration, yet often cannot exist alone; for example, the methyl radical $-CH_3$, or the carboxyl radical $-COOH$.

radicle part of a plant embryo that develops into the primary root. Usually it emerges from the seed before the embryonic shoot, or ◊plumule, its tip protected by a root cap, or calyptra, as it pushes through the soil. The radicle may form the basis of the entire root system, or it may be replaced by adventitious roots (positioned on the stem).

radio transmission and reception of radio waves. In radio transmission a microphone converts ◊sound waves (pressure variations in the air) into ◊electromagnetic waves that are then picked up by a receiving aerial and fed to a loudspeaker, which converts them back into sound waves.

The theory of electromagnetic waves was first developed by Scottish physicist James Clerk Maxwell 1864, given practical confirmation in the laboratory 1888 by German physicist Heinrich Hertz, and put to practical use by Italian inventor Guglielmo Marconi, who in 1901 achieved reception of a signal in Newfoundland transmitted from Cornwall, England.

To carry the transmitted electrical signal, an ◊oscillator produces a carrier wave of high frequency; different stations are allocated different transmitting carrier frequencies. A modulator superimposes the audiofrequency signal on the carrier. There are two main ways of doing this: ◊amplitude modulation (AM), used for long-and medium-wave broadcasts, in which the strength of the carrier is made to fluctuate in time with the audio signal; and ◊frequency modulation (FM), as used for VHF broadcasts, in which the frequency of the carrier is made to fluctuate. The transmitting aerial emits the modulated electromagnetic waves, which travel outwards from it.

In radio reception a receiving aerial picks up minute voltages in response to the waves sent out by a transmitter. A tuned circuit selects a particular frequency, usually by means of a variable ◊capacitor connected across a coil of wire. A demodulator disentangles the audio signal from the carrier, which is now discarded, having served its purpose. An amplifier boosts the audio signal for feeding to the loudspeaker. In a ◊superheterodyne receiver, the incoming signal is mixed with an internally-generated signal of fixed frequency so that the amplifier circuits can operate near their optimum frequncy.

radioactive decay process of continuous disintegration undergone by the nuclei of radioactive elements, such as radium and various isotopes of uranium and the transuranic elements. This changes the element's atomic number, thus transmuting one element into another, and is accompanied by the emission of radiation. Alpha and beta decay are the most common forms.

In **alpha decay** (the loss of a helium nucleus – two protons and two neutrons) the atomic number decreases by two; in **beta decay** (the loss of an electron) the atomic number increases by one. Certain lighter artificially created isotopes also undergo radioactive decay. The associated radiation consists of alpha rays, beta rays, or gamma rays (or a combination of these), and it takes place at a constant rate expressed as a specific half-life, which is the time taken for half of any mass of that particular isotope to decay completely. Less commonly occurring decay forms include heavy-ion emission, electron capture, and spontaneous fission (in each of these the atomic number decreases).

The original nuclide is known as the parent sub-

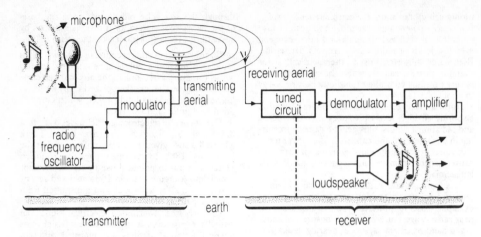

radio *Radio transmission and reception. The radio frequency oscillator generates rapidly varying electrical signals, which are sent to the transmitting aerial. In the aerial, the signals produce radio waves (the carrier wave), which spread out at the speed of light. The sound signal is added to the carrier wave by the modulator. When the radio waves fall on the receiving aerial, they induce an electrical current in the aerial. The electrical current is sent to the tuning circuit, which picks out the signal from the particular transmitting station desired. The demodulator separates the sound signal from the carrier wave and sends it, after amplification, to the loudspeaker.*

stance, and the product is a daughter nuclide (which may or may not be radioactive). The final product in all modes of decay is a stable element.

Radioactive Incident Monitoring Network (RIMNET) monitoring network at 46 (to be raised to about 90) Meteorological Office sites throughout the UK. It feeds into a central computer, and was installed 1989 to record contamination levels from nuclear incidents such as the ◊Chernobyl disaster.

radioactive tracer any of various radioactive ◊isotopes used in labelled compounds; see ◊tracer.

radioactive waste any waste that emits radiation in excess of the background level. See ◊nuclear waste.

radioactivity spontaneous alteration of the nuclei of radioactive atoms, accompanied by the emission of radiation. It is the property exhibited by the radioactive ◊isotopes of stable elements and all isotopes of radioactive elements, and can be either natural or induced. See ◊radioactive decay.

Radioactivity establishes an equilibrium in parts of the nuclei of unstable radioactive substances, ultimately to form a stable arrangement of nucleons (protons and neutrons); that is, a non-radioactive (stable) element. This is most frequently accomplished by the emission of ◊alpha particles (helium nuclei); ◊beta particles (electrons and positrons); or ◊gamma radiation (electromagnetic waves of very high frequency). It takes place either directly, through a one-step decay, or indirectly, through a number of decays that transmute one element into another. This is called a decay series or chain, and sometimes produces an element more radioactive than its predecessor.

The instability of the particle arrangements in the nucleus of a radioactive atom (the ratio of neutrons to protons and/or the total number of both) determines the lengths of the ◊half-lives of the isotopes of that atom, which can range from fractions of a second to billions of years. All isotopes of relative atomic mass 210 and greater are radio-active. Alpha, beta, and gamma radiation are ionizing in their effect and are therefore dangerous to body tissues, especially if a radioactive substance is ingested or inhaled.

RADIOACTIVITY: OUR EXPLODING BODIES

Our bodies contain tiny amounts of the radioactive isotopes potassium-40 and carbon-14. Around 38,000 atoms of potassium-40 and 1,200 of carbon-14 explode in your body each second. However, it has been calculated that the chance of this causing any serious damage to the body is negligible.

---

radio astronomy study of radio waves emitted naturally by objects in space, by means of a ◊radio telescope. Radio emission comes from hot gases (*thermal radiation*); electrons spiralling in magnetic fields (*synchrotron radiation*); and specific wavelengths (*lines*) emitted by atoms and molecules in space, such as the 21–cm/8–in line emitted by hydrogen gas.

Radio astronomy began 1932 when US astronomer Karl Jansky detected radio waves from the centre of our Galaxy, but the subject did not develop until after World War II. Radio astronomy has greatly improved our understanding of the evolution of stars, the structure of galaxies, and the origin of the universe. Astronomers have mapped the spiral structure of the Milky Way from the radio waves given out by interstellar gas, and they have detected many individual radio sources within our Galaxy and beyond.

Among radio sources in our Galaxy are the remains of ◊supernova explosions, such as the ◊Crab nebula and ◊pulsars. Short-wavelength radio waves have been detected from complex molecules in dense clouds of gas where stars are forming. Searches have been undertaken for signals from other civilizations in the Galaxy, so far without success.

Strong sources of radio waves beyond our Galaxy include ◊radio galaxies and ◊quasars. Their existence far off in the universe demonstrates how the universe has evolved with time. Radio astronomers have also detected weak *background radiation* thought to be from the ◊Big Bang explosion that marked the birth of the universe.

radio beacon radio transmitter in a fixed location, used in marine and aerial ◊navigation. Ships and aircraft pinpoint their positions by reference to continuous signals given out by two or more beacons.

radiocarbon dating or *carbon dating* method of dating organic materials (for example, bone or wood), used in archaeology. Plants take up carbon dioxide gas from the atmosphere and incorporate it into their tissues, and some of that carbon dioxide contains the radioactive isotope of carbon, carbon-14. This decays at a known rate (half of it decays every 5,730 years); the time elapsed since the plant died can therefore be measured in a laboratory. Animals take carbon-14 into their bodies from eating plant tissues and their remains can be similarly dated. After 120,000 years so little carbon-14 is left that no measure is possible (see ◊half-life).

Radiocarbon dating was first developed by US chemist Willard Libby 1949. The method yields reliable ages back to about 50,000 years, but its results require correction since Libby's assumption that the concentration of carbon-14 in the atmosphere was constant through time has subsequently been proved wrong. Radiocarbon dates from tree rings (see ◊dendrochronology) showed that material before 1000 BC had been exposed to greater concentrations of carbon-14. Now radiocarbon dates are calibrated against calendar dates obtained from tree rings, or, for earlier periods, against uranium/thorium dates obtained from coral. The carbon-14 content is determined by counting beta particles with either a proportional gas or a liquid scintillation counter for a period of time. A new advance, AMS (accelerator mass spectrometry), requires only tiny samples, and counts the atoms of carbon-14 directly, disregarding their decay.

radio, cellular portable telephone system; see ◊cellular phone.

radiochemistry chemical study of radioactive isotopes and their compounds (whether produced from naturally radioactive or irradiated materials) and their use in the study of other chemical processes.

When such isotopes are used in labelled compounds, they enable the biochemical and physiological functioning of parts of the living body to be observed. They can help in the testing of new drugs, showing where the drug goes in the body and how long it stays there. They are also useful in diagnosis – for example cancer, fetal abnormalities, and heart disease.

radio galaxy galaxy that is a strong source of electromagnetic waves of radio wavelengths. All galaxies, including our own, emit some radio waves, but radio galaxies are up to a million times more powerful.

In many cases the strongest radio emission comes not from the visible galaxy but from two clouds, invisible through an optical telescope, that can extend for millions of light years either side of the galaxy. This double structure at radio wavelengths is also shown by some ◊quasars, suggesting a close relationship between the two types of object. In both cases, the source of energy is thought to be a massive black hole at the centre. Some radio galaxies are thought to result from two galaxies in collision or recently merged.

radiography branch of science concerned with the use of radiation (particularly ◊X-rays) to produce images on photographic film or fluorescent screens. X-rays penetrate matter according to its nature, density, and thickness. In doing so they can cast shadows on photographic film, producing a radiograph. Radiography is widely used in medicine for examining bones and tissues and in industry for examining solid materials; for example, to check welded seams in pipelines.

radioisotope (contraction of *radioactive ◊isotope*) in physics, a naturally occurring or synthetic radioactive form of an element. Most radioisotopes are made by bombarding a stable element with neutrons in the core of a nuclear reactor. The radiations given off by radioisotopes are easy to detect (hence their use as ◊tracers), can in some instances penetrate substantial thicknesses of materials, and have profound effects (such as genetic ◊mutation) on living matter. Although dangerous, radioisotopes are used in the fields of medicine, industry, agriculture, and research.

Most natural isotopes of atomic mass below 208 are not radioactive. Those from 210 and up are all radioactive.

radioisotope scanning use of radioactive materials (radioisotopes or radionuclides) to pinpoint disease. It reveals the size and shape of the target organ and whether any part of it is failing to take up radioactive material, usually an indication of disease.

The speciality known as nuclear medicine makes use of the affinity of different chemical elements for certain parts of the body. Iodine, for instance, always makes its way to the thyroid gland. After being made radioactive, these materials can be given by mouth or injected, and then traced on scanners working on the Geiger-counter principle. The diagnostic record gained from radioisotope scanning is known as a *scintigram*.

radiometric dating method of dating rock by assessing the amount of ◊radioactive decay of naturally occurring ◊isotopes. The dating of rocks may be based on the gradual decay of uranium into lead. The ratio of the amounts of 'parent' to 'daughter' isotopes in a sample gives a measure of the time it has been decaying, that is, of its age. Different elements and isotopes are used depending on the isotopes present and the age of the rocks to be dated. Once-living matter can often be dated by ◊radiocarbon dating, employing the half-life of the isotope carbon-14, which is naturally present in organic tissue.

Radiometric methods have been applied to the decay of long-lived isotopes, such as potassium-40, rubidium-87, thorium-232, and uranium-238 which are found in rocks. These isotopes decay very slowly and this has enabled rocks as old as 3800 million years to be dated accurately. Carbon dating can be used for material between 100,000 and 1000 years old. *Potassium* dating is used for material more than 100,000 years old, *rubidium* for rocks more than 10 million years old, and *uranium* and *thorium* dating is suitable for rocks older than 20 million years.

radiosonde balloon carrying a compact package of meteorological instruments and a radio transmitter, used to 'sound', or measure, conditions in the atmosphere. The instruments measure temperature, pressure, and humidity, and the information gathered is transmitted back to observers on the ground. A radar target is often attached, allowing the balloon to be tracked.

radio telescope instrument for detecting radio waves from the universe in ◊radio astronomy. Radio telescopes usually consist of a metal bowl that collects and focuses radio waves the way a concave mirror collects and focuses light waves. Radio telescopes are much larger than optical telescopes, because the wavelengths they are detecting are much longer than the wavelength of light. The largest single dish is 305 m/1,000 ft across, at Arecibo, Puerto Rico.

A large dish such as that at ◊Jodrell Bank, England, can see the radio sky less clearly than a small optical telescope sees the visible sky. *Interferometry* is a technique in which the output from two dishes is combined to give better resolution of detail than with a single dish. *Very long baseline interferometry* (VBLI) uses radio telescopes spread across the world to resolve minute details of radio sources.

In *aperture synthesis*, several dishes are linked together to simulate the performance of a very large single dish. This technique was pioneered by Martin Ryle at Cambridge, England, site of a radio telescope consisting of eight dishes in a line 5 km/ 3 mi long. The ◊Very Large Array in New Mexico consists of 27 dishes arranged in a Y-shape, which simulates the performance of a single dish 27 km/ 17 mi in diameter. Other radio telescopes are shaped like long troughs, and some consist of simple rod-shaped aerials.

radiotherapy treatment of disease by ◊radiation from X-ray machines or radioactive sources. Radiation, which reduces the activity of dividing cells, is of special value for its effect on malignant tissues, certain nonmalignant tumours, and some diseases of the skin.

Generally speaking, the rays of the diagnostic X-ray machine are not penetrating enough to be efficient in treatment, so for this purpose more powerful machines are required, operating from 10,000 to over 30 million volts. The lower-voltage machines are similar to conventional X-ray machines; the higher-voltage ones may be of special design; for example, linear accelerators and betatrons. Modern radiotherapy is associated with fewer side effects than formerly, but radiotherapy to the head can cause temporary hair loss, and if the treatment involves the gut, diarrhoea and vomiting may occur. Much radiation now given uses synthesized ◊radioisotopes. Radioactive cobalt is

the most useful, since it produces gamma rays, which are highly penetrating, and it is used instead of very high-energy X-rays.

Similarly, certain radioactive substances may be administered to patients; for example, radioactive iodine for thyroid disease. Radium, formerly widely used for radiotherapy, has now been supplanted by more easily obtainable artificially produced substances.

radio wave electromagnetic wave possessing a long wavelength (ranging from about 10^{-3} to 10^4 m) and a low frequency (from about 10^5 to 10^{11} Hz). Included in the radio-wave part of the spectrum are ◊microwaves, used for both communications and for cooking; ultra high and very high frequency waves, used for television and FM (◊frequency modulation) radio communications; and short, medium, and long waves, used for AM (◊amplitude modulation) radio communications. Radio waves that are used for communications have all been modulated (see ◊modulation) to carry information. Stars emit radio waves, which may be detected and studied using ◊radio telescopes.

radium (Latin *radius* 'ray') white, radioactive, metallic element, symbol Ra, atomic number 88, relative atomic mass 226.02. It is one of the ◊alkaline-earth metals, found in nature in ◊pitchblende and other uranium ores. Of the 16 isotopes, the commonest, Ra-226, has a half-life of 1.622 years. The element was discovered and named in 1898 by Pierre and Marie Curie, who were investigating residues of pitchblende.

Radium decays in successive steps to produce radon (a gas), polonium, and finally a stable isotope of lead. The isotope Ra-223 decays through the uncommon mode of heavy-ion emission, giving off carbon-14 and transmuting directly to lead. Because radium luminesces, it was formerly used in paints that glowed in the dark; when the hazards of radioactivity became known its use was abandoned, but factory and dump sites remain contaminated and many former workers and neighbours contracted fatal cancers.

radius in biology, one of the two bones in the lower forearm of tetrapod (four-limbed) vertebrates.

radon colourless, odourless, gaseous, radioactive, nonmetallic element, symbol Rn, atomic number 86, relative atomic mass 222. It is grouped with the ◊inert gases and was formerly considered nonreactive, but is now known to form some compounds with fluorine. Of the 20 known isotopes, only three occur in nature; the longest half-life is 3.82 days.

Radon is the densest gas known and occurs in small amounts in spring water, streams, and the air, being formed from the natural radioactive decay of radium. Ernest Rutherford discovered the isotope Rn-220 in 1899, and Friedrich Dorn (1848–1916) in 1900; after several other chemists discovered additional isotopes, William Ramsay and R W Whytlaw-Gray isolated the element, which they named niton in 1908. The name radon was adopted in the 1920s.

railway method of transport in which trains convey passengers and goods along a twin rail track. Following the work of English steam pioneers such as

Scottish engineer James Watt, English engineer George Stephenson built the first public steam railway, from Stockton to Darlington, England, in 1825. This heralded extensive railway building in Britain, continental Europe, and North America, providing a fast and economical means of transport and communication. After World War II, steam engines were replaced by electric and diesel engines. At the same time, the growth of road building, air services, and car ownership destroyed the supremacy of the railways.

growth years Four years after building the first steam railway, Stephenson opened the first steam passenger line, inaugurating it with his locomotive *Rocket*, which achieved speeds of 50 kph/30 mph. The railway construction that followed resulted in 250 separate companies in Britain, which resolved into four systems 1921 and became the nationalized British Railways 1948, known as British Rail from 1965. In North America the growth of railways during the 19th century made shipping from the central and western territories economical and helped the North to win the American Civil War; US rail travel reached its peak in 1929. Railways were extended into Asia, the Middle East, Africa, and Latin America in the late 19th century and were used for troop and supply transport in both world wars.

gauge Railway tracks were at first made of wood but later of iron or steel, with ties wedging them apart and relatively parallel. The distance between the wheels is known as the gauge. Since much of the early development of the railway took place in Tyneside, the gauge of local coal wagons, 1.24 m/ 4 ft 8.5 in, was adopted 1824 for the Stockton–Darlington railway, and most other early railways followed suit. The main exception was the Great Western Railway (GWR) of Isambard Kingdom Brunel, opened 1841, with a gauge of 2.13 m/7 ft. The narrow gauge won legal backing in the UK 1846, but parts of GWR carried on with Brunel's broad gauge until 1892. British engineers building railways overseas tended to use the narrow gauge, and it became the standard in the USA from 1885. Other countries, such as Ireland and Finland, favoured the broad gauge. Although expensive, it offers a more comfortable journey.

decline of railways With the increasing use of private cars and government-encouraged road haulage after World War II, and the demise of steam, rising costs on the railways meant higher fares, fewer passengers, and declining freight traffic. In the UK many rural rail services closed down on the recommendations of the Beeching Report 1963, reducing the size of the network by more than 20% between 1965 and 1970, from a peak of 24,102 km/14,977 mi. In the 1970s, national railway companies began investing in faster

Blessings on Science, and her handmaid Steam/They make Utopia only half a dream.

On **railways** Charles Mackay (1814–1889)
Railways 1846

intercity services: in the UK, the diesel high-speed train (HST) was introduced. Elsewhere such trains run on specially built tracks; for example, the ◊Shinkansen (Japan) and ◊TGV (France) networks.

rain form of ◊precipitation in which separate drops of water fall to the Earth's surface from clouds. The drops are formed by the accumulation of fine droplets that condense from water vapour in the air. The condensation is usually brought about by rising and subsequent cooling of air.

Rain can form in three main ways—frontal (or cyclonic) rainfall, orographic (or relief) rainfall, and convectional rainfall.

Frontal rainfall takes place at the boundary, or ◊front, between a mass of warm air from the tropics and a mass of cold air from the poles. The water vapour in the warm air is chilled and condenses to form cloud and rain.

Orographic rainfall occurs when an airstream is forced to rise over a mountain range. The air becomes cooled and precipitation takes place. In the UK, the Pennine hills, which extend southwards from Northumbria to Derbyshire in N England, interrupt the path of the prevailing southwesterly winds, causing orographic rainfall. Their presence is partly responsible for the west of the UK being wetter than the east. The orographic effect can sometimes occur in large cities, when air rises over tall buildings.

Convectional rainfall, associated with hot climates, is brought about by rising and abrupt cooling of air that has been warmed by the extreme heat of the ground surface. The water vapour carried by the air condenses and so rain falls heavily. Convectional rainfall is usually accompanied by a thunderstorm.

rainbow arch in the sky displaying the seven colours of the ◊spectrum in bands. It is formed by the refraction, reflection, and dispersion of the Sun's rays through rain or mist. Its cause was discovered by German scientist Theodoric of Freiburg in the 14th century.

rainforest dense forest usually found on or near the ◊equator where the climate is hot and wet. Heavy rainfall results as the moist air brought by the converging tradewinds rises because of the heat. Over half the tropical rainforests are in Central and South America, the rest in SE Asia and Africa. They provide the bulk of the oxygen needed for plant and animal respiration. Tropical rainforest once covered 14% of the Earth's land surface, but are now being destroyed at an increasing rate as their valuable timber is harvested and the land cleared for agriculture, causing problems of ◊deforestation. Although by 1991 over 50% of the world's rainforest had been removed, they still comprise about 50% of all growing wood on the planet, and harbour at least 40% of the Earth's species (plants and animals).

Tropical rainforests are characterized by a great diversity of species, usually of tall broad-leafed evergreen trees, with many climbing vines and ferns, some of which are a main source of raw materials for medicines. A tropical forest, if properly preserved, can yield medicinal plants, oils (from cedar, juniper, cinnamon, sandalwood),

railways: chronology

1500s	Tramways—wooden tracks along which trolleys ran—were in use in mines.
1789	Flanged wheels running on cast-iron rails were first introduced; cars were still horse-drawn.
1804	Richard Trevithick built the first steam locomotive, and ran it on the track at the Pen-y-darren ironworks in South Wales.
1825	George Stephenson in England built the first public railway to carry steam trains—the Stockton and Darlington line—using his engine *Locomotion*.
1829	Stephenson designed his locomotive *Rocket*.
1830	Stephenson completed the Liverpool and Manchester Railway, the first steam passenger line. The first US-built locomotive, *Best Friend of Charleston*, went into service on the South Carolina Railroad.
1835	Germany pioneered steam railways in Europe, using *Der Adler*, a locomotive built by Stephenson.
1863	Robert Fairlie, a Scot, patented a locomotive with pivoting driving bogies, allowing tight curves in the track (this was later applied in the Garratt locomotives). London opened the world's first underground railway, powered by steam.
1869	The first US transcontinental railway was completed at Promontory, Utah, when the Union Pacific and the Central Pacific railroads met. George Westinghouse of the USA invented the compressed-air brake.
1879	Werner von Siemens demonstrated an electric train in Germany. Volk's Electric Railway along the Brighton seafront in England was the world's first public electric railway.
1883	Charles Lartique built the first monorail, in Ireland.
1885	The trans-Canada continental railway was completed, from Montréal in the east to Port Moody, British Columbia, in the west.
1890	The first electric underground railway opened in London.
1901	The world's longest-established monorail, the Wuppertal Schwebebahn, went into service in Germany.
1912	The first diesel locomotive took to the rails in Germany.
1938	The British steam locomotive *Mallard* set a steam-rail speed record of 203 kph/126 mph.
1941	Swiss Federal Railways introduced a gas-turbine locomotive.
1964	Japan National Railways inaugurated the 515 km/320 mi New Tokaido line between Osaka and Tokyo, on which the 210 kph/130 mph 'bullet' trains run.
1973	British Rail's High Speed Train (HST) set a diesel-rail speed record of 229 kph/142 mph.
1979	Japan National Railways' maglev test vehicle ML-500 attained a speed of 517 kph/321 mph.
1981	France's Train à Grande Vitesse (TGV) superfast trains began operation between Paris and Lyons, regularly attaining a peak speed of 270 kph/168 mph.
1987	British Rail set a new diesel-traction speed record of 238.9 kph/148.5 mph, on a test run between Darlington and York; France and the UK began work on the Channel Tunnel, a railway link connecting the two countries, running beneath the English Channel.
1988	The West German Intercity Experimental train reached 405 kph/252 mph on a test run between Würzburg and Fulda.
1990	A new rail-speed record of 515 kph/320 mph was established by a French TGV train, on a stretch of line between Tours and Paris.
1991	The British and French twin tunnels met 23 km/14 mi out to sea to form the Channel Tunnel.
1992	Trams returned to the streets of Manchester, England, after an absence of 40 years; towns in France and the USA follow suit.

spices, gums, resins (used in inks, lacquers, linoleum), tanning and dyeing materials, forage for animals, beverages, poisons, green manure, rubber, and animal products (feathers, hides, honey).

Other rainforests include montane, upper montane or cloud, mangrove, and subtropical.

Rainforests comprise some of the most complex and diverse ecosystems on the planet and help to regulate global weather patterns. When deforestation occurs, the microclimate of the mature forest disappears; soil erosion and flooding become major problems since rainforests protect the shallow tropical soils. Clearing of the rainforests may lead to a global warming of the atmosphere, and contribute to the ◊greenhouse effect. Deforestation also causes the salt level in the ground to rise to the surface, making the land unsuitable for farming or ranching.

RAM (acronym for *r*andom-*a*ccess *m*emory) in computing, a memory device in the form of a collection of integrated circuits (chips), frequently used in microcomputers. Unlike ◊ROM (read-only memory) chips, RAM chips can be both read from and written to by the computer, but their contents are lost when/the power is switched off. Microcomputers of the 1990s may have 16–32 megabytes of RAM.

ramjet simple jet engine (see ◊jet propulsion) used in some guided missiles. It only comes into operation at high speeds. Air is then 'rammed' into the combustion chamber, into which fuel is sprayed and ignited.

random access in computing, an alternative term for ◊direct access.

random number one of a series of numbers having no detectable pattern. Random numbers are used in ◊computer simulation and ◊computer games. It is impossible for an ordinary computer to generate true random numbers, but various techniques are available for obtaining pseudo-random numbers—close enough to true randomness for most purposes.

range check in computing, a ◊validation check applied to a numerical data item to ensure that its value falls in a sensible range.

rangefinder instrument for determining the range or distance of an object from the observer; used to focus a camera or to sight a gun accurately. A *rangefinder camera* has a rotating mirror or prism that alters the image seen through the viewfinder, and a secondary window. When the two images are brought together into one lens is sharply focussed.

(a) rate of reaction decreases with time

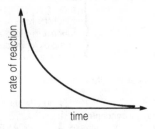

(b) concentration of reactant decreases with time

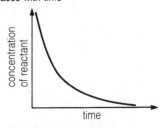

(c) concentration of product increases with time

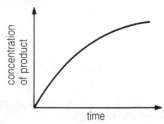

rate of reaction

rare-earth element alternative name for ◊lanthanide.

rare gas alternative name for ◊inert gas.

raster graphics computer graphics that are stored in the computer memory by using a map to record data (such as colour and intensity) for every ◊pixel that makes up the image. When transformed (enlarged, rotated, stretched, and so on), raster graphics become ragged and suffer loss of picture resolution, unlike ◊vector graphics. Raster graphics are typically used for painting applications, which allow the user to create artwork on a computer screen much as if they were painting on paper or canvas.

ratemeter instrument used to measure the count rate of a radioactive source. It gives a reading of the number of particles emitted from the source and captured by a detector in a unit of time (usually a second).

rate of reaction the speed at which a chemical reaction proceeds. It is usually expressed in terms of the concentration (usually in ◊moles per litre) of

a reactant consumed, or product formed, in unit time; so the units would be moles per litre per second (mol $l^{-1}{}_s{-}1$). The rate of a reaction may be affected by the concentration of the reactants, the temperature of the reactants, and the presence of a ◊catalyst. If the reaction is entirely in the gas state, the rate is affected by pressure, and, for solids, it is affected by the particle size.

During a reaction at constant temperature the concentration of the reactants decreases and so the rate of reaction decreases. These changes can be represented by drawing graphs.

The rate of reaction is at its greatest at the beginning of the reaction and it gradually slows down. For an ◊endothermic reaction (one that absorbs heat) increasing the temperature may produce large increases in the rate of reaction. A 10°C rise can double the rate while a 40°C rise can produce a 50–100–fold increase in the rate.

◊Collision theory is used to explain these effects. Increasing the concentration or the pressure of a gas means there are more particles per unit volume, therefore there are more collisions and more fruitful collisions. Increasing the temperature makes the particles move much faster, resulting in more collisions per unit time and more fruitful collisions; consequently the rate increases.

ratio measure of the relative size of two quantities or of two measurements (in similar units), expressed as a proportion. For example, the ratio of vowels to consonants in the alphabet is 5:21; the ratio of 500 m to 2 km is 500:2,000, or 1:4.

rationalized units units for which the defining equations conform to the geometry of the system. Equations involving circular symmetry contain the factor 2π; those involving spherical symmetry 4π. ◊SI units are rationalized, ◊c.g.s. units are not.

rational number in mathematics, any number that can be expressed as an exact fraction (with a denominator not equal to 0), that is, as a/b where a and b are integers. For example, $\frac{2}{1}$, $\frac{1}{4}$, $\frac{15}{4}$, $-\frac{3}{5}$ are all rational numbers, whereas π (which represents the constant 3.141592...) is not. Numbers such as π are called ◊irrational numbers.

Raunkiaer system method of plant classification devised by the Danish ecologist Christen Raunkiaer (1860–1938) whereby plants are divided into groups according to the position of their perennating (overwintering) buds in relation to the soil surface. For example, plants in cold areas, such as the tundra, generally have their buds protected below ground, whereas in hot, tropical areas they are above ground and freely exposed. This scheme is useful for comparing vegetation types in different parts of the world.

The main divisions are *phanerophytes* with buds situated well above the ground; *chamaephytes* with buds borne within 25 cm/10 in of the soil surface; *hemicryptophytes* with buds at or immediately below the soil surface; and *cryptophytes* with their buds either beneath the soil (*geophyte*) or below water (*hydrophyte*).

rayon any of various shiny textile fibres and fabrics made from ◊cellulose. It is produced by pressing whatever cellulose solution is used through very small holes and solidifying the resulting filaments. A common type is ◊viscose, which consists of

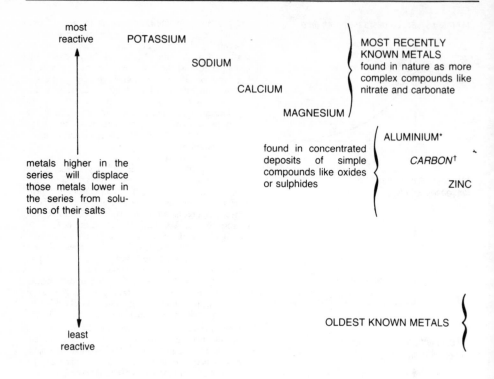

regenerated filaments of pure cellulose. Acetate and triacetate are kinds of rayon consisting of filaments of cellulose acetate and triacetate. Rayon was originally made in the 1930s from the Douglas fir tree.

rays original name for radiation of all types, such as ◊X-rays and gamma rays.

reactance property of an alternating current circuit that together with any ◊resistance makes up the ◊impedance (the total opposition of the circuit to the passage of a current).

The reactance of an inductance L is wL and that of the capacitance is $1/wC$, where $w = 2\pi f$, with f being the frequency of the alternating current. Reactance is measured in ◊ohms.

reaction in chemistry, the coming together of two or more atoms, ions, or molecules with the result that a ◊chemical change takes place. The nature of the reaction is portrayed by a chemical equation.

reaction force in physics, the equal and opposite force described by Newton's third law of motion that arises whenever one object applies a force (*action force*) to another. For example, if a magnet attracts a piece of iron, then that piece of iron will also attract the magnet with a force that is equal in magnitude but opposite in direction. When any object rests on the ground the downwards contact force applied to the ground always produces an equal, upwards reaction force.

reactivity series chemical series produced by arranging the metals in order of their ease of reaction with reagents such as oxygen, water, and acids. This arrangement aids the understanding of the properties of metals, helps to explain differences between them, and enables predictions to be made about a metal's behaviour, based on a knowledge of its position or properties.

real number in mathematics, any of the ◊rational numbers (which include the integers) or ◊irrational numbers. Real numbers exclude ◊imaginary numbers, found in ◊complex numbers of the general form $a + bi$ where $i = \sqrt{-1}$, although these do include a real component a.

real-time system in computing, a program that responds to events in the world as they happen, as, for example, an automatic pilot program in an aircraft must respond instantly in order to correct deviations from its course. Process control, robotics, games, and many military applications are examples of real-time systems.

receiver, radio component of a radio communication system that receives and processes radio waves. It detects and selects modulated radio waves (see ◊modulation) by means of an aerial and tuned circuit, and then separates the transmitted infor-

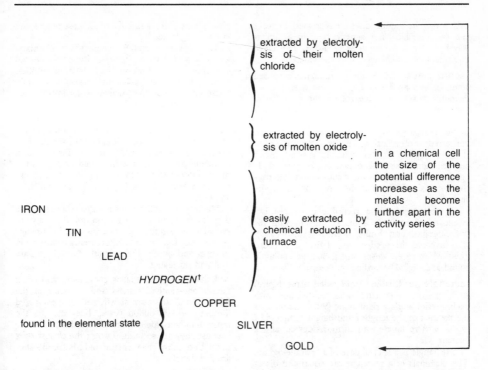

extracted by electrolysis of their molten chloride

extracted by electrolysis of molten oxide

easily extracted by chemical reduction in furnace

in a chemical cell the size of the potential difference increases as the metals become further apart in the activity series

IRON

TIN

LEAD

HYDROGEN†

COPPER

found in the elemental state

SILVER

GOLD

* The position of aluminium may appear anomalous as the oxide on its surface makes it unreactive.

†Carbon and hydrogen, although non-metals, are included to show that they will only reduce metals below them in the reactivity series.

reactivity series

mation from the carrier wave by a process that involves rectification (see ◊rectifier). The receiver device will usually also include the amplifiers that produce the audio signals (signals that are heard).

receptacle the enlarged end of a flower stalk to which the floral parts are attached. Normally the receptacle is rounded, but in some plants it is flattened or cup-shaped. The term is also used for the region on that part of some seaweeds which becomes swollen at certain times of the year and bears the reproductive organs.

recessive gene in genetics, an ◊allele (alternative form of a gene) that will show in the ◊phenotype (observed characteristics of an organism) only if its partner allele on the paired chromosome is similarly recessive. Such an allele will not show if its partner is dominant, that is if the organism is ◊heterozygous for a particular characteristic. Alleles for blue eyes in humans, and for shortness in pea plants are recessive. Most mutant alleles are recessive and therefore are only rarely expressed (see ◊haemophilia and ◊sickle cell disease).

reciprocal in mathematics, the result of dividing a given quantity into 1. Thus the reciprocal of 2 is $\frac{1}{2}$; of $\frac{2}{3}$ is $\frac{3}{2}$; of x^2 is $\frac{1}{x^2}$ or x^{-2}. Reciprocals are used to replace division by multiplication, since multiplying by the reciprocal of a number is the same as dividing by that number.

recombination in genetics, any process that recombines, or 'shuffles', the genetic material, thus increasing genetic variation in the offspring. The two main processes of recombination both occur during meiosis (reduction division of cells). One is ◊crossing over, in which chromosome pairs exchange segments; the other is the random reassortment of chromosomes that occurs when each gamete (sperm or egg) receives only one of each chromosome pair.

record in computing, a collection of related data items or *fields*; it usually forms part of a ◊file.

recording the process of storing information, or the information store itself. Sounds and pictures can be stored on discs or tape. The gramophone record or ◊compact disc stores music or speech as a spiral groove on a plastic disc and the sounds are reproduced by a record player. In ◊tape recording, sounds are stored as a magnetic pattern on plastic tape. The best-quality reproduction is achieved using digital audio tape.

In *digital recording* the signals picked up by the microphone are converted into precise numerical values by computer. These values, which represent the original sound wave form, are recorded on tape or compact disc. When it is played back by ◊laser, the exact values are retrieved. When the signal is fed via an amplifier to a loudspeaker, sound waves exactly like the original ones are recreated. Pictures can be recorded on magnetic tape in a similar way by using a ◊video tape recorder. The video equivalent of a compact disc is the video disc.

record player device for reproducing recorded sound stored as a spiral groove on a vinyl disc. A motor-driven turntable rotates the record at a constant speed, and a stylus or needle on the head of a pick-up is made to vibrate by the undulations in the record groove. These vibrations are then converted to electrical signals by a ◊transducer in the head (often a ◊piezoelectric crystal). After amplification, the signals pass to one or more loud-speakers, which convert them into sound. Alternative formats are ◊compact disc and magnetic ◊tape recording.

The pioneers of the record player were US scientist and inventor Thomas Edison, with his ◊phonograph, and Emile Berliner (1851–1929), who invented the predecessor of the vinyl record 1896. More recent developments are stereophonic sound and digital recording on compact disc.

rectangle quadrilateral (four-sided plane figure) with opposite sides equal and parallel and with each interior angle a right angle (90°). Its area A is the product of the length l and height h; that is, $A = l \times h$. A rectangle with all four sides equal is a ◊square.

A rectangle is a special case of a ◊parallelogram. The diagonals of a rectangle are equal and bisect each other.

rectifier in electrical engineering, a device used for obtaining one-directional current (DC) from an alternating source of supply (AC). (The process is necessary because almost all electrical power is generated, transmitted, and supplied as alternating current, but many devices, from television sets to electric motors, require direct current.) Types include plate rectifiers, thermionic ◊diodes, and ◊semiconductor diodes.

rectum lowest part of the digestive tract of animals, which stores faeces prior to elimination (defecation).

recursion in computing and mathematics, a technique whereby a ◊function or ◊procedure calls itself into use in order to enable a complex problem to be broken down into simpler steps. For example, a function that finds the factorial of a number n (calculates the product of all the whole numbers between 1 and n) would obtain its result by multiplying n by the factorial of $n - 1$.

recycling processing of industrial and household waste (such as paper, glass, and some metals and plastics) so that it can be reused. This saves expenditure on scarce raw materials, slows down the depletion of ◊nonrenewable resources, and helps to reduce pollution.

The USA recycles only around 13% of its waste, compared to around 33% in Japan. However, all US states encourage or require local recycling programmes to be set up. It was estimated 1992 that 4,000 cities collected waste from 71 million people for recycling. Most of these programmes were set up in 1989–92. Around 33% of newspapers, 22% of office paper, 64% of aluminium cans, 3% of

plastic containers, and 20% of all glass bottles and jars were recycled.

Most British recycling schemes are voluntary, and rely on people taking waste items to a central collection point. However, some local authorities, such as Leeds, now ask householders to separate waste before collection, making recycling possible on a much larger scale.

red blood cell or *erythrocyte* the most common type of blood cell, responsible for transporting oxygen around the body. It contains haemoglobin, which combines with oxygen from the lungs to form oxyhaemoglobin. When transported to the tissues, these cells are able to release the oxygen because the oxyhaemoglobin splits into its original constituents.

Mammalian erythrocytes are disc-shaped with a depression in the centre and no nucleus; they are manufactured in the bone marrow and, in humans, last for only four months before being destroyed in the liver and spleen. Those of other vertebrates are oval and nucleated.

red dwarf any star that is cool, faint, and small (about one-tenth the mass and diameter of the Sun). Red dwarfs burn slowly, and have estimated lifetimes of 100 billion years. They may be the most abundant type of star, but are difficult to see because they are so faint. Two of the closest stars to the Sun, ◊Proxima Centauri and ◊Barnard's star, are red dwarfs.

red giant any large bright star with a cool surface. It is thought to represent a late stage in the evolution of a star like the Sun, as it runs out of hydrogen fuel at its centre. Red giants have diameters between 10 and 100 times that of the Sun. They are very bright because they are so large, although their surface temperature is lower than that of the Sun, about 2,000–3,000K (1,700–2,700°C).

RED GIANT: SPREAD-OUT STARS

A red giant star is large enough to swallow up 100 million Suns, yet it contains no more matter than the Sun. The matter is spread over such a large space that it is not much denser than the best 'vacuum' that can be created in a laboratory. If a red giant were the size of an ordinary living room, its energy-generating core would be the size of this full stop.

---

redox reaction chemical change where one reactant is reduced and the other reactant oxidized. The reaction can only occur if both reactants are present and each changes simultaneously. For example, hydrogen reduces copper(II) oxide to copper while it is itself oxidized to water. The corrosion of iron and the reactions taking place in electric and electrolytic cells are just a few instances of redox reactions.

red shift in astronomy, the lengthening of the wavelengths of light from an object as a result of the object's motion away from us. It is an example of the ◊Doppler effect. The red shift in light from galaxies is evidence that the universe is expanding.

Lengthening of wavelengths causes the light to

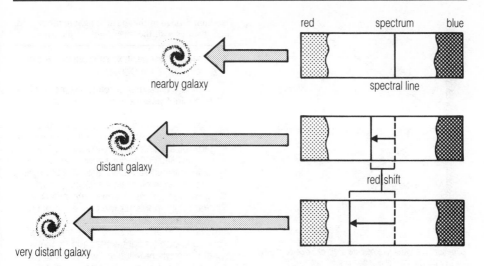

red   spectrum   blue

spectral line

nearby galaxy

distant galaxy

red shift

very distant galaxy

red shift *The red shift causes lines in the spectra of galaxies to be shifted towards the red end of the spectrum. More distant galaxies have greater red shifts than closer galaxies. The red shift indicates that distant galaxies are moving apart rapidly, as the universe expands.*

move or shift towards the red end of the ◊spectrum, hence the name. The amount of red shift can be measured by the displacement of lines in an object's spectrum. By measuring the amount of red shift in light from stars and galaxies, astronomers can tell how quickly these objects are moving away from us. A strong gravitational field can also produce a red shift in light; this is termed *gravitational red shift*.

Redstone rocket short-range US military missile, modified for use as a space launcher. Redstone rockets launched the first two flights of the ◊Mercury project. A modified Redstone, *Juno 1*, launched the first US satellite, *Explorer 1*, in 1958.

reducing agent substance that reduces another substance (see ◊reduction). In a redox reaction, the reducing agent is the substance that is itself oxidized. Strong reducing agents include hydrogen, carbon monoxide, carbon, and metals.

reduction in chemistry, the gain of electrons, loss of oxygen, or gain of hydrogen by an atom, ion, or molecule during a chemical reaction.

Reduction may be brought about by reaction with another compound, which is simultaneously oxidized (reducing agent), or electrically at the cathode (negative electrode) of an electric cell.

reed switch switch containing two or three thin iron strips, or reeds, inside a sealed glass tube. When a magnet is brought near the switch, it induces magnetism in the reeds, which then attract or repel each other according to the positions of their induced magnetic poles. Electric current is switched on or off as the reeds make or break contact.

A simple burglar alarm can be constructed by inserting a magnet into the closing edge of a door, and a reed switch into the part of the doorframe facing the magnet: when the door is closed the magnet remains close to the switch, keeping a circuit open; when the door is opened the magnet

moves away, closing the switch and setting off an alarm.

refining any process that purifies or converts something into a more useful form. Metals usually need refining after they have been extracted from their ores by such processes as ◊smelting. Petroleum, or crude oil, needs refining before it can be used; the process involves fractional ◊distillation, the separation of the substance into separate components or 'fractions'.

Electrolytic metal-refining methods use the principle of ◊electrolysis to obtain pure metals. When refining petroleum, or crude oil, further refinery processes after fractionation convert the heavier fractions into more useful lighter products. The most important of these processes is ◊cracking; others include ◊polymerization, hydrogenation, and reforming.

reflection the throwing back or deflection of waves, such as ◊light or ◊sound waves, when they hit a surface. The *law of reflection* states that the angle of incidence (the angle between the ray and a perpendicular line drawn to the surface) is equal to the angle of reflection (the angle between the reflected ray and a perpendicular to the surface).

reflex in animals, a very rapid automatic response to a particular stimulus. It is controlled by the ◊nervous system. A reflex involves only a few nerve cells, unlike the slower but more complex responses produced by the many processing nerve cells of the brain.

A *simple reflex* is entirely automatic and involves no learning. Examples of such reflexes include the sudden withdrawal of a hand in response to a painful stimulus, or the jerking of a leg when its kneecap is tapped. Sensory cells (receptors) in the knee send signals to the spinal cord along a sensory nerve cell. Within the spine a *reflex arc* switches the signals straight back to the muscles of the leg (effectors) via an

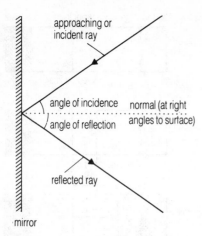

reflection *The law of reflection: the angle of incidence of a light beam equals the angle of reflection of the beam.*

intermediate nerve cell and then a motor nerve cell; contraction of the leg occurs, and the leg kicks upwards. Only three nerve cells are involved, and the brain is only aware of the response after it has taken place. Such reflex arcs are particularly common in lower animals, and have a high survival value, enabling organisms to take rapid action to avoid danger. In higher animals (those with a well-developed ◊central nervous system) the simple reflex can be modified by the involvement of the brain—for instance, humans can override the automatic reflex to withdraw a hand from a source of pain.

A *conditioned reflex* involves the modification of a reflex action in response to experience (learning). A stimulus that produces a simple reflex response becomes linked with another, possibly unrelated, stimulus. For example, a dog may salivate (a reflex action) when it sees its owner remove a tin-opener from a drawer because it has learned to associate that stimulus with the stimulus of being fed.

reflex angle an angle greater than 180° but less than 360°.

reflex camera camera that uses a mirror and prisms to reflect light passing through the lens into the viewfinder, showing the photographer the exact scene that is being shot. When the shutter button is released the mirror springs out of the way, allowing light to reach the film. The most common type is the single-lens reflex (◊SLR) camera. The twin-lens reflex (◊TLR) camera has two lenses: one has a mirror for viewing, the other is used for exposing the film.

refraction the bending of a wave of light, heat, or sound when it passes from one medium to another. Refraction occurs because waves travel at different velocities in different media. The ◊refractive index of a material indicates by how much light is bent.

refractive index measure of the refraction of a ray of light as it passes from one transparent medium to another. If the angle of incidence is i and the angle of refraction is r, the refractive index $n = \sin i/\sin r$. It is also equal to the speed of light in the first medium divided by the speed of light in the second, and it varies with the wavelength of the light.

REFRACTIVE INDEX: SLOWING LIGHT DOWN

Ordinary glass has a refractive index of 1.5, so light passing through it slows down to 200,000 m/64,000 ft per second. Diamond slows it further to 124,000 m/397,000 ft per second, while in the exotic gem rutile it only just breaks the 100,000 km/320,000 ft per second limit. With infrared radiation (the invisible light used to take night-time pictures) much bigger reductions in speed are possible. When light in this region passes through a lens made of germanium, it is slowed down by a factor of four; while the lowest speed known for light is reached in crystals of the element tellurium, where infrared radiation is slowed by a factor of 6.3.

refractory (of a material) able to resist high temperature, for example ◊ceramics made from clay, minerals, or other earthy materials. Furnaces are lined with refractory materials such as silica and dolomite.

Alumina (aluminium oxide) is an excellent refractory, often used for the bodies of spark plugs. Titanium and tungsten are often called refractory metals because they are temperature resistant. ◊Cermets are refractory materials made up of ceramics and metals.

refrigeration use of technology to transfer heat from cold to warm, against the normal temperature gradient, so that a body can remain substantially colder than its surroundings. Refrigeration

the knee-jerk reflex arc

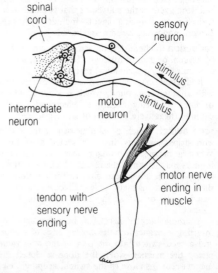

reflex

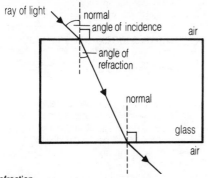

refraction

equipment is used for the chilling and deep freezing of food in ◊food technology, and in air conditioners and industrial processes.

It is commonly achieved by a vapour-compression cycle, in which a suitable chemical (the refrigerant) travels through a long circuit of tubing, during which it changes from a vapour to a liquid and back again. A compression chamber makes it condense, and thus give out heat. In another part of the circuit, called the evaporator coils, the pressure is much lower, so the refrigerant evaporates, absorbing heat as it does so. The evaporation process takes place near the central part of the refrigerator, which therefore becomes colder, while the compression process takes place near a ventilation grille, transferring the heat to the air outside.

regelation phenomenon in which water refreezes to ice after it has been melted by pressure at a temperature below the freezing point of water. Pressure makes an ice skate, for example, form a film of water that freezes once again after the skater has passed.

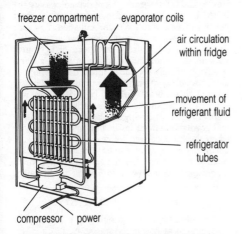

refrigeration *The circulating refrigerant fluid moves heat from the inside to the outside of a refrigerator. The refrigerant is pumped around a circuit of tubing which wraps around the freezer compartment and forms the radiator coil at the rear. The refrigerant evaporates when travelling around the freezer, extracting heat in the process. The compressor condenses the refrigerant back into liquid while it is travelling through the radiator coils, releasing heat.*

regeneration in biology, regrowth of a new organ or tissue after the loss or removal of the original. It is common in plants, where a new individual can often be produced from a 'cutting' of the original. In animals, regeneration of major structures is limited to lower organisms; certain lizards can regrow their tails if these are lost, and new flatworms can grow from a tiny fragment of an old one. In mammals, regeneration is limited to the repair of tissue in wound healing and the regrowth of peripheral nerves following damage.

register in computing, a memory location that can be accessed rapidly; it is often built into the computer's central processing unit. Some registers are reserved for special tasks—for example, an *instruction register* is used to hold the machine-code command that the computer is currently executing, while a *sequence-control register* keeps track of the next command to be executed. Other registers are used for holding frequently used data and for storing intermediate results.

regolith the surface layer of loose material that covers most rocks. It consists of eroded rocky material, volcanic ash, river alluvium, vegetable matter, or a mixture of these known as ◊soil.

Regulus or *Alpha Leonis* the brightest star in the constellation Leo, and the 21st brightest star in the sky. Regulus has a true luminosity 100 times that of the Sun, and is 69 light years from Earth.

rejuvenation in earth science, the renewal of a river's powers of downward erosion. It may be caused by a fall in sea level or a rise in land level, or by the increase in water flow that results when one river captures another (◊river capture).

relative in computing (of a value), variable and calculated from a base value. For example, a *relative address* is a memory location that is found by adding a variable to a base (fixed) address, and a *relative cell reference* locates a cell in a spreadsheet by its position relative to a base cell—perhaps directly to the left of the base cell or three columns to the right of the base cell. The opposite of relative is ◊absolute.

relative atomic mass the mass of an atom relative to one-twelfth the mass of an atom of carbon-12. It depends on the number of protons and neutrons in the atom, the electrons having negligible mass. If more than one ◊isotope of the element is present, the relative atomic mass is calculated by taking an average that takes account of the relative proportions of each isotope, resulting in values that are not whole numbers. The term *atomic weight*, although commonly used, is strictly speaking incorrect.

relative biological effectiveness (RBE) the relative damage caused to living tissue by different types of radiation. Some radiations do much more damage than others; alpha particles, for example, cause 20 times as much destruction as electrons (beta particles).

The RBE is defined as the ratio of the absorbed dose of a standard amount of radiation to the absorbed dose of 200 kV X-rays that produces the same amount of biological damage.

relative density or *specific gravity* the density (at 20°C/68°F) of a solid or liquid relative to

river features formed by rejuvenation

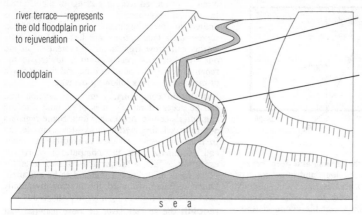

river terrace—represents the old floodplain prior to rejuvenation

floodplain

knick point —a sudden break of slope usually forming a waterfall or rapids

incised meander— steep sided meander formed as the river cuts down in its channel

s e a

rejuvenation

(divided by) the maximum density of water (at 4°C/ 39.2°F). The relative density of a gas is its density divided by the density of hydrogen (or sometimes dry air) at the same temperature and pressure.

relative humidity the concentration of water vapour in the air. It is expressed as the percentage that its moisture content represents of the maximum amount that the air could contain at the same temperature and pressure. The higher the temperature the more water vapour the air can hold.

relative molecular mass the mass of a molecule, calculated relative to one-twelfth the mass of an atom of carbon-12. It is found by adding the relative atomic masses of the atoms that make up the molecule. The term *molecular weight* is often used, but strictly this is incorrect.

When you are courting a nice girl an hour seems like a second. When you sit on a red hot cinder a second seems like an hour. That's relativity.

On **relativity** Albert Einstein *News Chronicle* 14 March 1949

relativity in physics, the theory of the relative rather than absolute character of motion and mass, and the interdependence of matter, time, and space, as developed by German physicist Albert Einstein in two phases:

special theory (1905) Starting with the premises that (1) the laws of nature are the same for all observers in unaccelerated motion, and (2) the speed of light is independent of the motion of its source, Einstein postulated that the time interval between two events was longer for an observer in whose frame of reference the events occur in different places than for the observer for whom they occur at the same place.

general theory of relativity (1915) The geometrical properties of space-time were to be conceived as modified locally by the presence of a body with

mass. A planet's orbit around the Sun (as observed in three-dimensional space) arises from its natural trajectory in modified space-time; there is no need to invoke, as Isaac Newton did, a force of ◊gravity coming from the Sun and acting on the planet. Einstein's theory predicted slight differences in the orbits of the planets from Newton's theory, which were observable in the case of Mercury. The new theory also said light rays should bend when they pass by a massive object, owing to the object's effect on local space-time. The predicted bending of starlight was observed during the eclipse of the Sun 1919, when light from distant stars passing close to the Sun was not masked by sunlight.

Einstein showed that for consistency with premises (1) and (2), the principles of dynamics as established by Newton needed modification; the most celebrated new result was the equation $E = mc^2$, which expresses an equivalence between mass (m) and ◊energy (E), c being the speed of light in a vacuum. Although since modified in detail, general relativity remains central to modern ◊astrophysics and ◊cosmology; it predicts, for example, the possibility of ◊black holes. General relativity theory was inspired by the simple idea that it is impossible in a small region to distinguish between acceleration and gravitation effects (as in a lift one feels heavier when the lift accelerates upwards), but the mathematical development of the idea is formidable. Such is not the case for the special theory, which a nonexpert can follow up to $E = mc^2$ and beyond.

EFFECTS OF RELATIVITY

The effects of high speed: the mass of a particle at rest is increased by 2% if it travels at 1/5 the speed of light; by 15% at 1/2 the speed of light; by 40% at 7/10 the speed of light; and by 66% at 8/10 the speed of light. The mass is more than doubled if the particle reaches 9/10 the speed of light.

RELATIVITY PUZZLE

According to Einstein's theory of relativity, nothing can travel faster than the speed of light. In fact nothing can even travel at the *same* speed as light, because mass increases with velocity and becomes infinite at the speed of light. How then does light manage the trick, and does a beam of light have infinite mass?

Perhaps light has no mass, so there is nothing to increase. Well, photons of light certainly have kinetic energy, and kinetic energy in turn implies some form of mass, so what is the answer? *See page 651 for the answer.*

relay in electrical engineering, an electromagnetic switch. A small current passing through a coil of wire wound around an iron core attracts an ◊armature whose movement closes a pair of sprung contacts to complete a secondary circuit, which may carry a large current or activate other devices. The solid-state equivalent is a thyristor switching device.

relay neuron nerve cell in the spinal cord, connecting motor neurons to sensory neurons. Relay neurons allow information to pass straight through the spinal cord, bypassing the brain. In humans such reflex actions, which are extremely rapid, cause the sudden removal of a limb from a painful stimulus.

rem acronym of *roentgen equivalent man* SI unit of radiation dose equivalent.

The rem has now been replaced in the SI system by the ◊sievert (one rem equals 0.01 sievert), but remains in common use.

remote sensing gathering and recording information from a distance. Space probes have sent back photographs and data about planets as distant as Neptune. In archaeology, surface survey techniques provide information without disturbing subsurface deposits.

Satellites such as *Landsat* have surveyed all the Earth's surface from orbit. Computer processing of data obtained by their scanning instruments, and the application of so-called false colours (generated by the computer), have made it possible to reveal surface features invisible in ordinary light. This has proved valuable in agriculture, forestry, and urban planning, and has led to the discovery of new deposits of minerals.

remote terminal in computing, a terminal that communicates with a computer via a modem (or acoustic coupler) and a telephone line.

REM sleep (acronym for *rapid-eye-movement sleep*) phase of sleep that recurs several times nightly in humans and is associated with dreaming. The eyes flicker quickly beneath closed lids.

renewable energy power from any source that replenishes itself. Most renewable systems rely on ◊solar energy directly or through the weather cycle as ◊wave power, ◊hydroelectric power, or wind power via ◊wind turbines, or solar energy collected by plants (alcohol fuels, for example). In addition, the gravitational force of the Moon can be harnessed through ◊tidal power stations, and the heat trapped in the centre of the Earth is used via ◊geothermal energy systems.

renewable resource natural resource that is replaced by natural processes in a reasonable amount of time. Soil, water, forests, plants, and animals are all renewable resources as long as they are properly conserved. Solar, wind, wave, and geothermal energies are based on renewable resources.

rennin (or *chymase*) enzyme found in the gastric juice of young mammals, used in the digestion of milk.

repellent anything whose smell, taste, or other properties discourages nearby creatures. *Insect repellent* is usually a chemical substance that keeps, for example, mosquitoes at bay; natural substances include citronella, lavender oil, and eucalyptus oils. A device that emits ultrasound waves is also claimed to repel insects and small mammals. The bitter-tasting denatonium saccharide may be added to medicines to prevent consumption by children, and to plastic garbage bags to repel foraging animals.

replication in biology, production of copies of the genetic material, DNA; it occurs during cell division (◊mitosis and ◊meiosis). Most mutations are caused by mistakes during replication.

reproduction in biology, process by which a living organism produces other organisms similar to itself. There are two kinds: ◊asexual reproduction and ◊sexual reproduction.

reproduction rate or *fecundity* in ecology, the rate at which a population or species reproduces itself.

reptile any member of a class (Reptilia) of vertebrates. Unlike amphibians, reptiles have hardshelled, yolk-filled eggs that are laid on land and from which fully formed young are born. Some snakes and lizards retain their eggs and give birth to live young. Reptiles are cold-blooded, produced from eggs, and the skin is usually covered with scales. The metabolism is slow, and in some cases (certain large snakes) intervals between meals may be months. Reptiles date back over 300 million years.

Many extinct forms are known, including the orders Pterosauria, Plesiosauria, Ichthyosauria, and Dinosauria. The chief living orders are the Chelonia (tortoises and turtles), Crocodilia (alligators and crocodiles), and Squamata, divided into three suborders: Lacertilia (lizards), Ophidia or Serpentes (snakes), and Amphisbaenia (worm lizards). The order Rhynchocephalia has one surviving species, the lizard-like tuatara of New Zealand.

research the primary activity in science, a combination of theory and experimentation directed towards finding scientific explanations of phenomena. It is commonly classified into two types: *pure research*, involving theories with little apparent relevance to human concerns; and *applied research*, concerned with finding solutions to

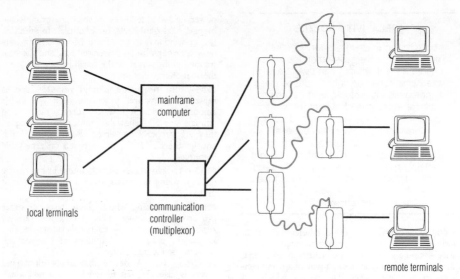

local terminals

mainframe computer

communication controller (multiplexor)

remote terminals

remote terminal

problems of social importance—for instance in medicine and engineering. The two types are linked in that theories developed from pure research may eventually be found to be of great value to society.

Scientific research is most often funded by government and industry, and so a nation's wealth and priorities are likely to have a strong influence on the kind of work undertaken.

In 1989 the European Community (EC) Council adopted a revised programme on research and technological development for the period 1990–94, requiring a total Community finance of 5,700 million ECUs, to be apportioned as follows: information and communications technology 2,221 million; industrial and materials technologies 888 million; life sciences and technologies 741 million; energy 814 million; human capacity and mobility 518 million; environment 518 million.

In the UK, government-funded research and development in 1989–90 was £4.8 billion, representing 3.8% of total central government expenditure. This amount was distributed between the Ministry of Defence (£2.2 billion), civil departments (£1 billion), and the science base (£1.6 billion), the latter comprising the combined total for the Research Councils, University Funding Council, and the Polytechnics and Colleges Funding Council.

The aim of research is the discovery of the equations which subsist between the elements of phenomena.

On **research** Ernst Mach *Popular Science Lectures* 1910

residual current device or *earth leakage circuit breaker* device that protects users of electrical equipment from electric shock by interrupting the electricity supply if a short circuit or current leakage occurs.

It contains coils carrying current to and from the electrical equipment. If a fault occurs, the currents become unbalanced and the residual current trips a switch. Residual current devices are used to protect household gardening tools as well as electrical equipment in industry.

residue in chemistry, a substance or mixture of substances remaining in the original container after the removal of one or more components by a separation process.

The non-volatile substance left in a container after ◊evaporation of liquid, the solid left behind after removal of liquid by filtration, and the substances left in a distillation flask after removal of components by ◊distillation, are all residues.

resin substance exuded from pines, firs, and other trees in gummy drops that harden in air. Varnishes are common products of the hard resins, and ointments come from the soft resins.

Rosin is the solid residue of distilled turpentine, a soft resin. The name 'resin' is also given to many synthetic products manufactured by polymerization; they are used in adhesives, plastics, and varnishes.

resistance in physics, that property of a substance that restricts the flow of electricity through it, associated with the conversion of electrical energy to heat; also the magnitude of this property. Resistance depends on many factors, such as the nature of the material, its temperature, dimensions, and thermal properties; degree of impurity; the nature and state of illumination of the surface; and the frequency and magnitude of the current. The SI unit of resistance is the ohm.

resistivity in physics, a measure of the ability of a material to resist the flow of an electric current. It is numerically equal to the ◊resistance of a sample of unit length and unit cross-sectional area, and its unit is the ohm metre. A good conductor has a low resistivity (1.7×10^{-8} ohm metres for

single force W

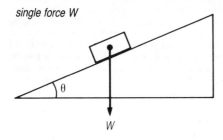

components of W

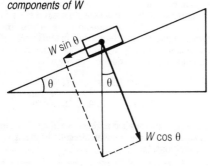

resolution of forces

copper); an insulator has a very high resistivity (10^{15} ohm metres for polyethane).

resistor in physics, any component in an electrical circuit used to introduce ◊resistance to a current. Resistors are often made from wire-wound coils or pieces of carbon. ◊Rheostats and ◊potentiometers are variable resistors.

resolution in computing, the number of dots per unit length in which an image can be reproduced on a screen or printer. A typical screen resolution for colour monitors is 75 dpi (dots per inch). A ◊laser printer will typically have a printing resolution of 300 dpi, and ◊dot matrix printers typically have resolutions from 60 to 180 dpi. Photographs in books and magazines have a resolution of 1,200 dpi or 2,400 dpi.

resolution of forces in mechanics, the division of a single force into two parts that act at right angles to each other. The two parts of a resolved force, called its **components**, have exactly the same effect on an object as the single force which they replace.

For example, the weight W of an object on a slope, tilted at an angle θ, can be resolved into two components: one acting at a right angle to the slope, equal to $W\cos θ$, and one acting parallel to and down the slope, equal to $W\sin θ$. The component acting down the slope (minus any friction force that may be acting in the opposite direction) is responsible for the acceleration of the object.

resonance rapid and uncontrolled increase in the size of a vibration when the vibrating object is subject to a force varying at its ◊natural frequency. In a trombone, for example, the length of the air column in the instrument is adjusted until it resonates with the note being sounded. Resonance effects are also produced by many electrical

circuits. Tuning a radio, for example, is done by adjusting the natural frequency of the receiver circuit until it coincides with the frequency of the radio waves falling on the aerial.

Resonance has many physical applications. Children use it to increase the size of the movement on a swing, by giving a push at the same point during each swing. Soldiers marching across a bridge in step could cause the bridge to vibrate violently if the frequency of their steps coincided with its natural frequency. Resonance was the cause of the collapse of the Tacoma Narrows bridge, USA, in 1940 when the frequency of the wind coincided with the natural frequency of the bridge.

respiration biochemical process whereby food molecules are progressively broken down (oxidized) to release energy in the form of ◊ATP. In most organisms this requires oxygen, but in some bacteria the oxidant is the nitrate or sulphate ion instead. In all higher organisms, respiration occurs in the ◊mitochondria. Respiration is also used to mean breathing, although this is more accurately described as a form of ◊gas exchange.

respiratory surface area used by an organism for the exchange of gases, for example the lungs, gills or, in plants, the leaf interior. The gases oxygen and carbon dioxide are both usually involved in respiration and photosynthesis. Although organisms have evolved different types of respiratory surface according to need, there are certain features in common. These include thinness and moistness, so that the gas can dissolve in a membrane and then diffuse into the body of the organism. In many animals the gas is then transported away from the surface and towards interior cells by the blood system.

response any change in an organism occurring as a result of a stimulus. There are many different types of response, some involving the entire organism, others only groups of cells or tissues. Examples include the muscular contractions in an animal, the movement of leaves towards the light, and the onset of hibernation by small mammals at the start of winter.

response time in computing, the delay between entering a command and seeing its effect.

rest mass in physics, the mass of a body when its velocity is zero. For subatomic particles, it is their mass at rest or at velocities considerably below that of light. According to the theory of ◊relativity, at very high velocities, there is a relativistic effect that increases the mass of the particle.

restriction enzyme bacterial ◊enzyme that breaks a chain of ◊DNA into two pieces at a specific point; used in ◊genetic engineering. The point along the DNA chain at which the enzyme can work is restricted to places where a specific sequence of base pairs occurs. Different restriction enzymes will break a DNA chain at different points. The overlap between the fragments is used in determining the sequence of base pairs in the DNA chain.

retina light-sensitive area at the back of the ◊eye connected to the brain by the optic nerve. It has several layers and in humans contains over a million rods and cones, sensory cells capable of converting light into nervous messages that pass down the optic nerve to the brain.

The *rod cells*, about 120 million in each eye, are distributed throughout the retina. They are sensitive to low levels of light, but do not provide detailed or sharp images, nor can they detect colour. The *cone cells*, about 6 million in number, are mostly concentrated in a central region of the retina called the *fovea*, and provide both detailed and colour vision. The cones of the human eye contain three visual pigments, each of which responds to a different primary colour (red, green, or blue). The brain can interpret the varying signal levels from the three types of cone as any of the different colours of the visible spectrum.

The image actually falling on the retina is highly distorted; research into the eye and the optic centres within the brain has shown that this poor quality image is processed to improve its quality. The retina can become separated from the wall of the eyeball as a result of a trauma, such as a boxing injury. It can be reattached by being 'welded' into place by a laser.

retinol or *vitamin A* fat-soluble chemical derived from β-carotene and found in milk, butter, cheese, egg yolk, and liver. Lack of retinol in the diet leads to the eye disease xerophthalmia.

retrovirus any of a family (*Retroviridae*) of ◊viruses containing the genetic material ◊RNA rather than the more usual ◊DNA.

For the virus to express itself and multiply within an infected cell, its RNA must be converted to DNA. It does this by using a built-in enzyme known as reverse transcriptase (since the transfer of genetic information from DNA to RNA is known as ◊transcription, and retroviruses do the reverse of this). Retroviruses include those causing ◊AIDS and some forms of leukemia. See ◊immunity.

reuse multiple use of a product (often a form of packaging), by returning it to the manufacturer or processor each time. Many such returnable items are sold with a deposit which is reimbursed if the item is returned. Reuse is usually more energy-and resource-efficient than ◊recycling unless there are large transport or cleaning costs.

reverberation in acoustics, the multiple reflections, or echoes, of sounds inside a building that merge and persist a short while (up to a few seconds) before fading away. At each reflection some of the sound energy is absorbed, causing the amplitude of the sound wave and the intensity of the sound to reduce a little.

Too much reverberation causes sounds to become confused and indistinct, and this is particularly noticeable in empty rooms and halls, and buildings such as churches and cathedrals where the hard, unfurnished surfaces do not absorb sound energy well. Where walls and surfaces absorb sound energy very efficiently, too little reverberation may cause a room or hall to sound dull or 'dead'. Reverberation is a key factor in the design of theatres and concert halls, and can be controlled by lining ceilings and walls with materials possessing specific sound-absorbing properties.

reverse osmosis the movement of solvent (liquid) through a semipermeable membrane from a more concentrated solution to a more dilute sol-ution. The solvent's direction of movement is opposite to that which it would experience during ◊osmosis, and is achieved by applying an external pressure to the solution on the more concentrated side of the membrane. The technique is used in desalination plants, when water (the solvent) passes from brine (a salt solution) into fresh water via a semipermeable filtering device.

reverse video alternative term for ◊*inverse video*.

reversible reaction chemical reaction that proceeds in both directions at the same time, as the product decomposes back into reactants as it is being produced. Such reactions do not run to completion, provided that no substance leaves the system. Examples include the manufacture of ammonia from hydrogen and nitrogen, and the oxidation of sulphur dioxide to sulphur trioxide.

$$N_2 + 3H_2 \rightleftharpoons 2NH_3$$

$$2SO_2 + O_2 \rightleftharpoons 2SO_3$$

The term is also applied to those reactions that can be made to go in the opposite direction by changing the conditions, but these run to completion because some of the substances escape from the reaction. Examples are the decomposition of calcium hydrogencarbonate on heating and the loss of water of crystallization by copper(II) sulphate pentahydrate.

$$Ca(HCO_3)_{2\,(aq)} \rightleftharpoons CaCO_{3\,(s)} + CO_{2\,(g)} + H_2O_{(l)}$$

$$CuSO_4.5H_2O_{(s)} \rightleftharpoons CuSO_{4\,(s)} + 5H_2O_{(g)}$$

Reynolds number number used in ◊fluid mechanics to determine whether fluid flow in a particular situation (through a pipe or around an aircraft body) will be turbulent or smooth. The Reynolds number is calculated using the flow velocity, viscosity of the fluid, and the dimensions of the flow channel. It is named after British engineer Osborne Reynolds.

RGB (abbreviation of *red–green–blue*) method of connecting a colour screen to a computer, involving three separate signals: red, green, and blue. All the colours displayed by the screen can be made up from these three component colours.

rhe unit of fluidity equal to the reciprocal of the ◊poise.

rhenium (Latin *Rhenus* 'Rhine') heavy, silver-white, metallic element, symbol Re, atomic number 75, relative atomic mass 186.2. It has chemical properties similar to those of manganese and a very high melting point (3,180°C/5,756°F), which makes it valuable as an ingredient in alloys.

It was identified by Walter Noddack and co-workers in Berlin in 1925, and named after the river Rhine.

rheostat in physics, a variable ◊resistor, usually consisting of a high-resistance wire-wound coil with a sliding contact. It is used to vary electrical resistance without interrupting the current (for example, when dimming lights). The circular type

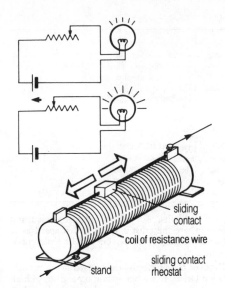

sliding contact

coil of resistance wire

sliding contact
stand   rheostat

rheostat *Rheostat used in a circuit to dim a bulb. Sliding the contact alters the length of resistance wire in the circuit, thus altering the current and dimming the bulb.*

in electronics (which can be used, for example, as the volume control of an amplifier) is also known as a ◊potentiometer.

rhesus factor ◊protein on the surface of red blood cells of humans, which is involved in the rhesus blood group system. Most individuals possess the main rhesus factor (Rh+); those without this factor (Rh−) produce ◊antibodies if they come into contact with it. The name comes from rhesus monkeys, in whose blood rhesus factors were first found.

If an Rh− mother carries an Rh+ fetus, she may produce antibodies if fetal blood crosses the ◊placenta. This is not normally a problem with the first infant because antibodies are only produced slowly. However, the antibodies continue to build up after birth, and a second Rh+ child may be attacked by antibodies passing from mother to fetus, causing the child to contract anaemia, heart failure, or brain damage. In such cases, the blood of the infant has to be changed for Rh− blood. Alternatively, the problem can be alleviated by giving the mother anti-Rh globulin just after the first pregnancy, preventing the formation of antibodies.

rhizoid hairlike outgrowth found on the ◊gametophyte generation of ferns, mosses and liverworts. Rhizoids anchor the plant to the substrate and can absorb water and nutrients. They may be composed of many cells, as in mosses, where they are usually brownish, or may be unicellular, as in liverworts, where they are usually colourless. Rhizoids fulfil the same functions as the ◊roots of higher plants but are simpler in construction.

rhizome or ◊rootstock horizontal underground plant stem. It is a ◊perennating organ in some species, where it is generally thick and fleshy, while in other species it is mainly a means of ◊vegetative reproduction, and is therefore long and slender,

with buds all along it that send up new plants. The potato is a rhizome that has two distinct parts, the tuber being the swollen end of a long, cordlike rhizome.

rhm (abbreviation of *roentgen–hour–metre*) the unit of effective strength of a radioactive source that produces gamma rays. It is used for substances for which it is difficult to establish radioactive disintegration rates.

rhodium (Greek *rhodon* 'rose') hard, silver-white, metallic element, symbol Rh, atomic number 45, relative atomic mass 102.905. It is one of the so-called platinum group of metals and is resistant to tarnish, corrosion, and acid. It occurs as a free metal in the natural alloy osmiridium and is used in jewellery, electroplating, and thermocouples.

Rhodium was discovered 1803 by English chemist William Wollaston (1766–1828) and named in 1804 after the red colour of its salts in solution.

rhombic sulphur allotropic form of sulphur. At room temperature, it is the stable allotrope (see ◊allotropy), unlike ◊monoclinic sulphur.

rhombus in geometry, an equilateral (all sides equal) ◊parallelogram. Its diagonals bisect each other at right angles, and its area is half the product of the lengths of the two diagonals. A rhombus whose internal angles are 90° is called a ◊square.

rhyolite ◊igneous rock, the fine-grained volcanic (extrusive) equivalent of granite.

rhythm method method of natural contraception that works by avoiding intercourse when the woman is producing egg cells (ovulating). The time of ovulation can be worked out by the calendar (counting days from the last period), by temperature changes, or by inspection of the cervical mucus. All these methods are unreliable because it is possible for ovulation to occur at any stage of the menstrual cycle.

ria long narrow sea inlet, usually branching and surrounded by hills. A ria is deeper and wider towards its mouth, unlike a ◊fjord. It is formed by the flooding of a river valley due to either a rise in sea level or a lowering of a landmass.

There are a number of rias in the UK—for example, in Salcombe and Dartmouth in Devon.

rib long, usually curved bone that extends laterally from the ◊spine in vertebrates. Most fishes and many reptiles have ribs along most of the spine, but in mammals they are found only in the chest area. In humans, there are 12 pairs of ribs. The ribs protect the lungs and heart, and allow the chest to expand and contract easily.

At the rear, each pair is joined to one of the vertebrae of the spine. The upper seven are joined by ◊cartilage directly to the breast bone (sternum). The next three are joined by cartilage to the end of the rib above. The last two ('floating ribs') are not attached at the front.

ribbon lake long, narrow lake found on the floor of a ◊glacial trough. A ribbon lake will often form in an elongated hollow carved out by a glacier, perhaps where it came across a weaker band of rock. Ribbon lakes can also form when water ponds up behind a terminal moraine or a landslide. The English Lake District is named after its many

Richter scale

magnitude	relative amount of energy released	examples
1		
2		
3		
4	1	Carlisle, England, 1979
5	30	San Francisco and New England, USA, 1979
		Wrexham, Wales, 1990
6	100	San Fernando, California, 1971
7	30,000	Chimbote, Peru, 1970
		San Francisco, 1989
8	1,000,000	Tangshan, China, 1976
		San Francisco, California, 1906
		Lisbon, Portugal, 1755
		Alaska, 1964
		Gansu, China, 1920

ribbon lakes, such as Lake Windermere and Coniston Water.

riboflavin or *vitamin B₂* ◊vitamin of the B complex whose absence in the diet causes stunted growth.

ribonucleic acid full name of ◊RNA.

ribosome in biology, the protein-making machinery of the cell. Ribosomes are located on the endoplasmic reticulum (ER) of eukaryotic cells, and are made of proteins and a special type of ◊RNA, ribosomal RNA. They receive messenger RNA (copied from the ◊DNA) and ◊amino acids, and 'translate' the messenger RNA by using its chemically coded instructions to link amino acids in a specific order, to make a strand of a particular protein.

Richter scale scale based on measurement of seismic waves, used to determine the magnitude of an ◊earthquake at its epicentre. The magnitude of an earthquake differs from its intensity, measured by the ◊Mercalli scale, which is subjective and varies from place to place for the same earthquake. The scale is named after US seismologist Charles Richter.

An earthquake's magnitude is a function of the total amount of energy released, and each point on the Richter scale represents a tenfold increase in energy over the previous point. The greatest earthquake ever recorded, in 1920 in Gansu, China, measured 8.6 on the Richter scale.

rickets defective growth of bone in children due to an insufficiency of calcium deposits. The bones, which do not harden adequately, are bent out of shape. It is usually caused by a lack of vitamin D and insufficient exposure to sunlight. Renal rickets, also a condition of malformed bone, is associated with kidney disease.

ridge of high pressure elongated area of high atmospheric pressure extending from an anticyclone. On a synoptic weather chart it is shown as a pattern of lengthened isobars. The weather under a ridge of high pressure is the same as that under an anticyclone.

The UK is often under the influence of a ridge of high pressure—in summer it originates from the Azores, and in winter from Scandinavia.

rift valley valley formed by the subsidence of a block of the Earth's ◊crust between two or more

parallel ◊faults. Rift valleys are steep-sided and form where the crust is being pulled apart, as at ◊ocean ridges, or in the Great Rift Valley of E Africa.

Rigel or *Beta Orionis* brightest star in the constellation Orion. It is a blue-white supergiant, with an estimated diameter 50 times that of the Sun. It is 900 light years from Earth, and is about 100,000 times more luminous than our Sun. It is the seventh brightest star in the sky.

right-angled triangle triangle in which one of the angles is a right angle (90°). It is the basic form of triangle for defining trigonometrical ratios (for example, sine, cosine, and tangent) and for which ◊Pythagoras' theorem holds true. The longest side of a right-angled triangle is called the hypotenuse; its area is equal to half the product of the lengths of the two shorter sides.

Any triangle constructed with its hypotenuse as the diameter of a circle, with its opposite vertex (corner) on the circumference, is a right-angled triangle. This is a fundamental theorem in geometry, first credited to the Greek mathematician Thales about 580 BC.

right ascension in astronomy, the coordinate on the ◊celestial sphere that corresponds to longitude on the surface of the Earth. It is measured in hours, minutes, and seconds eastwards from the point where the Sun's path, the ecliptic, once a year intersects the celestial equator; this point is called the *vernal equinox*.

right-hand rule in physics, a memory aid used to recall the relative directions of motion, magnetic field, and current in an electric generator. It was devised by English physicist John Fleming. See ◊Fleming's rules.

ring circuit household electrical circuit in which appliances are connected in series to form a ring with each end of the ring connected to the power supply.

ripple tank in physics, shallow water-filled tray used to demonstrate various properties of waves, such as reflection, refraction, diffraction, and interference, by programming and manipulating their movement.

RISC (acronym for *reduced instruction-set computer*) in computing, a microprocessor (processor on a single chip) that carries out fewer instructions

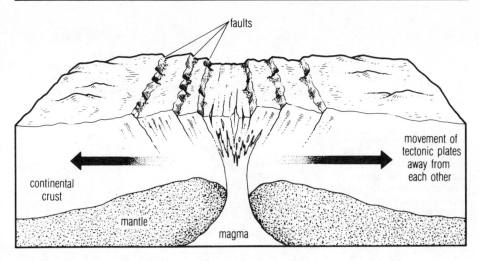

rift valley

than other (◊CISC) microprocessors in common use in the 1990s. Because of the low number of ◊machine-code instructions, the processor carries out those instructions very quickly.

RISC microprocessors became commercially available in the late 1980s, but are less widespread than traditional processors. The Archimedes range of computers, popular in schools, is based on RISC processors.

ritualization in ethology, a stereotype that occurs in certain behaviour patterns when these are incorporated into displays. For example, the exaggerated and stylized head toss of the goldeneye drake during courtship is a ritualization of the bathing movement used to wet the feathers; its duration and form have become fixed. Ritualization may make displays clearly recognizable, so ensuring that individuals mate only with members of their own species.

river long water course that flows down a slope along a channel. It originates at a point called its *source*, and enters a sea or lake at its *mouth*. Along its length it may be joined by smaller rivers called *tributaries*. A river and its tributaries are contained within a drainage basin.

river capture the diversion of the headwaters of one river into a neighbouring river. River capture occurs when a stream is carrying out rapid ◊headward erosion (backwards erosion at its source). Eventually the stream will cut into the course of a neighbouring river, causing the headwaters of that river to be diverted, or 'captured'.

The headwaters will then flow down to a lower level (often making a sharp bend, called an elbow of capture) over a steep slope, called a knickpoint. A waterfall will form here. ◊Rejuvenation then occurs, causing rapid downwards erosion.

river terrace part of an old ◊flood plain that has been left perched on the side of a river valley. It results from ◊rejuvenation, a renewal in the erosive powers of a river. River terraces are fertile and are often used for farming. They are also commonly chosen as sites for human settlement because they

are safe from flooding. Many towns and cities throughout the world have been built on terraces, including London, which is built on the terrace of the river Thames.

riveting method of joining metal plates. A hot metal pin called a rivet, which has a head at one end, is inserted into matching holes in two overlapping plates, then the other end is struck and formed into another head, holding the plates tight. Riveting is used in building construction, boilermaking, and shipbuilding.

RNA *ribonucleic acid* nucleic acid involved in the process of translating the genetic material ◊DNA into proteins. It is usually single-stranded, unlike the double-stranded DNA, and consists of a large number of nucleotides strung together, each of which comprises the sugar ribose, a phosphate group, and one of four bases (uracil, cytosine, adenine, or guanine). RNA is copied from DNA by the formation of ◊base pairs, with uracil taking the place of thymine.

RNA occurs in three major forms, each with a different function in the synthesis of protein molecules. *Messenger RNA* (mRNA) acts as the template for protein synthesis. Each ◊codon (a set of three bases) on the RNA molecule is matched up with the corresponding amino acid, in accordance with the ◊genetic code. This process (translation) takes place in the ribosomes, which are made up of proteins and *ribosomal RNA* (rRNA). *Transfer RNA* (tRNA) is responsible for combining with specific amino acids, and then matching up a special 'anticodon' sequence of its own with a codon on the mRNA. This is how the genetic code is translated.

Although RNA is normally associated only with the process of protein synthesis, it makes up the hereditary material itself in some viruses, such as ◊retroviruses.

road specially constructed route for wheeled vehicles to travel on. Reinforced tracks became necessary with the invention of wheeled vehicles in about 3000 BC and most ancient civilizations had

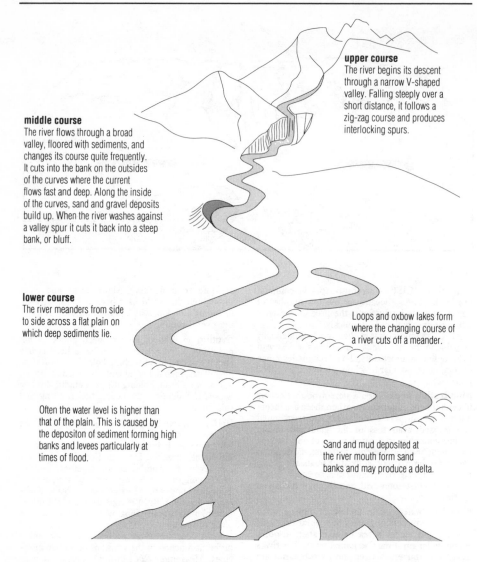

upper course
The river begins its descent through a narrow V-shaped valley. Falling steeply over a short distance, it follows a zig-zag course and produces interlocking spurs.

middle course
The river flows through a broad valley, floored with sediments, and changes its course quite frequently. It cuts into the bank on the outsides of the curves where the current flows fast and deep. Along the inside of the curves, sand and gravel deposits build up. When the river washes against a valley spur it cuts it back into a steep bank, or bluff.

lower course
The river meanders from side to side across a flat plain on which deep sediments lie.

Loops and oxbow lakes form where the changing course of a river cuts off a meander.

Often the water level is higher than that of the plain. This is caused by the depositon of sediment forming high banks and levees particularly at times of flood.

Sand and mud deposited at the river mouth form sand banks and may produce a delta.

river

some form of road network. The Romans developed engineering techniques that were not equalled for another 1,400 years.

Until the late 18th century most European roads were haphazardly maintained, making winter travel difficult. In the UK the turnpike system of collecting tolls created some improvement. The Scottish engineers Thomas Telford and John McAdam introduced sophisticated construction methods in the early 19th century. Recent developments have included durable surface compounds and machinery for rapid ground preparation.

The motorway is the most advanced form of road.

robot any computer-controlled machine that can be programmed to move or carry out work. Robots are often used in industry to transport materials or to perform repetitive tasks. For instance, robotic arms, fixed to a floor or workbench, may be used to paint machine parts or assemble electronic circuits. Other robots are designed to work in situations that would be dangerous to humans—for example, in defusing bombs or in space and deep-sea exploration. Some robots are equipped with sensors, such as touch sensors and video cameras, and can make be programmed to make simple decisions based on the sensory data received.

The first International Robot Olympics was held in Glasgow, Scotland in 1990. The world's fastest two-legged robot won a gold medal for completing a 3–m/9.8–ft course in less than a minute.

roche moutonnée outcrop of tough bedrock having one smooth side and one jagged side, found on the floor of a ◊glacial trough (U-shaped valley). It may be up to 40 m/132 ft high. A roche moutonnée is a feature of glacial erosion—as a glacier

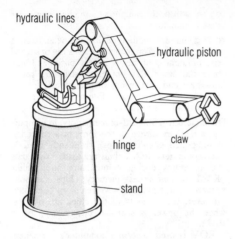

hydraulic lines

hydraulic piston

hinge

claw

stand

robot *The robot arm is the basic industrial robot used for assembly work.*

moved over its surface, ice and debris eroded its upstream side by corrasion, smoothing it and creating long scratches or striations. On the sheltered downstream side fragments of rock were plucked away by the ice, causing it to become steep and jagged.

rock constituent of the Earth's crust, composed of mineral particles and/or materials of organic origin consolidated into a hard mass as ◊igneous, ◊sedimentary, or ◊metamorphic rocks.

Igneous rock is made from magma (molten rock) that has solidified on or beneath the Earth's surface—for example, basalt, dolerite, granite, obsidian; *sedimentary rock* is formed by the compression of particles deposited by water, wind, or ice—for example, sandstone from sand particles, limestone from the remains of sea creatures; *metamorphic rock* is formed by changes in existing igneous or sedimentary rocks under high pressure or temperature, or chemical action—for example, marble is formed from limestone.

rocket projectile driven by the reaction of gases produced by a fast-burning fuel. Unlike jet engines, which are also reaction engines, modern rockets carry their own oxygen supply to burn their fuel and do not require any surrounding atmosphere. For warfare, rocket heads carry an explosive device.

Rockets have been valued as fireworks over the last seven centuries, but their intensive development as a means of propulsion to high altitudes, carrying payloads, started only in the interwar years with the state-supported work in Germany (primarily by Werner von Braun) and of Robert Hutchings Goddard (1882–1945) in the USA. Being the only form of propulsion available that can function in a vacuum, rockets are essential to exploration in outer space. ◊Multistage rockets have to be used, consisting of a number of rockets joined together.

Two main kinds of rocket are used: one burns liquid propellants, the other solid propellants. The fireworks rocket uses gunpowder as a solid propellant. The ◊space shuttle's solid rocket boosters use a mixture of powdered aluminium in a synthetic rubber binder. Most rockets, however, have liquid propellants, which are more powerful and easier to control. Liquid hydrogen and kerosene are common fuels, while liquid oxygen is the most common oxygen provider, or oxidizer. One of the biggest rockets ever built, the Saturn V moon rocket, was a three-stage design, standing 111 m/ 365 ft high, weighed more than 2,700 tonnes/3,000 tons on the launch pad, developed a takeoff thrust of some 3.4 million kg/7.5 million lb, and could place almost 140 tonnes/150 tons into low Earth orbit. In the early 1990s, the most powerful rocket system was the Soviet Energiya, capable of placing 100 tonnes/110 tons into low Earth orbit. The US space shuttle can put only 24 tonnes/26 tons into orbit. See ◊nuclear warfare and ◊missile.

rod a type of light-sensitive cell in the retina of most vertebrates. Rods are highly sensitive and provide only black and white vision. They are used when lighting conditions are poor and are the only type of visual cell found in animals active at night.

rodent any mammal of the worldwide order Rodentia, making up nearly half of all mammal species. Besides ordinary 'cheek teeth', they have a single front pair of incisor teeth in both upper and lower jaw, which continue to grow as they are worn down.

They are often subdivided into three suborders: Sciuromorpha, including primitive rodents, with squirrels as modern representatives; Myomorpha, rats and mice and their relatives; and Hystricomorpha, including the Old World and New World porcupines and guinea pigs.

roentgen or *röntgen* unit (symbol R) of radiation exposure, used for X-and gamma rays. It is defined in terms of the number of ions produced in one cubic centimetre of air by the radiation. Exposure to 1,000 roentgens gives rise to an absorbed dose of about 870 rads (8.7 grays), which is a dose equivalent of 870 rems (8.7 sieverts).

The annual dose equivalent from natural sources in the UK is 1,100 microsieverts.

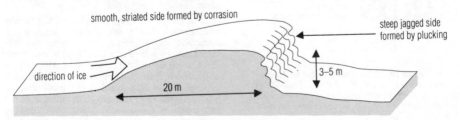

smooth, striated side formed by corrasion

steep jagged side formed by plucking

direction of ice

20 m

3–5 m

roche moutonnée

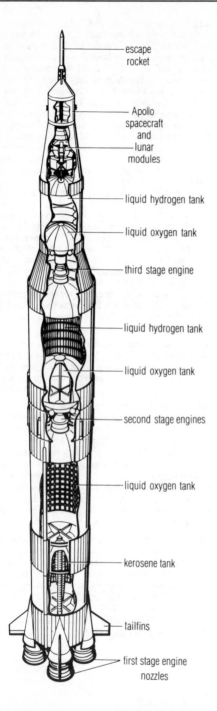

— escape rocket

— Apollo spacecraft and

— lunar modules

— liquid hydrogen tank

— liquid oxygen tank

— third stage engine

— liquid hydrogen tank

— liquid oxygen tank

— second stage engines

— liquid oxygen tank

— kerosene tank

— tailfins

— first stage engine nozzles

rocket *The three-stage Saturn V rocket used in the Apollo moonshots of the 1960s and 1970s. It stood 111 m/365 ft high, as tall as a 30–storey skyscraper, weighed 2,700 tonnes/3,000 tons when loaded with fuel, and developed a power equivalent to 50 Boeing 747 jumbo jets.*

rogue value in computing, another name for ◊data terminator.

rolling common method of shaping metal. Rolling is carried out by giant mangles, consisting of several sets, or stands, of heavy rollers positioned one above the other. Red-hot metal slabs are rolled into sheet and also (using shaped rollers) girders and rails. Metal sheets are often cold-rolled finally to impart a harder surface.

ROM (acronym for *r*ead-*o*nly *m*emory) in computing, a memory device in the form of a collection of integrated circuits (chips), frequently used in microcomputers. ROM chips are loaded with data and programs during manufacture and, unlike ◊RAM (random-access memory) chips, can subsequently only be read, not written to, by computer. However, the contents of the chips are not lost when the power is switched off, as happens in RAM.

ROM is used to form a computer's permanent store of vital information, or of programs that must be readily available but protected from accidental or deliberate change by a user. For example, a microcomputer ◊operating system is often held in ROM memory.

Roman numerals ancient European number system using symbols different from Arabic numerals (the ordinary numbers 1, 2, 3, 4, 5, and so on). The seven key symbols in Roman numerals, as represented today, are I (1), V (5), X (10), L (50), C (100), D (500) and M (1,000). There is no zero, and therefore no place-value as is fundamental to the Arabic system. The first ten Roman numerals are I, II, III, IV (or IIII), V, VI, VII, VIII, IX, and X. When a Roman symbol is preceded by a symbol of equal or greater value, the values of the symbols are added (XVI = 16). When a symbol is preceded by a symbol of less value, the values are subtracted (XL = 40). A horizontal bar over a symbol indicates a multiple of 1,000 ($\bar{X}$ = 10,000). Although addition and subtraction are fairly straightforward using Roman numerals, the absence of a zero makes other arithmetic calculations (such as multiplication) clumsy and difficult.

Although their role in mathematics is long obsolete, Roman numerals continue to enjoy a limited use as inscribed figures (for example, on clock faces, in the pagination of some written material, or as dates on buildings or in films).

röntgen alternative spelling for ◊roentgen, unit of X-and gamma-ray exposure.

root in biology, the part of a tooth that is not covered in enamel and is embedded in the jawbone.

It is also used to mean the point of origin of a nerve.

root in botany, the part of a plant that is usually underground, and whose primary functions are anchorage and the absorption of water and dissolved mineral salts. Roots usually grow downwards and towards water (that is, they are positively geotropic and hydrotropic; see ◊tropism). Plants such as epiphytic orchids, which grow above ground, produce aerial roots that absorb moisture from the atmosphere. Others, such as ivy, have climbing roots arising from the stems, which serve to attach the plant to trees and walls.

The absorptive area of roots is greatly increased by the numerous, slender root hairs formed near the tips. A calyptra, or root cap, protects the tip of the root from abrasion as it grows through the soil.

Symbiotic associations occur between the roots of certain ,plants, such as clover, and various bacteria that fix nitrogen from the air (see ◊nitrogen fixation). Other modifications of roots include ◊contractile roots, ◊pneumatophores, ◊taproots, and ◊prop roots.

ROOT AREA

In a study of a four-month-old rye plant, the root system was estimated to have 14 billion root hairs, with a total surface area of over 600 sq m/6,450 sq ft. This was 130 times more than the surface area of the plant above ground.

root in mathematics, a value that satisfies the equality of an equation. For example, $x = 0$ and $x = 5$ are roots of the equation $x^2 - 5x = 0$.

root cap cap at the tip of a growing root. It gives protection to the zone of actively dividing cells as the root pushes through the soil. As the cells are worn away by friction they are continually being replaced by the ◊meristem.

root directory in computing, the top directory in a ◊tree-and-branch filing system. It contains all the other directories.

root hair tubular outgrowth from a cell on the surface of a plant root. It is a delicate structure, which survives for a few days only and does not develop into a root. New root hairs are continually being formed near the root tip to replace the ones that are lost.

The majority of land plants possess root hairs, which greatly increase the surface area available for the absorption of water and mineral salts from the soil. The layer of the root's epidermis that produces root hairs is known as the *piliferous layer.*

root-mean-square (RMS) in mathematics, value obtained by taking the ◊square root of the mean of the squares of a set of values; for example the RMS value of four quantities a, b, c, and d is $\sqrt{[(a^2 + b^2 + c^2 + d^2)/4]}$.

For an alternating current (AC), the RMS value is equal to the peak value divided by the square root of 2.

root nodule a clearly visible swelling that develops in the roots of members of the bean family, the Leguminosae. The cells inside this tumerous growth have been invaded by the bacteria Rhizobium, a soil microbe capable of converting gaseous nitrogen into nitrate. The nodule is therefore an association between a plant and a bacterium, with both partners benefiting. The plant obtains nitrogen compounds while the bacterium obtains nutrition and shelter.

Nitrogen fixation by bacteria is one of the main ways by which the nitrogen in the atmosphere is cycled back into living things. The economic value of the process is so great that research has been carried out into the possibility of stimulating the

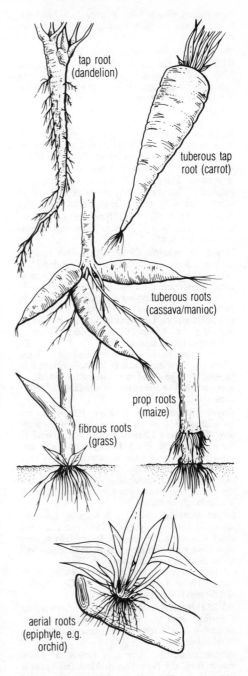

root *Types of root. Many flowers (dandelion) and vegetables (carrot) have swollen tap roots with smaller lateral roots. The tuberous roots of the cassava are swollen parts of an underground stem modified to store food. The fibrous roots of the grasses are all of equal size. Prop roots grow out from the stem and then grow down into the ground to support a heavy plant. Aerial roots grow from stems but do not grow into the ground; many absorb moisture from the air.*

formation of root nodules in crops such as wheat, which do not normally form an association with rhizobium.

rootstock another name for ◊rhizome.

ROSAT joint US/German/UK satellite launched 1990 to study cosmic sources of X-rays and extremely short ultraviolet wavelengths, named after Wilhelm Röntgen, the discoverer of X-rays.

rotifer any of the tiny invertebrates, also called 'wheel animalcules', of the phylum Rotifera. Mainly freshwater, some marine, rotifers have a ring of ◊cilia that carries food to the mouth and also provides propulsion. They are the smallest of multicellular animals, few reach 0.05 cm/0.02 in.

roughage alternative term for dietary ◊fibre, material of plant origin that cannot be digested by enzymes normally present in the human ◊gut.

Royal Botanic Gardens, Kew botanic gardens in Richmond, Surrey, England, popularly known as ◊Kew Gardens.

Royal Greenwich Observatory the national astronomical observatory of the UK, founded 1675 at Greenwich, E London, England, to provide navigational information for sailors. After World War II it was moved to Herstmonceux Castle, Sussex; in 1990 it was transferred to Cambridge. It also operates telescopes on La Palma in the Canary Islands, including the 4.2–m/165–in William Herschel Telescope, commissioned 1987.

The observatory was founded by King Charles II. The eminence of its work resulted in Greenwich Time and the Greenwich Meridian being adopted as international standards of reference 1884.

Royal Horticultural Society British society established 1804 for the improvement of horticulture. The annual Chelsea Flower Show, held in the grounds of the Royal Hospital, London, is also a social event, and another flower show is held at Vincent Square, London. There are gardens, orchards, and trial grounds at Wisley, Surrey, and the Lindley Library has one of the world's finest horticultural collections.

Royal Society oldest and premier scientific society in Britain, originating 1645 and chartered 1660; Christopher Wren and Isaac Newton were prominent early members. Its Scottish equivalent is the Royal Society of Edinburgh 1783.

The headquarters of the Royal Society is in Carlton House Terrace, London; the Royal Society of Edinburgh is in George Street.

RS-232 interface standard type of computer ◊interface used to connect computers to serial devices. It is used for modems, mice, screens, and serial printers.

rubber coagulated latex of a variety of plants, mainly from the New World. Most important is Para rubber, which derives from the tree *Hevea brasiliensis* of the spurge family. It was introduced from Brazil to SE Asia, where most of the world supply is now produced, the chief exporters being Peninsular Malaysia, Indonesia, Sri Lanka, Cambodia, Thailand, Sarawak, and Brunei. At about seven years the tree, which may grow to 20 m/ 60 ft, is ready for 'tapping'. Small incisions are made in the trunk and the latex drips into collecting cups. In pure form, rubber is white and has the formula $(C_5H_8)_n$.

Other sources of rubber are the Russian dandelion *Taraxacum koksagyz*, which grows in temperate climates and can yield about 45 kg/100 lb of rubber per tonne of roots, and guayule *Parthenium argentatum*, a small shrub of the Compositae family, which grows in SW USA and Mexico.

Early uses of rubber were limited by its tendency to soften on hot days and harden on colder ones, a tendency that was eliminated by Charles Goodyear's invention of ◊vulcanization 1839.

In the 20th century, world production of rubber has increased a hundredfold, and World War II stimulated the production of synthetic rubber to replace the supplies from Malaysian sources overrun by the Japanese. There are an infinite variety of synthetic rubbers adapted to special purposes, but economically foremost is SBR (styrene-butadiene rubber). Cheaper than natural rubber, it is preferable for some purposes; for example, on car tyres, where its higher abrasion-resistance is useful, and it is either blended with natural rubber or used alone for industrial moulding and extrusions, shoe soles, hoses, and latex foam.

rubidium (Latin *rubidus* 'red') soft, silver-white, metallic element, symbol Rb, atomic number 37, relative atomic mass 85.47. It is one of the ◊alkali metals, ignites spontaneously in air, and reacts violently with water. It is used in photoelectric cells and vacuum-tube filaments.

Rubidium was discovered spectroscopically by German physicists Robert Bunsen and Gustav Kirchhoff in 1861, and named after the red lines in its spectrum.

ruby the red transparent gem variety of the mineral ◊corundum Al_2O_3, aluminium oxide. Small amounts of chromium oxide, Cr_2O_3, substituting for aluminium oxide, give ruby its colour. Natural rubies are found mainly in Myanmar (Burma), but rubies can also be produced artificially and such synthetic stones are used in ◊lasers.

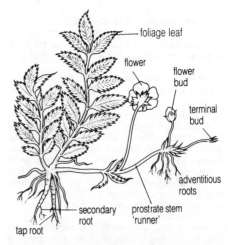

runner *A runner, or stolon, grows horizontally near the base of some plants, such as the strawberry. It produces roots along its length and new plants grow at these points.*

ruminant any even-toed hoofed mammal with a rumen, the 'first stomach' of its complex digestive system. Plant food is stored and fermented before being brought back to the mouth for chewing (chewing the cud) and then is swallowed to the next stomach. Ruminants include cattle, antelopes, goats, deer, and giraffes, all with a four-chambered stomach. Camels are also ruminants, but they have a three-chambered stomach.

runner in botany, aerial stem that produces new plants; also called a ◊*stolon*.

runoff the amount of water that flows off the land, either through streams or over the surface.

rust reddish-brown oxide of iron formed by the action of moisture and oxygen on the metal. It consists mainly of hydrated iron(III) oxide ($Fe_2O_3.H_2O$ and iron(III) hydroxide ($Fe(OH)_3$).

Paints that penetrate beneath any moisture, and plastic compounds that combine with existing rust to form a protective coating, are used to avoid rusting.

ruthenium hard, brittle, silver-white, metallic element, symbol Ru, atomic number 44, relative atomic mass 101.07. It is one of the so-called plati-num group of metals; it occurs in platinum ores as a free metal and in the natural alloy osmiridium. It is used as a hardener in alloys and as a catalyst; its compounds are used as colouring agents in glass and ceramics.

It was discovered in 1827 and named in 1828 after its place of discovery, the Ural mountains in Ruthenia (now part of the Ukraine). Pure ruthenium was not isolated until 1845.

rutherfordium name proposed by US scientists for the element currently known as ◊unnilquadium (atomic number 104), to honour New Zealand physicist Ernest Rutherford.

rutile TiO_2, titanium oxide, a naturally occurring ore of titanium. It is usually reddish brown to black, with a very bright (adamantine) surface lustre. It crystallizes in the tetragonal system. Rutile is common in a wide range of igneous and metamorphic rocks and also occurs concentrated in sands; the coastal sands of E and W Australia are a major source. It is also used as a pigment that gives a brilliant white to paint, paper, and plastics.

Rydberg constant in physics, a constant that relates atomic spectra to the ◊spectrum of hydrogen. Its value is 1.0977×10^7 per metre.

S

sabin unit of sound absorption, used in acoustical engineering. One sabin is the absorption of one square foot (0.093 square metre) of a perfectly absorbing surface (such as an open window).

saccharide another name for a sugar molecule.

saccharin or *ortho-sulpho benzimide* $C_7H_5NO_3S$ sweet, white, crystalline solid derived from coal tar and substituted for sugar. Since 1977 it has been regarded as potentially carcinogenic. Its use is not universally permitted and it has been largely replaced by other sweetening agents.

safety glass glass that does not splinter into sharp pieces when smashed. *Toughened glass* is made by heating a glass sheet and then rapidly cooling it with a blast of cold air; it shatters into rounded pieces when smashed. *Laminated glass* is a 'sandwich' of a clear plastic film between two glass sheets; when this is struck, it simply cracks, the plastic holding the glass in place.

safety lamp portable lamp designed for use in places where flammable gases such as methane may be encountered; for example, in coal mines. The electric head lamp used as a miner's working light has the bulb and contacts in protected enclosures. The flame safety lamp, now used primarily for gas detection, has the wick enclosed within a strong glass cylinder surmounted by wire gauzes. English chemist Humphrey Davy 1815 and English engineer George Stephenson each invented flame safety lamps.

Saffir–Simpson damage-potential scale scale of potential damage from wind and sea when a hurricane is in progress: 1 is minimal damage, 5 is catastrophic.

Sagittarius zodiac constellation in the southern hemisphere, represented as a centaur aiming a bow and arrow at neighbouring Scorpius. The Sun passes through Sagittarius from mid-Dec to mid-Jan, including the winter solstice, when it is farthest south of the equator. The constellation contains many nebulae and ◊globular clusters, and open ◊star clusters. Kaus Australis and Nunki are its brightest stars. The centre of our Galaxy, the Milky Way, is marked by the radio source Sagittarius A. In astrology, the dates for Sagittarius are between about 22 Nov and 21 Dec (see ◊precession).

Sahel (Arabic *sahil* 'coast') marginal area to the S of the Sahara, from Senegal to Somalia, where the desert is gradually encroaching. The desertification is partly due to climatic change but has also been caused by the pressures of a rapidly expanding population, which have led to overgrazing and the destruction of trees and scrub for fuelwood. In recent years many famines have taken place in the area.

St Elmo's fire bluish, flamelike electrical discharge that sometimes occurs above ships' masts and other pointed objects or about aircraft in stormy weather. Although high voltage, it is low current and therefore harmless. St Elmo (or St Erasmus) is the patron saint of sailors.

sal ammoniac former name for ◊ammonium chloride.

salicylic acid HOC_6H_4COOH the active chemical constituent of aspirin, an analgesic drug. The acid and its salts (salicylates) occur naturally in many plants; concentrated sources include willow bark and oil of wintergreen.

When purified, salicylic acid is a white solid that crystallizes into prismatic needles at 318°F/159°C. It is used as an antiseptic, in food preparation and dyestuffs, and in the preparation of aspirin.

saliva in vertebrates, a secretion from the salivary glands that aids the swallowing and digestion of food in the mouth. In mammals, it contains the enzyme amylase, which converts starch to sugar. The salivary glands of mosquitoes and other blood-sucking insects produce ◊anticoagulants.

Salmonella very varied group of bacteria. They can be divided into three broad groups. One of these causes typhoid and paratyphoid fevers, while a second group causes salmonella food poisoning, which is characterized by stomach pains, vomiting, diarrhoea, and headache. It can be fatal in elderly

	negative ions		
	NO_3^-	SO_4^{2-}	Cl^-
Na⁺	$NaNO_3$ sodium nitrate	Na_2SO_4 sodium sulphate	$NaCl$ sodium chloride
Mg²⁺	$Mg(NO_3)_2$ magnesium nitrate	$MgSO_4$ magnesium sulphate	$MgCl_2$ magnesium chloride
Fe³⁺	$Fe(NO_3)_3$ iron (III) nitrate	$Fe_2(SO_4)_3$ iron (III) sulphate	$FeCl_3$ iron (III) chloride

(positive ions)

salt

people, but others usually recover in a few days without antibiotics. Most cases are caused by contaminated animal products, especially poultry meat.

Human carriers of the disease may be well themselves but pass the bacteria on to others through unhygienic preparation of food. Domestic pets can also carry the bacteria while appearing healthy.

In 1989 the British government was forced to take action after it was claimed that nearly all English eggs were infected with salmonella. Many chickens were slaughtered and consumers were advised to hardboil eggs.

salt in chemistry, any compound containing a positive ion (cation) derived from a ◊metal or ammonia and a negative ion (anion) derived from an ◊acid or ◊nonmetal. If the negative ion has a replaceable hydrogen atom it is an *acid salt* (for example, sodium hydrogensulphate, $NaHSO_4$); if not, it is classed as a *normal salt* (for example sodium chloride, $NaCl$; potassium sulphate, K_2SO_4). *Common salt* is sodium chloride. Salts have the properties typical of ionic compounds.

Most inorganic salts readily dissolve in water to give an electrolyte (a solution that conducts electricity). The ions in a solid salt normally adopt a regular arrangement to form crystals. Some salts only form stable crystals as hydrates (when combined with water).

TINY CRYSTALS OF SALT

Crystals of salt (NaCl) are made up of small cubes of atoms. A crystal of rock salt contains 40,000 million million million atoms in each cubic centimetre, arranged in tiny cubes 0.00000002 cm/0.000000008 in across.

saltation (Latin *saltare* 'to leap') in biology, the idea that an abrupt genetic change can occur in an individual, which then gives rise to a new species. The idea has now been largely discredited, although the appearance of ◊polyploid individuals can be considered an example.

salt, common or *sodium chloride* NaCl white crystalline solid, found dissolved in sea water and as rock salt (halite) in large deposits and salt domes.

Common salt is used extensively in the food industry as a preservative and for flavouring, and in the chemical industry in the making of chlorine and sodium.

While common salt is an essential part of our diet, some medical experts believe that excess salt can lead to high blood pressure and increased risk of heart attacks.

Salt has historically been considered a sustaining substance, often taking on religious significance in ancient cultures. Roman soldiers were paid part of their wages as salt allowance (Latin *salerium argentinium*), hence the term 'salary'.

salt marsh wetland with halophytic vegetation (tolerant to sea water). Salt marshes develop around estuaries and on the sheltered side of sand and shingle spits. Salt marshes usually have a network of creeks and drainage channels by which tidal waters enter and leave the marsh.

Typical plants of European salt marshes include salicornia, or saltwort, which has fleshy leaves like a succulent; sea lavender, sea pink, and sea aster. Geese such as brent, greylag, and bean are frequent visitors to salt marshes in winter, feeding on plant material.

saltpetre former name for potassium nitrate (KNO_3), the compound used in making gunpowder (from about 1500). It occurs naturally, being deposited during dry periods in places with warm climates, such as India.

Salvarsan historical proprietory name for arsphenamine (technical name 3,3–diamino-4,4–dihydroxyarsenobenzene dichloride), the first specific antibacterial agent, discovered by German bacteriologist Paul Ehrlich in 1909. Because of its destructive effect on *Spirochaeta pallida*, it was used in the treatment of syphilis before the development of antibiotics.

sal volatile another name for ◊smelling salts.

Salyut (Russian 'salute') series of seven space stations launched by the USSR 1971–82. Salyut was cylindrical in shape, 15 m/50 ft long, and weighed 19 tonnes/21 tons. It housed two or three cosmonauts at a time, for missions lasting up to eight months.

Salyut 1 was launched 19 April 1971. It was occupied for 23 days in June 1971 by a crew of three, who died during their return to Earth when their ◊Soyuz ferry craft depressurized. *Salyut 2*, in 1973, broke up in orbit before occupation. The first fully successful Salyut mission was a 14–day visit to *Salyut 3* in July 1974. In 1984–85 a team of three cosmonauts endured a record 237–day flight in *Salyut 7*. In 1986, the Salyut series was superseded by ◊*Mir*, an improved design capable of being enlarged by additional modules sent up from Earth.

Crews observed Earth and the sky, and carried out processing of materials in weightlessness. The last in the series, *Salyut 7*, crashed to Earth in Feb 1991, scattering debris in Argentina.

samara in botany, a winged fruit, a type of ◊achene.

samarium hard, brittle, grey-white, metallic element of the ◊lanthanide series, symbol Sm, atomic number 62, relative atomic mass 150.4. It is widely distributed in nature and is obtained commercially from the minerals monzanite and bastnasite. It is used only occasionally in industry, mainly as a catalyst in organic reactions. Samarium was discovered by spectroscopic analysis of the mineral samarskite and named in 1879 by French chemist Paul Lecoq de Boisbaudran (1838–1912) after its source.

San Andreas fault geological fault line stretching for 1,125 km/700 mi in a NW–SE direction through the state of California, USA.

Two sections of the Earth's crust meet at the San Andreas fault, and friction is created as the coastal Pacific plate moves NW, rubbing against the American continental plate, which is moving slowly SE. The relative movement is only about 5 cm/2 in a year, which means that Los Angeles will reach San Francisco's latitude in 10 million years. The friction caused by this tectonic movement gives rise to periodic ◊earthquakes.

sand loose grains of rock, sized 0.0625–2.00 mm/0.0025–0.08 in in diameter, consisting chiefly of ◊quartz, but owing their varying colour to mixtures of other minerals. Sand is used in cement-making, as an abrasive, in glass-making, and for other purposes.

Sands are classified into marine, freshwater, glacial, and terrestrial. Some 'light' soils contain up to 50% sand. Sands may eventually consolidate into ◊sandstone.

sandbar ridge of sand built up by the currents across the mouth of a river or bay. A sandbar may be entirely underwater or it may form an elongated island that breaks the surface. A sandbar stretching out from a headland is a *sand spit*.

Coastal bars can extend across estuaries to form *bay bars*.

sandstone ◊sedimentary rocks formed from the consolidation of sand, with sand-sized grains (0.0625–2 mm/0.0025–0.08 in) in a matrix or cement. Their principal component is quartz. Sandstones are commonly permeable and porous, and may form freshwater ◊aquifers. They are mainly used as building materials.

Sandstones are classified according to the matrix or cement material (whether derived from clay or silt; for example, as calcareous sandstone, ferruginous sandstone, siliceous sandstone).

sap the fluids that circulate through ◊vascular plants, especially woody ones. Sap carries water and food to plant tissues. Sap contains alkaloids, protein, and starch; it can be milky (as in rubber trees), resinous (as in pines), or syrupy (as in maples).

saponification in chemistry, the ◊hydrolysis (splitting) of an ◊ester by treatment with a strong alkali, resulting in the liberation of the alcohol from which the ester had been derived and a salt of the constituent fatty acid. The process is used in the manufacture of soap.

sapphire deep-blue, transparent gem variety of the mineral ◊corundum Al_2O_3, aluminium oxide. Small amounts of iron and titanium give it its colour. A corundum gem of any colour except red (which is a ruby) can be called a sapphire; for example, yellow sapphire.

saprophyte in botany, an obsolete term for a ◊saprotroph, an organism that lives in dead or decaying matter.

saprotroph (formerly *saprophyte*) organism that feeds on the excrement or the dead bodies or tissues of others. They include most fungi (the rest being parasites); many bacteria and protozoa; animals such as dung beetles and vultures; and a few unusual plants, including several orchids. Saprotrophs cannot make food for themselves, so they are a type of ◊heterotroph. They are useful scavengers, and in sewage farms and refuse dumps break down organic matter into nutrients easily assimilable by green plants.

sarcoma malignant ◊tumour arising from the fat, muscles, bones, cartilage, or blood and lymph vessels, and connective tissues. Sarcomas are much less common than ◊carcinomas.

sard or *sardonyx* yellow or red-brown variety of ◊onyx.

satellite any small body that orbits a larger one, either natural or artificial. Natural satellites that orbit planets are called moons. The first *artificial satellite*, *Sputnik 1*, was launched into orbit around the Earth by the USSR 1957. Artificial satellites are used for scientific purposes, communications, weather forecasting, and military applications. The largest artificial satellites can be seen by the naked eye.

At any time, there are several thousand artificial satellites orbiting the Earth, including active satellites, satellites that have ended their working lives, and discarded sections of rockets. Artificial satellites eventually re-enter the Earth's atmosphere. Usually they burn up by friction, but sometimes debris falls to the Earth's surface, as with ◊*Skylab* and *Salyut 7*. The USA launched 23 nuclear-powered satellites 1961–77, of which four malfunctioned. The Strategic Defense Initiative (Star Wars) programme proposed to send as many as 100 nuclear reactors into space. From 1965, the USSR launched 39 nuclear reactors on orbiting satellites, of which six malfunctioned.

More than 70,000 pieces of space junk, ranging from disabled satellites to tiny metal fragments, are careering around the Earth. The amount of waste is likely to increase, as the waste particles in orbit are continually colliding and fragmenting further.

THE LARGEST SATELLITES

The Moon is the sixth largest satellite in the Solar System. It is 3,476 km/2,160 mi in diameter. The larger satellites are: Jupiter's moons, Io, Ganymede, and Callisto; Saturn's Titan; and Neptune's Triton.

satellite applications the uses to which artificial satellites are put. These include:

scientific experiments and observation Many astronomical observations are best taken above the disturbing effect of the atmosphere. Satellite observations have been carried out by *IRAS* (*Infrared Astronomical Satellite*, 1983) which made a complete infrared survey of the skies, and *Solar Max* 1980, which observed solar flares. The *Hipparcos* satellite, launched 1989, measured the positions of many

Sap flow in plants

Stephen Hales was a man of many scientific interests and, apparently, of settled habits. After scientific education at Cambridge, he was appointed vicar of Teddington, England in 1709, holding the post until he died in 1761. Perhaps he preferred a settled life so he could follow his many scientific interests. Amongst his investigations, he refined William Harvey's ideas on blood circulation in animals. This involved gruesome experiments on horses, dogs and frogs. In one, he attached a vertical tube 3.3 m/11 ft long to a horse's blood vessel to measure how high the blood rose—he was measuring blood pressure.

Circulation in plants?

In about 1724, Hales began studying the circulation of sap in plants. He showed how water is drawn in at the roots, transported to the leaves and then transpired. Several pieces of this jigsaw were already in place. After the microscope was developed in the mid 17th century, it was found that plants contained tubes running along the length of the stem, some full of water and some full of air. This suggested some sort of circulation system, comparable to that in animals. Common sense suggested that plants take in water through the roots, the moisture travelling to all parts of the plant. Around 1670 Italian Marcello Malpighi had shown that a plant's food substances were built up in the leaves. There must be a downward transport of these substances to other parts of the plant. Since these substances were sometimes stored in tubers, the downward flow must reach the roots.

The questions Hales addressed were: was there a closed circulation, as in animals, and what powered it? He undertook three crucial experiments. In the first, he established that the water was drawn upwards by the leaves. His notebook records:

'July 27 (1716). I fixed an apple branch . . . to a tube. I filled the tube with water, and then immersed the whole branch . . . into a vessel full of water. The water subsided 6 inches in the first two hours (being the filling of the sap vessels) and 6 inches the following night'.

Hales concluded that this showed the great drawing power of 'perspiration'. Evaporation from the leaves pulled the sap upwards, rather than pressure pushing from the roots. The next step was simple: a leafy plant in a closed vessel, and collecting the transpired liquid, confirmed that the material transpired into the air was water. Water was indeed flowing up the stem and being expired from the leaves.

Hales' apple branch

The third experiment was more complicated. Hales needed to discover whether the upward movement of the water was balanced by equal downward flow; water might be absorbed from the air by the leaves, making the flow a continuous circulation. His notebook records a beautifully simple and conclusive experiment:

'August 20 (1716). At 1 pm I took an apple branch nine foot long, 1 + ¾ inch diameter, with proportional lateral branches. I cemented it fast to one end of a vertical U tube; but first I cut away the bark, and last year's ringlet of wood, near the bottom of the branch. I then filled the tube with water, which was 22 feet long and ½ inch diameter. The water was very freely imbibed, at a rate of 3 + ½ inch in a minute. In half an hour's time I could plainly perceive the lower part of the cut to be moister than before; when at the same time the upper part of the wound looked white and dry.

'The water must necessarily ascend from the tube, through the innermost wood, because the last year's wood was cut away, for 3 inches all round the stem; and consequently, if the sap in its natural course descended by the last year's ringlet of wood, between that and the bark (as many have thought) the water should have descended by last year's wood, or the bark, and so have moistened the upper part of the gap; but, on the contrary, the lower part was moistened, and not the upper part.'

There was no sign of descending water, and hence no circulation of water within the plant. Hales had already shown that the sunflower transpires water 17 times faster than a man, bulk for bulk; if there was a circulation, it would have to be enormously fast.

Hales continued his work on plant physiology. He showed that plants take in part of the air (carbon dioxide) which is used in their nutrition. He measured growth rates. His interests were wide-ranging: from preservation of foods, water purification, and ventilation of ships and buildings, to the best way of supporting pie crusts.

In 1779 Dutch doctor Jan Ingen-Housz established two distinct processes in plants: respiration, in which oxygen is absorbed and carbon dioxide exhaled, as in animal respiration; and photosynthesis, in which carbon dioxide is taken in to make food, and oxygen is given out. It is the daily ebb and flow of these processes that transports water and food around a plant.

some environmental monitoring programmes using satellite remote sensing

programme	agency	status	objectives
POES: Polar-orbiting Operational Environment Satellites	NOAA	operational since 1970	weather observations
METEOSAT: Meteorology Satellite	ESA	operational since 1977	weather observations
LANDSAT: Land Remote Sensing Satellite	EOSAT	operational since 1972	vegetation, crop and land-use inventory
LAGEOS-1: Laser Geodynamics Satellite-1	NASA	operational since 1976	geodynamics, gravity field
SPOT-1: Système Probatoire d'Observation de la Terre-1	France	operational since 1986	land use, earth resources
IRS: Indian Remote Sensing Satellite (e.g. Rohini-2)	India	operational since 1981	Earth resources
MOS-1: Marine Observation Satellite-1	NASDA (Japan)	operational since 1987	state of sea surface and atmosphere
LAGEOS-2: Laser Geodynamics Satellite-2	NASA-PSN (Italy)	operational since 1988	geodynamics, gravity field
ERS-1: Earth Remote Sensing Satellite-1	ESA	launch 1990	imaging of oceans, ice fields, land areas
N-ROSS: Navy Remote Sensing System	US Navy	launch 1991	ocean topography, surface winds, ice extent
JERS-1: Japan Earth Remote Sensing Satellite-1	NASDA (Japan)	launch 1991	earth resources
TOPEX/POSEIDON: Ocean Topography Experiment	NASA-CNES (France)	start 1989, launch 1991	ocean surface topography
RADARSAT: Canadian Radar Satellite	Canada	start 1986, launch 1991	studies of Arctic ice, ocean studies, earth resources
GRM: Geopotential Research Mission	NASA	start 1986, launch 1992	measure global geoid and magnetic field
EOS: Earth Observing System/Polar-Orbiting Platforms	NASA	start 1989, launch 1994	long-term global earth observation
European Polar Orbiting Platform (Columbus)	ESA	planned	long-term comprehensive research, operational and commercial Earth observations
Rainfall Mission	NASA	start 1991, launch 1994	tropical precipitation measurements

representative Space Shuttle instruments

ATMOS: Atmospheric Trace Molecules Observed by Spectroscopy	NASA	current	atmospheric chemical composition
ACR: Active Cavity Radiometer	NASA	current	solar energy output
SUSIM: Solar Ultraviolet Spectral Irradiance Monitor	NASA	current	ultraviolet solar observations
MAPS: Measurement of Air Pollution from Shuttle	NASA	current/in development	tropospheric carbon monoxide

individual instruments for long-term global observations

Total Ozone Monitor	NASA	planned	monitor global ozone
Laser Ranger:	NASA	planned	continental motions
Scanning Radar Altimeter	NASA	planned	continental topography

CNES: Centre National d'Études Spatiales (France); EOSAT: EOSAT Company; ESA: European Space Agency; NASDA:Japan Space Agency; NOAA: National Oceanic and Atmospheric Administration (USA); NSF: National Science Foundation (USA); PSN: Piano Spaziale Nazionale (Italian National Space Plan); USGS: United States Geological Survey

largest natural planetary satellites

planet	satellite	diameter in km	mean distance from centre of primary in km/mi	orbital period in days	reciproca mass (planet = 1)
Jupiter	Ganymede	5,262/3,300	1,070,000	7.16	12,800
Saturn	Titan	5,150/3,200	1,221,800	15.95	4,200
Jupiter	Callisto	4,800/3,000	1,883,000	16.69	17,700
Jupiter	Io	3,630/2,240	421,600	1.77	21,400
Earth	Moon	3,476/2,160	384,400	27.32	81.3
Jupiter	Europa	3,138/1,900	670,900	3.55	39,700
Neptune	Triton	2,700/1,690	354,300	5.88	770

stars. In 1992, the COBE (Cosmic Background Explorer) satellite detected details of the Big Bang that mark the first stage in the formation of galaxies. Medical experiments have been carried out aboard crewed satellites, such as the Soviet *Mir* and the US *Skylab*.

reconnaissance, land resource, and mapping applications Apart from military use and routine mapmaking, the US *Landsat*, the French *SPOT*, and equivalent USSR satellites have provided much useful information about water sources and drainage, vegetation, land use, geological structures, oil and mineral locations, and snow and ice. *weather monitoring* The US NOAA series of satellites, and others launched by the European space agency, Japan, and India, provide continuous worldwide observation of the atmosphere.

navigation The US Global Positioning System, when complete in 1993, will feature 24 Navstar satellites that will enable users (including walkers and motorists) to find their position to within 100 m/328 ft. The US military will be able to make full use of the system, obtaining accuracy to within 1.5 m/4 ft 6ins. The Transit system, launched in the 1960s, with 12 satellites in orbit, locates users to within 100 m/328 ft.

communications A complete worldwide communications network is now provided by satellites such as the US-run ◊Intelsat system.

satellite television transmission of broadcast signals through artificial communications satellites. Mainly positioned in ◊geostationary orbit, satellites have been used since the 1960s to relay television pictures around the world. Higher-power satellites have more recently been developed to broadcast signals to cable systems or directly to people's homes.

Direct broadcasting began in the UK Feb 1989 with the introduction of Rupert Murdoch's Sky Television service; its rival British Satellite Broadcasting (BSB) was launched in April 1990, and they merged in Nov the same year.

saturated compound organic compound, such as propane, that contains only single covalent bonds. Saturated organic compounds can only undergo further reaction by ◊substitution reactions, as in the production of chloropropane from propane.

saturated fatty acid ◊fatty acid in which there are no double bonds in the hydrocarbon chain.

saturated solution in physics, a solution obtained when a solvent (liquid) can dissolve no more of a solute (usually a solid) at a particular temperature. Normally, a slight fall in temperature causes some

of the solute to crystallize out of solution. If this does not happen the phenomenon is called supercooling, and the solution is said to be *supersaturated*.

Saturn in astronomy, the second-largest planet in the Solar System, sixth from the Sun, and encircled by bright and easily visible equatorial rings. Viewed through a telescope it is ochre. Saturn orbits the Sun every 29.46 years at an average distance of 1,427,000,000 km/886,700,000 mi. Its equatorial diameter is 120,000 km/75,000 mi, but its polar diameter is 12,000 km/7,450 mi smaller, a result of its fast rotation and low density, the lowest of any planet.

Saturn spins on its axis every 10 hours 14 minutes at its equator, slowing to 10 hours 40 minutes at high latitudes. Its mass is 95 times that of Earth, and its magnetic field 1,000 times stronger. Saturn is believed to have a small core of rock and iron, encased in ice and topped by a deep layer of liquid hydrogen. There are 18 known moons, its largest being ◊Titan. The rings visible from Earth begin about 14,000 km/9,000 mi from the planet's cloudtops and extend out to about 76,000 km/47,000 mi. Made of small chunks of ice and rock (averaging 1 m/3 ft across), they are 275,000 km/170,000 mi rim to rim, but only 100 m/300 ft thick. The Voyager probes showed that the rings actually consist of thousands of closely spaced ringlets, looking like the grooves in a gramophone record.

Like Jupiter, Saturn's visible surface consists of swirling clouds, probably made of frozen ammonia at a temperature of –170°C/–274°F, although the markings in the clouds are not as prominent as Jupiter's. The space probes *Voyager 1* and *2* found winds reaching 1,800 kph/1,100 mph. The Voyagers photographed numerous small moons orbiting Saturn, taking the total to 18, more than for any other planet. The largest moon, Titan, has a dense atmosphere.

From Earth, Saturn's rings appear to be divided into three main sections. Ring A, the outermost, is separated from ring B, the brightest, by the Cassini division (named after its discoverer Italian astronomer Giovanni Cassini (1625–1712)), 3,000 km/2,000 mi wide; the innermost, transparent ring C is also called the Crepe Ring. Each ringlet of the rings is made of a swarm of icy particles like snowballs, a few centimetres to a few metres in diameter. Outside the A ring is the narrow and faint F ring, which the Voyagers showed to be twisted or braided. The rings of Saturn could be the remains of a shattered moon, or they may always have existed in their present form.

Saturn rocket family of large US rockets, developed by Wernher von Braun (1912–1977) for the ◊Apollo project. The two-stage Saturn IB was used for launching Apollo spacecraft into orbit around the Earth. The three-stage Saturn V sent Apollo spacecraft to the Moon, and launched the ◊*Skylab* space station. The liftoff thrust of a Saturn V was 3,500 tonnes. After Apollo and *Skylab*, the Saturn rockets were retired in favour of the ◊space shuttle.

savanna or *savannah* extensive open tropical grasslands, with scattered trees and shrubs. Savannas cover large areas of Africa, North and South America, and N Australia.

The name was originally given by Spaniards to the treeless plains of the tropical South American prairies. Most of North America's savannas have been built over.

scalar quantity in mathematics and science, a quantity that has magnitude but no direction, as distinct from a ◊vector quantity, which has a direction as well as a magnitude. Temperature, mass, and volume are scalar quantities.

scale in chemistry, ◊calcium carbonate deposits that form on the inside of a kettle or boiler as a result of boiling ◊hard water.

scandium silver-white, metallic element of the ◊lanthanide series, symbol Sc, atomic number 21, relative atomic mass 44.956. Its compounds are found widely distributed in nature, but only in minute amounts. The metal has little industrial importance.

Scandium is relatively more abundant in the Sun and other stars than on Earth. Scandium oxide (scandia) is used as a catalyst, in making crucibles and other ceramic parts, and scandium sulphate (in very dilute aqueous solution) is used in agriculture to improve seed germination.

The element was discovered and named in 1879 by Swedish chemist Lars Nilson (1840–1899) after Scandinavia, because it was found in the Scandinavian mineral euxenite.

scanner in computing, a device that can produce a digital image of a document for input and storage in a computer. It uses technology similar to that of a photocopier. Small scanners can be passed over the document surface by hand; larger versions have a flat bed, like that of a photocopier, on which the input document is placed and scanned.

Scanners are widely used to input graphics for use in ◊desktop publishing. If text is input with a scanner, the image captured is seen by the computer as a single digital picture rather than as separate characters. Consequently, the text cannot be processed by, for example, a word processor unless suitable optical character-recognition software is available to convert the image to its constituent characters. Scanners vary in their resolution, typical hand-held scanners ranging from 75 to 300 dpi. Types include flat-bed, drum, and overhead.

scanning electron microscope (SEM) electron microscope that produces three-dimensional images, magnified 10–200,000 times. A fine beam of electrons, focused by electromagnets, is moved, or scanned, across the specimen. Electrons reflected from the specimen are collected by a detector, giving rise to an electrical signal, which is then used to generate a point of brightness on a television-like screen. As the point moves rapidly over the screen, in phase with the scanning electron beam, an image of the specimen is built up.

The resolving power of an SEM depends on the size of the electron beam—the finer the beam, the greater the resolution. Present-day instruments typically have a resolution of 7–10 nm.

The first scanning electron picture was produced in 1935 by Max Knoll of the German company Telefunken, though the first commercial SEM (produced by the Cambridge Instrument Company in the UK) did not go on sale until 1965.

scanning transmission electron microscope (STEM) electron microscope that combines features of the ◊scanning electron microscope (SEM) and the ◊transmission electron microscope (TEM). First built in the USA in 1966, the microscope has both the SEM's contrast characteristics and lack of aberrations and the high resolution of the TEM. Magnifications of over 90 million times can be achieved, enough to image single atoms.

A fine beam of electrons, 0.3 nm in diameter, moves across the specimen, as in an SEM. However, because the specimen used is a thin slice, the beam also passes through the specimen (as in a TEM). The reflected electrons and those that penetrated the specimen are collected to form an electric signal, which is interpreted by computer to form an image on screen.

scanning tunnelling microscope (STM) microscope that produces a magnified image by moving a tiny tungsten probe across the surface of the specimen. The tip of the probe is so fine that it may consist of a single atom, and it moves so close to the specimen surface that electrons jump (or tunnel) across the gap between the tip and the surface.

The magnitude of the electron flow (current) depends on the distance from the tip to the surface, and so by measuring the current, the contours of the surface can be determined. These can be used to form an image on a computer screen of the surface, with individual atoms resolved. Magnifications up to 100 million times are possible.

The STM was invented 1981 by Gerd Binning from Germany and Heinrich Rohrer from Switzerland at the IBM Zurich Research Laboratory. With Ernst Ruska, who invented the transmission electron microscope in 1933, they were awarded the Nobel Prize for Physics in 1986.

In 1991, the Japanese electronics firm Hitachi used an STM to produce the smallest writing yet achieved. The message read 'Peace 91', followed by the initials of the Hitachi Central Research Laboratory. The letters were less than 1.5 nanometres (1.5–thousand-millionths of a metre) high, and were formed by removing individual sulphur atoms from the surface of a crystal of molybdenum disulphide.

scapolite group of white or greyish minerals, silicates of sodium, aluminium, and calcium, common in metamorphosed limestones and forming at high temperatures and pressures.

scapula or *shoulder blade* large bone forming part of the pectoral girdle, assisting in the articulation of the arm with the chest region. Its flattened shape allows a large region for the attachment of muscles.

scarp and dip in geology, the two slopes formed when a sedimentary bed outcrops as a landscape feature. The scarp is the slope that cuts across the bedding plane; the dip is the opposite slope which follows the bedding plane. The scarp is usually steep, while the dip is a gentle slope.

scatter diagram or *scattergram* a diagram whose purpose is to establish whether or not a relationship or ◊correlation exists between two variables; for example, between life expectancy and GNP (gross national product). Each observation is marked with a dot in a position that shows the value of both variables. The pattern of dots is then examined to see whether they show any underlying trend by means of a *line of best fit* (a straight line drawn so that its distance from the various points is as short as possible).

scattering in physics, the random deviation or reflection of a stream of particles or of a beam of radiation such as light.

Alpha particles scattered by a thin gold foil provided the first convincing evidence that atoms had very small, very dense, positive nuclei. From 1906 to 1908 New Zealand physicist Ernest Rutherford carried out a series of experiments from which he estimated that the closest approach of an alpha particle to a gold nucleus in a head-on collision was about 10^{-14} m. He concluded that the gold nucleus must be no larger than this. Most of the alpha particles fired at the gold foil passed straight through undeviated; however, a few were scattered in all directions and a very small fraction bounced back towards the source. This result so surprised Rutherford that he is reported to have commented: 'It was about as credible as if you had fired a 15–inch shell at a piece of tissue paper and it came back and hit you'.

Light is scattered from a rough surface, such as that of a sheet of paper, by random reflection from the varying angles of each small part of the surface. This is responsible for the dull, flat appearance of such surfaces and their inability to form images (unlike mirrors). Light is also scattered by particles suspended in a gas or liquid. The red and yellow colours associated with sunrises and sunsets are due to the fact that red light is scattered to a lesser extent than is blue light by dust particles in the atmosphere. When the Sun is low in the sky, its light passes through a thicker, more dusty layer of the atmosphere, and the blue light radiated by it is scattered away, leaving the red sunlight to pass through to the eye of the observer.

scent gland gland that opens onto the outer surface of animals, producing odorous compounds that are used for communicating between members of the same species (◊pheromones), or for discouraging predators.

schist ◊metamorphic rock containing ◊mica or another platy or elongate mineral, whose crystals are aligned to give a foliation (planar texture) known as schistosity. Schist may contain additional minerals such as ◊garnet.

schizocarp dry ◊fruit that develops from two or more carpels and splits, when mature, to form separate one-seeded units known as mericarps.

The mericarps may be dehiscent, splitting open to release the seed when ripe, as in *Geranium*, or indehiscent, remaining closed once mature, as in mallow *Malva* and plants of the Umbelliferae family, such as the carrot *Daucus carota* and parsnip *Pastinaca sativa*.

Schmidt Telescope reflecting telescope used for taking wide-angle photographs of the sky. Invented 1930 by Estonian astronomer Bernhard Schmidt (1879–1935), it has an added corrector lens to help focus the incoming light. Examples are the 1.2–m/ 48–in Schmidt telescope on ◊Mount Palomar and the UK Schmidt telescope, of the same size, at ◊Siding Spring Mountain.

science (Latin *scientia* 'knowledge') any systematic field of study or body of knowledge that aims, through experiment, observation, and deduction, to produce reliable explanation of phenomena, with reference to the material and physical world.

Activities such as healing, star-watching, and engineering have been practised in many societies since ancient times. Pure science, especially physics (formerly called natural philosophy), had traditionally been the main area of study for philosophers. The European scientific revolution between about 1650 and 1800 replaced speculative philosophy with a new combination of observation, experimentation, and rationality.

Today, scientific research involves an interaction among tradition, experiment and observation, and deduction. The subject area called *philosophy of science* investigates the nature of this complex interaction, and the extent of its ability to gain access to the truth about the material world. It has long been recognized that induction from observation cannot give explanations based on logic. In the 20th-century Austrian philosopher of science Karl Popper has described ◊scientific method as a rigorous experimental testing of a scientist's ideas or hypotheses (see ◊hypothesis). The origin and role of these ideas, and their interdependence with observation, have been examined, for example, by the US thinker Thomas S Kuhn, who places them in a historical and sociological setting. The *sociology of science* investigates how scientific theories and laws are produced, and questions the possibility of objectivity in any scientific endeavour. One controversial point of view is the replacement of scientific realism with scientific relativism, as proposed by US philosopher of science Paul K Feyerabend. Questions concerning the proper use of science and the role of science education are also restructuring this field of study.

Science is divided into separate areas of study, such as astronomy, biology, geology, chemistry, physics, and mathematics, although more recently attempts have been made to combine traditionally separate disciplines under such headings as ◊life sciences and ◊earth sciences. These areas are usually jointly referred to as the *natural sciences*. The *physical sciences* comprise mathematics, physics, and chemistry. The application of science for practical purposes is called *technology*. *Social science* is the systematic study of human behaviour, and includes such areas as anthropology, economics, psychology, and sociology. One area of contemporary debate is whether the social-science disciplines are actually sciences; that is, whether the study of human beings is capable of scientific precision or prediction in the same way as natural science is seen to be.

Progress in science depends on new techniques, new discoveries and new ideas, probably in that order.

On science Sidney Brenner *Nature* May 1980

scientific law in science, principles that are taken to be universally applicable.

Laws (for instance, ◊Boyle's law and ◊Newton's laws of motion) form the basic theoretical structure of the physical sciences, so that the rejection of a law by the scientific community is an almost inconceivable event. On occasion a law may be modified, as was the case when Einstein showed that Newton's laws of motion do not apply to objects travelling at speeds close to that of light.

scientific method in science, the belief that experimentation and observation, properly understood and applied, can avoid the influence of cultural and social values and so build up a picture of a reality independent of the observer.

Improved techniques and mechanical devices, which improve the reliability of measurements, may seem to support this theory; but the realization that observations of subatomic particles influence their behaviour has undermined the view that objectivity is possible in science (see ◊uncertainty principle).

scintillation counter instrument for measuring very low levels of radiation. The radiation strikes a scintillator (a device that emits a unit of light when a charged elementary particle collides with it), whose light output is 'amplified' by a ◊photomultiplier; the current pulses of its output are in turn counted or added by a scaler to give a numerical reading.

sclerenchyma plant tissue whose function is to strengthen and support, composed of thick-walled cells that are heavily lignified (toughened). On maturity the cell inside dies, and only the cell walls remain.

Sclerenchyma may be made up of one or two types of cells: *sclereids*, occurring singly or in small clusters, are often found in the hard shells of fruits and in seed coats, bark, and the stem cortex; *fibres*, frequently grouped in bundles, are elongated cells, often with pointed ends, associated with the vascular tissue (◊xylem and ◊phloem) of the plant.

Some fibres provide useful materials, such as flax from *Linum usitatissimum* and hemp from *Cannabis sativa*.

Scorpius zodiacal constellation in the southern hemisphere between Libra and Sagittarius, represented as a scorpion. The Sun passes briefly through Scorpius in the last week of Nov. The heart of the scorpion is marked by the red supergiant star Antares. Scorpius contains rich Milky Way star fields, plus the strongest ◊X-ray source in the sky, Scorpius X-1. In astrology, the dates for Scorpius are between about 24 Oct and 21 Nov (see ◊precession).

scrambling circuit in radiotelephony, a transmitting circuit that renders signals unintelligible unless received by the corresponding unscrambling circuit.

scree pile of rubble and sediment that collects at the foot of a mountain range or cliff. The rock fragments that form scree are usually broken off by the action of frost (◊freeze-thaw weathering). With time, the rock waste builds up into a heap or sheet of rubble that may eventually bury even the upper cliffs, and the growth of the scree then stops. Usually, however, erosional forces decompose the rock waste so that the scree stays restricted to lower slopes.

screen or *monitor* in computing, an output device on which the computer displays information for the benefit of the operator. The commonest type is the ◊cathode-ray tube (CRT), which is similar to a television screen. Portable computers often use ◊liquid crystal display (LCD) screens. These are harder to read than CRTs, but require less power, making them suitable for battery operation.

screen dump in computing, the process of making a printed copy of the current VDU screen display. The screen dump is sometimes stored as a data file instead of being printed immediately.

screw in construction, cylindrical or tapering piece of metal or plastic (or formerly wood) with a helical groove cut into it. Each turn of a screw moves it forward or backwards by a distance equal to the pitch (the spacing between neighbouring threads).

Its mechanical advantage equals $2 \, r/P$, where P is the pitch and r is the radius of the thread. Thus the mechanical advantage of a tapering wood screw, for example, increases as it is rotated into the wood.

scrolling in computing, the action by which data displayed on a VDU screen are automatically moved upwards and out of sight as new lines of data are added at the bottom.

scuba acronym for *s*elf-*c*ontained *u*nderwater *b*reathing *a*pparatus, another name for ◊aqualung.

scurvy disease caused by deficiency of vitamin C (ascorbic acid), which is contained in fresh vegetables and fruit. The signs are weakness and aching joints and muscles, progressing to bleeding of the gums and then other organs, and drying-up of the skin and hair. Treatment is by giving the vitamin.

sea breeze gentle coastal wind blowing off the sea towards the land. It is most noticeable in summer when the warm land surface heats the air above it and causes it to rise. Cooler air from the sea is drawn in to replace the rising air, so causing an onshore breeze. At night and in winter, air may move in the opposite direction, forming a ◊land breeze.

seafloor spreading growth of the ocean ◊crust outwards (sideways) from ocean ridges. The concept of seafloor spreading has been combined with that of continental drift and been incorporated into ◊plate tectonics.

Seafloor spreading was proposed by US geologist Harry Hess (1906–1969) in 1962, based on his observations of ocean ridges and the relative

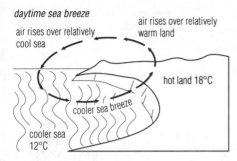

daytime sea breeze

air rises over relatively cool sea

air rises over relatively warm land

hot land 18°C

cooler sea breeze

cooler sea 12°C

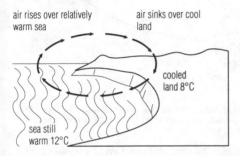

night-time land breeze

air rises over relatively warm sea

air sinks over cool land

cooled land 8°C

sea still warm 12°C

sea breeze

youth of all ocean beds. In 1963, British geophysicists Fred Vine and Drummond Matthews observed that the floor of the Atlantic Ocean was made up of rocks that could be arranged in strips, each strip being magnetized either normally or reversely (due to changes in the Earth's polarity when the North Pole becomes the South Pole and vice versa, termed ◊polar reversal). These strips were parallel and formed identical patterns on both sides of the ocean ridge. The inference was that each strip was formed at some stage in geological time when the magnetic field was polarized in a certain way. The seafloor magnetic-reversal patterns could be matched to dated magnetic reversals found in terrestrial rock. It could then be shown that new rock forms continuously and spreads away from the ocean ridges, with the oldest rock located farthest away from the midline. The observation was made independently in 1963 by Canadian geologist Lawrence Morley, studying an ocean ridge in the Pacific near Vancouver Island.

Confirmation came when sediments were discovered to be deeper further away from the oceanic ridge, because the rock there had been in existence longer and had had more time to accumulate sediment.

searching in computing, extracting a specific item from a large body of data, such as a file or table. The method used depends on how the data are organized. For example, a binary search, which requires the data to be in sequence, involves first deciding which half of the data contains the required item, then which quarter, then which eighth, and so on until the item is found.

search request in computing, a structured

request by a user for information from a ◊database. This may be a simple request for all the entries that have a single field meeting a certain condition. For example, a user searching a file of car registration details might request a list of all the records that have 'VOLKSWAGEN' in the ◊field recording the make of car. In more complex examples, the user may construct a search request using operators like AND, OR, NOT, CONTAINING, and BETWEEN.

An example of a search request using all these operators is:
> *CAR SEARCH*
> registration number *containing* XTW *and* make Volkswagen *and* model Polo *and* body hatchback *and* colour white or black *and* registered *between* 1989 *and* 1991

This search request would produce a list of all the white or black Volkswagen Polo hatchbacks registered between 1989 and 1991 that contained the letters XTW in their registration number.

season period of the year having a characteristic climate. The change in seasons is mainly due to the change in attitude of the Earth's axis in relation to the Sun, and hence the position of the Sun in the sky at a particular place. In temperate latitudes four seasons are recognized: spring, summer, autumn (fall), and winter. Tropical regions have two seasons—the wet and the dry. Monsoon areas around the Indian Ocean have three seasons: the cold, the hot, and the rainy.

The northern temperate latitudes have summer when the southern temperate latitudes have winter, and vice versa. During winter, the Sun is low in the sky and has less heating effect because of the oblique angle of incidence and because the sunlight has further to travel through the atmosphere. The differences between the seasons are more marked inland than near the coast, where the sea has a moderating effect on temperatures. In polar regions the change between summer and winter is abrupt; spring and autumn are hardly perceivable. In tropical regions, the belt of rain associated with the trade winds moves north and south with the Sun, as do the dry conditions associated with the belts of high pressure near the tropics. The monsoon's three seasons result from the influence of the Indian Ocean on the surrounding land mass of Asia in that area.

sea water the water of the seas and oceans, covering about 70% of the Earth's surface and comprising about 97% of the world's water (only about 3% is fresh water). Sea water contains a large amount of dissolved solids, the most abundant of which is sodium chloride (almost 3% by mass); other salts include potassium chloride, bromide, and iodide, magnesium chloride, and magnesium sulphate. It also contains a large amount of dissolved carbon dioxide, and thus acts as a carbon 'sink' that may help to reduce the greenhouse effect.

sebum oily secretion from the sebaceous glands that acts as a skin lubricant. Acne is caused by inflammation of the sebaceous glands and oversecretion of sebum.

sec or *s* abbreviation for *second*, a unit of time.

SECAM abbreviation for *sequential and*

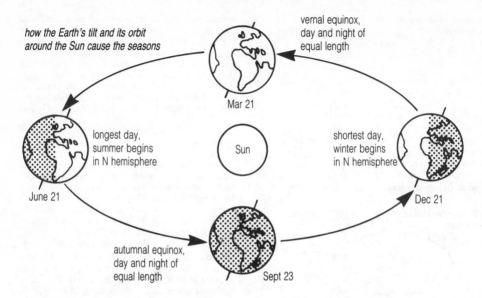

how the Earth's tilt and its orbit around the Sun cause the seasons

vernal equinox, day and night of equal length

Mar 21

longest day, summer begins in N hemisphere

Sun

shortest day, winter begins in N hemisphere

June 21

Dec 21

autumnal equinox, day and night of equal length

Sept 23

season *The cause of the seasons. As the Earth orbits the Sun, its axis of rotation always points in the same direction. This means that, during the northern hemisphere summer solstice (21 June), the Sun is overhead in the northern hemisphere. At the northern hemisphere winter solstice (22 December), the Sun is overhead in the southern hemisphere.*

memory, the coding system for colour-television broadcasting used in France and E Europe; see ◊television.

secant in trigonometry, the function of a given angle in a right-angled triangle, obtained by dividing the length of the hypotenuse (the longest side) by the length of the side adjacent to the angle. It is the ◊reciprocal of the ◊cosine (sec = 1/cos).

second basic SI unit (symbol sec or s) of time, one-sixtieth of a minute. It is defined as the duration of 9,192,631,770 periods of the radiation corresponding to the transition between two hyperfine levels of the ground state of the caesium-133 isotope. In mathematics, the second is a unit (symbol ″) of angular measurement, equalling one-sixtieth

of a minute, which in turn is one-sixtieth of a degree.

secondary emission in physics, an emission of electrons from the surface of certain substances when they are struck by high-speed electrons or other particles from an external source. See also ◊photomultiplier.

secondary growth or *secondary thickening* increase in diameter of the roots and stems of certain plants (notably shrubs and trees) that results from the production of new cells by the ◊cambium. It provides the plant with additional mechanical support and new conducting cells, the secondary ◊xylem and ◊phloem. Secondary growth is generally confined to ◊gymnosperms and, among the ◊angiosperms, to the dicotyledons. With just a few exceptions, the monocotyledons (grasses, lilies) exhibit only primary growth, resulting from cell division at the apical ◊meristems.

secondary sexual characteristic in biology, an external feature of an organism that is characteristic of its gender (male or female), but not the reproductive organs themselves. They include facial hair in men and breasts in women, combs in cockerels, brightly coloured plumage in many male birds, and manes in male lions. In many cases, they are involved in displays and contests for mates and have evolved by ◊sexual selection. Their development is stimulated by sex hormones.

secretin ◊hormone produced by the small intestine of vertebrates that stimulates the production of digestive secretions by the pancreas and liver.

secretion in biology, any substance (normally a fluid) produced by a cell or specialized gland, for example, sweat, saliva, enzymes, and hormones. The process whereby the substance is discharged from the cell is also known as secretion.

A

hypotenuse

β

B   adjacent   C

$$\text{sec(ant)}\ \beta = \frac{1}{\cos \beta} = \frac{\text{hypotenuse}}{\text{adjacent}} = \frac{AB}{BC}$$

secant *The secant of an angle is a function used in the mathematical study of the triangle. If the secant of angle B is known, then the hypotenuse can be found given the length of the adjacent side, or the adjacent side can be found from the hypotenuse.*

sector in computing, part of the magnetic structure created on a disc surface during ◊disc formatting so that data can be stored on it. The disc is first divided into circular tracks and then each circular track is divided into a number of sectors.

secular variable in astronomy, a star that, according to comparisons with ancient observations, has either increased or decreased substantially (and permanently) in brightness over the intervening centuries.

sedative any medication with the effect of lessening nervousness, excitement, or irritation. Sedatives will induce sleep in larger doses. Examples are ◊narcotics, barbiturates, and benzodiazepines.

sediment any loose material that has 'settled' — deposited from suspension in water, ice, or air, generally as the water current or wind speed decreases. Typical sediments are, in order of increasing coarseness, clay, mud, silt, sand, gravel, pebbles, cobbles, and boulders.

Sediments differ from sedimentary rocks in which deposits are fused together in a solid mass of rock by a process called ◊diagenesis. Pebbles are cemented into ◊conglomerates; sands become sandstones; muds become mudstones or shales; peat is transformed into coal.

sedimentary rock rock formed by the accumulation and cementation of deposits that have been laid down by water, wind, ice, or gravity. Sedimentary rocks cover more than two-thirds of the Earth's surface and comprise three major categories: clastic, chemically precipitated, and organic (or biogenic). Clastic sediments are the largest group and are composed of fragments of pre-existing rocks; they include clays, sands, and gravels. Chemical precipitates include some limestones and evaporated deposits such as gypsum and halite (rock salt). Coal, oil shale, and limestone made of fossil material are examples of organic sedimentary rocks.

Most sedimentary rocks show distinct layering (stratification), caused by alterations in composition or by changes in rock type. These strata may become folded or fractured by the movement of the Earth's crust, a process known as *deformation*.

Seebeck effect in physics, the generation of a voltage in a circuit containing two different metals, or semiconductors, by keeping the junctions between them at different temperatures. Discovered by the German physicist Thomas Seebeck (1770–1831), it is also called the thermoelectric effect, and is the basis of the ◊thermocouple. It is the opposite of the ◊Peltier effect (in which current flow causes a temperature difference between the junctions of different metals).

seed the reproductive structure of higher plants (◊angiosperms and ◊gymnosperms). It develops from a fertilized ovule and consists of an embryo and a food store, surrounded and protected by an outer seed coat, called the testa. The food store is contained either in a specialized nutritive tissue, the ◊endosperm, or in the ◊cotyledons of the embryo itself. In angiosperms the seed is enclosed within a ◊fruit, whereas in gymnosperms it is usually naked and unprotected, once shed from the female cone. Following ◊germination the seed develops into a new plant.

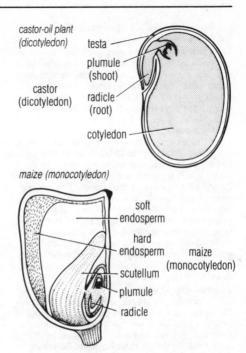

seed *The structure of seeds. The castor is a dicotyledon, a plant in which the developing plant has two leaves, developed from the cotyledon. In maize, a monocotyledon, there is a single leaf developed from the scutellum.*

Seeds may be dispersed from the parent plant in a number of different ways. Agents of dispersal include animals, as with ◊burs and fleshy edible fruits, and wind, where the seed or fruit may be winged or plumed. Water can disperse seeds or fruits that float, and various mechanical devices may eject seeds from the fruit, as in the pods of some leguminous plants (see ◊legume).

There may be a delay in the germination of some seeds to ensure that growth occurs under favourable conditions (see ◊after-ripening, ◊dormancy). Most seeds remain viable for at least 15 years if dried to about 5% water and kept at –20°C/–4°F, although 20% of them will not survive this process.

LARGEST AND SMALLEST SEEDS

The largest seeds in the world are those of the coco de mer coconut palm *Lodoicea maldivica*. Each weighs up to 18 kg/40 lb. The smallest seeds are those of tree-living orchids, of which 992 million seeds are needed to make up a gram (35 million to make up an ounce).

seed drill machine for sowing cereals and other seeds, developed by agriculturist Jethro Tull in England 1701, although simple seeding devices were known in Babylon 2000 BC.

The seed is stored in a hopper and delivered by tubes into furrows in the ground. The furrows are made by a set of blades, or coulters, attached to the front of the drill. A ◊harrow is drawn behind the drill to cover up the seeds.

seed plant any seed-bearing plant; also known as a *spermatophyte*. The seed plants are subdivided into two classes, the ◊angiosperms, or flowering plants, and the ◊gymnosperms, principally the cycads and conifers. Together, they comprise the major types of vegetation found on land.

Angiosperms are the largest, most advanced, and most successful group of plants at the present time, occupying a highly diverse range of habitats. There are estimated to be about 250,000 different species. Gymnosperms differ from angiosperms in their ovules which are borne unprotected (not within an ◊ovary) on the scales of their cones. The arrangement of the reproductive organs, and their more simplified internal tissue structure, also distinguishes them from the flowering plants. In contrast to the gymnosperms, the ovules of angiosperms are enclosed within an ovary and many species have developed highly specialized reproductive structures associated with ◊pollination by insects, birds, or bats.

seismic gap theory theory that aims to predict the location of ◊earthquakes. When records of past earthquakes are studied and plotted onto a map, it becomes possible to identify areas along a fault or plate margin where an earthquake should be due — such areas are called *seismic gaps*. According to the theory, an area that has not had an earthquake for some time will have a great deal of stress building up, which must eventually be released in the form of an earthquake.

Although the seismic gap theory can suggest areas that are likely to experience an earthquake, it does not enable scientists to predict when that earthquake will occur.

Research carried out in the vicinity of the ◊San Andreas fault in California has identified a seismic gap at the town of Parkfield, near San Francisco. Only time will tell whether the prediction will prove to be correct.

seismic wave energy wave generated by an ◊earthquake; it is responsible for the shaking of the ground and the conversion of soft deposits, such as clay, to a jellylike state (◊liquefaction).

The different speeds at which seismic waves move through the Earth have been used by scientists to deduce the Earth's layered internal structure.

seismograph instrument used to record the activity of an ◊earthquake. A heavy inert weight is suspended by a spring and attached to this is a pen that is in contact with paper on a rotating drum. During an earthquake the instrument frame and drum move, causing the pen to record a zigzag line on the paper; the pen does not move.

seismology study of earthquakes and how their shock waves travel through the Earth. By examining the global pattern of waves produced by an earthquake, seismologists can deduce the nature of the materials through which they have passed. This leads to an understanding of the Earth's internal structure.

On a smaller scale artificial earthquake waves, generated by explosions or mechanical vibrators, can be used to search for subsurface features in, for example, oil or mineral exploration. Earthquake waves from underground nuclear explosions can be distinguished from natural waves by their shorter wavelength and higher frequency.

selenium (Greek *Selene* 'Moon') grey, nonmetallic element, symbol Se, atomic number 34, relative atomic mass 78.96. It belongs to the sulphur group and occurs in several allotropic forms that differ in their physical and chemical properties. It is an essential trace element in human nutrition. Obtained from many sulphide ores and selenides, it is used as a red colouring for glass and enamel.

Because its electrical conductivity varies with the intensity of light, selenium is used extensively in photoelectric devices. It was discovered 1817 by Swedish chemist Jöns Berzelius and named after the Moon because its properties follow those of tellurium, whose name derives from Latin *Tellus* 'Earth'.

self-induction or *self-inductance* in physics, the creation of a counter emf (◊electromotive force) in a coil because of variations in the current flowing through it.

selva equatorial rainforest, such as that in the Amazon basin in South America.

semelparity in biology, the occurrence of a single act of reproduction during an organism's lifetime. Most semelparous species produce very large numbers of offspring when they do reproduce, and normally die soon afterwards. Examples include the Pacific salmon and the pine looper moth. Many plants are semelparous, or ◊monocarpic. Repeated reproduction is called ◊iteroparity.

semen fluid containing ◊sperm from the testes and secretions from various sex glands (such as the prostate gland) that are produced by male animals

a seismogram recorded by a seismograph

| quiet and stable before earthquake | first rumbles of earthquake | most violent shaking of earthquake | quiet again |

time—5 seconds approximately

seismograph

Earthquakes: advances in protection and prediction

1990 saw a number of earthquakes around the world, including the devastating Iranian earthquakes in June which measured 7.7 on the Richter scale and cost 50,000 lives. Earthquakes are caused by movement along faults, and so they are more frequent in regions that are known to have a great deal of fault activity— particularly areas close to the boundaries of the Earth's tectonic plates. Iran lies on the boundary between the Arabian and the Asian plates and so such an earthquake was not surprising.

The unexpected earthquake

Other earthquakes, however, occurred in regions not known for earthquake activity, such as Missouri in the USA on 26 Sept and the Welsh borders and Sheffield in the UK on 2 April and 8 Feb, respectively. Research conducted at the US Geological Survey's National Earthquake Information Center in Colorado, and the US Geological Survey Seismic Observatory in Virginia suggests that a big earthquake occurring in an area not traditionally prone to earthquakes could cause much more damage than one in the more earthquake-susceptible regions such as California. Historical statistics show that a big earthquake in the central and eastern part of the continent is about two-thirds as likely as one in California over the next 30 years. East of the Rockies the continental rocks are old and strong. They transmit shock waves much more efficiently. Should an earthquake disrupt them the effect will be felt over a much larger area than in the shattered faulted rocks of the west. Besides that, buildings of the east tend not to be designed with possible earthquake damage in mind.

Protection and . . .

A tall building whips back and forth with the vibration of an earthquake. If the dimensions of the building are such that the speed of the whipping action happens to be 'tuned' to the vibration of the earthquake, then the whipping builds up and the building collapses. The earthquake vibration depends largely on the nature of the surface on which the foundations are set, and so nowadays the vibration characteristics of the underlying soil are taken into account before a building is designed. A design feature that is under investigation by the University of California's Earthquake Engineering Research Center in San Francisco is that of 'base isolation'. The theory is that the building is separated from its foundations by bearings or rubber joints that absorb much of the earthquake shock before it reaches the building itself. Another idea, being pursued by the National Center for Earthquake Research at the State University of New York, involves encasing the building with wire mesh covered with cement. This would prevent the wall components, like bricks or blocks, from sliding sideways in relation to one another during the shaking.

More highly technological techniques are being tried out in Japan. These involve computerized sets of sensors in the basement that analyse the earthquake's vibration and instantaneously transmit the results to pistons and counterweights throughout the building. These continually adjust the building's loading and reduce the stresses in the structure.

. . . Prediction

The science of earthquake proofing is developing swiftly, but that of earthquake prediction less so. One development is the discovery, by Stanford University, that very low frequency radio waves passing through rocks change their amplitude a few hours before an earthquake. This may be due to electrical currents generated by pressure in the rocks, or it may be related to the opening of microscopic cracks as the rocks begin to fail. Possibly related to this is the discovery by Japanese researchers that electromagnetic radiation is emitted by rocks immediately prior to an earthquake. Up to now this has not been useful, as there was no way to tell the difference between this radiation and that produced by the Earth's ionosphere or by industry. Two researchers, Takahashi and Fijinawa, have managed to distinguish the horizontal electrical fields generated by ionospheric radiation from the vertical fields produced by the rocks, and have eliminated the artificial interference by siting their instruments far from cities and industry. Since their installation in March 1989 the instruments have recorded changes in radiation several hours before earthquakes on 5, 9 and 13 July 1990.

A National Center for Earthquake Engineering Research investigator at the State University of New York at Buffalo, monitors performance of a structure subjected to earthquake vibrations on the university's 'shake table'. The three-storey steel-frame structure is equipped with active tendon controls which give the model human-like ability to counter earthquake forces.

Dougal Dixon

during copulation. The secretions serve to nourish and activate the sperm cells, and prevent their clumping together.

semicircular canal one of three looped tubes that form part of the labyrinth in the inner ◊ear. They are filled with fluid and detect changes in the position of the head, contributing to the sense of balance.

semiconductor crystalline material with an electrical conductivity between that of metals (good) and insulators (poor). The conductivity of semiconductors can usually be improved by minute additions of different substances or by other factors. Silicon, for example, has poor conductivity at low temperatures, but this is improved by the application of light, heat, or voltage; hence silicon is used in ◊transistors, rectifiers, and ◊integrated circuits (silicon chips).

seminiferous tube in male vertebrates, one of a number of tightly packed, highly coiled tubes in the testis. The tubes are lined with germinal epithelium cells, from which ◊sperm are produced by cell division (mitosis and ◊meiosis) and in the midst of which the sperm mature, nourished and protected in the folds of large Sertoli cells. After maturation, the sperm are stored in a long tube called the epididymis.

Semtex plastic explosive, manufactured in the Czech Republic. It is safe to handle (it can only be ignited by a detonator) and difficult to trace, since it has no smell. It has been used by extremist groups in the Middle East and by the IRA in Northern Ireland.

0.5 kg of Semtex is thought to have been the cause of an explosion that destroyed a Pan-American Boeing 747 in flight over Lockerbie, Scotland, in Dec 1988, killing 270 people.

senescence in biology, the deterioration in physical and reproductive capacities associated with old age. See ◊ageing.

sense organ any organ that an animal uses to gain information about its surroundings. All sense organs have specialized receptors (such as light receptors in an eye) and some means of translating their response into a nerve impulse that travels to the brain. The main human sense organs are the eye, which detects light and colour (different wavelengths of light); the ear, which detects sound (vibrations of the air) and gravity; the nose, which detects some of the chemical molecules in the air; and the tongue, which detects some of the chemicals in food, giving a sense of taste. There are also many small sense organs in the skin, including pain sensors, temperature sensors, and pressure sensors, contributing to our sense of touch.

Research suggests that our noses may also be sensitive to magnetic forces, giving us an innate 'sense of direction'. This sense is well developed in other animals, as are a variety of senses that we do not share. Some animals can detect small electric discharges, underwater vibrations, minute vibrations of the ground, or sounds that are below (infrasound) or above (ultrasound) our range of hearing. Sensitivity to light varies greatly. Most mammals cannot distinguish different colours, whereas some birds can detect the polarization of

light. Many insects can see light in the ultraviolet range, which is beyond our spectrum, while snakes can form images of infrared radiation (radiant heat). In many animals, light is also detected by another organ, the pineal gland, which 'sees' light filtering through the skull, and measures the length of the day to keep track of the seasons.

sensitivity the ability of an organism, or part of an organism, to detect changes in the environment. Although all living things are capable of some sensitivity, evolution has led to the formation of highly complex mechanisms for detecting light, sound, chemicals, and other stimuli. It is essential to an animal's survival that it can process this type of information and make an appropriate response.

sensor in computing, a device designed to detect a physical state or measure a physical quantity, and produce an input signal for a computer. For example, a sensor may detect the fact that a printer has run out of paper or may measure the temperature in a kiln.

The signal from a sensor is usually in the form of an analogue voltage, and must therefore be converted to a digital signal, by means of an ◊analogue-to-digital converter, before it can be input.

sepal part of a flower, usually green, that surrounds and protects the flower in bud. The sepals are derived from modified leaves, and collectively are known as the ◊calyx.

In some plants, such as the marsh marigold *Caltha palustris*, where true ◊petals are absent, the sepals are brightly coloured and petal-like, taking over the role of attracting insect pollinators to the flower.

sequence-control register or *program counter* in computing, a special memory location used to hold the address of the next instruction to be fetched from the immediate access memory for execution by the computer (see ◊fetch-execute cycle). It is located in the control unit of the ◊central processing unit.

sequencing in biochemistry, determining the sequence of chemical subunits within a large molecule. Techniques for sequencing amino acids in proteins were established in the 1950s, insulin being the first for which the sequence was completed. Efforts are now being made to determine the sequence of base pairs within ◊DNA.

sequential file in computing, a file in which the records are arranged in order of a ◊key field and the computer can use a searching technique, like a ◊binary search, to access a specific record. See ◊file access.

sere plant ◊succession developing in a particular habitat. A *lithosere* is a succession starting on the surface of bare rock. A *hydrosere* is a succession in shallow freshwater, beginning with planktonic vegetation and the growth of pondweeds and other aquatic plants, and ending with the development of swamp. A *plagiosere* is the sequence of communities that follows the clearing of the existing vegetation.

serial device in computing, a device that communicates binary data by sending the bits that represent each character one by one along a single data line, unlike a ◊parallel device.

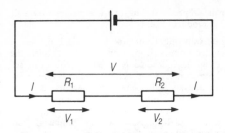

series circuit *Circuit diagram of two resistors, R_1 and R_2, connected in series. The current I flowing through each resistor is the same, but the potential difference across each is divided up in the ratio of their resistances ($V_1:V_2$ is the same as $R_1:R_2$). Such an arrangement is simple to wire, but if one of the components fails the whole circuit is broken and all the other components will also fail.*

serial file in computing, a file in which the records are not stored in any particular order and therefore a specific record can be accessed only by reading through all the previous records. See ◊file access.

series circuit electrical circuit in which the components are connected end to end, so that the current flows through them all one after the other.

The division of the ◊terminal voltage across each conductor is in the ratio of the resistances of each conductor. If the potential differences across two conductors of resistance R_1 and R_2, connected in series, are V_1 and V_2 respectively, then the ratio of those potential differences is given by the equation:

$$V_1/V_2 = R_1/R_2$$

The total resistance R of those conductors is given by:

$$R = R_1 + R_2$$

Compare ◊parallel circuit.

Serpens constellation of the equatorial region of the sky, represented as a serpent coiled around the body of Ophiuchus. It is the only constellation divided into two halves: **Serpens Caput**, the head (on one side of Ophiuchus), and **Serpens Cauda**, the tail (on the other side). Its main feature is the Eagle nebula.

serpentine group of minerals, hydrous magnesium silicate, $Mg_3Si_2O_5(OH)_4$, occurring in soft ◊metamorphic rocks and usually dark green. The fibrous form **chrysotile** is a source of ◊asbestos; other forms are **antigorite** and **lizardite** Serpentine minerals are formed by hydration of ultrabasic rocks during metamorphism. Rare snake-patterned forms are used in ornamental carving.

serum clear fluid that remains after blood clots. It is blood plasma with the anticoagulant proteins removed, and contains ◊antibodies and other proteins, as well as the fats and sugars of the blood. It can be produced synthetically, and is used to protect against disease.

servomechanism automatic control system used in aircraft, motor cars, and other complex machines. A specific input, such as moving a lever or joystick, causes a specific output, such as feeding current to an electric motor that moves, for example, the rudder of the aircraft. At the same time, the position of the rudder is detected and fed back to the central control, so that small adjustments can continually be made to maintain the desired course.

sessile in botany, a leaf, flower, or fruit that lacks a stalk and sits directly on the stem, as with the sessile acorns of certain oak trees. In zoology, it is an animal that normally stays in the same place, such as a barnacle or mussel. The term is also applied to the eyes of ◊crustaceans when these lack stalks and sit directly on the head.

set or **class** in mathematics, any collection of defined things (elements), provided the elements are distinct and that there is a rule to decide whether an element is a member of a set. It is usually denoted by a capital letter and indicated by curly brackets { }.

For example, L may represent the set that consists of all the letters of the alphabet. The symbol ϵ stands for 'is a member of'; thus $p \in L$ means that p belongs to the set consisting of all letters,

Venn diagram of two intersecting sets

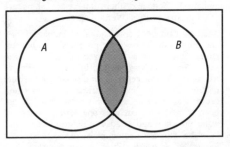

Venn diagram showing the whole numbers from 1 to 20 and the subsets of the prime and odd numbers

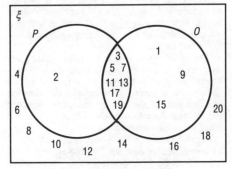

ξ = set of whole numbers from 1 to 20

O = set of odd numbers

P = set of prime numbers

the intersection of P and O ($P \cap O$) contains all the prime numbers that are also odd

set

and $4 \notin L$ means that 4 does not belong to the set consisting of all letters.

There are various types of sets. A *finite set* has a limited number of members, such as the letters of the alphabet; an *infinite set* has an unlimited number of members, such as all whole numbers; an *empty* or *null set* has no members, such as the number of people who have swum across the Atlantic Ocean, written as { } or Ø; a *single-element set* has only one member, such as days of the week beginning with M, written as {Monday}. *Equal sets* have the same members; for example, if W = {days of the week} and S = {Sunday, Monday, Tuesday, Wednesday, Thursday, Friday, Saturday}, it can be said that $W = S$. Sets with the same number of members are *equivalent sets*. Sets with some members in common are *intersecting sets*; for example, if R = {red playing cards} and F = {face cards}, then R and F share the members that are red face cards. Sets with no members in common are *disjoint sets*. Sets contained within others are *subsets*; for example, V = {vowels} is a subset of L = {letters of the alphabet}. Sets and their interrelationships are often illustrated by a ◊Venn diagram.

sewage disposal the disposal of human excreta and other waterborne waste products from houses, streets, and factories. Conveyed through sewers to sewage works, sewage has to undergo a series of treatments to be acceptable for discharge into rivers or the sea, according to various local laws and ordinances. Raw sewage, or sewage that has not been treated adequately, is one serious source of water pollution and a cause of ◊eutrophication.

In the industrialized countries of the West, most industries are responsible for disposing of their own wastes. Government agencies establish industrial waste-disposal standards. In most countries, sewage works for residential areas are the responsibility of local authorities. The solid waste (sludge) may be spread over fields as a fertilizer or, in a few countries, dumped at sea. A significant proportion of bathing beaches in densely populated regions have unacceptably high bacterial content, largely as a result of untreated sewage being discharged into rivers and the sea.

The use of raw sewage as a fertilizer (long practised in China) has the drawback that disease-causing microorganisms can survive in the soil and be transferred to people or animals by consumption of subsequent crops. Sewage sludge is safer, but may contain dangerous levels of heavy metals and other industrial contaminants.

In 1987, Britain dumped more than 4,700 tonnes of sewage sludge into the North Sea, and 4,200 tonnes into the Irish Sea and other coastal areas. Also dumped in British coastal waters, other than the Irish Sea, were 6,462 tonnes of zinc, 2,887 tonnes of lead, 1,306 tonnes of chromium, and 8 tonnes of arsenic. Dumped into the Irish Sea were 916 tonnes of zinc, 297 tonnes of lead, 200 tonnes of chromium, and 1 tonne of arsenic. Strict European rules will phase out sea dumping by 1998.

sex chromosome chromosome that differs between the sexes and that serves to determine the sex of the individual. In humans, the homogametic pair XX produces females, while the heterogametic XY produces males.

sex determination process by which the sex of an organism is determined. In many species, the sex of an individual is dictated by the two sex chromosomes (X and Y) it receives from its parents. In mammals, some plants, and a few insects, males are XY, and females XX; in birds, reptiles, some amphibians, and butterflies the reverse is the case. In bees and wasps, males are produced from unfertilized eggs, females from fertilized eggs. Environmental factors can affect some fish and reptiles, such as turtles, where sex is influenced by the temperature at which the eggs develop. In 1991 it was shown that maleness is caused by a single gene, 14 base pairs long, on the Y chromosome.

Most fish have a very flexible system of sex determination, which can be affected by external factors. For example, in wrasse all individuals develop into females, but the largest individual in each area or school changes sex to become the local breeding male.

DETERMINING THE SEX OF TURTLES

The sex of young turtles is determined by the temperature at which they develop. If the temperature in the nest is below 28°C/82°F, all the young will be males. If the temperature is above 33°C/91°F, all the eggs hatch into females. If the nest temperature falls between these two temperatures, roughly equal numbers of males and females will be hatched.

---

sex hormone steroid hormone produced and secreted by the gonads (testes and ovaries). Sex hormones control development and reproductive functions and influence sexual and other behaviour.

sex linkage in genetics, the tendency for certain characteristics to occur exclusively, or predominantly, in one sex only. Human examples include red-green colour blindness and haemophilia, both found predominantly in males. In both cases, these characteristics are ◊recessive and are determined by genes on the ◊X chromosome.

Since females possess two X chromosomes, any such recessive ◊allele on one of them is likely to be masked by the corresponding allele on the other. In males (who have only one X chromosome paired with a largely inert ◊Y chromosome) any gene on the X chromosome will automatically be expressed. Colour blindness and haemophilia can appear in females, but only if they are ◊homozygous for these traits, due to inbreeding, for example.

sextant navigational instrument for determining latitude by measuring the angle between some heavenly body and the horizon. It was invented by John Hadley (1682–1744) in 1730 and can be used only in clear weather.

When the horizon is viewed through the right-hand side *horizon glass*, which is partly clear and partly mirrored, the light from a star can be seen

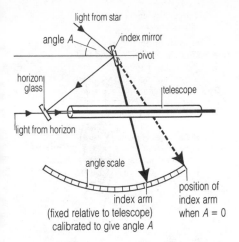

sextant *The geometry of the sextant. When the light from a star can be seen at the same time as light from the horizon, the angle A can be read from the position of the index arm on the angle scale.*

at the same time in the mirrored left-hand side by adjusting an *index mirror*. The angle of the star to the horizon can then be read on a calibrated scale.

sexual reproduction reproductive process in organisms that requires the union, or ◊fertilization, of gametes (such as eggs and sperm). These are usually produced by two different individuals, although self-fertilization occurs in a few ◊hermaphrodites such as tapeworms. Most organisms other than bacteria and cyanobacteria (◊blue-green algae) show some sort of sexual process. Except in some lower organisms, the gametes are of two distinct types called eggs and sperm. The organisms producing the eggs are called females, and those producing the sperm, males. The fusion of a male and female gamete produces a *zygote*, from which a new individual develops.

The alternatives to sexual reproduction are parthenogenesis and asexual reproduction by means of ◊spores.

sexual selection process similar to ◊natural selection but relating exclusively to success in finding a mate for the purpose of sexual reproduction and producing offspring. Sexual selection occurs when one sex (usually but not always the female) invests more effort in producing young than the other. Members of the other sex compete for access to this limited resource (usually males competing for the chance to mate with females).

Sexual selection often favours features that increase a male's attractiveness to females (such as the pheasant's tail) or enable males to fight with one another (such as a deer's antlers). More subtly, it can produce hormonal effects by which the male makes the female unreceptive to other males, causes the abortion of fetuses already conceived, or removes the sperm of males who have already mated with a female.

Seyfert galaxy galaxy whose small, bright centre is caused by hot gas moving at high speed around

a massive central object, possibly a ◊black hole. Almost all Seyferts are spiral galaxies. They seem to be closely related to ◊quasars, but are about 100 times fainter. They are named after their discoverer Carl Seyfert (1911–1960).

shackle unit of length, used at sea for measuring cable or chain. One shackle is 15 fathoms (90 ft/ 27 m).

shadoof or *shaduf* machine for lifting water, consisting typically of a long, pivoted wooden pole acting as a lever, with a weight at one end. The other end is positioned over a well, for example. The shadoof was in use in ancient Egypt and is still used in Arab countries today.

shadow area of darkness behind an opaque object that cannot be reached by some or all of the light coming from a light source in front. Its presence may be explained in terms of light rays travelling in straight lines and being unable to bend round obstacles. A point source of light produces an ◊umbra, a completely black shadow with sharp edges. An extended source of light produces both

female reproductive system

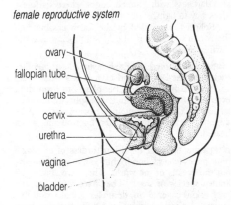

male reproductive system

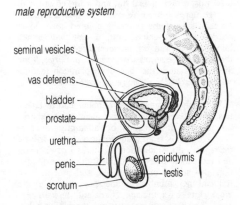

sexual reproduction *The human reproductive organs. In the female, gametes called ova are produced regularly in the ovaries after puberty. The Fallopian tubes carry the ova to the uterus or womb, in which a developing baby is held. In the male, sperm is produced inside the testes after puberty; about 10 million sperm cells are produced each day, enough to populate the world in six months. The sperm duct or vas deferens, a continuation of the epididymis, carries sperm to the urethra during ejaculation.*

reproductive organs in flowering plants

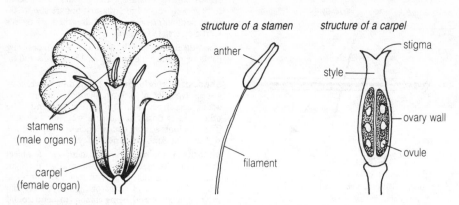

structure of a stamen

anther

filament

structure of a carpel

stigma

style

ovary wall

ovule

stamens
(male organs)

carpel
(female organ)

sexual reproduction in plants

a central umbra and a ◊penumbra, a region of semidarkness with blurred edges where darkness gives way to light.

◊Eclipses are caused by the Earth passing into the Moon's shadow or the Moon passing into the Earth's shadow.

shale fine-grained and finely layered ◊sedimentary rock composed of silt and clay. It is a weak rock, splitting easily along bedding planes to form thin, even slabs (by contrast, mudstone splits into irregular flakes). Oil shale contains kerogen, a solid bituminous material that yields ◊petroleum when heated.

shark any member of various orders of cartilaginous fishes (class Chondrichthyes), found throughout the oceans of the world. There are about 400 known species of shark. They have tough, usually grey, skin covered in denticles (small toothlike scales). A shark's streamlined body has side pectoral fins, a high dorsal fin, and a forked tail with a large upper lobe. Five open gill slits are visible on each side of the generally pointed head. Most sharks are fish-eaters, and a few will attack humans. They range from several feet in length to the great **white shark** *Carcharodon carcharias*, 9 m/ 30 ft long, and the harmless plankton-feeding **whale shark** *Rhincodon typus*, over 15 m/50 ft in length.

Relatively few attacks on humans lead to fatalities, and research suggests that the attacking sharks are not searching for food, but attempting to repel perceived rivals from their territory. Game fishing for 'sport', the eradication of sharks in swimming and recreation areas, and their industrial exploitation as a source of leather, oil, and protein have reduced their numbers. Some species, such as the great white shark, the tiger shark, and the hammerhead, are now endangered and their killing has been banned in US waters since July 1991. Other species will be protected by catch quotas.

Their eyes, though lacking acuity of vision or sense of colour, are highly sensitive to light. Their sense of smell is so acute that one-third of the brain is given up to interpreting its signals; they can detect blood in the water up to 1 km/1,100 yd

away. They also respond to electrical charges emanating from other animals.

The **basking shark** *Cetorhinus maximus* of temperate seas reaches 12 m/40 ft, but eats only marine organisms. The whale shark is the largest living fish. Sharks have remained virtually unchanged for millions of years.

shelf sea relatively shallow sea, usually no deeper than 200 m/650 ft, overlying the continental shelf around the coastlines. Most fishing and marine mineral exploitations are carried out in shelf seas.

shell the hard outer covering of a wide variety of invertebrates. The covering is usually mineralized, normally with large amounts of calcium. The shell of birds' eggs is also largely made of calcium.

SHF in physics, the abbreviation for **superhigh** ◊frequency.

point source of light

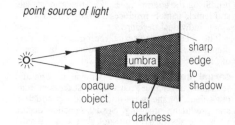

sharp edge to shadow

umbra

opaque object

total darkness

extended source of light

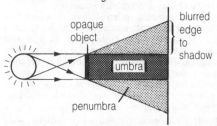

blurred edge to shadow

opaque object

umbra

penumbra

shadow

shield in geology, alternative name for ◊craton, the ancient core of a continent.

shield volcano broad flat ◊volcano formed at a ◊constructive margin between tectonic plates or over a hot spot. The magma (molten rock) associated with shield volcanoes is usually basalt—thin and free-flowing. An example is Mauna Loa in Hawaii. A ◊composite volcano, on the other hand, is formed at a destructive margin.

Shinkansen (Japanese 'new trunk line') fast railway network operated by Japanese Railways, on which the bullet trains run. The network, opened 1964, uses specially built straight and level track, on which average speeds of 160 kph/100 mph are attained.

The Shinkansen between Tokyo and Osaka carried 270,000 passengers a day by 1990.

ship large seagoing vessel. The Greeks, Phoenicians, Romans, and Vikings used ships extensively for trade, exploration, and warfare. The 14th century was the era of European exploration by sailing ship, largely aided by the invention of the compass. In the 15th century Britain's Royal Navy was first formed, but in the 16th–19th centuries Spanish and Dutch fleets dominated the shipping lanes of both the Atlantic and Pacific. The ultimate sailing ships, the fast US and British tea clippers, were built in the 19th century. Also in the 19th century, iron was first used for some shipbuilding instead of wood. Steam-propelled ships of the late 19th century were followed by compound engine and turbine-propelled vessels from the early 20th century.

The Greeks and Phoenicians built wooden ships, propelled by oar or sail. The Romans and Carthaginians built war galleys equipped with rams and several tiers of rowers. The oak ships of the Vikings were built for rough seas and propelled by oars and sail. The Crusader fleet of Richard the Lion-Heart was largely of sail. The invention of the compass in the 14th century led to exploration by sailing ship, especially by the Portuguese, resulting in the discovery of 'new worlds'. In the 15th century Henry VIII built the *Great Harry*, the first double-decked English warship. In the 16th century ships were short and high-sterned, and despite Pett's three-decker in the 17th century, English ships did not bear comparison with the Spanish and Dutch until the early 19th century. In the 1840s iron began replacing wood in shipbuilding, pioneered by British engineer Isambard Kingdom Brunel's *Great Britain* 1845.

The USA and Britain experimented with steam propulsion as the 19th century opened. The paddle-wheel-propelled *Comet* appeared 1812, the Canadian *Royal William* crossed the Atlantic 1833, and the English *Great Western* steamed from Bristol to New York 1838. Francis Pettit Smith first used the screw propeller in the *Archimedes* 1839, and after 1850 the paddle wheel was used mainly on inland waterways, especially the great rivers of the Americas. The introduction of the compound engine and turbine (the latter 1902) completed the revolution in propulsion until the advent of nuclear-powered vessels after World War II, chiefly submarines (which are considered boats, not ships).

More recently ◊hovercraft and ◊hydrofoil boats have been developed for specialized purposes, particularly as short-distance ferries—for example, the catamarans introduced 1991 by Hoverspeed cross the English Channel from Dover to Calais in 35 min, cruising at a speed of 35 knots (84.5 kph/52.5 mph). Sailing ships in automated form for cargo purposes, and ◊magnetohydrodynamics ships, were in development in the early 1990s.

shock absorber in technology, any device for absorbing the shock of sudden jarring actions or movements. Shock absorbers are used in conjunction with coil springs in most motor-vehicle suspension systems and are usually of the telescopic type, consisting of a piston in an oil-filled cylinder. The resistance to movement of the piston through the oil creates the absorbing effect.

shock wave narrow band or region of high pressure and temperature in which the flow of a fluid changes form subsonic to supersonic. Shock waves are produced when an object moves through a fluid at a supersonic speed. See ◊sonic boom.

shoot in botany, the parts of a ◊vascular plant growing above ground, comprising a stem bearing leaves, buds, and flowers. The shoot develops from the ◊plumule of the embryo.

shooting star another name for a ◊meteor.

short circuit direct connection between two points in an electrical circuit. Its relatively low resistance means that a large current flows through it, bypassing the rest of the circuit, and this may cause the circuit to overheat dangerously.

shrub perennial woody plant that typically produces several separate stems, at or near ground level, rather than the single trunk of most trees. A shrub is usually smaller than a tree, but there is no clear distinction between large shrubs and small trees.

shunt in electrical engineering, a conductor of very low resistance that is connected in parallel to an ◊ammeter in order to enable it to measure larger electric currents. Its low resistance enables it to act like a bypass, diverting the extra current through itself and away from the ammeter, and thereby reducing the instrument's sensitivity.

SI abbreviation for *Système International [d'Unités]* (French 'International System [of Metric Units]'); see ◊SI units.

sial in geochemistry and geophysics, the substance of the Earth's continental ◊crust, as distinct from the ◊sima of the ocean crust. The name, now used rarely, is derived from *si*lica and *al*umina, its two main chemical constituents.

sick building syndrome malaise diagnosed in the early 1980s among office workers and thought to be caused by such pollutants as formaldehyde (from furniture and insulating materials), benzene (from paint), and the solvent trichloroethene, concentrated in air-conditioned buildings. Symptoms include headache, sore throat, tiredness, colds, and flu. Studies have found that it can cause a 40% drop in productivity and a 30% rise in absenteeism.

Work on improving living conditions of astronauts showed that the causes were easily and inexpensively removed by potplants in which interaction is thought to take place between the plant and

ships: chronology

8000–7000 BC	Reed boats were developed in Mesopotamia and Egypt; dugout canoes were used in NW Europe.
4000–3000 BC	The Egyptians used single-masted square-rigged ships on the Nile.
1200 BC	The Phoenicians built keeled boats with hulls of wooden planks.
1st century BC	The Chinese invented the rudder.
AD 200	The Chinese built ships with several masts.
200–300	The Arabs and Romans developed fore-and-aft rigging that allowed boats to sail across the direction of wind.
800–900	Square-rigged Viking longboats crossed the North Sea to Britain, the Faroe Islands, and Iceland.
1090	The Chinese invented the magnetic compass.
1400–1500	Three-masted ships were developed in W Europe, stimulating voyages of exploration.
1620	Dutch engineer Cornelius Drebbel invented the submarine.
1776	US engineer David Bushnell built a hand-powered submarine, *Turtle*, with buoyancy tanks.
1777	The first boat with an iron hull was built in Yorkshire, England.
1783	Frenchman Jouffroy d'Abbans built the first paddle-driven steamboat.
1802	Scottish engineer William Symington launched the first stern paddle-wheel steamer, the *Charlotte Dundas*.
1807	The first successful steamboat, the *Clermont*, designed by US engineer and inventor Robert Fulton, sailed between New York and Albany.
1836	The screw propeller was patented, by Francis Pettit Smith in the UK.
1838	British engineer Isambard Kingdom Brunel's *Great Western*, the first steamship built for crossing the Atlantic, sailed from Bristol to New York in 15 days.
1845	*Great Britain*, also built by Isambard Kingdom Brunel, became the first propeller-driven iron ship to cross the Atlantic.
1845	The first clipper ship, *Rainbow*, was launched in the USA.
1863	*Plongeur*, the first submarine powered by an air-driven engine, was launched in France.
1866	The British clippers *Taeping* and *Ariel* sailed, laden with tea, from China to London in 99 days.
1886	German engineer Gottlieb Daimler built the first boat powered by an internal-combustion engine.
1897	English engineer Charles Parson fitted a steam turbine to *Turbinia*, making it the fastest boat of the time.
1900	Irish-American John Philip Holland designed the first modern submarine *Holland VI*, fitted with an electric motor for underwater sailing and an internal-combustion engine for surface travel; E Forlanini of Italy built the first hydrofoil.
1902	The French ship *Petit-Pierre* became the first boat to be powered by a diesel engine.
1955	The first nuclear-powered submarine, *Nautilus*, was built in the USA; the hovercraft was patented by British inventor Christopher Cockerell.
1959	The first nuclear-powered surface ship, the Soviet ice-breaker *Lenin*, was commissioned; the US *Savannah* became the first nuclear-powered merchant (passenger and cargo) ship.
1980	Launch of the first wind-assisted commercial ship for half a century, the Japanese tanker *Shin-Aitoku-Maru*.
1983	German engineer Ortwin Fries invented a hinged ship designed to bend into a V-shape in order to scoop up oil spillages in its jaws.
1989	*Gentry Eagle* set a record for the fastest crossing of the Atlantic in a power vessel, taking 2 days, 14 hours, and 7 minutes.
1990	*Hoverspeed Great Britain*, a wave-piercing catamaran, crossed the Atlantic in 3 days, 7 hours, and 52 minutes, setting a record for the fastest crossing by a passenger vessel. The world's largest car and passenger ferry, the *Silja Serenade*, entered service between Stockholm and Helsinki, carrying 2,500 passengers and 450 cars.
1992	Japanese propellerless ship *Yamato* driven by magnetohydrodynamics completes its sea trials. The ship uses magnetic forces to suck in and eject sea water like a jet engine.

microorganisms in its roots. Among the most useful are chrysanthemums (counteracting benzene), English ivy and the peace lily (trichloroethene), and the spider plant (formaldehyde).

sickle-cell disease hereditary chronic blood disorder common among people of black African descent; also found in the E Mediterranean, parts of the Persian Gulf, and in NE India. It is characterized by distortion and fragility of the red blood cells, which are lost too rapidly from the circulation. This often results in ◊anaemia.

People with this disease have abnormal red blood cells (sickle cells), containing a defective ◊haemoglobin. The presence of sickle cells in the blood, with or without accompanying anaemia, is called **sicklemia**. It confers a degree of protection against ◊malaria because fewer normal red blood cells are available to the parasites for infection.

sidereal period the orbital period of a planet around the Sun, or a moon around a planet, with reference to a background star. The sidereal period of a planet is in effect a 'year'. A ◊synodic period is a full circle as seen from Earth.

Siding Spring Mountain peak 400 km/250 mi NW of Sydney, site of the UK Schmidt Telescope, opened 1973, and the 3.9–m/154–in **Anglo-Australian Telescope**, opened 1975, which was the first big telescope to be fully computer-controlled. It is one of the most powerful telescopes in the southern hemisphere.

siemens SI unit (symbol S) of electrical conductance, the reciprocal of the ◊impedance of an electrical circuit. One siemens equals one ampere per volt. It was formerly called the mho or reciprocal ohm.

sievert SI unit (symbol Sv) of radiation dose equivalent. It replaces the rem (1 Sv equals 100 rem). Some types of radiation do more damage than others for the same absorbed dose; for example, the same absorbed dose of alpha radiation causes 20 times as much biological damage as the same dose of beta radiation. The equivalent dose in sieverts is equal to the absorbed dose of radiation in rays multiplied by the relative biological effectiveness. Humans can absorb up to 0.25 Sv without immediate ill effects; 1 Sv may produce radiation sickness; and more than 8 Sv causes death.

sight the detection of light by an ◊eye, which can form images of the outside world.

Sigma Octantis the star closest to the south celestial pole (see ◊celestial sphere), in effect the southern equivalent of ◊Polaris, although far less conspicuous. Situated just less than 1° from the south celestial pole in the constellation Octans, Sigma Octantis is 120 light years away.

signal-to-noise ratio ratio of the power of an electrical signal to that of the unwanted noise accompanying the signal. It is expressed in ◊decibels. In general, the higher the signal-to-noise ratio, the better. For a telephone, an acceptable ratio is 40 decibels; for television, the acceptable ratio is 50 decibels.

significant figures the figures in a number that, by virtue of their place value, express the magnitude of that number to a specified degree of accuracy. The final significant figure is rounded up if the following digit is greater than 5. For example, 5,463,254 to three significant figures is 5,460,000; 3.462891 to four significant figures is 3.463; 0.00347 to two significant figures is 0.0035.

silencer (North American *muffler*) device in the exhaust system of cars and motorbikes. Gases leave the engine at supersonic speeds, and the exhaust system and silencer are designed to slow them down, thereby silencing them.

Some silencers use baffle plates (plates with holes, which disrupt the airflow), others use perforated tubes and an expansion box (a large chamber that slows down airflow).

silica silicon dioxide, SiO_2, the composition of the most common mineral group, of which the most familiar form is quartz. Other silica forms are ◊chalcedony, chert, opal, tridymite, and cristobalite. Common sand consists largely of silica in the form of quartz.

silicate one of a group of minerals containing silicon and oxygen in tetrahedral units of SiO_4, bound together in various ways to form specific structural types. Silicates are the chief rock-forming minerals. Most rocks are composed, wholly or in part, of silicates (the main exception being limestones). Glass is a manufactured complex polysilicate material in which other elements (boron in borosilicate glass) have been incorporated.

Generally, additional cations are present in the structure, especially Al^{3+}, Fe^{2+}, Mg^{2+}, Ca^{2+}, Na^+, K^+, but quartz and other polymorphs of SiO_2 are also considered to be silicates; stishovite (a high pressure form of SiO_2) is a rare exception to the usual tetrahedral coordination of silica and oxygen.

In *orthosilicates*, the oxygens are all ionically bonded to cations such as Mg^{2+} or Fe^{2+} (as olivines), and are not shared between tetrahedra. All other silicate structures involve some degree of oxygen sharing between adjacent tetrahedra. For example, beryl is a *ring silicate* based on tetrahedra linked by sharing oxygens to form a circle. Pyroxenes are single *chain silicates*, with chains of linked tetrahedra extending in one direction through the structure; amphiboles are similar but have *double chains* if tetrahedra. In micas, which are *sheet silicates*, the tetrahedra are joined to form continuous sheets that are stacked upon one another. *Framework silicates*, such as feldspars and quartz, are based on three-dimensional frameworks of tetrahedra in which all oxygens are shared.

silicon (Latin *silicium* 'silica') brittle, nonmetallic element, symbol Si, atomic number 14, relative atomic mass 28.086. It is the second most abundant element (after oxygen) in the Earth's crust and occurs in amorphous and crystalline forms. In nature it is found only in combination with other elements, chiefly with oxygen in silica (silicon dioxide, SiO_2) and the silicates. These form the mineral ◊quartz, which makes up most sands, gravels, and beaches.

Pottery glazes and glassmaking are based on the use of silica sands and date from prehistory. Today the crystalline form of silicon is used as a deoxidizing and hardening agent in steel, and has become the basis of the electronics industry because of its ◊semiconductor properties, being used to make 'silicon chips' for microprocessors.

The element was isolated by Swedish chemist Jöns Berzelius in 1823, having been named in 1817 by Scottish chemist Thomas Thomson by analogy with boron and carbon because of its chemical resemblance to these elements.

silicon chip ◊integrated circuit with microscopically small electrical components on a piece of silicon crystal only a few millimetres square.

One silicon chip may contain more than a million components. A chip is mounted in a rectangular plastic package and linked via gold wires to metal pins, so that it can be connected to a printed circuit board for use in electronic devices, such as computers, calculators, television sets, car dashboards, and domestic appliances.

silk-screen printing or *serigraphy* method of ◊printing based on stencils. It can be used to print on most surfaces, including paper, plastic, cloth, and wood. An impermeable stencil (either paper or photographic) is attached to a finely meshed silk screen that has been stretched on a wooden frame, so that the ink passes through to the area beneath only where the image is required. The design can also be painted directly on the screen with varnish. A series of screens can be used to add successive layers of colour to the design.

The process was developed in the early 20th century for commercial use and adopted by many artists from the 1930s onwards.

sill sheet of igneous rock created by the intrusion of magma (molten rock) between layers of pre-existing rock. (A ◊dyke, by contrast, is formed when magma cuts *across* layers of rock.) An example of a sill in the UK is the Great Whin Sill, which forms the ridge along which Hadrian's Wall was built.

Sill is usually formed of *dolerite*, a rock that is extremely resistant to erosion and weathering, and often forms ridges in the landscape or cuts across rivers to create ◊waterfalls.

sillimanite aluminium silicate, Al_2SiO_5, a mineral that occurs either as white to brownish prismatic crystals or as minute white fibres. It is an indicator of high temperature conditions in metamorphic rocks formed from clay sediments. Andalusite, kyanite, and sillimanite are all polymorphs of Al_2SiO_5.

silt sediment intermediate in coarseness between clay and sand; its grains have a diameter of 0.002–0.02 mm/0.00008–0.0008 in. Silt is usually deposited in rivers, and so the term is often used generically to mean a river deposit, as in, the silting-up of a channel.

Silurian period of geological time 439–409 million years ago, the third period of the Palaeozoic era. Silurian sediments are mostly marine and consist of shales and limestone. Luxuriant reefs were built by coral-like organisms. The first land plants began to evolve during this period, and there were many ostracoderms (armoured jawless fishes). The first jawed fishes (called acanthodians) also appeared.

silver white, lustrous, extremely malleable and ductile, metallic element, symbol Ag (from Latin *argentum*), atomic number 47, relative atomic mass 107.868. It occurs in nature in ores and as a free metal; the chief ores are sulphides, from which the metal is extracted by smelting with lead. It is one of the best metallic conductors of both heat and electricity; its most useful compounds are the chloride and bromide, which darken on exposure to light and are the basis of photographic emulsions.

Silver is used ornamentally, for jewellery and tableware, for coinage, in electroplating, electrical contacts, and dentistry, and as a solder. It has been mined since prehistory; its name is an ancient non-Indo-European one, *silubr*, borrowed by the Germanic branch as *silber*.

sima in geochemistry and geophysics, the substance of the Earth's oceanic ◊crust, as distinct from the ◊sial of the continental crust. The name, now used rarely, is derived from *si*lica and *ma*gnesia, its two main chemical constituents.

simple harmonic motion (SHM) oscillatory or vibrational motion in which an object (or point) moves so that its acceleration towards a central point is proportional to its distance from it. A simple example is a pendulum, which also demonstrates another feature of SHM, that the maximum deflection is the same on each side of the central point.

A graph of the varying distance with respect to time is a sine curve, a characteristic of the oscillating current or voltage of an alternating current (AC), which is another example of SHM.

simultaneous equations in mathematics, one of two or more algebraic equations that contain two or more unknown quantities that may have a unique solution. For example, in the case of two linear equations with two unknown variables, such as (i) $x + 3y = 6$ and (ii) $3y − 2x = 4$, the solution will be those unique values of x and y that are valid for both equations. Linear simultaneous equations can be solved by using algebraic manipulation to

eliminate one of the variables, ◊coordinate geometry, or matrices (see ◊matrix).

For example, by using algebra, both sides of equation (i) could be multiplied by 2, which gives $2x + 6y = 12$. This can be added to equation (ii) to get $9y = 16$, which is easily solved: $y = {}^{16}/_9$. The variable x can now be found by inserting the known y value into either original equation and solving for x. Another method is by plotting the equations on a graph, because the two equations represent straight lines in coordinate geometry and the coordinates of their point of intersection are the values of x and y that are true for both of them. A third method of solving linear simultaneous equations involves manipulating matrices. If the equations represent either two parallel lines or the same line, then there will be no solutions or an infinity of solutions respectively.

sine in trigonometry, a function of an angle in a right-angled triangle which is defined as the ratio of the length of the side opposite the angle to the length of the hypotenuse (the longest side).

Various properties in physics vary sinusoidally; that is, they can be represented diagrammatically by a sine wave (a graph obtained by plotting values of angles against the values of their sines). Examples include ◊simple harmonic motion, such as the way alternating current (AC) electricity varies with time.

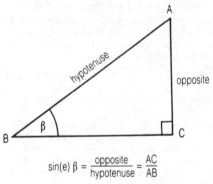

$$\sin(e)\ \beta = \frac{\text{opposite}}{\text{hypotenuse}} = \frac{AC}{AB}$$

sine

sine rule in trigonometry, a rule that relates the sides and angles of a triangle, stating that the ratio of the length of each side and the sine of the angle opposite is constant (twice the radius of the circumscribing circle). If the sides of a triangle are a, b, and c, and the angles opposite are A, B, and C, respectively, then the sine rule may be expressed as

$a/\sin A = b/\sin B = c/\sin C.$

single-sideband transmission radio-wave transmission using either the frequency band above the carrier wave frequency, or below, instead of both (as now).

singularity in astrophysics, the point at the centre of a ◊black hole at which it is predicted that the infinite gravitational forces will compress the infalling mass of the collapsing star to infinite density. It is a point in space-time at which the known laws of physics break down. It is also thought, in the

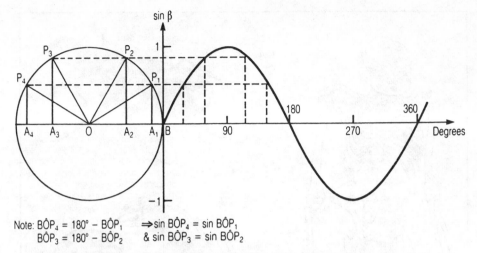

Note: $B\hat{O}P_4 = 180° - B\hat{O}P_1$ $\Rightarrow \sin B\hat{O}P_4 = \sin B\hat{O}P_1$
$B\hat{O}P_3 = 180° - B\hat{O}P_2$ $\&\ \sin B\hat{O}P_3 = \sin B\hat{O}P_2$

sine *(left) The sine of an angle; (right) constructing a sine wave. The sine of an angle is a function used in the mathematical study of the triangle. If the sine of angle β is known, then the hypotenuse can be found given the length of the opposite side, or the opposite side can be found from the hypotenuse. Within a circle of unit radius (left), the height P_1A_1 equals the sine of angle P_1OA_1. This fact and the equalities below the circle allow a sine curve to be drawn, as on the right.*

◊Big Bang theory of the origin of the universe, to be the point from which the expansion of the universe began.

sink hole funnel-shaped hollow in an area of limestone. A sink hole is usually formed by the enlargement of a joint, or crack, by ◊carbonation (the dissolving effect of water). It should not be confused with a swallow hole, or swallet, which is the opening through which a stream disappears underground when it passes onto limestone.

siphon tube in the form of an inverted U with unequal arms. When it is filled with liquid and the shorter arm is placed in a tank or reservoir, liquid flows out of the longer arm provided that its exit is below the level of the surface of the liquid in the tank.

It works because the weight of liquid in the short arm is less than that in the long arm. This produces an imbalance of pressure which tends to force the liquid from the tank up the short arm.

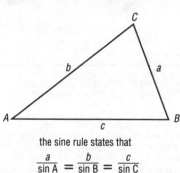

the sine rule states that

$$\frac{a}{\sin A} = \frac{b}{\sin B} = \frac{c}{\sin C}$$

or

$$\frac{\sin A}{a} = \frac{\sin B}{b} = \frac{\sin C}{c}$$

sine rule

siphonogamy plant reproduction in which a pollen tube grows to enable male gametes to pass to the ovary without leaving the protection of the plant. Siphonogamous reproduction is found in all angiosperms and most gymnosperms, and has enabled these plants to reduce their dependency on wet conditions for reproduction, unlike the zoidogamous plants (see ◊zoidogamy).

Sirius or *Dog Star* or *Alpha Canis Majoris* the brightest star in the sky, 8.6 light years from Earth in the constellation Canis Major. Sirius is a white star with a mass 2.3 times that of the Sun, a diameter 1.8 times that of the Sun, and a luminosity of 23 Suns. It is orbited every 50 years by a white dwarf, Sirius B, also known as the Pup.

sirocco hot, normally dry and dust-laden wind that blows from the deserts of N Africa across the Mediterranean into S Europe. It occurs mainly in the spring. The name 'sirocco' is also applied to any hot oppressive wind.

site of special scientific interest (SSSI) in the UK, land that has been identified as having animals, plants, or geological features that need to be protected and conserved. Numbers fluctuate, but there were over 5,000 SSSIs in 1991, covering about 6% of Britain. Although SSSIs enjoy some legal protection, this does not in practice always prevent damage or destruction; during 1989, for example, 44 SSSIs were so badly damaged that they were no longer worth protecting.

SI units (French *Système International d'Unités*) standard system of scientific units used by scientists worldwide. Originally proposed in 1960, it replaces the ◊m.k.s., ◊c.g.s., and ◊f.p.s. systems. It is based on seven basic units: the metre (m) for length, kilogram (kg) for mass, second (s) for time, ampere (A) for electrical current, kelvin (K) for temperature, mole (mol) for amount of substance, and candela (cd) for luminosity.

skeleton the rigid or semirigid framework that

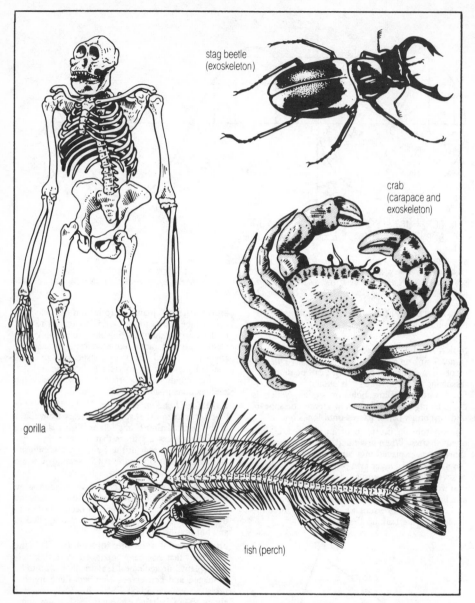

stag beetle
(exoskeleton)

crab
(carapace and
exoskeleton)

gorilla

fish (perch)

skeleton *The skeleton gives the body of an animal support and protection. Some skeletons are inside the body, like those of the gorilla and the perch. These are endoskeletons and are made of bone or cartilage. Skeletons outside the body are called exoskeletons. They are harder and more rigid than endoskeletons.*

supports an animal's body, protects its internal organs, and provides anchorage points for its muscles. The skeleton may be composed of bone and cartilage (vertebrates), chitin (arthropods), calcium carbonate (molluscs and other invertebrates), or silica (many protists).

It may be internal, forming an ◊**endoskeleton**, or external, forming an ◊**exoskeleton**. Another type of skeleton, found in invertebrates such as earthworms, is the **hydrostatic skeleton**. This gains partial rigidity from fluid enclosed within a body cavity. Because the fluid cannot be compressed, contraction of one part of the body results in extension of another part, giving peristaltic motion.

skin the covering of the body of a vertebrate. In mammals, the outer layer (epidermis) is dead and protective, and its cells are constantly being rubbed away and replaced from below. The lower layer (dermis) contains blood vessels, nerves, hair roots, and sweat and sebaceous glands, and is supported by a network of fibrous and elastic cells.

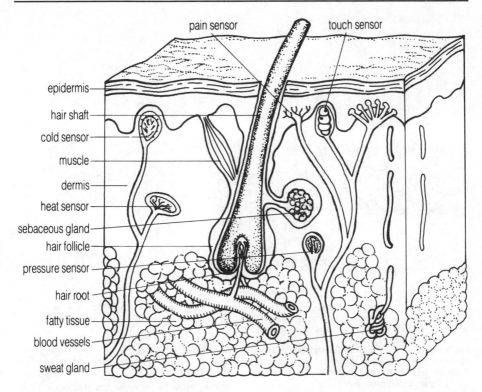

pain sensor   touch sensor

epidermis
hair shaft
cold sensor
muscle
dermis
heat sensor
sebaceous gland
hair follicle
pressure sensor
hair root
fatty tissue
blood vessels
sweat gland

skin *The skin of an adult man covers about 1.9 sq m/20 sq ft; a woman's skin covers about 1.6 sq m/17 sq ft. During our lifetime, we shed about 18 kg/40 lb of skin.*

Skin grafting is the repair of injured skin by placing pieces of skin, taken from elsewhere on the body, over the injured area. The field of medicine dealing with skin is called dermatology.

skull in vertebrates, the collection of flat and irregularly shaped bones (or cartilage) that enclose the brain and the organs of sight, hearing, and smell, and provide support for the jaws. In mammals, the skull consists of 22 bones joined by sutures. The floor of the skull is pierced by a large hole for the spinal cord and a number of smaller apertures through which other nerves and blood vessels pass.

The bones of the face include the upper jaw, enclose the sinuses, and form the framework for the nose, eyes, and roof of the mouth cavity. The lower jaw is hinged to the middle of the skull at its lower edge. The opening to the middle ear is located near the jaw hinge. The plate at the back of the head is jointed at its lower edge with the upper section of the spine. Inside, the skull has various shallow cavities into which fit different parts of the brain.

The human skull has evolved from robust to gracile in the past 5.5 million years; it exhibits ◊neoteny (infantilism), whereby the youthful features of ancient ◊human species are retained in the adult skulls of modern humans–probably to make room for the evolving and enlarging brain.

Skylab US space station, launched 14 May 1973, made from the adapted upper stage of a Saturn V rocket. At 75 tonnes/82.5 tons, it was the heaviest object ever put into space, and was 25.6 m/84 ft long. *Skylab* contained a workshop for carrying out experiments in weightlessness, an observatory for monitoring the Sun, and cameras for photographing the Earth's surface.

Damaged during launch, it had to be repaired by the first crew of astronauts. Three crews, each of three astronauts, occupied *Skylab* for periods of up to 84 days, at that time a record duration for human spaceflight. *Skylab* finally fell to Earth on 11 July 1979, dropping debris on Western Australia.

slag in chemistry, the molten mass of impurities that is produced in the smelting or refining of metals.

The slag produced in the manufacture of iron in a ◊blast furnace floats on the surface above the molten iron. It contains mostly silicates, phosphates, and sulphates of calcium. When cooled, the solid is broken up and used as a core material in the foundations of roads and buildings.

slaked lime technical name *calcium hydroxide* Ca(OH)₂ substance produced by adding water to quicklime (calcium oxide, CaO). Much heat is given out and the solid crumbles as it absorbs water. A solution of slaked lime is called ◊limewater.

slash and burn simple agricultural method whereby natural vegetation is cut and burned, and the clearing then farmed for a few years until the soil loses its fertility, whereupon farmers move on

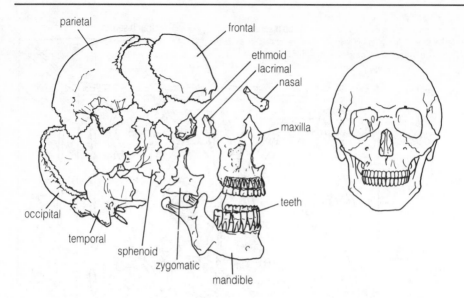

skull *The skull is a protective box for the brain, eyes, and hearing organs. It is also a framework for the teeth and flesh of the face. The cranium has eight bones: occipital, two temporal, two parietal, frontal, sphenoid, ethmoid. The face has 14 bones, the main ones being two maxillae, two nasal, two zygoma, two lacrimal, and the mandible.*

and leave the area to regrow. Although this is possible with a small, widely dispersed population, it becomes unsustainable with more people and is now a form of ◊deforestation.

slate fine-grained, usually grey metamorphic rock that splits readily into thin slabs along its ◊cleavage planes. It is the metamorphic equivalent of ◊shale.

Slate is highly resistant to atmospheric conditions and can be used for writing on with chalk (actually gypsum). Quarrying slate takes such skill and time that it is now seldom used for roof and sill material except in restoring historic buildings.

sleep state of reduced awareness and activity that occurs at regular intervals in most mammals and birds, though there is considerable variation in the amount of time spent sleeping. Sleep differs from hibernation in that it occurs daily rather than seasonally, and involves less drastic reductions in metabolism. The function of sleep is unclear. People deprived of sleep become irritable, uncoordinated, forgetful, hallucinatory, and even psychotic.

In humans, sleep is linked with hormone levels and specific brain electrical activity, including delta waves, quite different from the brain's waking activity. REM (rapid eye movement) phases, associated with dreams, occur at regular intervals during sleep, when the eyes move rapidly beneath closed lids.

In some species sleep may make animals inconspicuous at times when they might be vulnerable to predators.

slide rule mathematical instrument with pairs of logarithmic sliding scales, used for rapid calculations, including multiplication, division, and the extraction of square roots. It has been largely superseded by the electronic calculator.

It was invented 1622 by the English mathematician William Oughtred. A later version was devised by the French army officer Amédée Mannheim (1831–1906).

slime mould or *myxomycete* extraordinary organism that shows some features of ◊fungus and some of ◊protozoa. Slime moulds are not closely related to any other group, although they are often classed, for convenience, with the fungi. There are two kinds, cellular slime moulds and plasmodial slime moulds, differing in their complex life cycles.

Cellular slime moulds go through a phase of living as single cells, looking like amoebae, and feed by engulfing the bacteria found in rotting wood, dung, or damp soil. When a food supply is exhausted, up to 100,000 of these amoebae form into a colony resembling a single sluglike animal and migrate to a fresh source of bacteria. The colony then takes on the aspect of a fungus, and forms long-stalked fruiting bodies which release spores. These germinate to release amoebae, which repeat the life cycle.

Plasmodial slime moulds have a more complex life cycle involving sexual reproduction. They form a slimy mass of protoplasm with no internal cell walls, which slowly spreads over the bark or branches of trees.

SLR abbreviation for *single-lens reflex*, a type of ◊camera in which the image can be seen through the lens before a picture is taken.

A small mirror directs light entering the lens to the viewfinder. When a picture is taken the mirror moves rapidly aside to allow the light to reach the film. The SLR allows different lenses, such as close-up or zoom lenses, to be used because the photographer can see exactly what is being focused on.

slug obsolete unit of mass, equal to 14.6 kg/ 32.17 lb. It is the mass that will have an acceleration of one foot per second when under a force of one pound weight.

slurry form of manure composed mainly of liquids. Slurry is collected and stored on many farms, especially when large numbers of animals are kept in factory units. When slurry tanks are accidentally or deliberately breached, large amounts can spill into rivers, killing fish and causing ◊eutrophication. Some slurry is spread on fields as a fertilizer.

Slurry spills in the UK increased enormously in the 1980s and were by 1991 running at rates of over 4,000 a year. Tighter regulations were being introduced to curb this pollution.

smart card plastic card with an embedded microprocessor and memory. It can store, for example, personal data, identification, and bank-account details, to enable it to be used as a credit or debit card. The card can be loaded with credits, which are then spent electronically, and reloaded as needed. Possible other uses range from hotel door 'keys' to passports.

The smart card was invented by French journalist Juan Moreno in 1974. By the year 2000 it will be possible to make cards with as much computing power as the leading personal computers of 1990.

smart fluid or *electrorheological fluid* liquid suspension that solidifies to form a jellylike solid when a high-voltage electric field is applied across it and that returns to the liquid state when the field is removed. Most smart fluids are ◊zeolites or metals coated with polymers or oxides.

Applications being investigated include electrically operated clutches and vibration dampers.

smell sense that responds to chemical molecules in the air. It works by having receptors for particular chemical groups, into which the airborne chemicals must fit to trigger a message to the brain.

A sense of smell is used to detect food and to communicate with other animals (see ◊pheromone and ◊scent gland). Aquatic animals can sense chemicals in water, but whether this sense should be described as 'smell' or 'taste' is debatable. See also ◊nose.

smelling salts or *sal volatile* a mixture of ammonium carbonate, bicarbonate, and carbamate together with other strong-smelling substances, formerly used as a restorative for dizziness or fainting.

smelting processing a metallic ore in a furnace to produce the metal. Oxide ores such as iron ore are smelted with coke (carbon), which reduces the ore into metal and also provides fuel for the process.

A substance such as limestone is often added during smelting to facilitate the melting process and to form a slag, which dissolves many of the impurities present.

Smithsonian Institution academic organization in Washington, DC, USA, founded in 1846 with money left by British chemist and mineralogist James Smithson. The Smithsonian Institution undertakes scientific research but is also the parent organization of a collection of museums. These include the United States National Museum, the National Air and Space Museum, the National Gallery of Art, the Freer Gallery of Art, the National Portrait Gallery, and the National Collec-

tion of Fine Arts. The Smithsonian Institution also has a zoo and an astrophysical observatory.

smog natural fog containing impurities (unburned carbon and sulphur dioxide) from domestic fires, industrial furnaces, certain power stations, and internal-combustion engines (petrol or diesel). It can cause substantial illness and loss of life, particularly among chronic bronchitics, and damage to wildlife.

The London smog of 1952 killed 4,000 people from heart and lung diseases. The use of smokeless fuels, the treatment of effluent, and penalties for excessive smoke from poorly maintained and operated vehicles can be extremely effective in cutting down smog, as in London, but it still occurs in many cities throughout the world.

smokeless fuel fuel that does not give off any smoke when burned, because all the carbon is fully oxidized to carbon dioxide (CO_2). Natural gas, oil, and coke are smokeless fuels.

smoker or *hydrothermal vent* crack in the ocean floor, associated with an ◊ocean ridge, through which hot, mineral-rich ground water erupts into the sea, forming thick clouds of suspended material. The clouds may be dark or light, depending on the mineral content, thus producing 'white smokers' or 'black smokers'.

Sea water percolating through the sediments and crust is heated in the active area beneath and dissolves minerals from the hot rocks. As the charged water is returned to the ocean, the sudden cooling causes these minerals to precipitate from solution, so forming the suspension. The chemical-rich water around a smoker gives rise to colonies of bacteria, and these form the basis of food chains that can be sustained without sunlight and photosynthesis. Strange animals that live in such regions include huge tube worms 2 m/6 ft long, giant clams, and species of crab, anemone, and shrimp found nowhere else.

smooth muscle muscle capable of slow contraction over a period of time. Its presence in the wall of the alimentary canal allows slow rhythmic movements known as ◊peristalsis, which cause food to be mixed and forced along the gut. Smooth muscle has a microscopic structure distinct from other forms, and is not under conscious control.

snellen unit expressing the visual power of the eye.

Snell's law of refraction in optics, the law stating that for light rays passing from one transparent medium to another, the ratio of the sine of the angle of incidence and the sine of the angle of refraction is constant. The ratio for any pair of transparent media is called its ◊refractive index *n*. The law is named after its discoverer, Dutch physicist Willebrand Snell.

snout in earth science, the front end of a ◊glacier, representing the furthest advance of the ice at any one time. Deep cracks, or ◊crevasses, and ice falls are common.

Because the snout is the lowest point of a glacier it tends to be affected by the warmest weather. Considerable melting takes place, and so it is here that much of the rocky material carried by the glacier becomes deposited. Material dumped just ahead of the snout may form ◊terminal moraine.

The advance or retreat of the snout depends upon the glacier budget—the balance between ◊accumulation (the addition of snow and ice to the glacier) and ◊ablation (their loss by melting and evaporaton).

soap mixture of the sodium salts of various ◊fatty acids: palmitic, stearic, and oleic acid. It is made by the action of sodium hydroxide (caustic soda) or potassium hydroxide (caustic potash) on fats of animal or vegetable origin. Soap makes grease and dirt disperse in water in a similar manner to a ◊detergent.

Soap was mentioned by Galen in the 2nd century for washing the body, although the Romans seem to have washed with a mixture of sand and oil. Soap was manufactured in Britain from the 14th century, but better-quality soap was imported from Castile or Venice. The Soapmakers' Company, London, was incorporated 1638. Soap was taxed in England from the time of Cromwell in the 17th century to 1853.

soapstone compact, massive form of impure ◊talc.

social behaviour in zoology, behaviour concerned with altering the behaviour of other individuals of the same species. Social behaviour allows animals to live harmoniously in groups by establishing hierarchies of dominance to discourage disabling fighting. It may be aggressive or submissive (for example, cowering and other signals of appeasement), or designed to establish bonds (such as social grooming or preening).

The social behaviour of mammals and birds is generally more complex than that of lower organisms, and involves relationships with individually recognized animals. Thus, courtship displays allow individuals to choose appropriate mates and form the bonds necessary for successful reproduction. In the social systems of bees, wasps, ants, and termites, an individual's status and relationships with others are largely determined by its biological form, as a member of a caste of workers, soldiers, or reproductives; see ◊eusociality.

sociobiology study of the biological basis of all social behaviour, including the application of ◊population genetics to the evolution of behaviour. It builds on the concept of ◊inclusive fitness, contained in the notion of the 'selfish gene'. Contrary to some popular interpretations, it does not assume that all behaviour is genetically determined.

The New Zealand biologist W D Hamilton introduced the concept of inclusive fitness, which emphasizes that the evolutionary function of behaviour is to allow an organism to contribute as many of its own ◊alleles as it can to future generations: this idea is encapsulated in the British zoologist Richard Dawkins's notion of the 'selfish gene'.

soda ash former name for sodium carbonate (Na_2CO_3).

soda lime powdery mixture of calcium hydroxide and sodium hydroxide or potassium hydroxide, used in medicine and as a drying agent.

sodium soft, waxlike, silver-white, metallic element, symbol Na (from Latin *natrium*), atomic number 11, relative atomic mass 22.898. It is one of the ◊alkali metals and has a very low density, being light enough to float on water. It is the sixth most abundant element (the fourth most abundant metal) in the Earth's crust. Sodium is highly reactive, oxidizing rapidly when exposed to air and reacting violently with water. Its most familiar compound is sodium chloride (common salt), which occurs naturally in the oceans and in salt deposits left by dried-up ancient seas.

Other sodium compounds used industrially include sodium hydroxide (caustic soda, NaOH), sodium carbonate (washing soda, Na_2CO_3) and hydrogencarbonate (sodium bicarbonate, $NaHCO_3$), sodium nitrate (saltpetre, $NaNO_3$, used as a fertilizer), and sodium thiosulphate (hypo, $Na_2S_2O_3$, used as a photographic fixer). Thousands of tons of these are manufactured annually. Sodium metal is used to a limited extent in spectroscopy, in discharge lamps, and alloyed with potassium as a heat-transfer medium in nuclear reactors. It was isolated from caustic soda in 1807 by English chemist Humphry Davy.

sodium chloride or *common salt* or *table salt* NaCl white, crystalline compound found widely in nature. It is a typical ionic solid with a high melting point (801°C/1,474°F); it is soluble in water, insoluble in organic solvents, and is a strong electrolyte when molten or in aqueous solution. Found in concentrated deposits, it is widely used in the food industry as a flavouring and preservative, and in the chemical industry in the manufacture of sodium, chlorine, and sodium carbonate.

sodium hydrogencarbonate chemical name for ◊bicarbonate of soda.

sodium hydroxide or *caustic soda* NaOH the commonest ◊alkali. The solid and the solution are corrosive. It is used to neutralize acids, in the manufacture of soap, and in oven cleaners. It is prepared industrially from sodium chloride by the ◊electrolysis of concentrated brine.

soft-sectored disc another name for an unformatted blank disc; see ◊disc formatting.

software in computing, a collection of programs and procedures for making a computer perform a specific task, as opposed to ◊hardware, the physical components of a computer system. Software is created by programmers and is either distributed on a suitable medium, such as the ◊floppy disc, or built into the computer in the form of ◊firmware. Examples of software include ◊operating systems, ◊compilers, and application programs, such as payrolls. No computer can function without some form of software.

soft water water that contains very few dissolved metal ions such as calcium (Ca^{2+}) or magnesium (Mg^{2+}). It lathers easily with soap, and no ◊scale is formed inside kettles or boilers. It has been found that the incidence of heart disease is higher in soft-water areas.

soil loose covering of broken rocky material and decaying organic matter overlying the bedrock of the Earth's surface. Various types of soil develop under different conditions: deep soils form in warm wet climates and in valleys; shallow soils form in

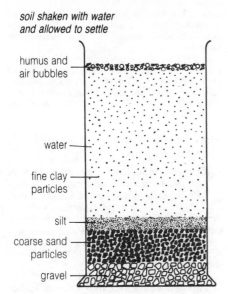

soil shaken with water and allowed to settle

- humus and air bubbles
- water
- fine clay particles
- silt
- coarse sand particles
- gravel

soil

clay

- small clay particle
- small air space

sand

- large air space
- large sand particle

cool dry areas and on slopes. **Pedology**, the study of soil, is significant because of the relative importance of different soil types to agriculture.

The organic content of soil is widely variable, ranging from zero in some desert soils to almost 100% in peats.

Soil Association pioneer British ecological organization founded 1946, which campaigns against pesticides and promotes organic farming.

soil creep gradual movement of soil down a slope. As each soil particle is dislodged by a raindrop it moves slightly further downhill. This eventually results in a mass downward movement of soil on the slope.

Evidence of soil creep includes the formation of terracettes (steplike ridges along the hillside), leaning walls and telegraph poles, and trees that grow in a curve to counteract progressive leaning.

soil depletion a decrease in soil quality over time. Causes include loss of nutrients caused by overfarming, erosion by wind, and chemical imbalances caused by acid rain.

soil erosion the wearing away and redistribution of the Earth's soil layer. It is caused by the action of water, wind, and ice, and also by improper methods of ◊agriculture. If unchecked, soil erosion results in the formation of deserts (◊desertification). It has been estimated that 20% of the world's cultivated topsoil was lost between 1950 and 1990.

If the rate of erosion exceeds the rate of soil formation (from rock and decomposing organic matter), then the land will decline and eventually become infertile. The removal of forests or other vegetation often leads to serious soil erosion, because plant roots bind soil, and without them the soil is free to wash or blow away, as in the American ◊dust bowl. The effect is worse on hillsides, and there has been devastating loss of soil where forests

have been cleared from mountainsides, as in Madagascar. Improved agricultural practices such as contour ploughing are needed to combat soil erosion. Windbreaks, such as hedges or strips planted with coarse grass, are valuable, and organic farming can reduce soil erosion by as much as 75%.

Soil degradation and erosion are becoming as serious as the loss of the rainforest. It is estimated that more than 10% of the world's soil lost a large amount of its natural fertility during the latter half of the 20th century. Some of the worst losses are in Europe, where 17% of the soil is damaged by human activity such as mechanized farming and fallout from acid rain. Mexico and Central America have 24% of soil highly degraded, mostly as a result of deforestation.

soil mechanics branch of engineering that studies the nature and properties of the soil. Soil is investigated during construction work to ensure that it has the mechanical properties necessary to support the foundations of dams, bridges, and roads.

sol ◊colloid of very small solid particles dispersed in a liquid that retains the physical properties of a liquid.

solar energy energy derived from the Sun's radiation. The amount of energy falling on just 1 sq km/0.3861 sq mi is about 4,000 megawatts, enough to heat and light a small town. In one second the Sun gives off 13 million times more energy than all the electricity used in the USA in one year. *Solar heaters* have industrial or domestic uses. They usually consist of a black (heat-absorbing) panel containing pipes through which air or water, heated by the Sun, is circulated, either by thermal ◊convection or by a pump.

Solar energy may also be harnessed indirectly using *solar cells* (photovoltaic cells) made of panels of ◊semiconductor material (usually silicon), which generate electricity when illuminated by sunlight. Although it is difficult to generate a high output from solar energy compared to sources such as nuclear or fossil fuels, it is a major nonpolluting and renewable energy source used as far north as Scandinavia as well as in the SW USA and in Mediterranean countries.

A solar furnace, such as that built in 1970 at Odeillo in the French Pyrénées, has thousands of mirrors to focus the Sun's rays; it produces uncontaminated intensive heat (up to 3,000°C/5,4000°F) for industrial and scientific or experimental purposes. The world's first solar power station connected to a national grid opened in 1991 at Adrano in Sicily. Scores of giant mirrors move to follow the Sun throughout the day, focusing the rays into a boiler. Steam from the boiler drives a conventional turbine. The plant generates up to 1 megawatt. A similar system, called Solar 1, has been built in the Mojave desert near Daggett, California, USA. It consists of 1818 computer-controlled mirrors arranged in circles around a 91 m/300 ft central boiler tower. Advanced schemes have been proposed that would use giant solar reflectors in space to harness ◊solar energy and beam it down to Earth in the form of ◊microwaves. Despite their low running costs, their high installation cost and low power output have meant that solar cells have found few applications outside space probes and artificial satellites. Solar heating is, however, widely used for domestic purposes in many parts of the world.

If sunbeams were weapons of war, we would have had solar energy long ago.

On **solar energy** George Porter in the *Observer* 26 Aug 1973

Solar Maximum Mission satellite launched by the US agency NASA 1980 to study solar activity, which discovered that the Sun's luminosity increases slightly when sunspots are most numerous. It was repaired in orbit by astronauts from the space shuttle 1984 and burned up in the Earth's atmosphere 1989.

solar pond natural or artificial 'pond', such as the Dead Sea, in which salt becomes more soluble in the Sun's heat. Water at the bottom becomes saltier and hotter, and is insulated by the less salty water layer at the top. Temperatures at the bottom reach about 100°C/212°F and can be used to generate electricity.

solar radiation radiation given off by the Sun, consisting mainly of visible light, ◊ultraviolet radiation, and ◊infrared radiation, although the whole spectrum of ◊electromagnetic waves is present, from radio waves to X-rays. High-energy charged particles such as ◊electrons are also emitted, especially from solar ◊flares. When these reach the Earth, they cause magnetic storms (disruptions of the Earth's magnetic field), which interfere with radio communications.

Solar System the Sun (a star) and all the bodies orbiting it: the nine ◊planets (Mercury, Venus, Earth, Mars, Jupiter, Saturn, Uranus, Neptune, and Pluto), their moons, the asteroids, and the

comets. It is thought to have formed from a cloud of gas and dust in space about 4.6 billion years ago. The Sun contains 99% of the mass of the Solar System.

In 1992 an object, designated 1992 QB1, was found orbiting in the outer Solar System beyond Pluto. Probably a ball of dirty ice about 200 km/ 125 mi across, it is the most remote member known in the Solar System. It orbits at between 37° and 59° further from the Sun than the Earth.

Men Very Easily Make Jugs Serve Useful Nocturnal Purposes

Mnemonic for the order of planets in the **Solar System**, from the Sun outwards: Mercury, Venus, Earth, Mars, Jupiter, Saturn, Uranus, Neptune, and Pluto

---

solar wind stream of atomic particles, mostly protons and electrons, from the Sun's corona, flowing outwards at speeds of between 300 kps/200 mps and 1,000 kps/600 mps.

The fastest streams come from 'holes' in the Sun's corona that lie over areas where no surface activity occurs. The solar wind pushes the gas of comets' tails away from the Sun, and 'gusts' in the solar wind cause geomagnetic disturbances and auroras on Earth.

SOLAR WIND

Every second, the Sun pumps more than a million tonnes of material into space as a stream of electrically-charged particles. This stream, called the solar wind, blows as far as Pluto, the most distant planet.

---

solder any of various alloys used when melted for joining metals such as copper, its common alloys (brass and bronze), and tin-plated steel, as used for making food cans.

Soft solders (usually alloys of tin and lead, sometimes with added antimony) melt at low temperatures (about 200°C/392°F), and are widely used in the electrical industry for joining copper wires. Hard (or brazing) solders, such as silver solder (an alloy of copper, silver, and zinc), melt at much higher temperatures and form a much stronger joint.

A necessary preliminary to making any solder joint is thorough cleaning of the surfaces of the metal to be joined (to remove oxide) and the use of a flux (to prevent the heat applied to melt the solder from reoxidizing the metal).

solenoid coil of wire, usually cylindrical, in which a magnetic field is created by passing an electric current through it (see ◊electromagnet). This field can be used to move an iron rod placed on its axis. Mechanical valves attached to the rod can be operated by switching the current on or off, so converting electrical energy into mechanical energy. Solenoids are used to relay energy from the battery of a car to the starter motor by means of the ignition switch.

solid in physics, a state of matter that holds its own shape (as opposed to a liquid, which takes up

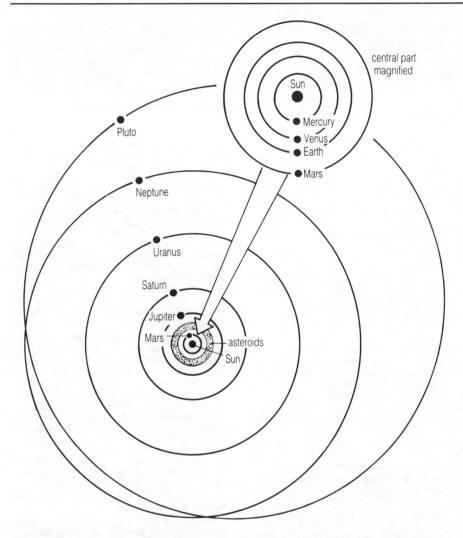

Solar System *Most of the objects in the Solar System lie close to the plane of the ecliptic. The planets are tiny compared to the Sun. If the Sun were the size of a basketball, the planet closest to the Sun, Mercury, would be the size of a mustard seed 15 m/48 ft from the Sun. The most distant planet, Pluto, would be a pinhead 1.6 km/1 mi away from the Sun. The Earth, which is the third planet out from the Sun, would be the size of a pea 32 m/100 ft from the Sun.*

the shape of its container, or a gas, which totally fills its container). According to ◊kinetic theory, the atoms or molecules in a solid are not free to move but merely vibrate about fixed positions, such as those in crystal lattices.

solidification change of state from liquid (or vapour) to solid that occurs at the freezing point (see ◊freezing) of a substance.

solid-state circuit electronic circuit where all the components (resistors, capacitors, transistors, and diodes) and interconnections are made at the same time, and by the same processes, in or on one piece of single-crystal silicon. The small size of this construction accounts for its use in electronics for space vehicles and aircraft.

solifluction the downhill movement of topsoil that has become saturated with water. Solifluction is common in periglacial environments (those bordering glacial areas) during the summer months, when the frozen topsoil melts to form an unstable soggy mass. This may then flow slowly downhill under gravity to form a *solifluction lobe* (a tonguelike feature).

Solifluction material, or head, is found at the bottom of chalk valleys in southern England; it is partly responsible for the rolling landscape typical of chalk scenery.

soliton solitary wave that maintains its shape and velocity, and does not widen and disperse in the normal way. The mathematical equations that sum up the behaviour of solitons are being used to further research in nuclear fusion and superconductivity.

comparative curves for copper(II) sulphate and potassium nitrate

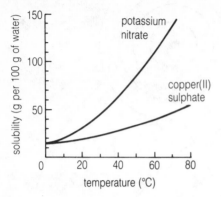

solubility curve

It is so named from a solitary wave seen on a canal by Scottish engineer John Scott Russell (1808–1882), who raced after it on his horse. Before he lost it, it had moved on for over a mile as a smooth, raised, and rounded form, rather than widening and dispersing.

solstice either of the points at which the Sun is farthest north or south of the celestial equator each year. The *summer solstice*, when the Sun is farthest north, occurs around June 21; the *winter solstice* around Dec 22.

solubility in physics, a measure of the amount of solute (usually a solid or gas) that will dissolve in a given amount of solvent (usually a liquid) at a particular temperature. Solubility may be expressed as grams of solute per 100 grams of solvent or, for a gas, in parts per million (ppm) of solvent.

solubility curve graph that indicates how the solubility of a substance varies with temperature. Most salts increase their solubility with an increase in temperature, as they dissolve endothermically. These curves can be used to predict which salt, and how much of it, will crystallize out of a mixture of salts.

solute substance that is dissolved in another substance (see ◊solution).

solution in chemistry, two or more substances mixed to form a single, homogenous phase. One of the substances is the *solvent* and the others (*solutes*) are said to be dissolved in it.

The constituents of a solution may be solid, liquid, or gaseous. The solvent is normally the substance that is present in greatest quantity; however, if one of the constituents is a liquid this is considered to be the solvent even if it is not the major substance.

solution in earth science, the dissolving in water of minerals within a rock. It may result in weathering (for example, when weakly acidic rainfall causes ◊carbonation) and erosion (when flowing water passes over rocks).

Solution commonly affects limestone and chalk, and may be responsible for forming features such as ◊sink holes.

Solvay process industrial process for the manufacture of sodium carbonate.

It is a multistage process in which carbon dioxide is generated from limestone and passed through ◊brine saturated with ammonia. Sodium hydrogen carbonate is isolated and heated to yield sodium carbonate. All intermediate by-products are recycled so that the only ultimate by-product is calcium chloride.

solvent substance, usually a liquid, that will dissolve another substance (see ◊solution). Although the commonest solvent is water, in popular use the term refers to low-boiling-point organic liquids, which are harmful if used in a confined space. They can give rise to respiratory problems, liver damage, and neurological complaints.

Typical organic solvents are petroleum distillates (in glues), xylol (in paints), alcohols (for synthetic and natural resins such as shellac), esters (in lacquers, including nail varnish), ketones (in cellulose lacquers and resins), and chlorinated hydrocarbons (as paint stripper and dry-cleaning fluids). The fumes of some solvents, when inhaled (glue-sniffing), affect mood and perception. In addition to damaging the brain and lungs, repeated inhalation of solvent from a plastic bag can cause death by asphyxia.

sonar (acronym for *so*und *na*vigation and *r*anging) method of locating underwater objects by the reflection of ultrasonic waves. The time taken for an acoustic beam to travel to the object and back to the source enables the distance to be found since the velocity of sound in water is known. Sonar devices, or *echo sounders*, were developed 1920.

The process is similar to that used in ◊radar. During World War I and after, the Allies developed and perfected an apparatus for detecting the presence of enemy U-boats beneath the sea surface by the use of ultrasonic echoes. It was originally named ASDIC, from the initials of the Allied Submarine Detection Investigation Committee responsible for its development, but in 1963 the name was changed to sonar.

sone unit of subjective loudness. A tone of 40 decibels above the threshold of hearing with a frequency of 1,000 hertz is defined as one sone; any sound that seems twice as loud as this has a value of two sones, and so on. A loudness of one sone corresponds to 40 ◊phons.

sonic boom noise like a thunderclap that occurs when an aircraft passes through the ◊sound barrier, or begins to travel faster than the speed of sound. It happens when the cone-shaped shock wave caused by the plane touches the ground.

sonoluminescence emission of light by a liquid that is subjected to high-frequency sound waves. The rapid changes of pressure induced by the sound cause minute bubbles to form in the liquid, which then collapse. Light is emitted at the final stage of the collapse, probably because it squeezes and heats gas inside the bubbles.

sorbic acid $CH_3CH{:}CHCH{:}CHCOOH$ tasteless acid found in the fruit of the mountain ash (genus *Sorbus*) and prepared synthetically. It is widely used in the preservation of food—for example, cider, wine, soft drinks, animal feeds, bread, and cheese.

sorting arranging data in sequence. When sorting a collection, or file, of data made up of several different ◊fields, one must be chosen as the **key field** used to establish the correct sequence. For example, the data in a company's mailing list might include fields for each customer's first names, surname, address, and telephone number. For most purposes the company would wish the records to be sorted alphabetically by surname; therefore, the surname field would be chosen as the key field.

The choice of sorting method involves a compromise between running time, memory usage, and complexity. Those used include **selection sorting**, in which the smallest item is found and exchanged with the first item, the second smallest exchanged with the second item, and so on; **bubble sorting**, in which adjacent items are continually exchanged until the data are in sequence; and **insertion sorting**, in which each item is placed in the correct position and subsequent items moved down to make a place for it.

sorus in ferns, a group of sporangia, the reproductive structures that produce ◊spores. They occur on the lower surface of fern fronds.

sound physiological sensation received by the ear, originating in a vibration (pressure variation in the air) that communicates itself to the air, and travels in every direction, spreading out as an expanding sphere. All sound waves in air travel with a speed dependent on the temperature; under ordinary conditions, this is about 330 m/1,070 ft per second. The pitch of the sound depends on the number of vibrations imposed on the air per second, but the speed is unaffected. The loudness of a sound is dependent primarily on the amplitude of the vibration of the air.

The lowest note audible to a human being has a frequency of about 20 ◊hertz (vibrations per second), and the highest one of about 15,000 Hz; the lower limit of this range varies little with the person's age, but the upper range falls steadily from adolescence onwards.

sound barrier concept that the speed of sound, or sonic speed (about 1,220 kph/760 mph at sea level), constitutes a speed limit to flight through the atmosphere, since a badly designed aircraft suffers severe buffeting at near sonic speed owing to the formation of shock waves. US test pilot Chuck Yeager first flew through the 'barrier' in 1947 in a Bell X-1 rocket plane. Now, by careful design, such aircraft as Concorde can fly at supersonic speed with ease, though they create in their wake a ◊sonic boom.

sound synthesis the generation of sound (usually music) by electronic ◊synthesizer.

soundtrack band at one side of a cine film on which the accompanying sound is recorded. Usually it takes the form of an optical track (a pattern of light and shade). The pattern is produced on the film when signals from the recording microphone are made to vary the intensity of a light beam. During playback, a light is shone through the track on to a photocell, which converts the pattern of light falling on it into appropriate electrical signals. These signals are then fed to loudspeakers to recreate the original sounds.

source language in computing, the language in which a program is written, as opposed to ◊machine code, which is the form in which the program's instructions are carried out by the computer. Source languages are classified as either ◊high-level languages or ◊low-level languages, according to whether each notation in the source language stands for many or only one instruction in machine code.

Programs in high-level languages are translated into machine code by either a ◊compiler or an ◊interpreter program. Low-level programs are translated into machine code by means of an ◊assembler program. The program, before translation, is called the **source program**; after translation into machine code it is called the **object program**.

source program in computing, a program written in a ◊source language.

souring change that occurs to wine on prolonged exposure to air. The ethanol in the wine is oxidized by the air (oxygen) to ethanoic acid. It is the presence of the ethanoic (acetic) acid that produces the sour taste.

$$CH_3CH_2OH_{(aq)} + O_{2(g)} = CH_3COOH_{(aq)} + H_2O_{(l)}$$

South African Astronomical Observatory national observatory of South Africa at Sutherland, founded 1973 after the merger of the Royal Observatory, Cape Town, and the Republic Observatory, Johannesburg, and operated by the Council for Scientific and Industrial Research of South Africa. Its main telescope is a 1.88–m/74–in reflector formerly at the Radcliffe Observatory, Pretoria.

Southern Cross popular name for the constellation ◊Crux.

southern lights common name for the aurora australis, coloured light in southern skies; see ◊aurora.

Soyuz (Russian 'union') Soviet series of spacecraft, capable of carrying up to three cosmonauts. Soyuz spacecraft consist of three parts: a rear section containing engines; the central crew compartment; and a forward compartment that gives additional room for working and living space. They are now used for ferrying crews up to space stations, though they were originally used for independent space flight.

Soyuz 1 crashed on its first flight April 1967, killing the lone pilot, Vladimir Komarov. The *Soyuz 11* 1971 had three deaths on re-entry. In 1975 the ◊Apollo–Soyuz test project resulted in a successful docking in orbit.

space or **outer space** the void that exists beyond Earth's atmosphere. Above 120 km/75 mi, very little atmosphere remains, so objects can continue to move quickly without extra energy. The space between the planets is not entirely empty, but filled with the tenuous gas of the ◊solar wind as well as dust specks.

space flight: chronology

1903   Russian scientist Konstantin Tsiolkovsky published the first practical paper on aeronautics.

1926   US engineer Robert Goddard launched the first liquid-fuel rocket.

1937–45 In Germany, Wernher von Braun developed the V2 rocket.

1957   4 Oct: The first space satellite, *Sputnik 1* (USSR, Russian 'fellow-traveller'), orbited the Earth at a height of 229–898 km/142–558 mi in 96.2 min.

3 Nov: *Sputnik 2* was launched carrying a dog, 'Laika'; it died on board seven days later.

1958   31 Jan: *Explorer 1*, the first US satellite, discovered the Van Allen radiation belts.

1961   12 April: the first crewed spaceship, *Vostok 1* (USSR), with Yuri Gagarin on board, was recovered after a single orbit of 89.1 min at a height of 142–175 km/88–109 mi.

1962   20 Feb: John Glenn in *Friendship 7* (USA) became the first American to orbit the Earth. *Telstar* (USA), a communications satellite, sent the first live television transmission between the USA and Europe.

1963   16–19 June: Valentina Tereshkova in *Vostok 1* (USSR) became the first woman in space.

1967   24 April: Vladimir Komarov was the first person to be killed in space research, when his ship, *Soyuz 1* (USSR), crash-landed on the Earth.

1969   20 July: Neil Armstrong of *Apollo 11* (USA) was the first person to walk on the Moon.

1970   10 Nov: *Luna 17* (USSR) was launched; its space probe, *Lunokhod*, took photographs and made soil analyses of the Moon's surface.

1971   19 April: *Salyut 1* (USSR), the first orbital spacec station, was established; it was later visited by the *Soyuz 11* crewed spacecraft.

1973   *Skylab 2*, the first US orbital space station, was established.

1975   15–24 July: *Apollo 18* (USA) and *Soyuz 19* (USSR) made a joint flight and linked up in space.

1979   The European Space Agency's satellite launcher, *Ariane 1*, was launched.

1981   12 April: The first reusable crewed spacecraft, the space shuttle *Columbia* (USA), was launched.

1986   Space shuttle *Challenger* (USA) exploded shortly after take-off, killing all seven crew members.

1988   US shuttle programme resumed with launch of *Discovery*. Soviet shuttle *Buran* was launched from the rocket *Energiya*. Soviet cosmonauts Musa Manarov and Vladimir Titov in space station *Mir* spent a record 365 days 59 min in space.

1990   April: Hubble Space Telescope (USA) was launched from Cape Canaveral.

1990   1 June: X-ray and ultraviolet astronomy satellite *ROSAT* (USA/Germany/UK) was launched from Cape Canaveral.

2 Dec: *Astro-1* ultraviolet observatory and the Broad Band X-ray Telescope were launched from the space shuttle *Columbia*.

Japanese television journalist Toyohiro Akiyama was launched with Viktor Afanasyev and Musa Manarov to the space station *Mir*.

1991   5 April: The Gamma Ray Observatory was launched from the space shuttle *Atlantis* to survey the sky at gamma-ray wavelengths.

1991   18 May: Astronaut Helen Sharman, the first Briton in space, was launched with Anatoli Artsebarsky and Sergei Krikalek to *Mir* space station, returning to Earth 26 May in *Soyuz TM-11* with Viktor Afanasyev and Musa Manarov. Manarov set a record for the longest time spent in space, 541 days, having also spent a year aboard *Mir* 1988.

1992   European satellite *Hipparcos*, launched 1989 to measure the position of 120,000 stars, failed to reach geostationary orbit and went into a highly elliptical orbit, swooping to within 500 km/308 mi of the Earth every ten hours. The mission was later retrieved.

16 May: Space shuttle *Endeavor* returned to Earth after its maiden voyage. During its mission, it circled the Earth 141 times and travelled 4 million km/2.5 million mi.

23 Oct: *LAGEOS II* (Laser Geodynamics Satellite) was released from the space shuttle *Columbia* into an orbit so stable that it will still be circling the Earth in billions of years.

SPACE PUZZLE 1

When a spacecraft re-enters the Earth's atmosphere, friction with air molecules heats up its surface. Unless special precautions are taken, the craft will eventually burn up on re-entry, just as meteorites burn up as 'shooting stars'.

However, in order to reach orbit a spacecraft must be accelerated to orbital speed through the same atmosphere. Why then does it not burn up on launch, and why do we only worry about re-entry? *See page 651 for the answer.*

Spacelab small space station built by the European Space Agency, carried in the cargo bay of the US space shuttle, in which it remains throughout each flight, returning to Earth with the shuttle. Spacelab consists of a pressurized module in which astronauts can work, and a series of pallets, open to the vacuum of space, on which equipment is mounted.

Spacelab is used for astronomy, Earth observation, and experiments utilizing the conditions of weightlessness and vacuum in orbit. The pressurized module can be flown with or without pallets, or the pallets can be flown on their own, in which case the astronauts remain in the shuttle's own crew compartment. All the sections of Spacelab can be reused many times. The first Spacelab mission, consisting of a pressurized module and pallets, lasted ten days Nov–Dec 1983.

space probe any instrumented object sent beyond Earth to collect data from other parts of the Solar System and from deep space. The first probe was the Soviet *Lunik 1*, which flew past the Moon 1959. The first successful planetary probe was the US *Mariner 2*, which flew past Venus 1962, using ◊transfer orbit. The first space probe to leave the Solar System was *Pioneer 10* 1983. Space probes include *Galileo, Giotto, Magellan, Mars Observer,*

space probe chronology

1959	13 Sept: *Luna 2* (USSR) hit the Moon, the first craft to do so.
	10 Oct: *Luna 3* photographed the far side of the Moon.
1962	14 Dec: *Mariner 2* (USA) flew past Venus; launch date 26 Aug 1962.
1964	31 July: *Ranger 7* (USA) hit the Moon, having sent back 4,316 pictures before impact.
1965	14 July: *Mariner 4* flew past Mars; launch date 28 Nov 1964.
1966	3 Feb: *Luna 9* achieved the first soft landing on the Moon, having transmitted 27 closeup panoramic photographs; launch date 31 Jan 1966.
	2 June: *Surveyor 1* (USA) landed softly on the Moon and returned 11,150 pictures; launch date 30 May 1965.
1971	13 Nov: *Mariner 9* entered orbit of Mars; launch date 30 May 1971.
1973	3 Dec: *Pioneer 10* (USA) flew past Jupiter; launch date 3 March 1972.
1974	29 March: *Mariner 10* flew past Mercury; launch date 3 Nov 1973.
1975	22 Oct: *Venera 9* (USSR) landed softly on Venus and returned its first pictures; launch date 8 June 1975.
1976	20 July: *Viking 1* (USA) first landed on Mars; launch date 20 Aug 1975.
	3 Sept: *Viking 2* transmitted data from the surface of Mars.
1977	20 Aug: *Voyager 2* (USA) launched. 5 Sept: *Voyager 1* launched.
1978	4 Dec: *Pioneer-Venus 1* (USA) orbited Venus; launch date 20 May 1978.
1979	5 March and 9 July: *Voyager 1* and *Voyager 2* encountered Jupiter, respectively.
1980	12 Nov: *Voyager 1* reached Saturn.
1981	25 Aug: *Voyager 2* flew past Saturn.
1982	1 March: *Venera 13* transmitted its first colour pictures of the surface of Venus; launch date 30 Oct 1981.
1983	10 Oct: *Venera 15* mapped the surface of Venus from orbit; launch date 2 June 1983.
1985	2 July: *Giotto* (European Space Agency) launched to Halley's comet.
1986	24 Jan: *Voyager 2* encountered Uranus.
	13–14 March: *Giotto* met Halley's comet, closest approach 596 km/370 mi, at a speed 50 times faster than that of a bullet.
1989	4 May: *Magellan* (USA) launched from space shuttle *Atlantis* on a 15–month cruise to Venus across 15 million km/9 million mi of space.
	25 Aug: *Voyager 2* reached Neptune (4,400 million km/2,700 million mi from Earth), approaching it to within 4,850 km/3,010 mi.
	18 Oct: *Galileo* (USA) launched from space shuttle *Atlantis* for six-year journey to Jupiter.
1990	10 Aug: *Magellan* arrived at Venus and transmitted its first pictures 16 Aug 1990.
	6 Oct: *Ulysses* (European Space Agency) launched from space shuttle *Discovery*, to study the Sun.
1991	29 Oct: *Galileo* made the closest-ever approach to an asteroid, Gaspra, flying within 1,600 km/990 mi.
1992	8 Feb: *Ulysses* flew past Jupiter at a distance of 380,000 km/236,000 mi from the surface, just inside the orbit of Io and closer than 11 of Jupiter's 16 moons.
	10 July: *Giotto* (USA) flew at a speed of 14 kms/8.5 mps to within 200 km/124 mi of comet Grigg-Skellerup, 12 light years (240 million km/150 mi) away from Earth.
	25 Sept: *Mars Observer* (USA) launched from Cape Canaveral, the first US mission to Mars for 17 years.
	10 Oct: *Pioneer-Venus 1* burned up in the atmosphere of Venus.

Ulysses, the ◊Moon probes, and the Mariner, Pioneer, Viking, and Voyager series. Japan launched its first space probe 1990.

space shuttle reusable crewed spacecraft. The first was launched 12 April 1981 by the USA. It was developed by NASA to reduce the cost of using space for commercial, scientific, and military purposes. After leaving its payload in space, the space-shuttle orbiter can be flown back to Earth to land on a runway, and is then available for reuse.

Four orbiters were built: *Columbia*, *Challenger*, *Discovery*, and *Atlantis*. *Challenger* was destroyed in a midair explosion just over a minute after its tenth launch 28 Jan 1986, killing all seven crew members, the result of a failure in one of the solid rocket boosters. Flights resumed with redesigned boosters in Sept 1988. A replacement orbiter, *Endeavor*, was built, which had its maiden flight in May 1992. At the end of the 1980s, an average of $375 million had been spent on each space-shuttle mission.

The USSR produced a shuttle of similar size and appearance to the US one. The first Soviet shuttle, *Buran*, was launched without a crew by the Energiya rocket 15 Nov 1988. In Japan, development of a crewless shuttle began 1986.

The space-shuttle orbiter, the part that goes into space, is 37.2 m/122 ft long and weighs 68 tonnes. Two to eight crew members occupy the orbiter's nose section, and missions last up to 30 days. In its cargo bay the orbiter can carry up to 29 tonnes of satellites, scientific equipment, ◊Spacelab, or military payloads. At launch, the shuttle's three main engines are fed with liquid fuel from a cylindrical tank attached to the orbiter; this tank is discarded shortly before the shuttle reaches orbit. Two additional solid-fuel boosters provide the main thrust for launch, but are jettisoned after two minutes.

space sickness or *space adaptation syndrome* feeling of nausea, sometimes accompanied by vomiting, experienced by about 40% of all astronauts during their first few days in space. It is akin to travel sickness, and is thought to be caused by confusion of the body's balancing mechanism, located in the inner ear, by weightlessness. The sensation passes after a few days as the body adapts.

SPACE PUZZLE 2

Two spacecraft are in orbit, one directly behind the other. The trailing craft decides to leave orbit, and fires its rocket motors directly away from the Earth. The crew then find themselves overtaking the leading craft. Should they be surprised?
See page 651 for the answer.

---

space station any large structure designed for human occupation in space for extended periods of time. Space stations are used for carrying out astronomical observations and surveys of Earth, as well as for biological studies and the processing of materials in weightlessness. The first space station was *Salyut 1* (see ◊Salyut), and the USA has launched ◊*Skylab*.

NASA plans to build a larger space station, to be called *Freedom*, in orbit during the 1990s, in cooperation with other countries, including the European Space Agency, which is building a module called *Columbus*, and Japan, also building a module.

Space isn't remote at all. It's only an hour's drive away if your car could go straight upwards.

On **space** Fred Hoyle in the *Observer* Sept 1979

space suit protective suit worn by astronauts and cosmonauts in space. It provides an insulated, air-conditioned cocoon in which people can live and work for hours at a time while outside the spacecraft. Inside the suit is a cooling garment that keeps the body at a comfortable temperature even during vigorous work. The suit provides air to breathe, and removes exhaled carbon dioxide and moisture. The suit's outer layers insulate the occupant from the extremes of hot and cold in space ($-150°C/-240°F$ in the shade to $+180°C/+350°F$ in sunlight), and from the impact of small meteorites. Some space suits have a jet-propelled backpack, which the wearer can use to move about.

space–time in physics, combination of space and time used in the theory of ◊relativity. When developing relativity, Albert Einstein showed that time was in many respects like an extra dimension (or direction) to space. Space and time can thus be considered as entwined into a single entity, rather than two separate things.

Space–time is considered to have four dimensions: three of space and one of time. In relativity theory, events are described as occurring at points in space–time. The **general theory of relativity** describes how space–time is distorted by the presence of material bodies, an effect that we observe as gravity.

spadix in botany, an ◊inflorescence consisting of a long, fleshy axis bearing many small, stalkless flowers. It is partially enclosed by a large bract or ◊spathe. A spadix is characteristic of plants belonging to the family Araceae, including the arum lily *Zantedeschia aethiopica*.

spark chamber electronic device for recording tracks of charged ◊subatomic particles, decay products, and rays. In combination with a stack of photographic plates, a spark chamber enables the point where an interaction has taken place to be located, to within a cubic centimetre. At its simplest, it consists of two smooth threadlike ◊electrodes that are positioned 1–2 cm/0.5–1 in apart, the space between being filled by an inert gas such as neon. Sparks jump through the gas along the ionized path created by the radiation. See ◊particle detector.

spark plug plug that produces an electric spark in the cylinder of a petrol engine to ignite the fuel mixture. It consists essentially of two electrodes insulated from one another. High-voltage (18,000 V) electricity is fed to a central electrode via the distributor. At the base of the electrode, inside the cylinder, the electricity jumps to another electrode earthed to the engine body, creating a spark. See also ◊ignition coil.

spathe in flowers, the single large bract surrounding the type of inflorescence known as a ◊spadix. It is sometimes brightly coloured and petal-like, as in the brilliant scarlet spathe of the flamingo plant *Anthurium andreanum* from South America; this serves to attract insects.

speciation emergence of a new species during evolutionary history. One cause of speciation is the geographical separation of populations of the parent species, followed by reproductive isolation, so that they no longer produce viable offspring unless they interbreed. Other causes are ◊assortative mating and the establishment of a ◊polyploid population.

species in biology, a distinguishable group of organisms that resemble each other or consist of a few distinctive types (as in ◊polymorphism), and that can all interbreed to produce fertile offspring. Species are the lowest level in the system of biological classification.

Related species are grouped together in a genus. Within a species there are usually two or more separate ◊populations, which may in time become distinctive enough to be designated subspecies or varieties, and could eventually give rise to new species through ◊speciation. Around 1.4 million species have been identified so far, of which 750,000 are insects, 250,000 are plants, and 41,000 are vertebrates. In tropical regions there are roughly two species for each temperate-zone species. It is estimated that one species becomes extinct every day through habitat destruction.

A **native** species is a species that has existed in that country at least from prehistoric times; a **naturalized** species is one known to have been introduced by humans from another country, but which now maintains itself; while an **exotic** species is one that requires human intervention to survive.

specific gravity alternative term for ◊relative density.

specific heat capacity in physics, quantity of heat required to raise unit mass (1 kg) of a substance by one kelvin (1°C) (see ◊kelvin scale). The unit of specific heat capacity in the SI system is the ◊joule per kilogram kelvin (J kg^{-1} K^{-1}).

specific latent heat in physics, the heat that changes the physical state of a unit mass (one kilogram) of a substance without causing any temperature change.

The *specific latent heat of fusion* of a solid substance is the heat required to change one kilogram of it from solid to liquid without any temperature change. The *specific latent heat of vaporization* of a liquid substance is the heat required to change one kilogram of it from liquid to vapour without any temperature change.

speckle interferometry technique whereby large telescopes can achieve high resolution of astronomical objects despite the adverse effects of the atmosphere through which light from the object under study must pass. It involves the taking of large numbers of images, each under high magnification and with short exposure times. The pictures are then combined to form the final picture. The technique was introduced by the French astronomer Antoine Labeyrie 1970.

spectacles pair of lenses fitted in a frame and worn in front of the ◊eyes to correct or assist defective vision. Common defects of the eye corrected by spectacle lenses are short-sightedness, or myopia (corrected by concave spherical lenses), long-sightedness, or hypermetropia (corrected by convex spherical lenses), and astigmatism (corrected by cylindrical lenses).

Spherical and cylindrical lenses may be combined in one lens. Bifocal spectacles correct vision both at a distance and for reading by combining two lenses of different curvatures in one piece of glass.

Spectacles are said to have been invented in the 13th century by a Florentine monk. Few people found the need for spectacles until printing was invented, when the demand for them increased rapidly. It is not known when spectacles were introduced into England, but in 1629 Charles I granted a charter to the Spectacle Makers' Guild. Using photosensitive glass, lenses can be produced that darken in glare and return to normal in ordinary light conditions. Lightweight plastic lenses are also common. The alternative to spectacles is contact lenses.

spectator ion in a chemical reaction that takes place in solution, an ion that remains in solution without taking part in the chemical change. For example, in the precipitation of barium sulphate from barium chloride and sodium sulphate, the sodium and chloride ions are spectator ions.

$$BaCl_{2\ (aq)} + Na_2SO_{4\ (aq)} = BaSO_{4\ (s)} + 2NaCl_{(aq)}$$

spectrometer in physics and astronomy, an instrument used to study the composition of light emitted by a source. The range, or ◊spectrum, of wavelengths emitted by a source depends upon its constituent elements, and may be used to determine its chemical composition.

The simpler forms of spectrometer analyse only visible light. A *collimator* receives the incoming rays and produces a parallel beam, which is then split into a spectrum by either a ◊diffraction grating or a prism mounted on a turntable. As the turntable is rotated each of the constituent colours of the beam may be seen through a *telescope*, and the angle at which each has been deviated may be measured on a circular scale. From this information the wavelengths of the colours of light can be calculated.

Spectrometers are used in astronomy to study the electromagnetic radiation emitted by stars and other celestial bodies. The spectral information gained may be used to determine their chemical composition, or to measure the ◊red shift of colours associated with the expansion of the universe and thereby calculate the speed with which distant stars are moving away from the Earth.

spectroscopy study of spectra (see ◊spectrum) associated with atoms or molecules in solid, liquid, or gaseous phase. Spectroscopy can be used to identify unknown compounds and is an invaluable tool in science, medicine, and industry (for example, in checking the purity of drugs).

Emission spectroscopy is the study of the characteristic series of sharp lines in the spectrum produced when an ◊element is heated. Thus an unknown mixture can be analysed for its component elements. Related is *absorption spectroscopy*, dealing with atoms and molecules as they absorb energy in a characteristic way. Again, dark lines can be used for analysis. More detailed structural information can be obtained using *infrared spectroscopy* (concerned with molecular vibrations) or *nuclear magnetic resonance (NMR) spectroscopy* (concerned with interactions between adjacent atomic nuclei).

spectrum (plural *spectra*) in physics, an arrangement of frequencies or wavelengths when electromagnetic radiations are separated into their constituent parts. Visible light is part of the ◊electromagnetic spectrum and most sources emit waves over a range of wavelengths that can be broken up or 'dispersed'; white light can be separated into red, orange, yellow, green, blue, indigo, and violet. The visible spectrum was first studied by English physicist Isaac Newton, who showed in 1672 how white light could be broken up into different colours.

There are many types of spectra, both emission and absorption, for radiation and particles, used in ◊spectroscopy. An incandescent body gives rise to a *continuous spectrum* where the dispersed radiation is distributed uninterruptedly over a range of wavelengths. An element gives a *line spectrum*—one or more bright discrete lines at characteristic wavelengths. Molecular gases give *band spectra* in which there are groups of close-packed lines shaded in one direction of wavelength. In an *absorption spectrum* dark lines or spaces replace the characteristic bright lines of the absorbing medium. The *mass spectrum* of an element is obtained from a mass spectrometer and shows the relative proportions of its constituent ◊isotopes.

speech recognition or *voice input* in computing, any technique by which a computer can understand ordinary speech. Spoken words are divided into 'frames', each lasting about one-thirtieth of a second, which are converted to a wave form. These are then compared with a series of stored frames to determine the most likely word. Research into speech recognition started in 1938, but the technology did not become sufficiently developed for commercial applications until the late 1980s.

There are three types: *separate word recognition* for distinguishing up to several hundred separately spoken words; *connected speech recognition* for speech in which there is a short pause between words; and *continuous speech recognition* for normal but carefully articulated speech.

speech synthesis or *voice output* computer-based technology for generating speech. A speech synthesizer is controlled by a computer, which supplies strings of codes representing basic speech sounds (phonemes); together these make up words. Speech-synthesis applications include children's toys, car and aircraft warning systems, and talking books for the blind.

speed the rate at which an object moves. The average speed v of an object may be calculated by dividing the distance s it has travelled by the time t taken to do so, and may be expressed as:

$$v = s/t$$

The usual units of speed are metres per second or kilometres per hour.

Speed is a scalar quantity in which direction of motion is unimportant (unlike the vector quantity ◊velocity, in which both magnitude and direction must be taken into consideration).

speed of light speed at which light and other ◊electromagnetic waves travel through empty space. Its value is 299,792,458 m/186,281 mi per second. The speed of light is the highest speed possible, according to the theory of ◊relativity, and its value is independent of the motion of its source and of the observer. It is impossible to accelerate any material body to this speed because it would require an infinite amount of energy.

SPEED OF LIGHT: QUICK AS A FLASH

Light takes: 0.14 seconds to travel around the world; 1.25 seconds to travel from the Moon to the Earth; 8 minutes 27 seconds to travel from the Sun; 6 hours to travel from Pluto; 4.3 years to travel from the nearest star, Proxima Centauri; 75,000 years to travel from the most distant stars in our galaxy; 160,000 years to travel from the nearest small galaxy, the Larger Magellanic Cloud; 2,200,000 years to travel from the nearest large galaxy, the Andromeda galaxy; 13 billion years to travel from quasars, the most distant objects in the Universe.

---

speed of sound speed at which sound travels through a medium, such as air or water. In air at a temperature of 0°C/32°F, the speed of sound is 331 m/1,087 ft per second. At higher temperatures, the speed of sound is greater; at 18°C/64°F it is 342 m/1,123 ft per second. It is greater in liquids and solids; for example, in water it is around 1,440 m/4,724 ft per second, depending on the temperature.

speleology scientific study of caves, their origin, development, physical structure, flora, fauna, folklore, exploration, mapping, photography, cave-diving, and rescue work. *Potholing*, which involves following the course of underground rivers or streams, has become a popular sport.

Speleology first developed in France in the late 19th century, where the Société de Spéléologie was founded in 1895.

sperm or *spermatozoon* in biology, the male ◊gamete of animals. Each sperm cell has a head capsule containing a nucleus, a middle portion containing ◊mitochondria (which provide energy), and a long tail (flagellum). See ◊sexual reproduction.

In most animals, the sperm are motile, and are propelled by a long flagellum, but in some (such as crabs and lobsters) they are nonmotile. Sperm cells are produced in the testes (see ◊testis). From there they pass through the sperm ducts via the seminal vesicles and the ◊prostate gland, which produce fluids called semen, that give the sperm cells energy and keep them moving after they leave the body. Hundreds of millions of sperm cells are contained in only a small amount of semen.

The term is sometimes applied to the motile male gametes (◊antherozoids) of lower plants.

SPERM: ENOUGH TO POPULATE THE WORLD?

A human male produces 1,000 sperm cells each second, or 86 million each day. This is enough to populate the whole world in less than two months.

---

spermatophore small capsule containing ◊sperm and other nutrients produced in invertebrates, newts, and cephalopods.

spermatophyte in botany, another name for a ◊seed plant.

spermicide any cream, jelly, pessary, or other preparation that kills the ◊sperm cells in semen. Spermicides are used for contraceptive purposes, usually in combination with a ◊condom or ◊diaphragm. Sponges impregnated with spermicide have been developed but are not yet in widespread use. Spermicide used alone is only 75% effective in preventing pregnancy.

sphalerite the chief ore of zinc, composed of zinc sulphide with a small proportion of iron, formula (Zn,Fe)S. It is brown with a nonmetallic lustre unless an appreciable amount of iron is present (up to 26% by weight). Sphalerite usually occurs in ore veins in limestones, where it is often associated with galena. It crystallizes in the cubic system but does not normally form perfect cubes.

sphere in mathematics, a perfectly round object with all points on its surface the same distance from the centre. This distance is the radius of the sphere. For a sphere of radius r, the volume $V = \frac{4}{3}\pi r^3$ and the surface area $A = 4\pi r^2$.

sphincter ring of muscle found at various points in the alimentary canal, which contracts and relaxes to control the movement of food. The *pyloric*

As swift as light

The speed of light in a vacuum, c, is a fundamental constant of nature. It occurs in many scientific equations, such as Einstein's famous $E = mc^2$. Numerous attempts have been made to determine c as accurately as possible.

First estimates

The first reasonable estimate was made around 1675 using observations by a young Danish astronomer, Ole Roemer, at the Royal Observatory in Paris. Roemer was studying the motion of Jupiter's moons, which take between 1.7 and 16.7 days to orbit Jupiter. The orbital time of a moon can be measured by timing its eclipse by Jupiter's shadow. Roemer noticed that the interval between successive eclipses gradually decreased when Jupiter and the Earth were approaching one another and increased when they were receding. He explained this by assuming that light travels through space at a finite though very great speed. Thus if Jupiter is approaching the Earth, the time the light from a moon takes to reach the Earth will gradually decrease as Jupiter and the Earth get closer, and the interval between eclipses will appear to decrease. Roemer was content to demonstrate that the speed of light is finite, but other investigators used his observations to calculate that the speed of light must be about 300,000 km/ 186,400 mi per second.

In 1728, British Astronomer Royal James Bradley (1693–1762) made another estimate using a different approach. Tradition has it that Bradley devised his method while sailing on the River Thames. He noticed how the apparent wind direction depended on the boat's movement and the windspeed. Bradley reasoned that if starlight was regarded as a wind blowing earthwards, its direction should appear to change throughout the year as the Earth moved round the Sun. By observing the change he was able to calculate and verify the speed of light.

Laboratory measurements

The first laboratory measurement was made by French physicist Armand Fizeau in 1849. He used a bright beam of light which travelled to a distant mirror and was reflected back along its path. Using a partly silvered mirror, he could look along the light path and see the reflected beam. A large square-toothed wheel in front of the mirror cut the beam into a sequence of flashes as it turned. The speed of rotation could be carefully controlled. With the wheel spinning, Fizeau could only see a flash of light if it passed through the gap between two teeth. He increased the speed of the wheel until the first gap coincided with the returning flash. It was then simple, using the known rotation speed and the dimensions of the teeth, to calculate how long it had taken the tooth to move out of the beam path. Fizeau's result, for the speed of light was 313,300 km/194,700 mi per second.

A year later another French physicist, Leon Foucault, improved on Fizeau's experiment. A beam of light was reflected by a rotating mirror, travelled to a distant mirror and then returned to the rotating mirror. The rotating mirror had turned a little, so the returned beam appeared to be in a slightly different position from when it set out. This change could be measured accurately with a micrometer and used to calculate the rotation speed of the mirror more accurately than in Fizeau's experiment. Foucault used a baseline of only 20 m/65 ft, but achieved a slightly more accurate result: 298,000 km/ 185,200 mi per second.

All done by mirrors

In 1926 US physicist Albert Michelson (1852–1931) refined Foucault's experiment. He used a much longer baseline, starting at the Mount Wilson Observatory and travelling to a mirror on nearby Mount San Antonio. The baseline was measured precisely by the US Coast and Geodetic Survey as 35,385.53 m/ 116,092.84 ft, accurate to within 3 mm/ 0.1 in. Michelson used a multi-sided octagonal mirror, which could be spun at 550 revolutions per second. The exact speed was measured by stroboscopic comparison with an electrically driven tuning fork.

Light from an electric arc passed through a narrow slit and was reflected from one facet of the rotating mirror. The beam travelled to the Mount San Antonio station, was reflected from a flat mirror and returned to Mount Wilson. For the reflection to be observed, the octagonal mirror had to have made one-eighth of a rotation while the light travelled along the baseline and returned. The round trip took 0.00023 seconds when the mirror spun at 528 revolutions per second. The calculated speed of light, after correction for slowing of the beam by air along its path, was 299,863 km/186,335 mi per second. In 1929 Michelson repeated the experiment inside a long excavated tube, to eliminate the uncertainty caused by variations in air density. His result was $c =$ 299,774 km/186,280 mi per second: within 0.006% of the modern value.

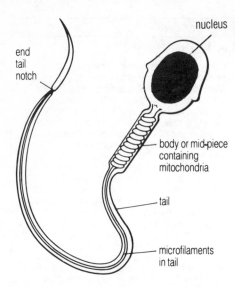

sperm *Only a single sperm is needed to fertilize a female egg, or ovum. Yet up to 500 million may start the journey towards the egg. Once a sperm has fertilized an egg, the egg's wall cannot be penetrated by other sperm. The unsuccessful sperm die after about three days.*

sphincter, at the base of the stomach, controls the release of the gastric contents into the duodenum. After release the sphincter contracts, closing off the stomach.

Spica or *Alpha Virginis* the brightest star in the constellation Virgo and the 16th brightest star in the sky. Spica has a true luminosity of over 1,500 times that of the Sun, and is 140 light years from Earth. It is also a spectroscopic ◊binary star, the components of which orbit each other every four days.

spicules, solar in astronomy, short-lived jets of hot gas in the upper ◊chromosphere of the Sun. Spiky in appearance, they move at high velocities along lines of magnetic force to which they owe their shape, and last for a few minutes each. Spicules appear to disperse material into the ◊corona.

spider any arachnid (eight-legged animal) of the order Araneae. There are about 30,000 known species. Unlike insects, the head and breast are merged to form the cephalothorax, connected to the abdomen by a characteristic narrow waist. There are eight legs, and usually eight simple eyes. On the undersurface of the abdomen are spinnerets, usually six, which exude a viscid fluid. This hardens on exposure to the air to form silky threads, used to make silken egg cases, silk-lined tunnels, or various kinds of webs and snares for catching prey that is then wrapped. The fangs of spiders inject substances to subdue and digest prey, the juices of which are then sucked into the stomach by the spider.

Spiders are found everywhere in the world except Antarctica. Species of interest include the zebra spider *Salticus scenicus*, a longer-sighted species which stalks its prey and has pads on its feet which enable it to walk even on glass; the poisonous tarantula and black widow; the cross spider *Araneus diadematus*, which spins webs of remarkable beauty; the only aquatic species of spider, the water spider *Argyroneta aquatica*, which fills a 'diving bell' home with air trapped on the hairs of the body; and the largest members of the group, the **bird-eating spider** genus *Mygale* of South America, with a body about 6–9 cm/2.4–3.5 in long and a leg-span of 30 cm/1 ft. Spider venom is a powerful toxin which paralyses its prey.

In Britain there are over 600 species of spider. One of the most familiar is the common garden spider *Araneus diadematus*, which spins webs of remarkable beauty. There are three species of house spider: *Tegenaria domestica*, *T. atrica*, both up to 2 cm/0.75 in long, and *T. parietina*, better known as the cardinal spider.

spikelet in botany, one of the units of a grass ◊inflorescence. It comprises a slender axis on which one or more flowers are borne.

Each individual flower or floret has a pair of scalelike bracts, the glumes, and is enclosed by a membranous lemma and a thin, narrow palea, which may be extended into a long, slender bristle, or **awn**.

spin in physics, the intrinsic ◊angular momentum of a subatomic particle, nucleus, atom, or molecule, which continues to exist even when the particle comes to rest. A particle in a specific energy state has a particular spin, just as it has a particular electric charge and mass. According to ◊quantum theory, this is restricted to discrete and indivisible values, specified by a spin ◊quantum number. Because of its spin, a charged particle acts as a small magnet and is affected by magnetic fields.

spinal cord major component of the ◊central nervous system in vertebrates.

It is enclosed by the bones of the ◊spine, and links the peripheral nervous system to the brain.

spine backbone of vertebrates. In most mammals, it contains 26 small bones called vertebrae, which enclose and protect the spinal cord (which links the peripheral nervous system to the brain). The spine connects with the skull, ribs, back muscles, and pelvis.

In humans, there are seven cervical **vertebrae**, in the neck; 12 thoracic, in the upper trunk; five lumbar, in the lower back; the sacrum (consisting of five rudimentary vertebrae fused together, joined to the hipbones); and the coccyx (four vertebrae, fused into a tailbone). The human spine has four curves (front to rear), which allow for the increased size of the chest and pelvic cavities, and permit springing, to minimize jolting of the internal organs.

spinel any of a group of 'mixed oxide' minerals consisting mainly of the oxides of magnesium and aluminium, $MgAl_2O_4$ and $FeAl_2O_4$. Spinels crystallize in the cubic system, forming octahedral crystals. They are found in high-temperature igneous and metamorphic rocks. The aluminium oxide spinel contains gem varieties, such as the ruby spinels of Sri Lanka and Myanmar (Burma).

spiracle in insects, the opening of a ◊trachea, through which oxygen enters the body and carbon dioxide is expelled. In cartilaginous fishes (sharks

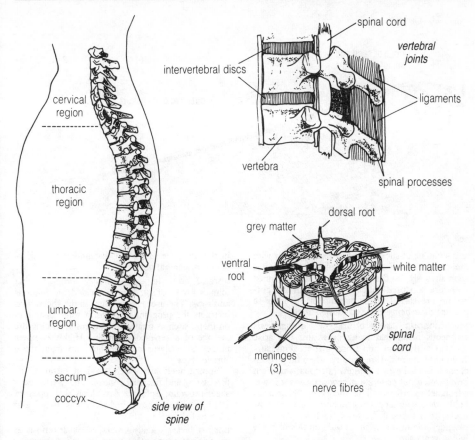

cervical
region

thoracic
region

lumbar
region

sacrum

coccyx

*side view of
spine*

spinal cord

*vertebral
joints*

intervertebral discs

ligaments

vertebra

spinal processes

dorsal root

grey matter

ventral
root

white matter

*spinal
cord*

meninges
(3)

nerve fibres

spine *The human spine extends every night during sleep. During the day, the cartilage discs between the vertebra are squeezed when the body is in a vertical position, standing or sitting, but at night, with pressure released, the discs swell and the spine lengthens by about 8 mm/0.3 in.*

and rays), the same name is given to a circular opening that marks the remains of the first gill slit.

In tetrapod vertebrates, the spiracle of early fishes has evolved into the Eustachian tube, which connects the middle ear cavity with the pharynx.

spiral a plane curve formed by a point winding round a fixed point from which it distances itself at regular intervals, for example the spiral traced by a flat coil of rope. Various kinds of spirals can be generated mathematically—for example, an equiangular or logarithmic spiral (in which a tangent at any point on the curve always makes the same angle with it) and an ◊involute. Spirals also occur in nature as a normal consequence of accelerating growth, such as the spiral shape of the shells of snails and some other molluscs.

spirits of salts former name for ◊hydrochloric acid.

spirochaete spiral-shaped bacterium. Some spirochaetes are free-living in water, others inhabit the intestines and genital areas of animals. The sexually transmitted disease syphilis is caused by a spirochaete.

spit ridge of sand or shingle projecting from the land into a body of water. It is deposited by waves

carrying material from one direction to another across the mouth of an inlet (◊longshore drift). Deposition in the brackish water behind a spit may result in the formation of a ◊salt marsh.

spleen organ in vertebrates, part of the lymphatic system, which helps to process ◊lymphocytes. It also regulates the number of red blood cells in circulation by destroying old cells, and stores iron. It is situated behind the stomach.

sponge any saclike simple invertebrate of the phylum Porifera, usually marine. A sponge has a hollow body, its cavity lined by cells bearing flagellae, whose whiplike movements keep water circulating, bringing in a stream of food particles. The body walls are strengthened with protein (as in the bath sponge) or small spikes of silica, or a framework of calcium carbonate.

spontaneous combustion burning that is not initiated by the direct application of an external source of heat. A number of materials and chemicals, such as hay and sodium chlorate, can react with their surroundings, usually by oxidation, to produce so much internal heat that combustion results.

Special precautions must be taken for the storage

Blakeney spit, Norfolk

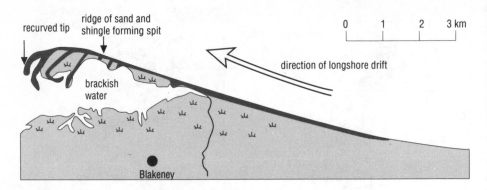

spit

and handling of substances that react violently with moisture or air. For example, phosphorus ignites spontaneously in the presence of air and must therefore be stored under water; sodium and potassium are stored under paraffin oil in order to prevent their being exposed to moisture.

spontaneous generation or *abiogenesis* erroneous belief that living organisms can arise spontaneously from non-living matter. This survived until the mid-19th century, when the French chemist Louis Pasteur demonstrated that a nutrient broth would not generate microorganisms if it was adequately sterilized. The theory of ◊biogenesis holds that spontaneous generation cannot now occur; it is thought, however, to have played an essential role in the origin of ◊life on this planet 4 billion years ago.

spooling in computing, the process in which information to be printed is stored temporarily in a file, the printing being carried out later. It is used to prevent a relatively slow printer from holding up the system at critical times, and to enable several computers or programs to share one printer.

sporangium structure in which ◊spores are produced.

spore small reproductive or resting body, usually consisting of just one cell. Unlike a ◊gamete, it does not need to fuse with another cell in order to develop into a new organism. Spores are produced by the lower plants, most fungi, some bacteria, and certain protozoa. They are generally light and easily dispersed by wind movements.

Plant spores are haploid and are produced by the sporophyte, following ◊meiosis; see ◊alternation of generations.

SHOOTING SPORES

Mushrooms tend to shoot out their spores at around mid-afternoon on dry days. The process involves static electricity, which builds up on the spores and rips them off their anchorages, launching them into the air.

sporophyte diploid spore-producing generation

in the life cycle of a plant that undergoes ◊alternation of generations.

spreadsheet in computing, a program that mimics a sheet of ruled paper, divided into columns and rows. The user enters values in the sheet, then instructs the program to perform some operation on them, such as totalling a column or finding the average of a series of numbers. Highly complex numerical analyses may be built up from these simple steps.

Spreadsheets are widely used in business for forecasting and financial control. The first spreadsheet program, Software Arts' VisiCalc, appeared 1979. The best known include Lotus 1–2–3 and Microsoft Excel.

spring device, usually a metal coil, that returns to its original shape after being stretched or compressed. Springs are used in some machines (such as clocks) to store energy, which can be released at a controlled rate. In other machines (such as engines) they are used to close valves.

In vehicle-suspension systems, springs are used to cushion passengers from road shocks. These springs are used in conjunction with ◊shock absorbers to limit their amount of travel. In bedding and upholstered furniture springs add comfort.

spring in geology, a natural flow of water from the ground, formed where the water table meets the ground's surface. The source of water is rain that has percolated through the overlying rocks. During its underground passage, the water may have dissolved mineral substances, which may then be precipitated at the spring (hence, a mineral spring).

A spring may be continuous or intermittent, and depends on the position of the water table and the topography (surface features).

sprite in computing, a graphics object made up of a pattern of ◊pixels (picture elements) defined by a computer programmer. Some ◊high-level languages and ◊applications programs contain routines that allow a user to define the shape, colours, and other characteristics of individual graphics objects. These objects can then be manipulated and combined to produce animated games or graphic screen displays.

	A	B	C	D	E	
			═══ COMPUTER SUPPLIES ORDER (SS) ═══			
1	DESCRIPTION		NUMBER	UNIT COST(£)	COST	
2						
3	FLOPPY DISCS		50	0.55	27.5	
4	INK RIBBONS		10	2.25	22.5	
5	INK JET CARTRIDGES		5	12.5	62.5	
6	PAPER (BOX 500 SHEETS)		12	3.75	45	
7						
8				TOTAL COST	£ 157.50	
9						
10						

spreadsheet

spur ridge of rock jutting out into a valley or plain. In mountainous areas rivers often flow around ◊interlocking spurs because they are not powerful enough to erode through the spurs. Spurs may be eroded away by large and powerful glaciers to form ◊truncated spurs.

Sputnik (Russian 'fellow traveller') series of ten Soviet Earth-orbiting satellites. *Sputnik 1* was the first artificial satellite, launched 4 Oct 1957. It weighed 84 kg/185 lb, with a 58 cm/23 in diameter, and carried only a simple radio transmitter which allowed scientists to track it as it orbited Earth. It burned up in the atmosphere 92 days later. Sputniks were superseded in the early 1960s by the Cosmos series.

Sputnik 2, launched 3 Nov 1957, weighed about 500 kg/1,100 lb including the dog Laika, the first living creature in space. Unfortunately, there was no way to return the dog to Earth, and it died in space. Later Sputniks were test flights of the ◊Vostok spacecraft.

sq abbreviation for *square* (measure).

SQL (abbreviation of *structured query language*) high-level computer language designed for use with ◊relational databases. Although it can be used by programmers in the same way as other languages, it is often used as a means for programs to communicate with each other. Typically, one program (called the 'client') uses SQL to request data from a database 'server'.

Although originally developed by IBM, SQL is now widely used on many types of computer.

square in geometry, a quadrilateral (four-sided) plane figure with all sides equal and each angle a right angle. Its diagonals bisect each other at right angles. The area A of a square is the length l of one side multiplied by itself ($A = l \times l$).

Also, any quantity multiplied by itself is also termed a square, represented by an ◊exponent of power 2; for example, $4 \times 4 = 4^2 = 16$ and $6.8 \times 6.8 = 6.8^2 = 46.24$.

An algebraic term is squared by doubling its exponent and squaring its coefficient if it has one; for example, $(x^2)2 = x^4$ and $(6y^3)^2 = 36y^6$. A number that has a whole number as its ◊square root is known as a *perfect square*; for example, 25, 144

and 54,756 are perfect squares (with roots of 5, 12 and 234, respectively).

square root in mathematics, a number that when squared (multiplied by itself) equals a given number. For example, the square root of 25 (written $\sqrt{25}$) is ± 5, because $5 \times 5 = 25$, and $(-5) \times (-5) = 25$. As an ◊exponent, a square root is represented by ½, for example, $16^{1/2} = 4$.

Negative numbers (less than 0) do not have square roots that are ◊real numbers. Their roots are represented by ◊complex numbers, in which the square root of -1 is given the symbol i (that is, $\pm i^2 = -1$). Thus the square root of -4 is $\sqrt{[(-1) \times 4]} = \sqrt{-1} \times \sqrt{4} = 2i$.

SRAM (acronym for *s*tatic *r*andom-*a*ccess *m*emory) computer memory device in the form of a silicon chip used to provide ◊immediate access memory. SRAM is faster but more expensive than ◊DRAM (dynamic random-access memory).

DRAM loses its contents unless they are read and rewritten every 2 milliseconds or so. This process is called *refreshing* the memory. SRAM does not require such frequent refreshing.

SSSI abbreviation for ◊*site of special scientific interest*.

stability measure of how difficult it is to move an object from a position of balance or ◊equilibrium with respect to gravity.

An object displaced from equilibrium does not remain in its new position if its weight, acting vertically downwards through its ◊centre of mass, no longer passes through the line of action of the contact force (the force exerted by the surface on which the object is resting), acting vertically upwards through the object's new base. If the lines of action of these two opposite but equal forces do not coincide they will form a couple and create a moment (see ◊moment of a force) that will cause the object either to return to its original rest position or to topple over into another position.

An object in *stable equilibrium* returns to its rest position after being displaced slightly. This form of equilibrium is found in objects that are difficult to topple over; these usually possess a relatively wide base and a low centre of mass—for example, a cone resting on its flat base on a

stable equilibrium

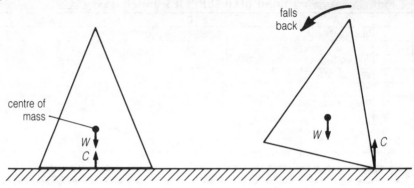

unstable equilibrium

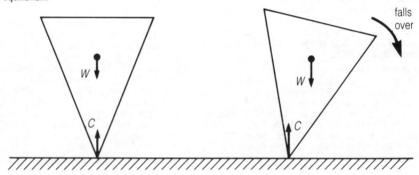

neutral equilibrium

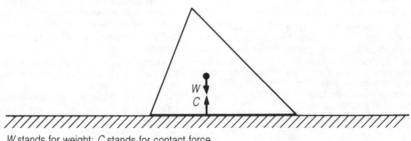

W stands for weight; *C* stands for contact force

stability

horizontal surface. When such an object is tilted slightly its centre of mass is raised and the line of action of its weight no longer coincides with that of the contact force exerted by its new, smaller base area. The moment created will tend to lower the centre of mass and so the cone will fall back to its original position.

An object in **unstable equilibrium** does not remain at rest if displaced, but falls into a new position; it does not return to its original rest posi-

tion. Objects possessing this form of equilibrium are easily toppled and usually have a relatively small base and a high centre of mass—for example, a cone balancing on its point, or apex, on a horizontal surface. When an object such as this is given the slightest push its centre of mass is lowered and the displacement of the line of action of its weight creates a moment. The moment will tend to lower the centre of mass still further and so the object will fall on to another position.

An object in *neutral equilibrium* stays at rest if it is moved into a new position—neither moving back to its original position nor on any further. This form of equilibrium is found in objects that are able to roll, such as a cone resting on its curved side placed on a horizontal surface. When such an object is rolled its centre of mass remains in the same position, neither rising nor falling, and the line of action of its weight continues to coincide with the contact force; no moment is created and so its equilibrium is maintained.

stabilizer one of a pair of fins fitted to the sides of a ship, especially one governed automatically by a ◊gyroscope mechanism, designed to reduce side-to-side rolling of the ship in rough weather.

stack in computing, a method of storing data in which the most recent item stored will be the first to be retrieved. The technique is commonly called 'last in, first out'.

Stacks are used to solve problems involving nested structures; for example, to analyse an arithmetical expression containing subexpressions in parentheses, or to work out a route between two points when there are many different paths.

stack in earth science, an isolated pillar of rock that has become separated from a headland by ◊coastal erosion. It is usually formed by the collapse of an arch. Further erosion will reduce it to a stump, which is exposed only at low tide.

Examples of stacks in the UK are the Needles, off the Isle of Wight, which are formed of chalk.

stain in chemistry, a coloured compound that will bind to other substances. Stains are used extensively in microbiology to colour microorganisms and in histochemistry to detect the presence and whereabouts in plant and animal tissue of substances such as fats, cellulose, and proteins.

stainless steel widely used ◊alloy of iron, chromium, and nickel that resists rusting. Its chromium content also gives it a high tensile strength. It is used for cutlery and kitchen fittings. Stainless steel was first produced in the UK 1913 and in Germany 1914.

stalactite and stalagmite cave structures formed by the deposition of calcite dissolved in ground water.

Stalactites grow downwards from the roofs or walls and can be icicle-shaped, straw-shaped, curtain-shaped, or formed as terraces.

Stalagmites grow upwards from the cave floor and can be conical, fir-cone-shaped, or resemble a stack of saucers. Growing stalactites and stalagmites may meet to form a continuous column from floor to ceiling.

Stalactites are formed when ground water, hanging as a drip, loses a proportion of its carbon dioxide into the air of the cave. This reduces the amount of calcite that can be held in solution, and a small trace of calcite is deposited. Successive drips build up the stalactite over many years. In stalagmite formation the calcite comes out of the solution because of agitation—the shock of a drop of water hitting the floor is sufficient to remove some calcite from the drop. The different shapes result from the splashing of the falling water.

stamen male reproductive organ of a flower. The

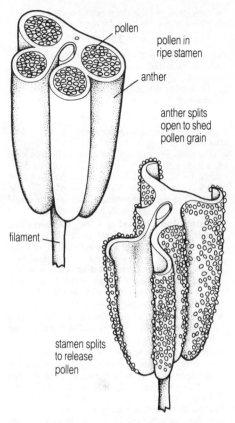

stamen *The stamen is the male reproductive organ of a flower. It has a thin stalk called a filament with an anther at the tip. The anther contains pollen sacs, which split to release tiny grains of pollen.*

stamens are collectively referred to as the ◊androecium. A typical stamen consists of a stalk, or filament, with an anther, the pollen-bearing organ, at its apex, but in some primitive plants, such as *Magnolia*, the stamen may not be markedly differentiated.

The number and position of the stamens are significant in the classification of flowering plants. Generally the more advanced plant families have fewer stamens, but they are often positioned more effectively so that the likelihood of successful pollination is not reduced.

stand-alone computer self-contained computer, usually a microcomputer, that is not connected to a network of computers and can be used in isolation from any other device.

standard atmosphere alternative term for ◊atmosphere, a unit of pressure.

standard deviation in statistics, a measure (symbol σ or s) of the spread of data. The deviation (difference) of each of the data items from the mean is found, and their values squared. The mean value of these squares is then calculated. The standard deviation is the square root of this mean.

If n is the number of items of data, x is the value

of each item, and $\bar{x}$ is the mean value, the standard deviation σ may be given by the formula:

$$\sigma = \sqrt{[\Sigma(x - \bar{x})^2/n]}$$

where Σ indicates that the differences between the value of each item of data and the mean should be summed.

To simplify the calculations, the formula may be rearranged to:

$$\sigma = \sqrt{[\Sigma x^2/n - \bar{x}^2]}$$

As a result, it becomes necessary only to calculate $\bar{x}$ and Σx^2.

For example, if the ages of a set of children were 4, 4.5, 5, 5.5, 6, 7, 9, and 11, Σx would be 52, $\bar{x}$ would be $52/n = 52/8 = 6.5$, and Σx^2 would be 378.5 ($= 4^2 + 4.5^2 + 5^2 + 5.5^2 + 6^2 + 7^2 + 9^2 + 11^2$). Therefore, the standard deviation σ would be $\sqrt{[^{378.5}/_8 - (6.5)^2]} = \sqrt{5.0625} = 2.25$.

standard form method of writing numbers often used by scientists, particularly for very large or very small numbers. The numbers are written with one digit before the decimal point and multiplied by a power of 10. The number of digits given after the decimal point depends on the accuracy required. For example, the ◊speed of light is 2.9979×10^8 m/1.8628×10^5 mi per second.

standard gravity acceleration due to gravity, generally taken as 9.81274 m/32.38204 ft per second per second. See ◊g scale.

standard illuminant any of three standard light intensities, A, B, and C, used for illumination when phenomena involving colour are measured. A is the light from a filament at 2,848K (2,575°C/4,667°F), B is noon sunlight, and C is normal daylight. B and C are defined with respect to A. Standardization is necessary because colours appear different when viewed in different lights.

standard model in physics, the modern theory of ◊elementary particles and their interactions. According to the standard model, elementary particles are classified as leptons (light particles, such as electrons), ◊hadrons (particles, such as neutrons and protons, that are formed from quarks), and gauge bosons. Leptons and hadrons interact by exchanging ◊gauge bosons, each of which is responsible for a different fundamental force: photons mediate the electromagnetic force, which affects all charged particles; gluons mediate the strong nuclear force, which affects quarks; gravitons mediate the force of gravity; and the weakons (intermediate vector bosons) mediate the weak nuclear force. See also ◊forces, fundamental, ◊quantum electrodynamics, and ◊quantum chromodynamics.

standard temperature and pressure (STP) in chemistry, a standard set of conditions for experimental measurements, to enable comparisons to be made between sets of results. Standard temperature is 0°C and standard pressure 1 atmosphere (101,325 Pa).

standard volume in physics, the volume occupied by one kilogram molecule (the molecular mass in kilograms) of any gas at standard temperature and pressure. Its value is approx 22.414 cubic metres.

standing crop in ecology, the total number of individuals of a given species alive in a particular area at any moment. It is sometimes measured as the weight (or ◊biomass) of a given species in a sample section.

standing wave in physics, a wave in which the positions of ◊nodes (positions of zero vibration) and antinodes (positions of maximum vibration) do not move. Standing waves result when two similar waves travel in opposite directions through the same space.

For example, when a sound wave is reflected back along its own path, as when a stretched string is plucked, a standing wave is formed. In this case the antinode remains fixed at the centre and the nodes are at the two ends. Water and ◊electromagnetic waves can form standing waves in the same way.

staphylococcus spherical bacterium that occurs in clusters. It is found on the skin and mucous membranes of humans and other animals. It can cause abscesses and systemic infections that may prove fatal.

star luminous globe of gas, mainly hydrogen and helium, which produces its own heat and light by nuclear reactions. Although stars shine for a very long time—many billions of years—they are not eternal, and have been found to change in appearance at different stages in their lives.

Stars are born when nebulae (giant clouds of dust and gas) contract under the influence of gravity. As each new star contracts, the temperature and pressure in its core rises. At about 10 million°C the temperature is hot enough for a nuclear reaction to begin (the fusion of hydrogen nuclei to form helium nuclei); vast amounts of energy are released, contraction stops, and the star begins to shine. Stars at this stage are called **main-sequence stars**—the Sun is such a star and is expected to remain at this stage for the next 5 billion years. Their surface temperatures range from 2,000°C/3,600°F to above 30,000°C/54,000°F and the corresponding colours range from red to blue-white.

The smallest mass possible for a star is about 8% that of the Sun (80 times the mass of the planet Jupiter), otherwise nuclear reactions do not occur. Objects with less than this critical mass shine only dimly, and are termed **brown dwarfs**.

When all the hydrogen at the core of a main-sequence star has been converted into helium, the star swells to become a **red giant**, about 100 times its previous size and with a cooler, redder surface. When, after this brief stage, the star can produce no more nuclear energy, its outer layers drift off into space to form a planetary nebula, and its core collapses in on itself to form a small and very dense body called a **white dwarf**. Eventually the white dwarf fades away, leaving a non-luminous **dark body**.

Some very large main-sequence stars do not end their lives as white dwarves—they pass through their life cycle quickly, becoming red ◊supergiants that eventually explode into brilliant **supernovae**. Part of the core remaining after such an explosion may collapse to form a small superdense star, which consists almost entirely of neutrons and is therefore called a **neutron star**. Neutron stars, called

pulsars, spin very quickly, giving off pulses of radio waves (rather as a lighthouse gives off flashes of light). If the collapsing core of the supernova is very massive it does not form a neutron star; instead it forms a **black hole**, a region so dense that its gravity not only draws in all nearby matter but also all radiation, including its own light.

See also ◊binary star, ◊Hertzsprung–Russell diagram, and ◊variable star.

HOTTEST STAR

In April 1992 the Hubble Space Telescope discovered the hottest star ever seen. The star, which lies in the centre of the Large Magellanic Cloud, has a surface temperature of 200,000°C/360,000°F—33 times that of the Sun.

---

stars: the top twenty brightest stars

common name	scientific name	apparent magnitude
Sirius	α Canis Majoris	−1.46
Canopus	α Carinae	−0.72
Rigil Kent	α Centauri	−0.27*
Arcturus	α Boötis	−0.4
Vega	α Lyrae	+0.03
Capella	α Aurigae	0.08
Rigel	β Orionis	0.12
Procyon	α Canis Minoris	0.38
Achernar	α Eridani	0.46
Betelgeuse	α Orionis	0.50**
Hadar	β Centauri	0.61
Altair	α Aquilae	0.77
Acrux	α Crucis	0.79*
Aldebaran	α Tauri	0.85**
Antares	α Scorpii	0.96**
Spica	α Virginis	0.98
Pollux	β Geminorum	1.14
Fomalhaut	α Piscis Austrini	1.16
Deneb	α Cygni	1.25
Mimosa	β Crucis	1.25

* combined magnitude of double star
** variable

starch widely distributed, high-molecular-mass ◊carbohydrate, produced by plants as a food store; main dietary sources are cereals, legumes, and tubers, including potatoes. It consists of varying proportions of two ◊glucose polymers (◊polysaccharides): straight-chain (amylose) and branched (amylopectin) molecules.

Purified starch is a white powder used to stiffen textiles and paper and as a raw material for making various chemicals. It is used in the food industry as a thickening agent. Chemical treatment of starch gives rise to a range of 'modified starches' with varying properties. Hydrolysis (splitting) of starch by acid or enzymes generates a variety of 'glucose syrups' or 'liquid glucose' for use in the food industry. Complete hydrolysis of starch with acid generates the ◊monosaccharide glucose only. Incomplete hydrolysis or enzymic hydrolysis yields a mixture of glucose, maltose and non-hydrolysed fractions called 'dextrins'.

star cluster group of related stars, usually held together by gravity. Members of a star cluster are thought to form together from one large cloud of gas in space. **Open clusters** such as the Pleiades contain from a dozen to many hundreds of young stars, loosely scattered over several light years. ◊Globular clusters are larger and much more densely packed, containing perhaps 100,000 old stars.

state change in science, a change in the physical state (solid, liquid, or gas) of a material. For instance, melting, boiling, evaporation, and their opposites, solidification and condensation, are changes of state. The former set of changes are brought about by heating or decreased pressure; the latter by cooling or increased pressure.

These changes involve the absorption or release of heat energy, called ◊latent heat, even though the temperature of the material does not change during the transition between states.

In the unusual change of state called **sublimation**, a solid changes directly to a gas without passing through the liquid state. For example, solid carbon dioxide (dry ice) sublimes to carbon dioxide gas.

states of matter forms (solid, liquid, or gas) in which material can exist. Whether a material is solid, liquid, or gas depends on its temperature and the pressure on it. The transition between states takes place at definite temperatures, called melting point and boiling point.

◊Kinetic theory describes how the state of a material depends on the movement and arrangement of its atoms or molecules. A hot ionized gas or ◊plasma is often called the fourth state of matter, but liquid crystals, ◊colloids, and glass also have a claim to this title.

state symbol symbol used in chemical equations to indicate the physical state of the substances present. The symbols are: (s) for solid, (l) for liquid, (g) for gas, and (aq) for aqueous.

static electricity ◊electric charge that is stationary, usually acquired by a body by means of electrostatic induction or friction. Rubbing different materials can produce static electricity, as seen in the sparks produced on combing one's hair or removing a nylon shirt. In some processes static electricity is useful, as in paint spraying where the parts to be sprayed are charged with electricity of opposite polarity to that on the paint droplets, and in ◊xerography.

statics branch of mechanics concerned with the behaviour of bodies at rest and forces in equilibrium, and distinguished from ◊dynamics.

statistical mechanics branch of physics in which the properties of large collections of particles are predicted by considering the motions of the constituent particles.

statistics branch of mathematics concerned with the collection and interpretation of data. For example, to determine the ◊mean age of the children in a school, a statistically acceptable answer might be obtained by calculating an average based on the ages of a representative sample, consisting, for example, of a random tenth of the pupils from each class. ◊Probability is the branch of statistics dealing with predictions of events.

staurolite silicate mineral, $(Fe,Mg)_2(Al,Fe)_9$ $Si_4O_{20}(OH)_2$. It forms brown crystals that may

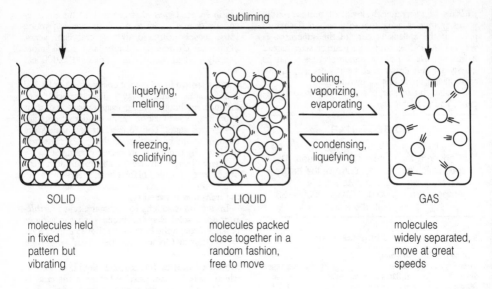

subliming

liquefying, melting

freezing, solidifying

boiling, vaporizing, evaporating

condensing, liquefying

SOLID

molecules held
in fixed
pattern but
vibrating

LIQUID

molecules packed
close together in a
random fashion,
free to move

GAS

molecules
widely separated,
move at great
speeds

state change

be twinned in the form of a cross. It is a useful indicator of medium grade (moderate temperature and pressure) in metamorphic rocks formed from clay sediments.

steady-state theory in astronomy, a rival theory to that of the ◊Big Bang, which claims that the universe has no origin but is expanding because new matter is being created continuously throughout the universe. The theory was proposed 1948 by Hermann Bondi, Thomas Gold (1920–   ), and Fred Hoyle, but was dealt a severe blow in 1965 by the discovery of ◊cosmic background radiation (radiation left over from the formation of the universe) and is now largely rejected.

steam in chemistry, a dry, invisible gas formed by vaporizing water. The visible cloud that normally forms in the air when water is vaporized is due to minute suspended water particles. Steam is widely used in chemical and other industrial processes and for the generation of power.

steam engine engine that uses the power of steam to produce useful work. It was the principal power source during the British Industrial Revolution in the 18th century. The first successful steam engine was built 1712 by English inventor Thomas Newcomen, and it was developed further by Scottish mining engineer James Watt from 1769 and by English mining engineer Richard Trevithick, whose high-pressure steam engine 1802 led to the development of the steam locomotive.

In Newcomen's engine, steam was admitted to a cylinder as a piston moved up, and was then condensed by a spray of water, allowing air pressure to force the piston downwards. James Watt improved Newcomen's engine in 1769 by condensing the steam outside the cylinder (thus saving energy formerly used to reheat the cylinder) and by using steam to force the piston upwards. Watt also introduced the **double-acting engine**, in which steam is alternately sent to each end of the cylinder. The **compound engine** (1781) uses the exhaust from one cylinder to drive the piston of another. A later development was the steam ◊turbine, still used today to power ships and generators in power stations. In other contexts, the steam engine was superseded by the ◊internal-combustion engine.

stearic acid $CH_3(CH_2)_{16}COOH$ saturated long-chain ◊fatty acid, soluble in alcohol and ether but not in water. It is found in many fats and oils, and is used to make soap and candles and as a lubricant. The salts of stearic acid are called stearates.

stearin mixture of stearic and palmitic acids, used to make soap.

steel alloy or mixture of iron and up to 1.7% carbon, sometimes with other elements, such as manganese, phosphorus, sulphur, and silicon. The USA, Russia, Ukraine, and Japan are the main steel producers. Steel has innumerable uses, including ship and automobile manufacture, sky-scraper frames, and machinery of all kinds.

Steels with only small amounts of other metals are called **carbon steels**. These steels are far stronger than pure iron, with properties varying with the composition. **Alloy steels** contain greater amounts of other metals. Low-alloy steels have less than 5% of the alloying material; high-alloy steels have more. Low-alloy steels containing up to 5% silicon with relatively little carbon have a high electrical resistance and are used in power transformers and motor or generator cores, for example. **Stainless steel** is a high-alloy steel containing at least 11% chromium. Steels with up to 20% tungsten are very hard and are used in high-speed cutting tools. About 50% of the world's steel is now made from scrap.

Steel is produced by removing impurities, such as carbon, from raw or pig iron, produced by a ◊blast furnace. The main industrial process is the ◊**basic–oxygen process**, in which molten pig iron and scrap steel is placed in a container lined with heat-resistant, alkaline (basic) bricks. A pipe or lance is lowered near to the surface of the molten metal and pure oxygen blown through it at high pressure. The surface of the metal is disturbed by the blast and the impurities are oxidized (burned out). The **open-hearth process** is an older steel-making method in which molten iron and limestone are placed in a shallow bowl or hearth (see ◊open-hearth furnace). Burning oil or gas is blown over the surface of the metal, and the impurities are oxidized. High-quality steel is made in an **electric furnace**. A large electric current flows through electrodes in the furnace, melting a charge of scrap steel and iron. The quality of the steel produced can be controlled precisely because the temperature of the furnace can be maintained exactly and there are no combustion by-products to contaminate the steel. Electric furnaces are also used to refine steel, producing the extra-pure steels used, for example, in the petrochemical industry.

The steel produced is cast into ingots, which can be worked when hot by hammering (forging) or pressing between rollers to produce sheet steel. Alternatively, the **continuous-cast process**, in which the molten metal is fed into an open-ended mould cooled by water, produces an unbroken slab of steel.

Stefan–Boltzmann law in physics, a law that relates the energy, E, radiated away from a perfect emitter (a ◊black body), to the temperature, T, of that body. It has the form $M = \sigma\, T^4$, where M is the energy radiated per unit area per second, T is the temperature, and σ is the **Stefan–Boltzmann constant**. Its value is 5.6697×10^{-8} W m^{-2} K^{-4}. The law was derived by Austrian physicists Joseph Stefan and Ludwig Boltzmann.

stem main supporting axis of a plant that bears the leaves, buds, and reproductive structures; it may be simple or branched. The plant stem usually grows above ground, although some grow underground, including ◊rhizomes, ◊corms, ◊rootstocks, and ◊tubers. Stems contain a continuous vascular system that conducts water and food to and from all parts of the plant.

The point on a stem from which a leaf or leaves arise is called a node, and the space between two successive nodes is the internode. In some plants, the stem is highly modified; for example, it may form a leaflike ◊cladode or it may be twining (as in many climbing plants), or fleshy and swollen to store water (as in cacti and other succulents). In plants exhibiting ◊secondary growth, the stem may become woody, forming a main trunk, as in trees, or a number of branches from ground level, as in shrubs.

stepper motor electric motor that can be precisely controlled by signals from a computer. The motor turns through a precise angle each time it receives a signal pulse from the computer. By varying the rate at which signal pulses are produced, the motor can be run at different speeds or turned through an exact angle and then stopped. Switching circuits can be constructed to allow the computer to reverse the direction of the motor.

By combining two or more motors, complex movement control becomes possible. For example, if stepper motors are used to power the wheels of a small vehicle a computer can manoeuver the vehicle in any direction.

Stepper motors are commonly used in small-scale applications where computer-controlled movement is required. In larger applications, where greater power is necessary, pneumatic or hydraulic systems are usually preferred.

step rocket another term for ◊multistage rocket.

steradian SI unit (symbol sr) of measure of solid (three-dimensional) angles, the three-dimensional equivalent of the ◊radian. One steradian is the angle at the centre of a sphere when an area on the surface of the sphere equal to the square of the sphere's radius is joined to the centre.

stereophonic sound system of sound reproduction using two complementary channels leading to two loudspeakers, which gives a more natural depth to the sound. Stereo recording began with the introduction of two-track magnetic tape in the 1950s. See ◊hi-fi.

sterilization any surgical operation to terminate the possibility of reproduction. In women, this is normally achieved by sealing or tying off the ◊Fallopian tubes (tubal ligation) so that fertilization can no longer take place. In men, the transmission of sperm is blocked by ◊vasectomy.

Sterilization is a safe alternative to ◊contraception and may be encouraged by governments to limit population growth or as part of a selective-breeding policy (see ◊eugenics).

sterilization the killing or removal of living organisms such as bacteria and fungi. A sterile environment is necessary in medicine, food processing, and some scientific experiments. Methods include heat treatment (such as boiling), the use of chemicals (such as disinfectants), irradiation with gamma rays, and filtration. See also ◊asepsis.

sterling silver ◊alloy containing 925 parts of silver and 75 parts of copper. The copper hardens the silver, making it more useful.

sternum large flat bone at the front of the chest, joined to the ribs. It gives protection to the heart and lungs. During open heart surgery the sternum must be cut to give access to the thorax.

steroid in biology, any of a group of cyclic, unsaturated alcohols (lipids without fatty acid components), which, like sterols, have a complex molecular structure consisting of four carbon rings. Steroids include the sex hormones, such as ◊testosterone, the corticosteroid hormones produced by the ◊adrenal gland, bile acids, and ◊cholesterol. The term is commonly used to refer to anabolic steroid, any hormone of the steroid group that stimulates tissue growth, used in sports to increase muscle bulk.

sterol any of a group of solid, cyclic, unsaturated alcohols, with a complex structure that includes four carbon rings; cholesterol is an example. Steroids are derived from sterols.

stigma in a flower, the surface at the tip of a ◊carpel that receives the ◊pollen. It often has short outgrowths, flaps, or hairs to trap pollen and may produce a sticky secretion to which the grains adhere.

stimulus any change in environmental factors, such as light, heat, or pressure, which can be detected by an organism's receptors.

stipule outgrowth arising from the base of a leaf or leaf stalk in certain plants. Stipules usually occur in pairs or fused into a single semicircular structure.

They may have a leaf-like appearance, as in goosegrass *Galium aparine*, be spiny, as in false acacia *Robina*, or look like small scales. In some species they are large, and contribute significantly to the photosynthetic area, as in the garden pea *Pisum sativum*.

Stirling engine A hot-air external combustion engine invented by Scottish priest Robert Stirling 1816. The engine operates by adapting to the fact that the air in its cylinders heats up when it is compressed and cools when it expands. The engine will operate on any fuel, is non-polluting, and relatively quiet. It was used fairly widely in the 19th century before the appearance of small, powerful, and reliable electric motors. Attempts have, however, been made in recent times to use Stirling's engine to power a variety of machines.

stokes cgs unit (symbol St) of kinematic viscosity (a liquid's resistance to flow).

Liquids with higher kinematic viscosity have higher turbulence than those with low kinematic viscosity. It is found by dividing the dynamic viscosity in ◊poise by the density of the liquid.

STOL (acronym for *s*hort *t*ake*o*ff and *l*anding) aircraft fitted with special devices on the wings (such as sucking flaps) that increase aerodynamic lift at low speeds. Small passenger and freight STOL craft may become common with the demand for small airports, especially in difficult terrain.

stolon in botany, a type of ◊runner.

stoma (plural *stomata*) in botany, a pore in the epidermis of a plant. Each stoma is surrounded by a pair of guard cells that are crescent-shaped when the stoma is open but can collapse to an oval shape, thus closing off the opening between them. Stomata allow the exchange of carbon dioxide and oxygen (needed for ◊photosynthesis and ◊respiration) between the internal tissues of the plant and the outside atmosphere. They are also the main route by which water is lost from the plant, and they can be closed to conserve water, the movements being controlled by changes in turgidity of the guard cells.

Stomata occur in large numbers on the aerial parts of a plant, and on the undersurface of leaves, where there may be as many as 45,000 per square centimetre.

stomach the first cavity in the digestive system of animals. In mammals it is a bag of muscle situated just below the diaphragm. Food enters it from the oesophagus, is digested by the acid and ◊enzymes secreted by the stomach lining, and then passes into the duodenum. Some plant-eating mammals have multichambered stomachs that harbour bacteria in one of the chambers to assist in the digestion of ◊cellulose. The gizzard is part of the stomach in birds.

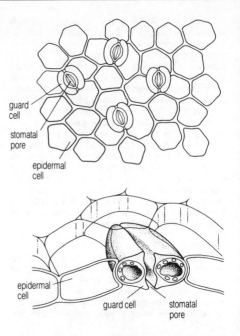

stoma *The stomata, tiny openings in the epidermis of a plant, are surrounded by pairs of crescent-shaped cells, called guard cells. The guard cells open and close the stoma by changing shape.*

stone (plural *stone*) imperial unit (abbreviation st) of mass. One stone is 14 pounds (6.35 kg).

storm surge abnormally high tide brought about by a combination of a severe atmospheric depression (very low pressure) over a shallow sea area, high spring tides, and winds blowing from the appropriate direction. A storm surge can cause severe flooding of lowland coastal regions and river estuaries.

Bangladesh is particularly prone to surges, being sited on a low-lying ◊delta where the Indian Ocean funnels into the Bay of Bengal. In May 1991, 125,000 people were killed there in such a disaster. In Feb 1953 more than 2,000 died when a North Sea surge struck the Dutch and English coasts.

STP abbreviation for ◊standard temperature and pressure.

strain measure of the extent to which a body is distorted when a deforming force (stress) is applied to it. It is a ratio of the extension or compression of that body (its length, area, or volume) to its original dimensions. For example, linear strain is the ratio of the change in length of a body to its original length.

strata (singular *stratum*) layers or ◊beds of sedimentary rock.

stratigraphy branch of geology that deals with the sequence of formation of ◊sedimentary rock layers and the conditions under which they were formed. Its basis was developed by William Smith, a British canal engineer.

Stratigraphy involves both the investigation of sedimentary structures to determine ancient geographies and environments, and the study of fossils for identifying and dating particular beds of rock.

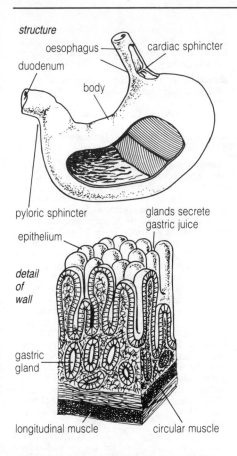

structure

oesophagus — cardiac sphincter

duodenum

body

pyloric sphincter — glands secrete gastric juice

epithelium

detail
of
wall

gastric
gland

longitudinal muscle — circular muscle

stomach *The human stomach can hold about 1.5 l/2.6 pt of liquid. The digestive juices are acidic enough to dissolve metal. To avoid damage, the cells of the stomach lining are replaced quickly—500,000 cells are replaced every minute, and the whole stomach lining every three days.*

Stratigraphy is a also an important tool in the interpretation of archaeological excavations. The basic principle of superimposition establishes that upper layers or deposits have accumulated later in time than lower ones, thus providing a relative chronology for the levels and the artefacts within them.

stratosphere that part of the atmosphere 10–40 km/6–25 mi from the Earth's surface, where the temperature slowly rises from a low of –55°C/–67°F to around 0°C/32°F. The air is rarefied and at around 25 km/15 mi much ◊ozone is concentrated.

streamlining shaping a body so that it offers the least resistance when travelling through a medium such as air or water. Aircraft, for example, must be carefully streamlined to reduce air resistance, or ◊drag.

High-speed aircraft must have swept-back wings, supersonic craft a sharp nose and narrow body.

strength of acids and bases in chemistry, the ability of ◊acids and ◊bases to dissociate in solution with water, and hence to produce a low or high ◊pH respectively.

A strong acid is fully dissociated in aqueous solution, whereas a weak acid is only partly dissociated. Since the ◊dissociation of acids generates hydrogen ions, a solution of a strong acid will have a high concentration of hydrogen ions and therefore a low pH. A strong base will have a high pH, whereas a weaker base will not dissociate completely and will have a pH of nearer 7.

stress and strain in the science of materials, measures of the deforming force applied to a body (stress) and of the resulting change in its shape (strain). For a perfectly elastic material, stress is proportional to strain (◊Hooke's law).

stridulatory organs in insects, organs that produce sound when rubbed together. Crickets rub their wings together, but grasshoppers rub a hind leg against a wing. Stridulation is thought to be used for attracting mates, but may also serve to mark territory.

string group of characters manipulated as a single object by the computer. In its simplest form a string may consist of a single letter or word—for example, the single word SMITH might be established as a string for processing by a computer. A string can also consist of a combination of words, spaces, and numbers—for example, 33 HIGH STREET ANYTOWN ALLSHIRE could be established as a single string.

Most high-level languages have a variety of string-handling ◊functions. For example, functions may be provided to read a character from any given position in a string or to count automatically the number of characters in a string.

string theory mathematical theory developed in the 1980s to explain the behaviour of ◊elementary particles; see ◊superstring theory.

strobilus in botany, a reproductive structure found in most ◊gymnosperms and some ◊pteridophytes, notably the club mosses. In conifers the strobilus is commonly known as a ◊cone.

stroboscope instrument for studying continuous periodic motion by using light flashing at the same frequency as that of the motion; for example, rotating machinery can be optically 'stopped' by illuminating it with a stroboscope flashing at the exact rate of rotation.

stromatolite mound produced in shallow water by mats of algae that trap mud particles. Another mat grows on the trapped mud layer and this traps another layer of mud and so on. The stromatolite grows to heights of a metre or so. They are uncommon today but their fossils are among the earliest evidence for living things—over 2,000 million years old.

strong nuclear force one of the four fundamental ◊forces of nature, the other three being the electromagnetic force, gravity, and the weak nuclear force. The strong nuclear force was first described by Japanese physicist Hideki Yukawa 1935. It is the strongest of all the forces, acts only over very small distances (within the nucleus of the atom), and is responsible for binding together ◊quarks to form ◊hadrons, and for binding together protons and neutrons in the atomic nucleus. The particle that is the carrier of the strong nuclear force is the ◊gluon, of which there are eight kinds, each with zero mass and zero charge.

strontium soft, ductile, pale-yellow, metallic element, symbol Sr, atomic number 38, relative atomic mass 87.62. It is one of the ◊alkaline-earth metals, widely distributed in small quantities only as a sulphate or carbonate. Strontium salts burn with a red flame and are used in fireworks and signal flares.

The radioactive isotopes Sr-89 and Sr-90 (half-life 25 years) are some of the most dangerous products of the nuclear industry; they are fission products in nuclear explosions and in the reactors of nuclear power plants. Strontium is chemically similar to calcium and deposits in bones and other tissues, where the radioactivity is damaging. The element was named in 1808 by English chemist Humphry Davy, who isolated it by electrolysis, after Strontian, a mining location in Scotland where it was first found.

structured programming in computing, the process of writing a program in small, independent parts. This makes it easier to control a program's development and to design and test its individual component parts. Structured programs are built up from units called *modules*, which normally correspond to single ◊procedures or ◊functions. Some programming languages, such as PASCAL and Modula-2, are better suited to structured programming than others.

strychnine $C_{21}H_{22}O_2N_2$ bitter-tasting, poisonous alkaloid. It is a poison that causes violent muscular spasms, and is usually obtained by powdering the seeds of plants of the genus *Strychnos* (for example *S. nux vomica*). Curare is a related drug.

style in flowers, the part of the ◊carpel bearing the ◊stigma at its tip. In some flowers it is very short or completely lacking, while in others it may be long and slender, positioning the stigma in the most effective place to receive the pollen.

Usually the style withers after fertilization but in certain species, such as traveller's joy *Clematis vitalba*, it develops into a long feathery plume that aids dispersal of the fruit.

subatomic particle in physics, a particle that is smaller than an atom. Such particles may be indivisible ◊elementary particles, such as the ◊electron and ◊quark, or they may be composites, such as the ◊proton, ◊neutron, and ◊alpha particle. See also ◊particle physics.

subduction zone region where two plates of the Earth's rigid lithosphere collide, and one plate descends below the other into the semiliquid asthenosphere. Subduction occurs along ocean trenches, most of which encircle the Pacific Ocean; portions of the ocean plate slide beneath other plates carrying continents.

Ocean trenches are usually associated with volcanic ◊island arcs and deep-focus earthquakes (more than 185 mi/300 km below the surface), both the result of disturbances caused by the plate subduction. The Aleutian Trench bordering Alaska is an example of an active subduction zone, which has produced the Aleutian Island arc.

subglacial beneath a glacier. Subglacial rivers are those that flow under a glacier: subglacial material is debris that has been deposited beneath glacier ice. Features formed subglacially include ◊drumlins and ◊eskers.

sublimation in chemistry, the conversion of a solid to vapour without passing through the liquid phase.

Some substances that do not sublime at atmospheric pressure can be made to do so at low pressures. This is the principle of freeze-drying, during which ice sublimes at low pressure.

submarine vessel capable of travelling and functioning underwater, used in research and military operations. On either side of the submarine's hull there are long, hollow ballast tanks. These are filled with water for submerging, and are emptied by blasts of compressed air for surfacing. Underwater the submarine is driven forward by its propeller, and downwards by hydroplanes, similar to the tail on an aeroplane. These are powered by electric motors and batteries in small submarines; larger ones are nuclear powered. Petrol or diesel engines are impractical underwater because they use up too much air and give out exhaust.

history The first underwater boat was constructed in 1620 for James I of England by the Dutch scientist Cornelius van Drebbel. In the 1760s, the American David Bushnell designed a submarine called Turtle, for attacking British ships, and in 1800, Robert Fulton designed a metal-hulled submarine called Nautilus, for Napoleon for the same purpose. John P Holland, an Irish immigrant to the USA, designed a submarine about 1875, which was used by both the US and British navies at the turn of the century.

The conventional submarine of World War I was driven by diesel engine on the surface and by battery-powered electric motors underwater. In both world wars submarines played a vital role. The first nuclear-powered submarine, the Nautilus, was launched by the USA 1954.

In oceanography, salvage, and pipe-laying smaller submarines called submersibles, are used.

subroutine in computing, a small section of a program that is executed ('called') from another part of the program. Subroutines provide a method of performing the same task at more than one point in the program, and also of separating the details of a program from its main logic. In some computer languages, subroutines are similar to ◊functions or ◊procedures.

substitution reaction in chemistry, the replacement of one atom or ◊functional group in an organic molecule by another.

substrate in biochemistry, a compound or mixture of compounds acted on by an enzyme. The term also refers to a substance such as ◊agar that provides the nutrients for the metabolism of microorganisms. Since the enzyme systems of microorganisms regulate their metabolism, the essential meaning is the same.

succession in ecology, a series of changes that occur in the structure and composition of the vegetation in a given area from the time it is first colonized by plants (*primary succession*), or after

	Group	Particle	Symbol	Charge	Mass (MeV)	Spin	Lifetime (sec)
elementary	quark	up	u	$\frac{2}{3}$	336	$\frac{1}{2}$	?
particle		down	d	$-\frac{1}{3}$	336	$\frac{1}{2}$	?
		(top)	t	$(\frac{2}{3})$	(<600,000)	$(\frac{1}{2})$	?
		bottom	b	$-\frac{1}{3}$	4,700	$\frac{1}{2}$	?
		strange	s	$-\frac{1}{3}$	540	$\frac{1}{2}$	?
		charm	c	$\frac{2}{3}$	1,500	$\frac{1}{2}$	?
	lepton	electron	e^-	-1	0·511	$\frac{1}{2}$	stable
		electron neutrino	ν_e	0	(0)	$\frac{1}{2}$	stable
		muon	μ^-	-1	105·66	$\frac{1}{2}$	$2\cdot2 \times 10^{-6}$
		muon neutrino	ν_μ	0	(0)	$\frac{1}{2}$	stable
		tau	τ^-	-1	1,784	$\frac{1}{2}$	$3\cdot4 \times 10^{-13}$
		tau neutrino	ν_τ	0	(0)	$\frac{1}{2}$	?
	gauge boson	photon	γ	0	0	1	stable
		graviton	g	0	(0)	2	stable
		gluon	g	0	0	1	?
		weakon	$W^\pm$	±1	81,000	1	?
			Z	0	94,000	1	?
hadron	meson	pion	π^+	1	139·57	0	$2\cdot6 \times 10^{-8}$
			π^0	0	134·96	0	$8\cdot3 \times 10^{-17}$
		kaon	K^+	1	493·67	0	$1\cdot2 \times 10^{-8}$
			K^0_s	0	497·67	0	$8\cdot9 \times 10^{-11}$
			K^0_L	0	497·67	0	$5\cdot18 \times 10^{-8}$
		psi	Ψ	0	3,100	1	$6\cdot3 \times 10^{-2}$
		upsilon	Y	0	9,460	1	$\sim1 \times 10^{-20}$
	baryon	nucleon:					
		proton	p	1	938·28	$\frac{1}{2}$	stable
		neutron	n	0	939·57	$\frac{1}{2}$	920
		hyperon:					
		lambda	Λ	0	1,115·6	$\frac{1}{2}$	$2\cdot63 \times 10^{-10}$
		sigma	Σ^+	1	1,189·4	$\frac{1}{2}$	$8\cdot0 \times 10^{-11}$
			Σ^-	-1	1,197·3	$\frac{1}{2}$	$1\cdot5 \times 10^{-10}$
			Σ^0	0	1,192·5	$\frac{1}{2}$	$5\cdot8 \times 10^{-20}$
		xi	Ξ^-	-1	1,321·3	$\frac{1}{2}$	$1\cdot64 \times 10^{-10}$
			Ξ^0	0	1,314·9	$\frac{1}{2}$	$2\cdot9 \times 10^{-10}$
		omega	Ω	-1	1,672·4	$\frac{3}{2}$	$8\cdot2 \times 10^{-11}$

? indicates that the particle's lifetime has yet to be determined
( ) indicates that the property has been deduced but not confirmed
MeV = million electron volts

subatomic particle

it has been disturbed by fire, flood, or clearing (*secondary succession*).

If allowed to proceed undisturbed, succession leads naturally to a stable ◊climax community (for example, oak and hickory forest or savannah grassland) that is determined by the climate and soil characteristics of the area.

succulent plant thick, fleshy plant that stores water in its tissues; for example, cacti.

sucrase enzyme capable of digesting sucrose into its constituent molecules of glucose and fructose. In mammals this action takes place within the wall of the intestine, the products of the reaction being liberated into the lumen. This is an example of intracellular digestion.

sucrose or *cane sugar* or *beet sugar* $C_{12}H_{22}O_{10}$ a sugar found in the pith of sugar cane and in sugar beets. It is popularly known as sugar.

Sucrose is a disaccharide sugar, each of its molecules being made up of two simple sugar (monosaccharide) units: glucose and fructose.

sulphate SO_4^{2-} salt or ester derived from sulphuric acid. Most sulphates are water soluble (the exceptions are lead, calcium, strontium, and barium sulphates), and require a very high temperature to decompose them.

The commonest sulphates seen in the laboratory are copper(II) sulphate ($CuSO_4$), iron(II) sulphate ($FeSO_4$), and aluminium sulphate ($Al_2(SO_4)_3$). The ion is detected in solution by using barium chloride or barium nitrate to precipitate the insoluble sulphate.

sulphide compound of sulphur and another element in which sulphur is the more electronegative element (see ◊electronegativity). Sulphides occur in a number of minerals. Some of the more volatile sulphides have extremely unpleasant odours (hydrogen sulphide smells of bad eggs).

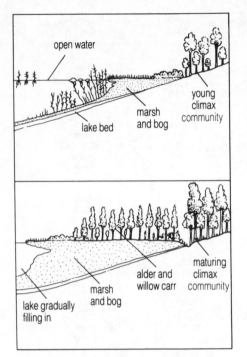

succession *The succession of plant types along a lake. As the lake gradually fills in, a mature climax community of trees forms inland from the shore. Extending out from the shore, a series of plant communities can be discerned with small, rapidly growing species closest to the shore.*

sulphite SO_3^{2-} salt or ester derived from sulphurous acid.

sulphonamide any of a group of compounds containing the chemical group sulphonamide (SO_2NH_2) or its derivatives, which were, and still are in some cases, used to treat bacterial diseases. Sulphadiazine ($C_{10}H_{10}N_4O_2S$) is an example.

Sulphonamide was the first commercially available antibacterial drug, the forerunner of a range of similar drugs. Toxicity and increasing resistance have limited their use chiefly to the treatment of urinary-tract infection.

sulphur brittle, pale-yellow, nonmetallic element, symbol S, atomic number 16, relative atomic mass 32.064. It occurs in three allotropic forms: two crystalline (called rhombic and monoclinic, following the arrangements of the atoms within the crystals) and one amorphous. It burns in air with a blue flame and a stifling odour. Insoluble in water but soluble in carbon disulphide, it is a good electrical insulator. Sulphur is widely used in the manufacture of sulphuric acid (used to treat phosphate rock to make fertilizers) and in making paper, matches, gunpowder and fireworks, in vulcanizing rubber, and in medicines and insecticides.

It is found abundantly in nature in volcanic regions combined with both metals and nonmetals, and also in its elemental form as a crystalline solid. It is a constituent of proteins, and has been known since ancient times.

sulphur dioxide SO_2 pungent gas produced by burning sulphur in air or oxygen. It is widely used for disinfecting food vessels and equipment, and as a preservative in some food products. It occurs in industrial flue gases and is a major cause of ◊acid rain.

sulphuric acid or *oil of vitriol* H_2SO_4 a dense, viscous, colourless liquid that is extremely corrosive. It gives out heat when added to water and can cause severe burns. Sulphuric acid is used extensively in the chemical industry, in the refining of petrol, and in the manufacture of fertilizers, detergents, explosives, and dyes. It forms the acid component of car batteries.

sulphurous acid H_2SO_3 solution of sulphur dioxide (SO_2) in water. It is a weak acid.

sulphur trioxide SO_3 colourless solid prepared by reacting sulphur dioxide and oxygen in the presence of a vanadium(V) oxide catalyst in the ◊contact process. It reacts violently with water to give sulphuric acid.

The violence of its reaction with water makes it extremely dangerous. In the contact process, it is dissolved in concentrated sulphuric acid to give oleum ($H_2S_2O_7$).

Sun the ◊star at the centre of the Solar System. Its diameter is 1,392,000 km/865,000 mi; its temperature at the surface is about 5,800K (5,530°C/ 9,980°F), and at the centre 15,000,000K (15,000,000°C/27,000,000°F). It is composed of about 70% hydrogen and 30% helium, with other elements making up less than 1%. The Sun's energy is generated by nuclear fusion reactions that turn hydrogen into helium at its centre. The gas core is far denser than mercury or lead on Earth. The Sun is about 4.7 billion years old, with a predicted lifetime of 10 billion years.

At the end of its life, it will expand to become a ◊red giant the size of Mars' orbit, then shrink to become a ◊white dwarf. The Sun spins on its axis every 25 days near its equator, but more slowly towards its poles. Its rotation can be followed by watching the passage of dark ◊sunspots across its disc. Sometimes bright eruptions called ◊flares occur near sunspots. Above the Sun's ◊photosphere lies a layer of thinner gas called the ◊chromosphere, visible only in special instruments or at eclipses. Tongues of gas called ◊prominences extend from the chromosphere into the corona, a

rhombic sulphur crystal   *monoclinic sulphur crystal*

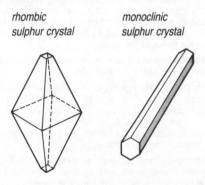

sulphur

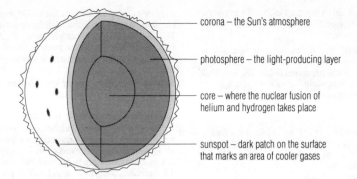

corona – the Sun's atmosphere

photosphere – the light-producing layer

core – where the nuclear fusion of helium and hydrogen takes place

sunspot – dark patch on the surface that marks an area of cooler gases

Sun

halo of hot, tenuous gas surrounding the Sun. Gas boiling from the corona streams outwards through the solar system, forming the ◊solar wind. Activity on the Sun, including sunspots, flares, and prominences, waxes and wanes during the **solar cycle**, which peaks every 11 years or so.

> *... in my studies of astronomy and philosophy I hold this opinion about the universe, that the Sun remains fixed in the centre of the circle of heavenly bodies, without changing its place; and the Earth, turning upon itself, moves round the Sun.*
>
> On the Sun Galileo Galilei, letter to Cristina di Lorena 1615

sunshine recorder device for recording the hours of sunlight during a day. The **Campbell-Stokes sunshine recorder** consists of a glass sphere that focuses the sun's rays on a graduated paper strip. A track is burned along the strip corresponding to the time that the Sun is shining.

sunspot dark patch on the surface of the Sun, actually an area of cooler gas, thought to be caused by strong magnetic fields that block the outward flow of heat to the Sun's surface. Sunspots consist of a dark central **umbra**, about 4,000K (3,700°C/6,700°F), and a lighter surrounding **penumbra**, about 5,500K (5,200°C/9,400°F). They last from several days to over a month, ranging in size from 2,000 km/1,250 mi to groups stretching for over 100,000 km/62,000 mi. The number of sunspots visible at a given time varies from none to over 100 in a cycle averaging 11 years.

superactinide any of a theoretical series of superheavy, radioactive elements, starting with atomic number 113, that extend beyond the ◊transactinide series in the periodic table. They do not occur in nature and none has yet been synthesized.

It is postulated that this series has a group of elements that have half-lives longer than those of the transactinide series. This group, centred on element 114, is referred to as the 'island of stability', based on the nucleon arrangement. The longer half-lives will, it is hoped, allow enough time for their chemical and physical properties to be studied when they have been synthesized.

supercomputer the fastest, most powerful type of computer, capable of performing its basic operations in picoseconds (thousand-billionths of a second), rather than nanoseconds (billionths of a second), like most other computers.

To achieve these extraordinary speeds, supercomputers use several processors working together and techniques such as cooling processors down to nearly ◊absolute zero temperature, so that their components conduct electricity many times faster than normal. Supercomputers are used in weather forecasting, fluid and aerodynamics. Manufacturers include Cray, Fujitsu, and NEC.

In 1992 Fujitsu announced the launch of the world's most powerful computer. It can perform 300 billion calculations a second.

superconductivity in physics, increase in electrical conductivity at low temperatures. The resistance of some metals and metallic compounds decreases uniformly with decreasing temperature until at a critical temperature (the superconducting point), within a few degrees of absolute zero (0 K/ –273.16°C/–459.67°F), the resistance suddenly falls to zero. The phenomenon was discovered by Dutch scientist Heike Kamerlingh-Onnes (1853–1926) in 1911.

In this superconducting state, an electric current will continue indefinitely after the magnetic field has been removed, provided that the material remains below the superconducting point. In 1986 IBM researchers achieved superconductivity with some ceramics at –243°C/–405°F; Paul Chu at the University of Houston, Texas, achieved superconductivity at –179°C/–290°F, a temperature that can be sustained using liquid nitrogen. In 1992, two Japanese researchers developed a material which becomes superconducting at around –103°C/ –153°F.

Some metals, such as platinum and copper, do not become superconductive; as the temperature decreases, their resistance decreases to a certain point but then rises again. Superconductivity can be nullified by the application of a large magnetic field. Superconductivity has been produced in a synthetic organic conductor that would operate at much higher temperatures, thus cutting costs.

supercooling in physics, the lowering in temperature of a ◊saturated solution without crystallization taking place, forming a supersaturated solution.

Usually crystallization rapidly follows the introduction of a small (seed) crystal or agitation of the supercooled solution.

superfluid fluid that flows without viscosity or friction and has a very high thermal conductivity. Liquid helium at temperatures below 2K (–271°C/–456°F) is a superfluid: it shows unexpected behaviour; for instance, it flows uphill in defiance of gravity and, if placed in a container, will flow up the sides and away.

supergiant the largest and most luminous type of star known, with a diameter of up to 1,000 times that of the Sun and absolute magnitudes of between –5 and –9.

superheterodyne receiver the most widely used type of radio receiver, in which the incoming signal is mixed with a signal of fixed frequency generated within the receiver circuits. The resulting signal, called the intermediate-frequency (i.f.) signal, has a frequency between that of the incoming signal and the internal signal. The intermediate frequency is near the optimum frequency of the amplifier to which the i.f. signal is passed. This arrangement ensures greater gain and selectivity. The superheterodyne system is also used in basic television receivers.

superior planet planet that is farther away from the Sun than the Earth is: that is, Mars, Jupiter, Saturn, Uranus, Neptune, and Pluto.

supernova the explosive death of a star, which temporarily attains a brightness of 100 million Suns or more, so that it can shine as brilliantly as a small galaxy for a few days or weeks. Very approximately, it is thought that a supernova explodes in a large galaxy about once every 100 years. Many supernovae remain undetected because of obscuring by interstellar dust—astronomers estimate some 50%.

Type I supernovae are thought to occur in ◊binary star systems in which gas from one star falls on to a ◊white dwarf, causing it to explode. *Type II* supernovae occur in stars ten or more times as massive as the Sun, which suffer runaway internal nuclear reactions at the ends of their lives, leading to explosions. These are thought to leave behind ◊neutron stars and ◊black holes. Gas ejected by such an explosion causes an expanding radio source, such as the ◊Crab nebula. Supernovae are thought to be the main source of elements heavier than hydrogen and helium.

The first supernova recorded (although not identified as such at the time) was in AD 185 in China. The last supernova seen in our Galaxy was in 1604, but many others have been seen since in other galaxies. In 1987 a supernova visible to the unaided eye occurred in the Large ◊Magellanic Cloud, a small neighbouring galaxy. Eta Carinae, an unusual star in the constellation Carina in the southern hemisphere, may become a supernova in a few hundred years.

The name 'supernova' was coined by US astronomers Fritz Zwicky and Walter Baade in 1934. Zwicky was also responsible for the division into types I and II.

superphosphate phosphate fertilizer made by treating apatite with sulphuric or phosphoric acid. The commoercial mixture contains largely mono-calcium phosphate. Single-superphosphate obtained from apatite and sulphuric acid contains 16–20% available phosphorus, as P_2O_5; triple-superphosphate, which contains 45–50% phosphorus, is made by treating apatite with phosphoric acid.

supersaturation in chemistry, the state of a solution that has a higher concentration of ◊solute than would normally be obtained in a ◊saturated solution.

Many solutes have a higher ◊solubility at high temperatures. If a hot saturated solution is cooled slowly, sometimes the excess solute does not come out of solution. This is an unstable situation and the introduction of a small solid particle will encourage the release of excess solute.

supersonic speed speed greater than that at which sound travels, measured in ◊Mach numbers. In dry air at 0°C/32°F, sound travels at about 1,170 kph/727 mph, but decreases its speed with altitude until, at 12,000 m/39,000 ft, it is only 1,060 kph/658 mph.

When an aircraft passes the ◊sound barrier, shock waves are built up that give rise to ◊sonic boom, often heard at ground level. US pilot Captain Charles Yeager was the first to achieve supersonic flight, in a Bell VS-1 rocket plane on 14 Oct 1947.

superstring theory in physics, a mathematical theory developed in the 1980s to explain the properties of ◊elementary particles and the forces between them (in particular, gravity and the nuclear forces) in a way that combines ◊relativity and ◊quantum theory. In string theory, the fundamental objects in the universe are not pointlike particles but extremely small stringlike objects. These objects exist in a universe of ten dimensions, although, for reasons not yet understood, only three space dimensions and one dimension of time are discernible.

There are many unresolved difficulties with superstring theory, but some physicists think it may be the ultimate 'theory of everything' that explains all aspects of the universe within one framework.

supersymmetry in physics, a theory that relates the two classes of elementary particle, the ◊fermions and the ◊bosons. According to supersymmetry, each fermion particle has a boson partner particle, and vice versa. It has not been possible to marry up all the known fermions with the known bosons, and so the theory postulates the existence of other, as yet undiscovered fermions, such as the photinos (partners of the photons), gluinos (partners of the gluons), and gravitinos (partners of the gravitons). Using these ideas, it has become possible to develop a theory of gravity—called supergravity—that extends Einstein's work and considers the gravitational, nuclear, and electromagnetic forces to be manifestations of an underlying superforce. Supersymmetry has been incorporated into the ◊superstring theory, and appears to be a crucial ingredient in the 'theory of everything' sought by scientists.

support environment in computing, a collection of programs (◊software) used to help people design and write other programs. At its simplest, this includes a ◊text editor (word-processing software)

surface areas of common three-dimensional shapes

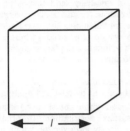

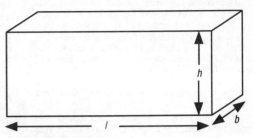

surface area of a *cube*
(faces are identical)
= 6 × area of each face
= 6 l^2

surface area of a *cuboid*
(opposite faces are identical)
= area of two end faces = area of two sides
  + area of top and base
= 2 lh + 2 hb + 2 lb
= 2(lh + hb × lb)

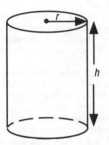

surface area of a *cylinder*
= area of a curved surface + area of top
and base
= 2π × (radius of cross-section × height)
  + 2π (radius of cross-section)^2
= $2\pi rh + 2\pi r^2$
= $2\pi r (h + r)$

surface area of a *cone*
= area of a curved surface + area of base
= π × (radius of cross-section × slant height)
  + π (radius of cross-section)^2
= $\pi r l + \pi r^2$
= $\pi r (l + r)$

surface area of a *sphere*
= 4π × radius^2
= $4\pi r^2$

surface area

and a ◊compiler for translating programs into executable form; but it can also include interactive debuggers for helping to locate faults, data dictionaries for keeping track of the data used, and rapid prototyping tools for producing quick, experimental mock-ups of programs.

suprarenal gland alternative name for the ◊adrenal gland.

surd an expression containing the root of an ◊irrational number that can never be exactly expressed—for example, $\sqrt{3} = 1.732050808. . . .$

surface area the area of the outer surface of a three-dimensional shape, or solid.

surface-area-to-volume ratio the ratio of an animal's surface area (the area covered by its skin) to its total volume. This is high for small animals, but low for large animals such as elephants.

The ratio is important for endothermic (warm-blooded) animals because the amount of heat lost by the body is proportional to its surface area, whereas the amount generated is proportional to its volume. Very small birds and mammals, such as hummingbirds and shrews, lose a lot of heat and need a high intake of food to maintain their body temperature. Elephants, on the other hand, are in danger of overheating, which is why they have no fur.

surface tension in physics, the property that causes the surface of a liquid to behave as if it were covered with a weak elastic skin; this is why a needle can float on water. It is caused by the exposed surface's tendency to contract to the smallest possible area because of unequal cohesive forces between ◊molecules at the surface. Allied phenomena include the formation of droplets, the concave profile of a meniscus, and the ◊capillary action by which water soaks into a sponge.

surge abnormally high tide; see ◊storm surge.

surgical spirit ◊ethanol to which has been added a small amount of methanol to render it unfit to drink. It is used to sterilize surfaces and to cleanse skin abrasions and sores.

susceptibility ratio of the intensity of magnetization produced in a material to the intensity of the magnetic field to which the material is exposed. It measures the extent to which a material is magnetized by an applied magnetic field. ◊Diamagnetic materials have small negative susceptibilities; ◊paramagnetic materials have small positive susceptibilities; ◊ferromagnetic materials have large positive susceptibilities. See also ◊magnetism.

suspension mixture consisting of small solid particles dispersed in a liquid or gas, which will settle on standing. An example is milk of magnesia, which is a suspension of magnesium hydroxide in water.

suspensory ligament in the ◊eye, a ring of fibre supporting the lens. The ligaments attach to the ciliary muscles, the circle of muscle mainly responsible for changing the shape of the lens during ◊accommodation. If the ligaments are put under tension, the lens becomes flatter, and therefore able to focus on objects in the far distance.

sustainable capable of being continued indefinitely. For example, the sustainable yield of a forest is equivalent to the amount that grows back. Environmentalists made the term a catchword, in advocating the sustainable use of resources.

sustained-yield cropping in ecology, the removal of surplus individuals from a ◊population of organisms so that the population maintains a constant size. This usually requires selective removal of animals of all ages and both sexes to ensure a balanced population structure. Taking too many individuals can result in a population decline, as in overfishing.

Excessive cropping of young females may lead to fewer births in following years, and a fall in population size. Appropriate cropping frequencies can be determined from an analysis of a ◊life table.

swallow hole or *swallet* hole, often found in limestone areas, through which a surface stream disappears underground. It will usually lead to an underground network of caves.

swamp region of low-lying land that is permanently saturated with water and usually overgrown with vegetation; for example, the everglades of Florida, USA. A swamp often occurs where a lake has filled up with sediment and plant material. The flat surface so formed means that runoff is slow, and the water table is always close to the surface. The high humus content of swamp soil means that good agricultural soil can be obtained by draining.

sweat gland ◊gland within the skin of mammals that produces surface perspiration. In primates, sweat glands are distributed over the whole body, but in most other mammals they are more localized; for example, in cats and dogs, they are restricted to the feet and around the face.

In humans, sweat glands occur in larger numbers in the male than the female and the odours produced are thought to be used in communicating sexual and social messages.

sweetener any chemical that gives sweetness to food. Caloric sweeteners are various forms of ◊sugar; noncaloric, or artificial, sweeteners are used by dieters and diabetics, and provide neither energy nor bulk.

Sweeteners are used to make highly processed foods attractive, whether sweet or savoury. Most of the noncaloric sweeteners do not have ◊E numbers, and some are banned from baby foods and for young children: thaumatin, aspartame, acesulfame-K, sorbitol, and mannitol. Cyclamate is banned in the UK and the USA; acesulfame-K is banned in the USA.

relative sweetness of various sweeteners

artificial sweetener (sucrose = 1.0)	relative sweetness
thaumatin	3,000
saccharin	300
aspartame	200
acesulfame-K	150
cyclamate	30

swim bladder thin-walled, air-filled sac found between the gut and the spine in bony fishes. Air enters the bladder from the gut or from surrounding capillaries (see ◊capillary), and changes of air pressure within the bladder maintain buoyancy whatever the water depth.

In evolutionary terms, the swim bladder of higher fishes is a derivative of the lungs present in all primitive fishes (not just lungfishes).

swing wing correctly *variable-geometry wing* aircraft wing that can be moved during flight to provide a suitable configuration for either low-speed or high-speed flight. The British engineer Barnes Wallis developed the idea of the swing wing, first used on the US-built Northrop X-4, and since used in several aircraft, including the US F-111, F-114, and the B-1, the European Tornado, and several Soviet-built aircraft. These craft have their wings projecting nearly at right angles for takeoff and landing and low-speed flight, and swung back for high-speed flight.

syenite grey, crystalline, plutonic (intrusive) ◊igneous rock, consisting of feldspar and hornblende; other minerals may also be present, including small amounts of quartz.

symbiosis any close relationship between two organisms of different species, and one where both partners benefit from the association. A well-known example is the pollination relationship between insects and flowers, where the insects feed on nectar and carry pollen from one flower to another. This is sometimes known as ◊mutualism. Symbiosis in a broader sense includes ◊commensalism and parasitism (see ◊parasite).

symbol letter or letters used to represent a chemical element, usually derived from the beginning of its English or Latin name. Symbols derived from English include B, boron; C, carbon; Ba, barium; and Ca, calcium. Those derived from Latin include Na, sodium (Latin *natrium*); Pb, lead (Latin *plumbum*); and Au, gold (Latin *aurum*).

symbolic address in computing, a symbol used in ◊assembly-language programming to represent the binary ◊address of a memory location.

symbolic processor computer purpose-built to run so-called symbol-manipulation programs rather than programs involving a great deal of numerical computation. They exist principally for the ◊artificial intelligence language ◊LISP, although some have also been built to run ◊PROLOG.

symmetry exact likeness in shape about a given line (axis), point, or plane. A figure has symmetry if one half can be rotated and/or reflected onto the other. (Symmetry preserves length, angle, but not necessarily orientation.) In a wider sense, symmetry

forms delta wing
with tailplane

fully
extended
position
(low speed)

swept-back
position
(high-speed)

swing wing *The swing-wing fighter aircraft forms a delta shape for supersonic flight. There are considerable engineering problems involved in hinging the wings, but there are several types of swing-wing craft now in use.*

exits if a change in the system leaves the essential features of the system unchanged; for example, reversing the sign of electric charges does not change the electrical behaviour of an arrangement of charges.

synapse junction between two ◊nerve cells, or between a nerve cell and a muscle (a neuromuscular junction), across which a nerve impulse is transmitted. The two cells are separated by a narrow gap called the ***synaptic cleft***. The gap is bridged by a chemical ◊neurotransmitter, released by the nerve impulse.

The threadlike extension, or ◊axon, of the transmitting nerve cell has a slightly swollen terminal point, the ***synaptic knob***. This forms one half of the synaptic junction and houses membrane-bound vesicles, which contain a chemical neurotransmitter. When nerve impulses reach the knob, the vesicles release the transmitter and this flows across the gap and binds itself to special receptors on the receiving cell's membrane. If the receiving cell is a nerve cell, the other half of the synaptic junction will be one or more extensions called ◊dendrites; these will be stimulated by the neurotransmitter to set up an impulse, which will then be conducted along the length of the nerve cell and on to its own axons. If the receiving cell is a muscle cell, it will be stimulated by the neurotransmitter to contract.

Synapsida group of mammal-like reptiles living 315–195 million years ago, whose fossil record is largely complete, and who were for a while the dominant land animals, before being replaced by the dinosaurs. The true mammals are their descendants.

synchrotron a type of particle ◊accelerator in which particles move, at increasing speed, around a hollow ring. The particles are guided around the ring by electromagnets, and accelerated by electric fields at points around the ring. Synchrotrons come in a wide range of sizes, the smallest being about a metre across while the largest is 27 km across. This huge machine is the ◊Large Electron Positron Collider (LEP) at the ◊CERN laboratories near Geneva in Switzerland. a synchrotron 6 km in circumference at ◊Fermilab accelerates protons and antiprotons to 1 TeV; it is called the Tevatron.

syncline geological term for a fold in the rocks of the Earth's crust in which the layers or ◊beds dip inwards, thus forming a trough-like structure with a sag in the middle. The opposite structure, with the beds arching upwards, is an ◊anticline.

synodic period the time taken for a planet or moon to return to the same position in its orbit as seen from the Earth; that is, from one ◊opposition to the next. It differs from the ◊sidereal period because the Earth is moving in orbit around the Sun.

synovial fluid viscous yellow fluid that bathes movable joints between the bones of vertebrates. It nourishes and lubricates the ◊cartilage at the end of each bone.

Synovial fluid is secreted by a membrane that links movably jointed bones. The same kind of fluid is found in bursae, the membranous sacs that buffer some joints, such as in the shoulder and hip region.

plane symmetry

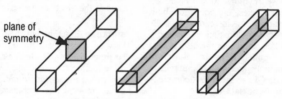

plane of
symmetry

cuboid has three planes of symmetry

square-based pyramid has four planes of symmetry

sphere has infinite number of planes of symmetry

axes of symmetry

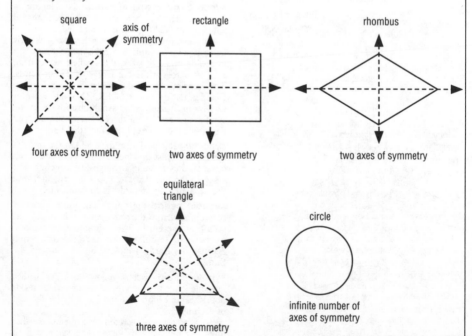

square

axis of
symmetry

rectangle

rhombus

four axes of symmetry

two axes of symmetry

two axes of symmetry

equilateral
triangle

circle

infinite number of
axes of symmetry

three axes of symmetry

symmetry

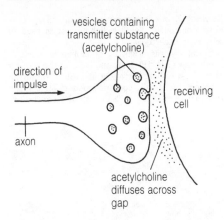

vesicles containing
transmitter substance
(acetylcholine)

direction of
impulse

receiving
cell

axon

acetylcholine
diffuses across
gap

synapse

synthesis in chemistry, the formation of a substance or compound from more elementary compounds. The synthesis of a drug can involve several stages from the initial material to the final product; the complexity of these stages is a major factor in the cost of production.

synthesizer device that uses electrical components to produce sounds. In *preset synthesizers*, the sound of various instruments is produced by a built-in computer-type memory. In *programmable synthesizers* any number of new instrumental or other sounds may be produced at the will of the performer. *Speech synthesizers* can break down speech into 128 basic elements (allophones), which are then combined into words and sentences, as in the voices of electronic teaching aids.

In preset synthesizers the memory triggers all the control settings required to produce the sound of a trumpet or violin. For example, the 'sawtooth' sound wave produced by a violin is artificially produced by an electrical tone generator, or oscillator, and then fed into an electrical filter set to have the resonances characteristic of a violin body. The first electronic sound synthesizer was developed 1964 by US engineer Robert Moog (1934–   ). Synthesizers are played via a keyboard. A drum machine is a specialized form of synthesizer.

synthetic any material made from chemicals. Since the 1900s, more and more of the materials used in everyday life are synthetics, including plastics (polythene, polystyrene), ◊synthetic fibres (nylon, acrylics, polyesters), synthetic resins, and synthetic rubber. Most naturally occurring organic substances are now made synthetically, especially pharmaceuticals.

Plastics are made mainly from petroleum chemicals by ◊polymerization, in which small molecules are joined to make very large ones.

synthetic fibre fibre made by chemical processes, unknown in nature. There are two kinds. One is made from natural materials that have been chemically processed in some way; ◊rayon, for example, is made by processing the cellulose in wood pulp. The other type is the true synthetic fibre, made entirely from chemicals. ◊Nylon was the original synthetic fibre, made from chemicals obtained from petroleum (crude oil).

Fibres are drawn out into long threads or filaments, usually by so-called spinning methods, melting or dissolving the parent material and then forcing it through the holes of a perforated plate, or spinneret.

syrinx the voice-producing organ of a bird. It is situated where the trachea divides in two and consists of vibrating membranes, a reverberating capsule, and numerous controlling muscles.

Système International d'Unités official French name for ◊SI units.

system flow chart type of ◊flow chart used to describe the flow of data through a particular computer system.

systems analysis in computing, the investigation of a business activity or clerical procedure, with a view to deciding if and how it can be computerized. The analyst discusses the existing procedures with the people involved, observes the flow of data through the business, and draws up an outline specification of the required computer system (see also ◊systems design).

Systems in use in the 1990s include Yourdon, SSADM (Structured Systems Analysis and Design Methodology), and Soft Systems Methodology.

systems design in computing, the detailed design of an ◊applications package. The designer breaks the system down into component programs, and designs the required input forms, screen layouts, and printouts. Systems design forms a link between systems analysis and ◊programming.

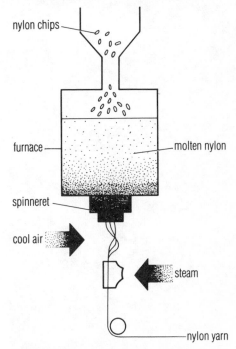

nylon chips

furnace

molten nylon

spinneret

cool air

steam

nylon yarn

synthetic *The manufacture of nylon fibre. Nylon chips are melted in an inert atmosphere and the molten liquid extruded through a spinneret. The fibres formed solidify by cooling in air, and are drawn to several times their original length.*

systems program in computing, a program that performs a task related to the operation and performance of the computer system itself. For example, a systems program might control the operation of the display screen, or control and organize backing storage. In contrast, an ◊applications program is designed to carry out tasks for the benefit of the computer user.

System X in communications, a modular, computer-controlled, digital switching system used in telephone exchanges.

System X was originally developed by the UK companies GEC, Plessey, and STC at the request of the Post Office, beginning in 1969. A prototype exchange was finally commissioned 1978, and the system launched 1980. STC left the consortium 1982.

t symbol for ◊*tonne*, ◊*ton*.

tachograph combined speedometer and clock that records a vehicle's speed (on a small card disc, magnetic disc, or tape) and the length of time the vehicle is moving or stationary. It is used to monitor a lorry driver's working hours.

taiga or *boreal forest* Russian name for the forest zone south of the ◊tundra, found across the northern hemisphere. Here, dense forests of conifers (spruces and hemlocks), birches, and poplars occupy glaciated regions punctuated with cold lakes, streams, bogs, and marshes. Winters are prolonged and very cold, but the summer is warm enough to promote dense growth.

The varied fauna and flora are in delicate balance because the conditions of life are so precarious. This ecology is threatened by mining, forestry, and pipeline construction.

talc $Mg_3Si_4O_{10}(OH)_2$, mineral, hydrous magnesium silicate. It occurs in tabular crystals, but the massive impure form, known as *steatite* or *soapstone*, is more common. It is formed by the alteration of magnesium compounds and usually found in metamorphic rocks. Talc is very soft, ranked 1 on the Mohs' scale of hardness. It is used in powdered form in cosmetics, lubricants, and as an additive in paper manufacture.

French chalk and potstone are varieties of talc. Soapstone has a greasy feel to it, and is used for carvings such as Inuit sculptures.

Tanegashima Space Centre Japanese rocket-launching site on a small island off S Kyushu.

Tanegashima is run by the National Space Development Agency (NASDA), responsible for the practical applications of Japan's space programme (research falls under a separate organization based at ◊Kagoshima Space Centre). NASDA, founded 1969, has headquarters in Tokyo; a tracking and testing station, the Tsukuba Space Centre, in E central Honshu; and an Earth observation centre near Tsukuba. By 1988 NASDA had put 20 satellites into orbit.

tangent in geometry, a straight line that touches a curve and gives the ◊gradient of the curve at the point of contact. At a maximum, minimum, or point of inflection, the tangent to a curve has zero gradient. Also, in trigonometry, a function of an acute angle in a right-angled triangle, defined as the ratio of the length of the side opposite the angle to the length of the side adjacent to it; a way of expressing the gradient of a line.

tannic acid or *tannin* $C_{14}H_{10}O_9$ yellow astringent substance, composed of several ◊phenol rings, occurring in the bark, wood, roots, fruits, and galls (growths) of certain trees, such as the oak. It precipitates gelatin to give an insoluble compound, used in the manufacture of leather from hides (tanning).

tantalum hard, ductile, lustrous, grey-white, metallic element, symbol Ta, atomic number 73, relative atomic mass 180.948. It occurs with niobium in tantalite and other minerals. It can be drawn into wire with a very high melting point and great tenacity, useful for lamp filaments subject to vibration. It is also used in alloys, for corrosion-resistant laboratory apparatus and chemical equipment, as a catalyst in manufacturing synthetic rubber, in tools and instruments, and in rectifiers and capacitors.

It was discovered and named 1802 by Swedish chemist Anders Ekeberg (1767–1813) after the mythological Greek character Tantalus.

tape recording, magnetic method of recording electric signals on a layer of iron oxide, or other magnetic material, coating a thin plastic tape. The electrical signals from the microphone are fed to the electromagnetic recording head, which magnetizes the tape in accordance with the frequency and amplitude of the original signal. The impulses may be audio (for sound recording), video (for television), or data (for computer). For playback, the tape is passed over the same, or another, head to convert magnetic into electrical signals, which are then amplified for reproduction. Tapes are easily demagnetized (erased) for reuse, and come in cassette, cartridge, or reel form.

tangent to a curve

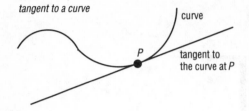

curve

P

tangent to
the curve at P

tangent to a circle

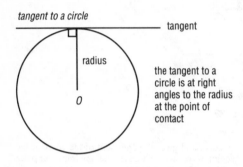

tangent

radius

O

the tangent to a
circle is at right
angles to the radius
at the point of
contact

tangent of an angle

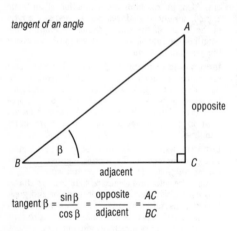

A

opposite

β

B

adjacent

C

$$\text{tangent } \beta = \frac{\sin \beta}{\cos \beta} = \frac{\text{opposite}}{\text{adjacent}} = \frac{AC}{BC}$$

tangent A tangent of a curve or circle is a straight line that touches the curve or circle at only one point. The tangent of an angle is a mathematical function used in the study of right-angled triangles. If the tangent of an angle β is known, then the length of the opposite side can be found given the length of the adjacent side, or vice versa.

tape streamer in computing, a backing storage device consisting of a continuous loop of magnetic tape. Tape streamers are largely used to store dumps (rapid backup copies) of important data files.

taproot in botany, a single, robust, main ◊root that is derived from the embryonic root, or ◊radicle, and grows vertically downwards, often to considerable depth. Taproots are often modified for food storage and are common in biennial plants such as the carrot *Daucus carota*, where they act as ◊perennating organs.

tar dark brown or black viscous liquid obtained by the destructive distillation of coal, shale, and wood. Tars consist of a mixture of hydrocarbons, acids, and bases. ◊Creosote and ◊paraffin are produced from wood tar. See also ◊coal tar.

tartaric acid $HCOO(CHOH)_2COOH$ organic acid present in vegetable tissues and fruit juices in the form of salts of potassium, calcium, and magnesium. It is used in carbonated drinks and baking powders.

tartrazine (E102) yellow food colouring produced synthetically from petroleum. Many people are allergic to foods containing it. Typical effects are skin disorders and respiratory problems. It has been shown to have an adverse effect on hyperactive children.

taste sense that detects some of the chemical constituents of food. The human ◊tongue can distinguish only four basic tastes (sweet, sour, bitter, and salty) but it is supplemented by the nose's sense of smell. What we refer to as taste is really a composite sense made up of both taste and smell.

tau ◊elementary particle with the same electric charge as the electron but a mass nearly double that of a proton. It has a lifetime of around 3×10^{-13} seconds and belongs to the ◊lepton family of particles—those that interact via the electromagnetic, weak nuclear, and gravitational forces, but not the strong nuclear force.

Taurus zodiacal constellation in the northern hemisphere near Orion, represented as a bull. The Sun passes through Taurus from mid-May to late June. Its brightest star is Aldebaran, seen as the bull's red eye. Taurus contains the Hyades and Pleiades open ◊star clusters, and the Crab nebula. In astrology, the dates for Taurus are between about 20 April and 20 May (see ◊precession).

tautomerism form of isomerism in which two interconvertible ◊isomers are in equilibrium. It is often specifically applied to an equilibrium between the keto (–C–C=O) and enol (–C=C–OH) forms of carbonyl compounds.

taxis (plural *taxes*) or *tactic movement* in botany, the movement of a single cell, such as a bacterium, protozoan, single-celled alga, or gamete, in response to an external stimulus. A movement directed towards the stimulus is described as positive taxis, and away from it as negative taxis. The alga *Chlamydomonas*, for example, demonstrates positive *phototaxis* by swimming towards a light source to increase the rate of photosynthesis. *Chemotaxis* is a response to a chemical stimulus, as seen in many bacteria that move towards higher concentrations of nutrients.

taxonomy another name for the ◊classification of living organisms.

TBT (abbreviation for ◊*tributyl tin*) chemical used in antifouling paints that has become an environmental pollutant.

T cell or *T lymphocyte* immune cell (see ◊immunity and ◊lymphocyte) that plays several roles in the body's defences. T cells are so called because they mature in the ◊thymus.

There are three main types of T cells: T helper cells (Th cells), which allow other immune cells to go into action; T suppressor cells (Ts cells), which stop specific immune reactions from occurring; and T cytotoxic cells (Tc cells), which kill cells that are cancerous or infected with viruses. Like ◊B cells, to which they are related, T cells have surface receptors that make them specific for particular antigens.

tear gas any of various volatile gases that produce irritation and tearing of the eyes, used by police against crowds and used in chemical warfare. The gas is delivered in pressurized, liquid-filled canisters or grenades, thrown by hand or launched from a specially adapted rifle. Gases (such as Mace) cause violent coughing and blinding tears, which pass when the victim breathes fresh air, and there are no lasting effects. Blister gases (such as mustard gas) and nerve gases are more harmful and may cause permanent injury or death.

tears salty fluid exuded by lachrymal glands in the eyes. The fluid contains proteins that are antibacterial, and also absorbs oils and mucus. Apart from cleaning and disinfecting the surface of the eye, the fluid supplies nutrients and gases to the cornea, which does not have a blood supply.

If insufficient fluid is produced, as sometimes happens in older people, the painful condition of 'dry-eye' results and the eye may be damaged.

technetium (Greek *technetos* 'artificial') silver-grey, radioactive, metallic element, symbol Tc, atomic number 43, relative atomic mass 98.906. It occurs in nature only in extremely minute amounts, produced as a fission product from uranium in ◊pitchblende and other uranium ores. Its longest-lived isotope, Tc-99, has a half-life of 216,000 years. It is a superconductor and is used as a hardener in steel alloys and as a medical tracer.

It was synthesized 1937 (named 1947) by Italian physicists Carlo Perrier and Emilio Segrè, who bombarded molybdenum with deuterons, looking to fill a missing spot in the ◊periodic table of the elements (at that time it was considered not to occur in nature). It was later isolated in large amounts from the fission-product debris of uranium fuel in nuclear reactors.

technology the use of tools, power, and materials, generally for the purposes of production. Almost every human process for getting food and shelter depends on complex technological systems, which have been developed over a 3–million-year period. Significant milestones include the advent of ◊steam engine 1712, the introduction of ◊electricity and the ◊internal-combustion engine in the mid-1800s, and recent developments in ◊communications, ◊electronics, and the nuclear and space industries. The *advanced technology* (highly automated and specialized) on which modern industrialized society depends is frequently contrasted with the *low technology* (labour-intensive and unspecialized) that characterizes some developing countries. ◊*Intermediate technology* is an attempt to adapt scientifically advanced inventions to less developed areas by using local materials and methods of manufacture.

power In human prehistory, the only power available was muscle power, augmented by primitive tools, such as the wedge or lever. The domestication of animals about 8500 BC and invention of the wheel about 300 BC paved the way for the water

mill (1st century BC) and later the windmill (12th century AD). Not until 1712 did an alternative source of power appear in the form of the first working steam engine, constructed by English inventor Thomas Newcomen; subsequent modifications improved its design. English chemist and physicist Michael Faraday's demonstration of the dynamo 1831 revealed the potential of the electrical motor, and in 1876 the German scientist Nicholas Otto introduced the four-stroke cycle used in the modern internal-combustion engine. The 1940s saw the explosion of the first atomic bomb and the subsequent development of the nuclear power industry. Latterly concern over the use of non-renewable power sources and the ◊pollution caused by the burning of fossil fuels has caused technologists to turn increasingly to exploring renewable sources of energy, in particular ◊solar energy, ◊wind energy, and ◊wave power.

materials The earliest materials used by humans were wood, bone, horn, shell, and stone. Metals were rare and/or difficult to obtain, although forms of bronze and iron were in use from 6000 BC and 1000 BC respectively. The introduction of the blast furnace in the 15th century enabled cast iron to be extracted, but this process remained expensive until English ironmaker Abraham Darby substituted coke for charcoal 1709, thus ensuring a plentiful supply of cheap iron at the start of the Industrial Revolution. Rubber, glass, leather, paper, bricks, and porcelain underwent similar processes of trial and error before becoming readily available. From the mid-1800s, entirely new materials, synthetics, appeared. First dyes, then plastic and the more versatile celluloid, and later drugs were synthesized, a process continuing into the 1980s with the growth of ◊genetic engineering, which enabled the production of synthetic insulin and growth hormones.

production The utilization of power sources and materials for production frequently lagged behind their initial discovery. The ◊lathe, known in antiquity in the form of a pole powered by a foot treadle, was not fully developed until the 18th century when it was used to produce objects of great precision, ranging from astronomical instruments to mass-produced screws. The realization that gears, cranks, cams, and wheels could operate in harmony to perform complex motion made mechanization possible. Early attempts at automation include Scottish engineer James Watt's introduction of the fly-ball governor into the steam engine 1769 to regulate the machine's steam supply automatically, and French textile maker Joseph Marie Jacquard's demonstration 1804 of how looms could be controlled automatically by punched cards. The first moving assembly line appeared 1870 in meat-packing factories in Chicago, USA, transferring to the motor industry 1913. With the perfection of the programmable electronic computer in the 1960s, the way lay open for fully automatic plants. The 1960s–90s saw extensive developments in the electronic and microelectronic industries (initially in the West but latterly Japan and the Pacific region have become primary producers) and in communications.

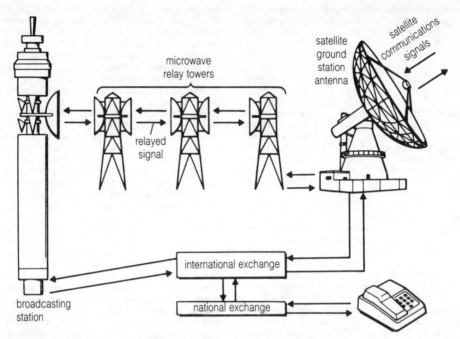

telecommunications *The international telecommunications system relies on microwave and satellite links for long-distance international calls. Cable links are increasingly made of optical fibres. The capacity of these links is enormous. The TDRS-C (tracking data and relay satellite communications) satellite, the world's largest and most complex satellite, can transmit in a single second the contents of a 20-volume encyclopedia, with each volume containing 1,200 pages of 2,000 words. A bundle of optical fibres, no thicker than a finger, can carry 10,000 phone calls—more than a copper wire as thick as an arm.*

Technology . . . the knack of so arranging the world that we don't have to experience it.

On **technology** Max Frisch, quoted in *The Image* D J Boorstin

tectonics in geology, the study of the movements of rocks on the Earth's surface. On a small scale tectonics involves the formation of ◊folds and ◊faults, but on a large scale ◊plate tectonics deals with the movement of the Earth's surface as a whole.

Teflon trade name for polytetrafluoroethene (PTFE), a tough, waxlike, heat-resistant plastic used for coating nonstick cookware and in gaskets and bearings.

tektite (from Greek *tektos* 'molten') small, rounded glassy stone, found in certain regions of the Earth, such as Australasia. Tektites are probably the scattered drops of molten rock thrown out by the impact of a large ◊meteorite.

telecommunications communications over a distance, generally by electronic means. Long-distance voice communication was pioneered 1876 by Scottish–US inventor Alexander Graham Bell, when he invented the telephone as a result of English chemist and physicist Michael Faraday's discovery of electromagnetism. Today it is possible to communicate with most countries by telephone cable, or by satellite or microwave link, with over 100,000 simultaneous conversations and several television channels being carried by the latest satellites. ◊Integrated Services Digital Network (ISDN) makes videophones and high-quality fax possible; the world's first large-scale centre of ISDN began operating in Japan 1988. ISDN is a system that transmits voice and image data on a single transmission line by changing them into digital signals. The chief method of relaying long-distance calls on land is microwave radio transmission.

The first mechanical telecommunications systems were the semaphore and heliograph (using flashes of sunlight), invented in the mid-19th century, but the forerunner of the present telecommunications age was the electric telegraph. The earliest practicable telegraph instrument was invented by William Cooke and Charles Wheatstone in Britain 1837 and used by railway companies. In the USA, Samuel Morse invented a signalling code, ◊Morse code, which is still used, and a recording telegraph, first used commercially between England and France 1851. As a result of German physicist Heirich Hertz's discoveries using electromagnetic waves, Italian inventor Gugliemo Marconi pioneered a 'wireless' telegraph, ancestor of the radio. He established wireless communication between England and France 1899 and across the Atlantic 1901. The modern telegraph uses teleprinters to send coded messages along telecommunications lines. Telegraphs are keyboard-operated machines that transmit a five-unit

Baudot code (see ◊baud). The receiving teleprinter automatically prints the received message.

The drawback to long-distance voice communication via microwave radio transmission is that the transmissions follow a straight line from tower to tower, so that over the sea the system becomes impracticable. A solution was put forward 1945 by the science-fiction writer Arthur C Clarke, when he proposed a system of communications satellites in an orbit 35,900 km/22,300 mi above the equator, where they would circle the Earth in exactly 24 hours, and thus appear fixed in the sky. Such a system is now in operation internationally, by ◊Intelsat. The satellites are called geostationary satellites (syncoms). The first to be successfully launched, by Delta rocket from Cape Canaveral, was *Syncom 2* in July 1963. Many such satellites are now in use, concentrated over heavy traffic areas such as the Atlantic, Indian, and Pacific oceans. Telegraphy, telephony, and television transmissions are carried simultaneously by high-frequency radio waves. They are beamed to the satellites from large dish antennae or Earth stations, which connect with international networks. Recent advances include the use of fibre-optic cables consisting of fine glass fibres for telephone lines instead of the usual copper cables. The telecommunications signals are transmitted along the fibres on pulses of laser light.

The first public telegraph line was laid between Paddington and Slough 1843. In 1980 the Post Office opened its first System X (all-electronic, digital) telephone exchange in London, a method already adopted in North America. In the UK

Goonhilly and Madley are the main Earth stations for satellite transmissions.

telegraphy transmission of coded messages along wires by means of electrical signals. The first modern form of telecommunication, it now uses printers for the transmission and receipt of messages. Telex is an international telegraphy network.

Overland cables were developed in the 1830s, but early attempts at underwater telegraphy were largely unsuccessful until the discovery of the insulating gum gutta-percha 1843 enabled a cable to be laid across the English Channel 1851. *Duplex telegraph* was invented in the 1870s, enabling messages to be sent in both directions simultaneously. Early telegraphs were mainly owned by the UK: 72% of all submarine cables were British-owned 1900.

telemetry measurement at a distance, in particular the systems by which information is obtained and sent back by instruments on board a spacecraft. See ◊remote sensing.

telephone instrument for communicating by voice over long distances, invented by Scottish–US inventor Alexander Graham Bell 1876. The transmitter (mouthpiece) consists of a carbon microphone, with a diaphragm that vibrates when a person speaks into it. The diaphragm vibrations compress grains of carbon to a greater or lesser extent, altering their resistance to an electric current passing through them. This sets up variable electrical signals, which travel along the telephone lines to the receiver of the person being called.

telecommunications: chronology

1794	Claude Chappe in France built a long-distance signalling system using semaphore.
1839	Charles Wheatstone and William Cooke devised an electric telegraph in England.
1843	Samuel Morse transmitted the first message along a telegraph line in the USA, using his Morse code of signals—short (dots) and long (dashes).
1858	The first transatlantic telegraph cable was laid.
1876	Alexander Graham Bell invented the telephone.
1877	Thomas Edison invented the carbon transmitter for the telephone.
1878	The first telephone exchange was opened at New Haven, Connecticut.
1884	The first long-distance telephone line was installed, between Boston and New York.
1891	A telephone cable was laid between England and France.
1892	The first automatic telephone exchange was opened, at La Porte, Indiana.
1894	Guglielmo Marconi pioneered wireless telegraphy in Italy, later moving to England.
1900	Reginald Fessenden in the USA first broadcast voice by radio.
1901	Marconi transmitted the first radio signals across the Atlantic.
1904	John Ambrose Fleming invented the thermionic valve.
1907	Charles Krumm introduced the forerunner of the teleprinter.
1920	Stations in Detroit and Pittsburgh began regular radio broadcasts.
1922	The BBC began its first radio transmissions, for the London station 2LO.
1932	The Post Office introduced the Telex in Britain.
1956	The first transatlantic telephone cable was laid.
1962	Telstar pioneered transatlantic satellite communications, transmitting live TV pictures.
1966	Charles Kao in England advanced the idea of using optical fibres for telecommunications transmissions.
1969	Live TV pictures were sent from astronauts on the Moon back to Earth.
1975	The Post Office announced Prestel, the world's first viewdata system, using the telephone lines to link a computer data bank with the TV screen.
1977	The first optical fibre cable was installed in California.
1984	First commercial cellphone service started in Chicago, USA.
1988	International Services Digital Network (ISDN), an international system for sending signals in digital format along optical fibres and coaxial cable, launched in Japan.
1989	The first transoceanic optical fibre cable, capable of carrying 40,000 simultaneous telephone conversations, was laid between Europe and the USA.
1991	ISDN introduced in the UK.
1992	Videophones, made possible by advances in image compression and the development of ISDN, introduced in the UK.

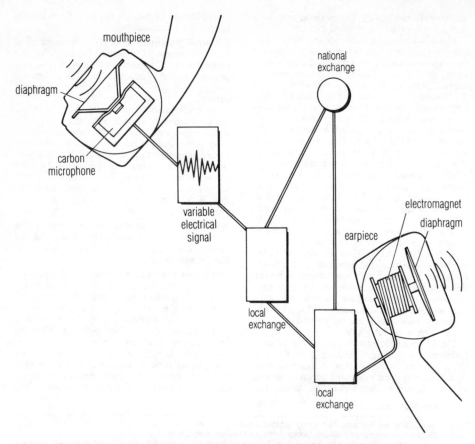

mouthpiece

diaphragm

carbon microphone

variable electrical signal

national exchange

electromagnet

diaphragm

earpiece

local exchange

local exchange

telephone *Sound vibrations are in the telephone converted to an electric signal and back again. The mouthpiece contains a carbon microphone that produces an electrical signal which varies in step with the spoken sounds. The signal is routed to the receiver via local or national exchanges. The earpiece contains an electromagnetic loudspeaker which reproduces the sounds by vibrating a diaphragm.*

There they cause the magnetism of an electromagnet to vary, making a diaphragm above the electromagnet vibrate and give out sound waves, which mirror those that entered the mouthpiece originally.

The standard instrument has a handset, which houses the transmitter (mouthpiece), and receiver (earpiece), resting on a base, which has a dial or push-button mechanism for dialling a telephone number. Some telephones combine a push-button mechanism and mouthpiece and earpiece in one unit. A cordless telephone is of this kind, connected to a base unit not by wires but by radio. It can be used at distances up to about 100 m/330 ft from the base unit. In 1988 Japan and in 1991 Britain introduced an ◊Integrated Services Digital Network (see ◊telecommunications), providing fast transfer of computerized information.

Mr Watson, come here; I want you.

Alexander Graham Bell, first complete sentence spoken over the **telephone** March 1876

telephoto lens photographic lens of longer focal length than normal that takes a very narrow view

and gives a large image through a combination of telescopic and ordinary photographic lenses.

teleprinter or **teletypewriter** transmitting and receiving device used in telecommunications to handle coded messages. Teleprinters are automatic typewriters keyed telegraphically to convert typed words into electrical signals (using a 5–unit Baudot code, see ◊baud) at the transmitting end, and signals into typed words at the receiving end.

telescope optical instrument that magnifies images of faint and distant objects; any device for collecting and focusing light and other forms of electromagnetic radiation. It is a major research tool in astronomy and is used to sight over land and sea; small telescopes can be attached to cameras and rifles. A telescope with a large aperture, or opening, can distinguish finer detail and fainter objects than one with a small aperture. The **refracting telescope** uses lenses, and the **reflecting telescope** uses mirrors. A third type, the **catadioptric telescope**, with a combination of lenses and mirrors, is used increasingly. See also ◊radio telescope.

In a refractor, light is collected by a ◊lens called

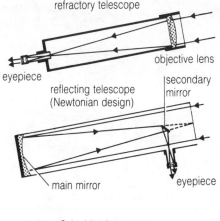

refractory telescope

objective lens

eyepiece

reflecting telescope
(Newtonian design)

secondary mirror

main mirror

eyepiece

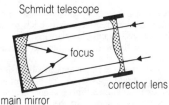

Schmidt telescope

focus

corrector lens

main mirror

telescope *Three kinds of telescope. The refracting telescope uses a large objective lens to gather light and form an image which the smaller eyepiece lens magnifies. A reflecting telescope uses a mirror to gather light. The Schmidt telescope uses a corrective lens to achieve a wide field of view. It is one of the most widely used tools of astronomy.*

the *object glass* or *objective*, which focuses light down a tube, forming an image magnified by an *eyepiece*. Invention of the refractor is attributed to a Dutch optician, Hans Lippershey 1608. The largest refracting telescope in the world, at ◊Yerkes Observatory, Wisconsin, USA, has an aperture of 102 cm/40 in.

In a reflector, light is collected and focused by a concave mirror. The first reflector was built about 1670 by English physicist and mathematician Isaac Newton. Large mirrors are cheaper to make and easier to mount than large lenses, so all the largest telescopes are reflectors. The largest reflector with a single mirror, 6 m/236 in, is at ◊Zelenchukskaya, Russia. Telescopes with larger apertures composed of numerous smaller segments have been built, such as the Keck Telescope on ◊Mauna Kea. A *multiple-mirror telescope* was installed on Mount Hopkins, Arizona 1979. It consists of six mirrors of 1.8 m/72 in aperture, which perform like a single 4.5 m/176 in mirror. ◊*Schmidt telescopes* are used for taking wide-field photographs of the sky. They have a main mirror plus a thin lens at the front of the tube to increase the field of view.

Large telescopes can now be placed in orbit above the distorting effects of the Earth's atmosphere. Telescopes in space have been used to study infrared, ultraviolet, and X-ray radiation that does not penetrate the atmosphere but carries much information about the births, lives, and deaths of stars and galaxies. The 2.4–m/94–in ◊Hubble

Space Telescope, launched 1990, can see the sky more clearly than can any telescope on Earth, despite an optical defect that impairs its performance.

TELESCOPES: NO PEEKING

Modern astronomers rarely look through their telescopes. For years most observations were recorded on photographic plates and, more recently, on electronic devices called charge-coupled devices, or as computer data to be analysed later. The computer screen provides a visual picture.

teletext broadcast system of displaying information on a television screen. The information — typically about news items, entertainment, sport, and finance — is constantly updated. Teletext is a form of ◊videotext, pioneered in Britain by the British Broadcasting Corporation (BBC) with Ceefax and by Independent Television with Teletext.

television (TV) reproduction at a distance by radio waves of visual images. For transmission, a television camera converts the pattern of light it takes in into a pattern of electrical charges. This is scanned line by line by a beam of electrons from an electron gun, resulting in variable electrical signals that represent the picture. These signals are combined with a radio carrier wave and broadcast as electromagnetic waves. The TV aerial picks up the wave and feeds it to the receiver (TV set). This separates out the vision signals, which pass to a cathode-ray tube where a beam of electrons is made to scan across the screen line by line, mirroring the action of the electron gun in the TV camera. The result is a recreation of the pattern of light that entered the camera. Twenty-five pictures are built up each second with interlaced scanning in Europe (30 in North America), with a total of 625 lines in Europe (525 lines in North America and Japan).

television channels In addition to transmissions received by all viewers, the 1970s and 1980s saw the growth of pay-television cable services, which are received only by subscribers, and of devices, such as those used in the Qube system (USA), which allow the viewers' opinions to be transmitted instantaneously to the studio via a response button, so that, for example, a home viewing audience can vote in a talent competition. The number of programme channels continues to increase, following the introduction of satellite-beamed TV signals.

Further use of TV sets has been brought about by ◊videotext and the use of video recorders to tape programmes for playback later or to play pre-recorded videocassettes, and by their use as computer screens and for security systems. Extended-definition television gives a clear enlargement from a microscopic camera and was first used 1989 in neurosurgery to enable medical students to watch brain operations.

history In 1873 it was realized that, since the electrical properties of the nonmetallic chemical element selenium vary according to the amount of

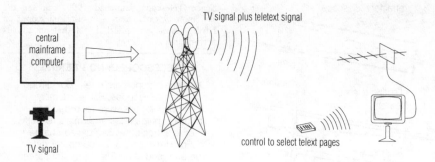

TV signal plus teletext signal

central mainframe computer

TV signal

control to select telext pages

teletext

light to which it is exposed, light could be converted into electrical impulses, making it possible to transmit such impulses over a distance and then reconvert them into light. The chief difficulty was seen to be the 'splitting of the picture' so that the infinite variety of light and shade values might be transmitted and reproduced. In 1908 Campbell-Swinton pointed out that cathode-ray tubes would best effect transmission and reception. Mechanical devices were used at the first practical demonstration of television, given by Scottish electrical engineer John Logie Baird in London 27 Jan 1926, and cathode-ray tubes were used experimentally in the UK from 1934.

colour television Baird gave a demonstration of colour TV in London 1928, but it was not until Dec 1953 that the first successful system was adopted for broadcasting, in the USA. This is called the ◊NTSC system, since it was developed by the National Television System Committee, and variations of it have been developed in Europe; for example, SECAM (sequential and memory) in France and Eastern Europe, and PAL (phase alternation by line) in most of Western Europe. The three differ only in the way colour signals are prepared for transmission, the scanning rate, and the number of lines used. When there was no agreement on a universal European system 1964, in 1967 the UK, West Germany, the Netherlands, and Switzerland adopted PAL while France and the USSR adopted SECAM. In 1989 the European Community agreed to harmonize TV channels from 1991, allowing any station to show programmes anywhere in the EC.

The method of colour reproduction is related to that used in colour photography and printing. It uses the principle that any colours can be made by mixing the primary colours red, green, and blue in appropriate proportions. In colour television the receiver reproduces only three basic colours: red, green, and blue. The effect of yellow, for example, is reproduced by combining equal amounts of red and green light, while white is formed by a mixture of all three basic colours. Signals indicate the amounts of red, green, and blue light to be generated at the receiver. To transmit each of these three signals in the same way as the single brightness signal in black and white television would need three times the normal band width and reduce the number of possible stations and programmes to one-third of that possible with monochrome

television. The three signals are therefore coded into one complex signal, which is transmitted as a more or less normal black and white signal and produces a satisfactory—or compatible—picture on black and white receivers. A fraction of each primary red, green, and blue signal is added together to produce the normal brightness, or luminance, signal. The minimum of extra colouring information is then sent by a special subcarrier signal, which is superimposed on the brightness signal. This extra colouring information corresponds to the hue and saturation of the transmitted colour, but without any of the fine detail of the picture. The impression of sharpness is conveyed only by the brightness signal, the colouring being added as a broad colour wash. The various colour systems differ only in the way in which the colouring information is sent on the subcarrier signal. The colour receiver has to amplify the complex signal and decode it back to the basic red, green, and blue signals; these primary signals are then applied to a colour cathode-ray tube.

The colour display tube is the heart of any colour receiver. Many designs of colour picture tubes have been invented; the most successful of these is known as the 'shadow mask tube'. It operates on similar electronic principles to the black and white television picture tube, but the screen is composed of a fine mosaic of over one million dots arranged in an orderly fashion. One-third of the dots glow red when bombarded by electrons, one-third glow green, and one-third blue. There are three sources of electrons, respectively modulated by the red, green, and blue signals. The tube is arranged so that the shadow mask allows only the red signals to hit red dots, the green signals to hit green dots, and the blue signals to hit blue dots. The glowing dots are so small that from a normal viewing distance the colours merge into one another and a picture with a full range of colours is seen.

high-definition television (HDTV) This system offers a significantly greater number of scanning lines, and therefore a clearer picture, than the 525/ 625 lines of established television systems. In 1989 the Japanese broadcasting station NHK and a consortium of manufacturers launched the Hi-Vision HDTV system, with 1,125 lines and a wide-screen format. In 1986, European electronics companies, research laboratories, and broadcasting authorities joined together to develop a European HDTV system, called HD-MAC, which uses 1,250 lines.

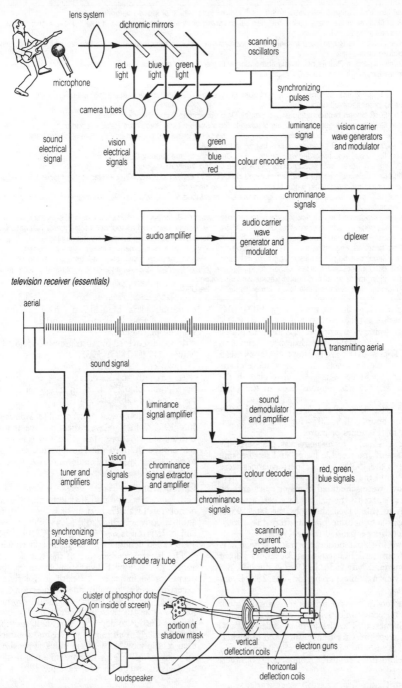

television transmitter (essentials)

lens system

dichromic mirrors

scanning oscillators

red light   blue light   green light

microphone

synchronizing pulses

camera tubes

sound electrical signal

vision electrical signals

luminance signal

vision carrier wave generators and modulator

green
blue
red

colour encoder

chrominance signals

audio amplifier

audio carrier wave generator and modulator

diplexer

television receiver (essentials)

aerial

transmitting aerial

sound signal

luminance signal amplifier

sound demodulator and amplifier

tuner and amplifiers

vision signals

chrominance signal extractor and amplifier

colour decoder

red, green, blue signals

chrominance signals

synchronizing pulse separator

scanning current generators

cathode ray tube

cluster of phosphor dots (on inside of screen)

portion of shadow mask

vertical deflection coils

electron guns

horizontal deflection coils

loudspeaker

television *Simplified block diagram of a complete colour television system—transmitting and receiving. The camera separates the picture into three colours—red, blue, green—by using filters and different camera tubes for each colour. The audio signal is produced separately from the video signal. Both signals are transmitted from the same aerial using a special coupling device called a diplexer. There are four sections in the receiver: the aerial, the tuners, the decoders, and the display. As in the transmitter, the audio and video signals are processed separately. The signals are amplified at various points.*

television: chronology

1878	William Crookes in England invented the Crookes tube, which produced cathode rays.
1884	Paul Nipkow in Germany built a mechanical scanning device, the Nipkow disc, a rotating disc with a spiral pattern of holes in it.
1897	Karl Ferdinand Braun, also in Germany, modified the Crookes tube to produce the ancestor of the TV receiver picture tube.
1906	Boris Rosing in Russia began experimenting with the Nipkow disc and cathode-ray tube, eventually succeeding in transmitting some crude TV pictures.
1923	Vladimir Zworykin in the USA invented the first electronic camera tube, the iconoscope.
1926	John Logie Baird demonstrated a workable TV system, using mechanical scanning by Nipkow disc.
1928	Baird demonstrated colour TV.
1929	The BBC began broadcasting experimental TV programmes using Baird's system.
1936	The BBC began regular broadcasting using Baird's system from Alexandra Palace, London.
1940	Experimental colour TV transmission began in the USA, using the present-day system of colour reproduction.
1953	Successful colour TV transmissions began in the USA.
1956	The first videotape recorder was produced in California by the Ampex Corporation.
1962	TV signals were transmitted across the Atlantic via the Telstar satellite.
1970	The first videodisc system was announced by Decca in Britain and AEG-Telefunken in Germany, but it was not perfected until the 1980s, when laser scanning was used for playback.
1973	The BBC and Independent Television in the UK introduced the world's first teletext systems, Ceefax and Oracle, respectively.
1975	Sony introduced their videocassette tape-recorder system, Betamax, for domestic viewers, six years after their professional U-Matic system. The UK Post Office (now British Telecom) announced their Prestel viewdata system.
1979	Matsushita in Japan developed a pocket-sized, flat-screen TV set, using a liquid-crystal display.
1986	Data broadcasting using digital techniques was developed; an enhancement of teletext was produced.
1989	The Japanese began broadcasting high-definition television; satellite television was introduced in the UK.
1990	The BBC introduced a digital stereo sound system (NICAM); MAC, a European system allowing greater picture definition, more data, and sound tracks, was introduced.
1992	All-digital high-definition television demonstrated in the USA.

However, the slow development of the HD-MAC system has meant that rival systems, such as PAL-plus, may triumph in Europe. Advances in digital technology also make the European situation unclear. Digital data compression techniques mean that digital television sets can receive more channels than analogue sets. An all-digital high-definition sustem was demonstrated in the USA 1992.

In the UK, 33% of the people watch TV for over three hours a day, 20% for over five hours, and 7% for 11 hours or more.

In the USA, TV technology was pioneered by David Sarnoff and Lee De Forest and sets became available in the 1930s, but few performances were televized until the late 1940s, when local and network shows were scheduled in major cities and, by coaxial cable, across the nation. Live performances gave way to videotaped shows by the late 1950s, and colour sets became popular from the 1960s. Cable networks expanded in the 1980s.

The world's first public television service was started from the BBC station at Alexandra Palace in N London, 2 Nov 1936. In 1990 in the UK, the average viewing time per person was 25.5 hours each week.

telex (acronym for *tele*printer *ex*change) international telecommunications network that handles telegraph messages in the form of coded signals. It uses ◊teleprinters for transmitting and receiving, and makes use of land lines (cables) and radio and satellite links to make connections between subscribers.

tellurium (Latin *Tellus* 'Earth') silver-white, semi-metallic (◊metalloid) element, symbol Te, atomic number 52, relative atomic mass 127.60. Chemically it is similar to sulphur and selenium, and it is considered as one of the sulphur group. It occurs naturally in telluride minerals, and is used in colouring glass blue–brown, in the electrolytic refining of zinc, in electronics, and as a catalyst in refining petroleum.

It was discovered by Austrian mineralogist Franz Müller (1740–1825) 1782, and named 1798 by German chemist Martin Klaproth.

Its strength and hardness are greatly increased by addition of 0.1% lead; in this form it is used for pipes and cable sheaths.

Telstar US communications satellite, launched 10 July 1962, which relayed the first live television transmissions between the USA and Europe. *Telstar* orbited the Earth in 158 minutes, and so had to be tracked by ground stations, unlike the geostationary satellites of today.

temperature state of hotness or coldness of a body, and the condition that determines whether or not it will transfer heat to, or receive heat from, another body according to the laws of ◊thermodynamics. It is measured in degrees Celsius (before 1948 called centigrade), kelvin, or Fahrenheit.

The normal temperature of the human body is about 36.9°C/98.4°F. Variation by more than a degree or so indicates ill-health, a rise signifying excessive activity (usually due to infection), and a decrease signifying deficient heat production (usually due to lessened vitality). To convert degrees Celsius to degrees Fahrenheit, multiply by ⁹⁄₅ and add 32; Fahrenheit to Celsius, subtract 32, then multiply by ⁵⁄₉. A useful quick approximation for converting Celsius to Fahrenheit is to double the Celsius and add 30, for example 12°C = 24 + 30 = 54°F.

temperature regulation the ability of an organism to control its internal body temperature; in warm-blooded animals this is known as ◊homeothermy.

Although some plants have evolved ways of

RECENT PROGRESS IN TELEVISION AND VIDEO TECHNOLOGY

High-definition TV: a confusing picture

Current television pictures in Europe are built from 625 horizontal scanning lines. On a small screen, the lines are too close together to be visible; on a large screen, especially a projected television picture, they become very noticeable. In the USA and Japan, television pictures consist of just 525 lines, and look even coarser on large screens.

Around 20 years ago, engineers with Japan's state broadcaster NHK started developing a system with 1,125 lines, giving a movie-quality picture. They also made the picture wider, with a 16:9 aspect ratio instead of conventional television's squarish 4:3. This is high-definition television (HDTV).

The HDTV signal contains at least four times more information than a conventional television signal, and cannot be transmitted in the channels allocated for conventional terrestrial broadcasts. Satellite transmitters have wider bandwidths, which is why Japan pushed ahead with satellite broadcasting. Satellites are now broadcasting HDTV programmes direct into Japanese homes.

European confusion

In 1986 European manufacturers and broadcasters cooperated to develop a native European HDTV system. Called HD-MAC, it uses 1,250 lines, twice the current number, to facilitate compatibility with existing television sets. It is not directly compatible with the PAL sets used in most of Europe, or the SECAM sets used in France. HD-MAC builds instead on a new 625-line system called MAC.

This was invented in the early 1980s in the research laboratories of the Independent Broadcasting Authority (now privatized as National Transcommunications Ltd). Through the 1980s, other European broadcasters and electronics companies helped perfect MAC. It was intended for satellite use, and designed to bridge the gap between 'old-fashioned' 4:3 625-line pictures, new 625-line widescreen pictures, and future 1,250-line HDTV widescreen pictures.

In 1986 the EC decreed that all European satellite broadcasters must use MAC, not PAL or SECAM. The aim was to provide Europe with an elegant upgrade path to widescreen and 1,250-line television. Unfortunately, Europe split the MAC standard into incompatible variants: D-MAC for the UK and D2-MAC for France, Germany and most other countries. This slowed development of the vital MAC receiver microchips. The EC's directive contained legal loopholes that allowed some broadcasters (including the UK's Sky and the German cable channels) to use PAL from Luxembourg's Astra satellite. In late 1992, the UK vetoed EC plans to subsidise satellite transmissions using the D2-MAC system. The veto effectively killed off MAC and HD-MAC. In February 1993, the EC said that it would not try to force broadcasters to use these systems.

Meanwhile, European broadcasters have been developing PALplus, a widescreen version of the existing PAL system. Television stations will transmit programmes in 'letterbox' format, with black borders at the top and bottom of the screen. New, widescreen television sets will expand this image to fill a full 16:9 screen, extracting signals hidden in the black borders to restore the clarity lost by expansion. The PALplus pictures still have only 625 lines, but look clearer (and wider) than existing PAL pictures.

Many people now question the need for 1,250-line HDTV pictures unless the screen is very large; and screen size is limited by the difficulty of making large cathode-ray tubes.

Digital systems

All existing television and HDTV systems are analogue-based. But the major electronics companies in Europe, Japan and the USA have been working on digital systems. The US government looks set to issue a digital standard for HDTV.

The key is data compression, which can reduce the number of digital bits needed to carry a picture by 30 or 40 times, without noticeable loss of quality. An existing television channel could carry at least one HDTV channel or several programmes of today's quality. So digital television gives viewers wider programme choice. The technology works, and should be available for domestic use by the late 1990s. Both terrestrial and satellite broadcasters can use it, creating market opportunities for a new kind of home video recorder.

Existing VCRs record pictures as analogue signals. Picture quality from the VHS and Video 8 formats has been improved by the Super-VHS and Hi 8 variants. These deliver quality matching or exceeding broadcast standards, so few users need the further improvement offered by switching to digital recording. Digital recording, however, will allow taping of a digital broadcast channel and subsequent decoding, watching either HDTV on a widescreen set or choosing between a selection of conventional-quality programmes, all taped together from the same channel.

Barry Fox

resisting extremes of temperature, sophisticated mechanisms for maintaining the correct temperature are found in multicellular animals. Such mechanisms may be behavioural, as when a lizard moves into the shade in order to cool down, or internal, as in mammals and birds, where temperature is regulated by the ◊medulla.

tempering heat treatment for improving the properties of metals, often used for steel alloys. The metal is heated to a certain temperature and then cooled suddenly in a water or oil bath.

temporary hardness hardness of water that is removed by boiling (see ◊hard water).

tendon or *sinew* cord of tough, fibrous connective tissue that joins muscle to bone in vertebrates. Tendons are largely composed of the protein collagen, and because of their inelasticity are very efficient at transforming muscle power into movement.

tendril in botany, a slender, threadlike structure that supports a climbing plant by coiling around suitable supports, such as the stems and branches of other plants. It may be a modified stem, leaf, leaflet, flower, leaf stalk, or stipule (a small append-

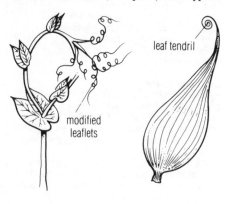

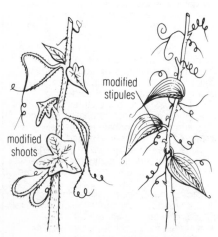

tendril *Tendrils are specially modifed leaves, shoots, or stems. They support the plant by twining around the stems of other plants nearby, as in the pea, or they may attach themselves to suitable surfaces by means of suckers, as in the virginia creeper.*

age on either side of the leaf stalk), and may be simple or branched. The tendrils of Virginia creeper *Parthenocissus quinquefolia* are modified flower heads with suckerlike pads at the end that stick to walls, while those of the grapevine *Vitis* grow away from the light and thus enter dark crevices where they expand to anchor the plant firmly.

tension reaction force set up in a body that is subjected to stress. In a stretched string or wire it exerts a pull that is equal in magnitude but opposite in direction to the stress being applied at its ends. Tension originates in the net attractive intermolecular force created when a stress causes the mean distance separating a material's molecules to become greater than the equilibrium distance. It is measured in newtons.

terbium soft, silver-grey, metallic element of the ◊lanthanide series, symbol Tb, atomic number 81, relative atomic mass 158.925. It occurs in gadolinite and other ores, with yttrium and ytterbium, and is used in lasers, semiconductors, and television tubes. It was named in 1843 by Swedish chemist Carl Mosander (1797–1858) for the town of Ytterby, Sweden, where it was first found.

terminal in computing, a device consisting of a keyboard and display screen (◊VDU)—or, in older systems, a teleprinter—to enable the operator to communicate with the computer. The terminal may be physically attached to the computer or linked to it by a telephone line (remote terminal). A 'dumb' terminal has no processor of its own, whereas an 'intelligent' terminal has its own processor and takes some of the processing load away from the main computer.

terminal moraine linear, slightly curved ridge of rocky debris deposited at the front end, or snout, of a glacier. It represents the furthest point of advance of a glacier, being formed when deposited material (till), which was pushed ahead of the snout as it advanced, became left behind as the glacier retreated.

A terminal moraine may be hundreds of metres in height; for example, the Franz Joseph glacier in New Zealand has terminal moraine that is over 400 m/1,320 ft high.

terminal velocity or *terminal speed* the maximum velocity that can be reached by an object moving through a fluid (gas or liquid). As the speed of the object increases so does the total magnitude of the forces resisting its motion. Terminal velocity is reached when the resistive forces exactly balance the applied force that has caused the object to accelerate; because there is now no resultant force, there can be no further acceleration.

For example, an object falling through air will reach a terminal velocity and cease to accelerate under the influence of gravity when the air resistance equals the object's weight. Parachutes are designed to increase air resistance so that the acceleration of a falling person or package ceases more rapidly, thereby limiting terminal velocity to a safe level.

terminal voltage the potential difference (pd) or voltage across the terminals of a power supply, such as a battery of cells. When the supply is not connected in circuit its terminal voltage is the same as its ◊electromotive force (emf); however, as soon as it begins to supply current to a circuit its terminal voltage falls because some electric potential energy is lost in driving current against the supply's own ◊internal resistance. As the current flowing in the circuit is increased the terminal voltage of the supply falls.

terrane in geology, a tract of land with a distinct geological character. The term *exotic terrane* is commonly used to describe a rock mass that has a very different history from others nearby. The exotic terranes of the Rocky Mountains represent old island chains that have been brought to the North American continent by the movements of plate tectonics, and welded to its edge during the formation of the ◊young fold mountains.

territorial behaviour in biology, any behaviour that serves to exclude other members of the same species from a fixed area or ◊territory. It may involve aggressively driving out intruders, marking the boundary (with dung piles or secretions from special scent glands), conspicuous visual displays, characteristic songs, or loud calls.

territory in animal behaviour, a fixed area from which an animal or group of animals excludes other members of the same species. Animals may hold territories for many different reasons; for example, to provide a constant food supply, to monopolize potential mates, or to ensure access to refuges or nest sites. The size of a territory depends in part on its function: some nesting and mating territories may be only a few square metres, whereas feeding territories may be as large as hundreds of square kilometres.

Tertiary period of geological time 65–1.64 million years ago, divided into five epochs: Palaeocene, Eocene, Oligocene, Miocene, and Pliocene. During the Tertiary, mammals took over all the ecological niches left vacant by the extinction of the dinosaurs, and became the prevalent land animals. The continents took on their present positions, and climatic and vegetation zones as we know them became established. Within the geological time column the Tertiary follows the Cretaceous period and is succeeded by the Quaternary period.

Terylene trade name for a polyester synthetic fibre produced by the chemicals company ICI. It is made by polymerizing ethylene glycol and terephthalic acid. Cloth made from Terylene keeps its shape after washing and is hard-wearing.

Terylene was the first wholly synthetic fibre invented in Britain. It was created by the chemist J R Whinfield of Accrington 1941. In 1942 the rights were sold to ICI (Du Pont in the USA) and bulk production began 1955. Since 1970 it has been the most widely produced synthetic fibre, often under the generic name polyester. In 1989 8.4 million tonnes were produced, constituting over 50% of world synthetic fibre output.

tesla SI unit (symbol T) of ◊magnetic flux density. One tesla represents a flux density of one ◊weber per square metre, or 10^4 ◊gauss. It is named after the Croatian engineer Nikola Tesla.

testa the outer coat of a seed, formed after fertilization of the ovule. It has a protective function and is usually hard and dry. In some cases the coat is adapted to aid dispersal, for example by being hairy. Humans have found uses for many types of testa, for example the fibre of the cotton seed.

test cross in genetics, a breeding experiment used to discover the genotype of an individual organism. By crossing with a double recessive of the same species, the offspring will indicate whether the test individual is homozygous or heterozygous for the characteristic in question. In peas, a tall plant under investigation would be crossed with a double recessive short plant with known genotype tt. The results of the cross will be all tall plants if the test plant is TT. If the individual is in fact Tt then there will be some short plants (genotype tt) among the offspring.

testis (plural *testes*) the organ that produces ◊sperm in male (and hermaphrodite) animals. In vertebrates it is one of a pair of oval structures that are usually internal, but in mammals (other than elephants and marine mammals), the paired testes (or testicles) descend from the body cavity during development, to hang outside the abdomen in a scrotal sac.

testosterone in vertebrates, hormone secreted chiefly by the testes, but also by the ovaries and the cortex of the adrenal glands. It promotes the development of secondary sexual characteristics in males. In animals with a breeding season, the onset of breeding behaviour is accompanied by a rise in the level of testosterone in the blood.

Synthetic or animal testosterone is used to treat inadequate development of male characteristics or (illegally) to aid athletes' muscular development. Like other sex hormones, testosterone is a ◊steroid.

Tethys Sea sea that once separated ◊Laurasia from ◊Gondwanaland. It has now closed up to become the Mediterranean, the Black, the Caspian, and the Aral seas.

tetrachloromethane or *carbon tetrachloride* CCl_4 chlorinated organic compound that is a very efficient solvent for fats and greases, and was at one time the main constituent of household dry-cleaning fluids and of fire extinguishers used with electrical and petrol fires. Its use became restricted after it was discovered to be carcinogenic and it has now been largely removed from educational and industrial laboratories.

tetraethyl lead $Pb(C_2H_5)_4$ compound added to leaded petrol as a component of ◊antiknock to increase the efficiency of combustion in car engines. It is a colourless liquid that is insoluble in water but soluble in organic solvents such as benzene, ethanol, and petrol.

tetrahedron (plural *tetrahedra*) in geometry, a solid figure (◊polyhedron) with four triangular faces; that is, a ◊pyramid on a triangular base. A regular tetrahedron has equilateral triangles as its faces.

In chemistry and crystallography, tetrahedra describe the shapes of some molecules and crystals; for example, the carbon atoms in a crystal of diamond are arranged in space as a set of interconnected regular tetrahedra.

tetrapod (Latin 'four-legged') type of ◊vertebrate. The group includes mammals, birds, reptiles, and amphibians. Birds are included because they

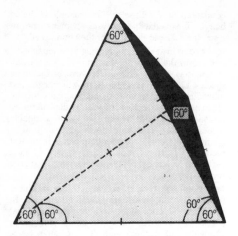

tetrahedron *A regular tetrahedron is a pyramid on a triangular base with all its sides equal in length.*

evolved from four-legged ancestors, the forelimbs having become modified to form wings. Even snakes are tetrapods, because they are descended from four-legged reptiles.

text editor in computing, a program that allows the user to edit text on the screen and to store it in a file. Text editors are similar to ◊word processors, except that they lack the ability to format text into paragraphs and pages and to apply different typefaces and styles.

textured vegetable protein manufactured meat substitute; see ◊TVP.

TGV *train à grande vitesse* (French 'high-speed train') French electrically powered train that provides the world's fastest rail service. Since it began operating 1981, it has carried more than 100 million passengers between Paris and Lyon, travelling at average speeds of 268 kph/167 mph and covering the 425 km/264 mi distance in two hours. In 1990 a second service, the Atlantique, was launched, running from Paris to Le Mans and Tours. A third is planned, linking Paris with the Channel Tunnel, Brussels, and Cologne. In 1990, a TGV broke the world speed record, reaching a speed of 515.3 kmh/320.2 mph (about half that of a passenger jet aircraft) on a stretch of line near Tours.

thallium (Greek *thallos* 'young green shoot') soft, bluish-white, malleable, metallic element, symbol Tl, atomic number 81, relative atomic mass 204.37. It is a poor conductor of electricity. Its compounds are poisonous and are used as insecticides and rodent poisons; some are used in the optical-glass and infrared-glass industries and in photoelectric cells.

Discovered spectroscopically 1861 by its green line, thallium was isolated and named by William Crookes later that year.

thallus any plant body that is not divided into true leaves, stems, and roots. It is often thin and flattened, as in the body of a seaweed, lichen, or liverwort, and the gametophyte generation (◊prothallus) of a fern.

Some flowering plants (◊angiosperms) that are adapted to an aquatic way of life may have a very simple plant body which is described as a thallus (for example, duckweed *Lemna*).

thaumatin naturally occurring, non-carbohydrate sweetener derived from the bacterium *Thaumatococcus danielli*. Its sweetness is not sensed as quickly as that of other sweeteners, and it is not as widely used in the food industry.

thaumatrope in photography, a disc with two different pictures at opposite ends of its surface. The images combine into one when rapidly rotated because of the persistence of visual impressions.

thebaine $C_{19}H_{21}NO_3$ highly poisonous extract of ◊opium.

theodolite instrument for the measurement of horizontal and vertical angles, used in surveying. It consists of a small telescope mounted so as to move on two graduated circles, one horizontal and the other vertical, while its axes pass through the centre of the circles. See also ◊triangulation.

theorem mathematical proposition that can be deduced by logic from a set of axioms (basic facts that are taken to be true without proof). Advanced mathematics consists almost entirely of theorems and proofs, but even at a simple level theorems are important.

theory in science, a set of ideas, concepts, principles, or methods used to explain a wide set of observed facts. Among the major theories of science are ◊relativity, ◊quantum theory, ◊evolution, and ◊plate tectonics.

A theory is a good theory if it satisfies two requirements: It must accurately describe a large class of observations on the basis of a model that contains only a few arbitrary elements, and it must make definite predictions about the results of future observations.'

On **theories** Stephen Hawking *A Brief History of Time* 1988

Theory of Everything (ToE) another name for ◊grand unified theory.

therm unit of energy defined as 10^5 British thermal units; equivalent to 1.055×10^8 joules. It is no longer in scientific use.

thermal capacity another name for ◊heat capacity

thermal conductivity in physics, the ability of a substance to conduct heat. Good thermal conductors, like good electrical conductors, are generally materials with many free electrons (such as metals).

Thermal conductivity is expressed in units of joules per second per metre per kelvin ($J\ s^{-1}\ m^{-1}\ K^{-1}$). For a block of material of cross-sectional area a and length l, with temperatures T_1 and T_2 at its end faces, the thermal conductivity λ equals $Hl/at(T_2 - T_1)$, where H is the amount of heat transferred in time t.

thermal decomposition breakdown of a compound into simpler substances by heating it; the reaction is irreversible. The catalytic ◊cracking of hydrocarbons is an example.

thermal dissociation reversible breakdown of a compound into simpler substances by heating it (see ◊dissociation). The splitting of ammonium chloride into ammonia and hydrogen chloride is an example. On cooling, they recombine to form the salt.

thermal expansion in physics, expansion that is due to a rise in temperature. It can be expressed in terms of linear, area, or volume expansion.

The coefficient of linear expansion α is the increase in unit length per degree temperature rise; area, or superficial, expansion β is the increase in unit area per degree; and volume, or cubic, expansion γ is the increase in unit volume per degree. As a close approximation, $\beta = 2\alpha$ and $\gamma = 3\alpha$.

THERMAL EXPANSION PUZZLE

Take a metal block and heat it. Obviously it expands, and the surface moves into what was free space. Now drill a hole right through the middle and heat the block again. In a similar manner, the surface of the hole will also move into free space as the block expands, and the hole will become smaller. On the other hand, expansion should affect all dimensions equally, so that the hole should have become larger. Which is true—does the hole become larger or smaller? *See page 651 for the answer.*

---

thermic lance cutting tool consisting of a tube of mild steel, enclosing tightly packed small steel rods and fed with oxygen. On ignition temperatures above 3,000°C/5,400°F are produced and the thermic lance becomes its own sustaining fuel. It rapidly penetrates walls and a 23–cm/9–in steel door can be cut through in less than 30 seconds.

thermionics branch of electronics dealing with the emission of electrons from matter under the influence of heat.

The *thermionic valve* (electron tube), used in telegraphy and telephony and in radio and radar, is a device using space conduction by thermionically emitted electrons from an electrically heated cathode. In most applications valves have been replaced by ◊transistors.

thermistor device whose electrical ◊resistance falls as temperature rises. The current passing through a thermistor increases rapidly as its temperature rises, and so they are used in electrical thermometers.

thermite process method used in incendiary devices and welding operations. It uses a powdered mixture of aluminium and (usually) iron oxide, which, when ignited, gives out enormous heat. The oxide is reduced to iron, which is molten at the high temperatures produced. This can be used to make a weld. The process was discovered 1895 by German chemist Hans Goldschmidt (1861–1923).

thermocouple electric temperature-measuring device consisting of a circuit having two wires made of different metals welded together at their ends.

A current flows in the circuit when the two junctions are maintained at different temperatures (◊Seebeck effect). The electromotive force generated—measured by a millivoltmeter—is proportional to the temperature difference.

thermodynamics branch of physics dealing with the transformation of heat into and from other forms of energy. It is the basis of the study of the efficient working of engines, such as the steam and internal-combustion engines. The three laws of thermodynamics are (1) energy can be neither created nor destroyed, heat and mechanical work being mutually convertible; (2) it is impossible for an unaided self-acting machine to convey heat from one body to another at a higher temperature; and (3) it is impossible by any procedure, no matter how idealized, to reduce any system to the ◊absolute zero of temperature (0K/–273°C) in a finite number of operations. Put into mathematical form, these laws have widespread applications in physics and chemistry.

thermography photographic recording of heat patterns. It is used medically as an imaging technique to identify 'hot spots' in the body—for example, tumours, where cells are more active than usual.

Thermography was developed in the 1970s and 1980s by the military to assist night vision by detecting the body heat of an enemy or the hot engine of a tank. It uses a photographic method

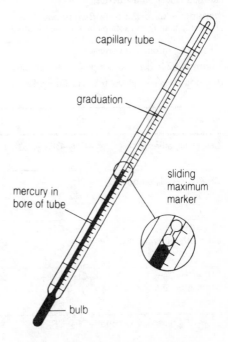

thermometer *Maximum and minimum thermometers are universally used in weather-reporting stations. The maximum thermometer, shown here, includes a magnet that fits tightly inside a capillary tube and is moved up it by the rising mercury. When the temperature falls, the magnet remains in position, thus enabling the maximum temperature to be recorded.*

(using infrared radiation) employing infrared-sensitive films.

thermoluminescence release in the form of light of stored energy from a substance heated by ◊irradiation. It occurs with most crystalline substances to some extent. It is used in archaeology to date pottery, and by geologists in studying terrestrial rocks and meteorites.

thermometer instrument for measuring temperature. There are many types, designed to measure different temperature ranges to varying degrees of accuracy. Each makes use of a different physical effect of temperature.

Expansion of a liquid is employed in common *liquid-in-glass thermometers*, such as those containing mercury or alcohol. The more accurate *gas thermometer* uses the effect of temperature on the pressure of a gas held at constant volume. A *resistance thermometer* takes advantage of the change in resistance of a conductor (such as a platinum wire) with variation in temperature. Another electrical thermometer is the ◊*thermocouple*. Mechanically, temperature change can be indicated by the change in curvature of a *bimetallic strip* (as commonly used in a ◊thermostat).

thermopile instrument for measuring radiant heat, consisting of a number of ◊thermocouples connected in series with alternate junctions exposed to the radiation. The current generated (measured by an ◊ammeter) is proportional to the radiation falling on the device.

thermoplastic or *thermosoftening plastic* type of ◊plastic that always softens on repeated heating. Thermoplastics include polyethylene (polyethene), polystyrene, nylon, and polyester.

Thermos trade name for a type of ◊vacuum flask.

thermoset type of ◊plastic that remains rigid when set, and does not soften with heating. Thermosets have this property because the long-chain polymer molecules cross-link with each other to give a rigid structure. Examples include Bakelite, resins, melamine, and urea-formaldehyde resins.

thermosphere layer in the Earth's ◊atmosphere above the mesosphere and below the exosphere. Its lower level is about 80 km/50 mi above the ground, but its upper level is undefined. The ionosphere is located in the thermosphere. In the thermosphere the temperature rises with increasing height to several thousand degrees Celsius. However, because of the thinness of the air, very little heat is actually present.

thermostat temperature-controlling device that makes use of feedback. It employs a temperature sensor (often a bimetallic strip) to operate a switch or valve to control electricity or fuel supply. Thermostats are used in central heating, ovens, and car engines.

At the required preset temperature (for example of a room or gas oven), the movement of the sensor switches off the supply of electricity to the room heater or gas to the oven. As the room or oven cools down, the sensor turns back on the supply of electricity or gas.

thiamine or *vitamin B_1* a water-soluble vitamin of the B complex. It is found in seeds and grain. Its absence from the diet causes the disease beriberi.

35 mm width of photographic film, the most popular format for the camera today. The 35–mm camera falls into two categories, the ◊SLR and the ◊rangefinder.

thorax in ◊tetrapod vertebrates, the part of the body containing the heart and lungs, and protected by the rib cage; in arthropods, the middle part of the body, between the head and abdomen.

In mammals the thorax is separated from the

monomer	polymer	name	uses
$CH_2=CH_2$ ethene	$+CH_2-CH_2 +_n$	poly(ethene), polythene	bottles, packaging, insulation, pipes
$CH_2=CH-CH_3$ propene	$+CH_2-CH +_n$ | CH_3	poly(propene), polypropylene	mouldings, film, fibres
$CH_2=CH-Cl$ chloroethene (vinyl chloride)	$+CH_2-CH +_n$ | Cl	polyvinylchloride (PVC), poly(chloroethene)	insulation, flooring, household fabric
$CH_2=CH-C_6H_5$ phenylethene (styrene)	$+CH_2-CH +_n$ | C_6H_5	polystyrene, poly(phenylethene)	insulation, packaging
$CF_2=CF_2$ tetrafluoroethene	$+CF_2-CF_2 +_n$ $(n = 1000+)$	poly(tetrafluoroethene) (PTFE)	high resistance to chemical and electrical reaction, low-friction applications

monomer I	monomer II	polymer name	uses
formaldehyde (methanal)	phenol	PF resins (Bakelites)	electrical fittings, radio cabinets
formaldehyde	urea	UF resins	electrical fittings, insulation, adhesives
formaldehyde	melamine	melamines	laminates for furniture

thermoset

abdomen by the muscular diaphragm. In insects the thorax bears the legs and wings. The thorax of spiders and crustaceans is fused with the head, to form the cephalothorax.

thorium dark-grey, radioactive, metallic element of the ◊actinide series, symbol Th, atomic number 90, relative atomic mass 232.038. It occurs throughout the world in small quantities in minerals such as thorite and is widely distributed in monazite beach sands. It is one of three fissile elements (the others are uranium and plutonium), and its longest-lived isotope has a half-life of 1.39 x 10^10 years. Thorium is used to strengthen alloys. It was discovered by Jöns Berzelius 1828 and was named by him after the Norse god Thor.

throat in human anatomy, the passage that leads from the back of the nose and mouth to the ◊trachea and ◊oesophagus. It includes the ◊pharynx and the ◊larynx, the latter being at the top of the trachea. The word 'throat' is also used to mean the front part of the neck, both in humans and other vertebrates; for example, in describing the plumage of birds. In engineering, it is any narrowing entry, such as the throat of a carburettor.

thulium soft, silver-white, malleable and ductile, metallic element, of the ◊lanthanide series, symbol Tm, atomic number 69, relative atomic mass 168.94. It is the least abundant of the rare-earth metals, and was first found in gadolinite and various other minerals. It is used in arc lighting.

The X-ray-emitting isotope Tm-170 is used in portable X-ray units. Thulium was named by French chemist Paul Lecoq de Boisbaudran 1886 after the northland, Thule.

thunderstorm severe storm of very heavy rain, thunder, and lightning. Thunderstorms are usually caused by the intense heating of the ground surface during summer. The warm air rises rapidly to form tall cumulonimbus clouds with a characteristic anvil-shaped top. Electrical charges accumulate in the clouds and are discharged to the ground as flashes of lightning. Air in the path of lightning becomes heated and expands rapidly, creating shock waves that are heard as a crash or rumble of thunder.

The rough distance between an observer and a lightning flash can be calculated by timing the number of seconds between the flash and the thunder. A gap of 3 seconds represents about a kilometre; 5 seconds represents about a mile.

thymus organ in vertebrates, situated in the upper chest cavity in humans. The thymus processes ◊lymphocyte cells to produce T-lymphocytes (T denotes 'thymus-derived'), which are responsible for binding to specific invading organisms and killing them or rendering them harmless.

The thymus reaches full size at puberty, and shrinks thereafter; the stock of T-lymphocytes is built up early in life, so this function diminishes in adults, but the thymus continues to function as an ◊endocrine gland, producing the hormone thymosin, which stimulates the activity of the T-lymphocytes.

thyristor type of ◊rectifier, an electronic device that conducts electricity in one direction only. The thyristor is composed of layers of ◊semiconductor material sandwiched between two electrodes called the anode and cathode. The current can be switched on by using a third electrode called the gate.

Thyristors are used to control mains-driven motors and in lighting dimmer controls.

thyroid ◊endocrine gland of vertebrates, situated in the neck in front of the trachea. It secretes several hormones, principally thyroxine, an iodine-containing hormone that stimulates growth, metabolism, and other functions of the body. The thyroid gland may be thought of as the regulator gland of the body's metabolic rate. If it is overactive, as in thyrotoxicosis, the sufferer feels hot and sweaty, has an increased heart rate, diarrhoea, and weight loss. Conversely, an underactive thyroid leads to myxoedema, a condition characterized by sensitivity to the cold, constipation, and weight gain. In infants, an underactive thyroid leads to cretinism, a form of mental retardation.

tibia the anterior of the pair of bones found between the ankle and the knee. In humans, the tibia is the shinbone.

tidal power station ◊hydroelectric power plant that uses the 'head' of water created by the rise and fall of the ocean tides to spin the water turbines. An example is located on the estuary of the river Rance in the Gulf of St Malo, Brittany, France, which has been in use since 1966.

tidal wave misleading name for a ◊tsunami.

tide rise and fall of sea level due to the gravitational forces of the Moon and Sun. High tide occurs at an average interval of 12 hr 24 min 30 sec. The highest or *spring tides* are at or near new and full Moon; the lowest or *neap tides* when the Moon is in its first or third quarter. Some seas, such as the Mediterranean, have very small tides.

Other factors affecting sea level are (1) a combination of naturally high tides with storm surge, as sometimes happens along the low-lying coasts of

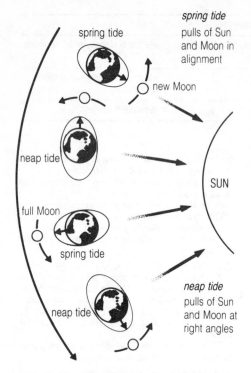

spring tide

spring tide

spring tide
pulls of Sun
and Moon in
alignment

new Moon

neap tide

SUN

full Moon

spring tide

neap tide
pulls of Sun
and Moon at
right angles

neap tide

neap tide

tide *The gravitational pull of the Moon is the main cause of the tides. Water on the side of the Earth nearest the Moon feels the Moon's pull and accumulates directly under the Moon. When the Sun and the Moon are in line, at new and full moon, the gravitational pull of Sun and Moon are in line and produce a high spring tide. When the Sun and Moon are at right angles, lower neap tides occur.*

Germany and the Netherlands; (2) the water walls created by typhoons and hurricanes, such as often hit Bangladesh; (3) underwater upheavals in the Earth's crust which may cause a ◊tsunami; and (4) global temperature change melting the polar ice caps.

Gravitational tides—the pull of nearby groups of stars—have been observed to affect the galaxies.

till or *boulder clay* deposit of clay, mud, gravel, and boulders left by a ◊glacier. It is unsorted, with all sizes of fragments mixed up together, and shows no stratification; that is, it does not form clear layers or ◊beds.

tilt-rotor aircraft type of vertical takeoff aircraft, also called a ◊convertiplane.

time continuous passage of existence, recorded by division into hours, minutes, and seconds. Formerly the measurement of time was based on the Earth's rotation on its axis, but this was found to be irregular. Therefore the second, the standard ◊SI unit of time, was redefined 1956 in terms of the Earth's annual orbit of the Sun, and 1967 in terms of a radiation pattern of the element caesium.

Universal time (UT), based on the Earth's actual rotation, was replaced by coordinated universal time (UTC) 1972, the difference between the two involving the addition (or subtraction) of leap

seconds on the last day of June or Dec. National observatories (in the UK until 1990 the Royal Greenwich Observatory) make standard time available, and the BBC broadcasts six pips at certain hours (five short, from second 55 to second 59, and one long, the start of which indicates the precise minute). Its computerized clock has an accuracy greater than 1 second in 4,000 years. From 1986 the term Greenwich Mean Time was replaced by UTC. However, the Greenwich meridian, adopted 1884, remains that from which all longitudes are measured, and the world's standard time zones are calculated from it.

For tribal man space was the uncontrollable mystery. For technological man it is time that occupies the same role.

On **time** Marshall McLuhan (1911–1980) *The Mechanical Bride* 1951

time-sharing in computing, a way of enabling several users to access the same computer at the same time. The computer rapidly switches between user ◊terminals and programs, allowing each user to work as if he or she had sole use of the system.

Time-sharing was common in the 1960s and 1970s before the spread of cheaper computers.

tin soft, silver-white, malleable and somewhat ductile, metallic element, symbol Sn (from Latin *stannum*), atomic number 50, relative atomic mass 118.69. Tin exhibits ◊allotropy, having three forms: the familiar lustrous metallic form above 55.8°F/13.2°C; a brittle form above 321.8°F/161°C; and a grey powder form below 55.8°F/13.2°C (commonly called tin pest or tin disease). The metal is quite soft (slightly harder than lead) and can be rolled, pressed, or hammered into extremely thin sheets; it has a low melting point. In nature it occurs rarely as a free metal. It resists corrosion and is therefore used for coating and plating other metals.

Tin and copper smelted together form the oldest desired alloy, bronze; since the Bronze Age (3,500 BC) that alloy has been the basis of both useful and decorative materials. The mines of Cornwall were the principal western source from then until the 19th century, when rich deposits were found in South America, Africa, and SE Asia. Tin is also alloyed with metals other than copper to make solder and pewter. It was recognized as an element by Antoine Lavoisier, but the name is very old and comes from the Germanic form *zinn*.

tin ore mineral from which tin is extracted, principally cassiterite, SnO_2. The world's chief producers are Malaysia, Thailand, and Bolivia.

The UK was a major producer in the 19th century but today only a few working mines remain, and production is small.

tinplate milled steel coated with tin, the metal used for most 'tin' cans. The steel provides the strength, and the tin provides the corrosion resistance, ensuring that the food inside is not contaminated. Tinplate may be made by ◊electroplating or by dipping in a bath of molten tin.

tissue in biology, any kind of cellular fabric that occurs in an organism's body. Several kinds of tissue can usually be distinguished, each consisting of cells of a particular kind bound together by cell walls (in plants) or extracellular matrix (in animals). Thus, nerve and muscle are different kinds of tissue in animals, as are ◊parenchyma and ◊sclerenchyma in plants.

tissue culture process by which cells from a plant or animal are removed from the organism and grown under controlled conditions in a sterile medium containing all the necessary nutrients. Tissue culture can provide information on cell growth and differentiation, and is also used in plant propagation and drug production. See also ◊meristem.

Titan in astronomy, largest moon of the planet Saturn, with a diameter of 5,150 km/3,200 mi and a mean distance from Saturn of 1,222,000 km/ 759,000 mi. It was discovered 1655 by Dutch mathematician and astronomer Christiaan Huygens, and is the second largest moon in the Solar System (Ganymede, of Jupiter, is larger).

Titan is the only moon in the Solar System with a substantial atmosphere (mostly nitrogen), topped with smoggy orange clouds that obscure the surface, which may be covered with liquid ethane lakes. Its surface atmospheric pressure is greater than Earth's. Radar signals suggest that Titan has dry land as well as oceans (among the planets, only Earth has both in the Solar System).

titanium strong, lightweight, silver-grey, metallic element, symbol Ti, atomic number 22, relative atomic mass 47.90. The ninth most abundant element in the Earth's crust, its compounds occur in practically all igneous rocks and their sedimentary deposits. It is very strong and resistant to corrosion, so it is used in building high-speed aircraft and spacecraft; it is also widely used in making alloys, as it unites with almost every metal except copper and aluminium. Titanium oxide is used in high-grade white pigments.

The element was discovered 1791 by English mineralogist William Gregor (1761–1817) and was named by German chemist Martin Klaproth 1796 after Titan, one of the giants of Greek mythology. It was not obtained in pure form until 1925.

titanium ore any mineral from which titanium is extracted, principally ilmenite (FeTiO$_3$) and rutile (TiO$_2$). Brazil, India, and Canada are major producers. Both these ore minerals are found either in rock formations or concentrated in heavy mineral sands.

Titan rocket family of US space rockets, developed from the Titan intercontinental missile. Two-stage Titan rockets launched the ◊Gemini crewed missions. More powerful Titans, with additional stages and strap-on boosters, were used to launch spy satellites and space probes, including the ◊Viking and ◊Voyager probes and ◊*Mars Observer*.

titration in analytical chemistry, a technique to find the concentration of one compound in a solution by determining how much of it will react with a known amount of another compound in solution.

One of the solutions is measured by ◊pipette into the reaction vessel. The other is added a little at a time from a ◊burette. The end-point of the

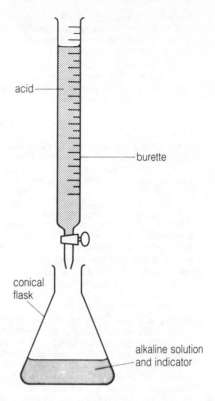

acid

burette

conical flask

alkaline solution and indicator

titration

reaction is determined with an ◊indicator or an electrochemical device.

TLR camera twin-lens reflex camera that has a viewing lens of the same angle of view and focal length mounted above and parallel to the taking lens.

TNT abbreviation for *trinitrotoluene* CH$_3$C$_6$H$_2$ (NO$_2$)$_3$ powerful high explosive. It is a yellow solid, prepared in several isomeric forms from ◊toluene by using sulphuric and nitric acids.

toadstool common name for many umbrella-shaped fruiting bodies of fungi. The term is normally applied to those that are inedible or poisonous.

tocopherol or *vitamin E* fat-soluble chemical found in vegetable oils. Deficiency of tocopherols leads to multiple adverse effects on health. In rats, vitamin E deficiency has been shown to cause sterility.

tog unit of measure of thermal insulation used in the textile trade; a light summer suit provides 1.0 tog.

The tog-value of an object is equal to ten times the temperature difference (in 8°C) between its two surfaces when the flow of heat is equal to one watt per square metre; one tog equals 0.645 ◊clo.

tokamak experimental machine designed by Soviet scientists to investigate controlled nuclear fusion. It consists of a doughnut-shaped chamber surrounded by electromagnets capable of exerting very powerful magnetic fields. The fields are generated to confine a very hot (millions of degrees) ◊plasma of ions and electrons, keeping it away from the chamber walls. See also ◊JET.

toluene or *methyl benzene* $C_6H_5CH_3$ colourless, inflammable liquid, insoluble in water, derived from petroleum. It is used as a solvent, in aircraft fuels, in preparing phenol (carbolic acid, used in making resins for adhesives, pharmaceuticals, and as a disinfectant), and the powerful high explosive ◊TNT.

tomography the obtaining of plane-section X-ray photographs, which show a 'slice' through any object. Crystal detectors and amplifiers can be used that have a sensitivity 100 times greater than X-ray film, and, in conjunction with a computer system, can detect, for example, the difference between a brain tumour and healthy brain tissue.

Godfrey Hounsfield was a leading pioneer in the development of this technique. In modern medical imaging there are several types, such as the ◊CAT scan (computerized axial tomography).

ton imperial unit of mass. The *long ton*, used in the UK, is 1,016 kg/2,240 lb; the *short ton*, used in the USA, is 907 kg/2,000 lb. The *metric ton* or *tonne* is 1,000 kg/2,205 lb.

tongue in tetrapod vertebrates, a muscular organ usually attached to the floor of the mouth. It has a thick root attached to a U-shaped bone (hyoid), and is covered with a ◊mucous membrane containing nerves and 'taste buds'. It directs food to the teeth and into the throat for chewing and swallowing. In humans, it is crucial for speech; in other animals, for lapping up water and for grooming, among other functions.

tonne the metric ton of 1,000 kg/2,204.6 lb; equivalent to 0.9842 of an imperial ◊ton.

tonsils in higher vertebrates, masses of lymphoid tissue situated at the back of the mouth and throat (palatine tonsils), and on the rear surface of the tongue (lingual tonsils). The tonsils contain many ◊lymphocytes and are part of the body's defence system against infection.

The adenoids are sometimes called pharyngeal tonsils.

tooth in vertebrates, one of a set of hard, bonelike structures in the mouth, used for biting and chewing food, and in defence and aggression. In humans, the first set (20 milk teeth) appear from age six months to two and a half years. The permanent ◊dentition replaces these from the sixth year onwards, the wisdom teeth (third molars) sometimes not appearing until the age of 25 or 30. Adults have 32 teeth: two incisors, one canine (eye tooth), two premolars, and three molars on each side of each jaw. Each tooth consists of an enamel coat (hardened calcium deposits), dentine (a thick, bonelike layer), and an inner pulp cavity, housing nerves and blood vessels. Mammalian teeth have roots surrounded by cementum, which fuses them into their sockets in the jawbones. The neck of the tooth is covered by the ◊gum, while the enamel-covered crown protrudes above the gum line.

The chief diseases of teeth are misplacements resulting from defect or disturbance of the tooth-germs before birth, eruption out of their proper places, and caries (decay).

topaz mineral, aluminium fluosilicate, $Al_2SiO_4(F,OH)_2$. It is usually yellow, but pink if it has been heated, and is used as a gemstone when transparent. It ranks 8 on the Mohs' scale of hardness.

topography the surface shape and aspect of the land, and its study. Topography deals with relief and contours, the distribution of mountains and valleys, the patterns of rivers, and all other features, natural and artificial, that produce the landscape. Such features are shown on *topographical*

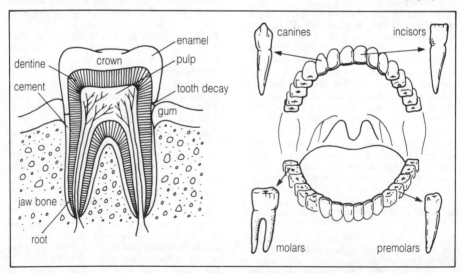

tooth *Adults have 32 teeth: two incisors, one canine, two premolars, and three molars on each side of each jaw. Each tooth has three parts: crown, neck, and root. The crown consists of a dense layer of mineral, the enamel, surrounding hard dentine with a soft centre, the pulp.*

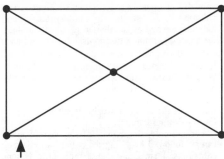

this figure is topologically equivalent to this one

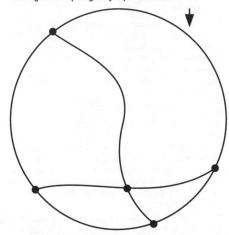

topology *Topology is often called 'rubber sheet geometry'. It is the branch of mathematics which studies the properties of a geometric figure which are unchanged when the figure is distorted; for example, the two coloured lines still intersect when the rectangle is distorted.*

maps (for example, those produced by the Ordnance Survey).

topology branch of geometry that deals with those properties of a figure that remain unchanged even when the figure is transformed (bent, stretched) — for example, when a square painted on a rubber sheet is deformed by distorting the sheet. Topology has scientific applications, as in the study of turbulence in flowing fluids. The map of the London Underground system is an example of the topological representation of a network; connectivity (the way the lines join together) is preserved, but shape and size are not.

The topological theory, proposed 1880, that only four colours are required in order to produce a map in which no two adjoining countries have the same colour; inspired extensive research, and was proved 1972 by Kenneth Appel and Wolfgang Haken.

tor isolated mass of rock, usually granite, left upstanding on a hilltop after the surrounding rock has been broken down. Weathering takes place along the joints in the rock, reducing the outcrop into a mass of rounded blocks.

tornado extremely violent revolving storm with swirling, funnel-shaped clouds, caused by a rising column of warm air propelled by strong wind. A tornado can rise to a great height, but with a diameter of only a few hundred metres or yards or less. Tornadoes move with wind speeds of 160–480 kph/100–300 mph, destroying everything in their path. They are common in central USA and Australia.

MOST DESTRUCTIVE TORNADO

The most destructive tornado on record occurred in Annapolis, Missouri, USA, on 18 March 1925. In three hours it tore through the town leaving a 300–m/980–ft wide trail of demolished buildings, uprooted trees, and overturned cars. The tornado killed 823 people and injured almost 3,000.

torque the turning effect of force on an object. A turbine produces a torque that turns an electricity generator in a power station. Torque is measured by multiplying the force by its perpendicular distance from the turning point.

torque converter device similar to a turbine, filled with oil, used in automatic transmission systems in motor vehicles and locomotives to transmit power (torque) from the engine to the gears.

torr unit of pressure equal to 1/760 of an ◊atmosphere, used mainly in high-vacuum technology.

One torr is equivalent to 133.322 pascals, and for practical purposes is the same as the millimetre of mercury. It is named after Italian physicist Evangelista Torricelli.

torsion in physics, the state of strain set up in a twisted material; for example, when a thread, wire, or rod is twisted, the torsion set up in the material tends to return the material to its original state. The *torsion balance*, a sensitive device for measuring small gravitational or magnetic forces, or electric charges, balances these against the restoring force set up by them in a torsion suspension.

torus circular ring with a D-shaped cross-section used to contain ◊plasma in nuclear fusion reactors such as the Joint European Torus (◊JET) reactor.

total internal reflection the complete reflection of a beam of light that occurs from the surface of an optically 'less dense' material. For example, a beam from an underwater light source can be reflected from the surface of the water, rather than escaping through the surface. Total internal reflection can only happen if a light beam hits a surface at an angle greater than the ◊critical angle for that particular pair of materials.

Total internal reflection is used as a means of reflecting light inside ◊prisms and ◊optical fibres. Light is contained inside an optical fibre not by the cladding around it, but by the ability of the internal surface of the glass-fibre core to reflect 100% of the light, thereby keeping it trapped inside the fibre.

touch sensation produced by specialized nerve endings in the skin. Some respond to light pressure, others to heavy pressure. Temperature detection may also contribute to the overall sensation of touch. Many animals, such as nocturnal ones, rely on touch more than humans do. Some have specialized organs of touch that project from the body, such as whiskers or antennae.

touch screen in computing, an input device allowing the user to communicate with the computer by touching a display screen with a finger. In this way, the user can point to a required ◊menu option or item of data. Touch screens are used less widely than other pointing devices such as the ◊mouse or ◊joystick.

Typically, the screen is able to detect the touch either because the finger presses against a sensitive membrane or because it interrupts a grid of light beams crossing the screen surface.

touch sensor in a computer-controlled ◊robot, a device used to give the robot a sense of touch, allowing it to manipulate delicate objects or move automatically about a room. Touch sensors provide the feedback necessary for the robot to adjust the force of its movements and the pressure of its grip. The main types include the strain gauge and the microswitch.

tourmaline hard, brittle mineral, a complex silicate of various metals, but mainly sodium aluminium borosilicate.

Small tourmalines are found in granites and gneisses. The common varieties range from black (schorl) to pink, and the transparent gemstones may be colourless (achroite), rose pink (rubellite), green (Brazilian emerald), blue (indicolite, verdelite, Brazilian sapphire), or brown (dravite).

toxicity tests tests carried out on new drugs, cosmetics, food additives, pesticides, and other synthetic chemicals to see whether they are safe for humans to use. They aim to identify potential toxins, carcinogens, teratogens, and mutagens.

Traditionally such tests use live animals such as rats, rabbits, and mice. Animal tests have become a target for criticism by ◊antivivisection groups, and alternatives have been sought. These include tests on human cells cultured in a test tube and on bacteria.

In Europe in 1990, 4,365 animals were used for testing cosmetic products, out of a total of 276,674 animals used to test products other than pharmaceuticals. Including pharmaceuticals, a total of around 3.2 million animals were used in experimentation. The US Office of Technology Assessment estimates that around 1.6 million animals are used annually in government research laboratories, of which 90% are rats or mice.

toxic waste dumped ◊hazardous substance.

toxin any chemical molecule that can damage the living body. In vertebrates, toxins are broken down by ◊enzyme action, mainly in the liver.

trace in computing, a method of checking that a computer program is functioning correctly by causing the changing values of all the ◊variables involved to be displayed while the program is running. In this way it becomes possible to narrow down the search for a bug, or error, in the program to the exact instruction that causes the variables to take unexpected values.

trace element chemical element necessary in minute quantities for the health of a plant or animal. For example, magnesium, which occurs in chlorophyll, is essential to photosynthesis, and iodine is needed by the thyroid gland of mammals for making hormones that control growth and body chemistry.

tracer in science, a small quantity of a radioactive ◊isotope (form of an element) used to follow the path of a chemical reaction or a physical or biological process. The location (and possibly concentration) of the tracer is usually detected by using a Geiger–Muller counter.

For example, the activity of the thyroid gland can be followed by giving the patient an injection containing a small dose of a radioactive isotope of iodine, which is selectively absorbed from the bloodstream by the gland.

trachea tube that forms an airway in air-breathing animals. In land-living ◊vertebrates, including humans, it is also known as the *windpipe* and runs from the larynx to the upper part of the chest. Its diameter is about 1.5 cm/0.6 in and its length 10 cm/4 in. It is strong and flexible, and reinforced by rings of ◊cartilage. In the upper chest, the trachea branches into two tubes: the left and right bronchi, which enter the lungs. Insects have a branching network of tubes called tracheae, which conduct air from holes (◊spiracles) in the body surface to all the body tissues. The finest branches of the tracheae are called tracheoles.

Some spiders also have tracheae but, unlike insects, they possess gill-like lungs (book lungs) and rely on their circulatory system to transport gases throughout the body.

tracheid cell found in the water-conducting tissue (◊xylem) of many plants, including gymnosperms (conifers) and pteridophytes (ferns). It is long and thin with pointed ends. The cell walls are thickened by ◊lignin, except for numerous small rounded areas, or pits, through which water and dissolved minerals pass from one cell to another. Once mature, the cell itself dies and only its walls remain.

track in computing, part of the magnetic structure created on a disc surface during ◊disc formatting so that data can be stored on it. The disc is first divided into circular tracks and then each circular track is divided into a number of sectors.

tracked vehicle vehicle, such as a tank or bulldozer, that runs on its own tracks (known as caterpillar tracks).

tractor in agriculture, a powerful motor vehicle, commonly having large rear wheels or caterpillar tracks, used for pulling farm machinery and loads. It is usually powered by a diesel engine and has a power-takeoff mechanism for driving machinery, and a hydraulic lift for raising and lowering implements.

In military usage, a *combat tractor* usually has two drivers (one forwards, one backwards), and can excavate 2 tonnes in a single action, so as, for example, to hide a Chieftain tank in 11 minutes. It can also cross rivers, operate in nuclear radiation, and clear minefields.

trade wind prevailing wind that blows towards the equator from the northeast and southeast. Trade winds are caused by hot air rising at the equator and the consequent movement of air from north

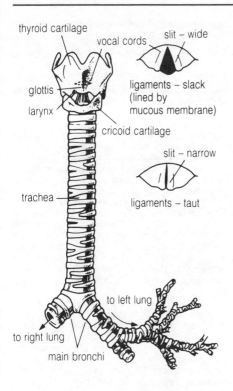

insect

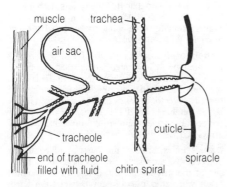

trachea *The human trachea, or windpipe. The larynx, or voice box, lies at the entrance to the trachea. The two vocal cords are membranes that normally remain open and still. When they are drawn together, the passage of air makes them vibrate and produce sounds.*

and south to take its place. The winds are deflected towards the west because of the Earth's west-to-east rotation. The unpredictable calms known as the ◊doldrums lie at their convergence.

The trade-wind belts move north and south about 5° with the seasons. The name is derived from the obsolete expression 'to blow *trade*' meaning consistently in a constant direction, which indicates the trade winds' importance to navigation in the days of cargo-carrying sailing ships.

traffic vehicles using public roads. In 1970 there were 100 million cars and lorries in use worldwide; in 1990 there were 550 million. One-fifth of the space in European and North American cities is taken up by cars. In 1989 UK road-traffic forecasts predicted that traffic demand would rise between 83% and 142% by the year 2025.

In 1988 there were 4,531 deaths from motor-vehicle traffic accidents in England and Wales. In 1991 in the UK there were about 10 million cars and lorries; the congestion they cause costs £2–5 billion per year. The government spends £6 billion per year on roads, and receives around £14 million from fuel tax and vehicle excise duty.

tramway transport system for use in cities, where wheeled vehicles run along parallel rails. Trams are powered either by electric conductor rails below ground or by conductor arms connected to overhead wires. Greater manoeuvrability is achieved with the ◊trolley bus, similarly powered by conductor arms overhead but without tracks.

Trams originated in collieries in the 18th century, and the earliest passenger system was in New York 1832. Tramways were widespread in Europe and the USA from the late 19th to the mid-20th century, and are still found in many European cities; in the Netherlands several neighbouring towns share an extensive tram network. They were phased out, especially in the USA and the UK, under pressure from the motor-transport lobby, but in the 1990s both trams and trolley buses were being revived in some areas. Both vehicles have the advantage of being nonpolluting to the local environment, though they require electricity generation, which is polluting at source.

Trams returned to Manchester in 1992 after an absence of 40 years. The Metrolink scheme connects two commuter railways by 3 km of track through the centre of the city. Two towns in France have bought back the tram and one town in the US. Another tramway in the UK is also under construct in Sheffield and around 40 other areas are considering plans.

transactinide element any of a series of nine radioactive, metallic elements with atomic numbers that extend beyond the ◊actinide series, those from 104 (rutherfordium) to 112 (unnamed). They are grouped because of their expected chemical similarities (all are bivalent), the properties differing only slightly with atomic number. All have half-lives of less than two minutes.

transcription in living cells, the process by which the information for the synthesis of a protein is transferred from the ◊DNA strand on which it is carried to the messenger ◊RNA strand involved in the actual synthesis.

It occurs by the formation of ◊base pairs when a single strand of unwound DNA serves as a template for assembling the complementary nucleotides that make up the new RNA strand.

transducer device that converts one form of energy into another. For example, a thermistor is a transducer that converts heat into an electrical voltage, and an electric motor is a transducer that converts an electrical voltage into mechanical energy. Transducers are important components in many types of ◊sensor, converting the physical quantity to be measured into a proportional voltage signal.

transfer orbit elliptical path followed by a spacecraft moving from one orbit to another, designed to save fuel although at the expense of a longer journey time.

Space probes travel to the planets on transfer orbits. A probe aimed at Venus has to be 'slowed down' relative to the Earth, so that it enters an elliptical transfer orbit with its perigee (point of closest approach to the Sun) at the same distance as the orbit of Venus; towards Mars, the vehicle has to be 'speeded up' relative to the Earth, so that it reaches its apogee (furthest point from the Sun) at the same distance as the orbit of Mars. *Geostationary transfer orbit* is the highly elliptical path followed by satellites to be placed in ◊geostationary orbit around the Earth (an orbit coincident with Earth's rotation). A small rocket is fired at the transfer orbit's apogee to place the satellite in geostationary orbit.

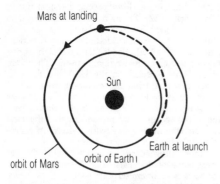

transfer orbit *The transfer orbit used by a spacecraft when travelling from Earth to Mars. The orbit is chosen to minimize the fuel needed by the spacecraft; the craft is in free fall for most of the journey.*

transformer device in which, by electromagnetic induction, an alternating current (AC) of one voltage is transformed to another voltage, without change of ◊frequency. Transformers are widely used in electrical apparatus of all kinds, and in particular in power transmission where high voltages and low currents are utilized.

A transformer has two coils, a primary for the input and a secondary for the output, wound on a common iron core. The ratio of the primary to the secondary voltages (and currents) is directly (and inversely) proportional to the number of turns in the primary and secondary coils.

transfusion intravenous delivery of blood or blood products (plasma, red cells) into a patient's circulation to make up for deficiencies due to disease, injury, or surgical intervention. Cross-matching is carried out to ensure the patient receives the right type of blood. Because of worries about blood-borne disease, self-transfusion with units of blood 'donated' over the weeks before an operation is popular.

Blood transfusion, first successfully pioneered in humans 1818, remained highly risky until the discovery of blood groups, by Austrian-born immunologist Karl Landsteiner 1900, indicated the need for compatibility of donated blood.

transgenic organism plant, animal, bacterium, or other living organism which has had a foreign gene added to it by means of ◊genetic engineering.

transistor solid-state electronic component, made of ◊semiconductor material, with three or more ◊electrodes, that can regulate a current passing through it. A transistor can act as an ◊amplifier, ◊oscillator, ◊photocell, or switch, and (unlike earlier thermionic valves) usually operates on a very small amount of power. Transistors commonly consist of a tiny sandwich of germanium or silicon, alternate layers having different electrical properties.

A crystal of pure germanium or silicon would act as an insulator (nonconductor). By introducing impurities in the form of atoms of other materials (for example, boron, arsenic, or indium) in minute amounts, the layers may be made either *n-type*, having an excess of electrons, or *p-type*, having a deficiency of electrons. This enables electrons to flow from one layer to another in one direction only.

Transistors have had a great impact on the electronics industry, and thousands of millions are now made each year. They perform many of the functions of the thermionic valve, but have the advantages of greater reliability, long life, compactness, and instantaneous action, no warming-up period being necessary. They are widely used in most electronic equipment, including portable radios and televisions, computers, and satellites, and are the basis of the ◊integrated circuit (silicon chip). They were invented at Bell Telephone Laboratories in the USA in 1948 by John Bardeen and Walter Brattain, developing the work of the US physicist William Shockley.

transistor–transistor logic (TTL) in computing, the type of integrated circuit most commonly used in building electronic products. In TTL chips the bipolar transistors are directly connected (usually collector to base). In mass-produced items, large numbers of TTL chips are commonly replaced by a small number of ◊uncommitted logic arrays (ULAs), or logic gate arrays.

transit in astronomy, the passage of a smaller object across the visible disc of a larger one. Transits of the inferior planets (Mercury and Venus) occur when they pass directly between the Earth and the Sun, and are seen as tiny dark spots against the Sun's disc.

Other forms of transit include the passage of a satellite (moon) or its shadow across the disc of Jupiter and the passage of planetary surface features across the central ◊meridian of that planet as seen from Earth. The passage of an object in the sky across the observer's meridian is also known as a transit.

transition metal any of a group of metallic elements that have incomplete inner electron shells (see ◊atom, electronic structure) and exhibit variable valency—for example, cobalt, copper, iron, and molybdenum. They are excellent conductors of electricity, and generally form highly coloured compounds.

translation in living cells, the process by which proteins are synthesized. During translation, the information coded as a sequence of nucleotides in messenger ◊RNA is transformed into a sequence of amino acids in a peptide chain. The process involves the 'translation' of the ◊genetic code. See also ◊transcription.

translation program in computing, a program that translates another program written in a high-level language or assembly language into the machine-code instructions that a computer can obey. See ◊assembler, ◊compiler, and ◊interpreter.

transmission electron microscope (TEM) the most powerful type of ◊electron microscope, with a resolving power ten times better than that of a ◊scanning electron microscope and a thousand times better than that of an optical microscope. A fine electron beam passes through the specimen, which must therefore be sliced extremely thinly—typically to about one-thousandth of the thickness of a sheet of paper (100 nanometres). The TEM can resolve objects 0.001 micrometres (0.04 millionth of an inch) apart, a gap that is 100,000 times smaller than the unaided eye can see.

A TEM consists of a tall evacuated column at the top of which is a heated filament that emits electrons. The electrons are accelerated down the column by a high voltage, (around 100,000 volts) and pass through the slice of specimen at a point roughly half-way down. Because the density of the specimen varies, the 'shadow' of the beam falls on a fluorescent screen near the bottom of the column and forms an image. A camera is mounted beneath the screen to record the image.

The electron beam is controlled by magnetic fields produced by electric coils, called electron lenses. One electron lens, called the condenser, controls the beam size and brightness before it strikes the specimen. Another electron lens, called the objective, focuses the beam on the specimen and magnifies the image about 50 times. Other electron lenses below the specimen then further magnify the image.

The *high voltage transmission electron*

microscope (HVEM) uses voltages of up to 3 million volts to accelerate the electron beam. The largest of these instruments is as tall as a three-storey building.

The first experimental TEM was built in 1931 by German scientists Max Knoll and Ernest Ruska of the Technische Hochschule, Berlin, Germany. They produced a picture of a platinum grid magnified 117 times. The first commercial electron microscope was built in England in 1936.

transparency in photography, a picture on slide film. This captures the original in a positive image (direct reversal) and can be used for projection or printing on positive-to-positive print material, for example by the Cibachrome or Kodak R-type process.

Slide film is usually colour but can be obtained in black and white.

transpiration the loss of water from a plant by evaporation. Most water is lost from the leaves through pores known as ◊stomata, whose primary function is to allow ◊gas exchange between the plant's internal tissues and the atmosphere. Transpiration from the leaf surfaces causes a continuous upward flow of water from the roots via the ◊xylem, which is known as the transpiration stream.

A single maize plant has been estimated to transpire 245 1/54 gal of water in one growing season.

transplant in medicine, the transfer of a tissue or organ from one human being to another or from one part of the body to another (skin grafting). In most organ transplants, the operation is for life-saving purposes, though the immune system tends to reject foreign tissue. Careful matching and immunosuppressive drugs must be used, but these are not always successful.

Corneal grafting, which may restore sight to a diseased or damaged eye, was pioneered 1905, and is the oldest successful human transplant procedure. Of the internal organs, kidneys were first transplanted successfully in the early 1950s and are the most readily received by the body. Recent transplantation also encompasses hearts, lungs, livers, pancreatic, bone, and bone-marrow tissue.

transpiration stream

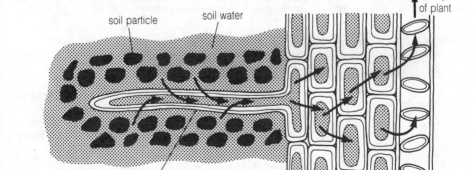

soil particle   soil water   to top of plant

root hair cell

arrows show direction of water flow   epidermal cell   cortex cells   xylem vessel

transpiration

Most transplant material is taken from cadaver donors, usually those suffering death of the ◊brainstem, or from frozen tissue banks. In rare cases, kidneys, corneas, and part of the liver may be obtained from living donors. Besides the shortage of donated material, the main problem facing transplant surgeons is rejection of the donated organ by the new body. The 1990 Nobel Prize for Medicine and Physiology was awarded to two US surgeons, Donnall Thomas and Joseph Murray, for their pioneering work on organ transplantation.

Under the UK transplant code of 1979 covering the use of material from a donor, two doctors (independent of the transplant team and clinically independent of each other) must certify that the donor is brain dead.

transputer in computing, a member of a family of microprocessors designed for parallel processing, developed in the UK by Inmos. In the circuits of a standard computer the processing of data takes place in sequence; in a transputer's circuits processing takes place in parallel, greatly reducing computing time for those programs that have been specifically written for it.

The transputer implements a special programming language called OCCAM, which Inmos based on CSP (communicating sequential processes),

developed by C A R Hoare of Oxford University Computing Laboratory.

transuranic element or *transuranium element* chemical element with an atomic number of 93 or more—that is, with a greater number of protons in the nucleus than has uranium. All transuranic elements are radioactive. Neptunium and plutonium are found in nature; the others are synthesized in nuclear reactions.

transverse wave ◊wave in which the displacement of the medium's particles is at right-angles to the direction of travel of the wave motion.

It is characterized by its alternating crests and troughs. Simple water waves, such as the ripples produced when a stone is dropped into a pond, are transverse waves, as are the waves on a vibrating string.

All ◊electromagnetic waves have a transverse-wave form; their electric and magnetic fields (rather than the particles of their medium) vibrate at right-angles to their direction of travel.

trapezium (North American *trapezoid*) in geometry, a four-sided plane figure (quadrilateral) with two of its sides parallel. If the parallel sides have lengths a and b and the perpendicular distance

transuranic elements

Atomic Number	Name	Symbol	Year discovered	Source of first preparation	Isotope identified	Half-life of first isotope identified
Actinide series						
93	neptunium	Np	1940	irradiation of uranium-238 with neutrons	Np-239	2.35 days
94	plutonium	Pu	1941	bombardment of uranium-238 with deuterons	Pu-238	86.4 years
95	americium	Am	1944	irradiation of plutonium-239 with neutrons	Am-241	458 years
96	curium	Cm	1944	bombardment of plutonium-239 with helium nuclei	Cm-242	162.5 days
97	berkelium	Bk	1949	bombardment of americium-241 with helium nuclei	Bk-243	4.5 hours
98	californium	Cf	1950	bombardment of curium-242 with helium nuclei	Cf-245	44 minutes
99	einsteinium	Es	1952	irradiation of uranium-238 with neutrons in first thermonuclear explosion	Es-253	20 days
100	fermium	Fm	1953	irradiation of uranium-238 with neutrons in first thermonuclear explosion	Fm-235	20 hours
101	mendelevium	Md	1955	bombardment of einsteinium-253 with helium nuclei	Md-256	76 minutes
102	nobelium	No	1958	bombardment of curium-246 with carbon nuclei	No-255	2.3 seconds
103	lawrencium	Lr	1961	bombardment of californium-252 with boron nuclei	Lr-257	4.3 seconds
transactinide elements						
104	unnilquadium* (also called rutherfordium or kurchatovium)	Unq	1969	bombardment of californium-249 with carbon-12 nuclei	Unq-257	3.4 seconds
105	unnilpentium* (also called hahnium or nielsbohrium)	Unp	1970	bombardment of californium-249 with nitrogen-15 nuclei	Unp-260	1.6 seconds
106	unnilhexium*	Unh	1974	bombardment of californium-249 with oxygen-18 nuclei	Unh-263	0.9 seconds
107	unnilseptium*	Uns	1977	bombardment of bismuth-209 with nuclei of chromium-54	Uns	102 milliseconds
108	unniloctium*	Uno	1984	bombardment of lead-208 with nuclei of iron-58	Uno-265	1.8 milliseconds
109	unnilennium*	Une	1982	bombardment of bismuth-209 with nuclei of iron-58	Une	3.4 milliseconds

* names for elements 104–109 are as proposed by the International Union for Pure and Applied Chemistry (IUPAC) 1980

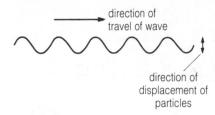

direction of
travel of wave

direction of
displacement of
particles

transverse wave

between them is *h* (the height of the trapezium), its area $A = h(a + b)/2$.

An isosceles trapezium has its sloping sides equal, and is symmetrical about a line drawn through the midpoints of its parallel sides.

travelator moving walkway, rather like a flat escalator.

tree perennial plant with a woody stem, usually a single stem or 'trunk', made up of ◊wood and protected by an outer layer of ◊bark. It absorbs water through a ◊root system. There is no clear dividing line between shrubs and trees, but sometimes a minimum height of 6 m/20 ft is used to define a tree.

A treelike form has evolved independently many times in different groups of plants. Among the ◊angiosperms, or flowering plants, most trees are ◊dicotyledons. This group includes trees such as oak, beech, ash, chestnut, lime, and maple, and they are often referred to as ◊broad-leaved trees because their leaves are broader than those of conifers, such as pine and spruce. In temperate regions angiosperm trees are mostly ◊deciduous (that is, they lose their leaves in winter), but in the tropics most angiosperm trees are evergreen. There are fewer trees among the ◊monocotyledons, but the palms and bamboos (some of which are treelike) belong to this group. The ◊gymnosperms include many trees and they are classified into four orders: Cycadales (including cycads and sago palms), Coniferales (the conifers), Ginkgoales (including only one living species, the ginkgo, or maidenhair tree), and Taxales (including yews). Apart from the ginkgo and the larches (conifers), most gymnosperm trees are evergreen. There are also a few living trees in the ◊pteridophyte group, known as tree ferns. In the swamp forests of the Carbonifer-

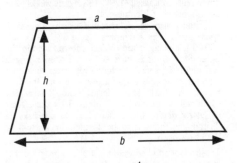

area of a trapezium $= \frac{1}{2}(a + b) \times h$

trapezium

ous era, 300 million years ago, there were giant treelike horsetails and club mosses in addition to the tree ferns. The world's oldest trees are found in the Pacific forest of North America, some more than 2,000 years old.

The great storm Oct 1987 destroyed some 15 million trees in Britain, and showed that large roots are less significant than those of 10 cm/4 in diameter or less. If enough of these are cut, the tree dies or falls.

tree-and-branch filing system in computing, a filing system where all files are stored within directories, like folders in a filing cabinet. These directories may in turn be stored within further directories. The root directory contains all the other directories and may be thought of as equivalent to the filing cabinet. Another way of picturing the system is as a tree with branches from which grow smaller branches, ending in leaves (individual files).

tree diagram in probability theory, a branching diagram consisting only of arcs and nodes (but not loops curving back on themselves), which is used to establish probabilities.

tree rings rings visible in the wood of a cut tree; see ◊annual rings.

tremor minor ◊earthquake.

triangle in geometry, a three-sided plane figure, the sum of whose interior angles is 180°. Triangles can be classified by the relative lengths of their sides. A *scalene triangle* has no sides of equal length; an *isosceles triangle* has at least two equal sides; an *equilateral triangle* has three equal sides (and three equal angles of 60°).

A right-angled triangle has one angle of 90°. If the length of one side of a triangle is *l* and the perpendicular distance from that side to the opposite corner is *h* (the height or altitude of the triangle), its area $A = (l \times h)/2$.

triangle of forces method of calculating the force produced by two other forces (the resultant). It is based on the fact that if three forces acting at a point can be represented by the sides of a triangle, the forces are in equilibrium. See ◊parallelogram of forces.

triangulation technique used in surveying and navigation to determine distances, using the properties of the triangle. To begin, surveyors measure a certain length exactly to provide a base line. From each end of this line they then measure the angle to a distant point, using a ◊theodolite. They now have a triangle in which they know the length of one side and the two adjacent angles. By simple trigonometry they can work out the lengths of the other two sides.

To make a complete survey of the region, they repeat the process, building on the first triangle.

Triassic period of geological time 245–208 million years ago, the first period of the Mesozoic era. The continents were fused together to form the world continent ◊Pangaea. Triassic sediments contain remains of early dinosaurs and other reptiles now extinct. By late Triassic times, the first mammals had evolved.

The climate was generally dry; desert sandstones are typical Triassic rocks.

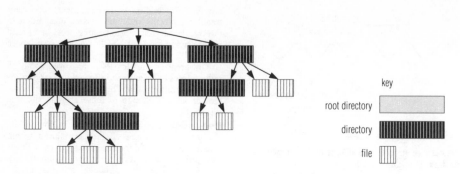

key

root directory

directory

file

tree-and-branch filing system

tribology the science of friction, lubrication, and lubricants. It studies the origin of frictional forces, the wearing of interacting surfaces, and the problems of efficient lubrication.

tributyl tin (TBT) chemical used in antifouling paints on ships' hulls and other submarine structures to deter the growth of barnacles. The tin dissolves in sea water and enters the food chain; the use of TBT has therefore been banned in many countries, including the UK.

triceratops any of a genus *Triceratops* of massive, horned dinosaurs of the order Ornithischia. They had three horns and a neck frill and were up to 8 m/25 ft long; they lived in the Cretaceous period.

trichloromethane technical name for ◊chloroform.

tricuspid valve flap of tissue situated on the right side of the ◊heart between the atrium and the ventricle. It prevents blood flowing backwards when the ventricle contracts.

As in all valves, its movements are caused by pressure changes during the beat rather than by any intrinsic muscular activity. As the valve snaps shut, a vibration passes through the chest cavity and is detectable as the first sound of the heartbeat.

triglyceride chemical name for ◊fat.

trigonometry branch of mathematics that solves problems relating to plane and spherical triangles. Its principles are based on the fixed proportions of sides for a particular angle in a right-angled triangle, the simplest of which are known as the ◊sine, ◊cosine, and ◊tangent (so-called trigonometrical ratios). It is of practical importance in navigation, surveying, and simple harmonic motion in physics.

Invented by Greek astronomer Hipparchus, trigonometry was developed by Ptolemy of Alexandria and was known to early Hindu and Arab mathematicians.

triiodomethane technical name for ◊iodoform.

trilobite any of a large class (Trilobita) of extinct, marine, invertebrate arthropods of the Palaeozoic era, with a flattened, oval body, 1–65 cm/0.4–26 in long. The hard-shelled body was divided by two deep furrows into three lobes.

Some were burrowers, others were swimming and floating forms. Their worldwide distribution, many species, and the immense quantities of their remains make them useful in geologic dating.

triode three-electrode thermionic ◊valve contain-ing an anode and a cathode (as does a ◊diode) with an additional negatively biased control grid. Small variations in voltage on the grid bias result in large variations in the current. The triode was commonly used in amplifiers but has now been almost entirely superseded by the ◊transistor.

triple bond three covalent bonds between adjacent atoms, as in the ◊alkynes (–C≡C–).

triploblastic in biology, having a body wall composed of three layers. The outer layer is the *ecto-derm*, the middle layer the *mesoderm*, and the inner layer the *endoderm*. This pattern of development is shown by most multicellular animals (including humans).

triticale cereal crop of recent origin that is a cross between wheat *Triticum* and rye *Secale*. It can produce heavy yields of high-protein grain, principally for use as animal feed.

tritium radioactive isotope of hydrogen, three times as heavy as ordinary hydrogen, consisting of one proton and two neutrons. It has a half-life of 12.5 years.

Triton in astronomy, the largest of Neptune's moons. It has a diameter of 2,700 km/1,680 mi, and orbits Neptune every 5.88 days in a retrograde (east to west) direction.

It is slightly larger than the planet Pluto, which it is thought to resemble in composition and appearance. Probably Triton was formerly a separate body like Pluto but was captured by Neptune. Triton was discovered in 1846 by British astronomer William Lassell (1799–1880) only weeks after the discovery of Neptune.

Triton's surface, as revealed by the *Voyager 2* space probe, has a temperature of 38K (–235°C/–391°F), making it the coldest known place in the Solar System. It is covered with frozen nitrogen and methane, some of which evaporates to form a tenuous atmosphere with a pressure only 0.00001 that of the Earth at sea level. Triton has a pink south polar cap, probably coloured by the effects of solar radiation on methane ice. Dark streaks on Triton are thought to be formed by geysers of liquid nitrogen.

Trojan horse in computing, a ◊virus program that appears to function normally but, while undetected by the normal user, causes damage to other files or circumvents security procedures. The earliest appeared in the UK in about 1988.

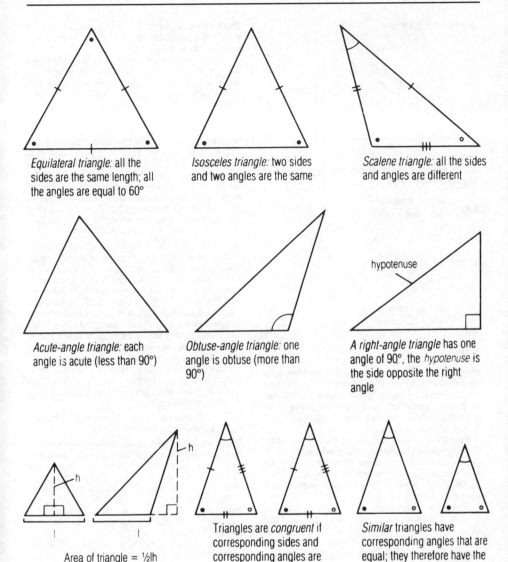

Equilateral triangle: all the sides are the same length; all the angles are equal to 60°

Isosceles triangle: two sides and two angles are the same

Scalene triangle: all the sides and angles are different

Acute-angle triangle: each angle is acute (less than 90°)

Obtuse-angle triangle: one angle is obtuse (more than 90°)

A right-angle triangle has one angle of 90°, the hypotenuse is the side opposite the right angle

Area of triangle = ½lh

Triangles are congruent if corresponding sides and corresponding angles are equal

Similar triangles have corresponding angles that are equal; they therefore have the same shape

triangle

trolley bus bus driven by electric power collected from overhead wires. It has greater manoeuvrability than a tram (see ◊tramway), but its obstructiveness in present-day traffic conditions led to its withdrawal in the UK.

trophic level in ecology, the position occupied by a species (or group of species) in a ◊food chain. The main levels are *primary producers* (photosynthetic plants), *primary consumers* (herbivores), *secondary consumers* (carnivores), and *decomposers* (bacteria and fungi).

tropical cyclone another term for ◊hurricane.

tropics the area between the tropics of Cancer and Capricorn, defined by the parallels of latitude approximately 23°30′ N and S of the equator. They are the limits of the area of Earth's surface in which the Sun can be directly overhead. The mean monthly temperature is over 20°C/68°F.

Climates within the tropics lie in parallel bands. Along the equator is the ◊intertropical convergence zone, characterized by high temperatures and year-round heavy rainfall. Tropical rainforests are found here. Along the tropics themselves lie the tropical high-pressure zones, characterized by descending dry air and desert conditions. Between these, the conditions vary seasonally between wet and dry, producing the tropical grasslands.

tropism or *tropic movement* the directional growth of a plant, or part of a plant, in response to an external stimulus such as gravity or light. If the

trigonometrical ratios

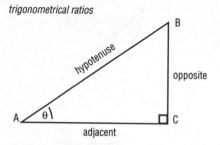

for any right-angled triangle with angle θ as shown the **trigonometrical ratios** are

$$\sin \theta = \frac{BC}{AB} = \frac{opposite}{hypotenuse}$$

$$\cos \theta = \frac{AC}{AB} = \frac{adjacent}{hypotenuse}$$

$$\tan \theta = \frac{BC}{AC} = \frac{opposite}{adjacent}$$

trigonometry

movement is directed towards the stimulus it is described as positive; if away from it, it is negative. *Geotropism* for example, the response of plants to gravity, causes the root (positively geotropic) to grow downwards, and the stem (negatively geotropic) to grow upwards.

Phototropism occurs in response to light, *hydrotropism* to water, *chemotropism* to a chemical stimulus, and *thigmotropism*, or *haptotropism*, to physical contact, as in the tendrils of climbing plants when they touch a support and then grow around it.

Tropic movements are the result of greater rate of growth on one side of the plant organ than the other. Tropism differs from a ◊nastic movement in being influenced by the direction of the stimulus.

troposphere lower part of the Earth's ◊atmosphere extending about 10.5 km/6.5 mi from the Earth's surface, in which temperature decreases with height to about –60°C/–76°F except in local layers of temperature inversion. The *tropopause* is the upper boundary of the troposphere, above which the temperature increases slowly with height within the atmosphere. All of the Earth's weather takes place within the troposphere.

troy system system of units used for precious metals and gems. The pound troy (0.37 kg) consists of 12 ounces (each of 120 carats) or 5,760 grains (each equal to 65 mg).

truncated spur blunt-ended ridge of rock jutting from the side of a glacial trough, or valley. As a glacier moves down a river valley it is unable to flow around the ◊interlocking spurs that project from either side, and so it erodes straight through them, shearing away their tips and forming truncated spurs.

truth table in electronics, a diagram showing the effect of a particular ◊logic gate on every combination of inputs.

Every possible combination of inputs and outputs for a particular gate or combination of gates is described, thereby defining their action in full. When logic value 1 is written in the table, it indicates a 'high' (or 'yes') input of perhaps 5 volts; logic value 0 indicates a 'low' (or 'no') input of 0 volts.

trypsin an enzyme in the vertebrate gut responsible for the digestion of protein molecules. It is secreted by the pancreas but in an inactive form known as trypsinogen. Activation into working trypsin occurs only in the small intestine, owing to the action of another enzyme enterokinase, secreted by the wall of the duodenum. Unlike the digestive enzyme pepsin, found in the stomach, trypsin does not require an acid environment.

tsunami (Japanese 'harbour wave') wave generated by an undersea ◊earthquake or volcanic eruption. In the open ocean it may take the form of several successive waves, rarely in excess of a metre in height but travelling at speeds of 650–800 kph/ 400–500 mph. In the coastal shallows tsunamis slow down and build up, producing towering waves that can sweep inland and cause great loss of life and property. In 1983 an earthquake in the Pacific caused tsunamis up to 3 m high, which killed over 100 people in Akita, Northern Japan.

Before each wave there may be a sudden, unexpected withdrawal of water from the beach. Used synonymously with tsunami, the popular term 'tidal wave' is misleading.

TTL abbreviation for ◊*transistor–transistor logic*, a family of integrated circuits.

tuber swollen region of an underground stem or root, usually modified for storing food. The potato is a *stem tuber*, as shown by the presence of terminal and lateral buds, the 'eyes' of the potato. *Root tubers*, for example dahlias, developed from ◊adventitious roots (growing from the stem, not from other roots) lack these. Both types of tuber can give rise to new individuals and so provide a means of ◊vegetative reproduction.

Unlike a bulb, a tuber persists for one season only; new tubers developing on a plant in the following year are formed in different places. See also ◊rhizome.

Tucana constellation of the southern hemisphere, represented as a toucan. It contains the second most prominent ◊globular cluster in the sky, 47 Tucanae, and the Small ◊Magellanic Cloud.

Tucana is one of the 11 constellations named by Johann Bayer (1572–1625) early in the 17th century to complement the 65 constellations delineated in ancient times.

tufa or *travertine* soft, porous, ◊limestone rock, white in colour, deposited from solution from carbonate-saturated ground water around hot springs and in caves.

tumour overproduction of cells in a specific area of the body, often leading to a swelling or lump. Tumours are classified as *benign* or *malignant* (see ◊cancer). Benign tumours grow more slowly, do not invade surrounding tissues, do not spread

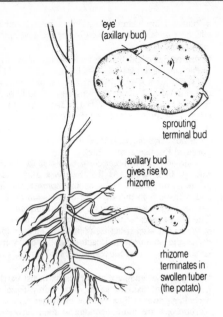

'eye'
(axillary bud)

sprouting
terminal bud

axillary bud
gives rise to
rhizome

rhizome
terminates in
swollen tuber
(the potato)

tuber *Tubers are produced underground from stems, as in the potato, or from roots, as in the dahlia. Tubers can grow into new plants.*

to other parts of the body, and do not usually recur after removal. However, benign tumours can be dangerous in areas such as the brain. The most familiar types of benign tumour are warts on the skin. In some cases, there is no sharp dividing line between benign and malignant tumours.

tundra region of high latitude almost devoid of trees, resulting from the presence of ◊permafrost. The vegetation consists mostly of grasses, sedges, heather, mosses, and lichens. Tundra stretches in a continuous belt across N North America and Eurasia.

The term was originally applied to part of N Russia, but is now used for all such regions.

tung oil oil used in paints and varnishes, obtained from trees of the genus *Aleurites*, family Euphorbiaceae, native to China.

tungsten (Swedish *tung sten* 'heavy stone') hard, heavy, grey-white, metallic element, symbol W (from German *Wolfram*), atomic number 74, relative atomic mass 183.85. It occurs in the minerals wolframite, scheelite, and hubertite. It has the highest melting point of any metal (6,170°F/3,410°C) and is added to steel to make it harder, stronger, and more elastic; its other uses include high-speed cutting tools, electrical elements, and thermionic couplings. Its salts are used in the paint and tanning industries.

Tungsten was first recognized in 1781 by Swedish chemist Karl Scheele in the ore scheelite. It was isolated in 1783 by Spanish chemists Fausto D'Elhuyar (1755–1833) and his brother Juan José (1754–1796).

tungsten ore either of the two main minerals, wolframite (FeMn)WO$_4$ and scheelite, CaWO$_4$, from which tungsten is extracted. Most of the

world's tungsten reserves are in China, but the main suppliers are Bolivia, Australia, Canada, and the USA.

Tunguska Event explosion at Tunguska, central Siberia, Russia, in June 1908, which devastated around 6,500 sq km/2,500 sq mi of forest. It is thought to have been caused by either a cometary nucleus or a fragment of ◊Encke's comet about 200 m/220 yards across. The magnitude of the explosion was equivalent to an atom bomb and produced a colossal shock wave; a bright falling object was seen 600 km/375 mi away and was heard up to 1,000 km/625 mi away.

An expedition to the site was made in 1927. The central area of devastation was occupied by trees that were erect but stripped of their branches. Further out, to a radius of 20 km/12 mi, trees were flattened and laid out radially.

tunicate any marine ◊chordate of the subphylum Tunicata (Urochordata), for example the seasquirt. Tunicates have transparent or translucent tunics made of cellulose. They vary in size from a few millimetres to 30 cm/1 ft in length, and are cylindrical, circular, or irregular in shape. There are more than a thousand species.

tunnel passageway through a mountain, under a body of water, or under ground. Tunnelling is a significant branch of civil engineering in both mining and transport. In the 19th century there were two major advances: the use of compressed air within underwater tunnels to balance the external pressure of water, and the development of the tunnel shield to support the face and assist excavation.

In recent years there have been notable developments in linings (for example, concrete segments and steel liner plates), and in the use of rotary diggers and cutters and explosives.

Major tunnels include:

Orange–Fish River (South Africa) 1975, longest irrigation tunnel, 82 km/51 mi;

Chesapeake Bay Bridge-Tunnel (USA) 1963, combined bridge, causeway, and tunnel structure, 28 km/17.5 mi;

St Gotthard (Switzerland–Italy) 1980, longest road tunnel, 16.3 km/10.1 mi;

Seikan (Japan) 1964–85, longest rail tunnel, Honshu–Hokkaido, under Tsugaru Strait, 53.9 km/33.5 mi, 23.3 km/14.5 mi undersea (however, a bullet-train service is no longer economical);

Simplon (Switzerland–Italy) 1906, longest rail tunnel on land, 19.8 km/12.3 mi;

Rogers Pass (Canada) 1989, longest tunnel in the western hemisphere, 35 km/22 mi long, through the Selkirk Mountains, British Columbia.

Plans for the *Channel tunnel* linking England and France were approved by the French and British governments in 1986, and work is under way with a schedule for completion in early 1993.

turbine engine in which steam, water, gas, or air (see ◊windmill) is made to spin a rotating shaft by pushing on angled blades, like a fan. Turbines are among the most powerful machines. Steam turbines are used to drive generators in power stations and ships' propellers; water turbines spin the generators in hydroelectric power plants; and gas

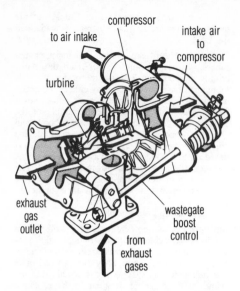

compressor

to air intake

intake air to compressor

turbine

exhaust gas outlet

wastegate boost control

from exhaust gases

turbocharger *The turbocharger increases the power of a car engine by forcing compressed air into the engine cylinders. The air is sucked in by a small turbine spun around by exhaust gases from the engine.*

turbines (as jet engines; see ◊jet propulsion) power most aircraft and drive machines in industry.

The high-temperature, high-pressure steam for **steam turbines** is raised in boilers heated by furnaces burning coal, oil, or gas, or by nuclear energy. A steam turbine consists of a shaft, or rotor, which rotates inside a fixed casing (stator). The rotor carries 'wheels' consisting of blades, or vanes. The stator has vanes set between the vanes of the rotor, which direct the steam through the rotor vanes at the optimum angle. When steam expands through the turbine, it spins the rotor by ◊reaction. The steam engine of Hero of Alexandria (130 BC), called the *aeolipile*, was the prototype of this type of turbine, called a *reaction turbine*. Modern development of the reaction turbine is largely due to English engineer Charles Parsons. Less widely used is the *impulse turbine*, patented by Carl Gustaf Patrick de Laval (1845–1913) 1882. It works by directing a jet of steam at blades on a rotor. Similarly there are reaction and impulse water turbines: impulse turbines work on the same principle as the water wheel and consist of sets of buckets arranged around the edge of a wheel; reaction turbines look much like propellers and are fully immersed in the water.

In a *gas turbine* a compressed mixture of air and gas, or vaporized fuel, is ignited, and the hot gases produced expand through the turbine blades, spinning the rotor. In the industrial gas turbine, the rotor shaft drives machines. In the jet engine, the turbine drives the compressor, which supplies the compressed air to the engine, but most of the power developed comes from the jet exhaust in the form of propulsive thrust.

turbocharger turbine-driven device fitted to engines to force more air into the cylinders, producing extra power. The turbocharger consists of

a 'blower', or ◊compressor, driven by a turbine, which in most units is driven by the exhaust gases leaving the engine.

turbofan jet engine of the type used by most airliners, so called because of its huge front fan. The fan sends air not only into the engine for combustion but also around the engine for additional thrust. This results in a faster and more fuel-efficient propulsive jet (see ◊jet propulsion).

turbojet jet engine that derives its thrust from a jet of hot exhaust gases. Pure turbojets can be very powerful but use a lot of fuel.

A single-shaft turbojet consists of a shaft (rotor) rotating in a casing. At the front is a multiblade ◊compressor, which takes in and compresses air and delivers it to one or more combustion chambers. Fuel (kerosene) is then sprayed in and ignited. The hot gases expand through a nozzle at the rear · of the engine after spinning a ◊turbine. The turbine drives the compressor. Reaction to the backward stream of gases produces a forward propulsive thrust.

turboprop jet engine that derives its thrust partly from a jet of exhaust gases, but mainly from a propeller powered by a turbine in the jet exhaust. Turboprops are more economical than turbojets but can be used only at relatively low speeds.

A turboprop typically has a twin-shaft rotor. One shaft carries the compressor and is spun by one turbine, while the other shaft carries a propeller and is spun by a second turbine.

turbulence irregular fluid (gas or liquid) flow, in which vortices and unpredictable fluctuations and motions occur. ◊Streamlining reduces the turbulence of flow around an object, such as an aircraft, and reduces drag. Turbulent flow of a fluid occurs when the ◊Reynolds number is high.

turgor the rigid condition of a plant caused by the fluid contents of a plant cell exerting a mechanical pressure against the cell wall. Turgor supports plants that do not have woody stems.

turpentine solution of resins distilled from the sap of conifers, used in varnish and as a paint solvent but now largely replaced by ◊white spirit.

turquoise mineral, hydrous basic copper aluminium phosphate. Blue-green, blue, or green, it is a gemstone. Turquoise is found in Iran, Turkestan, Mexico, and southwestern USA.

turtle small computer-controlled wheeled robot. The turtle's movements are determined by programs written by a computer user, typically using the high-level programming language ◊LOGO.

TVP (abbreviation for *t*exturized *v*egetable *p*rotein) meat substitute usually made from soya beans. In manufacture, the soya-bean solids (what remains after oil has been removed) are ground finely and mixed with a binder to form a sticky mixture. This is forced through a spinneret and extruded into fibres, which are treated with salts and flavourings, wound into hanks, and then chopped up to resemble meat chunks.

twilight period of faint light that precedes sunrise and follows sunset, caused by the reflection of light from the upper layers of the atmosphere. The limit of twilight is usually regarded as being when the Sun is 18° below the horizon. The length of twilight

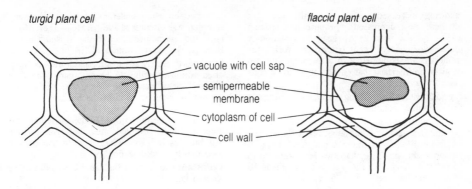

turgid plant cell

flaccid plant cell

vacuole with cell sap

semipermeable membrane

cytoplasm of cell

cell wall

turgor

depends on the latitude—at the tropics, it only lasts a few minutes; near the poles, it may last all night.

twin one of two young produced from a single pregnancy. Human twins may be genetically identical, having been formed from a single fertilized egg that split into two cells, both of which became implanted. Nonidentical twins are formed when two eggs are fertilized at the same time.

two's complement number system number system, based on the ◊binary number system, that allows both positive and negative numbers to be conveniently represented for manipulation by computer.

In the two's complement system the most significant column heading (the furthest to the left) is always taken to represent a negative number. For example, the four-column two's complement number 1101 stands for:

−8s	4s	2s	1s
1	1	0	1

It is therefore equivalent to the decimal number −3, since −8 + 4 + 1 = −3.

two-stroke cycle operating cycle for internal

floor turtle drawing a picture under computer control

turtle *The turtle is a wheeled robot that moves under computer control, tracing a line with the pen as it goes. It has been used to introduce children to computing, using a computer language called LOGO.*

combustion piston engines. The engine cycle is completed after just two strokes (movement up or down) of the piston, which distinguishes it from the more common ◊four-stroke cycle. Power mowers and lightweight motorcycles use two-stroke petrol engines, which are cheaper and simpler than four-strokes.

Most marine diesel engines are also two-strokes. In a typical two-stroke motorcycle engine, fuel mixture is drawn into the crankcase as the piston moves up on its first stroke to compress the mixture above it. Then the compressed mixture is ignited, and hot gases are produced, which drive the piston down on its second stroke. As it moves down, it uncovers an opening (port) that allows the fresh fuel mixture in the crankcase to flow into the combustion space above the piston. At the same time, the exhaust gases leave through another port.

type metal ◊alloy of tin, lead, and antimony, for making the metal type used by printers.

typesetting means by which text, or copy, is prepared for ◊printing, now usually carried out by computer. Text is keyed on a typesetting machine in a similar way to typing. Laser or light impulses are projected on to light-sensitive film that, when developed, can be used to make plates for printing.

typewriter keyboard machine that produces characters on paper. The earliest known typewriter design was patented by Henry Mills in England 1714. However, the first practical typewriter was built 1867 in Milwaukee, Wisconsin, USA, by Christopher Sholes, Carlos Glidden, and Samuel Soulé. By 1873 Remington and Sons, US gunmakers, produced under contract the first machines for sale and 1878 patented the first with lower-case as well as upper-case (capital) letters.

The first typewriter patented by Sholes included an alphabetical layout of keys, but Remington's first commercial typewriter had the QWERTY keyboard, designed by Sholes to slow down typists who were too fast for their mechanical keyboards. (Other layouts include the Dvorak keyboard developed by John Dvorak 1932, in which the most commonly used letters are evenly distributed between left and right, and under the strongest fingers.) Later developments include tabulators from about 1898, portable machines about 1907, the gradual introduction of electrical operation (allowing increased speed, since the keys are touched, not depressed), proportional spacing 1940, and the rotating typehead with stationary plates 1962. More recent typewriters work electronically, are equipped with a memory, and can be given an interface that enables them to be connected to a computer.

typhoon violently revolving storm, a ◊hurricane in the W Pacific Ocean.

tyrannosaurus any of a genus *Tyrannosaurus* of gigantic flesh-eating ◊dinosaurs, order Saurischia, which lived in North America and Asia about 70 million years ago. They had two feet, were up to 15 m/50 ft long, 6.5 m/20 ft tall, weighed 10 tonnes, and had teeth 15 cm/6 in long.

Only a few whole skeletons are known; the most complete was discovered 1989 in Hell Creek, Montana, and will be preserved in the Museum of the Rockies, Bozeman, Montana, USA.

tyre (North American *tire*) inflatable rubber hoop fitted round the rims of bicycle, car, and other road-vehicle wheels. The first pneumatic rubber tyre was patented in 1845 by the Scottish engineer Robert William Thomson (1822–73), but it was Scottish inventor John Boyd Dunlop of Belfast who independently reinvented pneumatic tyres for use with bicycles 1888–89. The rubber for car tyres is hardened by ◊vulcanization.

Tyuratam site of the ◊Baikonur Cosmodrome in Kazakhstan.

U

UHF (abbreviation for *ultra high frequency*) referring to radio waves of very short wavelength, used, for example, for television broadcasting.

UHT abbreviation for *ultra heat treated*; see ◊food technology.

ULA abbreviation for ◊*uncommitted logic array*, a type of integrated circuit.

ulna one of the two bones found in the lower limb of the tetrapod (four-limbed) vertebrate.

ultrabasic in geology, an igneous rock with a lower silica content than basic rocks (less than 45% silica).

ultrafiltration process by which substances in solution are separated on the basis of their molecular size. A solution is forced through a membrane with pores large enough to permit the passage of small solute molecules but not large ones.

Ultrafiltration is a vital mechanism in the vertebrate kidney: the cell membranes lining the Bowman's capsule act as semipermeable membranes allowing water and low-molecular-weight substances such as urea and salts to pass through into the urinary tubules but preventing the larger proteins from being lost from the blood.

ultrasonics study and application of the sound and vibrations produced by ultrasonic pressure waves (see ◊ultrasound).

The earliest practical application was to detect submarines during World War I, but recently the field of ultrasonics has greatly expanded. Frequencies above 80,000 Hz have been used to produce echoes as a means of measuring the depth of the sea and to detect flaws in metal; in medicine, high-frequency pressure waves are used to investigate various body organs (ultrasound scanning). High-power ultrasound has been used with focusing arrangements to destroy tissue deep in the body, and extremely high frequencies (1,000 MHz or more) are used in ultrasonic microscopes.

ultrasound pressure waves similar in nature to sound waves but occurring at frequencies above 20,000 Hz (cycles per second), the approximate

upper limit of human hearing (15–16 Hz is the lower limit). ◊Ultrasonics is concerned with the study and practical application of these phenomena.

The earliest practical application was to detect submarines during World War I, but recently the field of ultrasonics has greatly expanded. Frequencies above 80,000 Hz have been used to produce echoes as a means of measuring the depth of the sea or to detect flaws in metal, and in medicine, high-frequency pressure waves are used to investigate various body organs. Ultrasonic pressure waves transmitted through the body are absorbed and reflected to different degrees by different body tissues. By recording the 'echoes', a picture (sonogram) of the different structures being scanned can be built up. Ultrasound scanning is valued as a safe, noninvasive technique which often eliminates the need for exploratory surgery. Free of the risks of ionizing radiation, unlike X-rays and computerized axial tomography (◊CAT scan), it is especially valuable in obstetrics, where it has revolutionized fetal evaluation and diagnosis. High-power ultrasound has been used with focusing arrangements to destroy deep-lying tissue in the body, and extremely high frequencies of 1,000 MHz (megahertz) or more are used in ultrasonic microscopes.

ultraviolet astronomy study of cosmic ultraviolet emissions using artificial satellites. The USA has launched a series of satellites for this purpose, receiving the first useful data 1968. Only a tiny percentage of solar ultraviolet radiation penetrates the atmosphere, this being the less dangerous longer-wavelength ultraviolet. The dangerous shorter-wavelength radiation is absorbed by gases in the ozone layer high in the Earth's upper atmosphere.

The US Orbiting Astronomical Observatory (OAO) satellites provided scientists with a great deal of information regarding cosmic ultraviolet emissions. *OAO-1*, launched 1966, failed after only three days, although *OAO-2*, put into orbit 1968, operated for four years instead of the intended one year, and carried out the first ultraviolet observations of a supernova and also of Uranus.

OAO-3 (*Copernicus*), launched 1972, continued transmissions into the 1980s and discovered many new ultraviolet sources. The *International Ultraviolet Explorer (IUE)*, launched Jan 1978 and still operating in the early 1990s, observed all the main objects in the Solar System (including Halley's comet), stars, galaxies, and the interstellar medium.

ultraviolet radiation electromagnetic radiation invisible to the human eye, of wavelengths from about 400 to 4 nm (where the ◊X-ray range begins). Physiologically, ultraviolet radiation is extremely powerful, producing sunburn and causing the formation of vitamin D in the skin.

Ultraviolet rays are strongly germicidal and may be produced artificially by mercury vapour and arc lamps for therapeutic use. The radiation may be detected with ordinary photographic plates or films. It can also be studied by its fluorescent effect on certain materials. The desert iguana *Disposaurus dorsalis* uses it to locate the boundaries of its territory and to find food.

ULTRAVIOLET FISH

Several types of fish and some birds can detect ultraviolet radiation. On misty or dull days, ultraviolet penetrates the haze and reveals the Sun's position, enabling the animals to navigate by the Sun.

Ulysses space probe to study the Sun's poles, launched 1990 by a US space shuttle. It is a joint project by NASA and the European Space Agency. In Feb 1992, the gravity of Jupiter swung *Ulysses* on to a path that loops it first under the Sun's south pole in 1994 and then over the north pole in 1995 to study the Sun and solar wind at latitudes not observable from the Earth.

umbilical cord connection between the ◊embryo and the ◊placenta of placental mammals. It has one vein and two arteries, transporting oxygen and nutrients to the developing young, and removing waste products. At birth, the connection between the young and the placenta is no longer necessary. The umbilical cord drops off or is severed, leaving a scar called the navel.

umbra region of a ◊shadow that is totally dark because no light reaches it, and from which no part of the light source can be seen (compare ◊penumbra). In astronomy, it is a region of the Earth from which a complete ◊eclipse of the Sun or Moon can be seen.

uncertainty principle or *indeterminacy principle* in quantum mechanics, the principle that it is meaningless to speak of a particle's position, momentum, or other parameters, except as results of measurements; measuring, however, involves an interaction (such as a ◊photon of light bouncing off the particle under scrutiny), which must disturb the particle, though the disturbance is noticeable only at an atomic scale. The principle implies that one cannot, even in theory, predict the moment-to-moment behaviour of such a system.

It was established by German physicist Werner Heisenberg, and gave a theoretical limit to the precision with which a particle's momentum and position can be measured simultaneously: the more

accurately the one is determined, the more uncertainty there is in the other.

uncommitted logic array (ULA) or *gate array* in computing, a type of semicustomized integrated circuit in which the logic gates are laid down to a general-purpose design but are not connected to each other. The interconnections can then be set in place according to the requirements of individual manufacturers. Producing ULAs may be cheaper than using a large number of TTL (◊transistor-transistor logic) chips or commissioning a fully customized chip.

unconformity in geology, a break in the sequence of ◊sedimentary rocks. It is usually seen as an eroded surface, with the ◊beds above and below lying at different angles. An unconformity represents an ancient land surface, where exposed rocks were worn down by erosion and later covered in a renewed cycle of deposition.

UNEP acronym for *United Nations Environmental Programme.*

unicellular organism animal or plant consisting of a single cell. Most are invisible without a microscope but a few, such as the giant ◊amoeba, may be visible to the naked eye. The main groups of unicellular organisms are bacteria, protozoa, unicellular algae, and unicellular fungi or yeasts. Some become disease-causing agents, ◊pathogens.

unidentified flying object or *UFO* any light or object seen in the sky whose immediate identity is not apparent. Despite unsubstantiated claims, there is no evidence that UFOs are alien spacecraft. On investigation, the vast majority of sightings turn out to have been of natural or identifiable objects, notably bright stars and planets, meteors, aircraft, and satellites, or to have been perpetrated by pranksters. The term *flying saucer* was coined in 1947 and has been in use since.

unified field theory in physics, the theory that attempts to explain the four fundamental forces (strong nuclear, weak nuclear, electromagnetic, and gravity) in terms of a single unified force (see ◊particle physics).

Research was begun by Albert Einstein and, by 1971, a theory developed by US physicists Steven Weinberg and Sheldon Glashow, Pakistani physicist Abdus Salam, and others, had demonstrated the link between the weak and electromagnetic forces. The next stage is to develop a theory (called the ◊grand unified theory, or GUT) that combines the strong nuclear force with the electroweak force. The final stage will be to incorporate gravity into the scheme. Work on the ◊superstring theory indicates that this may be the ultimate 'theory of everything'.

uniformitarianism in geology, the principle that processes that can be seen to occur on the Earth's surface today are the same as those that have occurred throughout geological time. For example, desert sandstones containing sand-dune structures must have been formed under conditions similar to those present in deserts today. The principle was formulated by James Hutton and expounded by Charles Lyell, both Scottish geologists.

unit standard quantity in relation to which other quantities are measured. There have been many systems of units. Some ancient units, such as the day, the foot, and the pound, are still in use. ◊SI units, the latest version of the metric system, are widely used in science.

United Kingdom Infrared Telescope (UKIRT) 3.8–m/150–in reflecting telescope for observing at infrared wavelengths, opened in 1978 on ◊Mauna Kea and operated by the Royal Observatory, Edinburgh, Scotland.

universal indicator in chemistry, a mixture of ◊pH indicators, used to gauge the acidity or alkalinity of a solution. Each component changes colour at a different pH value, and so the indicator is capable of displaying a range of colours, according to the pH of the test solution, from red (at pH 1) to purple (at pH 13).

The pH of a substance may be found by adding a few drops of universal indicator and noting the colour, or by dipping in an absorbent paper strip that has been impregnated with the indicator.

universal joint flexible coupling used to join rotating shafts; for example, the propeller shaft in a car. In a typical universal joint the ends of the shafts to be joined are in U-shaped yokes. They dovetail into each other and pivot flexibly about an X-shaped spider. This construction allows side-to-side and up-and-down movement, while still transmitting rotary motion.

universal time (UT) another name for ◊Greenwich Mean Time. It is based on the rotation of the Earth, which is not quite constant. Since 1972, UT has been replaced by *coordinated universal time* (UTC), which is based on uniform atomic time; see ◊time.

universe all of space and its contents, the study of which is called cosmology. The universe is thought to be between 10 billion and 20 billion years old, and is mostly empty space, dotted with ◊galaxies for as far as telescopes can see. The most distant detected galaxies and ◊quasars lie 10 billion light years or more from Earth, and are moving farther apart as the universe expands. Several theories attempt to explain how the universe came into being and evolved, for example, the ◊Big Bang theory of an expanding universe originating in a single explosive event, and the contradictory ◊steady-state theory.

A lot of prizes have been awarded for showing that the universe is not as simple as we might have thought!

On the **universe** Stephen Hawking *A Brief History of Time* 1988

Apart from those galaxies within the ◊Local Group, all the galaxies we see display ◊red shifts in their spectra, indicating that they are moving away from us. The farther we look into space, the greater are the observed red shifts, which implies that the more distant galaxies are receding at ever greater speeds. This observation led to the theory of an expanding universe, first proposed by Edwin Hubble 1929, and to Hubble's law, which states that the speed with which one galaxy moves away from another is proportional to its distance from

it. Current data suggest that the galaxies are moving apart at a rate of 50–100 kps/30–60 mps for every million ◊parsecs of distance.

Unix multiuser ◊operating system designed for minicomputers but becoming increasingly popular on large microcomputers, workstations, mainframes, and supercomputers. It was developed by AT&T's Bell Laboratories in the USA during the late 1960s, using the programming language ◊C. It could therefore run on any machine with a C compiler, so ensuring its wide portability. Its wide range of functions and flexibility have made it widely used by universities and in commercial software.

unleaded petrol petrol manufactured without the addition of ◊antiknock. It has a slightly lower octane rating than leaded petrol, but has the advantage of not polluting the atmosphere with lead compounds. Many cars can be converted to running on unleaded petrol by altering the timing of the engine, and most new cars are designed to do so. Cars fitted with a ◊catalytic converter must use unleaded fuel.

The use of unleaded petrol has been standard in the USA for some years, and is increasing in the UK (encouraged by a lower rate of tax than that levied on leaded petrol).

unnilennium synthesized radioactive element of the ◊transactinide series, symbol Une, atomic number 109, relative atomic mass 266. It was first produced in 1982 at the Laboratory for Heavy Ion Research in Darmstadt, Germany, by fusing bismuth and iron nuclei; it took a week to obtain a single new, fused nucleus.

The element is (like unnilhexium, uniloctium, unnilpentium, unnilquadium, and unnilseptium) as yet unnamed; temporary identification was assigned until a name is approved by the International Union of Pure and Applied Chemistry.

UNNILENNIUM: THE RAREST ELEMENT

The rarest element of all is the synthesized element unnilennium. A single atom of this element was produced by German scientists in 1982. Unfortunately, the atom vanished by radioactive decay almost instantly.

---

unnilhexium synthesized radioactive element of the ◊transactinide series, symbol Unh, atomic number 106, relative atomic mass 263. It was first synthesized in 1974 by two institutions, each of which claims priority. The University of California at Berkeley bombarded californium with oxygen nuclei to get isotope 263; the Joint Institute for Nuclear Research in Dubna, Russia, bombarded lead with chromium nuclei to obtain isotopes 259 and 260.

uniloctium synthesized, radioactive element of the ◊transactinide series, symbol Uno, atomic number 108, relative atomic mass 265. It was first synthesized in 1984 by the Laboratory for Heavy Ion Research in Darmstadt, Germany.

unnilpentium synthesized, radioactive, metallic element of the ◊transactinide series, symbol Unp, atomic number 105, relative atomic mass 262. Six isotopes have been synthesized, each with very short (fractions of a second) half-lives. Two institutions claim to have been the first to produce it: the Joint Institute for Nuclear Research in Dubna, Russia, in 1967 (proposed name *nielsbohrium*); and the University of California at Berkeley, USA, who disputed the Soviet claim, in 1970 (proposed name *hahnium*).

unnilquadium synthesized, radioactive, metallic element, the first of the ◊transactinide series, symbol Unq, atomic number 104, relative atomic mass 262. It is produced by bombarding californium with carbon nuclei and has ten isotopes, the longest-lived of which, Unq-262, has a half-life of 70 seconds.

Two institutions claim to be the first to have synthesized it: the Joint Institute for Nuclear Research in Dubna, Russia, in 1964 (proposed name *kurchatovium*); and the University of California at Berkeley, USA, in 1969 (proposed name *rutherfordium*).

unnilseptium synthesized, radioactive element of the ◊transactinide series, symbol Uns, atomic number 107, relative atomic mass 262. It was first synthesized by the Joint Institute for Nuclear Research in Dubna, Russia, in 1976; in 1981 the Laboratory for Heavy Ion Research in Darmstadt, Germany, confirmed its existence.

unsaturated compound chemical compound in which two adjacent atoms are bonded by a double or triple covalent bond.

Examples are ◊alkenes and ◊alkynes, where the two adjacent atoms are both carbon, and ◊ketones, where the unsaturation exists between atoms of different elements. The laboratory test for unsaturated compounds is the addition of bromine water; if the test substance is unsaturated, the bromine water will be decolorized.

unsaturated solution solution that is capable of dissolving more solute than it already contains at the same temperature.

uraninite uranium oxide, UO_2, an ore mineral of uranium, also known as *pitchblende* when occurring in massive form. It is black or brownish black, very dense, and radioactive. It occurs in veins and as massive crusts, usually associated with granite rocks.

uranium hard, lustrous, silver-white, malleable and ductile, radioactive, metallic element of the ◊actinide series, symbol U, atomic number 92, relative atomic mass 238.029. It is the most abundant radioactive element in the Earth's crust, its decay giving rise to essentially all radioactive elements in nature; its final decay product is the stable element lead. Uranium combines readily with most elements to form compounds that are extremely poisonous. The chief ore is ◊pitchblende, in which the element was discovered by German chemist Martin Klaproth 1789; he named it after the planet Uranus, which had been discovered 1781.

Small amounts of certain compounds containing uranium have been used in the ceramics industry to make orange-yellow glazes and as mordants in dyeing; however, this practice was discontinued when the dangerous effects of radiation became known.

Uranium is one of three fissile elements (the others are thorium and plutonium). It was long considered to be the element with the highest atomic number to occur in nature. The isotopes U-238 and U-235 have been used to help determine the age of the Earth.

Uranium-238, which comprises about 99% of all naturally occurring uranium, has a half-life of 4.51×10^9 years. Because of its abundance, it is the isotope from which fissile plutonium is produced in breeder ◊nuclear reactors. The fissile isotope U-235 has a half-life of 7.13×10^8 years and comprises about 0.7% of naturally occurring uranium; it is used directly as a fuel for nuclear reactors and in the manufacture of nuclear weapons.

Many countries mine uranium; large deposits are found in Canada, the USA, Australia, and South Africa.

uranium ore material from which uranium is extracted, often a complex mixture of minerals. The main ore is uraninite (or pitchblende) UO_2, which is commonly found with sulphide minerals. The USA, Canada, and South Africa are the main producers in the West.

Uranus the seventh planet from the Sun, discovered by English astronomer William Herschel 1781. It is twice as far out as the sixth planet, Saturn. Uranus has a diameter of 50,800 km/31,600 mi and a mass 14.5 times that of Earth. It orbits the Sun in 84 years at an average distance of 2.9 billion km/1.8 billion mi. The spin axis of Uranus is tilted at 98°, so that one pole points towards the Sun, giving extreme seasons. It has 15 moons, and in 1977 was discovered to have thin rings around its equator.

Uranus has a peculiar magnetic field, whose axis is tilted at 60° to its axis of spin, and is displaced about one-third of the way from the planet's centre to its surface. Observations of the magnetic field show that the solid body of the planet rotates every 17.2 hours. Uranus spins from east to west, the opposite of the other planets, with the exception of Venus and possibly Pluto. The rotation rate of the atmosphere varies with latitude, from about 16 hours in mid-southern latitudes to longer than 17 hours at the equator.

The space probe *Voyager 2* detected eleven rings, composed of rock and dust, around the planet's equator, and found ten small moons in addition to the five visible from Earth. Titania, the largest moon, has a diameter of 1,580 km/980 mi. The rings are charcoal black, and may be debris of former 'moonlets' that have broken up.

urea $CO(NH_2)_2$ waste product formed in the mammalian liver when nitrogen compounds are broken down. It is excreted in urine. When purified, it is a white, crystalline solid. In industry it is used to make urea-formaldehyde plastics (or resins), pharmaceuticals, and fertilizers.

ureter tube connecting the kidney to the bladder. Its wall contains fibres of smooth muscle, whose contractions aid the movement of urine out of the kidney.

urethra in mammals, a tube connecting the bladder to the exterior. It carries urine and, in males, semen.

uric acid $C_5H_4N_4O_3$ nitrogen-containing waste substance, formed from the breakdown of food and body protein. It is only slightly soluble in water. Uric acid is the normal means by which most land animals that develop in a shell (birds, reptiles, insects, and land gastropods) deposit their waste products. The young are unable to get rid of their excretory products while in the shell and therefore store them in this insoluble form.

Humans and other primates produce some uric acid as well as urea, the normal nitrogen-waste product of mammals, adult amphibians, and many marine fishes. If formed in excess and not excreted, uric acid may be deposited in sharp crystals in the joints and other tissues, causing gout; or it may form stones (calculi) in the kidneys or bladder.

What is man, when you come to think upon him, but a minutely set, ingenious machine for turning, with infinite artfulness, the red wine of Shiraz into urine?

Isak Dinesen (Karen Blixen) (1885–1962)
Seven Gothic Tales 1934

urinary system system of organs that removes nitrogenous waste products and excess water from the bodies of animals. In vertebrates, it consists of a pair of kidneys, which produce urine; ureters, which drain the kidneys; and (in bony fishes, amphibians, some reptiles, and mammals) a blad-der, which stores the urine before its discharge. In mammals, the urine is expelled through the ure-thra; in other vertebrates, the urine drains into a common excretory chamber called a ⟡cloaca, and the urine is not discharged separately.

urine amber-coloured fluid made by the kidneys from the blood. It contains excess water, salts, pro-teins, waste products in the form of urea, a pig-ment, and some acid.

The kidneys pass it through two fine tubes (uret-ers) to the bladder, which may act as a reservoir for up to 0.7 l/1.5 pt at a time. In mammals, it then passes into the urethra, which opens to the outside by a sphincter (constricting muscle) under voluntary control. In reptiles and birds, nitrogenous wastes are discharged as an almost solid substance made mostly of ⟡uric acid, rather than urea.

Ursa Major (Latin 'Great Bear') third largest con-stellation in the sky, in the north polar region. Its seven brightest stars make up the familiar shape of the *Big Dipper* or *Plough*. The second star of the 'handle' of the dipper, called Mizar, has a companion star, Alcor. Two stars forming the far side of the 'bowl' act as pointers to the north pole star, Polaris.

Ursa Minor (Latin 'Little Bear') constellation in the northern sky. It is shaped like a dipper, with the north pole star Polaris at the end of the handle. It contains the orange subgiant Kochab, about 95 light years from Earth.

user interface in computing, the procedures and

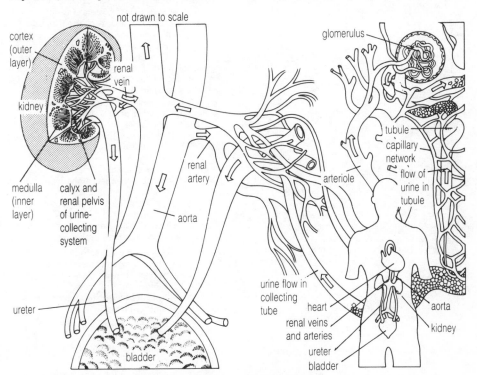

urinary system *The human urinary system. At the bottom right, the complete system in outline; on the left, the vessels linking a kidney to other organs in the body; at the top right, a detail of the network of vessels within a kidney.*

methods through which the user operates a program. These might include ◊menus, input forms, error messages, and keyboard procedures. A ◊graphical user interface (GUI or WIMP) is one that makes use of icons (small pictures) and allows the user to make menu selections with a mouse.

The study of the ways in which people interact with computers is a subbranch of ergonomics. It aims to make it easier for people to use computers effectively and comfortably, and has become a focus of research for many national and international programmes.

U-shaped valley another term for a ◊glacial trough, a valley formed by a glacier.

US Naval Observatory US government observatory in Washington, DC, which provides the nation's time service and publishes almanacs for navigators, surveyors, and astronomers. It contains a 66–cm/26–in refracting telescope opened 1873. A 1.55–m/61–in reflector for measuring positions of celestial objects was opened 1964 at Flagstaff, Arizona.

UTC abbreviation for *coordinated universal time*, the standard measurement of ◊time.

uterus hollow muscular organ of female mammals, located between the bladder and rectum, and connected to the Fallopian tubes above and the vagina below. The embryo develops within the uterus, and in placental mammals is attached to it after implantation via the ◊placenta and umbilical cord. The lining of the uterus changes during the ◊menstrual cycle. In humans and other higher primates, it is a single structure, but in other mammals it is paired.

The outer wall of the uterus is composed of smooth muscle, capable of powerful contractions (induced by hormones) during childbirth.

utility program in computing, a ◊systems program designed to perform a specific task related to the operation of the computer when requested to do so by the computer user. For example, a utility program might be used to complete a screen dump, format a disc, or convert the format of a data file so that it can be accessed by a different ◊applications program.

UV in physics, abbreviation for *ultraviolet*.

uvula muscular structure descending from the soft palate. Its length is variable (5–20 mm/0.2–0.8 inches in humans). It is raised during swallowing and is sometimes used in the production of speech sounds.

V

V Roman numeral for *five*; in physics, symbol for *volt*.

v in physics, symbol for *velocity*.

vaccine any preparation of modified viruses or bacteria that is introduced into the body, usually either orally or by a hypodermic syringe, to induce the specific ◊antibody reaction that produces ◊immunity against a particular disease.

In 1796, English physician Edward Jenner was the first to inoculate a child successfully with cowpox virus to produce immunity to smallpox. His method, the application of an infective agent to an abraded skin surface, is still used in smallpox inoculation.

In the UK, children are routinely vaccinated against diphtheria, tetanus, whooping cough, polio, measles, mumps, German measles, and tuberculosis.

vacuole in biology, a fluid-filled, membrane-bound cavity inside a cell. It may be a reservoir for fluids that the cell will secrete to the outside, or may be filled with excretory products or essential nutrients that the cell needs to store. In amoebae (single-cell animals), vacuoles are the sites of digestion of engulfed food particles. Plant cells usually have a large central vacuole for storage.

vacuum in general, a region completely empty of matter; in physics, any enclosure in which the gas pressure is considerably less than atmospheric pressure (101,325 pascals).

vacuum flask or *Dewar flask* or *Thermos flask* container for keeping things either hot or cold. It has two silvered glass walls with a vacuum between them, in a metal or plastic outer case. This design reduces the three forms of heat transfer: radiation (prevented by the silvering), conduction, and convection (both prevented by the vacuum). A vacuum flask is therefore equally efficient at keeping cold liquids cold, or hot liquids hot.

vagina the front passage in female mammals, linking the uterus to the exterior. It admits the penis during sexual intercourse, and is the birth canal down which the fetus passes during delivery.

valence electron in chemistry, an electron in the outermost shell of an ◊atom. It is the valence electrons that are involved in the formation of ionic and covalent bonds (see ◊molecule). The number of electrons in this outermost shell represents the maximum possible valency for many elements and matches the number of the group that the element occupies in the ◊periodic table of the elements.

valency in chemistry, the measure of an element's ability to combine with other elements, expressed as the number of atoms of hydrogen (or any other standard univalent element) capable of uniting with (or replacing) its atoms. The number of electrons in the outermost shell of the atom dictates the combining ability of an element.

The elements are described as uni-, di-, tri-, and tetravalent when they unite with one, two, three, and four univalent atoms respectively. Some elements have *variable valency*: for example, nitrogen and phosphorus have a valency of both three and five. The valency of oxygen is two: hence the formula for water, H_2O (hydrogen being univalent).

valency shell in chemistry, outermost shell of electrons in an ◊atom. It contains the ◊valence electrons. Elements with four or more electrons in their outermost shell can show variable valency. Chlorine can show valencies of 1, 3, 5, and 7 in different compounds.

validation in computing, the process of checking input data to ensure that it is complete, accurate, and reasonable. Although it would be impossible to guarantee that only valid data are entered into a computer, a suitable combination of validation checks should ensure that most errors are detected.

Common validation checks include:

character-type check Each input data item is checked to ensure that it does not contain invalid characters. For example, an input name might be checked to ensure that it contains only letters of

Turkish travels and English milkmaids

Today, during their first few months of life, infants are routinely vaccinated against diphtheria, tetanus, whooping cough and poliomyelitis. After one year, vaccination against measles, mumps and rubella is also recommended. Thanks to such preventive medicine and to proper nutrition, the common infectious diseases of childhood have largely disappeared from the developed world.

The regular use of vaccination began in 1796 as a result of the pioneering work of Edward Jenner, a British physician (1749–1823). In the 18th century, smallpox was one of the commonest and most deadly diseases. Most people contracted it, and a face completely unscarred by the disease was rare. In 17th-century London, some 10% of all deaths were due to smallpox.

Because the disease was so common, it was well known that an earlier, non-fatal attack of smallpox conferred immunity in any following epidemics. In Eastern countries, people were deliberately exposed to mild forms of the disease. This method was brought back to England in 1721 by Lady Mary Wortley Montague, wife of the British Ambassador to Turkey. She had her own children 'vaccinated' (the procedure was then called *variolation*), encountering much prejudice as a result.

Kill or cure?

In England, an epidemic of smallpox swept Gloucestershire in 1788, and variolation (first described in 1713) was widely practised. It involved scratching a vein in the arm of a healthy person, and working into it a small of amount of matter from a smallpox pustule taken from a person with a mild attack of the disease. This risky procedure had two major disadvantages. The inoculated subject, unless isolated, was likely to start a fresh smallpox epidemic; and if the dose was too virulent, the resulting disease was fatal.

Edward Jenner had a country medical practice, and was familiar with both human and animal diseases. He noted that milkmaids often caught the disease cowpox, and inquired further. He saw that milking was regularly done by both men and maidservants. The men, after changing the dressings on horses suffering from a disease called 'the grease', went on to milk the cows, thus infecting them at the same time. In due course, the cows became diseased, and in turn the milk maids who milked them caught cowpox.

Jenner wrote: 'Thus the disease makes its progress from the horse to the nipple of the cow, and from the cow to the human subject.' He went on: 'What makes the cowpox virus so extremely singular is that the person who has been thus affected is for ever secure from the infection of the smallpox.' Jenner then describes a great number of instances, in proof of his observations. Here are some of them.

Case I. Joseph Merret, the undergardener to the Earl of Berkeley, had been a farmer's servant in 1770, and sometimes helped with the milking. He also attended the horses. Merret caught cowpox. In April 1795, Jenner found that despite repeatedly inserting variolous matter into Merret's arm, it was impossible to infect him with smallpox. During the whole time his family had smallpox, Merret stayed with them, and remained perfectly healthy. Jenner was at pains to make sure that Merret had at no time previously caught smallpox.

Case II. Sarah Portlock nursed one of her own children who, in 1792, had accidentally caught smallpox. She considered herself safe from infection, for as a farmer's servant 27 years previously, she had contracted cowpox. She remained in the same room as her child, and as in the previous case, variolation produced no disease.

Jenner then performed the experiment that was to make him famous, and give medicine vaccination, one of the most powerful weapons against disease.

The first vaccination

Case XVII. On 14 May 1796, Jenner selected a healthy eight-year-old boy and inoculated him with cowpox, taken from a sore in the hand of a dairymaid. He inserted the infected matter in two incisions, each about 25 mm/1 in long. The boy showed only mild symptoms: on the seventh day he complained of a slight headache, became a little chilly and suffered loss of appetite.

Then on 1 July, Jenner inoculated the boy with variolous matter, inserting it in several slight punctures and incisions in both arms. No disease followed. The only symptoms the boy showed were those of someone who had recovered from smallpox, or had previously suffered from cowpox. Several months later, the boy was inoculated again with similar results.

Jenner published his results privately on the advice of Fellows of the Royal Society, who considered that he should not risk his reputation by presenting anything 'so much at variance with established knowledge'. However, in a few years, vaccination was a widespread practice.

valency shells

group number	I	II	III	IV	V	VI	VII
element	Na	Mg	Al	Si	P	S	Cl
atomic number	11	12	13	14	15	16	17
electron arrangement	2.8.1	2.8.2	2.8.3	2.8.4	2.8.5	2.8.6	2.8.7
valencies	1	2	3	4(2)	5(3)	6(2)	7(1)

the alphabet, or an input six-figure date might be checked to ensure it contains only numbers.

field-length check The number of characters in an input field is checked to ensure that the correct number of characters has been entered. For example, a six-figure date field might be checked to ensure that it does contain exactly six digits.

control-total check The arithmetic total of a specific field from a group of records is calculated—for example, the hours worked by a group of employees might be added together—and then input with the data to which it refers. The program recalculates the control total and compares it with the one entered to ensure that entry errors have not been made.

hash-total check An otherwise meaningless control total is calculated—for example, by adding together account numbers. Even though the total has no arithmetic meaning, it can still be used to check the validity of the input account numbers.

parity check Parity bits are added to binary number codes to ensure that each number in a set of data has the same ◊parity (that each binary number has an even number of 1s, for example). The binary numbers can then be checked to ensure that their parity remains the same. This check is often applied to data after it has been transferred from one part of the computer to another; for example, from a disc drive into the immediate-access memory.

check digit A digit is calculated from the digits of a code number and then added to that number as an extra digit. For example, in the ISBN (International Standard Book Number) 0 631 90057 8, the 8 is a check digit calculated from the book code number 063190057 and then added to it to make the full ISBN. When the full code number is input, the computer recalculates the check digit and compares it with the one entered. If the entered and calculated check digits do not match, the computer reports that an entry error of some kind has been made.

range check An input numerical data item is checked to ensure that its value falls in a sensible range. For example, an input two-digit day of the month might be checked to ensure that it is in the range 01 to 31.

valve device that controls the flow of a fluid. Inside a valve, a plug moves to widen or close the opening through which the fluid passes. The valve was invented by US radio engineer Lee de Forest (1873–1961).

Common valves include the cone or needle valve, the globe valve, and butterfly valve, all named after the shape of the plug. Specialized valves include the one-way valve, which permits fluid flow in one direction only, and the safety valve, which cuts off flow under certain conditions.

valve in animals, a structure for controlling the direction of the blood flow. In humans and other vertebrates, the contractions of the beating heart cause the correct blood flow into the arteries because a series of valves prevent back flow. Diseased valves, detected as 'heart murmurs', have decreased efficiency. The tendency for low-pressure venous blood to collect at the base of limbs under the influence of gravity is counteracted by a series of small valves within the veins. It was the existence of these valves that prompted the 17th-century physician William Harvey to suggest that the blood circulated around the body.

valve or **electron tube** in electronics, a glass tube containing gas at low pressure, which is used to control the flow of electricity in a circuit. Three or more metal electrodes are inset into the tube. By varying the voltage on one of them, called the **grid electrode**, the current through the valve can be controlled, and the valve can act as an amplifier. Valves have been replaced for most applications by ◊transistors. However, they are still used in high-power transmitters and amplifiers, and in some hi-fi systems.

vanadium silver-white, malleable and ductile, metallic element, symbol V, atomic number 23, relative atomic mass 50.942. It occurs in certain iron, lead, and uranium ores and is widely distributed in small quantities in igneous and sedimentary rocks. It is used to make steel alloys, to which it adds tensile strength.

Spanish mineralogist Andrés del Rio (1764–1849) and Swedish chemist Nils Sefström (1787–1845) discovered vanadium independently, the former 1801 and the latter 1831. Del Rio named it 'erythronium', but was persuaded by other chemists that he had not in fact discovered a new element; Sefström gave it its present name, after the Norse goddess of love and beauty, Vanadis (or Freya).

vanadium oxide or **vanadium pentoxide** V_2O_5 crystalline compound used as a catalyst in the ◊contact process for the manufacture of sulphuric acid.

Van.Allen radiation belts two zones of charged particles around the Earth's magnetosphere, discovered 1958 by US physicist James Van Allen. The atomic particles come from the Earth's upper atmosphere and the ◊solar wind, and are trapped by the Earth's magnetic field. The inner belt lies 1,000–5,000 km/620–3,100 mi above the equator, and contains ◊protons and ◊electrons. The outer belt lies 15,000–25,000 km/9,300–15,500 mi above the equator, but is lower around the magnetic poles. It contains mostly electrons from the solar wind.

The Van Allen belts are hazardous to astronauts, and interfere with electronic equipment on satellites.

van de Graaff generator electrostatic generator capable of producing a voltage of over a million volts. It consists of a continuous vertical conveyor

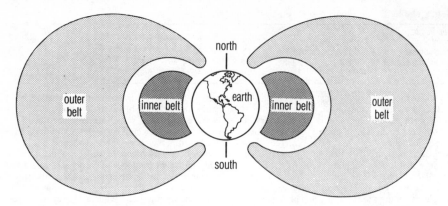

Van Allen radiation belts *The Van Allen belts of trapped charged particles are a hazard to spacecraft, affecting on-board electronics and computer systems. Similar belts have been discovered around the planets Mercury, Jupiter, Saturn, Uranus, and Neptune.*

belt that carries electrostatic charges (resulting from friction) up to a large hollow sphere supported on an insulated stand. The lower end of the belt is earthed, so that charge accumulates on the sphere. The size of the voltage built up in air depends on the radius of the sphere, but can be increased by enclosing the generator in an inert atmosphere, such as nitrogen.

van der Waals' law modified form of the ◊gas laws that includes corrections for the non-ideal behaviour of real gases (the molecules of ideal gases occupy no space and exert no forces on each other). It is named after Dutch physicist Johannes Diderik van der Waals (1837–1923).

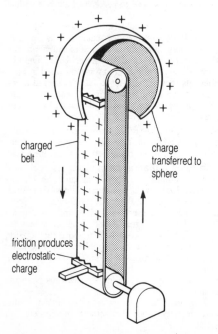

van de Graaff generator *The van de Graaff generator can produce more than a million volts. Such generators are used as particle accelerators for experiments with charged particles.*

The equation derived from the law states that:

$$(P + a/V^2)(V - b) = RT$$

where P, V, and T are the pressure, volume, and temperature (in kelvin) of the gas, respectively; R is the ◊gas constant; and a and b are constants for that particular gas.

Vanguard early series of US Earth-orbiting satellites and their associated rocket launcher. *Vanguard 1* was the second US satellite, launched 17 March 1958 by the three-stage Vanguard rocket. Tracking of its orbit revealed that Earth is slightly pear-shaped. The series ended Sept 1959 with *Vanguard 3*.

vapour one of the three states of matter (see also ◊solid and ◊liquid). The molecules in a vapour move randomly and are far apart, the distance between them, and therefore the volume of the vapour, being limited only by the walls of any vessel in which they might be contained. A vapour differs from a ◊gas only in that a vapour can be liquefied by increased pressure, whereas a gas cannot unless its temperature is lowered below its ◊critical temperature; it then becomes a vapour and may be liquefied.

vapour density density of a gas, expressed as the ◊mass of a given volume of the gas divided by the mass of an equal volume of a reference gas (such as hydrogen or air) at the same temperature and pressure. It is equal approximately to half the relative molecular weight (mass) of the gas.

vapour pressure pressure of a vapour given off by (evaporated from) a liquid or solid, caused by vibrating atoms or molecules continuously escaping from its surface. In an enclosed space, a maximum value is reached when the number of particles leaving the surface is in equilibrium with those returning to it; this is known as the *saturated vapour pressure*.

variable in mathematics, a changing quantity (one that can take various values), as opposed to a ◊constant. For example, in the algebraic expression $y = 4x^3 + 2$, the variables are x and y, whereas 4 and 2 are constants.

A variable may be dependent or independent. Thus if y is a ◊function of x, written $y = f(x)$, such that $y = 4x^3 + 2$, the domain of the function includes all values of the *independent variable x* while the range (or codomain) of the function is defined by the values of the *dependent variable y*.

In computer programming, variables are used to represent different items of data in the course of a program. A computer programmer will choose a symbol to represent each variable used in a program. The computer will then automatically assign a memory location to store the current value of each variable, and use the chosen symbol to identify this location. For example, the letter *P* might be chosen by a programmer to represent the price of an article. The computer would automatically reserve a memory location with the symbolic address *P* to store the price being currently processed.

Different programming languages place different restrictions on the choice of symbols used to represent variables. Some languages only allow a single letter followed, where required, by a single number. Other languages allow a much freer choice, allowing, for example, the use of the full word 'price' to represent the price of an article.

A *global variable* is one that can be accessed by any program instruction; a *local variable* is one that can only be accessed by the instructions within a particular subroutine.

variable-geometry wing technical name for what is popularly termed a ◊swing wing, a type of movable aircraft wing.

variable star in astronomy, a star whose brightness changes, either regularly or irregularly, over a period ranging from a few hours to months or even years. The ◊Cepheid variables regularly expand and contract in size every few days or weeks.

Stars that change in size and brightness at less precise intervals include *long-period variables*, such as the red giant Mira in the constellation Cetus (period about 330 days), and *irregular variables*, such as some red supergiants. *Eruptive variables* emit sudden outbursts of light. Some suffer flares on their surfaces, while others, such as a ◊nova, result from transfer of gas between a close pair of stars. A ◊supernova is the explosive death of a star. In an ◊*eclipsing binary*, the variation is due not to any change in the star itself, but to the periodic eclipse of a star by a close companion.

variance in statistics, the square of the ◊standard deviation, the measure of spread of data. Population and sample variance are denoted by σ^2 or s^2, respectively.

variation in biology, a difference between individuals of the same species, found in any sexually reproducing population. Variations may be almost unnoticeable in some cases, obvious in others, and can concern many aspects of the organism. Typically, variation in size, behaviour, biochemistry, or colouring may be found. The cause of the variation is genetic (that is, inherited), environmental, or more usually a combination of the two. The origins of variation can be traced to the recombination of the genetic material during the formation of the gametes, and, more rarely, to mutation.

variegation description of plant leaves or stems that exhibit patches of different colours. The term is usually applied to plants that show white, cream, or yellow on their leaves, caused by areas of tissue that lack the green pigment ◊chlorophyll. Variegated plants are bred for their decorative value, but they are often considerably weaker than the normal, uniformly green plant. Many will not breed true and require ◊vegetative reproduction.

The term is sometimes applied to abnormal patchy colouring of petals, as in the variegated petals of certain tulips, caused by a virus infection. A mineral deficiency in the soil may also be the cause of variegation.

varve in geology, a pair of thin sedimentary beds, one coarse and one fine, representing a cycle of thaw followed by an interval of freezing, in lakes of glacial regions.

Each couplet thus constitutes the sedimentary record of a year, and by counting varves in glacial lakes a record of absolute time elapsed can be determined. Summer and winter layers often are distinguished also by colour, with lighter layers representing summer deposition, and darker layers the result of dark clay settling from water while the lake was frozen.

vascular bundle strand of primary conducting tissue (a 'vein') in vascular plants, consisting mainly of water-conducting tissues, metaxylem and protoxylem, which together make up the primary ◊xylem, and nutrient-conducting tissue, ◊phloem. It extends from the roots to the stems and leaves. Typically the phloem is situated nearest to the epidermis and the xylem towards the centre of the

cross section through a young stem

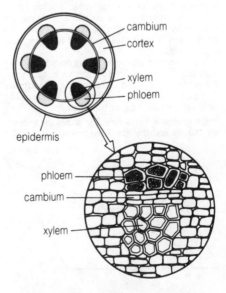

vascular bundle The fluid-carrying tissue of most plants is normally arranged in units called vascular bundles. The vascular tissue is of two types: xylem and phloem. The xylem carries water up through the plant; the phloem distributes food made in the leaves to all parts of the plant.

bundle. In plants exhibiting ◊secondary growth, the xylem and phloem are separated by a thin layer of vascular ◊cambium, which gives rise to new conducting tissues.

vascular plant plant containing vascular bundles. ◊Pteridophytes (ferns, horsetails, and club mosses), ◊gymnosperms (conifers and cycads), and ◊angiosperms (flowering plants) are all vascular plants.

vas deferens in male vertebrates, a tube conducting sperm from the testis to the urethra. The sperms are carried in a fluid secreted by various glands, and can be transported very rapidly when the smooth muscle in the wall of the vas deferens undergoes rhythmic contraction, as in sexual intercourse.

VDU abbreviation for ◊visual display unit.

vector graphics computer graphics that are stored in the computer memory by using geometric formulas. Vector graphics can be transformed (enlarged, rotated, stretched, and so on) without loss of picture resolution. It is also possible to select and transform any of the components of a vector-graphics display because each is separately defined in the computer memory. In these respects vector graphics are superior to ◊raster graphics. Vector graphics are typically used for drawing applications, allowing the user to create and modify technical diagrams such as designs for houses or cars.

vector quantity any physical quantity that has both magnitude and direction (such as the velocity or acceleration of an object) as distinct from ◊scalar quantity (such as speed, density, or mass), which has magnitude but no direction. A vector is represented either geometrically by an arrow whose length corresponds to its magnitude and points in an appropriate direction, or by a pair of numbers written vertically and placed within brackets $(x\ y)$.

Vectors can be added graphically by constructing a parallelogram of vectors (such as the ◊parallelogram of forces commonly employed in physics and engineering).

If two forces p and q are acting on a body, then the parallelogram of forces is drawn to determine the resultant force and direction r. p, q, and r are vectors. In technical writing, a vector is denoted by **bold** type, underlined, or overlined.

Vega or *Alpha Lyrae* brightest star in the constellation Lyra and the fifth brightest star in the sky.

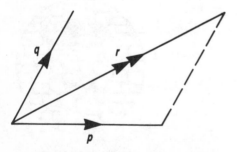

vector quantity *A parallelogram of vectors. Vectors can be added graphically using the parallelogram rule. According to the rule, the sum of vectors p and q is the vector r which is the diagonal of the parallelogram with sides p and q.*

It is a blue-white star, 25 light years from Earth, with a luminosity 50 times that of the Sun.

In 1983 the Infrared Astronomy Satellite (IRAS) discovered a ring of dust around Vega, possibly a disc from which a planetary system is forming.

vegetative reproduction type of ◊asexual reproduction in plants that relies not on spores, but on multicellular structures formed by the parent plant. Some of the main types are ◊stolons and runners, ◊gemmae, ◊bulbils, sucker shoots produced from roots (such as in the creeping thistle *Cirsium arvense*), ◊tubers, ◊bulbs, ◊corms, and ◊rhizomes. Vegetative reproduction has long been exploited in horticulture and agriculture, with various methods employed to multiply stocks of plants. See also ◊plant propagation.

vein in animals with a circulatory system, any vessel that carries blood from the body to the heart. Veins contain valves that prevent the blood from running back when moving against gravity. They always carry deoxygenated blood, with the exception of the veins leading from the lungs to the heart in birds and mammals, which carry newly oxygenated blood.

The term is also used more loosely for any system of channels that strengthens living tissues and supplies them with nutrients—for example, leaf veins (see ◊vascular bundle), and the veins in insects' wings.

Vela constellation of the southern hemisphere near Carina, represented as the sails of a ship. It contains large wisps of gas—called the Gum nebula after its discoverer, the Australian astronomer Colin Gum (1924–1960)—believed to be the remains of one or more ◊supernovae. Vela also contains the second optical ◊pulsar (a pulsar that flashes at a visible wavelength) to be discovered. Its four brightest stars are second-magnitude, one of them being Suhail, about 490 light years from Earth.

velocity speed of an object in a given direction. Velocity is a ◊vector quantity, since its direction is important as well as its magnitude (or speed).

The velocity at any instant of a particle travelling in a curved path is in the direction of the tangent to the path at the instant considered.

velocity ratio (VR), or *distance ratio* in a machine, the ratio of the distance moved by an effort force to the distance moved by the machine's load in the same time. It follows that the velocities of the effort and the load are in the same ratio. Velocity ratio has no units.

vena cava one of the large, thin-walled veins found just above the ◊heart, formed from the junction of several smaller veins. The *posterior vena cava* receives oxygenated blood returning from the lungs, and empties into the left atrium. The *anterior vena cava* collects deoxygenated blood returning from the rest of the body and passes it into the right side of the heart, from where it will be pumped into the lungs.

Venn diagram in mathematics, a diagram representing a ◊set or sets and the logical relationships between them. The sets are drawn as circles. An area of overlap between two circles (sets) contains elements that are common to both sets, and thus represents a third set. Circles that do not overlap represent sets with no elements in common (disjoint sets). The method is named after the British logician John Venn (1834–1923).

ventral surface the front of an animal. In vertebrates, the side furthest from the backbone; in invertebrates, the side closest to the ground. The positioning of the main nerve pathways on the ventral side is a characteristic of invertebrates.

ventricle in zoology, one of the chambers of the heart that forces blood into the arteries by muscular contraction. The term also refers to a space within the brain in which cerebrospinal fluid is produced.

Venturi tube device for measuring the rate of fluid flow through a pipe. It consists of a tube with a constriction (narrowing) in the middle of its length. The constriction causes a drop in pressure in the fluid flowing in the pipe. A pressure gauge attached to the constriction measures the pressure drop and this is used to find the rate of fluid flow.

Venturi tubes are also used to draw petrol into the carburettor of a motor car.

Venus second planet from the Sun. It orbits the Sun every 225 days at an average distance of 108.2 million km/67.2 million mi and can approach the Earth to within 38 million km/24 million mi, closer than any other planet. Its diameter is 12,100 km/ 7,500 mi and its mass is 0.82 that of Earth. Venus rotates on its axis more slowly than any other planet, once every 243 days and from east to west, the opposite direction to the other planets (except Uranus and possibly Pluto). Venus is shrouded by clouds of sulphuric acid droplets that sweep across the planet from east to west every four days. The atmosphere is almost entirely carbon dioxide, which traps the Sun's heat by the ◊greenhouse effect and raises the planet's surface temperature to 480°C/900°F, with an atmospheric pressure of 90 times that at the surface of the Earth.

The surface of Venus consists mainly of plains dotted with deep impact craters. The largest highland area is Aphrodite Terra near the equator, half the size of Africa. The highest mountains are on the N highland region of Ishtar Terra, where the massif of Maxwell Montes rises to 10,600 m/ 35,000 ft above the average surface level. The highland areas on Venus were formed by volcanoes, which may still be active.

VENUS: MORNING AND EVENING STAR

It was once thought that the 'evening star' and the 'morning star' were different bodies. The evening star was called Hesperus and the morning star Phosporous. It was eventually discovered that they were one and the same object— the planet Venus.

The first artificial object to hit another planet was the Soviet probe *Venera 3*, which crashed on Venus 1 March 1966. Later Venera probes parachuted down through the atmosphere and landed successfully on its surface, analysing surface material and sending back information and pictures. In Dec 1978 a US Pioneer Venus probe (see ◊Pioneer probes) went into orbit around the planet

and mapped most of its surface by radar, which penetrates clouds. In 1992 the US space probe Magellan mapped 95% of the planet's surface to a resolution of 100 m/ 330 ft.

verdigris green-blue coating of copper ethanoate that forms naturally on copper, bronze, and brass. It is an irritating, greenish, poisonous compound made by treating copper with ethanoic acid, and was formerly used in wood preservatives, antifouling compositions, and green paints.

verification in computing, the process of checking that data being input to a computer have been accurately copied from a source document. This may be done visually, by checking the original copy of the data against the copy shown on the VDU screen. A more thorough method is to enter the data twice, using two different keyboard operators, and then to check the two sets of input copies against each other. The checking is normally carried out by the computer itself, any differences between the two copies being reported for correction by one of the the the keyboard operators.

Where large quantities of data have to be input, a separate machine called a *verifier* may be used to prepare fully verified tapes or discs for direct input to the main computer.

vernal equinox spring ◊equinox.

vernalization the stimulation of flowering by exposure to cold. Certain plants will not flower unless subjected to low temperatures during their development. For example, winter wheat will flower in summer only if planted in the previous autumn. However, by placing partially germinated seeds in low temperatures for several days, the cold requirement can be supplied artificially, allowing the wheat to be sown in the spring.

vernier device for taking readings on a graduated scale to a fraction of a division. It consists of a short divided scale that carries an index or pointer and is slid along a main scale. It was invented by Pierre Vernier.

vertebral column the backbone, giving support to an animal and protecting the central nervous system. It is made up of a series of bones or vertebrae running from the skull to the tail with a central canal containing the nerve fibres of the spinal cord. In tetrapods the vertebrae show some specialization with the shape of the bones varying according to position. In the chest region the upper or thoracic vertebrae are shaped to form connections to the ribs. The backbone is only slightly flexible to give adequate rigidity to the animal structure.

vertebrate any animal with a backbone. The 41,000 species of vertebrates include mammals, birds, reptiles, amphibians, and fishes. They include most of the larger animals, but in terms of numbers of species are only a tiny proportion of the world's animals. The zoological taxonomic group Vertebrata is a subgroup of the ◊phylum *Chordata*.

vertex (plural *vertices*) in geometry, a point shared by three or more sides of a solid figure; the point farthest from a figure's base; or the point of intersection of two sides of a plane figure or the two rays of an angle.

vertical takeoff and landing craft (VTOL) aircraft that can take off and land vertically. ◊Helicopters, airships, and balloons can do this, as can a few fixed-wing aeroplanes, like the ◊convertiplane.

Very Large Array (VLA) largest and most complex single-site radio telescope in the world. It is located on the Plains of San Augustine, 80 km/ 50 mi west of Socorro, New Mexico. It consists of 27 dish antennae, each 25 m/82 ft in diameter, arranged along three equally spaced arms forming a Y-shaped array. Two of the arms are 21 km/13 mi long, and the third, to the north, is 19 km/ 11.8 mi long. The dishes are mounted on railway tracks enabling the configuration and size of the array to be altered as required.

Pairs of dishes can also be used as separate interferometers (see ◊radio telescope), each dish having its own individual receivers that are remotely controlled, enabling many different frequencies to be studied.

There are four standard configurations of antennae ranging from A (the most extended) through B and C to D. In the A configuration the antennae are spread out along the full extent of the arms and the VLA can map small, intense radio sources with high resolution. The smallest configuration, D, uses arms that are just 0.6 km/0.4 mi long for mapping larger sources. Here the resolution is lower, although there is greater sensitivity to fainter, extended fields of radio emission.

vestigial organ in biology, an organ that remains in diminished form after it has ceased to have any significant function in the adult organism. In humans, the appendix is vestigial, having once had a digestive function in our ancestors.

veterinary science the study, prevention, and cure of disease in animals. More generally, it covers animal anatomy, breeding, and relations to humans.

Professional bodies include the Royal College of Veterinary Surgeons 1844 in the UK and the American Veterinary Medical Association 1883 in the USA.

VHF (abbreviation for very *h*igh *f*requency) referring to radio waves that have very short wavelengths (10 m-1 m). They are used for interference-free ◊FM (frequency-modulated) transmissions. VHF transmitters have a relatively short range because the waves cannot be reflected over the horizon like longer radio waves.

video camera portable television camera that takes moving pictures electronically on magnetic tape. It produces an electrical output signal corresponding to rapid line-by-line scanning of the field of view. The output is recorded on video cassette and is played back on a television screen via a videotape recorder.

video cassette recorder (VCR) device for recording on and playing back video cassettes; see ◊video tape recorder.

video disc disc with pictures and sounds recorded on it, played back by laser. The video disc is a type of ◊compact disc.

The video disc (originated by Scottish inventor John Logie Baird 1928; commercially available from 1978) is chiefly used to provide commercial films for private viewing. Most systems use a 30 cm/12 in rotating vinyl disc coated with a

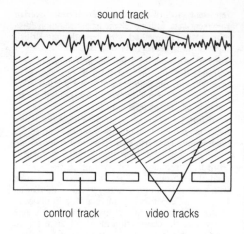

sound track

control track   video tracks

video tape recorder *The signal on a video tape is recorded diagonally across the tape to hold the maximum amount of information. If the signal was recorded in a straight line down the tape, it would take 33 km/20 mi of tape to make a one-hour recording.*

reflective material. Laser scanning recovers picture and sound signals from the surface where they are recorded as a spiral of microscopic pits.

video game electronic game played on a visual-display screen or, by means of special additional or built-in components, on the screen of a television set. The first commercially sold was a simple bat-and-ball game developed in the USA 1972, but complex variants are now available in colour and with special sound effects.

video tape recorder (VTR) device for recording pictures and sound on cassettes or spools of magnetic tape. The first commercial VTR was launched 1956 for the television broadcasting industry, but from the late 1970s cheaper models developed for home use, to record broadcast programmes for future viewing and to view rented or owned video cassettes of commercial films.

Video recording works in the same way as audio ◊tape recording: the picture information is stored as a line of varying magnetism, or track, on a plastic tape covered with magnetic material. The main difficulty—the huge amount of information needed to reproduce a picture—is overcome by arranging the video track diagonally across the tape. During recording, the tape is wrapped around a drum in a spiral fashion. The recording head rotates inside the drum. The combination of the forward motion of the tape and the rotation of the head produces a diagonal track. The audio signal accompanying the video signal is recorded as a separate track along the edge of the tape.

Two video cassette systems were introduced by Japanese firms in the 1970s. The Sony Betamax was technically superior, but Matsushita's VHS had larger marketing resources behind it and after some years became the sole system on the market. Super-VHS is an improved version of the VHS system, launched 1989, with higher picture definition and colour quality.

videotext system in which information (text and

simple pictures) is displayed on a television (video) screen. There are two basic systems, known as ◊teletext and ◊viewdata. In the teletext system information is broadcast with the ordinary television signals, whereas in the viewdata system information is relayed to the screen from a central data bank via the telephone network. Both systems require the use of a television receiver with special decoder.

viewdata system of displaying information on a television screen in which the information is extracted from a computer data bank and transmitted via the telephone lines. It is one form of ◊videotext. The British Post Office (now British Telecom) developed the world's first viewdata system, Prestel, 1975, and similar systems are now in widespread use in other countries. Viewdata users have access to an almost unlimited store of information, presented on the screen in the form of 'pages'.

Prestel has hundreds of thousands of pages, presenting all kinds of information, from local weather and restaurant menus to share prices and airport timetables. Since viewdata uses telephone lines, it can become a two-way interactive information system, making possible, for example, home banking and shopping. In contrast, the only user input allowed by the ◊teletext system is to select the information to be displayed.

Viking probes two US space probes to Mars, each one consisting of an orbiter and a lander. They were launched 20 Aug and 9 Sept 1975. They transmitted colour pictures, and analysed the soil. No definite signs of life were found.

The *Viking 1* lander touched down in the Chryse lowland area 20 July 1976; *Viking 2* landed in Utopia 3 Sept 1976.

villus plural *villi* small fingerlike projection extending into the interior of the small intestine and increasing the absorptive area of the intestinal wall. Digested food, including sugars and amino acids, pass into the villi and are carried away by the circulating blood.

Virgo zodiacal constellation, the second largest in the sky. It is represented as a maiden holding an ear of wheat. The Sun passes through Virgo from late Sept to the end of Oct. Virgo's brightest star is the first-magnitude ◊Spica. Virgo contains the nearest large cluster of galaxies to us, 50 million light years away, consisting of about 3,000 galaxies centred on the giant elliptical galaxy M87. Also in Virgo is the nearest ◊quasar, 3C 273, an estimated 3 billion light years distant. In astrology, the dates for Virgo are between about 23 Aug and 22 Sept (see ◊precession).

virion the smallest unit of a mature ◊virus.

virtual in computing, without physical existence. Some computers have ◊virtual memory, making their immediate-access memory seem larger than it is; some computers can also simulate *virtual devices*. For example, the Acorn A3000 and A5000 computers have only one floppy-disc drive but can behave as if they were equipped with two, using part of the ◊RAM to simulate the second drive. ◊Virtual reality is a computer simulation of a whole physical environment.

virtual memory in computing, a technique whereby a portion of the computer backing storage, or external, ◊memory is used as an extension of its immediate-access, or internal, memory. The contents of an area of the immediate-access memory are stored on, say, a hard disc while they are not needed, and brought back into main memory when required.

The process, called paging or segmentation, is controlled by the computer ◊operating system and is hidden from the programmer, to whom the computer's internal memory appears larger than it really is. The technique can be successfully implemented only if very fast backing store is available, so that 'pages' of memory can be rapidly switched into and out of the immediate-access memory.

virtual reality advanced form of computer simulation, in which a participant has the illusion of being part of an artificial environment. The participant views the environment through two tiny television screens (one for each eye) built into a visor. Sensors detect movements of the participant's head or body, causing the apparent viewing position to change. Gloves (datagloves) fitted with sensors may be worn, which allow the participant seemingly to pick up and move objects in the environment.

The technology is still under development but is expected to have widespread applications; for example, in military and surgical training, architecture, and home entertainment.

virus infectious particle consisting of a core of nucleic acid (DNA or RNA) enclosed in a protein shell. Viruses are acellular and able to function and reproduce only if they can invade a living cell to use the cell's system to replicate themselves. In the process they may disrupt or alter the host cell's own DNA. The healthy human body reacts by producing an antiviral protein, ◊interferon, which prevents the infection spreading to adjacent cells.

Many viruses mutate continuously so that the host's body has little chance of developing permanent resistance; others transfer between species, with the new host similarly unable to develop resistance. The viruses that cause ◊AIDS and Lassa fever are both thought to have 'jumped' to humans from other mammalian hosts.

Among diseases caused by viruses are canine distemper, chickenpox, common cold, herpes, influenza, rabies, smallpox, yellow fever, AIDS, and many plant diseases. Recent evidence implicates viruses in the development of some forms of cancer (see ◊oncogene).

Bacteriophages are viruses that infect bacterial cells.

Retroviruses are of special interest because they have an RNA genome and can produce DNA from this RNA by a process called reverse transcription.

Viroids, discovered 1971, are even smaller than viruses; they consist of a single strand of nucleic acid with no protein coat. They may cause stunting in plants and some rare diseases in animals, including humans. It is debatable whether viruses and viroids are truly living organisms, since they are incapable of an independent existence. Outside the cell of another organism they remain completely inert. The origin of viruses is also unclear, but it is believed that they are degenerate forms of life, derived from cellular organisms, or pieces of nucleic acid that have broken away from the genome of some higher organism and taken up a parasitic existence.

Antiviral drugs are difficult to develop because viruses replicate by using the genetic machinery of host cells, so that drugs tend to affect the host cell as well as the virus. Acyclovir (used against the herpes group of diseases) is one of the few drugs so far developed that is successfully selective in its action. It is converted to its active form by an enzyme that is specific to the virus, and it then specifically inhibits viral replication. Some viruses have shown developing resistance to the few antiviral drugs available.

Viruses have recently been found to be very abundant in seas and lakes, with between 5 and 10 million per millilitre of water at most sites tested, but up to 250 million per millilitre in one polluted lake. These viruses infect bacteria and, possibly, single-celled algae. They may play a crucial role in controlling the survival of bacteria and algae in the plankton.

SIZE OF VIRUSES

An average virus is only around 50 nm (50 millionths of a millimetre/2 millionths of an inch) across. Hundreds of thousands of viruses could fit on the full stop at the end of this sentence.

---

virus in computing, a piece of ◊software that can replicate itself and transfer itself from one computer to another, without the user being aware of it. Some viruses are relatively harmless, but others can damage or destroy data. They are written by anonymous programmers, often maliciously, and are spread along telephone lines or on ◊floppy discs. Antivirus software can be used to detect and destroy well-known viruses, but new viruses continually appear and these may bypass existing antivirus programs.

Computer viruses may be programmed to operate on a particular date, such as the Michelangelo Virus, which was triggered on 6 March 1992 (the anniversary of the birthday of Italian artist Michelangelo) and erased hard discs.

viscose yellowish, syrupy solution made by treating cellulose with sodium hydroxide and carbon disulphide. The solution is then regenerated as continuous filament for the making of ◊rayon and as cellophane.

viscosity in physics, the resistance of a fluid to flow, caused by its internal friction, which makes it resist flowing past a solid surface or other layers of the fluid. It applies to the motion of an object moving through a fluid as well as the motion of a fluid passing by an object.

Fluids such as pitch, treacle, and heavy oils are highly viscous; for the purposes of calculation, many fluids in physics are considered to be perfect, or nonviscous.

vision system computer-based device for interpreting visual signals from a video camera. Computer vision is important in robotics where sensory abilities would considerably increase the flexibility and usefulness of a robot.

Although some vision systems exist for recognizing simple shapes, the technology is still in its infancy.

visual display unit (VDU) computer terminal consisting of a keyboard for input data and a screen for displaying output. The oldest and the most popular type of VDU screen is the ◊cathode-ray tube (CRT), which uses essentially the same technology as a television screen. Other types use plasma display technology and ◊liquid crystal displays.

vitalism the idea that living organisms derive their characteristic properties from a universal life force. In the present century, this view is associated with the French philosopher Henri Bergson.

vitamin any of various chemically unrelated organic compounds that are necessary in small quantities for the normal functioning of the body. Many act as coenzymes, small molecules that enable ◊enzymes to function effectively. They are normally present in adequate amounts in a balanced diet. Deficiency of a vitamin will normally lead to a metabolic disorder ('deficiency disease'), which can be remedied by sufficient intake of the vitamin. They are generally classified as **water-soluble** (B and C) or **fat-soluble** (A, D, E, and K). See separate entries for individual vitamins, also ◊nicotinic acid, ◊folic acid, and ◊pantothenic acid.

Scurvy (the result of vitamin C deficiency) was observed at least 3,500 years ago, and sailors from the 1600s were given fresh sprouting cereals or citrus-fruit juice to prevent or cure it. The concept of scurvy as a deficiency disease, however, caused by the absence of a specific substance, emerged later. In the 1890s a Dutch doctor, Christiaan Eijkman, discovered that he could cure hens suffering from a condition like beriberi by feeding them on whole-grain, rather than polished, rice. In 1912 Casimir Funk, a Polish-born biochemist, had proposed the existence of what he called 'vitamines', but it was not fully established until about 1915 that several deficiency diseases were preventable and curable by extracts from certain foods. By then it was known that two groups of factors were involved, one being water-soluble and present, for example, in yeast, rice-polishings, and wheat germ, and the other being fat-soluble and present in egg yolk, butter, and fish-liver oils. The water-soluble substance, known to be effective against beriberi, was named vitamin B. The fat-soluble vitamin complex was at first called vitamin A. As a result of analytical techniques these have been subsequently separated into their various components, and others have been discovered.

Megavitamin therapy has yielded at best unproven effects; some vitamins (A, for example) are extremely toxic in high doses.

Other animals may also need vitamins, but not necessarily the same ones. For example, choline, which humans can synthesize, is essential to rats and some birds, which cannot produce sufficient for themselves.

recommended daily vitamin intake

vitamin	adult requirement	deficiency disease	vitamin function
water-soluble B₁ (thiamine)		beriberi	regulation of carbohydrate metabolism
B₂ (riboflavin)	1.5 mg	range of symptoms	converts food to energy
niacin	15–20 mg	pellagra	converts food to energy
B₆ (pyridoxine)	3 mg	nervous system and skin disorders	metabolism of amino acids
folate	200 microg	anaemia	for cell metabolism, particularly blood cells
B₁₂ (cyanocobalamin)	3 microg	pernicious anaemia	for cell metabolism, particularly nerve cells
C (ascorbic acid)	30–60 mg	scurvy	structure of connective tissue, absorption of iron
fat-soluble A (retinol)		blindness, infection	structure of skin and mucus-secreting tissue
D (cholecalciferol)	3–10 microg	rickets	transport of calcium from blood to bone
E (tocopherol)	10 mg	range of symptoms	protecting essential fatty acids from free radicals
K	100 microg	haemorrhaging	normal clotting of blood

[A vitamin is] a substance that makes you ill if you don't eat it.

On **vitamins** Albert Szent-Györgyi (1893–1986)

vitamin A another name for ◊retinol.

vitamin B₁ another name for ◊thiamine.

vitamin B₂ another name for ◊riboflavin.

vitamin B₆ another name for ◊pyridoxine.

vitamin B₁₂ another name for ◊cyanocobalamin.

vitamin C see ◊ascorbic acid.

vitamin D another name for ◊cholecalciferol.

vitamin E another name for ◊tocopherol.

vitamin H another name for ◊biotin.

vitamin K another name for ◊phytomenadione.

vitreous humour transparent jellylike substance behind the lens of the vertebrate ◊eye. It gives rigidity to the spherical form of the eye and allows light to pass through to the retina.

vitriol any of a number of sulphate salts. Blue, green, and white vitriols are copper, ferrous, and zinc sulphate, respectively. *Oil of vitriol* is sulphuric acid.

vivipary in animals, a method of reproduction in which the embryo develops inside the body of the female from which it gains nourishment (in contrast to ◊ovipary and ◊ovovivipary). Vivipary is best developed in placental mammals, but also occurs in some arthropods, fishes, amphibians, and reptiles that have placentalike structures. In plants, it is the formation of young plantlets or bulbils instead of flowers. The term also describes seeds that germinate prematurely, before falling from the parent plant.

Premature germination is common in mangrove trees, where the seedlings develop sizable spearlike roots before dropping into the swamp below; this prevents their being washed away by the tide.

vivisection literally, cutting into a living animal. Used originally to mean experimental surgery or dissection practised on a live subject, the term is often used by antivivisection campaigners to include any experiment on animals, surgical or otherwise.

Britain's 1876 Cruelty to Animals Act was the world's first legislation specifically to protect laboratory animals.

VLF in physics, the abbreviation for **very low** ◊frequency.

VLSI (abbreviation for **very large-scale integration**) in electronics, the early-1990s level of advanced technology in the microminiaturization of ◊integrated circuits, and an order of magnitude smaller than ◊LSI (large-scale integration).

vocal cords folds of tissue within a mammal's larynx, and a bird's syrinx. Air passing over them makes them vibrate, producing sounds. Muscles in the larynx change the pitch of the sound by adjusting the tension of the vocal cords.

voice sound produced through the mouth and by the passage of air between the ◊vocal cords. In humans the sound is much amplified by the hollow sinuses of the face, and is modified by the movements of the lips, tongue, and cheeks.

voice input in computing and electronics, an alternative name for ◊speech recognition.

voice output in computing and electronics, an alternative name for ◊speech synthesis.

vol abbreviation for **volume**.

volatile in chemistry, term describing a substance that readily passes from the liquid to the vapour phase. Volatile substances have a high ◊vapour pressure.

volatile memory in computing, ◊memory that loses its contents when the power supply to the computer is disconnected.

volcanic rock another name for ◊extrusive rock, igneous rock that has formed on the surface of the Earth.

volcano crack in the Earth's crust through which hot magma (molten rock) and gases well up. The magma becomes known as lava when it reaches the surface. A volcanic mountain, usually cone shaped with a crater on top, is formed around the opening, or vent, by the build-up of solidified lava and ashes (rock fragments). Most volcanoes arise on plate margins (see ◊plate tectonics), where the movements of plates generate magma or allow it to rise from the mantle beneath. However, a number are found far from plate-margin activity, on 'hot spots' where the Earth's crust is thin.

There are two main types of volcano:

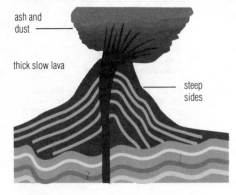

composite volcano

ash and dust

thick slow lava

steep sides

shield volcano

old volcanic lava

uneven surface

volcano

volt SI unit of electromotive force or electric potential, symbol V. A small battery has a potential of 1.5 volts; the domestic electricity supply in the UK is 240 volts (110 volts in the USA); and a high-tension transmission line may carry up to 765,000 volts.

The absolute volt is defined as the potential difference necessary to produce a current of one ampere through an electric circuit with a resistance of one ohm. It is named after the Italian scientist Alessandro Volta.

voltage commonly used term for ◊potential difference (pd).

voltage amplifier electronic device that increases an input signal in the form of a voltage or ◊potential difference, delivering an output signal that is larger than the input by a specified ratio.

voltmeter instrument for measuring potential difference (voltage). It has a high internal resistance (so that it passes only a small current), and is connected in parallel with the component across which potential difference is to be measured. A common type is constructed from a sensitive current-detecting moving-coil ◊galvanometer placed in series with a high-value resistor (multiplier). To measure an AC (◊alternating-current) voltage, the circuit must usually include a rectifier; however, a moving-iron instrument can be used to measure alternating voltages without the need for such a device.

volume in geometry, the space occupied by a three-dimensional solid object. A prism (such as a cube) or a cylinder has a volume equal to the area of the base multiplied by the height. For a pyramid or cone, the volume is equal to one-third of the area of the base multiplied by the perpendicular height. The volume of a sphere is equal to $\frac{4}{3}\pi r^3$, where r is the radius. Volumes of irregular solids may be calculated by the technique of ◊integration.

volumetric analysis procedure used for determining the concentration of a solution. A known volume of a solution of unknown concentration is reacted with a solution of known concentration (standard). The standard solution is delivered from a ◊burette so the volume added is known. This technique is known as ◊titration. Often an indicator is used to show when the correct proportions have reacted. This procedure is used for acid–base, ◊redox, and certain other reactions involving solutions.

Voskhod (Russian 'ascent') Soviet spacecraft used in the mid-1960s; it was modified from the single-seat Vostok, and was the first spacecraft capable of

Composite volcanoes, such as Stromboli and Vesuvius in Italy, are found at destructive plate margins (areas where plates are being pushed together), usually in association with island arcs and coastal mountain chains. The magma is mostly derived from plate material and is rich in silica. This makes a very stiff lava such as andesite, which solidifies rapidly to form a high, steep-sided volcanic mountain. The magma often clogs the volcanic vent, causing violent eruptions as the blockage is blasted free, as in the eruption of Mount St Helens, USA, in 1980. The crater may collapse to form a ◊caldera.

Shield volcanoes, such as Mauna Loa in Hawaii, are found along the rift valleys and ocean ridges of constructive plate margins (areas where plates are moving apart), and also over hot spots. The magma is derived from the Earth's mantle and is quite free-flowing. The lava formed from this magma — usually basalt — flows for some distance over the surface before it sets and so forms broad low volcanoes. The lava of a shield volcano is not ejected violently but simply flows over the crater rim.

Many volcanoes are submarine and occur along mid-ocean ridges. The chief terrestrial volcanic regions are around the Pacific rim (Cape Horn to Alaska); the central Andes of Chile (with the world's highest volcano, Guallatiri, 6,060 m/ 19,900 ft); North Island, New Zealand; Hawaii; Japan; and Antarctica. There are more than 1,300 potentially active volcanoes on Earth. Volcanism has helped shape other members of the Solar System, including the Moon, Mars, Venus, and Jupiter's moon Io.

volumes of common three-dimensional shapes

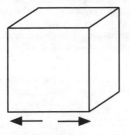

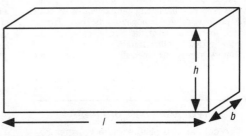

volume of a cube

= length³

= l^3

volume of a cuboid

= length × breadth × height

= $l \times b \times h$

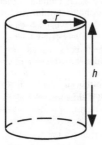

volume of a cylinder

= π × (radius of cross-section)² × height

= $\pi r^2 h$

volume of a cone

= $\frac{1}{3}$ π × (radius of cross-section)² × height

= $\frac{1}{2}\pi r^2 h$

volume of a sphere

= $\frac{4}{3}$ π radius³

= $\frac{4}{3}\pi r^3$

volume

carrying two or three cosmonauts. During *Voskhod* 2's flight 1965, Aleksi Leonov made the first space walk.

Vostok (Russian 'east') first Soviet spacecraft, used 1961–63. Vostok was a metal sphere 2.3 m/ 7.5 ft in diameter, capable of carrying one cosmonaut. It made flights lasting up to five days.

Vostok 1 carried the first person into space, Yuri Gagarin.

Voyager probes two US space probes, originally ◊Mariners. *Voyager 1*, launched 5 Sept 1977, passed Jupiter March 1979, and reached Saturn Nov 1980. *Voyager 2* was launched earlier, 20 Aug 1977, on a slower trajectory that took it past Jupiter

July 1979, Saturn Aug 1981, Uranus Jan 1986, and Neptune Aug 1989. Like the ◊Pioneer probes, the Voyagers are on their way out of the Solar System. Their tasks now include helping scientists to locate the position of the heliopause, the boundary at which the influence of the Sun gives way to the forces exerted by other stars.

Both Voyagers carry specially coded long-playing records called 'Sounds of Earth' for the enlightenment of any other civilizations that might find them.

VOYAGER 2 AT URANUS

To get the best pictures from *Voyager 2* during its Uranus flyby, mission control on Earth had to know its position to within 100 km/60 mi, even though the craft had travelled 4,954 million km/3,078 million mi. This accuracy is equivalent to a golfer hitting a hole in one on a fairway 2,520 km/1,560 miles long.

V-shaped valley river valley with a V-shaped cross-section. Such valleys are usually found near the source of a river, where the steeper gradient means that there is a great deal of ◊corrasion (grinding away by rock particles) along the stream bed and erosion cuts downwards more than it does sideways. However, a V-shaped valley may also be formed in the lower course of a river when its powers of downward erosion become renewed by a fall in sea level, a rise in land level, or the capture of another river (see ◊rejuvenation).

vulcanization technique for hardening rubber by heating and chemically combining it with sulphur. The process also makes the rubber stronger and more elastic. If the sulphur content is increased to as much as 30%, the product is the inelastic solid known as ebonite. More expensive alternatives to sulphur, such as selenium and tellurium, are used to vulcanize rubber for specialized products such as vehicle tyres. The process was discovered accidentally by US inventor Charles Goodyear 1839 and patented 1844.

Accelerators can be added to speed the vulcanization process, which takes from a few minutes for small objects to an hour or more for vehicle tyres. Moulded objects are often shaped and vulcanized simultaneously in heated moulds; other objects may be vulcanized in hot water, hot air, or steam.

vulcanology the study of ◊volcanoes and the geological phenomena that cause them.

Vulpecula small constellation in the northern hemisphere of the sky just south of Cygnus, represented as a fox. It contains a major planetary ◊nebula, the Dumbbell, and the first ◊pulsar (pulsating radio source) to be discovered.

W in physics, symbol for *watt*.

wadi in arid regions of the Middle East, a steep-sided valley containing an intermittent stream that flows in the wet season.

wafer in microelectronics, a 'superchip' some 8–10 cm/3–4 in in diameter, for which wafer-scale integration (WSI) is used to link the equivalent of many individual ◊silicon chips, improving reliability, speed, and cooling.

Waldsterben (German 'forest death') tree decline related to air pollution, common throughout the industrialized world. It appears to be caused by a mixture of pollutants; the precise chemical mix varies between locations, but it includes acid rain, ozone, sulphur dioxide, and nitrogen oxides.

Waldsterben was first noticed in the Black Forest of Germany during the late 1970s, and is spreading to many Third World countries, such as China. Despite initial hopes that Britain's trees had not been damaged, research has now shown them to be among the most badly affected in Europe.

Wallace line imaginary line running down the Lombok Strait in SE Asia, between the island of Bali and the islands of Lombok and Sulawesi. It was identified by English naturalist Alfred Russel Wallace as separating the S Asian (Oriental) and Australian biogeographical regions, each of which has its own distinctive animals.

Subsequently, others have placed the boundary between these two regions at different points in the Malay archipelago, owing to overlapping migration patterns.

wall pressure in plants, the mechanical pressure exerted by the cell contents against the cell wall. The rigidity (turgor) of a plant often depends on the level of wall pressure found in the cells of the stem. Wall pressure falls if the plant cell loses water.

Wankel engine rotary petrol engine developed by the German engineer Felix Wankel (1902–) in the 1950s. It operates according to the same stages as the ◊four-stroke petrol engine cycle, but these stages take place in different sectors of a figure-eight chamber in the space between the chamber walls and a triangular rotor. Power is produced once on every turn of the rotor. The Wankel engine is simpler in construction than the four-stroke piston petrol engine, and produces rotary power directly (instead of via a crankshaft). Problems with rotor seals have prevented its widespread use.

warning coloration in biology, an alternative term for ◊aposematic coloration.

washing soda $Na_2CO_3.10H_2O$ (chemical name *sodium carbonate decahydrate*) substance added to washing water to 'soften' it (see ◊hard water).

Washington Convention alternative name for ◊*CITES*, the international agreement that regulates trade in endangered species.

waste materials that are no longer needed and are discarded. Examples are household waste, industrial waste (which often contains toxic chemicals), medical waste (which may contain organisms that cause disease), and ◊nuclear waste (which is radioactive). By ◊recycling, some materials in waste can be reclaimed for further use. In 1990 the industrialized nations generated 2 billion tonnes of waste. In the USA, 40 tonnes of solid waste are generated annually per person.

There has been a tendency to increase the amount of waste generated per person in industrialized countries, particularly through the growth in packaging and disposable products, creating a 'throwaway society'.

In Britain, the average person throws away about ten times their own body weight in household refuse each year. Collectively the country generates about 50 million tonnes of waste per year.

waste disposal depositing waste. Methods of waste disposal vary according to the materials in the waste and include incineration, burial at designated sites, and dumping at sea. Organic waste can be treated and reused as fertilizer (see ◊sewage disposal). ◊Nuclear waste and ◊toxic waste is usually buried or dumped at sea, although this does not negate the danger.

Waste disposal is an increasing problem in the late 20th century. Environmental groups, such as Greenpeace and Friends of the Earth, are campaigning for more recycling, a change in lifestyle so that less waste (from packaging and containers to nuclear materials) is produced, and safer methods of disposal.

The USA burns very little of its rubbish as compared with other industrialized countries. Most of its waste, 80%, goes into land fills. Many of the country's landfill sites will have to close in the 1990s because they do not meet standards to protect ground water.

The industrial waste dumped every year by the UK in the North Sea includes 550,000 tonnes of fly ash from coal-fired power stations. The British government agreed 1989 to stop North Sea dumping from 1993, but dumping in the heavily polluted Irish Sea will continue. Industrial pollution is suspected of causing ecological problems, including an epidemic that killed hundreds of seals 1989.

The Irish Sea receives 80 tonnes of uranium a year from phosphate rock processing, and 300 million gallons of sewage every day, 80% of it untreated or merely screened. In 1988, 80,000 tonnes of hazardous waste were imported into the UK for processing, including 6,000 tonnes of ◊polychlorinated biphenyls.

watch portable timepiece. In the early 20th century increasing miniaturization, mass production, and, in World War I, the advantages of the wristband led to the watch moving from the pocket to the wrist. Watches were also subsequently made waterproof, antimagnetic, self-winding, and shock-resistant. In 1957 the electric watch was developed, and in the 1970s came the digital watch, which dispensed with all moving parts.

history Traditional mechanical watches with analogue dials (hands) are based on the invention by Peter Henlein (1480–1542) of the mainspring as the energy store. By 1675 the invention of the balance spring allowed watches to be made small enough to move from waist to pocket. By the 18th century pocket-watches were accurate, and by the 20th century wristwatches were introduced. In the 1950s battery-run electromagnetic watches were developed; in the 1960s electronic watches were marketed, which use the ◊piezoelectric oscillations of a quartz crystal to mark time and an electronic circuit to drive the hands. In the 1970s quartz watches without moving parts were developed — the solid-state watch with a display of digits. Some include a tiny calculator and such functions as date, alarm, stopwatch, and reminder beeps.

The watch mechanism was perfected in Britain by inventors such as Thomas Tompion (1639–1713) and Thomas Earnshaw (1749–1829), but throughout the 19th century watches remained articles of value.

An electric watch has no mainspring, the mechanism being kept in motion by the mutual attraction of a permanent magnet and an electromagnet, which pushes the balance wheel. In a digital watch the time is usually indicated by a ◊liquid crystal display.

water H₂O liquid without colour, taste, or odour. It is an oxide of hydrogen. Water begins to freeze at 0°C or 32°F, and to boil at 100°C or 212°F. When liquid, it is virtually incompressible; frozen, it expands by ¹⁄₁₁ of its volume. At 39.2°F/4°C, one cubic centimetre of water has a mass of one gram; this is its maximum density, forming the unit of specific gravity. It has the highest known specific heat, and acts as an efficient solvent, particularly when hot. Most of the world's water is in the sea; less than 0.01% is fresh water.

Water covers 70% of the Earth's surface and occurs as standing (oceans, lakes) and running (rivers, streams) water, rain, and vapour and supports all forms of Earth's life.

Water makes up 60–70% of the human body or about 40 litres of which 25 are inside the cells, 15 outside (12 in tissue fluid, and 3 in blood plasma). A loss of 4 litres may cause hallucinations; a loss of 8–10 litres may cause death. About 1.5 litres a day are lost through breathing, perspiration, and faeces, and the additional amount lost in urine is the amount needed to keep the balance between input and output. People cannot survive more than five or six days without water or two or three days in a hot environment.

A family of two adults and two children uses approximately 200 litres per day (UK figures). The British water industry was privatized 1989, and in 1991 the UK was taken to court for failing to meet EC drinking-water standards on nitrate and pesticide levels.

Water is H₂O, hydrogen two parts, oxygen one,/ but there is also a third thing, that makes it water/ and nobody knows what that is.

On **water** D H Lawrence (1885–1930) 'The Third Thing' *Pansies* 1929

water-borne disease disease associated with poor water supply. In the Third World four-fifths of all illness is caused by water-borne diseases, with diarrhoea being the leading cause of childhood death. Malaria, carried by mosquitoes dependent on stagnant water for breeding, affects 400 million people every year and kills 5 million. Polluted water is also a problem in industrialized nations, where industrial dumping of chemical, hazardous, and radioactive wastes causes a range of diseases from headache to cancer.

water cycle or *hydrological cycle* in ecology, the natural circulation of water through the ◊biosphere. Water is lost from the Earth's surface to the atmosphere either by evaporation from the surface of lakes, rivers, and oceans or through the transpiration of plants. This atmospheric water forms clouds that condense to deposit moisture on the land and sea as rain or snow. The water that collects on land flows to the ocean in streams and rivers.

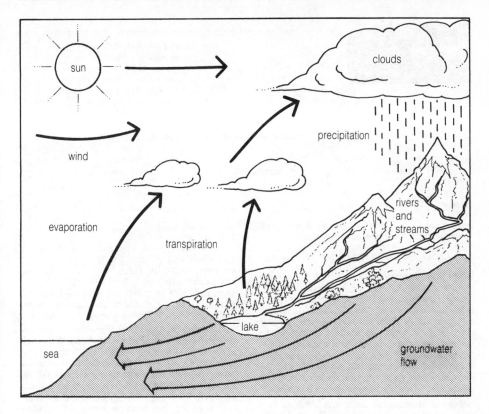

water cycle

WATER CYCLE: A ROUND TRIP

A drop of water may travel thousands of
kilometres or miles between the time it
evaporates and the time it falls to Earth
again as rain, sleet, or snow.

waterfall cascade of water in a river or stream. It
occurs when a river flows over a bed of rock that
resists erosion; weaker rocks downstream are worn
away, creating a steep, vertical drop and a plunge
pool into which the water falls. As the river ages,
continuing erosion causes the waterfall to retreat
upstream forming a deep valley, or ◊gorge.

water gas fuel gas consisting of a mixture of
carbon monoxide and hydrogen, made by passing
steam over red-hot coke. The gas was once the
chief source of hydrogen for chemical syntheses
such as the Haber process for making ammonia,
but has been largely superseded in this and other
reactions by hydrogen obtained from natural gas.

water glass common name for sodium metasilic-
ate (Na_2SiO_3). It is a colourless, jellylike substance
that dissolves readily in water to give a solution
used for preserving eggs and fireproofing porous
materials such as cloth, paper, and wood. It is also
used as an adhesive for paper and cardboard and
in the manufacture of soap and silica gel, a substance that absorbs moisture.

water mill machine that harnesses the energy in
flowing water to produce mechanical power, typically for milling (grinding) grain. Water from a
stream is directed against the paddles of a water
wheel to make it turn. Simple gearing transfers this
motion to the millstones. The modern equivalent
of the water wheel is the water turbine, used in
◊hydroelectric power plants.

Although early step wheels were used in ancient
China and Egypt, and parts of the Middle East,
the familiar vertical water wheel came into widespread use in Roman times. There were two types:
undershot, in which the wheel simply dipped into
the stream, and the more powerful **overshot**, in
which the water was directed at the top of the
wheel. The Domesday Book records over 7,000
water mills in Britain. Water wheels remained a
prime source of mechanical power until the development of a reliable steam engine in the 1700s,
not only for milling, but also for metalworking,
crushing and grinding operations, and driving
machines in the early factories. The two were combined to form paddlewheel steamboats in the 18th
century.

water of crystallization water chemically bonded
to a salt in its crystalline state. For example, in
copper(II) sulphate, there are five moles of water
per mole of copper sulphate: hence its formula is
$CuSO_4.5H_2O$. This water is responsible for the
colour and shape of the crystalline form. When the

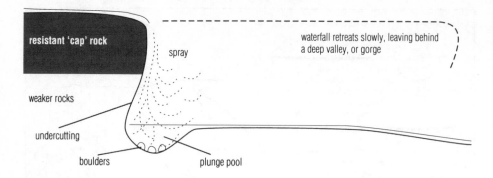

resistant 'cap' rock

spray

waterfall retreats slowly, leaving behind a deep valley, or gorge

weaker rocks

undercutting

boulders

plunge pool

waterfall

crystals are heated gently, the water is driven off as steam and a white powder is formed.

$$CuSO_4.5H_2O_{(s)} = CuSO_{4(s)} + 5H_2O_{(g)}$$

water pollution any addition to fresh or sea water that disrupts biological processes or causes a health hazard. Common pollutants include nitrate, pesticides, and sewage (see ◊sewage disposal), though a huge range of industrial contaminants, such as chemical byproducts and residues created in the manufacture of various goods, also enter water—legally, accidentally, and through illegal dumping.

In the UK, water pollution is controlled by the National Rivers Authority and, for large industrial plants, Her Majesty's Inspectorate of Pollution.

water softener any substance or unit that removes the hardness from water. Hardness is caused by the presence of calcium and magnesium ions, which combine with soap to form an insoluble scum, prevent lathering, and cause deposits to build up in pipes and cookware (kettle fur). A water softener replaces these ions with sodium ions, which are fully soluble and cause no scum.

water spout funnel-shaped column of water and cloud that is drawn by from the surface of the sea by a ◊tornado.

water supply distribution of water for domestic, municipal, or industrial consumption. Water supply in sparsely populated regions usually comes from underground water rising to the surface in natural springs, supplemented by pumps and wells. Urban sources are deep artesian wells, rivers, and reservoirs, usually formed from enlarged lakes or dammed and flooded valleys, from which water is conveyed by pipes, conduits, and aqueducts to filter beds. As water seeps through layers of shingle, gravel, and sand, harmful organisms are removed and the water is then distributed by pumping or gravitation through mains and pipes. Often other substances are added to the water, such as chlorine and fluorine; aluminium sulphate, a clarifying agent, is the most widely used chemical in water treatment. In towns, domestic and municipal (road washing, sewage) needs account for about 135 l/30 gal per head each day. In coastal desert areas, such as the Arabian peninsula, desalination plants remove salt from sea water. The Earth's waters, both fresh and saline, have been polluted by industrial and domestic chemicals, some of which are toxic and others radioactive.

In 1989 the regional water authorities of England and Wales were privatized to form ten water and sewerage companies. Following concern that some of the companies were failing to meet EC drinking-water standards on nitrate and pesticide levels, the companies were served with enforcement notices by the government Drinking Water Inspectorate.

water table the upper level of ground water (water collected underground in porous rocks). Water that is above the water table will drain downwards; a spring forms where the water table cuts the surface of the ground. The water table rises and falls in response to rainfall and the rate at which water is extracted, for example, for irrigation and industry.

In many irrigated areas the water table is falling due to the extraction of water. That below N China, for example, is sinking at a rate of 1 m/3 ft a year. Regions with high water tables and dense industrialization have problems with ◊pollution of the water table. In the USA, New Jersey, Florida, and Louisiana have water tables contaminated by both industrial ◊wastes and saline seepage from the ocean.

watt SI unit (symbol W) of power (the rate of expenditure or consumption of energy). One watt equals 0.00134 horsepower. A light bulb may use 40, 100, or 150 watts of power; an electric heater will use several kilowatts (thousands of watts). The watt is named after the Scottish engineer James Watt.

The absolute watt is defined as the power used when one joule of work is done in one second. In electrical terms, the flow of one ampere of current through a conductor whose ends are at a potential difference of one volt uses one watt of power (watts = volts × amperes).

wave in physics, a disturbance consisting of a series of oscillations that propagate through a medium (or space). There are two types: in a *longitudinal wave* (such as a ◊sound wave) the disturbance is parallel to the wave's direction of travel; in a *transverse wave* (such as an ◊electromagnetic wave) it is perpendicular. The medium only vibrates as the wave passes; it does not travel outward from the source with the waves.

A longitudinal wave is characterized by its alternating compressions and rarefactions. In the *compressions* the particles of the medium are pushed together (in a gas the pressure rises); in the *rarefactions* they are pulled apart (in a gas the pressure falls). A transverse wave is characterized by its alternating *crests* and *troughs*. Simple water waves, such as the ripples produced when a stone is dropped into a pond, are transverse waves, as are the waves on a vibrating string.

The speed *c*, frequency *f*, and wavelength λ of a wave are related by the equation $c = f\lambda$

wave in the oceans, a ridge or swell formed by wind or other causes. The power of a wave is determined by the strength of the wind and the distance of open water over which the wind blows (the fetch). Waves are the main agents of ◊coastal erosion and deposition: sweeping away or building up beaches, creating ◊spits and ◊berms, and wearing down cliffs by their hydraulic action and by the corrasion of the sand and shingle that they carry. A ◊tsunami (misleadingly called a 'tidal wave') is a type of freak wave.

As a wave approaches the shore it is forced to break as a result of friction with the sea bed. When it breaks on a beach, water and sediment are carried up the beach as *swash*; the water then drains back as *backwash*.

A *constructive wave* causes a net deposition of material on the shore because its swash tend be low and have crests that spill over gradually as they break. The backwash of a *destructive wave* is stronger than its swash, and therefore causes a net removal of material from the shore. Destructive waves are usually tall and have peaked crests that plunge downwards as they break, trapping air as they do so.

If waves strike a beach at an angle the beach material will be gradually moved along the shore (◊longshore drift), causing a deposition of material in some areas and erosion in others.

Atmospheric instability caused by the ◊greenhouse effect appears to be increasing the severity of Atlantic storms and the heights of the ocean waves. An increase of 20% in the heights of Atlantic waves has been recorded since the 1960s.

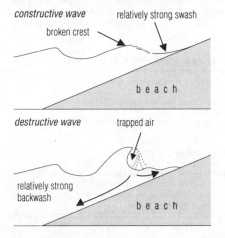

constructive wave   relatively strong swash

broken crest

b e a c h

destructive wave   trapped air

relatively strong backwash

b e a c h

wave

Freak or 'episodic' waves form under particular weather conditions at certain times of the year, travelling long distances in the Atlantic, Indian, and Pacific oceans. They are considered responsible for the sudden disappearance, without distress calls, of many ships.

Freak waves become extremely dangerous when they reach the shallow waters of the continental shelves at 100 fathoms (180 m/600 ft), especially when they meet currents: for example, the Agulhas Current to the east of South Africa, and the Gulf Stream in the N Atlantic. A wave height of 34 m/112 ft has been recorded.

wave-cut platform gently sloping rock surface found at the foot of a coastal cliff. Covered by water at high tide but exposed at low tide, it represents the last remnant of an eroded headland (see ◊coastal erosion).

waveguide hollow metallic tube containing a ◊dielectric used to guide a high-frequency electromagnetic wave (microwave) travelling within it. The wave is reflected from the internal surfaces of the guide. Waveguides are extensively used in radar systems.

wavelength the distance between successive crests of a ◊wave. The wavelength of a light wave determines its colour; red light has a wavelength of about 700 nanometres, for example. The complete range of wavelengths of electromagnetic waves is called the electromagnetic ◊spectrum.

wave power power obtained by harnessing the energy of water waves. Various schemes have been advanced since 1973, when oil prices rose dramatically and an energy shortage threatened. In 1974 the British engineer Stephen Salter developed the duck—a floating boom whose segments nod up and down with the waves. The nodding motion can be used to drive pumps and spin generators. Another device, developed in Japan, uses an oscillating water column to harness wave power. However, a major breakthrough will be required if wave power is ever to contribute significantly to the world's energy needs.

wax solid fatty substance of animal, vegetable, or mineral origin. Waxes are composed variously of ◊esters, ◊fatty acids, free ◊alcohols, and solid hydrocarbons.

Mineral waxes are obtained from petroleum and vary in hardness from the soft petroleum jelly (or petrolatum) used in ointments to the hard paraffin wax employed for making candles and waxed paper for drinks cartons.

Animal waxes include beeswax, the wool wax lanolin, and spermaceti from sperm-whale oil; they are used mainly in cosmetics, ointments, and polishes. Another animal wax is tallow, a form of suet obtained from cattle and sheep's fat, once widely used to make candles and soap. Sealing wax is made from lac or shellac, a resinous substance obtained from secretions of scale insects.

Vegetable waxes, which usually occur as a waterproof coating on plants that grow in hot, arid regions, include carnauba wax (from the leaves of the carnauba palm) and candelilla wax, both of which are components of hard polishes such as car waxes.

weak acid acid that only partially ionizes in aqueous solution (see ◊dissociation). Weak acids include ethanoic acid and carbonic acid. The pH of such acids lies between pH 3 and pH 6.

weak base base that only partially ionizes in aqueous solution (see ◊dissociation); for example, ammonia. The pH of such bases lies between pH 8 and pH 10.

weak nuclear force or **weak interaction** one of the four fundamental forces of nature, the other three being gravity, the electromagnetic force, and the strong force. It causes radioactive decay and other subatomic reactions. The particles that carry the weak force are called ◊weakons (or intermediate vector bosons) and comprise the positively and negatively charged W particles and the neutral Z particle.

weakon or **intermediate vector boson** in physics, a ◊gauge boson that carries the weak nuclear force, one of the fundamental forces of nature. There are three types of weakon, the positive and negative W particle and the neutral Z particle.

weapon any implement used for attack and defence, from simple clubs, spears, and bows and arrows in prehistoric times to machine guns and nuclear bombs in modern times. The first revolution in warfare came with the invention of ◊gunpowder and the development of cannons and shoulder-held guns. Many other weapons now exist, such as grenades, shells, torpedoes, rockets, and guided missiles. The ultimate in explosive weapons are the atomic (fission) and hydrogen (fusion) bombs. They release the enormous energy produced when atoms split or fuse together (see ◊nuclear warfare). There are also chemical and bacteriological weapons.

weather day-to-day variation of climatic and atmospheric conditions at any one place, or the state of these conditions at a place at any one time. Such conditions include humidity, precipitation, temperature, cloud cover, visibility, and wind. To a meteorologist the term 'weather' is limited to the state of the sky, precipitation, and visibility as affected by fog or mist. A region's ◊climate is derived from the average weather conditions over a long period of time. See also ◊meteorology.

Weather forecasts, in which the likely weather is predicted for a particular area, based on meteorological readings, may be short-range (covering a period of one or two days), medium-range (five to seven days), or long-range (a month or so). Readings from a series of scattered recording stations are collected and compiled on a weather map. Such a procedure is called synoptic forecasting. The weather map uses conventional symbols to show the state of the sky, the wind speed and direction, the kind of precipitation, and other details at each gathering station. Points of equal atmospheric pressure are joined by lines called isobars. The trends shown on such a map can be extrapolated to predict what weather is coming.

weathering process by which exposed rocks are broken down on the spot by the action of rain, frost, wind, and other elements of the weather. It differs from ◊erosion in that no movement or transportation of the broken-down material takes place. Two types of weathering are recognized: physical (or mechanical) and chemical. They usually occur together.

Physical weathering includes such effects as freeze–thaw (the splitting of rocks by the alternate freezing and thawing of water trapped in cracks) and exfoliation, or onion-skin weathering (flaking caused by the alternate expansion and contraction of rocks in response to extreme changes in temperature).

Chemical weathering is brought about by a chemical change in the rocks affected. The most common form is caused by rainwater that has absorbed carbon dioxide from the atmosphere and formed a weak carbonic acid. This then reacts with certain minerals in the rocks and breaks them down. Examples are the solution of caverns in limestone terrains, and the breakdown of feldspars in granite to form china clay or kaolin.

Although physical and chemical weathering normally occur together, in some instances it is difficult to determine which type is involved. For example, exfoliation, which produces rounded ◊inselbergs in arid regions, such as Ayers Rock in central Australia, may be caused by the daily physical expansion and contraction of the surface layers of the rock in the heat of the Sun, or by the chemical reaction of the minerals just beneath the surface during the infrequent rains of these areas.

weapons: chronology

13th century	Gunpowder brought to the West from China (where it was long in use but only for fireworks).
c. 1300	Guns invented by the Arabs, with bamboo muzzles reinforced with iron.
1346	Battle of Crécy in which gunpowder was probably used in battle for the first time.
1376	Explosive shells used in Venice.
17th century	Widespread use of guns and cannon in the Thirty Years' War and English Civil War.
1800	Henry Shrapnel invented shrapnel for the British army.
1862	Machine gun invented by Richard Gatling used against American Indians in the USA.
1863	TNT discovered by German chemist J Wilbrand.
1867	Dynamite patented by Alfred Nobel.
1915	Poison gas (chlorine) used for the first time by the Germans in World War I.
1916	Tanks used for the first time by the British at Cambrai.
1945	First test explosion and military use of atom bomb by the USA against Japan.
1954–73	Vietnam War, use of chemical warfare (defoliants and other substances) by the USA.
1983	Star Wars or Strategic Defense Initiative research announced by the USA to develop space laser and particle-beam weapons as a possible future weapons system in space.
1991	'Smart' weapons used by the USA and allied powers in the Gulf War; equipped with computers (using techniques such as digitized terrain maps) and laser guidance, they reached their targets with precision accuracy.

weathering

Physical weathering

temperature changes	weakening rocks by expansion and contraction
frost	wedging rocks apart by the expansion of water on freezing
unloading	the loosening of rock layers by release of pressure after the erosion and removal of those layers above

chemical weathering

carbonation	the breakdown of calcite by reaction with carbonic acid in rainwater
hydrolysis	the breakdown of feldspar into china clay by reaction with carbonic acid in rainwater
oxidation	the breakdown of iron-rich minerals due to rusting
hydration	the expansion of certain minerals due to the uptake of water

weber SI unit (symbol Wb) of ◊magnetic flux (the magnetic field strength multiplied by the area through which the field passes). One weber equals 10^8 ◊maxwells.

A change of flux at a uniform rate of one weber per second in an electrical coil with one turn produces an electromotive force of one volt in the coil. The unit is named after German physicist Wilhelm Weber.

wedge block of triangular cross-section that can be used as a simple machine. An axe is a wedge: it splits wood by redirecting the energy of the downward blow sideways, where it exerts the force needed to split the wood.

weedkiller or *herbicide* chemical that kills some or all plants. Selective herbicides are effective with cereal crops because they kill all broad-leaved plants without affecting grasslike leaves. Those that kill all plants include sodium chlorate and ◊paraquat; see also ◊Agent Orange. The widespread use of weedkillers in agriculture has led to a dramatic increase in crop yield but also to pollution of soil and water supplies and killing of birds and small animals, as well as creating a health hazard for humans.

weight the force exerted on an object by ◊gravity. The weight of an object depends on its mass—the amount of material in it—and the strength of the Earth's gravitational pull, which decreases with height. Consequently, an object weighs less at the top of a mountain than at sea level. On the Moon, an object has only one-sixth of its weight on Earth, because the pull of the Moon's gravity is one-sixth that of the Earth.

If the mass of a body is m kilograms and the gravitational field strength is g newtons per kilogram, its weight W in newtons is given by:

$$W = mg$$

weightlessness condition in which there is no gravitational force acting on a body, either because gravitational force is cancelled out by equal and opposite acceleration, or because the body is so far outside a planet's gravitational field that no force is exerted upon it.

Astronauts in space, though seemingly 'weightless', do not experience zero gravity because of the motion of their spacecraft.

weights and measures see under ◊c.g.s. system, ◊f.p.s. system, ◊m.k.s. system, ◊SI units.

welding joining pieces of metal (or nonmetal) at faces rendered plastic or liquid by heat or pressure (or both). The principal processes today are gas and arc welding, in which the heat from a gas flame or an electric arc melts the faces to be joined. Additional 'filler metal' is usually added to the joint.

Forge (or hammer) welding, employed by blacksmiths since early times, was the only method available until the late 19th century. Resistance welding is another electric method in which the weld is formed by a combination of pressure and resistance heating from an electric current. Recent developments include electric-slag, electron-beam, high-energy laser, and the still experimental radio-wave energy-beam welding processes.

Westerlies prevailing winds from the west that occur in both hemispheres between latitudes of about 35° and 60°. Unlike the ◊trade winds, they are very variable and produce stormy weather.

The Westerlies blow mainly from the SW in the northern hemisphere and the NW in the southern hemisphere, bringing moist weather to the W coast of the landmasses in these latitudes.

wetland permanently wet land area or habitat. Wetlands include areas of ◊marsh, fen, ◊bog, flood plain, and shallow coastal areas. Wetlands are extremely fertile. They provide warm, sheltered waters for fisheries, lush vegetation for grazing livestock, and an abundance of wildlife. Estuaries and seaweed beds are more than 16 times as productive as the open ocean.

The term is often more specifically applied to a naturally flooding area that is managed for agriculture or wildlife. A water meadow, where a river is expected to flood grazing land at least once a year thereby replenishing the soil, is a traditional example. In the UK, the Royal Society for the Protection of Birds (RSPB) manages 2,800 hectares/7,000 acres of wetland, using sluice gates and flood-control devices to produce sanctuaries for wading birds and wild flowers.

wetted perimeter the length of that part of a river's cross-section that is in contact with the water. The wetted perimeter is used to calculate a river's ◊hydraulic radius, a measure of its channel efficiency (ability to discharge water).

wheel and axle simple machine with a rope wound round an axle connected to a larger wheel with another rope attached to its rim. Pulling on the wheel rope (applying an effort) lifts a load attached to the axle rope. The velocity ratio of the machine (distance moved by load divided by distance moved by effort) is equal to the ratio of the wheel radius to the axle radius.

whirlwind rapidly rotating column of air, often

synonymous with a ◊tornado. On a smaller scale it produces the dust-devils seen in deserts.

white blood cell or *leucocyte* one of a number of different cells that play a part in the body's defences and give immunity against disease. Some (◊phagocytes and ◊macrophages) engulf invading microorganisms, others kill infected cells, while ◊lymphocytes produce more specific immune responses. White blood cells are colourless, with clear or granulated cytoplasm, and are capable of independent amoeboid movement. They occur in the blood, ◊lymph and elsewhere in the body's tissues.

Unlike mammalian red blood cells, they possess a nucleus. Human blood contains about 11,000 leucocytes to the cubic millimetre—about one to every 500 red cells.

White blood cell numbers may be reduced (leucopenia) by starvation, pernicious anaemia, and certain infections, such as typhoid and malaria. An increase in their numbers (leucocytosis) is a reaction to normal events such as digestion, exertion, and pregnancy, and to abnormal ones such as loss of blood, cancer, and most infections.

white dwarf small, hot ◊star, the last stage in the life of a star such as the Sun. White dwarfs have a mass similar to that of the Sun, but only 1% of the Sun's diameter, similar in size to the Earth. Most have surface temperatures of 8,000°C/14,400°F or more, hotter than the Sun. Yet, being so small, their overall luminosities may be less than 1% of that of the Sun. The Milky Way contains an estimated 50 billion white dwarfs.

White dwarfs consist of degenerate matter in which gravity has packed the protons and electrons together as tightly as is physically possible, so that a spoonful of it weighs several tonnes. White dwarfs are thought to be the shrunken remains of stars that have exhausted their internal energy supplies. They slowly cool and fade over billions of years.

WHITE DWARFS

A white dwarf cannot have a mass larger than 1.4 times that of the Sun. The lighter a white dwarf, the larger it is. The lightest white dwarfs weigh about 1/50 as much as the Sun and are 20 times the size of the Earth. Heavier white dwarfs are smaller, about the size of the Earth.

---

whiteout 'fog' of grains of dry snow caused by strong winds in temperatures of between –18°C/0°F and –1°C/30°F. The uniform whiteness of the ground and air causes disorientation in humans.

white spirit colourless liquid derived from petrol; it is used as a solvent and in paints and varnishes.

wide-angle lens photographic lens of shorter focal length than normal, taking in a wider angle of view.

wide area network in computing, a ◊network that connects computers distributed over a wide geographical area.

Wien's law in physics, a law of radiation stating that the wavelength carrying the maximum energy is inversely proportional to the body's absolute temperature: the hotter a body is, the shorter the wavelength. It has the form $\lambda_{max}T = $ constant, where λ_{max} is the wavelength of maximum intensity and T is the temperature. The law is named after German physicist Wilhelm Wien.

wilderness area of uncultivated and uninhabited land, which is usually located some distance from towns and cities.

In the USA wilderness areas are specially designated by Congress and protected by federal agencies.

wildlife trade international trade in live plants and animals, and in wildlife products such as skins, horns, shells, and feathers. The trade has made some species virtually extinct, and whole ecosystems (for example, coral reefs) are threatened. Wildlife trade is to some extent regulated by ◊CITES.

Species almost eradicated by trade in their products include many of the largest whales, crocodiles, marine turtles, and some wild cats. Until recently, some 2 million snake skins were exported from India every year. Populations of black rhino and African elephant have collapsed because of hunting for their horns and tusks (◊ivory), and poaching remains a problem in cases where trade is prohibited.

wild type in genetics, the naturally occurring gene for a particular character that is typical of most individuals of a given species, as distinct from new genes that arise by mutation.

wilting the loss of rigidity (◊turgor) in plants, caused by a decreasing wall pressure within the cells making up the supportive tissues. Wilting is most obvious in plants that have little or no wood.

WIMP (acronym for *w*indows, *i*cons, *m*enus, *p*ointing device) in computing, another name for ◊graphical user interface (GUI).

Winchester drive in computing, a small harddisc drive commonly used with microcomputers; *Winchester disc* has become synonymous with ◊hard disc.

wind the lateral movement of the Earth's atmosphere from high-pressure areas (anticyclones) to low-pressure areas (depression). Its speed is measured using an ◊anemometer or by studying its effects on, for example, trees by using the ◊Beaufort scale. Although modified by features such as land and water, there is a basic worldwide system of ◊trade winds, ◊Westerlies, ◊monsoons, and others.

A belt of low pressure (the ◊doldrums) lies along the equator. The trade winds blow towards this from the horse latitudes (areas of high pressure at about 30°N and 30°S of the equator), blowing from the NE in the northern hemisphere, and from the SE in the southern. The Westerlies (also from the horse latitudes) blow north of the equator from the SW and south of the equator from the NW.

Cold winds blow outwards from high-pressure areas at the poles. More local effects result from landmasses heating and cooling faster than the adjacent sea, producing onshore winds in the daytime and offshore winds at night.

The ◊monsoon is a seasonal wind of S Asia, blowing from the SW in summer and bringing the

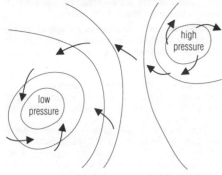

direction of air movement from a region of high atmospheric pressure to a region of low atmospheric pressure in the northern hemisphere

high pressure

low pressure

wind

rain on which crops depend. It blows from the NE in winter.

Famous or notorious warm winds include the *chinook* of the eastern Rocky Mountains, North America; the *föhn* of Europe's Alpine valleys; the *sirocco* (Italy)/*khamsin* (Egypt)/*sharav* (Israel), spring winds that bring warm air from the Sahara and Arabian deserts across the Mediterranean; and the *Santa Ana*, a periodic warm wind from the inland deserts that strikes the California coast. The dry northerly *bise* (Switzerland) and the *mistral*, which strikes the Mediterranean area of France, are unpleasantly cold winds.

Wind is caused by the trees waving their branches.

On **wind** Ogden Nash (1902–1971)

wind-chill factor or *wind-chill index* estimate of how much colder it feels when a wind is blowing. It is the sum of the temperature (in °F below zero) and the wind speed (in miles per hour). So for a wind of 15 mph at an air temperature of –5°F, the wind-chill factor is 20.

windmill mill with sails or vanes that, by the action of wind upon them, drive machinery for grinding corn or pumping water, for example. Wind turbines, designed to use wind power on a large scale, usually have a propeller-type rotor mounted on a tall shell tower. The turbine drives a generator for producing electricity.

Windmills were used in the East in ancient times, and in Europe they were first used in Germany and the Netherlands in the 12th century. The main types of traditional windmill are the *post mill*, which is turned around a post when the direction of the wind changes, and the *tower mill*, which has a revolving turret on top. It usually has a device (fantail) that keeps the sails pointing into the wind. In the USA windmills were used by the colonists and later a light type, with steel sails supported on a long steel girder shaft, was introduced for use on farms.

window in computing, a rectangular area on the screen of a ◊graphical user interface. A window is used to display data and can be manipulated in various ways by the computer user.

wind power the harnessing of wind energy to produce power. The wind has long been used as a source of energy: sailing ships and windmills are ancient inventions. After the energy crisis of the 1970s ◊wind turbines began to be used to produce electricity on a large scale. By the year 2000, 10% of Denmark's energy is expected to come from wind power.

wind tunnel test tunnel in which air is blown over, for example, a stationary model aircraft, motor vehicle, or locomotive to simulate the effects of movement. Lift, drag, and airflow patterns are observed by the use of special cameras and sensitive instruments. Wind-tunnel testing assesses aerodynamic design, preparatory to full-scale construction.

wind turbine windmill of advanced aerodynamic design connected to an electricity generator and used in wind-power installations. Wind turbines can be either large propeller-type rotors mounted on a tall tower, or flexible metal strips fixed to a vertical axle at top and bottom. In 1990, over 20,000 wind turbines were in use throughout the world, generating 1,600 megawatts of power.

The world's largest wind turbine is on Hawaii, in the Pacific Ocean. It has two blades 50 m/160 ft long on top of a tower 20 storeys high. An example of a propeller turbine is found at Tvind in Denmark and has an output of some 2 megawatts. Other machines use novel rotors, such as the 'egg-beater' design developed at Sandia Laboratories in New Mexico, USA.

The largest wind turbine on mainland Britain is at Richborough on the Kent coast. The three-bladed turbine, which is 35 m/115 ft across, produces 1 megawatt of power. Britain's largest vertical-axis wind turbine has two 24-m/80-ft blades and began operating in Dyfed, Wales, 1990.

wing in biology, the modified forelimb of birds and bats, or the membranous outgrowths of the ◊exoskeleton of insects, which give the power of flight. Birds and bats have two wings. Bird wings have feathers attached to the fused digits ('fingers') and forearm bones, while bat wings consist of skin stretched between the digits. Most insects have four wings, which are strengthened by wing veins.

The wings of butterflies and moths are covered with scales. The hind pair of a fly's wings are modified to form two knoblike balancing organs (halteres).

wireless original name for a radio receiver. In early experiments with transmission by radio waves, notably by Italian inventor Guglielmo Marconi in Britain, signals were sent in Morse code, as in telegraphy. Radio, unlike the telegraph, used no wires for transmission, and the means of communication was termed 'wireless telegraphy'.

wolfram alternative name for ◊tungsten

wolframite iron manganese tungstate, $(Fe,Mn)WO_4$, an ore mineral of tungsten. It is dark grey with a submetallic surface lustre, and often occurs in hydrothermal veins in association with ores of tin.

comparison of wing shapes

beetle

butterfly

bat

bird of prey

eagle

wing *Birds can fly because of the specialized shape of their wings: a rounded leading edge, flattened underneath and round on top. This aerofoil shape produces lift in the same way that an aircraft wing does. The outline of the wing is related to the speed of flight. Fast birds of prey have a streamlined shape. Larger birds, such as the eagle, have large wings with separated tip feathers which reduce drag and allow slow flight. Insect wings are not aerofoils. They push downwards to produce lift, in the same way that oars are used in water.*

womb common name for the ◊uterus.

wood the hard tissue beneath the bark of many perennial plants; it is composed of water-conducting cells, or secondary ◊xylem, and gains its hardness and strength from deposits of ◊lignin. *Hardwoods*, such as oak, and *softwoods*, such as pine, have commercial value as structural material and for furniture.

The central wood in a branch or stem is known as *heartwood* and is generally darker and harder than the outer wood; it consists only of dead cells. As well as providing structural support, it often contains gums, tannins, or pigments which may impart a characteristic colour and increased durability. The surrounding *sapwood* is the functional part of the xylem that conducts water.

The *secondary xylem* is laid down by the vascular ◊cambium which forms a new layer of wood annually, on the outside of the existing wood and visible as an ◊annual ring when the tree is felled; see ◊dendrochronology.

Commercial wood can be divided into two main types: hardwood, containing xylem vessels and obtained from angiosperms (for example, oak) and softwood, containing only ◊tracheids, obtained from gymnosperms (for example, pine). Although in general softwoods are softer than hardwoods, this is not always the case: balsa, the softest wood known, is a hardwood, while pitch pine, very dense and hard, is a softwood. A superhard wood is produced in wood-plastic combinations (WPC), in which wood is impregnated with liquid plastic (monomer) and the whole is then bombarded with gamma rays to polymerize the plastic.

woodland area in which trees grow more or less thickly; generally smaller than a forest. Temperate climates, with four distinct seasons a year, tend to support a mixed woodland habitat, with some conifers but mostly broad-leaved and deciduous trees, shedding their leaves in autumn and regrowing them in spring. In the Mediterranean region and parts of the southern hemisphere, the trees are mostly evergreen.

Temperate woodlands grow in the zone between the cold coniferous forest and the tropical forests of the hotter climates near the equator. They develop in areas where the closeness of the sea keeps the climate mild and moist.

Old woodland can rival tropical rainforest in the number of species it supports, but most of the species are hidden in the soil. A study in Oregon, USA 1991 found that the soil in a single woodland location contained 8,000 arthropod species (such as insects, mites, centipedes, and millipedes), compared with only 143 species of reptile, bird, and mammal in the forest above.

In England in 1900, about 2.5% of land was woodland, compared to about 3.4% in the 11th century. An estimated 33% of ancient woodland has been destroyed since 1945.

wood pitch by-product of charcoal manufacture, made from *wood tar*, the condensed liquid produced from burning charcoal gases. The wood tar is boiled to produce the correct consistency. It has been used since ancient times for caulking wooden ships (filling in the spaces between the hull planks to make them watertight).

wood pulp wood that has been processed into a pulpy mass of fibres. Its main use is for making paper, but it is also used in making ◊rayon and other cellulose fibres and plastics.

There are two methods of making wood pulp: mechanical and chemical. In the former, debarked logs are ground with water (to prevent charring) by rotating grindstones; the wood fibres are physically torn apart. In the latter, log chips are digested with chemicals (such as sodium sulphite). The chemicals dissolve the material holding the fibres together.

word in computing, a group of bits (binary digits) that a computer's central processing unit treats as a single working unit. The size of a word varies from one computer to another and, in general, increasing the word length leads to a faster and more powerful computer. In the late 1970s and early 1980s, most microcomputers were 8–bit machines. During the 1980s 16–bit microcomputers were introduced and 32–bit microcomputers are now available. Mainframe computers may be 32–bit or 64–bit machines.

word processor in computing, a program that allows the input, amendment, manipulation, storage, and retrieval of text; or a computer system that runs such software. Since word-processing programs became available to microcomputers, the method has been gradually replacing the typewriter for producing letters or other text.

Typical facilities include insert, delete, cut and paste, reformat, search and replace, copy, print, mail merge, and spelling check.

work in physics, a measure of the result of transferring energy from one system to another to cause an object to move. Work should not be confused with ◊energy (the capacity to do work, which is also measured in joules) or with ◊power (the rate of doing work, measured in joules per second).

Work is equal to the product of the force used and the distance moved by the object in the direction of that force. If the force is F newtons and the distance moved is d metres, then the work W is given by:

$$W=Fd$$

For example, the work done when a force of 10 newtons moves an object 5 metres against some sort of resistance is 50 joules (50 newton-metres).

WORK PUZZLE

Work is done when a force moves through a given distance. For example, if a weight is attached to a spring and released, the spring extends and its restoring force does work against the weight hanging from the extended spring. Thereafter no further work is done because the restoring force precisely matches the effect of gravity. In a similar way, if a weight is attached to the end of a fixed horizontal pole, no further work is necessary to maintain the pole in position because there is no further movement.

Suppose, however, that you hold a heavy weight at arm's length. Initially the position can be maintained, but gradually your arm will tire and you will be forced to stop. Since no movement occurs once the arm is extended, no mechanical work is being expended. What then is causing the arm to lose energy and become tired?

See page 651 for the answer.

---

workstation high-performance desktop computer with strong graphics capabilities, traditionally used for engineering (◊CAD and ◊CAM), scientific research, and desktop publishing. Frequently based on fast RISC (reduced instruction-set computer) chips, workstations generally offer more processing power than microcomputers (although the distinction between workstations and the more powerful microcomputer models is becoming increasingly blurred). Most workstations use Unix as their operating system, and have good networking facilities.

World Meteorological Organization agency, part of the United Nations since 1950, that promotes the international exchange of weather information through the establishment of a worldwide network of meteorological stations. It was founded as the International Meteorological Organization

1873, and its headquarters are now in Geneva, Switzerland.

World Wide Fund for Nature (WWF, formerly the *World Wildlife Fund*) international organization established 1961 to raise funds for conservation by public appeal. Projects include conservation of particular species, for example, the tiger and giant panda, and special areas, such as the Simen Mountains, Ethiopia.

The WWF has been criticized for investing in environmentally destructive companies, but the organization announced that this would cease. In 1990, it had 3.7 million members in 28 countries and an annual income of over £100 million. Its headquarters are in Gland, Switzerland.

World Wildlife Fund former and US name of the *World Wide Fund for Nature.*

worm any of various elongated limbless invertebrates belonging to several phyla. Worms include the ◊flatworms, such as flukes and tapeworms; the roundworms or ◊nematodes, such as the eelworm and the hookworm; the marine ribbon worms or nemerteans; and the segmented worms or ◊annelids.

In 1979, giant sea worms about 3 m/10 ft long, living within tubes created by their own excretions, were discovered in hydrothermal vents 2,450 m/ 8,000 ft beneath the Pacific NE of the Galápagos Islands.

The New Zealand flatworm *Artioposthia triangulata*, 15 cm/6 in long and weighing 2 g/0.07 oz, had by 1990 colonized every county of Northern Ireland and parts of Scotland. It can eat an earthworm in 30 minutes and so destroys soil fertility.

WORM (acronym for write *once* read *many* times) in computing, a storage device, similar to ◊CD-ROM. The computer can write to the disc directly, but cannot later erase or overwrite the same area. WORMs are mainly used for archiving and backup copies.

W particle in physics, an ◊elementary particle, one of the weakons responsible for transmitting the ◊weak nuclear force).

wrought iron fairly pure iron containing some beads of slag, widely used for construction work before the days of cheap steel. It is strong, tough, and easy to machine. It is made in a puddling furnace, invented by Henry Colt in England 1784. Pig iron is remelted and heated strongly in air with iron ore, burning out the carbon in the metal, leaving relatively pure iron and a slag containing impurities. The resulting pasty metal is then hammered to remove as much of the remaining slag as possible. It is still used in fences and grating.

wt abbreviation for *weight.*

WWF abbreviation for ◊*World Wide Fund for Nature* (formerly World Wildlife Fund).

WYSIWYG (acronym for *what you see is what you get*) in computing, a program that attempts to display on the screen a faithful representation of the final printed output. For example, a WYSIWYG ◊word processor would show actual page layout—line widths, page breaks, and the sizes and styles of type.

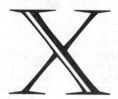

xanthophyll yellow pigment in plants that, like ◊chlorophyll, is responsible for the production of carbohydrates by photosynthesis.

X chromosome larger of the two sex chromosomes, the smaller being the ◊Y chromosome. These two chromosomes are involved in sex determination. Genes carried on the X chromosome produce the phenomenon of ◊sex linkage.

xenon (Greek *xenos* 'stranger') colourless, odourless, gaseous, non-metallic element, symbol Xe, atomic number 54, relative atomic mass 131.30. It is grouped with the ◊inert gases and was long believed not to enter into reactions, but is now known to form some compounds, mostly with fluorine. It is a heavy gas present in very small quantities in the air (about one part in 20 million).

Xenon is used in bubble chambers, light bulbs, vacuum tubes, and lasers. It was discovered in 1898 in a residue from liquid air by Scottish chemists William Ramsay and Morris Travers.

xerography dry, electrostatic method of producing images, without the use of negatives or sensitized paper, invented in the USA by Chester Carlson 1938 and applied in the Xerox ◊photocopier.

An image of the document to be copied is projected on to an electrostatically charged photoconductive plate. The charge remains only in the areas corresponding to its image. The latent image on the plate is then developed by contact with ink powder, which adheres only to the image, and is then usually transferred to ordinary paper or some other flat surface, and quickly heated to form a permanent print.

Applications include document copying, enlarging from microfilm, preparing printing masters for offset litho printing and dyeline machines, making X-ray pictures, and printing high-speed computer output.

xerophyte plant adapted to live in dry conditions. Common adaptations to reduce the loss of water by ◊transpiration include a reduction of leaf size, sometimes to spines or scales; a dense covering of hairs over the leaf to trap a layer of moist air (as in edelweiss); water storage cells; sunken stomata; and permanently rolled leaves or leaves that roll up in dry weather (as in marram grass). Many xerophytes have extensive root systems to take full advantage of occasional rains. Creosote bushes, for example, have roots in excess of 120 m/400 ft long. Many desert cacti are xerophytes.

X-ray band of electromagnetic radiation in the wavelength range 10^{-11} to 10^{-9} m (between gamma rays and ultraviolet radiation; see ◊electromagnetic waves). Applications of X-rays make use of their short wavelength (as in ◊X-ray diffraction) or their penetrating power (as in medical X-rays of internal body tissues). X-rays are dangerous and can cause cancer.

X-rays with short wavelengths pass through most body tissues, although dense areas such as bone

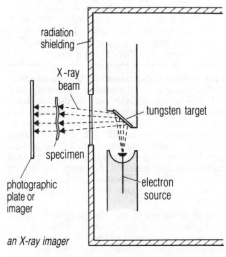

an X-ray imager

X-ray *An X-ray image. The X-rays are generated by high-speed electrons impinging on a tungsten target. The rays pass through the specimen and on to a photographic plate or imager.*

The bones in his wife's hands

On 8 November 1895, German physicist Wilhelm Conrad Röntgen (1845–1923) discovered X-rays—by accident. This key discovery in atomic physics transformed medical diagnosis, and later provided a powerful tool in cancer therapy.

Röntgen was investigating the effects of electricity discharged through gases at low pressures to produce a beam of cathode rays (or electrons). He used a Crookes tube: an improved vacuum tube invented by British chemist and physicist William Crookes (1832–1919). He directed a narrow beam of rays from the tube, which was covered with black cardboard, onto a screen in a darkened room, and noticed a faint light on a nearby bench, caused by fluorescence from another screen.

The unknown rays

Röntgen knew that cathode rays could only travel a few centimetres through air. This other screen was about a metre away; he had discovered a new phenomenon. Because their nature was unknown, he called the new rays X-rays. On 2 December, he made the first X-ray photograph, of his wife's hand. It shows that the bones in her hands were deformed by polyarthritis.

Röntgen announced his discovery on 28 December. He accurately described many of the properties of X-rays. They are produced at the walls of the discharge tube by cathode rays; like light, they travel in straight lines and cast shadows; most bodies are transparent to them in some degree; they can blacken photographic plates, and cause some substances to fluoresce; unlike cathode rays, they are not deflected by magnetic fields.

In 1901, Röntgen received the first Nobel Prize for Physics for his discovery. Initially, X-rays were used to examine the bones of the skeleton: effectively the start of radiology. Fractures could be seen clearly, as could embedded foreign bodies such as bullets or needles. For the first time, surgeons knew in advance of operating exactly what to do.

The barium meal

Solid objects and bones could be clearly visualized on photographic film, recently developed by George Eastman (1854–1932) in the USA. However, it was hard to distinguish the soft tissues of the body. US physician Walter Bradford Cannon (1871–1945), who researched digestive problems and pioneered the use of X-rays in studying the alimentary canal, invented the 'barium meal' in 1897. Barium salts are relatively impervious to X-rays; by swallowing a solution, the patient's digestive system is selectively thrown into relief on the X-ray film. By the 1930s, X-ray examination was routine medical practice, particularly in examining the lungs for tuberculosis.

As X-rays traverse living tissue, they also damage it. This danger to health was only understood much later, and both Röntgen and his assistant suffered X-ray poisoning. The destructive nature of radiation can be harnessed therapeutically. By focusing X-radiation on a tumour, its tissue is destroyed. When a cancer is not widely disseminated within the body, radiation therapy is a powerful weapon in the fight against cancer.

Traditional radiology now accounts for no more than half of medical photography. In tomography, an interesting extension of the medical uses for X-rays, they are used to photograph a selected plane of the human body. Tomography was first demonstrated successfully in 1928. In 1973, British engineer Godfrey Newbold Hounsfield (1919–) invented computerized axial tomography (CAT), in which a computer assembles a high-resolution X-ray picture of a 'slice' through the body (or head) of a patient from information provided by detectors rotating around the patient. CAT can detect very small changes in the density of living tissue. With Allan Macleod Cormack (1924–), the South African physicist who independently developed the mathematical basis for computer-assisted X-ray tomography, Hounsfield received the 1979 Nobel Prize for Physiology or Medicine.

From X-rays to radioactivity

Röntgen's discovery of X-rays led directly to the discovery of radioactivity by French physicist Antoine Henri Becquerel (1852–1908). In 1896 Becquerel was searching for X-rays in the fluorescence seen when certain salts absorb ultraviolet radiation: from the Sun, for example. He placed his test salt on a photographic plate wrapped in black paper, reasoning that any X-radiation given off by a fluorescing salt would fog the film. By chance, he left a test salt in a drawer near a wrapped plate. When the plate was developed, it had fogged.

The salt could not have fluoresced in the dark; some radiation other than X-rays must be responsible. His test salt was potassium uranyl sulphate, a salt of uranium; Becquerel had discovered radioactivity.

prevent their passage, showing up as white areas on X-ray photographs. The X-rays used in ◊radiotherapy have very short wavelengths that penetrate tissues deeply and destroy them.

X-rays were discovered by German experimental physicist Wilhelm Röntgen in 1895 and formerly called roentgen rays. They are produced when high-energy electrons from a heated filament cathode strike the surface of a target (usually made of tungsten) on the face of a massive heat-conducting anode, between which a high alternating voltage (about 100 kV) is applied.

X-ray astronomy detection of X-rays from intensely hot gas in the universe. Such X-rays are prevented from reaching the Earth's surface by the atmosphere, so detectors must be placed in rockets and satellites. The first celestial X-ray source, Scorpius X-1, was discovered by a rocket flight 1962.

Since 1970, special satellites have been orbited to study X-rays from the Sun, stars, and galaxies. Many X-ray sources are believed to be gas falling on to ◊neutron stars and ◊black holes.

Science without religion is lame, religion without science is blind.

Albert Einstein *Out of My Later Years* 1950

X-ray diffraction method of studying the atomic and molecular structure of crystalline substances by using ◊X-rays. X-rays directed at such substances spread out as they pass through the crystals owing to ◊diffraction (the slight spreading of waves around the edge of an opaque object) of the rays around the atoms. By using measurements of the position and intensity of the diffracted waves, it is possible to calculate the shape and size of the atoms in the crystal. The method has been used to study substances such as ◊DNA that are found in living material.

X-RAY DIFFRACTION: NEW STRUCTURES

Over 4,000 new chemical structures are determined by X-ray diffraction each year.

X-ray fluorescence spectrometry technique used to determine the major and trace elements in the chemical composition of such materials as ceramics, obsidian, and glass. A sample is bombarded with X-rays, and the wavelengths of the released energy, or fluorescent X-rays, are detected and measured. Different elements have unique wavelengths, and their concentrations can be estimated from the intensity of the released X-rays. This analysis may, for example, help an archaeologist in identifying the source of the material.

xylem tissue found in ◊vascular plants, whose main function is to conduct water and dissolved mineral nutrients from the roots to other parts of the plant. Xylem is composed of a number of different types of cell, and may include long, thin, usually dead cells known as ◊tracheids; fibres (schlerenchyma); thin-walled ◊parenchyma cells; and conducting vessels.

In most ◊angiosperms (flowering plants) water is moved through these vessels. Most ◊gymnosperms and ◊pteridophytes lack vessels and depend on tracheids for water conduction.

Non-woody plants contain only primary xylem, derived from the procambium, whereas in trees and shrubs this is replaced for the most part by secondary xylem, formed by ◊secondary growth from the actively dividing vascular ◊cambium. The cell walls of the secondary xylem are thickened by a deposit of ◊lignin, providing mechanical support to the plant; see ◊wood.

Yale lock trademark for a key-operated pin-tumbler cylinder lock invented by US locksmith Linus Yale Jr (1821–1868) 1865 and still widely used.

yard imperial unit (symbol yd) of length, equivalent to three feet (0.9144 m).

yardang ridge formed by wind erosion from a dried-up riverbed or similar feature, as in Chad, China, Peru, and North America. On the planet Mars yardangs occur on a massive scale.

Y chromosome smaller of the two sex chromosomes. In male mammals it occurs paired with the other type of sex chromosome (X), which carries far more genes. The Y chromosome is the smallest of all the mammalian chromosomes and is considered to be largely inert (that is, without direct effect on the physical body). See also ◊sex determination.

In humans, about one in 300 males inherits two Y chromosomes at conception, making him an XYY triploid. Few if any differences from normal XY males exist in these individuals, although at one time they were thought to be emotionally unstable and abnormally aggressive. In 1989 the gene determining that a human being is male was found to occur on the X as well as on the Y chromosome; however, it is not activated in the female.

yd abbreviation for *yard*.

year unit of time measurement, based on the orbital period of the Earth around the Sun. The *tropical year*, from one spring ◊equinox to the next, lasts 365.2422 days. It governs the occurrence of the seasons, and is the period on which the calendar year is based. The *sidereal year* is the time taken for the Earth to complete one orbit relative to the fixed stars, and lasts 365.2564 days (about 20 minutes longer than a tropical year). The difference is due to the effect of ◊precession, which slowly moves the position of the equinoxes. The *calendar year* consists of 365 days, with an extra day added at the end of Feb each leap year. *Leap years* occur in every year that is divisible by four, except that a century year is not a leap year unless it is divisible by 400. Hence 1900 was not a leap year, but 2000 will be.

A *historical year* begins on 1 Jan, although up to 1752, when the Gregorian ◊calendar was adopted in England, the civil or legal year began on 25 March. The English *fiscal/financial year* still ends on 5 April, which is 25 March plus the 11 days added under the reform of the calendar in 1752. The *regnal year* begins on the anniversary of the sovereign's accession; it is used in the dating of acts of Parliament.

The *anomalistic year* is the time taken by any planet in making one complete revolution from perihelion to perihelion; for the Earth this period is about 5 minutes longer than the sidereal year due to the gravitational pull of the other planets.

yeast one of various single-celled fungi (especially the genus *Saccharomyces*) that form masses of minute circular or oval cells by budding. When placed in a sugar solution the cells multiply and convert the sugar into alcohol and carbon dioxide. Yeasts are used as fermenting agents in baking, brewing, and the making of wine and spirits. Brewer's yeast *S. cerevisiae* is a rich source of vitamin B.

yeast artificial chromosome (YAC) fragment of ◊DNA from the human genome inserted into a yeast cell. The yeast replicates the fragment along with its own DNA. In this way the fragments are copied to be preserved in a gene library. YACs are characteristically between 250,000 and 1 million base pairs in length. A ◊cosmid works in the same way.

Yerkes Observatory astronomical centre in Wisconsin, USA, founded by George Hale 1897. It houses the world's largest refracting optical ◊telescope, with a lens of diameter 102 cm/40 in.

yield point or *elastic limit* the stress beyond which a material deforms by a relatively large amount for a small increase in stretching force. Beyond this stress, the material no longer obeys ◊Hooke's law.

yolk store of food, mostly in the form of fats and

proteins, found in the ◊eggs of many animals. It provides nourishment for the growing embryo.

yolk sac sac containing the yolk in the egg of most vertebrates. The term is also used for the membranous sac formed below the developing mammalian embryo and connected with the umbilical cord.

young fold mountain one of a range of mountains formed by the crumpling of the Earth's crust at a destructive margin (where two plates collide). For example, the Himalayas of central Asia, which have formed over the last 50 million years at the margin between the Indian and Eurasian plates. Other ranges of young fold mountains are the European Alps, the Andes of South America, and the Rockies of North America.

Fold mountains are formed when the continental crust of the two colliding plates is relatively buoyant; neither plate will therefore subduct, or descend, beneath the other. The resulting pressure between the plates causes the crustal material to become folded and uplifted.

Young fold mountains are very active areas. Earthquakes and landslides are common, and in the Andes range there are several active volcanoes, such as Nevade del Ruiz in Colombia.

ytterbium soft, lustrous, silvery, malleable, and ductile element of the ◊lanthanide series, symbol Yb, atomic number 70, relative atomic mass 173.04. It occurs with (and resembles) yttrium in gadolinite and other minerals, and is used in making steel and other alloys.

In 1878 Swiss chemist Jean-Charles de Marignac gave the name ytterbium (after the Swedish town of Ytterby, near where it was found) to what he believed to be a new element. French chemist Georges Urbain (1872–1938) discovered 1907 that this was in fact a mixture of two elements: ytterbium and lutetium.

yttrium silver-grey, metallic element, symbol Y, atomic number 39, relative atomic mass 88.905. It

the Himalayas are an example of young fold mountains

young fold mountains

is associated with and resembles the ◊rare-earth elements (◊lanthanides), occurring in gadolinite, xenotime, and other minerals. It is used in colour-television tubes and to reduce steel corrosion.

The name derives from the Swedish town of Ytterby, near where it was first discovered 1788. Swedish chemist Carl Mosander isolated the element 1843.

Z

Z in physics, the symbol for *impedance* (electricity and magnetism).

Zelenchukskaya site of the world's largest single-mirror optical telescope, with a mirror of 6 m/236 in diameter, in the Caucasus Mountains of Russia. At the same site is the RATAN 600 radio telescope, consisting of radio reflectors in a circle of 600 m/2,000 ft diameter. Both instruments are operated by the Academy of Sciences in St Petersburg.

zenith uppermost point of the celestial horizon, immediately above the observer; the ◊nadir is below, diametrically opposite. See ◊celestial sphere.

zeolite any of the hydrous aluminium silicates, also containing sodium, calcium, barium, strontium, or potassium, chiefly found in igneous rocks and characterized by a ready loss or gain of water. Zeolites are used as 'molecular sieves' to separate mixtures because they are capable of selective absorption. They have a high ion-exchange capacity and can be used to make petrol, benzene, and toluene from low-grade raw materials, such as coal and methanol.

Permutit is a synthetic zeolite used to soften hard water.

zidovudine (formerly *AZT*) antiviral drug used in the treatment of ◊AIDS. It is not a cure for AIDS but was thought to be effective in suppressing the causative virus (HIV).

Zidovudine was developed in the mid-1980s and approved for use by 1987. Taken every four hours, night and day, it reduces the risk of opportunistic infection and relieves many neurological complications. However, frequent blood monitoring is required to control anaemia, a potentially life-threatening side effect. Blood transfusions are often necessary, and the drug must be withdrawn if bone-marrow function is severely affected. Tests in 1993 led to doubts about its efficacy.

zinc (Germanic *zint* 'point') hard, brittle, bluish-white, metallic element, symbol Zn, atomic number 30, relative atomic mass 65.37. The principal ore is sphalerite or zinc blende (zinc sulphide, ZnS). Zinc is little affected by air or moisture at ordinary temperatures; its chief uses are in alloys such as brass and in coating metals (for example, galvanized iron). Its compounds include zinc oxide, used in ointments (as an astringent) and cosmetics, paints, glass, and printing ink.

Zinc has been used as a component of brass since the Bronze Age, but it was not recognized as a separate metal until 1746, when it was described by German chemist Andreas Sigismund Marggraf (1709–1782). The name derives from the shape of the crystals on smelting.

The zinc industry in Europe generates about 80,000 tons of zinc waste each year.

zinc chloride $ZnCl_2$ white, crystalline compound that is deliquescent and sublimes easily. It is used as a catalyst, as a dehydrating agent, and as a flux in soldering.

zinc ore mineral from which zinc is extracted, principally sphalerite $(Zn,Fe)S$, but also zincite, ZnO_2, and smithsonite, $ZnCO_3$, all of which occur in mineralized veins. Ores of lead and zinc often occur together, and are common worldwide; Canada, the USA, and Australia are major producers.

zinc oxide ZnO white powder, yellow when hot, that occurs in nature as the mineral zincite. It is used in paints and as an antiseptic in zinc ointment; it is the main ingredient of calamine lotion.

zinc sulphide ZnS yellow-white solid that occurs in nature as the mineral sphalerite (also called zinc blende). It is the principal ore of zinc, and is used in the manufacture of fluorescent paints.

zip fastener fastening device used in clothing, invented in the USA by Whitcomb Judson 1891, originally for doing up shoes. It has two sets of interlocking teeth, meshed by means of a slide that moves up and down.

zircon zirconium silicate, $ZrSiO_4$, a mineral that occurs in small quantities in a wide range of igneous, sedimentary, and metamorphic rocks. It is very durable and is resistant to erosion and weathering. It is usually coloured brown, but can be other colours, and when transparent may be used as a gemstone.

zirconium (Germanic *zircon*, from Persian *zargun* 'golden') lustrous, greyish-white, strong, ductile, metallic element, symbol Zr, atomic number 40, relative atomic mass 91.22. It occurs in nature as the mineral zircon (zirconium silicate), from which it is obtained commercially. It is used in some ceramics, alloys for wire and filaments, steel manufacture, and nuclear reactors, where its low neutron absorption is advantageous.

It was isolated 1824 by Swedish chemist Jöns Berzelius. The name was proposed by English chemist Humphry Davy 1808.

zodiac zone of the heavens containing the paths of the Sun, Moon, and planets. When this was devised by the ancient Greeks, only five planets were known, making the zodiac about 16° wide. In astrology, the zodiac is divided into 12 signs, each 30° in extent: Aries, Taurus, Gemini, Cancer, Leo, Virgo, Libra, Scorpio, Sagittarius, Capricorn, Aquarius, and Pisces. These do not cover the same areas of sky as the astronomical constellations.

zodiacal light cone-shaped light sometimes seen extending from the Sun along the ◊ecliptic, visible after sunset or before sunrise. It is due to thinly spread dust particles in the central plane of the Solar System. It is very faint, and requires a dark, clear sky to be seen.

zoetrope optical toy with a series of pictures on the inner surface of a cylinder. When the pictures are rotated and viewed through a slit, it gives the impression of continuous motion.

zoidogamy type of plant reproduction in which male gametes (antherozoids) swim in a film of water to the female gametes. Zoidogamy is found in algae, bryophytes, pteridophytes, and some gymnosperms (others use ◊siphonogamy).

zone system in photography, a system of exposure estimation invented by US photographer Ansel Adams that groups infinite tonal gradations into ten zones, zone 0 being black and zone 10 white. An ◊f-stop change in exposure is required from zone to zone.

zoo abbreviation for *zoological gardens*, a place where animals are kept in captivity. Originally created purely for visitor entertainment and education, zoos have become major centres for the breeding of endangered species of animals; a 1984 report identified 2,000 vertebrate species in need of such maintenance. The Arabian oryx has already

been preserved in this way; it was captured 1962, bred in captivity, and released again in the desert 1972, where it has flourished.

Notable zoos exist in New York, San Diego, Toronto, Chicago, London, Paris, Berlin, Moscow, and Beijing (Peking).

Henry I started a royal menagerie at Woodstock, Oxfordshire, later transferred to the Tower of London. The Zoological Society of London was founded 1826 by Stamford Raffles in Regent's Park, London, and in 1827 the gardens were opened to members. In 1831 William IV presented the royal menagerie to the Zoological Society; the public were admitted from 1848. The name 'zoo' dates from 1867. London Zoo currently houses some 8,000 animals of over 900 species. Threatened by closure 1991 because of falling income, the zoo was given a one-year extension to July 1992; the decision to close it was reversed Sept 1992 and it is now planned to transform the zoo into a conservation park, with Whipsnade as the national collection of animals. In 1991 the number of animals in Britain's zoos totalled 35,000.

zoology branch of biology concerned with the study of animals. It includes description of present-day animals, the study of evolution of animal forms, anatomy, physiology, embryology, behaviour, and geographical distribution.

. . . It next will be right/To describe each particular batch:/Distinguishing those that have feathers, and bite,/From those that have whiskers, and scratch.

On **zoology** Lewis Carroll (1832–1898) *The Hunting of the Snark* 1876

zoom lens photographic lens that, by variation of focal length, allows speedy transition from long shots to close-ups.

Z particle in physics, an ◊elementary particle, one of the weakons responsible for carrying the ◊weak nuclear force.

zwitterion ion that has both a positive and a negative charge, such as an ◊amino acid in neutral solution. For example, glycine contains both a basic amino group (NH_2) and an acidic carboxyl group (-COOH); when both these are ionized in aqueous solution, the acid group loses a proton to the amino group, and the molecule is positively charged at one end and negatively charged at the other.

zygote ◊ovum (egg) after ◊fertilization but before it undergoes cleavage to begin embryonic development.

Answers to the Puzzles:

capillarity puzzle

Paper itself consists of a mass of fibres with tiny spaces between them, which can act as capillaries. Indeed, very porous paper causes problems because the ink tends to spread sideways. The contact point between pen and paper can also act as a capillary in the same way that water can be drawn up between two sheets of glass. Interestingly, there is actually no reservoir in felt-tip pens. Instead, the tip extends right into the body of the pen and is so porous that it holds enough ink to act as its own reservoir.

friction puzzle

It is wrong to suppose that treadless tyres have a poorer grip. In fact they offer greater contact with the road surface and therefore give better grip, but only in dry weather and on dry roads. On wet roads the surface layer of water under the tyre dramatically reduces grip, and the purpose of treading is to remove the surface water by opening and closing under pressure as the tyre revolves, picking up the water and squeezing it out sideways.

genetics puzzle

Fortunately for history, nature has introduced a further shuffle to the process. During the separation of a chromosome pair to form reproductive cells there is an interchange of *bits* of chromosome between each half of the pair, so that separated single chromosomes already contain shuffled information. The result is that the only identical copies around are genetic twins, triplets, etc.

laser puzzle

The energy is not 'absorbed' in the same way in the laser. When light is absorbed by an object it causes atoms to vibrate, increasing kinetic energy and, therefore, temperature. The result is a possible melting of the target. In the laser, however, the energy is stored by raising the electrons to higher energy levels, and the laser effect occurs when all the electrons drop back at the same time. There may be some heating effect, but this will not be as great as the effect on the target.

magnetism puzzle

No. Magnetic fields exist only between poles, and since the north pole surrounds the south pole, the entire magnetic field exists only within the metal. The sphere will therefore behave simply as a normal, unmagnetized sphere.

relativity puzzle

Casual statements about relativity and the speed of light generally gloss over the meaning of 'mass'. Einstein's equations generally relate to rest mass, when an object is not moving, rather than to any kinetic energy associated with movement. Although photons have kinetic energy, their rest mass is zero and will remain so even at the speed of light.

space puzzle 1

The problem is one of acceleration and deceleration. On launch a spacecraft must be accelerated by on-board fuel all the way to orbit, and only approaches orbital speed at the end of launch. On re-entry, however, the craft is not slowed by prolonged fuel-burning but rather by the friction effect with the atmosphere. As the craft descends it will not slow significantly until the atmosphere becomes substantial. The effect is that the craft is travelling much faster at lower altitudes on re-entry than on launch, and frictional heat is therefore much more of a problem.

space puzzle 2

No! Just as skaters pull in their arms in order to spin faster, the spacecraft has moved into a slightly smaller orbit but has increased in speed, thereby causing it to overtake the leading craft. To be more precise, the spacecraft has maintained its angular momentum but reduced its orbital radius. The result, as with a skater, is a greater angular velocity. What the crew should have done is fire their motors in the direction of the lead craft. This would slow them down and cause them to drop into an even lower orbit. In fact, in order to manoeuvre, a spacecraft in orbit must sometimes fire its motors at right-angles to the intended direction of motion. Steering a spaceship is not like steering a car or a boat.

thermal expansion puzzle

The hole becomes larger because all dimensions increase, and the idea of 'moving into free space' is wrong. The effect was at one time used to put metal rims on wooden wagon wheels by heating them oversize and then letting them shrink to fit the wheel as they cooled.

work puzzle

Springs and fixed poles are static systems in which no chemical changes take place. The body, however, is a dynamic system in which muscles can either be tensed or relaxed, depending on their chemical state. The various substances involved are metabolized as they are used, and must therefore be continually replenished to maintain a muscle in a tensed state. Chemical energy is consumed in the process, the supply of the necessary chemicals dwindles, and the arm eventually tires.

Greek alphabet

A	α	alpha
B	β	beta
Γ	γ	gamma
Δ	δ	delta
E	ε	epsilon
Z	ζ	zeta
H	η	eta
Θ	θ	theta
I	ι	iota
K	κ	kappa
Λ	λ	lambda
M	μ	mu
N	ν	nu
Ξ	ξ	xi
O	o	omicron
Π	π	pi
P	ρ	rho
Σ	σ	sigma
T	τ	tau
Y	υ	upsilon
Φ	φ	phi
X	χ	chi
Ψ	ψ	psi
Ω	ω	omega

fundamental constants

constant	symbol	value in SI units
acceleration of free fall	g	9.80665 m s^{-2}
Avogadro's constant	N_A	6.02252×10^{23} mol^{-1}
Boltzmann's constant	$k = R/N_A$	1.380622×10^{23} J K^{-1}
electronic charge	e	1.602192×10^{-19} C
electronic rest mass	m_e	9.109558×10^{-31} kg
Faraday's constant	F	9.648670×10^{4} C mol^{-1}
gas constant	R	8.31434 J K^{-1} mol^{-1}
gravitational constant	G	6.664×10^{-11} N m^2 kg^{-2}
Loschmidt's number	N_L	2.68719×10^{25} m^{-3}
neutron rest mass	m_n	1.67492×10^{-27} kg
Planck's constant	h	6.626196×10^{-34} J s
proton rest mass	m_p	1.672614×10^{-27} kg
speed of light	c	2.99792458×10^{8} m s^{-1}
standard atmospheric pressure	P	1.01325×10^{5} Pa
Stefan–Boltzmann constant	σ	5.6697×10^{-8} W m^{-2} K^{-4}

SI units

quantity	SI unit	symbol
absorbed radiation dose	gray	Gy
amount of substance	mole*	mol
electric capacitance	farad	F
electric charge	coulomb	C
electric conductance	siemens	S
electric current	ampere*	A
energy or work	joule	J
force	newton	N
frequency	hertz	Hz
illuminance	lux	lx
inductance	henry	H
length	metre*	m
luminous flux	lumen	lm
luminous intensity	candela*	cd
magnetic flux	weber	Wb
magnetic flux density	tesla	T
mass	kilogram*	kg
plane angle	radian	rad
potential difference	volt	V
power	watt	W
pressure	pascal	Pa
radiation dose equivalent	sievert	Sv
radiation exposure	roentgen	r
radioactivity	becquerel	Bq
resistance	ohm	W
solid angle	steradian	sr
sound intensity	decibel	dB
temperature	°Celsius	°C
temperature, thermodynamic	kelvin*	K
time	second*	s

*SI base unit

SI prefixes

multiple	prefix	symbol	example
1,000,000,000,000,000,000 (10^{18})	exa-	E	Eg (exagram)
1,000,000,000,000,000 (10^{15})	peta-	P	PJ (petajoule)
1,000,000,000,000 (10^{12})	tera-	T	TV (teravolt)
1,000,000,000 (10^{9})	giga-	G	GW (gigawatt)
1,000,000 (10^{6})	mega-	M	MHz (megahertz)
1,000 (10^{3})	kilo-	k	kg (kilogram)
100 (10^{2})	hecto-	h	hm (hectometre)
10	deca-	da-	daN (decanewton)
1/10 (10^{-1})	deci-	d	dC (decicoulomb)
1/100 (10^{-2})	centi-	c	cm (centimetre)
1/1,000 (10^{-3})	milli-	m	mA (milliampere)
1/1,000,000 (10^{-6})	micro-	m	μF (microfarad)
1/1,000,000,000 (10^{-9})	nano-	n	nm (nanometre)
1/1,000,000,000,000 (10^{-12})	pico-	p	ps (picosecond)
1/1,000,000,000,000,000 (10^{-15})	femto-	f	frad (femtoradian)
1/1,000,000,000,000,000,000 (10^{-18})	atto-	a	aT (attotesla)

periodic table of the elements

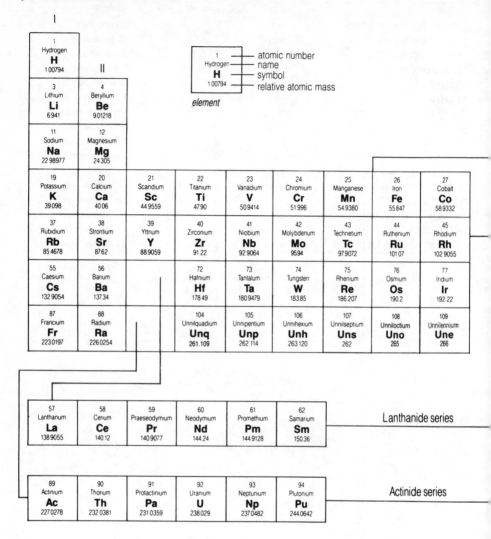

							0
							2 Helium **He** 4.00260
	III	IV	V	VI	VII		
	5 Boron **B** 10.81	6 Carbon **C** 12.011	7 Nitrogen **N** 14.0067	8 Oxygen **O** 15.9994	9 Fluorine **F** 18.99840		10 Neon **Ne** 20.179
	13 Aluminium **Al** 26.98154	14 Silicon **Si** 28.086	15 Phosphorus **P** 30.97376P	16 Sulphur **S** 32.06	17 Chlorine **Cl** 35.453		18 Argon **Ar** 39.948

28 Nickel **Ni** 58.70	29 Copper **Cu** 63.546	30 Zinc **Zn** 65.38	31 Gallium **Ga** 69.72	32 Germanium **Ge** 72.59	33 Arsenic **As** 74.9216	34 Selenium **Se** 78.96	35 Bromine **Br** 79.904	36 Krypton **Kr** 83.80
46 Palladium **Pd** 106.4	47 Silver **Ag** 107.868	48 Cadmium **Cd** 112.40	49 Indium **In** 114.82	50 Tin **Sn** 118.69	51 Antimony **Sb** 121.75	52 Tellurium **Te** 127.75	53 Iodine **I** 126.9045	54 Xenon **Xe** 131.30
78 Platinum **Pt** 195.09	79 Gold **Au** 196.9665	80 Mercury **Hg** 200.59	81 Thallium **Tl** 204.37	82 Lead **Pb** 207.37	83 Bismuth **Bi** 207.2	84 Polonium **Po** 210	85 Astatine **At** 211	86 Radon **Rn** 222.0176

63 Europium **Eu** 151.96	64 Gadolinium **Gd** 157.25	65 Terbium **Tb** 158.9254	66 Dysprosium **Dy** 162.50	67 Holmium **Ho** 164.9304	68 Erbium **Er** 167.26	69 Thulium **Tm** 168.9342	70 Ytterbium **Yb** 173.04	71 Lutetium **Lu** 174.97
95 Americium **Am** 243.0614	96 Curium **Cm** 247.0703	97 Berkelium **Bk** 247.0703	98 Californium **Cf** 251.0786	99 Einsteinium **Es** 252.0828	100 Fermium **Fm** 257.0951	101 Mendelevium **Me** 258.0986	102 Nobelium **No** 259.1009	103 Lawrencium **Lr** 260.1054

Nobel Prize for Chemistry

prizewinners	
1945	Artturi Virtanen (Finland): agriculture and nutrition, especially fodder preservation
1946	James Sumner (USA): crystallization of enzymes; John Northrop (USA) and Wendell Stanley (USA): preparation of pure enzymes and virus proteins
1947	Robert Robinson (UK): biologically important plant products, especially alkaloids
1948	Arne Tiselius (Sweden): electrophoresis and adsorption analysis, and discoveries concerning serum proteins
1949	William Giauque (USA): chemical thermodynamics, especially at very low temperatures
1950	Otto Diels (Germany) and Kurt Alder (Germany): discovery and development of diene synthesis
1951	Edwin McMillan (USA) and Glenn T. Seaborg (USA): chemistry of transuranic elements
1952	Archer Martin (UK) and Richard Synge (UK): invention of partition chromatography
1953	Hermann Staudinger (West Germany): discoveries in macromolecular chemistry
1954	Linus Pauling (USA): nature of chemical bonds, especially in complex substances
1955	Vincent Du Vigneaud (USA): investigations into biochemically important sulphur compounds, and the first synthesis of a polypeptide hormone
1956	Cyril Hinshelwood (UK) and Nikoly Semenov (USSR): mechanism of chemical reactions
1957	Alexander Todd (UK): nucleotides and nucleotide coenzymes
1958	Frederick Sanger (UK): structure of proteins, especially insulin
1959	Jaroslav Heyrovsky´ (Czechoslovakia): polarographic methods of chemical analysis
1960	Willard Libby (USA): radiocarbon dating in archaeology, geology, and geography
1961	Melvin Calvin (USA): assimilation of carbon dioxide by plants
1962	Max Perutz (UK) and John Kendrew (UK): structures of globular proteins
1963	Karl Ziegler (West Germany) and Giulio Natta (Italy): chemistry and technology of high polymers
1964	Dorothy Crowfoot Hodgkin (UK): crystallographic determination of the structures of biochemical compounds, notably penicillin and cyanocobalamin (vitamin B_{12})
1965	Robert Woodward (USA): organic synthesis
1966	Robert Mulliken (USA): molecular orbital theory of chemical bonds and structures
1967	Manfred Eigen (West Germany), Ronald Norrish (UK), and George Porter (UK): investigation of rapid chemical reactions by means of very short pulses of energy
1968	Lars Onsager (USA): discovery of reciprocal relations, fundamental for the thermodynamics of irreversible processes
1969	Derek Barton (UK) and Odd Hassel (Norway): concept and applications of conformation
1970	Luis Federico Leloir (Argentina): discovery of sugar nucleotides and their role in carbohydrate biosynthesis
1971	Gerhard Herzberg (Canada): electronic structure and geometry of molecules, particularly free radicals
1972	Christian Anfinsen (USA), Stanford Moore (USA), and William Stein (USA): amino-acid structure and biological activity of the enzyme ribonuclease
1973	Ernst Fischer (West Germany) and Geoffrey Wilkinson (UK): chemistry of organometallic sandwich compounds
1974	Paul Flory (USA): physical chemistry of macromolecules
1975	John Cornforth (Australia): stereochemistry of enzyme-catalysed reactions. Vladimir Prelog (Yugoslavia): stereochemistry of organic molecules and their reactions
1976	William N Lipscomb (USA): structure and chemical bonding of boranes (compounds of boron and hydrogen)
1977	Ilya Prigogine (USSR): thermodynamics of irreversible and dissipative processes
1978	Peter Mitchell (UK): biological energy transfer and chemiosmotic theory
1979	Herbert Brown (USA) and Georg Wittig (West Germany): use of boron and phosphorus compounds, respectively, in organic syntheses
1980	Paul Berg (USA): biochemistry of nucleic acids, especially recombinant-DNA. Walter Gilbert (USA) and Frederick Sanger (UK): base sequences in nucleic acids
1981	Kenichi Fukui (Japan) and Roald Hoffmann (USA): theories concerning chemical reactions
1982	Aaron Klug (UK): crystallographic electron microscopy: structure of biologically important nucleic-acid--protein complexes
1983	Henry Taube (USA): electron-transfer reactions in inorganic chemical reactions
1984	Bruce Merrifield (USA): chemical syntheses on a solid matrix
1985	Herbert A Hauptman (USA) and Jerome Karle (USA): methods of determining crystal structures
1986	Dudley Herschbach (USA), Yuan Lee (USA), and John Polanyi (Canada): dynamics of chemical elementary processes
1987	Donald Cram (USA), Jean-Marie Lehn (France), and Charles Pedersen (USA): molecules with highly selective structure-specific interactions
1988	Johann Deisenhofer (West Germany), Robert Huber (West Germany), and Hartmut Michel (West Germany): three-dimensional structure of the reaction centre of photosynthesis
1989	Sydney Altman (USA) and Thomas Cech (USA): discovery of catalytic function of RNA
1990	Elias James Corey (USA): new methods of synthesizing chemical compounds
1991	Richard R Ernst (Switzerland): improvements in the technology of nuclear magnetic resonance (NMR) imaging
1992	Rudolph A Marcus (USA): theoretical discoveries relating to reduction and oxidation reactions.

Nobel Prize for Medicine or Physiology

prizewinners

1962	Francis Crick (UK), James Watson (USA), and Maurice Wilkins (UK): discovery of the double-helical structure of DNA and of the significance of this structure in the replication and transfer of genetic information
1963	John Eccles (Australia), Alan Hodgkin (UK), and Andrew Huxley (UK): ionic mechanisms involved in the communication or inhibition of impulses across neuron-cell membranes
1964	Konrad Bloch (USA) and Feodor Lynen (West Germany): cholesterol and fatty-acid metabolism
1965	François Jacob (France), André Lwoff (France), and Jacques Monod (France): genetic control of enzyme and virus synthesis
1966	Peyton Rous (USA): discovery of tumour- inducing viruses; Charles Huggins (USA): hormonal treatment of prostatic cancer
1967	Ragnar Granit (Sweden), Haldan Hartline (USA), and George Wald (USA): physiology and chemistry of vision
1968	Robert Holley (USA), Har Gobind Khorana (USA), and Marshall Nirenberg (USA): interpretation of genetic code and its function in protein synthesis
1969	Max Delbruck (USA), Alfred Hershey (USA), and Salvador Luria (USA): replication mechanism and genetic structure of viruses
1970	Bernard Katz (UK), Ulf von Euler (Austria), and Julius Axelrod (USA): storage, release, and inactivation of neurotransmitters
1971	Earl Sutherland(USA): discovery of cyclic AMP, a chemical messenger that plays a role in the action of many hormones
1972	Gerald Edelman (USA) and Rodney Porter (UK): chemical structure of antibodies
1973	Karl von Frisch (Austria), Konrad Lorenz (Austria), and Nikolaas Tinbergen (Netherlands): animal behaviour patterns
1974	Albert Claude (USA), Christian de Duve (Belgium), and George Palade (USA): structural and functional organization of the cell
1975	David Baltimore (USA), Renato Dulbecco (USA), and Howard Temin (USA): interactions between tumour-inducing viruses and the genetic material of the cell
1976	Baruch Blumberg (USA) and Carleton Gajdusek (USA): new mechanisms for the origin and transmission of infectious diseases
1977	Roger Guillemin (USA) and Andrew Schally (USA): discovery of hormones produced by the hypothalamus region of the brain; Rosalyn Yalow (USA): radioimmunoassay techniques by which minute quantities of hormone may be detected
1978	Werner Arber (Switzerland), Daniel Nathans (USA), and Hamilton Smith (USA): discovery of restriction enzymes and their application to molecular genetics
1979	Allan Cormack (USA) and Godfrey Hounsfield (UK): development of the CAT (Computerized axial tomogrophy) scan
1980	Baruj Benacerraf (USA), Jean Dausset (France), and George Snell (USA): genetically determined structures on the cell surface that regulate immunological reactions
1981	Roger Sperry (USA): functional specialization of the brain's cerebral hemispheres; David Hubel (USA) and Torsten Wiesel (Sweden): visual perception
1982	Sune Bergström (Sweden), Bengt Samuelson (Sweden), and John Vane (UK): discovery of prostaglandins and related biologically reactive substances
1983	Barbara McClintock (USA): discovery of mobile genetic elements
1984	Niels Jerne (Denmark), Georges Köhler (West Germany), and César Milstein (UK): work on immunity and discovery of a technique for producing highly specific, monoclonal antibodies
1985	Michael Brown (USA) and Joseph L Goldstein (USA): regulation of cholesterol metabolism
1986	Stanley Cohen (USA) and Rita Levi-Montalcini (Italy): discovery of factors that promote the growth of nerve and epidermal cells
1987	Susumu Tonegawa (Japan): process by which genes alter to produce a range of different antibodies
1988	James Black (UK), Gertrude Elion (USA), and George Hitchings (USA): principles governing the design of new drug treatment
1989	Michael Bishop (USA) and Harold Varmus (USA): discovery of oncogenes, genes carried by viruses that can trigger cancerous growth in normal cells
1990	Joseph Murray (USA) and Donnall Thomas (USA): pioneering work in organ and cell transplants
1991	Erwin Neher (Germany) and Bert Sakmann (Germany): discovery of how gatelike structures (ion channels) regulate the flow of ions into and out of cells
1992	Edmond Fisher (USA) and Edwin Krebs (USA): isolating and describing the action of the enzyme responsible for reversible protein phosphorylation, a major biological control mechanism

Nobel Prize for Physics

prizewinners

1961	Robert Hofstadter (USA): scattering of electrons in atomic nuclei, and structure of protons and neutrons; Rudolf Mössbauer (Germany): resonance absorption of gamma radiation
1962	Lev Landau (USSR): theories of condensed matter, especially liquid helium
1963	Eugene Wigner (USA): discovery and application of symmetry principles in atomic physics; Maria Goeppert-Mayer (USA) and Hans Jensen (Germany): discovery of the shell-like structure of atomic nuclei
1964	Charles Townes (USA), Nikolai Basov (USSR), and Aleksandr Prokhorov (USSR): quantum electronics leading to construction of oscillators and amplifiers based on maser–laser principle
1965	Sin-Itiro Tomonaga (Japan), Julian Schwinger (USA), and Richard Feynman (USA): quantum electrodynamics
1966	Alfred Kastler (France): development of optical pumping, whereby atoms are raised to higher energy levels by illumination
1967	Hans Bethe (USA): theory of nuclear reactions, and discoveries concerning production of energy in stars
1968	Luis Alvarez (USA): elementary-particle physics, and discovery of resonance states, using hydrogen bubble chamber and data analysis
1969	Murray Gell-Mann (USA): classification of elementary particles, and study of their interactions
1970	Hannes Alfvén (Sweden): magnetohydrodynamics and its applications in plasma physics; Louis Néel (France): antiferromagnetism and ferromagnetism in solid-state physics
1971	Dennis Gabor (UK): invention and development of holography
1972	John Bardeen (USA), Leon Cooper (USA), and John Robert Schrieffer (USA): theory of superconductivity
1973	Leo Eskai (Japan) and Ivar Giaver (USA): tunnelling phenomena in semiconductors and superconductors; Brian Josephson (UK): theoretical predictions of the properties of a supercurrent through a tunnel barrier
1974	Martin Ryle (UK) and Antony Hewish (UK): development of radioastronomy, particularly aperture-synthesis technique, and the discovery of pulsars
1975	Aage Bohr (Denmark), Ben Mottelson (Denmark), and James Rainwater (USA): discovery of connection between collective motion and particle motion in atomic nuclei, and development of theory of nuclear structure
1976	Burton Richter (USA) and Samuel Ting (USA): discovery of the psi meson
1977	Philip Anderson (USA), Nevill Mott (UK), and John Van Vleck (USA): electronic structure of magnetic and disordered systems
1978	Pyotr Kapitza (USSR): low-temperature physics; Arno Penzias (Germany) and Robert Wilson (USA): discovery of cosmic background radiation
1979	Sheldon Glashow (USA), Abdus Salam (Pakistan), and Steven Weinberg (USA): unified theory of weak and electromagnetic fundamental forces, and prediction of the existence of the weak neutral current
1980	James W Cronin (USA) and Val Fitch (USA): violations of fundamental symmetry principles in the decay of neutral kaon mesons
1981	Nicolaas Bloemergen (USA) and Arthur Schawlow (USA): development of laser spectroscopy. Kai Siegbahn (Sweden): high-resolution electron spectroscopy
1982	Kenneth Wilson (USA): theoryof critical phenomena in connection with phase transitions
1983	Subrahmanyan Chandrasekhar (USA): theoretical studies of physical processes in connection with structure and evolution of stars; William Fowler (USA): nuclear reactions involved in the formation of chemical elements in the universe
1984	Carlo Rubbia (Italy) and Simon van der Meer (Netherlands): contributions to the discovery of the W and Z particles (weakons)
1985	Klaus von Klitzing (Germany): discovery of the quantized Hall effect
1986	Erns Ruska (Germany): electron optics, and design of the first electron microscope; Gerd Binnig (Germany) and Heinrich Rohrer (Switzerland): design of scanning tunnelling microscope
1987	Georg Bednorz (Germany) and Alex Müller (Switzerland): superconductivity in ceramic materials
1988	Leon M Lederman (USA), Melvin Schwartz (USA), and Jack Steinberger (Germany): neutrino-beam method, and demonstration of the doublet structure of leptons through discovery of muon neutrino
1989	Norman Ramsey (USA): measurement techniques leading to discovery of caesium atomic clock; Hans Dehmelt (USA) and Wolfgang Paul (Germany): ion-trap method for isolating single atoms
1990	Jerome Friedman (USA), Henry Kendall (USA), and Richard Taylor (Canada): experiments demonstrating that protons and neutrons are made up of quarks
1991	Pierre-Gilles de Gennes (France): work on disordered systems including polymers and liquid crystals; development of mathematical methods for studying the behaviour of molecules in a liquid on the verge of solidifying
1992	Georges Charpak (Poland): invention and development of detectors used in high-energy physics

Great Scientists

Anthropologists

Beattie, John Hugh Marshall 1915–1990
Blacking, John 1928–1990
Boas, Franz 1858–1942
Dart, Raymond 1893–1988
Frazer, James George 1854–1941
Harrisson, Tom 1911–1976
Heyerdahl, Thor 1914–
Kroeber, Alfred Louis 1876–1960
Lévi-Strauss, Claude 1908–1990
Malinowski, Bronislaw 1884–1942
Mead, Margaret 1901–1978
Montagu, Ashley 1905–
Morgan, Lewis Henry 1818–1881

Archaeologists

Catherwood, Frederick 1799–1854
Champollion, Jean François, le Jeune 1790–1832
Childe, Vere Gordon 1892–1957
Crawford, Osbert Guy Stanhope 1886–1957
Daniel, Glyn 1914–1986
Emery, Walter Bryan 1903–1971
Erim, Kenan Tevfig 1929–1990
Evans, Arthur John 1851–1941
Kenyon, Kathleen 1906–1978
Layard, Austen Henry 1817–1894
Leakey, Louis 1903–1972
Leakey, Mary 1913–
Leakey, Richard 1944–
Mariette, Auguste Ferdinand François
  1821–1881
Marsh, Othniel Charles 1831–1899
Pendlebury, John Devitt Stringfellow 1904–1941
Petrie, Flinders 1853–1942
Pitt-Rivers, Augustus Henry 1827–1900
Stukeley, William 1687–1765
Thomsen, Christian 1788–1865
Ventris, Michael 1922–1956
Wheeler, Mortimer 1890–1976
Woolley, Leonard 1880–1960

Astronomers, astronauts

Adams, John Couch 1819–1892
Airy, George Biddell 1801–1892
Aldrin, Edwin 'Buzz' 1930–
Anaximander c. 610–c. 546 BC
Aristarchus of Samos c. 320–c. 250 BC
Armstrong, Neil Alden 1930–
Baade, Walter 1893–1960
Bailly, Jean Sylvain 1736–1793
Bessel, Friedrich Wilhelm 1784–1846
Bode, Johann Elert 1747–1826
Bondi, Hermann 1919–
Bowditch, Nathaniel 1773–1838
Boyle, Charles, 4th Earl of Orrery 1676–1731
Bradley, James 1693–1762
Brahe, Tycho 1546–1601
Burnell, Jocelyn 1943–
Cannon, Annie Jump 1863–1941
Cassini, Giovanni Domenico 1625–1712
Chandrasekhar, Subrahmanyan 1910–
Clarke, Arthur C 1917–
Copernicus, Nicolaus 1473–1543
Eddington, Arthur Stanley 1882–1944

Eudoxus, of Cnidus c. 390–c. 340 BC
Flamsteed, John 1646–1719
Fowler, William 1911–
Gagarin, Yuri (Alexeyevich) 1934–1968
Galle, Johann Gottfried 1812–1910
Gamow, George 1904–1968
Glenn, John (Herschel), Jr 1921–
Hale, George Ellery 1868–1938
Halley, Edmond 1656–1742
Herschel, Caroline Lucretia 1750–1848
Herschel, John Frederick William 1792–1871
Herschel, William 1738–1822
Hewish, Antony 1924–
Hipparchus c. 190–c. 120 BC
Hoyle, Fred(erick) 1915–
Hubble, Edwin Powell 1889–1953
Huggins, William 1824–1910
Jansky, Karl Guthe 1905–1950
Kepler, Johannes 1571–1630
Korolev, Sergei Pavlovich 1906–1966
Kuiper, Gerard Peter 1905–1973
Laplace, Pierre Simon, Marquis de Laplace
  1749–1827
Leavitt, Henrietta Swan 1868–1921
Lemaître, Georges Edouard 1894–1966
Leonov, Aleksei Arkhipovich 1934–
Leverrier, Urbain Jean Joseph 1811–1877
Lippershey, Hans c. 1570–1619
Lovell, Bernard 1913–
Lowell, Percival 1855–1916
Maskelyne, Nevil 1732–1811
Messier, Charles 1730–1817
Olbers, Heinrich 1758–1840
Omar Khayyám c. 1050–1123
Oort, Jan Hendrik 1900–1992
Penston, Michael 1943–1990
Piazzi, Giuseppe 1746–1826
Ptolemy, (Claudius Ptolemaeus) c. 100–AD 170
Ryle, Martin 1918–1984
Schiaparelli, Giovanni 1835–1910
Secchi, Pietro Angelo 1818–1878
Shapley, Harlow 1885–1972
Sharman, Helen 1963–
Shepard, Alan (Bartlett) 1923–
Struve, Friedrich Georg Wilhelm 1793–1864
Tereshkova, Valentina Vladimirovna 1937–
Tombaugh, Clyde (William) 1906–
Van Allen, James (Alfred) 1914–
Whipple, Fred Lawrence 1906–
Young, John Watts 1930–
Zwicky, Fritz 1898–1974

Biologists

Beadle, George Wells 1903–1989
Berg, Paul 1926–
Brenner, Sidney 1927–
Calvin, Melvin 1911–
Carson, Rachel 1907–1964
Cori, Carl 1896–1984 and Gerty 1896–1957
Cousteau, Jacques Yves 1910–
Crick, Francis 1916–
Darwin, Charles Robert 1809–1882
Darwin, Erasmus 1731–1802
Dawkins, Richard 1941–

Delbruck, Max 1906–1981
Dobzhansky, Theodosius 1900–1975
Dubos, René Jules 1901–1981
Duve, Christian de 1917–
Fabre, Jean Henri Casimir 1823–1915
Fisher, Ronald Aylmer 1890–1962
Franklin, Rosalind 1920–1958
Frisch, Karl von 1886–1982
Gallo, Robert Charles 1937–
Galton, Francis 1822–1911
Gould, Stephen Jay 1941–
Haeckel, Ernst Heinrich 1834–1919
Haldane, J(ohn) B(urdon) S(anderson)
    1892–1964
Hamilton, William D 1936–
Hodgkin, Alan Lloyd 1914–
Hopkins, Frederick Gowland 1861–1947
Hounsfield, Godrey (Newbold) 1919–
Huxley, Andrew 1917–
Huxley, Julian 1887–1975
Huxley, Thomas Henry 1825–1895
Jacob, François 1920–
Jeffreys, Alec John 1950–
Katz, Bernard 1911–
Khorana, Har Gobind 1922–
Krebs, Hans 1900–1981
Lamarck, Jean Baptiste de 1744–1829
Lederberg, Joshua 1925–
Leeuwenhoek, Anton van 1632–1723
Lorenz, Konrad 1903–1989
Lysenko, Trofim Denisovich 1898–1976
Malpighi, Marcello 1628–1694
Maynard Smith, John 1920–
Mendel, Gregor Johann 1822–1884
Milstein, César 1927–
Monod, Jacques 1910–1976
Morgan, Thomas Hunt 1866–1945
Müller, Johannes Peter 1801–1858
Needham, Joseph 1900–
Northrop, John 1891–1987
Owen, Richard 1804–1892
Sanger, Frederick 1918–
Smith, John Maynard
Spallanzani, Lazzaro 1729–1799
Sutherland, Earl Wilbur Jr 1915–1974
Tatum, Edward Lawrie 1909–1975
von Gesner, Konrad 1516–1565
Waksman, Selman Abraham 1888–1973
Wallace, Alfred Russel 1823–1913
Warburg, Otto 1878–1976
Watson, James Dewey 1928–
Weismann, August 1834–1914
Whipple, George 1878–1976
Wilkins, Maurice Hugh Frederick 1916–
Wilson, Edward O 1929–
Wright, Sewall 1889–1988
Yersin, Alexandre Emile Jean 1863–1943

Botanists

Amici, Giovanni Batista 1786–1863
Banks, Joseph 1744–1820
Brown, Robert 1773–1858
Cesalpino, Andrea, Andrea 1519–1603
De Vries, Hugo 1848–1935
Gray, Asa 1810–1888
Hales, Stephen 1677–1761
Hofmeister, Wilhelm 1824–1877

Hooker, Joseph Dalton 1817–1911
Linnaeus, Carolus 1707–1778
Mee, Margaret 1909–1988
Ray, John 1627–1705
Solander, Daniel Carl 1736–1772
Sprengel, Christian Konrad 1750–1816
Tradescant, John 1570–1638

Chemists

Achard, Franz Karl 1753–1821
Adams, Roger 1889–1971
Andrews, John 1813–1885
Arrhenius, Svante August 1859–1927
Baekeland, Leo Hendrik 1863–1944
Bergius, Friedrich Karl Rudolph 1884–1949
Berthelot, Pierre Eugène Marcellin 1827–1907
Bertholet, Claude Louis 1748–1822
Berzelius, Jöns Jakob 1779–1848
Bloch, Konrad 1912–
Bosch, Carl 1874–1940
Buchner, Eduard 1860–1917
Bunsen, Robert Wilhelm von 1811–1899
Cannizzaro, Stanislao 1826–1910
Carothers, Wallace 1896–1937
Chardonnet, Hilaire Bernigaud 1839–1924
Chevreul, Michel-Eugène 1786–1889
Claude, Georges 1870–1960
Cleve, Per Teodor 1840–1905
Cornforth, John Warcup 1917–
Crookes, William 1832–1919
Curie, Marie (born Sklodovska) 1867–1934
Dalton, John 1766–1844
Dam, (Henrik) Carl (Peter) 1895–1976
Davy, Humphry 1778–1829
Dewar, James 1842–1923
Diels, Otto 1876–1954
Dodds, Charles 1899–1973
Doisy, Edward 1893–1986
Domagk, Gerhard 1895–1964
Dulong, Pierre 1785–1838
Edelman, Gerald Maurice 1929–
Eigen, Manfred 1927–
Faraday, Michael 1791–1867
Fischer, Emil Hermann 1852–1919
Fischer, Hans 1881–1945
Fröhlich, Herbert 1905–1991
Funk, Casimir 1884–1967
Galvani, Luigi 1737–1798 Geber, Latinized form
    of *Jabir* ibn Hayyan *c.* 721– *c.* 776
Gibbs, Josiah Willard 1839–1903
Gilbert, Walter 1932–
Glauber, Johann 1604–1668
Graham, Thomas 1805–1869
Grignard, François Auguste-Victor 1871–1935
Haber, Fritz 1868–1934
Hall, Charles 1863–1914
Haworth, Norman 1883–1950
Helmont, Jean Baptiste van 1577–1644
Henry, William 1774–1836
Hevesy, Georg von 1885–1966
Hinshelwood, Cyril Norman 1897–1967
Hodgkin, Dorothy Crowfoot 1910–
Hofmann, August Wilhelm von 1818–1892
Kekulé von Stradonitz, Friedrich August
    1829–1896
Kendall, Edward 1886–1972
Kendrew, John 1917–

Klaproth, Martin Heinrich 1743–1817
Kornberg, Arthur 1918–
Kuhn, Richard 1900–1967
Langmuir, Irving 1881–1957
Lavoisier, Antoine Laurent 1743–1794
Leblanc, Nicolas 1742–1806
Libby, Willard Frank 1908–1980
Liebig, Justus, Baron von 1803–1873
Lipmann, Fritz 1899–1986
Martin, Archer John Porter 1910–
Mendeleyev, Dmitri Ivanovich 1834–1907
Miller, Stanley 1930–
Mitchell, Peter 1920–1992
Moissan, Henri 1852–1907
Mond, Ludwig 1839–1909
Muller, Hermann Joseph 1890–1967
Müller, Paul 1899–1965
Nernst, (Walther) Hermann 1864–1941
Newlands, John Alexander Reina 1838–1898
Nobel, Alfred Bernhard 1833–1896
Northrop, John 1891–1987
Ochoa, Severo 1905–
Ostwald, Wilhelm 1853–1932
Paracelsus 1493–1541
Pauling, Linus Carl 1901–
Perey, Marguérite (Catherine) 1909–1975
Perkin, William Henry 1838–1907
Perutz, Max 1914–
Porter, George 1920–
Porter, Rodney Robert 1917–1985
Priestley, Joseph 1733–1804
Prigogine, Ilya 1917–
Proust, Joseph Louis 1754–1826
Prout, William 1785–1850
Ramsay, William 1852–1916
Raoult, François 1830–1901
Reichstein, Tadeus 1897–
Richards, Theodore 1868–1928
Robinson, Robert 1886–1975
Ruzicka, Leopold Stephen 1887–1976
Sabatier, Paul 1854–1951
Scheele, Karl Wilhelm 1742–1786
Seaborg, Glenn Theodore 1912–
Semenov, Nikoly 1896–1986
Smithson, James 1765–1829
Soddy, Frederick 1877–1956
Sørensen, Søren 1868–1939
Stahl, Georg Ernst 1660–1734
Stanley, Wendell 1904–1971
Staudinger, Hermann 1881–1965
Sumner, James 1887–1955
Svedberg, Theodor 1884–1971
Synge, Richard 1914–
Szent-Györgyi, Albert 1893–1986
Taube, Henry 1915–
Tiselius, Arne 1902–1971
Todd, Alexander, Baron Todd 1907–
Travers, Morris William 1872–1961
Urey, Harold Clayton 1893–1981
Vane, John 1927–
van't Hoff, Jacobus Henricus 1852–1911
Virtanen, Artturi Ilmari 1895–1973
Wald, George 1906–
Werner, Alfred 1866–1919
Wöhler, Friedrich 1800–1882
Wollaston, William 1766–1828
Woodward, Robert 1917–1979

Ziegler, Karl 1898–1973
Zsigmondy, Richard 1865–1929

Environmentalists, ecologists

Elton, Charles 1900–1991
Kelly, Petra 1947–1992
Muir, John 1838–1914
Porritt, Jonathon 1950–
Schumacher, Fritz (Ernst Friedrich) 1911–1977
Scott, Peter (Markham) 1909–1989

Geologists, Earth scientists

Agassiz, Jean Louis Rodolphe 1807–1873
Beaufort, Francis 1774–1857
Boucher de Crèvecoeur de Perthes, Jacques 1788–1868
Buffon, George Louis Leclerc, Comte de 1707–1778
Cuvier, Georges, Baron Cuvier 1769–1832
Humboldt, Alexander von 1769–1859
Hutton, James 1726–1797
Lyell, Charles 1797–1875
Maury, Matthew Fontaine 1806–1873
Mercator, Gerardus 1512–1594
Miller, William 1801–1880
Mohs, Friedrich 1773–1839
Murchison, Roderick 1792–1871
Osborn, Henry Fairfield 1857–1935
Richter, Charles Francis 1900–1985
Saussure, Horace de 1740–1799
Smith, William 1769–1839
Strabo c. 63 BC –AD 24
Wegener, Alfred Lothar 1880–1930
Werner, Abraham Gottlob 1750–1815

Inventors, manufacturers, engineers

Abel, Frederick Augustus 1827–1902
Ader, Clément 1841–1925
Albone, Dan 1860–1906
Appert, Nicolas 1750–1841
Arkwright, Richard 1732–1792
Armstrong, Edwin Howard 1890–1954
Armstrong, William George 1810–1900
Austin, Herbert, 1st Baron 1866–1941
Bailey, Donald Coleman 1901–1985
Baird, John Logie 1888–1946
Baker, Benjamin 1840–1907
Balfour, Eve 1898–1990
Batten, Jean 1909–1982
Baxter, George 1804–1867
Bazalgette, Joseph 1819–1890
Beeching, Richard, Baron Beeching 1913–1985
Bell, Alexander Graham 1847–1922
Bell, Patrick c. 1800–1869
Benz, Karl Friedrich 1844–1929
Bessemer, Henry 1813–1898
Birdseye, Clarence 1886–1956
Biro, Lazlo 1900–1985
Bishop, Ronald Eric 1903–1989
Blériot, Louis 1872–1936
Borlaug, Norman Ernest 1914–
Boulton, Matthew 1728–1809
Bourdon, Eugène 1808–1884
Boyd-Orr, John 1880–1971
Bramah, Joseph 1748–1814
Bridgewater, Francis Egerton, 3rd Duke of 1736–1803

Brindley, James 1716–1772
Brinell, Johann Auguste 1849–1925
Brunel, Isambard Kingdom 1806–1859
Brunel, Marc Isambard 1769–1849
Burroughs, William Steward 1857–1898
Campbell, Donald Malcolm 1921–1967
Carlson, Chester 1906–1968
Cartwright, Edmund 1743–1823
Carver, George Washington 1864–1943
Cayley, George 1773–1857
Charles, Jacques Alexandre César 1746–1823
Chrysler, Walter Percy 1875–1940
Cierva, Juan de la 1895–1936
Cockerell, Christopher 1910–
Cody, Samuel Franklin 1862–1913
Coke, Thomas William 1754–1842
Colt, Samuel 1814–1862
Cort, Henry 1740–1800
Coster, Laurens Janszoon 1370–1440
Courtauld, Samuel 1793–1881
Creed, Frederick George 1871–1957
Crompton, Samuel 1753–1827
Cugnot, Nicolas-Joseph 1728–1804
Culshaw, John 1924–1980
Curtiss, Glenn Hammond 1878–1930
Daimler, Gottlieb 1834–1900
Dalen, Nils 1869–1937
Daniell, John Frederic 1790–1845
Darby, Abraham 1677–1717
De Forest, Lee 1873–1961
De Havilland, Geoffrey 1882–1965
de Lesseps, Ferdinand, Vicomte Diesel, Rudolf 1858–1913
Doolittle, James Harold 1896–
Dornier, Claude 1884–1969
Dunlop, John Boyd 1840–1921
Duwez, Pol 1907–
Eastman, George 1854–1932
Edison, Thomas Alva 1847–1931
Edwards, George 1908–
Eiffel, (Alexandre) Gustave 1832–1923
Ericsson, John 1803–1889
Farman, Henry 1874–1958
Fender, Leo 1909–1991
Ferguson, Harry 1884–1960
Ferranti, Sebastian de 1864–1930
Fessenden, Reginald Aubrey 1866–1932
Fitch, John 1743–1798
Fleming, John Ambrose 1849–1945
Ford, Henry 1863–1947
Fraze, Ermal Cleon 1913–1989
Fulton, Robert 1765–1815
Fulton, Robert 1745–1815
Gatling, Richard Jordan 1818–1903
Giffard, Henri 1825–1882
Gillette, King Camp 1855–1932
Goddard, Robert Hutchings 1882–1945
Goodyear, Charles 1800–1860
Gutenberg, Johann c. 1400–1468
Hargreaves, James died 1778
Heinkel, Ernst 1888–1958
Hinkler, Herbert John Louis 1892–1933
Hooke, Robert 1635–1703
Hoover, William Henry 1849–1932
Howe, Elias 1819–1867
Hughes, David 1831–1900
Hume-Rothery, William 1899–1968

Issigonis, Alec 1906–1988
Ives, Frederic Eugene 1856–1937
Jacquard, Joseph Marie 1752–1834
Jacuzzi, Candido 1903–1986
Kaiser, Henry J 1882–1967
Kaplan, Viktor 1876–1934
Kay, John 1704–c. 1764
Kennelly, Arthur Edwin 1861–1939
Kliegl, John H 1869–1959 and Anton T 1872–1927
Lavrentiev, Mikhail 1900–
Lawes, John Bennet 1814–1900
Lesseps, Ferdinand, Vicomte de Lesseps 1805–1894
Lilienthal, Otto 1848–1896
Lindbergh, Charles A(ugustus) 1902–1974
Lumière, Auguste Marie 1862–1954 and Louis Jean 1864–1948
Lunardi, Vincenzo 1759–1806
McAdam, John Loudon 1756–1836
McCormick, Cyrus Hall 1809–1884
MacCready, Paul 1925–
Macintosh, Charles 1766–1843
Marconi, Guglielmo 1874–1937
Maxim, Hiram Stevens 1840–1916
Meikle, Andrew 1719–1811
Mergenthaler, Ottmar 1854–1899
Messerschmitt, Willy 1898–1978
Mitchell, R(eginald) J(oseph) 1895–1937
Montgolfier, Joseph Michel 1740–1810 and Étienne Jacques 1745–1799
Moon, William 1818–1894
Morse, Samuel (Finley Breese) 1791–1872
Murdock, William 1754–1839
Nasmyth, James 1808–1890
Newcomen, Thomas 1663–1729
Niepce, Joseph Nicéphore 1765–1833
Noyce, Robert Norton 1927–1990
Olds, Ransom Eli 1864–1950
Onassis, Aristotle 1906–1975
Otis, Elisha Graves 1811–1861
Otto, Nikolaus August 1832–1891
Page, Frederick Handley 1885–1962
Parsons, Charles Algernon 1854–1931
Paul, Les 1909–
Penney, William 1909–1991
Philips, Anton 1874–1951
Pilcher, Percy 1867–1899
Porsche, Ferdinand 1875–1951
Poulsen, Valdemar 1869–1942
Pullman, George 1831–1901
Ransome, Robert 1753–1830
Réaumur, Réné Antoine Ferchault de 1683–1757
Remington, Eliphalet 1793–1861
Remington, Philo 1816–1889
Rennie, John 1761–1821
Rickover, Hyman George 1900–1986
Rolls, Charles Stewart 1877–1910
Royce, (Frederick) Henry 1863–1933
Savery, Thomas c. 1650–1715
Scott, Douglas 1913–1990
Siemens, William 1823–1883
Sikorsky, Igor 1889–1972
Sinclair, Clive 1940–
Singer, Isaac Merit 1811–1875
Slater, Samuel 1768–1835
Smeaton, John 1724–1792

Smith, Keith Macpherson 1890–1955
Sommeiler, Germain 1815–1871
Sperry, Elmer Ambrose 1860–1930
Steinmetz, Charles 1865–1923
Stephenson, George 1781–1848
Stephenson, Robert 1803–1859
Stevenson, Robert 1772–1850
Swan, Joseph Wilson 1828–1914
Symington, William 1763–1831
Talbot, William Henry Fox 1800–1877
Telford, Thomas 1757–1834
Thomas, Seth 1785–1859
Thomson, Elihu 1853–1937
Todt, Fritz 1891–1942
Trevithick, Richard 1771–1833
Tsiolkovsky, Konstantin 1857–1935
Tull, Jethro 1674–1741
Tyndall, John 1820–1893
Vail, Alfred Lewis 1807–1859
van de Graaff, Robert Jemison 1901–1967
von Braun, Wernher 1912–1977
Wallis, Barnes (Neville) 1887–1979
Watson-Watt, Robert Alexander 1892–1973
Watt, James 1736–1819
Weber, Wilhelm Eduard 1804–1891
Westinghouse, George 1846–1914
Whitehead, Robert 1823–1905
Whitney, Eli 1765–1825
Whitten-Brown, Arthur 1886–1948
Whittle, Frank 1907–
Wickham, Henry 1846–1928
Young, Arthur 1741–1820
Zworykin, Vladimir Kosma 1889–1982

Mathematicians, computer scientists

Abel, Niels Henrik 1802–1829
Aiken, Howard 1900–
Alembert, Jean le Rond d' 1717–1783
Apollonius of Perga c. 260–c. 190 BC
Archimedes c. 287–212 BC
Babbage, Charles 1792–1871
Balmer, Johann Jakob 1825–1898
Barrow, Isaac 1630–1677
Bayes, Thomas 1702–1761
Boole, George 1815–1864
Byron, Augusta Ada 1815–1851
Cantor, Georg 1845–1918
Cauchy, Augustin Louis 1789–1857
Cayley, Arthur 1821–1895
Dedekind, Richard 1831–1916
Descartes, René 1596–1650
Diophantus lived c. 250
Dirichlet, Peter Gustav Lejeune 1805–1859
Eckert, John Presper Jr 1919–
Eratosthenes c. 276–194 BC
Euclid c. 330–c. 260 BC
Euler, Leonhard 1707–1783
Fermat, Pierre de 1601–1665
Fibonacci, Leonardo, or *Leonardo of Pisa*
   c. 1175– c. 1250
Fourier, Jean Baptiste Joseph 1768–1830
Galois, Évariste 1811–1832
Gauquelin, Michel 1928–1991
Gauss, Karl Friedrich 1777–1855
Germain, Sophie 1776–1831
Gödel, Kurt 1906–1978
Gunter, Edmund 1581–1626

Hamilton, William Rowan 1805–1865
Herapath, John 1790–1868
Hilbert, David 1862–1943
Hollerith, Herman 1860–1929
Khwārizmī, ibn-Mūsā al-Muhammad
   c. 780–c. 850
Kovalevsky, Sonja Vasilevna 1850–1891
Lagrange, Joseph Louis 1736–1813
Laplace, Pierre Simon, Marquis de Laplace
   1749–1827
Leibniz, Gottfried Wilhelm 1646–1716
Lighthill, James 1924–
Lobachevsky, Nikolai Ivanovich 1792–1856
Lorenz, Ludwig Valentine 1829–1891
Mandelbrot, Benoit B 1924–
Markov, Andrei 1856–1922
Mauchly, John William 1907–1980
Mersenne, Marin 1588–1648
Möbius, August Ferdinand 1790–1868
Napier, John 1550–1617
Noyce, Robert Norton 1927–1990
Oughtred, William 1575–1660
Pascal, Blaise 1623–1662
Pearson, Karl 1857–1936
Poincaré, Jules Henri 1854–1912
Poisson, Siméon Denis 1781–1840
Poncelet, Jean-Victor 1788–1867
Pythagoras c. 580–500 BC
Quetelet, Lambert Adolphe Jacques 1796–1874
Richard of Wallingford 1292–1335
Riemann, Georg Friedrich Bernhard 1826–1866
Rubik, Erno 1944–
Russell, Bertrand, 3rd Earl Russell 1872–1970
Shannon, Claude Elwood 1916–
Turing, Alan Mathison 1912–1954
Vernier, Pierre 1580–1637
Viète, François 1540–1603
Von Neumann, John 1903–1957
Wang, An 1920–1990
Whitehead, Alfred North 1861–1947
Wiener, Norbert 1894–1964

Medical figures

Abel, John Jacob 1857–1938
Abraham, Edward Penley 1913–
Addison, Thomas 1793–1863
Adrian, Edgar, 1st Baron Adrian 1889–1977
Alexander, Frederick Matthias 1869–1955
Allbutt, Thomas Clifford 1836–1925
Anderson, Elizabeth Garrett 1836–1917
Axelrod, Julius 1912–
Banting, Frederick Grant 1891–1941
Barker, Herbert 1869–1950
Barnard, Christiaan 1922–
Bayliss, William Maddock 1860–1924
Beaumont, William 1785–1853
Behring, Emil von 1854–1917
Best, Charles Herbert 1899–1978
Bichat, Marie François Xavier 1771–1802
Black, James 1924–
Blackwell, Elizabeth 1821–1910
Borelli, Giovanni Alfonso 1608–1679
Bovet, Daniel 1907–1992
Bright, Richard 1789–1858
Burnet, Macfarlane 1899–1985
Calmette, Albert 1863–1933
Cardano, Girolamo 1501–1576

Carrel, Alexis 1873–1944
Chain, Ernst Boris 1906–1979
Charcot, Jean-Martin 1825–1893
Colombo, Matteo Realdo *c.* 1516–1559
Cushing, Harvey Williams 1869–1939
Dale, Henry Hallett 1875–1968
Dick-Read, Grantly 1890–1959
Dodds, Charles 1899–1973
Doll, William Richard 1912–
Domagk, Gerhard 1895–1964
Donald, Ian 1910–1987
Dooley, Thomas Anthony 1927–1961
Dubos, René Jules 1901–1981
Ehrlich, Paul 1854–1915
Eijkman, Christiaan 1858–1930
Einthoven, Willem 1860–1927
Enders, John Franklin 1897–1985
Eustachio, Bartolommeo 1520–1574
Fabricius, Geronimo 1537–1619
Fernel, Jean François 1497–1558
Finsen, Niels Ryberg 1860–1904
Fixx, James 1932–1984
Fleming, Alexander 1881–1955
Florey, Howard Walter, Baron Florey 1898–1968
Forssmann, Werner 1904–1979
Fracastoro, Girolamo *c.* 1478–1553
Friedman, Maurice 1903–1991
Galen *c.* 130–*c.* 200
Gallo, Robert Charles 1937–
Garrod, Archibald Edward 1857–1937
Gibbon, John Heysham 1903–1974
Golgi, Camillo 1843–1926
Graaf, Regnier de 1641–1673
Gregg, Norman 1892–1966
Guérin, Camille 1872–1961
Haller, Albrecht von 1708–1777
Harvey, William 1578–1657
Helmholtz, Hermann Ludwig Ferdinand von
    1821–1894
Hench, Philip Showalter 1896–1965
Herophilus of Chalcedon *c.* 330–*c.* 260 BC
Hill, Austin Bradford 1897–1991
Hippocrates *c.* 460–*c.* 370 BC
Hodgkin, Thomas 1798–1856
Hunter, John 1728–1793
Ingenhousz, Jan 1730–1799
Isaacs, Alick 1921–1967
Jackson, John Hughlings 1835–1911
Jenner, Edward 1749–1823
Kitasato, Shibasaburo 1852–1931
Koch, Robert 1843–1910
Koller, Carl 1857–1944
Laënnec, René Théophile Hyacinthe 1781–1826
Levi-Montalcini, Rita 1909–
Linacre, Thomas *c.* 1460–1524
Lister, Joseph, 1st Baron Lister 1827–1912
Loewi, Otto 1873–1961
Ludwig, Karl Friedrich Wilhelm 1816–1895
McIndoe, Archibald 1900–1960
Manson, Patrick 1844–1922
Mayo, William James 1861–1939
Mechnikov, Ilya 1845–1916
Medawar, Peter (Brian) 1915–1987
Menninger, Karl Augustus 1893–1990
Mesmer, Friedrich Anton 1734–1815
Meyer, Alfred 1895–1990
Moniz, Antonio Egas 1874–1955

Morgagni, Giovanni Battista 1682–1771
Morton, William Thomas Green 1819–1868
Murray, Joseph E 1919–
Nicolle, Charles 1866–1936
Noguchi, Hideyo 1876–1928
Paracelsus 1493–1541
Paré, Ambroise 1509–1590
Parkinson, James 1755–1824
Pasteur, Louis 1822–1895
Pincus, Gregory Goodwin 1903–1967
Pirquet, Clemens von 1874–1929
Reed, Walter 1851–1902
Ross, Ronald 1857–1932
Sabin, Albert 1906–
Salk, Jonas Edward 1914–
Sanctorius, Sanctorius 1561–1636
Sanger, Margaret Higgins 1883–1966
Saunders, Cicely 1918–
Schweitzer, Albert 1875–1965
Semmelweis, Ignaz Philipp 1818–1865
Servetus, Michael 1511–1553
Sharpey-Schäfer, Edward Albert 1850–1935
Sherrington, Charles Scott 1857–1952
Simpson, James Young 1811–1870
Simpson, (Cedric) Keith 1907–1985
Sloane, Hans 1660–1753
Spock, Benjamin McLane 1903–
Starling, Ernest Henry 1866–1927
Steptoe, Patrick Christopher 1913–1988
Sydenham, Thomas 1624–1689
Tagliacozzi, Gaspare 1546–1599
Taussig, Helen Brooke 1898–1986
Vesalius, Andreas 1514–1564
Virchow, Rudolf Ludwig Carl 1821–1902
Wagner-Jauregg, Julius 1857–1940
Wassermann, August von 1866–1925
Weber, Ernst Heinrich 1795–1878
Yalow, Rosalyn Sussman 1921–

Physicists

Alhazen, Ibn al Haytham *c.* 965–1038
Alvarez, Luis Walter 1911–1988
Ampère, André Marie 1775–1836
Anderson, Carl David 1905–1991
Ångström, Anders Jonas 1814–1874
Appleton, Edward Victor 1892–1965
Aston, Francis William 1877–1945
Avogadro, Amedeo Conte di Quaregna
    1776–1856
Bainbridge, Kenneth Tompkins 1904–
Balmer, Johann Jakob 1825–1898
Bardeen, John 1908–1991
Basov, Nikolai Gennadievich 1912–
Becquerel, Antoine Henri 1852–1908
Bell, John 1928–1990
Bethe, Hans Albrecht 1906–
Biot, Jean 1774–1862
Black, Joseph 1728–1799
Blackett, Patrick Maynard Stuart,
    Baron Blackett 1897–1974
Bloch, Felix 1905–1983
Bohr, Aage 1922–
Bohr, Niels Henrik David 1885–1962
Boltzmann, Ludwig 1844–1906
Born, Max 1882–1970
Bose, Jagadis Chunder 1858–1937
Bose, Satyendra Nath 1894–1974

Bothe, Walther 1891–1957
Boyle, Robert 1627–1691
Bragg, William Henry 1862–1942
Brattain, Walter Houser 1902–1987
Brewster, David 1781–1868
Bridgman, Percy Williams 1882–1961
Broglie, Louis de, 7th Duc de Broglie 1892–1987
Broglie, Maurice de, 6th Duc de Broglie
  1875–1960
Cavendish, Henry 1731–1810
Chadwick, James 1891–1974
Chamberlain, Owen 1920–
Cherenkov, Pavel 1904–
Cherwell, Frederick Alexander Lindemann
  1886–1957
Chladni, Ernst Florens Friedrich 1756–1827
Clausius, Rudolf Julius Emanuel 1822–1888
Cockcroft, John Douglas 1897–1967
Compton, Arthur Holly 1892–1962
Cooper, Leon 1930–
Coulomb, Charles Auguste de 1736–1806
Curie, Marie 1867–1934
Davisson, Clinton Joseph 1881–1958
Debye, Peter 1884–1966
Dirac, Paul Adrien Maurice 1902–1984
Doppler, Christian Johann 1803–1853
Einstein, Albert 1879–1955
Eötvös, Roland von, Baron 1848–1919
Esaki, Leo 1925–
Fahrenheit, Gabriel Daniel 1686–1736
Faraday, Michael 1791–1867
Fermi, Enrico 1901–1954
Feynman, Richard Phillips 1918–1988
Fitzgerald, George 1851–1901
Fortin, Jean 1750–1831
Foucault, Jean Bernard Léon 1819–1868
Fowler, William 1911–
Franck, James 1882–1964
Frank, Ilya 1908–
Franklin, Benjamin 1706–1790
Fraunhofer, Joseph von 1787–1826
Fresnel, Augustin 1788–1827
Frisch, Otto 1904–1979
Gabor, Dennis 1900–1979
Galileo, properly Galileo Galilei 1564–1642
Gassendi, Pierre 1592–1655
Gauss, Karl Friedrich 1777–1855
Gay-Lussac, Joseph Louis 1778–1850
Geiger, Hans 1882–1945
Gell-Mann, Murray 1929–
Gilbert, William 1544–1603
Glaser, Donald Arthur 1926–
Glashow, Sheldon Lee 1932–
Goeppert-Mayer, Maria 1906–1972
Guillaume, Charles 1861–1938
Hahn, Otto 1879–1968
Hawking, Stephen 1942–
Heaviside, Oliver 1850–1925
Heisenberg, Werner Carl 1901–1976
Helmholtz, Hermann Ludwig Ferdinand von
  1821–1894
Henry, Joseph 1797–1878
Hertz, Heinrich 1857–1894
Hess, Victor 1883–1964
Hofstadter, Robert 1915–1990
Hooke, Robert 1635–1703
Huygens, Christiaan 1629–1695

Jeans, James Hopwood 1877–1946
Joliot-Curie, Irène 1897–1956 and Frédéric
  1900–1958
Josephson, Brian 1940–
Joule, James Prescott 1818–1889
Kamerlingh-Onnes, Heike 1853–1926
Kapitza, Peter 1894–1984
Kelvin, William Thomson, 1st Baron Kelvin
  1824–1907
Kirchhoff, Gustav Robert 1824–1887
Lamb, Willis 1913–
Landau, Lev Davidovich 1908–1968
Langevin, Paul 1872–1946
Langley, Samuel Pierpoint 1834–1906
Laue, Max Theodor Felix von 1879–1960
Lawrence, Ernest O(rlando) 1901–1958
Lebedev, Peter Nikolaievich 1866–1912
Leclanché, Georges 1839–1882
Lee Tsung-Dao 1926–
Lenard, Philipp Eduard Anton 1862–1947
Lodge, Oliver Joseph 1851–1940
Lorentz, Hendrik Antoon 1853–1928
Lorenz, Ludwig Valentine 1829–1891
Mach, Ernst 1838–1916
McMillan, Edwin Mattison 1907–
Mariotte, Edme 1620–1684
Maxwell, James Clerk 1831–1879
Mayer, Julius Robert von 1814–1878
Meitner, Lise 1878–1968
Michelson, Albert Abraham 1852–1931
Millikan, Robert Andrews 1868–1953
Morley, Edward 1838–1923
Moseley, Henry Gwyn-Jeffreys 1887–1915
Mott, Nevill Francis 1905–
Mulliken, Robert Sanderson 1896–1986
Newton, Isaac 1642–1727
Oersted, Hans Christian 1777–1851
Ohm, Georg Simon 1787–1854
Onsager, Lars 1903–1976
Oppenheimer, J Robert 1904–1967
Pauli, Wolfgang 1900–1958
Perrin, Jean 1870–1942
Petit, Alexis 1791–1820
Piccard, Auguste 1884–1962
Planck, Max 1858–1947
Popov, Alexander 1859–1905
Powell, Cecil Frank 1903–1969
Prokhorov, Aleksandr 1916–
Rabi, Isidor Isaac 1898–1988
Raman, Venkata 1888–1970
Rayleigh, John W Strutt, 3rd Baron 1842–1919
Reynolds, Osborne 1842–1912
Richardson, Owen Willans 1879–1959
Richter, Burton 1931–
Röntgen, Wilhelm Konrad 1845–1923
Rubbia, Carlo 1934–
Rumford, Benjamin Thompson, Count Rumford
  1753–1814
Rutherford, Ernest 1871–1937
Sagan, Carl 1934–
Sakharov, Andrei Dmitrievich 1921–1989
Salam, Abdus 1926–
Schrödinger, Erwin 1887–1961
Schwinger, Julian 1918–
Segrè, Emilio 1905–1989
Shockley, William 1910–1989
Snell, Willebrord 1581–1626

Sommerfeld, Arnold 1868–1951
Stark, Johannes 1874–1957
Stefan, Joseph 1835–1893
Stern, Otto 1888–1969
Stokes, George Gabriel 1819–1903
Street, J(abez) C(urry) 1906–1989
Szilard, Leo 1898–1964
Teller, Edward 1908–
Tesla, Nikola 1856–1943
Theodoric of Freiburg *c.* 1250–1310
Thomson, George Paget 1892–1975
Thomson, J(oseph) J(ohn) 1856–1940
Ting, Samuel 1936–
Torricelli, Evangelista 1608–1647
Townes, Charles 1915–
Ulam, Stanislaw Marcin 1909–1985
van der Waals, Johannes Diderik 1837–1923
Volta, Alessandro 1745–1827
Walton, Ernest 1903–
Watt, James 1736–1819
Weber, Ernst Heinrich 1795–1878
Weinberg, Steven 1933–
Wheatstone, Charles 1802–1875
Wien, Wilhelm 1864–1928
Wigner, Eugene Paul 1902–
Wilson, Charles Thomson Rees 1869–1959
Young, Thomas 1773–1829
Yukawa, Hideki 1907–1981
Zeeman, Pieter 1865–1943
Zernike, Frits 1888–1966

Psychiatrists, psychoanalysts

Adler, Alfred 1870–1937
Bettelheim, Bruno 1903–1990
Binet, Alfred 1857–1911
Bowlby, John 1907–1990
Breuer, Josef 1842–1925
Burt, Cyril Lodowic 1883–1971
Coué, Émile 1857–1926
de Bono, Edward 1933–
Ellis, Havelock 1859–1939
Eysenck, Hans Jurgen 1916–
Fechner, Gustav 1801–1887
Freud, Sigmund 1865–1939

Fromm, Erich 1900–1980
Gall, Franz Joseph 1758–1828
Gesell, Arnold Lucius 1880–1961
Goffman, Erving 1922–1982
Guthrie, Edwin R(ay) 1886–1959
James, William 1842–1910
Jung, Carl Gustav 1875–1961
Klein, Melanie 1882–1960
Krafft-Ebing, Baron Richard von 1840–1902
Laing, R(onald) D(avid) 1927–1989
Menninger, Karl Augustus 1893–1990
Myers, F(rederic) W(illiam) H(enry) 1843–1901
Pavlov, Ivan Petrovich 1849–1936
Perls, Laura (born Lore Posner) 1906–1990
Piaget, Jean 1896–1980
Reich, Wilhelm 1897–1957
Rhine, Joseph Banks 1895–1980
Rogers, Carl 1902–1987
Skinner, B F 1903–1990
Thorndike, Edward Lee 1874–1949
Watson, John Broadus 1878–1958
Wertheimer, Max 1880–1943
Wundt, Wilhelm Max 1832–1920

Zoologists

Adamson, Joy 1910–1985
Agassiz, Jean Louis Rodolphe 1807–1873
Baer, Karl Ernst von 1792–1876
Beebe, Charles 1877–1962
Bernard, Claude 1813–1878
Black, Davidson 1884–1934
Bordet, Jules 1870–1961
Cuvier, Georges, Baron Cuvier 1769–1832
Eccles, John Carew 1903–
Fallopius, Gabriel 1523–1562
Fossey, Dian 1938–1985
Hagenbeck, Carl 1844–1913
Hildegard of Bingen 1098–1179
Landsteiner, Karl 1868–1943
McClintock, Barbara 1902–1992
Marsh, Othniel Charles 1831–1899
Osborn, Henry Fairfield 1857–1935
Tinbergen, Niko(laas) 1907–
Wynne-Edwards, Vera 1906–

Thematic Indexes

Agriculture and food technology

additive
afforestation
agriculture
agrochemical
antioxidant
blast freezing
carbohydrate
cereal
cholesterol
colouring
crop rotation
deep freezing
diet
dough
E number
emulsifier
fertilizer
fibre, dietary
food
food irradiation
food technology
food test
forestry
freeze-drying
fruit
glycogen
harrow
herbicide
high-yield variety
honey
insecticide
irradiation
irrigation
lanolin
legume
margarine
milk
nut
organic farming
overfishing
pasteurization
plough
polyunsaturate
preservative
seed drill
set-aside scheme
slash and burn
slurry
sweetener
tractor
triticale
UHT
weedkiller

Archaeology

anthropology
archaeology
CAT scan
computerized axial tomography
dendrochronology
electron spin resonance
eolith

ethnoarchaeology
ethnography
ethnology
fission-track dating
flotation process
geochemical analysis
infrared absorption
  spectrometry
metal detector
neutron activation analysis
optical emission spectrometry
radiocarbon dating
remote sensing
stratigraphy
X-ray diffraction
X-ray fluorescence
  spectrometry

Astronomy

aberration of starlight
albedo
aphelion
apogee
asteroid
astrometry
astronomical unit
astronomy
astrophysics
binary star
black hole
brown dwarf
celestial sphere
Cepheid variable
comet
constellation
cosmic radiation
cosmology
double star
ecliptic
exobiology
galaxy
globular cluster
gravitational lensing
Hertzsprung–Russell diagram
Hubble's law
infrared astronomy
Jodrell Bank
light year
Local Group
luminosity
magnitude
Milky Way
moon
nadir
nebula
neutron star
nova
observatory
parallax
parsec
perigee
perihelion
precession
pulsar

radar astronomy
radio astronomy
red shift
red dwarf
red giant
Seyfert galaxy
shooting star
sidereal period
singularity
Solar System
space probe
speckle interferometry
star cluster
star
steady-state theory
Sun
supergiant
supernova
telescope
ultraviolet astronomy
universe
variable star
Very Large Array
white dwarf
X-ray astronomy
zenith
zodiac

Atmosphere and ocean

abyssal zone
air
air mass
anticyclone
atmosphere
atoll
barometer
bathyal zone
beach
Beaufort scale
Benguela current
climate
cloud
coastal erosion
continental shelf
coral
Coriolis effect
current
cyclone
deep-sea trench
doldrums
estuary
exosphere
fog
front
frost
greenhouse effect
gyre
Hadley cell
hail
Humboldt Current
hurricane
hydrological cycle
island
isobar

jet stream
lagoon
magnetic storm
Mariana Trench
Mediterranean climate
mesosphere
meteorology
Mid-Atlantic Ridge
monsoon
ocean trench
ocean
ocean ridge
ooze
rain
seafloor spreading
shelf sea
tide
tornado
trade wind
troposphere
tsunami
whirlwind
wind

Biology

aerobic
agar
allele
alternation of generations
anaerobic
archaebacteria
arthropod
bacillus
bacteria
basal metabolic rate
binary fission
biochemistry
biology
biosynthesis
blue-green algae
carnivore
cell
chemosynthesis
chromosome
coccus
crustacean
cytoplasm
decomposer
diploid
ectoplasm
embryo
endoplasm
eukaryote
fertilization
food chain
gamete
gas exchange
gene
genetics
genome
genotype
haploid
hermaphrodite
heterotroph
heterozygous
homozygous
karyotype

meiosis
metamorphosis
methanogenic bacteria
microorganism
nitrogen cycle
nucleus
ovulation
parthenogenesis
phenotype
protist
protozoa
replication
respiration
virus
zygote

Botany

angiosperm
annual plant
auxin
bark
berry
biennial plant
botany
bud
cell wall
chemotropism
chlorophyll
chloroplast
coleoptile
corm
cotyledon
cytokinin
deciduous
dendrochronology
dicotyledon
dormancy
drupe
endosperm
evergreen
flower
fruit
germination
gibberellin
gymnosperm
herbaceous plant
hydroponics
leaf
legume
monocotyledon
nitrogen fixation
periennial plant
petal
phloem
photosynthesis
phototropism
leaf
plant classification
pollen
root
seed
succulent plant
taxis
tropism
tuber
vegetative reproduction
xylem

Chemistry, general

absolute zero
acid
acid salt
acidic oxide
activation energy
activity series
addition reaction
affinity
alcohol
alkali metal
alkali
alkaline-earth metal
allotropy
alloy
amalgam
amphoteric
analysis
analytical chemistry
anhydrous
anion
anode
aqueous solution
aromatic compound
assay
atom
atomic number
Avogadro's hypothesis
Avogadro's number
base
basic oxide
binding energy
boiling point
bond
buffer
bunsen burner
burning
catalyst
cation
cell, electrolytic
change of state
Charles's law
chemical change
chemical equilibrium
chemical element
chemistry
colligative property
combustion
concentration
condensation
conservation of energy
conservation of mass
constant composition, law of
covalent bond
crystal
decomposition
deliquescence
density
dew point
diatomic molecule
dibasic acid
diffusion
dilution
dipole
displacement reaction
double decomposition

amine
antiknock
aromatic compound
azodye
buckminsterfullerene
buckyballs
carbohydrate
carboxyl group
carboxylic acid
chlorofluorocarbon
colloid
cross linking
detergent
disaccharide
drug
dye
dynamite
elastomer
ester
ethene
ether
ethyne
fatty acid
fluorocarbon
fusel oil
glyceride
halon
homologous series
hydrocarbon
isomer
kerosene
ketone
lipid
macromolecule
monosaccharide
narcotic
olefin
oligosaccharide
optical activity
paraffin
peptide
peptide bond
plastic
polyester
polymer
polymerization
polysaccharide
protein
saturated fatty acid
saturated compound
tautomerism
wax

Communications

Appleton layer
broadcasting
cable television
camcorder
cellular phone
citizens' band
closed-circuit television
coaxial cable
cybernetics
digital audiotape
digital compact cassette
Discman
Dolby system

electronic mail
Eutelsat
fax
film, photographic
four-colour process
gramophone
hi-fi
high-definition television
holography
interactive video
letterpress
liquid-crystal display
loudspeaker
magnetic tape
microphone
microwave
Mini Disc
Morse code
musical instrument digital
    interface
optical fibre
phonograph
photocopier
printing
projector
radio, cellular
radio
receiver, radio
record player
recording
satellite television
signal-to-noise ratio
soundtrack
stereophonic sound
superheterodyne receiver
tape recording, magnetic
telecommunications
telegraphy
teleprinter
teletext
television
telex
videocamera
videocassette recorder
videodisc
videotext
videotape recorder
viewdata
virtual reality
wireless

Computing

ADA
address
ALGOL
analogue computer
analytical engine
AND gate
application
arithmetic and logic unit
artificial intelligence
ASCII
assembly language
BASIC
baud
benchmark
binary number system

biological computer
bit
boot
bubble memory
bug
bus
byte
CAD
CAL
CD-ROM
central processing unit
chip
CMOS
COBOL
computer
computer graphics
computer game
CP/M
CPU
data
dBASE
desktop publishing
DIANE
digital computer
disc drive
DOS
dot matrix printer
DRAM
EPROM
expert system
file
flash memory
FORTRAN
fuzzy logic
gate
GIGO
GUI
hacking
hardware
high-level language
IBM
ink-jet printer
integrated circuit
interface
joystick
knowledge-based system
laptop computer
laser printer
light pen
logic gate
LOGO
Lotus 1–2–3
Macintosh
mainframe
megabyte
memory
microchip
microcomputer
MIDI
modem
MS-DOS
multimedia
NAND gate
NOR gate
NOT gate
object-oriented programming
OCR

square root
statistics
tangent
topology
trapezium
triangle
trigonometry
variable
vector quantity
Venn diagram

Medicine

acquired immune deficiency
  syndrome (AIDS)
anabolism
analgesic
antibiotic
anticoagulant
antiseptic
astigmatism
baldness
beriberi
bioengineering
blood
catabolism
coil
CAT scan
contraceptive
dialysis
diaphragm
disease
dopamine
drug
ECG
electrocardiogram
electroencephalogram
endorphin
endoscopy
genetic disease
grafting
heart–lung machine
heavy metal
HIV
hormone
immunity
jet lag
lymphocyte
magnetic resonance imaging
medicine
menopause
miscarriage
monoclonal antibody
narcotic
opium
pacemaker
pathogen
pellagra
pneumothorax
prostate gland
psychology
pulse
radiation sickness
radioisotope scanning
radiotherapy
retina
retrovirus
rhesus factor

rickets
scurvy
sebum
sedative
sick building syndrome
sterilization
transfusion
ultrasonics
ultrasound
veterinary science
vitamin
water-borne disease
zidovudine

Particle physics

accelerator
alpha particle
antimatter
artificial radioactivity
atomic structure
background radiation
baryon
becquerel
beta particle
boson
bubble chamber
charm
cloud chamber
cosmic radiation
critical mass
cyclotron
decay, radioactive
electron
elementary particle
Euratom
fast breeder
Fermilab
fermion
forces, fundamental particle
fusion
gamma radiation
gauge boson
Geiger counter
gluon
hadron
half-life
Higgs particle
hyperon
intermediate vector boson
ion
Large Electron Positron
  Collider
lepton
linear accelerator
meson
neutrino
neutron
nuclear energy
nuclear fusion
nuclear reaction
nuclear physics
particle physics
plasma
positron
proton
quantum chromodynamics
quantum electrodynamics

quantum theory
quark
relative biological effectiveness
scanning tunnelling microscope
sievert
soliton
strong nuclear force
subatomic particle
synchrotron
van de Graaff generator
W particle
weak nuclear force
weakon
Z particle

Photography

aperture
ASA
calotype
camera
camera obscura
charge-coupled device
Cibachrome
cine camera
daguerreotype
developing
dye-transfer print
exposure meter
film, photographic
f-number
focus
f-stop
hypo
ISO
negative/positive
orthochromatic
panchromatic
photogram
photography
Polaroid camera
rangefinder
reflex camera
SLR
telephoto lens
thaumatrope
35 mm
TLR camera
transparency
wide-angle lens
zoetrope
zone system
zoom lens

Physics

absolute zero
acoustics
analogue signal
Boltzmann constant
chain reaction
condensation number
critical mass
cryogenics
diffraction
dynamics
efficiency
elasticity
electric current